THE HUTCHISON SERIES IN MATHEMATICS

Elementary and Intermediate Algebra

fourth edition

Stefan Baratto
Clackamas Community College

Barry Bergman
Clackamas Community College

Don Hutchison
Clackamas Community College

The McGraw·Hill Companies

HUTCHISON'S ELEMENTARY AND INTERMEDIATE ALGEBRA, FOURTH EDITION

Published by McGraw-Hill, a business unit of The McGraw-Hill Companies, Inc., 1221 Avenue of the Americas, New York, NY 10020.

This book is printed on acid-free paper.

1 2 3 4 5 6 7 8 9 0 WDQ/WDQ 1 0 9 8 7 6 5 4 3 2 1 0

ISBN 978–0–07–338419–1
MHID 0–07–338419–4

ISBN 978–0–07–729216–4 (Annotated Instructor's Edition)
MHID 0–07–729216–2

Vice President, Editor-in-Chief: *Marty Lange*
Vice President, EDP: *Kimberly Meriwether David*
Director of Development: *Kristine Tibbetts*
Editorial Director: *Stewart K. Mattson*
Executive Editor: *David Millage*
Developmental Editor: *Adam Fischer*
Marketing Manager: *Victoria Anderson*
Senior Project Manager: *April R. Southwood*
Senior Production Supervisor: *Kara Kudronowicz*
Senior Media Project Manager: *Sandra M. Schnee*
Senior Designer: *David W. Hash*
Cover Designer: *John Joran*
(USE) Cover Image: *close-up of green and red gerbera flower, © Ron Evans/Red Cover/Getty Images, Inc.*
Senior Photo Research Coordinator: *Lori Hancock*
Supplement Producer: *Mary Jane Lampe*
Compositor: *MPS Limited*
Typeface: *10/12 New Times Roman*
Printer: *Worldcolor*

Library of Congress Cataloging-in-Publication Data

Baratto, Stefan.
Hutchison's elementary and intermediate algebra / Stefan Baratto, Barry Bergman.—4th ed.
p. cm.
Rev. ed. of: Elementary and intermediate algebra / Baratto, Bergaman, Hutchison. 3rd ed. 2008.
Includes index.
ISBN 978–0 07–338419–1—ISBN 0–07–338419–4 (hard copy : alk. paper) 1. Algebra–Textbooks. I. Bergman, Barry. II. Hutchison, Donald, 1948- Elementary and intermediate algebra. III. Title. IV. Title: Elementary and intermediate algebra.
QA152.3.H874 2011
512.9–dc22

2009034027

www.mhhe.com

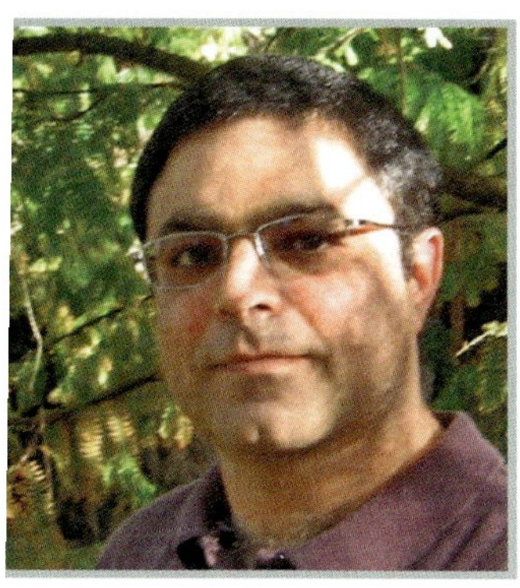

Stefan Baratto

Stefan began teaching math and science in New York City middle schools. He also taught math at the University of Oregon, Southeast Missouri State University, and York County Technical College. Currently, Stefan is a member of the mathematics faculty at Clackamas Community College where he has found a niche, delighting in the CCC faculty, staff, and students. Stefan's own education includes the University of Michigan (BGS, 1988), Brooklyn College (CUNY), and the University of Oregon (MS, 1996).

Stefan is currently serving on the AMATYC Executive Board as the organization's Northwest Vice President. He has also been involved with ORMATYC, NEMATYC, NCTM, and the State of Oregon Math Chairs group, as well as other local organizations. He has applied his knowledge of math to various fields, using statistics, technology, and web design. More personally, Stefan and his wife, Peggy, try to spend time enjoying the wonders of Oregon and the Pacific Northwest. Their activities include scuba diving, self-defense training, and hiking.

Barry Bergman

Barry has enjoyed teaching mathematics to a wide variety of students over the years. He began in the field of adult basic education and moved into the teaching of high school mathematics in 1977. He taught high school math for 11 years, at which point he served as a K-12 mathematics specialist for his county. This work allowed him the opportunity to help promote the emerging NCTM standards in his region.

In 1990, Barry began the next portion of his career, having been hired to teach at Clackamas Community College. He maintains a strong interest in the appropriate use of technology and visual models in the learning of mathematics.

Throughout the past 32 years, Barry has played an active role in professional organizations. As a member of OCTM, he contributed several articles and activities to the group's journal. He has presented at AMATYC, OCTM, NCTM, ORMATYC, and ICTCM conferences. Barry also served 4 years as an officer of ORMATYC and participated on an AMATYC committee to provide feedback to revisions of NCTM's standards.

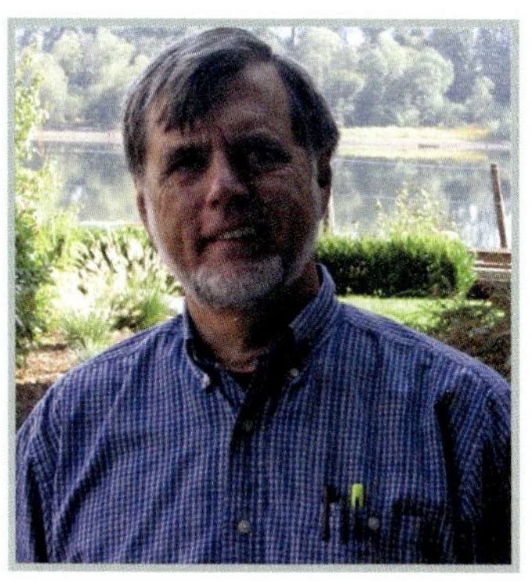

Don Hutchison

Don began teaching in a preschool while he was an undergraduate. He subsequently taught children with disabilities, adults with disabilities, high school mathematics, and college mathematics. Although each position offered different challenges, it was always breaking a challenging lesson into teachable components that he most enjoyed.

It was at Clackamas Community College that he found his professional niche. The community college allowed him to focus on teaching within a department that constantly challenged faculty and students to expect more. Under the guidance of Jim Streeter, Don learned to present his approach to teaching in the form of a textbook. Don has also been an active member of many professional organizations. He has been president of ORMATYC, AMATYC committee chair, and ACM curriculum committee member. He has presented at AMATYC, ORMATYC, AACC, MAA, ICTCM, and a variety of other conferences.

Above all, he encourages you to be involved, whether as a teacher or as a learner. Whether discussing curricula at a professional meeting or homework in a cafeteria, it is the process of communicating an idea that helps one to clarify it.

Dedication

We dedicate this text to the thousands of students who have helped us become better teachers, better communicators, better writers, and even better people. We read and respond to every suggestion we get—every one is invaluable. If you have any thoughts or suggestions, please contact us at

Stefan Baratto: sbaratto@clackamas.edu
Barry Bergman: bfbergman@gmail.com
Don Hutchison: donh@collegemathtext.com

Thank you all.

Message from the Authors

Dear Colleagues,

We believe the key to learning mathematics, at any level, is active participation. We have revised our textbook series to specifically emphasize GROWING MATH SKILLS through active learning. Students who are active participants in the learning process have a greater opportunity to construct their own mathematical ideas and make stronger connections to concepts covered in their course. This participation leads to better understanding, retention, success, and confidence.

In order to grow student math skills, we have integrated features throughout our textbook series that reflect our philosophy. Specifically, our *chapter-opening vignettes* and an array of section exercises relate to a singular topic or theme to engage students while identifying the relevance of mathematics.

Check Yourself exercises, which include optional calculator references, are designed to keep students actively engaged in the learning process. Our exercise sets include application problems as well as challenging and collaborative writing exercises to give students more opportunity to sharpen their skills.

Originally formatted as a work-text, this textbook allows students to make use of the margins where exercise answer space is available to further facilitate active learning. This makes the textbook more than just a reference. Many of these exercises are designed for insight to generate mathematical thought while reinforcing continual practice and mastery of topics being learned. Our hope is that students who use our textbook will grow their mathematical skills and become better mathematical thinkers as a result.

As we developed our series, we recognized that the use of technology should not be simply a supplement, but should be an essential element in learning mathematics. We understand that these "millennial students" are learning in different modes than just a few short years ago. Attending course lectures is not the only demand these students face—their daily schedules are pulling them in more directions than ever before. To meet the needs of these students, we have developed videos to better explain key mathematical concepts throughout the textbook. The goal of these videos is to provide students with a better framework—showing them how to solve a specific mathematical topic, regardless of their classroom environment (online or traditional lecture). The videos serve as refreshers or preparatory tools for classroom lecture, in several formats, including iPOD/MP3 format, to accommodate the different ways students access information.

Finally, with our series focus on growing math skills, we strongly believe that ALEKS® software can truly help students to remediate and grow their math skills given its adaptiveness. ALEKS is available to accompany our textbooks to help build proficiency. ALEKS has helped our own students to identify mathematical skills they have mastered and skills where remediation is required. Thank you for using our textbook. We look forward to learning of your success!

Stefan Baratto
Barry Bergman
Donald Hutchison

About the Cover

A flower symbolizes transformation and growth—a change from the ordinary to the spectacular. Similarly, students in an elementary and intermediate algebra course have the potential to grow their math skills to become stronger math students. Authors Stefan Baratto, Barry Bergman, and Don Hutchison help students *grow their mathematical skills*—guiding them through the stages to mathematical success!

"The Baratto/Bergman/Hutchison textbook gives the student a well-rounded foundation into many concepts of algebra, taking the student from prior knowledge, to guided practice, to independent practice, and then to assessment. Each chapter builds upon concepts learned in other chapters. Items such as Check Yourself exercises and Activities at the end of most chapters help the student to be more successful in many of the concepts taught."

– Karen Day, Elizabethtown Technical & Community College

With Baratto/Bergman/Hutchison!

Grow Your Mathematical Skills Through Better Conceptual Tools!

Stefan Baratto, Barry Bergman, and Don Hutchison know that students succeed once they have built a strong conceptual understanding of mathematics. ***"Make the Connection"*** chapter-opening vignettes help students to better understand mathematical concepts through everyday examples. Further reinforcing real-world mathematics, each vignette is accompanied by activities and exercises in the chapter to help students focus on the mathematical skills required for mastery.

- ***Make the Connection***
- ***Chapter-Opening Vignettes***
- ***Activities***
- ***Reading Your Text***
- ***Learning Objectives***
- ***Self-Tests***
- ***Cumulative Reviews***
- ***Group Activities***

Grow Your Mathematical Skills Through Better Exercises, Examples, and Applications!

A wealth of exercise sets is available for students at every level to actively involve them through the learning process in an effort to grow mathematical skills, including:

- ***Check Yourself Exercises***
- ***Application Exercises***
- ***End-of-Section Exercises***
- ***Summary Exercises***

Grow Your Mathematical Study Skills Through Better Active Learning Tools!

In an effort to meet the needs the "millennial student," we have made active-learning tools available to sharpen mathematical skills and build proficiency.

- ***ALEKS***
- ***MathZone***
- ***Conceptual Videos***
- ***Lecture Videos***

"This is a good book. The best feature, in my opinion is the readability of this text. It teaches through example and has students immediately check their own skills. This breaks up long text into small bits easier for students to digest."

– Robin Anderson, Southwestern Illinois College

Grow Your Mathematical Skills

"Make the Connection"*—*Chapter-Opening Vignettes provide interesting, relevant scenarios that will capture students' attention and engage them in the upcoming material. Exercises and *Activities* related to the *Opening Vignette* are available to utilize the theme most effectively for better mathematical comprehension (marked with an icon).

INTRODUCTION

We expect to use mathematics both in our careers and when making financial decisions. But, there are many more opportunities to use math, even when enjoying life's pleasures. For instance, we use math regularly when traveling.

When traveling to another country, you need to be able to convert currency, temperature, and distance. Even figuring out when to call home so that you do not wake up family and friends during the night is a computation.

The equation is a very old tool for solving problems and writing relationships clearly and accurately. In this chapter, you will learn to solve linear equations. You will also learn to write equations that accurately describe problem situations. Both of these skills will be demonstrated in many settings, including international travel.

From Arithmetic to Algebra

CHAPTER 1 OUTLINE

Source: Chapter 1

Activities are incorporated to promote *active learning* by requiring students to find, interpret, and manipulate real-world data. The activity in the chapter-opening vignette ties the chapter together by way of questions to sharpen student mathematical and conceptual understanding, highlighting the cohesiveness of the chapter. Students can complete the activities on their own, but they are best worked in small groups.

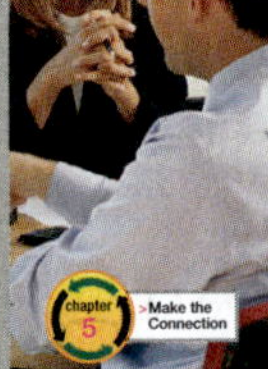

Activity 5 :: Wealth and Compound Interest

Suppose that when you were born, an uncle put \$500 in the bank for you. He never deposited money again, but the bank paid 5% interest on the money every year on your birthday. How much money was in the bank after 1 year? After 2 years? After 1 year, the amount is $\$500 + 500(0.05)$, which can be written as $\$500(1 + 0.05)$ because of the distributive property. Because $1 + 0.05 = 1.05$, the amount in the bank after 1 year was 500(1.05). After 2 years, this amount was again multiplied by 1.05. How much is in the bank today? Complete the chart.

Birthday	Computation	Amount
0 (Day of birth)		\$500
1	\$500(1.05)	
2	\$500(1.05)(1.05)	
3	\$500(1.05)(1.05)(1.05)	

Source: Chapter 5

NEW! Reading Your Text offers a brief set of exercises at the end of each section to assess students' knowledge of key vocabulary terms. These exercises are designed to encourage careful reading for greater conceptual understanding. *Reading Your Text* exercises address vocabulary issues, which students often struggle with in learning core mathematical concepts. Answers to these exercises are provided at the end of the book.

Reading Your Text

SECTION 2.5

(a) The vertical line test is a graphical test for identifying a ______________.

(b) A ______________ is a function if no vertical line passes through two or more points on its graph.

(c) The ______________ of a function is the set of inputs that can be substituted for the independent variable.

(d) The range of a function is the set of ______________ or y-values.

Source: Chapter 2 (Section 5)

Self-Tests appear in each chapter to provide students with an opportunity to check their progress and to review important concepts, as well as to provide confidence and guidance in preparing for exams. The answers to the *Self-Test* exercises are given at the end of the book.

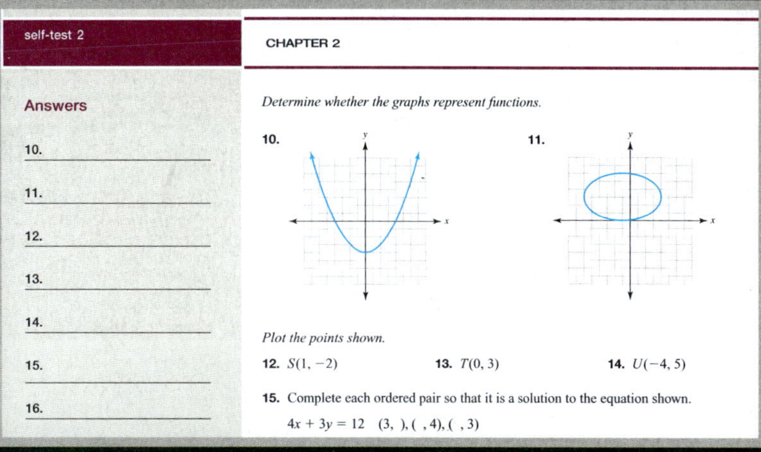

self-test 2

CHAPTER 2

Answers

10. ____
11. ____
12. ____
13. ____
14. ____
15. ____
16. ____

Determine whether the graphs represent functions.

10.

11.

Plot the points shown.

12. $S(1, -2)$ **13.** $T(0, 3)$ **14.** $U(-4, 5)$

15. Complete each ordered pair so that it is a solution to the equation shown.

$4x + 3y = 12$ $(3, \ )$, $(\ , 4)$, $(\ , 3)$

Source: Chapter 2

Cumulative Reviews are included, starting with Chapter 2. These reviews help students build on previously covered material and give them an opportunity to reinforce the skills necessary to prepare for midterm and final exams. These reviews assist students with the retention of knowledge throughout the course. The answers to these exercises are also given at the end of the book.

cumulative review chapters 0-4

Name ____

Section ____ Date ____

Answers

1. ____
2. ____
3. ____
4. ____
5. ____
6. ____

The following exercises are presented to help you review concepts from earlier chapters that you may have forgotten. This is meant as review material and not as a comprehensive exam. The answers are presented in the back of the text. If you have difficulty with any of these exercises, be certain to at least read through the summary related to that section.

Solve.

1. $3x - 2(x + 5) = 12 - 3x$

2. $2x - 7 < 3x - 5$

3. $x + 8 \geq 4x - 3$

4. $2x + 3(x - 2) = -4(x + 1) + 16$

Graph.

5. $5x + 7y = 35$

6. $2x + 3y < 6$

7. Solve the equation $P = P_0 + IRT$ for R.

8. Find the slope of the line connecting (4, 6) and (3, −1).

Source: Chapter 4

Group Activities offer practical exercises designed to grow student comprehension through group work. Group activities are great for instructors and adjuncts—bringing a more interactive approach to teaching mathematics.

Activity 2 :: Graphing with a Calculator

The graphing calculator is a tool that can be used to help you solve many different kinds of problems. This activity walks you through several features of the TI-83 or TI-84 Plus. By the time you complete this activity, you will be able to graph equations, change the viewing window to better accommodate a graph, or look at a table of values that represent some of the solutions for an equation. The first portion of this activity demonstrates how you can create the graph of an equation. The features described here can be found on most graphing calculators. See your calculator manual to learn how to get your particular calculator model to perform this activity.

Menus and Graphing

1. To graph the equation $y = 2x + 3$ on a graphing calculator, follow these steps.

a. Press the [Y =] key.

Grow Your Mathematical Skills with Better Worked Examples, Exercises, and Applications!

"Check Yourself" Exercises are a hallmark of the Hutchison series; they are designed to actively involve students in the learning process. Every example is followed by an exercise that encourages students to solve a problem similar to the one just presented and check, through practice, what they have just learned. Answers are provided at the end of the section for immediate feedback.

$$3 < x + 6$$
$$3 - 6 < x + 6 - 6$$
$$-3 < x$$

Subtract 6 from both sides.

The graph of the solution set is

Check Yourself 6

Solve and graph the solution set of the inequality

$4x - 5 < 5x - 9$

Some applications are solved by using an inequality instead of an equation. Example 7 illustrates such an application.

Source: Chapter 1 (Section 8)

End-of-Section Exercises enable students to evaluate their conceptual mastery through practice as they conclude each section. These comprehensive exercise sets are structured to highlight the progression in level, not only providing clarity for the student, but also making it easier for instructors to determine exercises for assignments. The *application exercises* that are now integrated into every section are a crucial component of this organization.

5.1 exercises

Boost *your* GRADE at ALEKS.com!

ALEKS

- Practice Problems
- Self-Tests
- NetTutor
- e-Professors
- Videos

Name ______

Section ______ Date ______

Answers

1. ______ 2. ______

Basic Skills | Challenge Yourself | Calculator/Computer | Career Applications | Above and Beyond

< Objective 1 >

Write each expression in simplest exponential form.

1. $x^4 \cdot x^5$ **2.** $x^7 \cdot x^9$

3. $x^5 \cdot x^3 \cdot x^2$ **4.** $x^8 \cdot x^4 \cdot x^7$

5. $3^5 \cdot 3^2$ **6.** $(-3)^4(-3)^6$

7. $(-2)^3(-2)^5$ **8.** $4^3 \cdot 4^4$

9. $4 \cdot x^2 \cdot x^4 \cdot x^7$ **10.** $3 \cdot x^3 \cdot x^5 \cdot x^8$

Source: Chapter 5 (Section 1)

Summary and Summary Exercises at the end of each chapter allow students to review important concepts. The *Summary Exercises* provide an opportunity for the student to practice these important concepts. The answers to odd-numbered exercises are provided in the answers appendix.

summary :: chapter 4

Definition/Procedure	Example	Reference
Graphing Systems of Linear Equations		**Section 4.1**
A **system of linear equations** is two or more linear equations considered together. A solution for a linear system in two variables is an ordered pair of real numbers (x, y) that satisfies both equations in the system. There are three solution te... addition method, and the subs...	The solution for the system $2x - y = 7$ ...	*p. 308*
Solving by the Graphing M... system on the same set of coo... will correspond to the point o... a system is called a **consisten...**		

summary exercises :: chapter 4

4.3 *Use the addition method to solve each system. If a unique solution does not exist, state whether the given system is inconsistent or dependent.*

15. $x + 2y = 7$, $x - y = 1$

16. $x + 3y = 14$, $4x + 3y = 29$

17. $3x - 5y = 5$, $-x + y = -1$

18. $x - 4y = 12$, $2x - 8y = 24$

19. $6x + 5y = -9$, $-5x + 4y = 32$

20. $3x + y = 8$, $-6x - 2y = -10$

21. $5x - y = -17$, $4x + 3y = -6$

22. $4x - 3y = 1$, $6x + 5y = 30$

23. $x - \frac{1}{2}y = 8$, $\frac{2}{?}x + \frac{3}{?}y = -2$

24. $\frac{1}{5}x - 2y = 4$, $\frac{3}{?}x + \frac{2}{?}y = -8$

Source: Chapter 4

Grow Your Mathematical Study Skills Through Better Active Learning Tools!

Tips for Student Success offers a resource to help students learn how to study, which is a problem many new students face, especially when taking their first exam in college mathematics. For this reason, Baratto/Bergman/Hutchison has incorporated *Tips for Student Success* boxes in the beginning of this textbook. The same suggestions made by great teachers in the classroom are now available to students outside of the classroom, offering extra direction to help improve understanding and further insight.

Tips for Student Success

Throughout this text, we present you with a series of class-tested techniques designed to improve your performance in this math class.

Become familiar with your syllabus

In the first class meeting, your instructor probably handed out a class syllabus. If you haven't done so already, you need to incorporate important information into your calendar and address book.

1. Write all important dates in your calendar. This includes homework due dates, quiz dates, test dates, and the date and time of the final exam. Never allow yourself to be surprised by a deadline!
2. Write your instructor's name, e-mail address, and office number in your address book. Also include the office hours. Make it a point to see your instructor early in the term. Although this is not the only person who can help clear up your confusion, your instructor is the most important person.
3. Make note of the other resources available to you. These include CDs, videotapes, Web pages, and tutoring.

Given all of these resources, it is important that you never let confusion or frustration mount. If you can't "get it" from the text, try another resource. All the resources are there specifically for you, so take advantage of them!

Source: Chapter 1 (Section 1)

Notes and Recalls accompany the step-by-step worked examples helping students **focus on information critical to their success.** ***Recall Notes*** give students a ***just-in-time*** reminder, reinforcing previously learned material through references.

NOTE

John Wallis (1616–1702), an English mathematician, was the first to fully discuss the meaning of 0, negative, and rational exponents. You will learn about rational exponents in Chapter 7.

RECALL

If two numbers have a product of 1, they must be reciprocals of each other.

Source: Chapter 5 (Section 2, page 495)

Cautions are integrated throughout the textbook to alert students to common mistakes and how to avoid them.

> CAUTION

This is different from

$(3c)^2 = [3 \cdot (4)]^2$
$= 12^2 = 144$

(a) $5a + 7b$

$5a + 7b =$

$=$

(b) $3c^2$

$3c^2 = 3 \cdot ($

$= 3 \cdot 1$

(c) $7(c + d)$

$7(c + d) =$

Source: Chapter 1 (Section 2, page 86)

grow your math skills with ALEKS®

Experience Student Success!

ALEKS® ALEKS is a unique online math tool that uses adaptive questioning and artificial intelligence to correctly place, prepare, and remediate students . . . all in one product! Institutional case studies have shown that **ALEKS has improved pass rates by over 20% versus traditional online homework, and by over 30% compared to using a text alone.**

By offering each student an individualized learning path, ALEKS directs students to work on the math topics that they are ready to learn. To help students keep pace in their course, instructors can correlate ALEKS to their textbook or syllabus in seconds.

To learn more about how ALEKS can be used to boost student performance, please visit www.aleks.com/highered/math or contact your McGraw-Hill representative.

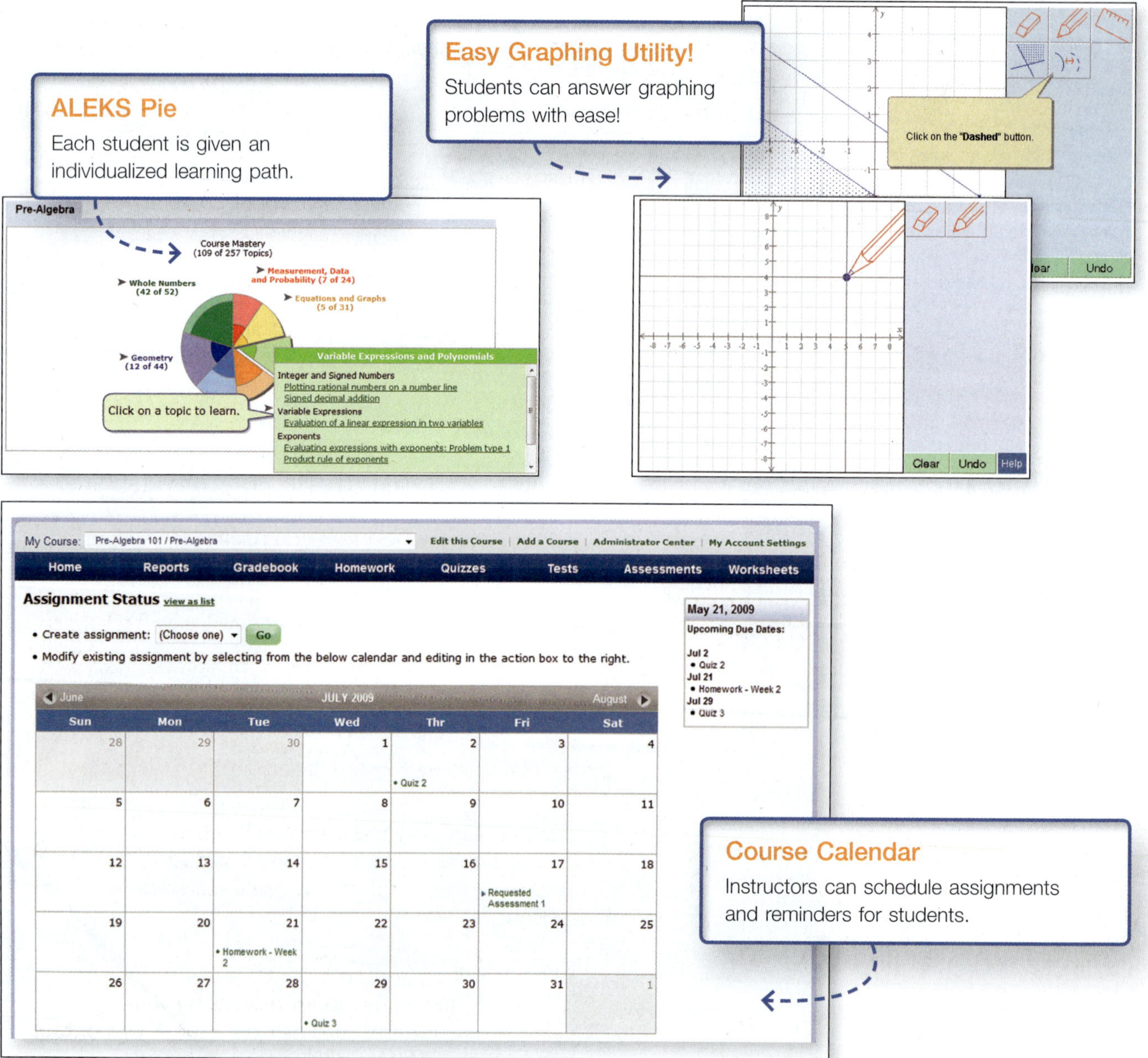

New ALEKS Instructor Module

Enhanced Functionality and Streamlined Interface Help to Save Instructor Time

ALEKS® The new ALEKS Instructor Module features enhanced functionality and a streamlined interface based on research with ALEKS instructors and homework management instructors. Paired with powerful assignment-driven features, textbook integration, and extensive content flexibility, the new ALEKS Instructor Module simplifies administrative tasks and makes ALEKS more powerful than ever.

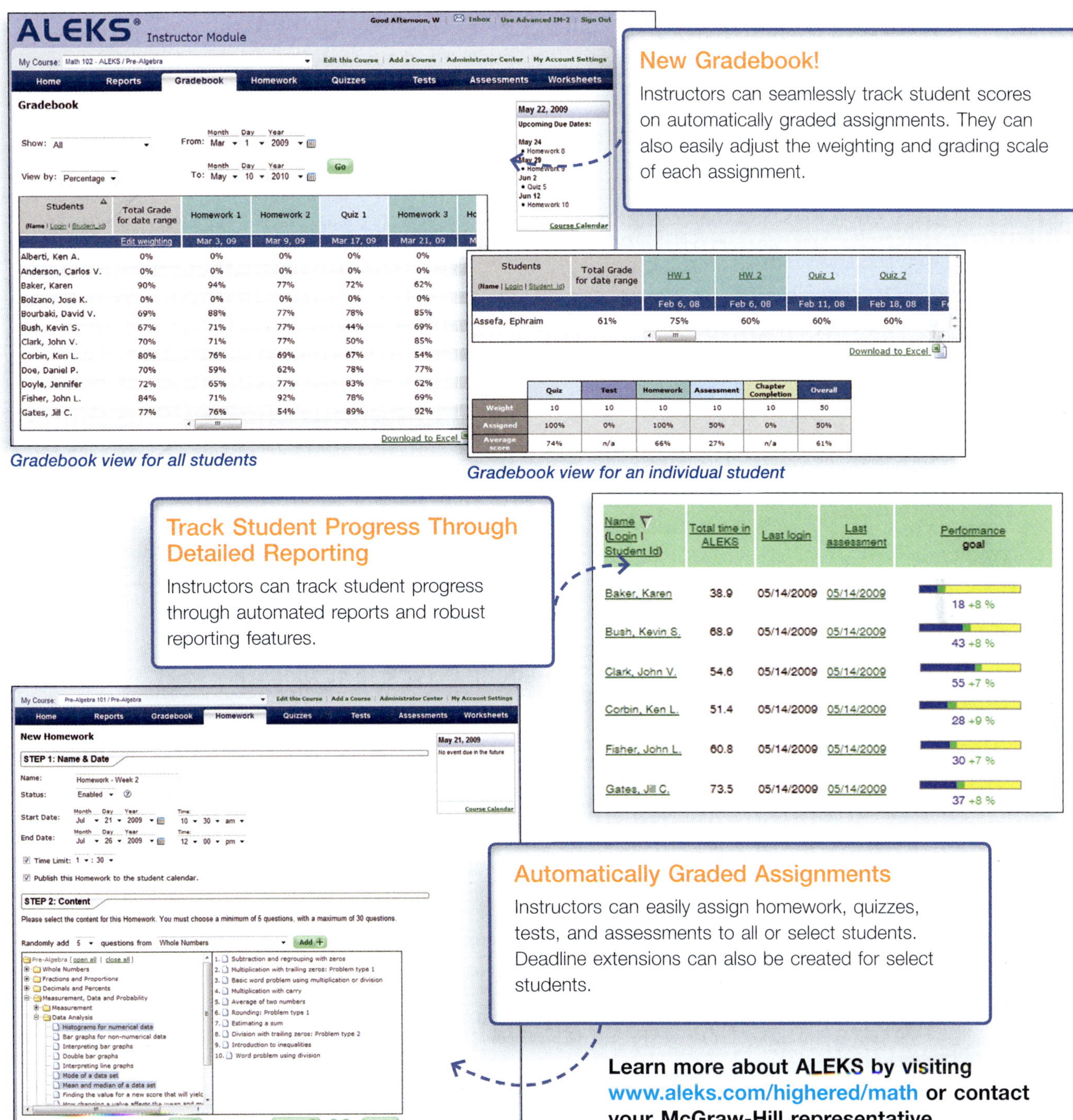

New Gradebook!

Instructors can seamlessly track student scores on automatically graded assignments. They can also easily adjust the weighting and grading scale of each assignment.

Gradebook view for all students

Gradebook view for an individual student

Track Student Progress Through Detailed Reporting

Instructors can track student progress through automated reports and robust reporting features.

Automatically Graded Assignments

Instructors can easily assign homework, quizzes, tests, and assessments to all or select students. Deadline extensions can also be created for select students.

Select topics for each assignment

Learn more about ALEKS by visiting www.aleks.com/highered/math or contact your McGraw-Hill representative.

360° Development Process

McGraw-Hill's 360° Development Process is an ongoing, never-ending, market-oriented approach to building accurate and innovative print and digital products. It is dedicated to continual large-scale and incremental improvement driven by multiple customer feedback loops and checkpoints. This is initiated during the early planning stages of our new products, and intensifies during the development and production stages, then begins again upon publication, in anticipation of the next edition.

A key principle in the development of any mathematics text is its ability to adapt to teaching specifications in a universal way. The only way to do so is by contacting those universal voices—and learning from their suggestions. We are confident that our book has the most current content the industry has to offer, thus pushing our desire for accuracy to the highest standard possible. In order to accomplish this, we have moved through an arduous road to production. Extensive and open-minded advice is critical in the production of a superior text.

Listening to you...

This textbook has been reviewed by over 300 teachers across the country. Our textbook is a commitment to your students, providing clear explanations, concise writing style, step-by-step learning tools, and the best exercises and applications in developmental mathematics. How do we know? You told us so!

Teachers *just like you* are saying great things about the Hutchison/Baratto/Bergman developmental mathematics series.

Acknowledgments and Reviewers

The development of this textbook series would never have been possible without the creative ideas and feedback offered by many reviewers. We are especially thankful to the following instructors for their careful review of the manuscript.

Symposia

Every year McGraw-Hill conducts a general mathematics symposium, which is attended by instructors from across the country. These events are an opportunity for editors from McGraw-Hill to gather information about the needs and challenges of instructors teaching these courses. This information helped to create the book plan for *Basic Mathematical Skills*. They also offer a forum for the attendees to exchange ideas and experiences with colleagues they might have not otherwise met.

Napa Valley Symposium

Antonio Alfonso, *Miami Dade College*

Lynn Beckett-Lemus, *El Camino College*

Kristin Chatas, *Washtenaw Community College*

Maria DeLucia, *Middlesex College*

Nancy Forrest, *Grand Rapids Community College*

Michael Gibson, *John Tyler Community College*

Linda Horner, *Columbia State College*

Matthew Hudock, *St. Phillips College*

Judith Langer, *Westchester Community College*

Kathryn Lavelle, *Westchester Community College*

Scott McDaniel, *Middle Tennessee State University*

Adelaida Quesada, *Miami Dade College*

Susan Schulman, *Middlesex College*

Stephen Toner, *Victor Valley College*

Chariklia Vassiliadis, *Middlesex County College*

Melanie Walker, *Bergen Community College*

Myrtle Beach Symposium

Patty Bonesteel, *Wayne State University*

Zhixiong Chen, *New Jersey City University*

Latonya Ellis, *Bishop State Community College*

Bonnie Filer-Tubaugh, *University of Akron*

Catherine Gong, *Citrus College*

Marcia Lambert, *Pitt Community College*

Katrina Nichols, *Delta College*

Karen Stein, *University of Akron*

Walter Wang, *Baruch College*

La Jolla Symposium

Darryl Allen, *Solano Community College*
Yvonne Aucoin, *Tidewater Community College*
Sylvia Carr, *Missouri State University*
Elizabeth Chu, *Suffolk County Community College*
Susanna Crawford, *Solano Community College*
Carolyn Facer, *Fullerton College*
Terran Felter, *Cal State Long Bakersfield*
Elaine Fitt, *Bucks County Community College*
John Jerome, *Suffolk County Community College*
Sandra Jovicic, *Akron University*
Carolyn Robinson, *Mt. San Antonio College*
Carolyn Shand-Hawkins, *Missouri State*

Manuscript Review Panels

Over 150 teachers and academics from across the country reviewed the various drafts of the manuscript to give feedback on content, design, pedagogy, and organization. This feedback was summarized by the book team and used to guide the direction of the text.

Reviewers of the Hutchison/Baratto/Bergman Developmental Mathematics Series

Board of Advisors

Timothy Brown, *South Georgia College*
Tony Craig, *Paradise Valley Community College*
Bruce Simmons, *Clackamas Community College*
Peter Williams, *California State University—San Bernardino*

Reviewers

Robin Anderson, *Southwestern Illinois College*
Nieves Angulo, *Hostos Community College*
Arlene Atchison, *South Seattle Community College*
Haimd Attarzadeh, *Kentucky Jefferson Community and Technical College*
Jody Balzer, *Milwaukee Area Technical College*
Rebecca Baranowski, *Estrella Mountain Community College*
Wayne Barber, *Chemeketa Community College*
Bob Barmack, *Baruch College*
Chris Bendixen, *Lake Michigan College*
Karen Blount, *Hood College*
Dr. Donna Boccio, *Queensborough Community College*
Dr. Steve Boettcher, *Estrella Mountain Community College*
Karen Bond, *Pearl River Community College—Poplarville*
Laurie Braga Jordan, *Loyola University-Chicago*
Kelly Brooks, *Pierce College*
Michael Brozinsky, *Queensborough Community College*
Amy Canavan, *Century Community and Technical College*
Faye Childress, *Central Piedmont Community College*
Kathleen Ciszewski, *University of Akron*
Bill Clarke, *Pikes Peak Community College*
Lois Colpo, *Harrisburg Area Community College*
Christine Copple, *Northwest State Community College*
Jonathan Cornick, *Queensborough Community College*
Julane Crabtree, *Johnson County Community College*
Carol Curtis, *Fresno City College*
Sima Dabir, *Western Iowa Tech Community College*
Reza Dai, *Oakton Community College*
Karen Day, *Elizabethtown Technical and Community College*
Mary Deas, *Johnson County Community College*
Anthony DePass, *St. Petersburg College-Ns*
Shreyas Desai, *Atlanta Metropolitan College*
Robert Diaz, *Fullerton College*
Michaelle Downey, *Ivy Tech Community College*
Ginger Eaves, *Bossier Parish Community College*
Azzam El Shihabi, *Long Beach City College*
Kristy Erickson, *Cecil College*
Steven Fairgrieve, *Allegany College of Maryland*
Jacqui Fields, *Wake Technical Community College*
Bonnie Filler-Tubaugh, *University of Akron*
Rhoderick Fleming, *Wake Tech Community College*
Matt Foss, *North Hennepin Community College*
Catherine Frank, *Polk Community College*
Matt Gardner, *North Hennepin Community College*
Judy Godwin, *Collin County Community College-Plano*
Lori Grady, *University of Wisconsin-Whitewater*
Brad Griffith, *Colby Community College*
Robert Grondahl, *Johnson County Community College*
Shelly Hansen, *Mesa State College*
Kristen Hathcock, *Barton County Community College*
Mary Beth Headlee, *Manatee Community College*
Kristy Hill, *Hinds Community College*
Mark Hills, *Johnson County Community College*
Sherrie Holland, *Piedmont Technical College*
Diane Hollister, *Reading Area Community College*
Denise Hum, *Canada College*

Byron D. Hunter, *College of Lake County*
Nancy Johnson, *Manatee Community College-Bradenton*
Joe Jordan, *John Tyler Community College-Chester*
Sandra Ketcham, *Berkshire Community College*
Lynette King, *Gadsden State Community College*
Jeff Koleno, *Lorain County Community College*
Donna Krichiver, *Johnson County Community College*
Indra B. Kshattry, *Colorado Northwestern Community College*
Patricia Labonne, *Cumberland County College*
Ted Lai, *Hudson County Community College*
Pat Lazzarino, *Northern Virginia Community College*
Richard Leedy, *Polk Community College*
Jeanine Lewis, *Aims Community College-Main Campus*
Michelle Christina Mages, *Johnson County Community College*
Igor Marder, *Antelope Valley College*
Donna Martin, *Florida Community College-North Campus*
Amina Mathias, *Cecil College*
Jean McArthur, *Joliet Junior College*
Carlea (Carol) McAvoy, *South Puget Sound Community College*
Tim McBride, *Spartanburg Community College*
Sonya McQueen, *Hinds Community College*
Maria Luisa Mendez, *Laredo Community College*
Madhu Motha, *Butler County Community College*
Shauna Mullins, *Murray State University*
Julie Muniz, *Southwestern Illinois College*
Kathy Nabours, *Riverside Community College*
Michael Neill, *Carl Sandburg College*
Nicole Newman, *Kalamazoo Valley Community College*
Said Ngobi, *Victor Valley College*
Denise Nunley, *Glendale Community College*
Deanna Oles, *Stark State College of Technology*
Staci Osborn, *Cuyahoga Community College-Eastern Campus*
Linda Padilla, *Joliet Junior College*
Karen D. Pain, *Palm Beach Community College*
George Pate, *Robeson Community College*
Margaret Payerle, *Cleveland State University-Ohio*
Jim Pierce, *Lincoln Land Community College*
Tian Ren, *Queensborough Community College*
Nancy Ressler, *Oakton Community College*
Bob Rhea, *J. Sargeant Reynolds Community College*
Minnie M. Riley, *Hinds Community College*
Melissa Rossi, *Southwestern Illinois College*
Anna Roth, *Gloucester County College*
Alan Saleski, *Loyola University-Chicago*
Lisa Sheppard, *Lorain County Community College*
Mark A. Shore, *Allegany College of Maryland*
Mark Sigfrids, *Kalamazoo Valley Community College*
Amber Smith, *Johnson County Community College*
Leonora Smook, *Suffolk County Community College-Brentwood*
Renee Starr, *Arcadia University*
Jennifer Strehler, *Oakton Community College*
Renee Sundrud, *Harrisburg Area Community College*
Harriet Thompson, *Albany State University*
John Thoo, *Yuba College*
Fred Toxopeus, *Kalamazoo Valley Community College*
Sara Van Asten, *North Hennepin Community College*
Felix Van Leeuwen, *Johnson County Community College*
Josefino Villanueva, *Florida Memorial University*
Howard Wachtel, *Community College of Philadelphia*
Dottie Walton, *Cuyahoga Community College Eastern Campus*
Walter Wang, *Baruch College*
Brock Wenciker, *Johnson County Community College*
Kevin Wheeler, *Three Rivers Community College*
Latrica Williams, *St. Petersburg College*
Paul Wozniak, *El Camino College*
Christopher Yarrish, *Harrisburg Area Community College*
Steve Zuro, *Joliet Junior College*

Finally, we are forever grateful to the many people behind the scenes at McGraw-Hill without whom we would still be on page 1. Most important, we give special thanks to all the students and instructors who will *grow* their *Math Skills!*

Supplements for the Student

McGraw-Hill's **MathZone** is a powerful Web-based tutorial for homework, quizzing, testing, and multimedia instruction. Also available in CD-ROM format, MathZone offers:

- **Practice exercises** based on the text and generated in an unlimited quantity for as much practice as needed to master any objective
- **Video** clips of classroom instructors showing how to solve exercises from the text, step by step
- **e-Professor** animations that take the student through step-by-step instructions, delivered on-screen and narrated by a teacher on audio, for solving exercises from the textbook; the user controls the pace of the explanations and can review as needed
- **NetTutor** offers personalized instruction by live tutors familiar with the textbook's objectives and problem-solving methods

Every assignment, exercise, video lecture, and e-Professor is derived from the textbook.

ALEKS Prep for Developmental Mathematics

ALEKS Prep for Beginning Algebra and Prep for Intermediate Algebra focus on prerequisite and introductory material for Beginning Algebra and Intermediate Algebra. These prep products can be used during the first 3 weeks of a course to prepare students for future success in the course and to increase retention and pass rates. Backed by two decades of National Science Foundation funded research, ALEKS interacts with students much like a human tutor, with the ability to precisely assess a student's preparedness and provide instruction on the topics the student is most likely to learn.

ALEKS Prep Course Products Feature:

- Artificial Intelligence Targets Gaps in Individual Students Knowledge
- Assessment and Learning Directed Toward Individual Students Needs
- Open Response Environment with Realistic Input Tools
- Unlimited Online Access-PC & Mac Compatible

Free trial at www.aleks.com/free_trial/instructor

Student's Solutions Manual

The *Student's Solutions Manual* provides comprehensive, worked-out solutions to the odd-numbered exercises in the Section Exercises, Summary Exercises, Self-Tests and the Cumulative Reviews. The steps shown in the solutions match the style of solved examples in the textbook.

grow your math skills

New Connect2Developmental Mathematics Video Series!

Available on DVD and the MathZone website, these innovative videos bring essential Developmental Mathematics concepts to life! The videos take the concepts and place them in a real world setting so that students make the connection from what they learn in the classroom to experiences outside the classroom. Making use of 3-D animations and lectures, Connect2Developmental Mathematics video series answers the age-old questions "Why is this important?" and "When will I ever use it?" The videos cover topics from Arithmetic and Basic Mathematics through the Algebra sequence, mixing student-oriented themes and settings with basic theory.

Video Lectures on Digital Video Disk

The video series is based on exercises from the textbook. Each presenter works through selected problems, following the solution methodology employed in the text. The video series is available on DVD or online as part of MathZone. The DVDs are closed-captioned for the hearing impaired, are subtitled in Spanish, and meet the Americans with Disabilities Act Standards for Accessible Design.

NetTutor

Available through MathZone, NetTutor is a revolutionary system that enables students to interact with a live tutor over the web. NetTutor's Web-based, graphical chat capabilities enable students and tutors to use mathematical notation and even to draw graphs as they work through a problem together. Students can also submit questions and receive answers, browse previously answered questions, and view previous sessions. Tutors are familiar with the textbook's objectives and problem-solving styles.

Supplements for the Instructor

McGraw-Hill's **MathZone** is a complete online tutorial and course management system for mathematics and statistics, designed for greater ease of use than any other management system. Available with selected McGraw-Hill textbooks, the system enables instructors to **create and share courses and assignments** with colleagues and adjuncts with only a few clicks of the mouse. All assignments, questions, e-Professors, online tutoring, and video lectures are directly tied to **text-specific** materials.

MathZone courses are customized to your textbook, but you can edit questions and algorithms, import your own content, and **create** announcements and due dates for assignments.

MathZone has **automatic grading** and reporting of easy-to-assign, algorithmically generated homework, quizzing, and testing. All student activity within **MathZone** is automatically recorded and available to you through a **fully integrated gradebook** that can be downloaded to Excel.

MathZone offers:

- **Practice exercises** based on the textbook and generated in an unlimited number for as much practice as needed to master any topic you study.
- **Videos** of classroom instructors giving lectures and showing you how to solve exercises from the textbook.
- **e-Professors** to take you through animated, step-by-step instructions (delivered via on-screen text and synchronized audio) for solving problems in the book, allowing you to digest each step at your own pace.
- **NetTutor,** which offers live, personalized tutoring via the Internet.

Instructor's Testing and Resource Online

Provides a wealth of resources for the instructor. Among the supplements is a **computerized test bank** utilizing Brownstone Diploma® algorithm-based testing software to create customized exams quickly. This user-friendly program enables instructors to search for questions by topic, format, or difficulty level; to edit existing questions or to add new ones; and to scramble questions and answer keys for multiple versions of a single test. Hundreds of text-specific, open-ended, and multiple-choice questions are included in the question bank. Sample chapter tests are also provided. CD available upon request.

Grow Your Knowledge with MathZone Reporting

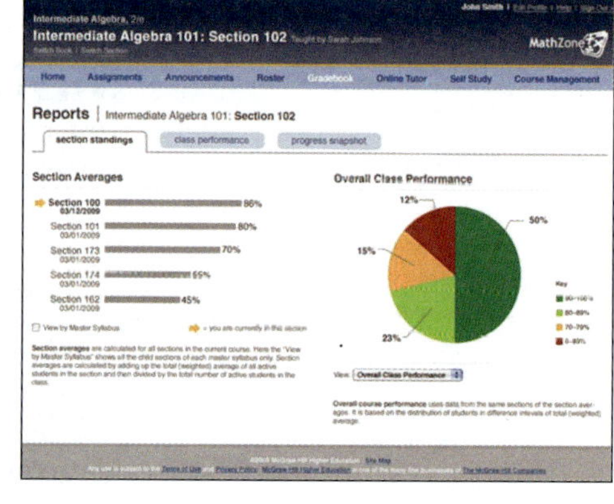

Visual Reporting

The new dashboard-like reports will provide the progress snapshot instructors are looking for to help them make informed decisions about their students.

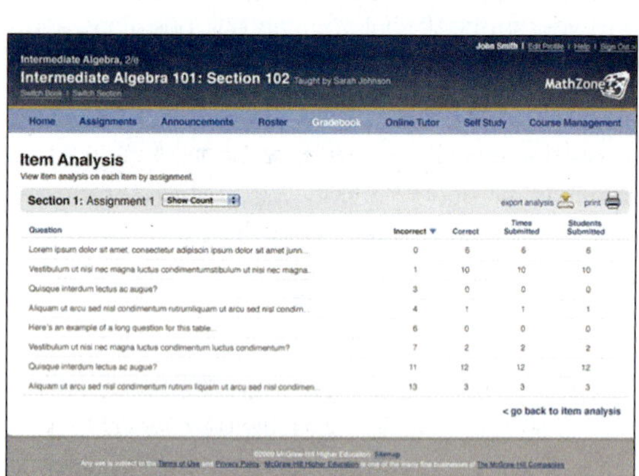

Item Analysis

Instructors can view detailed statistics on student performance at a learning objective level to understand what students have mastered and where they need additional help.

Managing Assignments for Individual Students

Instructors have greater control over creating individualized assignment parameters for individual students, special populations and groups of students, and for managing specific or ad hoc course events.

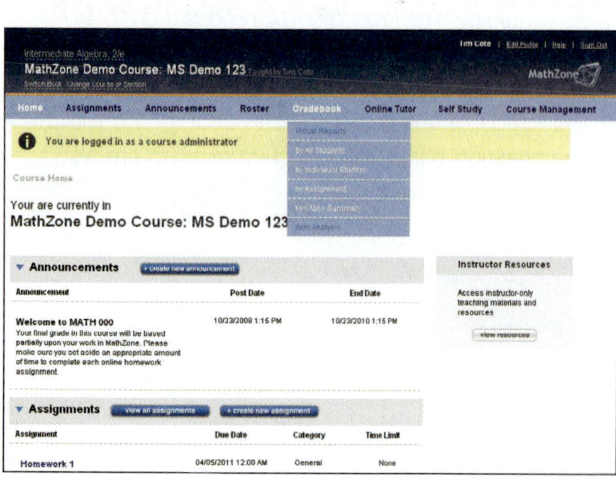

New User Interface

Designed by You! Instructors and students will experience a modern, more intuitive layout. Items used most commonly are easily accessible through the menu bar such as assignments, visual reports, and course management options.

grow your math skills

New ALEKS Instructor Module

The new ALEKS Instructor Module features enhanced functionality and a streamlined interface based on research with ALEKS instructors and homework management instructors. Paired with powerful assignment-driven features, textbook integration, and extensive content flexibility, the new ALEKS Instructor Module simplifies administrative tasks and makes ALEKS more powerful than ever. Features include:

Gradebook Instructors can seamlessly track student scores on automatically graded assignments. They can also easily adjust the weighting and grading scale of each assignment.

Course Calendar Instructors can schedule assignments and reminders for students.

Automatically Graded Assignments Instructors can easily assign homework, quizzes, tests, and assessments to all or select students. Deadline extensions can also be created for select students.

Set-Up Wizards Instructors can use wizards to easily set up assignments, course content, textbook integration, etc.

Message Center Instructors can use the redesigned Message Center to send, receive, and archive messages; input tools are available to convey mathematical expressions via email.

Baratto/Bergman/Hutchison Video Lectures on Digital Video Disk (DVD)

In the videos, qualified instructors work through selected problems from the textbook, following the solution methodology employed in the text. The video series is available on DVD or online as an assignable element of MathZone. The DVDs are closed-captioned for the hearing-impaired, are subtitled in Spanish, and meet the Americans with Disabilities Act Standards for Accessible Design. Instructors may use them as resources in a learning center, for online courses, and to provide extra help for students who require extra practice.

Annotated Instructor's Edition

In the *Annotated Instructor's Edition (AIE),* **answers to exercises and tests appear adjacent to each exercise set,** in a color used *only* for annotations.

Instructor's Solutions Manual

The *Instructor's Solutions Manual* provides comprehensive, worked-out solutions to all exercises in the Section Exercises, Summary Exercises, Self-Tests, and the Cumulative Reviews. The methods used to solve the problems in the manual are the same as those used to solve the examples in the textbook.

A commitment to accuracy

You have a right to expect an accurate textbook, and McGraw-Hill invests considerable time and effort to make sure that we deliver one. Listed below are the many steps we take to make sure this happens.

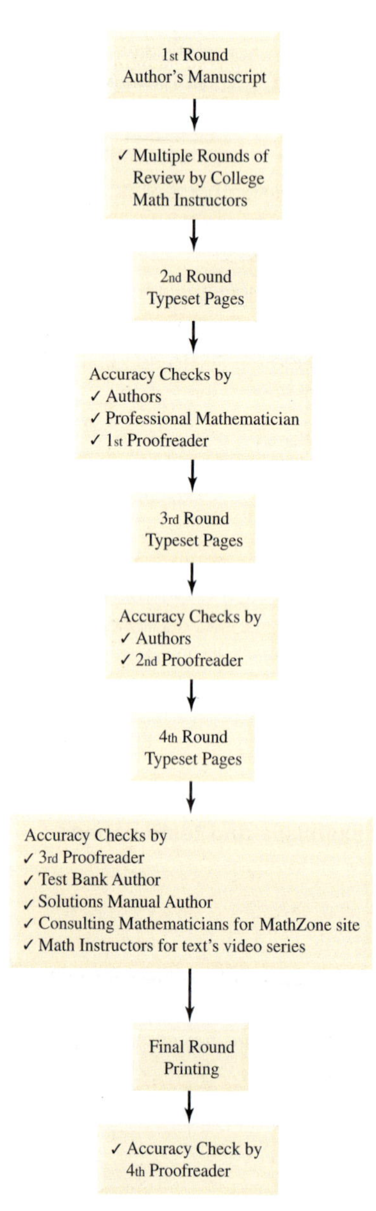

Our accuracy verification process

First Round

Step 1: Numerous **college math instructors** review the manuscript and report on any errors that they may find. Then the authors make these corrections in their final manuscript.

Second Round

Step 2: Once the manuscript has been typeset, the **authors** check their manuscript against the first page proofs to ensure that all illustrations, graphs, examples, exercises, solutions, and answers have been correctly laid out on the pages, and that all notation is correctly used.

Step 3: An outside, **professional, mathematician** works through every example and exercise in the page proofs to verify the accuracy of the answers.

Step 4: A **proofreader** adds a triple layer of accuracy assurance in the first pages by hunting for errors, then a second, corrected round of page proofs is produced.

Third Round

Step 5: The **author team** reviews the second round of page proofs for two reasons: (1) to make certain that any previous corrections were properly made, and (2) to look for any errors they might have missed on the first round.

Step 6: A **second proofreader** is added to the project to examine the new round of page proofs to double check the author team's work and to lend a fresh, critical eye to the book before the third round of paging.

Fourth Round

Step 7: A **third proofreader** inspects the third round of page proofs to verify that all previous corrections have been properly made and that there are no new or remaining errors.

Step 8: Meanwhile, in partnership with **independent mathematicians,** the text accuracy is verified from a variety of fresh perspectives:

- The **test bank author** checks for consistency and accuracy as he/she prepares the computerized test item file.
- The **solutions manual author** works every exercise and verifies his/her answers, reporting any errors to the publisher.
- A **consulting group of mathematicians,** who write material for the text's MathZone site, notifies the publisher of any errors they encounter in the page proofs.
- A video production company employing **expert math instructors** for the text's videos will alert the publisher of any errors they might find in the page proofs.

Final Round

Step 9: The **project manager,** who has overseen the book from the beginning, performs a **fourth proofread** of the textbook during the printing process, providing a final accuracy review.

⇒ What results is a mathematics textbook that is as accurate and error-free as is humanly possible. Our authors and publishing staff are confident that our many layers of quality assurance have produced textbooks that are the leaders in the industry for their integrity and correctness.

brief contents

contents

Chapter 8

Quadratic Functions

Chapter 9

Rational Expressions

Chapter 10

Exponential and Logarithmic Functions

Appendices

applications index

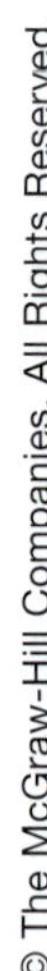

CHAPTER

Prealgebra Review

INTRODUCTION

Anthropologists and archeologists investigate modern human cultures and societies as well as cultures that existed so long ago that their characteristics must be inferred from objects found buried in lost cities or villages. With methods such as carbon dating, it has been established that large, organized cultures existed around 3000 B.C.E. in Egypt, 2800 B.C.E. in India, no later than 1500 B.C.E. in China, and around 1000 B.C.E. in the Americas.

Which is older, an object from 3000 B.C.E. or an object from 500 A.D.? An object from 500 A.D. is about 2,000 − 500 years old, or about 1,500 years old. But an object from 3000 B.C.E. is about 2,000 + 3,000 years old, or about 5,000 years old. Why subtract in the first case but add in the other? Because of the way years are counted before the common era (B.C.E.) and after the birth of Christ (A.D.), the B.C.E. dates must be considered as *negative* numbers.

Very early on, the Chinese accepted the idea that a number could be negative; they used red calculating rods for positive numbers and black for negative numbers. Hindu mathematicians in India worked out the arithmetic of negative numbers as long ago as 400 A.D., but western mathematicians did not recognize this idea until the sixteenth century. It would be difficult today to think of measuring things such as temperature, altitude, or money without using negative numbers.

CHAPTER 0 OUTLINE

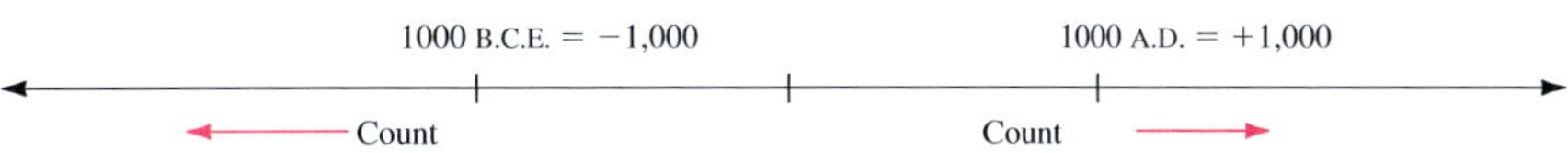

0.1 A Review of Fractions

< 0.1 Objectives >

1 > Simplify a fraction

2 > Multiply and divide fractions

3 > Add and subtract fractions

Tips for Student Success

Throughout this text, we present you with a series of class-tested techniques designed to improve your performance in this math class.

Become familiar with your textbook

Perform each task.

1. Use the Table of Contents to find the title of Section 5.1.
2. Use the Index to find the earliest reference to the term *factor.*
3. Find the answer to the first Check Yourself exercise in Section 0.1.
4. Find the answers to the Self-Test for Chapter 1.
5. Find the answers to the odd-numbered exercises in Section 0.1.

Now you know where some of the most important features of the text are. When you feel confused, think about using one of these features to help clear up your confusion.

We begin with certain assumptions about your previous mathematical learning. We assume you are reasonably comfortable using the basic operations of addition, subtraction, multiplication, and division with whole numbers.

We also assume you are familiar with and able to perform these operations on the most common type of fractions, *decimal fractions* or *decimals*.

Finally, we assume that you have worked with fractions and negative numbers in the past.

In this chapter, we review the basic operations and applications involving fractions and signed numbers. This is meant to be a brief review of these topics. If you need a more in-depth discussion of this content or any of the content discussed above, you should consider a course covering prealgebra material or a review of the text *Basic Mathematical Skills with Geometry* by Baratto, Bergman, and Hutchison in this same series.

The numbers used for counting are called the **natural numbers.** We write them as 1, 2, 3, 4, ... The three dots indicate that the pattern continues in the same way.

If we include zero in this group of numbers, we call them the **whole numbers.**

The **rational numbers** include all the whole numbers and all fractions, whether they are proper fractions such as $\frac{1}{2}$ and $\frac{2}{3}$ or improper fractions such as $\frac{7}{2}$ and $\frac{19}{5}$.

Every rational number can be written in fraction form $\frac{a}{b}$.

Interpreting fractions as a division statement allows you to avoid some common careless errors. Simply recall that the fraction bar represents division.

$$\frac{5}{8} = 5 \div 8$$

You can use this fact to understand some fraction basics.

$\frac{1}{6}$ is one-sixth of a whole, whereas

$\frac{6}{1}$ represents six "wholes" because this is $6 \div 1 = 6$.

Similarly, division by 0 is not defined, but you can have no parts of a whole.

$\frac{0}{3}$ means you have no thirds: $\frac{0}{3} = 0$.

On the other hand,

$\frac{3}{0} = 3 \div 0$ which does not exist. This expression has no meaning for us.

NOTE

$\frac{0}{3}$ means a whole is divided into three parts and you have none of them.

$\frac{3}{0}$ represents division by 0, which does not exist.

The number 1 has many different fraction forms. Any fraction in which the numerator and denominator are the same (and not zero) is another name for the number 1.

$$1 = \frac{2}{2} \qquad 1 = \frac{12}{12} \qquad 1 = \frac{257}{257}$$

To determine whether two fractions are equal or to find equivalent fractions, we use the **Fundamental Principle of Fractions.** The Fundamental Principle of Fractions states that multiplying the numerator and denominator of a fraction by the same number is the same as multiplying the fraction by 1. We express the principle in symbols here.

Property

The Fundamental Principle of Fractions

$$\frac{a}{b} = \frac{a \times c}{b \times c} \quad \text{or} \quad \frac{a \times c}{b \times c} = \frac{a}{b} \qquad c \neq 0$$

Example 1 Rewriting Fractions

Use the fundamental principle to write three fractional representations for each number.

(a) $\frac{2}{3}$

Multiplying the numerator and denominator by the same number is the same as multiplying by 1.

$\frac{2}{3} = \frac{2 \times 2}{3 \times 2} = \frac{4}{6}$ Multiply the numerator and denominator by 2.

$\frac{2}{3} = \frac{2 \times 3}{3 \times 3} = \frac{6}{9}$ Multiply the numerator and denominator by 3.

$\frac{2}{3} = \frac{2 \times 10}{3 \times 10} = \frac{20}{30}$

NOTE

Each representation is a numeral, or name, for the number. Each number has many names.

(b) 5

$5 = \frac{5 \times 2}{1 \times 2} = \frac{10}{2}$

$5 = \frac{5 \times 3}{1 \times 3} = \frac{15}{3}$

$5 = \frac{5 \times 100}{1 \times 100} = \frac{500}{100}$

Check Yourself 1

Use the fundamental principle to write three fractional representations for each number.

(a) $\frac{5}{8}$ **(b)** $\frac{4}{3}$ **(c)** 3

The fundamental principle can also be used to find the simplest fractional representation for a number. Fractions written in this form are said to be **simplified.**

Example 2 Simplifying Fractions

< Objective 1 >

RECALL

A prime number is any whole number greater than 1 that has only itself and 1 as factors.

Use the fundamental principle to simplify each fraction.

(a) $\frac{22}{55}$ **(b)** $\frac{35}{45}$ **(c)** $\frac{24}{36}$

In each case, we first write the numerator and denominator as a product of prime numbers.

(a) $\frac{22}{55} = \frac{2 \times 11}{5 \times 11}$

We then use the Fundamental Principle of Fractions to "remove" the common factor of 11.

$\frac{22}{55} = \frac{2 \times 11}{5 \times 11} = \frac{2}{5}$

(b) $\frac{35}{45} = \frac{5 \times 7}{3 \times 3 \times 5}$

Removing the common factor of 5 yields

$\frac{35}{45} = \frac{7}{3 \times 3} = \frac{7}{9}$

(c) $\frac{24}{36} = \frac{2 \times 2 \times 2 \times 3}{2 \times 2 \times 3 \times 3}$

Removing the common factor $2 \times 2 \times 3$ yields

$\frac{2}{3}$

NOTE

Often, we use the convention of "canceling" a factor that appears in both the numerator and denominator to prevent careless errors. In part (b),

$$\frac{5 \times 7}{3 \times 3 \times 5} = \frac{\cancel{5} \times 7}{3 \times 3 \times \cancel{5}} = \frac{7}{3 \times 3} = \frac{7}{9}$$

Check Yourself 2

Use the fundamental principle to simplify each fraction.

(a) $\frac{21}{33}$ (b) $\frac{15}{30}$ (c) $\frac{12}{54}$

NOTE

With practice, you will be able to simplify fractions mentally.

Fractions are often used in everyday situations. When solving an *application,* read the problem through carefully. Read the problem again and decide what you need to find and what you need to do. Then write out the problem completely and carefully. After completing the math work, be sure to answer the problem with a sentence.

Throughout this text, we use variations of this five-step process when working with applications. We will update this procedure after we introduce you to *algebra.*

Step by Step

Solving Applications

Step 1 Read the problem carefully to determine what you are being asked to find and what information is given in the application.

Step 2 Decide what you will do to solve the problem.

Step 3 Write down the complete (mathematical) statement necessary to solve the problem.

Step 4 Perform any calculations or other mathematics needed to solve the problem.

Step 5 Answer the question. Be sure to include units with your answer, when appropriate. Check to make certain that your answer is reasonable.

Example 3 Using Fractions in an Application

Jo, an executive vice president of information technology, already supervises 10 people and hires 2 more to fill out her staff. What fraction of her staff is new? Be sure to simplify your answer.

Step 1 We are being asked to find the fraction of Jo's staff that is new. We know that her staff consisted of 10 people and 2 new people were hired.

Step 2 First, we will figure out the size of her total staff. Then, we will figure out the fraction comparing the new people to the total staff.

Step 3 Total staff: 10 original people and 2 new people
$10 + 2$

We construct the ratio,

$$\frac{\text{New people}}{\text{Total staff}} = \frac{2}{10 + 2}$$

RECALL

We **cannot** simplify or "cancel" the twos in the sum.

$$\frac{2}{10 + 2} \neq \frac{\cancel{2}}{10 + \cancel{2}}$$

This is **incorrect.**

Step 4 $$\frac{2}{10 + 2} = \frac{2}{12} = \frac{1}{6}$$

Step 5 One-sixth of her staff is new.
This answer seems reasonable.

Check Yourself 3

There are 36 packaging machines in one division of Early Enterprises. At any given time, 4 of these machines are shut down for scheduled maintenance and service. What is the fraction of machines that are operating at one time? Be sure your answer is simplified.

When simplifying fractions, we are using the Fundamental Principle of Fractions, in reverse. In Example 3, we simplified the fraction in Step 4 by factoring a 2 from both the numerator and denominator. That quotient is equal to 1, which is the reason the numerator becomes 1 in this case.

$$\frac{2}{12} = \frac{2 \times 1}{2 \times 2 \times 3} \qquad \text{Prime factorization}$$

$$= \frac{2}{2} \times \frac{1}{2 \times 3} \qquad \text{The Fundamental Principle of Fractions}$$

$$= 1 \times \frac{1}{6} \qquad \frac{2}{2} = 1$$

$$= \frac{1}{6}$$

Usually, we write this step more simply:

$$\frac{2}{12} = \frac{\cancel{2}^1}{\cancel{2}_1 \times 6} = \frac{1}{6} \text{ or even } \frac{2}{12} = \frac{\cancel{2}^1}{\cancel{12}_6} = \frac{1}{6}$$

When multiplying fractions, we use the property

$$\frac{a}{b} \times \frac{c}{d} = \frac{a \times c}{b \times d}$$

We then write the numerator and denominator in factored form and simplify before multiplying.

Example 4 Multiplying Fractions

< Objective 2 >

RECALL

A product is the result of multiplication.

Find the product of the fractions.

$$\frac{9}{2} \times \frac{4}{3}$$

$$\frac{9}{2} \times \frac{4}{3} = \frac{9 \times 4}{2 \times 3}$$

$$= \frac{3 \times 3 \times 2 \times 2}{2 \times 3} = \frac{3 \times 2}{1}$$

$$= \frac{6}{1}$$ The denominator of 1 is not necessary.

$$= 6$$

Check Yourself 4

Multiply and simplify each pair of fractions.

(a) $\frac{3}{5} \times \frac{10}{7}$ **(b)** $\frac{12}{5} \times \frac{10}{6}$

RECALL

Multiplying or dividing a number by 1 leaves the number unchanged.

The process describing fraction multiplication gives us insight into a number of fraction operations and properties.

For instance, the Fundamental Principle of Fractions is easily explained with the multiplication property. When applying the Fundamental Principle of Fractions, all we are really doing is multiplying or dividing a given fraction by 1.

$$\frac{2}{3} = \frac{2}{3} \times 1$$

$$= \frac{2}{3} \times \frac{2}{2}$$ $1 = \frac{2}{2}$

$$= \frac{2 \times 2}{3 \times 2}$$ This is fraction multiplication.

$$= \frac{4}{6}$$

RECALL

We find a reciprocal by inverting the fraction.

Another property that arises from fraction multiplication allows us to rewrite a fraction as a product using both the numerator and the denominator. For example,

$$\frac{3}{4} = \frac{3 \times 1}{1 \times 4} = \frac{3}{1} \times \frac{1}{4} = 3 \times \frac{1}{4} \quad \text{and} \quad \frac{3}{4} = \frac{1 \times 3}{4 \times 1} = \frac{1}{4} \times \frac{3}{1} = \frac{1}{4} \times 3$$

To divide two fractions, the divisor is replaced with its **reciprocal;** then the fractions are multiplied.

$$\frac{a}{b} \div \frac{c}{d} = \frac{a}{b} \times \frac{d}{c} = \frac{a \times d}{b \times c}$$

Example 5 Dividing Fractions

NOTES

The divisor $\frac{5}{6}$ is inverted and becomes $\frac{6}{5}$.

The common factor of 3 is removed from the numerator and denominator. This is the same as dividing by $\frac{3}{3}$ or 1.

Find the quotient of the fractions.

$$\frac{7}{3} \div \frac{5}{6}$$

$$\frac{7}{3} \div \frac{5}{6} = \frac{7}{3} \times \frac{6}{5} = \frac{7 \times 6}{3 \times 5}$$

$$= \frac{7 \times 2 \times 3}{3 \times 5} = \frac{7 \times 2}{5} = \frac{14}{5}$$

Check Yourself 5

Find the quotient of the fractions.

$$\frac{9}{2} \div \frac{3}{5}$$

NOTE

In algebra, improper fractions are preferred to mixed numbers. However, mixed numbers are the preferred format when answering many application exercises.

When adding two fractions, we need to find the **least common denominator (LCD)** first. The least common denominator is the smallest number that both denominators evenly divide. The process of finding the LCD is outlined here.

Step by Step

To Find the Least Common Denominator

Step 1 Write the prime factorization for each of the denominators.

Step 2 Find all the prime factors that appear in any one of the prime factorizations.

Step 3 Form the product of those prime factors, using each factor the greatest number of times it occurs in any one factorization.

Example 6 Finding the Least Common Denominator (LCD)

Find the LCD of fractions with denominators 6 and 8.

Our first step in adding fractions with denominators 6 and 8 is to determine the least common denominator. Factor 6 and 8.

$6 = 2 \times 3$

$8 = 2 \times 2 \times 2$

Because 2 appears 3 times as a factor of 8, it is used 3 times in writing the LCD.

The LCD is $2 \times 2 \times 2 \times 3$, or 24.

Check Yourself 6

Find the LCD of fractions with denominators 9 and 12.

The process is similar if more than two denominators are involved.

Example 7 Finding the Least Common Denominator

Find the LCD of fractions with denominators 6, 9, and 15.

To add fractions with denominators 6, 9, and 15, we need to find the LCD. Factor the three numbers.

$6 = 2 \times 3$ — 2 and 5 appear only once in any one factorization.

$9 = 3 \times 3$ — 3 appears twice as a factor of 9.

$15 = 3 \times 5$

The LCD is $2 \times 3 \times 3 \times 5$, or 90.

Check Yourself 7

Find the LCD of fractions with denominators 5, 8, and 20.

To add two fractions, we use the property

$$\frac{a}{b} + \frac{c}{b} = \frac{a + c}{b}$$

Example 8 Adding Fractions

< Objective 3 >

RECALL

A sum is the result of addition.

RECALL

We use the LCD to write equivalent fractions.

$\frac{5}{8} = \frac{5 \times 3}{8 \times 3} = \frac{15}{24}$

Find the sum of the fractions.

$$\frac{5}{8} + \frac{7}{12}$$

The LCD of 8 and 12 is 24. Each fraction should be rewritten as a fraction with that denominator.

$$\frac{5}{8} = \frac{15}{24}$$ Multiply the numerator and denominator by 3.

$$\frac{7}{12} = \frac{14}{24}$$ Multiply the numerator and denominator by 2.

$$\frac{5}{8} + \frac{7}{12} = \frac{15}{24} + \frac{14}{24} = \frac{15 + 14}{24} = \frac{29}{24}$$ This fraction is simplified.

Check Yourself 8

Find the sum of the fractions.

(a) $\frac{4}{5} + \frac{7}{9}$ (b) $\frac{5}{6} + \frac{4}{15}$

To subtract two fractions, use the rule

$$\frac{a}{b} - \frac{c}{b} = \frac{a - c}{b}$$

Subtracting fractions is treated exactly like adding them, except the numerator becomes the difference of the two numerators.

Example 9 Subtracting Fractions

RECALL

The difference is the result of subtraction.

Find the difference.

$$\frac{7}{9} - \frac{1}{6}$$

The LCD is 18. We rewrite the fractions with that denominator.

$$\frac{7}{9} = \frac{14}{18}$$

$$\frac{1}{6} = \frac{3}{18}$$

$$\frac{7}{9} - \frac{1}{6} = \frac{14}{18} - \frac{3}{18} = \frac{14 - 3}{18} = \frac{11}{18}$$ This fraction is simplified.

Check Yourself 9

Find the difference $\frac{11}{12} - \frac{5}{8}$.

We present a final application of fraction arithmetic before concluding this section.

Example 10 A Crafts Application

A potter uses $\frac{2}{3}$ pound (lb) of clay when making a bowl. How many bowls can be made from 15 lb of clay?

Step 1 The question asks for the number of $\frac{2}{3}$-lb bowls that the potter can make from a 15-lb batch of clay.

Step 2 This is a division problem. We will divide to see how many full times $\frac{2}{3}$ goes into 15.

Step 3 $15 \div \frac{2}{3}$

Step 4 $15 \div \frac{2}{3} = 15 \times \frac{3}{2}$ Use the division property.

$= \frac{15 \times 3}{1 \times 2}$ Now multiply fractions: $15 = \frac{15}{1}$.

$= \frac{45}{2}$ or $22\frac{1}{2}$ Complete the computation.

RECALL

$\frac{45}{2} = 45 \div 2$

$= 22\frac{1}{2}$

Step 5 The potter can complete 22 (whole) bowls from a 15-lb batch.

Reasonableness

Because each bowl uses less than a pound of clay, we would expect to get more than 15 bowls.

Because each bowl uses more than a half-pound of clay, we would expect to get fewer than $15 \times 2 = 30$ bowls.

22 bowls is a reasonable answer.

Check Yourself 10

A student survey at a community college found that $\frac{3}{4}$ of the students held jobs while going to school. Of those who have jobs, $\frac{5}{6}$ reported working more than 20 hours per week. What fraction of those surveyed worked more than 20 hours per week?

Check Yourself ANSWERS

1. Answers will vary. **2.** **(a)** $\frac{7}{11}$; **(b)** $\frac{1}{2}$; **(c)** $\frac{2}{9}$ **3.** $\frac{8}{9}$
4. **(a)** $\frac{6}{7}$; **(b)** 4 **5.** $\frac{15}{2}$ **6.** 36 **7.** 40 **8.** **(a)** $\frac{71}{45}$; **(b)** $\frac{11}{10}$
9. $\frac{7}{24}$ **10.** $\frac{5}{8}$

Reading Your Text

We conclude each section with this feature.

The fill-in-the-blank exercises are designed to ensure that you understand some of the key vocabulary used in this section. You should base your answers on a careful reading of the section.

The answers are in the Answers section at the end of this text.

SECTION 0.1

(a) The numbers used for counting are called the __________ numbers.

(b) Multiplying the numerator and denominator of a fraction by the same number is the same as multiplying the fraction by __________ .

(c) A __________ is the result of multiplication.

(d) __________ fractions is treated exactly like adding them, except the numerator becomes the difference of the two numerators.

Basic Skills | Challenge Yourself | Calculator/Computer | Career Applications | Above and Beyond

0.1 exercises

Boost your GRADE at ALEKS.com!

ALEKS

- Practice Problems
- Self-Tests
- NetTutor
- e-Professors
- Videos

Name ______________

Section ________ Date ________

Use the Fundamental Principle of Fractions to write three fractional representations for each number.

1. $\frac{3}{7}$
2. $\frac{2}{5}$
3. $\frac{4}{9}$
4. $\frac{7}{8}$
5. $\frac{5}{6}$
6. $\frac{11}{13}$
7. $\frac{10}{17}$
8. $\frac{2}{7}$
9. $\frac{9}{16}$
10. $\frac{6}{11}$
11. $\frac{7}{9}$
12. $\frac{15}{16}$

< Objective 1 >

Use the Fundamental Principle of Fractions to write each fraction in simplest form.

13. $\frac{10}{15}$
14. $\frac{12}{15}$
15. $\frac{10}{14}$
16. $\frac{18}{60}$
17. $\frac{12}{18}$
18. $\frac{28}{35}$

Answers

1. ______________
2. ______________
3. ______________
4. ______________
5. ______________
6. ______________
7. ______________
8. ______________
9. ______________
10. ______________
11. ______________
12. ______________
13. ________ 14. ________
15. ________ 16. ________
17. ________ 18. ________

Answers

19. ____
20. ____
21. ____
22. ____
23. ____
24. ____
25. ____
26. ____
27. ____
28. ____
29. ____
30. ____
31. ____ 32. ____
33. ____ 34. ____
35. ____ 36. ____
37. ____ 38. ____
39. ____ 40. ____
41. ____ 42. ____
43. ____ 44. ____
45. ____ 46. ____

19. $\frac{35}{40}$

20. $\frac{28}{32}$

21. $\frac{11}{44}$

22. $\frac{10}{25}$

23. $\frac{11}{33}$

24. $\frac{18}{48}$

25. $\frac{24}{27}$

26. $\frac{27}{45}$

27. $\frac{32}{40}$

28. $\frac{17}{51}$

29. $\frac{75}{105}$

30. $\frac{62}{93}$

31. $\frac{24}{30}$

32. $\frac{48}{66}$

33. $\frac{105}{135}$ > Videos

34. $\frac{39}{91}$

< Objective 2 >

Multiply. Be sure to simplify each product.

35. $\frac{3}{7} \times \frac{4}{5}$

36. $\frac{2}{7} \times \frac{5}{9}$

37. $\frac{3}{4} \times \frac{7}{5}$

38. $\frac{3}{5} \times \frac{2}{7}$

39. $\frac{3}{5} \times \frac{5}{7}$

40. $\frac{6}{11} \times \frac{8}{6}$

41. $\frac{6}{13} \times \frac{4}{9}$ 

42. $\frac{5}{9} \times \frac{6}{11}$

43. $\frac{3}{11} \times \frac{7}{9}$

44. $\frac{7}{9} \times \frac{3}{5}$

45. $\frac{4}{21} \times \frac{7}{12}$

46. $\frac{5}{21} \times \frac{14}{25}$

Divide. Write each result in simplest form.

47. $\frac{1}{7} \div \frac{3}{5}$

48. $\frac{2}{5} \div \frac{1}{3}$

49. $\frac{2}{5} \div \frac{3}{4}$

50. $\frac{5}{8} \div \frac{3}{4}$

51. $\frac{8}{9} \div \frac{4}{3}$

52. $\frac{4}{7} \div \frac{6}{11}$

53. $\frac{7}{10} \div \frac{5}{9}$

54. $\frac{8}{9} \div \frac{11}{15}$

55. $\frac{8}{15} \div \frac{2}{5}$

56. $\frac{5}{27} \div \frac{15}{54}$ > Videos

57. $\frac{8}{21} \div \frac{24}{35}$

58. $\frac{9}{28} \div \frac{27}{35}$

Find the least common denominator (LCD) for fractions with the given denominators.

59. 30 and 50

60. 36 and 48

61. 48 and 80

62. 60 and 84

63. 3, 4, and 5

64. 3, 4, and 6

65. 8, 10, and 15

66. 6, 22, and 33

67. 5, 10, and 25

68. 8, 24, and 48

< Objective 3 >

Add. Write each result in simplest form.

69. $\frac{2}{5} + \frac{1}{4}$

70. $\frac{2}{3} + \frac{3}{10}$

71. $\frac{2}{5} + \frac{7}{15}$

72. $\frac{2}{3} + \frac{4}{5}$

73. $\frac{3}{8} + \frac{5}{12}$

74. $\frac{5}{36} + \frac{7}{24}$

75. $\frac{7}{30} + \frac{5}{18}$

76. $\frac{9}{14} + \frac{10}{21}$

77. $\frac{7}{15} + \frac{13}{18}$ > Videos

78. $\frac{12}{25} + \frac{19}{30}$

79. $\frac{1}{5} + \frac{1}{10} + \frac{1}{15}$

80. $\frac{1}{3} + \frac{1}{5} + \frac{1}{10}$

Answers

47. ______ 48. ______
49. ______ 50. ______
51. ______ 52. ______
53. ______ 54. ______
55. ______ 56. ______
57. ______ 58. ______
59. ______ 60. ______
61. ______ 62. ______
63. ______ 64. ______
65. ______ 66. ______
67. ______ 68. ______
69. ______ 70. ______
71. ______ 72. ______
73. ______ 74. ______
75. ______ 76. ______
77. ______ 78. ______
79. ______ 80. ______

Answers

81. __________

82. __________

83. __________

84. __________

85. __________

86. __________

87. __________

88. __________

89. __________

90. __________

91. __________

92. __________

93. __________

94. __________

95. __________

96. __________

97. __________

98. __________

Subtract. Write each result in simplest form.

81. $\frac{8}{9} - \frac{3}{9}$ **82.** $\frac{9}{10} - \frac{6}{10}$ **83.** $\frac{6}{7} - \frac{2}{7}$

84. $\frac{11}{12} - \frac{7}{12}$ **85.** $\frac{7}{8} - \frac{2}{3}$ > Videos **86.** $\frac{4}{9} - \frac{2}{5}$

87. $\frac{11}{18} - \frac{2}{9}$ **88.** $\frac{5}{6} - \frac{1}{4}$ **89.** $\frac{2}{3} - \frac{7}{11}$

90. $\frac{13}{18} - \frac{5}{12}$ **91.** $\frac{5}{42} - \frac{1}{36}$ **92.** $\frac{13}{18} - \frac{7}{15}$

Basic Skills | **Challenge Yourself** | Calculator/Computer | Career Applications | Above and Beyond

Determine whether each statement is **true** *or* **false.**

93. When adding two fractions, we add the numerators together and we add the denominators together.

94. When multiplying two fractions, we multiply the numerators together and we multiply the denominators together.

Complete each statement with **never, sometimes,** *or* **always.**

95. The least common denominator of three fractions is __________ the product of the three denominators.

96. To add two fractions with different denominators, we __________ rewrite the fractions so that they have the same denominator.

97. **Crafts** If a pancake recipe calls for $\frac{1}{3}$ cup of white flour, $\frac{1}{3}$ cup of wheat flour, and $\frac{1}{2}$ cup of soy flour, how much flour is in the recipe?

98. **Business and Finance** Deductions from your paycheck are made roughly as follows: $\frac{1}{8}$ for federal tax, $\frac{1}{20}$ for state tax, $\frac{1}{20}$ for Social Security, and $\frac{1}{40}$ for a savings withholding plan. What portion of your pay is deducted?

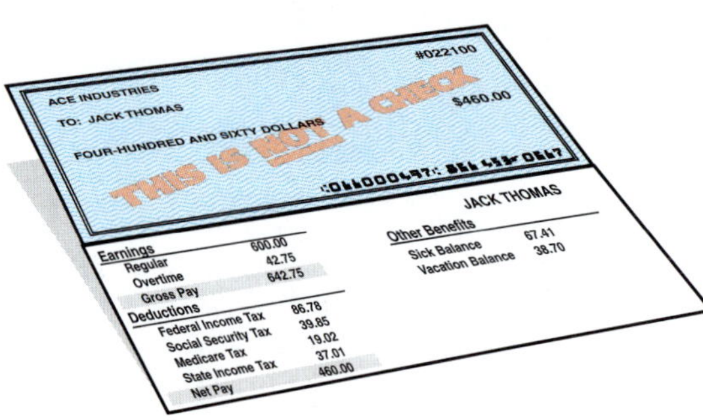

99. **Science and Medicine** Carol walked $\frac{3}{4}$ mile (mi) to the store, $\frac{1}{2}$ mi to a friend's house, and then $\frac{2}{3}$ mi home. How far did she walk?

100. **Geometry** Find the perimeter of, or the distance around, the accompanying figure by finding the sum of the lengths of the sides.

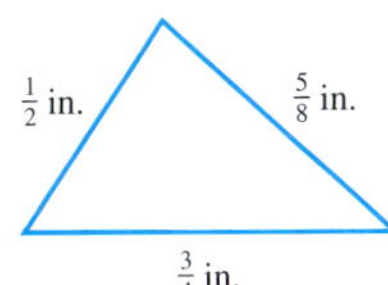

101. **Crafts** A hamburger that weighed $\frac{1}{4}$ pound (lb) before cooking weighed $\frac{3}{16}$ lb after cooking. How much weight was lost in cooking?

102. **Crafts** Geraldo has $\frac{3}{4}$ cup of flour. Biscuits use $\frac{5}{8}$ cup. Will he have enough left over for a small pie crust that requires $\frac{1}{4}$ cup? Explain.

Answers

99. ______

100. ______

101. ______

102. ______

Answers

1. $\frac{6}{14}, \frac{9}{21}, \frac{12}{28}$ **3.** $\frac{8}{18}, \frac{16}{36}, \frac{40}{90}$ **5.** $\frac{10}{12}, \frac{15}{18}, \frac{50}{60}$ **7.** $\frac{20}{34}, \frac{30}{51}, \frac{100}{170}$ **9.** $\frac{18}{32}, \frac{27}{48}, \frac{90}{160}$ **11.** $\frac{14}{18}, \frac{35}{45}, \frac{140}{180}$ **13.** $\frac{2}{3}$ **15.** $\frac{5}{7}$ **17.** $\frac{2}{3}$ **19.** $\frac{7}{8}$ **21.** $\frac{1}{4}$ **23.** $\frac{1}{3}$ **25.** $\frac{8}{9}$ **27.** $\frac{4}{5}$ **29.** $\frac{5}{7}$ **31.** $\frac{4}{5}$ **33.** $\frac{7}{9}$ **35.** $\frac{12}{35}$ **37.** $\frac{21}{20}$ **39.** $\frac{3}{7}$ **41.** $\frac{8}{39}$ **43.** $\frac{7}{33}$ **45.** $\frac{1}{9}$ **47.** $\frac{5}{21}$ **49.** $\frac{8}{15}$ **51.** $\frac{2}{3}$ **53.** $\frac{63}{50}$ **55.** $\frac{4}{3}$ **57.** $\frac{5}{9}$ **59.** 150 **61.** 240 **63.** 60 **65.** 120 **67.** 50 **69.** $\frac{13}{20}$ **71.** $\frac{13}{15}$ **73.** $\frac{19}{24}$ **75.** $\frac{23}{45}$ **77.** $\frac{107}{90}$ **79.** $\frac{11}{30}$ **81.** $\frac{5}{9}$ **83.** $\frac{4}{7}$ **85.** $\frac{5}{24}$ **87.** $\frac{7}{18}$ **89.** $\frac{1}{33}$ **91.** $\frac{23}{252}$ **93.** False **95.** sometimes **97.** $\frac{7}{6}$ cups or $1\frac{1}{6}$ cups **99.** $\frac{23}{12}$ mi **101.** $\frac{1}{16}$ lb

0.2 Real Numbers

< 0.2 Objectives >

1 > Identify integers
2 > Plot rational numbers on a number line
3 > Find the opposite of a number
4 > Find the absolute value of a number

In arithmetic, you learned to solve problems that involved working with numbers. In algebra, you will learn to use tools that will help you solve many new types of problems. Before we get there, we need more numbers. In this section, we expand our numbers beyond fractions and positive numbers. Let us look at some important sets of numbers.

NOTE

The set of three dots is called an *ellipsis* and indicates that a pattern continues.

The **natural numbers** are all the counting numbers 1, 2, 3, . . .

The **whole numbers** are the natural numbers together with zero.

We can represent whole numbers on a **number line.** Here is the number line.

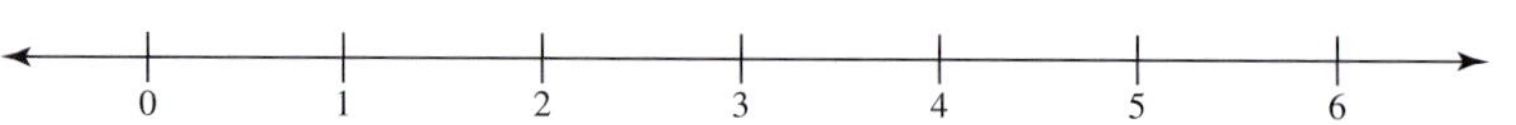

And here is the number line with the whole numbers 0, 1, 2, and 3 plotted.

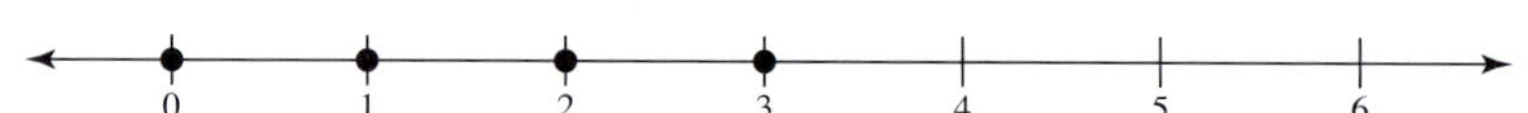

Now suppose you want to represent a temperature of 10 degrees below zero, a debt of $50, or an altitude 100 feet below sea level. These situations require a new set of numbers called *negative numbers.* We expand the number line to include negative numbers.

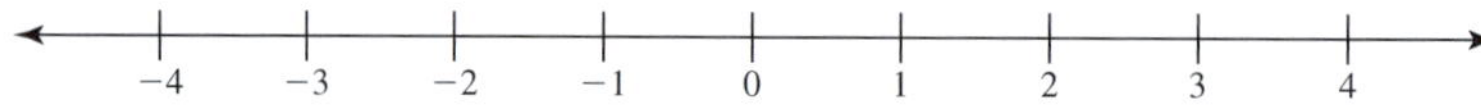

NOTE

Because −3 is to the left of 0, it is a negative number. Read −3 as "negative three."

Numbers to the right of (greater than) 0 on the number line are called **positive numbers.** Numbers to the left of (less than) 0 are called **negative numbers.** Zero is neither positive nor negative.

We indicate that a number is negative by placing a minus sign in front of the number. Positive numbers may be written with a plus sign, but are usually written with no sign at all.

Example 1 Identifying Real Numbers

RECALL

If no sign appears, a nonzero number is positive.

+6 is a positive number.

−9 is a negative number.

5 is a positive number.

0 is neither positive nor negative.

Check Yourself 1

Label each number as positive, negative, or neither.

(a) +3 (b) 7 (c) −5 (d) 0

The natural numbers, 0, and the negatives of natural numbers make up the set of integers.

Definition

Integers

The **integers** consist of the natural numbers, their negatives, and zero. We can represent the set of integers by

$\{\ldots, -3, -2, -1, 0, 1, 2, 3, \ldots\}$

NOTE

The arrowheads indicate that the number line extends forever in both directions.

Here we have a graphical representation of the set of integers.

The integers occur at the hash marks on the number line. Any plotted point that falls on one of the hash marks on the number line is an integer. This is true no matter how far in either direction we extend our number line.

Example 2 Identifying Integers

< Objective 1 >

Which numbers are integers?

$-3, \quad 5.3, \quad \frac{2}{3}, \quad 4$

Of these four numbers, only -3 and 4 are integers.

Check Yourself 2

Which numbers are integers?

$7, \quad 0, \quad \frac{4}{7}, \quad -5, \quad 0.2$

Definition

Rational Numbers

Any number that can be written as the ratio of two integers is called a **rational number.**

NOTE

6 is a rational number because it can be written as $\frac{6}{1}$.

Examples of rational numbers are $6, \frac{7}{3}, -\frac{15}{4}, 0, \frac{4}{1}$. On the number line, you can estimate the location of a rational number, as Example 3 illustrates.

Example 3 Plotting Rational Numbers

< Objective 2 >

Plot each rational number on the number line provided.

$\frac{2}{3}, \quad -3\frac{1}{4}, \quad \frac{27}{5}, \quad -1.445$

$\frac{2}{3}$ is between 0 and 1 (closer to one), so we plot that point on the number line shown here.

RECALL

Decimals are just a way of writing fractions when the denominator is a power of 10.

$-1.445 = -\frac{1,445}{1,000}$

$-3\frac{1}{4}$ is to the left of zero; it is $\frac{1}{4}$ farther than -3 from 0, so we plot this point, as well.

To find $\frac{27}{5}$ on a number line, we can do division, $\frac{27}{5} = 27 \div 5 = 5.4$, or write it as a mixed number $\frac{27}{5} = 5\frac{2}{5}$. Either way, we find the same point, farther than 5 units from 0 on the number line.

The point -1.445 is nearly halfway between -1 and -2, as shown here.

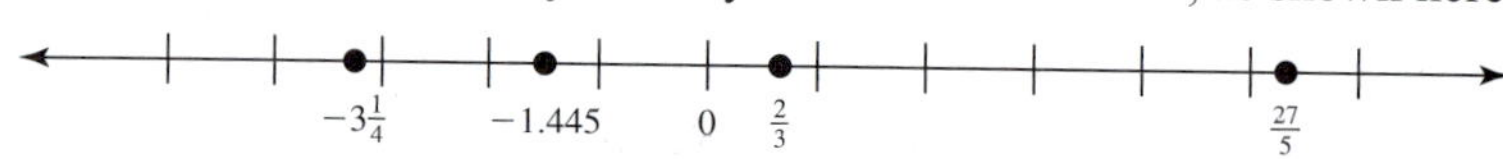

Check Yourself 3

Plot each rational number on the number line provided.

$-2\frac{1}{3}$, $\frac{37}{11}$, 5.66, $-\frac{1}{4}$

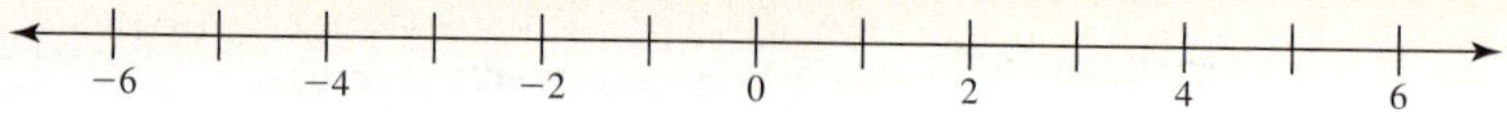

(number line from −6 to 6: −6, −4, −2, 0, 2, 4, 6)

One important property we can easily see on a number line is **order.** We say one number is **greater than** another if it is to the right on the number line. Similarly, the number on the left is **less than** the one on the right.

We use the symbols > and < to indicate order. The *inequality symbol* points to the smaller number. You should see how to use these symbols in the next example.

Example 4 Determining Order

> **CAUTION**

Because order is defined by position on the number line, you need to be careful when comparing two negative numbers.

(a) $6 > 3$

Six is *greater than* 3 because it is to the right of 3 on the number line.

(b) $2 < 5$

Two is *less than* 5; it is to the left of 5 on the number line.

(c) $-2 > -5$

-2 is to the right of -5 on the number line, so -2 is *greater than* -5.

Check Yourself 4

Fill in each blank with >, <, or = to make a true statement.

(a) -7 ____ 4 **(b)** $\frac{13}{4}$ ____ 3.25 **(c)** -12.08 ____ -12.2

Numbers that cannot be written as the ratio of two integers are called **irrational numbers.** Some examples of irrational numbers are $\sqrt{3}$, $\sqrt{7}$, and π. We will say more about these numbers later in the text. This diagram illustrates the relationships among the various sets of numbers.

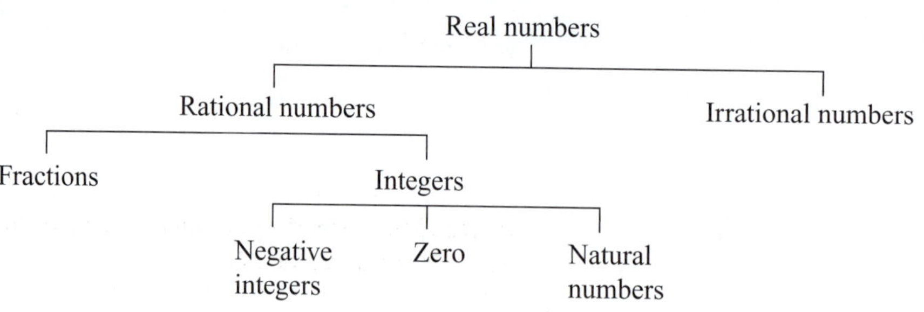

An important idea in our work with real numbers is the *opposite* of a number. Every number has an opposite.

Definition

Opposite of a Number

The **opposite** of a number corresponds to a point the same distance from 0 as the given number, but in the opposite direction.

Example 5 Writing the Opposite of a Real Number

< Objective 3 >

(a) The opposite of 5 is -5.

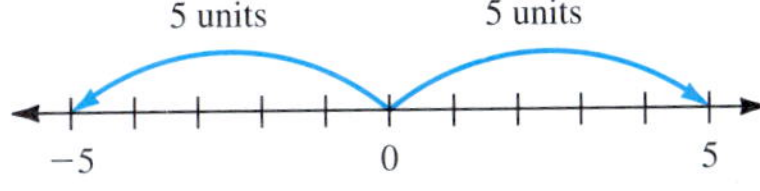

Both numbers are located 5 units from 0.

(b) The opposite of -3 is 3.

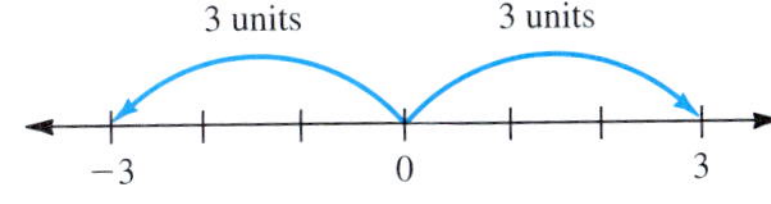

Both numbers correspond to points that are 3 units from 0.

Check Yourself 5

(a) What is the opposite of 8? **(b)** What is the opposite of -9?

NOTE

To represent the opposite of a number, place a minus sign in front of the number.

We write the opposite of 5 as -5. You can now think of -5 in two ways: as negative 5 and as the opposite of 5.

Using the same idea, we can write the opposite of a negative number. The opposite of -3 is $-(-3)$. Since we know from looking at the number line that the opposite of -3 is 3, this means that

$$-(-3) = 3$$

So the opposite of a negative number must be positive.

We summarize our results:

Property

The Opposite of a Real Number

1. The opposite of a positive number is negative.
2. The opposite of a negative number is positive.
3. The opposite of 0 is 0.

NOTE

The *magnitude* of a number is the same as its absolute value.

We also want to define the *absolute value,* or magnitude, of a real number.

Definition

Absolute Value

The **absolute value** of a real number is the distance (on the number line) between the number and 0.

Example 6 Finding Absolute Value

< Objective 4 >

NOTE

Both 5 and −5 have a magnitude of 5.

(a) The absolute value of 5 is 5.

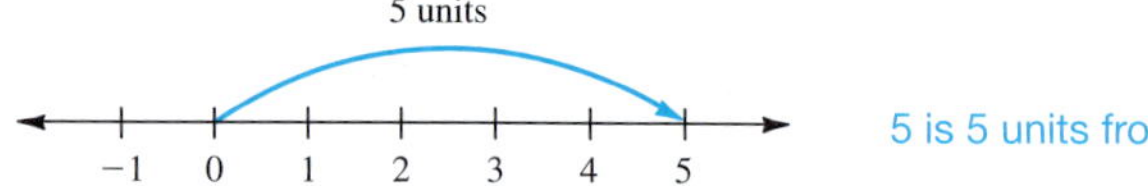

5 is 5 units from 0.

NOTE

|5| is read "the absolute value of 5."

(b) The absolute value of −5 is 5.

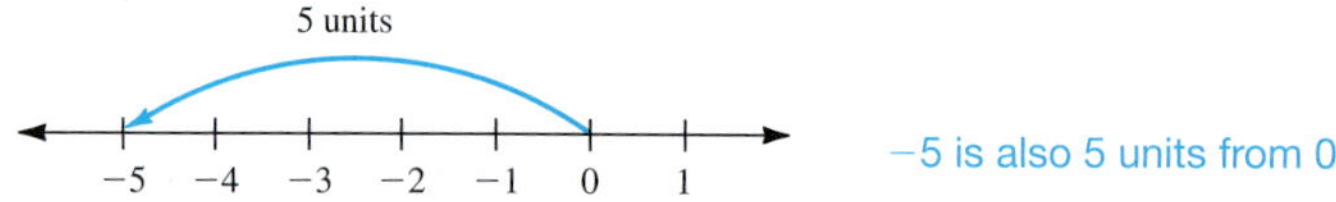

−5 is also 5 units from 0.

We usually write the absolute value of a number by placing vertical bars before and after the number. We can write

$|5| = 5$ and $|-5| = 5$

Check Yourself 6

Complete each statement.

(a) The absolute value of 9 is ________.

(b) The absolute value of −12 is ________.

(c) $|-6| =$ **(d)** $|15| =$

Example 7 Applying Real Numbers

RECALL

To arrange a set of numbers in *ascending order,* list them from least to greatest.

The elevations, in inches, of several points on a job site are shown below.

−18 27 −84 37 59 −13 4 92 49 66 −45

Arrange the elevations in ascending order.

Step 1 The question asks us to arrange the given numbers from least to greatest.

Steps 2 to 5 The number furthest left on the number line is −84, followed by −45, and so on.

NOTE

In this case, it makes sense to "combine" the remaining steps.

−84, −45, −18, −13, 4, 27, 37, 49, 59, 66, 92

Check Yourself 7

Several resistors were tested using an ohmmeter. Their resistance levels were entered into a table indicating their variance from 10,000 ohms (Ω). For example, if a resistor were to measure 9,900 Ω, it would be listed at −100.

Use their measured resistance to list the resistors in ascending order.

Resistor	#1	#2	#3	#4	#5	#6	#7
Variance (10,000 Ω)	+175	−60	−188	+10	+218	−65	−302

Check Yourself ANSWERS

1. **(a)** Positive; **(b)** positive; **(c)** negative; **(d)** neither **2.** 7, 0, -5

3.

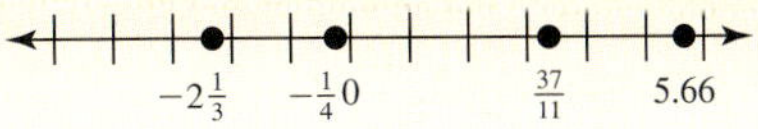

4. **(a)** $-7 < 4$; **(b)** $\frac{13}{4} = 3.25$; **(c)** $-12.08 > -12.2$ **5.** **(a)** -8; **(b)** 9

6. **(a)** 9; **(b)** 12; **(c)** 6; **(d)** 15 **7.** Resistors: #7, #3, #6, #2, #4, #1, and #5

Reading Your Text

SECTION 0.2

(a) The _______________ numbers are the natural numbers together with zero.

(b) We indicate that a number is _______________ by placing a minus sign in front of the number.

(c) The set of _______________ consists of the natural numbers, their negatives, and zero.

(d) The _______________ of a number is its absolute value.

0.2 exercises

Basic Skills | Challenge Yourself | Calculator/Computer | Career Applications | Above and Beyond

Name ____________________

Section ________ Date ________

Answers

1. ______ 2. ______
3. ______ 4. ______
5. ______ 6. ______
7. ______ 8. ______
9. ______ 10. ______
11. ______ 12. ______
13. ______ 14. ______
15. ______ 16. ______
17. ______ 18. ______
19. ______ 20. ______
21. ______ 22. ______
23. ______ 24. ______
25. ______ 26. ______
27. ______ 28. ______
29. ______ 30. ______
31. ______ 32. ______
33. ______ 34. ______

< Objectives 1, 3, and 4 >

Indicate whether each statement is **true** *or* **false.**

1. The opposite of -7 is 7.

2. The opposite of -10 is -10.

3. -9 is an integer.

4. 5 is an integer.

5. The opposite of -11 is 11.

6. The absolute value of -5 is 5.

7. $|-6| = -6$

8. $-(-30) = -30$

9. -12 is not an integer.

10. The opposite of -18 is 18.

11. $|-7| = -7$

12. The absolute value of -9 is -9.

13. $-(-8) = 8$

14. $\frac{2}{3}$ is not an integer.

15. $-|-15| = -15$

16. The absolute value of -3 is 3.

17. $\frac{3}{5}$ is an integer.

18. 0.7 is not an integer.

19. 0.15 is not an integer.

20. $|-9| = -9$

21. $\frac{5}{7}$ is not an integer.

22. 0.23 is not an integer.

23. $-(-7) = -7$

24. The opposite of 15 is -15.

Complete each statement.

25. The absolute value of -10 is _______.

26. $-(-12) =$ _______

27. $|-20| =$ _______ > Videos

28. The absolute value of -12 is _______.

29. The absolute value of -7 is _______.

30. The opposite of -9 is _______.

31. The opposite of 30 is _______.

32. $-|-15| =$ _______

33. $-(-6) =$ _______

34. The absolute value of 0 is _______.

35. $|50| =$ ______

36. The opposite of 18 is ______.

37. The absolute value of the opposite of 3 is ______.

38. The opposite of the absolute value of 3 is ______.

39. The opposite of the absolute value of -7 is ______.

40. The absolute value of the opposite of -7 is ______.

< Objective 2 >

Fill in each blank with $>$, $<$, or $=$ to make a true statement.

41. -5 ______ -9 

42. -15 ______ -10

43. -20 ______ -10

44. -15 ______ -14

45. $|3|$ ______ 3

46. $|-5|$ ______ $-(-5)$

47. -4 ______ $|-4|$

48. 7 ______ $|7|$

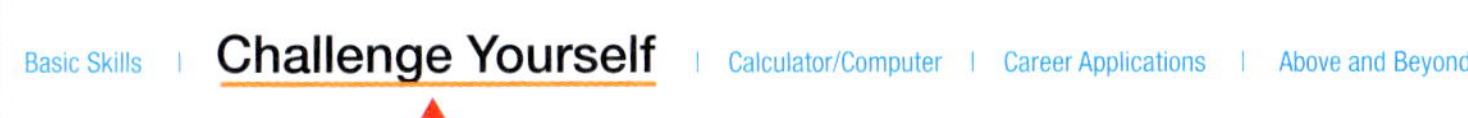

For exercises 49 to 52, use the numbers $-3, \frac{2}{3}, -1.5, 2,$ and 0.

49. Which of the numbers are integers? 

50. Which of the numbers are natural numbers?

51. Which of the numbers are whole numbers?

52. Which of the numbers are negative numbers?

For exercises 53 to 56, use the numbers $-2, -\frac{4}{3}, 3.5, 0,$ and 1.

53. Which of the numbers are integers?

54. Which of the numbers are natural numbers?

55. Which of the numbers are whole numbers?

56. Which of the numbers are negative numbers?

Answers

35. ______

36. ______

37. ______

38. ______

39. ______

40. ______

41. ______

42. ______

43. ______

44. ______

45. ______

46. ______

47. ______

48. ______

49. ______

50. ______

51. ______

52. ______

53. ______

54. ______

55. ______

56. ______

Answers

57. ______

58. ______

59. ______

60. ______

61. ______

62. ______

63. ______

64. ______

65. ______

66. ______

67. ______

68. ______

69. ______

70. ______

Complete each statement with **never, sometimes,** *or* **always.**

57. The opposite of a negative number is ________ negative.

58. The absolute value of a nonzero number is ________ positive.

59. The absolute value of a number is ________ equal to that number.

60. A rational number is ________ an integer.

Use a real number to represent each quantity. Be sure to include the appropriate sign and unit with each answer.

61. **Business and Finance** The withdrawal of $50 from a checking account

62. **Business and Finance** A $200 deposit into a savings account

63. **Science and Medicine** A 10°F temperature decrease in an hour

64. **Statistics** An eight-game losing streak by the local baseball team

65. **Social Science** A 25,000-person increase in a city's population

66. **Business and Finance** A country exported $90,000,000 more than it imported, creating a positive trade balance.

Use a real number to represent each quantity. Be sure to include the appropriate sign and unit with each answer.

67. **Agricultural Technology** The erosion of 4 in. of topsoil from an Iowa cornfield

68. **Agricultural Technology** The formation of 2.5 cm of new topsoil on the African savanna

Electrical Engineering Several 12-volt (V) batteries were tested using a voltmeter. The voltages were entered into a table indicating their variance from 12 V. Use this table to complete exercises 69–70.

Battery	#1	#2	#3	#4	#5
Variance (12 V)	+1	0	−1	−3	+2

69. Use their voltages to list the batteries in ascending order.

70. Which battery had the highest voltage measurement? What was its voltage measurement?

Basic Skills | Challenge Yourself | Calculator/Computer | Career Applications | **Above and Beyond**

71. **(a)** Every number has an opposite. The opposite of 5 is -5. In English, a similar situation exists for words. For example, the opposite of *regular* is *irregular*. Write the opposite of each word.

irredeemable, uncomfortable, uninteresting, uninformed, irrelevant, immoral

(b) Note that the idea of an opposite is usually expressed by a prefix such as *un-* or *ir-*. What other prefixes can be used to negate or change the meaning of a word to its opposite? List four words using these prefixes, and use the words in a sentence.

72. **(a)** What is the difference between positive integers and nonnegative integers?

(b) What is the difference between negative and nonpositive integers?

73. Simplify each expression.

(a) $-(-3)$ **(b)** $-(-(-3))$ **(c)** $-(-(-(-3)))$

(d) Use the results of parts (a), (b), and (c) to create a rule for simplifying expressions of this type.

(e) Use the rule created in part (d) to simplify $-(-(-(-(-(-(-7))))))$.

Answers

1. True **3.** True **5.** True **7.** False **9.** False **11.** False **13.** True **15.** True **17.** False **19.** True **21.** True **23.** False **25.** 10 **27.** 20 **29.** 7 **31.** -30 **33.** 6 **35.** 50 **37.** 3 **39.** -7 **41.** $>$ **43.** $<$ **45.** $=$ **47.** $<$ **49.** $-3, 2, 0$ **51.** 0, 2 **53.** $-2, 0, 1$ **55.** 0, 1 **57.** never **59.** sometimes **61.** $-\$50$ **63.** $-10°F$ **65.** $+25{,}000$ people **67.** -4 in. **69.** #4, #3, #2, #1, #5 **71.** Above and Beyond **73.** **(a)** 3; **(b)** -3; **(c)** 3; **(d)** Above and Beyond; **(e)** -7

Answers

71. ____________

72. ____________

73. ____________

0.3 Adding and Subtracting Real Numbers

< 0.3 Objectives >

1 > Add real numbers

2 > Use the commutative property of addition

3 > Use the associative property of addition

4 > Subtract real numbers

The number line can be used to demonstrate the sum of two real numbers. To add a positive number, we move to the right; to add a negative number, we move to the left.

Example 1 Finding the Sum of Real Numbers

< Objective 1 >

Find the sum $5 + (-2)$.

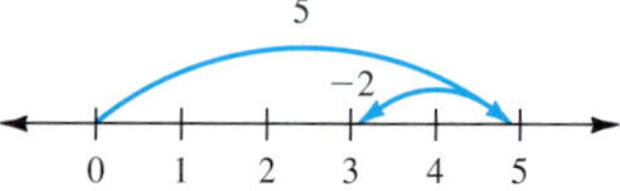

Begin 5 units *to the right* of 0. Then, to add -2, move 2 units *to the left.* We see that

$5 + (-2) = 3$

Check Yourself 1

Find the sum.

$9 + (-7)$

We can also use the number line to picture addition when two negative numbers are involved. Example 2 illustrates this approach.

Example 2 Finding the Sum of Real Numbers

NOTE

The sum of two positive numbers is positive, and the sum of two negative numbers is negative.

Find the sum $-2 + (-3)$.

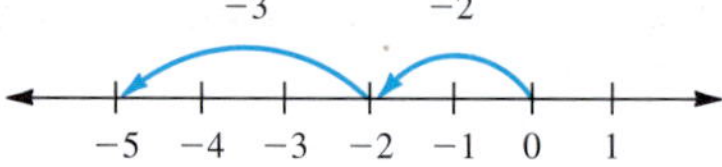

Begin 2 units *to the left* of 0 (because the first number is -2). Then move 3 more units *to the left* to add negative 3. We see that

$-2 + (-3) = -5$

Check Yourself 2

Find the sum.

$-7 + (-5)$

You may have noticed some patterns in the previous examples. These patterns let you do much of the addition mentally.

Property

To Add Real Numbers

RECALL

The magnitude of a number is given by its absolute value.

Case 1. If two numbers have the same sign, add their magnitudes. Give the sum the sign of the original numbers.

Case 2. If two numbers have different signs, subtract the smaller magnitude from the larger. Attach the sign of the number with the larger magnitude to the result.

Example 3 Finding the Sum of Real Numbers

Find each sum.

(a) $5 + 2 = 7$ The sum of two positive numbers is positive.

(b) $-2 + (-6) = -8$ Add the magnitudes of the two numbers ($2 + 6 = 8$). Give to the sum the sign of the original numbers.

Check Yourself 3

Find the sums.

(a) $6 + 7$ (b) $-8 + (-7)$

There are three important parts to the study of algebra. The first is the set of numbers, which we discuss in this chapter. The second is the set of operations, such as addition and multiplication. The third is the set of rules, which we call **properties.** Example 4 enables us to look at an important property of addition.

Example 4 Finding the Sum of Real Numbers

< Objective 2 >

Find each sum.

(a) $2 + (-7) = (-7) + 2 = -5$

(b) $-3 + (-4) = -4 + (-3) = -7$

In both cases the order in which we add the numbers does not affect the sum.

Check Yourself 4

Find the sums $-8 + 2$ and $2 + (-8)$. How do the results compare?

Property

The Commutative Property of Addition

The *order* in which we add two numbers does not change the sum. Addition is **commutative.** In symbols, for any numbers a and b,

$a + b = b + a$

What if we want to add more than two numbers? Another property of addition is helpful. Look at Example 5.

Example 5 Finding the Sum of Three Real Numbers

< Objective 3 >

Find the sum $2 + (-3) + (-4)$. First,

$$\underbrace{[2 + (-3)]} + (-4) \quad \text{Add the first two numbers.}$$
$$= -1 + (-4) \quad \text{Then add the third to that sum.}$$
$$= -5$$

Here is a second approach.

$$2 + \underbrace{[(-3) + (-4)]} \quad \text{This time, add the second and third numbers.}$$
$$= 2 + (-7) \quad \text{Then add the first number to that sum.}$$
$$= -5$$

Check Yourself 5

Show that $-2 + (-3 + 5) = [-2 + (-3)] + 5$

Do you see that it makes no difference which way we group numbers in addition? The final sum is the same.

Property

The Associative Property of Addition

The way we *group* numbers does not change the sum. Addition is **associative.** In symbols, for any numbers *a*, *b*, and *c*,

$(a + b) + c = a + (b + c)$

A number's opposite (or negative) is called its **additive inverse.** Use this rule to add opposite numbers.

Property

The Additive Inverse

The sum of any number and its additive inverse is 0. In symbols, for any number *a*,

$a + (-a) = 0$

Example 6 Finding the Sum of Additive Inverses

Find each sum.

(a) $6 + (-6) = 0$

(b) $-8 + 8 = 0$

Check Yourself 6

Find the sum.

$9 + (-9)$

So far we have looked only at the addition of integers. The process is the same if we want to add other types of real numbers.

Example 7 Finding the Sum of Real Numbers

Find each sum.

(a) $\frac{15}{4} + \left(-\frac{9}{4}\right) = \frac{6}{4} = \frac{3}{2}$ Subtract their magnitudes: $\frac{15}{4} - \frac{9}{4} = \frac{6}{4} = \frac{3}{2}$.

The sum is positive since $\frac{15}{4}$ has the larger magnitude.

(b) $-0.5 + (-0.2) = -0.7$ Add their magnitudes $(0.5 + 0.2 = 0.7)$. The sum is negative.

Check Yourself 7

Find each sum.

(a) $-\frac{5}{2} + \left(-\frac{7}{2}\right)$ **(b)** $5.3 + (-4.3)$

Now we turn our attention to the subtraction of real numbers. Subtraction is called the *inverse* operation to addition. This means that any subtraction problem can be written as a problem in addition.

Property

To Subtract Real Numbers

To subtract real numbers, add the first number and the *opposite* of the number being subtracted. In symbols, by definition

$a - b = a + (-b)$

Example 8 illustrates this property.

Example 8 Finding the Difference of Real Numbers

< Objective 4 >

(a) Subtract $5 - 3$.

$5 - 3 = 5 + (-3) = 2$ To subtract 3, we add the opposite of 3.

The opposite of 3

(b) Subtract $2 - 5$.

$2 - 5 = 2 + (-5) = -3$

The opposite of 5

NOTE

Use the subtraction property to add the opposite of 4, -4, to the value -3.

(c) Subtract $-3 - 4$.

$-3 - 4 = -3 + (-4) = -7$ -4 is the opposite of 4.

(d) Subtract $-10 - 15$.

$-10 - 15 = -10 + (-15) = -25$ -15 is the opposite of 15.

Check Yourself 8

Find each difference, using the definition of subtraction.

(a) $8 - 3$ **(b)** $7 - 9$ **(c)** $-5 - 9$ **(d)** $-12 - 6$

The subtraction rule works the same way when the number being subtracted is negative. Change the subtraction to addition and then replace the negative number being subtracted with its opposite, which is positive. Example 9 illustrates this principle.

Example 9 Subtracting Real Numbers

> CAUTION

Your graphing calculator can be used to simplify the kinds of problems we encounter in this section. The negation key is the [(−)] or the [+/−] found on the calculator. Do not confuse this with the subtraction key!

Simplify each expression.

(a) $5 - (-2) = 5 + (+2) = 5 + 2 = 7$ Change the subtraction to addition and replace −2 with its opposite, +2 or 2.

(b) $7 - (-8) = 7 + (+8) = 7 + 8 = 15$

(c) $-9 - (-5) = -9 + 5 = -4$

(d) $-12.7 - (-3.7) = -12.7 + 3.7 = -9$

(e) $-\frac{3}{4} - \left(-\frac{7}{4}\right) = -\frac{3}{4} + \frac{7}{4} = \frac{4}{4} = 1$

(f) Subtract −4 from −5. We write

$-5 - (-4) = -5 + 4 = -1$

Check Yourself 9

Subtract.

(a) 8 − (−2) **(b)** 3 − (−10) **(c)** −7 − (−2)
(d) −9.8 − (−5.8) **(e)** 7 − (−7)

> **NOTE**
>
> If your calculator is different from either of the ones we describe, refer to your manual, or ask your instructor for assistance.

The calculator can be a useful tool for checking arithmetic or performing complicated computations. In order to master your calculator, you should become familiar with some of the keys.

The first key is the subtraction key, [−]. This key is usually found in the right column of calculator keys along with the other "operation" keys such as addition, multiplication, and division.

The second key to find is the one for negative numbers. On graphing calculators, it usually looks like [(−)], whereas on scientific calculators, the key usually looks like [+/−]. In either case, the negative number key is usually found in the bottom row.

One very important difference between the two types of calculators is that when using a graphing calculator, you input the negative sign before keying in the number (as it is written). When using a scientific calculator, you input the negative sign button after keying in the number.

In Example 10, we illustrate this difference, while showing that subtraction remains the same.

Example 10 Subtracting with a Calculator

Use a calculator to find each difference.

(a) $-12.43 - 3.516$

> **NOTE**
>
> Graphing calculators usually have an [ENTER] key, whereas scientific calculators have an [=] key.

Graphing Calculator

[(−)] 12.43 [−] 3.516 [ENTER] The negative number sign comes before the number.

The display should read −15.946.

Scientific Calculator

12.43 [+/−] [−] 3.516 [=] The negative number sign comes after the number.

The display should read −15.946.

```
-12.43-3.516
          -15.946
23.56--4.7
            28.26
```

(b) $23.56 - (-4.7)$

Graphing Calculator

23.56 [–] [(–)] 4.7 [ENTER] 

The negative number key is pressed before the number.

The display should read 28.26.

Scientific Calculator

23.56 [–] 4.7 [+/–] [=]

The negative number key is pressed after the number.

The display should read 28.26.

Check Yourself 10

Use your calculator to find each difference.

(a) $-13.46 - 5.71$ **(b)** $-3.575 - (-6.825)$

Example 11 A Business and Finance Application

Oscar owns stock in four companies. This year, his holdings in Cisco went up \$2,250; his holdings in AT&T went down \$1,345; Chevron went down \$5,215; and IBM went down \$1,525.

How much did the total value of Oscar's holdings change during the year?

RECALL

We introduced this five-step problem-solving approach in Section 0.1.

Step 1 The question asks for the combined change in value of Oscar's holdings. We are given the amount each stock went up or down.

Step 2 To find the change in value, we add the increases and subtract the decreases.

Step 3 $\$2,250 - \$1,345 - \$5,215 - \$1,525$

Step 4 $\$2,250 - \$1,345 - \$5,215 - \$1,525 = -\$5,835$

Step 5 Oscar's holdings decreased in value by \$5,835 during the year.

Reasonableness

Oscar lost money on three stocks including over \$5,000 from one stock, so this answer seems reasonable.

Check Yourself 11

A bus with 15 people stopped at Avenue A. Nine people got off and 5 people got on. At Avenue B, 6 people got off and 8 people got on. At Avenue C, 4 people got off the bus and 6 people got on. How many people are now on the bus?

Check Yourself ANSWERS

1. 2 **2.** -12 **3.** **(a)** 13; **(b)** -15 **4.** $-6 = -6$ **5.** $0 = 0$
6. 0 **7.** **(a)** -6; **(b)** 1 **8.** **(a)** 5; **(b)** -2; **(c)** -14; **(d)** -18
9. **(a)** 10; **(b)** 13; **(c)** -5; **(d)** -4; **(e)** 14 **10.** **(a)** -19.17; **(b)** 3.25
11. 15 people

Reading Your Text

SECTION 0.3

(a) If two numbers have the same sign, add their ______________ and then give the sum the sign of the original numbers.

(b) The ______________ in which we add two numbers does not change their sum.

(c) Addition is ______________. In symbols, for any numbers a, b, and c, $(a + b) + c = a + (b + c)$.

(d) The sum of any number and its additive inverse is ______________.

Basic Skills | Challenge Yourself | Calculator/Computer | Career Applications | Above and Beyond

0.3 exercises

Boost your GRADE at ALEKS.com!

ALEKS

- Practice Problems
- Self-Tests
- NetTutor
- e-Professors
- Videos

Name ____________

Section ______ Date ______

< Objectives 1–4 >

Perform the indicated operation.

1. $6 + (-5)$

2. $3 + 9$

3. $11 + (-7)$

4. $-6 + (-7)$

5. $4 + (-6)$

6. $9 + (-2)$

7. $7 + 9$

8. $-7 + 11$

9. $(-11) + 5$

10. $5 + (-8)$

11. $-8 + (-7)$ > Videos

12. $8 + (-7)$

13. $-12 + 4$

14. $7 + (-7)$

15. $-9 + 10$

16. $-6 + 8$

17. $-4 + 4$

18. $5 + (-20)$

19. $7 + (-13)$

20. $0 + (-10)$

21. $-8 + 5$

22. $-7 + 3$

23. $6 + (-6)$

24. $-9 + 9$

25. $\frac{45}{16} - \frac{9}{16}$

26. $-\frac{35}{16} + \frac{17}{16}$

27. $\frac{29}{8} + \left(-\frac{17}{8}\right)$

28. $-\frac{81}{20} + \left(-\frac{107}{20}\right)$

29. $-\frac{73}{16} + \frac{119}{16}$

30. $-\frac{13}{8} - \left(-\frac{15}{4}\right)$

31. $4 + (-7) + (-5)$

32. $-7 + 8 + (-6)$

33. $-2 + (-6) + (-4)$

34. $12 + (-6) + (-4)$

Answers

1.	2.
3.	4.
5.	6.
7.	8.
9.	10.
11.	12.
13.	14.
15.	16.
17.	18.
19.	20.
21.	22.
23.	24.
25.	26.
27.	28.
29.	30.
31.	32.
33.	34.

Answers

35. ______ 36. ______
37. ______ 38. ______
39. ______ 40. ______
41. ______ 42. ______
43. ______ 44. ______
45. ______ 46. ______
47. ______ 48. ______
49. ______ 50. ______
51. ______ 52. ______
53. ______ 54. ______
55. ______ 56. ______
57. ______ 58. ______
59. ______ 60. ______
61. ______ 62. ______
63. ______ 64. ______
65. ______ 66. ______
67. ______ 68. ______
69. ______ 70. ______
71. ______ 72. ______

35. $-3 + (-7) + 5 + (-2)$

36. $7 + (-8) + (-9) + 10$

37. $11 - 13$

38. $7 - 5$

39. $9 - 3$

40. $4 - 9$

41. $-8 - 3$ > Videos

42. $-13 - 8$

43. $-12 - 8$

44. $9 - 15$

45. $-2 - (-3)$

46. $-9 - (-6)$

47. $-5 - (-5)$

48. $9 - (-7)$

49. $28 - (-22)$

50. $50 - (-25)$

51. $-15 - (-25)$ 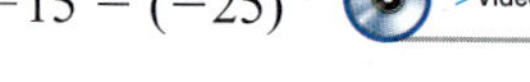

52. $-20 - (-30)$

53. $-25 - (-15)$

54. $-30 - (-20)$

55. $-(-20) - (-15)$

56. $18 - (-12)$

57. $48 - (-15)$

58. $-25 - (-30)$

59. $\frac{10}{2} - \left(-\frac{7}{2}\right)$

60. $-\frac{4}{2} - \frac{3}{2}$

61. $-\frac{7}{8} - \left(-\frac{19}{8}\right)$

62. $\frac{13}{4} - \left(-\frac{7}{4}\right)$

63. $-7 - (-5) - 6$

64. $-5 - (-8) - 10$

65. $-10 - 8 - (-7)$

66. $-5 - 8 - (-15)$

Basic Skills | **Challenge Yourself** | Calculator/Computer | Career Applications | Above and Beyond

Complete each statement with **never, sometimes,** *or* **always.**

67. The sum of two negative numbers is __________ negative.

68. The difference of two negative numbers is __________ negative.

69. The additive inverse of a negative number is __________ negative.

70. The sum of a number and its additive inverse is __________ zero.

Solve each application.

71. **Science and Medicine** The temperature in Chicago dropped from 18°F at 4 P.M. to −9°F at midnight. What was the drop in temperature?

72. **Business and Finance** Charley's checking account had \$175 deposited at the beginning of the month. After he wrote checks for the month, the account was \$95 *overdrawn.* What amount of checks did he write during the month?

73. **Technology** Micki entered the elevator on the 34th floor. From that point the elevator went up 12 floors, down 27 floors, down 6 floors, and up 15 floors before she got off. On what floor did she get off the elevator? > Videos

74. **Technology** A submarine dives to a depth of 500 ft below the ocean's surface. It then dives another 217 ft before climbing 140 ft. What is the depth of the submarine?

75. **Technology** A helicopter is 600 ft above sea level, and a submarine directly below it is 325 ft below sea level. How far apart are they?

76. **Business and Finance** Tom has received an overdraft notice from the bank telling him that his account is overdrawn by $142. How much must he deposit in order to have $625 in his account?

77. **Science and Medicine** At 9:00 A.M., Jose had a temperature of 99.8°. It rose another 2.5° before falling 3.7° by 1:00 P.M. What was his temperature at 1:00 P.M.?

78. **Business and Finance** Olga has $250 in her checking account. She deposits $52 and then writes a check for $77. What is her new balance?

Bal:	250
Dep:	52
CK # 1111:	77

79. **Statistics** Ezra's scores on five tests taken in a mathematics class were 87, 71, 95, 81, and 90. What was the difference between the highest and the lowest of his scores?

80. **Business and Finance** Aaron had $769 in his bank account on June 1. He deposited $125 and $986 during the month and wrote checks for $235, $529, and $712 during June. What was his balance at the end of the month?

Basic Skills | Challenge Yourself | **Calculator/Computer** | Career Applications | Above and Beyond

Use a calculator to find each difference.

81. $-11.392 - 13.491$

82. $-9.245 - 14.316$

83. $-7.259 - 4.235$

84. $-6.319 - 2.628$

85. $-18.271 - (-12.569)$

86. $-15.586 - (-9.874)$

87. $-17.346 - (-28.293)$

88. $-11.358 - (-23.145)$

Answers

73. ______
74. ______
75. ______
76. ______
77. ______
78. ______
79. ______
80. ______
81. ______
82. ______
83. ______
84. ______
85. ______
86. ______
87. ______
88. ______

Answers

89.

90.

91.

92.

93.

94.

Basic Skills | Challenge Yourself | Calculator/Computer | Career Applications | **Above and Beyond**

89. Complete the problem "4 − (−9) is the same as _______." Write an application problem that might be answered using this subtraction.

90. Explain the difference between the two phrases "7 less than a number" and "a number subtracted from 7." Use both symbols and English to explain the meanings of these phrases. Write some other ways of expressing subtraction in English.

91. Construct an example to show that subtraction of real numbers is *not* commutative.

92. Construct an example to show that subtraction of real numbers is *not* associative.

93. Do you think this statement is true?

$|a + b| = |a| + |b|$ for all numbers a and b

When we don't know whether such a statement is true, we refer to the statement as a **conjecture.** We may "test" the conjecture by substituting specific numbers for the letters.

Test the conjecture, using two positive numbers for a and b.

Test again, using a positive number for a and 0 for b.

Test again, using two negative numbers.

Now try using one positive number and one negative number.

Summarize your results in a rule that you feel is true.

94. If a represents a positive number and b represents a negative number, determine which expressions are positive and which are negative.

(a) $|b| + a$ **(b)** $b + (-a)$ **(c)** $(-b) + a$ **(d)** $-b + |-a|$

Answers

1. −11 **3.** 4 **5.** −2 **7.** 16 **9.** −6 **11.** −15 **13.** −8
15. 1 **17.** 0 **19.** −6 **21.** −3 **23.** 0 **25.** $\frac{9}{4}$ **27.** $\frac{3}{2}$
29. $\frac{23}{8}$ **31.** −8 **33.** −12 **35.** −7 **37.** −2 **39.** 6
41. −11 **43.** −20 **45.** 1 **47.** 0 **49.** 50 **51.** 10 **53.** −10
55. 35 **57.** 63 **59.** $\frac{17}{2}$ **61.** $\frac{3}{2}$ **63.** −8 **65.** −11
67. always **69.** never **71.** 27°F **73.** 28th floor **75.** 925 ft
77. 98.6° **79.** 24 points **81.** −24.883 **83.** −11.494
85. −5.702 **87.** 10.947 **89.** Above and Beyond
91. Above and Beyond **93.** Above and Beyond

0.4 Multiplying and Dividing Real Numbers

< 0.4 Objectives >

1 > Multiply real numbers
2 > Use the commutative property of multiplication
3 > Use the associative property of multiplication
4 > Use the distributive property
5 > Divide real numbers

Multiplication can be seen as repeated addition. That is, we can interpret

$3 \times 4 = 4 + 4 + 4 = 12$

We can use this interpretation together with the work of Section 0.3 to find the product of two real numbers.

Example 1 Finding the Product of Real Numbers

< Objective 1 >

NOTE

We use parentheses ( ) to indicate multiplication when negative numbers are involved.

Multiply.

(a) $(3)(-4) = (-4) + (-4) + (-4) = -12$

(b) $(4)\left(-\frac{1}{3}\right) = \left(-\frac{1}{3}\right) + \left(-\frac{1}{3}\right) + \left(-\frac{1}{3}\right) + \left(-\frac{1}{3}\right) = -\frac{4}{3}$

Check Yourself 1

Find the product by writing the expression as repeated addition.

(4)(−3)

Looking at the products we found by repeated addition in Example 1 should suggest our first rule for multiplying real numbers.

Property

To Multiply Real Numbers

Case 1 The product of two numbers with different signs is negative.

RECALL

The absolute value of a number is the same as its magnitude.

The rule is easy to use. To multiply two numbers with different signs, just multiply their absolute values and attach a minus sign to the product.

Example 2 Finding the Product of Real Numbers

Find each product.

$$(5)(-6) = -30$$
$$(10)(-12) = -120$$
$$(-7)\left(\frac{1}{8}\right) = -\frac{7}{8}$$
$$(1.5)(-0.3) = -0.45$$
$$\left(-\frac{5}{8}\right)\left(\frac{4}{15}\right) = -\left(\frac{\overset{1}{\cancel{5}}}{\underset{2}{\cancel{8}}} \times \frac{\overset{1}{\cancel{4}}}{\underset{3}{\cancel{15}}}\right) = -\frac{1}{6}$$

RECALL

$$-\frac{5}{8} \times \frac{4}{15} = -\frac{\cancel{5} \times \cancel{2} \times \cancel{2}}{\cancel{2} \times \cancel{2} \times 2 \times 3 \times \cancel{5}} = -\frac{1}{6}$$

Check Yourself 2

Find each product.

(a) (15)(−5) **(b)** (−0.8)(0.2) (c) $\left(-\frac{2}{3}\right)\left(\frac{6}{7}\right)$

The product of two negative numbers is harder to visualize. The pattern below may help you see how we can determine the sign of the product.

$$(3)(-2) = -6$$
$$(2)(-2) = -4$$
$$(1)(-2) = -2$$
$$(0)(-2) = 0$$
$$(-1)(-2) = 2$$
$$(-2)(-2) = 4$$

Do you see that the product *increases* by 2 each time the first factor *decreases* by 1?

RECALL

We already know that the product of two positive numbers is positive.

This suggests that the product of two negative numbers is positive, which is, in fact, the case. To extend our multiplication rule, we have the following.

Property

To Multiply Real Numbers

Case 2 The product of two numbers with the same sign is positive.

Example 3 Finding the Product of Real Numbers

Find each product.

$$8 \times 7 = 56$$
$$(-9)(-6) = 54$$
$$(-0.5)(-2) = 1$$

The numbers have the same sign, so the product is positive.

> **CAUTION**

(−8)(−6) tells you to multiply. The parentheses are *next to* one another. The expression −8 −6 tells you to subtract. The numbers are *separated* by the operation sign.

Check Yourself 3

Find each product.

(a) 5×7 **(b)** $(-8)(-6)$ **(c)** $(9)(-6)$ **(d)** $(-1.5)(-4)$

To multiply more than two real numbers, apply the multiplication rule repeatedly.

Example 4 Finding the Product of a Group of Real Numbers

NOTE

The original expression has an odd number of negative signs. Do you see that having an odd number of negative factors always results in a negative product?

Multiply.

$$\underbrace{(5)(-7)}(-3)(-2) \qquad (5)(-7) = -35$$
$$= \underbrace{(-35)(-3)}(-2) \qquad (-35)(-3) = 105$$
$$= (105)(-2)$$
$$= -210$$

Check Yourself 4

Find the product.

$(-4)(3)(-2)(5)$

In Section 0.3, we saw that the commutative and associative properties for addition can be extended to real numbers. The same is true for multiplication. Look at these examples.

Example 5 Using the Commutative Property of Multiplication

< Objective 2 >

Find each product.

$$(-5)(7) = (7)(-5) = -35$$
$$(-6)(-7) = (-7)(-6) = 42$$

The order in which we multiply does not affect the product.

Check Yourself 5

Show that $(-8)(-5) = (-5)(-8)$.

Property

The Commutative Property of Multiplication

The order in which we multiply does not change the product. Multiplication is *commutative*. In symbols, for any real numbers a and b,

$a \cdot b = b \cdot a$ The centered dot represents multiplication.

What about the way we group numbers in multiplication?

Example 6 Using the Associative Property of Multiplication

< Objective 3 >

NOTE

The symbols [ ] are called *brackets* and are used to group numbers in the same way as parentheses.

Multiply.

$$[(3)(-7)](-2) \qquad \text{or} \qquad (3)[(-7)(-2)]$$
$$= (-21)(-2) \qquad\qquad = (3)(14)$$
$$= 42 \qquad\qquad = 42$$

We group the first two numbers on the left and the second two numbers on the right. The product is the same in either case.

Check Yourself 6

Show that $[(2)(-6)](-3) = (2)[(-6)(-3)]$.

Property

The Associative Property of Multiplication

The way we *group* the numbers does not change the product. Multiplication is *associative*. In symbols, for any real numbers a, b, and c,

$(a \cdot b) \cdot c = a \cdot (b \cdot c)$

Another important property in mathematics is the **distributive property.** The distributive property involves addition and multiplication together. We can illustrate the property with an application.

RECALL

The area of a rectangle is the product of its length and width.

$A = L \cdot W$

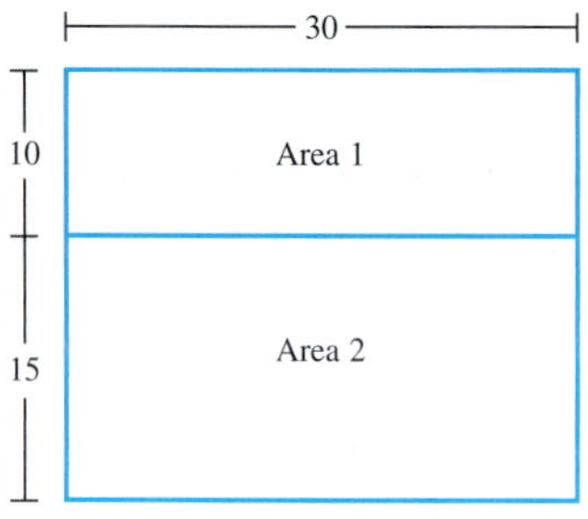

We can find the total area by multiplying the length by the overall width, which is found by adding the two widths.

or

We can find the total area as a sum of the two areas.

Length · Overall width

$\underbrace{30} \cdot \underbrace{(10 + 15)}$

$= 30 \cdot 25$

$= 750$

(Area 1) Length · Width, (Area 2) Length · Width

$30 \cdot 10 + 30 \cdot 15$

$= 300 + 450$

$300 + 450 = 750$

So

$30 \cdot (10 + 15) = 30 \cdot 10 + 30 \cdot 15$

This leads us to the following property.

Property

The Distributive Property

If a, b, and c are any numbers,

$a(b + c) = a \cdot b + a \cdot c$ and $(b + c)a = b \cdot a + c \cdot a$

Example 7 Using the Distributive Property

< Objective 4 >

Use the distributive property to simplify (remove the parentheses).

(a) $5(3 + 4)$

$5(3 + 4) = 5 \cdot 3 + 5 \cdot 4 = 15 + 20 = 35$

(b) $\frac{1}{3}(9 + 12) = \frac{1}{3} \cdot 9 + \frac{1}{3} \cdot 12$

$= 3 + 4 = 7$

NOTES

It is also true that

$5(3 + 4) = 5 \cdot (7) = 35$

It is also true that

$\frac{1}{3}(9 + 12) = \frac{1}{3}(21) = 7$

Check Yourself 7

Use the distributive property to simplify (remove the parentheses).

(a) $4(6 + 7)$ **(b)** $\frac{1}{5}(10 + 15)$

The distributive property can also be used to distribute multiplication over subtraction.

Example 8 Distributing Multiplication over Subtraction

Use the distributive property to remove the parentheses and simplify.

(a) $4(-3 - 6) = 4(-3) - 4(6) = -12 - 24 = -36$
(b) $-7(-3 - 2) = -7(-3) - (-7)(2) = 21 - (-14) = 21 + 14 = 35$

Check Yourself 8

Use the distributive property to remove the parentheses and simplify.

(a) $7(-3 - 4)$ **(b)** $-2(-4 - 3)$

We conclude our discussion of multiplication with a detailed explanation of why the product of two negative numbers must be positive.

Property

The Product of Two Negative Numbers

This argument shows why the product of two negative numbers is positive.

From our earlier work, we know that a number added to its opposite is 0. $5 + (-5) = 0$

Multiply both sides of the statement by -3. $(-3)[5 + (-5)] = (-3)(0)$

A number multiplied by 0 is 0, so on the right we have 0. $(-3)[5 + (-5)] = 0$

We can now use the distributive property on the left. $(-3)(5) + (-3)(-5) = 0$

We know that $(-3)(5) = -15$, so the statement becomes $-15 + (-3)(-5) = 0$

We now have a statement of the form $-15 + \square = 0$. This asks, What number must we add to -15 to get 0, where $\square$ is the value of $(-3)(-5)$? The answer is, of course, 15. This means that

$(-3)(-5) = 15$ The product must be positive.

It doesn't matter what numbers we use in the argument. The product of two negative numbers is always positive.

RECALL

We can interpret a fraction as a division problem.

$\frac{12}{3} = 12 \div 3$

Multiplication and division are related operations. So every division problem can be stated as an equivalent multiplication problem.

$8 \div 4 = 2$ because $8 = 4 \cdot 2$

$\frac{12}{3} = 4$ because $12 = 3 \cdot 4$

The operations are related, so the rules of signs for multiplication are also true for division.

Property

To Divide Real Numbers

Case 1 If two numbers have the same sign, the quotient is positive.

Case 2 If two numbers have different signs, the quotient is negative.

As you would expect, division with fractions or decimals uses the same rules for signs. Example 9 illustrates this concept.

Example 9 Dividing Real Numbers

< Objective 5 >

RECALL

$$\frac{3}{5} \cdot \frac{20}{9} = \frac{\cancel{3} \cdot 2 \cdot 2 \cdot \cancel{5}}{\cancel{5} \cdot \cancel{3} \cdot 3} = \frac{4}{3}$$

Divide.

$$\left(-\frac{3}{5}\right) \div \left(-\frac{9}{20}\right) = \frac{\overset{1}{\cancel{3}}}{\underset{1}{\cancel{5}}} \cdot \frac{\overset{4}{\cancel{20}}}{\underset{3}{\cancel{9}}} = \frac{4}{3}$$

The quotient of two negative numbers is positive, so we omit the negative signs and simply invert the divisor and multiply.

Check Yourself 9

Find each quotient.

(a) $-\frac{5}{8} \div \frac{3}{4}$ **(b)** $-4.2 \div (-0.6)$

As we discussed in Section 0.1, we must be careful when 0 is involved in a division problem. Remember that 0 divided by any nonzero number is 0. However, division *by* 0 is not allowed and is described as *undefined.*

Example 10 Dividing Real Numbers When Zero Is Involved

NOTE

An expression like $-9 \div 0$ has no meaning. There is no answer to the problem. Just write "undefined."

Divide.

(a) $0 \div 7 = 0$ **(b)** $\frac{0}{-4} = 0$

(c) $-9 \div 0$ is undefined. **(d)** $\frac{-5}{0}$ is undefined.

Check Yourself 10

Find each quotient, if possible.

(a) $\frac{0}{-7}$ **(b)** $\frac{-12}{0}$

You can use a calculator to confirm your results from Example 10, as we do in Example 11.

Example 11 Dividing with a Calculator

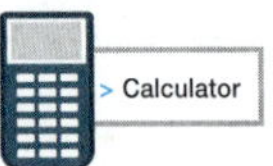


```
-12.567/0▮
```

Use your calculator to find each quotient.

(a) $\frac{-12.567}{0}$

The keystroke sequence on a graphing calculator

[(−)] 12.567 [÷] 0 [ENTER]

results in a "Divide by 0" error message. The calculator recognizes that it cannot divide by zero.

On a scientific calculator, 12.567 [+/−] [÷] 0 [=] results in an error message.

```
ERR:DIVIDE BY 0
1:Quit
2:Goto
```

```
-12.567/0
-10.992/-4.58
              2.4
```

(b) $-10.992 \div -4.58$

The keystroke sequence

[(–)] 10.992 [÷] [(–)] 4.58 [ENTER]

or 10.992 [+/–] [÷] 4.58 [+/–] [=]

yields 2.4.

Check Yourself 11

Find each quotient.

(a) $\dfrac{-31.44}{6.55}$ **(b)** $-23.6 \div 0$

Keep in mind, a calculator is a useful tool when performing computations. However, it is only a tool. The real work should be yours. You should not rely only on a calculator.

You need to have a good sense of whether an answer is reasonable, especially a calculator-derived answer. We all commit "typos" by pressing the wrong button now and again. You need to be able to look at a calculator answer and determine when it is unreasonable, indicating that you made a mistake entering the operation.

We recommend that you perform all computations by hand and then use the calculator to check your work.

Many students have difficulty applying the distributive property when negative numbers are involved. One key to applying the property correctly is to remember that the sign of a number "travels" with that number.

Example 12 Applying the Distributive Property with Negative Numbers

RECALL

We usually enclose negative numbers in parentheses in the middle of an expression to avoid careless errors.

We use brackets rather than nesting parentheses to avoid careless errors.

Evaluate each expression.

(a) $-7(3 + 6) = (-7) \cdot 3 + (-7) \cdot 6$ Apply the distributive property.

$= -21 + (-42)$ Multiply first, then add.

$= -63$

(b) $-3(5 - 6) = -3[5 + (-6)]$ First change the subtraction to addition.

$= (-3) \cdot 5 + (-3)(-6)$ Distribute the -3.

$= -15 + 18$ Multiply first, then add.

$= 3$

(c) $5(-2 - 6) = 5[-2 + (-6)]$

$= 5 \cdot (-2) + 5 \cdot (-6)$

$= -10 + (-30)$

$= -40$ The sum of two negative numbers is negative.

Check Yourself 12

Evaluate each expression.

(a) $-2(-3 + 5)$ **(b)** $4(-3 + 6)$ **(c)** $-7(-3 - 8)$

Recall that a negative sign indicates the opposite of the number that follows. For instance, we have already said that the opposite of 5 is -5, whereas the opposite of -5 is 5. This last instance can be translated as $-(-5) = 5$.

Also recall that any number must correspond to some point on the number line. That is, any nonzero number is either positive or negative. No matter how many negative signs a quantity has, you can always simplify it so that it is represented by a positive or a negative number (zero or one negative sign).

Example 13 Simplifying Real Numbers

> **NOTES**
>
> You should see a pattern emerge. An even number of negative signs gives a positive number, whereas an odd number of negative signs produces a negative number.
>
> In this text, we generally choose to write negative fractions with the sign outside the fraction, such as $-\frac{1}{2}$.

Simplify each expression.

(a) $-(-(-(-4)))$

The opposite of -4 is 4, so $-(-4) = 4$.
The opposite of 4 is -4, so $-(-(-4)) = -4$. The opposite of this last number, -4, is 4, so

$$-(-(-(-4))) = 4$$

(b) $-\frac{-3}{4}$

This is the opposite of $\frac{-3}{4}$ which is $\frac{3}{4}$, a positive number.

Check Yourself 13

Simplify each expression.

(a) $-(-(-(-(-(-(-12))))))$ **(b)** $-\frac{-2}{-3}$

Example 14 Solving an Application Involving Division

Bernal intends to purchase a new car for \$18,950. He will make a down payment of \$1,000 and agrees to make payments over a 48-month period. The total interest is \$8,546. What will his monthly payments be?

Step 1 We are trying to find the monthly payments.

Step 2 First, we subtract the down payment. Then we add the interest to that amount. Finally, we divide that total by the 48 months.

Step 3 \$18,950 − \$1,000 = \$17,950 Subtract the down payment.
\$17,950 + \$8,546 = \$26,496 Add the interest.

Step 4 \$26,496 ÷ 48 = \$552 Divide that total by the 48 months.

Step 5 The monthly payments are \$552, which seems reasonable.

Check Yourself 14

One \$13 bag of fertilizer covers 310 sq ft. What does it cost to cover 7,130 sq ft?

Check Yourself ANSWERS

1. $(-3) + (-3) + (-3) + (-3) = -12$ **2.** **(a)** -75; **(b)** -0.16; **(c)** $-\frac{4}{7}$ **3.** **(a)** 35; **(b)** 48; **(c)** -54; **(d)** 6 **4.** 120 **5.** $40 = 40$ **6.** $36 = 36$ **7.** **(a)** 52; **(b)** 5 **8.** **(a)** -49; **(b)** 14 **9.** **(a)** $-\frac{5}{6}$; **(b)** 7 **10.** **(a)** 0; **(b)** undefined **11.** **(a)** -4.8; **(b)** undefined **12.** **(a)** -4; **(b)** 12; **(c)** 77 **13.** **(a)** -12; **(b)** $-\frac{2}{3}$ **14.** \$299

Reading Your Text

SECTION 0.4

(a) ______________ can be seen as repeated addition.

(b) The product of two nonzero numbers with ______________ signs is always negative.

(c) The product of two nonzero numbers with the same sign is ______________.

(d) The _______ of a rectangle can be found by taking the product of its length and width.

0.4 exercises

Basic Skills | Challenge Yourself | Calculator/Computer | Career Applications | Above and Beyond

Boost *your* GRADE at ALEKS.com!

ALEKS

- Practice Problems
- Self-Tests
- NetTutor
- e-Professors
- Videos

Name ____________

Section ________ Date ________

Answers

1. ________ 2. ________
3. ________ 4. ________
5. ________ 6. ________
7. ________ 8. ________
9. ________ 10. ________
11. ________ 12. ________
13. ________ 14. ________
15. ________ 16. ________
17. ________ 18. ________
19. ________ 20. ________
21. ________ 22. ________
23. ________ 24. ________
25. ________ 26. ________
27. ________ 28. ________
29. ________ 30. ________

< Objectives 1–3 >

Multiply.

1. $7 \cdot 8$
2. $(6)(-12)$
3. $(4)(-3)$
4. $15 \cdot 5$
5. $(-8)(9)$
6. $(-8)(3)$
7. $(-5)\left(\frac{1}{3}\right)$
8. $(-12)(-2)$
9. $(-10)(0)$
10. $(10)(-10)$
11. $(-8)(-8)$
12. $(0)(-50)$
13. $(-4)\left(\frac{5}{7}\right)$ 

14. $(-25)(-8)$
15. $(-9)(-12)$ 

16. $(-3)(-27)$
17. $(-20)(1)$
18. $(1)(-30)$
19. $(-1.3)(6)$
20. $(-25)(5)$
21. $(-10)(-15)$
22. $(-2.4)(0.2)$ 

23. $\left(-\frac{7}{10}\right)\left(-\frac{5}{14}\right)$
24. $\left(-\frac{7}{20}\right)\left(\frac{10}{21}\right)$
25. $\left(\frac{3}{5}\right)\left(-\frac{10}{27}\right)$
26. $\left(-\frac{15}{4}\right)(0)$
27. $\left(-\frac{5}{8}\right)\left(-\frac{4}{15}\right)$
28. $\left(-\frac{8}{21}\right)\left(-\frac{7}{4}\right)$
29. $(-5)(3)(-8)$
30. $(4)(-3)(-5)$

31. $(-5)(-9)(-3)$

32. $(-7)(-5)(-2)$

33. $(2)(-5)(-3)(-5)$

34. $(-2)(-5)(-5)(-6)$

35. $(-4)(-3)(-6)(-2)$

36. $(-8)(3)(-2)(5)$

< Objective 4 >

Use the distributive property to remove parentheses and simplify.

37. $5(-6 + 9)$

38. $12(-5 + 9)$

39. $-8(-9 + 15)$

40. $-11(-8 + 3)$

41. $-4(-5 - 3)$

42. $-2(-7 - 11)$

43. $-4(-6 - 3)$

44. $-6(-3 - 2)$

< Objective 5 >

Divide.

45. $15 \div (-3)$

46. $\frac{90}{18}$

47. $\frac{54}{9}$

48. $-20 \div (-2)$

49. $\frac{-50}{5}$

50. $-36 \div 6$

51. $\frac{-24}{-3}$

52. $\frac{42}{-6}$

53. $\frac{90}{-6}$ > Videos

54. $70 \div (-10)$

55. $18 \div (-1)$

56. $\frac{-250}{-25}$

57. $\frac{0}{-9}$

58. $\frac{-12}{0}$

Answers

31. ______ 32. ______

33. ______ 34. ______

35. ______ 36. ______

37. ______ 38. ______

39. ______ 40. ______

41. ______

42. ______

43. ______

44. ______

45. ______

46. ______

47. ______

48. ______

49. ______

50. ______

51. ______

52. ______

53. ______

54. ______

55. ______ 56. ______

57. ______ 58. ______

Answers

59. ________
60. ________
61. ________
62. ________
63. ________
64. ________
65. ________
66. ________
67. ________
68. ________
69. ________
70. ________
71. ________
72. ________
73. ________
74. ________
75. ________
76. ________
77. ________
78. ________

59. $-180 \div (-15)$

60. $\dfrac{0}{-10}$

61. $-7 \div 0$

62. $\dfrac{-25}{-1}$

63. $\dfrac{-150}{6}$

64. $\dfrac{-80}{-16}$

65. $-45 \div (-9)$

66. $-\dfrac{2}{3} \div \dfrac{4}{9}$

67. $-\dfrac{8}{11} \div \dfrac{18}{55}$

68. $(-8) \div (-4)$

69. $\dfrac{7}{10} \div \left(-\dfrac{14}{25}\right)$

70. $\dfrac{6}{13} \div \left(-\dfrac{18}{39}\right)$

71. $\dfrac{-75}{15}$

72. $-\dfrac{5}{8} \div \left(-\dfrac{5}{16}\right)$

Basic Skills | **Challenge Yourself** | Calculator/Computer | Career Applications | Above and Beyond

Determine whether each statement is **true** *or* **false.**

73. A number divided by 0 is 0.

74. The product of 0 and any number is 0.

Complete each statement with **never, sometimes,** *or* **always.**

75. The product of three negative numbers is ____________ positive.

76. The quotient of a positive number and a negative number is ____________ negative.

77. **Science and Medicine** A patient lost 42 pounds (lb). If he lost 3 lb each week, how long has he been dieting?

78. **Business and Finance** Patrick worked all day mowing lawns and was paid \$9 per hour. If he had \$125 at the end of a 9-hour day, how much did he have before he started working?

79. Business and Finance A 4.5-lb can of food costs \$8.91. What is the cost per pound?

80. Business and Finance Suppose that you and your two brothers bought equal shares of an investment for a total of \$20,000 and sold it later for \$16,232. How much did each person lose?

81. Science and Medicine Suppose that the temperature outside is dropping at a constant rate. At noon, the temperature is 70°F and it drops to 58°F at 5:00 P.M. How much did the temperature change each hour? > Videos

82. Science and Medicine A chemist has 84 ounces (oz) of a solution. She pours the solution into test tubes. Each test tube holds $\frac{2}{3}$ oz. How many test tubes can she fill?

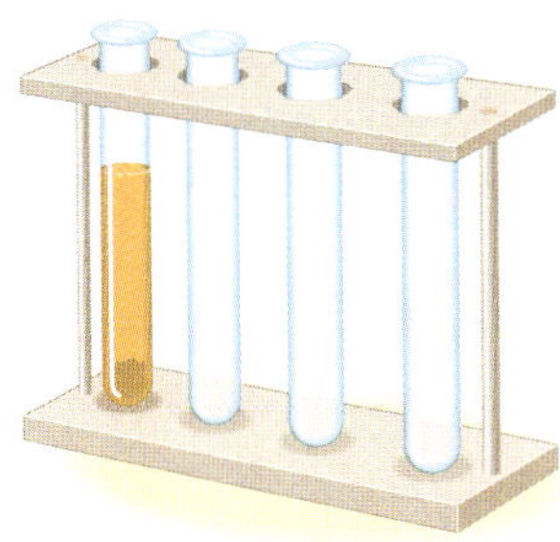

To evaluate an expression involving a fraction (indicating division), we evaluate the numerator and then the denominator. We then divide the numerator by the denominator as the last step. Using this approach, find the value of each expression.

83. $\frac{5 - 15}{2 + 3}$

84. $\frac{4 - (-8)}{2 - 5}$

85. $\frac{-6 + 18}{-2 - 4}$

86. $\frac{-4 - 21}{3 - 8}$

87. $\frac{(5)(-12)}{(-3)(5)}$

88. $\frac{(-8)(-3)}{(2)(-4)}$

Basic Skills | Challenge Yourself | **Calculator/Computer** | Career Applications | Above and Beyond

Divide by using a calculator. Round answers to the nearest thousandth.

89. $-5.634 \div 2.398$

90. $-2.465 \div 7.329$

91. $-18.137 \div (-5.236)$

92. $-39.476 \div (-17.629)$

93. $32.245 \div (-48.298)$

94. $43.198 \div (-56.249)$

Answers

79. ______

80. ______

81. ______

82. ______

83. ______

84. ______

85. ______

86. ______

87. ______

88. ______

89. ______

90. ______

91. ______

92. ______

93. ______

94. ______

Basic Skills | Challenge Yourself | Calculator/Computer | Career Applications | **Above and Beyond**

Answers

95. ______

96. ______

97. ______

98. ______

95. Create an example to show that the division of real numbers is *not* commutative.

96. Create an example to show that the division of real numbers is *not* associative.

97. Here is another conjecture to consider:

$|ab| = |a|\,|b|$ for all numbers a and b

(See the discussion in Section 0.3, following exercise 93, concerning testing a conjecture.) Test this conjecture for various values of a and b. Use positive numbers, negative numbers, and 0. Summarize your results in a rule.

98. Use a calculator (or mental calculations) to evaluate each expression.

$$\frac{5}{0.1}, \quad \frac{5}{0.01}, \quad \frac{5}{0.001}, \quad \frac{5}{0.0001}, \quad \frac{5}{0.00001}$$

In this series of problems, while the numerator is always 5, the denominator is getting smaller (and is getting closer to 0). As this happens, what is happening to the value of the fraction?

Write an argument that explains why $\frac{5}{0}$ could not have any finite value.

Answers

1. 56 **3.** -12 **5.** -72 **7.** $-\frac{5}{3}$ **9.** 0 **11.** 64 **13.** $-\frac{20}{7}$
15. 108 **17.** -20 **19.** -7.8 **21.** 150 **23.** $\frac{1}{4}$ **25.** $-\frac{2}{9}$
27. $\frac{1}{6}$ **29.** 120 **31.** -135 **33.** -150 **35.** 144 **37.** 15
39. -48 **41.** 32 **43.** 36 **45.** -5 **47.** 6 **49.** -10 **51.** 8
53. -15 **55.** -18 **57.** 0 **59.** 12 **61.** Undefined
63. -25 **65.** 5 **67.** $-\frac{20}{9}$ **69.** $-\frac{5}{4}$ **71.** -5 **73.** False
75. never **77.** 14 weeks **79.** \$1.98 **81.** -2.4°F **83.** -2
85. -2 **87.** 4 **89.** -2.349 **91.** 3.464 **93.** -0.668
95. Above and Beyond **97.** Above and Beyond

0.5 Exponents and Order of Operations

< 0.5 Objectives >

1 > Write a product of like factors in exponential form

2 > Evaluate numbers with exponents

3 > Use the order of operations

In Section 0.4, we mentioned that multiplication is a form for repeated addition. For example, an expression with repeated addition, such as

$3 + 3 + 3 + 3 + 3$

can be rewritten as

$5 \cdot 3$

Thus, multiplication is "shorthand" for repeated addition.

In algebra, we frequently have a number or variable that is repeated in an expression several times. For instance, we might have

$5 \cdot 5 \cdot 5$

To abbreviate this product, we write

$5 \cdot 5 \cdot 5 = 5^3$

> CAUTION
>
> Be careful: 5^3 is *not* the same as $5 \cdot 3$.
> $5^3 = 5 \cdot 5 \cdot 5 = 125$ and
> $5 \cdot 3 = 15$

This is called **exponential notation** or **exponential form.** The exponent or power, here 3, indicates the number of times that the factor or base, here 5, appears in a product.

$5 \cdot 5 \cdot 5 = 5^3$ (Exponent or power: 3; Factor or base: 5)

Example 1 Writing Expressions in Exponential Form

< Objective 1 >

(a) Write $3 \cdot 3 \cdot 3 \cdot 3$ in exponential form.

The number 3 appears 4 times in the product, so

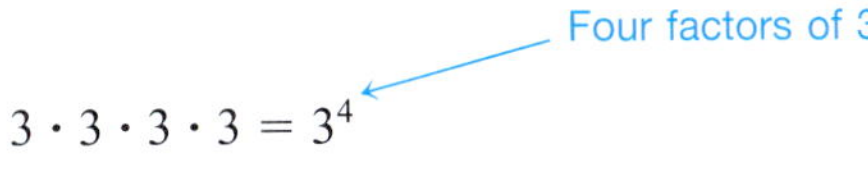

$3 \cdot 3 \cdot 3 \cdot 3 = 3^4$

This is read "3 to the fourth power."

(b) Write $10 \cdot 10 \cdot 10$ in exponential form.

Since 10 appears 3 times in the product, you can write

$10 \cdot 10 \cdot 10 = 10^3$

This is read "10 to the third power" or "10 cubed."

Check Yourself 1

Write in exponential form.

(a) $4 \cdot 4 \cdot 4 \cdot 4 \cdot 4 \cdot 4$ **(b)** $10 \cdot 10 \cdot 10 \cdot 10$

When evaluating a number raised to a power, it is important to note whether there is a sign attached to the number. Note that

$(-2)^4 = (-2)(-2)(-2)(-2) = 16$

whereas,

$-2^4 = -(2)(2)(2)(2) = -16$

Example 2 Evaluating Exponential Expressions

< Objective 2 >

NOTE

$-3^4 = -1 \cdot 3^4$
$= -1 \cdot 81$
$= -81$

whereas,

$(-3)^4 = (-3)(-3)(-3)(-3)$
$= 81$

Evaluate each expression.

(a) $(-3)^3 = (-3)(-3)(-3) = -27$

(b) $-3^3 = -(3)(3)(3) = -27$

(c) $(-3)^4 = (-3)(-3)(-3)(-3) = 81$

(d) $-3^4 = -(3)(3)(3)(3) = -81$

Check Yourself 2

Evaluate each expression.

(a) $(-4)^3$ **(b)** -4^3 **(c)** $(-4)^4$ **(d)** -4^4

NOTE

Most computer software, such as Excel, uses the caret, ^, when indicating that an exponent follows.

You can use a calculator to help you evaluate expressions containing exponents. If you have a graphing calculator, you can use the caret key, [^]. Enter the base, followed by the caret key, and then enter the exponent. Some calculators use a key labeled [y^x] instead of the caret key.

Remember, we use a calculator as an aid or tool, not a crutch. You should be able to evaluate each of these expressions by hand, if necessary.

Example 3 Evaluating Expressions with Exponents

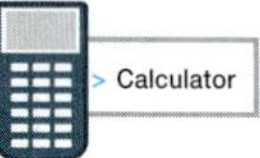

Use your calculator to evaluate each expression.

(a) $3^5 = 243$ Type 3 [^] 5 [ENTER]
or 3 [y^x] 5 [=]

(b) $2^{10} = 1{,}024$ 2 [^] 10 [ENTER]
or 2 [y^x] 10 [=]

Check Yourself 3

Use your calculator to evaluate each expression.

(a) 3^4 **(b)** 2^{16}

NOTE

To evaluate an expression, we find a number that is equal to the expression.

We used the word *expression* when discussing numbers taken to powers, such as 3^4. But what about something like $4 + 12 - 6$? We call *any* meaningful combination of numbers and operations an **expression.** When we evaluate an expression, we find a number that is equal to the expression. To evaluate an expression, we need to establish a set of rules that tell us the correct order in which to perform the operations. To see why, simplify the expression $5 + 2 \cdot 3$.

Method 1	or	Method 2
$5 + 2 \cdot 3$		$5 + 2 \cdot 3$
Add first		Multiply first
$= 7 \cdot 3$		$= 5 + 6$
$= 21$		$= 11$

> CAUTION

Only one of these results can be correct.

Since we get different answers depending on how we do the problem, the language of algebra would not be clear if there were no agreement on which method is correct. The following rules tell us the order in which operations should be done.

Step by Step

The Order of Operations

Parentheses, brackets, and fraction bars are all examples of grouping symbols.

- **Step 1** Evaluate all expressions inside grouping symbols.
- **Step 2** Evaluate all expressions involving exponents.
- **Step 3** Do any multiplication or division in order, working from left to right.
- **Step 4** Do any addition or subtraction in order, working from left to right.

Example 4 Evaluating Expressions

< Objective 3 >

NOTE

Method 2 in the previous discussion is the correct one.

(a) Evaluate $5 + 2 \cdot 3$.

There are no parentheses or exponents, so start with step 3: First multiply and then add.

$5 + 2 \cdot 3$ Multiply first

$= 5 + 6$ Then add

$= 11$

(b) Evaluate $10 \cdot 4 \div 2 \cdot 5$.

Perform the multiplication and division from left to right.

$10 \cdot 4 \div 2 \cdot 5$

$= 40 \div 2 \cdot 5$

$= 20 \cdot 5$

$= 100$

Check Yourself 4

Evaluate each expression.

(a) $20 - 3 \cdot 4$ **(b)** $9 + 6 \div 3$ **(c)** $10 \cdot 6 \div 3 \cdot 2$

Example 5 Evaluating Expressions

Evaluate $-5 \cdot 3^2$.

$-5 \cdot 3^2$ Evaluate the exponential expression first.

$= -5 \cdot 9$

$= -45$

Check Yourself 5

Evaluate $-4 \cdot 2^4$.

Modern calculators correctly interpret the order of operations as demonstrated in Example 6.

Example 6 Using a Calculator to Evaluate Expressions

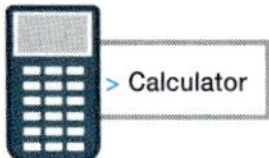

Use your scientific or graphing calculator to evaluate each expression.

(a) $24.3 + 6.2 \cdot 3.5$

When evaluating expressions by hand, you must consider the order of operations. In this case, the multiplication must be done before the addition. With a modern calculator, you need only enter the expression correctly. The calculator is programmed to follow the order of operations.

Entering 24.3 [+] 6.2 [×] 3.5 [ENTER]

yields the evaluation 46.

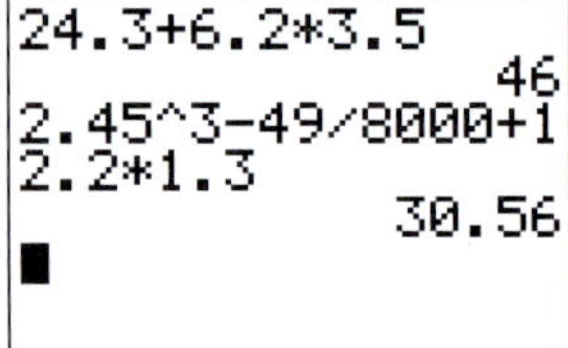

(b) $(2.45)^3 - 49 \div 8{,}000 + 12.2 \cdot 1.3$

As we mentioned earlier, some calculators use the caret (∧) to designate exponents. Others use the symbol x^y (or y^x).

Entering 2.45 [∧] 3 [−] 49 [÷] 8000 [+] 12.2 [×] 1.3 [ENTER]

or 2.45 [y^x] 3 [−] 49 [÷] 8000 [+] 12.2 [×] 1.3 [=]

yields 30.56.

Check Yourself 6

Use your calculator to evaluate each expression.

(a) $67.89 - 4.7 \cdot 12.7$ **(b)** $4.3 \cdot 55.5 - (3.75)^3 + 8{,}007 \div 1{,}600$

Operations inside grouping symbols are done first.

Example 7 Evaluating Expressions

Evaluate $(5 + 2) \cdot 3$.

Do the operation inside the parentheses as the first step.

$(5 + 2) \cdot 3 = 7 \cdot 3 = 21$

Add

Check Yourself 7

Evaluate $4(9 - 3)$.

The principle is the same when more than two "levels" of operations are involved.

Example 8 Evaluating Expressions

(a) Evaluate $4(-2 + 7)^3$.

Add inside the parentheses first.

$4(-2 + 7)^3 = 4(5)^3$ — Evaluate the exponential expression.

$= 4 \cdot 125$ — Multiply

$= 500$

(b) Evaluate $5(7 - 3)^2 - 10$.

Evaluate the expression inside the parentheses.

$5(7 - 3)^2 - 10 = 5(4)^2 - 10$ — Evaluate the exponential expression.

$= 5 \cdot 16 - 10$ — Multiply

$= 80 - 10 = 70$ — Subtract

Check Yourself 8

Evaluate.

(a) $4 \cdot 3^3 + 8 \cdot (-11)$ **(b)** $12 + 4(2 + 3)^2$

Parentheses and brackets are not the only types of grouping symbols. Example 9 demonstrates the fraction bar as a grouping symbol.

Example 9 Using the Order of Operations with Grouping Symbols

> CAUTION

You may not "cancel" the 2's, because the numerator is being added, not multiplied.

$\frac{2 + 14}{2}$ is incorrect!

Evaluate $3 + \frac{2 + 14}{2} \cdot 5$.

$3 + \frac{2 + 14}{2} \cdot 5 = 3 + \frac{16}{2} \cdot 5$ — The fraction bar acts as a grouping symbol.

$= 3 + 8 \cdot 5$ — We perform the division first because it precedes the multiplication.

$= 3 + 40$

$= 43$

Check Yourself 9

Evaluate $4 \cdot \frac{3^2 + 2 \cdot 3}{5} + 6$.

Many formulas require proper use of the order of operations. We conclude this chapter with one such application.

For obvious ethical reasons, children are rarely subjects in medical research. Nonetheless, when children are ill doctors sometimes determine that medication is necessary. Dosage recommendations for adults are based on research studies and the medical community believes that in most cases, children need smaller dosages than adults.

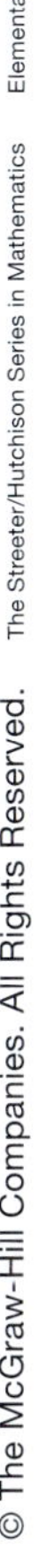

There are many formulas for determining the proper dosage for a child. The more complicated (and accurate) ones use a child's height, weight, or body mass. All of the formulas try to answer the question, "What fraction of an adult dosage should a child be given?"

Dr. Thomas Young constructed a conservative but simple model using only a child's age.

Example 10 An Allied Health Application

One formula for calculating the proper dosage of a medication for a child based on the recommended adult dosage and the child's age (in years) is Young's Rule.

$$\text{Child's dose} = \left(\frac{\text{Age}}{\text{Age} + 12}\right) \times \text{Adult dose}$$

Find the proper dose for a 3-year-old child if the recommended adult dose is 24 milligrams (mg), according to Young's Rule.

Step 1 We are being asked to use the formula to find the proper dosage for a child who is 3, given that an adult should take 24 mg.

Step 2 We need to evaluate the expression formed when the age of the child and the adult dosage are taken into account.

Step 3

$$\text{Child's dose} = \left(\frac{\text{Age}}{\text{Age} + 12}\right) \times \text{Adult dose}$$

$$= \left(\frac{(3)}{(3)+12}\right) \times (24 \text{ mg})$$

> **NOTE**
> The child's age is 3; an adult should take a 24-mg dose.

Step 4

$$\left(\frac{(3)}{(3)+12}\right) \times (24 \text{ mg}) = \frac{3}{15} \times 24 \text{ mg}$$

$$= \frac{1}{5} \times \frac{24}{1}$$

$$= \frac{24}{5}$$

$$\frac{24}{5} = 4\frac{4}{5} = 4.8$$

The fraction bar is a grouping symbol, so we add the numbers in the denominator first, and then we continue simplifying the fraction.

> **RECALL**
> $\frac{24}{5} = 24 \div 5$

Step 5 According to Young's Rule, the proper dose for a 3-year-old child is 4.8 mg.

Reasonableness

A 3-year-old is much younger than an adult, so we would expect the child's dose to be much smaller than the adult's dose.

Check Yourself 10

The approximate length of the belt pictured is given by

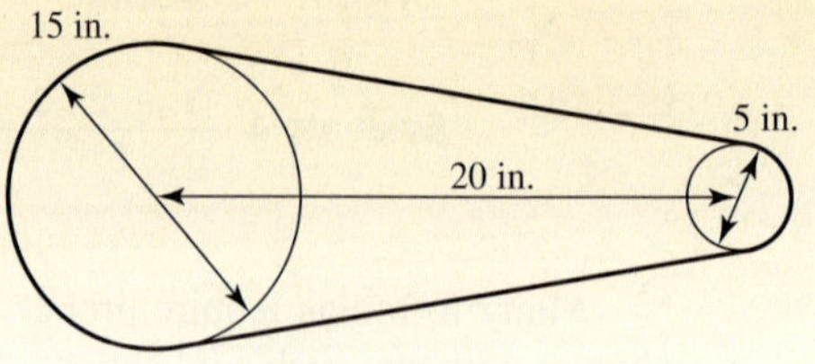

$$\frac{22}{7}\left(\frac{1}{2}\cdot 15 + \frac{1}{2}\cdot 5\right) + 2\cdot 21$$

Find the length of the belt.

> **NOTE**
> This formula uses an approximation of the formula for the *circumference*, or distance around, a circle.

29. 3×4^2

30. $(3 \times 4)^2$

31. $5 + 2^2$

32. $(5 + 2)^2$

33. $3^4 \times 2^4$

34. $(3 \times 2)^4$

< Objective 3 >

Evaluate each expression.

35. $4 + 3 \cdot 5$

36. $10 - 4 \cdot 2$

37. $(7 + 2) \cdot 6$

38. $(9 - 5) \cdot 3$

39. $-12 - 8 \div 4$

40. $10 + 20 \div 5$

41. $(12 - 8) \div 4$

42. $(10 + 20) \div 5$

43. $8 \cdot 7 + 2 \cdot 2$

44. $48 \div 8 - 14 \div 2$

45. $(7 \cdot 5) + 3 \cdot 2$

46. $48 \div (8 - 4) \div 2$

47. $3 \cdot 5^2$

48. $5 \cdot 2^3$

49. $(3 \cdot 5)^2$

50. $(5 \cdot 2)^3$

51. $4 \cdot 3^2 - 2$  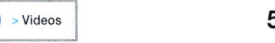

52. $3 \cdot 2^4 - 8$

53. $7(2^3 - 5)$

54. $3(7 - 3^2)$

55. $3 \cdot 2^4 - 26 \cdot 2$

56. $4 \cdot 2^3 - 15 \cdot 6$

57. $(2 \cdot 4)^2 - 8 \cdot 3$

58. $(3 \cdot 2)^3 - 7 \cdot 3$

Answers

29. ______ 30. ______

31. ______ 32. ______

33. ______ 34. ______

35. ______ 36. ______

37. ______ 38. ______

39. ______ 40. ______

41. ______ 42. ______

43. ______ 44. ______

45. ______ 46. ______

47. ______

48. ______

49. ______

50. ______

51. ______

52. ______

53. ______

54. ______

55. ______

56. ______

57. ______

58. ______

Answers

59. ____
60. ____
61. ____
62. ____
63. ____
64. ____
65. ____
66. ____
67. ____
68. ____
69. ____
70. ____
71. ____
72. ____
73. ____
74. ____
75. ____
76. ____
77. ____
78. ____
79. ____
80. ____

59. $5(3 + 4)^2$

60. $3(8 - 4)^2$

61. $(5 \cdot 3 + 4)^2$

62. $(3 \cdot 8 - 4)^2$

63. $5[3(2 + 5) - 5]$

64. $\dfrac{11 - (-9) + 6(8 - 2)}{2 + 3 \cdot 4}$

65. $-2[(3 - 5)^2 - (-4 + 2)^3 \cdot (8 \div 4 \cdot 2)]$

66. $5 \cdot 4 - 2^3$

67. $4(2 + 3)^2 - 125$

68. $8 + 2(3 + 3)^2$

69. $(4 \cdot 2 + 3)^2 - 25$

70. $8 + (2 \cdot 3 + 3)^2$

71. $[-20 - 4^2 + (-4)^2 + 2] \div 9$

72. $14 + 3 \cdot 9 - 28 \div 7 \cdot 2$

73. $4 \cdot 8 \div 2 - 5^2$

74. $-12 - 8 \div 4 \cdot 2$

75. $15 + 5 - 3 \cdot 2 + (-2)^3$

76. $-8 + 14 \div 2 \cdot 4 - 3$

Basic Skills | **Challenge Yourself** | Calculator/Computer | Career Applications | Above and Beyond

Determine whether each statement is **true** *or* **false.**

77. A negative number raised to an even power results in a positive number.

78. Exponential notation is shorthand for repeated addition.

Complete each statement with **never, sometimes,** *or* **always.**

79. Operations inside grouping symbols are ____________ done first.

80. In the order of operations, division is ____________ done before multiplication.

81. **Science and Medicine** Over the last 2,000 years, the earth's population has doubled approximately 5 times. Write the phrase "doubled 5 times" in exponential form.

82. **Geometry** The volume of a cube with each edge of length 9 inches (in.) is given by $9 \cdot 9 \cdot 9$. Write the volume, using exponential notation.

83. **Statistics** On an 8-hour trip, Jack drives $2\frac{3}{4}$ hr and Pat drives $2\frac{1}{2}$ hr. How much longer do they still need to drive?

84. **Statistics** A runner decides to run 20 miles (mi) each week. She runs $5\frac{1}{2}$ mi on Sunday, $4\frac{1}{4}$ mi on Tuesday, $4\frac{3}{4}$ mi on Wednesday, and $2\frac{1}{8}$ mi on Friday. How far does she need to run on Saturday to meet her goal?

Evaluate with a calculator. Round your answers to the nearest tenth.

85. $(1.2)^3 \div 2.0736 \cdot 2.4 + 1.6935 - 2.4896$

86. $(5.21 \cdot 3.14 - 6.2154) \div 5.12 - 0.45625$

87. $1.23 \cdot 3.169 - 2.05194 + (5.128 \cdot 3.15 - 10.1742)$

88. $4.56 + (2.34)^4 \div 4.7896 \cdot 6.93 \div 27.5625 - 3.1269 + (1.56)^2$

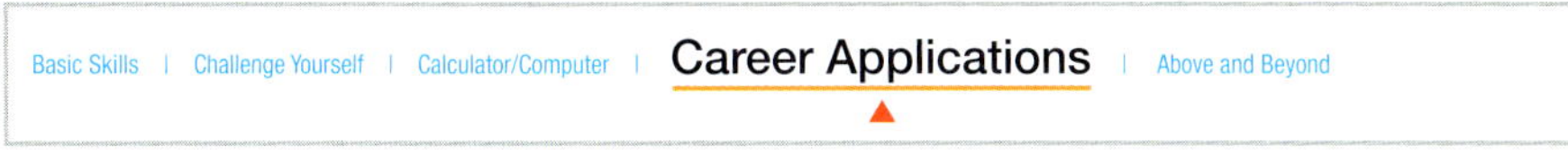

89. **Business and Finance** The interest rate on an auto loan was $12\frac{3}{8}\%$ in May and $14\frac{1}{4}\%$ in September. By how many percentage points did the interest rate increase between May and September?

90. **Manufacturing Technology** A $3\frac{3}{8}$-in. cut needs to be made in a piece of material. The cut rate is $\frac{3}{4}$ in. per minute. How many minutes does it take to make this cut?

91. **Manufacturing Technology** Peer's Pipe Fitters started July with 1,789 gallons (gal) of liquefied petroleum gas (LP) in its tank. After 21 working days, there were 676 gal left in the tank. How much gas was used on each working day, on average?

92. **Business and Finance** Three friends bought equal shares in an investment. Between them, they paid $21,000 for the shares. Later, they were able to sell their shares for only $17,232. How much did each person lose on the investment?

Answers

81. ______
82. ______
83. ______
84. ______
85. ______
86. ______
87. ______
88. ______
89. ______
90. ______
91. ______
92. ______

Basic Skills | Challenge Yourself | Calculator/Computer | Career Applications | **Above and Beyond**

Answers

93. ______

94. ______

93. Insert grouping symbols in the proper place so that the value of the expression $36 \div 4 + 2 - 4$ is 2.

94. Work with a small group of students.

Part 1: Write the numbers 1 through 25 on slips of paper and put the slips in a pile, face down. Each of you randomly draws a slip of paper until each person has five slips. Turn the papers over and write down the five numbers. Put the five papers back in the pile, shuffle, and then draw one more. This last number is the answer. The first five numbers are the problem. Your task is to arrange the first five into a computation, using all you know about the order of operations, so that the answer is the last number. Each number must be used and may be used only once. If you cannot find a way to do this, pose it as a question to the whole class. Is this guaranteed to work?

Part 2: Use your five numbers in a problem, each number being used and used only once, for which the answer is 1. Try this 9 more times with the numbers 2 through 10. You may find more than one way to do each of these. Surprising, isn't it?

Part 3: Be sure that when you successfully find a way to get the desired answer by using the five numbers, you can then write your steps, using the correct order of operations. Write your 10 problems and exchange them with another group to see if they get these same answers when they do your problems.

Answers

1. 3^5 **3.** 7^5 **5.** 8^6 **7.** $(-2)^3$ **9.** 9 **11.** 16 **13.** -512
15. -512 **17.** -25 **19.** 16 **21.** 32 **23.** 1,000
25. 1,000,000 **27.** 128 **29.** 48 **31.** 9 **33.** 1,296 **35.** 19
37. 54 **39.** -14 **41.** 1 **43.** 60 **45.** 41 **47.** 75 **49.** 225
51. 34 **53.** 21 **55.** -4 **57.** 40 **59.** 245 **61.** 361 **63.** 80
65. -72 **67.** -25 **69.** 96 **71.** -2 **73.** -9 **75.** 6
77. True **79.** always **81.** 2^5 **83.** $2\frac{3}{4}$ hr **85.** 1.2 **87.** 7.8
89. $1\frac{7}{8}\%$ **91.** 53 gal/day **93.** $36 \div (4 + 2) - 4$

Definition/Procedure	Example	Reference
A Review of Fractions		**Section 0.1**
Equivalent Fractions If the numerator and denominator of a fraction are both multiplied by some nonzero number, the result is a fraction that is equivalent to the original fraction.	$\frac{4 \cdot 3}{5 \cdot 3} = \frac{12}{15}$	p. 3
Simplifying Fractions A fraction is in simplest terms when the numerator and denominator have no common factor.	$\frac{9}{21} = \frac{3 \cdot 3}{3 \cdot 7} = \frac{3}{7}$	p. 4
Multiplying Fractions To multiply two fractions, multiply the numerators, then multiply the denominators. Simplification can be done before or after the multiplication.	$\frac{2}{3} \cdot \frac{5}{6} = \frac{10}{18} = \frac{5}{9}$	p. 6
Dividing Fractions To divide two fractions, invert the divisor (the second fraction), then multiply the fractions.	$\frac{3}{5} \div \frac{2}{7} = \frac{3}{5} \cdot \frac{7}{2} = \frac{21}{10}$	p. 6
Adding Fractions To add two fractions, find the LCD (least common denominator), rewrite the fractions with this denominator, then add the numerators.	$\frac{2}{5} + \frac{5}{8} = \frac{16}{40} + \frac{25}{40} = \frac{41}{40}$	p. 7
Subtracting Fractions To subtract two fractions, find the LCD, rewrite the fractions with this denominator, then subtract the numerators.	$\frac{2}{3} - \frac{1}{4} = \frac{8}{12} - \frac{3}{12} = \frac{5}{12}$	p. 8
Solving Applications Follow this step-by-step approach when solving applications. **Step 1** Read the problem carefully to determine what you are being asked to find and what information is given in the application. **Step 2** Decide what you will do to solve the problem. **Step 3** Write down the complete (mathematical) statement necessary to solve the problem. **Step 4** Perform any calculations or other mathematics needed to solve the problem. **Step 5** Answer the question. Be sure to include units with your answer, when appropriate. Check to make certain that your answer is reasonable.	A foundation requires 2,668 blocks. If a contractor has 879 blocks on hand, how many more blocks need to be ordered? **Step 1** We want to find out how many more blocks the contractor needs. The contractor has 879 blocks, but needs a total of 2,668 blocks. **Step 2** This is a subtraction problem. **Step 3** 2,668 − 879 **Step 4** 2,668 − 879 = 1,789 **Step 5** The contractor needs to order 1,789 blocks. *Reasonableness Check* 1,789 + 879 = 2,668	p. 4
Real Numbers		**Section 0.2**
Positive Numbers Numbers used to name points to the right of 0 on the number line. **Negative Numbers** Numbers used to name points to the left of 0 on the number line.	Negative numbers; Positive numbers −3 −2 −1 0 1 2 3 Zero is neither positive nor negative.	p. 16

Continued

Definition/Procedure	Example	Reference
Natural Numbers The counting numbers.	The natural numbers are $\{1, 2, 3, \ldots\}$	p. 16
Integers The set consisting of the natural numbers, their opposites, and 0.	The integers are $\{\ldots, -3, -2, -1, 0, 1, 2, 3, \ldots\}$	p. 17
Rational Number Any number that can be expressed as the ratio of two integers.	Rational numbers are $\frac{2}{3}, \frac{5}{1}, 0.234$	p. 17
Irrational Number Any number that is not rational.	Irrational numbers include $\sqrt{2}$ and π	p. 18
Real Numbers Rational and irrational numbers together.	All the numbers listed are real numbers.	p. 18
Opposites Two numbers are opposites if the points name the same distance from 0 on the number line, but in opposite directions.	5 units, 5 units; −5, 0, 5 The opposite of 5 is −5.	p. 19
The opposite of a positive number is negative.		p. 19
The opposite of a negative number is positive.	3 units, 3 units; −3, 0, 3 The opposite of −3 is 3.	p. 19
0 is its own opposite.		p. 19
Absolute Value The distance on the number line between the point named by a number and 0. The absolute value of a number is always positive or 0. The absolute value of a number is called its **magnitude.**	The absolute value of a number a is written $\|a\|$. $\|7\| = 7 \quad \|-8\| = 8$	p. 19

Operations on Real Numbers

Sections 0.3–0.4

Definition/Procedure	Example	Reference
To Add Real Numbers **1.** If two numbers have the same sign, add their magnitudes. Give to the sum the sign of the original numbers. **2.** If two numbers have different signs, subtract the smaller absolute value from the larger. Give to the result the sign of the number with the larger magnitude.	$5 + 8 = 13$ $-3 + (-7) = -10$ $5 + (-3) = 2$ $7 + (-9) = -2$	p. 27 p. 27
To Subtract Real Numbers To subtract real numbers, add the first number and the opposite of the number being subtracted.	$4 - (-2) = 4 + 2 = 6$ The opposite of −2	p. 29
To Multiply Real Numbers To multiply real numbers, multiply the absolute values of the numbers. Then attach a sign to the product according to the following rules: **1.** If the numbers have different signs, the product is negative. **2.** If the numbers have the same sign, the product is positive.	$5 \cdot 7 = 35$ $(-4)(-6) = 24$ $(8)(-7) = -56$	p. 37

Definition/Procedure	Example	Reference
To Divide Real Numbers To divide real numbers, divide the absolute values of the numbers. Then attach a sign to the quotient according to the following rules: **1.** If the numbers have the same sign, the quotient is positive. **2.** If the numbers have different signs, the quotient is negative.	$\frac{-8}{-2} = 4$ $27 \div (-3) = -9$ $\frac{-16}{8} = -2$	*p.* 42
The Properties of Addition and Multiplication		
The Commutative Properties If a and b are any numbers, then **1.** $a + b = b + a$ **2.** $a \cdot b = b \cdot a$	$3 + 4 = 4 + 3$ $7 = 7$	*p.* 39
The Associative Properties If a, b, and c are any numbers, then **1.** $a + (b + c) = (a + b) + c$ **2.** $a \cdot (b \cdot c) = (a \cdot b) \cdot c$	$3 \cdot (4 \cdot 5) = (3 \cdot 4) \cdot 5$ $3 \cdot (20) = (12) \cdot 5$ $60 = 60$	*p.* 39
The Distributive Property If a, b, and c are any numbers, then $a(b + c) = a \cdot b + a \cdot c$.	$2(5 + 3) = 2 \cdot 5 + 2 \cdot 3$ $2(8) = 10 + 6$ $16 = 16$	*p.* 40
Exponents and Order of Operations		Section 0.5
Notation Exponent ↓ $a^4 = \underbrace{a \cdot a \cdot a \cdot a}_{\text{4 factors}}$ ↑ Base The number or letter used as a factor, here a, is called the *base*. The *exponent,* which is written above and to the right of the base, tells us how many times the base is used as a factor.	$5^3 = 5 \cdot 5 \cdot 5$ $= 125$ $3^2 \cdot 7^3 = 3 \cdot 3 \cdot 7 \cdot 7 \cdot 7$	*p.* 51
The Order of Operations **Step 1** Do any operations within grouping symbols. **Step 2** Evaluate all expressions containing exponents. **Step 3** Do any multiplication or division in order, working from left to right. **Step 4** Do any addition or subtraction in order, working from left to right.	Operate inside grouping symbols. $5 + 3(6 - 4)^2$ Evaluate the exponential expression. $= 5 + 3 \cdot 2^2$ Multiply $= 5 + 3 \cdot 4$ Add $= 5 + 12$ $= 17$	*p.* 53

summary exercises :: chapter 0

This summary exercise set is provided to give you practice with each of the objectives of this chapter. Each exercise is keyed to the appropriate chapter section. When you are finished, you can check your answers to the odd-numbered exercises in the back of the text. If you have difficulty with any of these questions, go back and reread the examples from that section. The answers to the even-numbered exercises appear in the *Instructor's Manual.* Your instructor may give you guidelines on how best to use these exercises in your instructional setting.

0.1 *In exercises 1 to 3, write three fractional representations for each number.*

1. $\frac{5}{7}$

2. $\frac{3}{11}$

3. $\frac{4}{9}$

4. Use the fundamental principle to write the fraction $\frac{24}{64}$ in simplest form.

In exercises 5 to 12, perform the indicated operations. Write each answer in simplest form.

5. $\frac{7}{15} \times \frac{5}{21}$

6. $\frac{10}{27} \times \frac{9}{20}$

7. $\frac{5}{17} \div \frac{15}{34}$

8. $\frac{7}{15} \div \frac{14}{25}$

9. $\frac{7}{8} + \frac{15}{24}$

10. $\frac{5}{18} + \frac{7}{12}$

11. $\frac{11}{18} - \frac{2}{9}$

12. $\frac{11}{27} - \frac{5}{18}$

Solve each application.

13. Construction A kitchen measures $\frac{16}{3}$ by 4 yd. If you purchase linoleum that costs \$9 per square yard (you cannot purchase a partial square yard), how much will it cost to cover the floor?

14. Social Science The scale on a map uses 1 in. to represent 80 mi. If two cities are $\frac{11}{4}$ in. apart on the map, what is the actual distance between the cities?

15. Construction An 18-acre piece of land is to be subdivided into home lots that are each $\frac{3}{8}$ acre. How many lots can be formed?

16. Geometry Find the perimeter of the given figure.

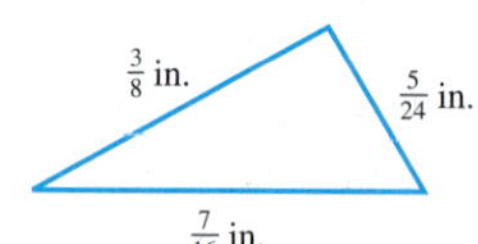

0.2 *Complete the statement.*

17. The absolute value of 12 is ________.

18. The opposite of -8 is ________.

19. $-|-3| =$ ________

20. $-(-20) =$ ________

21. $-|-4| =$ ________

22. $|-(-5)| =$ ________

23. The absolute value of -16 is ________.

24. The opposite of the absolute value of -9 is ________.

Complete the statement, using the symbol $<$, $>$, or $=$.

25. -3 ________ -1

26. -6 ________ $-|-6|$

27. $-|-7|$ ________ $-(-2)$

28. $-|-5|$ ________ $|-(-5)|$

0.3 *Simplify.*

29. $15 + (-7)$

30. $4 + (-9)$

31. $-23 - (-12)$

32. $\frac{5}{2} + \left(-\frac{4}{2}\right)$

33. $-\frac{9}{13} - \frac{4}{39}$

34. $5 + (-6) + (-3)$

35. $7 + (-4) + 8 + (-7)$

36. $-6 + 9 + 9 + (-5)$

37. $-35 + 30$

38. $-10 - 5$

39. $3 - (-2)$

40. $-7 - (-3)$

41. $\frac{23}{4} - \left(-\frac{3}{4}\right)$

42. $-3 - 2$

43. $8 - 12 - (-5)$

44. $-6 - 7 - (-18)$

45. $7 - (-4) - 7 - 4$

46. $-9 - (-6) - 8 - (-11)$

47. **Business and Finance** Jean deposited a check for \$625. She wrote two checks for \$69.74 and \$29.95, and used her debit card for a \$57.65 purchase. How much of her original deposit did she have left?

48. **Electrical Engineering** A certain electric motor spins at a rate of 5,400 rotations per minute (rpm). When a load is applied, the motor spins at 4,250 rpm. What is the change in rpm after loading?

0.4 *Multiply.*

49. $(-18)(-2)$

50. $(-10)(8)$

51. $(-5)(3)$

52. $\left(-\frac{3}{8}\right)\left(-\frac{4}{5}\right)$

53. $(-4)^2$

54. $(-2)(7)(-3)$

55. $(-6)(-5)(4)(-3)$

56. $(-9)(2)(-3)(1)$

Use the distributive property to remove parentheses and simplify.

57. $-4(8 - 7)$

58. $11(-15 + 4)$

59. $-8(5 - 2)$

60. $-4(-3 - 6)$

Divide.

61. $(-48) \div 12$

62. $\frac{-33}{-3}$

63. $-2 \div 0$

64. $-75 \div (-3)$

65. $-\frac{7}{9} \div \left(-\frac{2}{3}\right)$

66. $\left(-\frac{5}{11}\right) \div \frac{20}{33}$

67. $8 \div (-4)$

68. $(-12) \div (-1)$

69. **Business and Finance** An advertising agency lost a client who had been paying \$3,500 per month. How much revenue does the agency lose in a year?

70. **Social Science** A gambler lost \$180 over a 4-hr period. How much did the gambler lose per hour, on average?

0.5 *Write each expression in expanded form.*

71. 3^3

72. 5^4

73. 2^6

74. 4^5

Evaluate each expression.

75. $18 - 12 \div 2$

76. $(18 - 3) \cdot 5$

77. $6 \cdot 2^3$

78. $(5 \cdot 4)^2$

79. $5 \cdot 3^2 - 4$

80. $5(3^2 - 4)$

81. $5(4 - 2)^2$

82. $5 \cdot 4 - 2^2$

83. $(5 \cdot 4 - 2)^2$

84. $3(5 - 2)^2$

85. $3 \cdot 5 - 2^2$

86. $(3 \cdot 5 - 2)^2$

Statistics A professor grades a 20-question exam by awarding 5 points for each correct answer and subtracting 2 points for each incorrect answer. Points are neither added nor subtracted for answers left blank. Use this information to complete exercises 87 and 88.

87. Find the exam grade of a student who answers 14 questions correctly and 4 incorrectly, and leaves 2 questions blank.

88. Find the exam grade of a student who answers 17 questions correctly and 2 incorrectly, and leaves 1 question blank.

self-test 0

Name ____________

Section ________ Date ________

The purpose of this self-test is to help you assess your progress so that you can find concepts that you need to review before the next exam. Allow yourself about an hour to take this test. At the end of that hour, check your answers against those given in the back of this text. If you miss any, review the appropriate section until you have mastered that particular concept.

In exercises 1 and 2, use the fundamental principle to simplify each fraction.

1. $\frac{27}{99}$

2. $\frac{100}{64}$

Evaluate each expression. Write each answer in simplest form.

3. $\frac{4}{15} + \frac{7}{10}$

4. $\frac{9}{16} - \frac{3}{10}$

5. $13 + (-11) + (-5)$

6. $23 - 35$

7. $28 \div (-4)$

8. $(-44) \div (-11)$

9. $(-7)(5)$

10. $(-9)(-6)$

11. $-9 - 8 - (-5)$

12. $7 - 11 + 15$

13. $23 - 4 \times 12 \div 3 + |-4|$

14. $4 \cdot 5^2 - 35 + 21 - (-3)^3$

15. $\frac{3}{8} \times \frac{6}{11}$

16. $\frac{4}{5} \div \frac{2}{7}$

Fill in each blank with >, <, or = to make a true statement.

17. -7 ______ -5

18. $8 + (-3)^2$ ______ $8 - (-3)$

19. **Construction** A 14-acre piece of land is being developed into home lots. Each home site will be 0.35 acres and 2.8 acres will be used for roads. How many lots can be formed?

20. **Business and Finance** Michelle deposits $2,500 into her checking account each month. Each month, she pays her auto insurance ($200/mo), her auto loan ($250/mo), and her student loan ($275/mo). How much does she have left each month for other expenses?

Answers

1. ____________
2. ____________
3. ____________
4. ____________
5. ____________
6. ____________
7. ____________
8. ____________
9. ____________
10. ____________
11. ____________
12. ____________
13. ____________
14. ____________
15. ____________
16. ____________
17. ____________
18. ____________
19. ____________
20. ____________

CHAPTER

1

From Arithmetic to Algebra

INTRODUCTION

We expect to use mathematics both in our careers and when making financial decisions. But, there are many more opportunities to use math, even when enjoying life's pleasures. For instance, we use math regularly when traveling.

When traveling to another country, you need to be able to convert currency, temperature, and distance. Even figuring out when to call home so that you do not wake up family and friends during the night is a computation.

The equation is a very old tool for solving problems and writing relationships clearly and accurately. In this chapter, you will learn to solve linear equations. You will also learn to write equations that accurately describe problem situations. Both of these skills will be demonstrated in many settings, including international travel.

CHAPTER 1 OUTLINE

1.1 Transition to Algebra

< 1.1 Objectives >

1 > Introduce the concept of variables

2 > Identify algebraic expressions

3 > Translate from English to algebra

Tips for Student Success

Throughout this text, we present you with a series of class-tested techniques designed to improve your performance in this math class.

Become familiar with your syllabus

In the first class meeting, your instructor probably handed out a class syllabus. If you haven't done so already, you need to incorporate important information into your calendar and address book.

1. Write all important dates in your calendar. This includes homework due dates, quiz dates, test dates, and the date and time of the final exam. Never allow yourself to be surprised by a deadline!
2. Write your instructor's name, e-mail address, and office number in your address book. Also include the office hours. Make it a point to see your instructor early in the term. Although this is not the only person who can help clear up your confusion, your instructor is the most important person.
3. Make note of the other resources available to you. These include CDs, videotapes, Web pages, and tutoring.

Given all of these resources, it is important that you never let confusion or frustration mount. If you can't "get it" from the text, try another resource. All the resources are there specifically for you, so take advantage of them!

In arithmetic, you learned how to do calculations with numbers by using the basic operations of addition, subtraction, multiplication, and division.

In algebra, we still use numbers and the same four operations. However, we also use letters to represent numbers. Letters such as x, y, L, and W are called **variables** when they represent numerical values.

Here we see two rectangles whose lengths and widths are labeled with numbers.

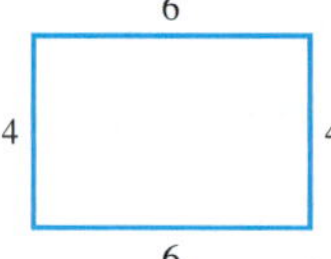

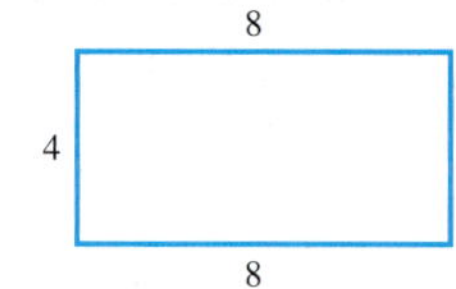

RECALL

In arithmetic:
- + denotes addition
- − denotes subtraction
- × denotes multiplication
- ÷ denotes division

If we want to represent the length and width of *any* rectangle, we can use the variables L for length and W for width.

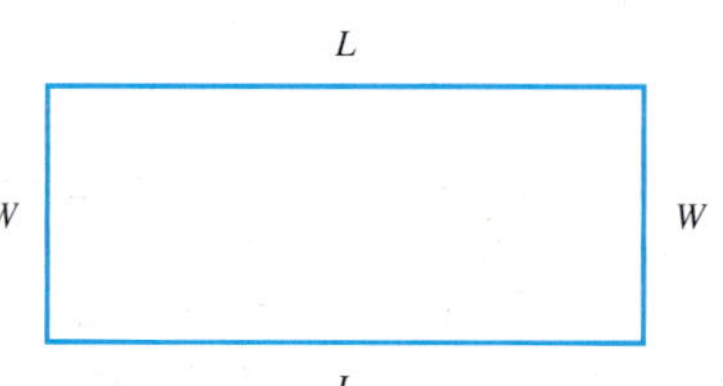

You are familiar with the four symbols (+, −, ×, ÷) used to indicate the fundamental operations of arithmetic.

Let's look at how these operations are indicated in algebra. We begin by looking at addition.

Definition

Addition $x + y$ means the *sum* of x and y, or x *plus* y.

Example 1 Writing Expressions That Indicate Addition

< Objective 1 >

(a) The *sum* of a and 3 is written as $a + 3$.

(b) L *plus* W is written as $L + W$.

(c) 5 *more than* m is written as $m + 5$.

(d) x *increased by* 7 is written as $x + 7$.

Check Yourself 1

Write, using symbols.

(a) The sum of y and 4
(b) a plus b
(c) 3 more than x
(d) n increased by 6

Now look at how subtraction is indicated in algebra.

Definition

Subtraction $x - y$ means the *difference* of x and y, or x *minus* y.

$x - y$ is not the same as $y - x$.

Example 2 Writing Expressions That Indicate Subtraction

(a) r *minus* s is written as $r - s$.

(b) The *difference* of m and 5 is written as $m - 5$.

(c) x *decreased by* 8 is written as $x - 8$.

(d) 4 *less than* a is written as $a - 4$.

(e) x subtracted from 5 is written as $5 - x$.

(f) 7 take away y is written as $7 - y$.

Check Yourself 2

Write, using symbols.

(a) w minus z
(b) The difference of a and 7
(c) y decreased by 3
(d) 5 less than b
(e) b subtracted from 8
(f) 4 take away x

You have seen that the operations of addition and subtraction are written exactly the same way in algebra as in arithmetic. This is not true for multiplication because the symbol × looks like the letter x, so we use other symbols to show multiplication to avoid confusion. Here are some ways to write multiplication.

Definition

Multiplication

NOTE

x and y are called the **factors** of the product xy.

A centered dot	$x \cdot y$	All these indicate the *product* of x and y, or x *times* y.
Writing the letters next to each other or separated only by parentheses	xy $x(y)$ $(x)(y)$	

Example 3 Writing Expressions That Indicate Multiplication

NOTE

You can place letters next to each other or numbers and letters next to each other to show multiplication. But you *cannot* place numbers side by side to show multiplication: 37 means the number thirty-seven, not 3 times 7.

(a) The product of 5 and a is written as $5 \cdot a$, $(5)(a)$, or $5a$. The last expression, $5a$, is the shortest and the most common way of writing the product.

(b) 3 times 7 can be written as $3 \cdot 7$ or $(3)(7)$.

(c) Twice z is written as $2z$.

(d) The product of 2, s, and t is written as $2st$.

(e) 4 more than the product of 6 and x is written as $6x + 4$.

Check Yourself 3

Write, using symbols.

(a) m times n

(b) The product of h and b

(c) The product of 8 and 9

(d) The product of 5, w, and y

(e) 3 more than the product of 8 and a

Before we move on to division, let's look at how we can combine the symbols we have learned so far.

Definition

Expression

An **expression** is a meaningful collection of numbers, variables, and symbols of operation.

Example 4 Identifying Expressions

< Objective 2 >

NOTE

Not every collection of symbols is an expression.

(a) $2m + 3$ is an expression. It means that we multiply 2 and m, then add 3.

(b) $x + \cdot + 3$ is not an expression. The three operations in a row have no meaning.

(c) $y = 2x - 1$ is not an expression, it is an *equation*. The equal sign is not an operation sign.

(d) $3a + 5b - 4c$ is an expression.

Check Yourself 4

Identify which are expressions and which are not.

(a) $7 - \cdot x$

(b) $6 + y = 9$

(c) $a + b - c$

(d) $3x - 5yz$

To write more complicated expressions in algebra, we need some "punctuation marks." Parentheses ( ) mean that an expression is to be thought of as a single quantity. Brackets [ ] are used in exactly the same way as parentheses in algebra. Look at the following example showing the use of these signs of grouping.

Example 5 Expressions with More Than One Operation

NOTES

This can be read as "3 times the quantity *a* plus *b*."

No parentheses are needed in part (b) since the 3 multiplies *only a*.

(a) 3 times the sum of a and b is written as

$3(a + b)$

The sum of *a* and *b* is a single quantity, so it is enclosed in parentheses.

(b) The sum of 3 times a and b is written as $3a + b$.

(c) 2 times the difference of m and n is written as $2(m - n)$.

(d) The product of s plus t and s minus t is written as $(s + t)(s - t)$.

(e) The product of b and 3 less than b is written as $b(b - 3)$.

Check Yourself 5

Write, using symbols.

(a) Twice the sum of p and q

(b) The sum of twice p and q

(c) The product of a and the quantity $b - c$

(d) The product of x plus 2 and x minus 2

(e) The product of x and 4 more than x

NOTE

In algebra the fraction form is usually used.

Now we look at the operation of division. In arithmetic, you see the division sign ÷, the long division symbol $\overline{)\ \ }$, and fraction notation. For example, to indicate the quotient when 9 is divided by 3, you could write

$$9 \div 3 \qquad \text{or} \qquad 3\overline{)9} \qquad \text{or} \qquad \frac{9}{3}$$

Definition

Division $\frac{x}{y}$ means x *divided* by y or the *quotient* of x and y.

Example 6 Writing Expressions That Indicate Division

< Objective 3 >

RECALL

The fraction bar acts as a grouping symbol.

(a) m divided by 3 is written as $\frac{m}{3}$.

(b) The quotient of a plus b, divided by 5 is written as $\frac{a + b}{5}$.

(c) The quantity p plus q divided by the quantity p minus q is written as $\frac{p + q}{p - q}$.

Check Yourself 6

Write, using symbols.

(a) r divided by s

(b) The quotient when x minus y is divided by 7

(c) The quantity a minus 2 divided by the quantity a plus 2

Notice that we can use many different letters to represent variables. In Example 6, the letters m, a, b, p, and q represented different variables. We often choose a letter that reminds us of what it represents, for example, L for *length* or W for *width*. These variables may be uppercase or lowercase letters, although lowercase is used more often.

Example 7 Writing Geometric Expressions

(a) *Length* times *width* is written $L \cdot W$.

(b) One-half of *altitude* times *base* is written $\frac{1}{2}\, a \cdot b$.

(c) *Length* times *width* times *height* is written $L \cdot W \cdot H$.

(d) Pi (π) times *diameter* is written πd.

Check Yourself 7

Write each geometric expression, using symbols.

(a) 2 times *length* plus two times *width*
(b) 2 times pi (π) times *radius*

Algebra can be used to model a variety of applications, such as the one shown in Example 8.

Example 8 Modeling Applications with Algebra

NOTE

We were asked to describe her pay given that her hours may vary.

Carla earns \$10.25 per hour in her job. Write an expression that describes her weekly gross pay in terms of the number of hours she works.

We represent the number of hours she works in a week by the variable h. Carla's pay is figured by taking the product of her hourly wage and the number of hours she works.

So, the expression

$$10.25h$$

describes Carla's weekly gross pay.

NOTE

The words "twice" and "doubled" indicate multiplication by 2.

Check Yourself 8

The specs for an engine cylinder call for the stroke length to be two more than twice the diameter of the cylinder. Write an expression for the stroke length of a cylinder based on its diameter.

We close this section by listing many of the common words used to indicate arithmetic operations.

Words Indicating Operations

The operations listed are usually indicated by the words shown.

Addition (+)	Plus, and, more than, increased by, sum
Subtraction (−)	Minus, from, less than, decreased by, difference, take away
Multiplication (·)	Times, of, by, product
Division (÷)	Divided, into, per, quotient

Check Yourself ANSWERS

1. **(a)** $y + 4$; **(b)** $a + b$; **(c)** $x + 3$; **(d)** $n + 6$
2. **(a)** $w - z$; **(b)** $a - 7$; **(c)** $y - 3$; **(d)** $b - 5$; **(e)** $8 - b$; **(f)** $4 - x$
3. **(a)** mn; **(b)** hb; **(c)** $8 \cdot 9$ or $(8)(9)$; **(d)** $5wy$; **(e)** $8a + 3$
4. **(a)** not an expression; **(b)** not an expression; **(c)** expression; **(d)** expression
5. **(a)** $2(p + q)$; **(b)** $2p + q$; **(c)** $a(b - c)$; **(d)** $(x + 2)(x - 2)$; **(e)** $x(x + 4)$
6. **(a)** $\frac{r}{s}$; **(b)** $\frac{x - y}{7}$; **(c)** $\frac{a - 2}{a + 2}$ **7.** **(a)** $2L + 2W$; **(b)** $2\pi r$ **8.** $2d + 2$

Reading Your Text

We conclude each section with this feature.

These fill-in-the-blank exercises are designed to ensure that you understand some of the key vocabulary used in this section. You should base your answers on a careful reading of the section.

The answers are in the Answers section at the end of this text.

SECTION 1.1

(a) In algebra, we use letters to represent numbers. We call these letters ______________.

(b) $x + y$ means the ______________ of x and y.

(c) $x \cdot y$, $(x)(y)$, and xy are all ways of indicating ______________ in algebra.

(d) An ______________ is a meaningful collection of numbers, variables, and symbols of operation.

1.1 exercises

Basic Skills | Challenge Yourself | Calculator/Computer | Career Applications | Above and Beyond

Name ____________

Section ________ Date ________

Answers

1. ________ 2. ________
3. ________ 4. ________
5. ________ 6. ________
7. ________ 8. ________
9. ________ 10. ________
11. ________ 12. ________
13. ________ 14. ________
15. ________ 16. ________
17. ________ 18. ________
19. ________ 20. ________
21. ________ 22. ________
23. ________ 24. ________
25. ________ 26. ________
27. ________ 28. ________

< Objectives 1 and 3 >

Write each phrase, using symbols.

1. The sum of c and d
2. a plus 7
3. w plus z
4. The sum of m and n
5. x increased by 5
6. 3 more than b
7. 10 more than y
8. m increased by 4
9. a minus b
10. s less than 5
11. 7 decreased by b
12. r minus 3
13. 6 less than r
14. x decreased by 3
15. w times z
16. The product of 3 and c
17. The product of 5 and t
18. 8 times a
19. The product of 8, m, and n
20. The product of 7, r, and s
21. The product of 8 and the quantity m plus n
22. The product of 5 and the sum of a and b
23. Twice the sum of x and y
24. 3 times the sum of m and n
25. The sum of twice x and y
26. The sum of 3 times m and n
27. Twice the difference of x and y
28. 3 times the difference of c and d

29. The quantity a plus b times the quantity a minus b

30. The product of x plus y and x minus y

31. The product of m and 3 less than m

32. The product of a and 7 less than a

33. 5 divided by x

34. The quotient when b is divided by 8

35. The sum of a and b, divided by 7

36. The quantity x minus y, divided by 9

37. The difference of p and q, divided by 4

38. The sum of a and 5, divided by 9

39. The sum of a and 3, divided by the difference of a and 3

40. The difference of m and n, divided by the sum of m and n

Write each phrase, using symbols. Use the variable x to represent the number in each case.

41. 5 more than a number

42. A number increased by 8

43. 7 less than a number

44. A number decreased by 8

45. 9 times a number

46. Twice a number

47. 6 more than 3 times a number

48. 5 times a number, decreased by the sum of the number and 3

49. Twice the sum of a number and 5

50. 3 times the difference of a number and 4 

51. The product of 2 more than a number and 2 less than that same number

52. The product of 5 less than a number and 5 more than that same number

53. The quotient of a number and 7

Answers

29. ______
30. ______
31. ______
32. ______
33. ______
34. ______
35. ______
36. ______
37. ______
38. ______
39. ______
40. ______
41. ______ 42. ______
43. ______ 44. ______
45. ______ 46. ______
47. ______
48. ______
49. ______
50. ______
51. ______
52. ______
53. ______

Answers

54. ____
55. ____
56. ____
57. ____
58. ____
59. ____
60. ____
61. ____
62. ____
63. ____
64. ____
65. ____
66. ____
67. ____
68. ____
69. ____
70. ____
71. ____
72. ____
73. ____
74. ____
75. ____
76. ____

54. A number divided by the sum of the number and 7

55. The sum of a number and 5, divided by 8

56. The quotient when 7 less than a number is divided by 3

57. 6 more than a number divided by 6 less than that same number 

58. The quotient when 3 less than a number is divided by 3 more than that same number

Write each geometric expression, using symbols.

59. Four times the length of a side s

60. $\frac{4}{3}$ times π times the cube of the radius r 

61. π times the radius r squared times the height h

62. Twice the length L plus twice the width W

63. One-half the product of the height h and the sum of two unequal sides b_1 and b_2

64. Six times the length of a side s squared

< Objective 2 >

Identify which are expressions and which are not.

65. $2(x + 5)$

66. $4 - (x + 3)$

67. $4 + \div m$

68. $6 + a = 7$

69. $2b = 6$

70. $x(y + 3)$

71. $2a(3b + 5)$

72. $4x + \cdot 7$

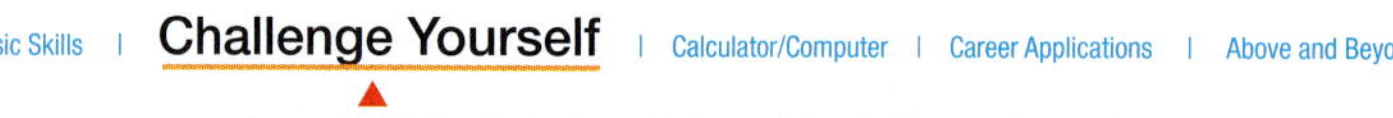

Determine whether each statement is **true** *or* **false.**

73. The phrase "7 more than x" indicates addition.

74. A product is the result of dividing two numbers.

Complete each statement with **never, sometimes,** *or* **always.**

75. An expression is ________ an equation.

76. A number written next to a letter ________ indicates multiplication.

77. **Number Problem** Two numbers have a sum of 35. If one number is x, express the other number in terms of x.

78. **Science and Medicine** It is estimated that the earth is losing 4,000 species of plants and animals every year. If S represents the number of species living last year, how many species are on the earth this year?

79. **Business and Finance** The simple interest earned when a principal P is invested at a rate r for a time t is calculated by multiplying the principal by the rate by the time. Write an expression for the interest earned.

80. **Science and Medicine** The kinetic energy of a particle of mass m is found by taking one-half of the product of the mass and the square of the velocity v. Write an expression for the kinetic energy of a particle.

81. **Business and Finance** Four hundred tickets were sold for a school play. The tickets were of two types: general admission and student. There were x general admission tickets sold. Write an expression for the number of student tickets sold.

82. **Business and Finance** Nate has $375 in his bank account. He wrote a check for x dollars for a concert ticket. Write an expression that represents the remaining money in his account.

Basic Skills | Challenge Yourself | Calculator/Computer | **Career Applications** | Above and Beyond

83. **Construction Technology** K Jones Manufacturing produces hex bolts and carriage bolts. They sold 284 more hex bolts than carriage bolts last month. Write an expression that describes the number of carriage bolts they sold last month.

84. **Allied Health** The standard dosage given to a patient is equal to the product of the desired dose D and the available quantity Q divided by the available dose H. Write an expression for the standard dosage.

85. **Information Technology** Mindy is the manager of the help desk at a large cable company. She notices that, on average, her staff can handle 50 calls/hr. Last week, during a thunderstorm, the call volume increased from 65 calls/hr to 150 calls/hr.

To figure out the average number of customers in the system, she needs to take the quotient of the average rate of customer arrivals (the call volume) a and the difference of the average rate at which customers are served h and the average rate of customer arrivals a. Write an expression for the average number of customers in the system.

86. **Electrical Engineering** Electrical power P is the product of voltage V and current I. Express this relationship algebraically.

Answers

77. ______

78. ______

79. ______

80. ______

81. ______

82. ______

83. ______

84. ______

85. ______

86. ______

Basic Skills | Challenge Yourself | Calculator/Computer | Career Applications | Above and Beyond

Answers

87. ________________

88. ________________

87. Rewrite each algebraic expression using English phrases. Exchange papers with another student to edit your writing. Be sure the meaning in English is the same as in algebra. These expressions are not complete sentences, so your English does not have to be in complete sentences. Here is an example.

Algebra: $2(x - 1)$

English: We could write "double 1 less than a number." Or we might write "a number diminished by 1 and then multiplied by 2."

(a) $n + 3$ **(b)** $\frac{x + 2}{5}$ **(c)** $3(5 + a)$ **(d)** $3 - 4n$ **(e)** $\frac{x + 6}{x - 1}$

88. Use the Internet to find the origins of the symbols $+$, $-$, $\times$, and $\div$. Summarize your findings.

Note: We provide a brief tutorial on searching the Internet in Appendix A.

Answers

1. $c + d$ **3.** $w + z$ **5.** $x + 5$ **7.** $y + 10$ **9.** $a - b$ **11.** $7 - b$
13. $r - 6$ **15.** wz **17.** $5t$ **19.** $8mn$ **21.** $8(m + n)$
23. $2(x + y)$ **25.** $2x + y$ **27.** $2(x - y)$ **29.** $(a + b)(a - b)$
31. $m(m - 3)$ **33.** $\frac{5}{x}$ **35.** $\frac{a + b}{7}$ **37.** $\frac{p - q}{4}$ **39.** $\frac{a + 3}{a - 3}$
41. $x + 5$ **43.** $x - 7$ **45.** $9x$ **47.** $3x + 6$ **49.** $2(x + 5)$
51. $(x + 2)(x - 2)$ **53.** $\frac{x}{7}$ **55.** $\frac{x + 5}{8}$ **57.** $\frac{x + 6}{x - 6}$ **59.** $4s$
61. $\pi r^2 h$ **63.** $\frac{1}{2}h(b_1 + b_2)$ **65.** Expression **67.** Not an expression
69. Not an expression **71.** Expression **73.** True **75.** never
77. $35 - x$ **79.** Prt **81.** $400 - x$ **83.** $H - 284$ **85.** $\frac{a}{h - a}$
87. Above and Beyond

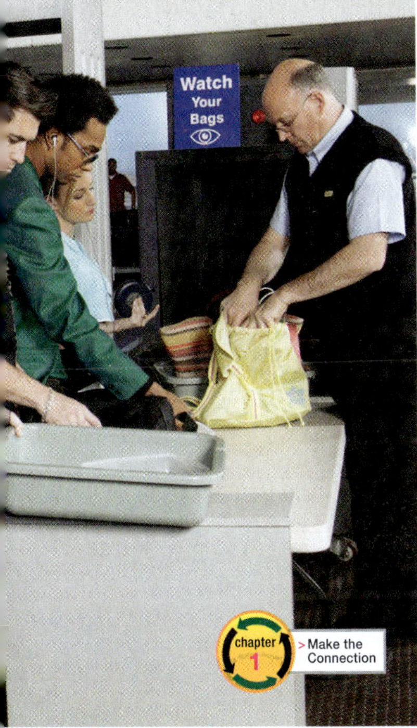

Activity 1 :: Monetary Conversions

Each activity in this text is designed to either enhance your understanding of the topics of the chapter, provide you with a mathematical extension of those topics, or both. The activities can be undertaken by one student, but they are better suited for a small group project. Occasionally it is only through discussion that different facets of the activity become apparent.

In the opener to this chapter, we discussed international travel and using exchange rates to acquire local currency. In this activity, we use these exchange rates to explore the idea of variables. Recall that a **variable** is a symbol used to represent an unknown quantity or a quantity that varies.

Currency exchange rates are published on a daily basis by many sources such as *Yahoo!Finance* and the *Wall Street Journal*. For instance, on May 20, 2006, the exchange rate for trading US$ for CAN$ was 1.1191. This means that US$1 is equivalent to CAN$1.1191. That is, if you exchanged $100 of U.S. money, you would have received $111.91 in Canadian dollars.

CAN$ = Exchange rate × US$

Activity

1. Choose a country that you would like to visit. Use a search engine to find the exchange rate between US$ and the currency of your chosen country.
2. If you are visiting for only a short time, you may not need too much money. Determine how much of the local currency you will receive in exchange for US$250.
3. If you stay for an extended period, you will need more money. How much would you receive in exchange for US$900?

Here, we treated the amount (US$) as a *variable.* This quantity varied, depending on our needs. If we visit Canada and let x = the amount exchanged in US$ and y = the amount received in CAN$, then, using the exchange rate previously given, we have the equation

$$y = 1.1191x$$

You may ask, "Isn't the amount of Canadian money received (y) a variable, too?" The answer is yes; in fact, all three quantities are variables. The exchange rate varies on a daily basis. For example, according to *Yahoo!Finance,* the exchange rate for US-CAN currency was 1.372 on December 14, 2001. If we let r = the exchange rate, then we can write the conversion equation as

$$y = rx$$

4. Consider the country you chose to visit above. Find the exchange rate for another date and repeat exercises 2 and 3 for this other exchange rate.
5. Choose another nation that you would like to visit. Repeat exercises 1–3 for this country.

This data set is provided for your convenience. We encourage you to find more current data on the Internet.

Data Set

Currency	US$	Yen (¥)	Euro (€)	CAN$	U.K. (£)	Aust$
1 US$	1	111.705	0.7833	1.1191	0.5327	1.3181
1 Yen (¥)	0.008952	1	0.007012	0.010018	0.004769	0.0118
1 Euro (€)	1.2766	142.6026	1	1.4286	0.6801	1.6827
1 CAN$	0.8936	99.8213	0.7	1	0.476	1.1779
1 U.K. (£)	1.8772	209.6924	1.4705	2.1007	1	2.4744
1 Aust$	0.7586	84.745	0.5943	0.849	0.4041	1

Source: Yahoo!Finance; 5/20/06.

1. We chose to visit Canada and will use the 5/20/06 exchange rate of 1.1191 from the sample data set.

2. Exchange rate × US$ = CAN$

   (1.1191) · (US$250) = CAN$279.775

   We would receive $279.78 in Canadian dollars for $250 in U.S. money (round Canadian money to two decimal places).

3. (1.1191) · (US$900) = CAN$1,007.19

4. Had we visited Canada on 12/14/01, we would have received an exchange rate of 1.372.

   (1.372) · (US$250) = CAN$343

   (1.372) · (US$900) = CAN$1,234.80

5. We choose to visit Japan. The 5/20/06 exchange rate was 111.705 Yen (¥) for each US$.

   (111.705) · (US$250) = ¥27,926.25

   (111.705) · (US$900) = ¥100,534.5

   We would receive 27,926 yen for US$250, and 100,535 yen for US$900.

1.2 Evaluating Algebraic Expressions

< 1.2 Objectives >

1 > Evaluate algebraic expressions given any real-number value for the variables

2 > Use a graphing calculator to evaluate algebraic expressions

Tips for Student Success

Working Together

How many of your classmates do you know? Whether you are by nature gregarious or shy, you have much to gain by getting to know your classmates.

1. It is important to have someone to call when you miss class or if you are unclear on an assignment.
2. Working with another person is almost always beneficial to both people. If you don't understand something, it helps to have someone to ask about it. If you do understand something, nothing cements that understanding more than explaining the idea to another person.
3. Sometimes we need to commiserate. If an assignment is particularly frustrating, it is reassuring to find out that it is also frustrating for other students.
4. Have you ever thought you had the right answer, but it doesn't match the answer in the text? Frequently the answers are equivalent, but that's not always easy to see. A different perspective can help you see that. Occasionally there is an error in a textbook (here we are talking about *other* textbooks). In such cases it is wonderfully reassuring to find that someone else has the same answer as you do.

In applying algebra to problem solving, you often want to find the value of an algebraic expression when you know certain values for the letters (or variables) in the expression. Finding the value of an expression is called *evaluating the expression* and uses the following steps.

Step by Step

To Evaluate an Algebraic Expression

Step 1 Replace each variable with its given number value.

Step 2 Do the necessary arithmetic operations, following the rules for order of operations.

Example 1 Evaluating Algebraic Expressions

< Objective 1 >

Suppose that $a = 5$ and $b = 7$.

(a) To evaluate $a + b$, we replace a with 5 and b with 7.

$a + b = (5) + (7) = 12$

(b) To evaluate $3ab$, we again replace a with 5 and b with 7.

$3ab = 3 \cdot (5) \cdot (7) = 105$

Check Yourself 1

If $x = 6$ and $y = 7$, evaluate.

(a) $y - x$ **(b)** $5xy$

Some algebraic expressions require us to follow the rules for the order of operations.

Example 2 Evaluating Algebraic Expressions

Evaluate each expression if $a = 2$, $b = 3$, $c = 4$, and $d = 5$.

(a) $5a + 7b$

$5a + 7b = 5 \cdot (2) + 7 \cdot (3)$ Multiply first

$= 10 + 21 = 31$ Then add

(b) $3c^2$

$3c^2 = 3 \cdot (4)^2$ Evaluate the power.

$= 3 \cdot 16 = 48$ Then multiply

> CAUTION
>
> This is different from
>
> $(3c)^2 = [3 \cdot (4)]^2$
>
> $= 12^2 = 144$

(c) $7(c + d)$

$7(c + d) = 7[(4) + (5)]$ Add inside the brackets.

$= 7 \cdot 9 = 63$ Multiply

(d) $5a^4 - 2d^2$

$5a^4 - 2d^2 = 5 \cdot (2)^4 - 2 \cdot (5)^2$ Evaluate the powers.

$= 5 \cdot 16 - 2 \cdot 25$ Multiply

$= 80 - 50 = 30$ Subtract

Check Yourself 2

If $x = 3$, $y = 2$, $z = 4$, and $w = 5$, evaluate each expression.

(a) $4x^2 + 2$ **(b)** $5(z + w)$ **(c)** $7(z^2 - y^2)$

To evaluate algebraic expressions when a fraction bar is used, do the following: Start by doing all the work in the numerator, then do the work in the denominator. Divide the numerator by the denominator as the last step.

Example 3 Evaluating Algebraic Expressions

If $p = 2$, $q = 3$, and $r = 4$, evaluate.

(a) $\dfrac{8p}{r}$

Replace p with 2 and r with 4.

$\dfrac{8p}{r} = \dfrac{8 \cdot (2)}{(4)} = \dfrac{16}{4} = 4$ Divide as the last step.

(b) $\dfrac{7q + r}{p + q}$

$\dfrac{7q + r}{p + q} = \dfrac{7 \cdot (3) + (4)}{(2) + (3)}$ Evaluate the top and bottom separately.

$= \dfrac{21 + 4}{2 + 3}$

$= \dfrac{25}{5} = 5$

Check Yourself 3

Evaluate each expression if $c = 5$, $d = 8$, and $e = 3$.

(a) $\frac{6c}{e}$ (b) $\frac{4d + e}{c}$ (c) $\frac{10d - e}{d + e}$

A calculator or computer can be used to evaluate an algebraic expression. We demonstrate this in Example 4.

Example 4 Using a Calculator to Evaluate an Expression

< Objective 2 >

Use a calculator to evaluate each expression for the given variable values.

(a) $\frac{4x + y}{z}$ if $x = 2$, $y = 1$, and $z = 3$

RECALL

Graphing calculators usually use an [ENTER] key instead of an [=] key.

Begin by writing the expression with the values substituted for the variables.

$$\frac{4x + y}{z} = \frac{4(2) + (1)}{(3)}$$

Then, enter the numerical expression into a calculator.

[(] 4 [×] 2 [+] 1 [)] [÷] 3 [ENTER]

Remember to enclose the entire numerator in parentheses.

The display should read 3.

```
(4*2+1)/3
                3
(7*2-6)/(3*-2-2)
               -1
```

(b) $\frac{7x - y}{3z - x}$ if $x = 2$, $y = 6$, and $z = -2$

Again, we begin by substituting:

$$\frac{7x - y}{3z - x} = \frac{7(2) - (6)}{3(-2) - 2}$$

Then, we enter the expression into a calculator.

[(] 7 [×] 2 [−] 6 [)] [÷] [(] 3 [×] [(−)] 2 [−] 2 [)] [ENTER]

The display should read −1.

Check Yourself 4

Use a calculator to evaluate each expression if $x = 2$, $y = -6$, and $z = 5$.

(a) $\frac{2x + y}{z}$ (b) $\frac{4y - 2z}{3x}$

> CAUTION

A calculator follows the correct order of operations when evaluating an expression. If we omit the parentheses in Example 4(b) and enter

7 [×] 2 [−] 6 [÷] 3 [×] [(−)] 2 [−] 2 [ENTER]

the calculator will interpret our input as $7 \cdot 2 - \frac{6}{3} \cdot (-2) - 2$, which is not what we wanted.

Whether working with a calculator or pencil and paper, you must remember to take care both with signs and with the order of operations.

Example 5 Evaluating Expressions

> **RECALL**
>
> Always follow the rules for the order of operations. Multiply first, then add.

Evaluate $5a + 4b$ if $a = -2$ and $b = \frac{3}{4}$.

Replace a with -2 and b with $\frac{3}{4}$.

$$
\begin{aligned}
5a + 4b &= 5(-2) + 4\left(\frac{3}{4}\right) \\
&= -10 + 3 \\
&= -7
\end{aligned}
$$

Check Yourself 5

Evaluate $3x + 5y$ if $x = -2$ and $y = -\frac{4}{5}$.

We follow the same rules no matter how many variables are in the expression.

Example 6 Evaluating Expressions

Evaluate each expression if $a = -4$, $b = 2$, $c = -5$, and $d = 6$.

(a) $7a - 4c$

This becomes $-(-20)$, or $+20$.

$$
\begin{aligned}
7a - 4c &= 7(-4) - 4(-5) \\
&= -28 + 20 \\
&= -8
\end{aligned}
$$

(b) $7c^2$

Evaluate the power first, then multiply by 7.

$$
\begin{aligned}
7c^2 = 7(-5)^2 &= 7 \cdot 25 \\
&= 175
\end{aligned}
$$

> **> CAUTION**
>
> When a squared variable is replaced by a negative number, square the negative.
>
> $(-5)^2 = (-5)(-5) = 25$
>
> The exponent applies to -5!
>
> $-5^2 = -(5 \cdot 5) = -25$
>
> The exponent applies only to 5!

(c) $b^2 - 4ac$

$$
\begin{aligned}
b^2 - 4ac &= (2)^2 - 4(-4)(-5) \\
&= 4 - 4(-4)(-5) \\
&= 4 - 80 \\
&= -76
\end{aligned}
$$

(d) $b(a + d)$

Add inside the brackets first.

$$
\begin{aligned}
b(a + d) &= (2)[(-4) + (6)] \\
&= 2(2) \\
&= 4
\end{aligned}
$$

Check Yourself 6

Evaluate if $p = -4$, $q = 3$, and $r = -2$.

(a) $5p - 3r$ **(b)** $2p^2 + q$ **(c)** $p(q + r)$
(d) $-q^2$ **(e)** $(-q)^2$

We will look at one more example that involves a fraction. Remember that the fraction bar is a grouping symbol. This means that you should do the required operations first in the numerator and then in the denominator. Divide as the last step.

Example 7 Evaluating Expressions

Evaluate each expression if $x = 4$, $y = -5$, $z = 2$, and $w = -3$.

(a) $\dfrac{z - 2y}{x}$

$$\frac{z - 2y}{x} = \frac{(2) - 2(-5)}{(4)} = \frac{2 + 10}{4} = \frac{12}{4} = 3$$

(b) $\dfrac{3x - w}{2x + w}$

$$\frac{3x - w}{2x + w} = \frac{3(4) - (-3)}{2(4) + (-3)} = \frac{12 + 3}{8 + (-3)} = \frac{15}{5} = 3$$

Check Yourself 7

Evaluate if $m = -6$, $n = 4$, and $p = -3$.

(a) $\dfrac{m + 3n}{p}$ **(b)** $\dfrac{4m + n}{m + 4n}$

The process of evaluating expressions has many common applications.

Example 8 An Application of Evaluating an Expression

A car is advertised for rent at a cost of \$59 per day plus 20 cents per mile. The total cost can be found by evaluating the expression

$59d + 0.20m$

in which d represents the number of days and m the number of miles. Find the total cost for a 3-day rental if 250 miles are driven.

$59(3) + 0.20(250)$
$= 177 + 50$
$= 227$

The total cost is \$227.

Check Yourself 8

The cost to hold a wedding reception at a certain cultural arts center is \$195 per hour plus \$27.50 per guest. The total cost can be found by evaluating the expression

$195h + 27.50g$

in which h represents the number of hours and g the number of guests. Find the total cost for a 4-hour reception with 220 guests.

Check Yourself ANSWERS

1. (a) 1; **(b)** 210 **2. (a)** 38; **(b)** 45, **(c)** 84 **3. (a)** 10; **(b)** 7; **(c)** 7
4. (a) −0.4; **(b)** −5.67 **5.** −10 **6. (a)** −14; **(b)** 35; **(c)** −4; **(d)** −9; **(e)** 9
7. (a) −2; **(b)** −2 **8.** \$6,830

Reading Your Text

SECTION 1.2

(a) To evaluate an algebraic expression, first replace each ______________ with its given number value.

(b) Finding the value of an expression is called ______________ the expression.

(c) To evaluate an algebraic expression, you must follow the rules for the order of ______________.

(d) When a squared variable is replaced by a negative number, ______________ the negative as well.

Graphing Calculator Option

Using the Memory Feature to Evaluate Expressions

The memory features of a graphing calculator are a great aid when you need to evaluate several expressions, using the same variables and the same values for those variables.

Your graphing calculator can store variable values for many different variables in different memory spaces. Using these memory spaces saves a great deal of time when evaluating expressions.

Evaluate each expression if $a = 4.6$, $b = -\frac{2}{3}$, and $c = 8$. Round your results to the nearest hundredth.

(a) $a + \frac{b}{ac}$

(b) $b - b^2 + 3(a - c)$

(c) $bc - a^2 - \frac{ab}{c}$

(d) $a^2b^3c - ab^4c^2$

Begin by entering each variable's value into a calculator memory space. When possible, use the memory space that has the same name as the variable you are saving.

Step 1 Type the value associated with one variable.

Step 2 Press the store key, STO▸, the green alphabet key to access the memory names, ALPHA, and the key indicating which memory space you want to use.

Note: By pressing ALPHA, you are accessing the green letters above selected keys. These letters name the variable spaces.

Step 3 Press ENTER.

Step 4 Repeat until every variable value has been stored in an individual memory space.

NOTE

We use the TI-84 Plus model graphing calculator throughout this text. If you have a different model, consult your instructor or the instruction manual.

In the example above, we store 4.6 in **Memory A**, $-\frac{2}{3}$ in **Memory B**, and 8 in **Memory C**.

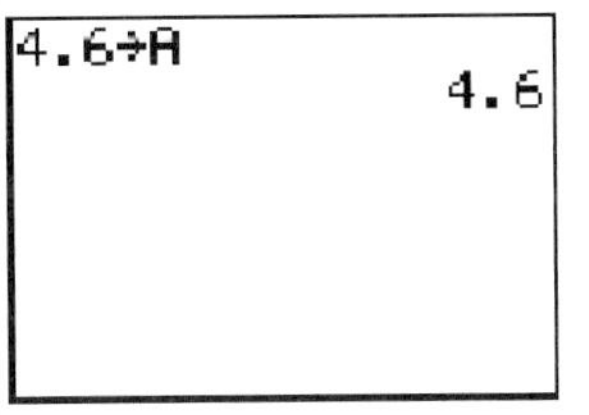

Memory A is with the MATH key.

```
4.6→A
            4.6
-2/3→B
  -.6666666667
```

Memory B is with the APPS key.

Divide to from a fraction.

```
4.6→A
            4.6
-2/3→B
  -.6666666667
8→C
              8
```

Memory C is with the PRGM key.

You can use the variables in the memory spaces rather than type in the numbers. Access the memory spaces by pressing ALPHA before pressing the key associated with the memory space. This will save time and make careless errors much less likely.

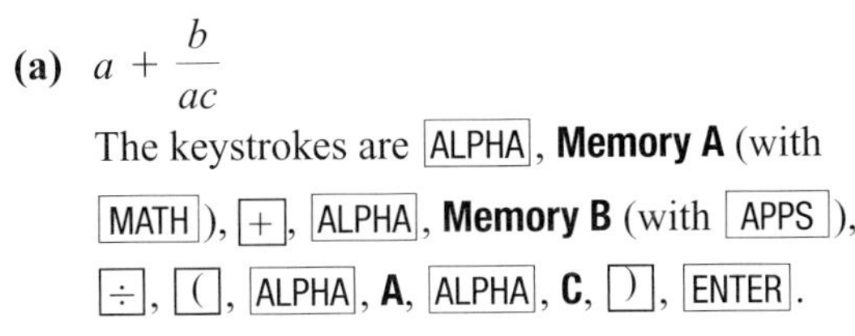

(a) $a + \frac{b}{ac}$

The keystrokes are ALPHA, **Memory A** (with MATH), +, ALPHA, **Memory B** (with APPS), ÷, (, ALPHA, **A**, ALPHA, **C**,), ENTER.

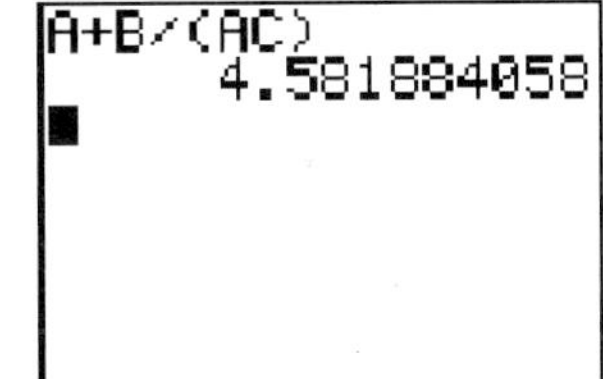

$a + \frac{b}{ac} = 4.58$, to the nearest hundredth.

Note: Because the fraction bar is a grouping symbol, you must remember to enclose the denominator in parentheses.

(b) $b - b^2 + 3(a - c)$

```
A+B/(AC)
    4.581884058
B-B²+3(A-C)
   -11.31111111
```

$b - b^2 + 3(a - c) = -11.31$

Use x^2 to square a value.

(c) $bc - a^2 - \frac{ab}{c}$

```
A+B/(AC)
    4.581884058
B-B²+3(A-C)
   -11.31111111
BC-A²-(AB)/C
         -26.11
```

$bc - a^2 - \frac{ab}{c} = -26.11$

(d) $a^2b^3c - ab^4c^2$

```
    4.581884058
B-B²+3(A-C)
   -11.31111111
BC-A²-(AB)/C
         -26.11
A²B^3C-AB^4C²
   -108.3101235
```

$a^2b^3c - ab^4c^2 = -108.31$

Use the caret key, ^, for general exponents.

Graphing Calculator Check

Evaluate each expression if $x = -8.3$, $y = \frac{5}{4}$, and $z = -6$. Round your results to the nearest hundredth.

(a) $\frac{xy}{z} - xz$

(b) $5(z - y) + \frac{x}{x - z}$

(c) $x^2y^5z - (x + y)^2$

(d) $\frac{-2(x + z)^2}{y^3z}$

Answers

(a) -48.07 **(b)** -32.64 **(c)** $-1{,}311.12$ **(d)** 34.90

Note: Throughout this text, we will provide additional graphing-calculator material. This material is optional. The authors will not assume that students have learned this, but we feel that students using a graphing calculator will benefit from these materials.

The screen shots and key commands are from the TI-84 Plus model from Texas Instruments. Most calculator models are fairly similar in how they handle memory. If you have a different model, consult your instructor or the instruction manual.

Basic Skills | Challenge Yourself | Calculator/Computer | Career Applications | Above and Beyond

1.2 exercises

Boost *your* GRADE at ALEKS.com!

ALEKS

- Practice Problems
- Self-Tests
- NetTutor
- e-Professors
- Videos

Name ______

Section ______ Date ______

< Objective 1 >

Evaluate each expression if $a = -2$, $b = 5$, $c = -4$, *and* $d = 6$.

1. $3c - 2b$

2. $4c - 2b$

3. $7c + 6b$

4. $7a - 2c$

5. $-b^2 + b$

6. $(-c)^2 + 5c$

7. $3a^2$

8. $6c^2$

9. $c^2 - 2d$

10. $3a^2 + 4c$

11. $2a^2 + 3b^2$

12. $4b^2 - 2c^2$

13. $2(c - d)$

14. $5(b - c)$

15. $4(2a - d)$

16. $6(3c - d)$

17. $a(b + 3c)$

18. $c(3a - d)$

19. $\dfrac{6d}{c}$

20. $\dfrac{3a}{5b}$

21. $\dfrac{3d + 2c}{b}$

22. $\dfrac{2b + 3d}{2a}$

23. $\dfrac{2b - 3a}{c + 2d}$ > Videos

24. $\dfrac{3d - 2b}{5a + d}$

25. $b^2 - d^2$

26. $d^2 - b^2$

27. $(b - d)^2$

28. $(d - b)^2$

29. $(d - b)(d + b)$

30. $(c - a)(c + a)$

31. $c^3 - a^3$

32. $c^3 + a^3$

33. $(c - a)^3$

34. $(c + a)^3$

35. $(d - b)(d^2 + db + b^2)$

36. $(c + a)(c^2 - ac + a^2)$ > Videos

37. $b^2 + a^2$

38. $d^2 - a^2$

39. $(b + a)^2$

40. $(d - a)^2$

41. $a^2 + 2ad + d^2$

42. $d^2 - 2ad + a^2$

Evaluate each expression if $x = -2$, $y = -3$, *and* $z = 4$.

43. $x^2 - 2y^2 + z^2$

44. $4yz + 6xy$

45. $2xy - (x^2 - 2yz)$

46. $3yz - 6xyz + x^2y^2$

47. $2y(z^2 - 2xy) + yz^2$

48. $-z - (-2x - yz)$

Answers

1. ______	2. ______
3. ______	4. ______
5. ______	6. ______
7. ______	8. ______
9. ______	10. ______
11. ______	12. ______
13. ______	14. ______
15. ______	16. ______
17. ______	18. ______
19. ______	20. ______
21. ______	22. ______
23. ______	24. ______
25. ______	26. ______
27. ______	28. ______
29. ______	30. ______
31. ______	32. ______
33. ______	34. ______
35. ______	36. ______
37. ______	38. ______
39. ______	40. ______
41. ______	42. ______
43. ______	44. ______
45. ______	46. ______
47. ______	48. ______

Answers

49. ______

50. ______

51. ______

52. ______

53. ______

54. ______

55. ______

56. ______

57. ______

58. ______

59. ______

60. ______

61. ______

Decide whether the given numbers make the statement **true** *or* **false.**

49. $x - 7 = 2y + 5; x = 22, y = 5$

50. $3(x - y) = 6; x = 5, y = -3$

51. $2(x + y) = 2x + y; x = -4, y = -2$ > Videos

52. $x^2 - y^2 = x - y; x = 4, y = -3$

Basic Skills | **Challenge Yourself** | Calculator/Computer | Career Applications | Above and Beyond

Determine whether each statement is **true** *or* **false.**

53. When evaluating an expression that has a fraction bar, dividing the numerator by the denominator is the first step.

54. The value of w^2 will be nonnegative, no matter what number is used to replace w.

Complete each statement with **never, sometimes,** *or* **always.**

55. When n is replaced with a number, the value of $-n^2$ is ________ positive.

56. When x is replaced with a number, the value of $-5x$ is ________ negative.

57. **Technology** The formula for the total resistance in a parallel circuit is $R_T = \dfrac{R_1R_2}{R_1 + R_2}$. Find the total resistance if $R_1 = 9$ ohms (Ω) and $R_2 = 15\ \Omega$.

58. **Geometry** The formula for the area of a triangle is given by $A = \dfrac{1}{2}ab$, where a is the altitude (or height) and b is the length of the base. Find the area of a triangle if $a = 4$ centimeters (cm) and $b = 8$ cm.

59. **Geometry** The perimeter of a rectangle of length L and width W is given by the formula $P = 2L + 2W$. Find the perimeter when $L = 10$ inches (in.) and $W = 5$ in.

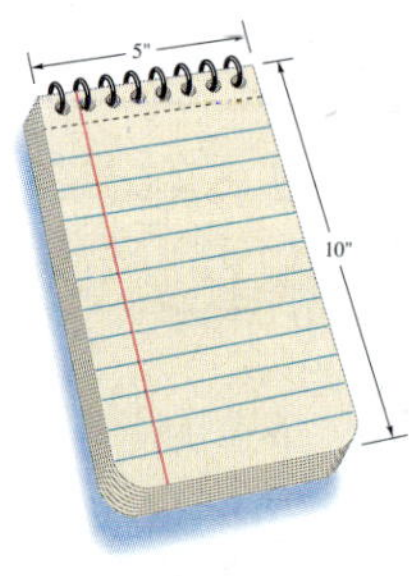

60. **Business and Finance** The simple interest I on a principal of P dollars at interest rate r for time t, in years, is given by $I = Prt$. Find the simple interest on a principal of \$6,000 at 4% for 3 years. (*Note:* 4% = 0.04.)

61. **Business and Finance** Use the formula $P = \dfrac{I}{r \cdot t}$ to find the principal if the total interest earned was \$150 and the rate of interest was 4% for 2 years.

62. **Business and Finance** Use the formula $r = \dfrac{I}{P \cdot t}$ to find the rate of interest if \$5,000 earns \$1,500 interest in 6 years.

63. **Science and Medicine** The formula that relates Celsius and Fahrenheit temperatures is $F = \dfrac{9}{5}C + 32$. If the temperature is $-10°C$, what is the Fahrenheit temperature?

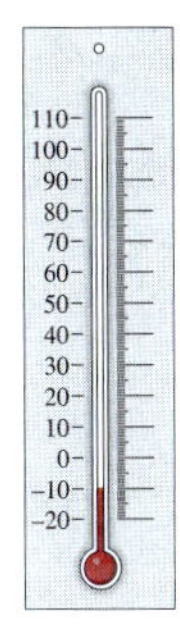

64. **Geometry** If the area of a circle whose radius is r is given by $A = \pi r^2$, where $\pi = 3.14$, find the area when $r = 3$ meters (m).

65. **Business and Finance** A local telephone company offers a long-distance telephone plan that charges \$5.25 per month and \$0.08 per minute of calling time. The expression $0.08t + 5.25$ represents the monthly long-distance bill for a customer who makes t minutes (min) of long-distance calling on this plan. Find the monthly bill for a customer who makes 173 min of long-distance calls on this plan.

66. **Science and Medicine** The speed of a model car as it slows down is given by $v = 20 - 4t$, where v is the speed in meters per second (m/s) and t is the time in seconds (s) during which the car has slowed. Find the speed of the car 1.5 s after it has begun to slow.

< Objective 2 >

Use a calculator to evaluate each expression if $x = -2.34$, $y = -3.14$, and $z = 4.12$. Round your answer to the nearest tenth.

67. $x + yz$

68. $y - 2z$

69. $y^2 - 2x^2$

70. $x^2 + y^2$

71. $\dfrac{xy}{z - x}$

72. $\dfrac{y^2}{zy}$

73. $\dfrac{2x + y}{2x + z}$

74. $\dfrac{y^2z^2}{xy}$

Use a calculator to evaluate the expression $x^2 - 4x^3 + 3x$ for each given value.

75. $x = 3$

76. $x = 12$

77. $x = 27$

78. $x = 48$

Answers

62. ______
63. ______
64. ______
65. ______
66. ______
67. ______
68. ______
69. ______
70. ______
71. ______
72. ______
73. ______
74. ______
75. ______
76. ______
77. ______
78. ______

Answers

79. ______

80. ______

81. ______

82. ______

83. ______

84. ______

85. ______

Basic Skills | Challenge Yourself | Calculator/Computer | **Career Applications** | Above and Beyond

79. **Allied Health** The concentration, in micrograms per milliliter (μg/mL), of an antihistamine in a patient's bloodstream can be approximated using the expression $-2t^2 + 13t + 1$, in which t is the number of hours since the drug was administered. Approximate the concentration of the antihistamine 1 hour after being administered. > Videos

80. **Allied Health** Use the expression given in exercise 79 to approximate the concentration of the antihistamine 3 hours after being administered.

81. **Electrical Engineering** Evaluate $\frac{rT}{5{,}252}$ for $r = 1{,}180$ and $T = 3$ (round to the nearest thousandth).

82. **Mechanical Engineering** The kinetic energy (in joules) of a particle is given by $\frac{1}{2}mv^2$. Find the kinetic energy of a particle if its mass is 60 kg and its velocity is 6 m/s.

Basic Skills | Challenge Yourself | Calculator/Computer | Career Applications | **Above and Beyond**

83. Write an English interpretation of each algebraic expression or equation.

(a) $(2x^2 - y)^3$ **(b)** $3n = \frac{n-1}{2}$ **(c)** $(2n + 3)(n - 4)$

84. Is $a^n + b^n = (a + b)^n$? Try a few numbers and decide whether this is true for all numbers, for some numbers, or never true. Write an explanation of your findings and give examples.

85. **(a)** Evaluate the expression $4x(5 - x)(6 - x)$ for $x = 0, 1, 2, 3, 4,$ and 5. Complete the table below.

Value of x	0	1	2	3	4	5
Value of expression						

(b) For which value of x does the expression value appear to be largest?

(c) Evaluate the expression for $x = 1.5, 1.6, 1.7, 1.8, 1.9, 2.0, 2.1, 2.2, 2.3, 2.4,$ and 2.5. Complete the table.

Value of x	1.5	1.6	1.7	1.8	1.9	2.0	2.1	2.2	2.3	2.4	2.5
Value of expression											

(d) For which value of x does the expression value appear to be largest?

(e) Continue the search for the value of x that produces the greatest expression value. Determine this value of x to the nearest hundredth.

86. Work with other students on this exercise.

Part 1: Evaluate the three expressions $\frac{n^2 - 1}{2}$, n, $\frac{n^2 + 1}{2}$, using odd values of n: 1, 3, 5, 7, etc. Make a chart like the one below and complete it.

n	$a = \frac{n^2-1}{2}$	$b = n$	$c = \frac{n^2+1}{2}$	a^2	b^2	c^2
1						
3						
5						
7						
9						
11						
13						
15						

Part 2: The numbers a, b, and c that you get in each row have a surprising relationship to each other. Complete the last three columns and work together to discover this relationship. You may want to find out more about the history of this famous number pattern.

87. In exercise 86 you investigated the numbers obtained by evaluating the following expressions for odd positive integer values of n: $\frac{n^2 - 1}{2}$, n, $\frac{n^2 + 1}{2}$. Work with other students to investigate what three numbers you get when you evaluate for a *negative* odd value. Does the pattern you observed before still hold? Try several negative odd numbers to test the pattern.

Have no fear of fractions—does the pattern work with fractions? Try even integers. Is there a pattern for the three numbers obtained when you begin with even integers?

88. Enjoyment of patterns in art, music, and language is common to all cultures, and many cultures also delight in and draw spiritual significance from patterns in numbers. One such set of patterns is that of the "magic" square. One of these squares appears in a famous etching by Albrecht Dürer, who lived from 1471 to 1528 in Europe. He was one of the first artists in Europe to use geometry to give perspective, a feeling of three dimensions, in his work. The magic square in his work is this one:

16	3	2	13
5	10	11	8
9	6	7	12
4	15	14	1

Why is this square "magic"? It is magic because every row, every column, and both diagonals add to the same number. In this square there are 16 spaces for the numbers 1 through 16.

Answers

86. ______

87. ______

88. ______

Answers

89. ______

Part 1: What number does each row and column add to?

Write the square that you obtain by adding −17 to each number. Is this still a magic square? If so, what number does each column and row add to? If you add 5 to each number in the original magic square, do you still have a magic square? You have been studying the operations of addition, multiplication, subtraction, and division with integers and with rational numbers. What operations can you perform on this magic square and still have a magic square? Try to find something that will not work. Use algebra to help you decide what will work and what won't. Write a description of your work and explain your conclusions.

Part 2: Here is the oldest published magic square. It is from China, about 250 B.C.E. Legend has it that it was brought from the River Lo by a turtle to Emperor Yii, who was a hydraulic engineer.

4	9	2
3	5	7
8	1	6

Check to make sure that this is a magic square. Work together to decide what operation might be done to every number in the magic square to make the sum of each row, column, and diagonal the *opposite* of what it is now. What would you do to every number to cause the sum of each row, column, and diagonal to equal zero?

89. Use the Internet to research magic squares such as the one appearing in Dürer's work (see the previous exercise).

Note: We provide a brief tutorial on searching the Internet in Appendix A.

Answers

1. −22 **3.** 2 **5.** −20 **7.** 12 **9.** 4 **11.** 83 **13.** −20 **15.** −40 **17.** 14 **19.** −9 **21.** 2 **23.** 2 **25.** −11 **27.** 1 **29.** 11 **31.** −56 **33.** −8 **35.** 91 **37.** 29 **39.** 9 **41.** 16 **43.** 2 **45.** −16 **47.** −72 **49.** True **51.** False **53.** False **55.** never **57.** 5.625 Ω **59.** 30 in. **61.** $1,875 **63.** 14°F **65.** $19.09 **67.** −15.3 **69.** −1.1 **71.** 1.1 **73.** 14.0 **75.** −90 **77.** −77,922 **79.** 12 μg/mL **81.** 0.674 **83.** Above and Beyond **85.** **(a)** 0, 80, 96, 72, 32, 0; **(b)** 2; **(c)** 94.5, 95.744, 96.492, 96.768, 96.596, 96, 95.004, 93.632, 91.908, 89.856, 87.5; **(d)** 1.8; **(e)** 1.81 **87.** Above and Beyond **89.** Above and Beyond

1.3 Adding and Subtracting Algebraic Expressions

< 1.3 Objectives >

1 > Combine like terms

2 > Add algebraic expressions

3 > Subtract algebraic expressions

To find the perimeter of (or the distance around) a rectangle, we add 2 times the length and 2 times the width. In the language of algebra, this can be written as

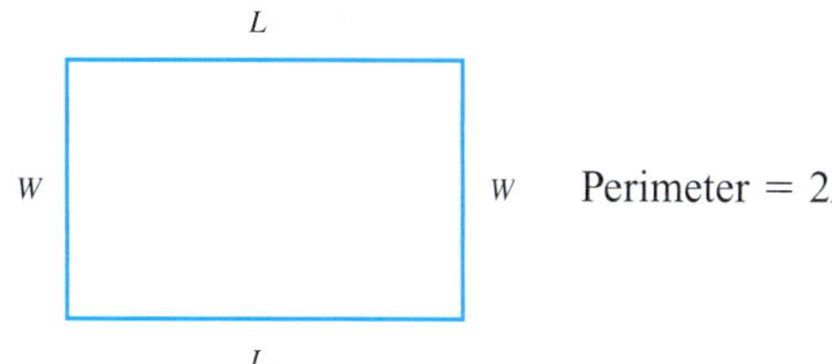

Perimeter $= 2L + 2W$

We call $2L + 2W$ an **algebraic expression,** or more simply an **expression.** As we discussed in Section 1.1, an expression allows us to write a mathematical idea in symbols. It can be thought of as a meaningful collection of letters, numbers, and operation symbols.

Some expressions are

1. $5x^2$

2. $3a + 2b$

3. $4x^3 - 2y + 1$

4. $3(x^2 + y^2)$

NOTE

If a variable has no exponent, it is raised to the power 1.

In algebraic expressions, the addition and subtraction signs break the expressions into smaller parts called *terms.*

Definition

Term — A **term** can be written as a number or the product of a number and one or more variables and their exponents.

In an expression, each sign (+ or −) is a part of the term that follows the sign.

Example 1 Identifying Terms

(a) $5x^2$ has one term.

(b) $3a + 2b$ has two terms: $3a$ and $2b$.

(Term: $3a$; Term: $2b$)

(c) $4x^3 - 2y + 1$ has three terms: $4x^3$, $-2y$, and 1.

(Term: $4x^3$; Term: $-2y$; Term: $+1$)

(d) $x - y$ has two terms: x and $-y$.

(e) (3)(2) is a term because we can write the product as the number 6.

NOTE

Each term "owns" the sign that precedes it.

NOTE

The numerical coefficient is usually referred to as the *coefficient*.

Check Yourself 1

List the terms of each expression.

(a) $2b^4$ (b) $5m + 3n$ (c) $2s^2 - 3t - 6$

Note that a term in an expression may have any number of factors. For instance, $5xy$ is a term. It has factors of 5, x, and y. The number-factor of a term is called the **numerical coefficient.** For the term $5xy$, the numerical coefficient is 5.

Example 2 Identifying the Numerical Coefficient

(a) $4a$ has the numerical coefficient 4.

(b) $6a^3b^4c^2$ has the numerical coefficient 6.

(c) $-7m^2n^3$ has the numerical coefficient -7.

(d) x has the numerical coefficient 1 since $x = 1 \cdot x$.

(e) $(4)(2)x^2$ has the numerical coefficient 8 because we can write the expression as $8x^2$.

Check Yourself 2

Give the numerical coefficient for each term.

(a) $8a^2b$ (b) $-5m^3n^4$ (c) y

If terms contain exactly the *same letters* (or variables) raised to the *same powers,* they are called **like terms.**

Example 3 Identifying Like Terms

(a) The following are like terms.

$6a$ and $7a$
$5b^2$ and b^2
$10x^2y^3z$ and $-6x^2y^3z$

Each pair of terms has the same letters, with matching letters raised to the same power—the numerical coefficients can be any number.

(b) The following are *not* like terms.

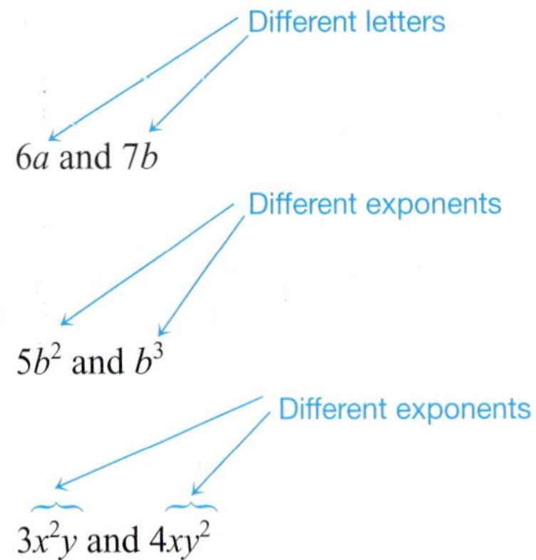

$6a$ and $7b$

$5b^2$ and b^3

$3x^2y$ and $4xy^2$

Check Yourself 3

Circle the like terms.

$5a^2b$ ab^2 a^2b $-3a^2$ $4ab$ $3b^2$ $-7a^2b$

NOTES

Here we use the distributive property.

You don't have to write all this out—just do it mentally!

Like terms of an expression can always be combined into a single term. Consider the following:

$$\underbrace{2x}_{x+x} + \underbrace{5x}_{x+x+x+x+x} = \underbrace{7x}_{x+x+x+x+x+x+x}$$

Rather than having to write out all those x's, try

$2x + 5x = (2 + 5)x = 7x$

In the same way,

$9b + 6b = (9 + 6)b = 15b$

and $10a - 4a = (10 - 4)a = 6a$

This leads us to the following procedure for combining like terms.

Step by Step

Combining Like Terms

To combine like terms, do the following steps.

Step 1 Add or subtract the numerical coefficients.

Step 2 Attach the common variables.

Example 4 Combining Like Terms

< Objective 1 >

RECALL

When any factor is multiplied by 0, the product is 0.

Combine like terms.

(a) $8m + 5m = (8 + 5)m = 13m$

(b) $5pq^3 - 4pq^3 = 1pq^3 = pq^3$

(c) $7a^3b^2 - 7a^3b^2 = 0a^3b^2 = 0$

Check Yourself 4

Combine like terms.

(a) $6b + 8b$
(b) $12x^2 - 3x^2$
(c) $8xy^3 - 7xy^3$
(d) $9a^2b^4 - 9a^2b^4$

Here are some expressions involving more than two terms. The idea is the same.

Example 5 Combining Like Terms

RECALL

The distributive property can be used over any number of like terms.

Combine like terms.

(a) $5ab - 2ab + 3ab$
$= (5 - 2 + 3)ab = 6ab$

(b) $\underbrace{8x - 2x} + 5y$ Only like terms can be combined.
$= 6x + 5y$

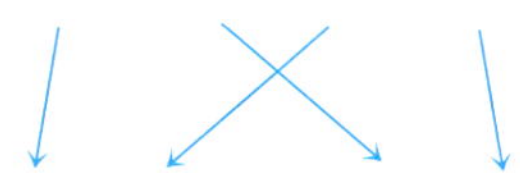

(c) $5m + 8n + 4m - 3n$ Rearrange the order of the terms using the associative and commutative properties of addition.
$= (5m + 4m) + (8n - 3n)$
$= 9m + 5n$

NOTE

With practice you will be doing this mentally rather than writing out these steps.

(d) $4x^2 + 2x - 3x^2 + x$
$= (4x^2 - 3x^2) + (2x + x)$
$= x^2 + 3x$

> CAUTION

Be careful when moving terms. Remember that they own the signs in front of them.

As these examples illustrate, combining like terms often means changing the grouping and the order in which the terms are written. Again all this is possible because of the properties of addition that we introduced in Section 0.3.

Check Yourself 5

Combine like terms.

(a) $4m^2 - 3m^2 + 8m^2$ (b) $9ab + 3a - 5ab$
(c) $4p + 7q + 5p - 3q$

Addition is always a matter of combining like quantities (two apples plus three apples, four books plus five books, and so on). If you keep that basic idea in mind, adding expressions is easy. It is just a matter of combining like terms. Suppose that you want to add

$5x^2 + 3x + 4$ and $4x^2 + 5x - 6$

Parentheses are sometimes used in adding, so for the sum of these expressions, we can write

$(5x^2 + 3x + 4) + (4x^2 + 5x - 6)$

Now what about the parentheses? You can use the following rule.

Property

Removing Grouping Symbols When Adding

When adding two expressions, if a plus sign (+) or nothing at all appears in front of parentheses, just remove the parentheses. No other changes are necessary.

NOTES

Just remove the parentheses. No other changes are necessary.

We use the associative and commutative properties to reorder and regroup.

Here we use the distributive property. For example, $5x^2 + 4x^2 = 9x^2$

Now we return to the addition.

$(5x^2 + 3x + 4) + (4x^2 + 5x - 6)$
$= 5x^2 + 3x + 4 + 4x^2 + 5x - 6$

Like terms Like terms Like terms

Collect like terms. (*Remember:* Like terms have the same variables raised to the same power.)

$= (5x^2 + 4x^2) + (3x + 5x) + (4 - 6)$

Combine like terms for the result:

$= 9x^2 + 8x - 2$

Alternatively, we could perform the addition in a vertical format. When using this method, be certain to align like terms in each column. In a vertical format the same addition looks like this.

$$\begin{array}{r} 5x^2 + 3x + 4 \\ + \; 4x^2 + 5x - 6 \\ \hline 9x^2 + 8x - 2 \end{array}$$

Much of this work can be done mentally. You can then write the sum directly by locating like terms and combining. Example 6 illustrates this property.

Example 6 Combining Like Terms

< Objective 2 >

Add $3x - 5$ and $2x + 3$.

Write the sum.

$(3x - 5) + (2x + 3)$
$= 3x - 5 + 2x + 3 = 5x - 2$

Like terms Like terms

Check Yourself 6

Add $6x^2 + 2x$ and $4x^2 - 7x$.

Subtracting expressions requires another rule for removing signs of grouping.

Property

Removing Grouping Symbols When Subtracting

When subtracting expressions, if a minus sign (−) appears in front of a set of parentheses, the parentheses can be removed by changing the sign of each term inside the parentheses.

When applying this rule, we are actually distributing the negative. This is illustrated in Example 7.

Example 7 Removing Parentheses

NOTE

This uses the distributive property,

$-(2x + 3y) = (-1)(2x + 3y)$
$= -2x - 3y$

In each case, remove the parentheses.

(a) $-(2x + 3y) = -2x - 3y$ Change each sign when removing the parentheses.

(b) $m - (5n - 3p) = m \underbrace{- 5n + 3p}_{\text{Sign changes}}$

(c) $2x - (-3y + z) = 2x \underbrace{+ 3y - z}_{\text{Sign changes}}$

Check Yourself 7

Remove the parentheses.

(a) $-(3m + 5n)$ **(b)** $-(5w - 7z)$
(c) $3r - (2s - 5t)$ **(d)** $5a - (-3b - 2c)$

Subtracting expressions is now a matter of using the previous rule when removing the parentheses and then combining the like terms.

Example 8 Subtracting Expressions

< Objective 3 >

RECALL

The expression following *from* is written first in the problem.

(a) Subtract $5x - 3$ from $8x + 2$.

Write

$(8x + 2) - (5x - 3)$
$= 8x + 2 \underbrace{- 5x + 3}_{\text{Sign changes}}$
$= 3x + 5$ Combine like terms: $8x - 5x = 3x$ and $2 + 3 = 5$.

(b) Subtract $4x^2 - 8x + 3$ from $8x^2 + 5x - 3$.

Write

$(8x^2 + 5x - 3) - (4x^2 - 8x + 3)$
$= 8x^2 + 5x - 3 \underbrace{- 4x^2 + 8x - 3}_{\text{Sign changes}}$
$= 4x^2 + 13x - 6$

RECALL

Combine like terms:

$8x^2 - 4x^2 = 4x^2$
$5x + 8x = 13x$
$-3 - 3 = -6$

Check Yourself 8

(a) Subtract $7x + 3$ from $10x - 7$.
(b) Subtract $5x^2 - 3x + 2$ from $8x^2 - 3x - 6$.

Example 9 demonstrates a business and finance application of some of the ideas presented in this section.

Example 9 A Business and Finance Application

S-Bar Electronics, Inc., sells a certain server for \$1,410. It pays the manufacturer \$849 for each server, and there are \$4,500 per week in fixed costs associated with the servers. Find an equation that represents the profit S-Bar Electronics earns by buying and selling these servers.

NOTE

A business can compute the profit it earns on a product by subtracting the costs associated with the product from the revenue earned by that product. We write

$P = R - C$

Let x be the number of servers bought and sold during the week.

Then, the revenue earned by S-Bar from these servers can be modeled by the formula

$R = 1{,}410x$

The cost can be modeled with the formula

$C = 849x + 4{,}500$

The profit can be modeled by the difference between the revenue and the cost.

$P = 1{,}410x - (849 + 4{,}500)$

$P = 1{,}410x - 849x - 4{,}500$

Simplify the given profit formula.

The like terms are $1{,}410x$ and $-849x$. We combine these to give a simplified formula

$P = 561x - 4{,}500$

NOTE

A negative profit means the company suffered a loss.

Check Yourself 9

S-Bar Electronics, Inc., also sells a 19-in. flat-screen monitor for \$799 each. The monitors cost S-Bar \$489 each. Additionally, there are weekly fixed costs of \$3,150 associated with the sale of the monitors. We can model the profits earned on the sale of y monitors in one week with the formula

$P = 799y - 489y - 3{,}150$

Simplify the profit formula.

Check Yourself ANSWERS

1. **(a)** $2b^4$; **(b)** $5m, 3n$; **(c)** $2s^2, -3t, -6$ **2.** **(a)** 8; **(b)** -5; **(c)** 1
3. The like terms are $5a^2b$, a^2b, and $-7a^2b$. **4.** **(a)** $14b$; **(b)** $9x^2$; **(c)** xy^3; **(d)** 0
5. **(a)** $9m^2$; **(b)** $4ab + 3a$; **(c)** $9p + 4q$ **6.** $10x^2 - 5x$
7. **(a)** $-3m - 5n$; **(b)** $-5w + 7z$; **(c)** $3r - 2s + 5t$; **(d)** $5a + 3b + 2c$
8. **(a)** $3x - 10$; **(b)** $3x^2 - 8$ **9.** $P = 310y - 3{,}150$

Reading Your Text

SECTION 1.3

(a) If a variable appears without an exponent, it is understood to be raised to the ______________ power.

(b) A ______________ can be written as a number or the product of a number and one or more variables and their exponents.

(c) A term may have any number of ______________.

(d) In the term $5xy$, the factor 5 is called the ______________.

Basic Skills | Challenge Yourself | Calculator/Computer | Career Applications | Above and Beyond

1.3 exercises

Name ______________

Section ________ Date ________

List the terms of each expression.

1. $5a + 2$

2. $7a - 4b$

3. $5x^4$

4. $3x^2$

5. $3x^2 + 3x - 7$

6. $2a^3 - a^2 + a$

Circle the like terms in each group of terms.

7. $5ab, 3b, 3a, 4ab$

8. $9m^2, 8mn, 5m^2, 7m$

9. $4xy^2, 2x^2y, 5x^2, -3x^2y, 5y, 6x^2y$ > Videos

10. $8a^2b, 4a^2, 3ab^2, -5a^2b, 3ab, 5a^2b$

< Objective 1 >

Combine the like terms.

11. $6p + 9p$

12. $6a^2 + 8a^2$

13. $7b^3 + 10b^3$

14. $7rs + 13rs$

15. $21xyz + 7xyz$

16. $4n^2m + 11n^2m$

17. $9z^2 - 3z^2$

18. $7m - 6m$

19. $5a^3 - 5a^3$

20. $9xy - 13xy$

21. $16p^2q - 17p^2q$

22. $7cd - 7cd$

23. $6p^2q - 21p^2q$

24. $8r^3s^2 - 17r^3s^2$

25. $10x^2 - 7x^2 + 3x^2$

26. $13uv + 5uv - 12uv$

27. $-6c + 3d + 5c$

28. $5m^2 - 3m + 6m^2$

29. $4x + 4y - 7x - 5y$

30. $7a - 4a^2 - 13a + 9a^2$

31. $2a - 7b - 3 + 2a - 3b + 2$

32. $5p^2 - 2p - 8 - 7p^2 + 5p + 6$

Remove the parentheses in each expression, and simplify where possible.

33. $-(2a + 3b)$

34. $-(7x - 4y)$

35. $5a - (2b - 3c)$

36. $7x - (4y + 3z)$

37. $3x - (4y + 5x)$

38. $10m - (3m - 2n)$

39. $5p - (-3p + 2q)$

40. $8d - (-7c - 2d)$

Answers

1. ______ 2. ______
3. ______ 4. ______
5. ______ 6. ______
7. ______ 8. ______
9. ______
10. ______
11. ______ 12. ______
13. ______ 14. ______
15. ______ 16. ______
17. ______ 18. ______
19. ______ 20. ______
21. ______ 22. ______
23. ______ 24. ______
25. ______ 26. ______
27. ______ 28. ______
29. ______ 30. ______
31. ______
32. ______
33. ______ 34. ______
35. ______ 36. ______
37. ______ 38. ______
39. ______ 40. ______

Answers

41. ______ 42. ______
43. ______ 44. ______
45. ______ 46. ______
47. ______
48. ______
49. ______
50. ______
51. ______ 52. ______
53. ______ 54. ______
55. ______
56. ______
57. ______ 58. ______
59. ______ 60. ______
61. ______ 62. ______
63. ______ 64. ______
65. ______
66. ______
67. ______ 68. ______
69. ______ 70. ______
71. ______ 72. ______
73. ______ 74. ______
75. ______
76. ______
77. ______
78. ______
79. ______
80. ______

< Objective 2 >

Add.

41. $6a - 5$ and $3a + 9$

42. $9x + 3$ and $3x - 4$

43. $-7p^2 + 9p$ and $4p^2 - 5p$

44. $2m^2 + 3m$ and $6m^2 - 8m$

45. $3x^2 - 2x$ and $-5x^2 + 2x$

46. $3p^2 + 5p$ and $-7p^2 - 5p$

47. $2x^2 + 5x - 3$ and $3x^2 - 7x + 4$

48. $4d^2 - 8d + 7$ and $5d^2 - 6d - 9$

49. $2b^2 + 8$ and $5b + 8$

50. $5p - 2$ and $4p^2 - 7p$

51. $8y^3 - 5y^2$ and $5y^2 - 2y$

52. $9x^4 - 2x^2$ and $2x^2 + 3$

53. $3x^2 - 7x^3$ and $-5x^2 + 4x^3$

54. $9m^3 - 2m$ and $-6m - 4m^3$

55. $4x^2 - 2 + 7x$ and $5 - 8x - 6x^2$

56. $5b^3 - 8b + 2b^2$ and $3b^2 - 7b^3 + 5b$

< Objective 3 >

Subtract.

57. $x + 2$ from $3x - 5$

58. $x - 2$ from $3x + 5$

59. $3m^2 - 2m$ from $4m^2 - 5m$

60. $9a^2 - 5a$ from $11a^2 - 10a$

61. $6y^2 + 5y$ from $4y^2 + 5y$

62. $9x^2 - 2x$ from $6x^2 - 2x$

63. $x^2 - 4x - 3$ from $3x^2 - 5x - 2$

64. $3x^2 - 2x + 4$ from $5x^2 - 8x - 3$

65. $3a + 7$ from $8a^2 - 9a$

66. $3x^3 + x^2$ from $4x^3 - 5x$

67. $2p - 5p^2$ from $-9p^2 + 4p$

68. $7y - 3y^2$ from $3y^2 - 2y$

69. $x^2 - 5 - 8x$ from $3x^2 - 8x + 7$

70. $4x - 2x^2 + 4x^3$ from $4x^3 + x - 3x^2$

Perform the indicated operations.

71. Subtract $3b + 2$ from the sum of $4b - 2$ and $5b + 3$.

72. Subtract $5m - 7$ from the sum of $2m - 8$ and $9m - 2$.

73. Subtract $5x^2 + 7x - 6$ from the sum of $2x^2 - 3x + 5$ and $3x^2 + 5x - 7$.

74. Subtract $4x^2 - 5x - 3$ from the sum of $x^2 - 3x - 7$ and $2x^2 - 2x + 9$.

75. Subtract $2x^2 - 3x$ from the sum of $4x^2 - 5$ and $2x - 7$.

76. Subtract $5a^2 - 3a$ from the sum of $3a - 3$ and $5a^2 + 5$.

77. Subtract the sum of $3y^2 - 3y$ and $5y^2 + 3y$ from $2y^2 - 8y$.

78. Subtract the sum of $3y^3 + 7y^2$ and $5y^3 - 7y^2$ from $4y^3 - 5y^2$.

79. $[(9x^2 - 3x + 5) - (3x^2 + 2x - 1)] - (x^2 - 2x - 3)$

80. $[(5x^2 + 2x - 3) - (-2x^2 + x - 2)] - (2x^2 + 3x - 5)$

Basic Skills | **Challenge Yourself** | Calculator/Computer | Career Applications | Above and Beyond

Determine whether each statement is **true** *or* **false.**

81. For two terms to be *like terms,* the numerical coefficients must match.

82. The key property that allows like terms to be combined is the distributive property.

Complete each statement with **never, sometimes,** *or* **always.**

83. Like terms can ______________ be combined.

84. When adding two expressions, the terms can ______________ be rearranged.

85. **GEOMETRY** A rectangle has sides of $8x + 9$ and $6x - 7$. Find an expression that represents its perimeter.

86. **GEOMETRY** A triangle has sides $4x + 7$, $6x + 3$, and $2x - 5$. Find an expression that represents its perimeter. > Videos

87. **BUSINESS AND FINANCE** The cost of producing x units of an item is $C = 150 + 25x$. The revenue for selling x units is $R = 90x - x^2$. The profit is given by the revenue minus the cost. Find an expression that represents profit.

88. **BUSINESS AND FINANCE** The revenue for selling y units is $R = 3y^2 - 2y + 5$, and the cost of producing y units is $C = y^2 + y - 3$. Find an expression that represents profit.

89. **CONSTRUCTION** A wooden beam is $(3y^2 + 3y - 2)$ meters (m) long. If a piece $(y^2 - 8)$ m is cut, find an expression that represents the length of the remaining piece of beam.

90. **CONSTRUCTION** A steel girder is $(9y^2 + 6y - 4)$ m long. Two pieces are cut from the girder. One has length $(3y^2 + 2y - 1)$ m and the other has length $(4y^2 + 3y - 2)$ m. Find the length of the remaining piece.

91. **GEOMETRY** Find an expression for the perimeter of the given triangle.

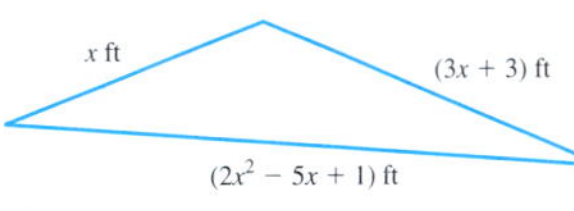

92. **GEOMETRY** Find an expression for the perimeter of the given rectangle.

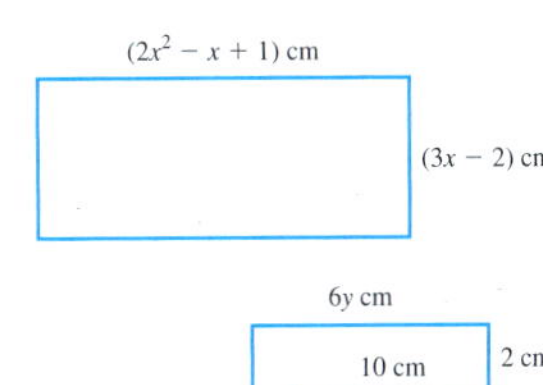

93. **GEOMETRY** Find an expression for the perimeter of the given figure.

6y cm
2 cm
10 cm
3y cm
8y cm
10 cm
5 cm
(5y + 2) cm

94. **GEOMETRY** Find the perimeter of the accompanying figure.

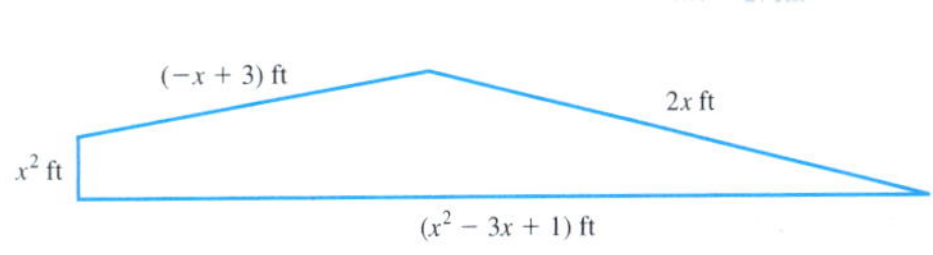

Answers

81. ______________

82. ______________

83. ______________

84. ______________

85. ______________

86. ______________

87. ______________

88. ______________

89. ______________

90. ______________

91. ______________

92. ______________

93. ______________

94. ______________

Answers

95. ______

96. ______

97. ______

98. ______

99. ______

100. ______

101. ______

102. ______

103. ______

104. ______

105. ______

106. ______

Basic Skills | Challenge Yourself | **Calculator/Computer** | Career Applications | Above and Beyond

Using your calculator, evaluate each expression for the given values of the variables. Round your answer to the nearest tenth.

95. $7x^2 - 5y^3$ for $x = 7.1695$ and $y = 3.128$

96. $2x^2 + 3y + 5x$ for $x = 3.61$ and $y = 7.91$

97. $4x^2y + 2xy^2 - 5x^3y$ for $x = 1.29$ and $y = 2.56$

98. $3x^3y - 4xy + 2x^2y^2$ for $x = 3.26$ and $y = 1.68$

99. **Mechanical Engineering** A primary beam can support a load of $54p$. A second beam is added that can support a load of $32p$. What is the total load that the two beams can support? > Videos

100. **Mechanical Engineering** Two objects are spinning on the same axis. The moment of inertia of the first object is $\frac{6^3}{12}b$. The moment of inertia of the second object is given by $\frac{30^3}{36}b$. The total moment of inertia is given by the sum of the moments of inertia of the two objects. Write a simplified expression for the total moment of inertia for the two objects described.

101. **Allied Health** A person's body mass index (BMI) can be calculated using their height h, in inches, and their weight w, in pounds, with the formula $\frac{703w}{h^2}$

Compute the BMI of a 69-inch, 190 pound man (to the nearest tenth).

102. **Allied Health** A person's body mass index (BMI) can be calculated using their height h, in centimeters, and their weight w, in kilograms, with the formula $\frac{10{,}000w}{h^2}$

Compute the BMI of a 160-cm, 70-kg woman (to the nearest tenth).

103. Does replacing each occurrence of the variable y in $3y^5 - 7y^4 + 3y$ with its opposite result in the opposite of the polynomial? Why or why not?

104. Write a paragraph explaining the difference between n^2 and $2n$.

105. Complete the explanation "x^3 and $3x$ are not the same because. . . ."

106. Complete the statement "$x + 2$ and $2x$ are different because. . . ."

107. Write an English phrase for each algebraic expression.

(a) $2x^3 + 5x$ **(b)** $(2x + 5)^3$ **(c)** $6(n + 4)^2$

108. Work with another student to complete this exercise. Place >, <, or = in the blanks in these statements.

1^2_____2^1

2^3_____3^2

3^4_____4^3

4^5_____5^4

Write an algebraic statement for the pattern of numbers. Do you think this is a pattern that continues? Add more examples and extend the pattern to the general case by writing the pattern in algebraic notation. Write a short paragraph stating your conjecture.

109. Compute and fill in the blanks.

Case 1: $1^2 - 0^2 =$ _____

Case 2: $2^2 - 1^2 =$ _____

Case 3: $3^2 - 2^2 =$ _____

Case 4: $4^2 - 3^2 =$ _____

Based on the pattern you see in these four cases, predict the value of case 5: $5^2 - 4^2$. Compute case 5 to check your prediction. Write an expression for case *n*. Describe in words the pattern that you see in this exercise.

Answers

107. ____________________

108. ____________________

109. ____________________

Answers

1. $5a, 2$ **3.** $5x^4$ **5.** $3x^2, 3x, -7$ **7.** $5ab, 4ab$
9. $2x^2y, -3x^2y, 6x^2y$ **11.** $15p$ **13.** $17b^3$ **15.** $28xyz$ **17.** $6z^2$
19. 0 **21.** $-p^2q$ **23.** $-15p^2q$ **25.** $6x^2$ **27.** $-c + 3d$
29. $-3x - y$ **31.** $4a - 10b - 1$ **33.** $-2a - 3b$ **35.** $5a - 2b + 3c$
37. $-2x - 4y$ **39.** $8p - 2q$ **41.** $9a + 4$ **43.** $-3p^2 + 4p$
45. $-2x^2$ **47.** $5x^2 - 2x + 1$ **49.** $2b^2 + 5b + 16$ **51.** $8y^3 - 2y$
53. $-3x^3 - 2x^2$ **55.** $-2x^2 - x + 3$ **57.** $2x - 7$ **59.** $m^2 - 3m$
61. $-2y^2$ **63.** $2x^2 - x + 1$ **65.** $8a^2 - 12a - 7$ **67.** $-4p^2 + 2p$
69. $2x^2 + 12$ **71.** $6b - 1$ **73.** $-5x + 4$ **75.** $2x^2 + 5x - 12$
77. $-6y^2 - 8y$ **79.** $5x^2 - 3x + 9$ **81.** False **83.** always
85. $28x + 4$ **87.** $-x^2 + 65x - 150$ **89.** $(2y^2 + 3y + 6)$ m
91. $(2x^2 - x + 4)$ ft **93.** $(22y + 29)$ cm **95.** 206.8 **97.** 6.5
99. $86p$ **101.** 28.1 **103.** Above and Beyond
105. Above and Beyond **107.** Above and Beyond
109. Above and Beyond

1.4 Solving Equations by Adding and Subtracting

< 1.4 Objectives >

1 > Determine whether a given number is a solution for an equation

2 > Use the addition property to solve equations

3 > Translate words to equation symbols

4 > Solve application problems

Tips for Student Success

Don't procrastinate!

1. Do your math homework while you're still fresh. If you wait until too late at night, your tired mind will have much greater difficulty understanding the concepts.
2. Do your homework the day it is assigned. The more recent the explanation, the easier it is to recall.
3. When you've finished your homework, try reading the next section through one time. This will give you a sense of direction when you next encounter the material. This works whether you are in a lecture or lab setting.

Remember that, in a typical math class, you are expected to do 2 or 3 hours of homework for each weekly class hour. This means 2 or 3 hours per night. Schedule the time and stick to your schedule.

In this chapter you will work with one of the most important tools of mathematics—the equation. The ability to recognize and solve various types of equations is probably the most useful algebraic skill you will learn. We will continue to build upon the methods of this chapter throughout the remainder of the text. To start, we describe what we mean by an *equation.*

Definition

Equation An **equation** is a mathematical statement that two expressions are equal.

Some examples are $3 + 4 = 7$, $x + 3 = 5$, $P = 2L + 2W$. As you can see, an equal sign (=) separates the two equal expressions. These expressions are usually called the *left side* and the *right side* of the equation.

$$x + 3 = 5$$

Left side Equals Right side

NOTE

An equation such as

$x + 3 = 5$

is called a **conditional equation** because it can be either true or false depending on the value of the variable.

An equation may be either true or false. For instance, $3 + 4 = 7$ is true because both sides name the same number. What about an equation such as $x + 3 = 5$ that has a letter or variable on one side? Any number can replace x in the equation. However, only one number will make this equation a true statement.

$$\text{If} \quad x = \begin{cases} 1 & 1 + 3 = 5 \text{ is false} \\ 2 & 2 + 3 = 5 \text{ is true} \\ 3 & 3 + 3 = 5 \text{ is false} \end{cases}$$

The number 2 is called a *solution* (or *root*) of the equation $x + 3 = 5$ because substituting 2 for x gives a true statement. 2 is the only solution to this equation.

Definition

Solution A **solution** to an equation is any value for the variable that makes the equation a true statement.

Example 1 **Verifying a Solution**

< Objective 1 >

NOTE

Until the left side equals the right side, we place a question mark over the equal sign.

RECALL

Always apply the rules for the order of operations. Multiply first; then add or subtract.

(a) Is 3 a solution for the equation $2x + 4 = 10$?

To find out, replace x with 3 and evaluate $2x + 4$ on the left.

Left side		Right side
$2 \cdot (3) + 4$	$\stackrel{?}{=}$	10
$6 + 4$	$\stackrel{?}{=}$	10
10	$=$	10

Since $10 = 10$ is a true statement, 3 is a solution of the equation.

(b) Is $\frac{5}{3}$ a solution of the equation $3x - \frac{2}{3} = 2x + 1$?

To find out, replace x with $\frac{5}{3}$ and evaluate each side separately.

Left side		Right side
$3 \cdot \left(\frac{5}{3}\right) - \frac{2}{3}$	$\stackrel{?}{=}$	$2 \cdot \left(\frac{5}{3}\right) + 1$
$\frac{15}{3} - \frac{2}{3}$	$\stackrel{?}{=}$	$\frac{10}{3} + \frac{3}{3}$
$\frac{13}{3}$	$=$	$\frac{13}{3}$

Because the two sides name the same number, we have a true statement, and $\frac{5}{3}$ is a solution.

Check Yourself 1

For the equation

$2x - 1 = x + 5$

(a) Is 6 a solution? **(b)** Is $\frac{8}{3}$ a solution?

NOTE

The equation $x^2 = 9$ is an example of a **quadratic equation.** We will learn to solve them in Chapters 6 and 8.

You may be wondering whether an equation can have more than one solution. It certainly can. For instance,

$x^2 = 9$

has two solutions. They are 3 and -3 because

$(3)^2 = 9$ and $(-3)^2 = 9$

In this chapter, we will generally work with *linear equations*. These are equations that can be put into the form

$ax + b = 0$

in which the variable is x, a and b are numbers, and a is not equal to 0. In a linear equation, the variable can appear only to the first power. No other power (x^2, x^3, etc.) can appear. Linear equations are also called **first degree equations.** The *degree* of an equation in one variable is the highest power to which the variable is raised. So, in the equation $5x^4 - 9x^2 + 7x - 2 = 0$, the highest power to which the x is raised is four. Therefore, it is a fourth-degree equation.

Property

Solutions for Linear Equations

Linear equations in one variable that can be written in the form

$ax + b = 0 \qquad a \neq 0$

have exactly one solution.

Example 2 Identifying Expressions and Equations

Label each statement as an expression, a linear equation, or a nonlinear equation. Recall that an equation is a statement in which an equal sign separates two expressions.

(a) $4x + 5$ is an expression.

(b) $2x + 8 = 0$ is a linear equation.

(c) $3x^2 - 9 = 0$ is a nonlinear equation.

(d) $5x = 15$ is a linear equation.

(e) $\dfrac{3}{x} + 2 = 0$ is a nonlinear equation.

NOTE

There can be no variable in the denominator of a linear equation.

Check Yourself 2

Label each as an expression, a linear equation, or a nonlinear equation.

(a) $2x^2 = 8$ **(b)** $2x - 3 = 0$ **(c)** $5x - 10$

(d) $2x + 1 = 7$ **(e)** $5 - \dfrac{6}{x} = 2x$

You can find the solution for an equation such as $x + 3 = 8$ by guessing the answer to the question "What plus 3 is 8?" Here the answer to the question is 5, which is also the solution for the equation. But for more complicated equations we need something more than guesswork. A better method is to transform the given equation to an *equivalent equation* whose solution can be found by inspection.

Definition

Equivalent Equations

Equations that have exactly the same solutions are called **equivalent equations.**

The following are all equivalent equations:

$2x + 3 = 5 \qquad 2x = 2 \qquad \text{and} \qquad x = 1$

They all have the same solution, 1. We say that a linear equation is *solved* when it is transformed to an equivalent equation of the form

$x = \square$

The variable is alone on one side.

The other side is some number, the solution.

NOTE

In some cases we write the equation in the form

$\square = x$

The number is the solution when the variable is isolated on either the left or the right.

The addition property of equality is the first property you need to transform an equation to an equivalent form.

Property

The Addition Property of Equality

If $\quad a = b$

then $\quad a + c = b + c$

In words, adding the same quantity to both sides of an equation gives an equivalent equation.

An equation is a statement that the two sides are equal. Adding the same quantity to both sides does not change the equality or "balance."

In Example 3 we apply this idea to solve an equation.

Example 3 Using the Addition Property to Solve an Equation

< Objective 2 >

NOTE

To check, replace x with 12 in the original equation:

$$x - 3 = 9$$
$$(12) - 3 \stackrel{?}{=} 9$$
$$9 = 9 \quad \text{True}$$

Because we have a true statement, 12 is the solution.

Solve.

$x - 3 = 9$

Remember that our goal is to isolate x on one side of the equation. Because 3 is being subtracted from x, we can add 3 to both sides of the equation.

$$\begin{aligned} x - 3 &= 9 \\ +3 & +3 \\ \hline x &= 12 \end{aligned}$$

Adding 3 "undoes" the subtraction and leaves x alone on the left.

Because 12 is the solution for the equivalent equation $x = 12$, it is the solution for our original equation.

Check Yourself 3

Solve and check.

$x - 5 = 4$

The addition property also allows us to add a negative number to both sides of an equation. This is really the same as subtracting the same quantity from both sides.

Example 4 Using the Addition Property to Solve an Equation

Solve.

$$x + 2 = \frac{11}{2}$$

In this case, 2 is *added* to x on the left. We can use the addition property to subtract 2 from both sides. This "undoes" the addition and leaves the variable x alone on one side of the equation.

NOTE

Because subtraction is defined in terms of addition, we can add *or* subtract the same quantity from both sides of the equation.

$$\begin{aligned} x + 2 &= \frac{11}{2} \\ -2 & -\frac{4}{2} \\ \hline x &= \frac{7}{2} \end{aligned}$$

We subtracted 2 from each side.

$$-\frac{4}{2} = -2$$

The solution is $\frac{7}{2}$. To check, replace x with $\frac{7}{2}$.

$$\left(\frac{7}{2}\right) + 2 = \frac{11}{2} \quad \text{True}$$

Check Yourself 4

Solve and check.

$$x + 6 = \frac{11}{3}$$

What if the equation has a variable term on both sides? You can use the addition property to add or subtract a term involving the variable to get the desired result.

Example 5 Using the Addition Property to Solve an Equation

Solve.

$5x = 4x + 7$

We start by subtracting $4x$ from both sides of the equation. Do you see why? Remember that an equation is solved when we have an equivalent equation of the form $x = \square$.

> **NOTE**
>
> Subtracting $4x$ is the same as adding $-4x$.

$$\begin{aligned} 5x &= 4x + 7 \\ -4x & \quad -4x \\ \hline x &= 7 \end{aligned}$$

Subtracting $4x$ from both sides *removes* $4x$ from the right.

To check: Since 7 is a solution for the equivalent equation $x = 7$, it should be a solution for the original equation. To find out, replace x with 7.

$$5 \cdot (7) \stackrel{?}{=} 4 \cdot (7) + 7$$
$$35 \stackrel{?}{=} 28 + 7$$
$$35 = 35 \quad \text{True}$$

Check Yourself 5

Solve and check.

$7x = 6x + 3$

Recall that addition can be set up either in a vertical format such as

$$\begin{array}{r} 256 \\ +192 \\ \hline 448 \end{array}$$

or in a horizontal format

$256 + 192 = 448$

When we use the addition property to solve an equation, the same choices are available. In our examples to this point we have used the vertical format. In Example 6 we use the horizontal format. In the remainder of this text, we assume that you are familiar with both formats.

Example 6 Using the Addition Property to Solve an Equation

Solve.

$7x - 8 = 6x$

We want all variables on *one* side of the equation. If we choose the left, we subtract $6x$ from both sides of the equation. This removes $6x$ from the right:

$$7x - 8 - 6x = 6x - 6x$$
$$x - 8 = 0$$

We want the variable alone, so we add 8 to both sides. This isolates x on the left.

$$x - 8 + 8 = 0 + 8$$
$$x = 8$$

We leave it to you to check that 8 is the solution.

Check Yourself 6

Solve and check.

$9x + 3 = 8x$

Often an equation has more than one variable term *and* more than one number. You have to apply the addition property twice to solve such equations.

Example 7 Using the Addition Property to Solve an Equation

Solve.

$5x - 7 = 4x + 3$

We would like the variable terms on the left, so we start by subtracting $4x$ to remove that term from the right side of the equation:

$$\begin{array}{rcl} 5x - 7 &=& 4x + 3 \\ -4x & & -4x \\ \hline x - 7 &=& 3 \end{array}$$

Now, to isolate the variable, we add 7 to both sides to undo the subtraction on the left:

$$\begin{array}{rcl} x - 7 &=& 3 \\ +7 & & +7 \\ \hline x &=& 10 \end{array}$$

The solution is 10. To check, replace x with 10 in the original equation:

$$5 \cdot (10) - 7 \stackrel{?}{=} 4 \cdot (10) + 3$$
$$43 = 43 \quad \text{True}$$

NOTE

You could just as easily have added 7 to both sides and *then* subtracted $4x$. The result would be the same. In fact, some students prefer to combine the two steps.

Check Yourself 7

Solve and check.

(a) $4x - 5 = 3x + 2$ **(b)** $6x + 2 = 5x - 4$

RECALL

By *simplify,* we mean to combine all like terms.

When solving an equation, you should always simplify each side as much as possible before using the addition property.

Example 8 Simplifying an Equation

Solve $5 + 8x - 2 = 2x - 3 + 5x$.

Like terms Like terms

$5 + 8x - 2 = 2x - 3 + 5x$

Notice that like terms appear on both sides of the equation. We start by combining the numbers on the left (5 and -2). Then we combine the like terms ($2x$ and $5x$) on the right. We have

$3 + 8x = 7x - 3$

Now we can apply the addition property, as before:

$$\begin{array}{rcl} 3 + 8x &=& 7x - 3 \\ -7x &=& -7x \\ \hline 3 + x &=& -3 \\ -3 & & -3 \\ \hline x &=& -6 \end{array}$$

Subtract $7x$.

Subtract 3 to isolate x.

The solution is -6. To check, always return to the original equation. That catches any possible errors in simplifying. Replacing x with -6 gives

$$5 + 8(-6) - 2 \stackrel{?}{=} 2(-6) - 3 + 5(-6)$$
$$5 - 48 - 2 \stackrel{?}{=} -12 - 3 - 30$$
$$-45 = -45 \quad \text{True}$$

Check Yourself 8

Solve and check.

(a) $3 + 6x + 4 = 8x - 3 - 3x$ (b) $5x + 21 + 3x = 20 + 7x - 2$

We may have to apply some of the properties discussed in Section 0.4 in solving equations. Example 9 illustrates the use of the distributive property to clear an equation of parentheses.

Example 9 Using the Distributive Property and Solving Equations

RECALL

$2(3x + 4) = 2(3x) + 2(4)$
$= 6x + 8$

Solve.

$2(3x + 4) = 5x - 6$

Applying the distributive property on the left gives

$6x + 8 = 5x - 6$

We can then proceed as before.

$$\begin{array}{rcrl} 6x + 8 & = & 5x - 6 & \\ -5x & = & -5x & \text{Subtract } 5x. \\ \hline x + 8 & = & -6 & \\ -8 & & -8 & \text{Subtract 8.} \\ \hline x & = & -14 & \end{array}$$

The solution is -14. We leave it to you to check this result.

Remember: Always return to the original equation to check.

Check Yourself 9

Solve and check each equation.

(a) $4(5x - 2) = 19x + 4$ (b) $3(5x + 1) = 2(7x - 3) - 4$

Given an expression such as

$-2(x - 5)$

we use the distributive property to create the equivalent expression

$-2x + 10$

The distribution of a negative number is shown in Example 10.

Example 10 Distributing a Negative Number

Solve each equation.

(a)
$$\begin{array}{rcrl} -2(x - 5) & = & -3x + 2 & \text{Distribute the } -2. \\ -2x + 10 & = & -3x + 2 & \\ +3x & & +3x & \text{Add } 3x. \\ \hline x + 10 & = & 2 & \\ -10 & = & -10 & \text{Subtract 10.} \\ \hline x & = & -8 & \text{The solution is } -8. \end{array}$$

(b)
$$\begin{array}{rcrl} -3(3x + 5) & = & -5(2x - 2) & \text{Distribute the } -3. \\ -9x - 15 & = & -5(2x - 2) & \text{Distribute the } -5. \\ -9x - 15 & = & -10x + 10 & \\ +10x & & +10x & \text{Add } 10x. \\ \hline x - 15 & = & 10 & \\ +15 & & +15 & \text{Add 15.} \\ \hline x & = & 25 & \text{The solution is 25.} \end{array}$$

RECALL

Return to the original equation to check your solution.

Check:

$$-3[3(25) + 5] \stackrel{?}{=} -5[2(25) - 2]$$
$$-3(75 + 5) \stackrel{?}{=} -5(50 - 2) \quad \text{Follow the order of operations.}$$
$$-3(80) \stackrel{?}{=} -5(48)$$
$$-240 = -240 \quad \text{True}$$

Check Yourself 10

Solve each equation.

(a) $-2(x - 3) = -x + 5$ (b) $-4(2x - 1) = -3(3x + 2)$

The main reason for learning how to set up and solve algebraic equations is so that we can use them to solve word problems. In fact, algebraic equations were *invented* to make solving word problems much easier. The first word problems that we know about are over 4,000 years old. They were literally "written in stone," on Babylonian tablets, about 500 years before the first algebraic equation made its appearance.

Before algebra, people solved word problems primarily by **substitution,** which is a method of finding unknown values by using trial and error in a logical way. Example 11 shows how to solve a word problem by using substitution.

Example 11 Solving a Word Problem by Substitution

NOTE

Consecutive integers are integers that follow each other, such as 8 and 9.

The sum of two consecutive integers is 37. Find the two integers.

If the two integers were 20 and 21, their sum would be 41. Since that's more than 37, the integers must be smaller. If the integers were 15 and 16, the sum would be 31. More trials yield that the sum of 18 and 19 is 37.

Check Yourself 11

The sum of two consecutive integers is 91. Find the two integers.

Most word problems are not so easily solved by substitution. For more complicated word problems, we use a five-step procedure. Using this step-by-step approach allows you to organize your work. Organization is a key to solving word problems.

Step by Step

To Solve Word Problems

Step 1 Read the problem carefully. Then reread it to decide what you are asked to find.

Step 2 Choose a letter to represent one of the unknowns in the problem. Then represent all other unknowns of the problem with expressions that use the same letter.

Step 3 Translate the problem to the language of algebra to form an equation.

Step 4 Solve the equation.

Step 5 Answer the question and include units in your answer, when appropriate. Check your solution by returning to the original problem.

The third step is usually the hardest. We must translate words to the language of algebra. Before we look at a complete example, the following table may help you review that translation step.

RECALL

We discussed these translations in Section 1.1. You might find it helpful to review that section before going on.

Translating Words to Algebra

Words	Algebra
The sum of x and y	$x + y$
3 plus a	$3 + a$ or $a + 3$
5 more than m	$m + 5$
b increased by 7	$b + 7$
The difference of x and y	$x - y$
4 less than a	$a - 4$
s decreased by 8	$s - 8$
The product of x and y	$x \cdot y$ or xy
5 times a	$5 \cdot a$ or $5a$
Twice m	$2m$
The quotient of x and y	$\frac{x}{y}$
a divided by 6	$\frac{a}{6}$
One-half of b	$\frac{b}{2}$ or $\frac{1}{2}b$

Now let's look at some typical examples of translating phrases to algebra.

Example 12 Translating Statements

< Objective 3 >

Translate each English expression to an algebraic expression.

(a) The sum of a and 2 times b

$a + 2b$

Sum — 2 times b

(b) 5 times m, increased by 1

$5m + 1$

5 times m — Increased by 1

(c) 5 less than 3 times x

$3x - 5$

3 times x — 5 less than

(d) The product of x and y, divided by 3

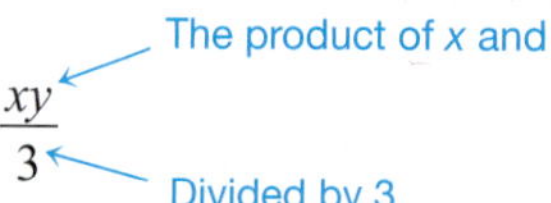

$\frac{xy}{3}$

Check Yourself 12

Translate to algebra.

(a) 2 more than twice x
(b) 4 less than 5 times n
(c) The product of twice a and b
(d) The sum of s and t, divided by 5

Now let's work through a complete example. Although this problem could be solved by substitution, it is presented here to help you practice the five-step approach.

Example 13 Solving an Application

< Objective 4 >

The sum of a number and 5 is 17. What is the number?

Step 1 *Read carefully.* You must find the unknown number.

Step 2 *Choose letters or variables.* Let x represent the unknown number. There are no other unknowns.

Step 3 *Translate.*

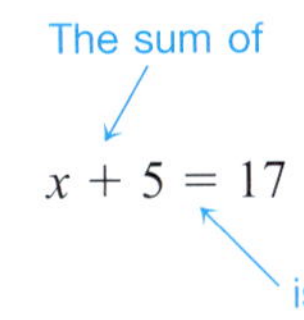

$$x + 5 = 17$$

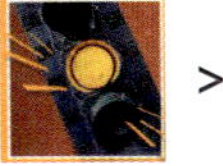

Always return to the *original problem* to check your result and *not* to the equation in step 3. This helps prevent possible errors!

Step 4 *Solve.*

$$x + 5 = 17 \qquad \text{Subtract 5.}$$
$$x + 5 - 5 = 17 - 5$$
$$x = 12$$

Step 5 *Answer.* The number is 12.

Check. Is the sum of 12 and 5 equal to 17? Yes $(12 + 5 = 17)$.

Check Yourself 13

The sum of a number and 8 is 35. What is the number?

Of course, there are many applications that require us to use the addition property to solve an equation. Consider the consumer application in the next example.

Example 14 A Consumer Application

An appliance store is having a sale on washers and dryers. They are charging $999 for a washer and dryer combination. If the washer sells for $649, how much is someone paying for the dryer as part of the combination?

Step 1 *Read carefully.* We are asked to find the cost of a dryer in this application.

Step 2 *Choose letters or variables.* Let d represent the cost of a dryer as part of the washer-dryer combination. This is the only unknown quantity in the problem.

Step 3 *Translate.*

$$d + 649 = 999$$

The washer costs $649.
Together, they cost $999.

RECALL

Always answer an application with a full sentence.

Step 4 *Solve.*

$$d + 649 = 999$$
$$d + 649 - 649 = 999 - 649 \qquad \text{Subtract 649 to isolate the variable.}$$
$$d = 350$$

Step 5 *Answer.* The dryer costs $350 as part of this combination.

Check. A $649 washer and a $350 dryer cost a total of $649 + $350 = $999.

Check Yourself 14

Of 18,540 votes cast in the school board election, 11,320 went to Carla. How many votes did her opponent Marco receive? Who won the election?

Let m be the number of votes Marco received and solve the equation $11{,}320 + m = 18{,}540$ in order to answer the questions.

Check Yourself ANSWERS

1. **(a)** 6 is a solution; **(b)** $\frac{8}{3}$ is not a solution.

2. **(a)** nonlinear equation; **(b)** linear equation; **(c)** expression; **(d)** linear equation; **(e)** nonlinear equation

3. 9 **4.** $-\frac{7}{3}$ **5.** 3 **6.** -3 **7.** **(a)** 7; **(b)** -6

8. **(a)** -10; **(b)** -3 **9.** **(a)** 12; **(b)** -13 **10.** **(a)** 1; **(b)** -10

11. 45 and 46 **12.** **(a)** $2x + 2$; **(b)** $5n - 4$; **(c)** $2ab$; **(d)** $\frac{s + t}{5}$

13. The equation is $x + 8 = 35$. The number is 27.

14. Marco received 7,220 votes; Carla won the election.

Reading Your Text

SECTION 1.4

(a) You should do your math homework while you are still ______________.

(b) An equation is a mathematical statement that two ______________ are equal.

(c) A ______________ of an equation is any value for the variable that makes the equation a true statement.

(d) Linear equations in one variable that can be written as $ax + b = 0$ $(a \neq 0)$ have exactly ______________ solution.

Basic Skills | Challenge Yourself | Calculator/Computer | Career Applications | Above and Beyond

1.4 exercises

Boost *your* GRADE at ALEKS.com!

ALEKS

- Practice Problems
- Self-Tests
- NetTutor
- e-Professors
- Videos

Name ______

Section ______ Date ______

< Objective 1 >

Is the number shown in parentheses a solution for the given equation?

1. $x + 7 = 12 \quad (5)$

2. $x + 2 = 11 \quad (8)$

3. $x - 15 = 6 \quad (-21)$

4. $x - 11 = 5 \quad (16)$

5. $5 - x = 2 \quad (4)$

6. $10 - x = 7 \quad (3)$

7. $8 - x = 5 \quad (-3)$

8. $5 - x = 6 \quad (-3)$

9. $3x + 4 = 13 \quad (8)$

10. $5x + 6 = 31 \quad (5)$

11. $4x - 5 = 7 \quad (3)$

12. $4x - 5 = 1 \quad \left(\frac{3}{2}\right)$

13. $5 - 2x = 10 \quad \left(-\frac{5}{2}\right)$

14. $4 - 5x = 9 \quad (-2)$

15. $6 - \frac{3}{4}x = 4 + \frac{1}{4}x \quad (2)$

16. $5x + 4 = 2x + 10 \quad (4)$

17. $x + 3 + 2x = 5 + x + 8 \quad (5)$

> Videos

18. $5x - 3 + 2x = 3 + x - 12 \quad (-2)$

19. $\frac{3}{4}x = 18 \quad (20)$

20. $\frac{3}{5}x = 24 \quad (40)$

21. $\frac{3}{5}x + 5 = 11 \quad (10)$

22. $\frac{2}{3}x + 8 = -12 \quad (-6)$

Label each statement as an expression, a linear equation, or a nonlinear equation.

23. $2x + 1 = 9$

24. $5x - 11$

25. $7x + 2x + 8 - 3$

26. $x + 5 = 13$

27. $3x + 5 = 9$

28. $12x^2 - 5x + 2 = 5$

< Objective 2 >

Solve each equation and check your results.

29. $x + 7 = 9$

30. $x - 4 = 6$

31. $x - 8 = 3$

32. $x + 11 = 15$

33. $x - 8 = -10$

34. $x - 8 = -11$

Answers

1. ______ 2. ______
3. ______ 4. ______
5. ______ 6. ______
7. ______ 8. ______
9. ______ 10. ______
11. ______ 12. ______
13. ______ 14. ______
15. ______ 16. ______
17. ______ 18. ______
19. ______ 20. ______
21. ______ 22. ______
23. ______
24. ______
25. ______
26. ______
27. ______
28. ______
29. ______ 30. ______
31. ______ 32. ______
33. ______ 34. ______

Answers

35. ______
36. ______
37. ______
38. ______
39. ______
40. ______
41. ______
42. ______
43. ______ 44. ______
45. ______ 46. ______
47. ______ 48. ______
49. ______ 50. ______
51. ______ 52. ______
53. ______ 54. ______
55. ______ 56. ______
57. ______ 58. ______
59. ______ 60. ______
61. ______ 62. ______
63. ______ 64. ______
65. ______
66. ______
67. ______
68. ______

35. $11 = x + 5$

36. $x + 7 = 0$

37. $4x = 3x + 4$

38. $7x = 6x - 8$

39. $11x = 10x - 10$

40. $2(x + 3) = x + 6$

41. $4x - 10 = 5(x - 2)$

42. $\frac{2}{3}x + \frac{7}{8} = \frac{5}{3}x + \frac{1}{8}$

43. $\frac{9}{5}x + \frac{5}{6} = \frac{4}{5}x + \frac{1}{6}$

44. $3x - 2 = 2x + 1$

45. $5x - 7 = 4x - 3$

46. $8x + 5 = 7x - 2$

47. $\frac{5}{3}x = 9 + \frac{2}{3}x$

48. $\frac{4}{7}x = 8 - \frac{3}{7}x$ > Videos

49. $3 + 6x + 2 = 3x + 11 + 2x$

50. $6x - 3 + 2x = 7x + 8$

51. $4x + 7 + 3x = 5x + 13 + x$

52. $5x + 9 + 4x = 9 + 8x - 7$

53. $4(3x + 4) = 11x - 2$

54. $2(5x - 3) = 9x + 7$

55. $3(7x + 2) = 5(4x + 1) + 17$ > Videos

56. $5(5x + 3) = 3(8x - 2) + 4$

57. $\frac{5}{4}x - 1 = \frac{1}{4}x + 7$

58. $\frac{7}{5}x + 3 = \frac{2}{5}x - 8$

59. $\frac{9}{2}x - \frac{3}{4} = \frac{7}{2}x + \frac{5}{4}$

60. $\frac{11}{3}x + \frac{1}{6} = \frac{8}{3}x + \frac{19}{6}$

61. $0.56x = 9 - 0.44x$

62. $8 - 0.37x = 5 + 0.63x$

63. $0.12x + 0.53x - 8 = -0.92x + 0.57x + 4$

64. $0.71x + 6 + 0.35x = 0.25x - 11 - 0.19x$

< Objective 3 >

In exercises 65 to 70, translate each English statement to an algebraic equation. Let x represent the number in each case.

65. 3 more than a number is 7.

66. 5 less than a number is 12.

67. 7 less than 3 times a number is twice that same number.

68. 4 more than 5 times a number is 6 times that same number.

69. 2 times the sum of a number and 5 is 18 more than that same number.

70. 3 times the sum of a number and 7 is 4 times that same number.

71. Which equation is equivalent to $8x + 5 = 9x - 4$?

(a) $17x = -9$ (b) $x = -9$ (c) $8x + 9 = 9x$ (d) $9 = 17x$

72. Which equation is equivalent to $5x - 7 = 4x - 12$?

(a) $9x = 19$ (b) $9x - 7 = -12$ (c) $x = -18$ (d) $x - 7 = -12$

73. Which equation is equivalent to $12x - 6 = 8x + 14$?

(a) $4x - 6 = 14$ (b) $x = 20$ (c) $20x = 20$ (d) $4x = 8$

74. Which equation is equivalent to $7x + 5 = 12x - 10$?

(a) $5x = -15$ (b) $7x - 5 = 12x$ (c) $-5 = 5x$ (d) $7x + 15 = 12x$

Basic Skills | **Challenge Yourself** | Calculator/Computer | Career Applications | Above and Beyond

Determine whether each statement is **true** *or* **false.**

75. Every linear equation with one variable has exactly one solution.

76. Isolating the variable on the right side of the equation will result in a negative solution.

77. If we add the same number to both sides of an equation, we always obtain an equivalent equation.

78. The equations $x^2 = 9$ and $x = 3$ are equivalent equations.

Complete each statement with **never, sometimes,** *or* **always.**

79. An equation __________ has one solution.

80. If a first-degree equation has a variable term on both sides, we must _______ use the addition property to solve the equation.

< Objective 4 >

Solve each word problem. Be sure to show the equation you use for the solution.

81. **Number Problem** The sum of a number and 7 is 33. What is the number?

82. **Number Problem** The sum of a number and 15 is 22. What is the number?

83. **Number Problem** The sum of a number and −15 is 7. What is the number?

Answers

69. ____________

70. ____________

71. ____________

72. ____________

73. ____________

74. ____________

75. ____________

76. ____________

77. ____________

78. ____________

79. ____________

80. ____________

81. ____________

82. ____________

83. ____________

Answers

84. ______

85. ______

86. ______

87. ______

88. ______

89. ______

90. ______

91. ______

92. ______

93. ______

84. **Number Problem** The sum of a number and −8 is 17. What is the number?

85. **Social Science** In an election, the winning candidate has 1,840 votes. If the total number of votes cast was 3,260, how many votes did the losing candidate receive?

86. **Business and Finance** Mike and Stefanie work at the same company and make a total of $2,760 per month. If Stefanie makes $1,400 per month, how much does Mike earn every month?

87. **Business and Finance** A washer-dryer combination costs $650. If the washer costs $360, what does the dryer cost?

88. **Technology** You have $2,350 saved for the purchase of a new computer that costs $3,675. How much more must you save?

89. **Construction Technology** K Jones Manufacturing produces hex bolts and carriage bolts. They sold 284 more hex bolts than carriage bolts last month. If they sold 2,680 carriage bolts, how many hex bolts did they sell?

90. **Electronics Technology** Berndt Electronics earns a marginal profit of $560 each on the sale of a particular server. If other costs involved amount to $4,500, will they earn a net profit of $5,000 on the sale of 15 servers? > Videos

91. **Engineering Technology** The specifications for an engine cylinder of a particular ship call for the stroke length to be two more than twice the diameter of the cylinder. Write an expression for the required stroke length given a cylinder's diameter d.

92. **Engineering Technology** Use your answer to exercise 91 to determine the required stroke length if the cylinder has a diameter of 52 in.

93. An algebraic equation is a complete sentence. It has a subject, a verb, and a predicate. For example, $x + 2 = 5$ can be written in English as "Two more than a number is five" or "A number added to two is five." Write an English version of each equation. Be sure to write complete sentences and that the sentences express the same idea as the equations. Exchange sentences with

another student and see if your interpretations of each other's sentences result in the same equation.

(a) $2x - 5 = x + 1$ **(b)** $2(x + 2) = 14$

(c) $n + 5 = \frac{n}{2} - 6$ **(d)** $7 - 3a = 5 + a$

94. Complete the explanation in your own words: "The difference between $3(x - 1) + 4 - 2x$ and $3(x - 1) + 4 = 2x$ is . . . ."

95. "I make \$2.50 an hour more in my new job." If x = the amount I used to make per hour and y = the amount I now make, which of the equations say the same thing as the previous statement? Explain your choices by translating the equation to English and comparing with the original statement.

(a) $x + y = 2.50$ **(b)** $x - y = 2.50$ **(c)** $x + 2.50 = y$

(d) $2.50 + y = x$ **(e)** $y - x = 2.50$ **(f)** $2.50 - x = y$

96. "The river rose 4 feet above flood stage last night." If a = the river's height at flood stage and b = the river's height now (the morning after), which of the equations say the same thing as the previous statement? Explain your choices by translating the equations to English and comparing the meaning with the original statement.

(a) $a + b = 4$ **(b)** $b - 4 = a$ **(c)** $a - 4 = b$

(d) $a + 4 = b$ **(e)** $b + 4 = a$ **(f)** $b - a = 4$

97. "Surprising Results!" Work with other students to try this experiment. Each person should do the six steps mentally, not telling anyone else what his or her calculations are.

(a) Think of a number. **(b)** Add 7.

(c) Multiply by 3. **(d)** Add 3 more than the original number.

(e) Divide by 4. **(f)** Subtract the original number.

What number do you end up with? Compare your answer with everyone else's. Does everyone have the same answer? Make sure that everyone followed the directions accurately. How do you explain the results? Algebra makes the explanation clear. Work together to do the problem again, using a variable for the number. Make up another series of computations that give "surprising results."

98. **(a)** Do you think this is a linear equation in one variable?

$$3(2x + 4) = 6(x + 2)$$

(b) What happens when you use the properties of this section to solve the equation?

(c) Pick *any* number to substitute for x in this equation. Now try a different number to substitute for x in the equation. Try yet another number to substitute for x in the equation. Summarize your findings.

(d) Can this equation be called *linear in one variable?* Refer to the definition as you explain your answer.

Answers

94. ______

95. ______

96. ______

97. ______

98. ______

Answers

99. ________

99. **(a)** Do you think this is a linear equation in one variable?

$$4(3x - 5) = 2(6x - 8) - 3$$

(b) What happens when you use the properties of this section to solve the equation?

(c) Do you think it is possible to find a solution for this equation?

(d) Can this equation be called *linear in one variable?* Refer to the definition as you explain your answer.

Answers

1. Yes **3.** No **5.** No **7.** No **9.** No **11.** Yes **13.** Yes
15. Yes **17.** Yes **19.** No **21.** Yes **23.** Linear equation
25. Expression **27.** Linear equation **29.** 2 **31.** 11
33. -2 **35.** 6 **37.** 4 **39.** -10 **41.** 0 **43.** $-\frac{2}{3}$
45. 4 **47.** 9 **49.** 6 **51.** 6 **53.** -18 **55.** 16
57. 8 **59.** 2 **61.** 9 **63.** 12 **65.** $x + 3 = 7$
67. $3x - 7 = 2x$ **69.** $2(x + 5) = x + 18$ **71.** (c) **73.** (a)
75. True **77.** True **79.** sometimes **81.** 26; $x + 7 = 33$
83. 22; $x - 15 = 7$ **85.** 1,420; $1{,}840 + x = 3{,}260$
87. \$290; $x + 360 = 650$ **89.** 2,964 hex bolts **91.** $2d + 2$
93. Above and Beyond **95.** Above and Beyond **97.** Above and Beyond
99. Above and Beyond

1.5 Solving Equations by Multiplying and Dividing

< 1.5 Objectives >

1 > Use the multiplication property to solve equations

2 > Use the multiplication property to solve applications

In this section we look at a different type of equation. What if we want to solve an equation like

$6x = 18$

The addition property that you just learned does not help. We need a second property for solving such equations.

Property

The Multiplication Property of Equality

If $a = b$ then $ac = bc$ with $c \neq 0$

In words, multiplying both sides of an equation by the same nonzero number produces an equivalent equation.

We now work through some examples, using this second rule.

Example 1 Solving Equations by Using the Multiplication Property

< Objective 1 >

NOTE

$$\frac{1}{6}(6x) = \left(\frac{1}{6} \cdot 6\right)x$$
$$= 1 \cdot x \quad \text{or } x$$

Solve.

$6x = 18$

Here the variable x is multiplied by 6. So we apply the multiplication property and multiply both sides by $\frac{1}{6}$. Keep in mind that we want an equation of the form

$x = \square$

$$\frac{1}{6}(6x) = \frac{1}{6}(18)$$

We can now simplify.

$1 \cdot x = 3 \qquad \text{or} \qquad x = 3$

The solution is 3. To check, replace x with 3.

$$6 \cdot (3) \stackrel{?}{=} 18$$
$$18 = 18 \quad \text{True}$$

Check Yourself 1

Solve and check.

$8x = 32$

In Example 1 we solved the equation by multiplying both sides by the reciprocal of the coefficient of the variable. Example 2 illustrates a slightly different approach to solving an equation by using the multiplication property.

Example 2 Solving Equations by Using the Multiplication Property

NOTE

Because division is defined in terms of multiplication, we can also divide both sides of an equation by the same nonzero number.

Solve.

$5x = -35$

The variable x is multiplied by 5. We *divide* both sides by 5 to "undo" that multiplication.

$$\frac{5x}{5} = \frac{-35}{5}$$

$$x = -7$$

The right side simplifies to −7. Be careful with the rules for signs.

The solution is −7.

We leave it to you to check the solution.

Check Yourself 2

Solve and check.

$7x = -42$

Example 3 Solving Equations by Using the Multiplication Property

RECALL

Dividing by −9 and multiplying by $-\frac{1}{9}$ produce the same result—they are the same operation.

Solve.

$-9x = 54$

In this case, x is multiplied by −9, so we divide both sides by −9 to isolate x on the left:

$$\frac{-9x}{-9} = \frac{54}{-9}$$

$$x = -6$$

The solution is −6. To check:

$$(-9)(-6) \stackrel{?}{=} 54$$

$$54 = 54 \quad \text{True}$$

Check Yourself 3

Solve and check.

$-10x = -60$

Example 4 illustrates the use of the multiplication property when there are fractions in an equation.

Example 4 Solving Equations by Using the Multiplication Property

RECALL

$\frac{x}{3} = \frac{1}{3}x$

(a) Solve $\frac{x}{3} = 6$.

Here x is *divided* by 3. We use multiplication to isolate x.

$$3\left(\frac{x}{3}\right) = 3 \cdot (6)$$

$$x = 18$$

This leaves x alone on the left because

$$3\left(\frac{x}{3}\right) = \frac{3}{1} \cdot \frac{x}{3} = \frac{x}{1} = x$$

The solution is 18.

To check:

$$\frac{(18)}{3} \stackrel{?}{=} 6$$

$$6 = 6 \quad \text{True}$$

(b) Solve $\frac{x}{5} = -9$.

$$5\left(\frac{x}{5}\right) = 5(-9)$$

$$x = -45$$

Because x is divided by 5, we multiply both sides by 5.

The solution is -45. To check, we replace x with -45:

$$\frac{(-45)}{5} \stackrel{?}{=} -9$$

$$-9 = -9 \quad \text{True}$$

The solution is verified.

Check Yourself 4

Solve and check.

(a) $\frac{x}{7} = 3$ (b) $\frac{x}{4} = -8$

When the variable is multiplied by a fraction that has a numerator other than 1, there are two approaches to finding the solution.

Example 5 Solving Equations by Using Reciprocals

Solve.

$$\frac{3}{5}x = 9$$

One approach is to multiply by 5 as the first step.

$$5\left(\frac{3}{5}x\right) = 5 \cdot (9)$$

$$3x = 45$$

Now we divide by 3.

$$\frac{3x}{3} = \frac{45}{3}$$

$$x = 15$$

To check the solution 15, substitute 15 for x.

$$\frac{3}{5} \cdot (15) \stackrel{?}{=} 9$$

$$9 = 9 \quad \text{True}$$

A second approach combines the multiplication and division steps and is generally a bit more efficient. We multiply by $\frac{5}{3}$.

$$\frac{5}{3}\left(\frac{3}{5}x\right) = \frac{5}{3} \cdot (9)$$

$$x = \frac{5}{\cancel{3}_1} \cdot \frac{\cancel{9}^3}{1} = 15$$

So $x = 15$, as before.

NOTE

We multiply by $\frac{5}{3}$ because it is the reciprocal of $\frac{3}{5}$, and the product of a number and its reciprocal is 1.

$$\left(\frac{5}{3}\right)\left(\frac{3}{5}\right) = 1$$

Check Yourself 5

Solve and check.

$\frac{2}{3}x = 18$

You may sometimes have to simplify an equation before applying the methods of this section. Example 6 illustrates this procedure.

Example 6 Combining Like Terms and Solving Equations

Solve and check.

$3x + 5x = 40$

Using the distributive property, we combine the like terms on the left to write

$8x = 40$

We now proceed as before.

$\frac{8x}{8} = \frac{40}{8}$ Divide by 8.

$x = 5$

The solution is 5. To check, we return to the original equation. Substituting 5 for x yields

$$3 \cdot (5) + 5 \cdot (5) \stackrel{?}{=} 40$$
$$15 + 25 \stackrel{?}{=} 40$$
$$40 = 40 \quad \text{True}$$

The solution is verified.

Check Yourself 6

Solve and check.

$7x + 4x = -66$

As with the addition property, there are many applications that require us to use the multiplication property.

Example 7 An Application Involving the Multiplication Property

< Objective 2 >

On her first day on the job in a photography lab, Samantha processed all of the film given to her. The next day, her boss gave her four times as much film to process. Over the two days, she processed 60 rolls of film. How many rolls did she process on the first day?

Step 1 We want to find the number of rolls Samantha processed on the first day.

Step 2 Let x be the number of rolls Samantha processed on her first day and solve the equation $x + 4x = 60$ to answer the question.

Step 3 $x + 4x = 60$

Step 4 $5x = 60$ Combine like terms first.

$\frac{1}{5}(5x) = \frac{1}{5}(60)$ Multiply by $\frac{1}{5}$ to isolate the variable.

$x = 12$

Step 5 Samantha processed 12 rolls of film on her first day.

Check: $4 \times 12 = 48$; $12 + 48 = 60$.

RECALL

Always use a sentence to give the answer to an application.

NOTE

The yen (¥) is the monetary unit of Japan.

Check Yourself 7

On a recent trip to Japan, Marilyn exchanged $1,200 and received 139,812 yen. What exchange rate did she receive?

Let x be the exchange rate and solve the equation $1{,}200x = 139{,}812$ to answer the question (to the nearest hundredth).

Check Yourself ANSWERS

1. 4 **2.** −6 **3.** 6 **4.** **(a)** 21; **(b)** −32

5. 27 **6.** −6 **7.** She received 116.51 yen for each dollar.

Reading Your Text

SECTION 1.5

(a) Multiplying both sides of an equation by the same ____________ number yields an equivalent equation.

(b) $\frac{5}{3}$ is the ____________ of $\frac{3}{5}$.

(c) To check a solution, we return to the ____________ equation.

(d) The product of a number and its ____________ is always 1.

1.5 exercises

Basic Skills | Challenge Yourself | Calculator/Computer | Career Applications | Above and Beyond

Name ______________

Section ________ Date ________

Answers

1. ______ 2. ______
3. ______ 4. ______
5. ______ 6. ______
7. ______ 8. ______
9. ______ 10. ______
11. ______ 12. ______
13. ______ 14. ______
15. ______ 16. ______
17. ______ 18. ______
19. ______ 20. ______
21. ______ 22. ______
23. ______ 24. ______
25. ______ 26. ______
27. ______ 28. ______
29. ______ 30. ______
31. ______ 32. ______
33. ______ 34. ______
35. ______ 36. ______
37. ______ 38. ______
39. ______ 40. ______

< Objective 1 >

Solve and check.

1. $5x = 20$ **2.** $6x = 30$

3. $7x = 42$ **4.** $6x = -42$

5. $63 = 9x$ **6.** $66 = 6x$

7. $4x = -16$ **8.** $-3x = 27$

9. $-9x = 72$ 

10. $-9x = 90$

11. $6x = -54$ **12.** $-7x = 49$

13. $-4x = -12$ **14.** $15 = -9x$

15. $-21 = 24x$ **16.** $-7x = -35$

17. $-6x = -54$ **18.** $-4x = -24$

19. $\frac{x}{4} = 2$ **20.** $\frac{x}{3} = 2$

21. $\frac{x}{5} = 3$ **22.** $\frac{x}{8} = 5$

23. $6 = \frac{x}{7}$ **24.** $6 = \frac{x}{3}$

25. $\frac{x}{5} = -4$ **26.** $\frac{x}{5} = -7$

27. $-\frac{x}{3} = 8$ 

28. $-\frac{x}{4} = -3$

29. $\frac{2}{3}x = 6$ **30.** $\frac{4}{5}x = 10$

31. $\frac{3}{4}x = -16$ **32.** $\frac{7}{8}x = -21$

33. $-\frac{2}{5}x = 10$ **34.** $-\frac{5}{6}x = -15$

35. $5x + 4x = 36$ **36.** $8x - 3x = -50$

37. $\frac{7}{9}x - 5 = \frac{3}{9}x + 11$ **38.** $\frac{4}{11}x - 9 + \frac{2}{11}x = 18 - \frac{3}{11}x$

39. $4(x + 5) + 7x = -3(x - 2)$ **40.** $-2(x - 3) + 10 = -4(5 - 4x)$

Certain equations involving decimal fractions can be solved by the methods of this section. For instance, to solve $2.3x = 6.9$, we simply use the multiplication property to divide both sides of the equation by 2.3. This isolates x on the left as desired. Use this idea to solve each equation.

41. $3.2x = 12.8$

42. $5.1x = -15.3$

43. $-4.5x = 13.5$

44. $-8.2x = -32.8$

45. $1.3x + 2.8x = 12.3$

46. $2.7x + 5.4x = -16.2$

47. $9.3x - 6.2x = 12.4$

48. $12.5x - 7.2x = -21.2$

Translate each statement to an equation. Let x represent the number in each case.

49. 6 times a number is 72.

50. Twice a number is 36.

51. A number divided by 7 is equal to 6.

52. A number divided by 5 is equal to -4.

53. $\frac{1}{3}$ of a number is 8.

54. $\frac{1}{5}$ of a number is 10.

55. $\frac{3}{4}$ of a number is 18.

56. $\frac{2}{7}$ of a number is 8.

57. Twice a number, divided by 5, is 12. 

58. 3 times a number, divided by 4, is 36.

Basic Skills | **Challenge Yourself** | Calculator/Computer | Career Applications | Above and Beyond

< Objective 2 >

Determine whether each statement is **true** *or* **false.**

59. To isolate x in the equation $\frac{3}{4}x = 9$, we can simply add $-\frac{3}{4}$ to both sides.

60. Dividing both sides of an equation by 5 is the same as multiplying both sides by $\frac{1}{5}$.

Complete each statement with **never, sometimes,** *or* **always.**

61. To solve a linear equation, we ________ must use the multiplication property.

Answers

41. ________
42. ________
43. ________
44. ________
45. ________
46. ________
47. ________
48. ________
49. ________
50. ________
51. ________
52. ________
53. ________
54. ________
55. ________
56. ________
57. ________
58. ________
59. ________
60. ________
61. ________

Answers

62. ______

63. ______

64. ______

65. ______

66. ______

67. ______

68. ______

69. ______

70. ______

71. ______

72. ______

62. If we want to obtain an equivalent linear equation by multiplying both sides by a number, that number can ________ be zero.

Solve each application.

63. **Statistics** Three-fourths of the theater audience left in disgust. If 87 angry patrons walked out, how many were there originally?

64. **Business and Finance** A mechanic charged \$45 an hour plus \$225 for parts to replace the ignition coil on a car. If the total bill was \$450, how many hours did the repair job take?

65. **Business and Finance** A call to Phoenix, Arizona, from Dubuque, Iowa, costs 55 cents for the first minute and 23 cents for each additional minute or portion of a minute. If Barry has \$6.30 in change, how long can he talk?

66. **Number Problem** The sum of 4 times a number and 14 is 34. Find the number.

67. **Number Problem** If 6 times a number is subtracted from 42, the result is 24. Find the number.

68. **Number Problem** When a number is divided by -6, the result is 3. Find the number.

69. **Geometry** Suppose that the circumference of a tree measures 9 ft 2 in., or 110 in. To find the diameter of the tree at that point, we must solve the equation

$$110 = 3.14d$$

Find the diameter of the tree to the nearest inch. (*Note:* 3.14 is an approximation for π.)

70. **Geometry** Suppose that the circumference of a circular swimming pool is 88 ft. Find the diameter of the pool by solving the equation to the nearest foot.

$$88 = 3.14d$$

71. **Problem Solving** While traveling in Europe, Susan noticed that the distance to the city she was heading to was 200 kilometers (km). She knew that to estimate this distance in miles she could solve the equation

> Videos

$$200 = \frac{8}{5}x$$

What was the equivalent distance in miles?

72. **Problem Solving** Aaron was driving a rental car while traveling in France, and saw a sign indicating a speed limit of 95 km/h. To approximate this speed in miles per hour, he used the equation

$$95 = \frac{8}{5}x$$

What is the corresponding speed, rounded to the nearest mile per hour?

Basic Skills | Challenge Yourself | Calculator/Computer | **Career Applications** | Above and Beyond

73. Automotive Technology One horsepower (hp) estimate of an engine is given by the formula

$$\text{hp} = \frac{d^2n}{2.5}$$

in which d is the diameter of the cylinder bore (in cm) and n is the number of cylinders.

Find the number of cylinders in a 194.4-hp engine if its cylinder bore has a 9-cm diameter.

74. Automotive Technology The horsepower (hp) of a diesel engine is calculated using the formula

$$\text{hp} = \frac{P \cdot L \cdot A \cdot N}{33{,}000}$$

in which P is the average pressure (in pounds per square inch), L is the length of the stroke (in feet), A is the area of the piston (in square inches), and N is the number of strokes per minute.

Determine the average pressure of a 144-hp diesel engine if its stroke length is $\frac{1}{3}$ ft, its piston area is 9 in.^{2}, and it completes 8,000 strokes per minute.

75. Manufacturing Technology The pitch of a gear is given by the number of teeth divided by the working diameter of the gear. Write an equation for the gear pitch p in terms of the number of teeth t and its diameter d.

76. Manufacturing Technology Use your answer to exercise 75 to determine the number of teeth needed for a gear with a working diameter of $6\frac{1}{4}$ in. to have a pitch of 4.

Answers

73. ______

74. ______

75. ______

76. ______

Answers

1. 4 **3.** 6 **5.** 7 **7.** -4 **9.** -8 **11.** -9
13. 3 **15.** $-\frac{7}{8}$ **17.** 9 **19.** 8 **21.** 15 **23.** 42
25. -20 **27.** -24 **29.** 9 **31.** $-\frac{64}{3}$ **33.** -25
35. 4 **37.** 36 **39.** -1 **41.** 4 **43.** -3 **45.** 3
47. 4 **49.** $6x = 72$ **51.** $\frac{x}{7} = 6$ **53.** $\frac{1}{3}x = 8$ **55.** $\frac{3}{4}x = 18$
57. $\frac{2x}{5} = 12$ **59.** False **61.** sometimes **63.** 116 patrons
65. 26 min **67.** 3 **69.** 35 in. **71.** 125 mi **73.** 6 cylinders
75. $p = \frac{t}{d}$

1.6 Combining the Rules to Solve Equations

< 1.6 Objectives >

1 > Combine addition and multiplication to solve equations

2 > Solve equations involving fractions

3 > Solve applications

In Section 1.4, we solved equations by using the addition property, which allowed us to solve equations such as $x + 3 = 9$. Then, in Section 1.5, we solved equations by using the multiplication property, which allowed us to solve equations such as $5x = 32$. Now, we will solve equations that require us to use both the addition and multiplication properties.

In Example 1, we check to see whether a given value for the variable is a solution to a given equation.

Example 1 Checking a Solution

Test to see if -3 is a solution for

$5x + 6 = 2x - 3$

-3 is a solution because replacing x with -3 gives

$$5(-3) + 6 \stackrel{?}{=} 2(-3) - 3$$
$$-15 + 6 \stackrel{?}{=} -6 - 3$$
$$-9 = -9 \qquad \text{A true statement}$$

Check Yourself 1

Test to see if 7 is a solution for the equation

$5x - 15 = 2x + 6$

In Example 2, we apply the addition and multiplication properties to find the solution of a linear equation.

Example 2 Applying the Properties of Equality

< Objective 1 >

RECALL

Why did we add 5? We added 5 because it is the *opposite* of -5, and the resulting equation has the variable term on the left and the constant term on the right.

We choose $\frac{1}{3}$ because $\frac{1}{3}$ is the *reciprocal* of 3 and $\frac{1}{3} \cdot 3 = 1$

Solve.

$3x - 5 = 4$

We start by using the addition property to add 5 to both sides of the equation.

$$3x - 5 + 5 = 4 + 5$$
$$3x = 9$$

Now we want to get the x-term alone on the left with a coefficient of 1 (we call this *isolating* the variable). To do this, we use the multiplication property and multiply both sides by $\frac{1}{3}$.

$$\frac{1}{3}(3x) = \frac{1}{3}(9)$$

So $x = 3$.

Because any application of the addition or multiplication property leads to an equivalent equation, all of these equations have the same solution, 3.

To check this result, we replace x with 3 in the original equation:

$$3(3) - 5 \stackrel{?}{=} 4$$
$$9 - 5 \stackrel{?}{=} 4$$
$$4 = 4 \quad \text{A true statement}$$

You may prefer a slightly different approach in the last step of the previous solution. From the equation $3x = 9$, the multiplication property can be used to *divide* both sides of the equation by 3. Then

$$\frac{3x}{3} = \frac{9}{3}$$
$$x = 3$$

The result is the same.

Check Yourself 2

Solve and check.

$4x - 7 = 17$

The steps involved in using the addition and multiplication properties to solve an equation are the same if more terms are involved in an equation.

Example 3 Applying the Properties of Equality

Solve.

$5x - 11 = 2x - 7$

Our objective is to use the properties of equality to isolate x on one side of an equivalent equation. We begin by adding 11 to both sides.

$$5x - 11 + 11 = 2x - 7 + 11$$
$$5x = 2x + 4$$

We continue by adding $-2x$ to (or subtracting $2x$ from) both sides.

$$5x + (-2x) = 2x + (-2x) + 4$$
$$3x = 4$$

To isolate x, we now multiply both sides by $\frac{1}{3}$.

$$\frac{1}{3}(3x) = \frac{1}{3}(4)$$
$$x = \frac{4}{3}$$

We leave it to you to check this result.

NOTES

Again, adding 11 leaves us with the constant term on the right.

If you prefer, write

$5x - 2x = 2x - 2x + 4$

Again,

$3x = 4$

This is the same as dividing both sides by 3. So

$\frac{3x}{3} = \frac{4}{3}$

$x = \frac{4}{3}$

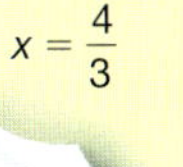

Check Yourself 3

Solve and check.

$7x - 12 = 2x - 9$

Both sides of an equation should be simplified as much as possible *before* the addition and multiplication properties are applied. If like terms are involved on one side (or on both sides) of an equation, they should be combined before an attempt is made to isolate the variable. Example 4 illustrates this approach.

Example 4 Applying the Properties of Equality with Like Terms

NOTE

There are like terms on both sides of the equation.

Solve.

$8x + 2 - 3x = 8 + 3x + 2$

We combine the like terms $8x$ and $-3x$ on the left and the like terms 8 and 2 on the right as our first step. We then have

$5x + 2 = 3x + 10$

We can now solve as before.

$$5x + 2 - 2 = 3x + 10 - 2 \quad \text{Subtract 2 from both sides.}$$
$$5x = 3x + 8$$

Then

$$5x - 3x = 3x - 3x + 8 \quad \text{Subtract } 3x \text{ from both sides.}$$
$$2x = 8$$
$$\frac{2x}{2} = \frac{8}{2} \quad \text{Divide both sides by 2.}$$
$$x = 4$$

The solution is 4.

Check:

RECALL

Return to the original equation to check your solution.

Always follow the order of operations when evaluating an expression.

$$8(4) + 2 - 3(4) \stackrel{?}{=} 8 + 3(4) + 2$$
$$32 + 2 - 12 \stackrel{?}{=} 8 + 12 + 2 \quad \text{Multiply first, then add and subtract.}$$
$$22 = 22 \quad \text{True}$$

Check Yourself 4

Solve and check.

$7x - 3 - 5x = 10 + 4x + 3$

If there are parentheses on one or both sides of an equation, the parentheses should be removed by applying the distributive property as the first step. Like terms should then be combined before isolating the variable. Consider Example 5.

Example 5 Applying the Properties of Equality with Parentheses

Solve.

$x + 3(3x - 1) = 4(x + 2) + 4$

First, apply the distributive property to remove the parentheses on the left and right sides.

$x + 9x - 3 = 4x + 8 + 4$

Combine like terms on each side of the equation.

$10x - 3 = 4x + 12$

RECALL

To isolate x, we must get x alone on one side with a coefficient of 1.

Now, isolate the variable x on the left side.

$$10x - 3 + 3 = 4x + 12 + 3 \quad \text{Add 3 to both sides.}$$
$$10x = 4x + 15$$
$$10x - 4x = 4x - 4x + 15 \quad \text{Subtract } 4x \text{ from both sides.}$$
$$6x = 15$$
$$\frac{6x}{6} = \frac{15}{6} \quad \text{Divide both sides by 6.}$$
$$x = \frac{5}{2}$$

The solution is $\frac{5}{2}$. Again, this should be checked by returning to the original equation.

RECALL

The LCM of a set of denominators is also called the **least common denominator (LCD)**.

Check Yourself 5

Solve and check.

$x + 5(x + 2) = 3(3x - 2) + 18$

To solve an equation involving fractions, the first step is to multiply both sides of the equation by the **least common multiple (LCM)** of all denominators in the equation. This clears the equation of fractions, and we can proceed as before.

Example 6 Applying the Properties of Equality with Fractions

< Objective 2 >

Solve.

$$\frac{x}{2} - \frac{2}{3} = \frac{5}{6}$$

First, multiply each side by 6, the least common multiple of 2, 3, and 6.

$$6\left(\frac{x}{2} - \frac{2}{3}\right) = 6\left(\frac{5}{6}\right)$$
$$6\left(\frac{x}{2}\right) - 6\left(\frac{2}{3}\right) = 6\left(\frac{5}{6}\right) \quad \text{Apply the distributive property.}$$
$$\overset{3}{\cancel{6}}\left(\frac{x}{\underset{1}{\cancel{2}}}\right) - \overset{2}{\cancel{6}}\left(\frac{2}{\underset{1}{\cancel{3}}}\right) = \overset{1}{\cancel{6}}\left(\frac{5}{\underset{1}{\cancel{6}}}\right) \quad \text{Simplify}$$
$$3x - 4 = 5$$

NOTE

The equation is now cleared of fractions.

Next, isolate the variable x on the left side.

$$3x = 9$$
$$x = 3$$

The solution 3, should be checked as before by returning to the original equation.

Check Yourself 6

Solve.

$$\frac{x}{4} - \frac{4}{5} = \frac{19}{20}$$

Be sure that the distributive property is applied properly so that *every term* of the equation is multiplied by the LCM.

Example 7 Applying the Properties of Equality with Fractions

Solve.

$$\frac{2x-1}{5}+1=\frac{x}{2}$$

First, multiply each side by 10, the LCM of 5 and 2.

$$10\left(\frac{2x-1}{5}+1\right)=10\left(\frac{x}{2}\right)$$ Apply the distributive property on the left. Simplify.

$$\overset{2}{\cancel{10}}\left(\frac{2x-1}{\underset{1}{\cancel{5}}}\right)+10(1)=\overset{5}{\cancel{10}}\left(\frac{x}{\underset{1}{\cancel{2}}}\right)$$ Next, isolate x. Here we isolate x on the right side.

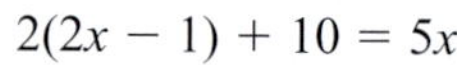

$$2(2x-1)+10=5x$$
$$4x-2+10=5x$$
$$4x+8=5x$$
$$8=x$$

NOTE

$4x$ is subtracted from both sides of the equation.

The solution is 8. We return to the original equation and follow the order of operations to check this result.

$$\frac{2(8)-1}{5}+1\overset{?}{=}\frac{(8)}{2}$$
$$\frac{16-1}{5}+1\overset{?}{=}4$$ The fraction bar is a grouping symbol.
$$\frac{15}{5}+1\overset{?}{=}4$$
$$3+1\overset{?}{=}4$$
$$4=4$$ True

Check Yourself 7

Solve and check.

$$\frac{3x+1}{4}-2=\frac{x+1}{3}$$

Conditional Equations, Identities, and Contradictions

1. An equation that is true for only particular values of the variable is called a **conditional equation.** For example, a linear equation that can be written in the form

   $ax+b=0$

   in which $a \neq 0$ is a conditional equation. We illustrated this case in our previous examples and exercises.

2. An equation that is true for all possible values of the variable is called an **identity.** In this case, *both* a and b are 0, so we get the equation $0 = 0$. This is the case if both sides of the equation reduce to the same expression (a true statement).

3. An equation that is never true, no matter what the value of the variable, is called a **contradiction.** For example, if a is 0 but b is 4, a contradiction results. This is the case if the equation simplifies to a false statement.

Example 8 illustrates the second and third cases.

Example 8 Identities and Contradictions

NOTE

By adding 6 to both sides of this equation, we have $0 = 0$.

(a) Solve $2(x - 3) - 2x = -6$.

Apply the distributive property to remove the parentheses.

$$2x - 6 - 2x = -6$$

$$-6 = -6 \quad \text{A \textit{true} statement}$$

Because the two sides simplify to the true statement $-6 = -6$, the original equation is an *identity,* and the solution set is the set of all real numbers. This is sometimes written as $\mathbb{R}$, which is read, "the set of all real numbers."

NOTE

This agrees with the definition of a contradiction. Subtracting 3 from both sides yields $0 = 1$.

(b) Solve $3(x + 1) - 2x = x + 4$.

Again, apply the distributive property.

$$3x + 3 - 2x = x + 4$$

$$x + 3 = x + 4$$

$$3 = 4 \quad \text{A \textit{false} statement}$$

Because the two sides reduce to the false statement $3 = 4$, the original equation is a contradiction. There are no values of the variable that can satisfy the equation. There is no solution. We sometimes use **empty set** or **null set** notation to write this: $\varnothing$ or $\{\ \}$.

Check Yourself 8

Determine whether each equation is a conditional equation, an identity, or a contradiction.

(a) $2(x + 1) - 3 = x$ **(b)** $2(x + 1) - 3 = 2x + 1$
(c) $2(x + 1) - 3 = 2x - 1$

NOTE

An **algorithm** is a step-by-step process for problem solving.

An organized step-by-step procedure is the key to an effective equation-solving strategy. The following algorithm summarizes our work in this section and gives you guidance in approaching the problems that follow.

Step by Step

Solving Linear Equations in One Variable

Step 1 Remove any grouping symbols by applying the distributive property.

Step 2 Multiply both sides of the equation by the LCM required to clear the equation of fractions or decimals.

Step 3 Combine any like terms that appear on either side of the equation.

Step 4 Apply the addition property of equality to write an equivalent equation with the variable term on *one side* of the equation and the constant term on the *other side.*

Step 5 Apply the multiplication property of equality to write an equivalent equation with the variable isolated on one side of the equation with coefficient 1.

Step 6 State the answer and check the solution in the *original* equation.

Note: If the equation derived in step 5 is always true, the original equation is an *identity.* If the equation is always false, the original equation is a *contradiction.* If you find a unique solution, the equation is *conditional*.

When you are solving an equation for which a calculator is recommended, it is often easiest to do all calculations as the last step. For more complex equations, it is usually best to calculate at each step.

Example 9 Solving Equations Using a Calculator

Solve the equation.

$$5(x - 3.25) + \frac{3}{4} = 2{,}110.75$$

Following the steps of the algorithm, we get

$$5x - 16.25 + \frac{3}{4} = 2{,}110.75$$ Remove parentheses.

$$20x - 65 + 3 = 8{,}443$$ Multiply by the LCD, 4.

$$20x = 8{,}443 + 62$$ Isolate the variable.

$$x = \frac{8{,}505}{20}$$

Now, we use a calculator to simplify the expression on the right.

$x = 425.25$

Check Yourself 9

Solve the equation for x.

$$7(x + 4.3) - \frac{3}{5} = 467$$

We first solved problems involving consecutive integers in Section 1.4. We use the following properties to solve these problems algebraically.

Property

Consecutive Integers

If x is an integer, then $x + 1$ is the next **consecutive integer,** $x + 2$ is the next, and so on.

If x is an odd integer, the next **consecutive odd integer** is $x + 2$, and the next is $x + 4$.

If x is an even integer, the next **consecutive even integer** is $x + 2$, and the next is $x + 4$.

NOTE

Consecutive integers are integers that follow one another, such as 10, 11, and 12.

We use this idea in Example 10.

Example 10 Solving an Application

< Objective 3 >

The sum of two consecutive integers is 41. What are the two integers?

Step 1 We want to find the two consecutive integers.

Step 2 Let x be the first integer. Then $x + 1$ must be the next.

Step 3 The first integer, The second integer

$$x + x + 1 = 41$$

The sum, Is

RECALL

We use the five-step method to solve word problems that we introduced in Section 1.4.

Step 4 $x + x + 1 = 41$

$2x + 1 = 41$

$2x = 40$

$x = 20$

Step 5 The first integer (x) is 20, and the next integer ($x + 1$) is 21.

The sum of the two integers 20 and 21 is 41.

Check Yourself 10

The sum of three consecutive integers is 51. What are the three integers?

Sometimes algebra is used to reconstruct missing information. Example 11 does just that with some election information.

Example 11 Solving an Application

There were 55 more yes votes than no votes on an election measure. If 735 votes were cast in all, how many yes votes were there? How many no votes?

Step 1 We want to find the number of yes votes and the number of no votes.

Step 2 Let x be the number of no votes. Then

$\underbrace{x + 55}$

55 more than x

is the number of yes votes.

Step 3 $x + \underbrace{x + 55} = 735$

No votes Yes votes Total votes cast

Step 4 $x + x + 55 = 735$

$2x + 55 = 735$

$2x = 680$

$x = 340$

Step 5 No votes (x) = 340

Yes votes ($x + 55$) = 395

Thus, 340 no votes plus 395 yes votes equal 735 total votes. The solution checks.

Check Yourself 11

Francine earns $120 per month more than Rob. If they earn a total of $2,680 per month, what are their monthly salaries?

Similar methods allow you to solve a variety of word problems. Example 12 includes three unknown quantities but uses the same basic solution steps.

Example 12 Solving an Application

Juan worked twice as many hours as Jerry. Marcia worked 3 h more than Jerry. If they worked a total of 31 h, find out how many hours each worked.

Step 1 We want to find the hours each worked, so there are three unknowns.

NOTE

There are other choices for x, but choosing the smallest quantity usually gives the easiest equation to write and solve.

Step 2 Let x be the hours that Jerry worked.

Juan worked twice Jerry's hours.

Then $2x$ is Juan's hours worked.

Marcia worked 3 h more than Jerry worked.

And $x + 3$ is Marcia's hours.

Step 3 Jerry Juan Marcia

$$x + 2x + x + 3 = 31$$

Sum of their hours

Step 4

$$\begin{aligned} x + 2x + x + 3 &= 31 \\ 4x + 3 &= 31 \\ 4x &= 28 \\ x &= 7 \end{aligned}$$

Step 5

Jerry's hours $(x) = 7$

Juan's hours $(2x) = 14$

Marcia's hours $(x + 3) = 10$

The sum of their hours $(7 + 14 + 10)$ is 31, and the solution is verified.

Check Yourself 12

Lucy jogged twice as many miles as Paul but 3 mi less than Isaac. If the three ran a total of 23 mi, how far did each person run?

Many applied problems involve the use of *percents.* The idea of *percent* is a useful way of naming parts of a whole. We can think of a percent as a fraction whose denominator is 100. Thus, 15% would be written as $\frac{15}{100}$ and represents 15 parts out of 100. A percentage can also be expressed as a decimal by converting the fractional representation to a decimal. So 15% is written as 0.15. Examples 13 and 14 illustrate some uses of percents in applications.

Example 13 Solving an Application

Marzenna inherits \$5,000 and invests part of her money in bonds at 4% and the remaining in savings at 3%. What amount has she invested at each rate if she receives \$180 in interest for 1 year?

Step 1 We want to find the amount invested at each rate, so there are two unknowns.

Step 2 Let x be the amount invested at 4%.

\$5,000 was the total amount of money invested.

So $5{,}000 - x$ is the amount invested at 3%.

$0.04x$ is the amount of interest from the 4% investment.

$0.03(5{,}000 - x)$ is the amount of interest from the 3% investment.

\$180 is the total interest for the year.

Step 3 $$0.04x + 0.03(5{,}000 - x) = 180$$

Step 4
$$\begin{aligned}
0.04x + 0.03(5{,}000) - 0.03x &= 180\\
0.04x + 150 - 0.03x &= 180\\
0.04x - 0.03x &= 180 - 150\\
0.01x &= 30\\
x &= \frac{30}{0.01}\\
x &= 3{,}000
\end{aligned}$$

NOTE

The check is left to you.

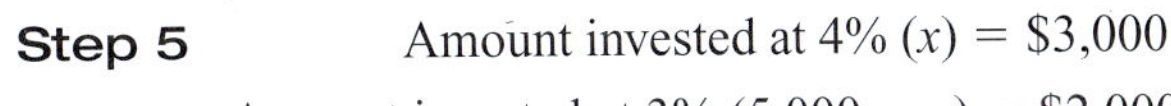

Step 5 Amount invested at 4% $(x) = \$3{,}000$

Amount invested at 3% $(5{,}000 - x) = \$2{,}000$

Check Yourself 13

Greg received an $8,000 bonus. He invested some of it in bonds at 2% and the rest in savings at 5%. If he receives $295 interest for 1 year, how much was invested at each rate?

Example 14 Solving an Application

Tony earns a take-home pay of $592 per week. If his deductions for taxes, retirement, union dues, and a medical plan amount to 26% of his wages, what is his weekly pay before the deductions?

Step 1 We want to find his weekly pay before deductions (gross pay).

Step 2 Let x = gross pay.

Since 26% of his gross pay is deducted from his weekly salary, the amount deducted is $0.26x$.

$592 is Tony's take-home pay (net pay).

Step 3 Net pay = Gross pay − Deductions

$$\$592 = x - 0.26x$$

Step 4
$$\begin{aligned}
592 &= 0.74x\\
\frac{592}{0.74} &= x\\
800 &= x
\end{aligned}$$

Step 5 So Tony's weekly pay before deductions is $800.

Check Yourself 14

Joan gives 10% of her take-home pay to the church. This amounts to $90 per month. In addition, her paycheck deductions are 25% of her gross monthly income. What is her gross monthly income?

Check Yourself ANSWERS

1. $5(7) - 15 \stackrel{?}{=} 2(7) + 6$ **2.** 6 **3.** $\frac{3}{5}$ **4.** -8 **5.** $-\frac{2}{3}$

$35 - 15 \stackrel{?}{=} 14 + 6$

$20 = 20$

A true statement

6. 7 **7.** 5 **8.** **(a)** conditional, $\{1\}$; **(b)** contradiction, $\{\ \}$; **(c)** identity, $\mathbb{R}$ **9.** 62.5 **10.** The equation is $x + x + 1 + x + 2 = 51$. The integers are 16, 17, and 18. **11.** The equation is $x + x + 120 = 2{,}680$. Rob's salary is \$1,280, and Francine's is \$1,400. **12.** Paul: 4 mi; Lucy: 8 mi; Isaac: 11 mi **13.** \$3,500 invested at 2% and \$4,500 at 5% **14.** \$1,200.00

Reading Your Text

SECTION 1.6

(a) Given an equation such as $3x = 9$, the multiplication property can be used to ______________ both sides of the equation by 3.

(b) ______________ by $\frac{1}{3}$ is the same as dividing by 3.

(c) We can check a solution by ______________ values into the original equation.

(d) Both sides of an equation should be ______________ as much as possible before using the addition and multiplication properties.

Basic Skills | Challenge Yourself | Calculator/Computer | Career Applications | Above and Beyond

1.6 exercises

< Objectives 1 and 2 >

Solve and check each equation.

1. $3x + 1 = 13$

2. $3x - 1 = 17$

3. $3x - 2 = 7$

4. $5x + 3 = 23$

5. $4 - 7x = 18$

6. $7 - 4x = -5$

7. $3 - 4x = -9$

8. $5 - 4x = 25$

9. $\frac{x}{2} + 1 = 5$

10. $\frac{x}{3} - 2 = 3$

11. $\frac{3}{4}x + 8 = 32$

12. $\frac{5}{6}x - 9 = 16$

13. $5x = 2x + 9$

14. $7x = 18 - 2x$

15. $9x + 2 = 3x + 38$

16. $4(2x - 1) = 2(3x - 2)$

17. $4x - 8 = x - 14$

18. $6x - 5 = 3x - 29$

19. $5(3x + 4) = 10(x + 2)$

20. $\frac{4}{3}x - 7 = 11 - \frac{5}{3}x$

21. $5x + 4 = 7x - 8$

22. $2x + 23 = 6x - 5$

23. $6x + 7 - 4x = 8 + 7x - 26$

24. $7x - 2 - 3x = 5 + 8x + 13$

25. $6x - 3 + 5x + 11 = 8x - 12$

26. $3x + 3 + 8x - 9 = 7x + 5$

27. $5(8 - x) = 3x$

28. $7x = 7(6 - x)$

29. $7(2x - 1) - 5x = x + 25$

30. $9(3x + 2) - 10x = 12x - 7$

31. $2(2x - 1) = 3(x + 1)$

32. $3(3x - 1) = 4(3x + 1)$

33. $8x - 3(2x - 4) = 17$

34. $7x - 4(3x + 4) = 9$

35. $7(3x + 4) = 8(2x + 5) + 13$

36. $-4(2x - 1) + 3(3x + 1) = 9$

37. $9 - 4(3x + 1) = 3(6 - 3x) - 9$

38. $13 - 4(5x + 1) = 3(7 - 5x) - 15$

39. $5.3x - 7 = 2.3x + 5$ > Videos

40. $9.8x + 2 = 3.8x + 20$

Name ________________

Section ________ Date ________

Answers

1.	2.
3.	4.
5.	6.
7.	8.
9.	10.
11.	12.
13.	14.
15.	16.
17.	18.
19.	20.
21.	22.
23.	24.
25.	26.
27.	28.
29.	30.
31.	32.
33.	34.
35.	36.
37.	38.
39.	40.

Answers

41. ______ 42. ______
43. ______ 44. ______
45. ______ 46. ______
47. ______ 48. ______
49. ______ 50. ______
51. ______ 52. ______
53. ______
54. ______
55. ______
56. ______
57. ______
58. ______
59. ______
60. ______
61. ______
62. ______
63. ______
64. ______
65. ______
66. ______
67. ______
68. ______
69. ______
70. ______

Solve each equation.

41. $\frac{2x}{3} - \frac{5}{3} = 3$

42. $\frac{3x}{4} + \frac{1}{4} = 4$

43. $\frac{x}{6} + \frac{x}{5} = 11$

44. $\frac{x}{6} - \frac{x}{8} = 1$

45. $\frac{2x}{5} - \frac{x}{3} = \frac{7}{15}$

46. $\frac{2x}{7} - \frac{3x}{5} = \frac{6}{35}$

47. $\frac{x}{5} - \frac{x-7}{3} = \frac{1}{3}$

48. $\frac{x}{6} + \frac{3}{4} = \frac{x-1}{4}$

49. $\frac{5x-3}{4} - 2 = \frac{x}{3}$

50. $\frac{6x-1}{5} - \frac{2x}{3} = 3$

51. $\frac{3x-2}{3} - \frac{2x-5}{5} = \frac{7}{15}$ > Videos

52. $\frac{4x}{7} - \frac{2x-3}{3} = \frac{19}{21}$

Classify each equation as a conditional equation, an identity, or a contradiction and give the solution.

53. $3(x - 1) = 2x + 3$

54. $2(x + 3) = 2x + 6$

55. $3(x - 1) = 3x + 3$

56. $2(x + 3) = x + 5$

57. $3(x - 1) = 3x - 3$

58. $2(2x - 1) = 3x - 4$

59. $3x - (x - 3) = 2(x + 1) + 2$

60. $5x - (x + 4) = 4(x - 2) + 4$ > Videos

61. $\frac{x}{2} - \frac{x}{3} = \frac{x}{6}$

62. $\frac{3x}{4} - \frac{2x}{3} = \frac{1}{6}$

Translate each statement to an equation. Let x represent the number in each case.

63. 3 more than twice a number is 7.

64. 5 less than 3 times a number is 25.

65. 7 less than 4 times a number is 41.

66. 10 more than twice a number is 44.

67. 5 more than two-thirds a number is 21.

68. 3 less than three-fourths of a number is 24.

69. 3 times a number is 12 more than that number.

70. 5 times a number is 8 less than that number.

Basic Skills | **Challenge Yourself** | Calculator/Computer | Career Applications | Above and Beyond

Determine whether each statement is **true** *or* **false.**

71. An equation that is never true, no matter what value is substituted, is called an identity.

72. A conditional equation can be an identity.

Complete each statement with **never, sometimes,** *or* **always.**

73. To solve a linear equation, we _________ must use both the addition property and the multiplication property.

74. We should _________ check a possible solution by substituting it into the original equation.

< Objective 3 >

Solve each word problem.

75. **Number Problem** The sum of twice a number and 16 is 24. What is the number?

76. **Number Problem** 3 times a number, increased by 8, is 50. Find the number.

77. **Number Problem** 5 times a number, minus 12, is 78. Find the number.

78. **Number Problem** 4 times a number, decreased by 20, is 44. What is the number?

79. **Number Problem** The sum of two consecutive integers is 71. Find the two integers.

80. **Number Problem** The sum of two consecutive integers is 145. Find the two integers.

81. **Number Problem** The sum of three consecutive integers is 90. What are the three integers?

82. **Number Problem** If the sum of three consecutive integers is 93, find the three integers.

83. **Number Problem** The sum of two consecutive even integers is 66. What are the two integers? (*Hint:* Consecutive even integers such as 10, 12, and 14 can be represented by x, $x + 2$, $x + 4$, and so on.)

84. **Number Problem** If the sum of two consecutive even integers is 110, find the two integers.

85. **Number Problem** If the sum of two consecutive odd integers is 52, what are the two integers? (*Hint:* Consecutive odd integers such as 21, 23, and 25 can be represented by x, $x + 2$, $x + 4$, and so on.)

86. **Number Problem** The sum of two consecutive odd integers is 88. Find the two integers.

Answers

71. ____________
72. ____________
73. ____________
74. ____________
75. ____________
76. ____________
77. ____________
78. ____________
79. ____________
80. ____________
81. ____________
82. ____________
83. ____________
84. ____________
85. ____________
86. ____________

Answers

87. ______

88. ______

89. ______

90. ______

91. ______

92. ______

93. ______

94. ______

95. ______

96. ______

97. ______

98. ______

99. ______

100. ______

87. **Number Problem** 4 times an integer is 9 more than 3 times the next consecutive integer. What are the two integers?

88. **Number Problem** 4 times an even integer is 30 less than 5 times the next consecutive even integer. Find the two integers.

89. **Social Science** In an election, the winning candidate had 160 more votes than the loser. If the total number of votes cast was 3,260, how many votes did each candidate receive?

90. **Business and Finance** Jody earns \$140 more per month than Frank. If their monthly salaries total \$2,760, what amount does each earn?

91. **Business and Finance** A washer-dryer combination costs \$950. If the washer costs \$90 more than the dryer, what does each appliance cost?

92. **Problem Solving** Yan Ling is 1 year less than twice as old as his sister. If the sum of their ages is 14 years, how old is Yan Ling?

93. **Problem Solving** Diane is twice as old as her brother Dan. If the sum of their ages is 27 years, how old are Diane and her brother?

94. **Business and Finance** Patrick has invested \$15,000 in two bonds; one bond yields 4% annual interest, and the other yields 3% annual interest. How much is invested in each bond if the combined yearly interest from both bonds is \$545?

95. **Business and Finance** Johanna deposited \$21,000 in two banks. One bank gives $2\frac{1}{2}\%$ annual interest, and the other gives $3\frac{1}{4}\%$ annual interest. How much did she deposit in each bank if she received a total of \$615 in annual interest?

96. **Business and Finance** Tonya takes home \$1,080 per week. If her deductions amount to 28% of her wages, what is her weekly pay before deductions?

97. **Business and Finance** Sam donates 5% of his net income to charity. This amounts to \$190 per month. His payroll deductions are 24% of his gross monthly income. What is Sam's gross monthly income?

98. **Business and Finance** The Randolphs used 12 more gallons (gal) of fuel oil in October than in September and twice as much oil in November as in September. If they used 132 gal for the 3 months, how much was used during each month?

99. While traveling in South America, Richard noted that temperatures were given in degrees Celsius. Wondering what the temperature 95°F would correspond to, he found that he could answer this if he could solve the equation

$$95 = \frac{9}{5}C + 32$$

What was the corresponding temperature?

100. While traveling in England, Marissa noted an outdoor thermometer showing 20°C. To convert this to degrees Fahrenheit, she solved the equation

$$20 = \frac{5}{9}(F - 32)$$

What was the Fahrenheit temperature?

Basic Skills | Challenge Yourself | Calculator/Computer | **Career Applications** | Above and Beyond

101. **Allied Health** The internal diameter d (in mm) of an endotracheal tube for a child is calculated using the formula

$$d = \frac{t + 16}{4}$$

in which t is the child's age (in years).

How old is a child who requires an endotracheal tube with an internal diameter of 7 mm?

102. **Construction Technology** The number of studs, s, required to build a wall (with studs spaced 16 inches on center) is equal to one more than $\frac{3}{4}$ times the length of the wall, w, in feet. We model this with the formula

$$s = \frac{3}{4}w + 1$$

If a contractor uses 22 studs to build a wall, how long is the wall?

103. **Information Technology** A compression program reduces the size of files by 36%. If a compressed folder has a size of 11.2 MB, how large was it before compressing?

> Videos

104. **Agricultural Technology** A farmer harvested 2,068 bushels of barley. This amounted to 94% of his bid on the futures market. How many bushels did he bid to sell on the futures market?

Basic Skills | Challenge Yourself | Calculator/Computer | Career Applications | **Above and Beyond**

105. Complete this statement in your own words: "You can tell that an equation is a linear equation when. . . ."

106. What is the common characteristic of equivalent equations?

107. What is meant by a *solution* to a linear equation?

108. Define **(a)** identity and **(b)** contradiction.

109. Why does the multiplication property of equality not include multiplying both sides of the equation by 0?

110. Maxine lives in Pittsburgh, Pennsylvania, and pays 8.33 cents per kilowatt-hour (kWh) for electricity. During the 6 months of cold winter weather, her household uses about 1,500 kWh of electric power per month. During the two hottest summer months, the usage is also high because the family uses electricity to run an air conditioner. During these summer months, the usage is 1,200 kWh per month; the rest of the year, usage averages 900 kWh per month.

Answers

101. ______

102. ______

103. ______

104. ______

105. ______

106. ______

107. ______

108. ______

109. ______

110. ______

Answers

111.

112.

(a) Write an expression for the total yearly electric bill.

(b) Maxine is considering spending \$2,000 for more insulation for her home so that it is less expensive to heat and to cool. The insulation company claims that "with proper installation the insulation will reduce your heating and cooling bills by 25%." If Maxine invests the money in insulation, how long will it take her to get her money back in savings on her electric bill? Write to her about what information she needs to answer this question. Give her your opinion about how long it will take to save \$2,000 on heating bills, and explain your reasoning. What is your advice to Maxine?

111. Solve each equation. Express each solution as a fraction.

(a) $2x + 3 = 0$ **(b)** $4x + 7 = 0$ **(c)** $6x - 1 = 0$

(d) $5x - 2 = 0$ **(e)** $-3x + 8 = 0$ **(f)** $-5x - 9 = 0$

(g) Based on these problems, express the solution to the equation

$$ax + b = 0$$

where a and b represent real numbers and $a \neq 0$.

112. You are asked to solve an equation, but one number is missing. It reads

$$\frac{5x - ?}{4} = \frac{9}{2}$$

The solution is 4. What is the missing number?

Answers

1. 4 **3.** 3 **5.** −2 **7.** 3 **9.** 8 **11.** 32 **13.** 3 **15.** 6 **17.** −2 **19.** 0 **21.** 6 **23.** 5 **25.** $-\frac{20}{3}$ **27.** 5 **29.** 4 **31.** 5 **33.** $\frac{5}{2}$ **35.** 5 **37.** $-\frac{4}{3}$ **39.** 4 **41.** 7 **43.** 30 **45.** 7 **47.** 15 **49.** 3 **51.** $\frac{2}{9}$ **53.** Conditional; 6 **55.** Contradiction; { } **57.** Identity; $\mathbb{R}$ **59.** Contradiction; { } **61.** Identity; $\mathbb{R}$ **63.** $2x + 3 = 7$ **65.** $4x - 7 = 41$ **67.** $\frac{2}{3}x + 5 = 21$ **69.** $3x = x + 12$ **71.** False **73.** sometimes **75.** 4 **77.** 18 **79.** 35, 36 **81.** 29, 30, 31 **83.** 32, 34 **85.** 25, 27 **87.** 12, 13 **89.** 1,550 votes, 1,710 votes **91.** Washer: \$520; dryer: \$430 **93.** 18 years old, 9 years old **95.** \$12,000 at $3\frac{1}{4}\%$; \$9,000 at $2\frac{1}{2}\%$ **97.** \$5,000 **99.** 35°C **101.** 12 yr old **103.** 17.5 MB **105.** Above and Beyond **107.** A value for which the original equation is true **109.** Multiplying by 0 would always give 0 = 0. **111. (a)** $-\frac{3}{2}$; **(b)** $-\frac{7}{4}$; **(c)** $\frac{1}{6}$; **(d)** $\frac{2}{5}$; **(e)** $\frac{8}{3}$; **(f)** $-\frac{9}{5}$; **(g)** $-\frac{b}{a}$

1.7 Literal Equations and Their Applications

< 1.7 Objectives >

1 > Solve a literal equation for any variable

2 > Solve applications involving geometric figures

3 > Solve mixture problems

4 > Solve motion problems

Formulas are extremely useful tools in any field in which mathematics is applied. Formulas are simply equations that express a relationship between two or more letters or variables. You are no doubt familiar with many formulas, such as

$A = \frac{1}{2}bh$ The area of a triangle

$I = Prt$ Interest

$V = \pi r^2 h$ The volume of a cylinder

A formula is also called a **literal equation** because it involves several letters or variables. For instance, our first formula or literal equation, $A = \frac{1}{2}bh$, involves the three variables A (for area), b (for base), and h (for height).

Unfortunately, formulas are not always given in the form needed to solve a particular problem. In such cases, we use algebra to change the formula to a more useful equivalent equation, solved for a particular variable. The steps used in the process are very similar to those you used in solving linear equations. Let's consider an example.

Example 1 Solving a Literal Equation for a Variable

< Objective 1 >

Suppose we know the area A and the base b of a triangle and want to find its height h.

We are given

$$A = \frac{1}{2}bh$$

We need to find an equivalent equation with h, the unknown, by itself on one side and everything else on the other side. We can think of $\frac{1}{2}b$ as the **coefficient** of h.

We can remove the two *factors* of that coefficient, $\frac{1}{2}$ and b, separately.

$$2A = 2\left(\frac{1}{2}bh\right)$$ Multiply both sides by 2 to clear the equation of fractions.

or

$$2A = bh$$

$$\frac{2A}{b} = \frac{bh}{b}$$ Divide by b to isolate h.

$$\frac{2A}{b} = h$$

or

$$h = \frac{2A}{b}$$ Reverse the sides to write h on the left.

RECALL

A coefficient is the factor by which a variable is multiplied.

NOTE

$$2\left(\frac{1}{2}bh\right) = \left(2 \cdot \frac{1}{2}\right)(bh)$$
$$= 1 \cdot bh$$
$$= bh$$

NOTE

Here,  means an expression containing all the numbers or letters *other than h*.

We now have the height h in terms of the area A and the base b. This is called **solving the equation for *h*** and means that we are rewriting the formula as an equivalent equation of the form

$$h = \square$$

Check Yourself 1

Solve $V = \frac{1}{3}Bh$ for h.

You have already learned the methods needed to solve most literal equations or formulas for some specified variable. As Example 1 illustrates, the rules you learned in Section 1.6 are applied in exactly the same way as they were applied to equations with one variable.

You may have to apply both the addition and the multiplication properties when solving a formula for a specified variable. Example 2 illustrates this situation.

Example 2 Solving a Literal Equation

NOTE

This is a linear equation in two variables. You will see this again in Chapter 2.

(a) Solve $y = mx + b$ for x.

Remember that we want to end up with x alone on one side of the equation. Start by subtracting b from both sides to undo the addition on the right.

$$y = mx + b$$

$$y - b = mx + b - b$$

$$y - b = mx$$

If we divide both sides by m, then x will be alone on the right side.

$$\frac{y - b}{m} = \frac{mx}{m}$$

$$\frac{y - b}{m} = x$$

or

$$x = \frac{y - b}{m}$$

(b) Solve $3x + 2y = 12$ for y.

Begin by isolating the y-term.

$$\begin{array}{rcl} 3x + 2y &=& 12 \\ -3x & & -3x \\ \hline 2y &=& -3x + 12 \end{array}$$

Then, isolate y by dividing by its coefficient.

$$\frac{2y}{2} = \frac{-3x + 12}{2}$$

$$y = \frac{-3x + 12}{2}$$

RECALL

Dividing by 2 is the same as multiplying by $\frac{1}{2}$.

Often, in a situation like this, we use the distributive property to separate the terms on the right-hand side of the equation.

$$y = \frac{-3x + 12}{2}$$

$$= \frac{-3x}{2} + \frac{12}{2}$$

$$= -\frac{3}{2}x + 6 \qquad \frac{-3x}{2} = -\frac{3}{2}x$$

NOTE

v and v_0 represent distinct quantities.

Check Yourself 2

(a) Solve $v = v_0 + gt$ for t.

(b) Solve $4x - 3y = 8$ for x.

Let's summarize the steps illustrated by our examples.

Step by Step

Solving a Formula or Literal Equation

Step 1 Remove any grouping symbols by applying the distributive property.

Step 2 Multiply both sides of the equation by the LCM required to clear the equation of fractions or decimals.

Step 3 Combine any like terms that appear on either side of the equation.

Step 4 Apply the addition property of equality to write an equivalent equation with the variable term on *one side* of the equation and the constant term on the *other side*.

Step 5 Apply the multiplication property of equality to write an equivalent equation with the variable isolated on one side of the equation with coefficient 1.

NOTE

These are the same steps used to solve **any** linear equation.

Here is one more example, using these steps.

Example 3 Solving a Literal Equation for a Variable

NOTE

$A = P + Prt$ is a formula for the *amount* of money in an account after interest has been earned.

Solve $A = P + Prt$ for r.

$$A = P + Prt$$

$$A - P = P - P + Prt \qquad \text{Subtracting } P \text{ from both sides leaves the term involving } r \text{ alone on the right.}$$

$$A - P = Prt$$

$$\frac{A - P}{Pt} = \frac{Prt}{Pt} \qquad \text{Dividing both sides by } Pt \text{ isolates } r \text{ on the right.}$$

$$\frac{A - P}{Pt} = r$$

or

$$r = \frac{A - P}{Pt}$$

Check Yourself 3

Solve $2x + 3y = 6$ for y.

Now we look at an application that requires us to solve a literal equation.

Example 4 Using a Literal Equation

Suppose that the amount in an account, 3 years after a principal of \$5,000 was invested, is \$6,050. What was the interest rate?

From Example 3,

$$A = P + Prt$$

> **NOTE**
>
> Do you see the advantage of having the equation solved for the desired variable?

in which A is the amount in the account, P is the principal, r is the interest rate, and t is the time in years that the money has been invested. By the result of Example 3 we have

$$r = \frac{A - P}{Pt}$$

and we can substitute the known values in the second equation:

$$r = \frac{(6{,}050) - (5{,}000)}{(5{,}000)(3)}$$

$$= \frac{1{,}050}{15{,}000} = 0.07 = 7\%$$

The interest rate was 7%.

Check Yourself 4

Suppose that the amount in an account, 4 years after a principal of \$3,000 was invested, is \$3,720. What was the interest rate?

In subsequent applications, we use the five-step process first described in Section 1.4. As a reminder, here are those steps.

Step by Step

To Solve Word Problems

> **NOTE**
>
> Part of checking a solution is making certain that it is **reasonable.**

Step 1 Read the problem carefully. Then reread it to decide what you are asked to find.

Step 2 Choose a letter to represent one of the unknowns in the problem. Then represent all other unknowns of the problem with expressions that use the same letter.

Step 3 Translate the problem to the language of algebra to form an equation.

Step 4 Solve the equation.

Step 5 Answer the question and include units in your answer, when appropriate. Check your solution by returning to the original problem.

Example 5 Solving a Geometry Application

< Objective 2 >

> **NOTE**
>
> When an application involves geometric figures, draw a sketch of the problem, including the labels you assigned in step 2.

Length $3x - 1$

Width x

The length of a rectangle is 1 centimeter (cm) less than 3 times the width. If the perimeter is 54 cm, find the dimensions of the rectangle.

Step 1 You want to find the dimensions (the width and length).

Step 2 Let x be the width.

Then $3x - 1$ is the length.

(3 times the width; 1 less than)

Step 3 To write an equation, we use this formula for the perimeter of a rectangle:

$P = 2W + 2L$ or $2W + 2L = P$

So

$$2x + 2(3x - 1) = 54$$

(Twice the width; Twice the length; Perimeter)

NOTE

Be sure to return to the original statement of the problem when checking your result.

Step 4 Solve the equation.

$$2x + 2(3x - 1) = 54$$
$$2x + 6x - 2 = 54$$
$$8x = 56$$
$$x = 7$$

Step 5 The width x is 7 cm, and the length, $3x - 1$, is 20 cm.

Check: We look at the two conditions specified in this problem.

The relationship between the length and the width
20 is 1 less than 3 times 7, so this condition is met.

The perimeter of a rectangle
The sum of twice the width and twice the length is $2(7) + 2(20) = 14 + 40 = 54$, which checks.

Check Yourself 5

The length of a rectangle is 5 inches (in.) more than twice the width. If the perimeter of the rectangle is 76 in., what are the dimensions of the rectangle?

RECALL

π is used to represent an irrational number.
$\pi \approx 3.14$

One reason you might need to *manipulate* a geometric formula is because it is sometimes easier to measure the *output* of a formula.

For instance, the formula for the circumference of a circle is

$C = 2\pi r$

However, in practice, we might be able to measure the circumference of a round object directly, but not its radius. But if we wanted to compute the area (or volume) of this object, we would need to know its radius.

Example 6 Solving a Geometry Application

Poplar trees often have a round trunk. You use a tape measure to find the circumference of one poplar tree. Its circumference is approximately 8.8 inches.

(a) Find the radius of the trunk, to the nearest tenth of an inch.

We are asked to find the radius of this tree trunk.

We begin with the formula for the circumference of a circle and solve for the radius, r.

$$\frac{C}{2\pi} = \frac{2\pi r}{2\pi}$$

$$\frac{C}{2\pi} = r \quad \text{or} \quad r = \frac{C}{2\pi}$$

Now we can substitute in the circumference, 8.8 inches.

$$r = \frac{C}{2\pi}$$
$$= \frac{(8.8)}{2\pi}$$
$$\approx 1.4$$

The radius is approximately 1.4 in.

NOTE

```
8.8/(2π)
        1.400563499
Ans→R
        1.400563499
```

Be sure to place parentheses around the denominator.

Recall that you can store this value if you want to use it later.

```
π(1.4)²(420)
        2586.159072
πR²*420
        2588.241347
```

If you use the stored value for the radius, you get 2,588 in.[3], which is more precise.

(b) The trunk of this particular poplar tree is 35 feet high (420 in.). The volume of the trunk, in cubic inches, is given by the formula

$$V = \pi r^2 h$$

in which r is the radius and h is the height (both in inches).

Find the volume of this poplar trunk, to the nearest cubic inch.

We use the radius found in part (a) along with the height, in inches.

$$\begin{aligned} V &= \pi r^2 h \\ &= \pi(1.4)^2(420) \\ &\approx 2{,}586 \end{aligned}$$

The volume is approximately 2,586 in.^{3}.

Check Yourself 6

The circumference of a telephone pole measures approximately 31.4 in.

(a) Find the radius of a telephone pole, to the nearest inch.

(b) Find the volume, to the nearest cubic inch, if the telephone pole is 40 feet (480 inches) tall.

We use parentheses often when solving *mixture problems.* Mixture problems involve combining things that have different values, rates, or strengths. Look at Example 6.

Example 7 Solving a Mixture Problem

< Objective 3 >

Four hundred tickets were sold for a school play. General admission tickets were \$4, while student tickets were \$3. If the total ticket sales were \$1,350, how many of each type of ticket were sold?

NOTE

We subtract x, the number of general admission tickets, from 400, the total number of tickets, to find the number of student tickets.

Step 1 You want to find the number of each type of ticket sold.

Step 2 Let x be the number of general admission tickets.

Then $400 - x$ student tickets were sold.

400 tickets were sold in all.

Step 3 The sales value for each kind of ticket is found by multiplying the price of the ticket by the number sold.

Value of general admission tickets:	$4x$	\$4 for each of the x tickets
Value of student tickets:	$3(400 - x)$	\$3 for each of the $400 - x$ tickets

So to form an equation, we have

$$4x + 3(400 - x) = 1{,}350$$

Value of general admission tickets — Value of student tickets — Total value

Step 4 Solve the equation.

$$\begin{aligned} 4x + 3(400 - x) &= 1{,}350 \\ 4x + 1{,}200 - 3x &= 1{,}350 \\ x + 1{,}200 &= 1{,}350 \\ x &= 150 \end{aligned}$$

Step 5 This shows that 150 general admission and 250 student tickets were sold. We leave the check to you.

Check Yourself 7

Beth bought 40¢ stamps and 15¢ stamps at the post office. If she purchased 60 stamps at a cost of $19, how many of each kind did she buy?

Many of the problems encountered by small businesses can be treated as mixture problems.

Example 8 A Small-Business Application

A coffee reseller wishes to mix two types of coffee beans. The Kona bean wholesales for $4.50 per pound; the Sumatran bean wholesales for $3.25 per pound. If she wishes to mix 200 pounds for a wholesale price of $4.00 per pound, how many pounds of each type of coffee should she include in the mix?

Step 1 We are asked to find the correct amount of each coffee bean so that her mixture contains 200 pounds of beans and wholesales for $4.00 per pound.

Step 2 Let x be the number of pounds of Kona beans needed. Then, $200 - x$ gives the amount of Sumatran beans needed.

Step 3 We set up the problem: Each pound of Kona beans costs $4.50 per pound and each pound of Sumatran beans costs $3.25 per pound. The total cost of the mixture is given by the expression

$$4.50\underbrace{x}_{\text{Kona}} + 3.25\underbrace{(200 - x)}_{\text{Sumatran}}$$

The total mixture will be 200 pounds and will cost $4.00 per pound.

$$4.00 \times 200 = 800$$

We set these two expressions equal to each other.

$$4.50x + 3.25(200 - x) = 800$$

Step 4 $4.50x + 3.25(200 - x) = 800$

$4.50x + 650 - 3.25x = 800$ Use the distributive property to remove the parentheses.

$1.25x + 650 = 800$ Combine like terms.

$1.25x = 150$ Subtract 650 from both sides to isolate the x-term.

$x = 120$ Divide both sides by 1.25 to isolate the variable.

Step 5 She needs 120 pounds of Kona beans and

$200 - x = 200 - (120) = 80$

80 pounds of Sumatran beans.

Check Yourself 8

Minh splits his $20,000 investment between two funds. At the end of a year, one fund grows by 3.25% and the other grows 4.5%. If the total earnings on his investment came to $793.75, how much did he invest in each fund?

Another common application is the *motion problem*. Motion problems involve a distance traveled, a rate (or speed), and an amount of time. To solve a motion problem, we need a relationship between these three quantities.

> CAUTION

Be careful to make your units consistent. If a rate is given in *miles per hour,* then the time must be given in *hours* and the distance in *miles.*

Suppose you travel at a rate of 50 miles per hour (mi/h) on a highway for 6 hours (h). How far (what distance) will you have gone? To find the distance, you multiply:

$$(50 \text{ mi/h})(6 \text{ h}) = 300 \text{ mi}$$

Speed or rate ↑ Time ↑ Distance ↑

Property

Motion Problems

If r is the rate, t is the time, and d is the distance traveled, then

$$d = r \cdot t$$

We apply this relationship in Example 9.

Example 9 Solving a Motion Problem

< Objective 4 >

On Friday morning Ricardo drove from his house to the beach in 4 h. When coming back Sunday afternoon, heavy traffic slowed his speed by 10 mi/h, and the trip took 5 h. What was his average speed (rate) in each direction?

Step 1 We want the speed or rate in each direction.

Step 2 Let x be Ricardo's speed to the beach. Then $x - 10$ is his return speed.

It is always a good idea to sketch the given information in a motion problem. Here we have

Going → x mi/h for 4 h

Returning ← $x - 10$ mi/h for 5 h

Step 3 Since we know that the distance is the same each way, we can write an equation using the fact that the product of the rate and the time each way must be the same. So

Distance (going) = Distance (returning)

Time · rate (going) = Time · rate (returning)

$$4x = 5(x - 10)$$

Time · rate (going) ↑ Time · rate (returning) ↑

A chart or table can help summarize the given information, especially when stumped about how to proceed. We begin with an "empty" table.

	Rate	Time	Distance
Going			
Returning			

Next, we fill the table with the information given in the problem.

	Rate	Time	Distance
Going	x	4	
Returning	$x - 10$	5	

Now we fill in the missing information. Here we use the fact that $d = rt$ to complete the table.

	Rate	Time	Distance
Going	x	4	$4x$
Returning	$x - 10$	5	$5(x - 10)$

From here we set the two distances equal to each other and solve as before.

NOTE

x was his rate going; $x - 10$, his rate returning.

Step 4 Solve.

$4x = 5(x - 10)$

$4x = 5x - 50$ Use the distributive property to remove the parentheses.

$-x = -50$ Subtract $5x$ from both sides to isolate the x-term.

$x = 50$ Divide both sides by -1 to isolate the variable.

Step 5 So Ricardo's rate going to the beach was 50 mi/h, and his rate returning was 40 mi/h.

To check, you should verify that the product of the time and the rate is the same in each direction.

Check Yourself 9

A plane made a flight (with the wind) between two towns in 2 h. Returning against the wind, the plane's speed was 60 mi/h slower, and the flight took 3 h. What was the plane's speed in each direction?

Example 10 illustrates another way of using the distance relationship.

Example 10 Solving a Motion Problem

Katy leaves Las Vegas, Nevada, for Los Angeles, California, at 10 A.M., driving at 50 mi/h. At 11 A.M. Jensen leaves Los Angeles for Las Vegas, driving at 55 mi/h along the same route. If the cities are 260 mi apart, at what time will they meet?

Step 1 Let's find the time that Katy travels until they meet.

Step 2 Let x be Katy's time.

Then $x - 1$ is Jensen's time. Jensen left 1 h later!

Again, you should draw a sketch of the given information.

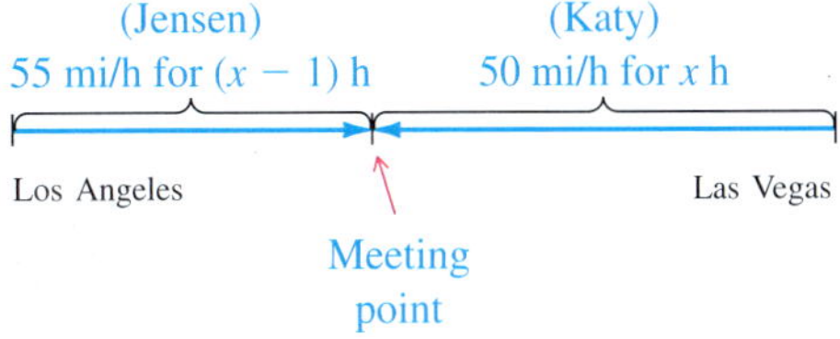

Step 3 To write an equation, we again need the relationship $d = rt$. From this equation, we can write

Katy's distance $= 50x$

Jensen's distance $= 55(x - 1)$

As before, we can use a table to solve.

	Rate	Time	Distance
Katy	50	x	$50x$
Jensen	55	$x - 1$	$55(x - 1)$

From the original problem, the sum of those distances is 260 mi, so

$$50x + 55(x - 1) = 260$$

Step 4
$$\begin{aligned} 50x + 55(x - 1) &= 260 \\ 50x + 55x - 55 &= 260 \\ 105x - 55 &= 260 \\ 105x &= 315 \\ x &= 3 \end{aligned}$$

NOTE

Be sure to answer the question asked in the problem.

Step 5 Finally, since Katy left at 10 A.M., the two will meet at 1 P.M. We leave the check of this result to you.

Check Yourself 10

At noon a jogger leaves one point, running at 8 mi/h. One hour later a bicyclist leaves the same point, traveling at 20 mi/h in the opposite direction. At what time will they be 36 mi apart?

Check Yourself ANSWERS

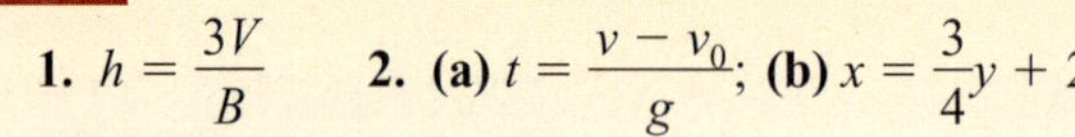

1. $h = \dfrac{3V}{B}$ **2.** **(a)** $t = \dfrac{v - v_0}{g}$; **(b)** $x = \dfrac{3}{4}y + 2$

3. $y = \dfrac{6 - 2x}{3}$ or $y = -\dfrac{2}{3}x + 2$ **4.** The interest rate was 6%.

5. The width is 11 in.; the length is 27 in. **6.** **(a)** 5 in.; **(b)** 37,699 in.^{3}

7. 40 at 40¢ and 20 at 15¢ **8.** \$8,500 at 3.25% and \$11,500 at 4.5%

9. 180 mi/h with the wind and 120 mi/h against the wind **10.** At 2 P.M.

Reading Your Text

SECTION 1.7

(a) A ______________ is also called a literal equation because it involves several letters or variables.

(b) A ______________ is the factor by which a variable is multiplied.

(c) Always return to the ______________ equation or statement when checking your result.

(d) In a motion problem, the ______________ traveled is found by taking the product of the rate of travel (speed) and the time traveled.

1.7 exercises

Name ____________

Section ________ Date ________

< Objective 1 >

Solve each literal equation for the indicated variable.

1. $P = 4s$ (for s) Perimeter of a square

2. $V = Bh$ (for B) Volume of a prism

3. $E = IR$ (for R) Voltage in an electric circuit

4. $I = Prt$ (for r) Simple interest

5. $V = LWH$ (for H) Volume of a rectangular solid

6. $V = \pi r^2 h$ (for h) Volume of a cylinder

7. $A + B + C = 180$ (for B) Measure of angles in a triangle

8. $P = I^2R$ (for R) Power in an electric circuit

9. $ax + b = 0$ (for x) Linear equation in one variable

10. $y = mx + b$ (for m) Slope-intercept form for a line 

11. $s = \frac{1}{2}gt^2$ (for g) Distance

12. $K = \frac{1}{2}mv^2$ (for m) Energy

13. $x + 5y = 15$ (for y) Linear equation in two variables

14. $2x + 3y = 6$ (for x) Linear equation in two variables

15. $P = 2L + 2W$ (for L) Perimeter of a rectangle

16. $ax + by = c$ (for y) Linear equation in two variables

17. $V = \frac{KT}{P}$ (for T) Volume of a gas

18. $V = \frac{1}{3}\pi r^2 h$ (for h) Volume of a cone

19. $x = \frac{a + b}{2}$ (for b) Average of two numbers

Answers

1. ______ 2. ______

3. ______ 4. ______

5. ______ 6. ______

7. ______

8. ______ 9. ______

10. ______

11. ______

12. ______

13. ______

14. ______

15. ______

16. ______

17. ______

18. ______

19. ______

Answers

20. ______

21. ______

22. ______

23. ______

24. ______

25. ______

26. ______

27. ______

28. ______

29. ______

30. ______

31. ______

32. ______

33. ______

34. ______

35. ______

36. ______

20. $D = \frac{C - s}{n}$ (for s) Depreciation

21. $F = \frac{9}{5}C + 32$ (for C) Celsius/Fahrenheit conversion

22. $A = P + Prt$ (for t) Amount at simple interest

23. $S = 2\pi r^2 + 2\pi rh$ (for h) Total surface area of a cylinder

24. $A = \frac{1}{2}h(B + b)$ (for b) Area of a trapezoid > Videos

< Objectives 2–4 >

25. **Geometry** A rectangular solid has a base with length 8 centimeters (cm) and width 5 cm. If the volume of the solid is 120 cm^3, find the height of the solid. (See exercise 5.) > Videos

26. **Geometry** A cylinder has a radius of 4 inches (in.). If the volume of the cylinder is 144π in.^{3}, what is the height of the cylinder? (See exercise 6.)

27. **Business and Finance** A principal of \$2,000 was invested in a savings account for 4 years. If the interest earned for the period was \$240, what was the interest rate? (See exercise 4.)

28. **Geometry** If the perimeter of a rectangle is 60 feet (ft) and the width is 12 ft, find its length. (See exercise 15.)

29. **Statistics** The high temperature in New York for a particular day was reported at 77°F. How would the same temperature have been given in degrees Celsius? (See exercise 21.) > Make the Connection

30. **Geometry** Rose's garden is in the shape of a trapezoid. If the height of the trapezoid is 16 meters (m), one base is 20 m, and the area is 224 m^2, find the length of the other base. (See exercise 24.)

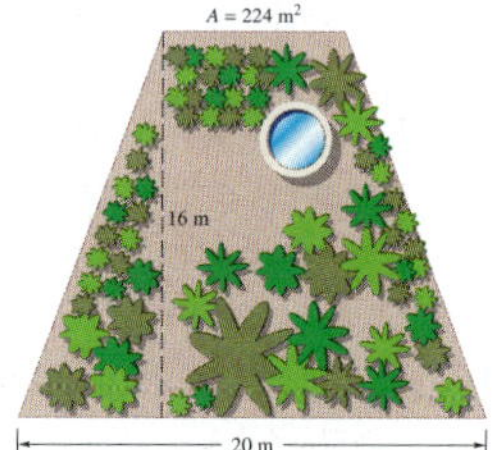

Translate each statement to an equation. Let x represent the number in each case.

31. Twice the sum of a number and 6 is 18.

32. The sum of twice a number and 4 is 20.

33. 3 times the difference of a number and 5 is 21.

34. The difference of 3 times a number and 5 is 21.

35. The sum of twice an integer and 3 times the next consecutive integer is 48.

36. The sum of 4 times an odd integer and twice the next consecutive odd integer is 46.

Determine whether each statement is **true** *or* **false.**

37. Another name for *formula* is *literal equation.*

38. The formula for the area of a rectangle is $P = 2L + 2W$.

39. The key relationship in motion problems is $d = rt$.

40. When solving for a variable in a formula, we use the same steps used in solving linear equations.

Solve each word problem.

41. Number Problem One number is 8 more than another. If the sum of the smaller number and twice the larger number is 46, find the two numbers.

42. Number Problem One number is 3 less than another. If 4 times the smaller number minus 3 times the larger number is 4, find the two numbers.

43. Number Problem One number is 7 less than another. If 4 times the smaller number plus 2 times the larger number is 62, find the two numbers.

44. Number Problem One number is 10 more than another. If the sum of twice the smaller number and 3 times the larger number is 55, find the two numbers. > Videos

45. Number Problem Find two consecutive integers such that the sum of twice the first integer and 3 times the second integer is 28. (*Hint:* If x represents the first integer, $x + 1$ represents the next consecutive integer.)

46. Number Problem Find two consecutive odd integers such that 3 times the first integer is 5 more than twice the second. (*Hint:* If x represents the first integer, $x + 2$ represents the next consecutive odd integer.)

47. Geometry The length of a rectangle is 1 inch (in.) more than twice its width. If the perimeter of the rectangle is 74 in., find the dimensions of the rectangle.

48. Geometry The length of a rectangle is 5 centimeters (cm) less than 3 times its width. If the perimeter of the rectangle is 46 cm, find the dimensions of the rectangle. > Videos

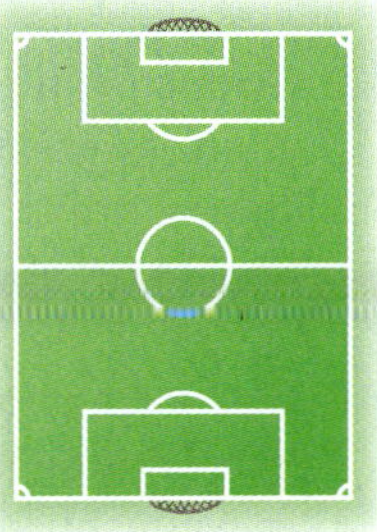

49. Geometry The length of a rectangular garden is 5 m more than 3 times its width. The perimeter of the garden is 74 m. What are the dimensions of the garden?

50. Geometry The length of a rectangular playing field is 5 ft less than twice its width. If the perimeter of the playing field is 230 ft, find the length and width of the field.

Answers

37. ______
38. ______
39. ______
40. ______
41. ______
42. ______
43. ______
44. ______
45. ______
46. ______
47. ______
48. ______
49. ______
50. ______

Answers

51. ________
52. ________
53. ________
54. ________
55. ________
56. ________
57. ________
58. ________
59. ________
60. ________
61. ________
62. ________
63. ________

51. **Geometry** The base of an isosceles triangle is 3 cm less than the length of the equal sides. If the perimeter of the triangle is 36 cm, find the length of each of the sides.

52. **Geometry** The length of one of the equal legs of an isosceles triangle is 3 in. less than twice the length of the base. If the perimeter is 29 in., find the length of each of the sides.

53. **Business and Finance** Tickets for a play cost \$14 for the main floor and \$9 in the balcony. If the total receipts from 250 tickets were \$3,000, how many of each type of ticket were sold?

54. **Business and Finance** Tickets for a basketball tournament were \$6 for students and \$9 for nonstudents. Total sales were \$10,500, and 250 more student tickets were sold than nonstudent tickets. How many of each type of ticket were sold? > Videos

55. **Problem Solving** Maria bought 50 stamps at the post office in 27¢ and 42¢ denominations. If she paid \$18 for the stamps, how many of each denomination did she buy?

56. **Business and Finance** A bank teller had a total of 125 \$10 bills and \$20 bills to start the day. If the value of the bills was \$1,650, how many of each denomination did he have?

57. **Business and Finance** Tickets for a train excursion were \$120 for a sleeping room, \$80 for a berth, and \$50 for a coach seat. The total ticket sales were \$8,600. If there were 20 more berth tickets sold than sleeping room tickets and 3 times as many coach tickets as sleeping room tickets, how many of each type of ticket were sold?

58. **Business and Finance** Admission for a college baseball game is \$6 for box seats, \$5 for the grandstand, and \$3 for the bleachers. The total receipts for one evening were \$9,000. There were 100 more grandstand tickets sold than box seat tickets. Twice as many bleacher tickets were sold as box seat tickets. How many tickets of each type were sold?

59. **Science and Medicine** Patrick drove 3 h to attend a meeting. On the return trip, his speed was 10 mi/h less, and the trip took 4 h. What was his speed each way?

60. **Science and Medicine** A bicyclist rode into the country for 5 h. In returning, her speed was 5 mi/h faster and the trip took 4 h. What was her speed each way?

61. **Science and Medicine** A car leaves a city and goes north at a rate of 50 mi/h at 2 P.M. One hour later a second car leaves, traveling south at a rate of 40 mi/h. At what time will the two cars be 320 mi apart? > Videos

62. **Science and Medicine** A bus leaves a station at 1 P.M., traveling west at an average rate of 44 mi/h. One hour later a second bus leaves the same station, traveling east at a rate of 48 mi/h. At what time will the two buses be 274 mi apart?

63. **Science and Medicine** At 8:00 A.M., Catherine leaves on a trip at 45 mi/h. One hour later, Max decides to join her and leaves along the same route, traveling at 54 mi/h. When will Max catch up with Catherine?

64. **Science and Medicine** Martina leaves home at 9 A.M., bicycling at a rate of 24 mi/h. Two hours later, John leaves, driving at the rate of 48 mi/h. At what time will John catch up with Martina?

65. If the temperature in Madrid is given as 35°C, what is the corresponding temperature in degrees Fahrenheit? chapter 1 > Make the Connection

66. What temperature in degrees Celsius is equivalent to 59°F? 

67. **Statistics and Mathematics** Mika leaves Boston for Baltimore at 10:00 A.M., traveling at 45 mi/h. One hour later, Hiroko leaves Baltimore for Boston on the same route, traveling at 50 mi/h. If the two cities are 425 mi apart, when will Mika and Hiroko meet?

68. **Statistics and Mathematics** A train leaves town A for town B, traveling at 35 mi/h. At the same time, a second train leaves town B for town A at 45 mi/h. If the two towns are 320 mi apart, how long will it take for the two trains to meet?

69. **Electrical Engineering** Resistance R (in ohms, Ω) is given by the formula

$$R = \frac{V^2}{D}$$

in which D is the power dissipation (in watts) and V is the voltage. Determine the power dissipation when 13.2 volts pass through a 220-Ω resistor.

70. **Mechanical Engineering** In a planetary gear, the size and number of teeth must satisfy the equation

$$Cx = By(F - 1)$$

Calculate the number of teeth y needed if $C = 9$ in., $x = 14$ teeth, $B = 2$ in., and $F = 8$.

71. **Allied Health** Yohimbine is used to reverse the effects of xylazine in deer. The recommended dose is 0.125 mg per kilogram of a deer's weight.

(a) Write a formula that expresses the required dosage level d for a deer of weight w.

(b) How much yohimbine should be administered to a 15-kg fawn?

(c) What size deer requires a 5.0-mg dosage?

72. **Electronics Technology** Temperature sensors output voltage, which varies with respect to temperature. For a particular sensor, the output voltage V for a given Celsius temperature C is given by

$$V = 0.28C + 2.2$$

(a) Determine the output voltage at 0°C.

(b) Determine the output voltage at 22°C.

(c) Determine the temperature if the sensor outputs 14.8 V.

(d) At what temperature is there no voltage output (two decimal places)?

Answers

64. ______

65. ______

66. ______

67. ______

68. ______

69. ______

70. ______

71. ______

72. ______

Answers

73. ____________

74. ____________

75. ____________

76. ____________

Basic Skills | Challenge Yourself | Calculator/Computer | Career Applications | **Above and Beyond**

73. There is a universally agreed on *order of operations* used to simplify expressions. Explain how the order of operations is used in solving equations. Be sure to use complete sentences.

74. Here is a common mistake in solving equations.

The equation: $2(x - 2) = x + 3$

First step in solving: $2x - 2 = x + 3$

Write a clear explanation of what error has been made. What could be done to avoid this error?

75. Here is another very common mistake.

The equation: $6x - (x + 3) = 5 + 2x$

First step in solving: $6x - x + 3 = 5 + 2x$

Write a clear explanation of what error has been made and what could be done to avoid the mistake.

76. Write an algebraic equation for the English statement "Subtract 5 from the sum of x and 7 times 3 and the result is 20." Compare your equation with those of other students. Did you all write the same equation? Are all the equations correct even though they don't look alike? Do all the equations have the same solution? What is wrong? The English statement is *ambiguous.* Write another English statement that leads correctly to more than one algebraic equation. Exchange with another student and see if she or he thinks the statement is ambiguous. Notice that the algebra is *not* ambiguous!

Answers

1. $s = \frac{P}{4}$ **3.** $R = \frac{E}{I}$ **5.** $H = \frac{V}{LW}$ **7.** $B = 180 - A - C$
9. $x = -\frac{b}{a}$ **11.** $g = \frac{2s}{t^2}$ **13.** $y = \frac{15 - x}{5}$ or $y = -\frac{1}{5}x + 3$
15. $L = \frac{P - 2W}{2}$ or $L = \frac{P}{2} - W$ **17.** $T = \frac{PV}{K}$ **19.** $b = 2x - a$
21. $C = \frac{5}{9}(F - 32)$ or $C = \frac{5(F - 32)}{9}$ **23.** $h = \frac{S - 2\pi r^2}{2\pi r}$ or $h = \frac{S}{2\pi r} - r$
25. 3 cm **27.** 3% **29.** 25°C **31.** $2(x + 6) = 18$
33. $3(x - 5) = 21$ **35.** $2x + 3(x + 1) = 48$ **37.** True **39.** True
41. 10, 18 **43.** 8, 15 **45.** 5, 6 **47.** 12 in., 15 in. **49.** 8 m, 29 m
51. Legs: 13 cm; base: 10 cm **53.** \$14-tickets: 150; \$9-tickets: 100
55. 20 27¢ stamps; 30 42¢ stamps
57. 60 coach, 40 berth, 20 sleeping room
59. going 40 mi/h, returning 30 mi/h **61.** 6 P.M. **63.** 2 P.M. **65.** 95°F
67. 3 P.M. **69.** 0.792 watt **71.** **(a)** $d = 0.125w$; **(b)** 1.875 mg; **(c)** 40 kg
73. Above and Beyond **75.** Above and Beyond

1.8

Solving Linear Inequalities

< 1.8 Objectives >

1 > Use inequality notation
2 > Graph the solution set of a linear inequality
3 > Use the addition property to solve a linear inequality
4 > Use the multiplication property to solve a linear inequality

Tips for Student Success

Preparing for a test

Preparing for a test begins on the first day of class. Everything you do in class and at home is part of that preparation. In fact, if you attend class every day, take good notes, and keep up with the homework, then you will already be prepared and will not need to "cram" for your exam.

Instead of cramming, here are a few things to focus on in the days before a scheduled test.

1. Study for your exam, but finish studying 24 hours before the test. Make certain to get some good rest before taking a test.
2. Study for the exam by going over the homework and class notes. Write down all of the problem types, formulas, and definitions that you think might give you trouble on the test.
3. The last item before you finish studying is to take the notes you made in step 2 and transfer the most important ideas to a 3×5 (index) card. You should complete this step a full 24 hours before your exam.
4. One hour before your exam, review the information on the 3×5 card you made in step 3. You will be surprised at how much you remember about each concept.
5. The biggest obstacle for many students is to believe that they can be successful on a test. You can overcome this obstacle easily enough. If you have been completing the homework and keeping up with the classwork, then you should perform quite well on the test. Truly anxious students are often surprised to score well on an exam. These students attribute a good test score to blind luck when it is not luck at all. This is the first sign that you "get it." Enjoy the success!

As pointed out earlier in this chapter, an equation is a statement that two expressions are equal. In algebra, an **inequality** is a statement that one expression is less than or greater than another. The inequality symbols are used when writing inequalities.

Example 1 Reading the Inequality Symbol

< Objective 1 >

$5 < 8$ is an inequality read "5 is less than 8."

$9 > 6$ is an inequality read "9 is greater than 6."

RECALL

The "arrowhead" always points toward the smaller quantity.

Check Yourself 1

Fill in each blank with the symbol $<$ or $>$.

(a) 12 ______ 8 **(b)** 20 ______ 25

Just as was the case with equations, inequalities that involve variables may be either true or false depending on the value that we give to the variable. For instance, consider the inequality

$x < 6$

$$\text{If } x = \begin{cases} 3 & 3 < 6 \text{ is true} \\ 6 & 6 < 6 \text{ is false} \\ -10 & -10 < 6 \text{ is true} \\ 8 & 8 < 6 \text{ is false} \end{cases}$$

Therefore, 3 and -10 are both *solutions* of the inequality $x < 6$; they make the inequality a true statement. You should see that 6 and 8 are *not* solutions.

Recall from Section 1.4 that a *solution of an equation* is any value for the variable that makes the equation a true statement. Similarly, the **solution of an inequality** is a value for the variable that makes the inequality a true statement.

In the discussion describing $x < 6$, above, there is more than one solution. We have also seen equations with more than one solution. To talk clearly about this type of problem, we define a term for all of the solutions of an equation or inequality in one variable. In Chapter 2, we will expand this definition to include equations and inequalities with more than one variable.

NOTE

Since there are so many solutions (an infinite number, in fact), we certainly do not want to try to list them all! A convenient way to show the solutions of an inequality is with a number line.

RECALL

The equation $x^2 = 9$ has two solutions.

Identities have an infinite number of solutions.

Definition

Solution Set

The **solution set** of an equation or inequality in one variable is the set of all values for the variable that make the equation or inequality a true statement.

That is, the solution set is the set of all solutions to an equation or inequality.

Example 2 Graphing Inequalities

< Objective 2 >

To graph the solution set of the inequality $x < 6$, we want to include all real numbers that are "less than" 6. This means all numbers *to the left* of 6 on the number line.

We then start at 6 and draw an arrow extending left, as shown:

Note: The parenthesis at 6 means that we do not include 6 in the solution set (6 is not less than itself). The colored arrow shows all the numbers in the solution set, with the arrowhead indicating that the solution set continues infinitely to the left.

NOTE

The colored arrow indicates the direction of the *solutions*.

Check Yourself 2

Graph the solution set of $x < -2$.

Two other symbols are used in writing inequalities. They are used with inequalities such as

$x \geq 5$ and $x \leq 2$

$x \geq 5$ is a combination of the two statements $x > 5$ and $x = 5$. It is read "x is greater than or equal to 5." The solution set includes 5 in this case.

The inequality $x \leq 2$ combines the statements $x < 2$ and $x = 2$. It is read "x is less than or equal to 2."

Example 3 Graphing Inequalities

NOTE

The bracket means that we include 5 in the solution set.

The solution set of $x \geq 5$ is graphed as

Check Yourself 3

Graph each solution set.

(a) $x \leq -4$ **(b)** $x \geq 3$

We have looked at graphs of the solution sets of some simple inequalities, such as $x < 8$ or $x \geq 10$. Now we will look at more complicated inequalities, such as

$2x - 3 < x + 4$

Fortunately, the methods used to solve this type of inequality are very similar to those we used earlier in this chapter to solve linear equations in one variable. Here is our first property for inequalities.

Property

The Addition Property of Inequality

If $a < b$ then $a + c < b + c$

In words, adding the same quantity to both sides of an inequality gives an equivalent inequality.

Equivalent inequalities have the same solution set.

Example 4 Solving Inequalities

< Objective 3 >

NOTE

The inequality is solved when an equivalent inequality has the form

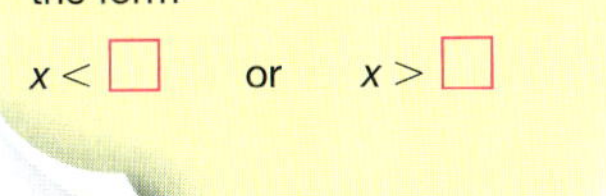

Solve and graph the solution set of $x - 8 < 7$.

To solve $x - 8 < 7$, add 8 to both sides of the inequality by the addition property.

$$x - 8 < 7$$

$$x - 8 + 8 < 7 + 8 \quad \text{Add 8 to both sides.}$$

$$x < 15 \quad \text{The inequality is solved.}$$

The graph of the solution set is

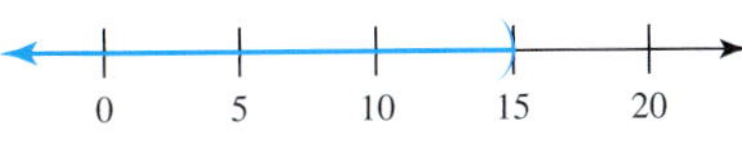

Check Yourself 4

Solve and graph the solution set of

$x - 9 > -3$

As with equations, the addition property allows us to *subtract* the same quantity from both sides of an inequality.

Example 5 Solving Inequalities

Solve and graph the solution set of $4x - 2 \geq 3x + 5$.

First, we subtract $3x$ from both sides of the inequality.

> **NOTE**
>
> We subtracted $3x$ and then added 2 to both sides. If these steps are done in the other order, the result is the same.

$$4x - 2 \geq 3x + 5$$

$$4x - 3x - 2 \geq 3x - 3x + 5 \qquad \text{Subtract } 3x \text{ from both sides.}$$

$$x - 2 \geq 5$$

$$x - 2 + 2 \geq 5 + 2 \qquad \text{Now we add 2 to both sides.}$$

$$x \geq 7$$

The graph of the solution set is

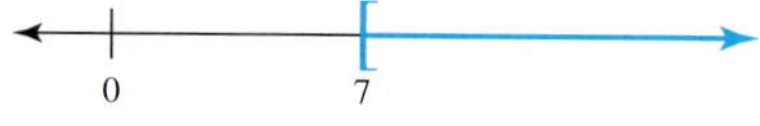

Check Yourself 5

Solve and graph the solution set.

$7x - 8 \leq 6x + 2$

Note that $x < 3$ is the same as $3 > x$. In our next example, we graph an inequality in which the variable is on the right side.

Example 6 Solving an Inequality

Solve and graph the solution set of the inequality

$2x + 3 < 3x + 6$

The coefficient of x is larger on the right side of the inequality than on the left side. Therefore, we isolate the variable on the right side.

$$2x - 2x + 3 < 3x - 2x + 6 \qquad \text{Subtract } 2x \text{ from both sides.}$$

$$3 < x + 6 \qquad \text{Subtract 6 from both sides.}$$

$$3 - 6 < x + 6 - 6$$

$$-3 < x$$

The graph of the solution set is

Check Yourself 6

Solve and graph the solution set of the inequality

$4x - 5 < 5x - 9$

Some applications are solved by using an inequality instead of an equation. Example 7 illustrates such an application.

Example 7 Solving an Inequality Application

Mohammed needs a mean score of 92 or higher on four tests to get an A. So far his scores are 94, 89, and 88. What scores on the fourth test will get him an A?

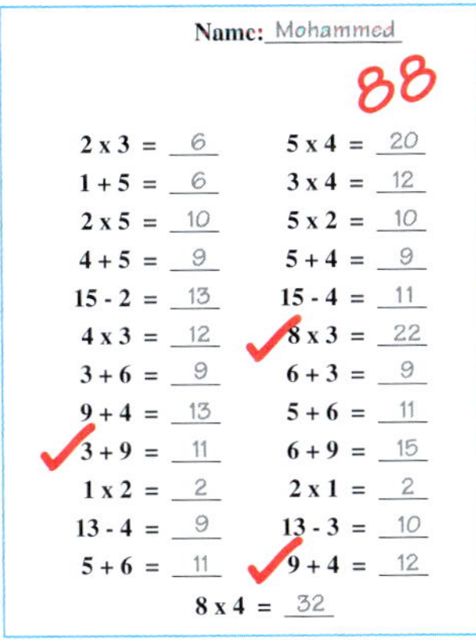

NOTES

What do you need to find?

Assign a letter to the unknown.

Write an inequality.

Solve the inequality.

Step 1 We are looking for the scores that will, when combined with the other scores, give Mohammed an A.

Step 2 Let x represent a fourth-test score that will get him an A.

Step 3 The inequality has the mean on the left side, which must be greater than or equal to the 92 on the right.

$$\frac{94 + 89 + 88 + x}{4} \geq 92$$

Step 4 First, multiply both sides by 4:

$$94 + 89 + 88 + x \geq 368$$

Then add the test scores:

$$183 + 88 + x \geq 368$$

$$271 + x \geq 368$$

Subtract 271 from both sides:

$$x \geq 97$$

Step 5 Mohammed needs to earn a 97 or above to earn an A.

To check the solution, we find the mean of the four test scores, 94, 89, 88, and 97.

$$\frac{94 + 89 + 88 + (97)}{4} = \frac{368}{4} = 92$$

Check Yourself 7

Felicia needs a mean score of at least 75 on five tests to get a passing grade in her health class. On her first four tests she has scores of 68, 79, 71, and 70. What scores on the fifth test will give her a passing grade?

As with equations, we need a rule for multiplying on both sides of an inequality. Here we have to be a bit careful. There is a difference between the multiplication property for inequalities and the one for equations. Look at the following:

$2 < 7$ A true inequality

Multiply both sides by 3.

$$2 < 7$$
$$3 \cdot 2 < 3 \cdot 7$$
$$6 < 21 \qquad \text{A true inequality}$$

Start again, but multiply both sides by −3.

$$2 < 7 \qquad \text{The original inequality}$$
$$(-3)(2) < (-3)(7)$$
$$-6 < -21 \qquad \textit{Not} \text{ a true inequality}$$

NOTE

When both sides of an inequality are multiplied by the same *negative* number, it is necessary to *reverse the direction* of the inequality to give an equivalent inequality.

Let's try something different.

$$2 < 7 \qquad \text{Change the direction of the inequality}$$
$$(-3)(2) > (-3)(7) \qquad < \text{ becomes } >.$$
$$-6 > -21 \qquad \text{This is now a true inequality.}$$

This suggests that multiplying both sides of an inequality by a negative number changes the direction of the inequality.

Property

The Multiplication Property of Inequality

If $a < b$ then $ac < bc$ if $c > 0$
and $ac > bc$ if $c < 0$

In words, multiplying both sides of an inequality by the same *positive* number gives an equivalent inequality. Multiplying both sides of an inequality by the same *negative* number gives an equivalent inequality if we also reverse the direction of the inequality sign.

As with equations, this rule applies to division, as well.

- Dividing both sides of an inequality by the same *positive* number gives an equivalent inequality.

  If $a < b$, then $\frac{a}{c} < \frac{b}{c}$ if $c > 0$.

- When dividing both sides of an inequality by the same *negative* number we must reverse the direction of the inequality sign to get an equivalent inequality.

  If $a < b$, then $\frac{a}{c} > \frac{b}{c}$ if $c < 0$.

Example 8 **Solving and Graphing Inequalities**

< Objective 4 >

NOTE

Multiplying both sides of the inequality by $\frac{1}{5}$ is the same as dividing both sides by 5:

$$\frac{5x}{(5)} < \frac{30}{(5)}$$

(a) Solve and graph the solution set of $5x < 30$.

Multiplying both sides of the inequality by $\frac{1}{5}$ gives

$$\frac{1}{5}(5x) < \frac{1}{5}(30)$$

Simplifying, we have

$$x < 6$$

The graph of the solution set is

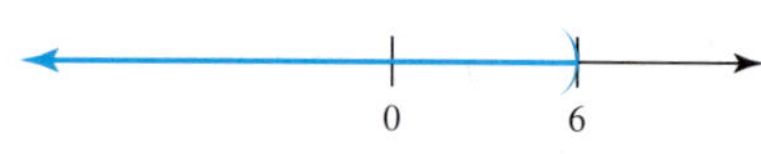

(b) Solve and graph the solution set of $-4x \geq 28$.

In this case we want to multiply both sides of the inequality by $-\frac{1}{4}$ to convert the coefficient of x to 1 on the left.

$$\left(-\frac{1}{4}\right)(-4x) \leq \left(-\frac{1}{4}\right)(28)$$

Reverse the direction of the inequality because you are multiplying by a negative number!

or $\quad x \leq -7$

The graph of the solution set is

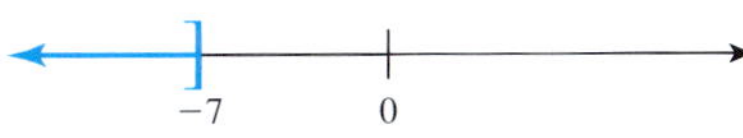

Check Yourself 8

Solve and graph the solution sets.

(a) $7x > 35$ **(b)** $-8x \leq 48$

Example 9 illustrates the use of the multiplication property when fractions are involved in an inequality.

Example 9 Solving and Graphing Inequalities

(a) Solve and graph the solution set of

$$\frac{x}{4} > 3$$

Here we multiply both sides of the inequality by 4. This isolates x on the left.

$$4\left(\frac{x}{4}\right) > 4(3)$$

$$x > 12$$

The graph of the solution set is

(b) Solve and graph the solution set of

$$-\frac{x}{6} \geq -3$$

In this case, we multiply both sides of the inequality by -6:

$$(-6)\left(-\frac{x}{6}\right) \leq (-6)(-3)$$

$$x \leq 18$$

NOTE

We reverse the direction of the inequality because we are multiplying by a negative number.

The graph of the solution set is

Check Yourself 9

Solve and graph the solution set of each inequality.

(a) $\frac{x}{5} \leq 4$ **(b)** $-\frac{x}{3} < -7$

We summarize our work of this and the previous sections by looking at the step-by-step procedure for solving an inequality in one variable. Note that the steps are nearly identical to those given to solve an equation in Section 1.6.

Step by Step

Solving a Linear Inequality in One Variable

Step 1 Remove any grouping symbols by applying the distributive property.

Step 2 Multiply both sides of the equation by the LCM to clear the inequality of fractions or decimals.

Step 3 Combine any like terms that appear on either side of the inequality.

Step 4 Apply the addition property of inequalities to write an equivalent inequality with the variable term on one side of the inequality and the constant term on the other.

Step 5 Apply the multiplication property to write an equivalent inequality with the variable isolated on one side of the inequality. Be sure to reverse the direction of the inequality if you multiply or divide by a negative number.

You should see the similarities and differences between equations and inequalities from the problems in the next example. Study them carefully and then complete Check Yourself 10 on your own.

Example 10 Solving and Graphing Inequalities

(a) Solve and graph the solution set of $5x - 3 < 2x$.

First, add 3 to both sides to undo the subtraction on the left.

$$5x - 3 < 2x$$

$$5x - 3 + 3 < 2x + 3 \quad \text{Add 3 to both sides to undo the subtraction.}$$

$$5x < 2x + 3$$

Now subtract $2x$, so that only the number remains on the right.

$$5x < 2x + 3$$

$$5x - 2x < 2x - 2x + 3 \quad \text{Subtract } 2x \text{ to isolate the number on the right.}$$

$$3x < 3$$

Next *divide* both sides by 3.

$$\frac{3x}{3} < \frac{3}{3}$$

$$x < 1$$

RECALL

The multiplication property also allows us to divide both sides by a nonzero number.

The graph of the solution set is

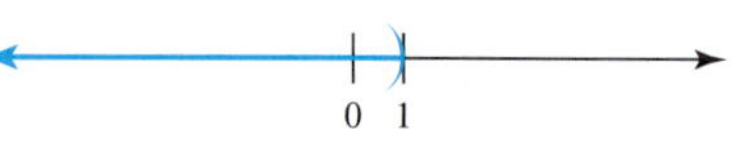

(b) Solve and graph the solution set of $2 - 5x < 7$.

$$2 - 5x < 7$$

$$2 - 2 - 5x < 7 - 2 \quad \text{Subtract 2.}$$

$$-5x < 5$$

$$\frac{-5x}{-5} > \frac{5}{-5} \quad \text{Divide by } -5\text{. Be sure to reverse the direction of the inequality.}$$

or $\quad x > -1$

The graph of the solution set is

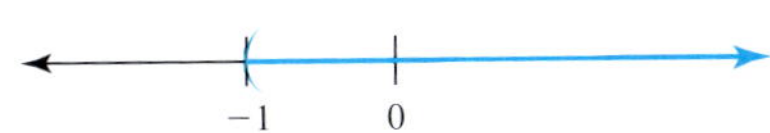

(c) Solve and graph the solution set of $5x - 5 \geq 3x + 4$.

$$5x - 5 \geq 3x + 4$$

$$5x - 5 + 5 \geq 3x + 4 + 5 \quad \text{Add 5.}$$

$$5x \geq 3x + 9$$

$$5x - 3x \geq 3x - 3x + 9 \quad \text{Subtract } 3x.$$

$$2x \geq 9$$

$$\frac{2x}{2} \geq \frac{9}{2} \quad \text{Divide by 2.}$$

$$x \geq \frac{9}{2}$$

The graph of the solution set is

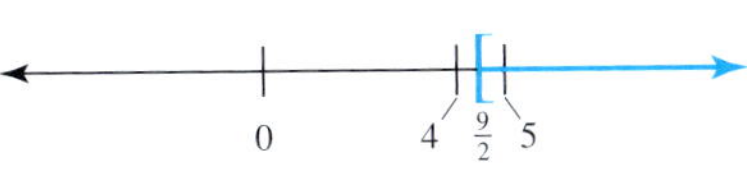

RECALL

Place $\frac{9}{2}$ on a number line in between 4 and 5.

(d) Solve and graph the solution set of $x + 2 < \frac{5}{2}x - 1$.

$$2(x + 2) < 2\left(\frac{5}{2}x - 1\right) \quad \text{Multiply by the LCD.}$$

$$2x + 4 < 5x - 2$$

$$2x + 4 - 4 < 5x - 2 - 4 \quad \text{Subtract 4.}$$

$$2x < 5x - 6$$

$$2x - 5x < 5x - 5x - 6 \quad \text{Subtract } 5x.$$

$$-3x < -6$$

Divide by -3, and reverse the direction of the inequality.

$$\frac{-3x}{-3} > \frac{-6}{-3}$$

$$x > 2$$

The graph of the solution set is

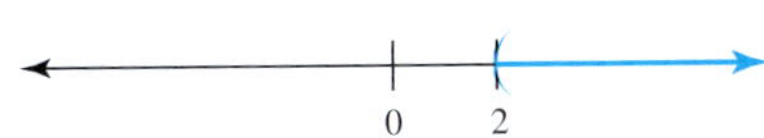

Check Yourself 10

Solve each inequality and graph each solution set.

(a) $4x + 9 \geq x$ **(b)** $5 - 6x < 41$

(c) $8x + 3 < 4x - 13$ **(d)** $5x + 12 \geq 10x - 8$

So far, we have represented our solution sets by graphing them on a number line. In Chapter 2, you will learn to present these solution sets algebraically by using *set-builder* and *interval* notations.

Check Yourself ANSWERS

1. (a) >; (b) < 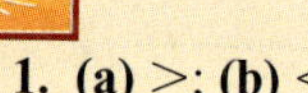 2. 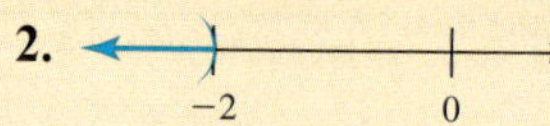

3. (a) ; (b)

4. $\{x \mid x > 6\}$

5. $\{x \mid x \le 10\}$

6. $\{x \mid x > 4\}$

7. She needs a score of 87 or greater.

8. (a)

(b)

9. (a)

(b)

10. (a)

(b)

(c)

(d)

Reading Your Text 

SECTION 1.8

(a) $9 > 6$ is read "9 is ____________ than 6."

(b) Adding the same quantity to both sides of an inequality yields an ____________ inequality.

(c) Multiplying both sides of an inequality by a ____________ number yields an equivalent inequality.

(d) Multiplying both sides of an inequality by a ____________ number yields an equivalent inequality only if we also reverse the direction of the inequality sign.

Basic Skills | Challenge Yourself | Calculator/Computer | Career Applications | Above and Beyond

1.8 exercises

< Objective 1 >

Complete the statements, using the symbol < or >.

1. 5 ______ 10

2. 9 ______ 8

3. 7 ______ -2

4. 0 ______ -5

5. 0 ______ 4

6. -10 ______ -5

7. -2 ______ -5 > Videos

8. -4 ______ -11

Write each inequality in words.

9. $x < 3$

10. $x \le -5$

11. $x \ge -4$

12. $x < -2$

13. $-5 \le x$

14. $2 < x$

< Objective 2 >

Graph the solution set of each inequality.

15. $x > 2$

16. $x < -3$

17. $x < 6$

18. $x > 4$

19. $x > 1$

20. $x < -2$

21. $x < 8$

22. $x > 3$

23. $x > -5$

24. $x < -2$

25. $x \ge 9$

26. $x \ge 0$ > Videos

Name ______________

Section ________ Date ________

Answers

1. ______ 2. ______
3. ______ 4. ______
5. ______ 6. ______
7. ______ 8. ______
9. ______
10. ______
11. ______
12. ______
13. ______
14. ______
15. ______
16. ______
17. ______
18. ______
19. ______
20. ______
21. ______
22. ______
23. ______
24. ______
25. ______
26. ______

Answers

27.
28.
29.
30.
31.
32.
33.
34.
35.
36.
37.
38.
39.
40.
41.
42.
43.
44.
45.
46.

27. $x < 0$

28. $x \le -3$

< Objectives 3 and 4 >

Solve and graph the solution set of each inequality.

29. $x - 8 \le 3$

30. $x + 5 \le 4$

31. $x + 8 \ge 10$

32. $x - 11 > -14$

33. $5x < 4x + 7$ 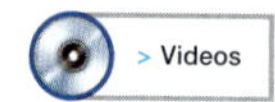

34. $8x \ge 7x - 4$

35. $6x - 8 \le 5x$

36. $3x + 2 > 2x$

37. $8x + 1 \ge 7x + 9$

38. $5x + 2 \le 4x - 6$

39. $7x + 5 < 6x - 4$

40. $8x - 7 > 7x + 3$

41. $\frac{3}{4}x - 5 \ge 7 - \frac{1}{4}x$

42. $\frac{7}{8}x + 6 < 3 - \frac{1}{8}x$

43. $11 + 0.63x > 9 - 0.37x$

44. $0.54x + 0.12x + 9 \le 19 - 0.34x$

45. $3x \le 9$

46. $5x > 20$

47. $5x > -35$

48. $6x \le -18$

49. $-6x \ge 18$

50. $-9x < 45$

51. $-2x < -12$

52. $-12x \ge -48$

53. $\frac{x}{4} > 5$

54. $\frac{x}{3} \le -3$

55. $-\frac{x}{2} \ge -3$

56. $-\frac{x}{4} < 5$

57. $\frac{2x}{3} < 6$

58. $\frac{3x}{4} \ge -9$

59. $5x > 3x + 8$

60. $4x \le x - 9$

61. $5x - 2 < 3x$ 

62. $7x + 3 \ge 2x$

63. $3 - 2x > 5$

64. $5 - 3x \le 17$

65. $2x \ge 5x + 18$

66. $3x < 7x - 28$

67. $\frac{1}{3}x - 5 \le \frac{5}{3}x + 11$

68. $\frac{3}{7}x + 6 \ge -\frac{12}{7}x - 9$

Answers

47. ______
48. ______
49. ______
50. ______
51. ______
52. ______
53. ______
54. ______
55. ______
56. ______
57. ______
58. ______
59. ______
60. ______
61. ______
62. ______
63. ______
64. ______
65. ______
66. ______
67. ______
68. ______

Answers

69. ______
70. ______
71. ______
72. ______
73. ______
74. ______
75. ______
76. ______
77. ______
78. ______
79. ______
80. ______
81. ______
82. ______
83. ______
84. ______
85. ______
86. ______
87. ______

69. $0.34x + 21 \geq 19 - 1.66x$

70. $-1.57x - 15 \geq 1.43x + 18$

71. $7x - 5 < 3x + 2$

72. $5x - 2 \geq 2x - 7$

73. $5x + 7 > 8x - 17$

74. $4x - 3 \leq 9x + 27$

75. $3x - 2 \leq 5x + 3$ 

76. $2x + 3 > 8x - 2$

Translate each statement to an inequality. Let x represent the number in each case.

77. 5 more than a number is greater than 3.

78. 3 less than a number is less than or equal to 5.

79. 4 less than twice a number is less than or equal to 7. 

80. 10 more than a number is greater than negative 2.

81. 4 times a number, decreased by 15, is greater than that number.

82. 2 times a number, increased by 28, is less than or equal to 6 times that number.

Determine whether each statement is **true** *or* **false.**

83. A linear inequality in one variable can have an infinite number of solutions.

84. The statement $x < 5$ has the same solution set as the statement $5 < x$.

85. The solution set of $3 \geq x$ is the same as the solution set of $x \leq 3$.

86. If we add a negative number to both sides of an inequality, we must reverse the direction of the inequality symbol.

Complete each statement with **never, sometimes,** *or* **always.**

87. Adding the same quantity to both sides of an inequality ______ gives an equivalent inequality. > Videos

88. We can ______________ solve an inequality just by using the addition property of inequality.

89. When both sides of an inequality are multiplied by a negative number, the direction of the inequality symbol is ______________ reversed.

90. If the graph of the solution set for an inequality extends infinitely to the right, the solution set ______________ includes the number 0.

Match each inequality on the right with a statement on the left.

91. x is nonnegative. **(a)** $x \geq 0$

92. x is negative. **(b)** $x \geq 5$

93. x is no more than 5. **(c)** $x \leq 5$

94. x is positive. **(d)** $x > 0$

95. x is at least 5. **(e)** $x < 5$

96. x is less than 5. **(f)** $x < 0$

97. **STATISTICS** There are fewer than 1,000 wild giant pandas left in the bamboo forests of China. Write an inequality expressing this relationship.

98. **STATISTICS** Let C represent the amount of Canadian forest and M represent the amount of Mexican forest. Write an inequality showing the relationship of the forests of Mexico and Canada if Canada contains at least 9 times as much forest as Mexico.

99. **STATISTICS** To pass a course with a grade of B or better, Liza must have an average of 80 or more. Her grades on three tests are 72, 81, and 79. Write an inequality representing the scores that Liza must get on the fourth test to obtain a B average or better for the course.

100. **STATISTICS** Sam must average 70 or more in his summer course in order to obtain a grade of C. His first three test grades were 75, 63, and 68. Write an inequality representing the scores that Sam must get on the last test in order to earn a C grade. > Videos

101. **BUSINESS AND FINANCE** Juanita is a salesperson for a manufacturing company. She may choose to receive \$500 or 5% commission on her sales as payment for her work. Write an inequality representing the amounts she needs to sell to make the 5% offer a better deal.

102. **BUSINESS AND FINANCE** The cost for a long-distance telephone call is \$0.24 for the first minute and \$0.11 for each additional minute or portion thereof. The total cost of the call cannot exceed \$3. Write an inequality representing the number of minutes a person could talk without exceeding \$3.

103. **BUSINESS AND FINANCE** Samantha's financial aid stipulates that her tuition not exceed \$1,500 per semester. If her local community college charges a \$45 service fee plus \$290 per course, what is the greatest number of courses for which Samantha can register?

Answers

88. ______________

89. ______________

90. ______________

91. ______________

92. ______________

93. ______________

94. ______________

95. ______________

96. ______________

97. ______________

98. ______________

99. ______________

100. ______________

101. ______________

102. ______________

103. ______________

Answers

104. ______

105. ______

106. ______

107. ______

108. ______

109. ______

110. ______

111. ______

112. ______

113. ______

114. ______

104. **Statistics** Nadia is taking a mathematics course in which five tests are given. To get a B, a student must average at least 80 on the five tests. Nadia scored 78, 81, 76, and 84 on the first four tests. What score on the last test will earn her at least a B?

105. **Geometry** The width of a rectangle is fixed at 40 cm, and the perimeter can be no greater than 180 cm. Find the maximum length of the rectangle.

106. **Business and Finance** The women's soccer team can spend at most \$900 for its annual awards banquet. If the restaurant charges a \$75 setup fee and \$24 per person, at most how many people can attend?

107. **Business and Finance** Joyce is determined to spend no more than \$125 on clothes. She wants to buy two pairs of identical jeans and a blouse. If she spends \$29 on the blouse, what is the maximum amount she can spend on each pair of jeans?

108. **Business and Finance** Ben earns \$750 per month plus 4% commission on all his sales over \$900. Find the minimum sales that will allow Ben to earn at least \$2,500 per month.

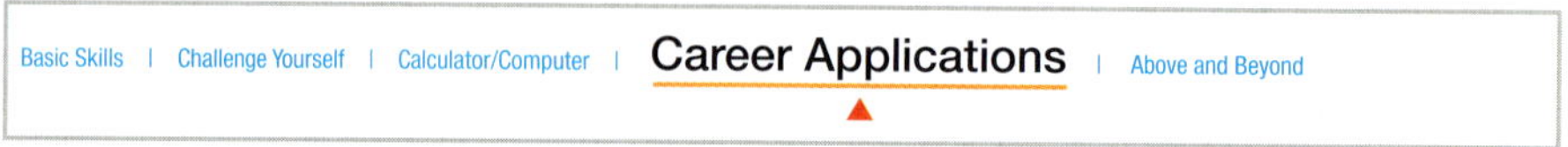

109. **Construction Technology** Pressure-treated wooden studs can be purchased for \$4.97 each. How many studs can be bought if a project's budget allots no more than \$250 for studs?

110. **Electronics Technology** Berndt Electronics earns a marginal profit of \$560 each on the sale of a particular server. If other costs involved amount to \$4,500, then how many servers does the company need to sell in order to earn a net profit of at least \$12,000?

Basic Skills | Challenge Yourself | Calculator/Computer | Career Applications | **Above and Beyond**

111. If an inequality simplifies to $7 > -5$, what is the solution set and why?

112. If an inequality simplifies to $7 < -5$, what is the solution set and why?

113. You are the office manager for a small company and need to acquire a new copier for the office. You find a suitable one that leases for \$250 per month from the copy machine company. It costs 2.5¢ per copy to run the machine. You purchase paper for \$3.50 per ream (500 sheets). If your copying budget is no more than \$950 per month, is this machine a good choice? Write a brief recommendation to the purchasing department. Use equations and inequalities to explain your recommendation.

114. Nutritionists recommend that, for good health, no more than 30% of our daily intake of calories come from fat. Algebraically, we can write this as $f \le 0.30(c)$, where f = calories from fat and c = total calories for the day. But this does not mean that everything we eat must meet this requirement.

For example, if you eat $\frac{1}{2}$ cup of Ben and Jerry's vanilla ice cream for dessert after lunch, you are eating a total of 250 calories, of which 150 are from fat. This amount is considerably more than 30% from fat, but if you are careful about what you eat the rest of the day, you can stay within the guidelines.

Set up an inequality based on your normal caloric intake. Solve the inequality to find how many calories in fat you could eat over the day and still have no more than 30% of your daily calories from fat. The American Heart Association says that to maintain your weight, your daily caloric intake should be 15 calories for every pound. You can compute this number to estimate the number of calories a day you normally eat. Do some research in your grocery store or library to determine what foods satisfy the requirements for your diet for the rest of the day. There are 9 calories in every gram of fat; many food labels give the amount of fat only in grams.

115. Your aunt calls to ask your help in making a decision about buying a new refrigerator. She says that she found two that seem to fit her needs, and both are supposed to last at least 14 years, according to *Consumer Reports.* The initial cost for one refrigerator is $712, but it uses only 88 kilowatt-hours (kWh) per month. The other refrigerator costs $519 and uses an estimated 100 kWh/ month. You do not know the price of electricity per kilowatt-hour where your aunt lives, so you will have to decide what, in cents per kilowatt-hour, will make the first refrigerator cheaper to run for its 14 years of expected usefulness. Write your aunt a letter, explaining what you did to calculate this cost, and tell her to make her decision based on how the kilowatt-hour rate she has to pay in her area compares with your estimation.

Answers

115. ________

Answers

1. $<$ **3.** $>$ **5.** $<$ **7.** $>$ **9.** x is less than 3
11. x is greater than or equal to -4 **13.** -5 is less than or equal to x

15. 
17. 
19. 
21. 
23. 
25. 
27. 
29. 
31. 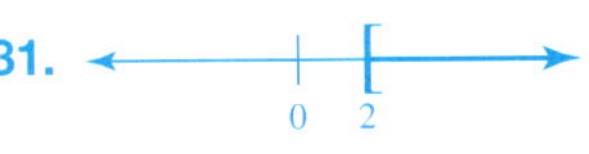
33. 
35. 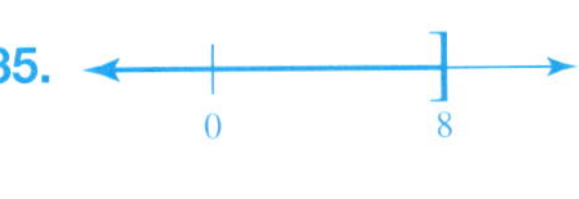
37. 
39. 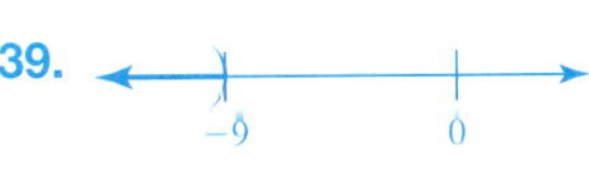
41. 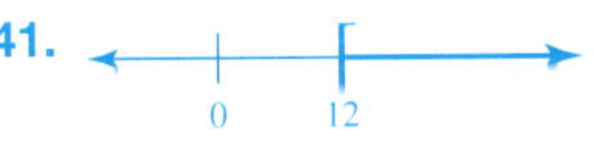
43. 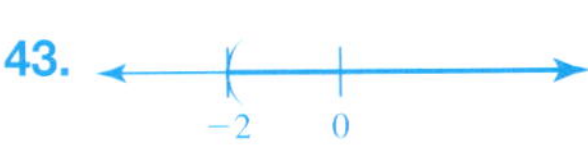
45. 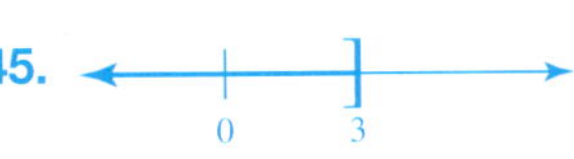

47. 

49. 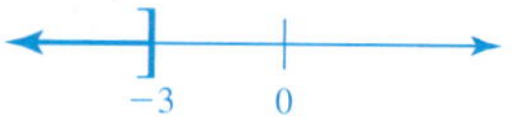

51. 

53. 

55. 

57. 

59. 

61. 

63. 

65. 

67. 

69. 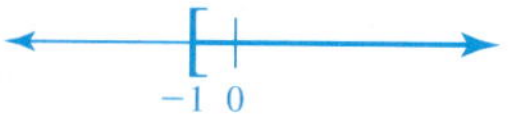

71. 

73. 

75. 

77. $x + 5 > 3$ **79.** $2x - 4 \leq 7$ **81.** $4x - 15 > x$ **83.** True
85. True **87.** always **89.** always **91.** (a) **93.** (c) **95.** (b)
97. $P < 1{,}000$ **99.** $x \geq 88$ **101.** Sales > \$10,000 **103.** 5 courses
105. 50 cm **107.** \$48 **109.** No more than 50 studs
111. Above and Beyond **113.** Above and Beyond

Definition/Procedure	Example	Reference
Transition to Algebra		**Section 1.1**
Addition $x + y$ means the **sum** of x **and** y or x **plus** y. Some other words indicating addition are *more than* and *increased by.*	The sum of x and 5 is $x + 5$. 7 more than a is $a + 7$. b increased by 3 is $b + 3$.	p. 73
Subtraction $x - y$ means the **difference** of x **and** y or x **minus** y. Some other words indicating subtraction are *less than* and *decreased by.*	The difference of x and 3 is $x - 3$. 5 less than p is $p - 5$. a decreased by 4 is $a - 4$.	p. 73
Multiplication $x \cdot y$, $(x)(y)$, xy — All these mean the **product** of x and y or x **times** y.	The product of m and n is mn. The product of 2 and the sum of a and b is $2(a + b)$.	p. 74
Division $\frac{x}{y}$ means x **divided by** y or the **quotient** when x is divided by y.	n divided by 5 is $\frac{n}{5}$. The sum of a and b, divided by 3, is $\frac{a+b}{3}$.	p. 75
Evaluating Algebraic Expressions		**Section 1.2**
To Evaluate an Algebraic Expression: **Step 1** Replace each variable by the given number value. **Step 2** Do the necessary arithmetic operations. (Be sure to follow the rules for the order of operations.)	Evaluate $\frac{4a-b}{2c}$ if $a = -6$, $b = 8$, and $c = -4$. $\frac{4a-b}{2c} = \frac{4(-6)-8}{2(-4)}$ $= \frac{-24-8}{-8}$ $= \frac{-32}{-8} = 4$	p. 85
Adding and Subtracting Algebraic Expressions		**Section 1.3**
Term A number or the product of a number and one or more variables and their exponents.	$3x^2y$ is a term.	p. 99
Like Terms Terms that contain exactly the same variables raised to the same powers.	$4a^2$ and $3a^2$ are like terms. $5x^2$ and $2xy^2$ are not like terms.	p. 100
Combining Like Terms **Step 1** Add or subtract the numerical coefficients. **Step 2** Attach the common variables.	$5a + 3a = 8a$ $7xy - 3xy = 4xy$	p. 101

Continued

Definition/Procedure	Example	Reference
Solving Algebraic Equations		**Sections 1.4–1.6**
Equation A statement that two expressions are equal.	$3x - 5 = 7$ is an equation.	*p.* 110
Solution A value for the variable that will make an equation a true statement.	4 is a solution for the equation because $3 \cdot 4 - 5 \stackrel{?}{=} 7$ $12 - 5 \stackrel{?}{=} 7$ $7 = 7$ True	*p.* 111
Equivalent Equations Equations that have exactly the same solutions.		*p.* 112
Writing Equivalent Equations There are two basic properties that will yield equivalent equations.		
Addition Property If $a = b$, then $a + c = b + c$. Adding (or subtracting) the same quantity on each side of an equation gives an equivalent equation.	If $x = y + 3$, then $x + 2 = y + 5$.	*p.* 112
Multiplication Property If $a = b$, then $ac = bc, c \neq 0$. Multiplying (or dividing) both sides of an equation by the same nonzero number gives an equivalent equation.	$5x = 20$ and $x = 4$ are equivalent equations.	*p.* 127
Solving Linear Equations We say that an equation is "solved" when we have an equivalent equation of the form $x = \square$ or $\square = x$ where $\square$ is some number The steps of solving a linear equation are as follows: **Step 1** Remove any grouping symbols by applying the distributive property. **Step 2** Multiply both sides of the equation by the LCM required to clear the equation of fractions or decimals. **Step 3** Combine any like terms that appear on either side of the equation. **Step 4** Apply the addition property of equality to write an equivalent equation with the variable term on *one side* of the equation and the constant term on the *other side.* **Step 5** Apply the multiplication property of equality to write an equivalent equation with the variable isolated on one side of the equation with coefficient 1. **Step 6** State the answer and check the solution in the *original* equation.	Solve: $3(x - 2) + 4x = 3x + 14$ $3x - 6 + 4x = 3x + 14$ $7x - 6 = 3x + 14$ $\quad + 6 \qquad + 6$ $7x = 3x + 20$ $-3x \qquad -3x$ $4x = 20$ $\frac{4x}{4} = \frac{20}{4}$ $x = 5$	*p.* 141
Literal Equations and Their Applications		**Section 1.7**
Literal Equation An equation that involves more than one letter or variable.	$a = \frac{2b + c}{3}$ is a literal equation.	*p.* 153

Definition/Procedure	Example	Reference
Solving Literal Equations **Step 1** Remove any grouping symbols by applying the distributive property. **Step 2** Multiply both sides of the equation by the LCM required to clear the equation of fractions or decimals. **Step 3** Combine any like terms that appear on either side of the equation. **Step 4** Apply the addition property of equality to write an equivalent equation with the variable term on *one side* of the equation and the constant term on the *other side.* **Step 5** Apply the multiplication property of equality to write an equivalent equation with the variable isolated on one side of the equation with coefficient 1.	Solve for b: $a = \frac{2b+c}{3}$ $3a = 3\left(\frac{2b+c}{3}\right)$ $3a = 2b + c$ $\underline{\quad - c \quad - c}$ $3a - c = 2b$ $\frac{3a-c}{2} = b$	p. 155
Applying Equations		
Using Equations to Solve Word Problems Follow these steps. **Step 1** Read the problem carefully. Then reread it to decide what you are asked to find. **Step 2** Choose a letter to represent one of the unknowns in the problem. Then represent each of the unknowns with an expression that uses the same letter. **Step 3** Translate the problem to the language of algebra to form an equation. **Step 4** Solve the equation. **Step 5** Answer the question and include units in your answer, when appropriate. Check your solution by returning to the original problem.		p. 156
Inequalities **Inequality** A statement that one quantity is less than (or greater than) another. Four symbols are used: $a < b$ ↑ a is less than b. $a > b$ ↑ a is greater than b. $a \le b$ ↑ a is less than or equal to b. $a \ge b$ ↑ a is greater than or equal to b.	$4 < 9$ $-1 > -6$ $2 \le 2$ $3 \ge -4$	Section 1.8 p. 169
Graphing Inequalities To graph $x < a$, we use a parenthesis and an arrow pointing left. [number line: arrow pointing left to parenthesis at a] To graph $x \ge b$, we use a bracket and an arrow pointing right. [number line: bracket at b with arrow pointing right]	$x < 6$ is graphed [number line: 0, 6; parenthesis at 6, arrow left] $x \ge 5$ [number line: 0, 5; bracket at 5, arrow right]	p. 170

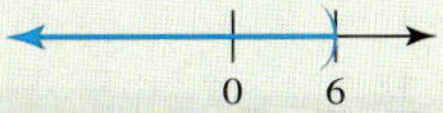

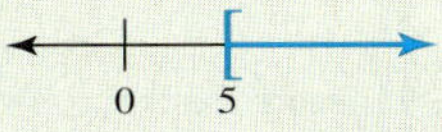

Continued

Definition/Procedure	Example	Reference
Solving Inequalities An inequality is "solved" when it is in the form $x < \square$ or $x > \square$. Proceed as in solving equations by using the following properties.	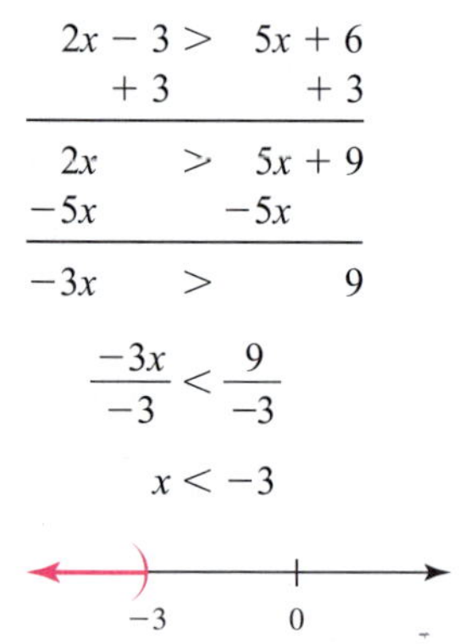	
Addition Property If $a < b$, then $a + c < b + c$. Adding (or subtracting) the same quantity to both sides of an inequality gives an equivalent inequality.		p. 171
Multiplication Property If $a < b$, then $ac < bc$ when $c > 0$ and $ac > bc$ when $c < 0$. Multiplying both sides of an inequality by the same *positive number* gives an equivalent inequality. When both sides of an inequality are multiplied by the same *negative number, you must reverse the direction* of the inequality to give an equivalent inequality.		p. 174

summary exercises :: chapter 1

This summary exercise set is provided to give you practice with each of the objectives of this chapter. Each exercise is keyed to the appropriate chapter section. When you are finished, you can check your answers to the odd-numbered exercises in the back of the text. If you have difficulty with any of these questions, go back and reread the examples from that section. The answers to the even-numbered exercises appear in the *Instructor's Solutions Manual.* Your instructor will give you guidelines on how best to use these exercises in your instructional setting.

1.1 *Write, using symbols.*

1. 8 more than y

2. c decreased by 10

3. The product of 8 and a

4. 5 times the product of m and n

5. The product of x and 7 less than x

6. 3 more than the product of 17 and x

7. The quotient when a plus 2 is divided by a minus 2

8. The product of 6 more than a number and 6 less than the same number

9. The quotient of 9 and a number

10. The product of a number and 3 more than twice the same number

1.2 *Evaluate the expressions if* $x = -3$, $y = 6$, $z = -4$, *and* $w = 2$.

11. $3x + w$

12. $5y - 4z$

13. $x + y - 3z$

14. $5z^2$

15. $5(x^2 - w^2)$

16. $\frac{6z}{2w}$

17. $\frac{2x - 4z}{y - z}$

18. $\frac{y(x - w)^2}{x^2 - 2xw + w^2}$

19. $-4x^2 - 2zw^2 + 4z$

20. $3x^3w^2 + xy^2$

1.3 *List the terms of the expressions.*

21. $4a^3 - 3a^2$

22. $5x^2 - 7x + 3$

Circle like terms.

23. $5m^2, -3m, -4m^2, 5m^3, m^2$

24. $4ab^2, 3b^2, -5a, ab^2, 7a^2, -3ab^2, 4a^2b$

Combine like terms.

25. $9x + 7x$

26. $2x + 5x$

27. $9xy - 6xy$

28. $5ab^2 + 2ab^2$

29. $7a + 3b + 12a - 2b$

30. $-3x + 2y - 5x - 7y$

31. $5x^3 + 17x^2 - 2x^3 - 8x^2$

32. $3a^3 + 5a^2 + 4a - 2a^3 - 3a^2 - a$

33. Subtract $4a^3$ from the sum of $2a^3$ and $12a^3$.

34. Subtract the sum of $3x^2$ and $5x^2$ from $15x^2$.

Write an expression for each exercise.

35. **Construction** If x feet (ft) is cut off the end of a board that is 37 ft long, how much is left?

36. **Business and Finance** Sergei has 25 nickels and dimes in his pocket. If x of these are dimes, how many of the coins are nickels?

37. **Geometry** The length of a rectangle is 4 meters (m) more than the width. Write an expression for the length of the rectangle.

38. **Number Problem** A number is 7 less than 6 times the number n. Write an expression for the number.

39. **Construction** A 25-ft plank is cut into two pieces. Write expressions for the length of each piece.

40. **Business and Finance** Bernie has d dimes and q quarters in his pocket. Write an expression for the amount of money (in dollars) that Bernie has in his pocket.

41. **Geometry** Find the perimeter of the given rectangle.

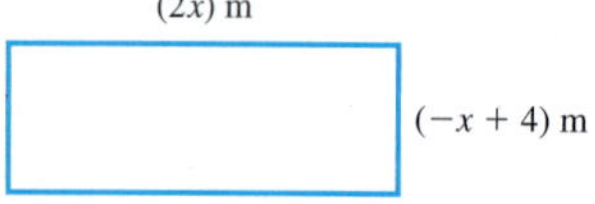

42. **Geometry** If the length of a building is x m and the width is $\frac{x}{6}$ m, what is the perimeter of the building?

1.4 *Determine whether the number shown in parentheses is a solution for the given equation.*

43. $5x - 3 = 7 \quad (2)$

44. $5x - 8 = 3x + 2 \quad (4)$

45. $7x - 2 = 2x + 8 \quad (2)$

46. $\frac{2}{3}x - 2 = 10 \quad (21)$

Solve each equation and check your results.

47. $x + 3 = 5$

48. $x - 9 = 3$

49. $5x = 4x - 5$

50. $4x - 9 = 3x$

51. $9x - 7 = 8x - 6$

52. $3 + 4x - 1 = x - 7 + 2x$

53. $4(2x + 3) = 7x + 5$

54. $5(5x - 3) = 6(4x + 1)$

1.5–1.6 *Solve each equation and check your results.*

55. $5x = 35$

56. $7x = -28$

57. $-9x = 36$

58. $-9x = -63$

59. $\frac{2}{3}x = 18$

60. $\frac{7}{8}x = 28$

61. $7x + 8 = 3x$

62. $3 - 5x = -17$

63. $4x - 7 = 2x$

64. $2 - 4x = 5$

65. $\frac{x}{3} - 5 = 1$

66. $\frac{3}{4}x - 2 = 7$

67. $7x + 4 = 2x + 6$

68. $9x - 8 = 7x - 3$

69. $2x + 7 = 4x - 5$

70. $3x - 15 = 7x - 10$

71. $\frac{10}{3}x - 5 = \frac{4}{3}x + 7$

72. $\frac{11}{4}x - 15 = 5 - \frac{5}{4}x$

73. $3.7x + 8 = 1.7x + 16$

74. $2.4x + 6 - 1.2x = 9 - 1.8x + 12$

75. $5(3x - 1) - 6x = 3x - 2$

76. $5x + 2(3x - 4) = 14x - 7$

77. $8x - 5(x + 3) = -10$

78. $3(2x - 5) - 2(x - 3) = 11$

79. $\frac{2x}{3} - \frac{x}{4} = 5$

80. $\frac{3x}{4} - \frac{2x}{5} = 7$

81. $\frac{x}{2} - \frac{x + 1}{3} = \frac{1}{6}$

82. $\frac{x + 1}{5} - \frac{x - 6}{3} = \frac{1}{3}$

1.7 *Solve for the indicated variable.*

83. $V = LWH$ (for W)

84. $P = 2L + 2W$ (for L)

85. $ax + by = c$ (for y)

86. $A = \frac{1}{2}bh$ (for h)

87. $A = P + Prt$ (for t)

88. $m = \frac{n - p}{q}$ (for p)

Solve each word problem.

89. Number Problem The sum of 3 times a number and 7 is 25. What is the number?

90. Number Problem 5 times a number, decreased by 8, is 32. Find the number.

91. Number Problem If the sum of two consecutive integers is 85, find the two integers.

92. Problem Solving Larry is 2 years older than Susan, while Nathan is twice as old as Susan. If the sum of their ages is 30 years, find each of their ages.

93. Science and Medicine Lisa left Friday morning, driving on the freeway to visit friends for the weekend. Her trip took 4 h. When she returned on Sunday, heavier traffic slowed her average speed by 6 mi/h, and the trip took $4\frac{1}{2}$ h. What was her average speed in each direction, and how far did she travel each way?

94. Science and Medicine At 9 A.M., David left New Orleans, Louisiana, for Tallahassee, Florida, averaging 47 mi/h. Two hours later, Gloria left Tallahassee for New Orleans along the same route, driving 5 mi/h faster than David. If the two cities are 391 mi apart, at what time will David and Gloria meet?

95. Business and Finance A firm producing running shoes finds that its fixed costs are $3,900 per week, and its variable cost is $21 per pair of shoes. If the firm can sell the shoes for $47 per pair, how many pairs of shoes must be produced and sold each week for the company to break even?

1.8 *Graph the solution sets.*

96. $x > 5$

97. $x \le -4$

98. $x \ge 9$

99. $x < 0$

100. $x - 2 \le 9$

101. $5x > 4x - 3$

102. $4x \ge -12$

103. $-\frac{x}{3} \ge 5$

104. $2x \le 8x - 3$

105. $7 - 6x > 15$

106. $5x - 2 \le 4x + 5$

107. $4x - 2 < 7x + 16$

self-test 1

Name ____________________

Section ________ Date ________

Answers

1. ________ 2. ________

3. ________ 4. ________

5. ________ 6. ________

7. ________ 8. ________

9. ________ 10. ________

11. ________ 12. ________

13. ________ 14. ________

15. ________ 16. ________

17. ________ 18. ________

19. ________ 20. ________

21. ____________________

22. ____________________

23. ____________________

24. 

25. 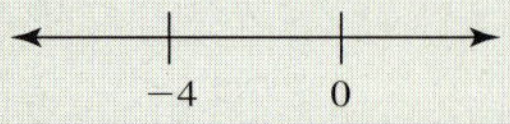

The purpose of this self-test is to help you assess your progress so that you can find concepts that you need to review before the next exam. Allow yourself about an hour to take this test. At the end of that hour, check your answers against those given in the back of this text. If you miss any, go back to the appropriate section to reread the examples until you have mastered that particular concept.

Write in symbols.

1. The sum of x and y

2. The difference m minus n

3. The product of a and b

4. The quotient when p is divided by 3 less than q

5. 5 less than c

6. The product of 3 and the quantity $2x$ minus $3y$

7. 3 times the difference of m and n

Evaluate when $x = -4$.

8. $-4x - 12$

9. $3x^2 + 2x - 4$

Evaluate each expression if $a = -2$, $b = 6$, and $c = -4$.

10. $4a - c$

11. $\dfrac{3a - 4b}{a + c}$

Combine like terms.

12. $8a - 3b - 5a + 2b$

13. $7x^2 - 3x + 2 - (5x^2 - 3x - 6)$

Tell whether the number shown in parentheses is a solution for the given equation.

14. $7x - 3 = 25 \quad (5)$

15. $8x - 3 = 5x + 9 \quad (4)$

Solve each equation and check your results.

16. $7x - 12 = 6x$

17. $\dfrac{4}{5}x = 24$

18. $5x - 3(x - 5) = 19$

19. $\dfrac{x - 5}{3} = \dfrac{5}{4}$

Solve for the indicated variable.

20. $V = \dfrac{1}{3}Bh$ (for B)

Solve each word problem.

21. 5 times a number, decreased by 7, is 28. What is the number?

22. Jan is twice as old as Juwan, while Rick is 5 years older than Jan. If the sum of their ages is 35 years, find each of their ages.

23. At 10 A.M., Sandra left her house on a business trip and drove an average of 45 mi/h. One hour later, Adam discovered that Sandra had left her briefcase behind, and he began driving at 55 mi/h along the same route. When will Adam catch up with Sandra?

Solve and graph the solution set of each inequality.

24. $x - 5 \le 9$

25. $5 - 3x > 17$

Functions and Graphs

INTRODUCTION

Math is used in so many places that, although we try to provide our readers with a variety of applications, we can touch on only a few of the settings and fields in which mathematics is applied.

Though the methods learned in introductory algebra have not changed, the technology associated with "doing mathematics" is different. Today, the power of math comes from the use of functions to model applications. We can concentrate on understanding the function model precisely because the tools and technology enhance our experience with "doing mathematics."

In Activity 2, we introduce you to many of the features of graphing calculators. If you have not had the opportunity to use a graphing calculator, we suggest that you work through the activity in this chapter. If you have had experience with a graphing calculator, you will undoubtedly agree that it is a very helpful tool for examining and understanding the function model.

CHAPTER 2 OUTLINE

2.1 Sets and Set Notation

< 2.1 Objectives >

1 > Write a set, using the roster method
2 > Write a set, using set-builder notation
3 > Write a set, using interval notation
4 > Plot the elements of a set on a number line
5 > Describe the solution set of an inequality
6 > Find the union and intersection of sets

For his birthday, Jacob received a jacket, a ticket to a play, some candy, and a pen. We could call this collection of gifts "Jacob's presents." Such a collection is called a **set.** The things in the set are called **elements** of the set. We can write the set as {jacket, ticket, candy, pen}. The braces tell us where the set begins and ends. Every person could have a set that describes the presents she or he received on their last birthday.

RECALL

We first introduced these *empty-set* notations in Section 1.6.

What if I received no presents on my last birthday? What would my set look like? It would be the set { }, which we call the **empty set.** Sometimes the symbol ∅ is used to indicate the empty set.

Many sets can be written in *roster form,* as was the case with Jacob's presents. The set of prime numbers less than 15 can be written in roster form as {2, 3, 5, 7, 11, 13}. In Example 1, we list some sets in roster form. **Roster form** is a list enclosed in braces.

Example 1 Listing the Elements of a Set

< Objective 1 >

Use the roster form to list the elements of each set described.

(a) The set of all factors of 12

The set of factors is {1, 2, 3, 4, 6, 12}.

(b) The set of all integers with an absolute value less than 4

The set of integers is {−3, −2, −1, 0, 1, 2, 3}.

Check Yourself 1

Use the roster form to list the elements of each set described.

(a) The set of all factors of 18 **(b)** The set of all even prime numbers

Each set that we examined had a limited number of elements. If we need to indicate that a set continues in some pattern, we use three dots, called an *ellipsis,* to indicate that the set continues with the pattern it started.

Example 2 Listing the Elements of a Set

Use the roster form to list the elements of each set described.

(a) The set of all natural numbers less than 100

The set {1, 2, 3, . . . , 98, 99} indicates that we continue increasing the numbers by 1 until we get to 99.

(b) The set of all positive multiples of 4

The set $\{4, 8, 12, 16, \ldots\}$ indicates that we continue counting by fours forever. (There is no indicated stopping point.)

(c) The set of all integers

$\{\ldots, -2, -1, 0, 1, 2, \ldots\}$ indicates that we continue forever in both directions.

Check Yourself 2

Use the roster form to list the elements of each set described.

(a) The set of all natural numbers between 200 and 300
(b) The set of all positive multiples of 3
(c) The set of all even numbers

NOTE

A statement such as

$1 < x < 2$

is called a *compound inequality.* It says that x is greater than 1 and also that x is less than 2.

Not all sets can be described using the roster form. What if we want to describe all the real numbers between 1 and 2? We could not list that set of numbers. Yet another way that we can describe the elements of a set is with **set-builder notation.** To describe the aforementioned set using this notation, we write

$\{x \mid 1 < x < 2\}$

We read this as "the set of all x, where x is between 1 and 2." Note that neither 1 nor 2 is included in this set.

Example 3 further illustrates this idea.

Example 3 Using Set-Builder Notation

< Objective 2 >

Use set-builder notation for each set described.

(a) The set of all real numbers less than 100

We write $\{x \mid x < 100\}$.

(b) The set of all real numbers greater than -4 but less than or equal to 9

$\{x \mid -4 < x \le 9\}$

The symbol $\le$ is a combination of the symbols $<$ and $=$. When we write $x \le 9$, we are indicating that either x is equal to 9 or it is less than 9.

Check Yourself 3

Use set-builder notation for each set described.

(a) The set of all real numbers greater than -2
(b) The set of all real numbers between 3 and 10 (inclusive)

Another notation that can be used to describe a set is called **interval notation.** For example, all the real numbers between 1 and 2 would be written as (1, 2). Note that the parentheses are used since neither 1 nor 2 is included in the set.

Interval notation should feel familiar based on your work graphing the solution set of an inequality on a number line in Section 1.8. You are simply "removing" the number line from the notation.

(1, 2)

With number line Without number line

Example 4 Using Interval Notation

< Objective 3 >

Use interval notation to represent each set described.

(a) The set of real numbers between 4 and 5

We write (4, 5).

(b) The set of real numbers greater than −3 but less than or equal to 9

We write (−3, 9].

A square bracket is used at 9 to indicate that 9 is included in the interval while a parenthesis is used at −3 because −3 is not part of the interval.

(c) The set of all real numbers greater than or equal to 45

We write [45, ∞).

The positive infinity symbol ∞ does not indicate a number. It is used to show that the interval includes all real numbers greater than or equal to 45.

(d) The set of all real numbers less than 15

We write (−∞, 15).

The negative infinity symbol −∞ is used to show that the interval includes all real numbers less than 15.

NOTE

Again, looking at interval notation in terms of a number line, ∞ means that we would shade in the number line as far as it goes: [45, ∞).

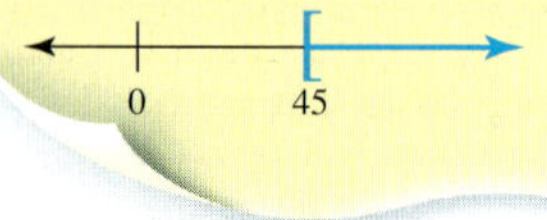

Check Yourself 4

Use interval notation to describe each set.

(a) The set of all real numbers less than 75
(b) The set of all real numbers between 5 and 10
(c) The set of all real numbers greater than 60
(d) The set of all real numbers greater than or equal to 23 but less than or equal to 38

Sets of numbers can also be represented graphically. In Example 5, we look at the connection between sets and their graphs.

Example 5 Plotting the Elements of a Set on a Number Line

< Objective 4 >

Plot the elements of each set on the number line.

(a) {−2, 1, 5}

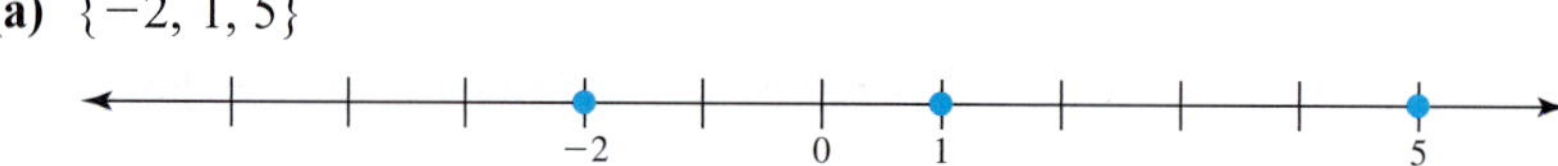

(b) $\{x \mid x < 3\}$

Note that the blue line and blue arrow indicate that we continue forever in the negative direction. The parenthesis at 3 indicates that the 3 is not part of the graph.

(c) $\{x \mid -2 < x < 5\}$

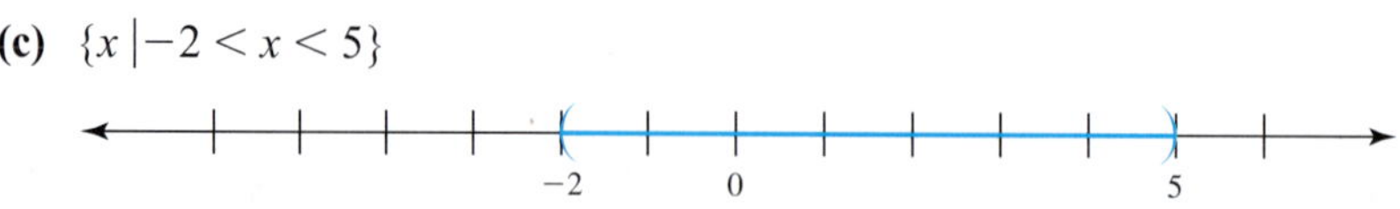

The parentheses indicate that the numbers −2 and 5 are not part of the set.

(d) $\{x \mid x \geq -2\}$

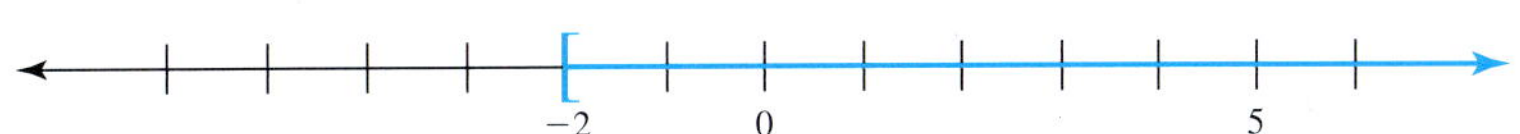

The bracket indicates that -2 is part of the set that is graphed.

Check Yourself 5

Plot the elements of each set on a number line.

(a) $\{-5, -3, 0\}$ **(b)** $\{x \mid -3 < x < -1\}$ **(c)** $\{x \mid x \leq 5\}$

This table summarizes the different ways of describing a set.

Basic Set Notation (*a* and *b* represent any real numbers)

Set	Set-Builder Notation	Interval Notation	Graph
All real numbers greater than b	$\{x \mid x > b\}$	(b, ∞)	
All real numbers less than or equal to b	$\{x \mid x \leq b\}$	$(-\infty, b]$	
All real numbers greater than a and less than b	$\{x \mid a < x < b\}$	(a, b)	
All real numbers greater than or equal to a and less than b	$\{x \mid a \leq x < b\}$	$[a, b)$	

Example 6 Using Set Notation

Express the set represented by each graph in both set-builder and interval notation.

(a)

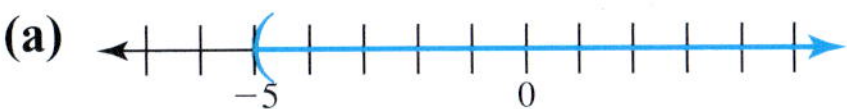

In set-builder notation the set is $\{x \mid x > -5\}$.

In interval notation it is $(-5, \infty)$.

(b)

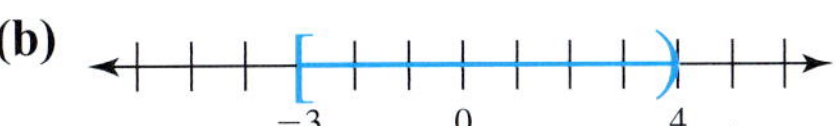

In set-builder notation the set is $\{x \mid -3 \leq x < 4\}$.

In interval notation it is $[-3, 4)$.

Check Yourself 6

Express the set represented by each graph in both set-builder and interval notation.

(a)

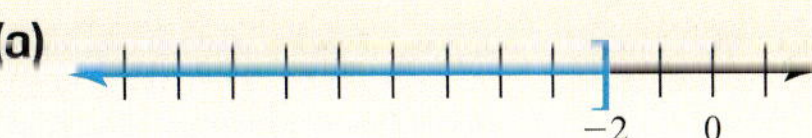

(b)

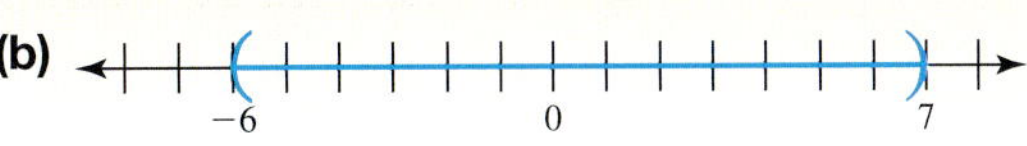

In Section 1.8, you learned to solve inequalities and to graph their *solution sets*. The language and notation of sets allow us to present the solution set of an inequality in other ways.

Example 7 Solving and Graphing Inequalities

< Objective 5 >

Solve each inequality. Represent each solution set using set-builder notation, interval notation, and with a graph, as appropriate.

(a) $5(x - 2) \geq -8$

Applying the distributive property on the left yields

$5x - 10 \geq -8$

Solving as before yields

$$5x - 10 + 10 \geq -8 + 10 \quad \text{Add 10.}$$
$$5x \geq 2$$
$$\text{or} \quad x \geq \frac{2}{5} \quad \text{Divide by 5.}$$

The graph of the solution set, $\left\{x \mid x \geq \frac{2}{5}\right\}$, is

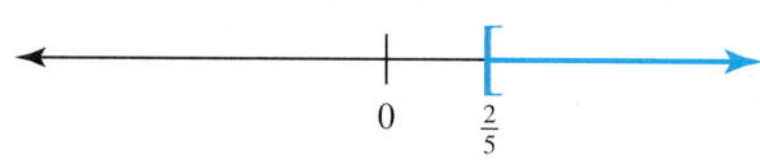

We write the solution set using interval notation as $\left[\frac{2}{5}, \infty\right)$.

(b) $-3(x + 2) \geq 5 - 3x$

$$-3x - 6 \geq 5 - 3x \quad \text{Apply the distributive property.}$$
$$0 - 6 \geq 5 \quad \text{Add } 3x \text{ to both sides.}$$
$$0 \geq 11 \quad \text{Add 6 to both sides.}$$

This is a false statement, so no real number satisfies this inequality or the original inequality. Thus, the solution set is the empty set, written { }. The graph of the solution set contains no points at all. (You might say that it is pointless!)

NOTE

When the answer is the empty set, we neither graph the solution nor use interval notation.

(c) $3(x + 2) > 3x - 4$

$$3x + 6 > 3x - 4 \quad \text{Apply the distributive property.}$$
$$6 > -4 \quad \text{Subtract } 3x \text{ from both sides.}$$

This is a true statement for all values of x, so this inequality and the original inequality are true for all real numbers.

The graph of the solution set $\{x \mid x \in \mathbb{R}\}$ is every point on the number line.

NOTE

Interval notation for the set of all real numbers is $(-\infty, \infty)$. We do not usually graph the solution set if it is the entire number line.

Check Yourself 7

Solve each inequality. Represent each solution set using set-builder notation, interval notation, and with a graph, as appropriate.

(a) $4(x + 3) < 9$ **(b)** $-2(4 - x) \leq 5 + 2x$ **(c)** $-4(x - 1) < 3 - 4x$

There are occasions when we need to combine sets. There are two commonly used operations to accomplish this: *union* and *intersection.*

Definition

Union and Intersection of Sets

The **union** of two sets *A* and *B*, written $A \cup B$, is the set of all elements that belong to either *A* or *B* or to both.

The **intersection** of two sets *A* and *B*, written $A \cap B$, is the set of all elements that belong to both *A* and *B*.

Example 8 Finding Union and Intersection

< Objective 6 >

Let $A = \{1, 3, 5\}$, $B = \{3, 5, 9\}$, and $C = \{9, 11\}$. List the elements in each of the following sets.

(a) $A \cup B$

This is the set of elements that are in *A* or *B* or in both.

$A \cup B = \{1, 3, 5, 9\}$

(b) $A \cap B$

This is the set of elements common to *A* and *B*.

$A \cap B = \{3, 5\}$

(c) $A \cup C$

This is the set of elements in *A* or *C* or in both.

$A \cup C = \{1, 3, 5, 9, 11\}$

(d) $A \cap C$

This is the set of elements common to *A* and *C*.

$A \cap C = \{\ \}$ or $\varnothing$ since there are no elements in common.

RECALL

$\{\ \}$ or $\varnothing$ is the symbol for the empty set.

Check Yourself 8

Let $A = \{2, 4, 7\}$, $B = \{4, 7, 10\}$, and $C = \{8, 12\}$. Find

(a) $A \cup B$ **(b)** $A \cap B$ **(c)** $A \cup C$ **(d)** $A \cap C$

Check Yourself ANSWERS

1. **(a)** $\{1, 2, 3, 6, 9, 18\}$; **(b)** $\{2\}$ 2. **(a)** $\{201, 202, 203, \ldots, 298, 299\}$; **(b)** $\{3, 6, 9, 12, \ldots\}$; **(c)** $\{\ldots, -6, -4, -2, 0, 2, 4, 6, \ldots\}$
3. **(a)** $\{x \mid x > -2\}$; **(b)** $\{x \mid 3 \le x \le 10\}$
4. **(a)** $(-\infty, 75)$; **(b)** $(5, 10)$; **(c)** $(60, \infty)$; **(d)** $[23, 38]$
5. **(a)** −5 −3 0
   (b) −3 −1 0
   (c) 0 5
6. **(a)** $\{x \mid x \le -2\}$, $(-\infty, -2]$; **(b)** $\{x \mid -6 < x < 7\}$, $(-6, 7)$
7. **(a)** $\left\{x \mid x < -\frac{3}{4}\right\}$, $\left(-\infty, -\frac{3}{4}\right)$, $-\frac{3}{4}$ 0
   (b) $\{x \mid x \in \mathbb{R}\}$, $(-\infty, \infty)$; **(c)** $\{\ \}$
8. **(a)** $\{2, 4, 7, 10\}$; **(b)** $\{4, 7\}$; **(c)** $\{2, 4, 7, 8, 12\}$; **(d)** $\varnothing$

Reading Your Text

SECTION 2.1

(a) The objects in a set are called the ____________ of the set.

(b) The symbol $\varnothing$ is often used to represent the ____________ set.

(c) The notation $\{x \mid x < 0\}$ is an example of ____________ notation.

(d) The notation in which the set of real numbers between 0 and 1 is written as (0, 1) is called ____________ notation.

2.1 exercises

Boost *your* GRADE at ALEKS.com!

ALEKS®

- Practice Problems
- Self-Tests
- NetTutor
- e-Professors
- Videos

Name ________________

Section ________ Date ________

< Objective 1 >

Use the roster method to list the elements of each set.

1. The set of all the days of the week

2. The set of all months of the year that have 31 days

3. The set of all factors of 18

4. The set of all factors of 24

5. The set of all prime numbers less than 30

6. The set of all prime numbers between 20 and 40

7. The set of all negative integers greater than -6

8. The set of all positive integers less than 6

9. The set of all even whole numbers less than 13

10. The set of all odd whole numbers less than 14

11. The set of integers greater than 2 and less than 7

12. The set of integers greater than 5 and less than 10

13. The set of integers greater than -4 and less than -1

14. The set of integers greater than -8 and less than -3

15. The set of integers between -5 and 2, inclusive 

16. The set of integers between 1 and 4

17. The set of odd whole numbers

Answers

1. ________________
2. ________________
3. ________________
4. ________________
5. ________________
6. ________________
7. ________________
8. ________________
9. ________________
10. ________________
11. ________________
12. ________________
13. ________________
14. ________________
15. ________________
16. ________________
17. ________________

Answers

18. ______

19. ______

20. ______

21. ______

22. ______

23. ______

24. ______

25. ______

26. ______

27. ______

28. ______

29. ______

30. ______

31. ______

32. ______

33. ______

34. ______

35. ______

36. ______

18. The set of even whole numbers

19. The set of all even whole numbers less than 100

20. The set of all odd whole numbers less than 100

21. The set of all positive multiples of 5

22. The set of all positive multiples of 6

< Objectives 2 and 3 >

Use set-builder notation and interval notation for each set described.

23. The set of all real numbers greater than 10

24. The set of all real numbers less than 25

25. The set of all real numbers greater than or equal to -5

26. The set of all real numbers less than or equal to -3 

27. The set of all real numbers greater than or equal to 2 and less than or equal to 7

28. The set of all real numbers greater than -3 and less than -1

29. The set of all real numbers between -4 and 4, inclusive

30. The set of all real numbers between -8 and 3, inclusive

< Objective 4 >

Plot the elements of each set on a number line.

31. $\{-2, -1, 0, 4\}$

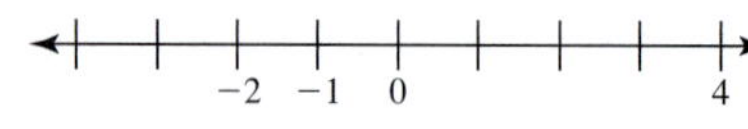

32. $\{-5, -1, 2, 3, 5\}$

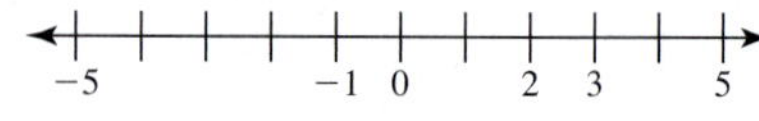

33. $\{x \mid x > 4\}$

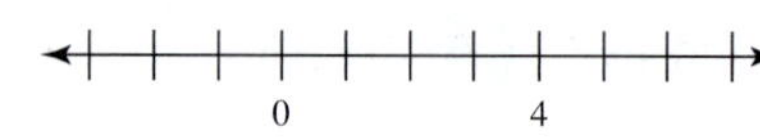

34. $\{x \mid x > -1\}$

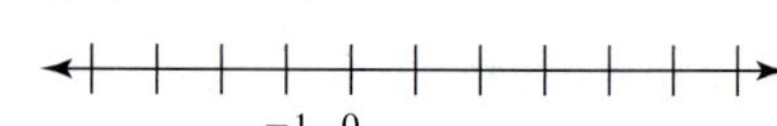

35. $\{x \mid x \geq -3\}$

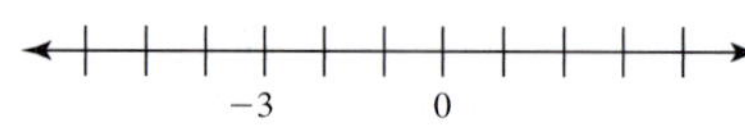

36. $\{x \mid x \leq 6\}$

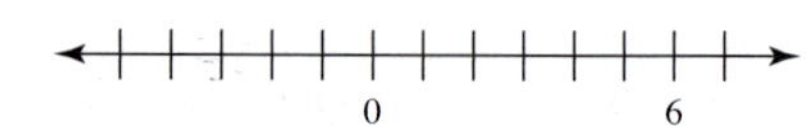

37. $\{x \mid 2 < x < 7\}$

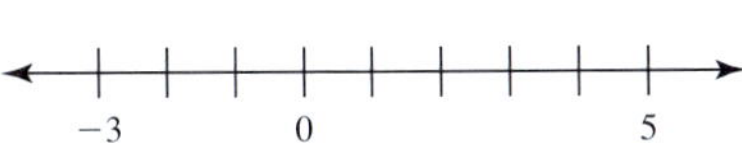

38. $\{x \mid 4 < x < 8\}$

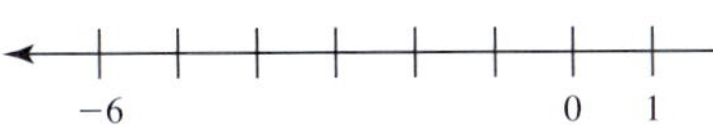

39. $\{x \mid -3 < x \le 5\}$

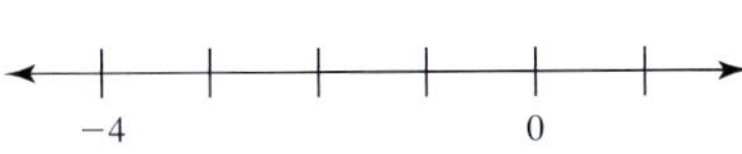

40. $\{x \mid -6 < x \le 1\}$

41. $\{x \mid -4 \le x < 0\}$

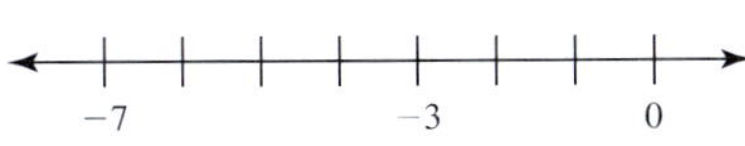

42. $\{x \mid -5 \le x < 2\}$

43. $\{x \mid -7 \le x \le -3\}$

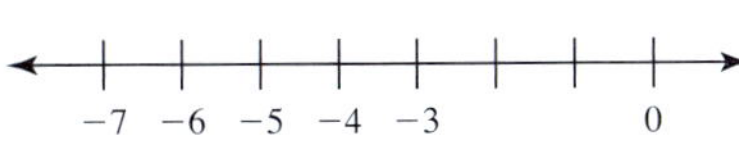

44. $\{x \mid -1 \le x \le 4\}$

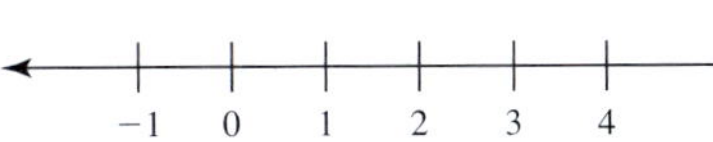

45. The set of all integers between -7 and -3, inclusive

46. The set of all integers between -1 and 4, inclusive

< Objective 5 >

Solve each inequality. Represent each solution set using set-builder notation, interval notation, and with a graph, as appropriate.

47. $3(x - 3) < 3$

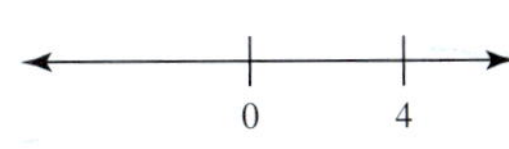

48. $3 - 2x \le 2$

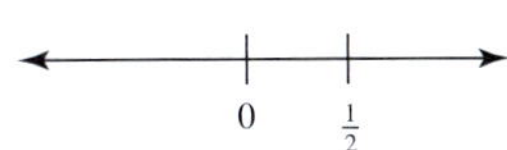

49. $-2(4x - 5) < 16$

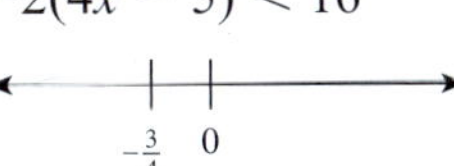

50. $5x + 4 \le 2x + 4$

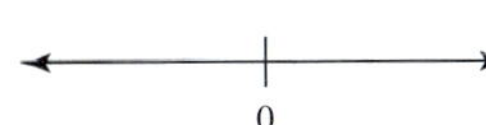

51. $-2(4 - x) \le 7 + 2x$

52. $6(x + 3) \ge 4 + 6x$

53. $-2(5 - x) \le -3(x + 2) + 5x$

54. $3(x + 5) \le 6(x + 2) - 3x$

Answers

37. ______
38. ______
39. ______
40. ______
41. ______
42. ______
43. ______
44. ______
45. ______
46. ______
47. ______
48. ______
49. ______
50. ______
51. ______
52. ______
53. ______
54. ______

Answers

55. ______

56. ______

57. ______

58. ______

59. ______

60. ______

61. ______

62. ______

63. ______

64. ______

65. ______

66. ______

67. ______

68. ______

69. ______

70. ______

71. ______

72. ______

73. ______

74. ______

75. ______

76. ______

77. ______

78. ______

Use set-builder notation and interval notation to describe each graphed set.

55. −4 −3 −2 −1 0 1 2 3 4

56. −3 −2 −1 0 1 2 3

57. **58.**

59. −3 −2 −1 0 1 2 3

60. −5 −4 −3 −2 −1 0 1 2

61. −3 −2 −1 0 1 2 3 4

62. −2 −1 0 1 2 3 4

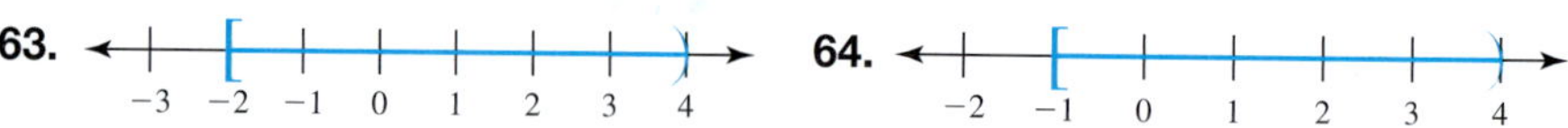

63. **64.**

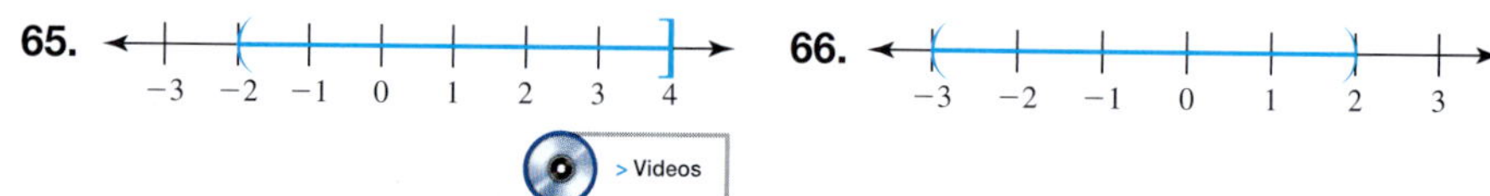

65. **66.**

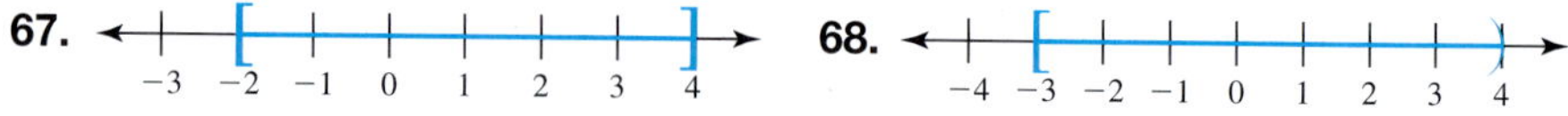

67. **68.**

Determine whether each statement is **true** *or* **false.**

69. The set of all even primes is finite.

70. We can list all the real numbers between 3 and 4 in roster form.

Complete each statement with **never, sometimes,** *or* **always.**

71. The intersection of two nonempty sets is ______ empty.

72. The union of two nonempty sets is ______ empty.

< Objective 6 >

In exercises 73 to 82, $A = \{x \mid x \text{ is an even natural number less than } 10\}$, $B = \{1, 3, 5, 7, 9\}$, *and* $C = \{1, 2, 3, 4, 5\}$. *List the elements in each set.*

73. $A \cup B$

74. $A \cap B$

75. $B \cap \varnothing$

76. $C \cup A$

77. $A \cup \varnothing$

78. $B \cup C$

79. $B \cap C$ > Videos

80. $C \cap A$

81. $(A \cup C) \cap B$

82. $A \cup (C \cap B)$

Basic Skills | Challenge Yourself | Calculator/Computer | Career Applications | **Above and Beyond**

83. Use the Internet to research the origin of the use of sets in mathematics.

chapter 2 > Make the Connection

Answers

1. {Monday, Tuesday, Wednesday, Thursday, Friday, Saturday, Sunday}
3. {1, 2, 3, 6, 9, 18} **5.** {2, 3, 5, 7, 11, 13, 17, 19, 23, 29}
7. {−5, −4, −3, −2, −1} **9.** {0, 2, 4, 6, 8, 10, 12} **11.** {3, 4, 5, 6}
13. {−3, −2} **15.** {−5, −4, −3, −2, −1, 0, 1, 2}
17. {1, 3, 5, 7, . . .} **19.** {0, 2, 4, 6, . . . , 96, 98}
21. {5, 10, 15, 20, . . .} **23.** $\{x \mid x > 10\}$; $(10, \infty)$
25. $\{x \mid x \geq -5\}$; $[-5, \infty)$ **27.** $\{x \mid 2 \leq x \leq 7\}$; $[2, 7]$
29. $\{x \mid -4 \leq x \leq 4\}$; $[-4, 4]$

31. 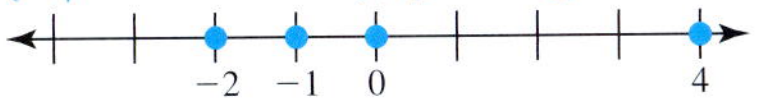

33. 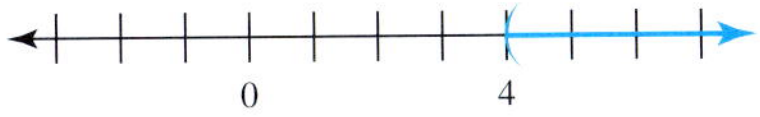

35. 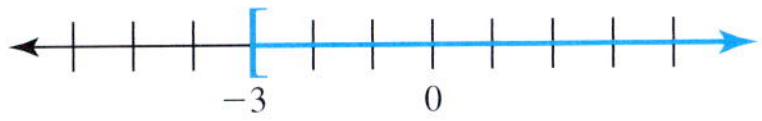

37. 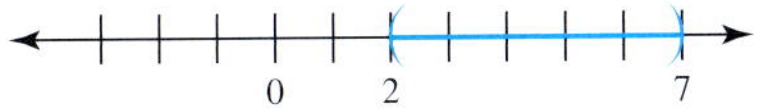

39. 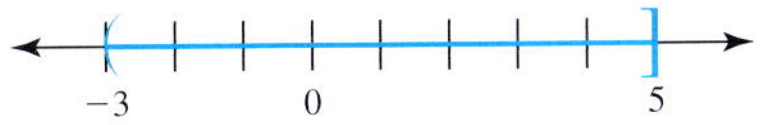

41. 

43. 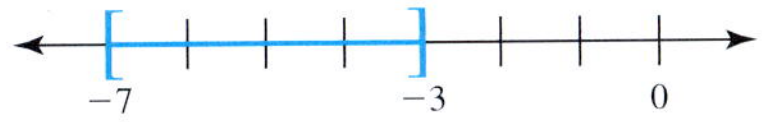

45. 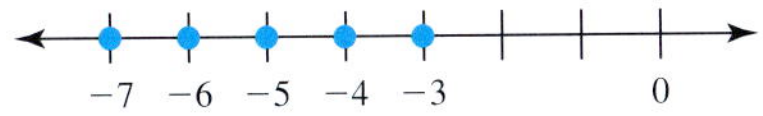

47. $\{x \mid x < 4\}$; $(-\infty, 4)$;

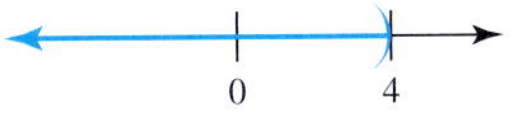

49. $\left\{x \mid x > -\frac{3}{4}\right\}$; $\left(-\frac{3}{4}, \infty\right)$;

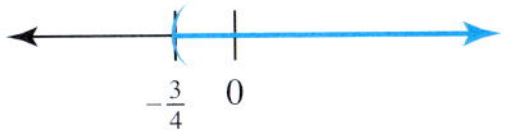

51. $\{x \mid x \in \mathbb{R}\}$; $(-\infty, \infty)$ **53.** $\{x \mid x \in \mathbb{R}\}$; $(-\infty, \infty)$
55. $\{x \mid x \leq 1\}$; $(-\infty, 1]$ **57.** $\{x \mid x > -2\}$; $(-2, \infty)$
59. $\{x \mid -2 \leq x \leq 2\}$; $[-2, 2]$ **61.** $\{x \mid -3 < x < 2\}$; $(-3, 2)$
63. $\{x \mid -2 \leq x < 4\}$; $[-2, 4)$ **65.** $\{x \mid -2 < x \leq 4\}$; $(-2, 4]$
67. $\{x \mid -2 \leq x \leq 4\}$; $[-2, 4]$ **69.** True **71.** sometimes
73. {1, 2, 3, 4, 5, 6, 7, 8, 9} **75.** ∅ **77.** {2, 4, 6, 8} **79.** {1, 3, 5}
81. {1, 3, 5} **83.** Above and Beyond

Answers

79. ____

80. ____

81. ____

82. ____

83. ____

Activity 2 :: Graphing with a Calculator

The graphing calculator is a tool that can be used to help you solve many different kinds of problems. This activity walks you through several features of the TI-83 or TI-84 Plus. By the time you complete this activity, you will be able to graph equations, change the viewing window to better accommodate a graph, or look at a table of values that represent some of the solutions for an equation. The first portion of this activity demonstrates how you can create the graph of an equation. The features described here can be found on most graphing calculators. See your calculator manual to learn how to get your particular calculator model to perform this activity.

Menus and Graphing

1. To graph the equation $y = 2x + 3$ on a graphing calculator, follow these steps.

a. Press the [Y =] key.

```
Plot1 Plot2 Plot3
\Y1=
\Y2=
\Y3=
\Y4=
\Y5=
\Y6=
\Y7=
```

b. Type $2x + 3$ at the Y_1 prompt. (This represents the first equation. You can type up to 10 separate equations.) Use the [X, T, θ, *n*] key for the variable.

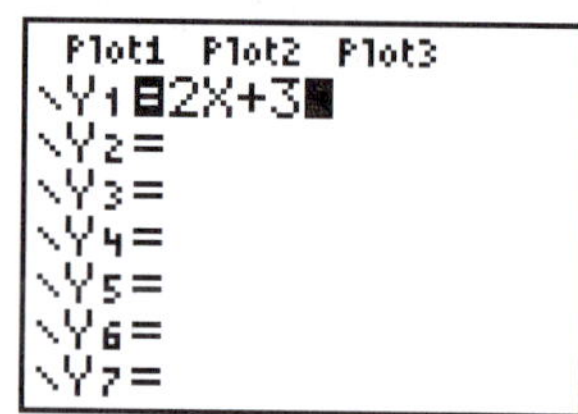

c. Press the [GRAPH] key to see the graph.

d. Press the [TRACE] key to display the equation. Once you have selected the [TRACE] key, you can use the left and right arrows of the calculator to move the cursor along the line. Experiment with this movement. Look at the coordinates at the bottom of the display screen as you move along the line.

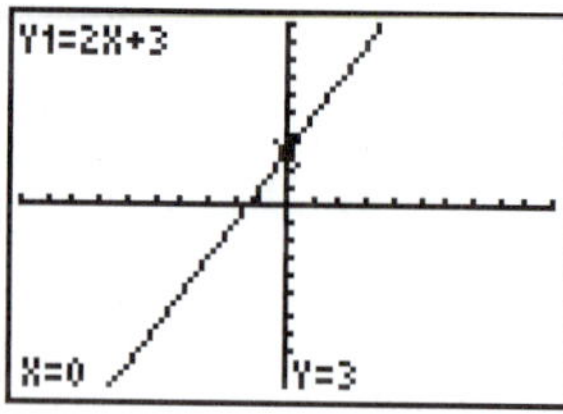

NOTE

Be sure the window is the standard window to see the same graph displayed.

Frequently, we can learn more about an equation if we look at a different section of the graph than the one offered on the display screen. The portion of the graph displayed is called the **window.** The second portion of the activity explains how this window can be changed.

2. Press the WINDOW key. The **standard** graphing screen is shown.

Xmin = left edge of screen
Xmax = right edge of screen
Xscl = scale given by each tick mark on x-axis
Ymin = bottom edge of screen
Ymax = top edge of screen
Yscl = scale given by each tick mark on y-axis
Xres = resolution (do not alter this)

```
WINDOW
 Xmin=-10
 Xmax=10
 Xscl=1
 Ymin=-10
 Ymax=10
 Yscl=1
 Xres=1
```

Note: To turn the scales off, enter a 0 for Xscl or Yscl. Do this when the intervals used are very large.

By changing the values for Xmin, Xmax, Ymin, and Ymax, you can adjust the viewing window. Change the viewing window so that Xmin = 0, Xmax = 40, Ymin = 0, and Ymax = 10. Again, press GRAPH. Notice that the tick marks along the x-axis are now much closer together. Changing Xscl from 1 to 5 will improve the display. Try it.

Sometimes we can learn something important about a graph by zooming in or zooming out. The third portion of this activity discusses this calculator feature.

3. a. Press the ZOOM key. There are 10 options. Use the ▼ key to scroll down.

```
ZOOM MEMORY
1:ZBox
2:Zoom In
3:Zoom Out
4:ZDecimal
5:ZSquare
6:ZStandard
7↓ZTrig
```

```
ZOOM MEMORY
4↑ZDecimal
5:ZSquare
6:ZStandard
7:ZTrig
8:ZInteger
9:ZoomStat
0:ZoomFit
```

b. Selecting the first option, ZBox, allows the user to enlarge the graph within a specified rectangle.

i. Graph the equation $y = x^2 + x - 1$ in the standard window. *Note:* To type in the exponent, use the x^2 key or the ∧ key.

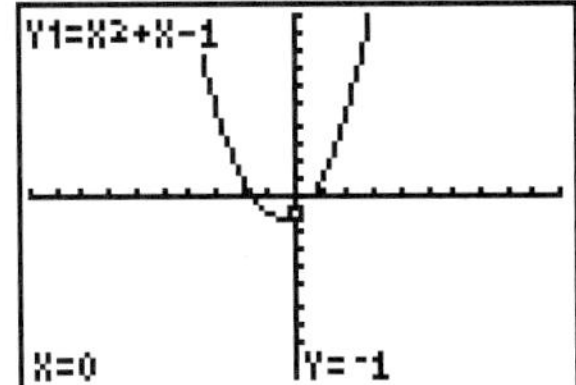

ii. When ZBox is selected, a blinking "+" cursor will appear in the graph window. Use the arrow keys to move the cursor to where you would like a corner of the screen to be; then press the ENTER key.

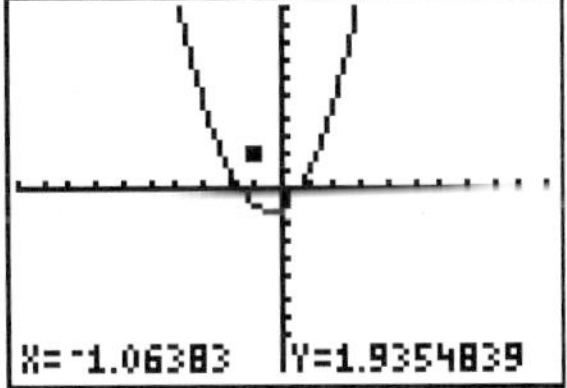

iii. Use the arrow keys to trace out the box containing the desired portion of the graph. Do not press the ENTER key until you have reached the diagonal corner and a full box is on your screen.

After using the down arrow

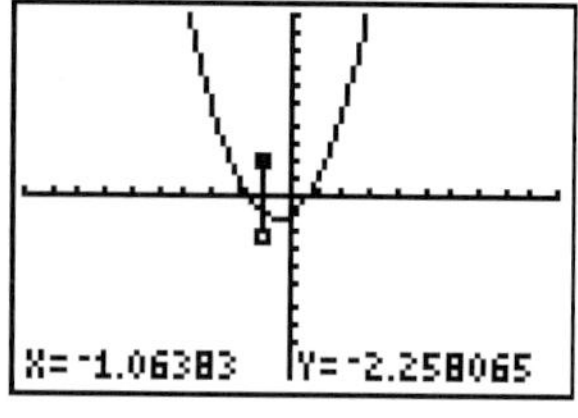

After using the right arrow

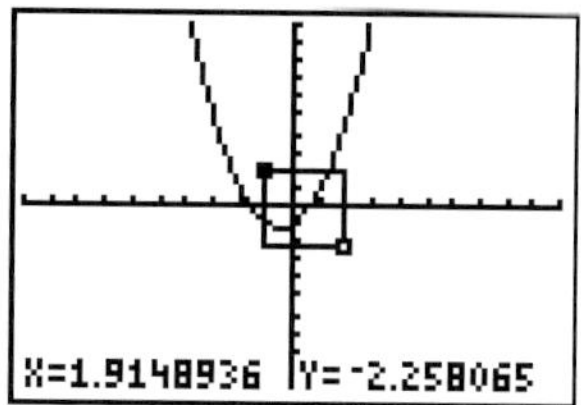

After pressing the ENTER *key a second time*

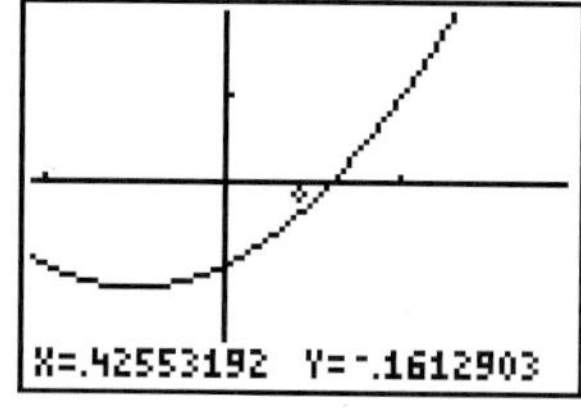

Now the desired portion of a graph can be seen more clearly.

The Zbox feature is especially useful when analyzing the roots (x-intercepts) of an equation.

c. Another feature that allows us to focus is Zoom In. Select the Zoom In option on the Zoom menu. Place the cursor in the center of the portion of the graph you are interested in and press the ENTER key. The window will reset with the cursor at the center of a zoomed-in view.

d. Zoom Out works like Zoom In, except that it sets the view larger (that is, it zooms out) to enable you to see a larger portion of the graph.

e. ZStandard sets the window to the standard window. This is a quick and convenient way to reset the viewing window.

f. ZSquare recalculates the view so that one horizontal unit is the same length as one vertical unit. This is sometimes necessary to get an accurate view of a graph because the width of the calculator screen is greater than its height.

4. Home Screen This is where all the basic computations take place. To get to the home screen from any other screen, press 2nd, Mode. This accesses the QUIT feature. To clear the home screen of calculations, press the CLEAR key (once or twice).

5. Tables The final feature that we look at here is the TABLE. Enter the equation $y = 2x + 3$ into the Y = menu. Then press 2nd, WINDOW to access the TBLSET menu. Set the table as shown here and press 2nd, GRAPH to access the TABLE feature. You will see the screens shown here.

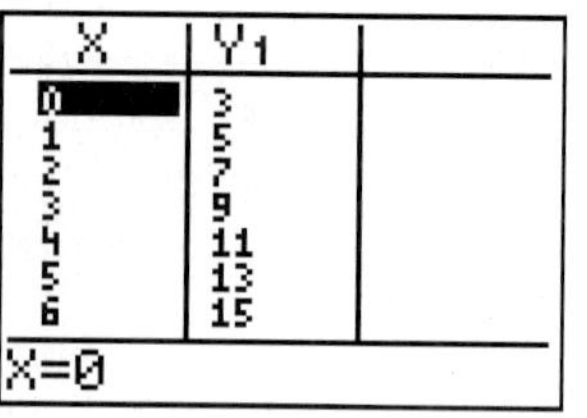

2.2 Solutions of Equations in Two Variables

< 2.2 Objectives >

1 > Identify solutions for an equation in two variables

2 > Use ordered-pair notation to write solutions for equations in two variables

RECALL

An equation is a statement that two expressions are equal.

We discussed finding solutions for equations in Section 1.4. Recall that a solution is a value for the variable that "satisfies" the equation, or makes the equation a true statement. For example, we know that 4 is a solution of the equation

$$2x + 5 = 13$$

because when we replace x with 4, we have

$$2(4) + 5 \stackrel{?}{=} 13$$

$$8 + 5 \stackrel{?}{=} 13$$

$$13 = 13 \quad \text{A true statement}$$

We now want to consider **equations in two variables.** In fact, in this chapter we will study equations of the form $Ax + By = C$, where A and B are not both 0. Such equations are called **linear equations in two variables,** and are said to be in **standard form.** An example is

$$x + y = 5$$

What does a solution look like? It is not going to be a single number, because there are two variables. Here a solution is a pair of numbers—one value for each of the variables x and y. Suppose that x has the value 3. In the equation $x + y = 5$, you can substitute 3 for x.

NOTE

An equation in two variables "pairs" two numbers, one for x and one for y.

$$(3) + y = 5$$

Solving for y gives

$$y = 2$$

So the pair of values $x = 3$ and $y = 2$ satisfies the equation because

$$(3) + (2) = 5$$

That pair of numbers is a *solution* for the equation in two variables.

Property

Equation in Two Variables

An **equation in two variables** is an equation for which *every* solution is a pair of values.

How many such pairs are there? Choose any value for x (or for y). You can always find the other *paired* or *corresponding* value in an equation of this form. We say that there are an *infinite* number of pairs that satisfy the equation. Each of these pairs is a solution. We find some other solutions for the equation $x + y = 5$ in Example 1.

Example 1 Solving for Corresponding Values

< Objective 1 >

For the equation $x + y = 5$, find **(a)** y if $x = 5$ and **(b)** x if $y = 4$.

(a) If $x = 5$,

$$(5) + y = 5, \quad \text{so} \quad y = 0$$

(b) If $y = 4$,

$$x + (4) = 5, \quad \text{so} \quad x = 1$$

So the pairs $x = 5$, $y = 0$ and $x = 1$, $y = 4$ are both solutions.

Check Yourself 1

You are given the equation $2x + 3y = 26$.

(a) If $x = 4$, $y = ?$ **(b)** If $y = 0$, $x = ?$

To simplify writing the pairs that satisfy an equation, we use **ordered-pair notation.** The numbers are written in parentheses and are separated by a comma. For example, we know that the values $x = 3$ and $y = 2$ satisfy the equation $x + y = 5$. So we write the pair as

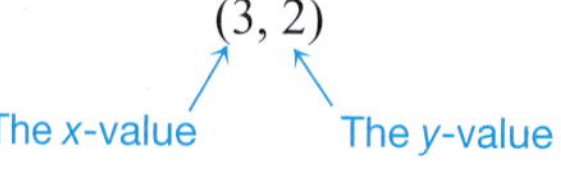

> CAUTION

(3, 2) means $x = 3$ and $y = 2$. (2, 3) means $x = 2$ and $y = 3$. (3, 2) and (2, 3) are entirely different. That's why we call them *ordered pairs.*

The first number of the pair is *always* the value for x and is called the **x-coordinate.** The second number of the pair is *always* the value for y and is the **y-coordinate.**

Using ordered-pair notation, we can say that (3, 2), (5, 0), and (1, 4) are all *solutions* for the equation $x + y = 5$. Each pair gives values for x and y that satisfy the equation.

Example 2 Identifying Solutions of Two-Variable Equations

< Objective 2 >

Which of the ordered pairs (2, 5), (5, −1), and (3, 4) are solutions for the equation $2x + y = 9$?

(a) To check whether (2, 5) is a solution, let $x = 2$ and $y = 5$ and see if the equation is satisfied.

$2x + y = 9$ Substitute 2 for x and 5 for y.

$$2(2) + (5) \stackrel{?}{=} 9$$
$$4 + 5 \stackrel{?}{=} 9$$
$$9 = 9 \quad \text{A true statement}$$

So (2, 5) is a solution for the equation $2x + y = 9$.

NOTE

(2, 5) is a solution because a *true statement* results.

(b) For (5, −1), let $x = 5$ and $y = -1$.

$$2(5) + (-1) \stackrel{?}{=} 9$$
$$10 - 1 \stackrel{?}{=} 9$$
$$9 = 9 \quad \text{A true statement}$$

So (5, −1) is a solution for $2x + y = 9$.

(c) For (3, 4), let $x = 3$ and $y = 4$. Then

$$2(3) + (4) \stackrel{?}{=} 9$$
$$6 + 4 \stackrel{?}{=} 9$$
$$10 = 9 \quad \textit{Not} \text{ a true statement}$$

So (3, 4) is *not* a solution for the equation.

Check Yourself 2

Which of the ordered pairs (3, 4), (4, 3), (1, −2), and (0, −5) are solutions for the equation

$3x - y = 5$

Equations such as those seen in Examples 1 and 2 are said to be in **standard form.**

Definition

Standard Form of Linear Equation

A linear equation in two variables is in **standard form** if it is written as

$Ax + By = C$ in which A and B are not both 0.

Note, for example, that if $A = 1$, $B = 1$, and $C = 5$, we have

$$(1)x + (1)y = (5)$$
$$x + y = 5$$

which is the equation in Example 1.

It is possible to view an equation in one variable as a two-variable equation. For example, if we have the equation $x = 2$, we can view this in standard form as

$$1x + 0y = 2$$

and we may search for ordered-pair solutions. The key is this: If the equation contains only one variable (in this case x), then the missing variable (in this case y) can take on any value. Consider Example 3.

Example 3 Identifying Solutions of One-Variable Equations

Which of the ordered pairs (2, 0), (0, 2), (5, 2), (2, 5), and (2, −1) are solutions for the equation $x = 2$?

A solution is any ordered pair in which the x-coordinate is 2. That makes (2, 0), (2, 5), and (2, −1) solutions for the given equation.

Check Yourself 3

Which of the ordered pairs (3, 0), (0, 3), (3, 3), (−1, 3), and (3, −1) are solutions for the equation $y = 3$?

Remember that when an ordered pair is presented, the first number is always the x-coordinate and the second number is always the y-coordinate.

Example 4 Completing Ordered-Pair Solutions

Complete the ordered pairs (9,), (, −1), (0,), and (, 0) so that each is a solution for the equation $x - 3y = 6$.

(a) The first number, 9, appearing in (9,) represents the x-value. To complete the pair (9,), substitute 9 for x and then solve for y.

$$(9) - 3y = 6$$
$$-3y = -3$$
$$y = 1$$

The ordered pair (9, 1) is a solution for $x - 3y = 6$.

(b) To complete the pair (, −1), let y be −1 and solve for x.

$$x - 3(-1) = 6$$
$$x + 3 = 6$$
$$x = 3$$

The ordered pair (3, −1) is a solution for the equation $x - 3y = 6$.

(c) To complete the pair (0,), let x be 0.

$$(0) - 3y = 6$$
$$-3y = 6$$
$$y = -2$$

So (0, −2) is a solution.

(d) To complete the pair (, 0), let y be 0.

$$x - 3(0) = 6$$
$$x - 0 = 6$$
$$x = 6$$

Then (6, 0) is a solution.

Check Yourself 4

Complete the ordered pairs so that each is a solution for the equation $2x + 5y = 10$.

(10,), (, 4), (0,), and (, 0)

Example 5 Finding Some Solutions of a Two-Variable Equation

Find four solutions for the equation

$$2x + y = 8$$

In this case the values used to form the solutions are *up to you.* You can assign any value for x (or for y). We demonstrate with some possible choices.

NOTE

Generally, you want to pick values for x (or for y) so that the resulting equation in one variable is easy to solve.

Solution with $x = 2$:

$$2x + y = 8$$
$$2(2) + y = 8$$
$$4 + y = 8$$
$$y = 4$$

The ordered pair (2, 4) is a solution for $2x + y = 8$.

Solution with $y = 6$:

$$2x + y = 8$$
$$2x + (6) = 8$$
$$2x = 2$$
$$x = 1$$

So (1, 6) is also a solution for $2x + y = 8$.

Solution with $x = 0$:

$$2x + y = 8$$
$$2(0) + y = 8$$
$$y = 8$$

NOTE

The solutions (0, 8) and (4, 0) have special significance when graphing. They are also easy to find!

And (0, 8) is a solution.

Solution with $y = 0$:

$$2x + y = 8$$
$$2x + (0) = 8$$
$$2x = 8$$
$$x = 4$$

So (4, 0) is a solution.

Check Yourself 5

Find four solutions for $x - 3y = 12$.

Each variable in a two-variable equation plays a different role. The variable for which the equation is solved is called the **dependent variable** because its value depends on what value is given the other variable, which is called the **independent variable.** Generally we use x for the independent variable and y for the dependent variable.

In applications, different letters tend to be used for the variables. These letters are selected to help us see what they stand for, so h is used for height, A is used for area, and so on.

We close this section with an application from the field of medicine.

Example 6 An Allied Health Application

NOTE

The value of the independent variable d, which represents the number of days, can only be a positive integer. We call the set of all possible values for the independent variable the *domain.*

For a particular patient, the weight (w), in grams, of a uterine tumor is related to the number of days (d) of chemotherapy treatment by the equation

$$w = -1.75d + 25$$

(a) What was the original size of the tumor?
The original size of the tumor is the value of w when $d = 0$.
Substituting 0 for d in the equation gives

$$w = -1.75(0) + 25$$
$$= 25$$

The tumor was originally 25 grams.

(b) How many days of chemotherapy are required to eliminate the tumor?
The tumor will be eliminated when the weight (w) is 0. So

$$(0) = -1.75d + 25$$
$$-25 = -1.75d$$
$$d \approx 14.3$$

NOTE

We round **up** in this case, because the tumor will be eliminated on the 15th day.

It will take about 14.3 days to eliminate the tumor. Because the domain for d is the set of positive integers, we answer the original question by saying it will take 15 days to eliminate the tumor.

Check Yourself 6

For a particular patient, the weight (w), in grams, of a uterine tumor is related to the number of days (d) of chemotherapy treatment by the equation

$w = -1.6d + 32$

(a) Find the original size of the tumor.
(b) Determine the number of days of chemotherapy required to eliminate the tumor.

Check Yourself ANSWERS

1. (a) $y = 6$; (b) $x = 13$ 2. $(3, 4)$, $(1, -2)$, and $(0, -5)$ are solutions.
3. $(0, 3)$, $(3, 3)$, and $(-1, 3)$ are solutions.
4. $(10, -2)$, $(-5, 4)$, $(0, 2)$, and $(5, 0)$
5. $(6, -2)$, $(3, -3)$, $(0, -4)$, and $(12, 0)$ are four possibilities.
6. (a) 32 grams; (b) 20 days

Reading Your Text

SECTION 2.2

(a) An equation in two variables is an equation for which every _____________ is a pair of values.

(b) Given an equation such as $x + y = 5$, there are an _____________ number of solutions.

(c) To simplify writing the pairs that satisfy an equation, we use _____________ notation.

(d) When an equation in two variables is solved for y, we say that y is the _____________ variable.

Basic Skills | Challenge Yourself | Calculator/Computer | Career Applications | Above and Beyond

2.2 exercises

< Objectives 1-2 >

Determine which of the ordered pairs are solutions for the given equation.

1. $x + y = 6$ $(4, 2), (-2, 4), (0, 6), (-3, 9)$

2. $x - y = 10$ $(11, 1), (11, -1), (10, 0), (5, 7)$

3. $2x - y = 8$ $(5, 2), (4, 0), (0, 8), (6, 4)$

4. $x + 5y = 20$ $(10, -2), (10, 2), (20, 0), (25, -1)$

5. $4x + y = 8$ $(2, 0), (2, 3), (0, 2), (1, 4)$

6. $x - 2y = 8$ $(8, 0), (0, 4), (5, -1), (10, -1)$

7. $2x - 3y = 6$ $(0, 2), (3, 0), (6, 2), (0, -2)$

8. $6x + 2y = 12$ $(0, 6), (2, 6), (2, 0), (1, 3)$

9. $3x - 2y = 12$ $(4, 0), \left(\frac{2}{3}, -5\right), (0, 6), \left(5, \frac{3}{2}\right)$

10. $3x + 4y = 12$ $(-4, 0), \left(\frac{2}{3}, \frac{5}{2}\right), (0, 3), \left(\frac{2}{3}, 2\right)$

11. $3x + 5y = 15$ $(0, 3), \left(1, \frac{12}{5}\right), (5, 3)$

12. $y = 2x - 1$ $(0, -2), (0, -1), \left(\frac{1}{2}, 0\right), (3, -5)$

13. $x = 3$ $(3, 5), (0, 3), (3, 0), (3, 7)$ > Videos

14. $y = 7$ $(0, 7), (3, 7), (-1, -4), (7, 7)$

Name ____________

Section ______ Date ______

Answers

1. ____________
2. ____________
3. ____________
4. ____________
5. ____________
6. ____________
7. ____________
8. ____________
9. ____________
10. ____________
11. ____________
12. ____________
13. ____________
14. ____________

2.2 exercises

Answers

15. ______
16. ______
17. ______
18. ______
19. ______
20. ______
21. ______
22. ______
23. ______
24. ______
25. ______
26. ______
27. ______
28. ______
29. ______
30. ______
31. ______
32. ______
33. ______
34. ______
35. ______
36. ______
37. ______
38. ______

Complete the ordered pairs so that each is a solution for the given equation.

15. $x + y = 12$ $(4, \ ), (\ , 5), (0, \ ), (\ , 0)$

16. $x - y = 7$ $(\ , 4), (15, \ ), (0, \ ), (\ , 0)$

17. $3x + y = 9$ $(3, \ ), (\ , 9), (\ , -3), (0, \ )$

18. $x + 4y = 12$ $(0, \ ), (\ , 2), (8, \ ), (\ , 0)$

19. $5x - y = 15$ $(\ , 0), (2, \ ), (4, \ ), (\ , -5)$

20. $x - 3y = 9$ $(0, \ ), (12, \ ), (\ , 0), (\ , -2)$

21. $4x - 2y = 16$ $(\ , 0), (\ , -6), (2, \ ), (\ , 6)$ > Videos

22. $2x + 5y = 20$ $(0, \ ), (5, \ ), (\ , 0), (\ , 6)$

23. $y = 3x + 9$ $(\ , 0), \left(\frac{2}{3}, \ \right), (0, \ ), \left(-\frac{2}{3}, \ \right)$

24. $6x + 8y = 24$ $(0, \ ), \left(\ , \frac{3}{4}\right), (\ , 0), \left(-\frac{2}{3}, \ \right)$

25. $y = 3x - 4$ $(0, \ ), (\ , 5), (\ , 0), \left(\frac{5}{3}, \ \right)$

26. $y = -2x + 5$ $(0, \ ), (\ , 5), \left(\frac{3}{2}, \ \right), (\ , 1)$

Find four solutions for each equation. Note: *Your answers may vary from those shown in the answer section.*

27. $x - y = 10$ > Videos

28. $x + y = 18$

29. $2x - y = 6$

30. $4x - 2y = 8$

31. $x + 4y = 8$

32. $x + 3y = 12$

33. $5x - 2y = 10$

34. $2x + 7y = 14$

35. $y = 2x + 3$

36. $y = 5x - 8$

37. $x = -5$

38. $y = 8$

Determine whether each statement is **true** *or* **false.**

39. The ordered pair (a, b) means the same thing as the ordered pair (b, a).

40. An equation in two variables has exactly two solutions.

41. For any number k, $(0, k)$ is a solution for the equation $y = k$.

42. For any number h, $(0, h)$ is a solution for the equation $x = h$.

43. **Business and Finance** When an employee produces x units per hour, the hourly wage in dollars is given by $y = 0.75x + 8$. What are the hourly wages for the following number of units: 2, 5, 10, 15, and 20?

44. **Science and Medicine** Celsius temperature readings can be converted to Fahrenheit readings by using the formula $F = \frac{9}{5}C + 32$. What is the Fahrenheit temperature that corresponds to each of the following Celsius temperatures: -10, 0, 15, 100?

45. **Geometry** The area of a square is given by $A = s^2$. What is the area of the squares whose sides are 4 cm, 11 cm, 14 cm, and 17 cm?

46. **Business and Finance** When x units are sold, the price of each unit is given by $p = 4x + 15$. Find the unit price in dollars when the following quantities are sold: 1, 5, 10, 12.

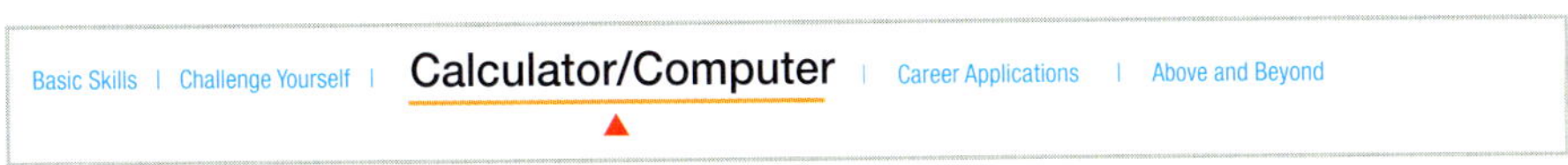

47. Given $y = 3.12x - 14.79$, use the TABLE utility on a graphing calculator to complete the ordered pairs.

(10,), (20,), (30,), (40,), (50,)

48. Given $y = -16x^2 + 90x + 23$, use the TABLE utility on a graphing calculator to complete the ordered pairs.

(1.5,), (2.5,), (3.5,), (4.5,), (5.5,)

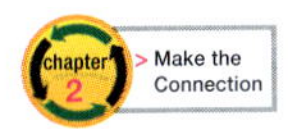

Basic Skills | Challenge Yourself | Calculator/Computer | **Career Applications** | Above and Beyond

49. **Construction Technology** The number of studs s (16 inches on center) required to build a wall that is L feet long is given by the formula

$$s = \frac{3}{4}L + 1$$

Determine the number of studs required to build walls of length 12 ft, 20 ft, and 24 ft.

Answers

39. ______
40. ______
41. ______
42. ______
43. ______
44. ______
45. ______
46. ______
47. ______
48. ______
49. ______

Answers

50. ______

51. ______

52. ______

53. ______

54. ______

55. ______

56. ______

57. ______

58. ______

59. ______

60. ______

50. Manufacturing Technology The number of board feet b of lumber in a 2" × 6" board of length L (in feet) is given by the equation

$$b = \frac{8.25}{144}L$$

Determine the number of board feet in 2" × 6" boards of length 12 ft, 16 ft, and 20 ft.

51. Allied Health The recommended dosage d (in mg) of the antibiotic ampicillin sodium for children weighing less than 40 kg is given by the linear equation $d = 7.5w$, in which w represents the child's weight (in kg).

> Videos

(a) Determine the dosage for a 30-kg child.

(b) What is the weight of a child who requires a 150-mg dose?

52. Allied Health The recommended dosage d (in μg) of neupogen (medication given to bone-marrow transplant patients) is given by the linear equation $d = 8w$, in which w is the patient's weight (in kg).

(a) Determine the dosage for a 92-kg patient.

(b) What is the weight of a patient who requires a 250-μg dose?

Basic Skills | Challenge Yourself | Calculator/Computer | Career Applications | **Above and Beyond**

An equation in three variables has an ordered triple as a solution. For example, (1, 2, 2) *is a solution to the equation* $x + 2y - z = 3$. *Complete the ordered-triple solutions for each equation.*

53. $x + y + z = 0$ $\quad (2, -3, \ )$

54. $2x + y + z = 2$ $\quad (\ , -1, 3)$

55. $x + y + z = 0$ $\quad (1, \ , 5)$

56. $x + y - z = 1$ $\quad (4, \ , 3)$

57. $2x + y + z = 2$ $\quad (-2, \ , 1)$

58. $x + y - z = 1$ $\quad (-2, 1, \ )$

59. You now have had practice solving equations with one variable and equations with two variables. Compare equations with one variable to equations with two variables. How are they alike? How are they different?

60. Each of the following sentences describes pairs of numbers that are related. After completing the sentences in parts (a) to (g), write two of your own sentences in (h) and (i).

(a) The *number of hours you work* determines the *amount you are* ________.

(b) The *number of gallons of gasoline* you put in your car determines *the amount you* ________.

(c) The *amount of the* ________ in a restaurant is related to *the amount of the tip.*

(d) The *sales amount of a purchase in a store* determines ____________.

(e) The *age of an automobile* is related to ________________.

(f) The *amount of electricity you use in a month* determines ____________.

(g) The *cost of food for a family* is related to ________________.

Think of two more:

(h) ____.

(i) ____.

Answers

1. (4, 2), (0, 6), (−3, 9) **3.** (5, 2), (4, 0), (6, 4) **5.** (2, 0), (1, 4)
7. (3, 0), (6, 2), (0, −2) **9.** (4, 0), $\left(\frac{2}{3}, -5\right)$, $\left(5, \frac{3}{2}\right)$ **11.** (0, 3), $\left(1, \frac{12}{5}\right)$
13. (3, 5), (3, 0), (3, 7) **15.** 8, 7, 12, 12 **17.** 0, 0, 4, 9 **19.** 3, −5, 5, 2
21. 4, 1, −4, 7 **23.** −3, 11, 9, 7 **25.** −4, 3, $\frac{4}{3}$, 1
27. (0, −10), (10, 0), (5, −5), (12, 2) **29.** (0, −6), (3, 0), (6, 6), (9, 12)
31. (8, 0), (−4, 3), (0, 2), (4, 1) **33.** (0, −5), (4, 5), (−6, −20), (2, 0)
35. (0, 3), (1, 5), (2, 7), (3, 9) **37.** (−5, 0), (−5, 1), (−5, 2), (−5, 3)
39. False **41.** True **43.** \$9.50, \$11.75, \$15.50, \$19.25, \$23
45. 16 cm^2, 121 cm^2, 196 cm^2, 289 cm^2 **47.** 16.41, 47.61, 78.81, 110.01, 141.21
49. 10 studs, 16 studs, 19 studs **51.** **(a)** 225 mg; **(b)** 20 kg **53.** 1
55. −6 **57.** 5 **59.** Above and Beyond

2.3 The Cartesian Coordinate System

< 2.3 Objectives >

1 > Identify plotted points

2 > Plot ordered pairs

3 > Scale the axes

In Section 2.2, we used ordered pairs to write solutions to equations in two variables. The next step is to graph those ordered pairs as points in a plane.

Since there are two numbers (one for x and one for y), we need two number lines: one line drawn horizontally, the other drawn vertically. Their point of intersection (at their respective zero points) is called the **origin.** The horizontal line is called the ***x*-axis,** and the vertical line is called the ***y*-axis.** Together the lines form the **rectangular** or **Cartesian coordinate system.**

The axes (pronounced "axees") divide the plane into four regions called **quadrants,** which are numbered (usually by Roman numerals) counterclockwise from the upper right.

NOTE

This system is called the **Cartesian coordinate system,** named in honor of its inventor, René Descartes (1596–1650), a French mathematician and philosopher.

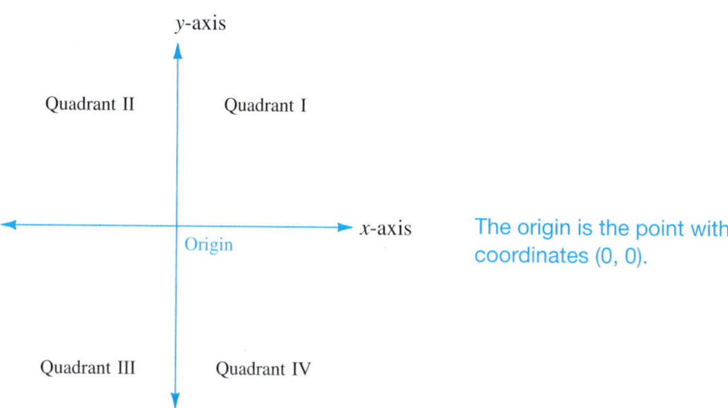

The origin is the point with coordinates (0, 0).

We now want to establish correspondences between ordered pairs of numbers (x, y) and points in the plane.

For any ordered pair,

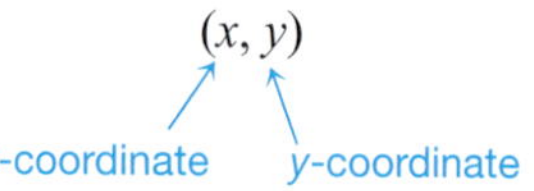

the following are true:

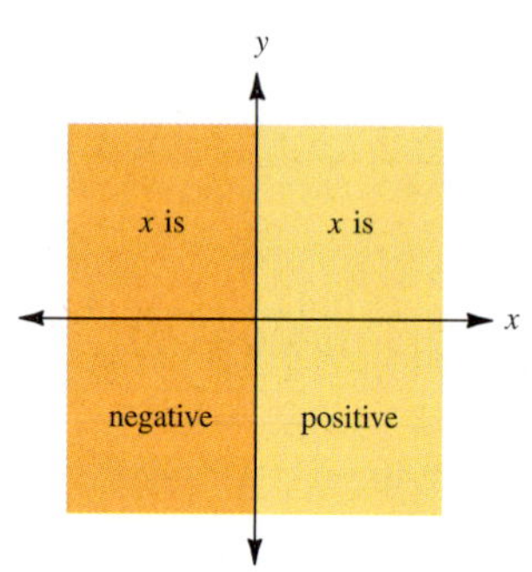

1. If the x-coordinate is

   Positive, the point corresponding to that pair is located x units to the *right* of the y-axis.

   Negative, the point is x units to the *left* of the y-axis.

   Zero, the point is on the y-axis.

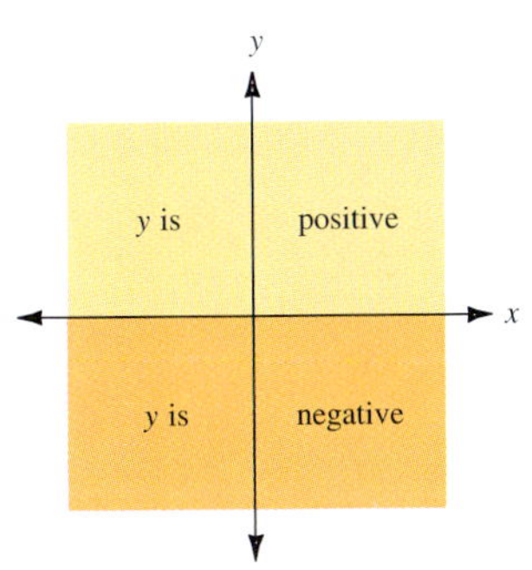

2. If the y-coordinate is

Positive, the point is y units *above* the x-axis.

Negative, the point is y units *below* the x-axis.

Zero, the point is on the x-axis.

Example 1 illustrates how to use these guidelines to match coordinates with points in the plane.

Example 1 Identifying the Coordinates for a Given Point

< Objective 1 >

Give the coordinates of each point shown. Assume that each tick mark represents 1 unit.

NOTE

The x-coordinate gives the *horizontal* distance from the y-axis. The y-coordinate gives the *vertical* distance from the x-axis.

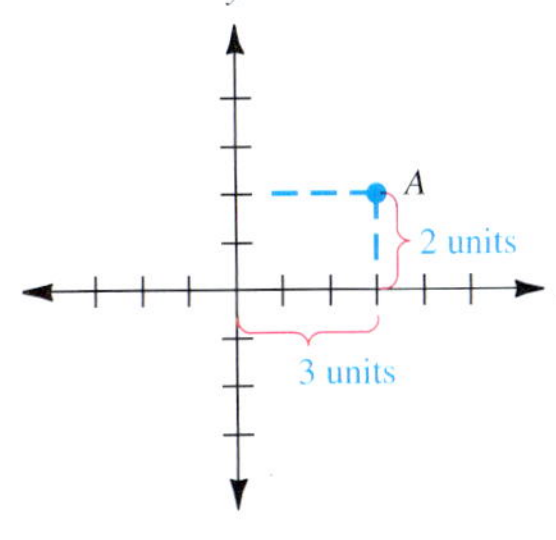

(a) Point A is 3 units to the *right* of the y-axis and 2 units *above* the x-axis. Point A has coordinates (3, 2).

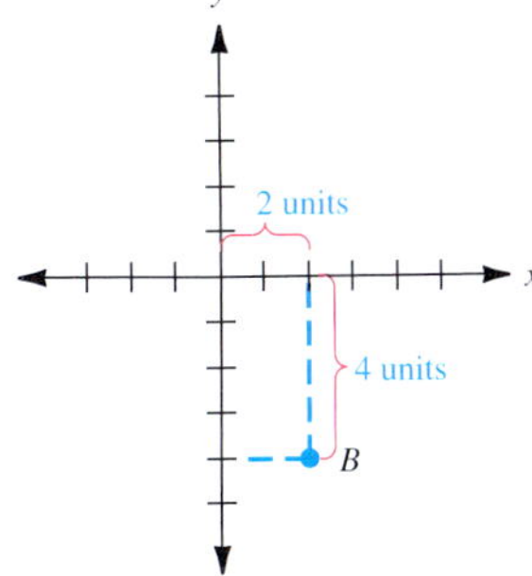

(b) Point B is 2 units to the *right* of the y-axis and 4 units *below* the x-axis. Point B has coordinates $(2, -4)$.

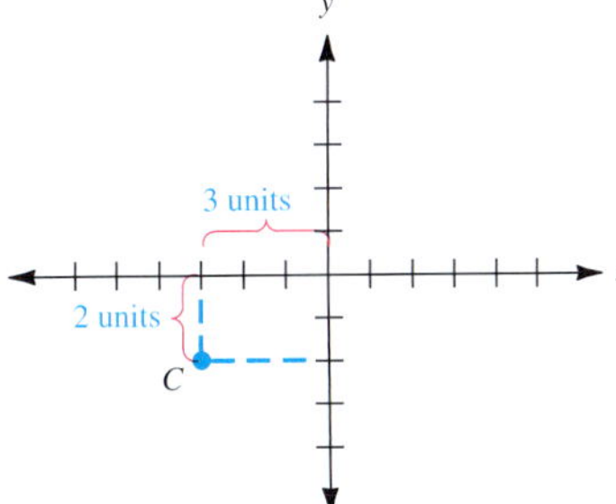

(c) Point C is 3 units to the *left* of the y-axis and 2 units *below* the x-axis. Point C has coordinates $(-3, -2)$.

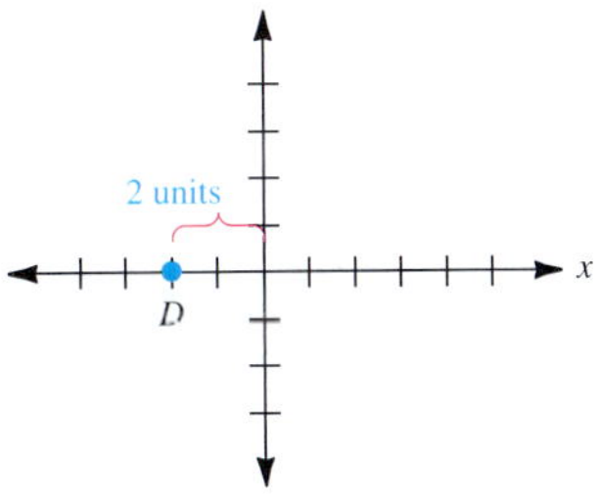

(d) Point D is 2 units to the *left* of the y-axis and *on* the x-axis. Point D has coordinates $(-2, 0)$.

Check Yourself 1

Give the coordinates of points *P*, *Q*, *R*, and *S*.

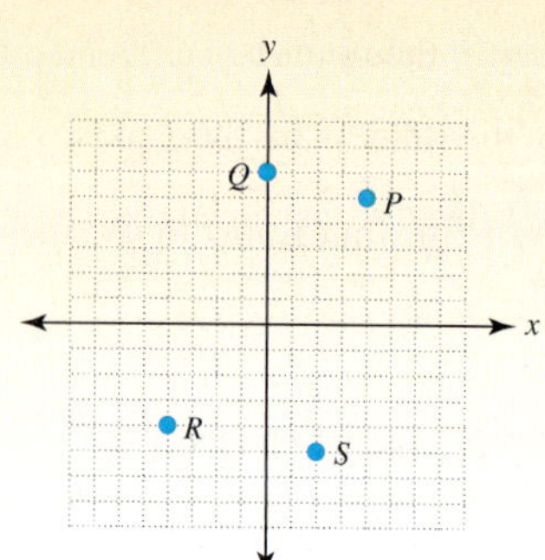

NOTE

Graphing individual points is sometimes called **point plotting.**

Reversing the process used in Example 1 allows us to graph (or plot) a point in the plane, given the coordinates of the point. You can use the following steps.

Step by Step

To Graph a Point in the Plane

Step 1 Start at the origin.
Step 2 Move right or left according to the value of the x-coordinate.
Step 3 Move up or down according to the value of the y-coordinate.

Example 2 Graphing Points

< Objective 2 >

(a) Graph the point corresponding to the ordered pair (4, 3).

Move 4 units to the right on the x-axis. Then move 3 units up from the point where you stopped on the x-axis. This locates the point corresponding to (4, 3).

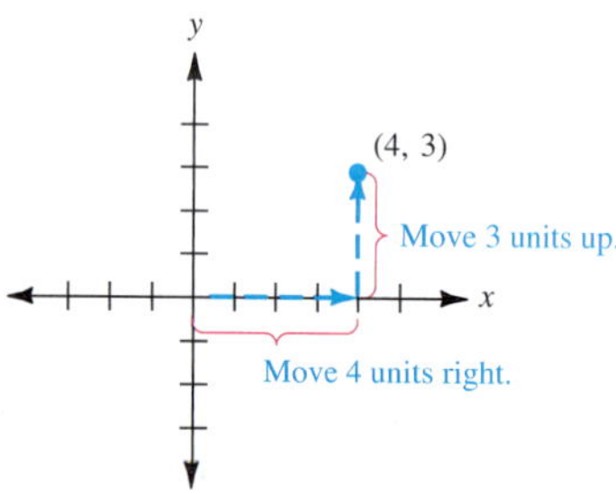

(b) Graph the point corresponding to the ordered pair (−5, 2).

In this case move 5 units *left* (because the x-coordinate is negative) and then 2 units *up*.

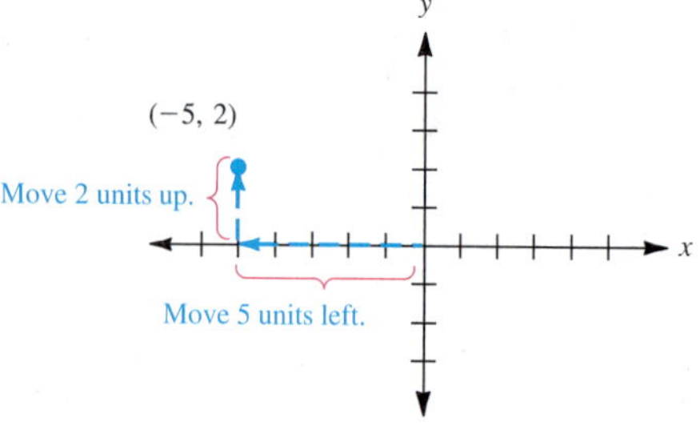

(c) Graph the point corresponding to $(-4, -2)$.

Here move 4 units *left* and then 2 units *down* (the y-coordinate is also negative).

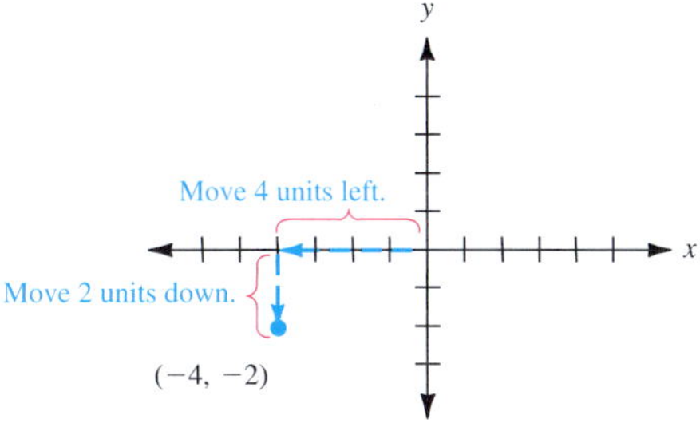

NOTE

Any point on an axis has 0 as one of its coordinates.

(d) Graph the point corresponding to $(0, -3)$.

There is *no* horizontal movement because the x-coordinate is 0. Move 3 units *down*.

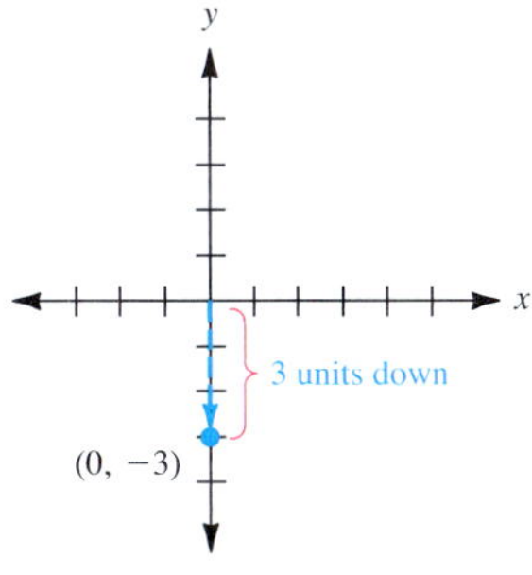

(e) Graph the point corresponding to $(5, 0)$.

Move 5 units *right*. The desired point is on the x-axis because the y-coordinate is 0.

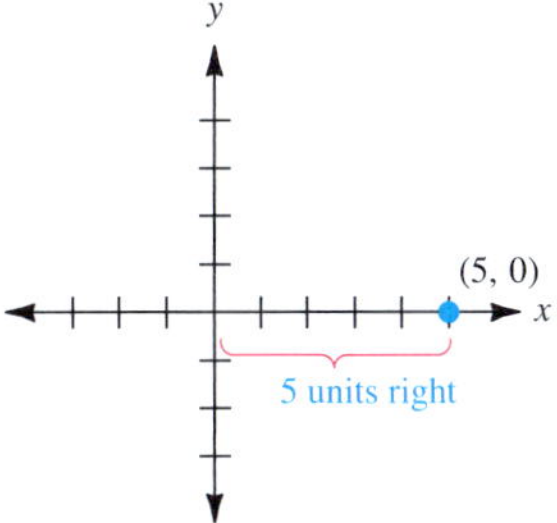

Check Yourself 2

Graph the points corresponding to $M(4, 3)$, $N(-2, 4)$, $P(-5, -3)$, and $Q(0, -3)$.

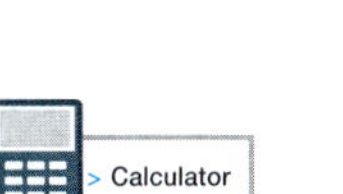

NOTE

The same decisions must be made when you use a graphing calculator. When graphing this kind of relation on a calculator, you must decide on an appropriate viewing window.

It is not necessary, or even desirable, to always use the same scale on both the x- and y-axes. For example, if we were plotting ordered pairs in which the first value represented the age of a used car and the second value represented the number of miles driven, it would be necessary to have a different scale on the two axes. If not, the following extreme cases could happen.

Assume that the cars range in age from 1 to 15 years. The cars have mileages from 2,000 to 150,000 miles (mi). If we used the same scale on both axes, 0.5 in. between each two counting numbers, how large would the paper have to be on which the points were plotted? The horizontal axis would have to be $15(0.5) = 7.5$ in. The vertical axis would have to be $150{,}000(0.5) = 75{,}000$ in. $= 6{,}250$ feet (ft) $=$ almost 1.2 mi long!

So what do we do? We simply use a different, but clearly marked, scale on the axes. In this case, we mark the horizontal axis in 5's with gridlines every unit, and we mark the vertical axis in 50,000's with gridlines every 10,000 units. Additionally, all the numbers are positive, so we only need the first quadrant, in which x and y are both always positive. We could scale the axes like this:

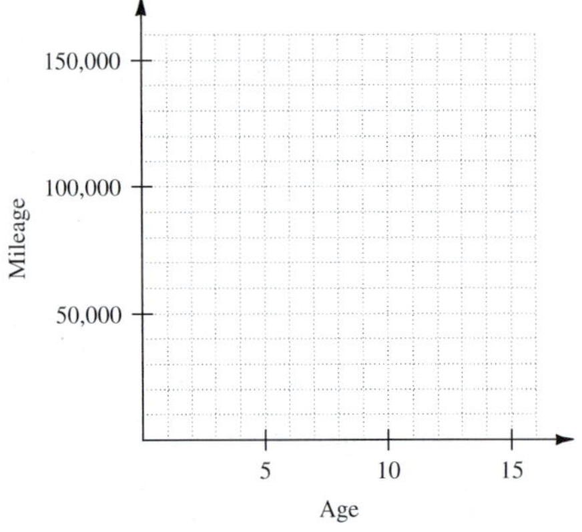

Example 3 Scaling the Axes

< Objective 3 >

A survey of residents in a large apartment building was recently taken. The following points represent ordered pairs in which the first number is the number of years of education a person has had, and the second number is his or her year 2009 income (in thousands of dollars). Estimate, and interpret, each ordered pair represented.

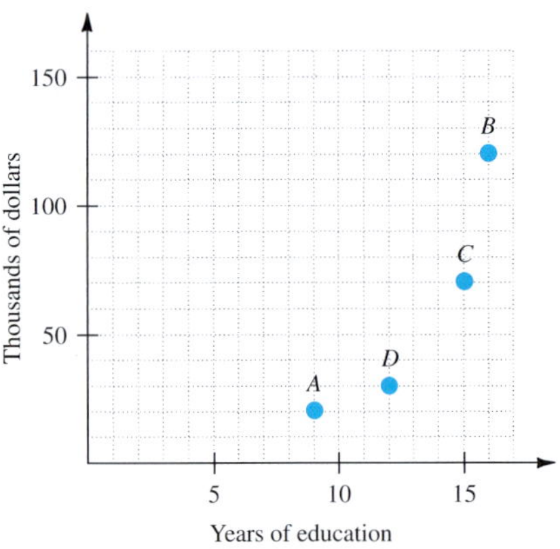

Point A is (9, 20), B is (16, 120), C is (15, 70), and D is (12, 30). Person A completed 9 years of education and made \$20,000 in 2009. Person B completed 16 years of education and made \$120,000 in 2009. Person C had 15 years of education and made \$70,000. Person D had 12 years and made \$30,000.

Note that there is no obvious "relation" that would allow one to predict income from years of education, but you might suspect that in most cases, more education results in more income.

Check Yourself 3

Each six months, Armand records his son's weight. The following points represent ordered pairs in which the first number represents his son's age and the second number represents his son's weight. For example, point *A* indicates that when his son was 1 year old, the boy weighed 14 pounds. Estimate and interpret each ordered pair represented.

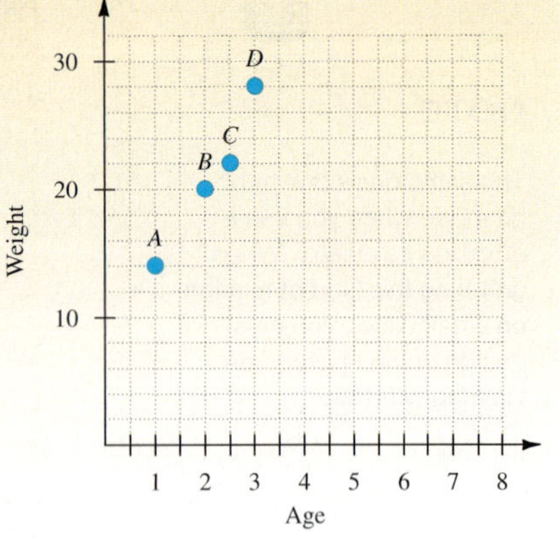

Here is an application from the field of manufacturing.

Example 4 A Graphing Application

A computer-aided design (CAD) operator has located three corners of a rectangle. The corners are at (5, 9), (−2, 9), and (5, 2). Find the location of the fourth corner.

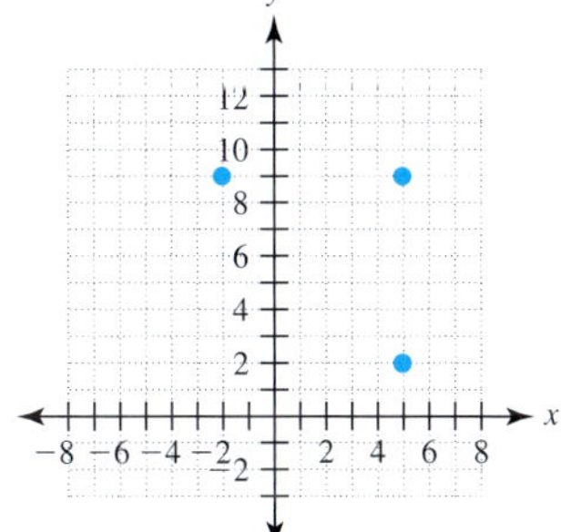

We plot the three indicated points on graph paper.

The fourth corner must lie directly underneath the point (−2, 9), so the x-coordinate must be −2. The corner must lie on the same horizontal as the point (5, 2), so the y-coordinate must be 2. Therefore, the coordinates of the fourth corner must be (−2, 2).

Check Yourself 4

A CAD operator has located three corners of a rectangle. The corners are at (−3, 4), (6, 4), and (−3, −7). Find the location of the fourth corner.

Check Yourself ANSWERS

1. $P(4, 5)$, $Q(0, 6)$, $R(-4, -4)$, and $S(2, -5)$

2.

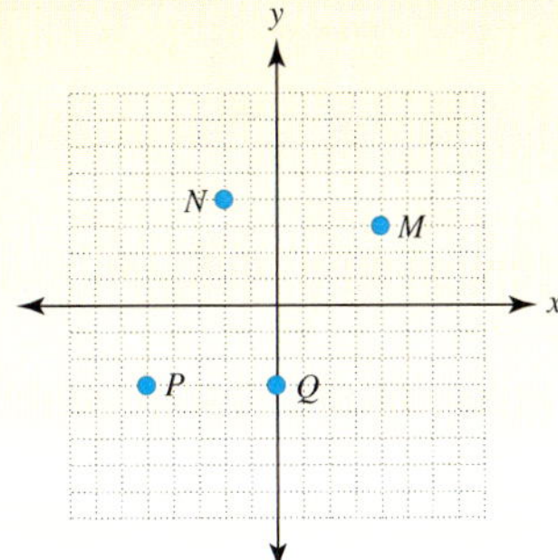

3. $A(1, 14)$, $B(2, 20)$, $C\left(\frac{5}{2}, 22\right)$, and $D(3, 28)$; The first number in each ordered pair represents the age in years. The second number represents the weight in pounds.

4. (6, −7)

Reading Your Text

SECTION 2.3

(a) In the rectangular coordinate system the horizontal line is called the ______________.

(b) In the rectangular coordinate system the vertical line is called the ______________.

(c) To graph a point we start at the ______________.

(d) Every ordered pair is either in one of the ______________ or on one of the axes.

2.3 exercises

Basic Skills | Challenge Yourself | Calculator/Computer | Career Applications | Above and Beyond

Name ______________________

Section __________ Date __________

Answers

1. ______________________
2. ______________________
3. ______________________
4. ______________________
5. ______________________
6. ______________________
7. ______________________
8. ______________________
9. ______________________
10. ______________________
11. ______________________
12. ______________________
13. ______________________
14. ______________________
15. ______________________
16. ______________________

< Objective 1 >

Give the coordinates of the points graphed below.

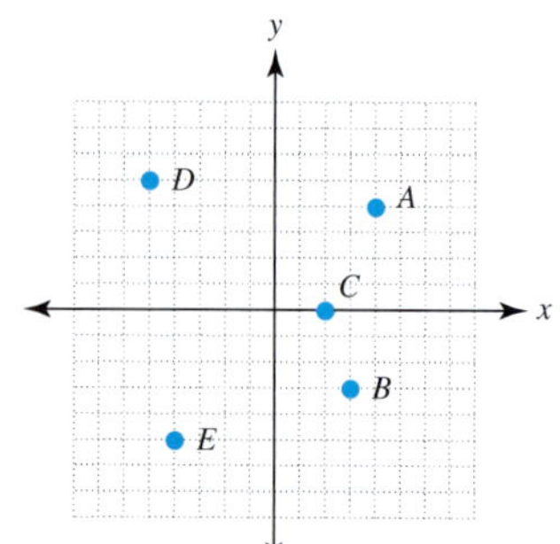

1. *A*
2. *B*
3. *C* 

4. *D*
5. *E* > Videos

Give the coordinates of the points graphed below.

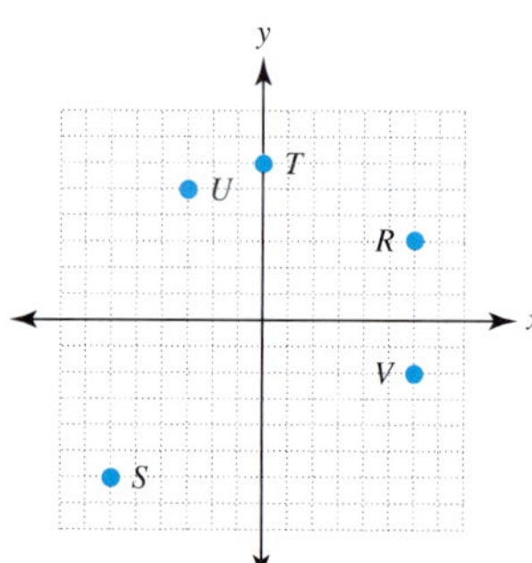

6. *R*
7. *S*
8. *T*
9. *U*
10. *V*

< Objective 2 >

Plot each point on a rectangular coordinate system.

11. $M(5, 3)$

12. $N(0, -3)$

13. $P(-4, 5)$

14. $Q(5, 0)$

15. $R(-4, -6)$

16. $S(-4, -3)$

17. $F(-3, -1)$

18. $G(4, 3)$

19. $H(4, -3)$

20. $I(-3, 0)$

21. $J(-5, 3)$ 

22. $K(0, 4)$

Give the quadrant in which each point is located or the axis on which the point lies.

23. $(4, 5)$

24. $(-3, 2)$

25. $(-6, -8)$

26. $(2, -4)$

27. $(5, 0)$

28. $(-1, 11)$

29. $(-4, 7)$

30. $(-3, -7)$

31. $(0, -4)$

32. $(-3, 0)$

33. $\left(5\frac{3}{4}, -3\right)$

34. $\left(-3, 4\frac{2}{3}\right)$

Answers

17. ______
18. ______
19. ______
20. ______
21. ______
22. ______
23. ______
24. ______
25. ______
26. ______
27. ______
28. ______
29. ______
30. ______
31. ______
32. ______
33. ______
34. ______

Answers

35. ______________________

36. ______________________

< Objective 3 >

35. A company has kept a record of the number of items produced by an employee as the number of days on the job increases. In the graph, points correspond to an ordered-pair relationship in which the first number represents days on the job and the second number represents the number of items produced. Estimate each ordered pair represented.

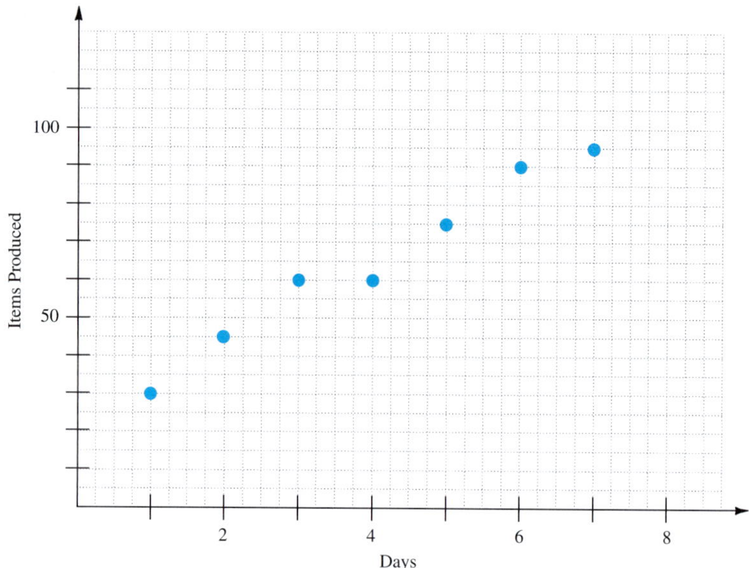

36. In the graph, points correspond to an ordered-pair relationship between height and age in which the first number represents age and the second number represents height. Estimate each ordered pair represented.

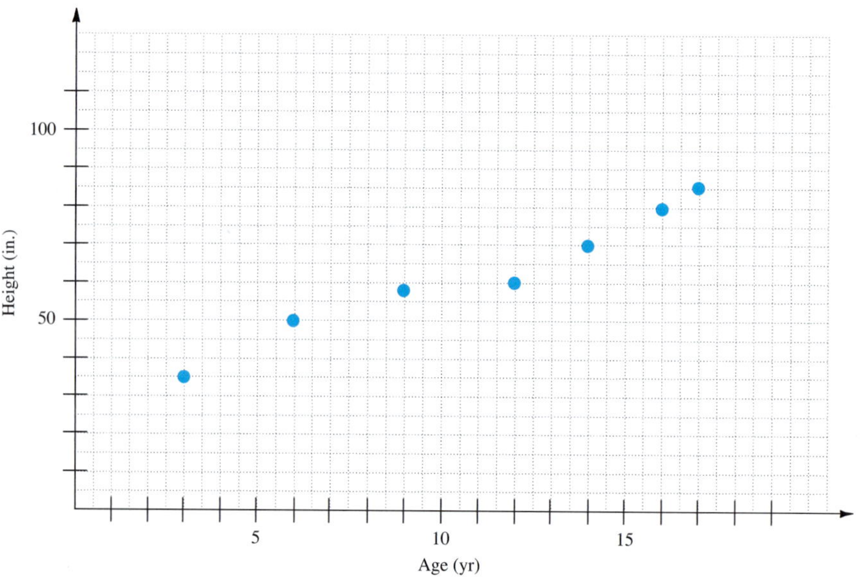

37. An unidentified company has kept a record of the number of hours devoted to safety training and the number of work hours lost due to on-the-job accidents. In the graph, the points correspond to an ordered-pair relationship in which the first number represents hours in safety training and the second number represents hours lost by accidents. Estimate each ordered pair represented.

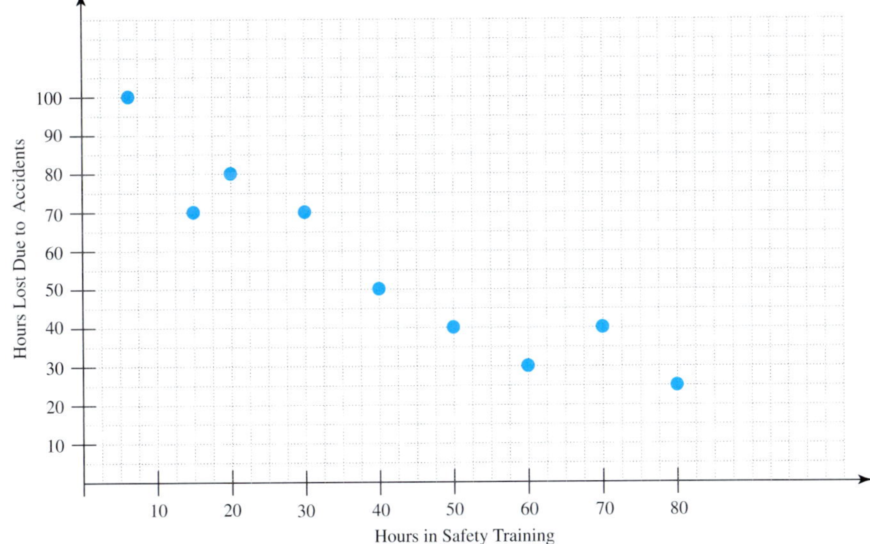

38. In the graph, points correspond to an ordered-pair relationship between the age of a person and the annual average number of visits to doctors and dentists for a person that age. The first number represents the age, and the second number represents the number of visits. Estimate each ordered pair represented.

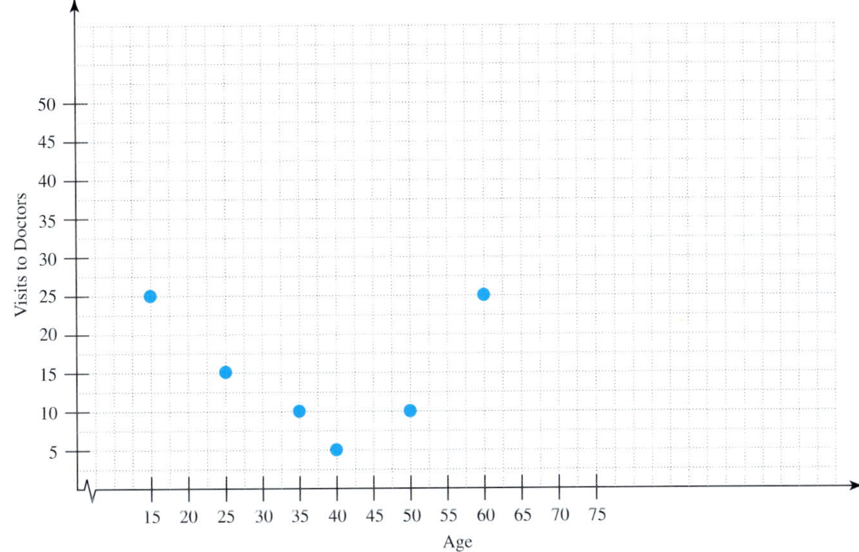

Answers

37. ______

38. ______

Basic Skills | **Challenge Yourself** | Calculator/Computer | Career Applications | Above and Beyond

Answers

39.

40.

41.

42.

43.

44.

45.

Complete each statement with **never, sometimes,** *or* **always.**

39. In the plane, a point on an axis __________ has a coordinate equal to zero.

40. The ordered pair (a, b) is __________ equal to the ordered pair (b, a).

41. If, in the ordered pair (a, b), a and b have different signs, then the point (a, b) is __________ in the second quadrant.

42. If $a \neq b$, then the ordered pair (a, b) is __________ equal to the ordered pair (b, a).

43. **Business and Finance** A plastics company is sponsoring a plastics recycling contest for the local community. The focus of the contest is on collecting plastic milk, juice, and water jugs. The company will award \$200, plus the current market price of the jugs collected, to the group that collects the most jugs in a single month. The number of jugs collected and the amount of money won can be represented as an ordered pair.

(a) In April, group A collected 1,500 pounds (lb) of jugs to win first place. The prize for the month was \$350. If x represents the pounds of jugs and y represents the amount of money that the group won, graph the point that represents the winner for April.

(b) In May, group B collected 2,300 lb of jugs to win first place. The prize for the month was \$430. Graph the point that represents the May winner on the same grid you used in part (a).

44. **Science and Medicine** The table gives the average temperature y (in degrees Fahrenheit) for the first 6 months of the year x. The months are numbered 1 through 6, with 1 corresponding to January. Plot the data given in the table.

> Videos

x	1	2	3	4	5	6
y	4	14	26	33	42	51

45. **Business and Finance** The table gives the total salary of a salesperson y for each of the four quarters of the year x. Plot the data given in the table.

x	1	2	3	4
y	\$6,000	\$5,000	\$8,000	\$9,000

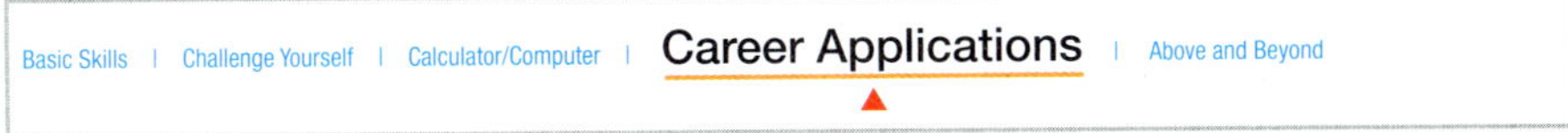

46. **Electronics** A solenoid uses an applied electromagnetic force to cause mechanical force. Typically, a wire conductor is coiled and current is applied, creating an electromagnet. The magnetic field induced by the energized coil attracts a piece of iron, creating mechanical movement.

Plot the force y (in newtons) for each applied voltage x (in volts) of a solenoid shown in the table.

> Videos

x	5	10	15	20
y	0.12	0.24	0.36	0.49

47. **Mechanical Engineering** Plot the temperature and pressure relationship of a coolant as described in the table.

Temperature (°F)	−10	10	30	50	70	90
Pressure (psi)	4.6	14.9	28.3	47.1	71.1	99.2

48. **Mechanical Engineering** Use the graph in exercise 47 to answer each question.

(a) Predict the pressure when the temperature is 60°F.

(b) At what temperature would you expect the coolant to be if the pressure reads 37 psi?

49. **Allied Health** Plot the baby's weight w (in pounds) recorded at well-baby checkups at the ages x (in months), as described in the table.

Age (months)	0	0.5	1	2	7	9
Weight (pounds)	7.8	7.14	9.25	12.5	20.25	21.25

Basic Skills | Challenge Yourself | Calculator/Computer | Career Applications | **Above and Beyond**

50. Graph points with coordinates $(-1, 3)$, $(0, 0)$, and $(1, -3)$. What do you observe? Can you give the coordinates of another point with the same property?

51. Graph points with coordinates (1, 5), (1, 3), and (3, 1). What do you observe? Can you give the coordinates of another point with the same property?

Answers

46. ________

47. ________

48. ________

49. ________

50. ________

51. ________

Answers

52. ______

53. ______

54. ______

55. ______

56. ______

52. Although high employment is a measure of a country's economic vitality, economists worry that periods of low unemployment will lead to inflation. Look at the table.

Year	Unemployment Rate (%)	Inflation Rate (%)
1965	4.5	1.6
1970	4.9	5.7
1975	8.5	9.1
1980	7.1	13.5
1985	7.2	3.6
1990	5.5	5.4
1995	5.6	2.5
2000	3.8	3.2

Plot the figures in the table with unemployment rates on the x-axis and inflation rates on the y-axis. What do these plots tell you? Do higher inflation rates seem to be associated with lower unemployment rates? Explain.

53. We mentioned that the Cartesian coordinate system was named for the French philosopher and mathematician René Descartes. What philosophy book is Descartes most famous for?

54. What characteristic is common to all points on the x-axis? On the y-axis?

55. How would you describe a rectangular coordinate system? Explain what information is needed to locate a point in a coordinate system.

56. Some newspapers have a special day that they devote to automobile want ads. Use this special section or the Sunday classified ads from your local newspaper to find all the want ads for a particular automobile model. Make a list of the model year and asking price for 10 ads, being sure to get a variety of ages for this model. After collecting the information, make a plot of the age and the asking price for the car.

Describe your graph, including an explanation of how you decided which variable to put on the vertical axis and which on the horizontal axis. What trends or other information does the graph portray?

Answers

1. (4, 4) **3.** (2, 0) **5.** (−4, −5) **7.** (−6, −6) **9.** (−3, 5)

11.

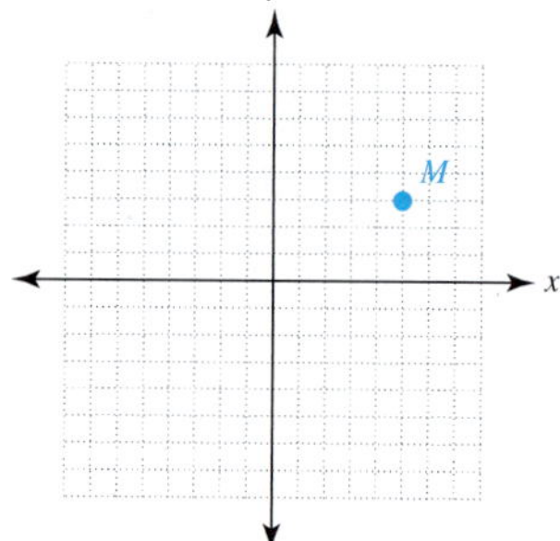

13.

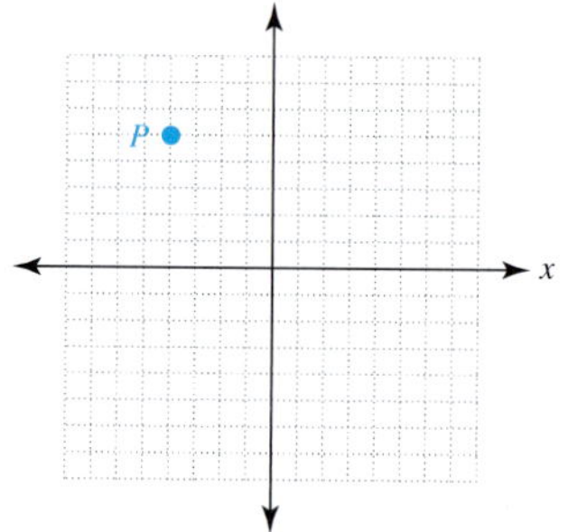

15.

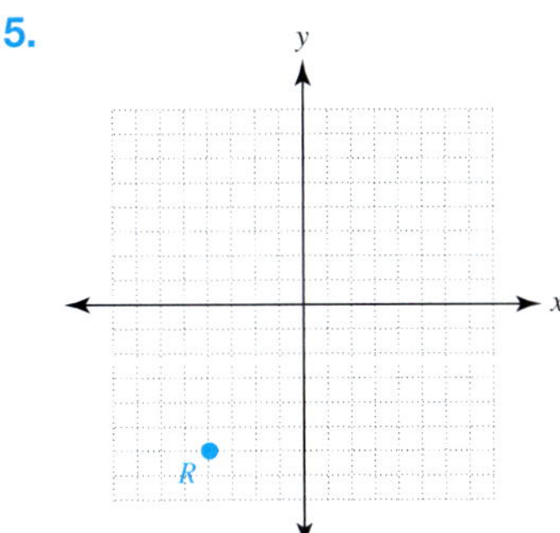

17.

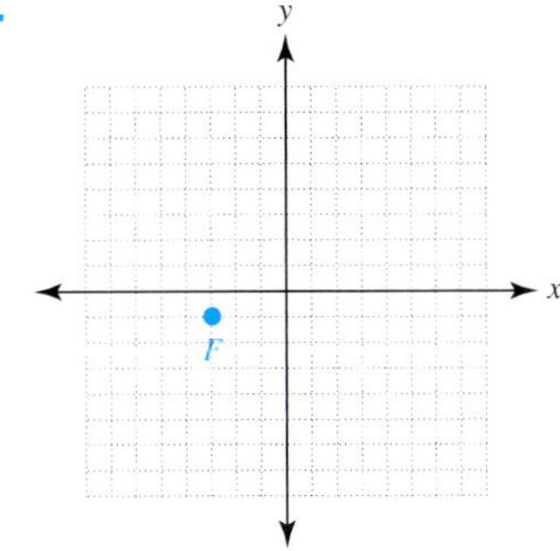

19.

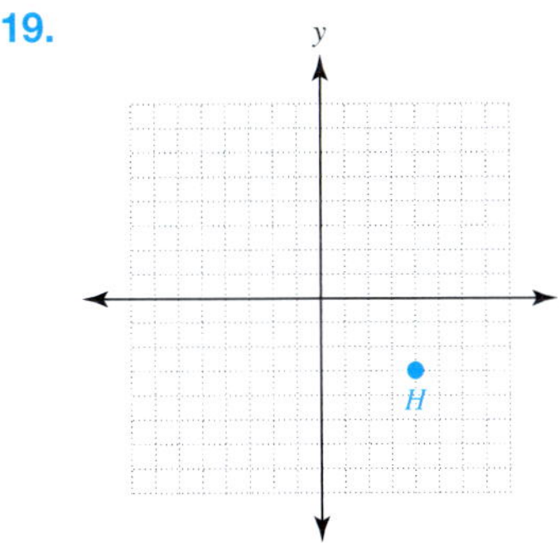

21.

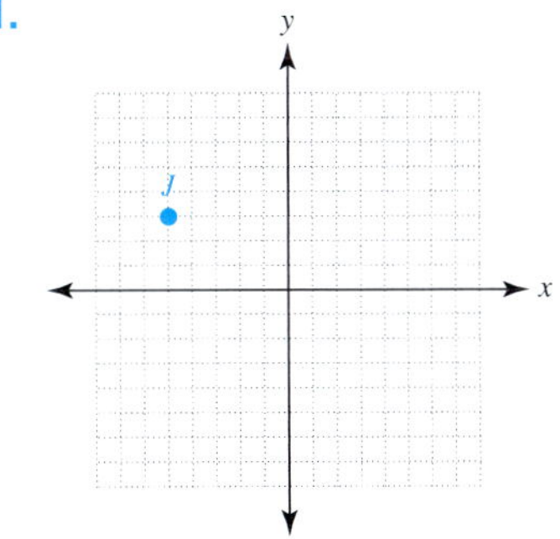

23. I **25.** III **27.** x-axis **29.** II **31.** y-axis **33.** IV
35. (1, 30), (2, 45), (3, 60), (4, 60), (5, 75), (6, 90), (7, 95)
37. (7, 100), (15, 70), (20, 80), (30, 70), (40, 50), (50, 40), (60, 30), (70, 40), (80, 25) **39.** always **41.** sometimes

43.

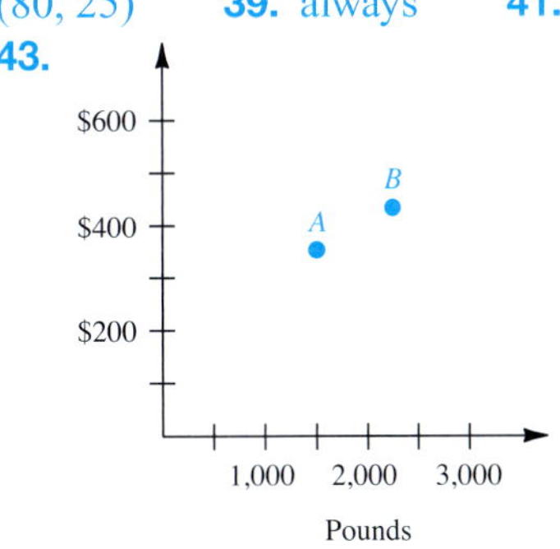

45.

47.

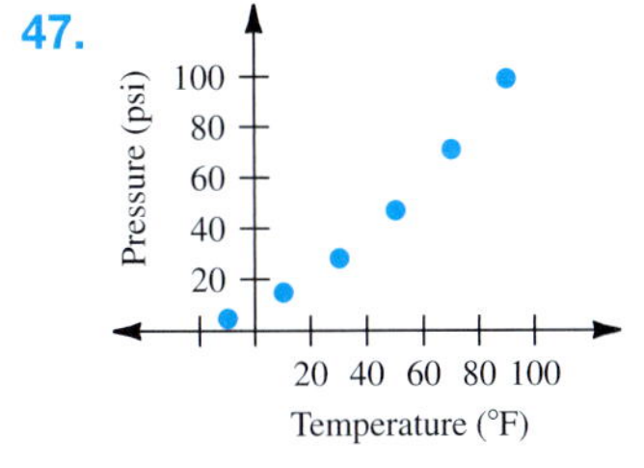

49.

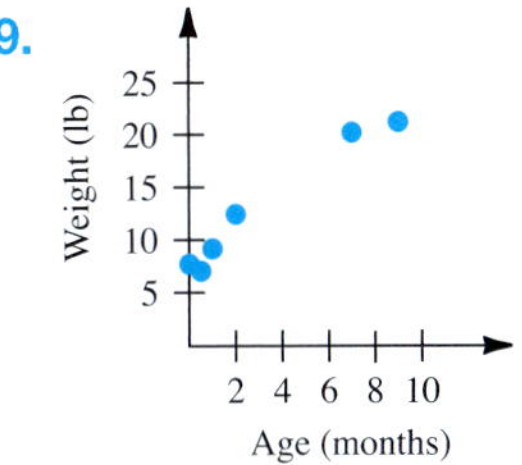

51. The points lie on a line; (3, 7) **53.** Above and Beyond
55. Above and Beyond

2.4 Relations and Functions

< 2.4 Objectives >

1 > Identify the domain and range of a relation
2 > Identify a function, using ordered pairs
3 > Evaluate a function
4 > Determine whether a relation is a function
5 > Write an equation as a function

In Section 2.2, we introduced the concept of ordered pairs. We now turn our attention to sets of ordered pairs.

Definition

Relation A set of ordered pairs is called a **relation.**

We usually denote a relation with a capital letter. Given

$A = \{$(Jane Trudameier, 123-45-6789),
(Jacob Smith, 987-65-4321),
(Julia Jones, 111-22-3333)$\}$

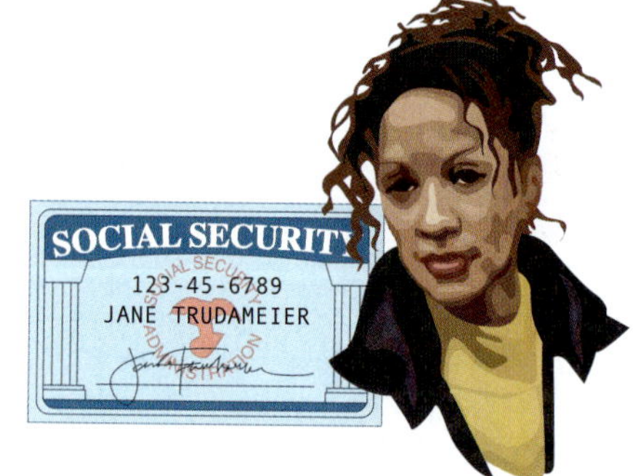

we have a relation, which we call A. In this case, there are three ordered pairs in the relation A.

Within this relation, there are two interesting sets. The first is the set of names, which happens to be the set of first elements. The second is the set of Social Security numbers, which is the set of second elements. Each of these sets has a name.

Definition

Domain The set of first elements in a relation is called the **domain** of the relation.

Example 1 Finding the Domain of a Relation

< Objective 1 >

Find the domain of each relation.

(a) $A = \{$(Ben Bender, 58), (Carol Clairol, 32), (David Duval, 29)$\}$
The domain of A is $\{$Ben Bender, Carol Clairol, David Duval$\}$.

(b) $B = \left\{\left(5, \frac{1}{2}\right), (-4, -5), (-12, 10), (-16, \pi)\right\}$
The domain of B is $\{5, -4, -12, -16\}$.

Check Yourself 1

Find the domain of each relation.

(a) $A = \{$(Secretariat, 10), (Seattle Slew, 8), (Charismatic, 5), (Gallant Man, 7)$\}$

(b) $B = \left\{\left(-\frac{1}{2}, \frac{3}{4}\right), (0, 0), (1, 5), (\pi, \pi)\right\}$

Definition

Range The set of second elements in a relation is called the **range** of the relation.

Example 2 Finding the Range of a Relation

Find the range for each relation.

(a) $A = \{$(Ben Bender, 58), (Carol Clairol, 32), (David Duval, 29)$\}$

The range of A is $\{58, 32, 29\}$.

(b) $B = \left\{\left(5, \frac{1}{2}\right), (-4, -5), (-12, 10), (-16, \pi), (-16, 1)\right\}$

The range of B is $\left\{\frac{1}{2}, -5, 10, \pi, 1\right\}$.

Check Yourself 2

Find the range of each relation.

(a) $A = \{$(Secretariat, 10), (Seattle Slew, 8), (Charismatic, 5), (Gallant Man, 7)$\}$

(b) $B = \left\{\left(-\frac{1}{2}, \frac{3}{4}\right), (0, 0), (1, 5), (\pi, \pi)\right\}$

The set of ordered pairs $B = \{(-2, 1), (-1, 1), (0, 3), (4, 3)\}$ can be represented in the following table:

x	y
−2	1
−1	1
0	3
4	3

The same set of ordered pairs can also be presented as a mapping.

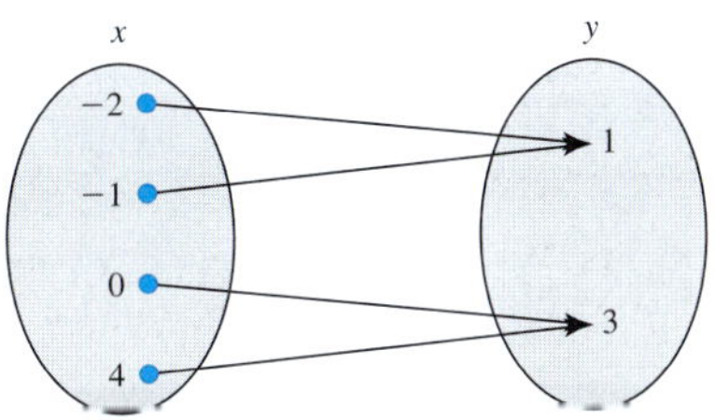

Note that, in this mapping, no x-value (domain element) is mapped to two different y-values (range elements). That leads to our definition of a function.

Definition

Function

A **function** is a set of ordered pairs in which no element of the domain is paired with more than one element of the range.

Example 3 Identifying a Function

< Objective 2 >

For each table of values, decide whether the relation is a function.

(a)

x	y
−2	1
−1	1
1	3
2	3

(b)

x	y
−5	−2
−1	3
−1	6
2	8

(c)

x	y
−3	1
−1	0
0	2
2	4

Part (a) represents a function. No two first coordinates are equal. Part (b) is not a function because −1 appears as a first coordinate with two different second coordinates. Part (c) is a function.

Check Yourself 3

For each table of values below, decide whether the relation is a function.

(a)

x	y
−3	0
−1	1
1	2
3	3

(b)

x	y
−2	−2
−1	−2
1	3
2	3

(c)

x	y
−2	0
−1	1
0	2
0	3

Next we look at another way to represent functions. Rather than being given a set of ordered pairs or a table, we may instead be given a rule or equation from which we must generate ordered pairs. To generate ordered pairs, we need to recall how to evaluate an expression, first introduced in Section 1.2, and apply the order-of-operations rules that you reviewed in Section 0.5.

We have seen that variables can be used to represent numbers whose values are unknown. By using addition, subtraction, multiplication, division, and exponentiation, these numbers and variables form expressions such as

$3 + 5 \qquad 7x - 4 \qquad x^2 - 3x - 4 \qquad x^4 - x^2 + 2$

If a specific value is given for the variable, we **evaluate the expression.**

Example 4 Evaluating Expressions

Evaluate the expression $x^4 - 2x^2 + 3x + 4$ for the indicated value of x.

(a) $x = 0$

Substituting 0 for x in the expression yields

$$\begin{aligned}(0)^4 - 2(0)^2 + 3(0) + 4 &= 0 - 0 + 0 + 4\\ &= 4\end{aligned}$$

(b) $x = 2$

Substituting 2 for x in the expression yields

$$(2)^4 - 2(2)^2 + 3(2) + 4 = 16 - 8 + 6 + 4$$
$$= 18$$

(c) $x = -1$

Substituting -1 for x in the expression yields

$$(-1)^4 - 2(-1)^2 + 3(-1) + 4 = 1 - 2 - 3 + 4$$
$$= 0$$

Check Yourself 4

Evaluate the expression $2x^3 - 3x^2 + 3x + 1$ for the indicated value of x.

(a) $x = 0$ **(b)** $x = 1$ **(c)** $x = -2$

We could design a machine whose purpose would be to crank out the value of an expression for each given value of x. We could call this machine something simple such as f, our **function machine.** Our machine might look like this.

For example, if we put -1 into the machine, the machine would substitute -1 for x in the expression, and 5 would come out the other end because

$$2(-1)^3 + 3(-1)^2 - 5(-1) - 1 = -2 + 3 + 5 - 1 = 5$$

Note that, with this function machine, an input of -1 will always result in an output of 5. One of the most important aspects of a function machine is that each input has a unique output.

NOTE

Two distinct input elements can have the same output. However, each input element can only be associated with exactly one output element.

In fact, the idea of the function machine is very useful in mathematics. Your graphing calculator can be used as a function machine. You can enter the expression into the calculator as Y_1 and then evaluate Y_1 for different values of x.

Generally, in mathematics, we do not write $Y_1 = 2x^3 + 3x^2 - 5x - 1$. Instead, we write $f(x) = 2x^3 + 3x^2 - 5x - 1$, which is read "$f$ of x is equal to. . . ." Instead of calling f a function machine, we say that f is a function of x. The greatest benefit of this notation is that it lets us easily note the input value of x along with the output of the function. Instead of "the value of Y_1 is 155 when $x = 4$," we can write $f(4) = 155$.

Example 5 Evaluating a Function

< Objective 3 >

Given $f(x) = x^3 + 3x^2 - x + 5$, find

(a) $f(0)$

Substituting 0 for x in the above expression, we get

$$(0)^3 + 3(0)^2 - (0) + 5 = 5$$

NOTE

$f(x)$ is just another name for y. The advantage of the $f(x)$ notation is seen here. It allows us to indicate the value for which we are evaluating the function.

(b) $f(-3)$

Substituting -3 for x in the above expression, we get

$$\begin{aligned}(-3)^3 + 3(-3)^2 - (-3) + 5 &= -27 + 27 + 3 + 5 \\ &= 8\end{aligned}$$

(c) $f\left(\frac{1}{2}\right)$

Substituting $\frac{1}{2}$ for x in the earlier expression, we get

$$\begin{aligned}\left(\frac{1}{2}\right)^3 + 3\left(\frac{1}{2}\right)^2 - \left(\frac{1}{2}\right) + 5 &= \frac{1}{8} + 3\left(\frac{1}{4}\right) - \frac{1}{2} + 5 \\ &= \frac{1}{8} + \frac{3}{4} - \frac{1}{2} + 5 \\ &= \frac{1}{8} + \frac{6}{8} - \frac{4}{8} + 5 \\ &= \frac{3}{8} + 5 \\ &= 5\frac{3}{8} \text{ or } \frac{43}{8}\end{aligned}$$

Check Yourself 5

Given $f(x) = 2x^3 - x^2 + 3x - 2$, find

(a) $f(0)$ **(b)** $f(3)$ **(c)** $f\left(-\frac{1}{2}\right)$

We can rewrite the relationship between x and $f(x)$ in Example 5 as a series of ordered pairs.

$f(x) = x^3 + 3x^2 - x + 5$

From this we found that

$f(0) = 5, \quad f(-3) = 8, \quad \text{and} \quad f\left(\frac{1}{2}\right) = \frac{43}{8}$

NOTE

Because $y = f(x)$, $(x, f(x))$ is another way of writing (x, y).

There is an ordered pair, which we could write as $(x, f(x))$, associated with each of these. Those three ordered pairs are

$(0, 5), \quad (-3, 8), \quad \text{and} \quad \left(\frac{1}{2}, \frac{43}{8}\right)$

Example 6 Finding Ordered Pairs

Given the function $f(x) = 2x^2 - 3x + 5$, find the ordered pair $(x, f(x))$ associated with each given value for x.

(a) $x = 0$

$f(0) = 2(0)^2 - 3(0) + 5 = 5$

The ordered pair is $(0, 5)$.

(b) $x = -1$

$f(-1) = 2(-1)^2 - 3(-1) + 5 = 10$

The ordered pair is $(-1, 10)$.

(c) $x = \frac{1}{4}$

$$f\left(\frac{1}{4}\right) = 2\left(\frac{1}{4}\right)^2 - 3\left(\frac{1}{4}\right) + 5 = \frac{35}{8}$$

The ordered pair is $\left(\frac{1}{4}, \frac{35}{8}\right)$.

Check Yourself 6

Given $f(x) = 2x^3 - x^2 + 3x - 2$, find the ordered pair associated with each given value of x.

(a) $x = 0$ **(b)** $x = 3$ **(c)** $x = -\frac{1}{2}$

We began this section by defining a relation as a set of ordered pairs. In Example 7, we will determine which relations can be modeled by a function machine.

Example 7 Modeling with a Function Machine

< Objective 4 >

Determine which relations can be modeled by a function machine.

(a) The set of all possible ordered pairs in which the first element is a U.S. state and the second element is a U.S. Senator from that state.

We cannot model this relation with a function machine. Because there are two senators from each state, each input does not have a unique output. In the picture, New Jersey is the input, but New Jersey has two different senators.

(b) The set of all ordered pairs in which the input is the year and the output is the U.S. Open golf champion of that year.

This relation can be modeled with the function machine. Each input has a unique output. In the picture, an input of 2000 gives an output of Tiger Woods. For any input year, there will be exactly one U.S. Open golf champion.

(c) The set of all ordered pairs in the relation R, when
$R = \{(1, 3), (2, 5), (2, 7), (3\ -4)\}$

This relation cannot be modeled with a function machine. An input of 2 results in two different outputs, 5 and 7.

(d) The set of all ordered pairs in the relation S, when
$S = \{(-1, 3), (0, 3), (3, 5), (5, -2)\}$

This relation can be modeled with a function machine. Each input has a unique output.

Check Yourself 7

Determine which relations can be modeled by a function machine.

(a) The set of all ordered pairs in which the first element is a U.S. city and the second element is the mayor of that city
(b) The set of all ordered pairs in which the first element is a street name and the second element is a U.S. city in which a street of that name is found
(c) The relation $A = \{(-2, 3), (-4, 9), (9, -4)\}$
(d) The relation $B = \{(1, 2), (3, 4), (3, 5)\}$

NOTE

We begin graphing functions in Section 2.5 and continue in Chapter 3.

If we are working with an equation in x and y, we may wish to rewrite the equation as a function of x. This is particularly useful if we want to use a graphing calculator to find y for a given x, or to view a graph of the equation.

Example 8 Writing Equations as Functions

< Objective 5 >

Rewrite each linear equation as a function of x. Use $f(x)$ notation in the final result.

(a) $y = 3x - 4$

We note that y is already isolated. Simply replace y with $f(x)$.

$f(x) = 3x - 4$

(b) $2x - 3y = 6$

We first solve for y.

$$-3y = -2x + 6$$

$$y = \frac{-2x + 6}{-3} \qquad \text{y has been isolated.}$$

$$y = \frac{2}{3}x - 2$$

$$f(x) = \frac{2}{3}x - 2 \qquad \text{Now replace y with f(x).}$$

Check Yourself 8

Rewrite each linear equation as a function of x. Use $f(x)$ notation in the final result.

(a) $y = -2x + 5$ **(b)** $3x + 5y = 15$

One benefit of having a function written in $f(x)$ form is that it makes it fairly easy to substitute values for x. Sometimes it is useful to substitute nonnumeric values for x.

Example 9 Substituting Nonnumeric Values for x

Let $f(x) = 2x + 3$. Evaluate f as indicated.

(a) $f(a)$

Substituting a for x in the equation, we see that

$$f(a) = 2a + 3$$

(b) $f(2 + h)$

Substituting $2 + h$ for x in the equation, we get

$$f(2 + h) = 2(2 + h) + 3$$

Distributing the 2 and then simplifying, we have

$$f(2 + h) = 4 + 2h + 3$$
$$= 2h + 7$$

Check Yourself 9

Let $f(x) = 4x - 2$. Evaluate f as indicated.

(a) $f(b)$ **(b)** $f(4 + h)$

The TABLE feature on a graphing calculator can also be used to evaluate a function. Example 10 illustrates this feature.

Example 10 Using a Graphing Calculator to Evaluate a Function

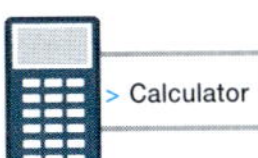

Evaluate the function $f(x) = 3x^3 + x^2 - 2x - 5$ for each x in the set $\{-6, -5, -4, -3, -2\}$.

1. Enter the function into a Y= screen.
2. Find the table setup screen.
3. Start the table at -6 with a change of 1.
4. View the table.

The table should look something like this.

X	Y_1	
−6	−605	
−5	−345	
−4	−173	
−3	−71	
−2	−21	
−1	−5	
0	−5	

X=−6

The Y_1 column is the function value for each value of x.

NOTE

Although we assumed that the graphing calculator was a TI, most such calculators have similar capability.

Check Yourself 10

Evaluate the function $f(x) = 2x^3 - 3x^2 - x + 2$ for each x in the set $\{-5, -4, -3, -2, -1, 0, 1\}$.

Check Yourself ANSWERS

1. **(a)** The domain of A is {Secretariat, Seattle Slew, Charismatic, Gallant Man}; **(b)** the domain of B is $\left\{-\frac{1}{2}, 0, 1, \pi\right\}$.
2. **(a)** The range of A is $\{10, 8, 5, 7\}$; **(b)** the range of B is $\left\{\frac{3}{4}, 0, 5, \pi\right\}$.
3. **(a)** Function; **(b)** function; **(c)** not a function
4. **(a)** 1; **(b)** 3; **(c)** -33
5. **(a)** -2; **(b)** 52; **(c)** -4
6. **(a)** $(0, -2)$; **(b)** $(3, 52)$; **(c)** $\left(-\frac{1}{2}, -4\right)$
7. **(a)** Function; **(b)** not a function; **(c)** function; **(d)** not a function
8. **(a)** $f(x) = -2x + 5$; **(b)** $f(x) = -\frac{3}{5}x + 3$
9. **(a)** $4b - 2$; **(b)** $4h + 14$
10.

X	Y_1	
−5	−318	
−4	−170	
−3	−76	
−2	−24	
−1	−2	
0	2	
1	0	

X=−5

Reading Your Text

SECTION 2.4

(a) A set of ordered pairs is called a ________________.

(b) The set of all first elements in a relation is called the ________________ of the relation.

(c) The set of second elements of a relation is called the ________________ of the relation.

(d) In a function of the form $y = f(x)$, x is called the ________________ variable, and y is called the dependent variable.

Basic Skills | Challenge Yourself | Calculator/Computer | Career Applications | Above and Beyond

2.4 exercises

Boost *your* GRADE at ALEKS.com!

ALEKS

- Practice Problems
- Self-Tests
- NetTutor
- e-Professors
- Videos

Name ______________

Section ________ Date ________

< Objective 1 >

Find the domain and range of each relation.

1. $A = \{(\text{Colorado}, 21), (\text{Edmonton}, 5), (\text{Calgary}, 18), (\text{Vancouver}, 17)\}$

2. $F = \left\{\left(\text{St. Louis}, \frac{1}{2}\right), \left(\text{Denver}, -\frac{3}{4}\right), \left(\text{Green Bay}, \frac{7}{8}\right), \left(\text{Dallas}, -\frac{4}{5}\right)\right\}$

3. $G = \left\{(\text{Chamber}, \pi), (\text{Testament}, 2\pi), \left(\text{Rainmaker}, \frac{1}{2}\right), (\text{Street Lawyer}, 6)\right\}$

4. $C = \{(\text{John Adams}, -16), (\text{John Kennedy}, -23), (\text{Richard Nixon}, -5),$ $(\text{Harry Truman}, -11)\}$

5. $\{(1, 2), (3, 4), (5, 6), (7, 8), (9, 10)\}$

6. $\{(2, 3), (3, 5), (4, 7), (5, 9), (6, 11)\}$

7. $\{(1, 2), (1, 3), (1, 4), (1, 5), (1, 6)\}$

8. $\{(3, 4), (3, 6), (3, 8), (3, 9), (3, 10)\}$

9. $\{(-1, 3), (-2, 4), (-3, 5), (4, 4), (5, 6)\}$ 

10. $\{(-2, 4), (1, 4), (-3, 4), (5, 4), (7, 4)\}$

11. **Business and Finance** The Dow Jones Industrial Averages over a 5-day period are displayed in the table. List this information as a set of ordered pairs, using the day of the week as the domain.

Day	1	2	3	4	5
Average	9,274	9,096	8,814	8,801	8,684

Answers

1. ______________
2. ______________
3. ______________
4. ______________
5. ______________
6. ______________
7. ______________
8. ______________
9. ______________
10. ______________
11. ______________

Answers

12. ________

13. ________

14. ________

15. ________

16. ________

17. ________

18. ________

19. ________

20. ________

21. ________

22. ________

23. ________

24. ________

25. ________

26. ________

12. Business and Finance In the snack department of the local supermarket, candy costs \$2.16 per pound. For 1 to 5 lb, write the cost of candy as a set of ordered pairs.

< Objective 2 >

Write a set of ordered pairs that describes each situation. Give the domain and range of each relation.

13. The first element is an integer between -3 and 3. The second coordinate is the cube of the first coordinate.

14. The first element is a positive integer less than 6. The second coordinate is the sum of the first coordinate and -2.

15. The first element is the number of hours worked—10, 20, 30, 40; the second coordinate is the salary at \$9 per hour.

16. The first coordinate is the number of toppings on a pizza (up to four); the second coordinate is the price of the pizza, which is \$9 plus \$1 per topping.

< Objective 3 >

Evaluate each function for the values specified.

17. $f(x) = x^2 - x - 2$; find **(a)** $f(0)$, **(b)** $f(-2)$, and **(c)** $f(1)$.

18. $f(x) = x^2 - 7x + 10$; find **(a)** $f(0)$, **(b)** $f(5)$, and **(c)** $f(-2)$.

19. $f(x) = 3x^2 + x - 1$; find **(a)** $f(-2)$, **(b)** $f(0)$, and **(c)** $f(1)$.

20. $f(x) = -x^2 - x - 2$; find **(a)** $f(-1)$, **(b)** $f(0)$, and **(c)** $f(2)$.

21. $f(x) = x^3 - 2x^2 + 5x - 2$; find **(a)** $f(-3)$, **(b)** $f(0)$, and **(c)** $f(1)$.

22. $f(x) = -2x^3 + 5x^2 - x - 1$; find **(a)** $f(-1)$, **(b)** $f(0)$, and **(c)** $f(2)$.

23. $f(x) = -3x^3 + 2x^2 - 5x + 3$; find **(a)** $f(-2)$, **(b)** $f(0)$, and **(c)** $f(3)$.

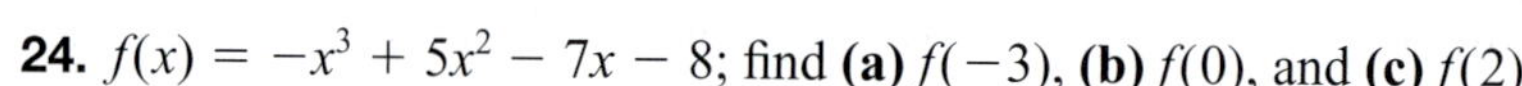

24. $f(x) = -x^3 + 5x^2 - 7x - 8$; find **(a)** $f(-3)$, **(b)** $f(0)$, and **(c)** $f(2)$.

25. $f(x) = 2x^3 + 4x^2 + 5x + 2$; find **(a)** $f(-1)$, **(b)** $f(0)$, and **(c)** $f(1)$.

26. $f(x) = -x^3 + 2x^2 - 7x + 9$; find **(a)** $f(-2)$, **(b)** $f(0)$, and **(c)** $f(2)$.

< Objective 4 >

In exercises 27 to 34, determine which of the relations are also functions.

27. $\{(1, 6), (2, 8), (3, 9)\}$

28. $\{(2, 3), (3, 4), (5, 9)\}$

29. $\{(-1, 4), (-2, 5), (-3, 7)\}$

30. $\{(-2, 1), (-3, 4), (-4, 6)\}$

> Videos

31. $\{(1, 3), (1, 2), (1, 1)\}$

32. $\{(2, 4), (2, 5), (3, 6)\}$

33. $\{(-3, 5), (6, 3), (6, 9)\}$

34. $\{(4, -4), (2, 8), (4, 8)\}$

Decide whether the relation, shown as a table of values, is a function.

35.

x	y
3	1
−2	4
5	3
−7	4

36.

x	y
−2	3
1	4
5	6
2	−1

37.

x	y
2	3
4	2
2	−5
−6	−3

38.

> Videos

x	y
1	5
3	−6
1	−5
−2	−9

39.

x	y
−1	2
3	6
6	2
−9	4

40.

x	y
4	−6
2	3
−7	1
−3	−6

< Objective 5 >

Rewrite each equation as a function of x. Use f(x) notation in the final result.

41. $y = -3x + 2$

42. $y = 5x + 7$

43. $y = 4x - 8$

44. $y = -7x - 9$

Answers

27. ______

28. ______

29. ______

30. ______

31. ______

32. ______

33. ______

34. ______

35. ______

36. ______

37. ______

38. ______

39. ______

40. ______

41. ______

42. ______

43. ______

44. ______

Answers

45. ______

46. ______

47. ______

48. ______

49. ______

50. ______

51. ______

52. ______

53. ______

54. ______

55. ______

56. ______

57. ______

58. ______

59. ______

60. ______

61. ______

62. ______

63. ______

64. ______

45. $3x + 2y = 6$

46. $4x + 3y = 12$

47. $-2x + 6y = 9$

48. $-3x + 4y = 11$

49. $-5x - 8y = -9$

50. $4x - 7y = -10$

Basic Skills | **Challenge Yourself** | Calculator/Computer | Career Applications | Above and Beyond

Complete each statement with **never, sometimes,** *or* **always.**

51. The domain of a relation ________ consists of the set of all first coordinates of the ordered pairs of the relation.

52. When evaluating a function at a particular x-value, we ________ obtain two y-values.

If $f(x) = 5x - 1$, *find*

53. $f(a)$

54. $f(2r)$

55. $f(x + 1)$ 

56. $f(a - 2)$

57. $f(x + h)$

58. $\dfrac{f(x + h) - f(x)}{h}$

If $g(x) = -3x + 2$, *find*

59. $g(m)$

60. $g(5n)$

61. $g(x + 2)$

62. $g(s - 1)$

Solve each application.

63. **Business and Finance** The marketing department of a company has determined that the profit for selling x units of a product is approximated by the function

$$f(x) = 50x - 600$$

Find the profit in selling 2,500 units.

64. **Business and Finance** The inventor of a new product believes that the cost of producing the product is given by the function

$C(x) = 1.75x + 7{,}000$ where x = units produced

What would be the cost of producing 2,000 units of the product?

65. **Business and Finance** A phone company has two different rates for calls made at different times of the day. These rates are given by the function

$$C(x) = \begin{cases} 24x + 33 & \text{between 5 P.M. and 11 P.M.} \\ 36x + 52 & \text{between 8 A.M. and 5 P.M.} \end{cases}$$

where x is the number of minutes of a call and C is the cost of a call in cents.

(a) What is the cost of a 10-minute call at 10:00 A.M.?
(b) What is the cost of a 10-minute call at 10:00 P.M.?

66. **Statistics** The number of accidents in 1 month involving drivers x years of age can be approximated by the function

$f(x) = 2x^2 - 125x + 3{,}000$

Find the number of accidents in 1 month that involved **(a)** 17-year-olds and **(b)** 25-year-olds.

67. **Science and Medicine** The distance x (in feet) that a car will skid on a certain road surface after the brakes are applied is a function of the car's velocity v (in miles per hour). The function can be approximated by

$x = f(v) = 0.017v^2$

How far will the car skid if the brakes are applied at **(a)** 55 mi/h? **(b)** 70 mi/h?

68. **Science and Medicine** An object is thrown upward with an initial velocity of 128 ft/s. Its height h in feet after t seconds is given by the function

$h(t) = -16t^2 + 128t$

What is the height of the object at **(a)** 2 s? **(b)** 4 s? **(c)** 6 s?

69. **Science and Medicine** Suppose that the weight (in pounds) of a baby boy x months old is predicted, for his first 10 months, by the function

$f(x) = 1.5x + 8.3$

(a) Find the predicted weight at the age of 4 months.
(b) Find the predicted weight at the age of 8 months.

70. **Science and Medicine** Suppose that the height (in inches) of a baby boy x months old is predicted, for his first 10 months, by the function

$f(x) = x + 21.3$

(a) Find the predicted height at the age of 4 months.
(b) Find the predicted height at the age of 8 months.

Answers

65. ______
66. ______
67. ______
68. ______
69. ______
70. ______

Answers

71. ________

72. ________

73. ________

74. ________

75. ________

76. ________

77. ________

78. ________

Basic Skills | Challenge Yourself | **Calculator/Computer** | Career Applications | Above and Beyond

Use your graphing calculator to evaluate the given function for each value in the given set.

71. $f(x) = 3x^2 - 5x + 7$; $\{-5, -4, -3, -2, -1, 0, 1, 2, 3, 4, 5\}$

72. $f(x) = 4x^3 - 7x^2 + 9$; $\{-3, -2, -1, 0, 1, 2, 3\}$

73. $f(x) = 2x^3 - 4x^2 + 5x - 9$; $\{-4, -3, -2, -1, 0, 1, 2, 3, 4\}$

74. $f(x) = -3x^4 + 5x^2 - 7x - 15$; $\{-3, -2, -1, 0, 1, 2, 3\}$

Basic Skills | Challenge Yourself | Calculator/Computer | **Career Applications** | Above and Beyond

75. Manufacturing Technology The pitch of a 6-in. gear is given by the number of teeth divided by 6.

(a) Write a function to describe this relationship.

(b) What is the pitch of a 6-in. gear with 30 teeth?

76. Allied Health Dimercaprol (BAL) is used to treat arsenic poisoning in mammals. The recommended dose is 4 milligrams (mg) per kilogram (kg) of the animal's weight.

(a) Construct a function for the dosage in terms of an animal's weight.

(b) How much BAL must be administered to a 5-kg cat?

(c) What size cow requires a 1,450-mg dose of BAL?

77. Construction Technology The cost of building a house is \$90 per square foot plus \$12,000 for the foundation.

(a) Give the cost of building a house as a function of the area of the house.

(b) How much does it cost to build an 1,800-ft^2 house?

78. Mechanical Engineering A computer-aided design (CAD) operator has located 3 corners of a rectangle, at (5, 9), (−2, 9), and (5, 2). Give the coordinates of the fourth corner.

Answers

1. Domain: {Colorado, Edmonton, Calgary, Vancouver}; Range: {21, 5, 18, 17}

3. Domain: {Chamber, Testament, Rainmaker, Street Lawyer}; Range: $\left\{\pi, 2\pi, \frac{1}{2}, 6\right\}$ **5.** Domain: {1, 3, 5, 7, 9}; Range: {2, 4, 6, 8, 10} **7.** Domain: {1}; Range: {2, 3, 4, 5, 6} **9.** Domain: $\{-3, -2, -1, 4, 5\}$; Range: {3, 4, 5, 6} **11.** {(1, 9,274), (2, 9,096), (3, 8,814), (4, 8,801), (5, 8,684)} **13.** $\{(-2, -8), (-1, -1), (0, 0), (1, 1), (2, 8)\}$; Domain: $\{-2, -1, 0, 1, 2\}$; Range: $\{-8, -1, 0, 1, 8\}$ **15.** {(10, 90), (20, 180), (30, 270), (40, 360)}; Domain: {10, 20, 30, 40}; Range: {90, 180, 270, 360} **17.** **(a)** -2; **(b)** 4; **(c)** -2 **19.** **(a)** 9; **(b)** -1; **(c)** 3 **21.** **(a)** -62; **(b)** -2; **(c)** 2 **23.** **(a)** 45; **(b)** 3; **(c)** -75 **25.** **(a)** -1; **(b)** 2; **(c)** 13 **27.** Function **29.** Function **31.** Not a function **33.** Not a function **35.** Function **37.** Not a function **39.** Function **41.** $f(x) = -3x + 2$ **43.** $f(x) = 4x - 8$ **45.** $f(x) = -\frac{3}{2}x + 3$ **47.** $f(x) = \frac{1}{3}x + \frac{3}{2}$ **49.** $f(x) = -\frac{5}{8}x + \frac{9}{8}$ **51.** always **53.** $5a - 1$ **55.** $5x + 4$ **57.** $5x + 5h - 1$ **59.** $-3m + 2$ **61.** $-3x - 4$ **63.** \$124,400 **65.** **(a)** \$4.12; **(b)** \$2.73 **67.** **(a)** 51.425 ft; **(b)** 83.3 ft **69.** **(a)** 14.3 lb; **(b)** 20.3 lb **71.** 107, 75, 49, 29, 15, 7, 5, 9, 19, 35, 57 **73.** $-221, -114, -51, -20, -9, -6, 1, 24, 75$ **75.** **(a)** $P(t) = \frac{t}{6}$; **(b)** 5 **77.** **(a)** $C(x) = 90x + 12{,}000$; **(b)** \$174,000

2.5

Tables and Graphs

< 2.5 Objectives >

1 > Use the vertical line test

2 > Identify the domain and range from the graph of a relation

3 > Read function values from a table

4 > Read function values from a graph

In Section 2.4, we defined a function in terms of ordered pairs. A set of ordered pairs can be specified in several ways; here are the most common.

Property

Ordered Pairs

1. We can present ordered pairs in a list or table.
2. We can give a rule or equation that will generate ordered pairs.
3. We can use a graph to indicate ordered pairs. The graph can show distinct ordered pairs, or it can show all the ordered pairs on a line or curve.

We have already seen functions presented as lists of ordered pairs, in tables, and as rules or equations. We now look at graphs of the ordered pairs from Example 3 in Section 2.4 to introduce the **vertical line test,** which is a graphical test for identifying a function.

RECALL

If a grid has no numeric labels, each mark represents one unit. In this text, each such grid represents x-values from -8 to 8 and y-values from -8 to 8.

(a) As a set of ordered pairs, the relation is $\{(-2, 1), (-1, 1), (1, 3), (2, 3)\}$. Recall that this relation does represent a function.

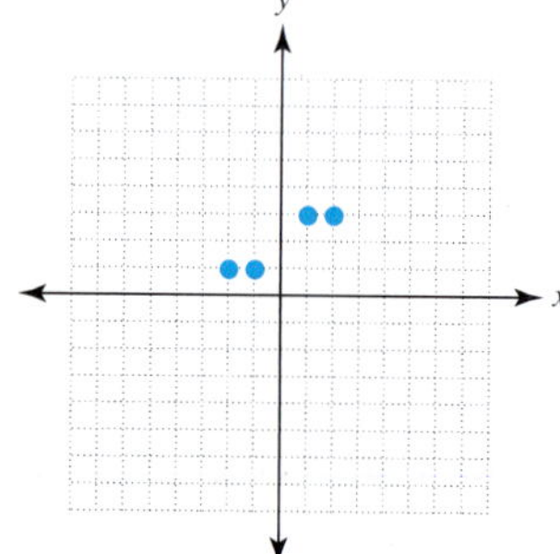

(b) As a set of ordered pairs, the relation is $\{(-5, -2), (-1, 3), (-1, 6), (2, 8)\}$. Recall that this relation does *not* represent a function.

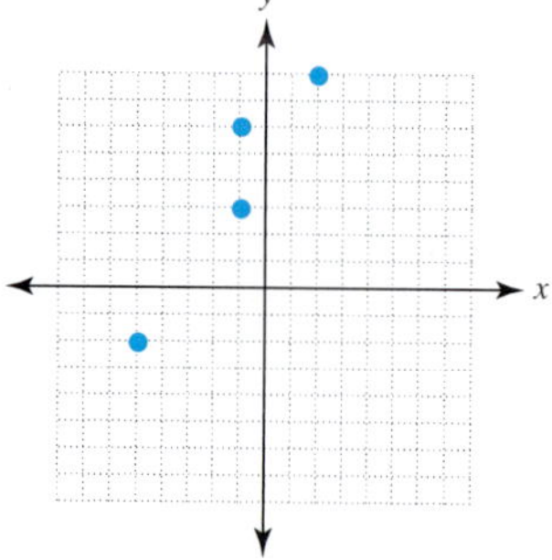

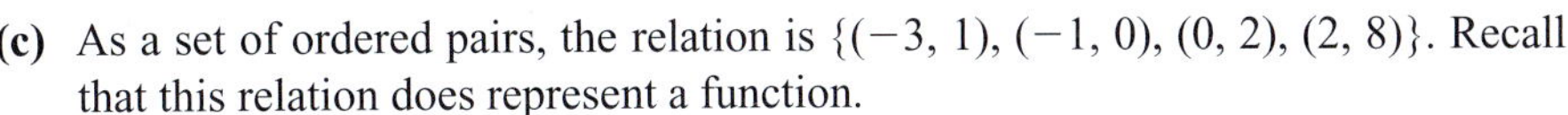

(c) As a set of ordered pairs, the relation is $\{(-3, 1), (-1, 0), (0, 2), (2, 8)\}$. Recall that this relation does represent a function.

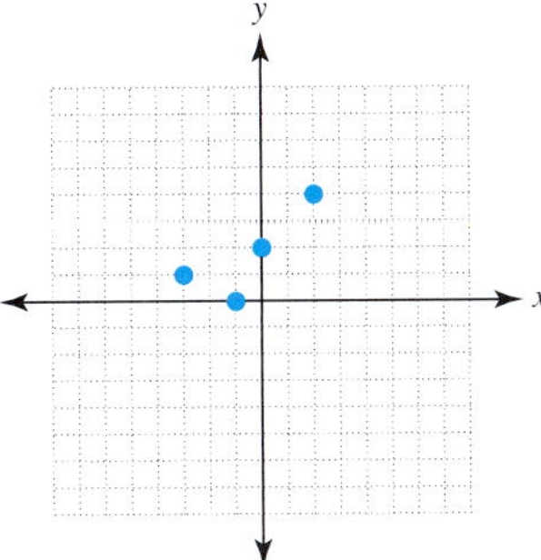

Notice that in the graphs of relations (a) and (c), there is no vertical line that can pass through two different points of the graph. In relation (b), a vertical line can pass through the two points that represent the ordered pairs $(-1, 3)$ and $(-1, 6)$. This leads to the following test.

Property

Vertical Line Test

A relation is a function if no vertical line can pass through two or more points on its graph.

Example 1 Identifying a Function

< Objective 1 >

For each set of ordered pairs, plot the related points on the provided axes. Then use the vertical line test to determine which of the sets is a function.

(a) $\{(0, -1), (2, 3), (2, 6), (4, 2), (6, 3)\}$

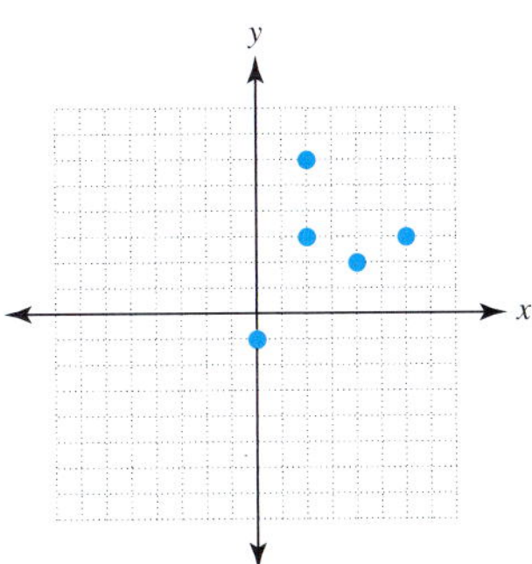

Because a vertical line can be drawn through the points (2, 3) and (2, 6), the relation does not pass the vertical line test. That is, if the input is 2, the output is *both* 3 and 6. This is not a function.

(b) $\{(1, 1), (2, 0), (3, 3), (4, 3), (5, 3)\}$

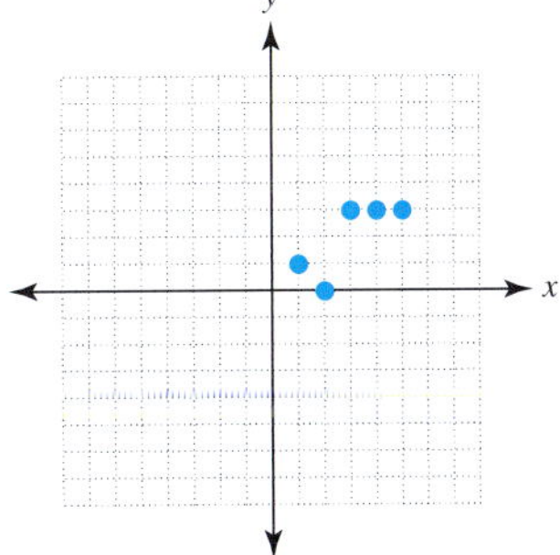

This is a function. Although a horizontal line can be drawn through several points, no vertical line passes through more than one point.

Check Yourself 1

For each set of ordered pairs, plot the related points. Then use the vertical line test to determine which of the sets is a function.

(a) $\{(-2, 4), (-1, 4), (0, 4), (1, 3), (5, 5)\}$
(b) $\{(-3, -1), (-1, -3), (1, -3), (1, 3)\}$

By studying the graph of a relation, we can also determine the domain and range, as shown in Example 2. Recall that the domain is the set of x-values that appear in the ordered pairs, while the range is the set of y-values.

Example 2 Identifying Functions, Domain, and Range

< Objective 2 >

Determine whether the given graph is the graph of a function. Also provide the domain and range in each case.

(a)

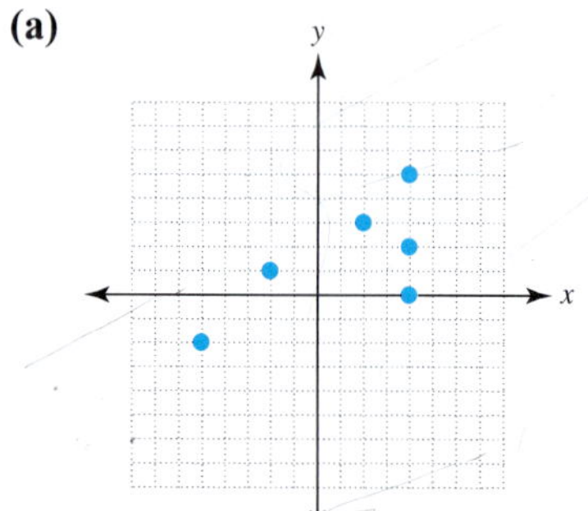

This is not a function. A vertical line at $x = 4$ passes through three points. The domain D of this relation is

$D = \{-5, -2, 2, 4\}$

and the range R is

$R = \{-2, 0, 1, 2, 3, 5\}$

(b)

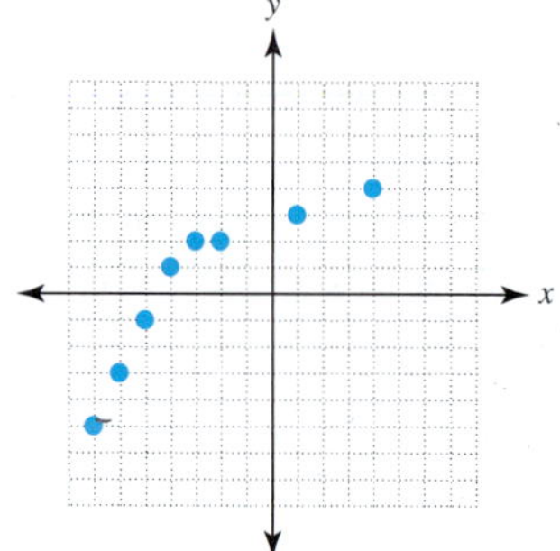

This is a function. No vertical line passes through more than one point. The domain is

$D = \{-7, -6, -5, -4, -3, -2, 1, 4\}$

and the range is

$R = \{-5, -3, -1, 1, 2, 3, 4\}$

Check Yourself 2

Determine whether the given graph is the graph of a function. Also provide the domain and range in each case.

(a)

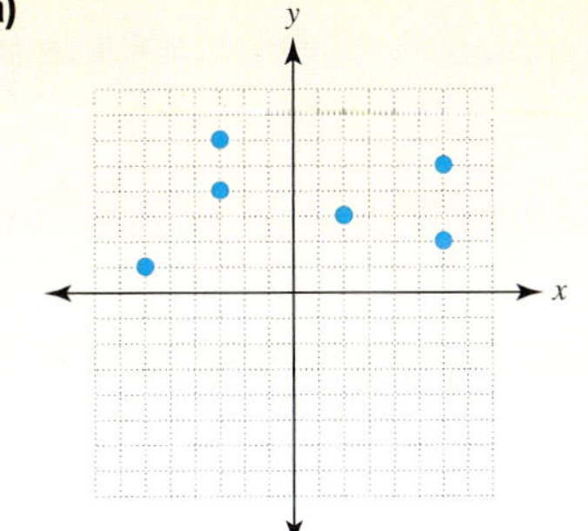

(b)

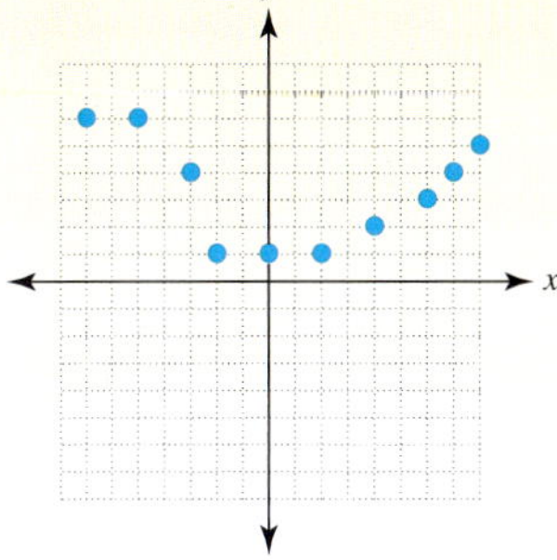

The graphs presented in this section have all depicted relations that are **finite** sets of ordered pairs. We now consider graphs composed of line segments, lines, or curves. Each such graph represents an **infinite** collection of points. The vertical line test can be used to decide whether the relation is a function, and we can name the domain and range.

Example 3 Identifying Functions, Domain, and Range

Determine whether the given graph is the graph of a function. Also provide the domain and range in each case.

NOTE

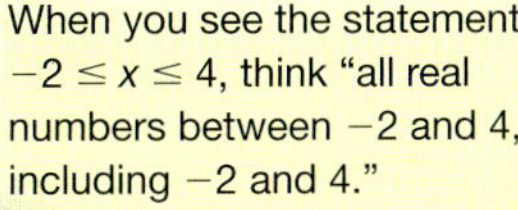

When you see the statement $-2 \le x \le 4$, think "all real numbers between -2 and 4, including -2 and 4."

(a)

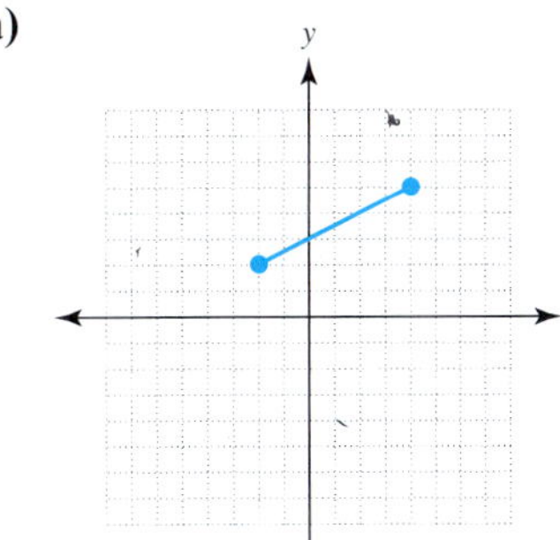

Because no vertical line will pass through more than one point, this is a function. The x-values that are used in the ordered pairs go from -2 to 4, inclusive. By using set-builder notation, we write the domain as

$$D = \{x \mid -2 \le x \le 4\}$$

The y-values that are used go from 2 to 5, inclusive. The range is

$$R = \{y \mid 2 \le y \le 5\}$$

(b)

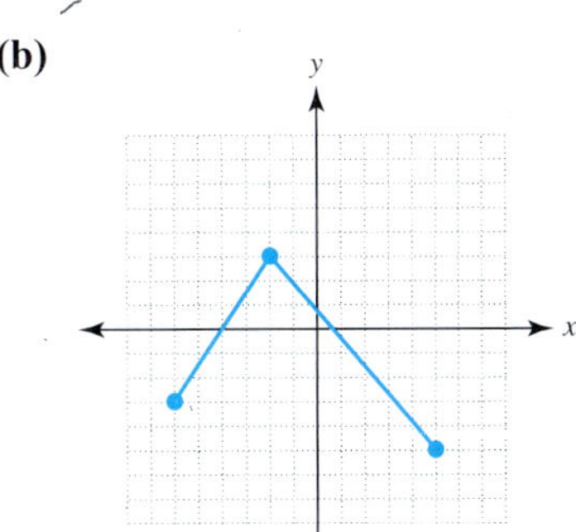

RECALL

When the endpoints are included, we use the "less than or equal to" symbol $\le$.

The relation graphed here is a function. The x-values run from -6 to 5, so

$$D = \{x \mid -6 \le x \le 5\}$$

The y-values go from -5 to 3, so

$$R = \{y \mid -5 \le y \le 3\}$$

Check Yourself 3

Determine whether the given graph is the graph of a function. Also provide the domain and range in each case.

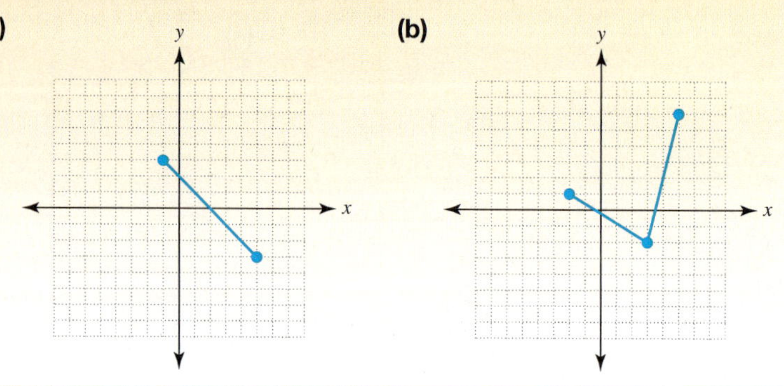

In Example 4, we consider the graphs of some common curves.

Example 4 Identifying Functions, Domain, and Range

Determine whether the given graph is the graph of a function. Also provide the domain and range in each case.

(a)

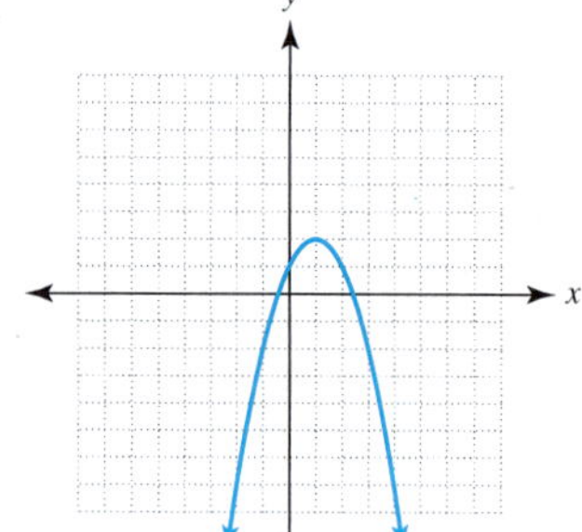

> **NOTE**
>
> Depicted here is a curve called a *parabola.*

> **RECALL**
>
> $\mathbb{R}$ is the symbol for the set of all real numbers.

Since no vertical line will pass through more than one point, this is a function. Note that the arrows on the ends of the graph indicate that the pattern continues indefinitely. The x-values that are used in this graph therefore consist of all real numbers. The domain is

$$D = \{x \mid x \text{ is a real number}\}$$

or simply $D = \mathbb{R}$.

The y-values, however, are never higher than 2. The range is the set of all real numbers less than or equal to 2. So

$$R = \{y \mid y \leq 2\}$$

(b)

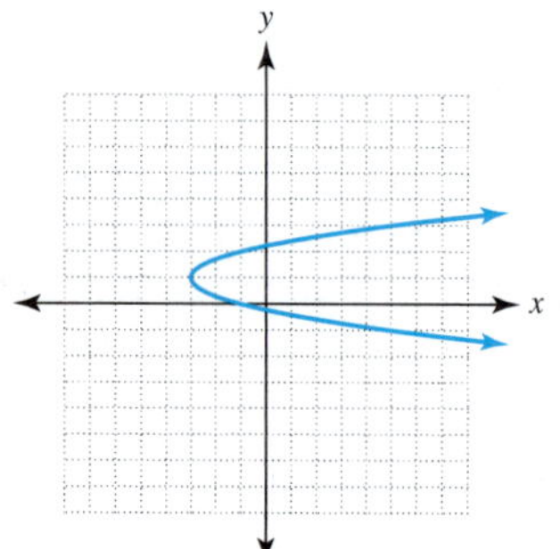

> **NOTE**
>
> This curve is also a parabola.

This relation is not a function. A vertical line drawn anywhere to the right of -3 will pass through two points. The x-values that are used begin at -3 and continue indefinitely to the right, so

$D = \{x \mid x \geq -3\}$

The y-values consist of all real numbers, so

$R = \{y \mid y \text{ is a real number}\}$

or simply $R = \mathbb{R}$.

(c)

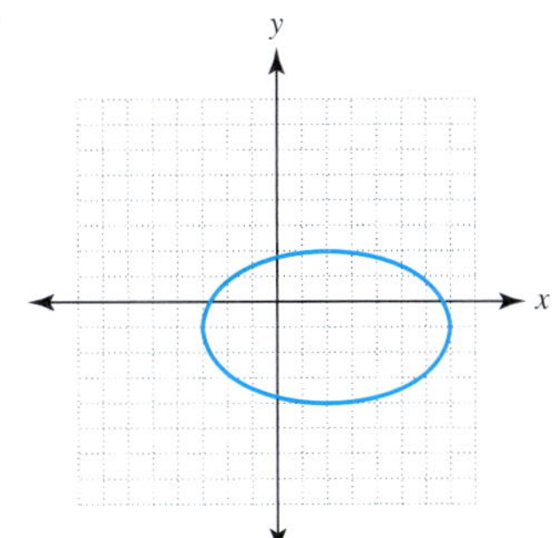

NOTE

This curve is called an *ellipse*.

This relation is not a function. A vertical line drawn anywhere between -3 and 7 will pass through two points. The x-values that are used run from -3 to 7, inclusive. Thus,

$D = \{x \mid -3 \leq x \leq 7\}$

The y-values used in the ordered pairs go from -4 to 2, inclusive, so

$R = \{y \mid -4 \leq y \leq 2\}$

Check Yourself 4

Determine whether the given graph is the graph of a function. Also provide the domain and range in each case.

(a)

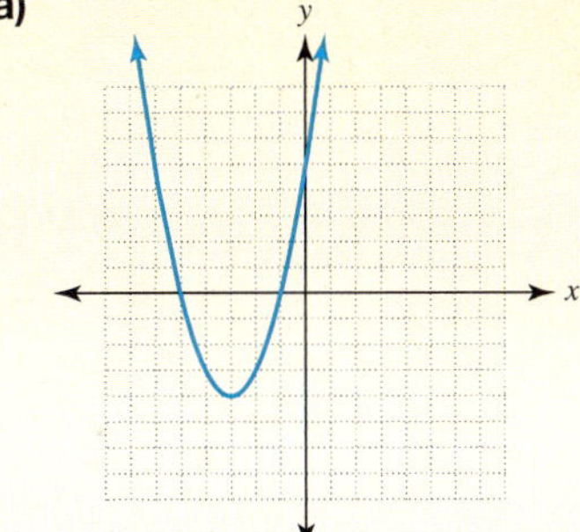

(b)

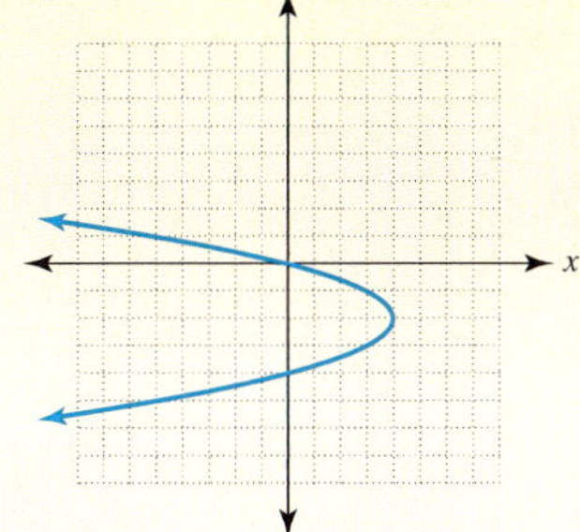

(c)

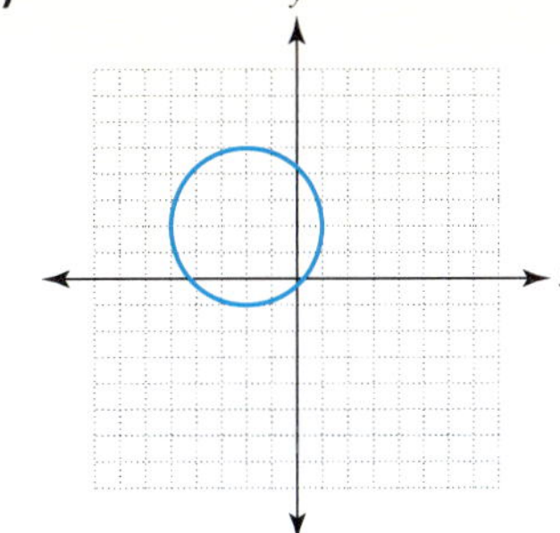

An important skill in working with functions is that of reading tables and graphs. If we are given a function f in either of these forms, our goals here are twofold.

1. Given x, we want to find $f(x)$.
2. Given $f(x)$, we want to find x.

Example 5 illustrates.

Example 5 Reading Values from a Table

< Objective 3 >

Suppose we have the functions f and g, as shown.

x	$f(x)$
−4	8
0	6
2	−4
1	−2

x	$g(x)$
−2	5
1	0
4	−4
8	−2

NOTE

Think of the x-values as "input" values and the $f(x)$ values as "outputs."

(a) Find $f(0)$.

This means that 0 is the input value (a value for x). We want to know what f does to 0. Looking in the table, we see that the output value is 6. So $f(0) = 6$.

(b) Find $g(4)$.

We are given $x = 4$, and we want $g(x)$. In the table we find $g(4) = -4$.

(c) Find x, given that $f(x) = -4$.

Now we are given the output value of -4. We ask, what x-value results in an output value of -4? The answer is 2. So $x = 2$.

(d) Find x, given that $g(x) = -2$.

Since the output is given as -2, we look in the table to find that when $x = 8$, $g(x) = -2$. So $x = 8$.

Check Yourself 5

Use the functions in Example 5 to find

(a) $f(1)$ **(b)** $g(-2)$
(c) x, given that $f(x) = 8$ **(d)** x, given that $g(x) = 5$

In Example 6 we consider the same goals, given the graph of a function: (1) given x, find $f(x)$; and (2) given $f(x)$, find x.

Example 6 Reading Values from a Graph

< Objective 4 >

Given the graph of f shown, find the desired values.

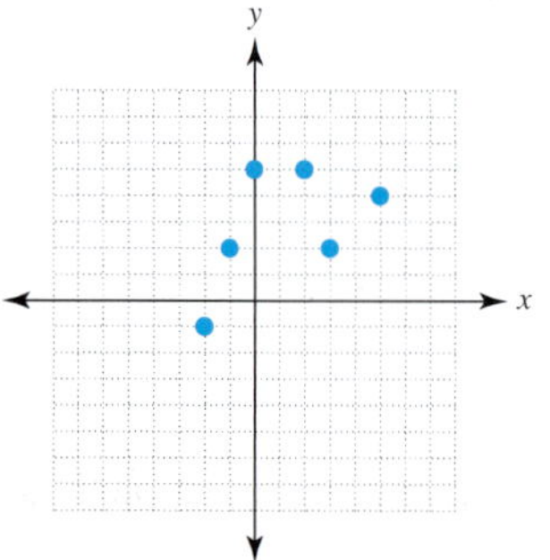

(a) Find $f(2)$.

Since 2 is the x-value, we move to 2 on the x-axis and then search vertically for a plotted point. We find (2, 5), which tells us that an input of 2 results in an output of 5. Thus, $f(2) = 5$.

(b) Find $f(-1)$.

Since $x = -1$, we move to -1 on the x-axis. We note the point $(-1, 2)$, so $f(-1) = 2$.

(c) Find all x such that $f(x) = 2$.

Now we are told that the output value is 2, so we move up to 2 on the y-axis and search horizontally for plotted points. There are two: $(-1, 2)$ and $(3, 2)$. So the desired x-values are -1 and 3.

(d) Find all x such that $f(x) = 4$.

We move to 4 on the y-axis and search horizontally. We find one point: (5, 4). So $x = 5$.

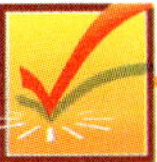

Check Yourself 6

Given the graph of *f* shown, find the desired values.

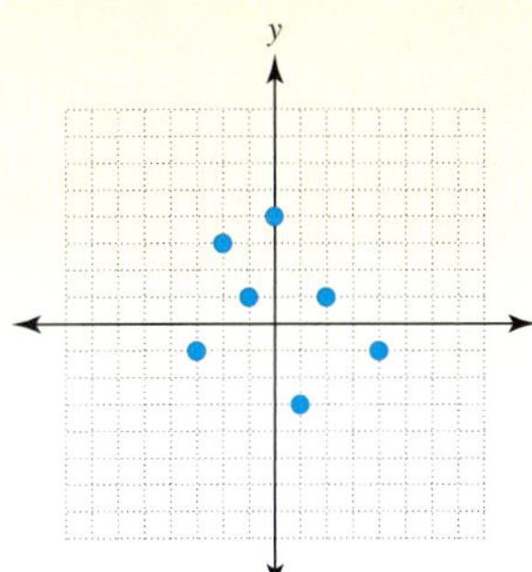

(a) Find $f(1)$.
(b) Find $f(-3)$.
(c) Find all x such that $f(x) = 1$.
(d) Find all x such that $f(x) = 4$.

Example 7 deals with graphs that represent **infinite** collections of points.

Example 7 Reading Values from a Graph

(a) Given the graph of f shown, find the desired values.

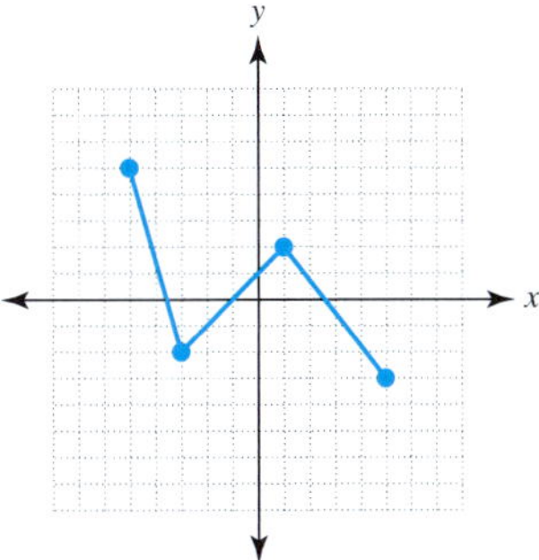

(i) Find $f(-1)$.

Since $x = -1$, we move to -1 on the x-axis. There we find the point $(-1, 0)$. So $f(-1) = 0$.

(ii) Find all x such that $f(x) = -1$.

We are given the output -1, so we move to -1 on the y-axis and search horizontally for plotted points. There are three of them, and we must estimate the coordinates for a couple of these. One point is exactly $(-2, -1)$, one is approximately $(-3.3, -1)$, and one is approximately $(3.5, -1)$. So the desired x-values are -3.3, -2, and 3.5.

(b) Given the graph of f shown, find the desired values.

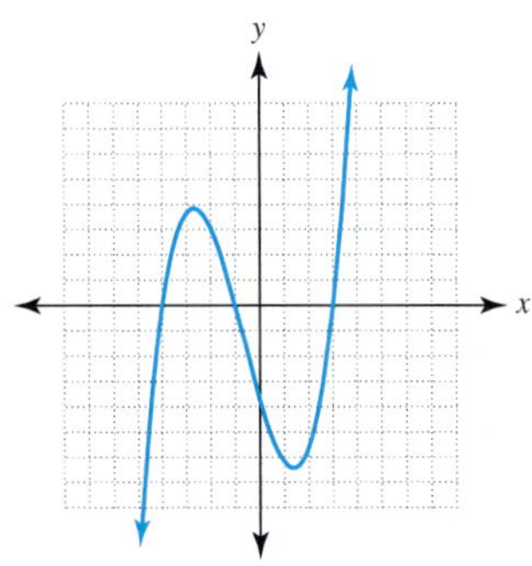

(i) Find $f(-3)$.

Since $x = -3$, we move to -3 on the x-axis. We search vertically and estimate a plotted point at approximately $(-3, 3.7)$. So $f(-3) \approx 3.7$.

(ii) Find all x such that $f(x) = 0$.

Since the output (y-value) is 0, we look for points with a y-coordinate of 0. There are three: $(-4, 0)$, $(-1, 0)$, and $(3, 0)$. So the desired x-values are -4, -1, and 3.

Check Yourself 7

Given the graph of f shown, find the desired values.

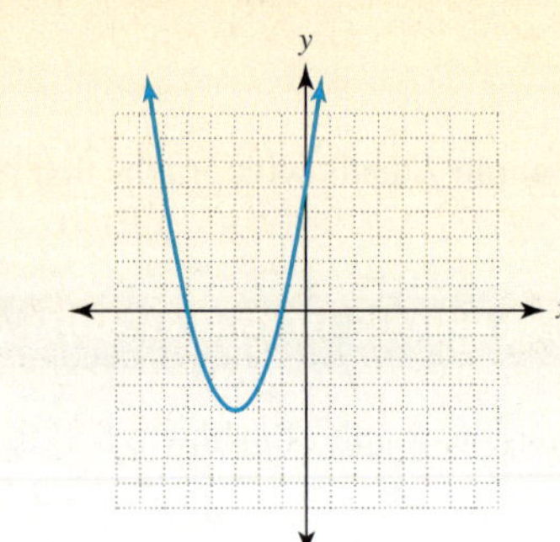

(a) Find $f(-4)$.

(b) Find all x such that $f(x) = 0$.

At this point, you may be wondering how the concept of function relates to anything outside the study of mathematics. A function is a relation that yields a single output (y-value) each time a specific input (x-value) is given. Any field in which predictions are made is building on the idea of functions. Here are a few examples:

- A physicist looks for the relationship that uses a planet's mass to predict its gravitational pull.
- An economist looks for the relationship that uses the tax rate to predict the employment rate.
- A business marketer looks for the relationship that uses an item's price to predict the number that will be sold.

- A college board looks for the relationship between tuition costs and the number of students enrolled at the college.
- A biologist looks for the relationship that uses temperature to predict a body of water's nutrient level.

In your future study of mathematics, you will see functions applied in areas such as these. In those applications, you should find that you put to good use the basic skills developed here: (1) given x, find $f(x)$; and (2) given $f(x)$, find x.

Check Yourself ANSWERS

1. **(a)** Is a function; **(b)** is not a function
2. **(a)** Is not a function; $D = \{-6, -3, 2, 6\}$; $R = \{1, 2, 3, 4, 5, 6\}$;
   (b) is a function; $D = \{-7, -5, -3, -2, 0, 2, 4, 6, 7, 8\}$; $R = \{1, 2, 3, 4, 5, 6\}$
3. **(a)** Is a function; $D = \{x \mid -1 \le x \le 5\}$; $R = \{y \mid -3 \le y \le 3\}$;
   (b) is a function; $D = \{x \mid -2 \le x \le 5\}$; $R = \{y \mid -2 \le y \le 6\}$
4. **(a)** Is a function; $D = \mathbb{R}$; $R = \{y \mid y \ge -4\}$;
   (b) is not a function; $D = \{x \mid x \le 4\}$; $R = \mathbb{R}$;
   (c) is not a function; $D = \{x \mid -5 \le x \le 1\}$; $R = \{y \mid -1 \le y \le 5\}$
5. **(a)** -2; **(b)** 5; **(c)** -4; **(d)** -2
6. **(a)** -3; **(b)** -1; **(c)** -1 and 2; **(d)** 0
7. **(a)** -3; **(b)** -5 and -1

Reading Your Text

SECTION 2.5

(a) The vertical line test is a graphical test for identifying a ______________.

(b) A ______________ is a function if no vertical line passes through two or more points on its graph.

(c) The ______________ of a function is the set of inputs that can be substituted for the independent variable.

(d) The range of a function is the set of ______________ or y-values.

2.5 exercises

Basic Skills | Challenge Yourself | Calculator/Computer | Career Applications | Above and Beyond

Boost *your* GRADE at ALEKS.com!

ALEKS®

- Practice Problems
- Self-Tests
- NetTutor
- e-Professors
- Videos

Name ____________

Section ________ Date ________

Answers

1. ____________
2. ____________
3. ____________
4. ____________
5. ____________
6. ____________
7. ____________
8. ____________
9. ____________
10. ____________

< Objective 1 >

For each set of ordered pairs, plot the related points. Then use the vertical line test to determine which sets are functions.

1. $\{(-3, 1), (-1, 2), (-2, 3), (1, 4)\}$

2. $\{(2, 2), (1, 1), (3, 3), (4, 5)\}$

3. $\{(-1, 1), (2, 2), (3, 4), (5, 6)\}$

4. $\{(1, 4), (-1, 5), (0, 2), (2, 3)\}$

5. $\{(1, 2), (1, 3), (2, 1), (3, 1)\}$

6. $\{(-1, 1), (3, 4), (-1, 2), (5, 3)\}$

< Objective 2 >

Determine whether the relation is a function. Also provide the domain and the range.

> Videos

7.

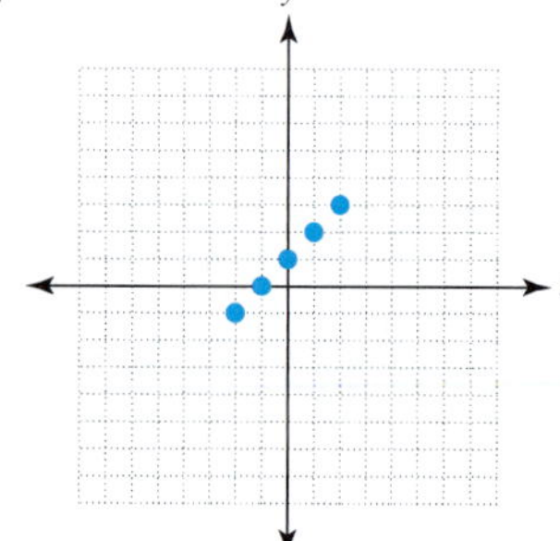

8.

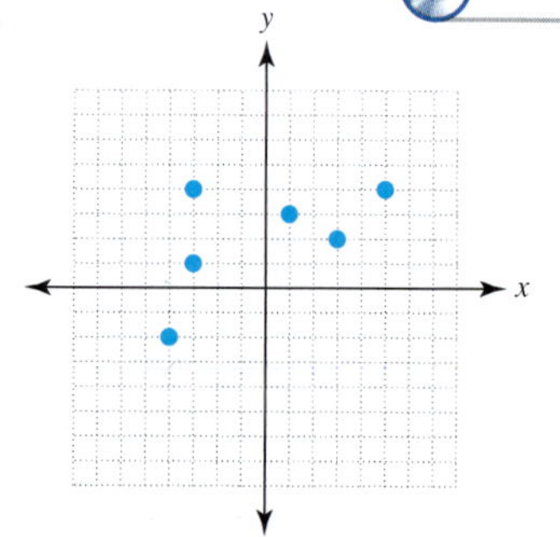

9.

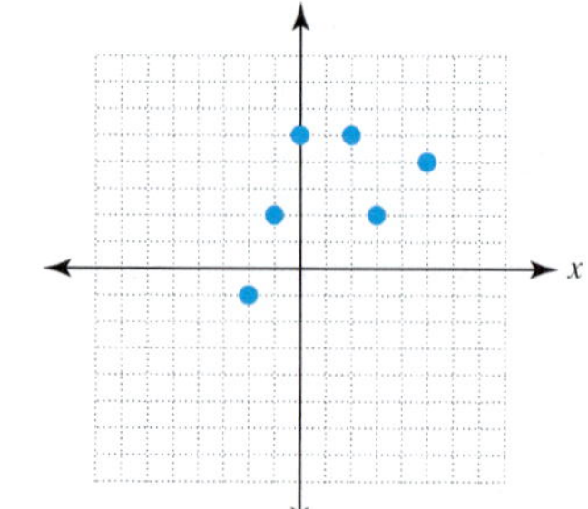

10.

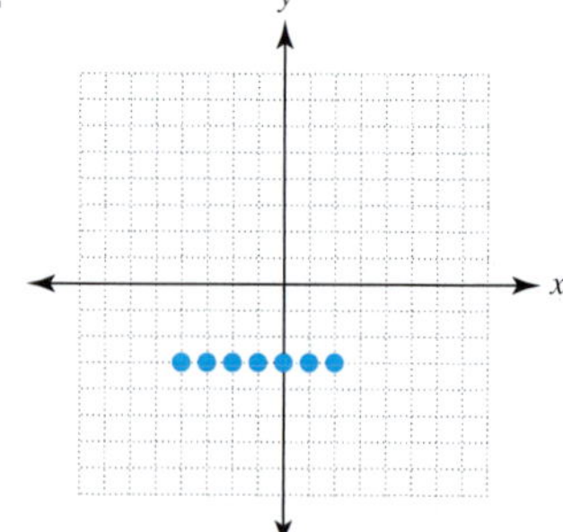

11.

12.

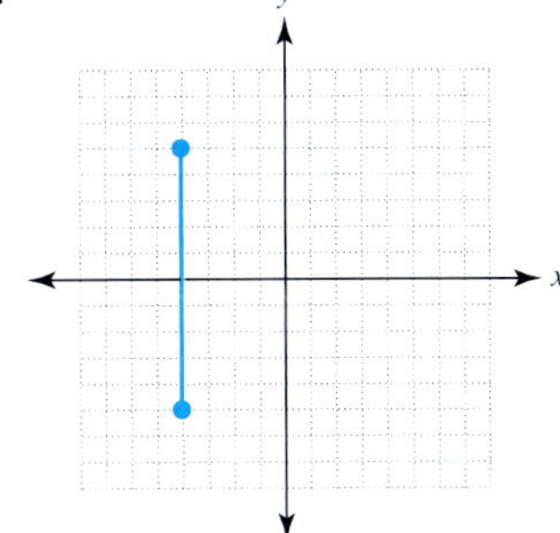

13.

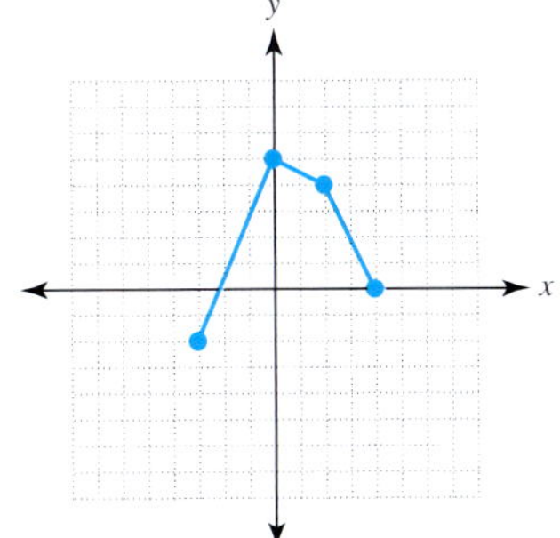

14.

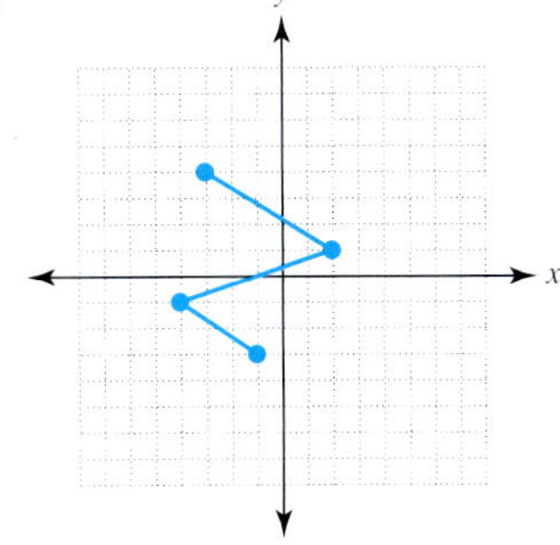

15.

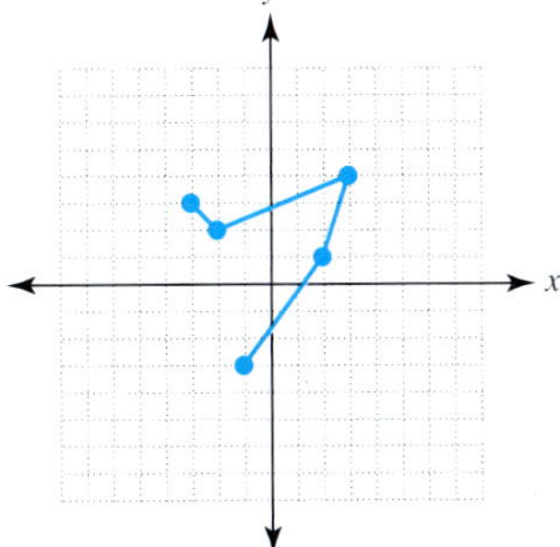

16.

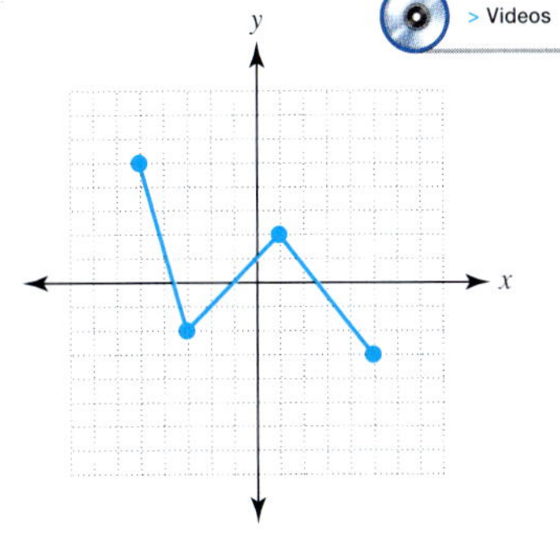

17.

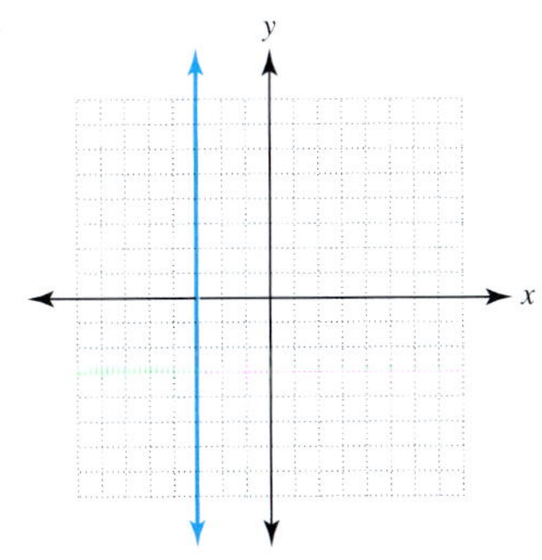

18.

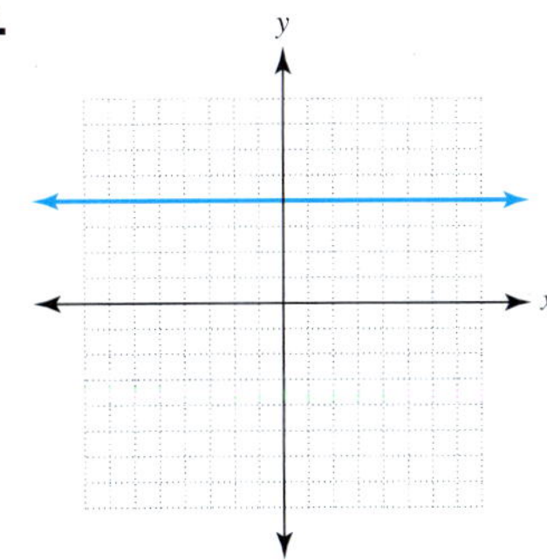

Answers

11. ______________________

12. ______________________

13. ______________________

14. ______________________

15. ______________________

16. ______________________

17. ______________________

18. ______________________

Answers

19. ____________________

20. ____________________

21. ____________________

22. ____________________

23. ____________________

24. ____________________

25. ____________________

26. ____________________

19.

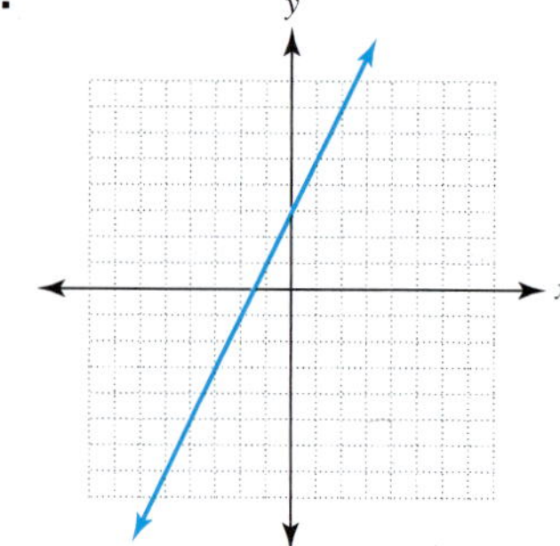

20.

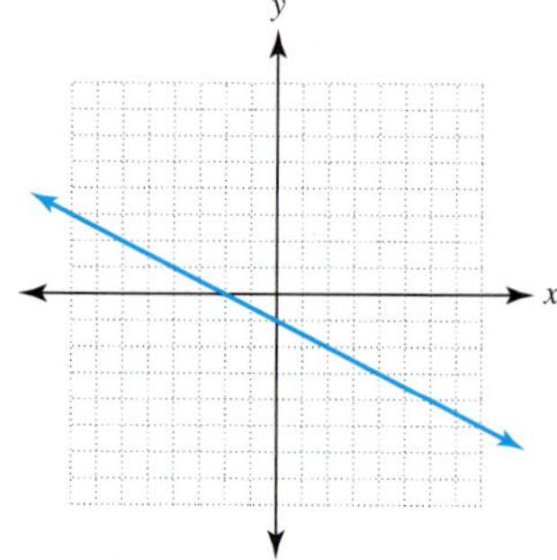

21.

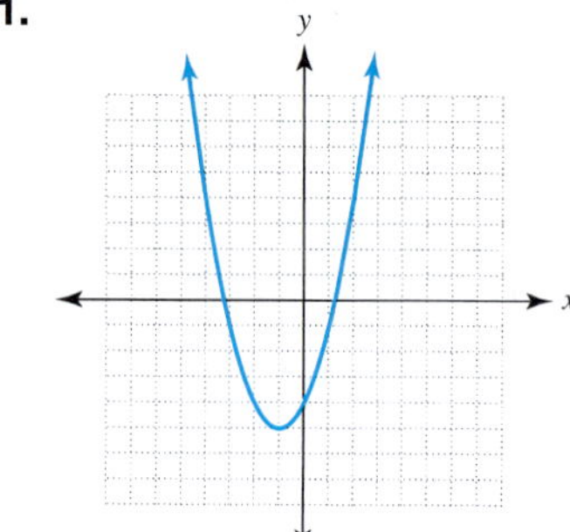

22.

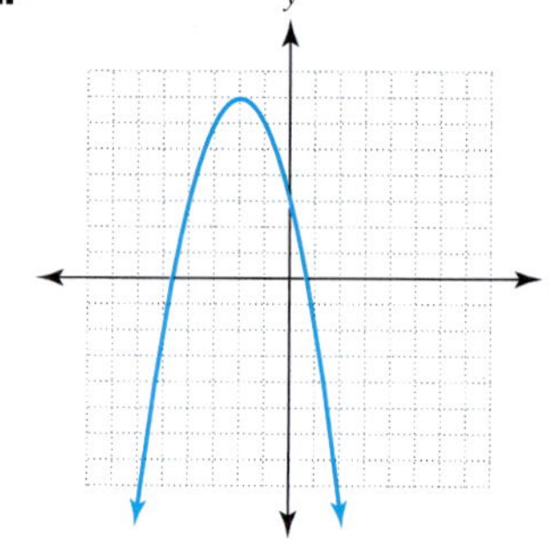

23.

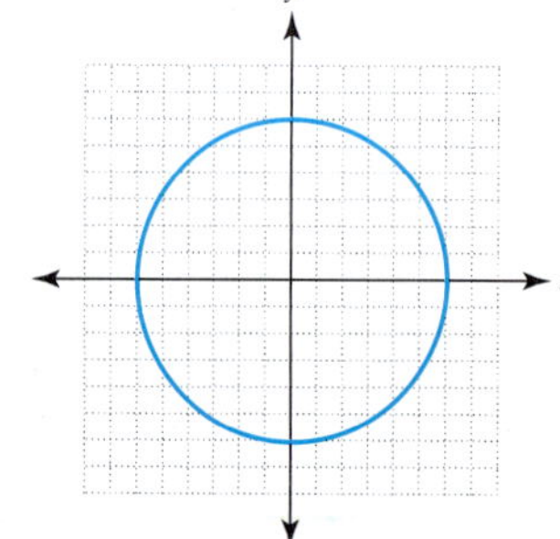

24.

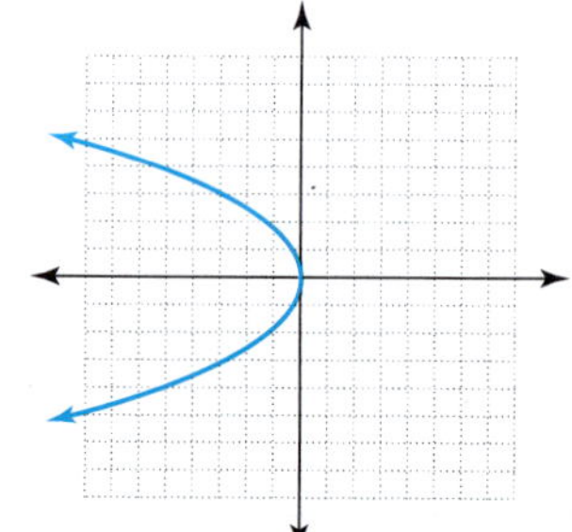

25.

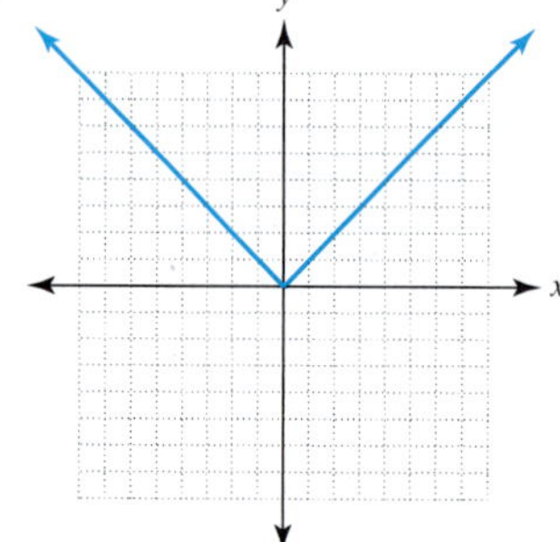

26.

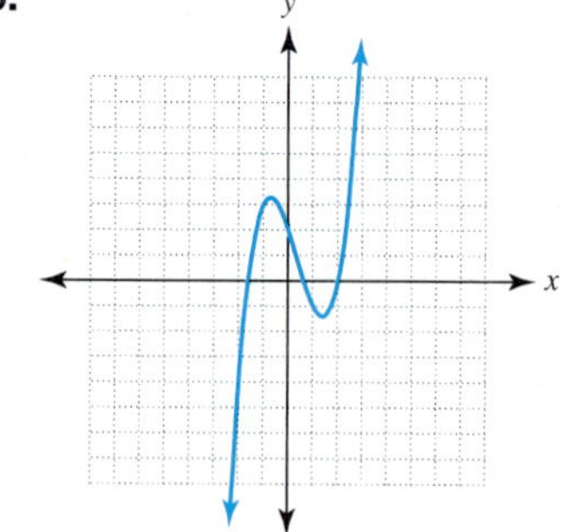

27.

28.

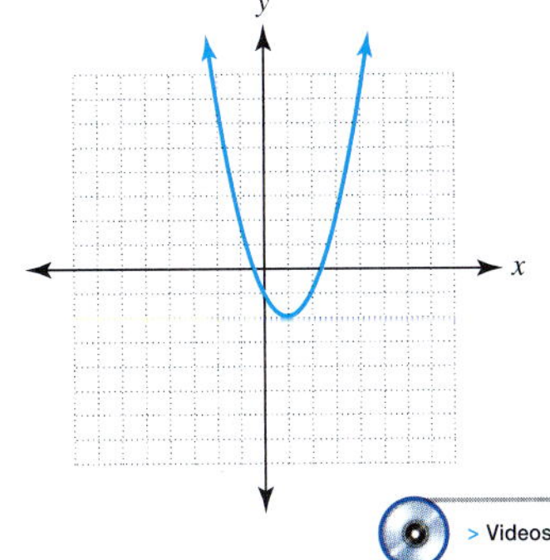

29.

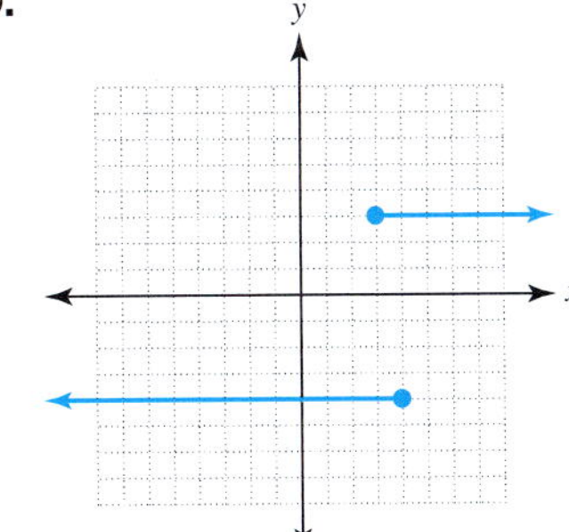

30.

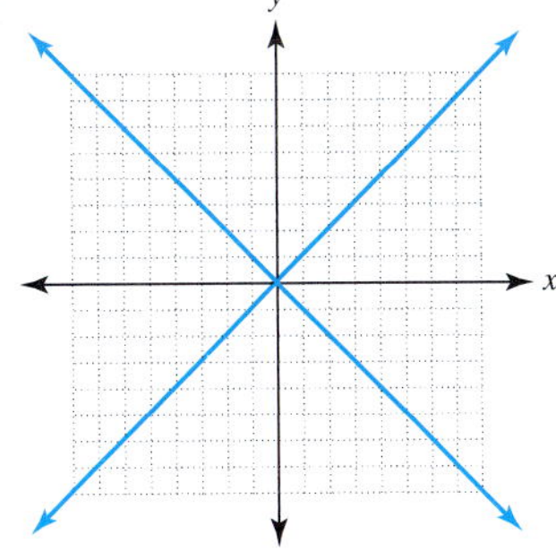

Basic Skills | **Challenge Yourself** | Calculator/Computer | Career Applications | Above and Beyond

Complete each statement with **never, sometimes,** *or* **always.**

31. If a vertical line passes through two points on the graph of a relation, the relation is ______________ a function.

32. If a horizontal line passes through two points on the graph of a relation, the relation is ______________ a function.

33. If the graph of a relation is a line that is not vertical, the relation is ______________ a function.

34. If the graph of a relation is a circle, the relation is ______________ a function.

< Objective 3 >

For exercises 35 to 48, use the tables to find the desired values.

x	$f(x)$
-3	-8
-1	2
2	4
5	-3

x	$g(x)$
-6	1
0	3
1	3
4	5

Answers

27. ______________

28. ______________

29. ______________

30. ______________

31. ______________

32. ______________

33. ______________

34. ______________

Answers

35. __________
36. __________
37. __________
38. __________
39. __________
40. __________
41. __________
42. __________
43. __________
44. __________
45. __________
46. __________
47. __________
48. __________
49. __________
50. __________

x	$h(x)$
-4	3
-2	7
3	5
7	-4

x	$k(x)$
-5	2
-3	-4
0	2
6	-4

35. $f(5)$

36. $g(-6)$

37. $h(3)$

38. $k(-5)$

39. All x such that $f(x) = -8$

40. All x such that $g(x) = 1$

41. All x such that $g(x) = 3$

42. All x such that $k(x) = 2$

43. $k(0)$

44. $g(4)$

45. $g(1)$

46. $h(-4)$

47. All x such that $k(x) = -4$

48. All x such that $h(x) = 3$

< Objective 4 >

For exercises 49 to 54, use the given graphs to find, or estimate, the desired values.

49.

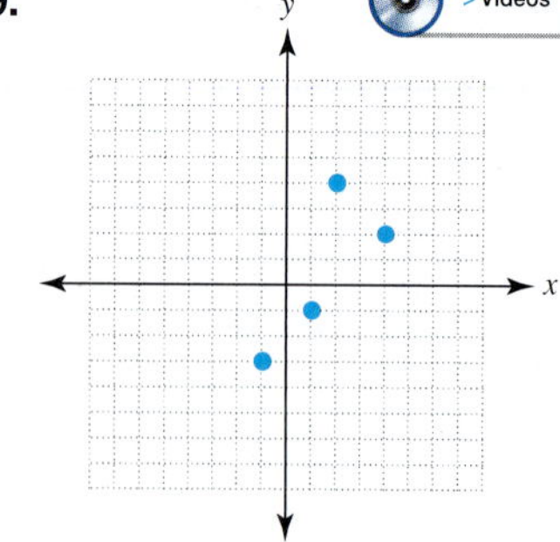

> Videos

(a) Find $f(2)$.
(b) Find $f(-1)$.
(c) Find all x such that $f(x) = 2$.
(d) Find all x such that $f(x) = -1$.

50.

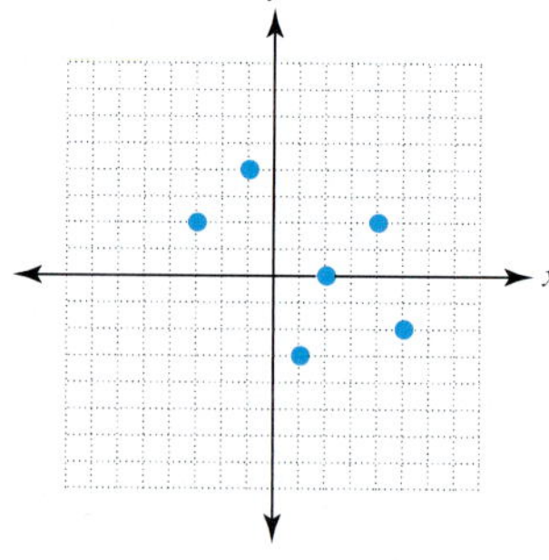

(a) Find $f(2)$.
(b) Find $f(-1)$.
(c) Find all x such that $f(x) = 2$.
(d) Find all x such that $f(x) = -3$.

51.

(a) Find $f(3)$.

(b) Find $f(4)$.

(c) Find all x such that $f(x) = 1$.

(d) Find all x such that $f(x) = 4$.

52.

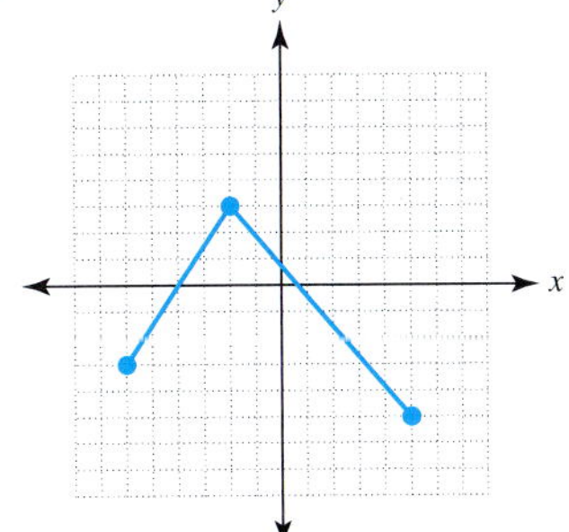

(a) Find $f(-2)$.

(b) Find $f(-5)$.

(c) Find all x such that $f(x) = 0$.

(d) Find all x such that $f(x) = -2$.

53.

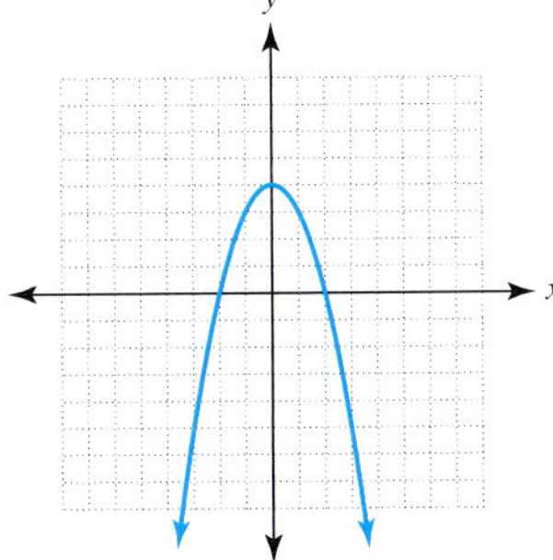

(a) Find $f(3)$.

(b) Find $f(0)$.

(c) Find all x such that $f(x) = 0$.

(d) Find all x such that $f(x) = -2$.

54.

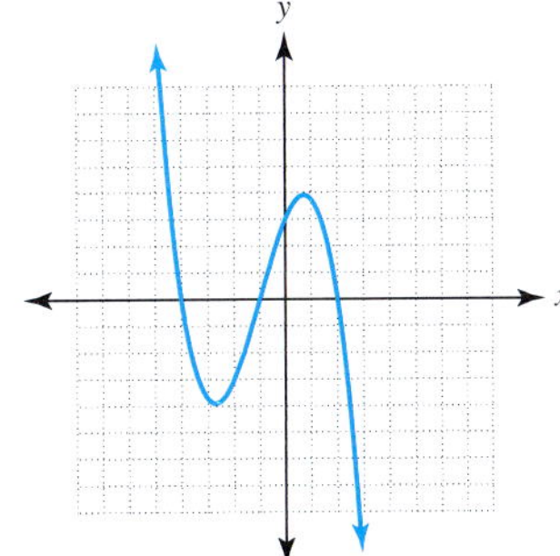

(a) Find $f(2)$.

(b) Find $f(-2)$.

(c) Find all x such that $f(x) = 3$.

(d) Find all x such that $f(x) = 5$.

Answers

51. ______

52. ______

53. ______

54. ______

Answers

1.

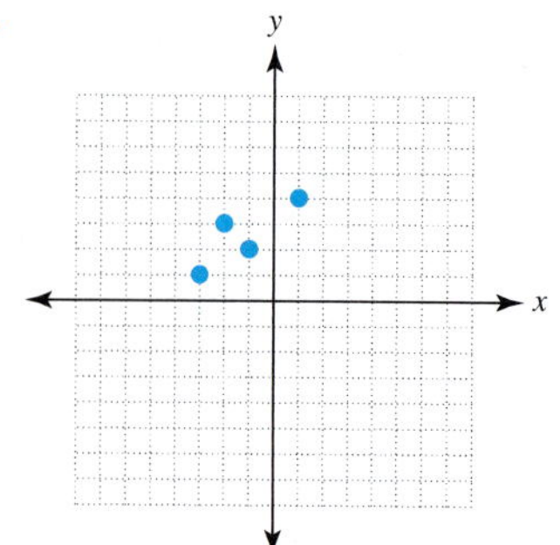

Function

3.

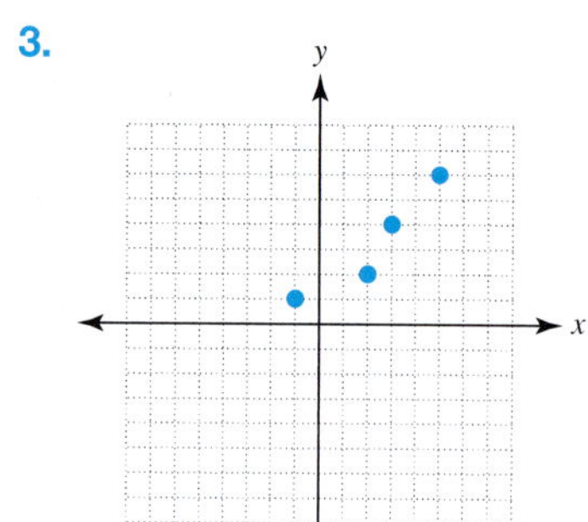

Function

5.

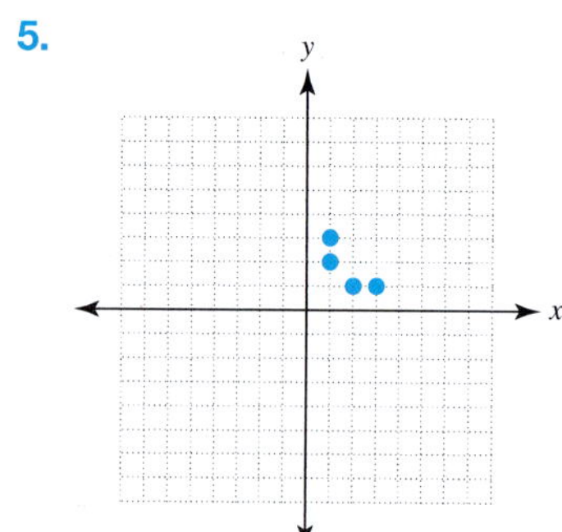

Not a function

7. Function; $D = \{-2, -1, 0, 1, 2\}$; $R = \{-1, 0, 1, 2, 3\}$
9. Function; $D = \{-2, -1, 0, 2, 3, 5\}$; $R = \{-1, 2, 4, 5\}$
11. Function; $D = \{x \mid -4 \le x \le 3\}$; $R = \{-2\}$
13. Function; $D = \{x \mid -3 \le x \le 4\}$; $R = \{y \mid -2 \le y \le 5\}$
15. Not a function; $D = \{x \mid -3 \le x \le 3\}$; $R = \{y \mid -3 \le y \le 4\}$
17. Not a function; $D = \{-3\}$; $R = \mathbb{R}$ **19.** Function; $D = \mathbb{R}$; $R = \mathbb{R}$
21. Function; $D = \mathbb{R}$; $R = \{y \mid y \ge -5\}$
23. Not a function; $D = \{x \mid -6 \le x \le 6\}$; $R = \{y \mid -6 \le y \le 6\}$
25. Function; $D = \mathbb{R}$; $R = \{y \mid y \ge 0\}$ **27.** Function; $D = \mathbb{R}$; $R = \{y \mid y \ge 3\}$
29. Not a function; $D = \mathbb{R}$; $R = \{-4, 3\}$ **31.** never **33.** always
35. -3 **37.** 5 **39.** -3 **41.** 0, 1 **43.** 2 **45.** 3 **47.** $-3, 6$
49. **(a)** 4; **(b)** -3; **(c)** 4; **(d)** 1 **51.** **(a)** -2; **(b)** 2; **(c)** $-2, 3.7$; **(d)** 4.5
53. **(a)** -5; **(b)** 4; **(c)** $-2, 2$; **(d)** $-2.5, 2.5$

summary :: chapter 2

Definition/Procedure	Example	Reference
Sets and Set Notation		Section 2.1
Set A set is a collection of objects classified together.	$A = \{2, 3, 4, 5\}$ is a set.	p. 198
Elements The elements are the objects in a set.	2 is an element of set A.	p. 198
Roster Form A set is said to be in roster form if the elements are listed and enclosed in braces.	$S = \{2, 4, 6, 8\}$ is in roster form.	p. 198
Set-Builder Notation $\{x \mid x > a\}$ is read "the set of all x, where x is greater than a." $\{x \mid x < a\}$ is read "the set of all x, where x is less than a." $\{x \mid a < x < b\}$ is read "the set of all x, where x is greater than a and less than b."	$\{x \mid x < 4\}$ is written in set-builder notation.	p. 199
Interval Notation (a, ∞) is read "all real numbers greater than a." $(-\infty, b)$ is read "all real numbers less than b." (a, b) is read "all real numbers greater than a and less than b." $[a, b]$ is read "all real numbers greater than or equal to a and less than or equal to b."	$(-4, 5]$ is written in interval notation.	p. 199
Plotting the Elements of a Set on a Number Line $\{x \mid x < a\}$ indicates the set of all points on the number line to the left of a. We plot those points by using a parenthesis at a (indicating that a is not included), then a bold line to the left.	$\{x \mid x < 4\}$ 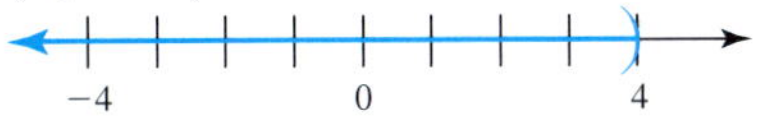 The parenthesis indicates every number below the marked value (here it is 4).	p. 200
$\{x \mid x \geq a\}$ indicates the set of all points on the number line to the right of, and including, a. We plot those points by using a bracket at a (indicating that a is included), then a bold line to the right.	$\{x \mid x \geq -3\}$ 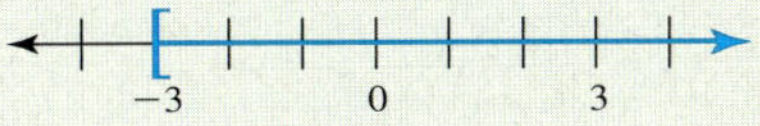 The bracket indicates every number at or above the indicated value (-3).	p. 201
$\{x \mid a \leq x < b\}$ indicates the set of all points on the number line between a and b, including a. We plot those points by using an opening bracket at a and a closing parenthesis at b, then a bold line in between.	$\{x \mid 3 \leq x < 10\}$ 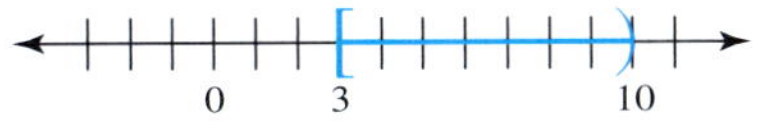 This notation indicates every number between 3 and 10, including 3 but not including 10.	p. 201
Set Operations **Union** $A \cup B$ is the set of elements in A or B or in both. **Intersection** $A \cap B$ is the set of elements in both A and B.	$A = \{2, 3, 4, 6\}$ and $B = \{3, 4, 7, 8\}$ $A \cup B = \{2, 3, 4, 6, 7, 8\}$ $A \cap B = \{3, 4\}$	p. 203

Continued

Definition/Procedure	Example	Reference
Solutions of Equations in Two Variables		**Section 2.2**
Solutions of Linear Equations Pairs of values that satisfy the equation. Solutions for linear equations in two variables are written as *ordered pairs.* An ordered pair has the form (x, y) — x-coordinate, y-coordinate	If $2x - y = 10$, then (6, 2) is a solution for the equation, because substituting 6 for x and 2 for y gives a true statement.	p. 214
The Cartesian Coordinate System		**Section 2.3**
The Rectangular Coordinate System A system formed by two perpendicular axes that intersect at a point called the **origin.** The horizontal line is called the ***x*-axis.** The vertical line is called the ***y*-axis.**		p. 224
Graphing Points from Ordered Pairs The coordinates of an ordered pair allow you to associate a point in the plane with every ordered pair. To graph a point in the plane: **Step 1** Start at the origin. **Step 2** Move right or left according to the value of the x-coordinate: to the right if x is positive or to the left if x is negative. **Step 3** Then move up or down according to the value of the y-coordinate: up if y is positive or down if y is negative.	To graph the point corresponding to (2, 3): 	p. 226

Definition/Procedure	Example	Reference
Relations and Functions		**Section 2.4**
Ordered Pair Given two related values x and y, we write the pair of values as (x, y).	$(1, 4)$ is an ordered pair.	p. 238
Relation A relation is a set of ordered pairs.	The set $\{(1, 4), (2, 5), (1, 6)\}$ is a relation.	p. 238
Domain The domain is the set of all first elements of a relation.	The domain is $\{1, 2\}$.	p. 238
Range The range is the set of all second elements of a relation.	The range is $\{4, 5, 6\}$.	p. 239
Function A function is a set of ordered pairs (a relation) in which no two first elements are equal.	$\{(1, 2), (2, 3), (3, 4)\}$ is a function. $\{(1, 2), (2, 3), (2, 4)\}$ is *not* a function.	p. 240
Tables and Graphs		**Section 2.5**
Graph The graph of a relation is the set of points in the plane that correspond to the ordered pairs of the relation.	y x	p. 254
Vertical Line Test The vertical line test is used to determine, from the graph, whether a relation is a function. If a vertical line meets the graph of a relation in two or more points, the relation is *not* a function. If no vertical line passes through two or more points on the graph of a relation, it is the graph of a function.	y x A relation—*not* a function	p. 255

Continued

Definition/Procedure	Example	Reference
Reading Values from Graphs For a specific value of x, let's call it a, we can find $f(a)$ with the following algorithm: **1.** Draw a vertical line through a on the x-axis. **2.** Find the point of intersection of that line with the graph. **3.** Draw a horizontal line through the graph at that point. **4.** Find the intersection of the horizontal line with the y-axis. **5.** $f(a)$ is that y-value.	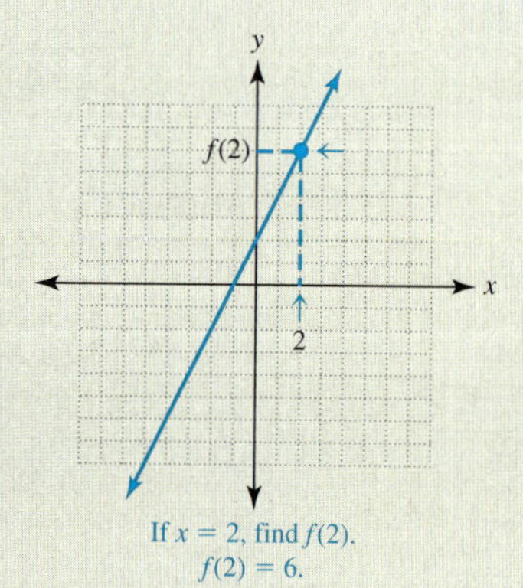 If $x = 2$, find $f(2)$. $f(2) = 6$.	p. 261
If given the function value, we find the x-value associated with it as follows: **1.** Find the given function value on the y-axis. **2.** Draw a horizontal line through that point. **3.** Find every point on the graph that intersects the horizontal line. **4.** Draw a vertical line through each of those points of intersection. **5.** The x-values are the points of intersection of the vertical lines and the x-axis.	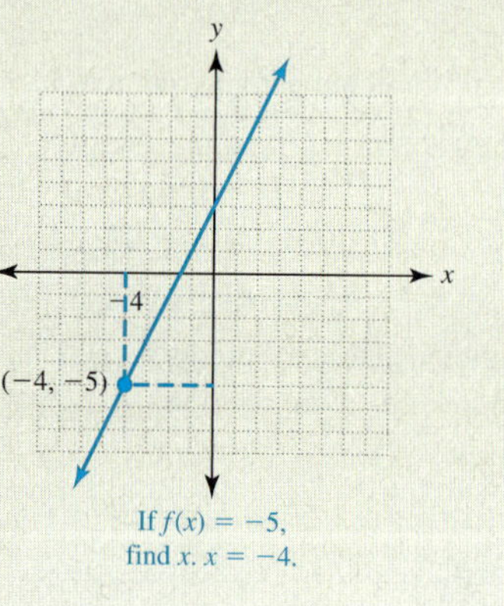 If $f(x) = -5$, find x. $x = -4$.	p. 261

summary exercises :: chapter 2

This summary exercise set is provided to give you practice with each of the objectives of this chapter. Each exercise is keyed to the appropriate chapter section. When you are finished, you can check your answers to the odd-numbered exercises in the back of the text. If you have difficulty with any of these questions, go back and reread the examples from that section. The answers to the even-numbered exercises appear in the *Instructor's Solutions Manual.* Your instructor will give you guidelines on how best to use these exercises in your instructional setting.

2.1 *Use the roster method to list the elements of each set.*

1. The set of all factors of 3

2. The set of all positive integers less than 7

3. The set of integers greater than -2 and less than 4

4. The set of integers between -4 and 3, inclusive

5. The set of all odd whole numbers less than 4

6. The set of all integers greater than -2 and less than 3

Use set-builder notation and interval notation to represent each set described.

7. The set of all real numbers greater than 9

8. The set of all real numbers greater than -2 and less than 4

9. The set of all real numbers less than or equal to -5

10. The set of all real numbers between -4 and 3, inclusive

Plot the elements of each set on a number line.

11. $\{x \mid x \geq 1\}$

12. $\{x \mid x \leq -2\}$

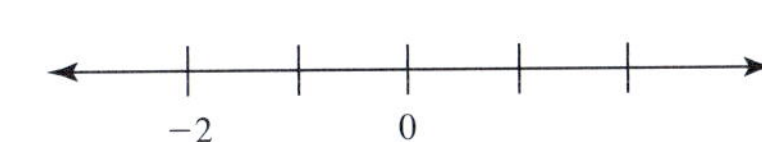

13. $\{x \mid -2 \leq x < 3\}$

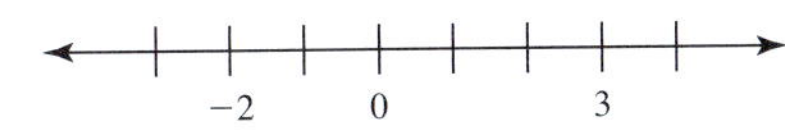

14. $\{x \mid -7 < x \leq -1\}$

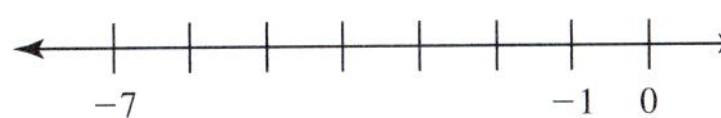

Use set-builder notation and interval notation to describe each set.

15.

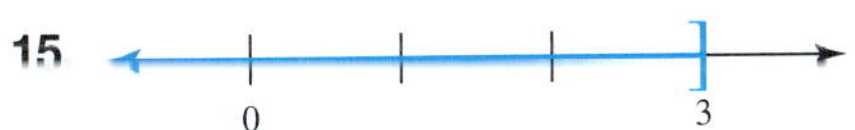

16.

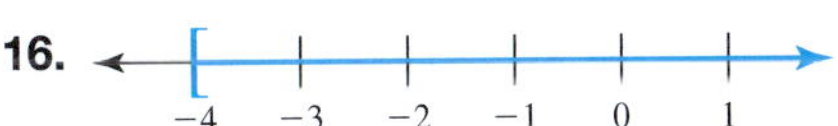

17.

18. 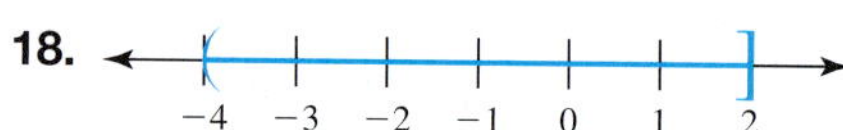

19. 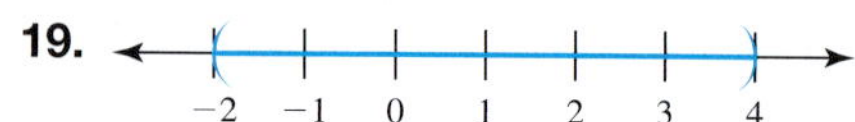

20. 

In exercises 21 to 24, $A = \{1, 5, 7, 9\}$ *and* $B = \{2, 5, 9, 11, 15\}$*. List the elements in each set.*

21. $A \cup B$

22. $A \cap B$

23. $B \cup \varnothing$

24. $A \cap \varnothing$

2.2 *Determine which of the ordered pairs are solutions for the given equations.*

25. $x - y = 6$ (6, 0), (3, 3), (3, −3), (0, −6)

26. $2x + 3y = 6$ (3, 0), (6, 2), (−3, 4), (0, 2)

2.3 *Give the coordinates of the labeled points on the graph.*

27. A

28. B

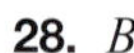

29. E

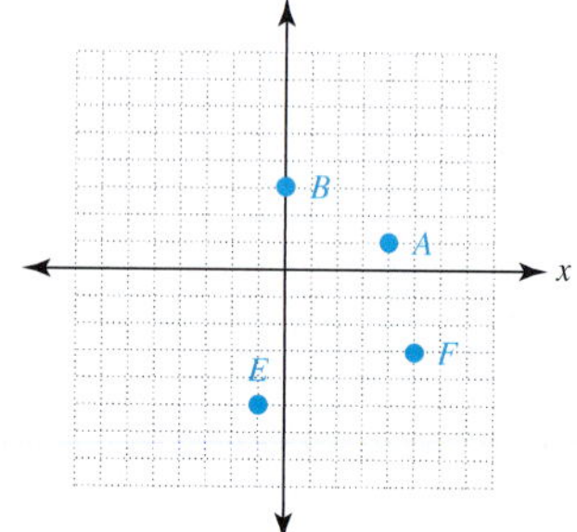

30. F

Plot the points with the given coordinates.

31. $P(4, 0)$

32. $Q(5, 4)$

33. $T(-2, 4)$

34. $U(4, -2)$

Give the quadrant in which each point is located or the axis on which the point lies.

35. (3, 6)

36. (−7, 5)

37. (−1, −6)

38. (−7, 8)

39. (−5, 0)

40. (0, −5)

Plot each point.

41. $(-1, 4)$ **42.** $(-1.25, 3.5)$ **43.** $(6, 3)$ **44.** $(-5, -2)$

2.4 *Find the domain and range of each relation.*

45. A = {(Maine, 5), (Massachusetts, 13), (Vermont, 7), (Connecticut, 11)}

46. B = {(John Wayne, 1969), (Art Carney, 1974), (Peter Finch, 1976), (Marlon Brando, 1972)}

47. C = {(Dean Smith, 65), (John Wooden, 47), (Denny Crum, 42), (Bob Knight, 41)}

48. E = {(Don Shula, 328), (George Halas, 318), (Tom Landry, 250), (Chuck Noll, 193)}

49. $\{(3, 5), (4, 6), (1, 2), (8, 1), (7, 3)\}$

50. $\{(-1, 3), (-2, 5), (3, 7), (1, 4), (2, -2)\}$

51. $\{(1, 3), (1, 5), (1, 7), (1, 9), (1, 10)\}$

52. $\{(2, 4), (-1, 4), (-3, 4), (1, 4), (6, 4)\}$

Determine which relations are also functions.

53. $\{(1, 3), (2, 4), (5, -1), (-1, 3)\}$

54. $\{(-2, 4), (3, 6), (1, 5), (0, 1)\}$

55. $\{(1, 2), (0, 4), (1, 3), (2, 5)\}$

56. $\{(1, 3), (2, 3), (3, 3), (4, 3)\}$

57.

x	y
-3	2
-1	1
0	3
1	4
3	5

58.

x	y
-1	3
0	2
1	3
2	4
3	5

59.

x	y
-2	3
-1	4
0	-1
1	5
-2	13

60.

x	y
-3	-4
1	0
2	3
1	5
5	2

Evaluate each function for the value specified.

61. $f(x) = x^2 - 3x + 5$; find **(a)** $f(0)$, **(b)** $f(-1)$, and **(c)** $f(1)$.

62. $f(x) = -2x^2 + x - 7$; find **(a)** $f(0)$, **(b)** $f(2)$, and **(c)** $f(-2)$.

63. $f(x) = x^3 - x^2 - 2x + 5$; find **(a)** $f(-1)$, **(b)** $f(0)$, and **(c)** $f(2)$.

64. $f(x) = -x^2 + 7x - 9$; find **(a)** $f(-3)$, **(b)** $f(0)$, and **(c)** $f(1)$.

65. $f(x) = 3x^2 - 5x + 1$; find **(a)** $f(-1)$, **(b)** $f(0)$, and **(c)** $f(2)$.

66. $f(x) = -x^3 + 3x - 5$; find **(a)** $f(2)$, **(b)** $f(0)$, and **(c)** $f(1)$.

Rewrite each equation as a function of x. Use f(x) notation in the final result.

67. $y = -2x + 5$

68. $y = 3x + 2$

69. $2x + 3y = 6$

70. $4x + 2y = 8$

71. $-3x + 4y = 12$

72. $-2x - 5y = -10$

Let $f(x) = -\frac{3}{4}x + 2$. *Evaluate, as indicated.*

73. $f(t)$

74. $f(x + 4)$

75. $f(x + h)$

76. $\frac{f(x + h) - f(x)}{h}$

If $f(x) = 3x - 2$, *find the following:*

77. $f(a)$

78. $f(x - 1)$

79. $f(x + h)$

80. $\frac{f(x + h) - f(x)}{h}$

2.5 *For each set of ordered pairs, plot the related points. Then use the vertical line test to determine which sets are functions.*

81. $\{(-3, -3), (-2, -2), (2, 2), (3, 3)\}$

82. $\{(-4, -4), (-2, 4), (2, 3), (0, 4)\}$

83. $\{(-2, 1), (-2, 3), (0, 1), (1, 2)\}$

84. $\{(0, 5), (1, 6), (1, -2), (3, 4)\}$

Use the vertical line test to determine whether the given graph represents a function. Find the domain and range of the relation.

85.

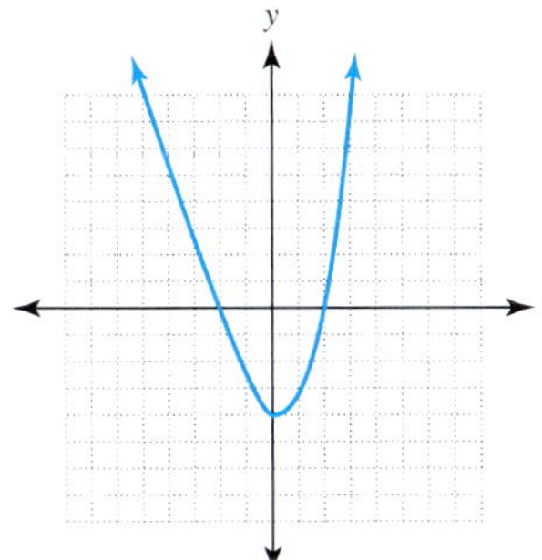

86.

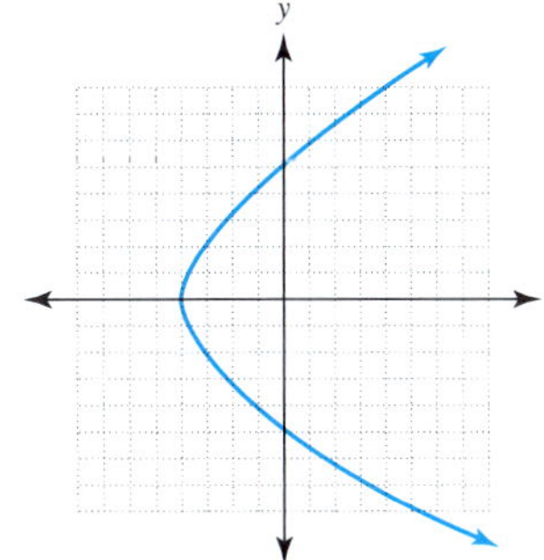

87.

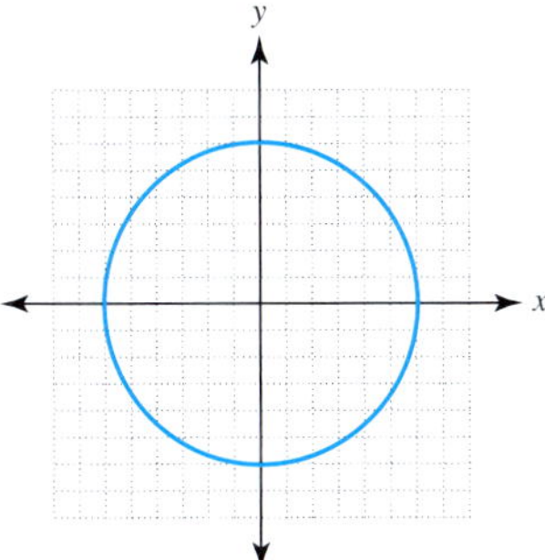

88.

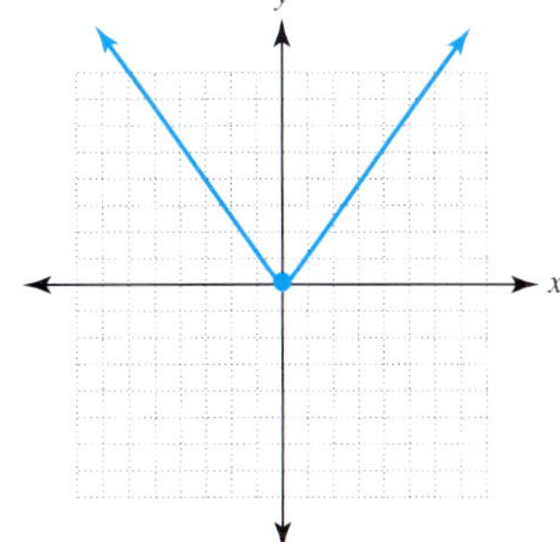

89. Use the given graph to answer parts (a) through (f). Estimate values where necessary.

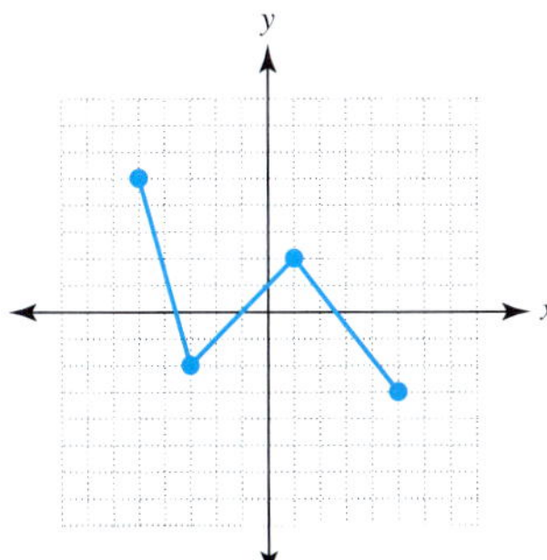

(a) Find $f(-1)$.

(b) Find $f(5)$.

(c) Find all x such that $f(x) = 5$.

(d) Find all x such that $f(x) = -3$.

(e) Find all x such that $f(x) = -1$.

(f) Find all x such that $f(x) = 2$.

90. Use the given graph to answer parts (a) through (f). Estimate values where necessary.

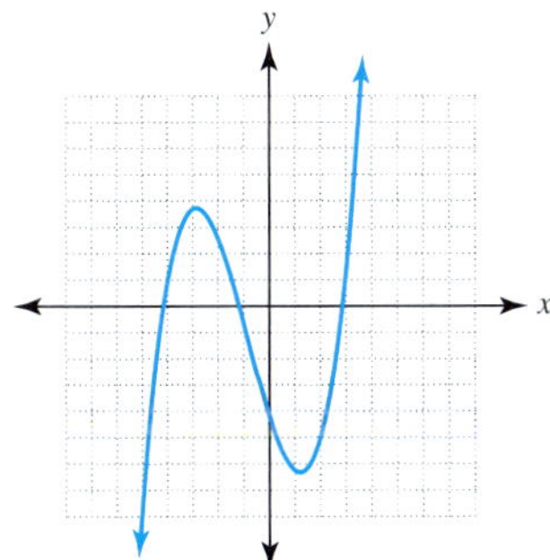

(a) Find $f(-3)$.

(b) Find $f(2)$.

(c) Find all x such that $f(x) = 7$.

(d) Find all x such that $f(x) = -8$.

(e) Find all x such that $f(x) = 3$.

(f) Find all x such that $f(x) = -4$.

91. Use the given table to answer parts (a) through (f).

x	$f(x)$
−5	−2
−2	0
4	−2
8	5
12	12

(a) Find $f(-5)$.

(b) Find $f(12)$.

(c) Find all x such that $f(x) = -2$.

(d) Find all x such that $f(x) = 0$.

(e) Find all x such that $f(x) = 5$.

(f) Find all x such that $f(x) = 12$.

92. Use the given table to answer parts (a) through (f).

x	$f(x)$
−3	0
−2	−3
0	−3
3	−2
7	0

(a) Find $f(-3)$.

(b) Find $f(0)$.

(c) Find $f(3)$.

(d) Find all x such that $f(x) = -3$.

(e) Find all x such that $f(x) = -2$.

(f) Find all x such that $f(x) = 0$.

self-test 2

The purpose of this self-test is to help you assess your progress so that you can find concepts that you need to review before the next exam. Allow yourself about an hour to take this test. At the end of that hour, check your answers against those given in the back of this text. If you miss any, go back to the appropriate section to reread the examples until you have mastered that particular concept.

1. Plot the elements of the set $\{x \mid -5 < x \leq 3\}$ on a number line.

2. **(a)** Use set-builder notation to describe the set pictured below.
(b) Describe the set using interval notation.

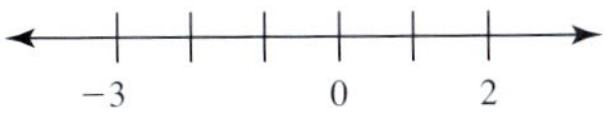

3. Determine which of the ordered pairs are solutions to the given equation.

$4x - y = 16 \quad (4, 0), (3, -1), (5, 4)$

Give the coordinates of the points graphed below.

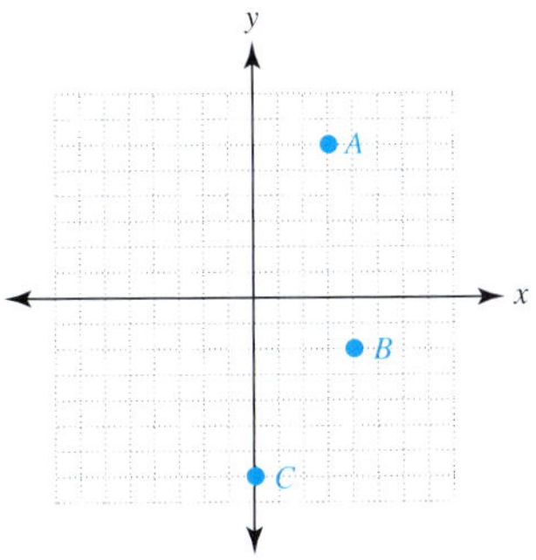

4. *A*

5. *B*

6. *C*

7. For each set of ordered pairs, identify the domain and range.

(a) $\{(1, 6), (-3, 5), (2, 1), (4, -2), (3, 0)\}$
(b) {(United States, 101), (Germany, 65), (Russia, 63), (China, 50)}

8. For each relation shown, determine whether the given relation is a function and identify its domain and range.

(a) $\{(2, 5), (-1, 6), (0, 2), (-4, 5)\}$

(b)

x	y
−3	2
0	4
1	7
2	0
−3	1

9. Plot the given points on a graph and use the vertical line test to determine whether the graph represents a function.

$\{(-1, 2), (0, 1), (2, 2), (3, -4)\}$

Name ______________________

Section ________ Date ________

Answers

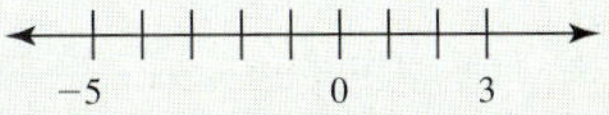

1. ______________________

2. ______________________

3. ______________________

4. ______________________

5. ______________________

6. ______________________

7. ______________________

8. ______________________

9. ______________________

Answers

10. ______

11. ______

12. ______

13. ______

14. ______

15. ______

16. ______

17. ______

18. ______

19. ______

20. ______

Determine whether the graphs represent functions.

10.

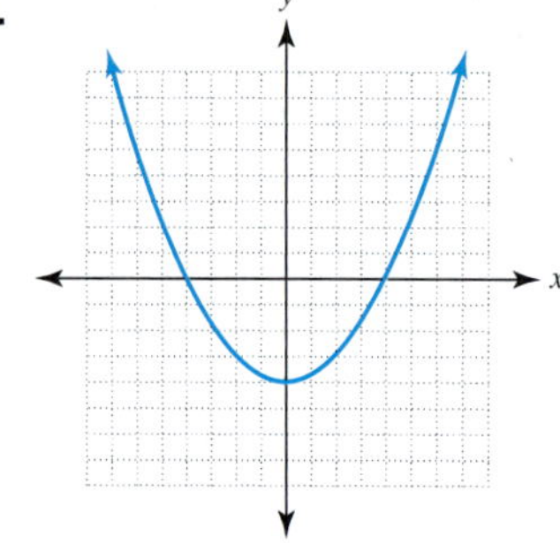

11.

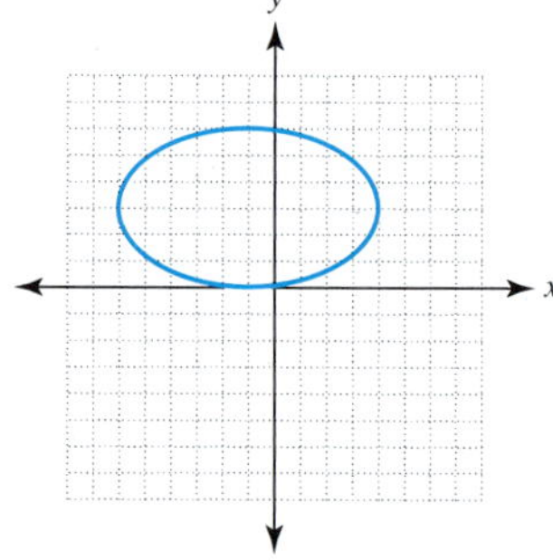

Plot the points shown.

12. $S(1, -2)$ **13.** $T(0, 3)$ **14.** $U(-4, 5)$

15. Complete each ordered pair so that it is a solution to the equation shown.

$4x + 3y = 12 \quad (3,\ ), (\ , 4), (\ , 3)$

16. If $f(x) = x^2 - 5x + 6$, find **(a)** $f(0)$; **(b)** $f(-1)$; and **(c)** $f(1)$.

17. If $f(x) = -3x^2 - 2x + 3$, find **(a)** $f(-1)$; and **(b)** $f(-2)$.

18. Graph the function $f(x) = -2x + 3$.

19. If $A = \{1, 2, 5\}$, and $B = \{3, 5, 7\}$, find

(a) $A \cup B$ **(b)** $A \cap B$

20. Use the table to find the desired values.

x	y
−5	3
−3	5
0	−1
1	9
4	2
5	3

(a) $f(-5)$ **(b)** $f(4)$

(c) Values of x such that $f(x) = 9$

(d) Values of x such that $f(x) = 3$

cumulative review chapters 0-2

Name ____________

Section ______ Date ______

We offer the following exercises to help you review concepts from earlier chapters. This is meant as review material and not as a comprehensive exam. The answers are presented in the back of the text. We provide section references for each concept along with the answers in the back of this text. If you have difficulty with any of these exercises, be certain to at least read through the summary related to that section.

Use the Fundamental Principle of Fractions to simplify each fraction.

1. $\frac{56}{88}$

2. $\frac{132}{110}$

Perform the indicated operations. Write each answer in simplest form.

3. $2 \cdot 3^2 - 8 \cdot 2$

4. $5(7 - 3)^2$

5. $|12 - 5|$

6. $|12| - |5|$

7. $(-7) + (-9)$

8. $\frac{17}{3} + \left(-\frac{5}{3}\right)$

9. $(-7)(-9)$

10. $(-3.2)(5)$

11. $\frac{0}{-13}$

12. $8 - 12 \div 2 \cdot 3 + 5$

13. $5 - 4^2 \div (-8) \cdot 2$

14. $\frac{4}{9} \times \frac{27}{36}$

15. $\frac{3}{4} + \frac{5}{6}$

16. $\frac{5}{6} \div \frac{25}{21}$

Evaluate each expression if $x = -2$, $y = 3$, and $z = 5$.

17. $3x - y$

18. $4x^2 - y$

19. $\frac{5z - 4x}{2y + z}$

20. $-y^2 - 8x$

Simplify and combine like terms.

21. $7x - 3y + 2(4x - 3y)$

22. $6x^2 - (5x - 4x^2 + 7) - 8x + 9$

Answers

1. ______ 2. ______

3. ______ 4. ______

5. ______ 6. ______

7. ______ 8. ______

9. ______ 10. ______

11. ______

12. ______

13. ______

14. ______

15. ______

16. ______

17. ______

18. ______

19. ______

20. ______

21. ______

22. ______

Answers

23. ____

24. ____

25. ____

26. ____

27. ____

28. ____

29. ____

30. ____

31. ____

32. ____

33. ____

34. ____

35. ____

36. ____

37. ____

38. ____

39. ____

40. ____

Solve each equation.

23. $12x - 3 = 10x + 5$

24. $\frac{x-2}{3} - \frac{x+1}{4} = 5$

25. $4(x - 1) - 2(x - 5) = 14$

Solve each inequality.

26. $7x + 5 \le 4x - 7$

27. $-5 \le 2x + 1 < 7$

Solve each equation for the indicated variable.

28. $I = Prt$ (for r)

29. $A = \frac{1}{2}bh$ (for h)

30. $ax + by = c$ (for y)

31. $P = 2L + 2W$ (for W)

Use the graph shown to determine.

32. $f(-3)$

33. $f(0)$

34. Value of x for which $f(x) = 3$

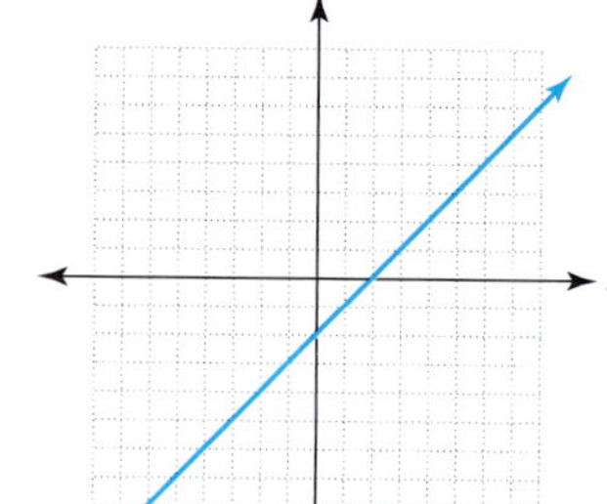

Solve each word problem. Be sure to show the equation used for the solution.

35. If 4 times a number decreased by 7 is 45, find that number.

36. The sum of two consecutive integers is 85. What are those two integers?

37. If 3 times an odd integer is 12 more than the next consecutive odd integer, what is that integer?

38. Michelle earns \$120 more per week than Dmitri. If their weekly salaries total \$720, how much does Michelle earn?

39. The length of a rectangle is 2 centimeters (cm) more than 3 times its width. If the perimeter of the rectangle is 44 cm, what are the dimensions of the rectangle?

40. One side of a triangle is 5 in. longer than the shortest side. The third side is twice the length of the shortest side. If the triangle perimeter is 37 in., find the length of each leg.

CHAPTER

3

Graphing Linear Functions

INTRODUCTION

Linear models describe many situations that we encounter in our other classes and careers. For instance, many people earn a paycheck based on the number of hours worked. In fact, many business and finance applications are best modeled with linear functions.

In this chapter, we learn to build, graph, and describe linear functions. We use properties such as *rate-of-change* to describe important ideas such as *marginal profit*. Using technology and real-world data, we look to you to make these powerful models your own. Doing so will ensure that you can use the math you learn in later settings.

CHAPTER 3 OUTLINE

3.1 Graphing Linear Functions

< 3.1 Objectives >

1 > Graph a linear equation by plotting points
2 > Graph horizontal and vertical lines
3 > Graph a linear equation using the intercept method
4 > Solve a linear equation for *y* and graph the result
5 > Write a linear equation using function notation

In Section 2.2, you learned to use ordered pairs to write the solutions of equations in two variables. In Section 2.3, we graphed ordered pairs in the Cartesian plane. Putting these ideas together helps us graph certain equations. Example 1 illustrates one approach to finding the graph of a linear equation.

Example 1 Graphing a Linear Equation

< Objective 1 >

Graph $x + 2y = 4$.

Step 1 Find some solutions for $x + 2y = 4$. To find solutions, we choose any convenient values for x, say $x = 0$, $x = 2$, and $x = 4$. Given these values for x, we can substitute and then solve for the corresponding value for y.

When $x = 0$, we have

$x + 2y = 4$

$(0) + 2y = 4$ Substitute $x = 0$.

$2y = 4$

$y = 2$ Divide both sides by 2.

Therefore, $(0, 2)$ is a solution.

When $x = 2$, we have

$(2) + 2y = 4$ Substitute $x = 2$.

$2y = 2$ Subtract 2 from both sides.

$y = 1$ Divide both sides by 2.

So, $(2, 1)$ is a solution.

When $x = 4$, $y = 0$, so $(4, 0)$ is a solution.

A handy way to show this information is in a table.

x	y
0	2
2	1
4	0

NOTES

We find *three* solutions for the equation. We will point out why shortly.

A table is a convenient way to display the information. It is the same as writing (0, 2), (2, 1), and (4, 0).

Step 2 We now graph the solutions found in step 1.

$x + 2y = 4$

x	y
0	2
2	1
4	0

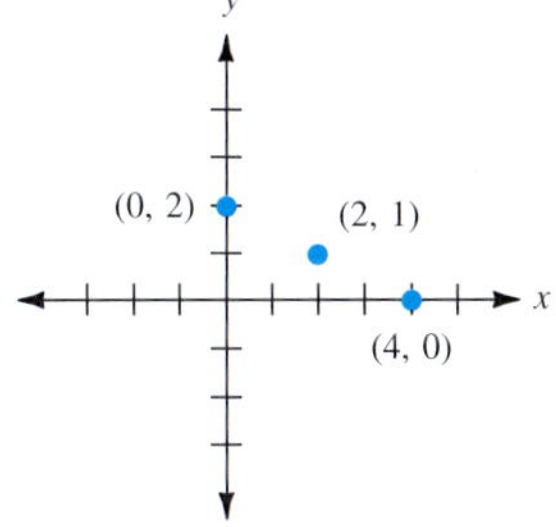

What pattern do you see? It appears that the three points lie on a straight line, which is the case.

Step 3 Draw a straight line through the three points graphed in step 2.

NOTE

Arrowheads on the end of the line mean that the line extends infinitely in each direction.

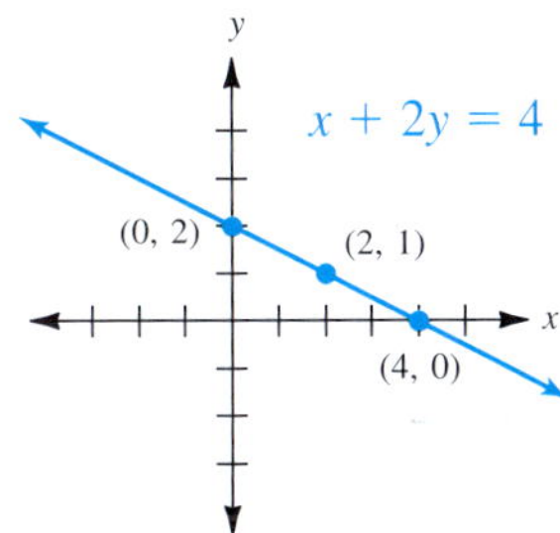

NOTE

A graph is a "picture" of the solutions for the given equation.

The line shown is the **graph** of the equation $x + 2y = 4$. It represents *all* the ordered pairs that are solutions (an infinite number) for that equation.

Every ordered pair that is a solution is plotted as a point on this line. Any point on the line represents a pair of numbers that is a solution for the equation.

Note: Why did we suggest finding *three* solutions in step 1? Two points determine a line, so technically you need only two. The third point that we find is a check to catch any possible errors.

Check Yourself 1

Graph $2x - y = 6$, using the steps shown in Example 1.

As mentioned in Section 2.2, an equation that can be written in the form

$Ax + By = C$ where A and B are not both 0

is called a linear equation in two variables **in standard form.** The graph of this equation is a *line.* That is why we call it a *linear* equation.

The steps of graphing follow.

Step by Step

To Graph a Linear Equation

Step 1 Find at least three solutions for the equation, and put your results in tabular form.

Step 2 Graph the solutions found in step 1.

Step 3 Draw a straight line through the points determined in step 2 to form the graph of the equation.

Example 2 Graphing a Linear Equation

> **NOTE**
>
> Let $x = 0$, 1, and 2, and substitute to determine the corresponding y-values. Again the choices for x are simply convenient. Other values for x would serve the same purpose.

Graph $y = 3x$.

Step 1 Some solutions are

x	y
0	0
1	3
2	6

Step 2 Graph the points.

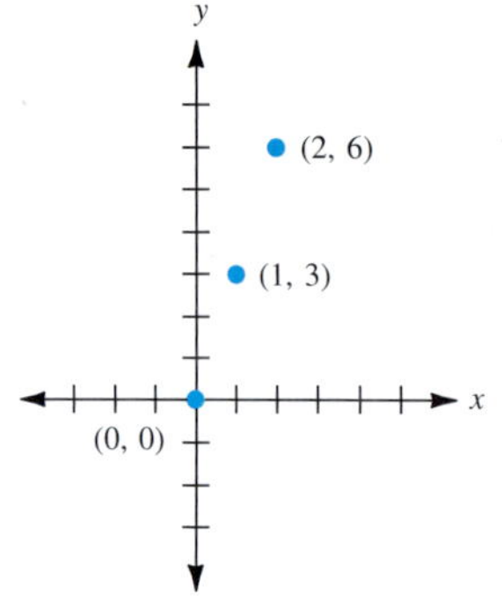

> **NOTE**
>
> Connecting any two of these points produces the same line.

Step 3 Draw a line through the points.

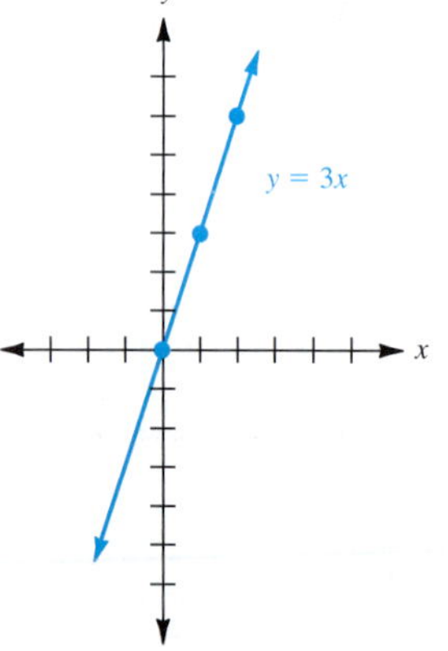

Check Yourself 2

Graph the equation $y = -2x$ after completing the table of values.

x	y
0	
1	
2	

We now work through another example of graphing a line from its equation.

Example 3 Graphing a Linear Equation

Graph $y = 2x + 3$.

Step 1 Some solutions are

x	y
0	3
1	5
2	7

Step 2 Graph the points corresponding to these values.

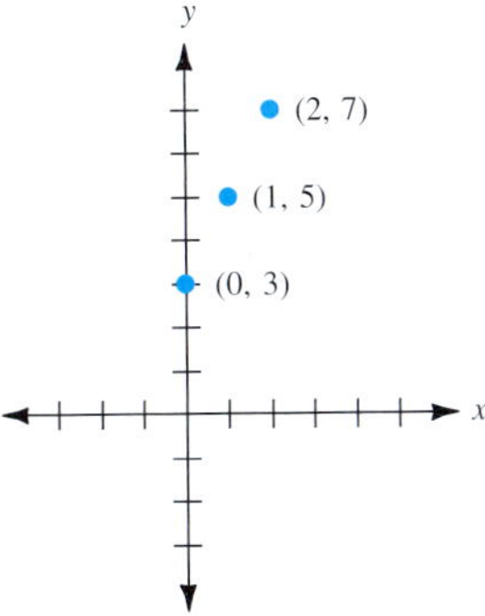

Step 3 Draw a line through the points.

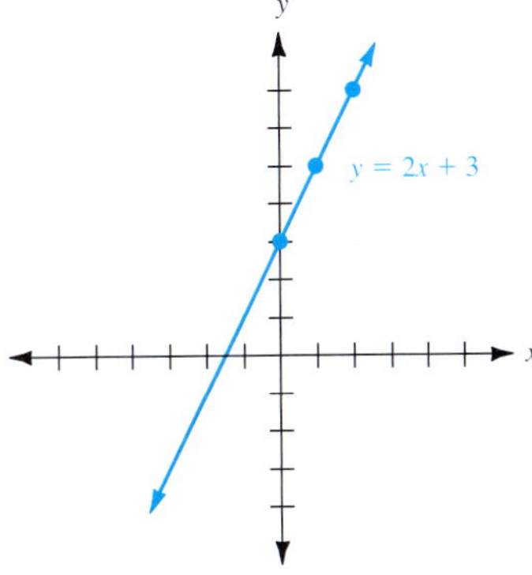

Check Yourself 3

Graph the equation $y = 3x - 2$ after completing the table of values.

x	y
0	
1	
2	

In graphing equations, particularly when fractions are involved, a careful choice of values for x can simplify the process. Consider Example 4.

Example 4 Graphing a Linear Equation

Graph

$$y = \frac{3}{2}x - 2$$

As before, we want to find solutions for the given equation by picking convenient values for x. Note that in this case, choosing *multiples of* 2, the denominator of the x coefficient, avoids fractional values for y making it much easier to plot these solutions. For instance, here we might choose values of -2, 0, and 2 for x.

Step 1

If $x = -2$:

$$y = \frac{3}{2}x - 2$$
$$= \frac{3}{2}(-2) - 2$$
$$= -3 - 2 = -5$$

If $x = 0$:

$$y = \frac{3}{2}x - 2$$
$$= \frac{3}{2}(0) - 2$$
$$= 0 - 2 = -2$$

If $x = 2$:

$$y = \frac{3}{2}x - 2$$
$$= \frac{3}{2}(2) - 2$$
$$= 3 - 2 = 1$$

In tabular form, the solutions are

x	y
-2	-5
0	-2
2	1

NOTE

Suppose we do *not* choose a multiple of 2, say, $x = 3$. Then

$$y = \frac{3}{2}(3) - 2$$
$$= \frac{9}{2} - 2$$
$$= \frac{5}{2}$$

$\left(3, \frac{5}{2}\right)$ is still a valid solution, but we must graph a point with fractional coordinates.

Step 2 Graph the points determined in step 1.

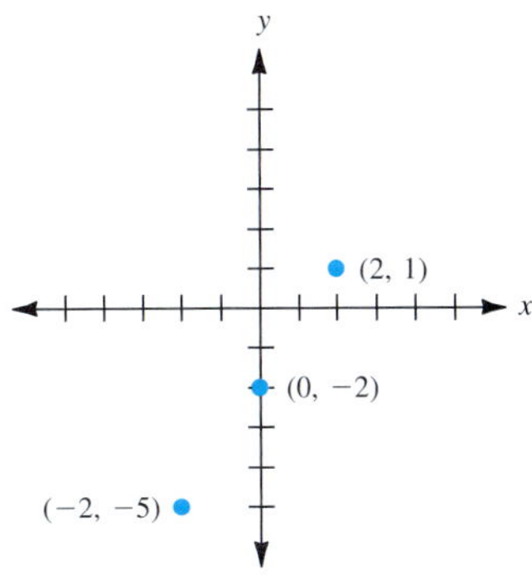

Step 3 Draw a line through the points.

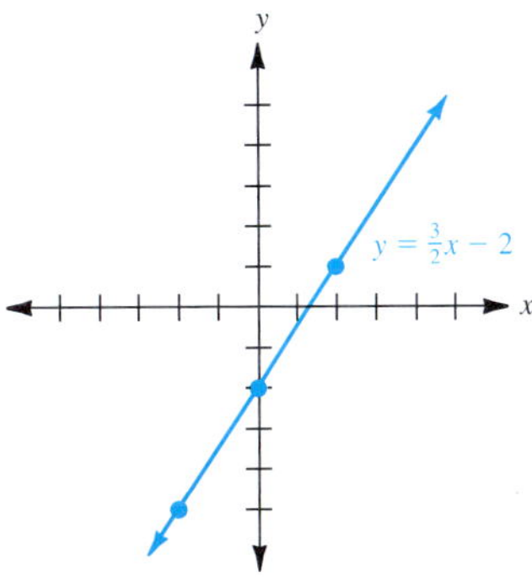

Check Yourself 4

Graph the equation $y = -\frac{1}{3}x + 3$ after completing the table of values.

x	y
−3	
0	
3	

Some special cases of linear equations are illustrated in Example 5.

Example 5 Graphing Special Equations

< Objective 2 >

NOTE

We cannot write $x = 3$ so that y is a function of x. Therefore, this equation does not represent a function.

(a) Graph $x = 3$.

The equation $x = 3$ is equivalent to $1 \cdot x + 0 \cdot y = 3$. Let's look at some solutions.

If $y = 1$:	If $y = 4$:	If $y = -2$:
$x + 0 \cdot (1) = 3$	$x + 0 \cdot (4) = 3$	$x + 0(-2) = 3$
$x = 3$	$x = 3$	$x = 3$

In tabular form,

x	y
3	1
3	4
3	−2

What do you observe? The variable x has the value 3, regardless of the value of y. Look at the graph.

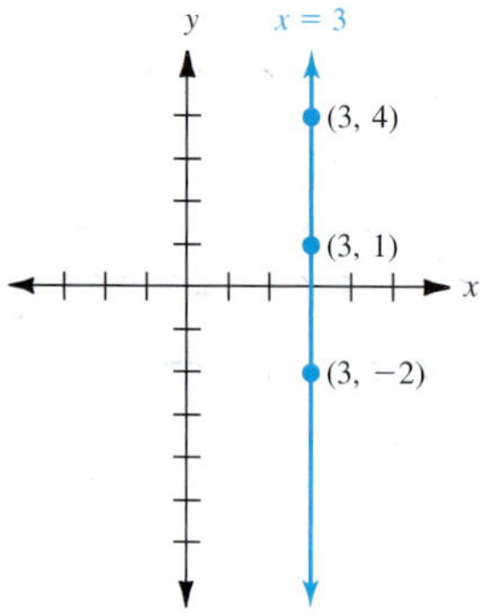

The graph of $x = 3$ is a vertical line crossing the x-axis at (3, 0).

Note that graphing (or plotting) points in this case is not really necessary. Simply recognize that the graph of $x = 3$ *must* be a vertical line (parallel to the y-axis) that intercepts the x-axis at (3, 0).

(b) Graph $y = 4$.

Since $y = 4$ is equivalent to $0 \cdot x + 1 \cdot y = 4$, any value for x paired with 4 for y will form a solution. A table of values might be

x	y
−2	4
0	4
2	4

Here is the graph.

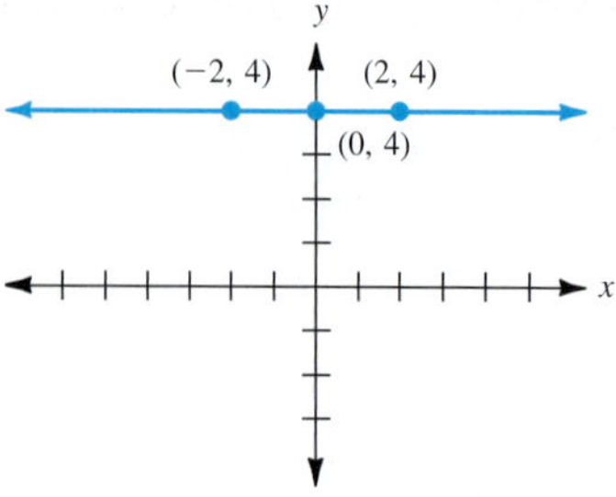

NOTE

A horizontal line represents the graph of a *constant function.* In this case, the function is written as $f(x) = 4$.

This time the graph is a horizontal line that crosses the y-axis at (0, 4). Again graphing the points is not required. The graph of $y = 4$ *must* be horizontal (parallel to the x-axis) and intercepts the y-axis at (0, 4).

Check Yourself 5

(a) Graph the equation $x = -2$.
(b) Graph the equation $y = -3$.

We call the function $f(x) = b$ a constant function because the y-value does not change, even as the input x changes. On the graph, the height of the line does not change so we think of it as constant.

The vertical line produced by the linear equation $x = a$ does not represent a function. We cannot write this equation so that y is a function of x because the one x-value is a and this "maps" to every real-number y.

This property box summarizes our work in Example 5.

Property

Vertical and Horizontal Lines

1. The graph of $x = a$ is a *vertical line* crossing the x-axis at $(a, 0)$.
2. The graph of $y = b$ is a *horizontal line* crossing the y-axis at $(0, b)$.

To simplify the graphing of certain linear equations, some students prefer the **intercept method** of graphing. This method makes use of the fact that the solutions that are easiest to find are those with an x-coordinate or a y-coordinate of 0. For instance, let's graph the equation

$$4x + 3y = 12$$

First, let $x = 0$ and solve for y.

$$4x + 3y = 12$$
$$4(0) + 3y = 12$$
$$3y = 12$$
$$y = 4$$

So $(0, 4)$ is one solution. Now let $y = 0$ and solve for x.

$$4x + 3y = 12$$
$$4x + 3(0) = 12$$
$$4x = 12$$
$$x = 3$$

A second solution is $(3, 0)$.

The two points corresponding to these solutions can now be used to graph the equation.

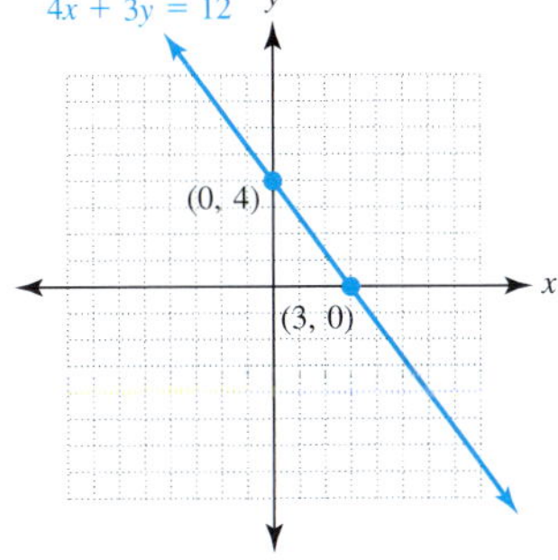

NOTE

With practice, this all can be done mentally, which is the big advantage of this method.

RECALL

Only two points are needed to graph a line. A third point is used as a check.

NOTE

The intercepts are the points where the line intersects the x- and y-axes. Here, the x-intercept has coordinates $(3, 0)$, and the y-intercept has coordinates $(0, 4)$.

The point (3, 0) is called the ***x*-intercept,** and the point (0, 4) is the ***y*-intercept** of the graph. Using these points to draw the graph gives the name to this method. Here is another example of graphing by the intercept method.

Example 6 Using the Intercept Method to Graph a Line

< Objective 3 >

Graph $3x - 5y = 15$, using the intercept method.

To find the x-intercept, let $y = 0$.

$$3x - 5 \cdot (0) = 15$$

$$x = 5$$

The x-value of the intercept

To find the y-intercept, let $x = 0$.

$$3 \cdot (0) - 5y = 15$$

$$y = -3$$

The y-value of the intercept

So (5, 0) and (0, −3) are solutions for the equation, and we can use the corresponding points to graph the equation.

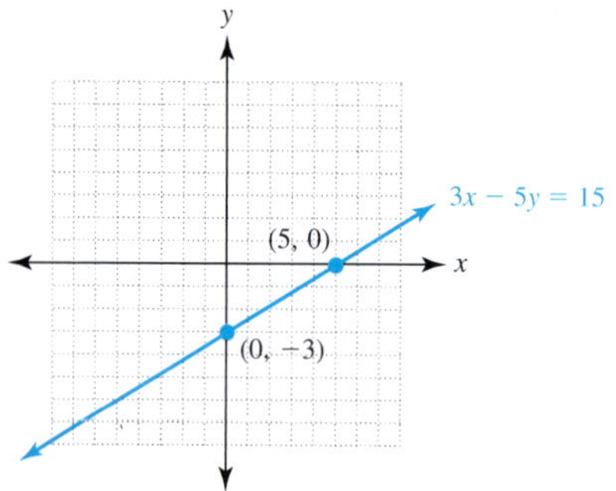

Check Yourself 6

Graph $4x + 5y = 20$, using the intercept method.

NOTE

Finding a third "checkpoint" is always a good idea.

This all looks quite easy, and for many equations it is. What are the drawbacks? For one, you don't have a third checkpoint, and it is possible for errors to occur. You can, of course, still find a third point (other than the two intercepts) to be sure your graph is correct. A second difficulty arises when the x- and y-intercepts are very close to each other (or are actually the same point—the origin). For instance, if we have the equation

$$3x + 2y = 1$$

the intercepts are $\left(\frac{1}{3}, 0\right)$ and $\left(0, \frac{1}{2}\right)$. It is hard to draw a line accurately through these intercepts, so choose other solutions farther away from the origin for your points.

We summarize the steps of graphing by the intercept method for appropriate equations.

Step by Step

Graphing a Line by the Intercept Method

Step 1 To find the x-intercept: Let $y = 0$, then solve for x.
Step 2 To find the y-intercept: Let $x = 0$, then solve for y.
Step 3 Graph the x- and y-intercepts.
Step 4 Draw a straight line through the intercepts.

A third method of graphing linear equations involves **solving the equation for *y*.** The reason we use this extra step is that it often makes it much easier to find solutions for the equation. Here is an example.

Example 7 Graphing a Linear Equation

< Objective 4 >

RECALL

Solving for y means that we want to leave y isolated on the left.

RECALL

We can write this equation in function form,

$f(x) = -\frac{2}{3}x + 2$

NOTE

Again, to pick convenient values for x, we suggest you look at the equation carefully. Here, for instance, picking multiples of 3 for x makes the work much easier.

Graph $2x + 3y = 6$.

Rather than finding solutions for the equation in this form, we solve for y.

$$2x + 3y = 6 \qquad \text{Subtract } 2x.$$
$$3y = 6 - 2x \qquad \text{Divide by 3.}$$
$$y = \frac{6 - 2x}{3}$$

We have solved for y. However, we will have reason in the coming sections to write this in a different form:

$$y = 2 - \frac{2}{3}x \qquad \text{We distributed the division.}$$
$$y = 2 + \left(-\frac{2}{3}x\right)$$
$$y = -\frac{2}{3}x + 2$$

Now find your solutions by picking convenient values for x.

If $x = -3$:

$$y = -\frac{2}{3}x + 2$$
$$= -\frac{2}{3}(-3) + 2$$
$$= 2 + 2 = 4$$

So $(-3, 4)$ is a solution.

If $x = 0$:

$$y = -\frac{2}{3}x + 2$$
$$= -\frac{2}{3}(0) + 2$$
$$= 0 + 2 = 2$$

So $(0, 2)$ is a solution.

If $x = 3$:

$$y = -\frac{2}{3}x + 2$$
$$= -\frac{2}{3}(3) + 2$$
$$= -2 + 2 = 0$$

So (3, 0) is a solution.

We can now plot the points that correspond to these solutions and form the graph of the equation as before.

x	y
−3	4
0	2
3	0

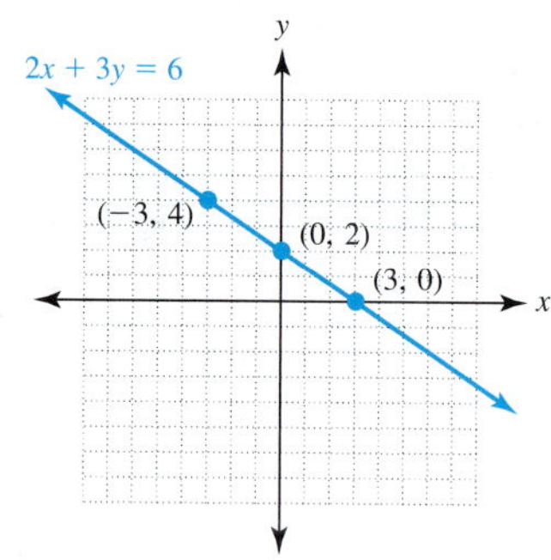

Check Yourself 7

Graph the equation $5x + 2y = 10$. Solve for y to determine solutions.

x	y
0	
2	
4	

Many students find it easier to keep themselves organized by using function notation when working with linear equations. In Chapter 2, you learned that when we solve a two-variable equation for y, we can write y as a function of x. In this case, we write

$$y = f(x)$$

One advantage to writing an equation with function notation is that it allows us to see the value we are using for x when evaluating the function.

Example 8 Graphing a Linear Function

< Objective 5 >

> **RECALL**
>
> To write y as a function of x, solve the equation for y and replace y with $f(x)$.

Rewrite the equation shown so that y is a function of x. Graph the function.

$$x - 2y = 6$$

Begin by solving the equation for y.

$$x - 2y = 6$$
$$x - 6 = 2y \quad \text{Subtract 6 and add } 2y \text{ to both sides.}$$
$$2y = x - 6 \quad \text{Switch sides so that the } y\text{-term is on the left.}$$

NOTE

Choosing even numbers for your inputs (or x-values) guarantees whole-number outputs.

$$y = \frac{x - 6}{2} \qquad \text{Divide both sides by 2.}$$

$$y = \frac{1}{2}x - 3 \qquad \text{Remember to use distribution.}$$

$$f(x) = \frac{1}{2}x - 3$$

Now we evaluate the function at three points to graph it.

$$f(0) = \frac{1}{2}(0) - 3 = -3 \qquad (0, -3)$$

$$f(2) = \frac{1}{2}(2) - 3 = 1 - 3 = -2 \qquad (2, -2)$$

$$f(4) = \frac{1}{2}(4) - 3 = 2 - 3 = -1 \qquad (4, -1)$$

Finally, we plot the three points and draw the line through them.

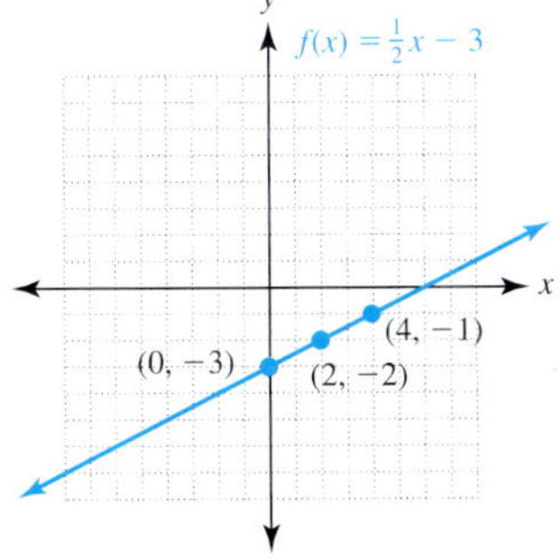

Check Yourself 8

Rewrite the equation shown so that y is a function of x, and graph the function.

$6x + 2y = 4$

One important reason to solve a linear equation for y is so that we can analyze it with a graphing calculator. In order to enter an equation into the [Y=] menu, we need to isolate y because your calculator needs to work with functions.

Example 9 Using a Graphing Calculator

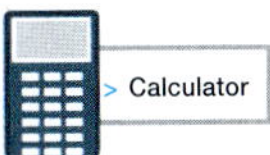

RECALL

A graphing calculator needs to "think" of the equation as a function so it must look like "y is a function of x."

Use a graphing calculator to graph the equation

$$2x + 3y = 6$$

In Example 7, we solved this equation for y to form the equivalent equation

$$y = -\frac{2}{3}x + 2$$

Enter the right side of the equation into the [Y=] menu in the Y_1 field, and then press the [GRAPH] key.

NOTE

A good way to enter fractions is to enclose them in parentheses.

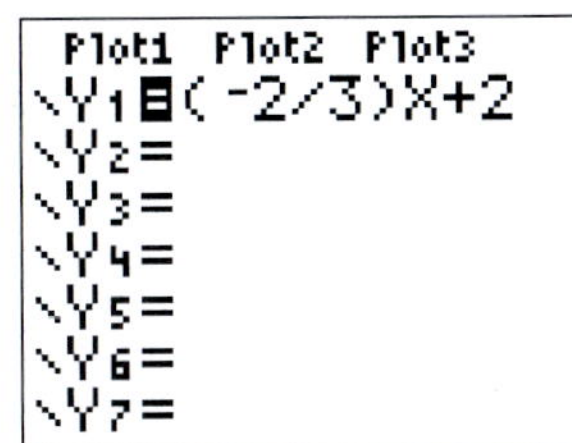

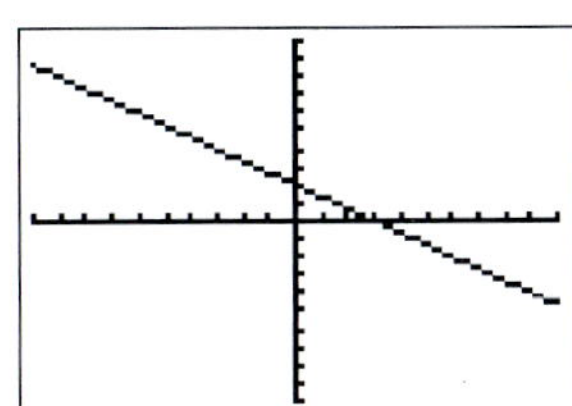

Check Yourself 9

Use a graphing calculator to graph the equation

$5x + 2y = 10$

When we scale the axes, it is important to include numbers on the axes at convenient grid lines. If we set the axes so that part of an axis is removed, we include a mark to indicate this. Both of these situations are illustrated in Example 10.

Example 10 Graphing in Nonstandard Windows

NOTE

In business, the constant, 2,500, is called the **fixed cost.** The slope, 45, is referred to as the **marginal cost.**

The cost, y, to produce x CD players is given by the equation $y = 45x + 2,500$. Graph the cost equation, with appropriately scaled and set axes.

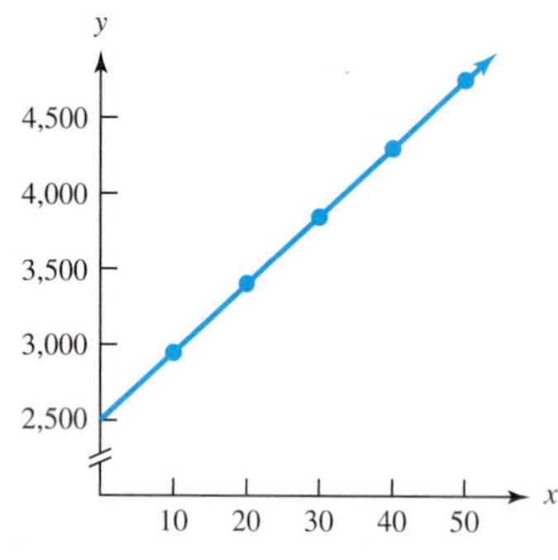

We removed part of the y-axis.

NOTE

We can also graph this with a graphing calculator.

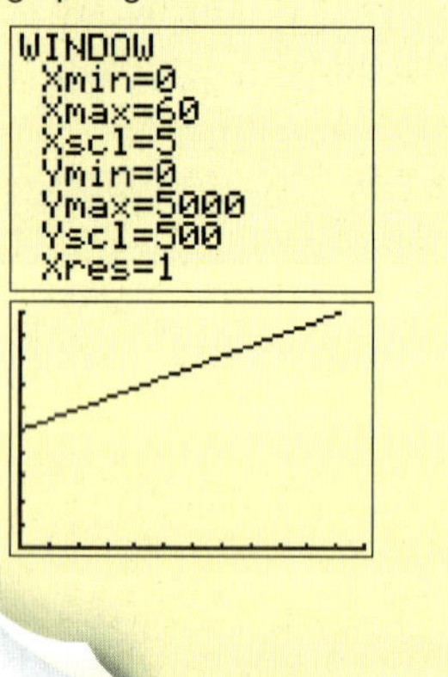

The y-intercept is (0, 2,500). We find more points to plot by creating a table.

x	10	20	30	40	50
y	2,950	3,400	3,850	4,300	4,750

Check Yourself 10

Graph the cost equation given by $y = 60x + 1,200$, with appropriately scaled and set axes.

Here is an application from the field of medicine.

Example 11 A Health Sciences Application

NOTE

The domain (the set of possible values for A) is restricted to positive integers.

The arterial oxygen tension (P_aO_2), in millimeters of mercury (mm Hg), of a patient can be estimated based on the patient's age (A), in years. If the patient is lying down, the equation $P_aO_2 = 103.5 - 0.42A$ is used to determine arterial oxygen tension. Draw the graph of this equation, using appropriately scaled and set axes.

We begin by creating a table. Using a calculator here is very helpful.

A	0	10	20	30	40	50	60	70	80
P_aO_2	103.5	99.3	95.1	90.9	86.7	82.5	78.3	74.1	69.9

RECALL

We can use the table feature to determine a reasonable viewing window.

X	Y1
0	103.5
20	95.1
40	86.7
60	78.3
80	69.9
100	61.5
120	53.1

X=0

We can also graph this with a graphing calculator.

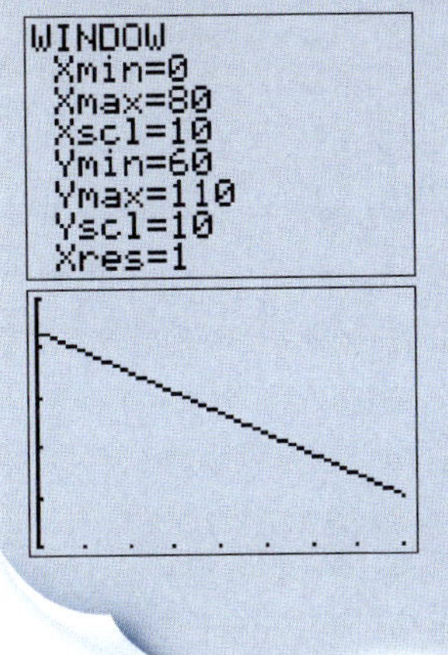

Seeing these values allows us to decide upon the vertical axis scaling. We scale from 60 to 110, and include a mark to show a break in the axis. We estimate the locations of these coordinates, and draw the line.

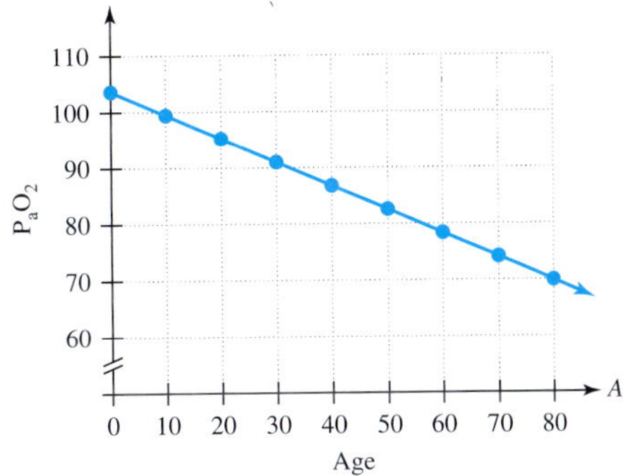

Check Yourself 11

The arterial oxygen tension (P_aO_2), in millimeters of mercury (mm Hg), of a patient can be estimated based on the patient's age (A), in years. If the patient is seated, the equation $P_aO_2 = 104.2 - 0.27A$ is used to approximate arterial oxygen tension. Draw the graph of this equation, using appropriately scaled and set axes.

Check Yourself ANSWERS

1.

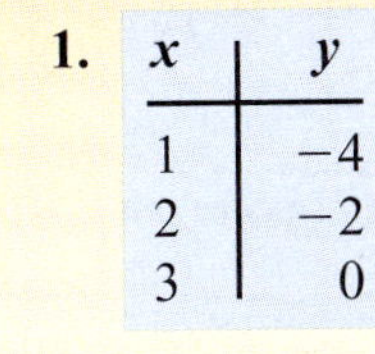

x	y
1	−4
2	−2
3	0

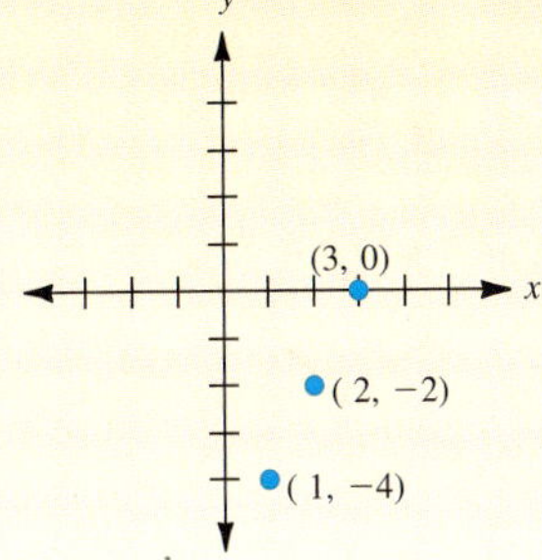

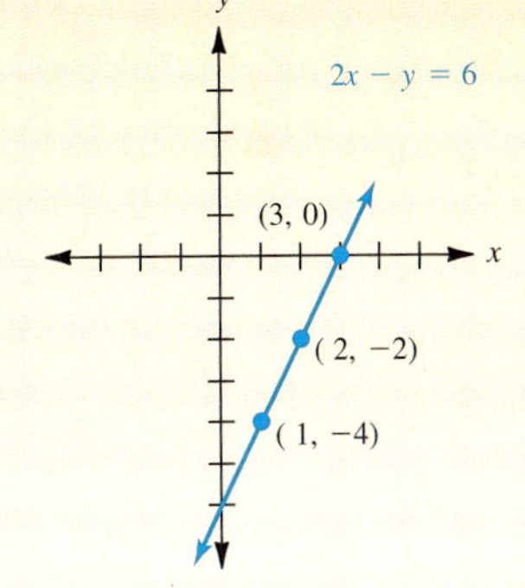

2.

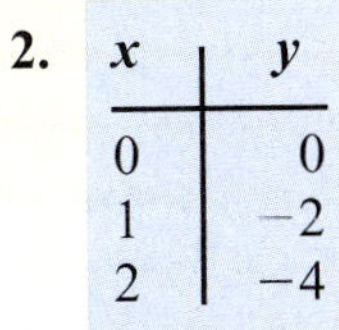

x	y
0	0
1	−2
2	−4

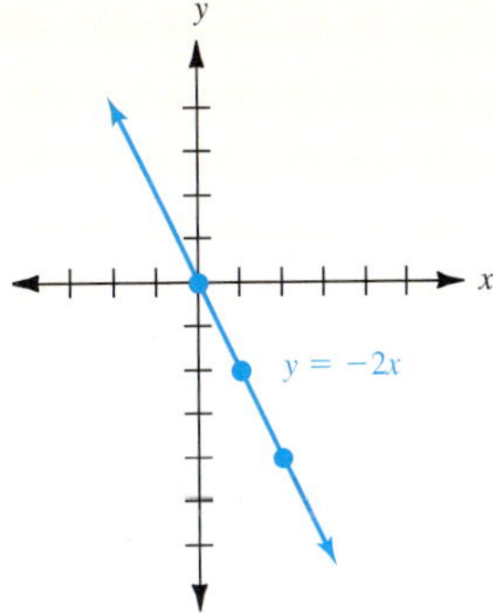

3.

x	y
0	−2
1	1
2	4

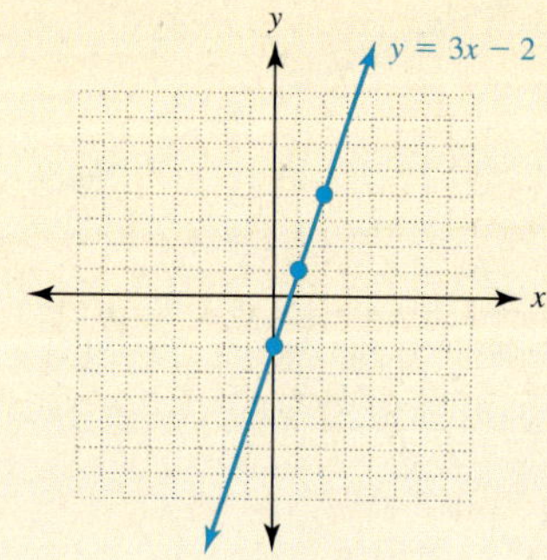

4.

x	y
−3	4
0	3
3	2

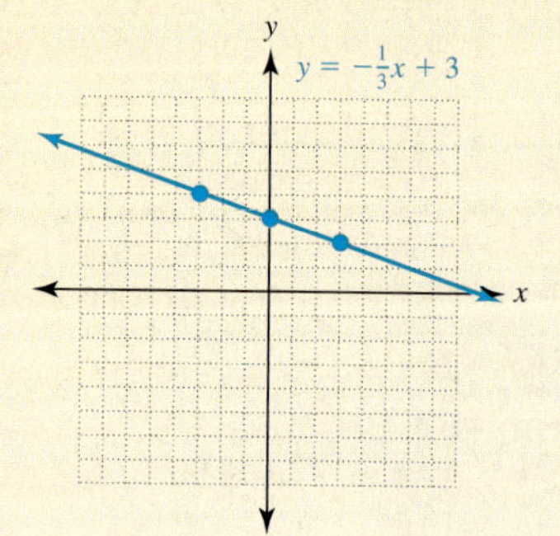

5. (a)

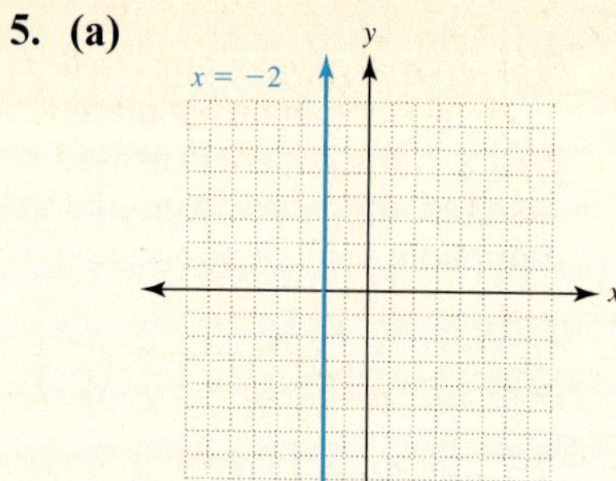

(b)

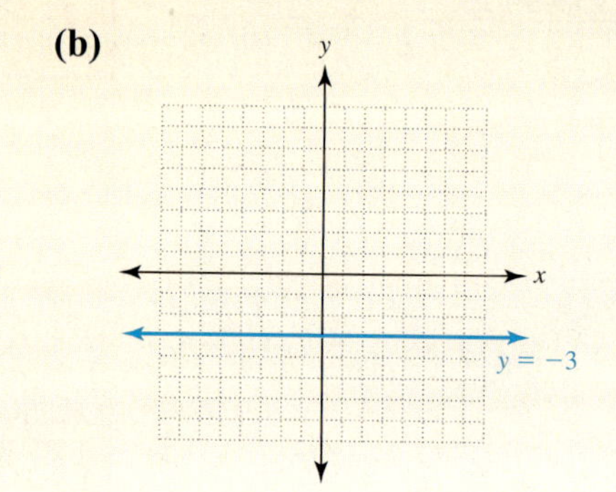

6.

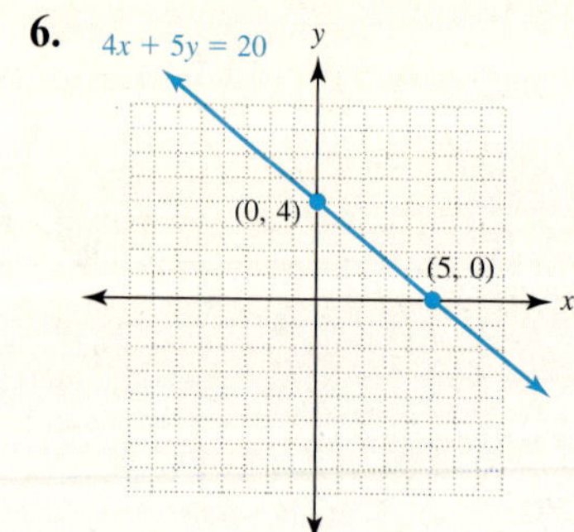

7.

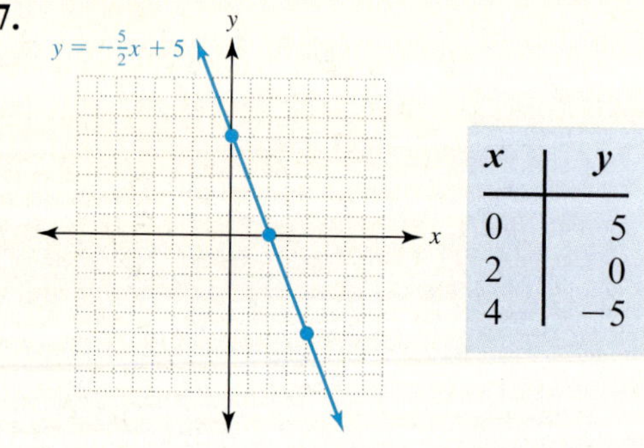

x	y
0	5
2	0
4	−5

8. $f(x) = -3x + 2$

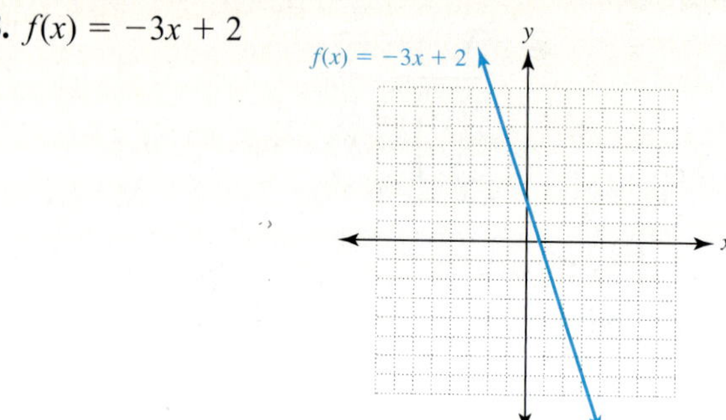

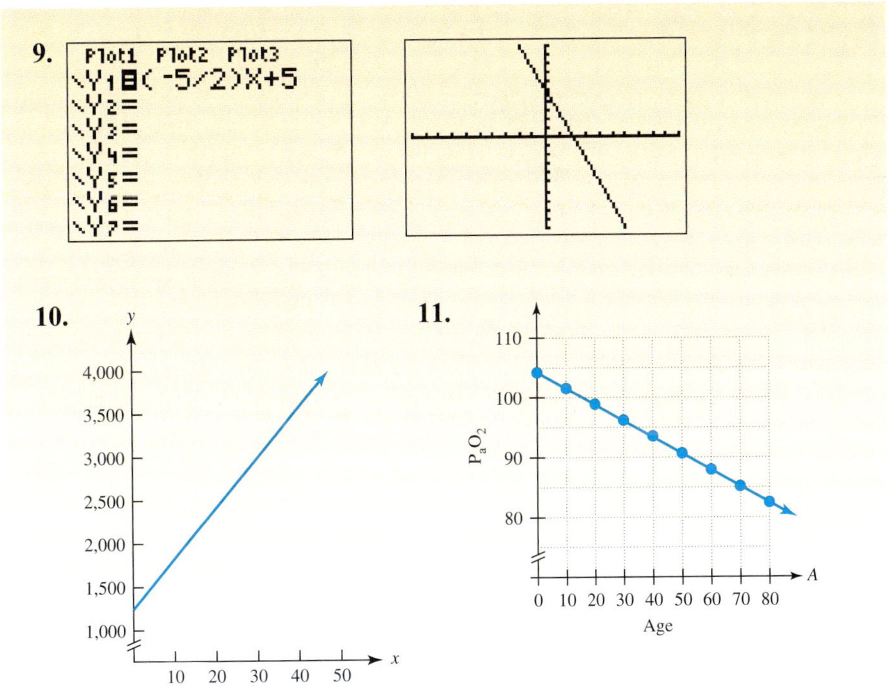

Reading Your Text

SECTION 3.1

(a) A graph is a picture of the ______________ for a given equation.

(b) The graph of $x = a$ is a ______________ line.

(c) A horizontal line represents the graph of a ______________ function.

(d) The x-coordinate is ______________ at the y-intercept of a graph.

Activity 3 ::
Linear Regression: A Graphing Calculator Activity

In Section 3.4, you will learn to build functions that model real-world phenomena. One of the more powerful features of graphing calculators is that they can create *regression equations* to approximate a data set. We will describe how to use Texas Instruments calculators, the TI-83 and TI-84 Plus, to plot data, find a linear regression equation, and use that equation. See your instructor or your calculator manual to learn the steps necessary to get your particular calculator model to perform these functions.

Scatter Plots

We begin by putting together a *scatter plot.* At its most basic level, a scatter plot is simply a set of points on the same graph. Of course, in order to be useful, the points should all be related in some way.

The data that we will use relate the amount of time (in hours) each of 13 students spent studying for an exam and their grades on the exam.

Study Time, x	0	1	2	4	4	5	5	5	6	6	7	7	8
Exam Grade, y	50	51	72	52	74	78	81	74	86	93	84	92	94

We want to enter the data into our calculator so that we can create a scatter plot.

1. Clear any existing data from the lists you will use.
   We will use Lists 1 and 2, so our first step is to make sure these lists are empty. We do this by accessing the statistics menu and clearing the lists.
   STAT **4:ClrList**
   "ClrList" will appear on the home screen.
   Then, tell the calculator to clear Lists 1 and 2.
   2nd [L1] , 2nd [L2] ENTER
   Note: On the Texas Instruments models, [L1] and [L2] are the second functions of the **1** and **2** number keys.

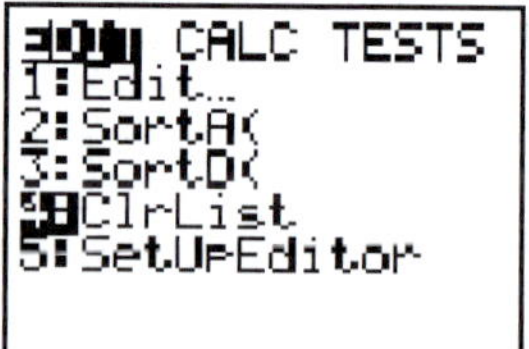

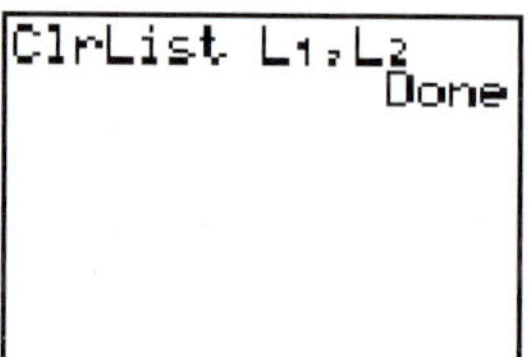

2. Enter the data into the lists.
   Access the lists by choosing the edit option from the statistics menu.
   STAT **1:Edit**
   Then, enter the x-values in the first list, pressing ENTER after each one.
   After entering the x-values, use the right-arrow key ▶ to move to the second list and enter the y-values.
   Note: It is surprisingly easy to make a mistake when entering data into the lists. You should double-check that you entered the data correctly and that the y-values that you enter are on the same line as the corresponding x-values.

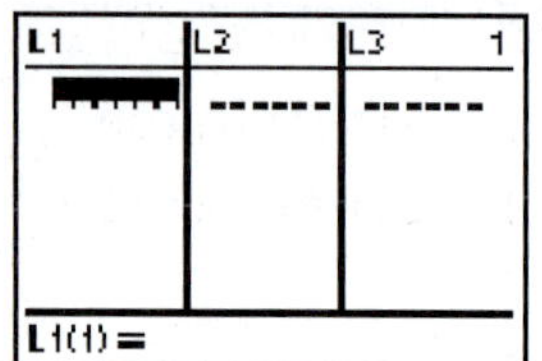

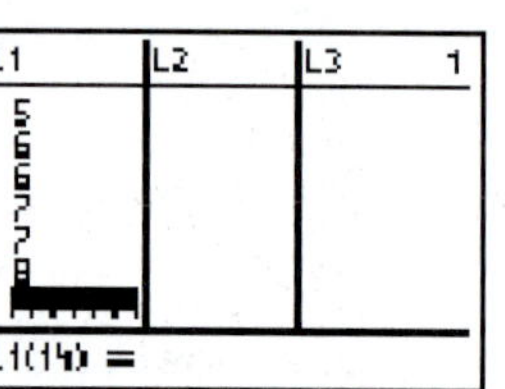

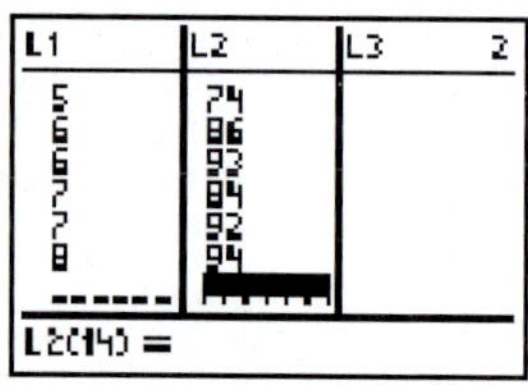

3. Create a scatter plot from the data.
Clear any equations from the function, or [Y=], menu.
Then, access the StatPlot menu, it is the second function of the [Y=] key. Select the first plot.
[2nd] [STAT PLOT] **1: Plot 1**
Select the **On** option and make sure the **Type** selected is the scatter plot, as shown in the figure to the right.

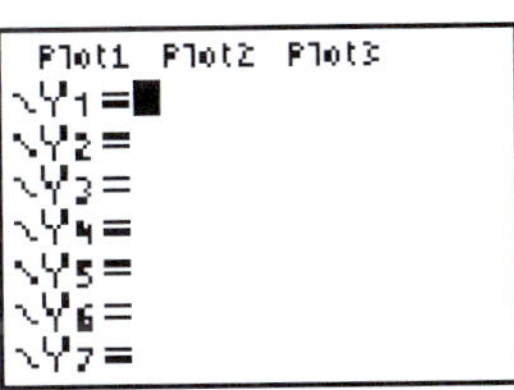

4. View the scatter plot.
Let the calculator choose an appropriate viewing window by using the ZoomStat feature.
[ZOOM] **9:ZoomStat**
You can use the window menu, [WINDOW], to modify the viewing window to improve your graph, if you wish. Press [GRAPH] to see the scatter plot when you are done.

NOTE

In the window menu, we increased the **Yscl** value to 10. **Yscl** gives the space between "tick marks" on the y-axis.

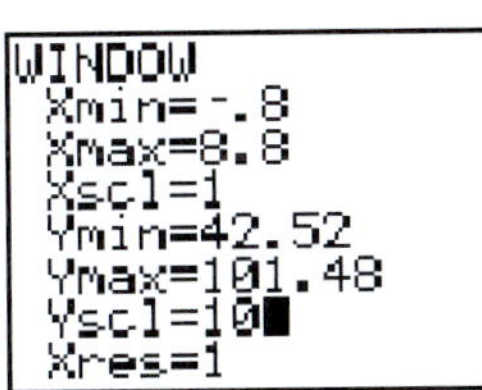

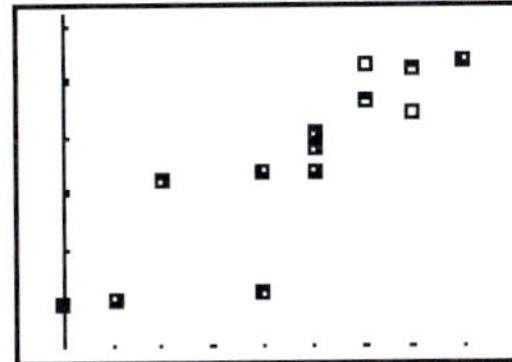

Regression Analysis

In this chapter, you will learn to construct a linear equation based on two points. Your calculator can accomplish the much more intense task of creating the *best* linear function to fit a larger set of data points.

1. Set your calculator to perform data analysis.
Access the statistics menu and set up the editor; you will need to enter the command in the home screen when it comes up:
[STAT] **5:SetUpEditor** [ENTER]
You also need to turn the calculator's diagnostics program on. You can do this by going to the *catalog* menu. The catalog menu is a complete listing of every function programmed into your calculator. The catalog menu is the second function of the **0** key.
[2nd] [CATALOG]
Move down the list until you reach **DiagnosticOn.** Press [ENTER] to send it to the home screen and press [ENTER] again to make it work.
We are now ready to perform the regression analysis.

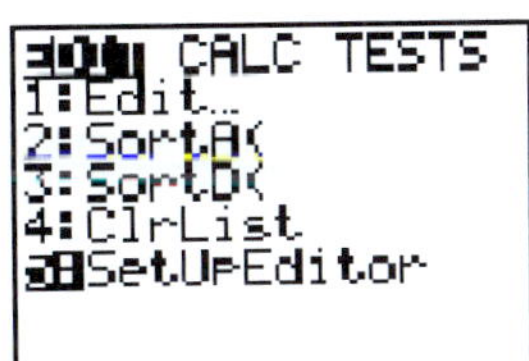

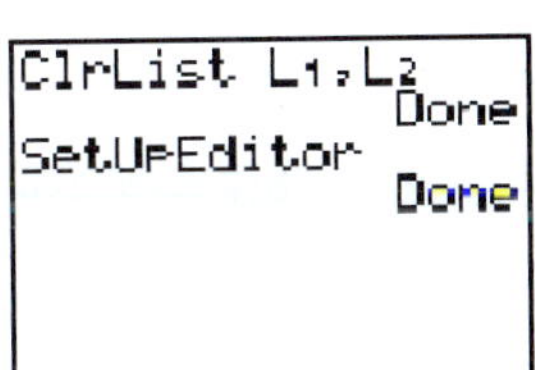

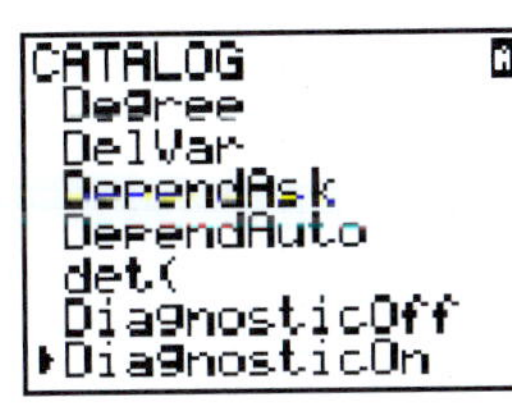

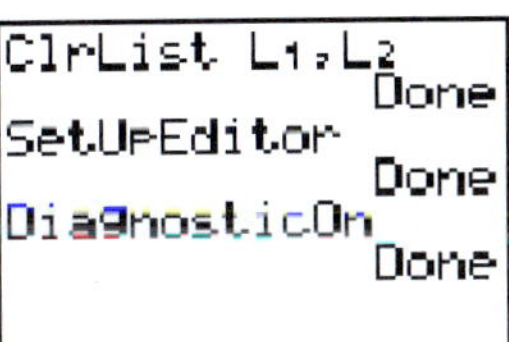

2. Perform a regression analysis on the data.
   Access the regression options by moving to the **CALC** submenu of the statistics menu. Then select the linear regression model.
   STAT ▶
   Note: This brings you to the **CALC** submenu of the statistics menu.
   4:LinReg(ax+b) ENTER

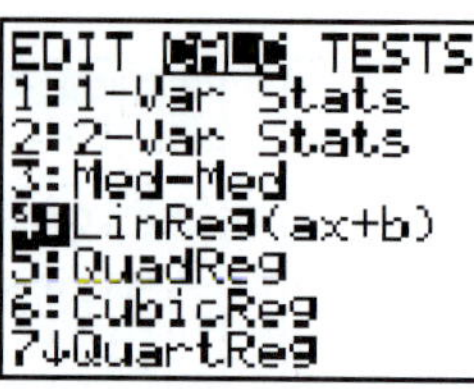

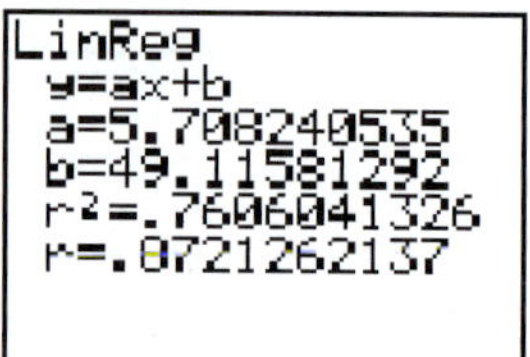

> **NOTE**
>
> Your calculator constructs a linear regression model by finding the line that minimizes the vertical distance between that line and the data set's y-values.

We will learn about the slope of a line beginning with the next section. After completing Sections 3.2 and 3.3, you should read this activity again. We briefly describe the information your calculator gives you.

The calculator is modeling a linear equation $y = ax + b$, so it is calling the slope a and the y-intercept is $(0, b)$ (in this equation, the number that multiplies x is the slope). In this case, the calculator is showing the line that best fits the student study data as

$y = 5.7x + 49.1$ (to one decimal place)

In the context of this application, a slope of 5.7 indicates that each additional hour of studying increased a student's exam score by 5.7 points. The y-intercept tells us that a student who did not study at all could expect to receive a 49.1 on the exam. *Note:* $\mathbf{r}^2$ and **r** are used to measure the validity of the model. The closer **r** is to 1 or −1, the better the model; the closer **r** is to 0, the worse the model.

3. Graph the linear regression model on the scatter plot.
   We command the calculator to paste the linear regression model into the function menu, Y=. The calculator has saved the regression model in a variables menu.
   Y= VARS **5:Statistics . . .** ▶ ▶ **1:RegEQ** GRAPH

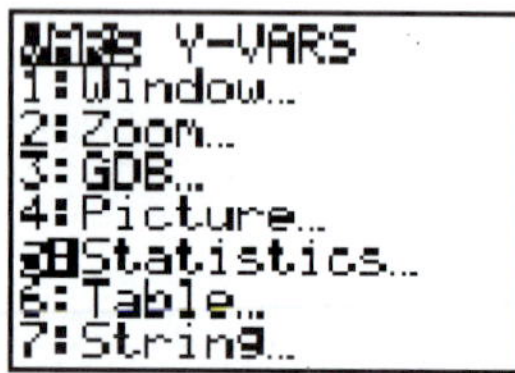

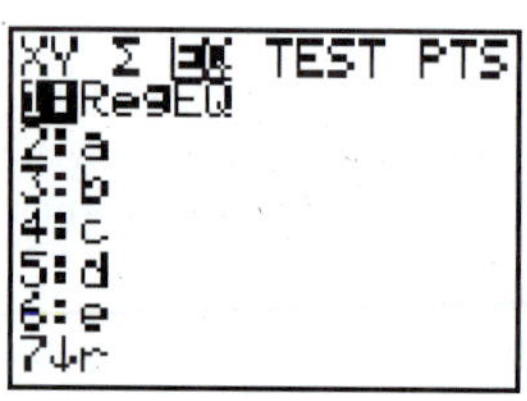

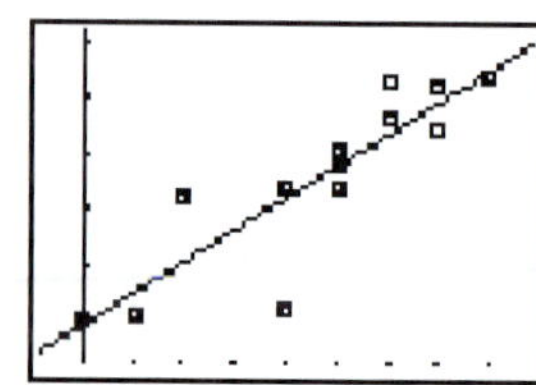

Exercise

The table gives the total acreage devoted to wheat in the United States over a recent 5-year period (all figures are in millions of acres).

Year	1	2	3	4	5
Acreage Planted, x	65.8	62.7	62.6	59.6	60.4
Acreage Harvested, y	59.0	53.8	53.1	48.6	45.8

Source: Farm Service Agency; U.S. Department of Agriculture (Aug, 2003).

(a) Create a scatter plot relating the acres planted and harvested.
(b) Perform a regression analysis on the data.
(c) Give the slope and y-intercept (one decimal place of accuracy) and interpret them in the context of this application.
(d) Graph the regression equation in the same window with the scatter plot.

Answers

(a)

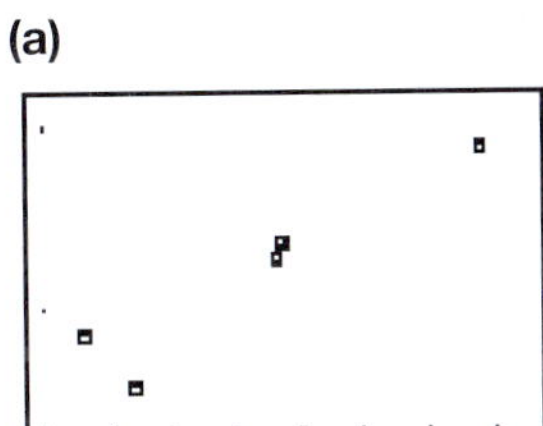

(b)

(c)

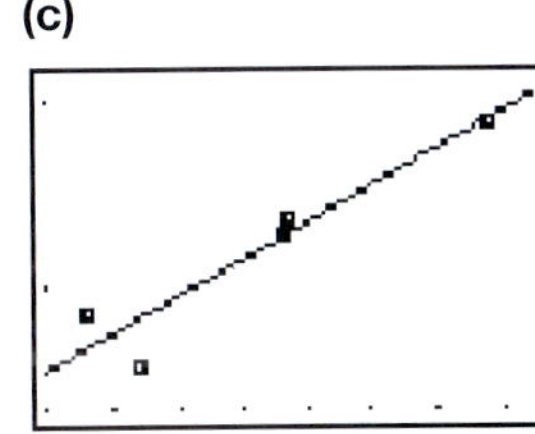

(c) The slope is approximately 2, which means that for each additional acre planted, we expect to harvest two additional acres of wheat. The y-intercept is $(0, -71.8)$, which claims that if we planted no wheat, we would harvest a negative amount of wheat. This is, of course, not true. This means that our model is not valid near $x = 0$.

3.1 exercises

Basic Skills | Challenge Yourself | Calculator/Computer | Career Applications | Above and Beyond

Boost *your* GRADE at ALEKS.com!

ALEKS®

- Practice Problems
- Self-Tests
- NetTutor
- e-Professors
- Videos

Name ______________

Section ________ Date ________

Answers

1. ______________
2. ______________
3. ______________
4. ______________
5. ______________
6. ______________
7. ______________
8. ______________
9. ______________
10. ______________
11. ______________
12. ______________

< Objectives 1–3 >

Graph each equation.

1. $x + y = 6$

2. $x - y = 5$

3. $x - y = -3$ 

4. $x + y = -3$

5. $3x + y = 6$

6. $x - 2y = 6$

7. $3x + y = 0$

8. $2x - y = 4$

9. $x + 4y = 8$

10. $2x - 3y = 6$

11. $y = 3x$

12. $y = -4x$

13. $y = 2x - 1$

14. $y = 2x + 5$

15. $y = -3x + 1$

16. $y = -3x - 3$

17. $y = \frac{1}{5}x$

18. $y = -\frac{1}{4}x$

19. $y = \frac{2}{3}x - 3$

20. $y = \frac{3}{4}x + 2$

21. $x = -3$ 

22. $y = -3$

23. $y = 1$

24. $x = 4$

25. $x - 2y = 4$

26. $6x + y = 6$

Answers

13. ______
14. ______
15. ______
16. ______
17. ______
18. ______
19. ______
20. ______
21. ______
22. ______
23. ______
24. ______
25. ______
26. ______

Answers

27. ______

28. ______

29. ______

30. ______

31. ______

32. ______

33. ______

34. ______

35. ______

36. ______

37. ______

38. ______

39. ______

40. ______

27. $5x + 2y = 10$

28. $2x + 3y = 6$

29. $3x + 5y = 15$

30. $4x + 3y = 12$

< Objectives 4 and 5 >

Solve each equation for y, write the equation in function form, and graph the function.

31. $x + 3y = 6$

32. $x - 2y = 6$

33. $3x + 4y = 12$

34. $2x - 3y = 12$

35. $5x - 4y = 20$

36. $7x + 3y = 21$

Basic Skills | **Challenge Yourself** | Calculator/Computer | Career Applications | Above and Beyond

Complete each statement with **never, sometimes,** *or* **always.**

37. If the ordered pair (x, y) is a solution to an equation in two variables, then the point (x, y) is ______ on the graph of the equation.

38. If the graph of a linear equation $Ax + By = C$ passes through the origin, then C ______ equals zero.

39. If the ordered pair (x, y) is *not* a solution to an equation in two variables, then the point (x, y) is ______ on the graph of the equation.

40. The graph of a horizontal line ______ passes through the origin.

Write an equation that describes each relationship between x and y.

41. y is twice x.

42. y is 3 times x.

43. y is 3 more than x.

44. y is 2 less than x.

45. y is 3 less than 3 times x.

46. y is 4 more than twice x.

47. The difference of x and the product of 4 and y is 12.

48. The difference of twice x and y is 6.

Graph each pair of equations on the same grid. Give the coordinates of the point where the lines intersect.

49. $x + y = 4$
$x - y = 2$

50. $x - y = 3$
$x + y = 5$

51. **Business and Finance** The function $f(x) = 0.10x + 200$ describes the amount of winnings a group earns for collecting plastic jugs in a recycling contest. Sketch the graph of the line.

52. **Business and Finance** In exercise 51, the contest sponsor will award a prize only if the winning group in the contest collects 100 lb of jugs or more. Use your graph to determine the minimum prize possible.

53. **Business and Finance** A high school class wants to raise some money by recycling newspapers. They decide to rent a truck for a weekend and to collect the newspapers from homes in the neighborhood. The market price for recycled newsprint is currently \$15 per ton. The function $f(x) = 15x - 100$ describes the amount of money the class will make, where $f(x)$ is the amount of money made in dollars, x is the number of tons of newsprint collected, and 100 is the cost in dollars to rent the truck.

(a) Draw a graph that represents the relationship between newsprint collected and money earned.

(b) The truck is costing the class \$100. How many tons of newspapers must the class collect to break even on this project?

(c) If the class members collect 16 tons of newsprint, how much money will they earn?

(d) Six months later the price of newsprint is \$17 dollars per ton, and the cost to rent the truck has risen to \$125. Construct a function describing the amount of money the class might make at that time.

54. **Business and Finance** The cost of producing x items is given by $C(x) = mx + b$, where b is the fixed cost and m is the marginal cost (the cost of producing one additional item).

(a) If the fixed cost is \$40 and the marginal cost is \$10, write the cost function.

Answers

41. ______

42. ______

43. ______

44. ______

45. ______

46. ______

47. ______

48. ______

49. ______

50. ______

51. ______

52. ______

53. ______

54. ______

Answers

55. ______

56. ______

57. ______

(b) Graph the cost function.

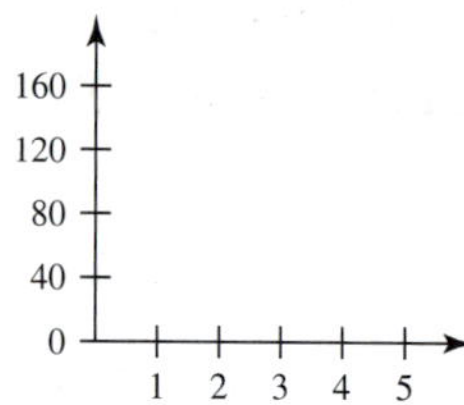

(c) The revenue generated from the sale of x items is given by $R(x) = 50x$. Graph the revenue function on the same set of axes as the cost function.

(d) How many items must be produced for the revenue to equal the cost (the break-even point)?

55. Business and Finance A car rental agency charges \$12 per day and 8¢ per mile for the use of a compact automobile. The cost of the rental C and the number of miles driven per day s are related by the equation

$C = 0.08s + 12$

Graph the relationship between C and s. Be sure to select appropriate scaling for the C and s axes.

56. Business and Finance A bank has this structure for charges on checking accounts: the monthly charges consist of a fixed amount of \$8 and an additional charge of 4¢ per check. The monthly cost of an account C and the number of checks written per month n are related by the equation

$C = 0.04n + 8$

Graph the relationship between C and n. 

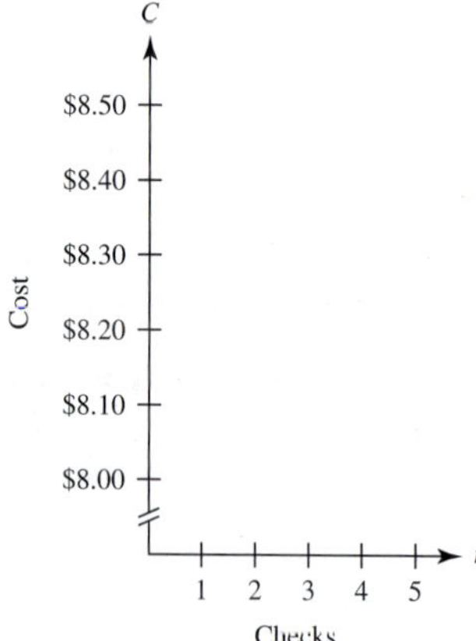

57. Business and Finance A college has tuition charges based on a pattern: tuition is \$35 per credit-hour plus a fixed student fee of \$75.

(a) Write a linear function describing the relationship between the total tuition charge T and the number of credit-hours taken h.

(b) Graph the relationship between T and h.

58. **Business and Finance** A salesperson's weekly salary is based on a fixed amount of \$200 plus 10% of the total amount of weekly sales.

(a) Write an equation that shows the relationship between the weekly salary S and the amount of weekly sales x (in dollars).

(b) Graph the relationship between S and x.

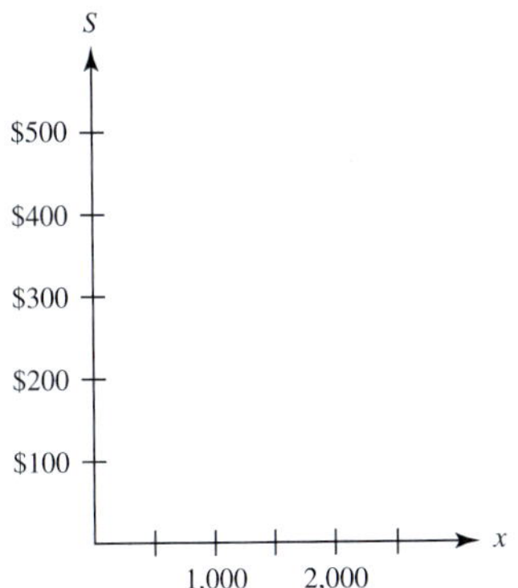

Basic Skills | Challenge Yourself | **Calculator/Computer** | Career Applications | Above and Beyond

59. Use a graphing calculator to draw the graph for the equation you created in exercise 53, part (d). Choose a window that shows results from $x = 0$ to $x = 20$. Sketch the graph you see on your screen, and indicate the viewing window that you chose. chapter 3 > Make the Connection

60. Use a graphing calculator to draw the graphs of the equations you created in exercise 54, parts (a) and (c). Choose a window that shows results from $x = 0$ to $x = 4$. Sketch what you see on your screen, and indicate the viewing window that you chose.

61. Use a graphing calculator to draw the graph of the equation given in exercise 55. Choose a window that shows results from $s = 0$ to $s = 300$. Sketch the graph you see on your screen, and indicate the viewing window that you chose. chapter 3 > Make the Connection

62. Use a graphing calculator to draw the graph for the equation you created in exercise 58. Choose a window that shows results from $x = 0$ to $x = 3{,}000$. Sketch the graph you see on your screen, and indicate the viewing window that you chose.

Answers

58. ______

59. ______

60. ______

61. ______

62. ______

Answers

63.

64.

65.

66.

67.

68.

69.

70.

Basic Skills | Challenge Yourself | Calculator/Computer | **Career Applications** | Above and Beyond

63. **Allied Health** The weight w (in kg) of a uterine tumor is related to the number of days d of chemotherapy treatment by the function $w(d) = -1.75d + 25$. Sketch a graph of the weight of a tumor in terms of the number of days of treatment.

64. **Mechanical Engineering** The force that a coil exerts on an object is related to the distance that the coil is pulled from its natural (at-rest) position. The formula to describe this is $F = kx$. Graph this relationship for a coil for which $k = 72$ pounds per foot.

65. **Construction Technology** The number of studs s (16 inches on center) required to build a wall that is L feet long is given by the formula

$$s = \frac{3}{4}L + 1$$

Graph the equation with appropriately scaled axes.

66. **Manufacturing Technology** The number of board feet b of lumber in a 2" × 6" board of length L (in feet) is given by the equation

$$b = \frac{8.25}{144}L$$

Graph the equation with appropriately scaled axes.

Basic Skills | Challenge Yourself | Calculator/Computer | Career Applications | **Above and Beyond**

In each exercise, graph both functions on the same set of axes and report what you observe about the graphs.

67. $f(x) = 2x$ and $g(x) = 2x + 1$

68. $f(x) = 3x + 1$ and $g(x) = 3x - 1$

69. $f(x) = 2x$ and $g(x) = -\frac{1}{2}x$

70. $f(x) = \frac{1}{3}x + \frac{7}{3}$ and $g(x) = -3x + 2$

71. Consider the equation $y = 2x + 3$.

(a) Complete the table of values, and plot the resulting points.

Point	x	y
A	5	
B	6	
C	7	
D	8	
E	9	

(b) As the x-coordinate changes by 1 (for example, as you move from point A to point B), how much does the corresponding y-coordinate change?

(c) Is your answer to part (b) the same if you move from B to C? from C to D? from D to E?

(d) Describe the "growth rate" of the line, using these observations. Complete the statement: When the x-value grows by 1 unit, the y-value ________.

72. Describe how answers to parts (b), (c), and (d) would change if you were to repeat exercise 71 using $y = 2x + 5$.

73. Describe how answers to parts (b), (c), and (d) would change if you were to repeat exercise 71 using $y = 3x - 2$.

74. Describe how answers to parts (b), (c), and (d) would change if you were to repeat exercise 71 using $y = 3x - 4$.

75. Describe how answers to parts (b), (c), and (d) would change if you were to repeat exercise 71 using $y = -4x + 50$.

76. Describe how answers to parts (b), (c), and (d) would change if you were to repeat exercise 71 using $y = -4x + 40$.

Answers

71. ________

72. ________

73. ________

74. ________

75. ________

76. ________

Answers

1.

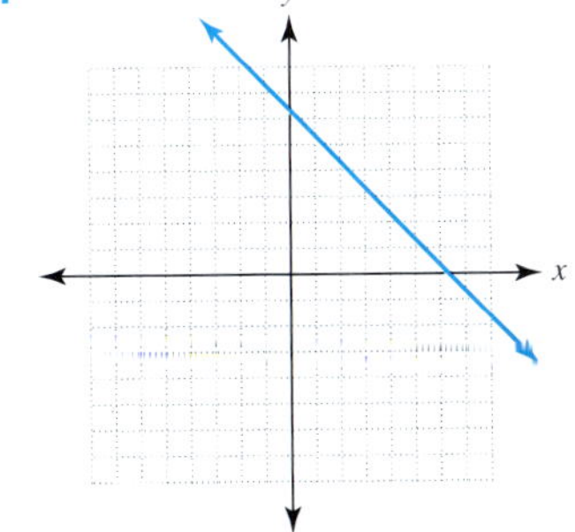

3.

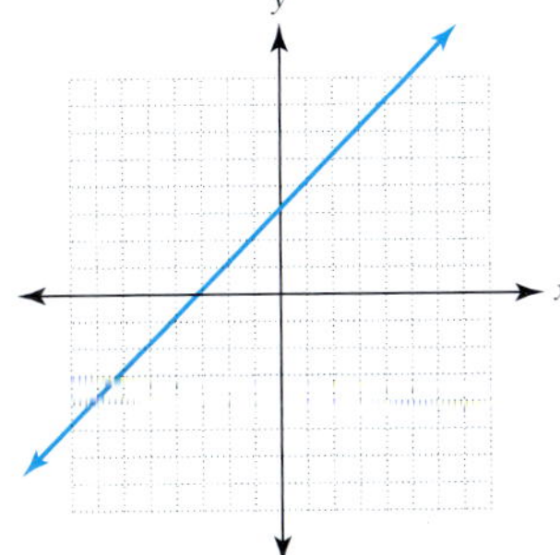

5.

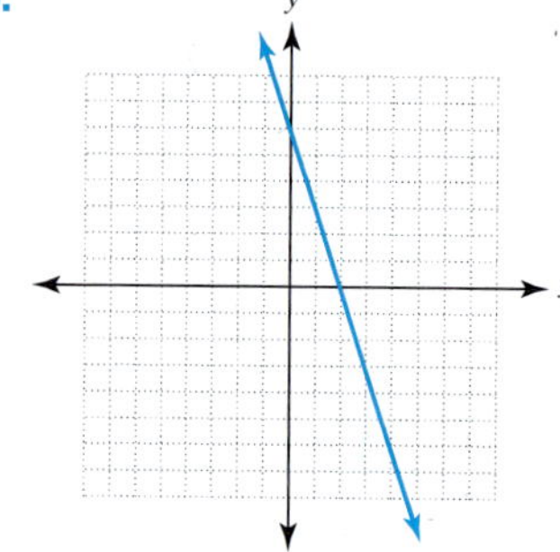

7.

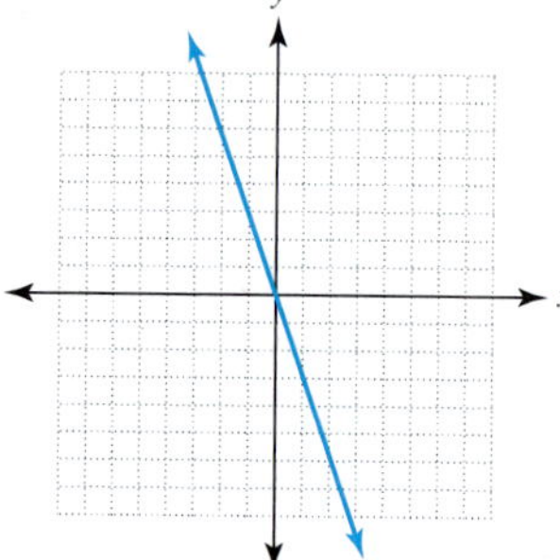

9.

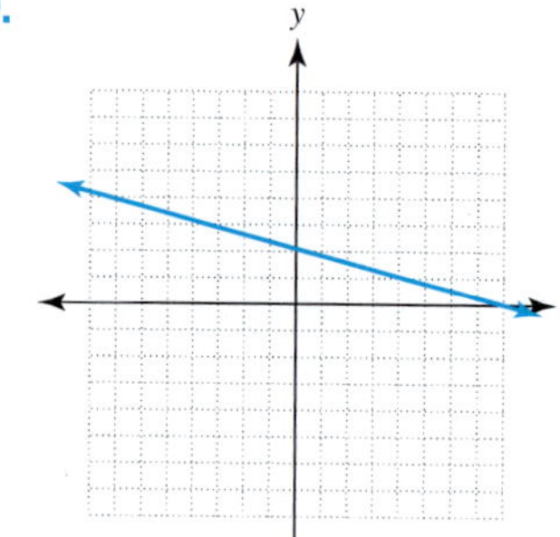

11.

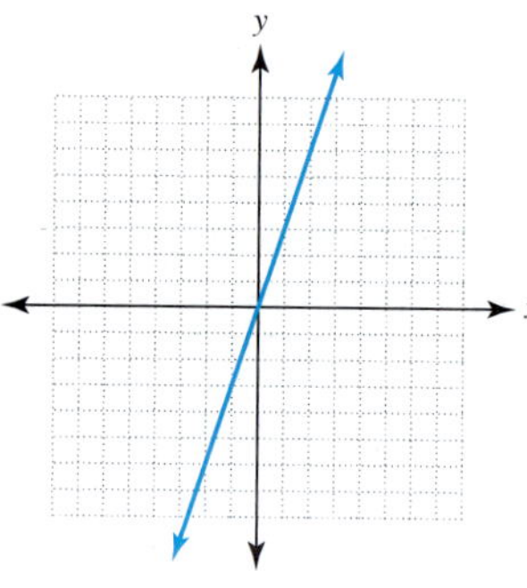

13.

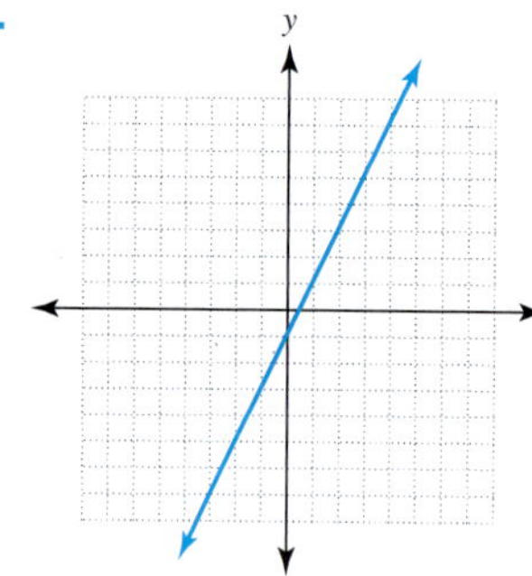

15.

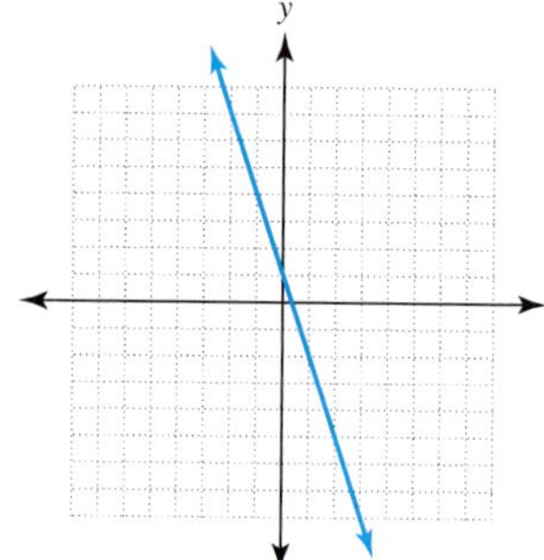

17.

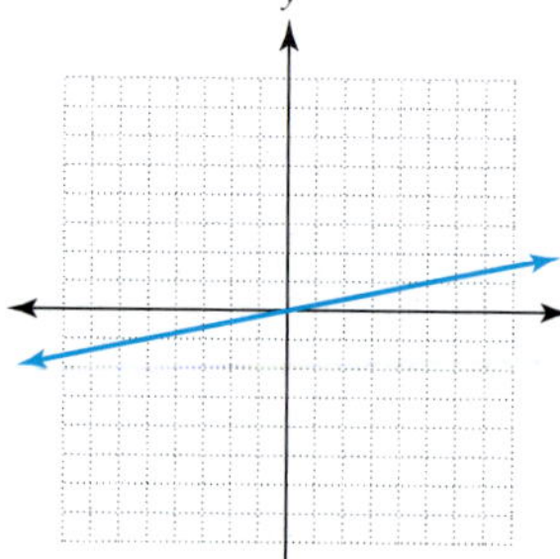

19.

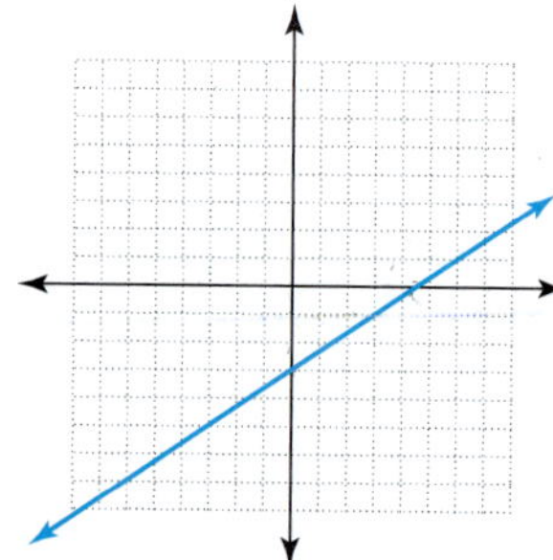

21.

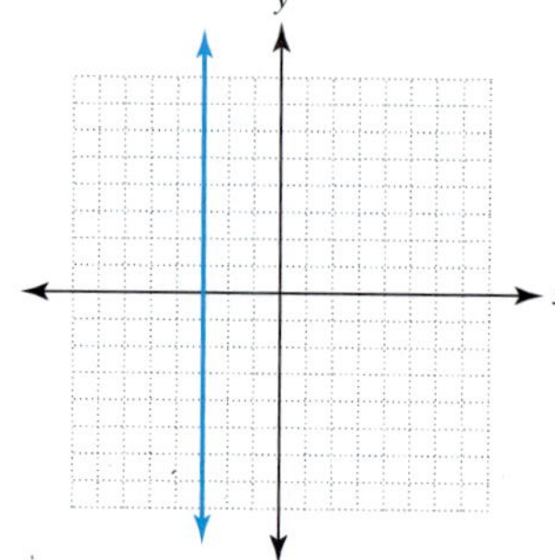

23.

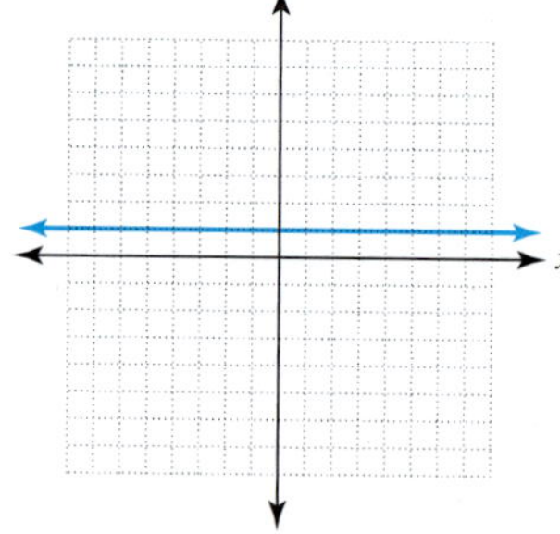

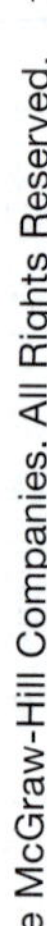

25.

27.

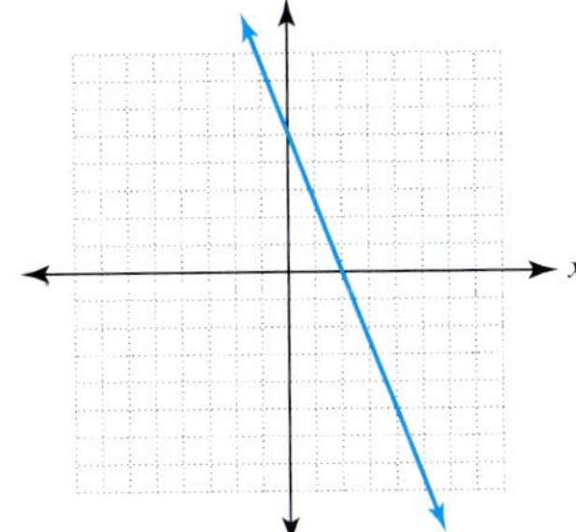

29.

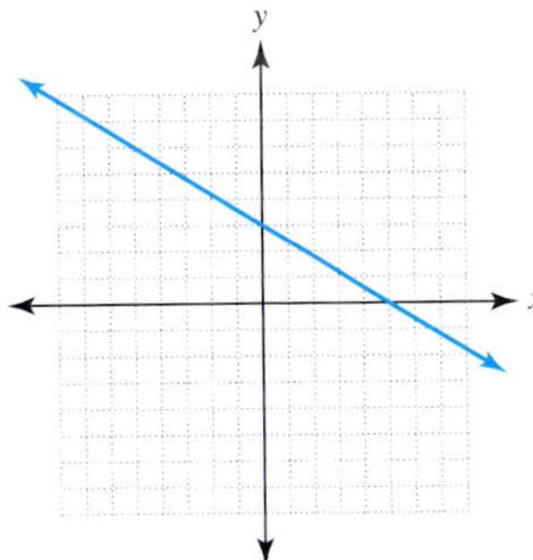

31.

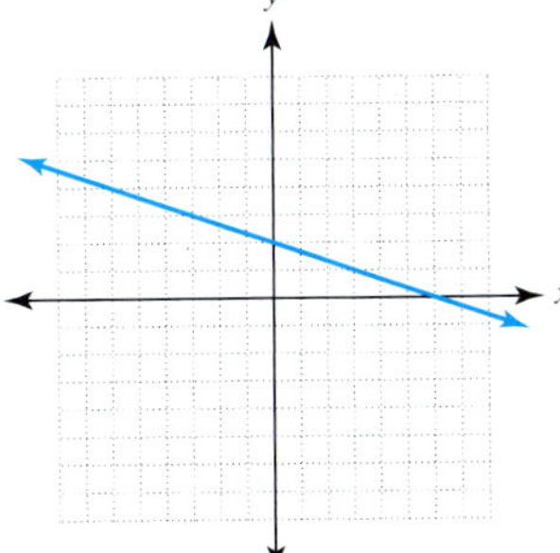

$f(x) = -\frac{1}{3}x + 2$

33.

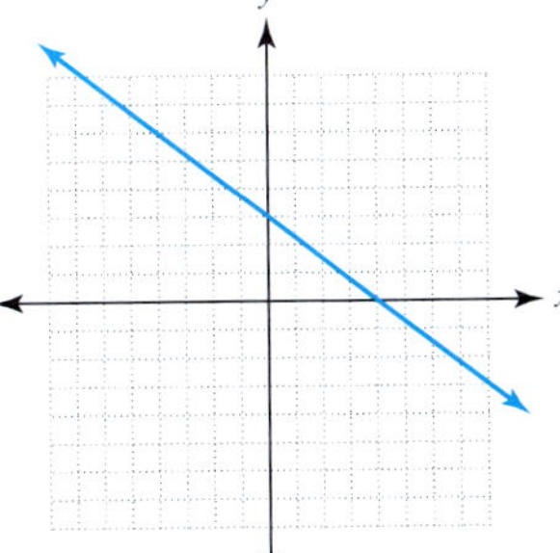

$f(x) = -\frac{3}{4}x + 3$

35.

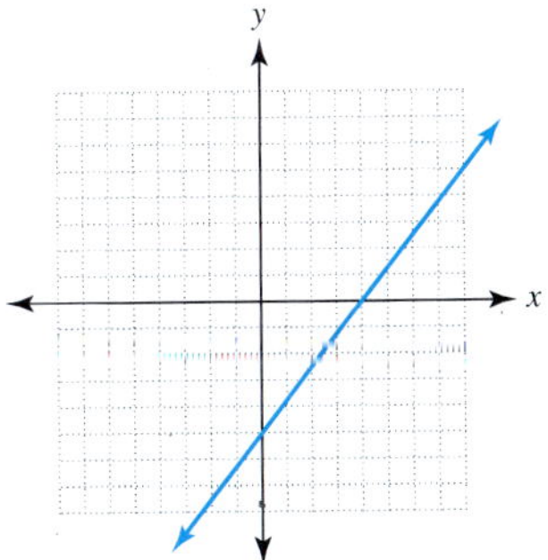

$f(x) = \frac{5}{4}x - 5$

37. always **39.** never **41.** $y = 2x$ **43.** $y = x + 3$
45. $y = 3x - 3$ **47.** $x - 4y = 12$ **49.** $(3, 1)$
51.

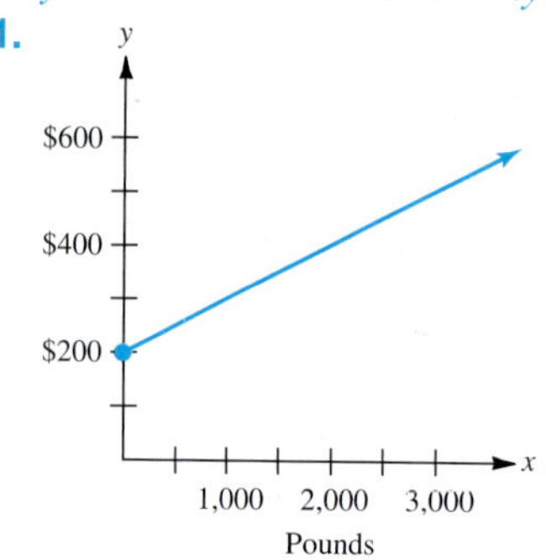

53. **(a)**

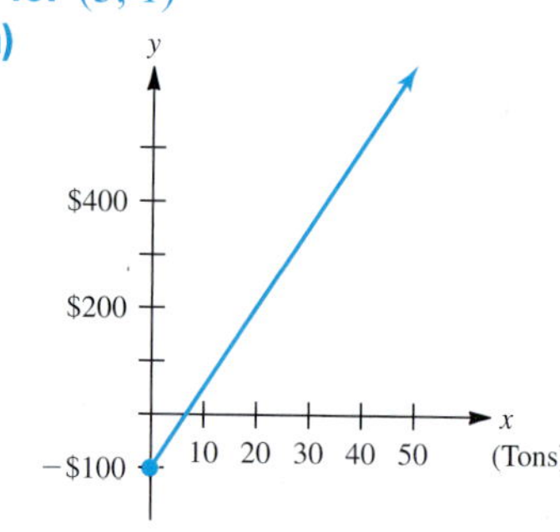

(b) $\frac{100}{15}$ or ≈ 7 tons; **(c)** \$140;
(d) $f(x) = 17x - 125$

55.

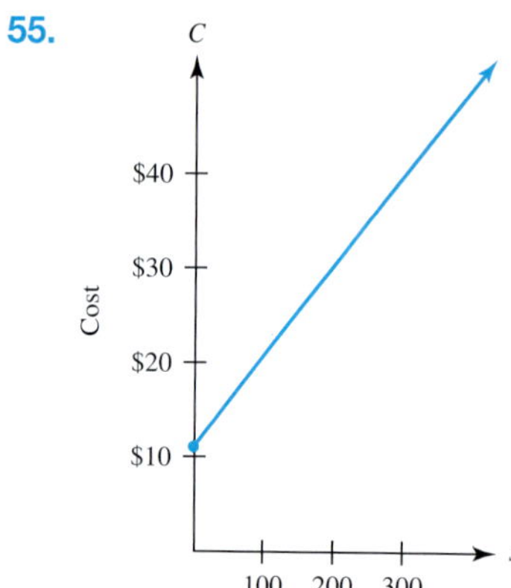

57. **(a)** $T(h) = 35h + 75$; **(b)**

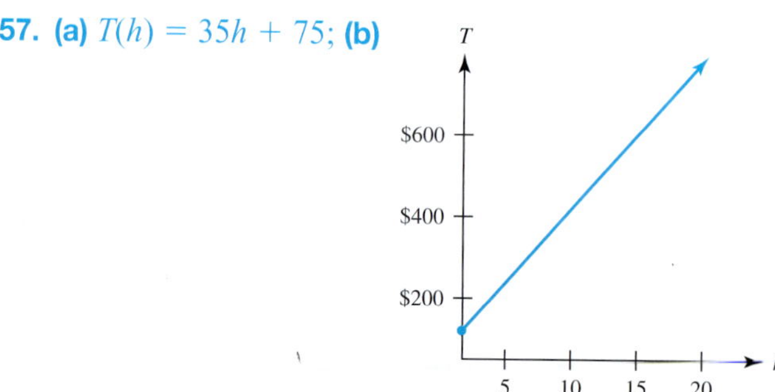

59. **61.**

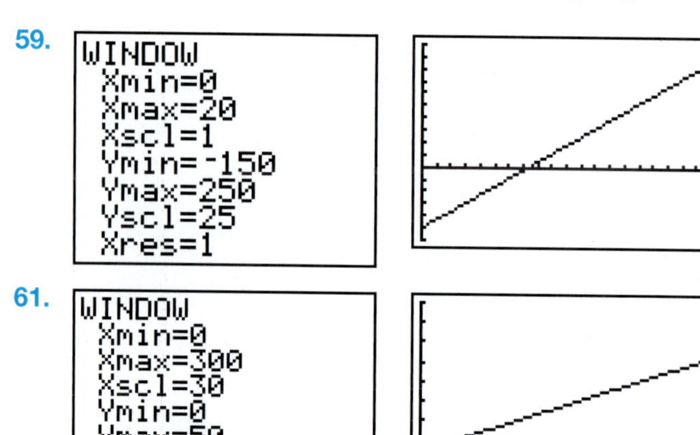

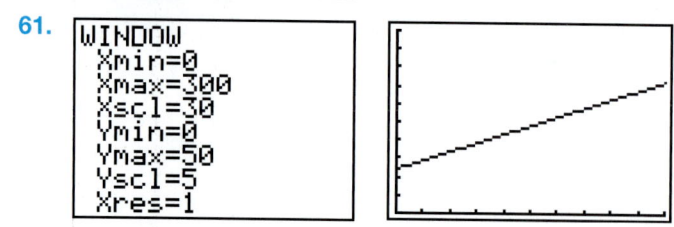

63.

65.

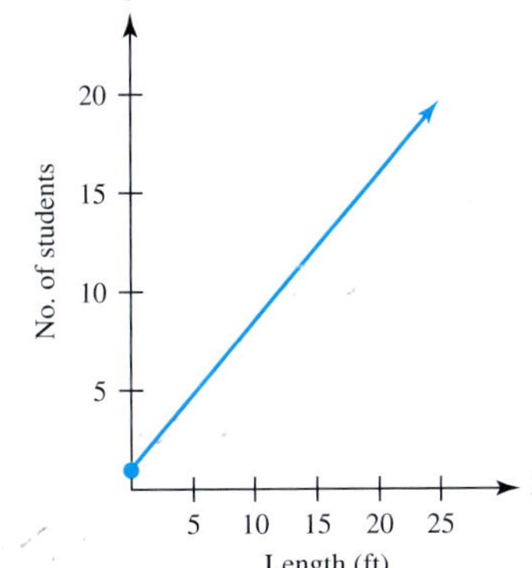

67.

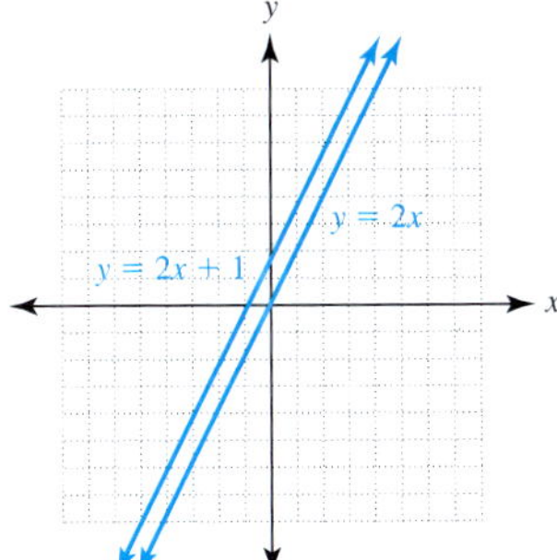

Parallel lines

69.

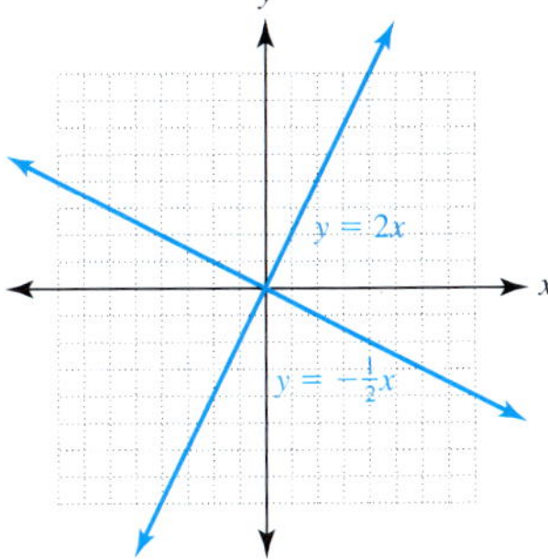

Perpendicular lines

71. (a) 13, 15, 17, 19, 21;

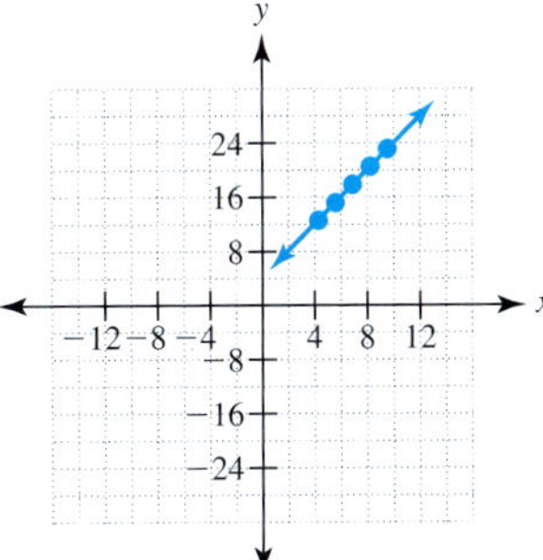

(b) Increases by 2; **(c)** Yes; **(d)** Grows by 2 units

73. (b) Increases by 3; **(c)** Yes; **(d)** Grows by 3 units

75. (b) Decreases by 4; **(c)** Yes; **(d)** Decreases by 4 units

3.2

The Slope of a Line

< 3.2 Objectives >

1 > Find the slope of a line

2 > Find the slopes and y-intercepts of horizontal and vertical lines

3 > Find the slope and y-intercept of a line, given an equation

4 > Write the equation of a line given the slope and y-intercept

5 > Graph linear equations, using the slope of a line

On the coordinate system below, plot a random point.

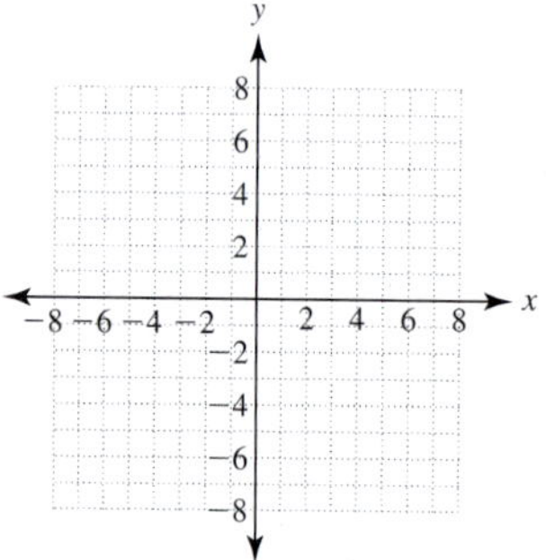

How many different lines can you draw through that point? Hundreds? Thousands? Millions? Actually, there is no limit to the number of different lines that pass through that point.

On the coordinate system below, plot two distinct points.

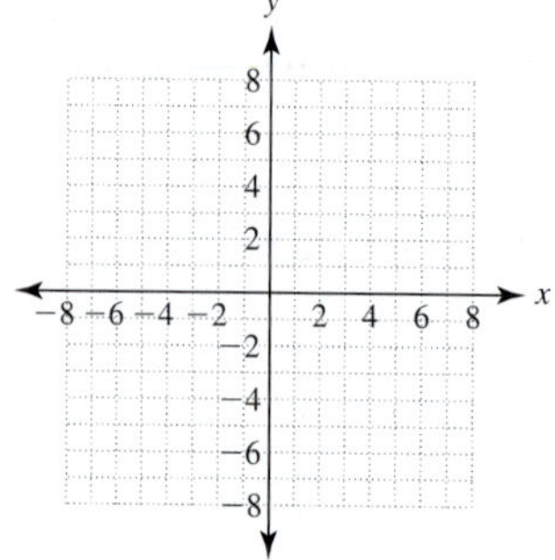

Now, how many different (straight) lines can you draw through those points? Only one! Two points are enough to define the line.

In Section 3.3, we will see how we can find the equation of a line if we are given two of its points. The first part of finding that equation is finding the **slope** of the line, which is a way of describing the *steepness* of a line.

Let us assume that the two points selected were $(-2, -3)$ and $(3, 7)$.

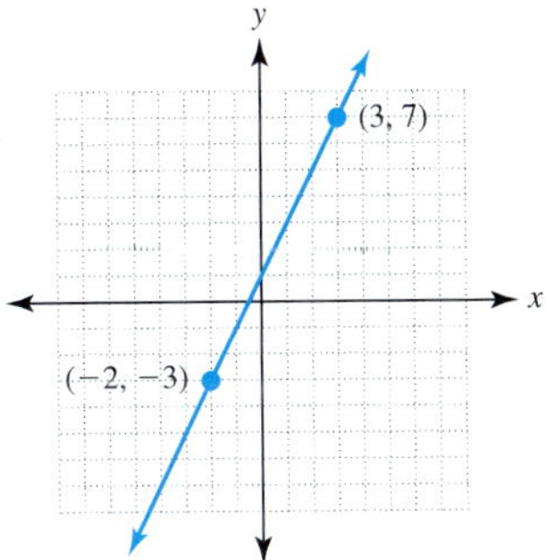

When moving between these two points, we go up 10 units and over 5 units.

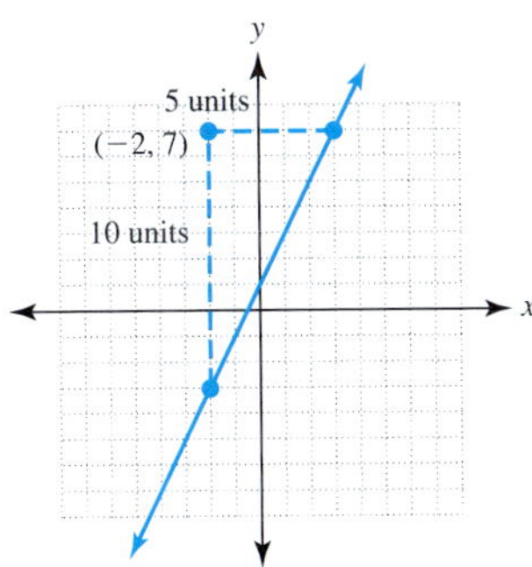

We refer to the 10 units as the *rise.* The 5 units is called the *run.* The slope is found by dividing the rise by the run. In this case, we have

$$\frac{\text{Rise}}{\text{Run}} = \frac{10}{5} = 2$$

NOTE

The difference $x_2 - x_1$ is called the **run.** The difference $y_2 - y_1$ is the **rise.**

Note that $x_1 \neq x_2$, or $x_2 - x_1 \neq 0$, ensures that the denominator is nonzero, so that the slope is defined.

The slope of this line is 2. This means that for any two points on the line, the rise (the change in the y-value) is twice as much as the run (the change in the x-value).

We now proceed to a more formal look at the process of finding the slope of the line through two given points.

To define a formula for slope, choose any two distinct points on the line, say, P with coordinates (x_1, y_1) and Q with coordinates (x_2, y_2). As we move along the line from P to Q, the x-value, or coordinate, changes from x_1 to x_2. That change in x, also called the **horizontal change,** is $x_2 - x_1$. Similarly, as we move from P to Q, the corresponding change in y, called the **vertical change,** is $y_2 - y_1$. The *slope* is then defined as the ratio of the vertical change to the horizontal change. The letter m is used to represent the slope, which we now define.

Definition

Slope of a Line

The **slope** of a line through two distinct points $P(x_1, y_1)$ and $Q(x_2, y_2)$ is given by

$$m = \frac{\text{Change in } y}{\text{Change in } x} = \frac{y_2 - y_1}{x_2 - x_1}$$

where $x_1 \neq x_2$.

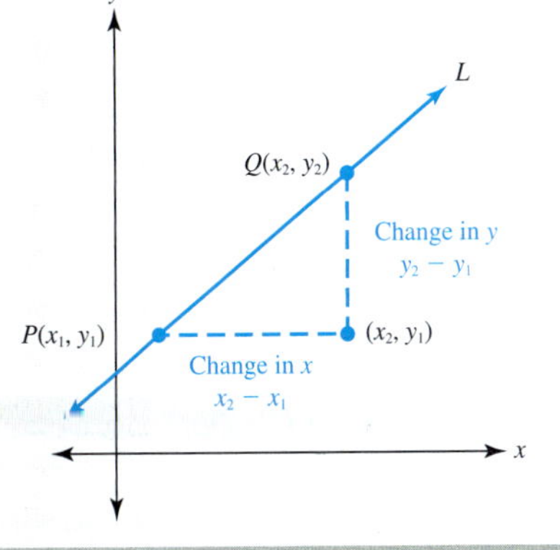

This definition provides the numerical measure of "steepness" that we want. If a line "rises" as we move from left to right, its slope is positive—the steeper the line, the larger the numerical value of the slope. If the line "falls" from left to right, its slope is negative.

Example 1 Finding the Slope

< Objective 1 >

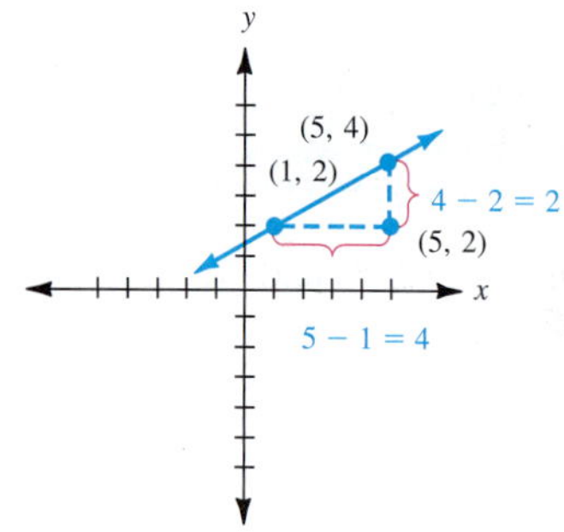

(a) Find the slope of the line containing points with coordinates (1, 2) and (5, 4).

Let $P(x_1, y_1) = (1, 2)$ and $Q(x_2, y_2) = (5, 4)$. Using the formula for the slope of a line gives

$$m = \frac{y_2 - y_1}{x_2 - x_1} = \frac{(4) - (2)}{(5) - (1)} = \frac{2}{4} = \frac{1}{2}$$

Note: We would have found the same slope if we had reversed P and Q and subtracted in the other order. In that case, $P(x_1, y_1) = (5, 4)$ and $Q(x_2, y_2) = (1, 2)$, so

$$m = \frac{(2) - (4)}{(1) - (5)} = \frac{-2}{-4} = \frac{1}{2}$$

It makes no difference which point is labeled (x_1, y_1) and which is (x_2, y_2)—the slope is the same. You must simply stay with your choice once it is made and *not* reverse the order of the subtraction in your calculations.

(b) Find the slope of the line containing points with the coordinates (−1, −2) and (3, 6).

Again, applying the definition, we have

$$m = \frac{(6) - (-2)}{(3) - (-1)} = \frac{6 + 2}{3 + 1} = \frac{8}{4} = 2$$

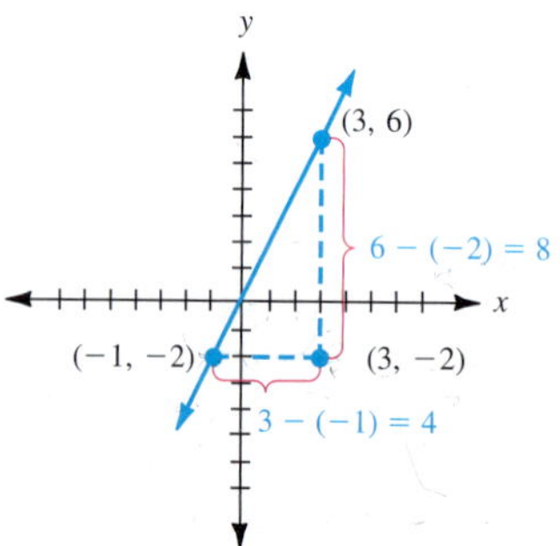

The figure below compares the slopes found in parts (a) and (b). Line l_1, from part (a), had slope $\frac{1}{2}$. Line l_2, from part (b), had slope 2. Do you see the idea of slope measuring steepness? The greater the value of a positive slope, the more steeply the line is inclined upward.

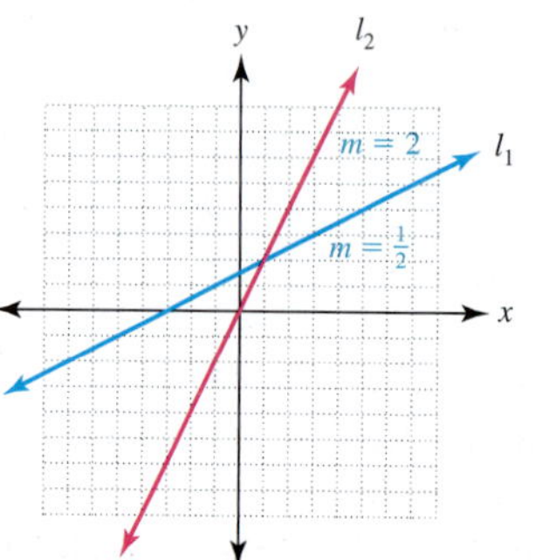

Check Yourself 1

(a) Find the slope of the line containing the points (2, 3) and (5, 5).

(b) Find the slope of the line containing the points (−1, 2) and (2, 7).

(c) Graph both lines on the same set of axes. Compare the lines and their slopes.

We now look at lines with a negative slope.

Example 2 Finding the Slope

Find the slope of the line containing points with coordinates (−2, 3) and (1, −3).

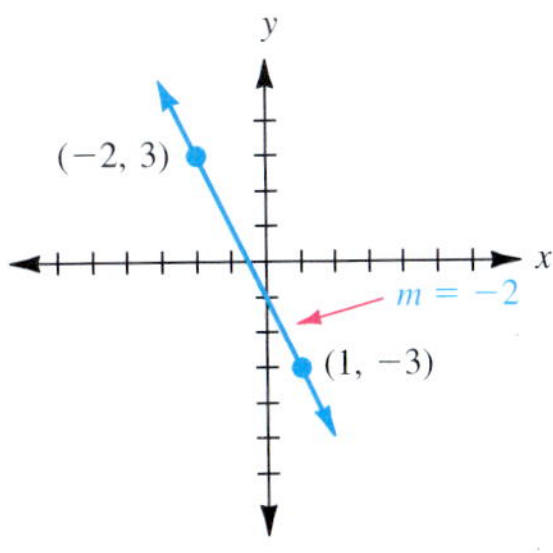

By the definition,

$$m = \frac{(-3) - (3)}{(1) - (-2)} = \frac{-6}{3} = -2$$

This line has a *negative* slope. The line *falls* as we move from left to right.

Check Yourself 2

Find the slope of the line containing points with coordinates (−1, 3) and (1, −3).

We have seen that lines with positive slope rise from left to right, and lines with negative slope fall from left to right. What about lines with a slope of 0? A line with a slope of 0 is especially important in mathematics.

Example 3 Finding the Slope

< Objective 2 >

Find the slope of the line containing points with coordinates (−5, 2) and (3, 2).

By the definition,

$$m = \frac{(2) - (2)}{(3) - (-5)} = \frac{0}{8} = 0$$

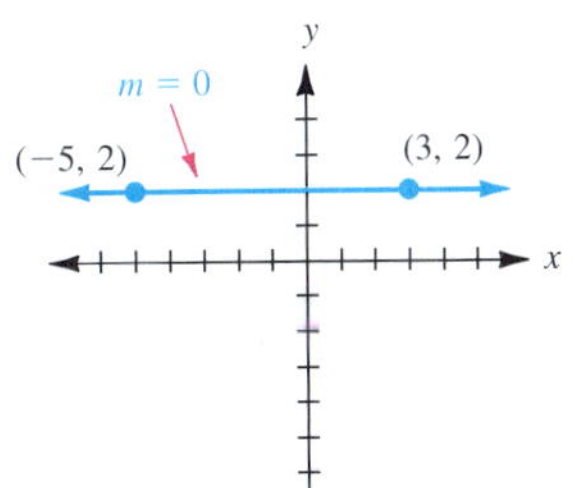

The slope of the line is 0. That is the case for any horizontal line. Since any two points on the line have the same y-coordinate, the vertical change $y_2 - y_1$ is always 0, and so the resulting slope is 0.

You should recall that this is the graph of $y = 2$.

Check Yourself 3

Find the slope of the line containing points with coordinates (−2, −4) and (3, −4).

Since division by 0 is undefined, it is possible to have a line with an undefined slope.

Example 4 Finding the Slope

Find the slope of the line containing points with coordinates (2, −5) and (2, 5).

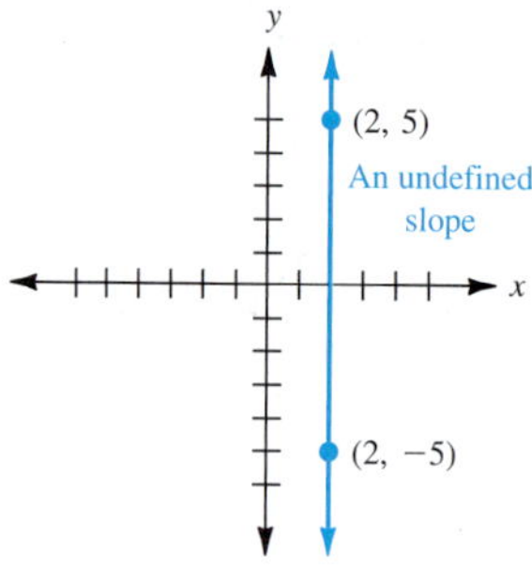

By the definition,

$$m = \frac{(5) - (-5)}{(2) - (2)} = \frac{10}{0}$$

Remember that division by 0 is undefined.

We say the vertical line has an undefined slope. On a vertical line, any two points have the same x-coordinate. This means that the horizontal change $x_2 - x_1$ is 0, and since division by 0 is undefined, the slope of a vertical line is always undefined.

You should recall that this is the graph of $x = 2$.

Check Yourself 4

Find the slope of the line containing points with the coordinates (−3, −5) and (−3, 2).

This sketch summarizes our results from Examples 1 through 4.

NOTE

As the slope gets closer to 0, the line gets "flatter."

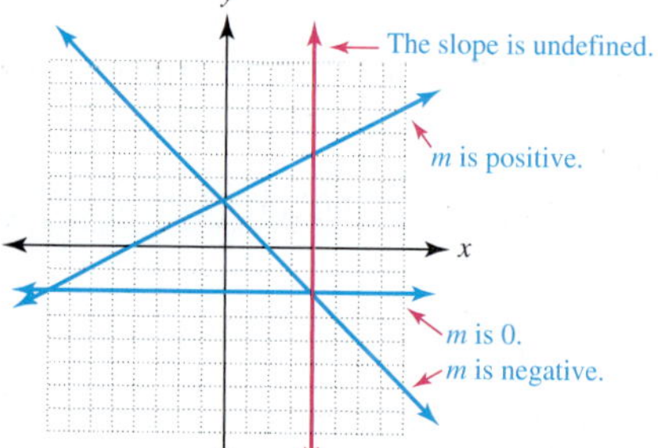

Four lines are illustrated in the figure. Note that

1. The slope of a line that rises from left to right is positive.
2. The slope of a line that falls from left to right is negative.
3. The slope of a horizontal line is 0.
4. A vertical line has an undefined slope.

We now want to consider finding the equation of a line when its slope and y-intercept are known. Suppose that the y-intercept is $(0, b)$. That is, the point at which the line crosses the y-axis has coordinates $(0, b)$. Look at the sketch.

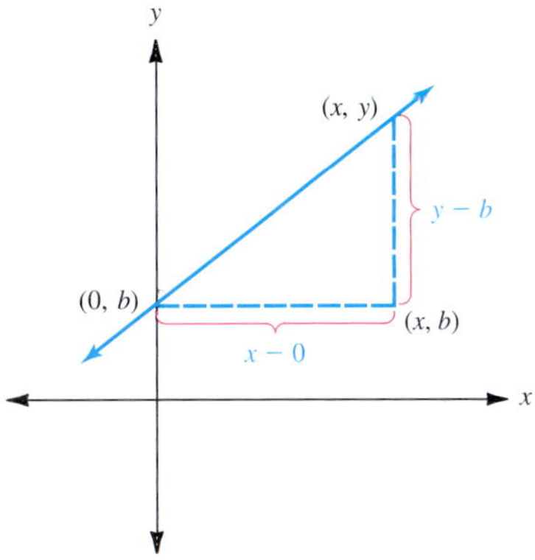

Now, using any other point (x, y) on the line and using our definition of slope, we can write

$$m = \frac{\overbrace{y - b}^{\text{Change in } y}}{\underbrace{x - 0}_{\text{Change in } x}}$$

or $\quad m = \dfrac{y - b}{x}$

Multiplying both sides by x, we have

$mx = y - b$

Finally, adding b to both sides gives

$$mx + b = y$$

or $\quad y = mx + b$

We can summarize the above discussion as follows:

Property

The Slope-Intercept Form for a Line

A linear function with slope m and y-intercept $(0, b)$ is expressed in *slope-intercept form* as

$y = mx + b \quad$ or $\quad f(x) = mx + b \quad$ (using function notation)

In this form, the equation is *solved* for y. The coefficient of x gives you the slope of the line, and the constant term gives the y-intercept.

Example 5 **Finding the Slope and y-Intercept**

< Objective 3 >

(a) Find the slope and y-intercept for the graph of the equation

$$y = \underset{m}{3}x + \underset{b}{4}$$

The graph has slope 3 and y-intercept $(0, 4)$.

NOTE

You briefly encountered this idea in Activity 3. You might want to review Activity 3 after completing this section.

(b) Find the slope and y-intercept for the graph of the equation

$$y = \underset{\underset{m}{\uparrow}}{-\frac{2}{3}}x - \underset{\underset{b}{\uparrow}}{5}$$

The slope of the line is $-\frac{2}{3}$; the y-intercept is $(0, -5)$.

Check Yourself 5

Find the slope and y-intercept for the graph of each of these equations.

(a) $y = -3x - 7$ **(b)** $y = \frac{3}{4}x + 5$

As Example 6 illustrates, we may have to solve for y as the first step in determining the slope and the y-intercept for the graph of an equation.

Example 6 Finding the Slope and y-Intercept

Find the slope and y-intercept for the graph of the equation

$$3x + 2y = 6$$

First, we solve the equation for y.

$$\begin{aligned} 3x + 2y &= 6 && \text{Subtract } 3x \text{ from both sides.} \\ 2y &= -3x + 6 && \text{Divide each term by 2.} \\ y &= -\frac{3}{2}x + 3 && \text{In function form, we have } f(x) = -\frac{3}{2}x + 3. \end{aligned}$$

NOTE

If we write the equation as

$$y = \frac{-3x + 6}{2}$$

it is more difficult to identify the slope and the y-intercept.

The equation is now in slope-intercept form. The slope is $-\frac{3}{2}$, and the y-intercept is $(0, 3)$.

Check Yourself 6

Find the slope and y-intercept for the graph of the equation

$2x - 5y = 10$

As we mentioned earlier, knowing certain properties of a line (namely, its slope and y-intercept) allows us to write the equation of the line by using the slope-intercept form. Example 7 illustrates this approach.

Example 7 Writing the Equation of a Line

< Objective 4 >

(a) Write the equation of a line with slope 3 and y-intercept $(0, 5)$.

We know that $m = 3$ and $b = 5$. Using the slope-intercept form, we have

$$y = \underset{\underset{m}{\nwarrow}}{3}x + \underset{\underset{b}{\nwarrow}}{5}$$

which is the desired equation.

(b) Write the equation of a line with slope $-\frac{3}{4}$ and y-intercept $(0, -3)$.

We know that $m = -\frac{3}{4}$ and $b = -3$. In this case,

$$y = \overset{m}{-\frac{3}{4}}x + \overset{b}{(-3)}$$

or $\quad y = -\frac{3}{4}x - 3$

which is the desired equation.

Check Yourself 7

Write the equation of a line with the properties

(a) Slope -2 and y-intercept $(0, 7)$

(b) Slope $\frac{2}{3}$ and y-intercept $(0, -3)$

We can also use the slope and y-intercept of a line in drawing its graph.

Example 8 Graphing a Line

< Objective 5 >

(a) Graph the line with slope $\frac{2}{3}$ and y-intercept $(0, 2)$.

Because the y-intercept is $(0, 2)$, we begin by plotting this point. The horizontal change (or run) is 3, so we move 3 units to the right *from that y-intercept.* The vertical change (or rise) is 2, so we move 2 units up to locate another point on the desired graph. Note that we have located that second point at $(3, 4)$. The final step is to draw a line through that point and the y-intercept.

NOTE

$m = \frac{2}{3} = \frac{\text{Rise}}{\text{Run}}$

The line rises from left to right because the slope is positive.

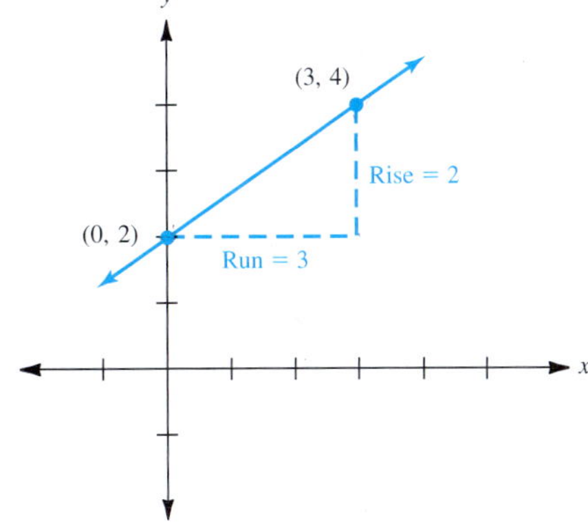

The equation of this line is $y = \frac{2}{3}x + 2$.

(b) Graph the line with slope -3 and y-intercept $(0, 3)$.

As before, first we plot the intercept point. In this case, we plot $(0, 3)$. The slope is -3, which we interpret as $\frac{-3}{1}$. Because the rise is negative, we go down rather than up. We move 1 unit in the horizontal direction, then 3 units down in the vertical direction. We plot this second point, (1, 0), and connect the two points to form the line.

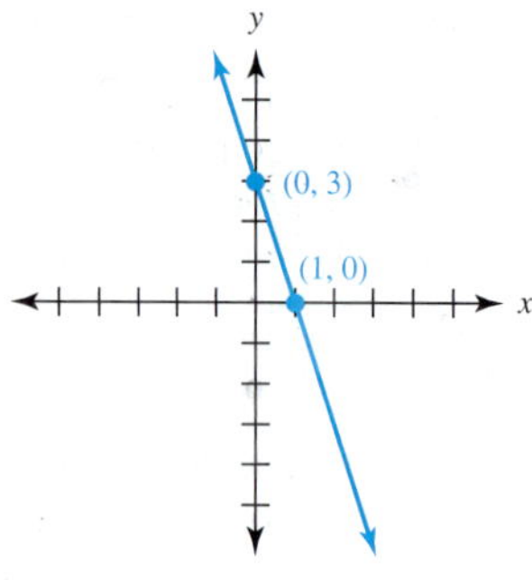

The equation of this line is $y = -3x + 3$.

Check Yourself 8

Graph the equation of a line with slope $\frac{3}{5}$ and y-intercept $(0, -2)$.

A line can certainly pass through the origin, as Example 9 demonstrates. In such cases, the y-intercept is (0, 0).

Example 9 Graphing a Line

Graph the line associated with the equation $y = -3x$ and the line associated with the equation $y = -3x - 3$.

In the first case, the slope is -3 and the y-intercept is (0, 0). We begin with the point (0, 0). From there, we move down 3 units and to the right 1 unit, arriving at the point $(1, -3)$. Now we draw a line through those two points. On the same axes, we draw the line with slope -3 through the intercept $(0, -3)$. Note that the two lines are parallel to each other.

NOTE

Nonvertical **parallel lines** have the same slope.

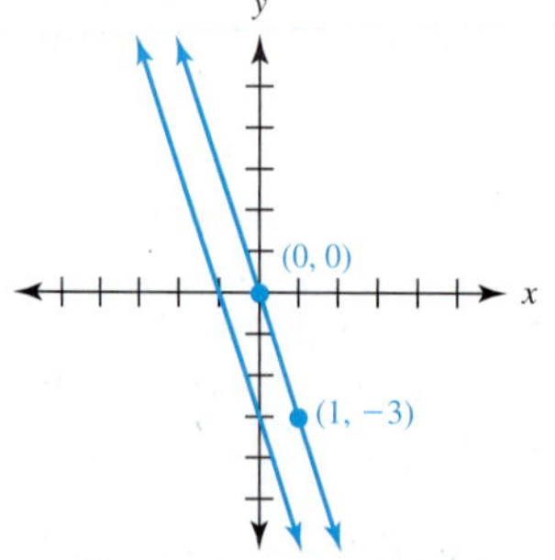

Check Yourself 9

Graph the line associated with the equation $y = \frac{7}{2}x$.

We summarize graphing with the slope-intercept form with the following algorithm.

Step by Step

Graphing by Using the Slope-Intercept Form

Step 1 Write the original equation of the line in slope-intercept form $y = mx + b$.

Step 2 Determine the slope m and the y-intercept $(0, b)$.

Step 3 Plot the y-intercept at $(0, b)$.

Step 4 Use m (the change in y over the change in x) to determine a second point on the desired line.

Step 5 Draw a line through the two points determined above to complete the graph.

You have now seen two methods for graphing lines: the slope-intercept method (this section) and the intercept method (Section 3.1). When you graph a linear equation, you should first decide which is the appropriate method.

Example 10 Selecting an Appropriate Graphing Method

Decide which of the two methods for graphing lines—the intercept method or the slope-intercept method—is more appropriate for graphing equations **(a)**, **(b)**, and **(c)**.

(a) $2x - 5y = 10$

Because both intercepts are easy to find, you should choose the intercept method to graph this equation.

(b) $2x + y = 6$

This equation can be quickly graphed by either method. As it is written, you might choose the intercept method. It can, however, be rewritten as $y = -2x + 6$, in which case the slope-intercept method is more appropriate.

(c) $y = \frac{1}{4}x - 4$

Since the equation is in slope-intercept form, that is the more appropriate method to choose.

Check Yourself 10

Which would be more appropriate for graphing each equation, the intercept method or the slope-intercept method?

(a) $x + y = -2$ **(b)** $3x - 2y = 12$ **(c)** $y = -\frac{1}{2}x - 6$

When working with applications, we are frequently asked to interpret the slope of a function as its *rate of change.* We will explore this more fully in Sections 3.3 and 3.4.

In short, the slope represents the change in the output, y or $f(x)$, when the input x is increased by one unit.

Graphically, the slope of a line is the change in the line's height when x increases by one unit. To remind you, a constant function has a slope equal to zero because the height of a horizontal line does not change when the input x increases by one unit.

In business applications, the slope of a linear function often correlates to the idea of *margin.* We learned about *marginal revenue, marginal cost,* and *marginal profit* in Chapter 2 and again in Section 3.1.

We conclude this section with an application from the field of electronics.

Example 11 An Electronics Application

The accompanying graph depicts the relationship between the position of a linear potentiometer (variable resistor) and the output voltage of some DC source. Consider the potentiometer to be a slider control, possibly to control volume of a speaker or the speed of a motor.

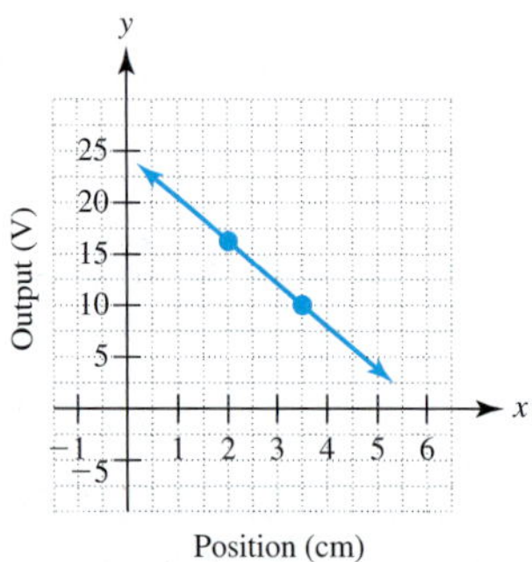

The linear position of the potentiometer is represented on the x-axis, and the resulting output voltage is represented on the y-axis. At the 2-cm position, the output voltage measured with a voltmeter is 16 VDC. At a position of 3.5 cm, the measured output was 10 VDC.

What is the slope of the resulting line?

We see that we have two ordered pairs: (2, 16) and (3.5, 10). Using our formula for slope, we have

$$m = \frac{16 - 10}{2 - 3.5} = \frac{6}{-1.5} = -4$$

The slope is -4.

Check Yourself 11

The same potentiometer described in Example 11 is used in another circuit. This time, though, when at position 0 cm, the output voltage is 12 volts. At position 5 cm, the output voltage is 3 volts. Draw a graph using the new data and determine the slope.

Check Yourself ANSWERS

1. (a) $m = \frac{2}{3}$; **(b)** $m = \frac{5}{3}$; **(c)**

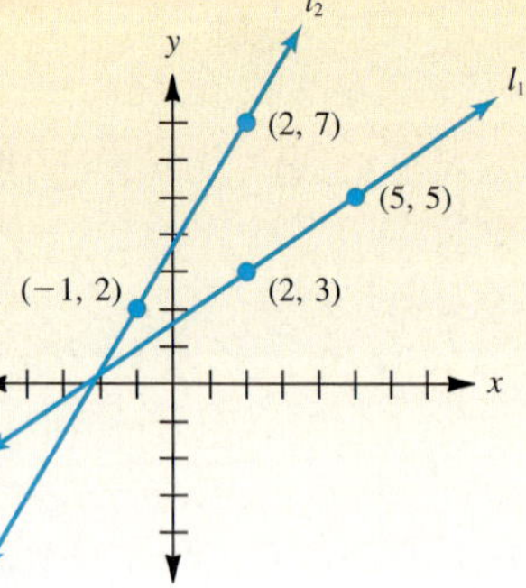

2. $m = -3$ **3.** $m = 0$ **4.** m is undefined

5. **(a)** $m = -3$, y-intercept: $(0, -7)$; **(b)** $m = \frac{3}{4}$, y-intercept: $(0, 5)$

6. $y = \frac{2}{5}x - 2$; $m = \frac{2}{5}$; y-intercept: $(0, -2)$

7. **(a)** $y = -2x + 7$; **(b)** $y = \frac{2}{3}x - 3$

8. 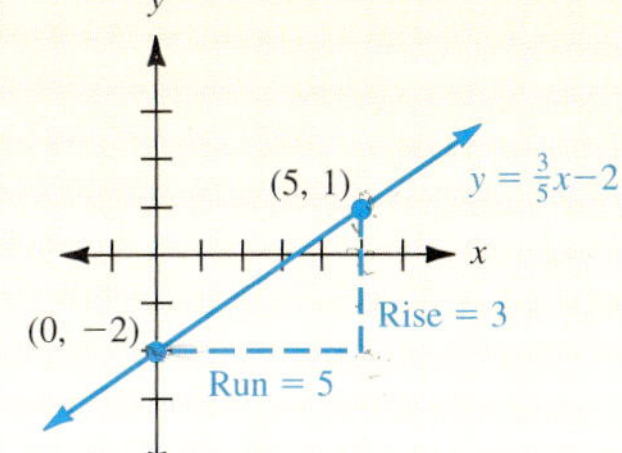

9. 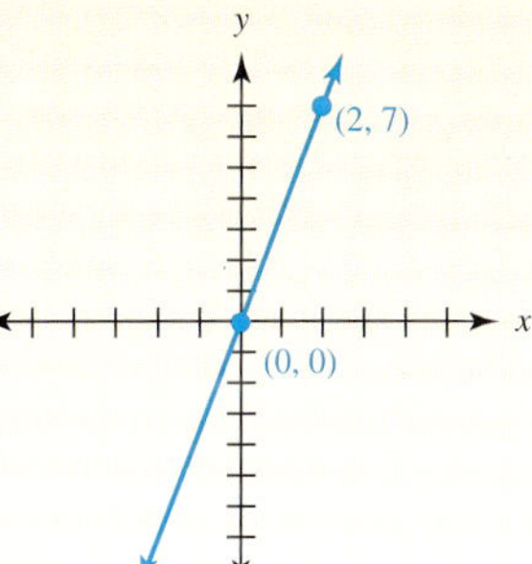

10. **(a)** either; **(b)** intercept; **(c)** slope-intercept

11. The slope is $-\frac{9}{5}$.

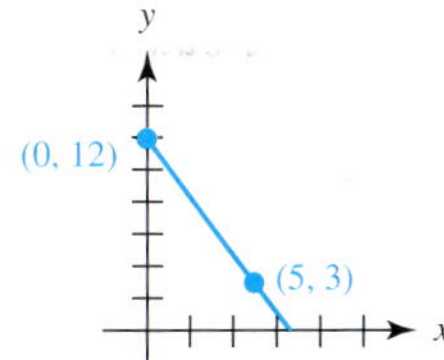

Reading Your Text

SECTION 3.2

(a) The ______________ of a line describes its steepness.

(b) The slope is defined as the ratio of the vertical change to the ______________ change.

(c) The change in the x-values between two points is called the run. The change in the y-values is called the ______________.

(d) Lines with ______________ slope fall from left to right.

3.2 exercises

Boost *your* GRADE at ALEKS.com!

ALEKS®

- Practice Problems
- Self-Tests
- NetTutor
- e-Professors
- Videos

Name ______________

Section ________ Date ________

Answers

1. ________ 2. ________
3. ________ 4. ________
5. ________ 6. ________
7. ________ 8. ________
9. ________ 10. ________
11. ________ 12. ________
13. ________ 14. ________
15. ________ 16. ________
17. ________ 18. ________
19. ________
20. ________
21. ________
22. ________
23. ________
24. ________
25. ________
26. ________

Basic Skills | Challenge Yourself | Calculator/Computer | Career Applications | Above and Beyond

< Objectives 1 and 2 >

Find the slope of the line through each pair of points.

1. $(5, 7)$ and $(9, 11)$

2. $(4, 9)$ and $(8, 17)$

3. $(-3, -1)$ and $(2, 3)$

4. $(-3, 2)$ and $(0, 17)$

5. $(-2, 3)$ and $(3, 7)$

6. $(-2, -5)$ and $(1, -4)$

7. $(-3, 2)$ and $(2, -8)$

8. $(-6, 1)$ and $(2, -7)$

9. $(3, -2)$ and $(5, -5)$ 

10. $(-2, 4)$ and $(3, 1)$

11. $(5, -4)$ and $(5, 2)$

12. $(-2, 8)$ and $(6, 8)$

13. $(-4, -2)$ and $(3, 3)$

14. $(-5, -3)$ and $(-5, 2)$

15. $(-2, -6)$ and $(8, -6)$

16. $(-5, 7)$ and $(2, -2)$

17. $(-1, 7)$ and $(2, 3)$

18. $(-3, -5)$ and $(2, -2)$

< Objective 3 >

Find the slope and y-intercept of the line represented by each equation.

19. $y = 3x + 5$

20. $y = -7x + 3$

21. $y = -3x - 6$

22. $y = 5x - 2$

23. $y = \frac{3}{4}x + 1$

24. $y = -5x$

25. $y = \frac{2}{3}x$

26. $y = -\frac{3}{5}x - 2$

Write each equation in function form. Give the slope and y-intercept of each function.

27. $4x + 3y = 12$

28. $5x + 2y = 10$

29. $y = 9$

30. $2x - 3y = 6$

31. $3x - 2y = 8$ 

32. $x = -3$

< Objective 4 >

Write the equation of the line with given slope and y-intercept. Then graph each line, using the slope and y-intercept.

33. Slope 3; y-intercept: (0, 5)

34. Slope -2; y-intercept: (0, 4)

35. Slope -4; y-intercept: (0, 5)

36. Slope 5; y-intercept: (0, -2)

37. Slope $\frac{1}{2}$; y-intercept: (0, -2)

38. Slope $-\frac{2}{5}$; y-intercept: (0, 6)

39. Slope $\frac{4}{3}$; y-intercept: (0, 0)

40. Slope $\frac{2}{3}$; y-intercept: (0, -2)

41. Slope $\frac{3}{4}$; y-intercept: (0, 3)

42. Slope -3; y-intercept: (0, 0)

Answers

27. __________
28. __________
29. __________
30. __________
31. __________
32. __________
33. __________
34. __________
35. __________
36. __________
37. __________
38. __________
39. __________
40. __________
41. __________
42. __________

Basic Skills | **Challenge Yourself** | Calculator/Computer | Career Applications | Above and Beyond

Answers

43. ______
44. ______
45. ______
46. ______
47. ______
48. ______
49. ______
50. ______
51. ______
52. ______
53. ______
54. ______
55. ______
56. ______
57. ______
58. ______

Complete each statement with **never, sometimes,** *or* **always.**

43. The slope of a line through the origin is ______ zero.

44. A line with an undefined slope is ______ the same as a line with a slope of zero.

45. Lines ______ have exactly one x-intercept.

46. The y-intercept of a line through the origin is ______ zero.

In which quadrant(s) are there no solutions for each equation?

47. $y = 4x + 5$

48. $y = 3x + 2$

49. $y = -x + 1$

50. $y = -7x + 3$

51. $y = -2x - 5$

52. $y = -5x - 7$

53. $y = 5$

54. $x = -2$

In exercises 55 to 62, match the graph with one of the equations below.

(a) $y = 2x$, **(b)** $y = x + 1$, **(c)** $y = -x + 3$, **(d)** $y = 2x + 1$,

(e) $y = -3x - 2$, **(f)** $y = \frac{2}{3}x + 1$, **(g)** $y = -\frac{3}{4}x + 1$, **(h)** $y = -4x$

55.

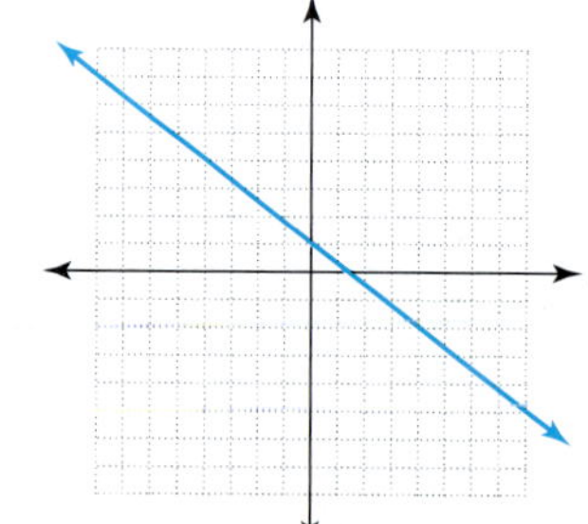

56.

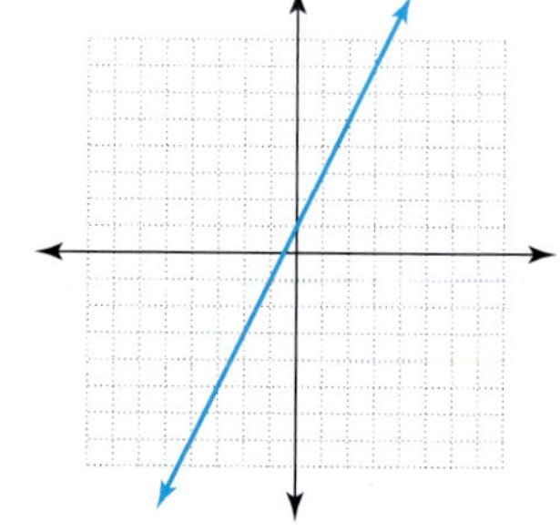

57.

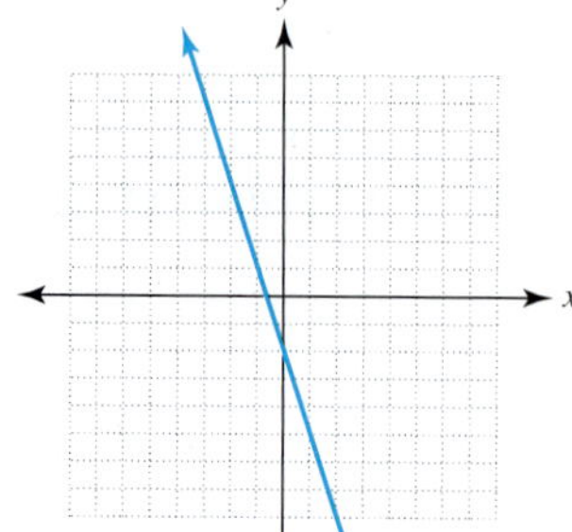

58.

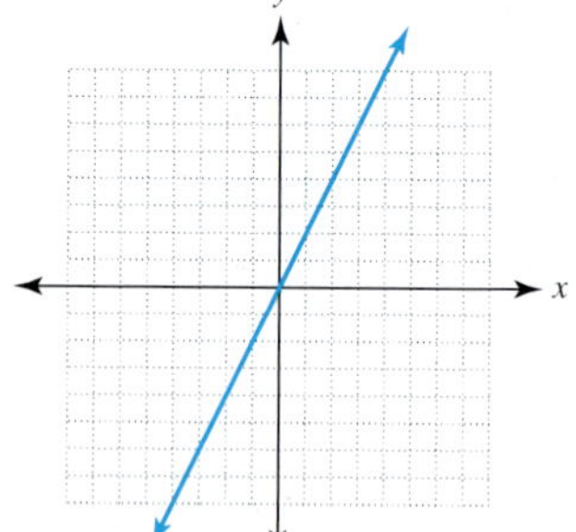

59.

60.

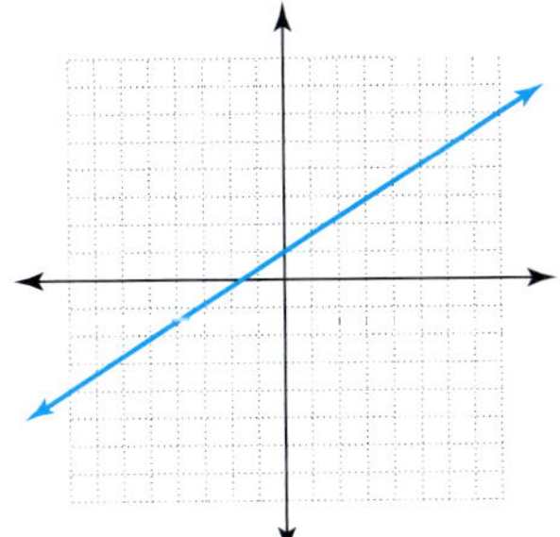

61.

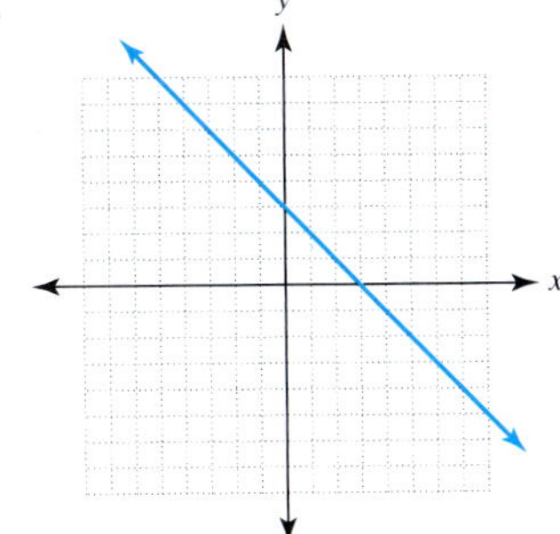

62.

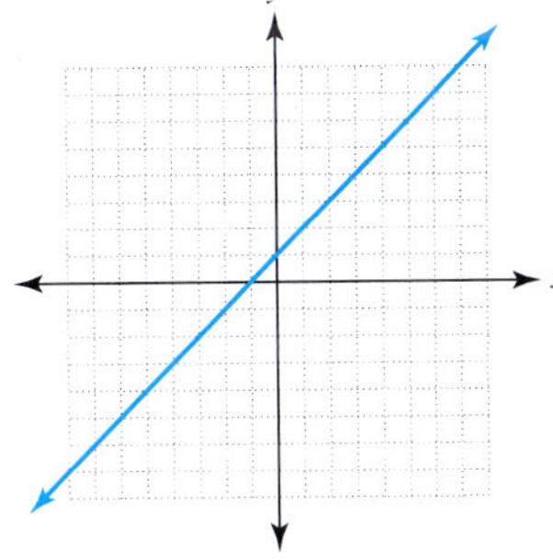

< Objective 5 >

In exercises 63 to 66, solve each equation for y, then graph each equation.

63. $2x + 5y = 10$ 

64. $5x - 3y = 12$

65. $x + 7y = 14$

66. $-2x - 3y = 9$

In exercises 67 to 74, use the graph to determine the slope of each line.

67.

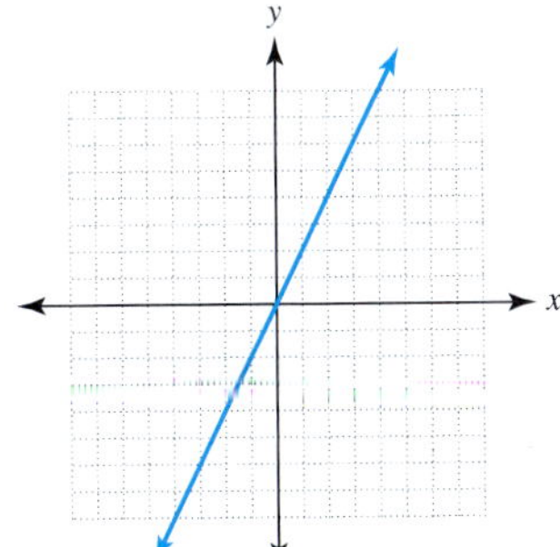

68.

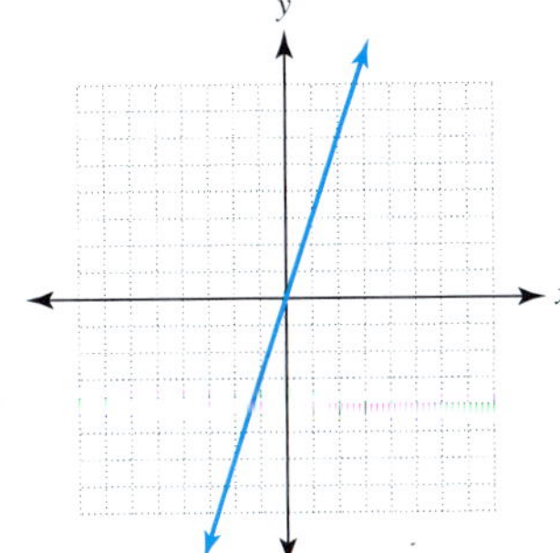

Answers

59. ____________

60. ____________

61. ____________

62. ____________

63. ____________

64. ____________

65. ____________

66. ____________

67. ____________

68. ____________

Answers

69. ____________________

70. ____________________

71. ____________________

72. ____________________

73. ____________________

74. ____________________

75. ____________________

76. ____________________

77. ____________________

78. ____________________

69.

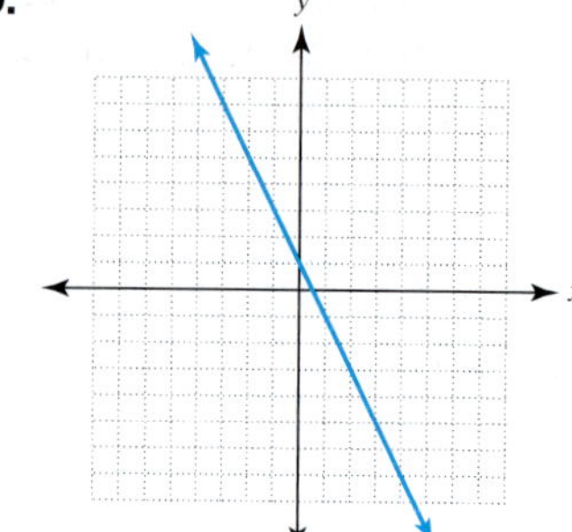

70.

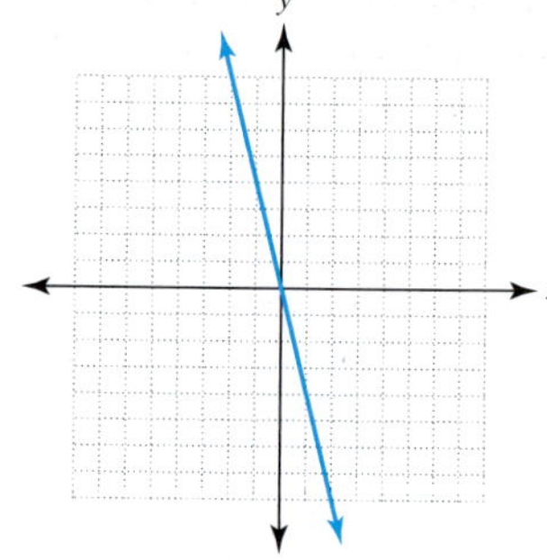

71.

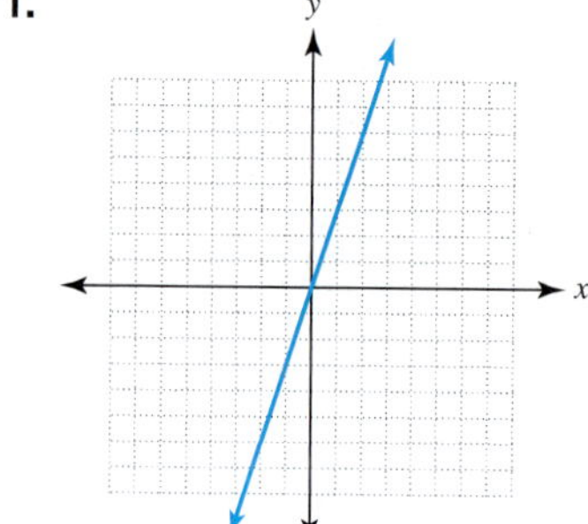

72.

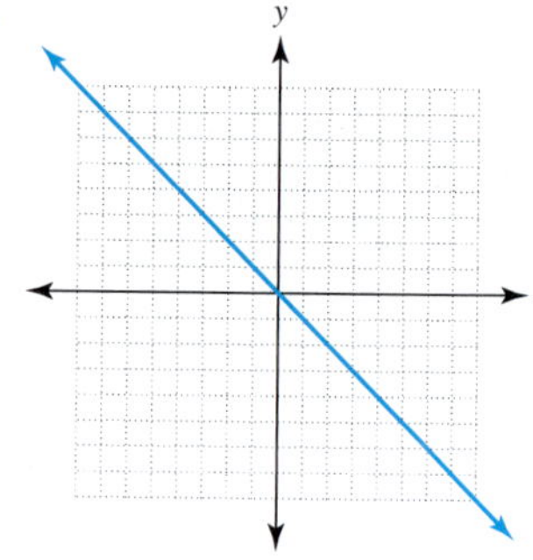

73.

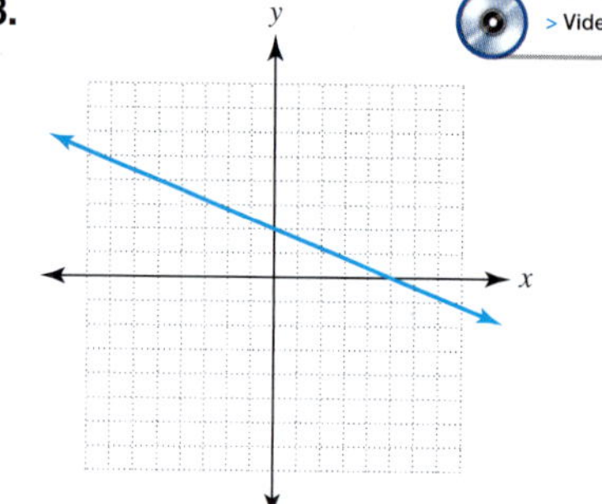

> Videos

74.

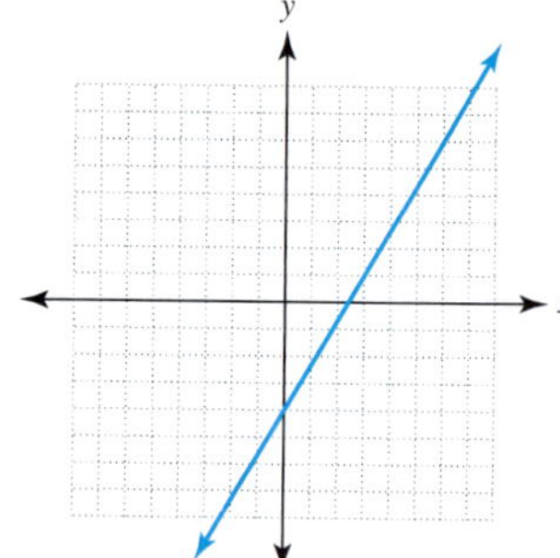

75. **BUSINESS AND FINANCE** We used the equation $y = 0.10x + 200$ to describe the award money in a recycling contest. What are the slope and the y-intercept for this equation? What does the slope of the line represent in the equation? What does the y-intercept represent?

76. **BUSINESS AND FINANCE** We used the equation $y = 15x - 100$ to describe the amount of money a high school class might earn from a paper drive. What are the slope and y-intercept for this equation?

77. **BUSINESS AND FINANCE** In the equation in exercise 76, what does the slope of the line represent? What does the y-intercept represent?

78. **CONSTRUCTION** A roof rises 8.75 feet (ft) in a horizontal distance of 15.09 ft. Find the slope of the roof to the nearest hundredth.

79. **Science and Medicine** An airplane covered 15 miles (mi) of its route while decreasing its altitude by 24,000 ft. Find the slope of the line of descent that was followed. (1 mi = 5,280 ft) Round to the nearest hundredth.

80. **Science and Medicine** Driving down a mountain, Tom finds that he has descended 1,800 ft in elevation by the time he is 3.25 mi horizontally away from the top of the mountain. Find the slope of his descent to the nearest hundredth.

81. **Business and Finance** In 1960, the cost of a soft drink was 20¢. By 2002, the cost of the same soft drink had risen to $1.50. During this time period, what was the annual rate of change of the cost of the soft drink?

82. **Science and Medicine** On a certain February day in Philadelphia, the temperature at 6:00 A.M. was 10°F. By 2:00 P.M. the temperature was up to 26°F. What was the hourly rate of temperature change?

Basic Skills | Challenge Yourself | Calculator/Computer | **Career Applications** | Above and Beyond

83. **Allied Health** The recommended dosage d (in mg) of the antibiotic ampicillin sodium for children weighing less than 40 kg is given by the linear equation $d = 7.5w$, in which w represents the child's weight (in kg). Sketch a graph of this equation.

84. **Allied Health** The recommended dosage d (in μg) of neupogen (medication given to bone-marrow transplant patients) is given by the linear equation $d = 8w$, in which w is the patient's weight (in kg). Sketch a graph of this equation.

Mechanical Engineering The graph shows the bending moment of a wood beam at various points x feet from the left end of the beam. Use the graph to complete exercises 85 and 86.

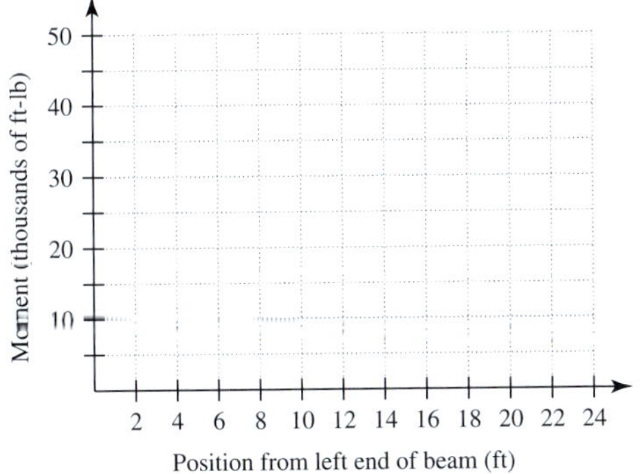

Answers

79. ______

80. ______

81. ______

82. ______

83. ______

84. ______

Answers

85. ____

86. ____

87. ____

88. ____

89. ____

90. ____

91. ____

92. ____

93. ____

94. ____

85. Determine the slope of the moment graph for points between 0 and 4 feet from the left end of the beam.

86. Determine the slope of the moment graph for points between 4 and 11 feet and between 11 and 19 feet.

Basic Skills | Challenge Yourself | Calculator/Computer | Career Applications | **Above and Beyond**

87. Complete the statement: "The difference between undefined slope and zero slope is. . . ."

88. Complete the statement: "The slope of a line tells you. . . ."

89. On two occasions last month, Sam Johnson rented a car on a business trip. Both times it was the same model from the same company, and both times it was in San Francisco. On both occasions he dropped the car at the airport booth and just got the total charge, not the details. Sam now has to fill out an expense account form and needs to know how much he was charged per mile and the base rate. All Sam knows is that he was charged \$210 for 625 mi on the first occasion and \$133.50 for 370 mi on the second trip. Sam has called accounting to ask for help. Plot these two points on a graph, and draw the line that goes through them. What question does the slope of the line answer for Sam? How does the y-intercept help? Write a memo to Sam, explaining the answers to his questions and how a knowledge of algebra and graphing has helped you find the answers.

90. On the same graph, sketch each line.

$y = 2x - 1$ and $y = 2x + 3$

What do you observe about these graphs? Will the lines intersect?

91. Repeat Exercise 90, using

$y = -2x + 4$ and $y = -2x + 1$

92. On the same graph, sketch each line.

$y = \frac{2}{3}x$ and $y = -\frac{3}{2}x$

What do you observe concerning these graphs? Find the product of the slopes of these two lines.

93. Repeat Exercise 92, using

$y = \frac{4}{3}x$ and $y = -\frac{3}{4}x$

94. Based on Exercises 92 and 93, write the equation of a line that is perpendicular to

$y = \frac{3}{5}x$

Answers

1. 1 **3.** $\frac{4}{5}$ **5.** $\frac{4}{5}$ **7.** -2 **9.** $-\frac{3}{2}$ **11.** Undefined
13. $\frac{5}{7}$ **15.** 0 **17.** $-\frac{4}{3}$ **19.** Slope 3; y-intercept: $(0, 5)$
21. Slope -3; y-intercept: $(0, -6)$ **23.** Slope $\frac{3}{4}$; y-intercept: $(0, 1)$
25. Slope $\frac{2}{3}$; y-intercept: $(0, 0)$ **27.** $f(x) = -\frac{4}{3}x + 4$; slope: $-\frac{4}{3}$; y-intercept: $(0, 4)$ **29.** $f(x) = 9$; slope: 0; y-intercept: $(0, 9)$
31. $f(x) = \frac{3}{2}x - 4$; slope: $\frac{3}{2}$; y-intercept: $(0, -4)$

33.

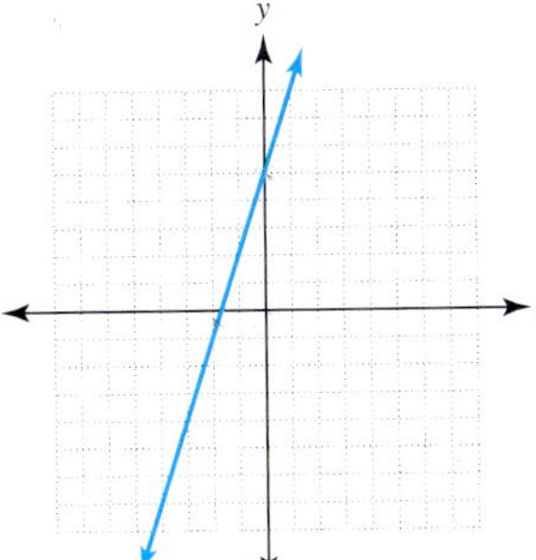

$y = 3x + 5$

35.

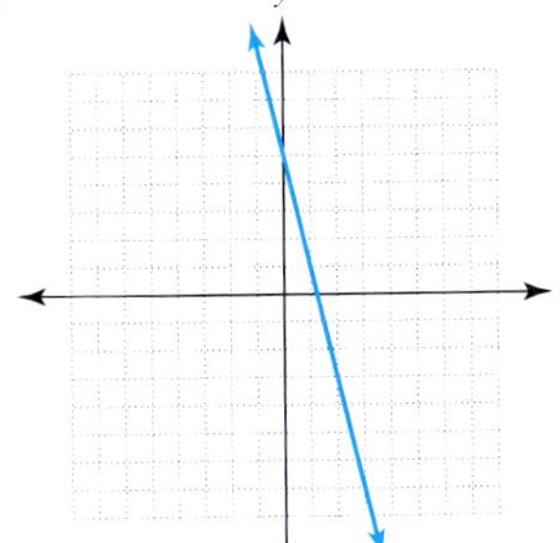

$y = -4x + 5$

37.

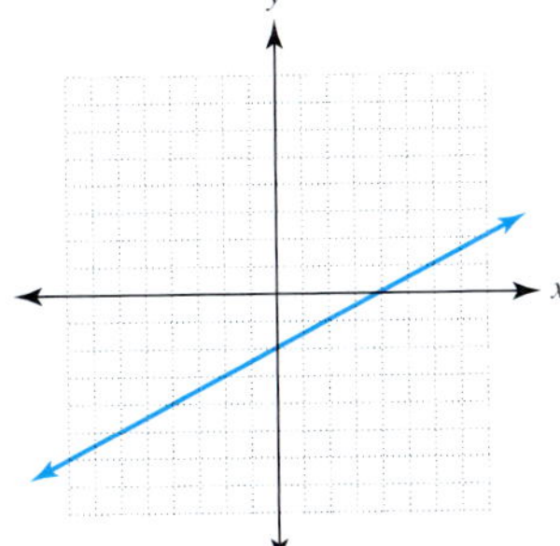

$y = \frac{1}{2}x - 2$

39.

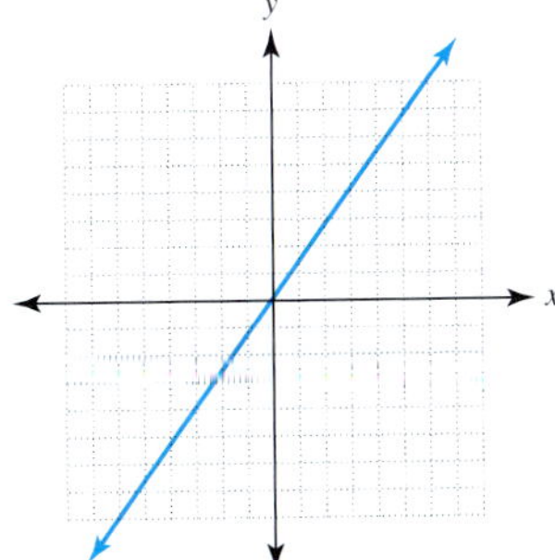

$y = \frac{4}{3}x$

41. 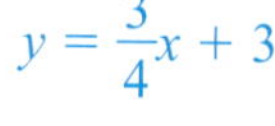

$y = \frac{3}{4}x + 3$

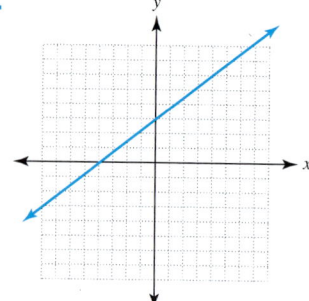

43. sometimes **45.** sometimes **47.** IV **49.** III **51.** I
53. III and IV **55.** (g) **57.** (e) **59.** (h) **61.** (c)

63.

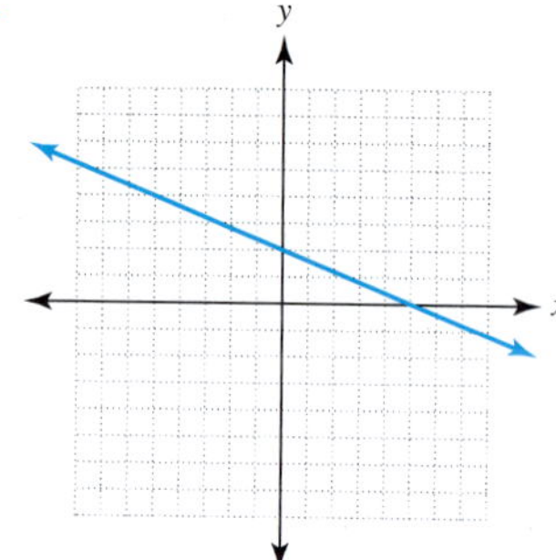

$y = -\frac{2}{5}x + 2$

65.

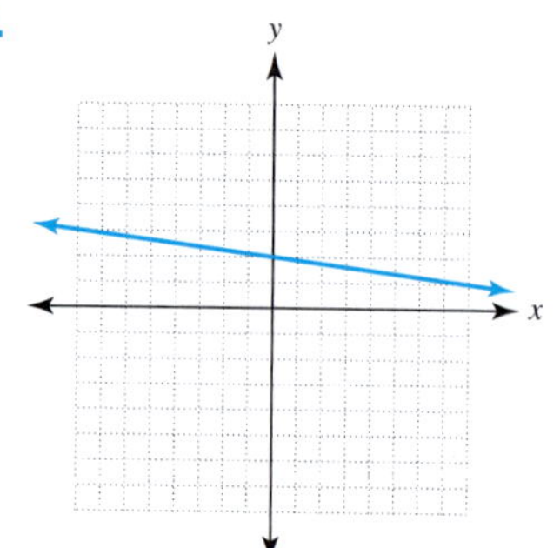

$y = -\frac{1}{7}x + 2$

67. 2 **69.** -2 **71.** 3 **73.** $-\frac{2}{5}$

75. Slope: 0.10, market price per pound; y-intercept: (0, 200), the minimum $200 award

77. Slope represents price of newsprint; y-intercept represents cost of the truck

79. -0.30 **81.** 3.10 ¢/yr

83.

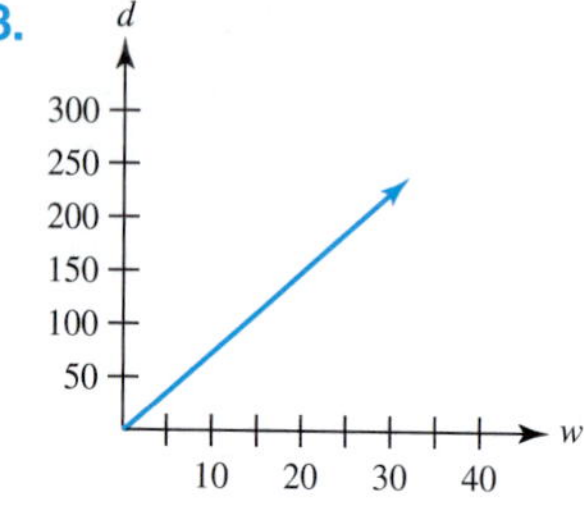

85. 6,250 ft-lb per foot

87. Above and Beyond **89.** Above and Beyond

91.

$y = -2x + 1$ $y = -2x + 4$

Parallel lines; no

93.

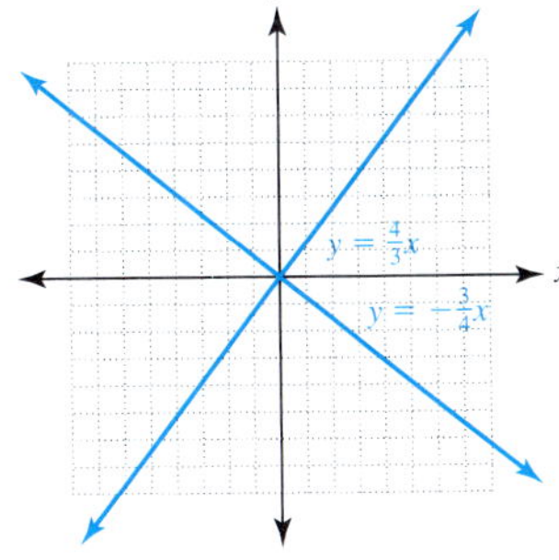

Perpendicular lines; -1

3.3 Forms of Linear Equations

< 3.3 Objectives >

1 > Use the equations of lines to determine whether two lines are parallel, perpendicular, or neither

2 > Write the equation of a line, given a slope and a point on the line

3 > Write the equation of a line, given two points

4 > Write the equation of a line satisfying given geometric conditions

Recall that the form

$$Ax + By = C$$

in which A and B cannot both be zero, is called the **standard form for a linear equation.** In Section 3.2 we determined the slope of a line from two ordered pairs. We then used the slope to write the equation of a line.

In this section, we will see that the slope-intercept form of a line clearly indicates whether the graphs of two lines are parallel, perpendicular, or neither. We will make frequent use of the following definitions.

Definition

Parallel Lines and Perpendicular Lines

When two lines have the same slope, we say they are **parallel lines.**

When two lines meet at right angles, we say they are **perpendicular lines.**

NOTE

We assume that neither line is vertical. We will discuss the special case involving a vertical line shortly.

Algebraically, the slopes of the two lines can be written as m_1 and m_2. For parallel lines, it will always be the case that

$$m_1 = m_2$$

For perpendicular lines, it will always be the case that the two slopes will be negative reciprocals. Algebraically, we write

$$m_1 = -\frac{1}{m_2}$$

Note that, by multiplying both sides by m_2, we can also write this as

$$m_1 \cdot m_2 = -1$$

Example 1 illustrates this concept.

Example 1 Verifying That Two Lines Are Perpendicular

< Objective 1 >

Show that the graphs of $3x + 4y = 4$ and $-4x + 3y = 12$ are perpendicular lines.

First, we solve each equation for y.

$$3x + 4y = 4$$
$$4y = -3x + 4$$
$$y = -\frac{3}{4}x + 1$$

Note that the slope of the line is $-\frac{3}{4}$. We can say $m_1 = -\frac{3}{4}$.

$$-4x + 3y = 12$$
$$3y = 4x + 12$$
$$y = \frac{4}{3}x + 4$$

The slope of the line is $\frac{4}{3}$. We can say $m_2 = \frac{4}{3}$.

We now look at the product of the two slopes: $-\frac{3}{4} \cdot \frac{4}{3} = -1$. Any two lines whose slopes have a product of -1 are perpendicular lines. These two lines are perpendicular.

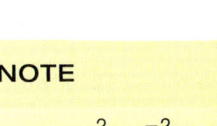

Check Yourself 1

Show that the graphs of the equations

$-3x + 2y = 4$ and $2x + 3y = 9$

are perpendicular lines.

In Example 2, we review how the slope-intercept form can be used in graphing a line.

Example 2 Graphing the Equation of a Line

Graph the line $2x + 3y = 3$.

Solving for y, we find the slope-intercept form for this equation is

$$y = -\frac{2}{3}x + 1$$

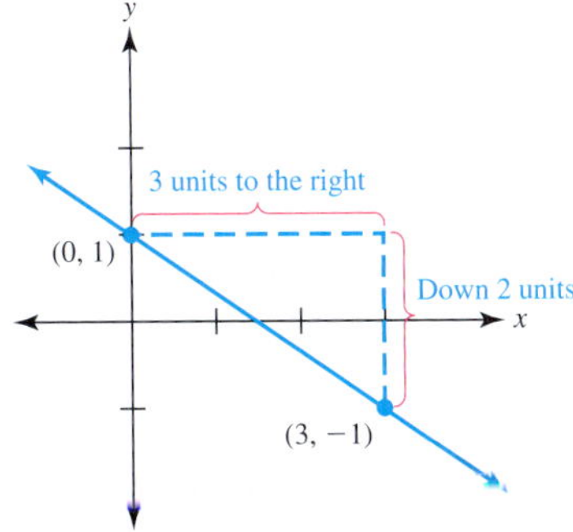

To graph the line, plot the y-intercept at (0, 1). Because the slope m is equal to $-\frac{2}{3}$, we move from (0, 1) to the right 3 units and then *down* 2 units, to locate a second point on the graph of the line, here $(3, -1)$. We can now draw a line through the two points to complete the graph.

NOTE

We treat $-\frac{2}{3}$ as $\frac{-2}{+3}$ to move to the right 3 units and down 2 units.

Check Yourself 2

Graph the line with equation

$3x - 4y = 8$

Hint: First rewrite the equation in slope-intercept form.

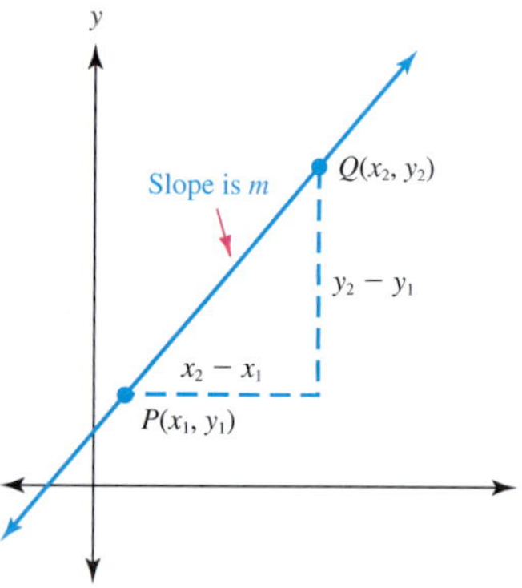

From the definition of slope, we can find another useful form for the equation of a line.

Recall that slope is defined as the change in y divided by the change in x. We write

$$m = \frac{y_2 - y_1}{x_2 - x_1}$$

Multiplying both sides by the LCD, we get

$$m(x_2 - x_1) = y_2 - y_1$$

or $\quad y_2 - y_1 = m(x_2 - x_1)$

This last equation is called the **point-slope form** for the equation of a line. All points lying on the line satisfy this equation. We state the general result.

Property

Point-Slope Form for the Equation of a Line

The equation of a line with slope m that passes through point (x_1, y_1) is given by

$$y - y_1 = m(x - x_1)$$

Example 3 Finding the Equation of a Line

< Objective 2 >

Write the equation for the line that passes through point $(3, -1)$ with a slope of 3.

Letting $(x_1, y_1) = (3, -1)$ and $m = 3$, we use the point-slope form to get

$$y - (-1) = 3[x - (3)]$$

or $\quad y + 1 = 3x - 9$

We can write the final result in slope-intercept form as

$$y = 3x - 10$$

Check Yourself 3

Write the equation of the line that passes through point $(-2, 4)$ with a slope of $\frac{3}{2}$. Write your result in slope-intercept form.

Since we know that two points determine a line, it is natural that we should be able to write the equation of a line passing through two given points. Using the point-slope form together with the slope formula allows us to write such an equation.

Example 4 Finding the Equation of a Line

< Objective 3 >

Write the equation of the line passing through (2, 4) and (4, 7).

First, we find m, the slope of the line. Here

$$m = \frac{7 - 4}{4 - 2} = \frac{3}{2}$$

Now we apply the point-slope form with $m = \frac{3}{2}$ and $(x_1, y_1) = (2, 4)$.

$$y - (4) = \frac{3}{2}[x - (2)]$$

$$y - 4 = \frac{3}{2}x - 3$$

We write the result in slope-intercept form.

$$y = \frac{3}{2}x + 1$$

NOTE

We could just as well choose to let

$(x_1, y_1) = (4, 7)$

The resulting equation is the same in either case. Take time to verify this for yourself.

Check Yourself 4

Write the equation of the line passing through (−2, 5) and (1, 3). Write your result in slope-intercept form.

A line with slope zero is a horizontal line. A line with an undefined slope is vertical. Example 5 illustrates the equations of such lines.

Example 5 Finding the Equation of a Line

< Objective 4 >

(a) Find the equation of a line passing through (7, −2) with a slope of 0.

We could find the equation by letting $m = 0$. Substituting into the slope-intercept form, we can solve for b.

$$y = mx + b$$

$$-2 = 0(7) + b$$

$$-2 = b$$

So $y = 0x - 2$, or $y = -2$

It is far easier to remember that any line with a zero slope is a horizontal line and has the form

$$y = b$$

The value for b is always the y-coordinate for the given point.

Note that, for any horizontal line, all of the points have the same y-value. Look at the graph of the line $y = -2$. Three points have been labeled.

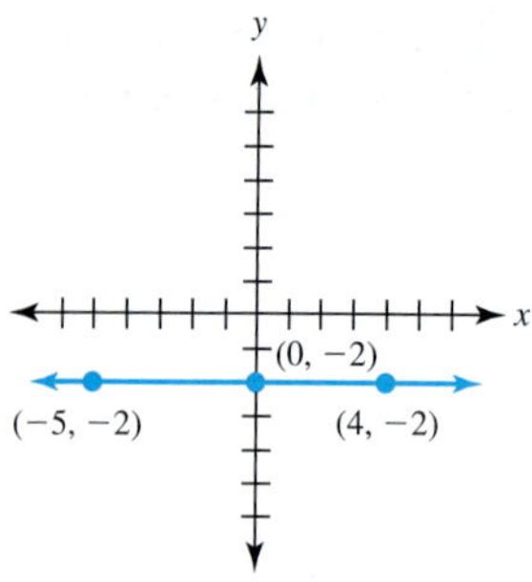

(b) Find the equation of a line with undefined slope passing through (4, −5).

A line with undefined slope is vertical. It always has the form $x = a$, where a is the x-coordinate for the given point. The equation is

$x = 4$

Note that, for any vertical line, all of the points have the same x-value. Look at the graph of the line $x = 4$. Three points have been labeled.

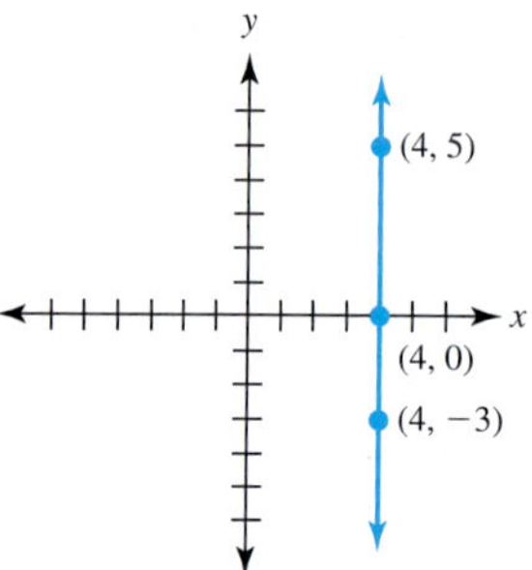

Check Yourself 5

(a) Find the equation of a line with zero slope that passes through point (−3, 5).
(b) Find the equation of a line passing through (−3, −6) with undefined slope.

There are alternative methods for finding the equation of a line through two points. Example 6 shows such an approach.

Example 6 Finding the Equation of a Line

NOTE

We could, of course, use the point-slope form seen earlier.

Write the equation of the line through points (−2, 3) and (4, 5).

First, we find m, as before.

$$m = \frac{(5) - (3)}{(4) - (-2)} = \frac{2}{6} = \frac{1}{3}$$

We now make use of the slope-intercept form, but in a different manner.

Using $y = mx + b$, and the slope just calculated, we can immediately write

$$y = \frac{1}{3}x + b$$

> **NOTE**
>
> We substitute these values because the line must pass through (−2, 3).

Now, if we substitute a known point for x and y, we can solve for b. We may choose either of the two given points. Using $(-2, 3)$, we have

$$3 = \frac{1}{3}(-2) + b$$

$$3 = -\frac{2}{3} + b$$

$$3 + \frac{2}{3} = b$$

$$b = \frac{11}{3}$$

Therefore, the equation of the desired line is

$$y = \frac{1}{3}x + \frac{11}{3}$$

Check Yourself 6

Repeat the Check Yourself 4 exercise, using the technique illustrated in Example 6.

We now know that we can write the equation of a line once we have been given a point on the line and the slope of that line. In some applications, the slope may not be given directly but through specified parallel or perpendicular lines instead.

Example 7 Finding the Equation of a Parallel Line

Find the equation of the line passing through $(-4, -3)$ and parallel to the line determined by $3x + 4y = 12$.

First, we find the slope of the given parallel line, as before.

> **NOTE**
>
> The slope of the given line is $-\frac{3}{4}$, the coefficient of x.

$$3x + 4y = 12$$

$$4y = -3x + 12$$

$$y = -\frac{3}{4}x + 3$$

The slopes of two parallel lines is the same. Because the slope of the desired line must also be $-\frac{3}{4}$, we can use the point-slope form to write the required equation.

> **NOTE**
>
> The line must pass through $(-4, -3)$, so let $(x_1, y_1) = (-4, -3)$

$$y - y_1 = m(x - x_1)$$

$$y - (-3) = -\frac{3}{4}[x - (-4)]$$ $m = -\frac{3}{4}$ is the slope; $(-4, -3)$ is a point on the line.

We simplify this to its slope-intercept form, $y = mx + b$.

$$y - (-3) = -\frac{3}{4}[x - (-4)]$$

$$y + 3 = -\frac{3}{4}(x + 4)$$ Simplify the signs.

$$y + 3 = -\frac{3}{4}x - 3$$ Distribute to remove the parentheses.

$$y = -\frac{3}{4}x - 6$$ Subtract 3 from both sides.

Check Yourself 7

Find the equation of the line passing through (2, −5) and parallel to the line determined by $4x - y = 9$.

Example 8 Finding the Equation of a Perpendicular Line

Find the equation of the line passing through $(3, -1)$ and perpendicular to the line $3x - 5y = 2$.

First, find the slope of the perpendicular line.

$$3x - 5y = 2$$
$$-5y = -3x + 2$$
$$y = \frac{3}{5}x - \frac{2}{5}$$

The slope of the perpendicular line is $\frac{3}{5}$. Recall that the slopes of perpendicular lines are negative reciprocals. The slope of our line is the negative reciprocal of $\frac{3}{5}$. It is therefore $-\frac{5}{3}$.

Using the point-slope form, we have the equation

$$y - (-1) = -\frac{5}{3}[x - (3)]$$
$$y + 1 = -\frac{5}{3}x + 5$$
$$y = -\frac{5}{3}x + 4$$

Check Yourself 8

Find the equation of the line passing through (5, 4) and perpendicular to the line with equation $2x - 5y = 10$.

There are many applications of linear equations. Here is just one of many typical examples.

Example 9 A Business and Finance Application

NOTE

In applications, it is common to use letters other than x and y. In this case, we use C to represent the **cost**.

In producing a new product, a manufacturer predicts that the number of items produced x and the cost in dollars C of producing those items will be related by a linear equation.

Suppose that the cost of producing 100 items is \$5,000 and the cost of producing 500 items is \$15,000. Find the linear equation relating x and C.

To solve this problem, we must find the equation of the line passing through points (100, 5,000) and (500, 15,000). Even though the numbers are considerably larger than we have encountered thus far in this section, the process is exactly the same.

First, we find the slope:

$$m = \frac{15{,}000 - 5{,}000}{500 - 100} = \frac{10{,}000}{400} = 25$$

We can now use the point-slope form as before to find the desired equation.

$$C - 5{,}000 = 25(x - 100)$$
$$C - 5{,}000 = 25x - 2{,}500$$
$$C = 25x + 2{,}500$$

To graph the equation we have just derived, we must choose the scaling on the x- and C-axes carefully to get a "reasonable" picture. Here we choose increments of 100 on the x-axis and 2,500 on the C-axis since those seem appropriate for the given information.

NOTE

The change in scaling "distorts" the slope of the line.

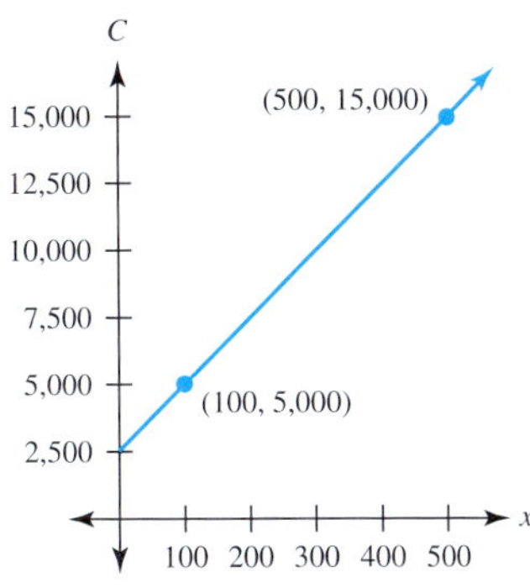

Check Yourself 9

A company predicts that the value in dollars, *V*, and the time that a piece of equipment has been in use, *t*, are related by a linear equation. If the equipment is valued at $1,500 after 2 years and at $300 after 10 years, find the linear equation relating *t* and *V*.

Earlier, we mentioned that when working with applications, we are frequently asked to interpret the slope of a function as its *rate of change*.

In short, the slope represents the change in the output, y or $f(x)$, when the input x is increased by one unit. We ask for such an interpretation in the next example, from the health sciences field.

Example 10 An Allied Health Application

A person's body mass index (BMI) can be calculated using his or her height h, in inches, and weight w, in pounds, with the formula

$$\text{BMI} = \frac{703w}{h^2}$$

NOTE

$69^2 = 4{,}761$

In the case of a 69-inch man, his height remains constant over many years, but his weight might vary, so we can model his body mass index as a function of his weight w.

$$\text{B}(w) = \frac{703}{4{,}761}w$$

(a) Find the body mass index of a 190-lb, 69-in. man (to the nearest tenth). We use the model above with $w = 190$.

$$\begin{aligned} B(190) &= \frac{703}{4{,}761}(190) \\ &= \frac{133{,}570}{4{,}761} \\ &\approx 28.1 \end{aligned}$$

(b) Determine the slope of this function.

The slope is $\dfrac{703}{4{,}761} \approx 0.15$

(c) Interpret the slope of this function in the context of the application.

The input of this function is the man's weight, which is given in pounds. Therefore, the slope can be interpreted as "for each additional pound that the man weighs, his body mass index increases by 0.15."

Check Yourself 10

Using the metric system, a person's body mass index (BMI) can be calculated using his or her height h, in centimeters, and weight w, in kilograms, with the formula

$$\text{BMI} = \frac{10{,}000w}{h^2}$$

In the case of a 160-cm woman, we can model her body mass index as a function of her weight w.

$$B(w) = \frac{25}{64}w$$

(a) Find the body mass index of a 160-cm, 70-kg woman (to the nearest tenth).
(b) Determine the slope of this function.
(c) Interpret the slope of this function in the context of the application.

NOTE

$$\frac{10{,}000}{h^2} = \frac{10{,}000}{160^2} = \frac{10{,}000}{25{,}600} = \frac{25}{64}$$

Check Yourself ANSWERS

1. $m_1 = \frac{3}{2}$ and $m_2 = -\frac{2}{3}$; $(m_1)(m_2) = -1$ **2.**

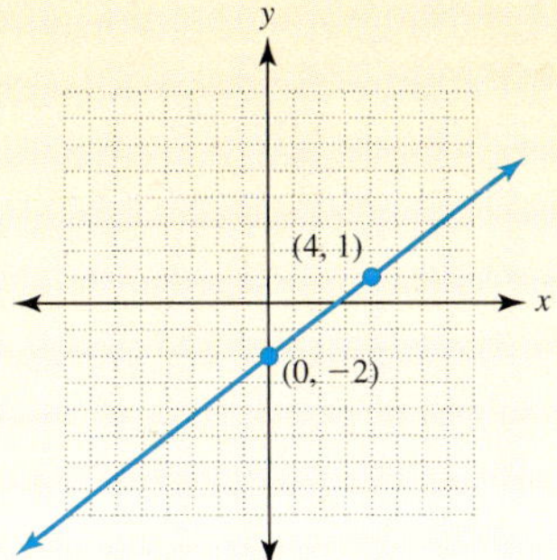

3. $y = \frac{3}{2}x + 7$ **4.** $y = -\frac{2}{3}x + \frac{11}{3}$ **5.** **(a)** $y = 5$; **(b)** $x = -3$

6. $y = -\frac{2}{3}x + \frac{11}{3}$ **7.** $y = 4x - 13$ **8.** $y = -\frac{5}{2}x + \frac{33}{2}$

9. $V = -150t + 1{,}800$ **10.** **(a)** $\frac{875}{32} \approx 27.3$; **(b)** $\frac{25}{64} \approx 0.39$;
(c) Each additional kilogram increases her BMI by 0.39.

Reading Your Text

SECTION 3.3

(a) Two (nonvertical) lines are ______________ if, and only if, their slopes are equal.

(b) Two (nonvertical) lines are ______________ if and only if their slopes are negative reciprocals.

(c) A vertical line has a slope that is ______________.

(d) A horizontal line has a slope that is ______________.

3.3 exercises

Basic Skills | Challenge Yourself | Calculator/Computer | Career Applications | Above and Beyond

Name ______________

Section ________ Date ________

Answers

1. ______
2. ______
3. ______
4. ______
5. ______
6. ______
7. ______ 8. ______
9. ______ 10. ______
11. ______
12. ______
13. ______
14. ______
15. ______
16. ______
17. ______
18. ______

< Objective 1 >

Determine whether each pair of lines is **parallel, perpendicular,** *or* **neither.**

1. L_1 through $(-2, -3)$ and $(4, 3)$; L_2 through $(3, 5)$ and $(5, 7)$

2. L_1 through $(-2, 4)$ and $(1, 8)$; L_2 through $(-1, -1)$ and $(-5, 2)$

3. L_1 through $(7, 4)$ and $(5, -1)$; L_2 through $(8, 1)$ and $(-3, -2)$

4. L_1 through $(-2, -3)$ and $(3, -1)$; L_2 through $(-3, 1)$ and $(7, 5)$

5. L_1 with equation $x - 3y = 6$; L_2 with equation $3x + y = 3$ 

6. L_1 with equation $2x + 4y = 8$; L_2 with equation $4x + 8y = 10$

7. Find the slope of any line parallel to the line through points $(-2, 3)$ and $(4, 5)$.

8. Find the slope of any line perpendicular to the line through points $(0, 5)$ and $(-3, -4)$.

9. A line passing through $(-1, 2)$ and $(4, y)$ is parallel to a line with slope 2. What is the value of y?

10. A line passing through $(2, 3)$ and $(5, y)$ is perpendicular to a line with slope $\frac{3}{4}$. What is the value of y?

< Objective 2 >

Write the equation of the line passing through each of the given points with the indicated slope. Give your results in slope-intercept form, where possible.

11. $(0, -5)$, slope $\frac{5}{4}$

12. $(0, -4)$, slope $-\frac{3}{4}$

13. $(1, 3)$, slope 5

14. $(-1, 2)$, slope 3

15. $(-2, -3)$, slope -3

16. $(1, 3)$, slope -2

17. $(5, -3)$, slope $\frac{2}{5}$ 

18. $(4, 3)$, slope 0

19. (1, −4), slope undefined

20. (2, −5), slope $\frac{1}{4}$

21. (5, 0), slope $-\frac{4}{5}$

22. (−3, 4), slope undefined

< Objective 3 >

Write the equation of the line passing through each of the given pairs of points. Write your result in slope-intercept form, where possible.

23. (2, 3) and (5, 6)

24. (3, −2) and (6, 4)

25. (−2, −3) and (2, 0)

26. (−1, 3) and (4, −2)

27. (−3, 2) and (4, 2)

28. (−5, 3) and (4, 1)

29. (2, 0) and (0, −3)

30. (2, −3) and (2, 4)

31. (0, 4) and (−2, −1)

32. (−4, 1) and (3, 1)

< Objective 4 >

Write the equation of the line L satisfying the given geometric conditions.

33. L has slope 4 and y-intercept (0, −2).

34. L has slope $-\frac{2}{3}$ and y-intercept (0, 4).

35. L has x-intercept (4, 0) and y-intercept (0, 2).

36. L has x-intercept (−2, 0) and slope $\frac{3}{4}$.

37. L has y-intercept (0, 4) and a 0 slope.

38. L has x-intercept (−2, 0) and an undefined slope.

39. L passes through (2, 3) with a slope of −2.

40. L passes through (−2, −4) with a slope of $-\frac{3}{2}$.

Answers

19. ____
20. ____
21. ____
22. ____
23. ____
24. ____
25. ____
26. ____
27. ____
28. ____
29. ____
30. ____
31. ____
32. ____
33. ____
34. ____
35. ____
36. ____
37. ____
38. ____
39. ____
40. ____

Answers

41. ______

42. ______

43. ______

44. ______

45. ______

46. ______

47. ______

48. ______

49. ______

50. ______

51. ______

52. ______

53. ______

54. ______

55. ______

56. ______

57. ______

58. ______

41. L has y-intercept (0, 3) and is parallel to the line with equation $y = 3x - 5$.

42. L has y-intercept (0, −3) and is parallel to the line with equation $y = \frac{2}{3}x + 1$.

43. L has y-intercept (0, 4) and is perpendicular to the line with equation $y = -2x + 1$.

44. L has y-intercept (0, 2) and is parallel to the line with equation $y = -1$.

45. L has y-intercept (0, 3) and is parallel to the line with equation $y = 2$.

46. L has y-intercept (0, 2) and is perpendicular to the line with equation $2x - 3y = 6$.

47. L passes through (−4, 5) and is parallel to the line $y = -4x + 5$.

48. L passes through (−4, 3) and is parallel to the line with equation $y = -2x + 1$.

49. L passes through (3, 2) and is parallel to the line with equation $y = \frac{4}{3}x + 4$.

50. L passes through (−2, −1) and is perpendicular to the line with equation $y = 3x + 1$.

51. L passes through (3, −1) and is perpendicular to the line with equation $y = -\frac{2}{3}x + 5$.

52. L passes through (−4, 2) and is perpendicular to the line with equation $y = 4x + 5$.

53. L passes through (−2, 1) and is parallel to the line with equation $x + 2y = 4$.

54. L passes through (−3, 5) and is parallel to the x-axis.

Basic Skills | **Challenge Yourself** | Calculator/Computer | Career Applications | Above and Beyond

Determine whether each statement is **true** *or* **false.**

55. If two nonvertical lines are parallel, then they have the same slope.

56. If two lines are perpendicular, with slopes m_1 and m_2, then the product of the slopes is 1.

Complete each statement with **never, sometimes,** *or* **always.**

57. Given two points of a line, we can ______ determine the equation of the line.

58. Given a nonvertical line, the slope of a line perpendicular to it will ______ be zero.

A four-sided figure (quadrilateral) is a parallelogram if the opposite sides have the same slope. If the adjacent sides are perpendicular, the figure is a rectangle. In exercises 59 to 62, for each quadrilateral ABCD, determine whether it is a parallelogram; then determine whether it is a rectangle.

59. $A(0, 0), B(2, 0), C(2, 3), D(0, 3)$

60. $A(-3, 2), B(1, -7), C(3, -4), D(-1, 5)$

61. $A(0, 0), B(4, 0), C(5, 2), D(1, 2)$

62. $A(-3, -5), B(2, 1), C(-4, 6), D(-9, 0)$

63. **Science and Medicine** A temperature of 10°C corresponds to a temperature of 50°F. Also, 40°C corresponds to 104°F. Find the linear equation relating F and C.

64. **Business and Finance** In planning for a new item, a manufacturer assumes that the number of items produced x and the cost in dollars C of producing these items are related by a linear equation. Projections are that 100 items will cost $10,000 to produce and that 300 items will cost $22,000 to produce. Find the equation that relates C and x.

65. **Business and Finance** Mike bills a customer at the rate of $65 per hour plus a fixed service call charge of $75.

(a) Write an equation that will allow you to compute the total bill for any number of hours x that it takes to complete a job.

(b) What will the total cost of a job be if it takes 3.5 hours to complete?

(c) How many hours would a job have to take if the total bill were $247.25?

66. **Business and Finance** Two years after an expansion, a company had sales of $42,000. Four years later (six years after the expansion) the sales were $102,000. Assuming that the sales in dollars S and the time t in years are related by a linear equation, find the equation relating S and t.

Use the graph to determine the slope and y-intercept of the line.

67.

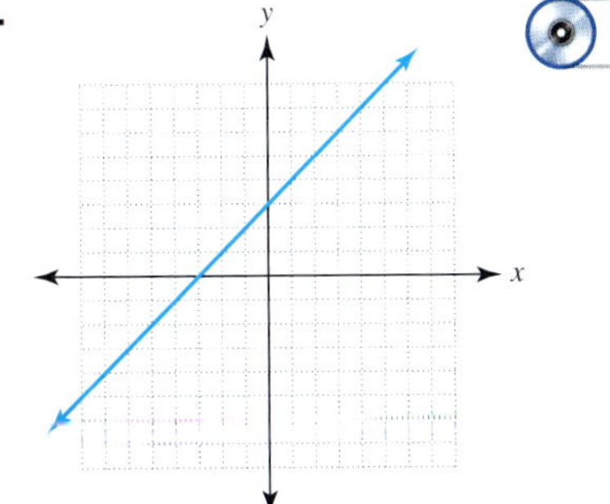

> Videos

68.

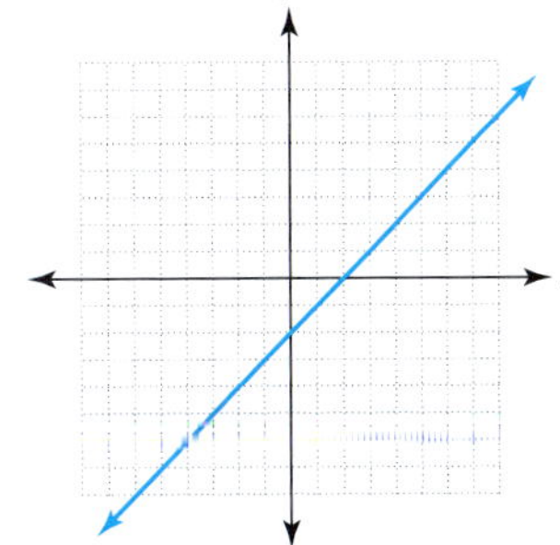

Answers

59. ______

60. ______

61. ______

62. ______

63. ______

64. ______

65. ______

66. ______

67. ______

68. ______

Answers

69. ______

70. ______

71. ______

72. ______

73. ______

74. ______

75. ______

76. ______

69.

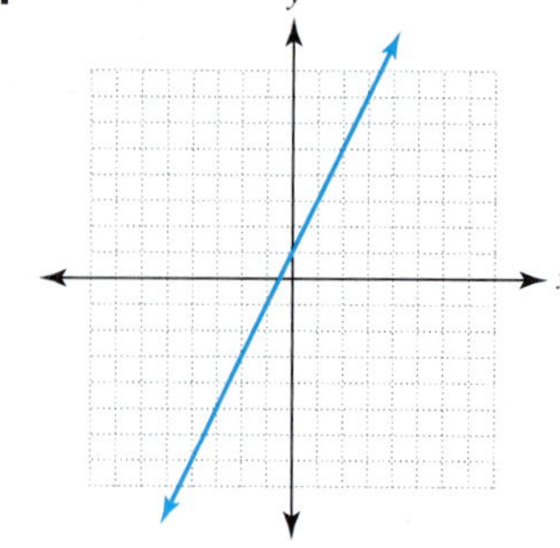

70.

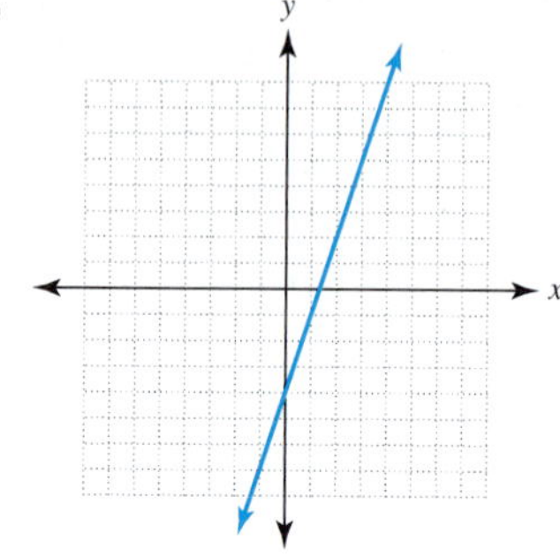

71.

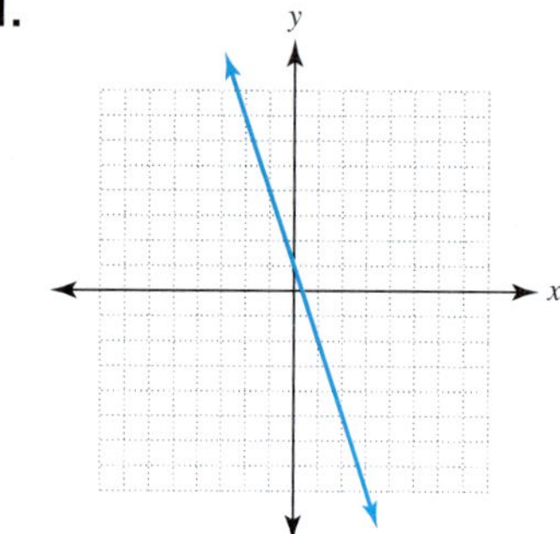

72.

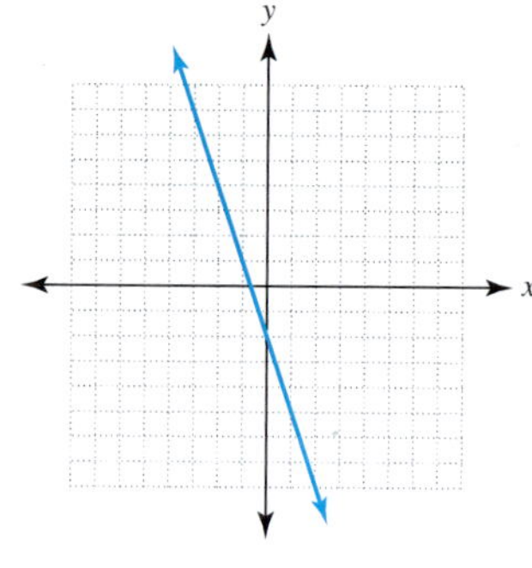

73.

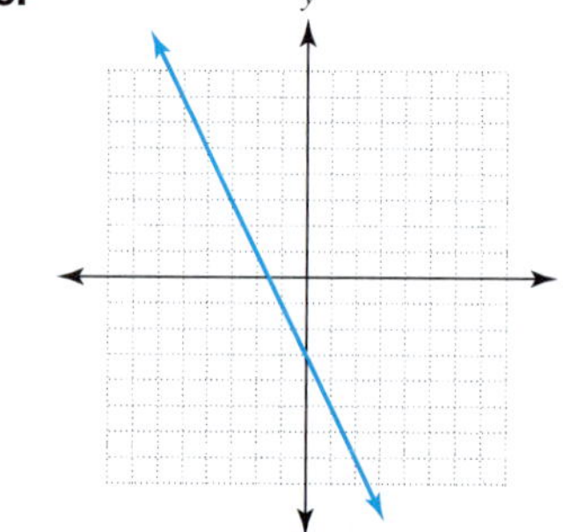

74.

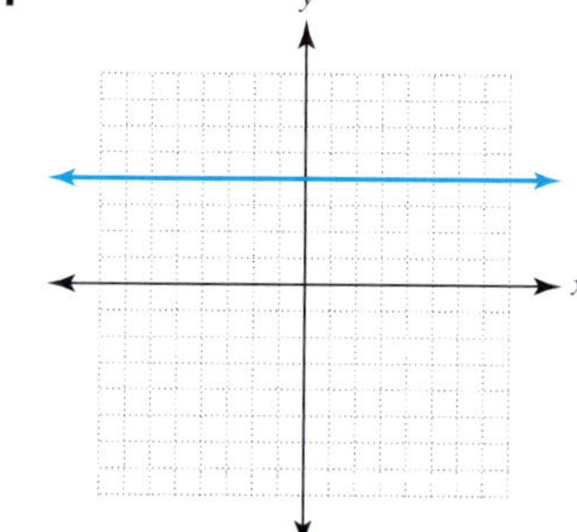

Basic Skills | Challenge Yourself | **Calculator/Computer** | Career Applications | Above and Beyond

75. Use a graphing calculator to graph the equations on the same screen.

$y = 0.5x + 7 \qquad y = 0.5x + 3 \qquad y = 0.5x - 1 \qquad y = 0.5x - 5$

Use the standard viewing window. Describe the results.

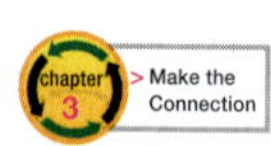

76. Use a graphing calculator to graph the equations on the same screen.

$y = \frac{2}{3}x \qquad y = -\frac{3}{2}x$

Use the standard viewing window first, and the regraph using a Zsquare utility on the calculator. Describe the results.

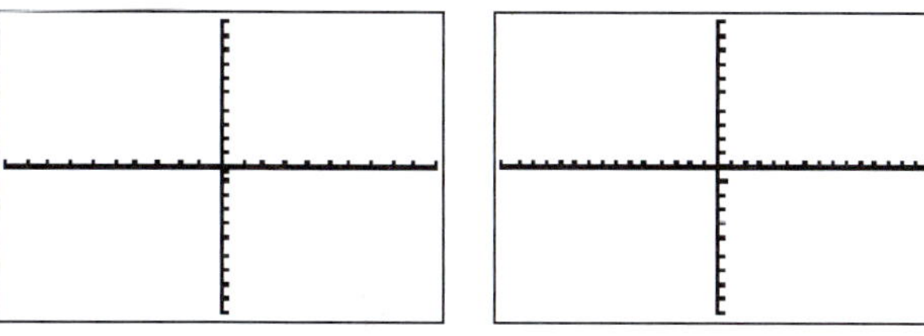

The lines appear perpendicular in the second graph.

Basic Skills | Challenge Yourself | Calculator/Computer | Career Applications | Above and Beyond

77. Agricultural Technology The yield Y (in bushels per acre) for a cornfield is estimated from the amount of rainfall R (in inches) using the formula

$$Y = \frac{43{,}560}{8{,}000}R$$

(a) Find the slope of the line described by this equation (to the nearest tenth).

(b) Interpret the slope in the context of this application.

78. Agricultural Technology During one summer period, the growth of corn plants follows a linear pattern approximated by the equation

$$h = 1.77d + 24.92$$

in which h is the height (in inches) of the corn plants and d is the number of days that have passed.

(a) Find the slope of the line described by this equation.

(b) Interpret the slope in the context of this application.

Allied Health The arterial oxygen tension (P_aO_2, in mm Hg) of a patient can be estimated based on the patient's age A (in years). The equation used depends on the position of the patient. Use this information to complete exercises 79 and 80.

79. If a patient is lying down, the arterial oxygen tension can be approximated using the formula

$$P_aO_2 = 103.5 - 0.42A$$

(a) Determine the slope of this formula.

(b) Interpret the slope in the context of this application.

80. If a patient is seated, the arterial oxygen tension can be approximated using the formula

$$P_aO_2 = 104.2 - 0.27A$$

(a) Determine the slope of this formula.

(b) Interpret the slope in the context of this application.

Answers

77. ______

78. ______

79. ______

80. ______

Answers

1. Parallel **3.** Neither **5.** Perpendicular **7.** $\frac{1}{3}$ **9.** 12

11. $y = \frac{5}{4}x - 5$ **13.** $y = 5x - 2$ **15.** $y = -3x - 9$ **17.** $y = \frac{2}{5}x - 5$

19. $x = 1$ **21.** $y = -\frac{4}{5}x + 4$ **23.** $y = x + 1$ **25.** $y = \frac{3}{4}x - \frac{3}{2}$

27. $y = 2$ **29.** $y = \frac{3}{2}x - 3$ **31.** $y = \frac{5}{2}x + 4$ **33.** $y = 4x - 2$

35. $y = -\frac{1}{2}x + 2$ **37.** $y = 4$ **39.** $y = -2x + 7$ **41.** $y = 3x + 3$

43. $y = \frac{1}{2}x + 4$ **45.** $y = 3$ **47.** $y = -4x - 11$ **49.** $y = \frac{4}{3}x - 2$

51. $y = \frac{3}{2}x - \frac{11}{2}$ **53.** $y = -\frac{1}{2}x$ **55.** True **57.** always

59. Yes; yes **61.** Yes; no **63.** $F = \frac{9}{5}C + 32$

65. **(a)** $C = 65x + 75$; **(b)** \$302.50; **(c)** 2.65 h **67.** Slope 1, y-intercept (0, 3)
69. Slope 2, y-intercept (0, 1) **71.** Slope -3, y-intercept (0, 1)
73. Slope -2, y-intercept (0, -3) **75.** The lines are parallel.

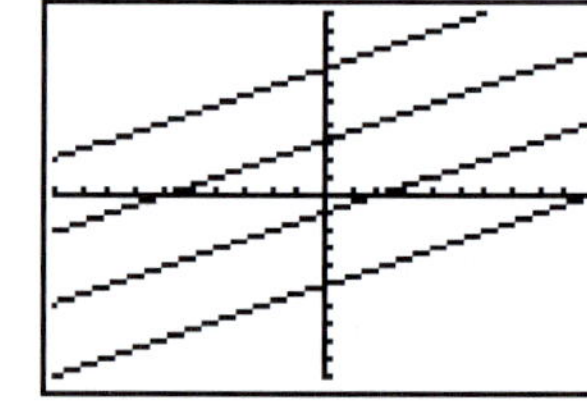

77. **(a)** 5.4; **(b)** Each additional inch of rainfall yields an additional 5.4 bushels per acre. **79.** **(a)** -0.42; **(b)** Each additional year of age reduces the arterial oxygen tension by 0.42 mm Hg.

3.4 Rate of Change and Linear Regression

< 3.4 Objectives >

1 > Construct a linear function to model an application

2 > Construct a linear function based on two data points

3 > Find the input necessary to produce a given function value

4 > Use regression analysis to produce a linear model based on a data set

In this section, we bring together the two main ideas that we presented in Chapters 2 and 3: Functions and Linear Equations. This will allow us to understand these powerful tools in real-world settings.

We begin by taking another look at the slope of a linear equation. Recall that we defined the slope of a line as a measure of its steepness. The question that we want to answer is, "Given an application, what are the properties represented by the slope?"

Consider the linear equation

RECALL

To graph the equation, locate the y-intercept, (0, 4), and use the slope to locate a second point such as (2, 5). Then, graph the line.

$$y = \frac{1}{2}x + 4$$

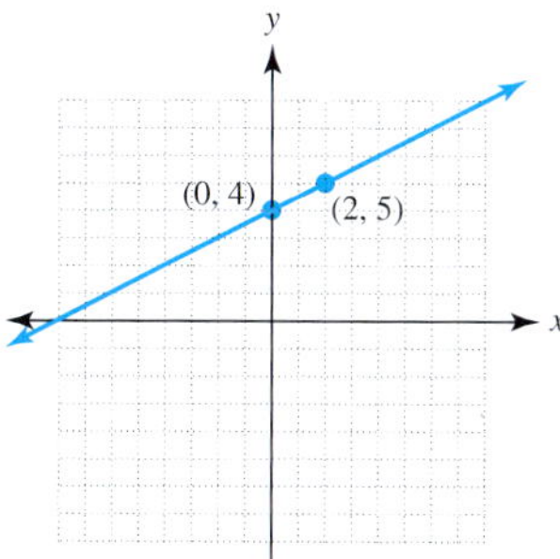

The slope of this line is $\frac{1}{2}$, which means that

$$m = \frac{\text{Change in } y}{\text{Change in } x} = \frac{y_2 - y_1}{x_2 - x_1} = \frac{1}{2}$$

In other words, if x increases by 2 units, then y increases by one unit.

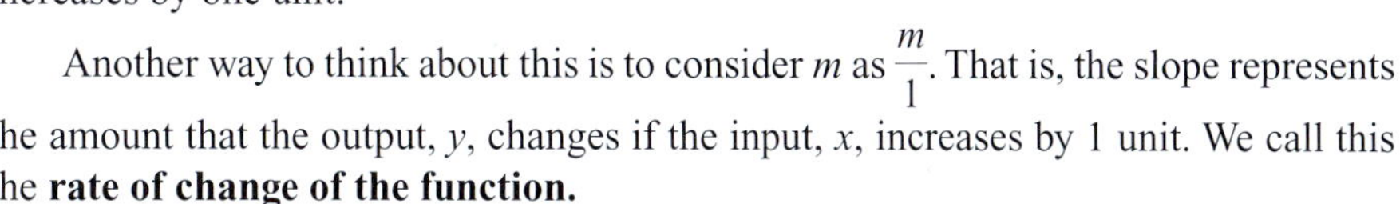

Another way to think about this is to consider m as $\frac{m}{1}$. That is, the slope represents the amount that the output, y, changes if the input, x, increases by 1 unit. We call this the **rate of change of the function.**

In the example above, this means that if x increases by one unit, then y increases by $\frac{1}{2}$ unit.

This is a powerful way of interpreting the slope of a linear model.

Example 1 Constructing a Function

< Objective 1 >

A store charges \$99.95 for a certain calculator. If we are interested in the revenue from the sales of this calculator, then the quantity that varies is the number of calculators it sells.

We begin by identifying this quantity and representing it with a variable. Let x be the number of calculators sold.

Because the store's revenue depends on the number of calculators sold and is computed by multiplying the number of calculators sold by the price of each, we identify, and name, a function to describe this relationship.

Let R represent the revenue from the sale of x calculators.

RECALL

This does not mean R times x.

The relationship we have is written using function notation as follows.

$R(x) = 99.95x$ Revenue is determined by multiplying the number of calculators sold by the price of each one.

This function is read, "R of x is 99.95 times x." We say "R is a function of x."

The function above is a linear function. The y-intercept is (0, 0) because the store earns no revenue if it does not sell any calculators.

The slope of the function is 99.95. This means that each additional calculator sold increases the revenue \$99.95.

NOTE

The input, or independent variable, represents the number of calculators sold.

The output, or function value, gives the revenue.

Consider the question, "How much is the store's revenue if it sells 10 calculators?" In notation, we say that we are trying to find $R(10)$.

$R(10) = 99.95(10) = 999.50$ We replace x with 10 everywhere it appears.

The store earns \$999.50 in revenue from the sale of 10 calculators.

Check Yourself 1

A store sells coffee by the pound. It charges \$6.99 for each pound of coffee beans.

(a) Construct a function that models its revenue from the sale of x pounds of coffee.

(b) Use the function created in (a) to determine its revenue if it sells 32.5 pounds of coffee.

In retail applications, it is common for revenue models to have the origin as the y-intercept because a business would need to sell something in order to earn revenue.

On the other hand, most businesses incur costs independent of how much they sell. In fact, there are two types of costs that we will focus on: *fixed cost* and *variable cost*.

The fixed cost represents the cost of running a business and having that business available. Fixed cost might include the cost of a lease, insurance costs, energy costs, and some labor costs, to name a few.

Marginal cost represents the cost of each item being sold. For a retail store, this is usually the *wholesale price* of an item. For instance, if the store in Example 1 bought the calculators from the manufacturer for \$64.95 each, then this is their wholesale price and represents the marginal cost associated with the calculators.

The variable cost for a product is the product of the marginal cost and the number of items sold.

Example 2 Modeling a Cost Function

A store purchases a graphing-calculator model at a wholesale price of \$64.95 each. Additionally, the store has a weekly fixed cost of \$450 associated with the sale of these graphing calculators.

(a) Construct a function to model the cost of selling these calculators.

Let x represent the number of these graphing calculators that the store buys. Let C represent the cost of purchasing x calculators.

Then, we construct the cost function:

$$C(x) = \underbrace{64.95}_{\text{Marginal cost}}x + \underbrace{450}_{\text{Fixed cost}}$$

NOTE

The slope of the function is given by the marginal cost, 64.95, because each additional calculator increases the cost \$64.95.

(b) Find the cost to the store if it sells 35 calculators in one week.

$$C(35) = 64.95(35) + 450$$
$$= 2{,}723.25$$

It costs the store \$2,723.25 to sell 35 calculators in one week.

(c) What is the additional cost if the store were to sell 36 calculators?

The slope gives the cost of selling one more unit. Therefore, selling one more calculator costs the store an additional \$64.95.

Check Yourself 2

A store purchases coffee beans at a wholesale price of \$4.50 per pound (lb). Its daily fixed cost, associated with the sale of coffee beans, is \$60.

(a) Construct a function to model the cost of coffee-bean sales.

(b) Find the cost if the store sells 40 lb of coffee beans in a day.

(c) What is the additional cost to the store if it were to sell 41 lb of coffee?

In many cases, there isn't an explicit or obvious function to use. For instance, consider the question, "How will an increase in spending on advertising affect sales?" We might think that sales will increase, but we do not know by what amount.

We saw in Section 3.3 that we can construct a linear model if we have two points. Before moving to more complicated examples in which we need to use technology, we review the techniques for building a linear model from two points.

Example 3 Building a Linear Model

< Objective 2 >

A small financial services company spent \$40,000 on advertising one month. It earned \$325,000 in profits that month. The following month, the company increased its advertising budget to \$55,000 and saw its profits increase to \$445,000.

(a) Use this information to determine two points and model the profits as a linear function of the advertising budget.

The company is interested in how its advertising spending affects its profits. As such, the independent, or input, variable is the amount of the advertising budget.

Let x be the amount spent on advertising in a month and let P be the profits that month.

We use thousands, as our unit, to keep the numbers reasonably small. This gives the points (40, 325) and (55, 445), because $P(40) = 325$ and $P(55) = 445$.

NOTE

Forty thousand dollars spent on advertising led to profits of 325 thousand dollars.

Similarly, 55 thousand dollars in advertising gave the company 445 thousand dollars in profits.

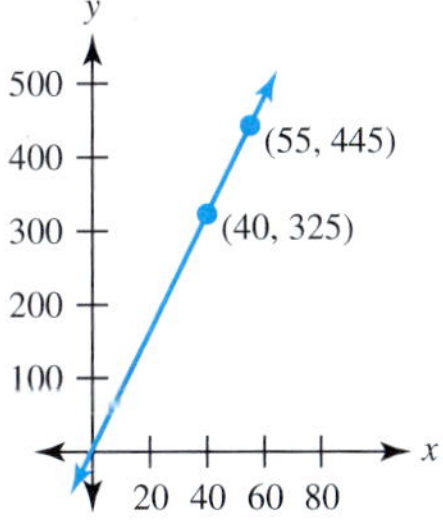

To construct a model, we find the slope of the line through our two points. We then use the point-slope equation for a line to construct the function.

$$m = \frac{y_2 - y_1}{x_2 - x_1} = \frac{445 - 325}{55 - 40} = \frac{120}{15} = 8$$

We choose to use the first point, (40, 325), in the point-slope formula with the slope $m = 8$. Recall that the point-slope formula for a line, given a point and the slope, is

$$y - y_1 = m(x - x_1)$$

Substitute into this formula:

$$y - (325) = (8)(x - 40)$$

$y - 325 = 8x - 320$ Distribute the 8 to remove the parentheses on the right.

$y = 8x + 5$ Add 325 to both sides.

$P(x) = 8x + 5$ Write the model using function notation.

NOTE

Choosing the other point, (55, 445), leads to the same function.

$y - 445 = 8(x - 55)$
$y - 445 = 8x - 440$
$y = 8x + 5$

(b) Interpret the slope as the rate of change of this function in the context of this application.

The slope is 8. This means that each time x increases by 1, y increases by 8. In the context of this application, the company can expect to see profits increase by $8,000 for each $1,000 increase in advertising spending.

RECALL

The y-intercept occurs where the x-value is 0.

In this case, $x = 0$ corresponds to the company spending $0 on advertising.

(c) Interpret the y-intercept in the context of this application.

The y-intercept is (0, 5). This means that the model predicts that the company would earn $5,000 in profits if it did not spend anything on advertising.

Check Yourself 3

At an underwater depth of 30 ft, the atmospheric pressure is approximately 28.2 pounds per square inch (psi). At 80 feet, the pressure increases to approximately 50.7 psi.

(a) Construct a linear function modeling the pressure underwater as a function of the depth.

(b) Give the slope of the function, to the nearest hundredth. Interpret the slope in the context of this application.

(c) Give the y-intercept, to the nearest tenth, and interpret it in the context of this application.

Once we have built a model to describe a situation, we can use the model in several different ways. We can examine different outputs based on changes to the input. Or, we can predict an input that would lead to a desired output. We do both in the next example.

Example 4 A Business and Finance Model

< Objective 3 >

In Example 3, we modeled the profits earned by a financial services company as a function of its advertising budget: $P(x) = 8x + 5$, in thousands.

(a) Use this model to predict the company's profits if it budgets $50,000 to advertising.

Because our units are thousands of dollars, we are being asked to find $P(50)$:

$P(50) = 8(50) + 5$ Replace x with 50 and evaluate.

$= 405$ Remember to follow the order of operations.

The company would expect to earn $405,000 in profits if it budgets $50,000 to advertising.

(b) How much would it need to budget to advertising in order for the profits to reach $500,000?

Do you see how this differs from the problem in (a)? This time, we are being given the profit, or output, of the function and being asked to find the appropriate input.

That is, we want to find x so that $P(x) = 500$.

We need to solve the equation

$$8x + 5 = 500$$ Because $P(x) = 500$, we set the expression $8x + 5$ equal to 500.

$$8x = 495$$ Subtract 5 from both sides.

$$x = \frac{495}{8}$$ Divide both sides by 8.

$$= 61.875$$

NOTE

Because x is in thousands, we have $61.875 \times 1{,}000 = 61{,}875$.

Therefore, in order to earn a $500,000 profit, the company should invest $61,875 in advertising.

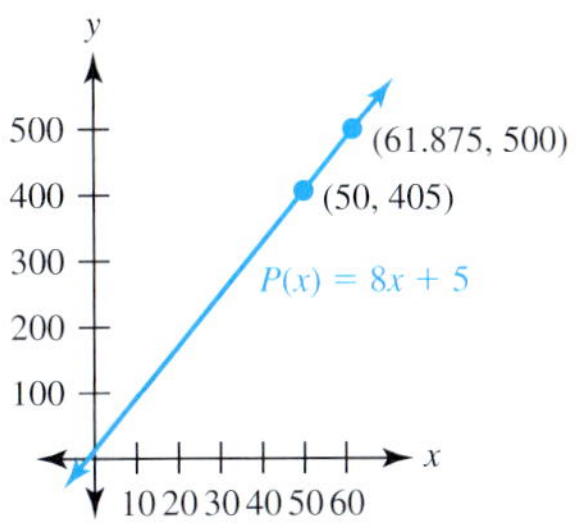

Check Yourself 4

In Check Yourself 3, you modeled atmospheric pressure as a function of underwater depth: $P(x) = 0.045x + 14.7$.

(a) What is the approximate pressure felt by a diver at a depth of 130 ft?

(b) At what depth is the pressure 60 psi (to the nearest foot)?

Of course, a company does not usually have a nice model demonstrating its profits as a function of its advertising. More likely, a company might spend $40,000 one month and see a $325,000 return, but might earn $300,000 the next time they spend that much in advertising. There are many factors that might influence a company's profits, and advertising is just one of them.

When modeling an application, it is much more likely that the data set does not form a straight line, even when the underlying phenomenon is basically linear. This is especially true when the data being studied relate to human health, such as children's heights or drug studies. In these situations, each subject is unique because each is a person. No two people respond exactly the same to a medication dosage.

NOTE

You will learn to construct regression models by hand when you study calculus. Until then, we use technology to do the computations.

We will use the *regression analysis* techniques that you learned in Activity 3 before the exercises in Section 3.1. In that activity, you learned to enter data into a graphing calculator to create a scatter plot and to find the linear function that *best fits* the data.

Example 5 Linear Regression

< Objective 4 >

The table below gives the 2005 population (in millions) and CO_2 emissions (in teragrams) for selected nations.

Nation	Population	Emissions
Australia	20	384
Austria	8	80
Belarus	10	55
Bulgaria	8	55
Canada	32	583
Czech Republic	10	126
Denmark	5	52
Finland	5	57
France	61	417
Germany	82	873
Italy	59	493
United Kingdom	60	558

Source: Statistics Division; United Nations (DYB 2005).

(a) Use a graphing calculator to create a scatter plot, perform a regression analysis, and graph the best-fit linear model on the scatter plot.

We follow the techniques learned in Activity 3 to create each screen.

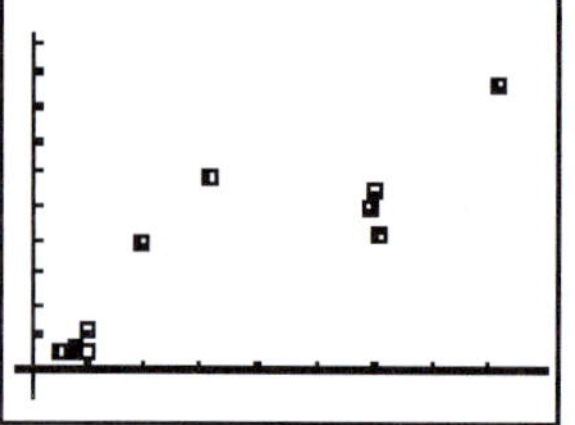

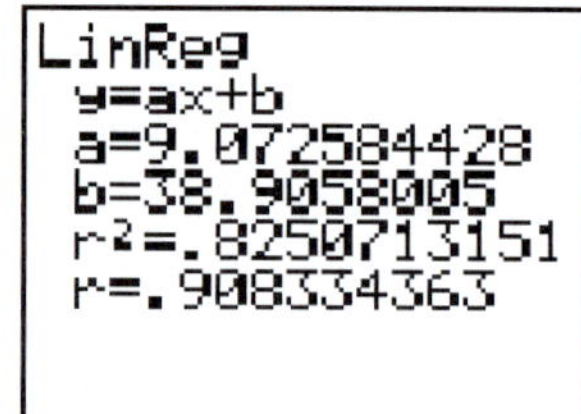

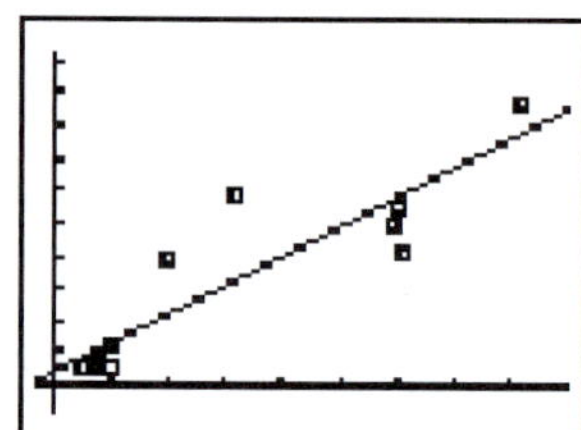

(b) Write the equation of the line-of-best-fit for the data, the linear regression model (round to the nearest tenth). What is the slope? Interpret the slope in the context of this application.

In the regression screen, *a* represents the slope of the line.

$y = 9.1x + 38.9$

The slope is approximately 9.1. We interpret this to mean that an increase of one million people leads to an increase of 9.1 teragrams of CO_2 emissions.

Check Yourself 5

The table below gives the age (in months) and weight (in pounds) for a set of 10 boys.

Age, x	12	12	13	15	15	16	19	20	20	24
Weight, y	19	25	24	21	26	26	29	26	31	33

Source: Adapted from U.S. Center for Disease Control and Prevention data.

(a) Use a graphing calculator to create a scatter plot, perform a regression analysis, and graph the best-fit linear model on the scatter plot.

(b) Write the equation of the line-of-best-fit for the data, the linear regression model (round to the nearest tenth). What is the slope? Interpret the slope in the context of this application.

Looking at Example 4 suggests how we can expand on Example 5. In Example 4, you evaluated the linear function at important points that were not part of the original model. We can use the line-of-best-fit in the same way.

Example 6 Using the Line-of-Best-Fit

Use the linear function constructed in Example 5 to answer each question. Use the model rounded to the nearest tenth.

(a) Estimate the 2005 CO_2 emissions (in teragrams) of a country if its population is 50 million people.

Use the function $f(x) = 9.1x + 38.9$, from Example 5, with $x = 50$.

$$f(50) = 9.1(50) + 38.9$$
$$= 493.9$$

We expect a country of 50 million people to emit approximately 493.9 teragrams of CO_2.

(b) In 2005, the United States emitted approximately 6,064 teragrams of CO_2. Use the line-of-best-fit from Example 5 (to the nearest tenth) to estimate the population of a nation that emits 6,064 teragrams of CO_2.

This time, we are being given the output, $f(x) = 6{,}064$, and asked to find the input that would produce that result.

NOTE

The U.S. Census Bureau estimates the nation's 2005 population at 296 million.

$$9.1x + 38.9 = 6{,}064$$
$$9.1x = 6{,}025.1 \quad \text{Subtract 38.9 from both sides.}$$
$$x = 662.1 \quad \text{Divide both sides by 9.1; round to one decimal place.}$$

We would expect a nation that emits 6,064 teragrams of CO_2 to have a population near 662 million people.

Check Yourself 6

In Check Yourself 5, you constructed a model for a boy's weight, in pounds, based on his age, in months. Use that model, accurate to one decimal place, to answer each question.

(a) Estimate the weight of an 18-month-old boy.

(b) Estimate the age (to the nearest month) of a boy who weighs 25 lb.

Check Yourself ANSWERS

1. **(a)** $R(x) = 6.99x$; **(b)** \$227.18
2. **(a)** $C(x) = 4.5x + 60$; **(b)** \$240; **(c)** \$4.50
3. **(a)** $P(x) = 0.45x + 14.7$; **(b)** The slope is approximately 0.45; the pressure increases approximately 0.45 psi for each additional foot of depth underwater; **(c)** The y-intercept is approximately (0, 14.7). At the surface (depth is 0 ft), the atmospheric pressure is approximately 14.7 psi.
4. **(a)** 73.2 psi; **(b)** 101 ft

5. (a)

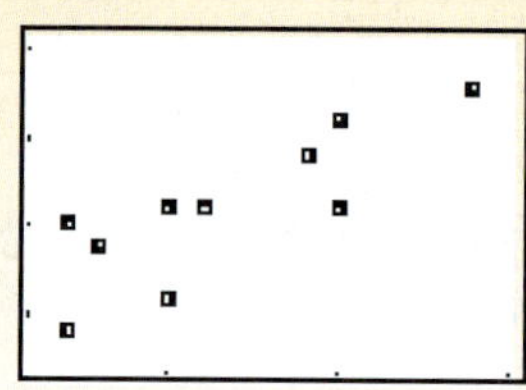

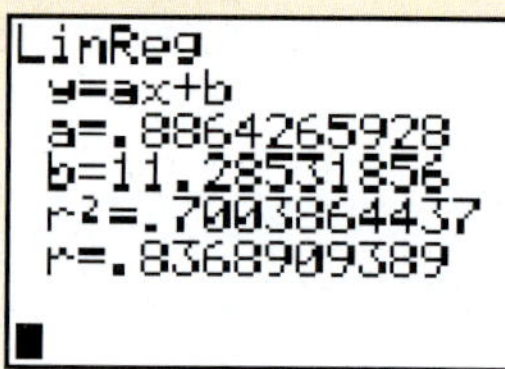

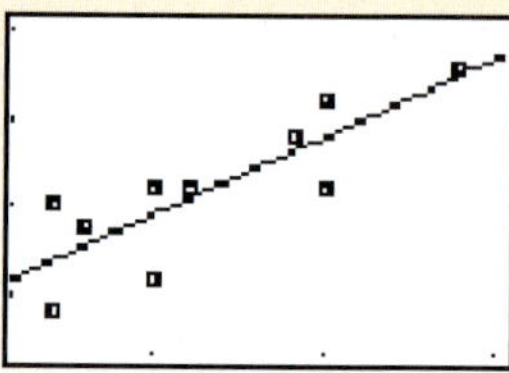

(b) $y = 0.9x + 11.3$; the slope is approximately 0.9, which means that for each month that a boy ages, we expect him to gain an additional 0.9 lb.

6. (a) 27.5 lb; **(b)** 15 months

Reading Your Text

SECTION 3.4

(a) The slope of a line represents the change in y when x is increased by ____________ unit.

(b) The ____________ of a linear function is called the rate of change of the function.

(c) The y-intercept occurs where the x-coordinate is ____________.

(d) We use regression analysis to find the ____________ that best fits a data set.

Basic Skills | Challenge Yourself | Calculator/Computer | Career Applications | Above and Beyond

3.4 exercises

< Objective 1 >

Model each relationship with a linear function.

1. In the snack department of a local supermarket, candy costs $1.58 per pound.

2. A cheese pizza costs $11.50. Each topping costs an additional $1.25.

3. The perimeter of a square is a function of the length of a side.

4. The temperature, in degrees Celsius, is a function of the temperature, in degrees Fahrenheit. *Hint:* You can find this on the Internet, or look in some cookbooks or science books. You can even build it using

   $0°C = 32°F$; $100°C = 212°F$.

Use the functions constructed in exercises 1 to 4 and function notation to answer each question.

5. How much does it cost to purchase 7 pounds of candy (exercise 1)?

6. How much does a pizza with 3 toppings cost (exercise 2)?

7. What is the perimeter of a square if the length of a side is 18 cm (exercise 3)?

8. What is the Celsius equivalent of 65°F (exercise 4)?

< Objective 3 >

9. How much candy can be purchased for $12 (exercise 1)?

10. How many pizza toppings can you get for $14 (exercise 2)?

11. What is the length of a side of a square if its perimeter is 42 ft (exercise 3)?

12. What is the Fahrenheit equivalent of 13°C (exercise 4)?

< Objective 2 >

13. **Science and Medicine** At 3 months old, a kitten weighed 4 lb. It reached 9 lb by the time it was 8 months old.

    (a) Construct a linear function modeling the kitten's weight as a function of its age.

    (b) Give the slope of the function. Interpret the slope in the context of this application.

14. **Science and Medicine** A young girl weighed 25 lb at 24 months old. When she reached 30 months old, she weighed 27 lb.

    (a) Construct a linear function modeling the girl's weight as a function of her age.

    (b) Give the slope of the function. Interpret the slope in the context of this application.

Name ____________

Section ________ Date ________

Answers

1. ____________
2. ____________
3. ____________
4. ____________
5. ____________
6. ____________
7. ____________
8. ____________
9. ____________
10. ____________
11. ____________
12. ____________
13. ____________
14. ____________

Answers

15.

16.

17.

15. **Information Technology** In AAC audio format, one song measured 3:13 (3 minutes 13 seconds or 193 seconds) and was 3.0 megabytes (MB) in size. A second song was 4:53 (293 seconds) long and took 4.2 MB.

(a) Construct a linear function modeling the size of a song as a function of length (round to three decimal places).

(b) Give the slope of the function. Interpret the slope in the context of this application.

(c) How much space would be required to store a 6:22 song (one decimal place)?

(d) How long would a song be if it required 4.6 MB (to the nearest second)?

16. **Social Science** A driver used 10.3 gal of gas driving 327 mi. The same driver drove 152 mi and used 5.4 gal.

(a) Construct a linear function modeling the gas used as a function of miles driven (round to three decimal places).

(b) Give the slope of the function. Interpret the slope in the context of this application.

(c) How much gas would be required to drive 225 mi (one decimal place)?

(d) How far can the driver go on 12 gal of gas (to the nearest mile)?

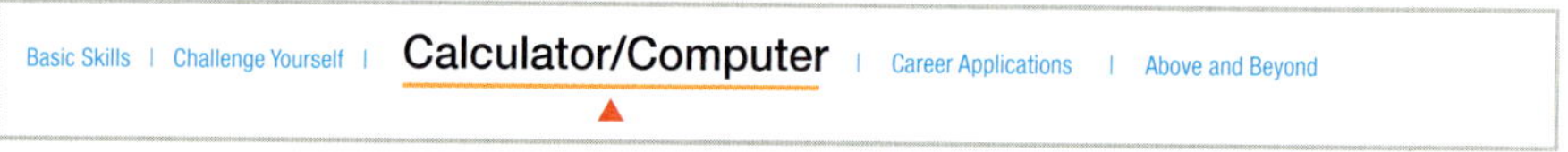

< Objective 4 >

17. **Science and Medicine** The table below gives the age (in months) and weight (in pounds) for a set of 10 girls.

Age	24	24	25	26	26	28	32	32	33	36
Weight	23	29	28	32	30	26	27	35	33	35

Source: Adapted from U.S. Center for Disease Control and Prevention data.

(a) Use a graphing calculator to create a scatter plot, perform a regression analysis, and graph the best-fit linear model on the scatter plot.

(b) Write the equation of the line-of-best-fit for the data, the linear regression model (round to the nearest tenth).

(c) What is the slope? Interpret the slope in the context of this application.

18. **Social Science** The table below gives the 2005 population (in millions) and CO_2 emissions (in teragrams) for selected nations.

Nation	Population	Emissions
Croatia	4	23
Greece	11	110
Ireland	4	46
Japan	128	1,288
Netherlands	16	181
Portugal	11	66
Russian Federation	143	1,698
Spain	44	352
Turkey	72	242
United States	296	6,064

Source: Statistics Division; United Nations (DYB 2005).

(a) Use a graphing calculator to create a scatter plot, perform a regression analysis, and graph the best-fit linear model on the scatter plot.

(b) Write the equation of the line-of-best-fit for the data, the linear regression model (round to the nearest tenth).

(c) What is the slope? Interpret the slope in the context of this application.

(d) How does the slope compare to that found in Example 5? Provide a reason for this discrepancy.

19. **Statistics** A brief review of ten syndicated news columns showed the number of words and the number of characters (including punctuation but not spaces) in the fourth paragraph of each column.

Words	53	90	52	27	22	49	25	44	87	98
Characters	281	510	324	142	119	233	128	225	435	417

(a) Use a graphing calculator to create a scatter plot, perform a regression analysis, and graph the best-fit linear model on the scatter plot.

(b) Write the equation of the line-of-best-fit for the data, the linear regression model (round to the nearest tenth).

(c) What is the slope? Interpret the slope in the context of this application.

(d) How many characters would you expect if a paragraph had 75 words?

(e) If a paragraph required 200 characters, how many words would you expect it to have?

Answers

18. ______

19. ______

Answers

20. ____

21. ____

22. ____

20. Science and Medicine Each of the Great Lakes contains many islands. The table below compares the number of islands in each lake to the total area of the lake's islands (in thousands of acres).

Lake	Islands	Area
Superior	41	390
Michigan	21	96
Huron	66	979
Erie	7	25
Ontario	16	82

Source: U.S. National Oceanic and Atmospheric Administration (1980).

(a) Use a graphing calculator to create a scatter plot, perform a regression analysis, and graph the best-fit linear model on the scatter plot.

(b) Write the equation of the line-of-best-fit for the data, the linear regression model (round to the nearest tenth).

(c) What is the slope? Interpret the slope in the context of this application.

(d) How much area would you expect 30 islands to require in a lake similar to a Great Lake?

(e) In a lake similar to a Great Lake, if islands made up 500,000 acres, how many islands would you expect?

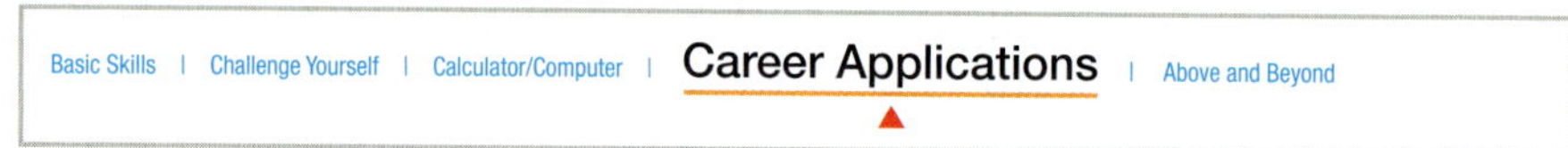

21. Allied Health Dimercaprol (BAL) is used to treat arsenic poisoning in mammals. The recommended dose is 4 mg per kg of the animal's weight.

(a) Construct a linear function describing the relationship between the recommended dose and the animal's weight.

(b) How much BAL must be administered to a 5-kg cat?

(c) What size cow requires a 1,450-mg dose of BAL?

22. Allied Health Yohimbine is used to reverse the effects of xylazine in deer. The recommended dose is 0.125 mg per kg of the deer's weight.

(a) Express the recommended dosage as a linear function of a deer's weight.

(b) How much yohimbine should be administered to a 15-kg fawn?

(c) What size deer requires a 5.0-mg dose of yohimbine?

23. **Allied Health** An abdominal tumor originally weighed 32 g. Every day, chemotherapy treatment reduces the size of the tumor by 2.33 g.

 (a) Express the size of the tumor as a linear function of the number of days spent in chemotherapy.

 (b) How much does the tumor weigh after 5 days of treatment?

 (c) How many days of chemotherapy are required to eliminate the tumor?

24. **Allied Health** A brain tumor originally weighs 41 g. Every day of chemotherapy treatment reduces the size of the tumor by 0.83 g.

 (a) Express the size of the tumor as a linear function of the number of days spent in chemotherapy.

 (b) How much does the tumor weigh after 2 weeks of treatment?

 (c) How many days of chemotherapy are required to eliminate the tumor?

25. **Mechanical Engineering** The input force required to lift an object with a two-pulley system is equal to one-half the object's weight plus eight pounds (to overcome friction).

 (a) Express the input force required as a linear function of an object's weight.

 (b) Report the input force required to lift a 300-lb object.

 (c) How much weight can be lifted with an input force of 650 lb?

26. **Mechanical Engineering** The pitch of a 6-in. gear is given by the number of teeth the gear has divided by six.

 (a) Express the pitch as a linear function of the number of teeth.

 (b) Report the pitch of a 6-in. gear with 30 teeth.

 (c) How many teeth does a 6-in. gear have if its pitch is 8?

27. **Mechanical Engineering** The working depth of a gear (in inches) is given by 2.157 divided by the pitch of the gear.

 (a) Express the depth of a gear as a function of its pitch. (This function is not linear.)

 (b) What is the working depth of a gear that has a pitch of 3.5 (round your result to the nearest hundredth of an inch)?

28. **Mechanical Engineering** Use exercises 26 and 27 to determine the working depth of a 6-in. gear with 42 teeth (to the nearest hundredth of an inch).

Electronics A temperature sensor outputs voltage at a certain temperature. The output voltage varies linearly with respect to temperature. For a particular sensor, the function describing the voltage output V for a given Celsius temperature x is given by

$$V(x) = 0.28x + 2.2$$

29. Determine the output voltage if $x = 0°C$.

30. Evaluate $V(22°C)$.

31. Determine the temperature if the sensor puts out 7.8 V.

Answers

23. ____________

24. ____________

25. ____________

26. ____________

27. ____________

28. ____________

29. ____________

30. ____________

31. ____________

Answers

32. ______

33. ______

34. ______

35. ______

36. ______

37. ______

38. ______

39. ______

Basic Skills | Challenge Yourself | Calculator/Computer | Career Applications | **Above and Beyond**

Extrapolation and Interpolation In Exercise 13, you modeled a kitten's weight (in pounds) based on its age (in months).

$W(x) = x + 1$

This model gives the weight of a kitten as 1 more than its age.

32. How much should a 7-month-old kitten weigh?

33. According to the model, how much should the kitten weigh when it is 5 years old (60 months)?

34. Write a paragraph giving your interpretations of the answers to exercises 32 and 33.

The extrapolation problem, above, has difficulty with making predictions based on data-derived models. Every model should be accompanied by a *domain* stating the input values for which the model is valid. Making predictions outside the given data is called *extrapolation*.

For instance, in the kitten model, the domain might be $3 \le x \le 12$, which means that the model could be used on kittens at least 3 months old but not older than a year. This would make sense because as they become cats, their growth rates (and weight gain) slows.

On the other hand, exercise 32 asks you to *interpolate*. This means that you are making a prediction based on an input (7 months) that is between the extremes of your data. That is, your data points were for a 3-month-old and an 8-month-old kitten.

We can usually extrapolate near to our data. For instance, it might be safe to predict the weight of a 10-month-old kitten, but a 5-year-old cat will not weigh 61 pounds!

In exercise 17, you modeled a young girl's weight as a function of her age, based on 10 girls between 24 months old and 36 months old.

$W(x) = 0.6x + 12.9$

35. According to the model, how much should a 32-month-old girl weigh?

36. According to the model, how much should a 40-month-old girl weigh?

37. According to the model, how much should a 50-year-old (600 months) woman weigh?

38. Which of the predictions above are interpolations and which are extrapolations?

39. Write a paragraph interpreting your predictions in exercises 35–37.

Answers

1. $C(x) = 1.58x$ **3.** $P(s) = 4s$ **5.** \$11.06 **7.** 72 cm **9.** 7.6 lb **11.** 10.5 ft **13. (a)** $W(x) = x + 1$; **(b)** 1; the kitten gains one pound per month. **15. (a)** $f(x) = 0.012x + 0.684$; **(b)** 0.012; each second requires 0.012 MB of space. **(c)** 5.3 MB; **(d)** 5.26 or 326 s

17. (a)

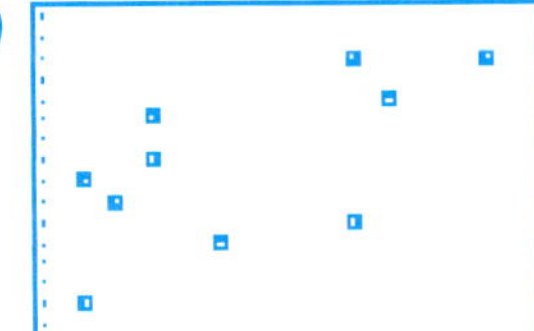

LinReg
y=ax+b
a=.5901442308
b=12.921875
r²=.4092666911
r=.6397395494

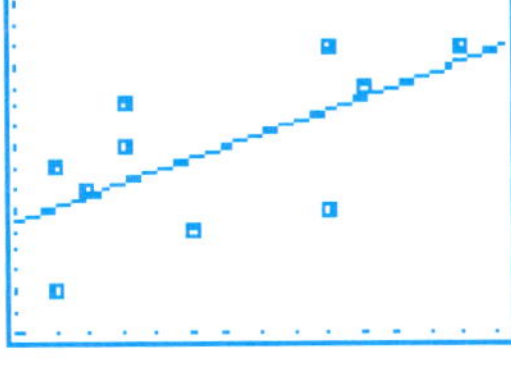

(b) $y = 0.6x + 12.9$; **(c)** 0.6; a young girl's weight increases about 0.6 lb for every month she ages.

19. (a)

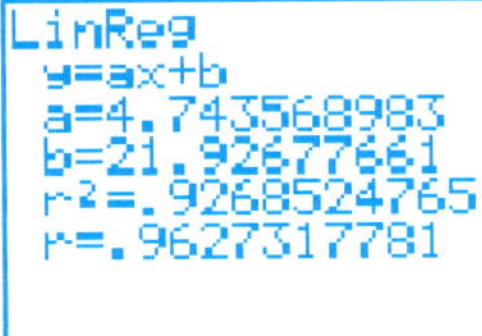

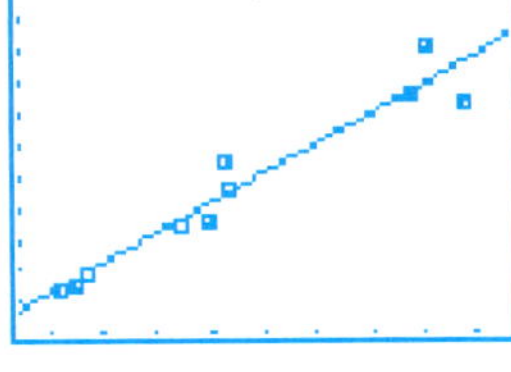

(b) $y = 4.7x + 21.9$; **(c)** 4.7; each additional word leads to approximately 4.7 additional characters in a paragraph; **(d)** 374.4 characters; **(e)** 37.9 words
21. (a) $d(x) = 4x$; **(b)** 20 mg; **(c)** 362.5 kg
23. (a) $W(x) = -2.33x + 32$; **(b)** 20.35 g; **(c)** 14 days
25. (a) $F(x) = \frac{1}{2}x + 8$; **(b)** 158 lb; **(c)** 1,284 lb
27. (a) $D(p) = \frac{2.157}{p}$; **(b)** 0.62 in. **29.** 2.2 V **31.** 20°C **33.** 61 lb
35. 32.1 lb **37.** 372.9 lb **39.** Above and Beyond

3.5 Graphing Linear Inequalities in Two Variables

< 3.5 Objectives >

1 > Graph a linear inequality in two variables

2 > Graph a region defined by linear inequalities

What does the solution set look like when we have an inequality in two variables? We will see that it is a set of ordered pairs best represented by a shaded region. The general form for a linear inequality in two variables is

$$Ax + By < C$$

in which A and B cannot both be 0. The symbol $<$ can be replaced with $>$, $\le$, or $\ge$. Some examples are

$$y < -2x + 6 \qquad x - 2y \le 4 \qquad \text{and} \qquad 2x - 3y \ge x + 5y$$

As was the case with an equation, the solution set of a linear inequality is a set of ordered pairs of real numbers. However, the solution set for a linear inequality consists of an entire region in the plane. We call this region a **half-plane.**

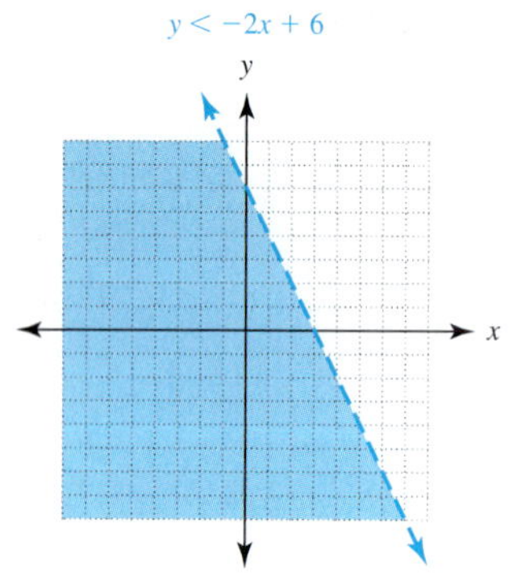

To determine such a solution set, we start with the first inequality listed above. To graph the solution set of

$$y < -2x + 6$$

we begin by writing the corresponding linear equation

$$y = -2x + 6$$

Note that the graph of $y = -2x + 6$ is simply a straight line.

To graph the solution set of $y < -2x + 6$, we must include all ordered pairs that satisfy that inequality. For instance if $x = 1$, we have

$$y < -2(1) + 6$$

$$y < 4$$

So we want to include all points of the form $(1, y)$, where $y < 4$. Of course, since $(1, 4)$ is *on* the corresponding line, this means that we want all points *below* the line along the vertical line $x = 1$. The result is similar for any choice of x, and our solution set contains all of the points below the line $y = -2x + 6$. We can graph the solution set as the shaded region shown. We have the following definition.

NOTES

The line is dashed to indicate that points on the line are *not* included.

We call the graph of the equation

$Ax + By = C$

the **boundary line** of the half-planes.

Definition

Solution Set of an Inequality

In general, the solution set of an inequality of the form

$$Ax + By < C \qquad \text{or} \qquad Ax + By > C$$

can be represented by a half-plane either above or below the corresponding line determined by

$$Ax + By = C$$

How do we decide which half-plane represents the desired solution set? The use of a **test point** provides an easy answer. Choose any point *not* on the line. Then substitute the coordinates of that point into the given inequality. If the coordinates satisfy the inequality (result in a true statement), then shade the region or half-plane that includes the test point; if not, shade the opposite half-plane. Example 1 illustrates the process.

Example 1 Graphing a Linear Inequality

< Objective 1 >

RECALL

The graph of $x - 2y = 4$ is shown below.

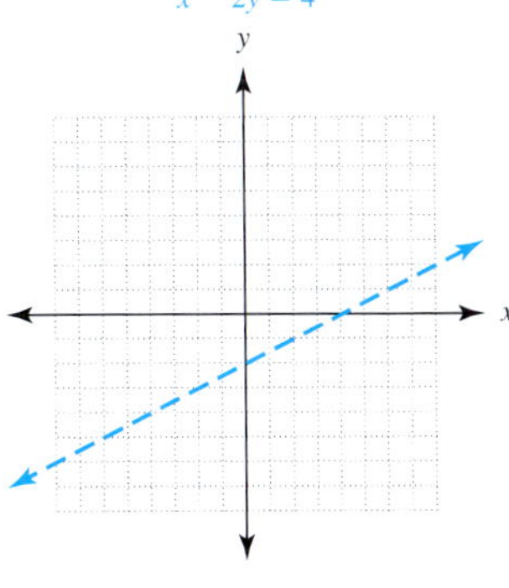

NOTE

Because we have a strict inequality, $x - 2y < 4$, the boundary line does not include solutions. In this case, we use a dashed line.

Graph the linear inequality

$$x - 2y < 4$$

First, we graph the corresponding equation

$$x - 2y = 4$$

to find the boundary line. To determine which half-plane is part of the solution set, we need a test point *not* on the line. As long as the line *does not pass through the origin,* we can use (0, 0) as a test point. It provides the easiest computation.

Here letting $x = 0$ and $y = 0$, we have

$$(0) - 2(0) \overset{?}{<} 4$$
$$0 < 4$$

Because this is a true statement, we shade the half-plane that includes the origin (the test point), as shown.

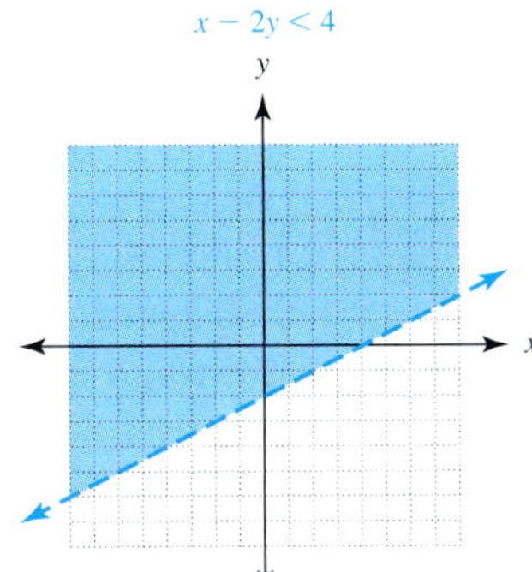

Check Yourself 1

Graph the solution set of $3x + 4y > 12$.

The graphs of some linear inequalities include the boundary line. That is the case whenever equality is included with the inequality statement, as illustrated in Example 2.

Example 2 Graphing a Linear Inequality

Graph the inequality

$$2x + 3y \geq 6$$

First, we graph the boundary line, here corresponding to $2x + 3y = 6$. This time we use a solid line because equality is included in the original statement.

Again, we choose a convenient test point not on the line. As before, the origin provides the simplest computation.

Substituting $x = 0$ and $y = 0$, we have

$$2(0) + 3(0) \overset{?}{\geq} 6$$
$$0 \geq 6$$

NOTE

A solid boundary line means that points on the line are solutions.

This occurs when the inequality symbol is either $\geq$ or $\leq$.

This is a *false* statement. Hence, the graph consists of all points on the *opposite* side of the origin. The graph is the upper half-plane shown.

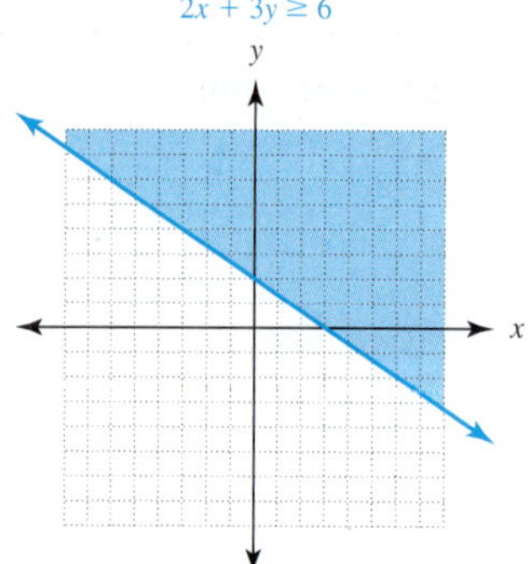

Check Yourself 2

Graph the solution set of $x - 3y \leq 6$.

Example 3 Graphing a Linear Inequality

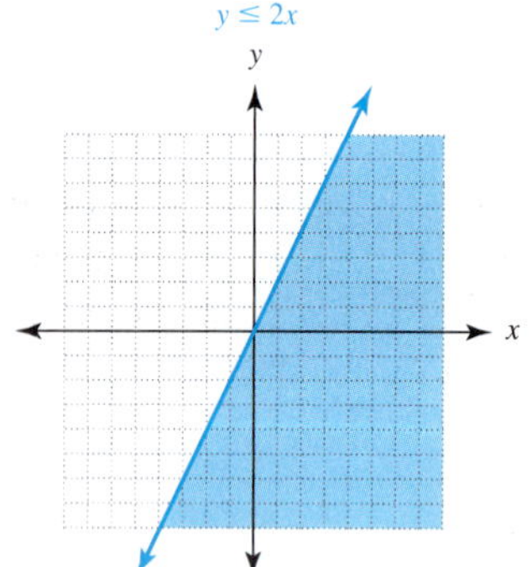

Graph the solution set of

$$y \leq 2x$$

We proceed as before by graphing the boundary line (it is solid since equality is included). The only difference between this and previous examples is that we *cannot use the origin* as a test point. Do you see why?

Choosing (1, 1) as our test point gives the statement

$$(1) \overset{?}{\leq} 2(1)$$

$$1 \leq 2$$

Because the statement is *true,* we shade the half-plane that *includes* the test point (1, 1).

NOTE

The choice of (1, 1) is arbitrary. We simply want *any* point *not* on the line.

Check Yourself 3

Graph the solution set of $3x + y > 0$.

We now consider a special case of graphing linear inequalities in the rectangular coordinate system.

Example 4 Graphing a Linear Inequality

NOTE

Here we specify the rectangular coordinate system to indicate we want a two-dimensional graph.

Graph the solution set of $x > 3$ in the rectangular coordinate system.

First, we draw the boundary line (a dashed line because equality is not included) corresponding to

$$x = 3$$

We can choose the origin as a test point in this case. It results in the false statement

$$0 > 3$$

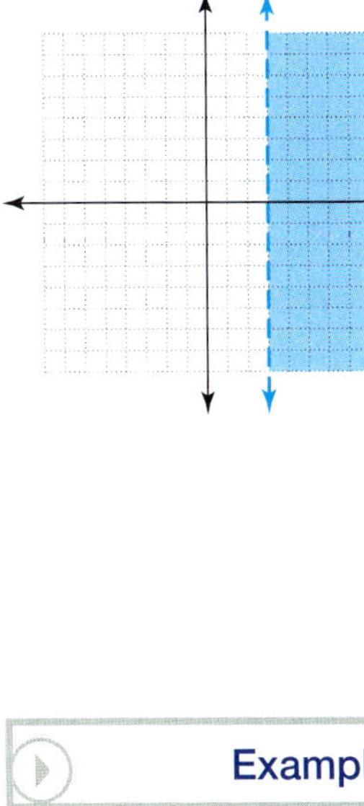

We then shade the half-plane *not* including the origin. In this case, the solution set is represented by the half-plane to the right of the vertical boundary line.

As you may have observed, in this special case choosing a test point is not really necessary. Because we want values of x that are *greater than* 3, we want those ordered pairs that are to the *right* of the boundary line.

Check Yourself 4

Graph the solution set of

$$y \le 2$$

in the rectangular coordinate system.

Applications of linear inequalities often involve more than one inequality condition. Consider Example 5.

Example 5 Graphing a Region Defined by Linear Inequalities

< Objective 2 >

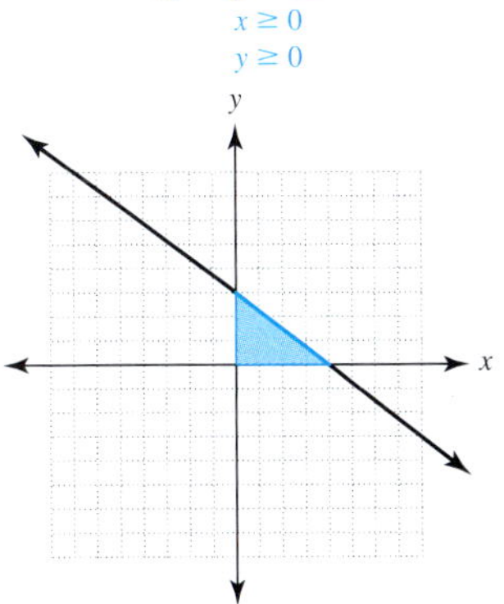

Graph the region satisfying the conditions.

$$3x + 4y \le 12$$
$$x \ge 0$$
$$y \ge 0$$

The solution set in this case must satisfy *all three conditions.* As before, the solution set of the first inequality is graphed as the half-plane *below* the boundary line. The second and third inequalities mean that x and y must also be nonnegative. Therefore, our solution set is restricted to the first quadrant (and the appropriate segments of the x- and y-axes), as shown.

Check Yourself 5

Graph the region satisfying the conditions.

$$3x + 4y < 12$$
$$x \ge 0$$
$$y \ge 0$$

Here is an algorithm summarizing our work in graphing linear inequalities in two variables.

Step by Step

To Graph a Linear Inequality

Step 1 Replace the inequality symbol with an equality symbol to form the equation of the boundary line of the solution set.

Step 2 Graph the boundary line. Use a dashed line if equality is not included ($<$ or $>$). Use a solid line if equality is included ($\le$ or $\ge$).

Step 3 Choose any convenient test point *not* on the boundary line.

Step 4 If the inequality is *true* for the test point, shade the half-plane that *includes* the test point. If the inequality is *false* for the test point, shade the half-plane that does *not include* the test point.

Check Yourself ANSWERS

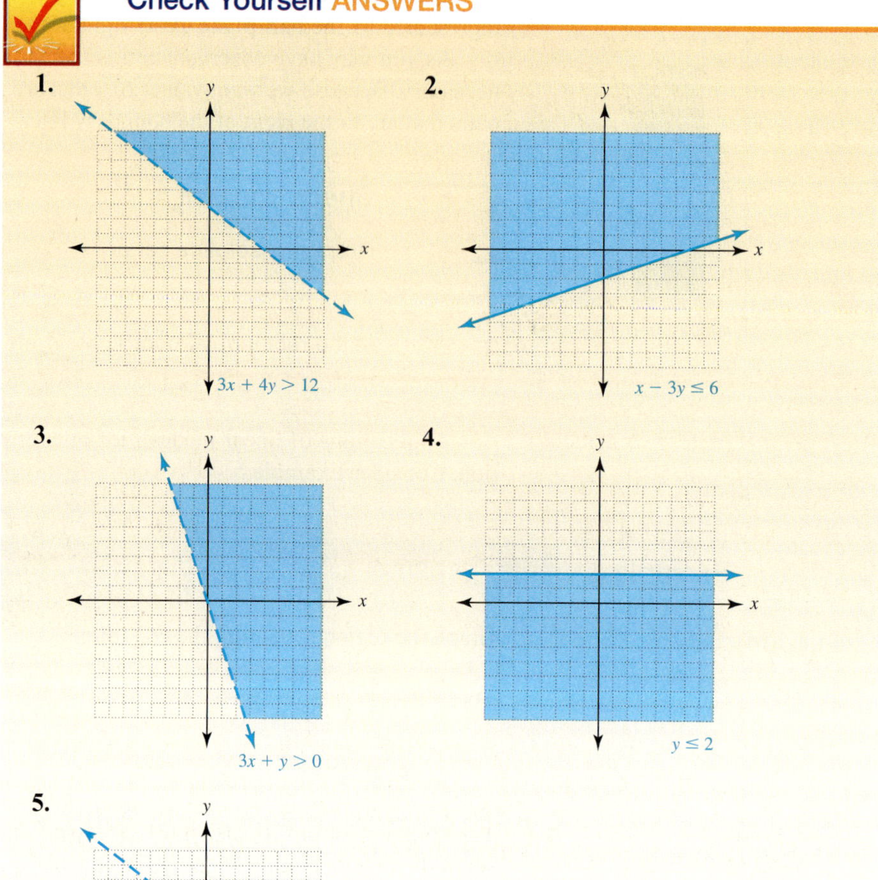

5.

y

x

$3x + 4y < 12$
$x \ge 0$
$y \ge 0$

Reading Your Text

SECTION 3.5

(a) In the case of linear inequalities, the solution set consists of all the points in an entire region of the plane, called a _______________.

(b) To decide which region represents the solution set for an inequality, we use a _______________ point.

(c) A _______________ boundary line means that the points on the line are solutions to the inequality.

(d) If the inequality is _______________ for the test point, shade the half-plane that does *not* include the test point.

Basic Skills | Challenge Yourself | Calculator/Computer | Career Applications | Above and Beyond

3.5 exercises

< Objective 1 >

Graph the solution set of each linear inequality.

1. $x + y < 4$ 

2. $x + y \geq 6$

3. $x - y \geq 3$

4. $x - y < 5$

5. $y \geq 2x + 1$

6. $y < 3x - 4$

7. $2x + 3y < 6$ 

8. $3x - 4y \geq 12$

9. $x - 4y > 8$

10. $2x + 5y \leq 10$

11. $y \geq 3x$

12. $y \leq -2x$

13. $x - 2y > 0$

14. $x + 4y \leq 0$

Name ____________

Section ________ Date ________

Answers

1. ____________
2. ____________
3. ____________
4. ____________
5. ____________
6. ____________
7. ____________
8. ____________
9. ____________
10. ____________
11. ____________
12. ____________
13. ____________
14. ____________

Answers

15. ______
16. ______
17. ______
18. ______
19. ______
20. ______
21. ______
22. ______
23. ______
24. ______
25. ______
26. ______
27. ______
28. ______

15. $x < 3$

16. $y < -2$

17. $y > 3$

18. $x \leq -4$

19. $3x - 6 \leq 0$

20. $-2y > 6$

21. $0 < x < 1$

22. $-2 \leq y \leq 1$

23. $1 \leq x \leq 3$

24. $1 < y < 5$

< Objective 2 >

Graph the region satisfying each set of conditions.

25. $0 \leq x \leq 3$
$2 \leq y \leq 4$

26. $1 \leq x \leq 5$
$0 \leq y \leq 3$

27. $x + 2y \leq 4$
$x \geq 0$
$y \geq 0$

28. $2x + 3y \leq 6$
$x \geq 0$
$y \geq 0$

Basic Skills | **Challenge Yourself** | Calculator/Computer | Career Applications | Above and Beyond

Determine whether each statement is **True** *or* **False.**

29. If a test point satisfies a linear inequality, then we shade the half-plane that contains the test point.

30. A dashed boundary line means that the points on that line are solutions for the inequality.

Complete each statement with **never, sometimes,** *or* **always.**

31. When graphing a linear inequality, there is ____________ a straight-line boundary.

32. When graphing a linear inequality, the point (0, 0) is ______________ on the boundary line.

33. **BUSINESS AND FINANCE** A manufacturer produces a standard model and a deluxe model of a 13-in. television set. The standard model requires 12 h to produce, while the deluxe model requires 18 h. The labor available is limited to 360 h per week.

If x represents the number of standard model sets produced per week and y represents the number of deluxe models, draw a graph of the region representing the feasible values for x and y. Keep in mind that the values for x and y must be nonnegative since they represent a quantity of items. (This will be the solution set for the system of inequalities.)

34. **BUSINESS AND FINANCE** A manufacturer produces standard record turntables and CD players. The turntables require 10 h of labor to produce while CD players require 20 h. Let x represent the number of turntables produced and y the number of CD players.

If the labor hours available are limited to 300 h per week, graph the region representing the feasible values for x and y.

35. **BUSINESS AND FINANCE** A hospital food service department can serve at most 1,000 meals per day. Patients on a normal diet receive 3 meals per day, and patients on a special diet receive 4 meals per day. Write a linear inequality that describes the number of patients that can be served per day and draw its graph.

36. **BUSINESS AND FINANCE** The movie and TV critic for the local radio station spends 3 to 7 h daily reviewing movies and less than 4 h reviewing TV shows. Let x represent the time (in hours) watching movies and y represent the time spent watching TV. Write two inequalities that model the situation, and graph their intersection.

Answers

29. ______

30. ______

31. ______

32. ______

33. ______

34. ______

35. ______

36. ______

Answers

37. ______

38. ______

39. ______

40. ______

41. ______

42. ______

Write an inequality for the shaded region shown in each figure.

37.

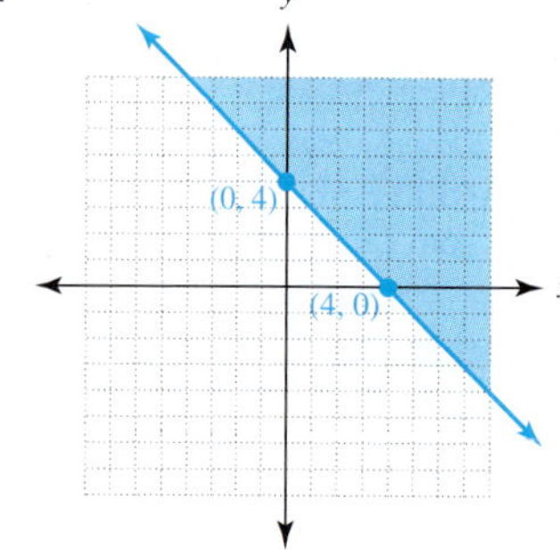

38.

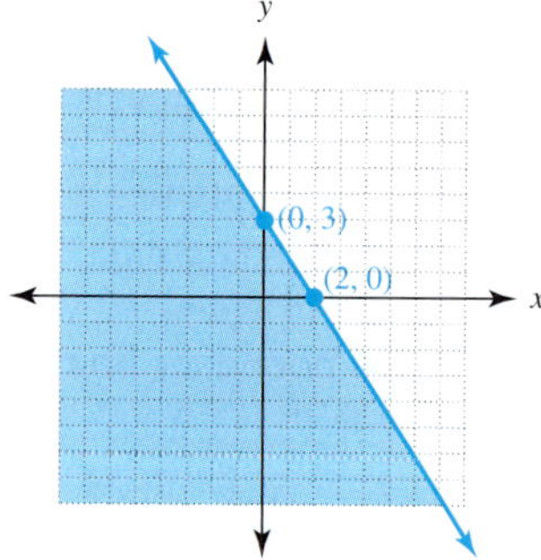

39.

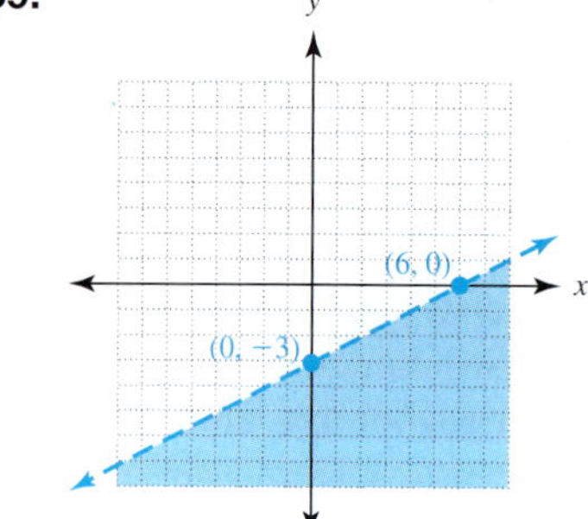

40.

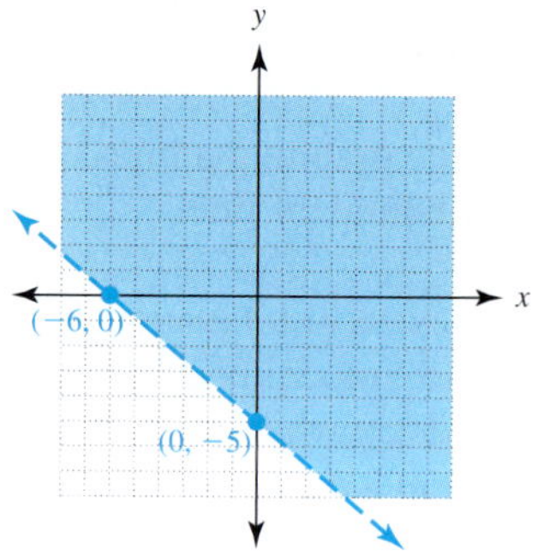

Basic Skills | Challenge Yourself | Calculator/Computer | **Career Applications** | Above and Beyond

41. Manufacturing Technology A manufacturer produces two-slice toasters and four-slice toasters. The two-slice toasters require 8 hours to produce, and the four-slice toasters require 10 hours to produce. The manufacturer has 400 hours of labor available each week.

(a) Write a linear inequality to represent the number of each type of toaster the manufacturer can produce in a week (use x for the two-slice toasters and y for the four-slice toasters).

(b) Graph the inequality (in the first quadrant).

(c) Is it feasible to produce 20 two-slice toasters and 30 four-slice toasters in the same week?

> Videos

42. Manufacturing Technology A certain company produces standard clock radios and deluxe clock radios. It costs the company \$15 to produce each standard clock radio and \$20 to produce each deluxe model. The company's budget limits production costs to \$3,000 per day.

(a) Write a linear inequality to represent the number of each type of clock radio that the company can produce in a day (use x for the standard model and y for the deluxe model).

(b) Graph the inequality (in the first quadrant).

(c) Is it feasible to produce 80 of each type of clock radio in the same day?

Basic Skills | Challenge Yourself | Calculator/Computer | Career Applications | **Above and Beyond**

43. Assume that you are working only with the variable x. Describe the set of solutions for the statement $x > -1$.

44. Now, assume that you are working in two variables x and y. Describe the set of solutions for the statement $x > -1$.

Answers

1.

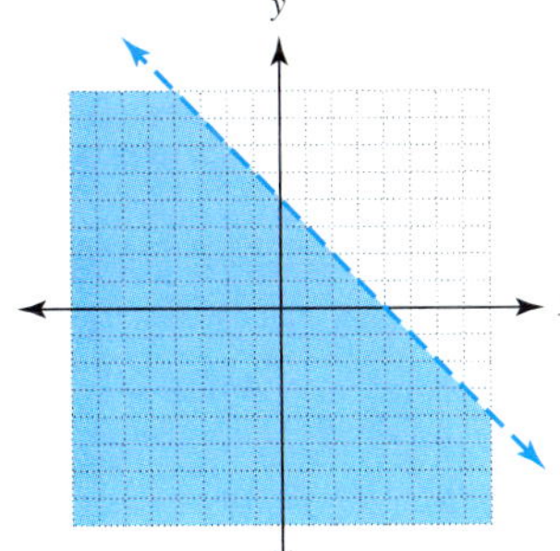

3.

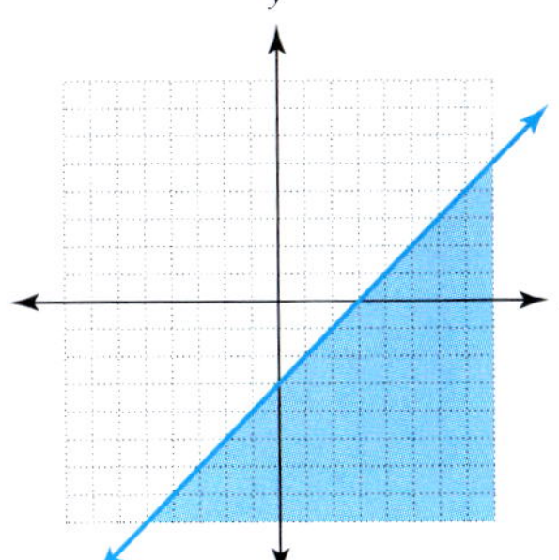

5.

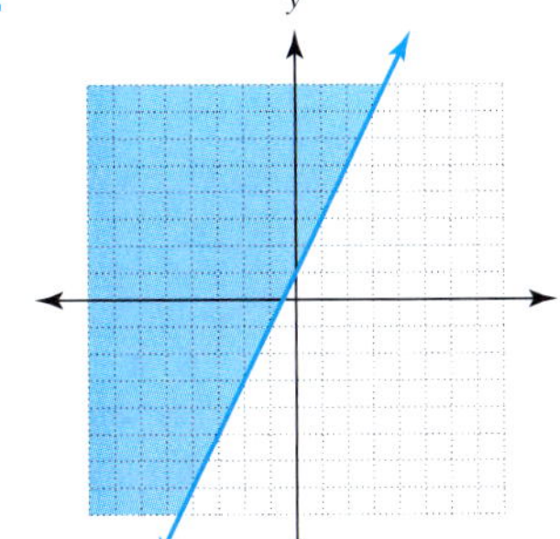

7.

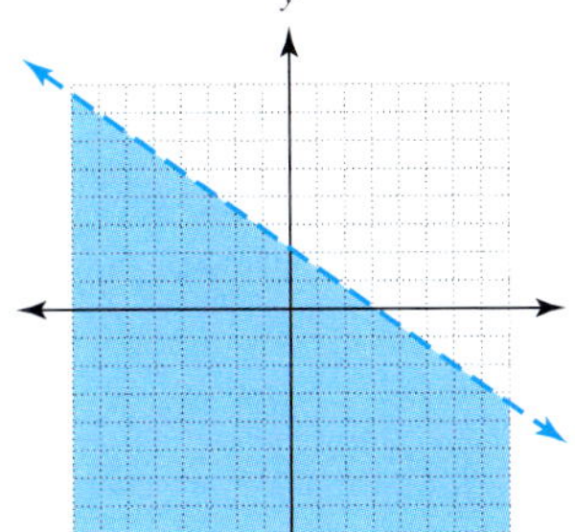

9.

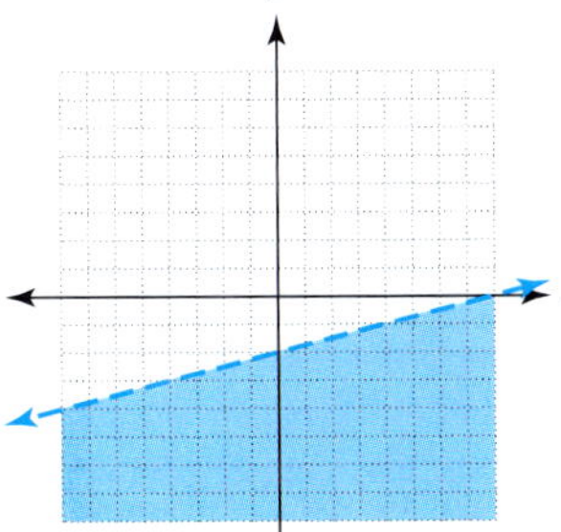

11.

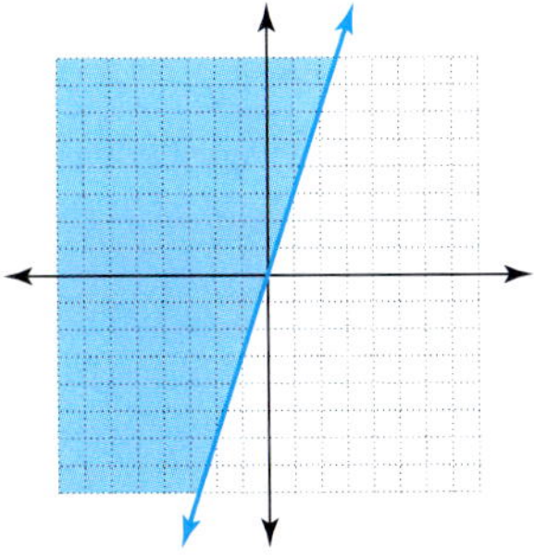

13.

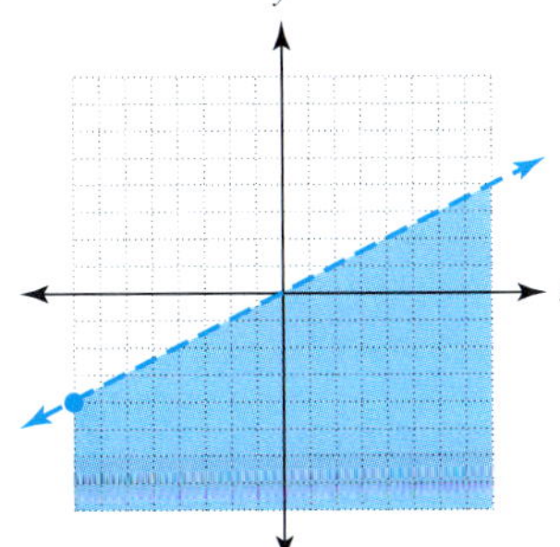

15.

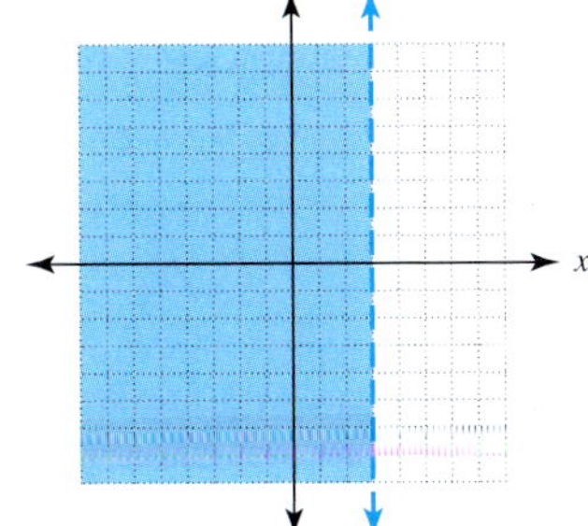

Answers

43. ____________

44. ____________

17.

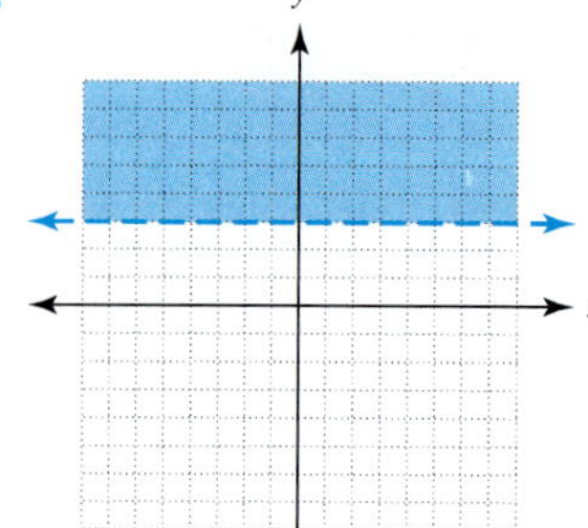

19.

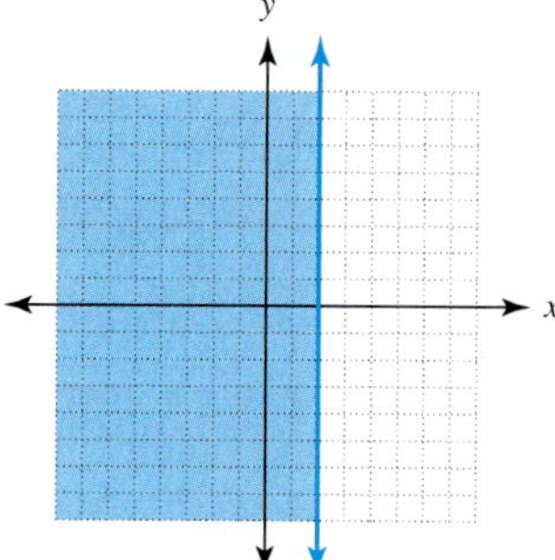

21.

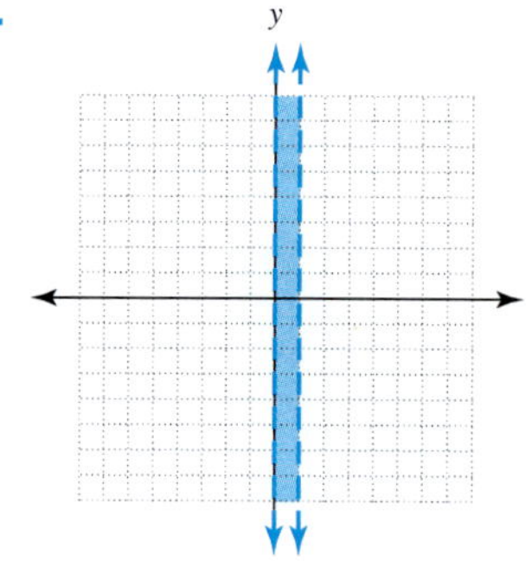

23.

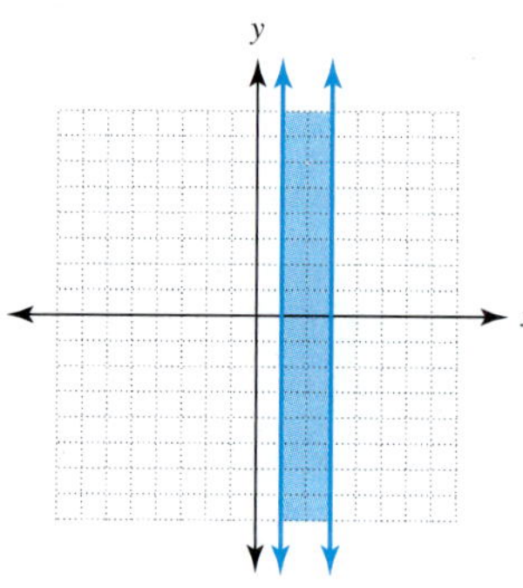

25.

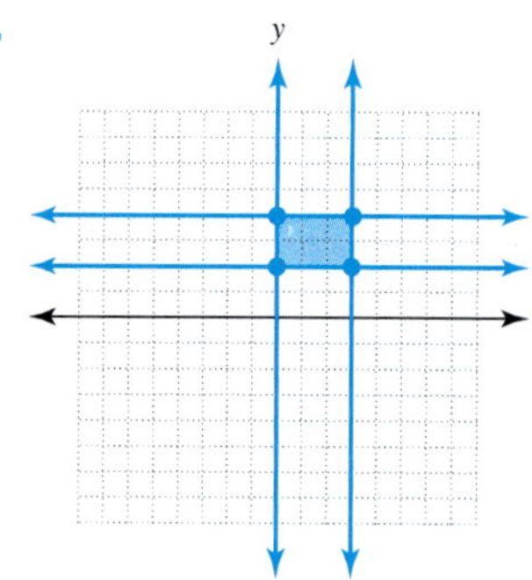

27.

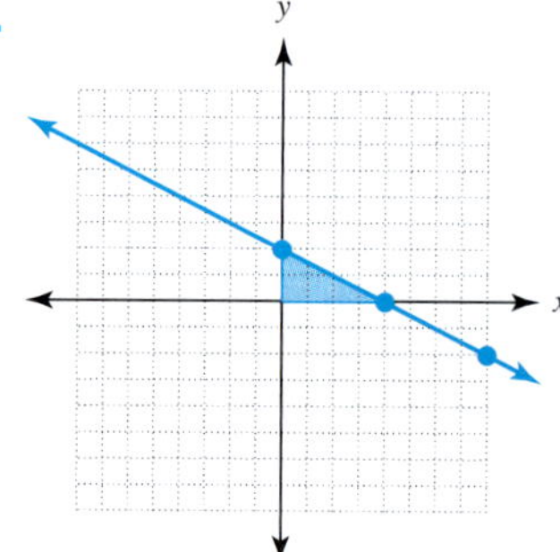

29. True **31.** always

33.

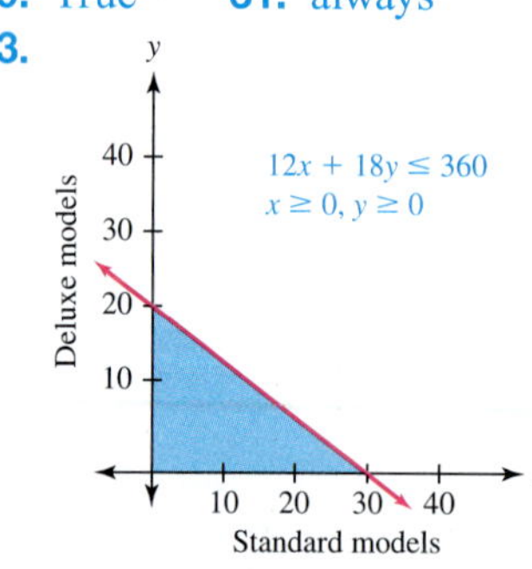

35.

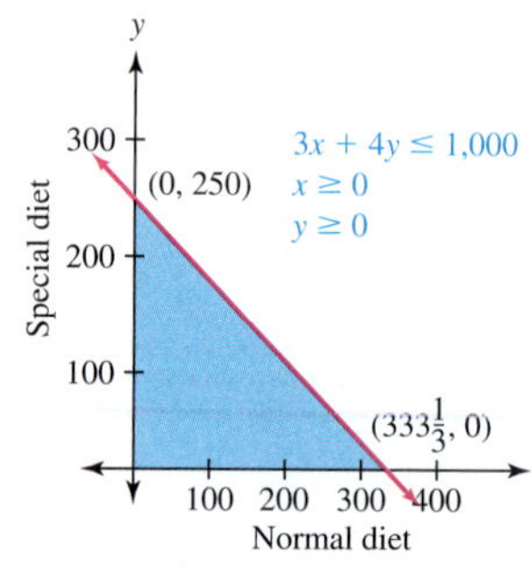

37. $y \geq -x + 4$ **39.** $y < \frac{1}{2}x - 3$ **41.** **(a)** $8x + 10y \leq 400$;

(b)

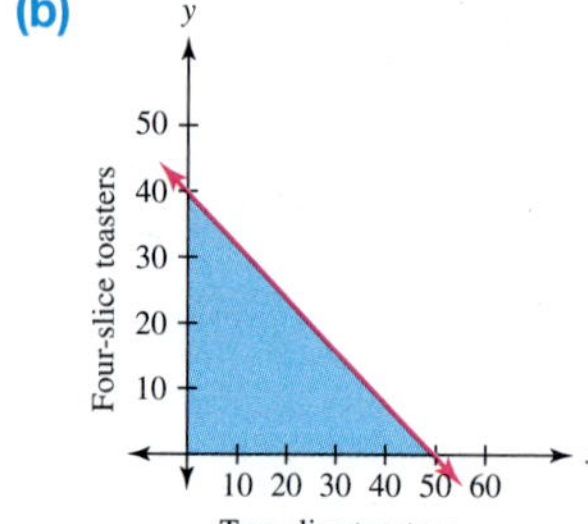

(c) No **43.** Above and Beyond

Definition/Procedure	Example	Reference

Graphing Linear Functions

Section 3.1

Linear Equation An equation that can be written in the form

$$Ax + By = C$$

in which A and B are not both 0.

Example: $2x - 3y = 4$ is a linear equation.

Reference: p. 287

Graphing Linear Equations

Step 1 Find at least three solutions for the equation, and put your results in tabular form.

Step 2 Graph the solutions found in step 1.

Step 3 Draw a straight line through the points determined in step 2 to form the graph of the equation.

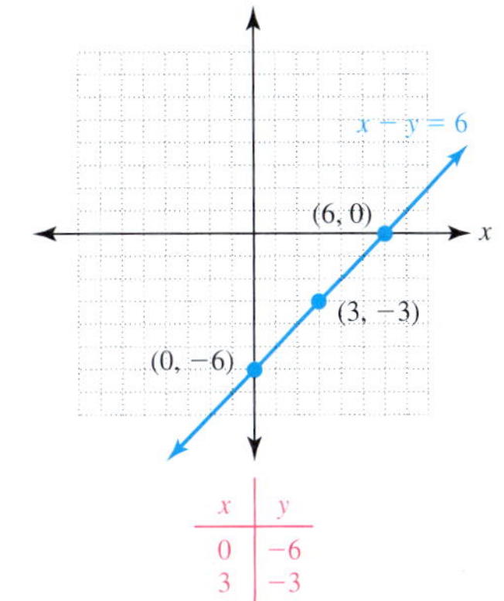

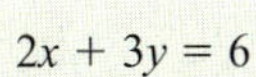

x	y
0	−6
3	−3
6	0

Reference: p. 287

Writing Linear Equations as Functions

Step 1 Solve the equation for the dependent variable y.

Step 2 Replace y with $f(x)$.

Example: $2x + 3y = 6$

Step 1 $3y = -2x + 6$

$$y = -\frac{2}{3}x + 2$$

Step 2 $f(x) = -\frac{2}{3}x + 2$

Reference: p. 296

The Slope of a Line

Section 3.2

Slope The slope of a line gives a numerical measure of the steepness of the line. The slope m of a line containing the distinct points in the plane $P(x_1, y_1)$ and $Q(x_2, y_2)$ is given by

$$m = \frac{y_2 - y_1}{x_2 - x_1} \quad \text{where } x_2 \neq x_1.$$

Example: To find the slope of the line through $(-2, -3)$ and $(4, 6)$,

$$m = \frac{(6) - (-3)}{(4) - (-2)}$$

$$= \frac{6 + 3}{4 + 2}$$

$$= \frac{9}{6} = \frac{3}{2}$$

Reference: p. 319

Continued

Definition/Procedure	Example	Reference

Slopes and Lines

- The slope of a line that rises from left to right is positive.
- The slope of a line that falls from left to right is negative.
- The slope of a horizontal line is 0.
  - The equation of the horizontal line with y-intercept $(0, b)$ is $y = b$.
- The slope of a vertical line is undefined.
  - The equation of the vertical line with x-intercept $(a, 0)$ is $x = a$.

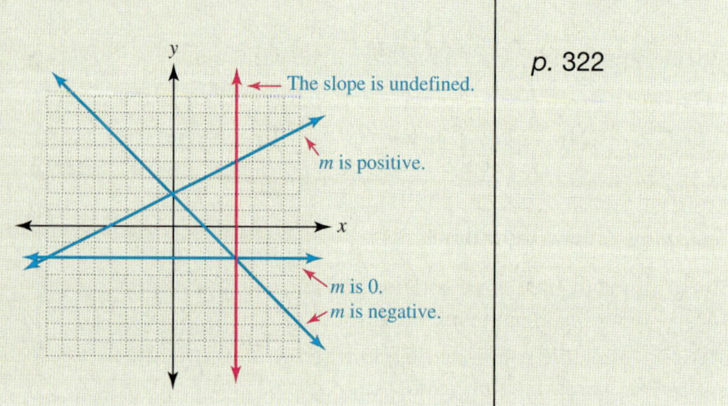

p. 322

The Slope-Intercept Form The slope-intercept form for the equation of a line is

$y = mx + b$

in which the line has slope m and y-intercept $(0, b)$.

Slope-Intercept and Graphing

Step 1 Write the equation of the line in slope-intercept form.

Step 2 Find the slope and y-intercept.

Step 3 Plot the y-intercept.

Step 4 Plot a second point based on the slope of the line.

Step 5 Draw a line through the two points.

For the equation

$y = \frac{2}{3}x - 3$

the slope m is $\frac{2}{3}$ and b, which determines the y-intercept, is -3.

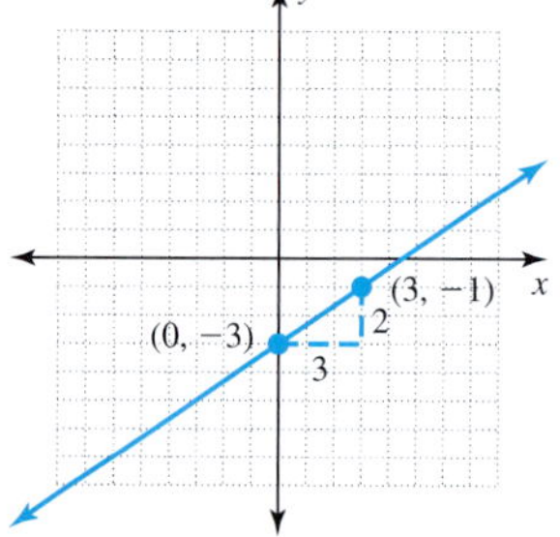

p. 323

Forms of Linear Equations

Section 3.3

Parallel Lines Two lines are **parallel** if and only if they have the same slope, so

$m_1 = m_2$

or both are vertical.

$y = 3x - 5$ and
$y = 3x + 2$ are parallel.

Parallel lines

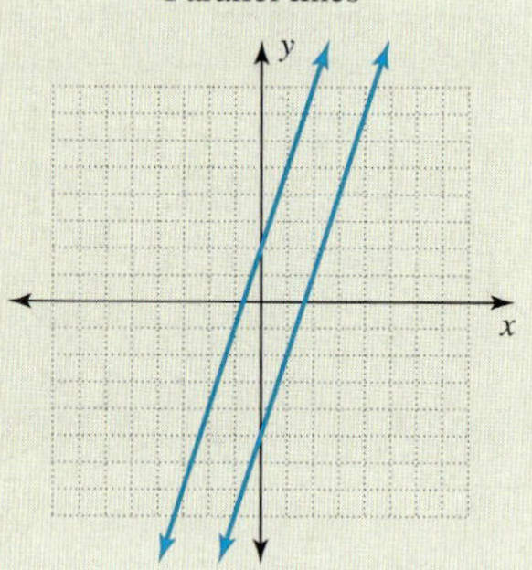

p. 340

Definition/Procedure	Example	Reference
Perpendicular Lines Two lines are **perpendicular** if and only if their slopes are negative reciprocals, that is, when $m_1 \cdot m_2 = -1$ or if one is vertical and the other horizontal.	$y = 5x + 2$ and $y = -\frac{1}{5}x - 3$ are perpendicular. Perpendicular lines 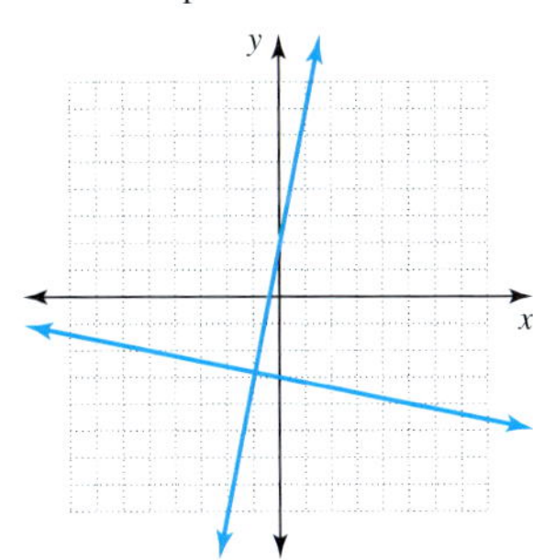	p. 340
The Point-Slope Form The equation of a line with slope m that passes through the point (x_1, y_1) is $y - y_1 = m(x - x_1)$	The line with slope $\frac{1}{3}$ passing through (4, 3) has the equation $y - 3 = \frac{1}{3}(x - 4)$ 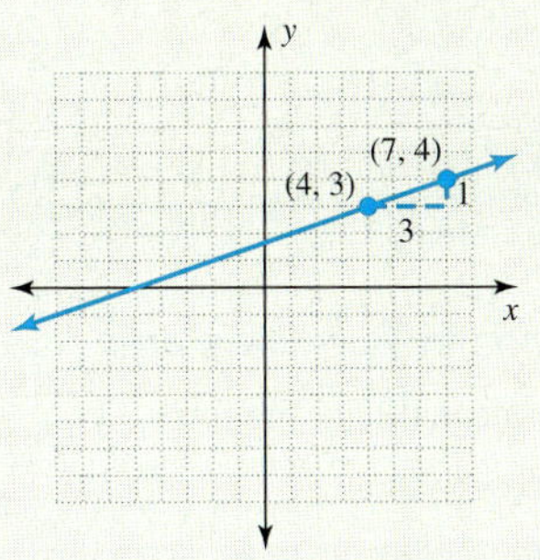	p. 342
Rate of Change and Linear Regression		**Section 3.4**
Rate of Change The rate of change of a linear function is equal to its slope. It represents the change in the output when the input is increased by 1.	Consider the cost model, $C(x) = 12x + 250$ The rate of change of this function is 12, which means that the cost increases by \$12 for each additional unit produced.	p. 357

Continued

Definition/Procedure	Example	Reference

Linear Regression

Step 1 Enter the x- and y-values into your calculator's lists.

Step 2 Create a scatter plot from the data.

Step 3 Perform a regression analysis on the data.

Step 4 Graph the line-of-best-fit on the scatter plot.

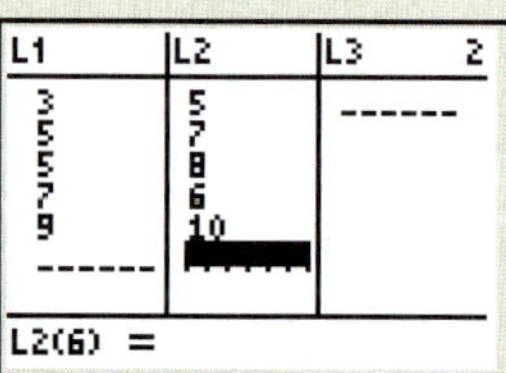
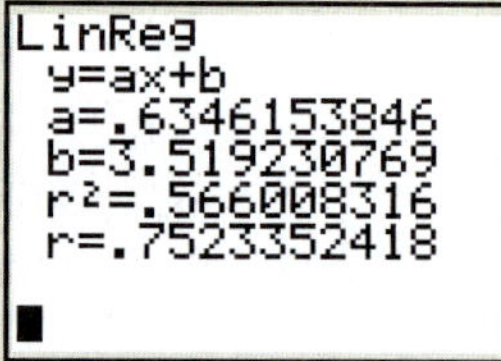
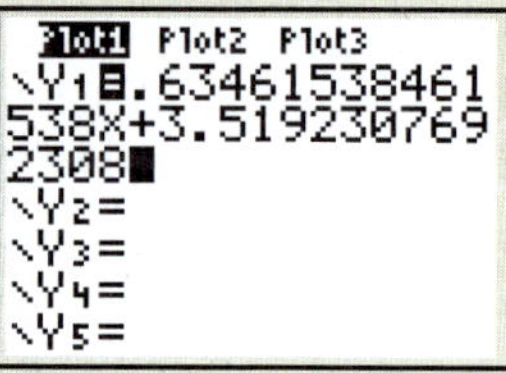
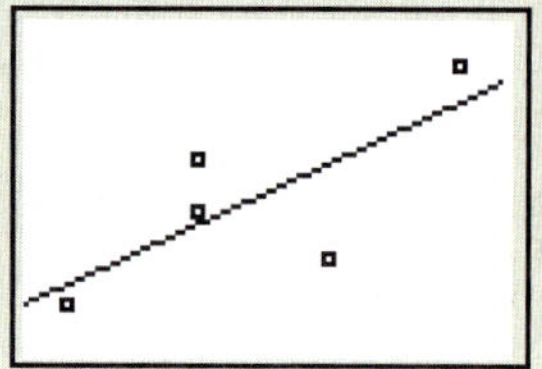

p. 362

Graphing Linear Inequalities in Two Variables

In general, the solution set of an inequality of the form

$Ax + By < C$ or $Ax + By > C$

will be a **half-plane** either above or below the **boundary line** determined by

$Ax + By = C$

The boundary line is included in the graph if equality is included in the statement of the original inequality. Such a line is solid. The boundary line is dashed if it is not included in the graph.

To graph

$x - 2y < 4$

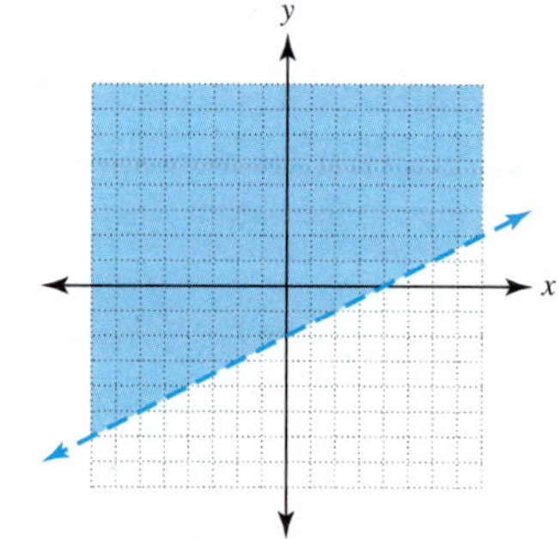

Section 3.5
p. 372

Definition/Procedure	Example	Reference
Graphing Linear Inequalities **Step 1** Replace the inequality symbol with an equality symbol to form the equation of the boundary line of the solution set. **Step 2** Graph the boundary line. Use a dashed line if equality is not included ($<$ or $>$). Use a solid line if equality is included ($\le$ or $\ge$). **Step 3** Choose any convenient test point *not* on the boundary line. **Step 4** If the inequality is *true* for the test point, shade the half-plane *including* the test point. If the inequality is *false* for the test point, shade the half-plane *not including* the test point.		*p.* 375

summary exercises :: chapter 3

This summary exercise set is provided to give you practice with each of the objectives of this chapter. Each exercise is keyed to the appropriate chapter section. When you are finished, you can check your answers to the odd-numbered exercises in the back of the text. If you have difficulty with any of these questions, go back and reread the examples from that section. The answers to the even-numbered exercises appear in the *Instructor's Solutions Manual.* Your instructor will give you guidelines on how best to use these exercises in your instructional setting.

3.1 *Graph each equation.*

1. $x + y = 5$

2. $x - y = 6$

3. $y = 5x$

4. $y = -3x$

5. $y = \frac{3}{2}x$

6. $y = 3x + 2$

7. $y = -2x + 4$

8. $y = -3x + 4$

9. $y = \frac{2}{3}x + 2$

10. $3x - y = 3$

11. $2x + y = 6$

12. $3x + 2y = 12$

13. $3x - 4y = 12$

14. $x = -5$

15. $y = -2$

16. $5x - 3y = 15$

17. $3x + 4y = 12$

18. $2x + y = 6$

19. $3x + 2y = 6$

20. $-4x - 5y = 20$

3.2 *Find the slope of the line through each pair of points.*

21. (3, 4) and (5, 8)

22. (−2, 3) and (1, −6)

23. (−2, 5) and (2, 3)

24. (−5, −2) and (1, 2)

25. (−2, 6) and (5, 6)

26. (−3, 2) and (−1, −3)

27. (−3, −6) and (5, −2)

28. (−6, −2) and (−6, 3)

Find the slope and y-intercept of the line represented by each equation.

29. $y = 2x + 5$

30. $y = -4x - 3$

31. $y = -\frac{3}{4}x$

32. $y = \frac{2}{3}x + 3$

33. $2x + 3y = 6$

34. $5x - 2y = 10$

35. $y = -3$

36. $x = 2$

Write the equation of the line with the given slope and y-intercept.

37. Slope 2, *y*-intercept (0, 3)

38. Slope $\frac{3}{4}$, *y*-intercept (0, −2)

39. Slope $-\frac{2}{3}$, *y*-intercept (0, 2)

3.3 *In exercises 40 to 43, are the pairs of lines* **parallel, perpendicular,** *or* **neither?**

40. L_1 through (−3, −2) and (1, 3)
L_2 through (0, 3) and (4, 8)

41. L_1 through (−4, 1) and (2, −3)
L_2 through (0, −3) and (2, 0)

42. L_1 with equation $x + 2y = 6$
L_2 with equation $x + 3y = 9$

43. L_1 with equation $4x - 6y = 18$
L_2 with equation $2x - 3y = 6$

Write an equation of the line passing through each point with the indicated slope. Give your result in slope-intercept form, where possible.

44. $(0, -5)$, $m = \frac{2}{3}$

45. $(0, -3)$, $m = 0$

46. $(2, 3)$, $m = 3$

47. $(4, 3)$, m is undefined

48. $(3, -2)$, $m = \frac{5}{3}$

49. $(-2, -3)$, $m = 0$

50. $(-2, -4)$, $m = -\frac{5}{2}$

51. $(-3, 2)$, $m = -\frac{4}{3}$

52. $\left(\frac{2}{3}, -5\right)$, $m = 0$

53. $\left(-\frac{5}{2}, -1\right)$, m is undefined

Write an equation of the line L satisfying each set of conditions.

54. L passes through $(-3, -1)$ and $(3, 3)$.

55. L passes through $(2, 3)$ and $(-2, -5)$.

56. L has slope $\frac{3}{4}$ and y-intercept $(0, 3)$.

57. L passes through $(4, -3)$ with a slope of $-\frac{5}{4}$.

58. L has y-intercept $(0, -4)$ and is parallel to the line with equation $3x - y = 6$.

59. L passes through $(-5, 2)$ and is perpendicular to the line with equation $5x - 3y = 15$.

60. L passes through $(2, -1)$ and is perpendicular to the line with equation $3x - 2y = 5$.

61. L passes through the point $(-5, -2)$ and is parallel to the line with equation $4x - 3y = 9$.

3.4 BUSINESS AND FINANCE It costs a lunch cart \$1.75 to make each gyro. The portion of the cart's fixed cost attributable to gyros comes to \$30 per day. Use this information to answer exercises 62–64.

62. Construct a linear function to model the cart's gyro costs.

63. How much does it cost to make 35 gyros in one day?

64. How many gyros can the cart make if it can spend \$150 making gyros?

Business and Finance The lunch cart earns a profit of \$2.75 on each gyro by selling them for \$4.50 each. The fixed cost associated with gyros reduces profits by \$30 per day. Use this information to complete exercises 65–67.

65. Construct a linear function to model the cart's gyro profits.

66. How much profit does the cart make by selling 35 gyros in one day?

67. How many gyros do they need to sell if they want to earn \$100 in gyro profits?

Statistics On a 63-mile trip, a driver used two gallons of gas. The same driver used 8 gal on a 252-mi trip. Use this information to complete exercises 68–72.

68. Construct a linear model for the gas used as a function of the miles driven.

69. What is the rate of change of the function constructed in exercise 68?

70. Interpret the rate of change in the context of this application

71. What is the y-intercept of the function constructed in exercise 68?

72. Interpret the y-intercept in the context of this application.

Social Science A survey of public school libraries and media centers provided data comparing the state's expenditures for library materials (per student) to the number of books acquired during the year (per 100 students). The data for five states are shown in the table below. Use this information to complete exercises 73–78.

State	Expenditures	Acquisitions
Arizona	\$15.30	121
Georgia	\$14.20	76
Minnesota	\$15.20	111
Ohio	\$10.90	75
Virginia	\$16.20	88

Source: National Center for Education Statistics (AY2003–04).

73. Create a scatter plot of the data and include the line-of-best-fit on your graph.

74. What is the equation of the best-fit line (two decimal places of accuracy)?

75. What is the slope of the best-fit line?

76. Interpret the slope in the context of this application.

77. How many books would you expect to be acquired (per 100 students) if a state's per-student expenditures were \$17 (to the nearest whole number)?

78. What expenditures should policy makers approve if they wanted their state's libraries to acquire 100 books (per 100 students) in a given year (to the nearest cent)?

3.5 *Graph the solution set for each linear inequality.*

79. $y < 2x + 1$

80. $y \geq -2x + 3$

81. $3x + 2y \geq 6$

82. $3x - 5y < 15$

83. $y < -2x$

84. $4x - y \geq 0$

85. $y \geq -3$

86. $x < 4$

self-test 3

Name ____________

Section ________ Date ________

The purpose of this self-test is to help you assess your progress so that you can find concepts that you need to review before the next exam. Allow yourself about an hour to take this test. At the end of that hour, check your answers against those given in the back of this text. If you miss any, go back to the appropriate section to reread the examples until you have mastered that particular concept.

Find the slope of the line through each pair of points.

1. $(-3, 5)$ and $(2, 10)$

2. $(4, 9)$ and $(-3, 6)$

Write the equation of the line with the given slope and y-intercept. Then graph each line.

3. Slope -3; y-intercept $(0, 6)$

4. Slope $\frac{2}{5}$; y-intercept $(0, -3)$

Crafts A cookbook recommends that you should roast a 10-lb stuffed turkey for 4 hr and an 18-lb stuffed bird for 6 hr. Use this information to complete exercises 5 and 6.

5. Construct a linear model for roasting times as a function of the size of a stuffed turkey.

6. According to the model, for how long should you roast 16-lb stuffed turkey?

Graph each equation.

7. $x + y = 4$

8. $y = 3x$

9. $y = \frac{3}{4}x - 4$

10. $x + 3y = 6$

11. $2x + 5y = 10$

12. $y = -4$

Graph each inequality.

13. $5x + 6y \le 30$

14. $x + 3y > 6$

15. $4x - 8 \le 0$

16. $2y + 4 > 0$

Answers

1. ____________
2. ____________
3. ____________
4. ____________
5. ____________
6. ____________
7. ____________
8. ____________
9. ____________
10. ____________
11. ____________
12. ____________
13. ____________
14. ____________
15. ____________
16. ____________

Answers

17. ______

18. ______

19. ______

20. ______

21. ______

22. ______

23. ______

24. ______

25. ______

Find the slope and y-intercept of the line represented by each equation.

17. $y = -5x - 9$ **18.** $6x + 5y = 30$ **19.** $y = 5$

Write an equation of the line L satisfying the given set of conditions.

20. L has slope 5 and y-intercept $(0, -2)$.

21. L passes through $(-5, 4)$ and $(-2, -8)$.

22. L has y-intercept $(0, 3)$ and is parallel to the line given by $4x - y = 9$.

23. L passes through the point $(-6, -2)$ and is perpendicular to the line given by $2x - 5y = 10$.

Science and Medicine The high and low temperatures at five locations were recorded one day.

Low	39°F	42°F	54°F	64°F	66°F
High	70°F	77°F	77°F	79°F	75°F

Source: National Weather Service (Oct. 14, 2008).

24. Construct a scatter plot of the data and include the line-of-best-fit.

25. Find the equation of the line-of-best-fit (round to two decimal places).

cumulative review chapters 0-3

Name ______________

Section ________ Date ________

We offer the following exercises to help you review concepts from earlier chapters. This is meant as review material and not as a comprehensive exam. The answers are presented in the back of the text. We provide section references for each concept along with the answers in the back of this text. If you have difficulty with any of these exercises, be certain to at least read through the summary related to that section.

Perform the indicated operations. Write each answer in simplest form.

1. $\frac{5}{6} - \left(\frac{2}{3} + \frac{1}{2}\right)$

2. $\frac{7}{15} \times \left(\frac{5}{6} \div \frac{7}{12}\right)$

3. $2^3 - |-8| \div (-4) \cdot 2 + 5$

4. $4^2 + (-16 \div 4 \cdot 2)$

Evaluate each expression if $x = -1$, $y = 3$, and $z = -2$.

5. $-4x^2 + 3y + 2z$

6. $\dfrac{2z - 3y^2 - 2}{y^2 + 2z}$

Simplify the given expression.

7. $9x - 5y - (3x - 8y)$

8. $2x^2 - 4x - (-3x + x^2) - (4 + x^2)$

9. $-4x^2 + 7x - 4 - (-7x^2 + 11x) - (9x^2 - 5x - 6)$

10. $7 - 5x + 2x^2 + 2(9 - 5x^2)$

Solve each equation.

11. $5x - 3(2x - 6) + 9 = -2(x - 5) + 6$

12. $\frac{4}{5}x - 2 = 3 + \frac{3}{4}x$

13. $\dfrac{x + 1}{3} - \dfrac{2x + 3}{4} = \dfrac{1}{6}$

14. $2x(x - 3) - 9 = 2x^2$

Solve the equation for the indicated variable.

15. $F = \frac{9}{5}C + 32$ (for C)

16. $V = \frac{1}{3}\pi r^2 h$ (for h)

Answers

1. ______________
2. ______________
3. ______________
4. ______________
5. ______________
6. ______________
7. ______________
8. ______________
9. ______________
10. ______________
11. ______________
12. ______________
13. ______________
14. ______________
15. ______________
16. ______________

Answers

17. ______
18. ______
19. ______
20. ______
21. ______ 22. ______
23. ______
24. ______
25. ______
26. ______
27. ______
28. ______
29. ______
30. ______
31. ______
32. ______
33. ______ 34. ______
35. ______
36. ______
37. ______
38. ______
39. ______
40. ______

Solve and graph the solution set for each inequality.

17. $4x - 7 < 9$

18. $6x + 4 > 3x - 8$

19. $4 \le 2x - 6 \le 12$

20. $x - 5 < -3$ or $x - 5 > 2$

Find the slope of the line through each pair of points.

21. $(6, -4)$ and $(-2, 12)$

22. $(4, 5)$ and $(7, 5)$

Find the slope and the y-intercept of the line represented by the given equation.

23. $y = -4x + 9$

24. $2x - 5y = 10$

25. $y = 9$

26. $x = 7$

Write an equation of the line L that satisfies the given conditions.

27. L has slope 5 and y-intercept of $(0, -6)$.

28. L passes through $(-4, 9)$ and $(6, 8)$.

29. L has y-intercept $(0, 6)$ and is parallel to the line with the equation $2x + 3y = 6$.

30. L passes through the point $(2, 4)$ and is perpendicular to the line with the equation $4x - 5y = 20$.

31. L has x-intercept $(2, 0)$ and y-intercept $(0, -3)$.

32. L has slope 3 and passes through the point $(-2, 4)$.

Solve each problem.

33. If one-third of a number is added to 3 times the number, the result is 30. Find the number.

34. Two more than 4 times a number is 30. Find the number.

35. On a particular flight, the cost of a coach ticket is one-half the cost of a first-class ticket. If the total cost of the tickets is \$1,350, how much does each ticket cost?

36. The length of one side of a triangle is twice that of the second and 4 less than that of the third. If the perimeter is 64 meters (m), find the length of each of the sides.

37. Graph the solution set for the inequality

$$2x + 3y < 6$$

BUSINESS AND FINANCE A shipping company charged \$50.52 to ship a 5-lb package across the country overnight. It charged \$70.27 to ship a 10-lb package overnight between the same addresses.

38. Construct a linear model for the cost of shipping as a function of a package's weight.

39. How much would you expect it to cost to ship a 12-lb package?

40. Interpret the slope of the model in the context of the application.

CHAPTER

4

Systems of Linear Equations

INTRODUCTION

Although agriculture is not typically thought of as a high-tech industry, technology has long been an important element of farming. In the industrial revolution, a lot of time and energy was spent assuring that farms were supplied with equipment to increase productivity.

In the computer information era, agriculture has again benefited greatly. Whether it is computer-operated watering systems or market analysis, computers and mathematics play an important role in agronomy.

CHAPTER 4 OUTLINE

4.1 Graphing Systems of Linear Equations

< 4.1 Objectives >

1 > Solve a system by graphing

2 > Use slopes to identify consistent systems

In Section 1.8, we defined a solution set as "the set of all values for the variable that make the equation a true statement." For the equation

$$2x - 3x + 5 = x - 7$$

the solution set is $\{6\}$. This tells us that 6 is the only value for the variable x that makes the equation a true statement.

When we studied equations in two variables in Chapter 2, we found that a solution to a two-variable equation is an ordered pair. Given the equation

$$y = 2x - 5$$

one possible solution to the equation is the ordered pair $(1, -3)$. There are an infinite number of other possible solutions which form a line when graphed.

In this chapter, we introduce a topic that has many applications in chemistry, business, economics, and physics. Each of these areas has occasion to solve systems of equations.

Definition

Systems of Equations

A **system of equations** is a set of two or more related equations.

Our goal in this chapter is to solve linear systems of equations.

Definition

Solutions for Systems of Equations

A **solution** for a system of equations in two variables is an ordered pair of real numbers (x, y) that satisfies all of the equations in the system.

Over the course of this chapter, we will look at different ways in which a linear system of equations can be solved. Our first method is a graphical method of solving a system.

Example 1 Solving a System by Graphing

< Objective 1 >

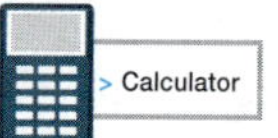

Solve the system by graphing.

$$2x + y = 4$$
$$x - y = 5$$

We graph the lines corresponding to the two equations of the system.

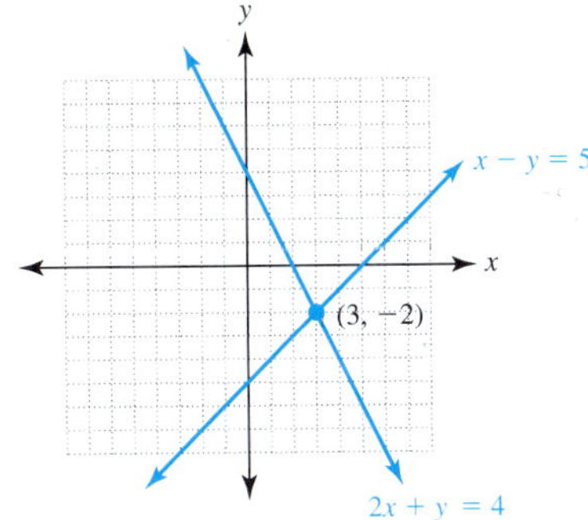

Each equation has an infinite number of solutions (ordered pairs) corresponding to points on a line. The point of intersection, here $(3, -2)$, is the *only* point lying on both lines, so $(3, -2)$ is the only ordered pair satisfying both equations and $(3, -2)$ is the solution for the system. The solution set is $\{(3, -2)\}$.

NOTES

Solve each equation for y and then graph.

$Y_1 = -2x + 4$

and

$Y_2 = x - 5$

We can *approximate* the solution by tracing the curves near their intersection.

Because there are two variables in the equations, we are searching for ordered pairs. We are looking for all of the ordered pairs that make *both* equations true.

Check Yourself 1

Solve the system by graphing.

$3x - y = 2$

$x + y = 6$

The definition of a *solution for a system of equations* states that a solution must satisfy all of the equations in the system. It is always a good idea to check a solution to a system, but it is especially important to do so when using a graphing approach.

Example 2 Checking the Solution to a System of Equations

> CAUTION

The difficulty with determining a solution exactly by graphing makes it especially important that you check solutions found using this method.

In Example 1, we found that $(3, -2)$ is a solution to the system of equations

$2x + y = 4$

$x - y = 5$

Check this result.

We check the solution to the system by checking that it is a solution to each equation, individually. Begin by substituting 3 for x and -2 for y into the first equation and seeing if the result is true.

$2x + y = 4$ Always use the original equation to check a result.

$2(3) + (-2) \stackrel{?}{=} 4$ Substitute $x = 3$ and $y = -2$.

$6 - 2 \stackrel{?}{=} 4$

$4 = 4$ True

Then, check the result using the second equation.

$x - y = 5$

$(3) - (-2) \stackrel{?}{=} 5$ Substitute into the second equation.

$5 = 5$ True

Because $(3, -2)$ checks as a solution in both equations, it is a solution to the system of equations.

NOTE

Remember to check the solution in *both* equations.

Check Yourself 2

Check that (2, 4) is a solution to the system of equations from Check Yourself 1.

$$3x - y = 2$$
$$x + y = 6$$

We put these last two ideas together into a single example.

Example 3 Solving a System of Equations

RECALL

Solve each equation for y to graph it. $5x + 2y = 5$ can be rewritten as

$$y = -\frac{5}{2}x + \frac{5}{2}$$

$3x + y = 2$ is equivalent to $y = -3x + 2$.

Solve the system by graphing and check the solution.

$$5x + 2y = 5$$
$$3x + y = 2$$

We begin by graphing the equations.

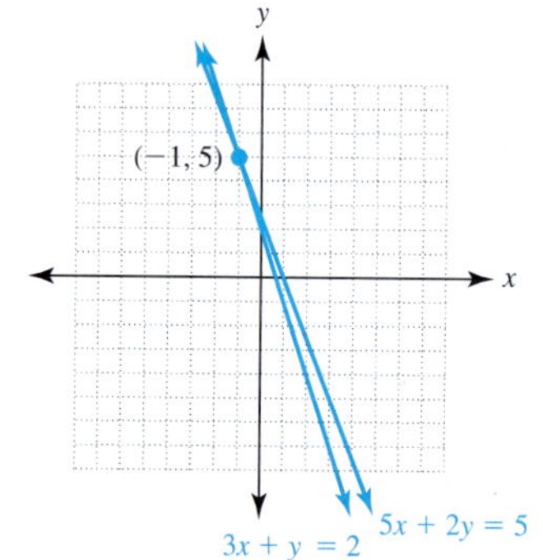

It looks like the graphed lines intersect at $(-1, 5)$. To be certain, we check that this is a solution to the system. We check the solution by substituting the x- and y-value in each equation.

First Equation

$5x + 2y = 5$ The first equation

$5(-1) + 2(5) \stackrel{?}{=} 5$ Substitute $x = -1$ and $y = 5$.

$-5 + 10 \stackrel{?}{=} 5$ Follow the order of operations.

$5 = 5$ True

Second Equation

$3x + y = 2$ The second equation

$3(-1) + (5) \stackrel{?}{=} 2$ Substitute $x = -1$ and $y = 5$.

$-3 + 5 \stackrel{?}{=} 2$ Follow the order of operations.

$2 = 2$ True

The solution $(-1, 5)$ checks in both equations so the solution set for the given system of equations is $\{(-1, 5)\}$.

Check Yourself 3

Solve the system by graphing and check your solution.

$$5x - 2y = 7$$
$$x + y = 7$$

In the previous examples, the two lines are nonparallel and intersect at only one point. Each system has a unique solution corresponding to that point. Such a system is called a **consistent system.** In the next example, we examine a system representing two lines that have no point of intersection.

Example 4 Solving a System by Graphing

Solve the system by graphing.

$$2x - y = 4$$
$$6x - 3y = 18$$

The lines corresponding to the two equations are graphed here.

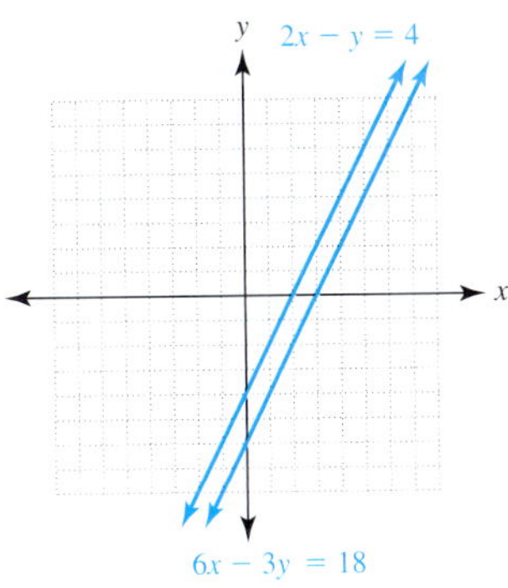

The lines are distinct and parallel. There is no point at which they intersect, so the system has **no solution.** We call such a system an **inconsistent system.**

Check Yourself 4

Solve the system, if possible.

$$3x - y = 1$$
$$6x - 2y = 13$$

Sometimes the equations in a system have the same graph.

Example 5 Solving a System by Graphing

Solve the system by graphing.

$$2x - y = 2$$
$$4x - 2y = 4$$

The equations are graphed, as follows.

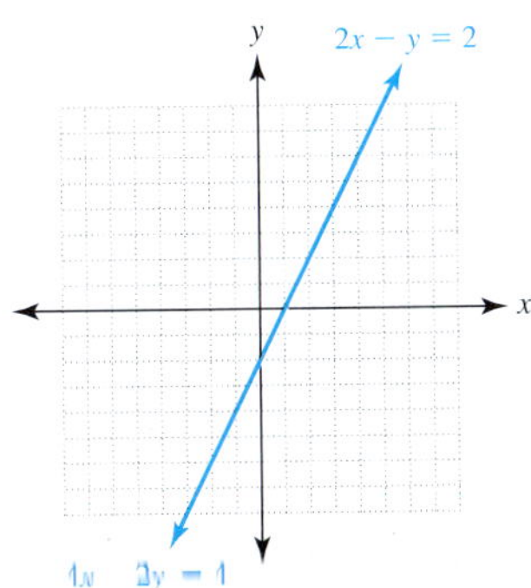

Both equations graph the same line, so they have an **infinite number of solutions** in common. We call such a system a **dependent system.**

Check Yourself 5

Solve the system by graphing.

$$6x - 3y = 12$$
$$y = 2x - 4$$

You have now seen the three possible types of solutions to a system of two linear equations. There will be a single solution (a consistent system), an infinite number of solutions (a dependent system), or no solution (an inconsistent system).

Note that, for both the dependent system and the inconsistent system, the slopes of the two lines in the system must be the same. (Do you see why that is true?) Given any two lines with different slopes, they will intersect at exactly one point. This idea is used in Example 6.

Example 6 Identifying the Type of a System

< Objective 2 >

For each system, determine the number of solutions, and identify the type of system.

(a) $y = 2x - 5$

$y = 2x + 9$

Both lines have a slope of 2, but different y-intercepts. We have two distinct parallel lines, and therefore there are no solutions. The system is inconsistent.

(b) $y = 3x + 7$

$y = -\frac{1}{3}x + 2$

These lines are perpendicular. There is one solution. The system is consistent.

(c) $2x - 3y = 7$

$3x + 5y = 2$

The lines have different slopes. The slopes are $\frac{2}{3}$ and $-\frac{3}{5}$. There is a single solution. The system is consistent.

(d) $y = \frac{2}{3}x - 6$

$2x - 3y = 12$

Both lines have a slope of $\frac{2}{3}$, but different y-intercepts. There are no solutions. The system is inconsistent.

NOTE

Solving $2x - 3y = 12$ for y gives $y = \frac{2}{3}x - 4$.

Check Yourself 6

For each system, determine the number of solutions, and identify the type of system.

(a) $y = 2x - 1$
$y = 3x + 7$

(b) $y = -3x - 2$
$y = -\frac{1}{3}x + 4$

(c) $6x - 3y = 4$
$-2x + y = 9$

(d) $y = \frac{1}{2}x - 4$
$x - y = 6$

Check Yourself ANSWERS

1.

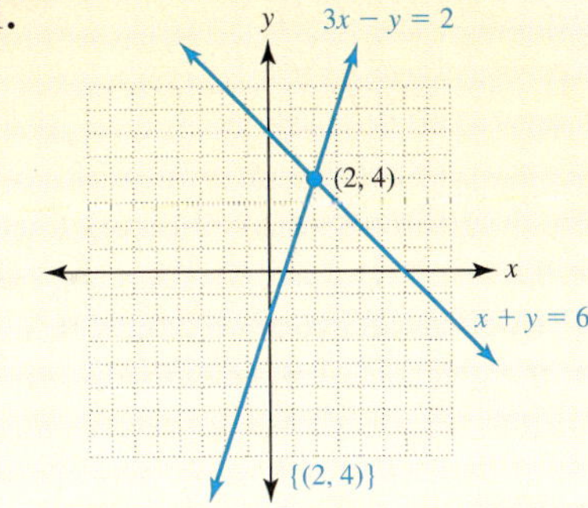

2. $3x - y = 2$ $\qquad x + y = 6$

$3(2) - (4) \stackrel{?}{=} 2$ $\qquad (2) + (4) \stackrel{?}{=} 6$

$6 - 4 \stackrel{?}{=} 2$ $\qquad 6 = 6$ True

$2 = 2$ True

3. $\{(3, 4)\}$

4.

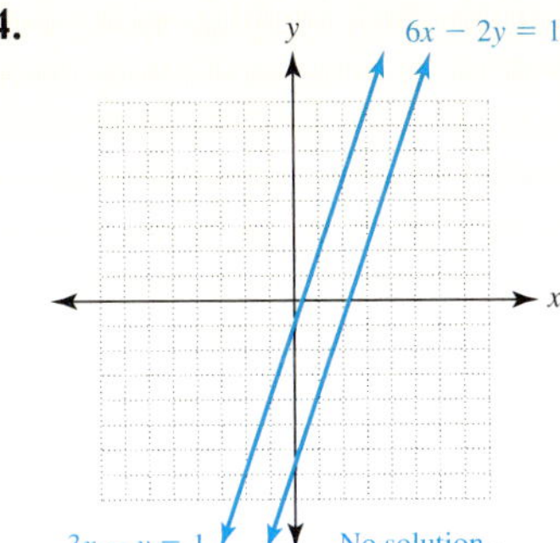

5.

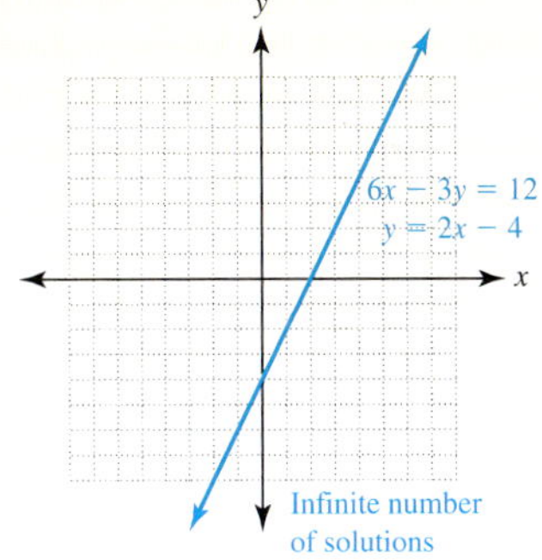

6. **(a)** one solution, consistent; **(b)** one solution, consistent; **(c)** no solutions, inconsistent; **(d)** one solution, consistent

Reading Your Text

SECTION 4.1

(a) A system of equations is a set of two or more ______________ equations.

(b) A solution for a system of equations in two variables is an __________ of real numbers that satisfies all of the equations in the system.

(c) A system that has a unique solution corresponding to only one point is called a ______________ system.

(d) A system having no solution is called an ______________ system.

Graphing Calculator Option

Solving a System of Equations

A graphing calculator can help us solve a system of equations. In order to use a graphing calculator, we must first solve each equation for y. Note that we do not actually need to put the equations into slope-intercept form.

After graphing both lines in a system, we find a good viewing window and use the calculator's **intersect** utility to find the point of intersection. If the lines do not intersect at a *nice* point, the calculator will give us an estimate of the coordinates.

Consider the system

$$37x + 15y = 2{,}531$$
$$45x + 29y = 3{,}946$$

We begin by solving each equation for y.

$$37x + 15y = 2{,}531$$
$$15y = 2{,}531 - 37x$$
$$y = \frac{2{,}531 - 37x}{15}$$

There is no need to write the equation in slope-intercept form.

$$45x + 29y = 3{,}946$$
$$29y = 3{,}946 - 45x$$
$$y = \frac{3{,}946 - 45x}{29}$$

It is important to remember to place the entire numerator in parentheses when entering these functions into a calculator.

Enter the functions into a graphing calculator and graph them on the same set of axes.

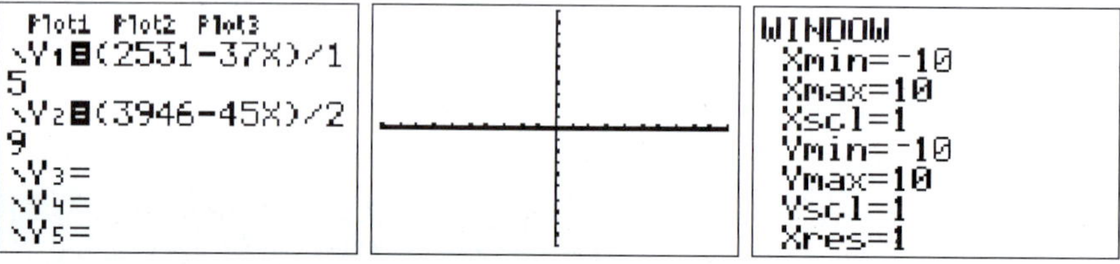

The graphs do not show on the standard (default) graphing window. This is because the graphs are outside this small range. That is, the y-values are not between -10 and 10 when x is in that range.

We can use the **TABLE** utility to find an appropriate viewing window.

X	Y1	Y2
0	168.73	136.07
10	144.07	120.55
20	119.4	105.03
30	94.733	89.517
40	70.067	74
50	45.4	58.483
60	20.733	42.966

X=30

When $x = 0$, the y-value of the first equation is approximately 169 and 136 in the second equation. Therefore, we need our window to include these y-values in order to see the graphs if the y-axis is part of our viewing window. To simplify our tasks, we set the graphs in the first quadrant and see what happens.

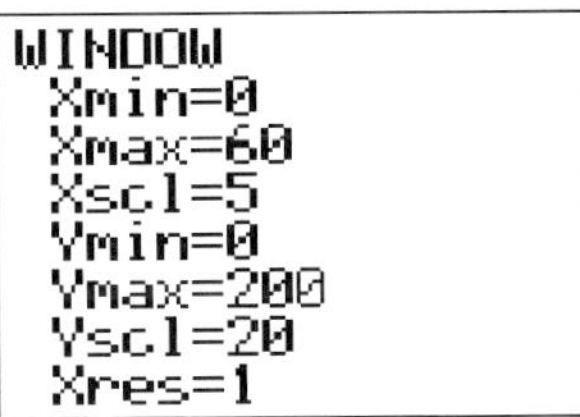

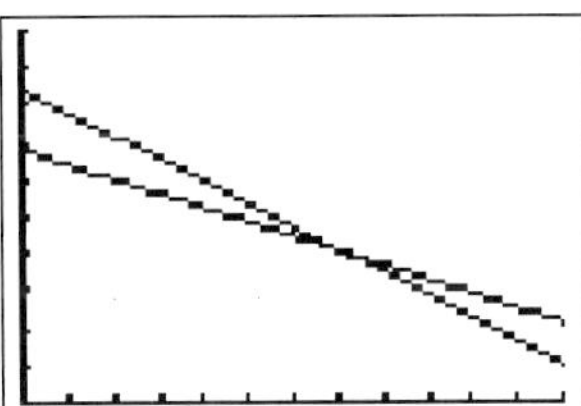

We can see the point of intersection on the screen, so there is no reason to modify the viewing window.

Next, we look for the point of intersection.

On the TI-84 Plus, we begin by opening the **CALC** menu. It is the second function above the TRACE key. Then, select the **intersect** utility.

2nd [CALC] **5:intersect**

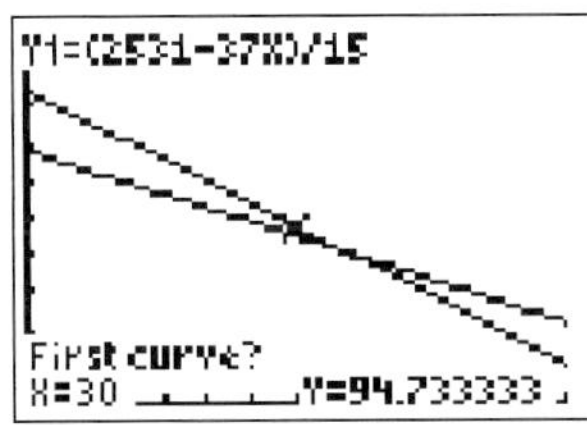

You must tell the calculator which graphs to examine and you must provide a guess. Simply press ENTER for each curve and for the guess and the calculator will approximate the intersection point.

ENTER ENTER ENTER

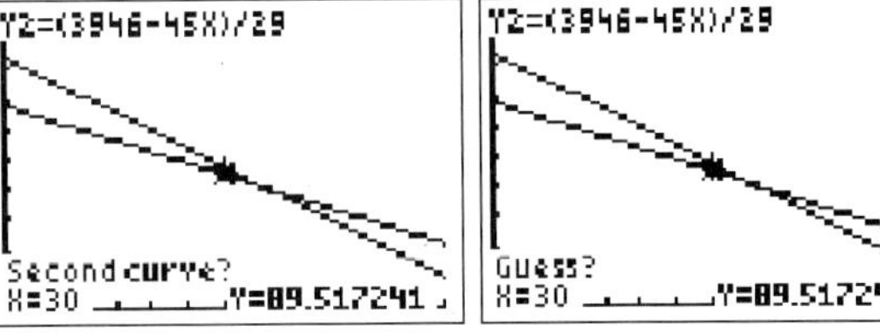

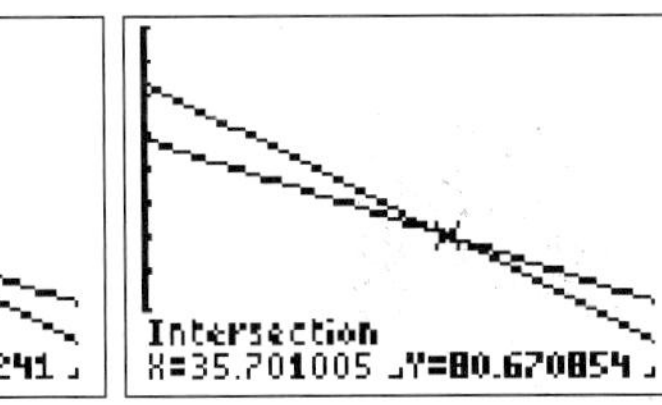

Note: You need to use the left/right arrows to move the cursor near the point of intersection when responding to the **Guess?** prompt if there is more than one intersection point on the screen. Similarly, you would need to use the up/down arrows to cycle to the correct curves if there are more than two functions graphed on your screen.

The final window gives the intersection point. We see that the solution for the system, to the nearest hundredth, is (35.70, 80.67).

Graphing Calculator Check

Use a graphing calculator to solve each system (round your results to the nearest hundredth).

(a) $19x + 83y = 4{,}587$
$36x + 51y = 4{,}229$

(b) $28x + 14y = 3{,}757$
$8x - 7y = -91$

(c) $3x + 5y = 10\sqrt{2}$
$x - 3y = 15$

(d) $x^2 - y = 6$
$x + 2y = 3\pi$

Note: This is not a linear system, but the methods are the same. In this case, there are two solutions.

Answers

(a) $\{(57.98, 41.99)\}$

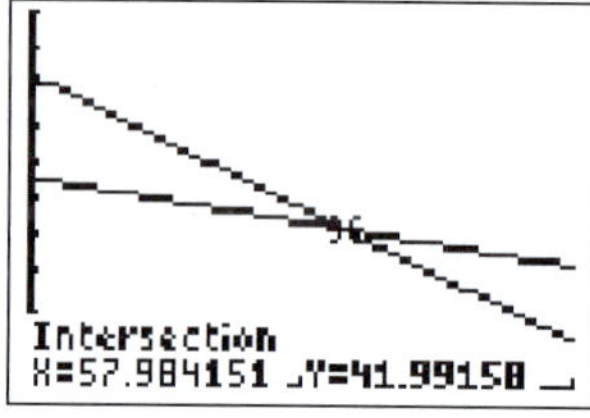

(b) $\{(81.25, 105.86)\}$

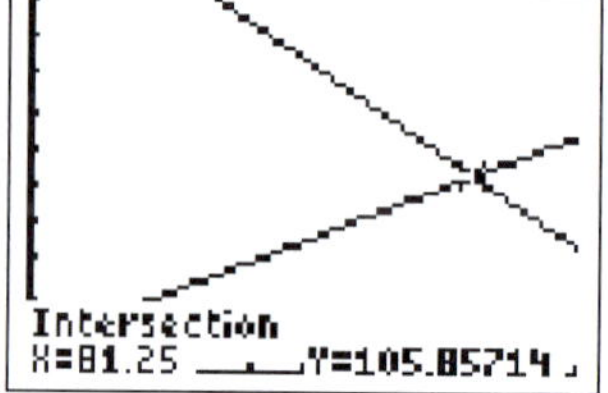

(c) $\{(8.39, -2.20)\}$

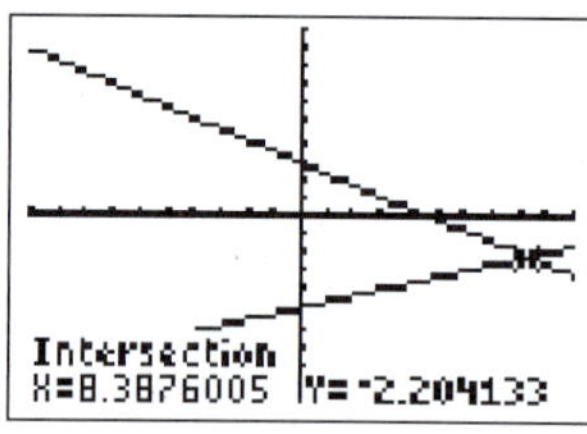

(d) $\{(-3.53, 6.48), (3.03, 3.20)\}$

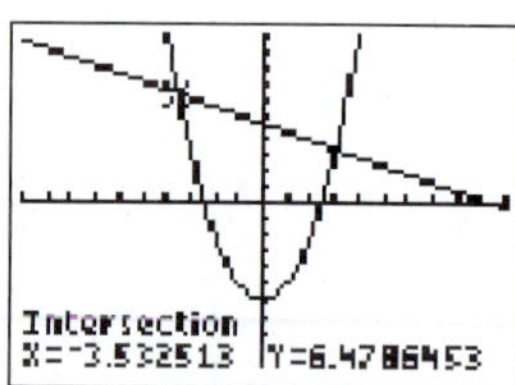

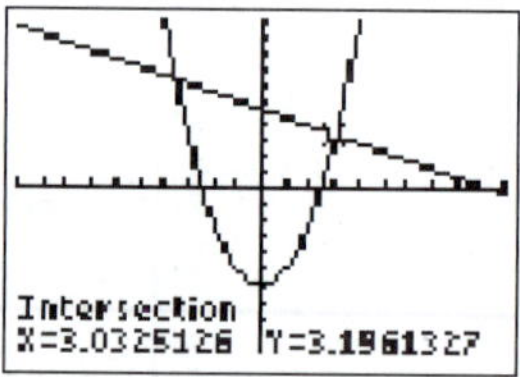

Basic Skills | Challenge Yourself | Calculator/Computer | Career Applications | Above and Beyond

4.1 exercises

< Objective 1 >

Graph each system of equations and then solve the system.

1. $x + y = 6$
$x - y = 4$

2. $x - y = 8$
$x + y = 2$

3. $x + y = 5$
$-x + y = 7$

4. $x + y = 7$
$-x + y = 3$

5. $x + 2y = 4$
$x - y = 1$

> Videos

6. $3x + y = 6$
$x + y = 4$

7. $3x - y = 21$
$3x + y = 15$

8. $x - 2y = -2$
$x + 2y = 6$

9. $x + 3y = 12$
$2x - 3y = 6$

10. $2x - y = 4$
$2x - y = 6$

11. $3x + 2y = 12$
$y = 3$

12. $5x - y = 11$
$2x - y = 8$

Name ____________

Section ________ Date ________

Answers

1. ____________
2. ____________
3. ____________
4. ____________
5. ____________
6. ____________
7. ____________
8. ____________
9. ____________
10. ____________
11. ____________
12. ____________

Answers

13. ______

14. ______

15. ______

16. ______

17. ______

18. ______

19. ______

20. ______

21. ______

22. ______

23. ______

24. ______

13. $x - y = 4$
$2x - 2y = 8$

14. $2x - y = 8$
$x = 2$

15. $x - 4y = -4$
$x + 2y = 8$

16. $4x + y = -7$
$-2x + y = 5$

17. $3x - 2y = 6$
$2x - y = 5$

> Videos

18. $4x + 3y = 12$
$x + y = 2$

19. $3x - y = 3$
$3x - y = 6$

> Videos

20. $3x - 6y = 9$
$x - 2y = 3$

21. $2y = 3$
$x - 2y = -3$

22. $x + y = -6$
$-x + 2y = 6$

23. $x = -5$
$y = 3$

24. $x = -3$
$y = 5$

< Objective 2 >

Determine whether each system is **consistent, inconsistent,** *or* **dependent.**

25. $y = 3x + 7$
$y = 7x - 2$

26. $y = 2x - 5$
$y = -2x + 9$

27. $y = 7x - 1$
$y = 7x + 8$

28. $y = -5x + 9$
$y = -5x - 11$

29. $3x + 4y = 12$
$9x - 5y = 10$

30. $2x - 4y = 11$
$-8x + 16y = 15$

31. $7x - 2y = 5$
$14x - 4y = 10$

32. $3x + 2y = 8$
$6x - 4y = -12$

Basic Skills | **Challenge Yourself** | Calculator/Computer | Career Applications | Above and Beyond

Complete each statement with **never, sometimes,** *or* **always.**

33. A linear system ________ has at least one solution.

34. If the graphs of two linear equations in a system have different slopes, the system ________ has exactly one solution.

35. If the graphs of two linear equations in a system have equal slopes, the system ________ has exactly one solution.

36. If the graphs of two linear equations in a system have equal slopes and equal y-intercepts, the system ________ has an infinite number of solutions.

Basic Skills | Challenge Yourself | **Calculator/Computer** | Career Applications | Above and Beyond

Use a graphing calculator to solve each exercise. Estimate your answer to the nearest hundredth. You may need to adjust the viewing window to see the point of intersection.

37. $88x + 57y = 1{,}909$
$95x + 48y = 1{,}674$

38. $32x + 45y = 2{,}303$
$29x - 38y = 1{,}509$

39. $25x - 65y = 5{,}312$
$-21x + 32y = 1{,}256$

40. $-27x + 76y = 1{,}676$
$56x - 2y = -678$

41. $15x + 20y = 79$
$7x + 5y = 115$

42. $23x - 31y = 1{,}915$
$15x + 42y = 1{,}107$

Answers

25. ________
26. ________
27. ________
28. ________
29. ________
30. ________
31. ________
32. ________
33. ________
34. ________
35. ________
36. ________
37. ________
38. ________
39. ________
40. ________
41. ________
42. ________

Answers

43. ______

44. ______

45. ______

46. ______

47. ______

48. ______

Basic Skills | Challenge Yourself | Calculator/Computer | **Career Applications** | Above and Beyond

43. Allied Health A medical lab technician needs to determine how much 15% hydrochloric acid (HCl) solution to mix with 5% HCl to produce 50 mL of 9% solution. Use the system of equations in which x is the amount of 15% solution to solve the application graphically.

$$x + y = 50$$
$$15x + 5y = 450$$

44. Allied Health A medical lab technician needs to determine how much 6-molar (M) copper sulfate ($CuSO_4$) solution to mix with 2-M $CuSO_4$ solution to produce 200 mL of a 3-M solution. Use the system of equations shown in which x is the amount of 6-M solution to solve the application graphically.

$$x + y = 200$$
$$6x + 2y = 600$$

45. Construction Technology The beam shown in the figure is 15 feet long and has the load indicated on each end. Graphically solve the system of equations shown in order to determine the point at which the beam balances.

$$x + y = 15$$
$$80x = 120y$$

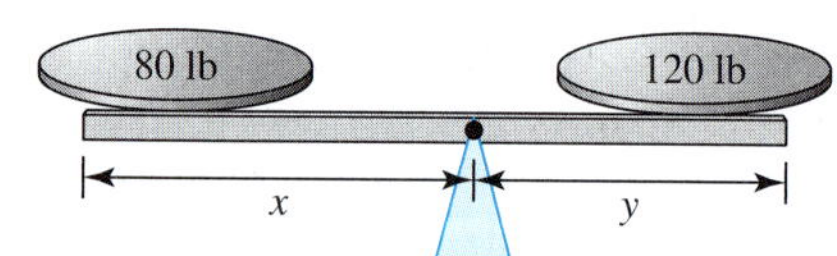

46. Mechanical Engineering For a plating bath, 10,000 L of 13% electrolyte solution is required. You have 8% and 16% solutions in stock. Solve the system of equations graphically, in which x represents the amount of 8% solution to use, to solve the application.

$$x + y = 10{,}000$$
$$0.08x + 0.16y = 1{,}300$$

Basic Skills | Challenge Yourself | Calculator/Computer | Career Applications | **Above and Beyond**

47. Find values for m and b so that (1, 2) is the solution to the system.

$$mx + 3y = 8$$
$$-3x + 4y = b$$

48. Find values for m and b so that $(-3, 4)$ is the solution to the system.

$$5x + 7y = b$$
$$mx + y = 22$$

49. Complete each statement in your own words.

"To solve an equation means to . . . ."

"To solve a system of equations means to . . . ."

50. A system of equations such as the one below is sometimes called a *2-by-2* system of linear equations.

$$3x + 4y = 1$$
$$x - 2y = 6$$

Explain this term.

51. Complete this statement in your own words: "All the points on the graph of the equation $2x + 3y = 6$ . . . ." Exchange statements with other students. Do you agree with other students' statements?

52. Does a system of linear equations always have a solution? How can you tell without graphing that a system of two equations will be graphed as two parallel lines? Give some examples to explain your reasoning.

53. Suppose we have the linear system

$$Ax + By = C$$
$$Dx + Ey = F$$

(a) Write the slope of the line determined by the first equation.

(b) Write the slope of the line determined by the second equation.

(c) What must be true about the given coefficients in order to guarantee that the system is consistent?

Answers

49. ______

50. ______

51. ______

52. ______

53. ______

Answers

1. 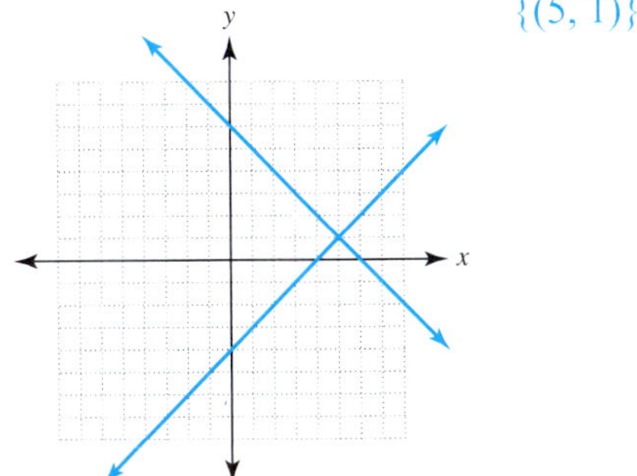

{(5, 1)}

3. 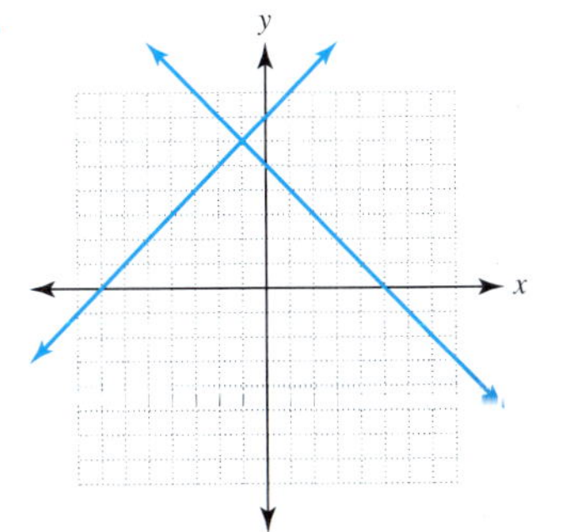

{(−1, 6)}

5. 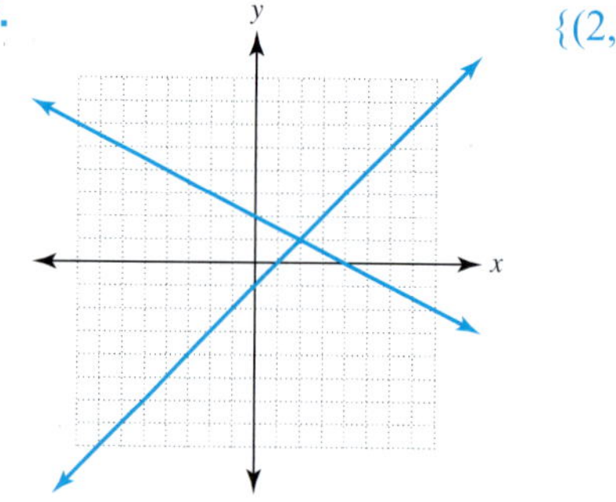

$\{(2, 1)\}$

7. 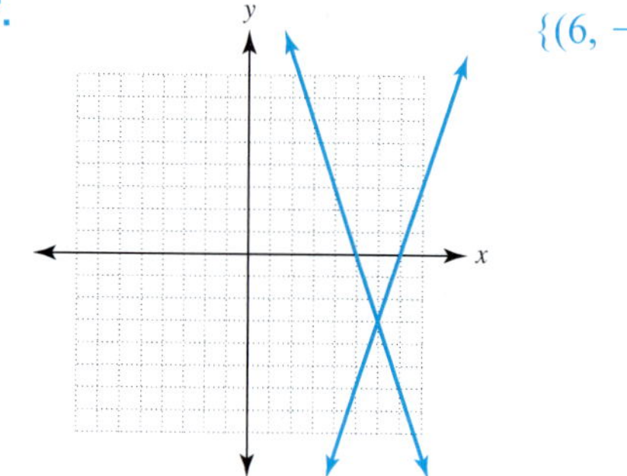

$\{(6, -3)\}$

9. 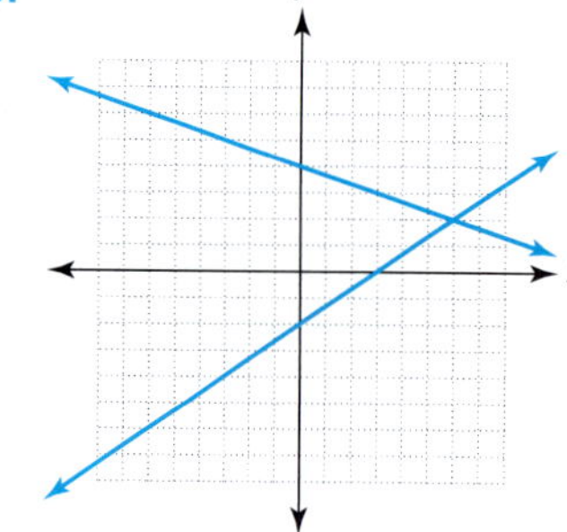

$\{(6, 2)\}$

11. 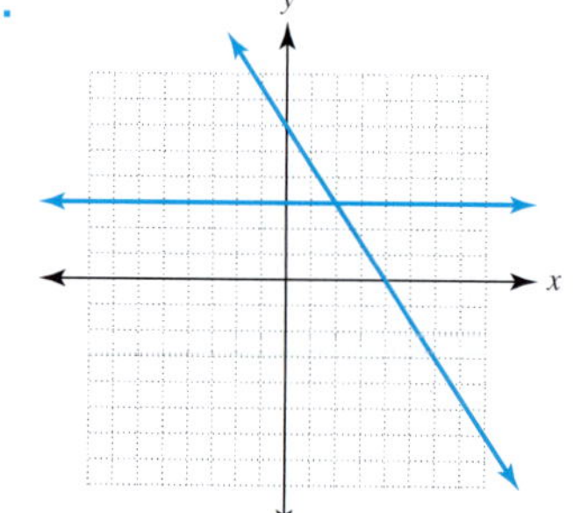

$\{(2, 3)\}$

13. 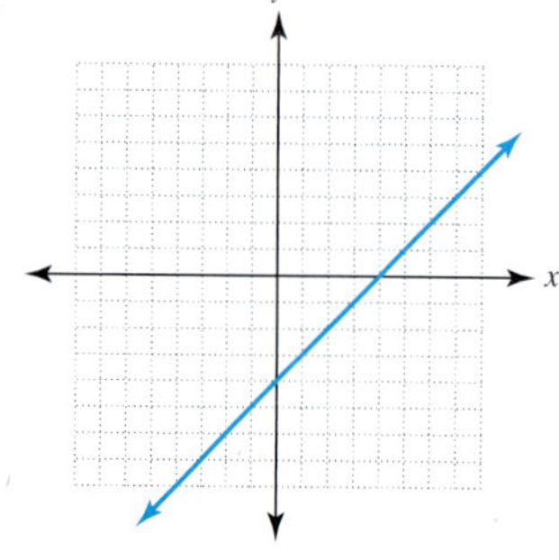

Infinite number of solutions, dependent system

15. 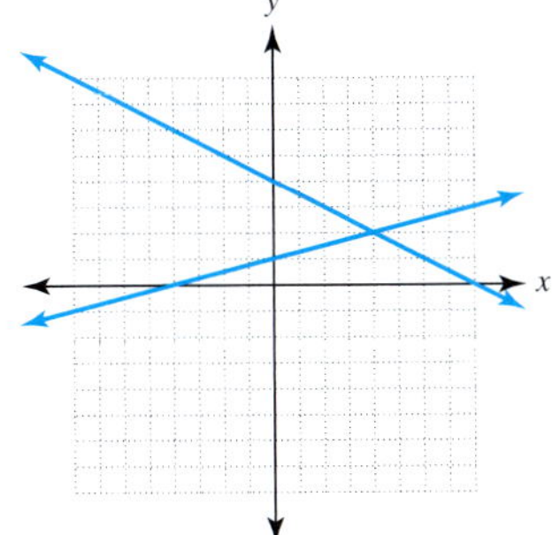

{(4, 2)}

17. 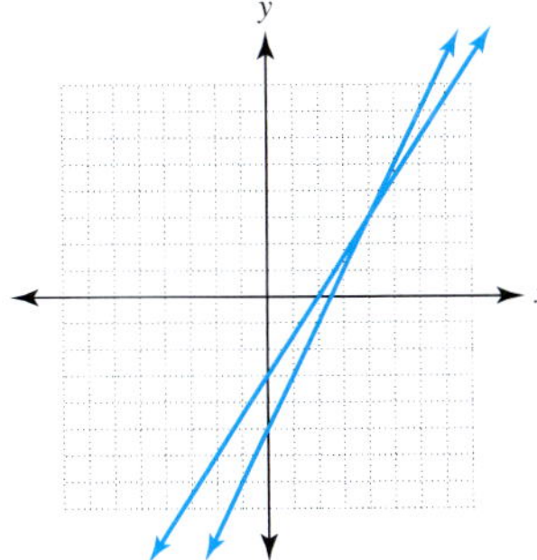

{(4, 3)}

19. 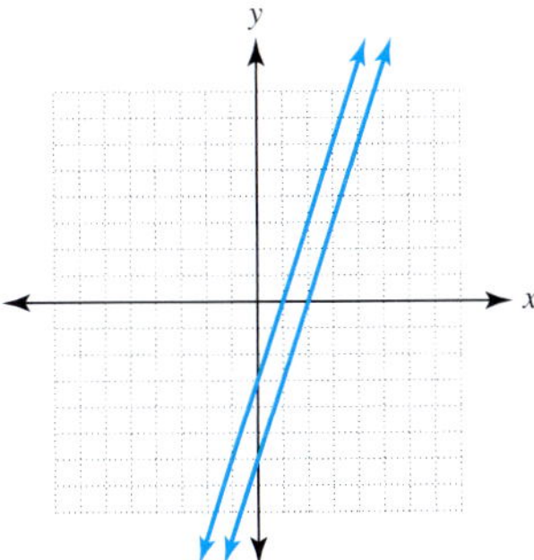

No solutions, inconsistent system

21. 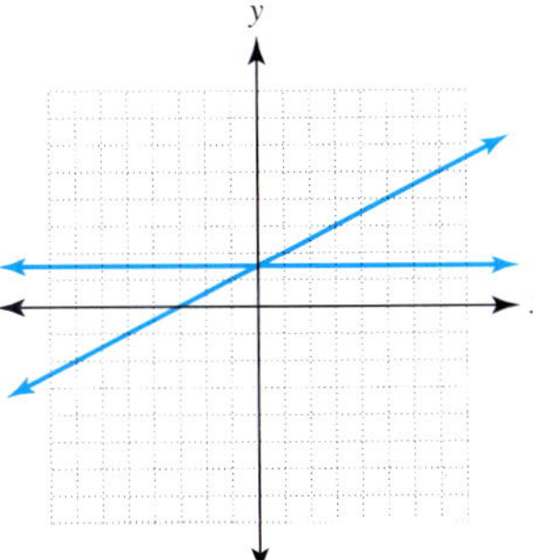

$\left\{\left(0, \frac{3}{2}\right)\right\}$

23. 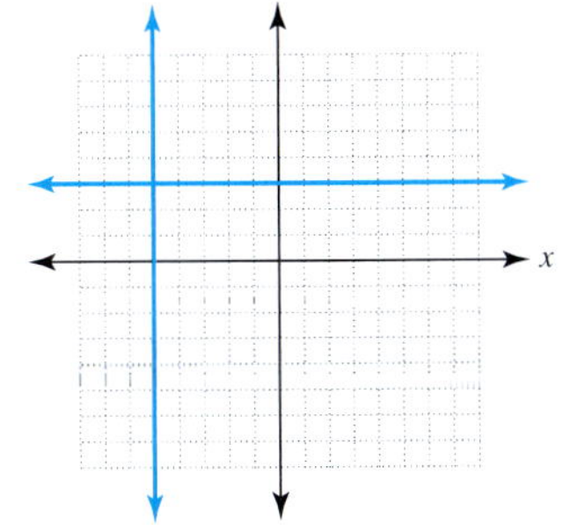

{(−5, 3)}

25. Consistent **27.** Inconsistent **29.** Consistent **31.** Dependent
33. sometimes **35.** never **37.** $\{(3.18, 28.58)\}$
39. $\{(-445.35, -253.01)\}$ **41.** $\{(29.31, -18.03)\}$
43. (20, 30); 20 mL of 15%, 30 mL of 5%

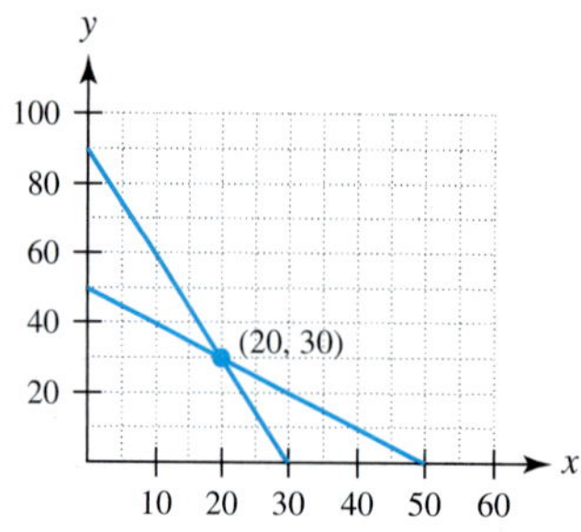

45. $x = 9$ ft, $y = 6$ ft

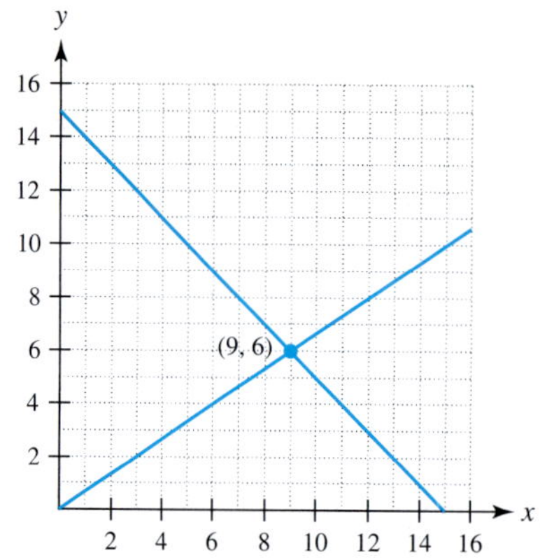

47. $m = 2$; $b = 5$ **49.** Above and Beyond **51.** Above and Beyond
53. **(a)** $-\dfrac{A}{B}$; **(b)** $-\dfrac{D}{E}$; **(c)** $AE - BD \neq 0$

Activity 4 :: Agricultural Technology

Nutrients and Fertilizers

When growing crops, it is not enough just to till the soil and plant seeds. The soil must be properly prepared before planting. Each crop takes nutrients out of the soil that must be replenished. Some of this is done with crop rotations (each crop takes some nutrients out of the soil while replenishing other nutrients), but maintaining proper nutrient levels ultimately requires that some additional nutrients be added. This is done through fertilizers.

The three most vital nutrients are nitrogen, phosphorus, and potassium. Three different fertilizer mixes are available:

Urea: Contains 46% nitrogen

Growth: Contains 16% nitrogen, 48% phosphorus, and 12% potassium

Restorer: Contains 21% phosphorus and 62% potassium

A soil test shows that a field requires 115 pounds of nitrogen, 78 pounds of phosphorus, and 61 pounds of potassium per acre. We need to determine how many pounds of each type of fertilizer to use on the field.

Solution

1. Let x equal the number of pounds of urea used. How many pounds of each nutrient are in a batch of urea?
2. Let y equal the number of pounds of the growth blend used. How many pounds of each nutrient are in a batch?
3. Let z equal the number of pounds of the soil restorer used. How many pounds of each nutrient are in a batch?
4. Create an equation for the amount of nitrogen in x pounds of urea, y pounds of growth blend, and z pounds of soil restorer.
5. Create similar equations for phosphorus and potassium.
6. Solve this system of equations to find the amount of each type of fertilizer required.

4.2 Solving Equations in One Variable Graphically

< 4.2 Objectives >

1 > Rewrite a linear equation in one variable as $f(x) = g(x)$

2 > Find and interpret the point of intersection of $f(x)$ and $g(x)$

3 > Solve a linear equation in one variable by writing it as the functional equality $f(x) = g(x)$

In Chapter 1, we learned to use algebraic methods to solve linear equations in one variable. It is interesting that our work with systems of equations in Section 4.1 leads us to a graphical approach to solving linear equations in one variable.

The techniques presented here are not meant to replace algebraic methods. But, they should be seen as powerful, alternative approaches to solving a variety of equations.

In this section, you will learn to use graphs to find an approximate solution to a problem. In such cases, it is often handy, but not necessary, to have access to a graphing calculator.

In our first example, we solve a straightforward linear equation. While the graphing approach may seem to be a bit much, once you master it, you will find it helpful.

NOTE

Visual learners should find this approach particularly helpful.

Example 1 Solving a Linear Equation Graphically

< Objective 1 >

Graphically solve the equation.

$$2x - 6 = 0$$

Step 1 Let each side of the equation represent a function of x.

$$f(x) = 2x - 6$$
$$g(x) = 0$$

Step 2 Graph the two functions on the same set of axes.

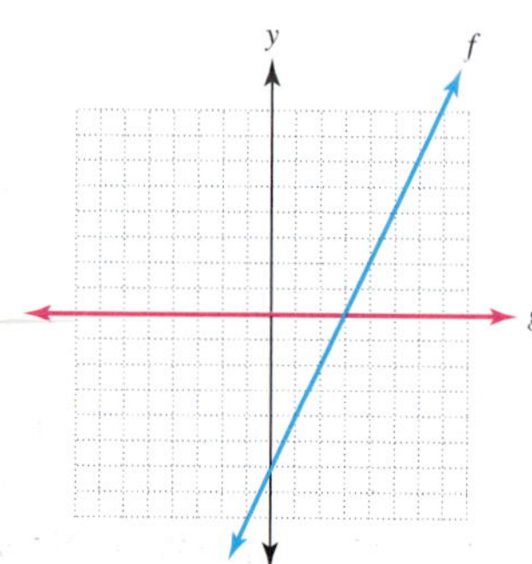

The graph of $y = g(x)$ is simply the x-axis.

NOTE

This is a one-variable equation. We are interested in values of x that make this true.

NOTE

We ask the question, When is the graph of f equal to the graph of g? Specifically, for what values of x does this occur?

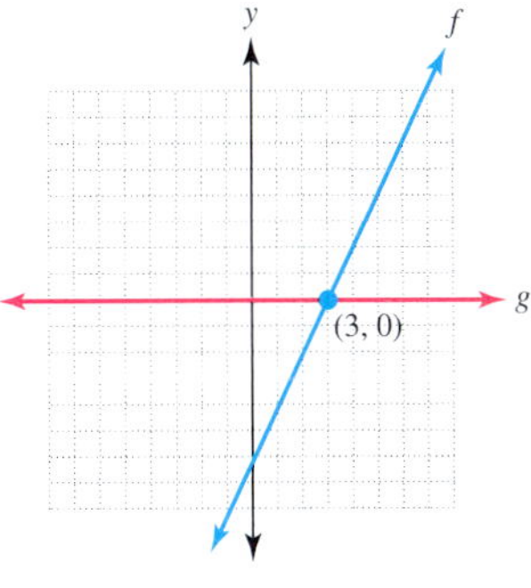

Step 3 Find the point of intersection of the two graphs. The x-coordinate of this point represents the solution to the original equation.

The two lines intersect on the x-axis at the point (3, 0). Again, because we are solving an equation in one variable (x), we are interested only in x-values. Thus, the solution is $x = 3$, and the solution set is $\{3\}$.

It is always a good idea to check your work, and it is especially important when you use graphical methods to solve a problem.

We check our solution by substituting it back into the original equation.

Check

$2x - 6 = 0$ The original equation

$2(3) - 6 \stackrel{?}{=} 0$ Substitute $x = 3$ into the original equation.

$6 - 6 \stackrel{?}{=} 0$

$0 = 0$ True!

Check Yourself 1

Graphically solve the equation.

$-3x + 6 = 0$

The same three-step process is used for solving any equation. In Example 2, we look for a point of intersection that is *not* on the x-axis.

Example 2 Solving a Linear Equation Graphically

< Objective 2 >

Graphically solve the equation.

$2x - 6 = -3x + 4$

Step 1 Let each side of the equation represent a function of x.

$f(x) = 2x - 6$

$g(x) = -3x + 4$

Step 2 Graph the two functions on the same set of axes.

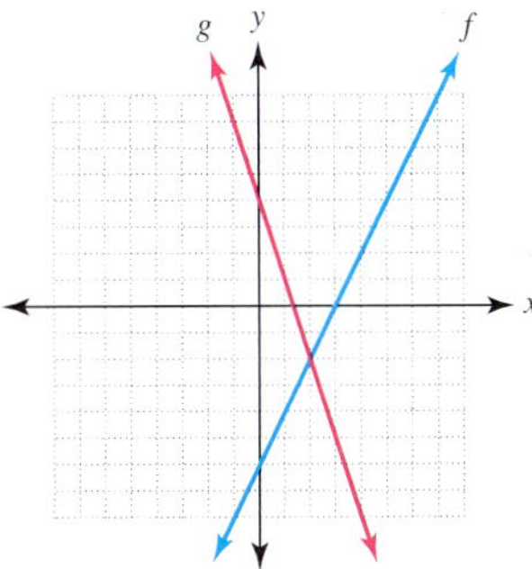

Step 3 Find the point of intersection of the two graphs. Because we want the x-coordinate of this point, we suggest the following: Draw a vertical line from the point of intersection (2, −2) to the x-axis, marking a

RECALL

We are only interested in the x-value of the intersection point.

point there. This is done to emphasize that we are interested only in the x-value: 2. The solution set for the original equation is $\{2\}$.

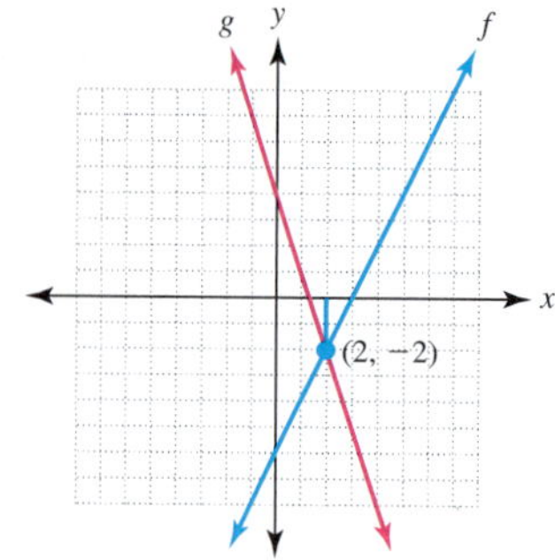

We leave it to you to check that this result is a solution.

Check Yourself 2

Graphically solve the equation.

$-2x - 5 = x - 2$

This algorithm summarizes our work in graphically solving a linear equation.

Step by Step

Solving a Linear Equation Graphically

Step 1 Let each side of the equation represent a function of x.

Step 2 Graph the two functions on the same set of axes.

Step 3 Find the point of intersection of the two graphs. Draw a vertical line from the point of intersection to the x-axis, marking a point there. The x-value at the indicated point represents the solution to the original equation.

We often apply some algebra even when we are taking a graphical approach. Consider Example 3.

Example 3 **Solving a Linear Equation Graphically**

< Objective 3 >

Solve the equation graphically.

$2(x + 3) = -3x - 4$

Use the distributive property to rid the left side of parentheses.

$2x + 6 = -3x - 4$

Now let $f(x) = 2x + 6$

$g(x) = -3x - 4$

Graphing both lines, we get

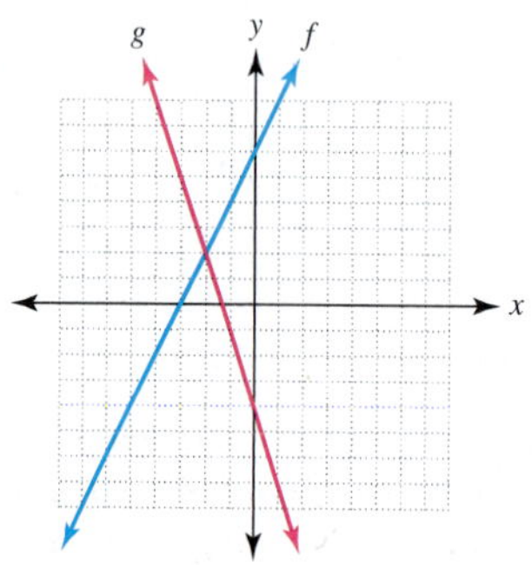

The point of intersection is $(-2, 2)$. Draw a vertical line to the x-axis and mark a point. The desired x-value is -2. The solution set for the original equation is $\{-2\}$.

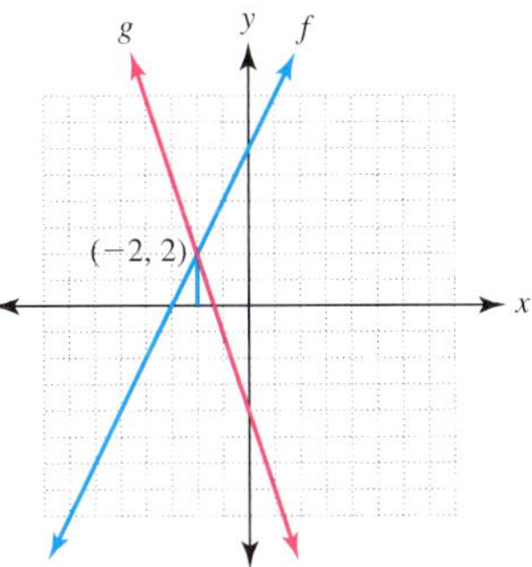

As before, we should check that our proposed solution is correct.

Check

$2(x + 3) = -3x - 4$	The original equation
$2[(-2) + 3] \stackrel{?}{=} -3(-2) - 4$	Substitute $x = -2$ into the original equation.
$2(1) \stackrel{?}{=} 6 - 4$	Remember to follow the correct order of operations.
$2 = 2$	True!

Check Yourself 3

Graphically solve the equation and check your result.

$3(x - 2) = -4x + 1$

A graphing calculator can certainly be used to solve equations in this manner. Using such a tool, we do not need to apply algebraic ideas such as the distributive property. We now demonstrate this with the same equation seen in Example 3.

Example 4 Solving a Linear Equation with a Graphing Calculator

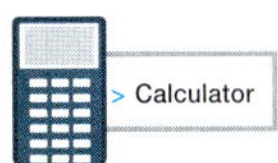

Use a graphing calculator to solve the equation.

$2(x + 3) = -3x - 4$

As before, let each side define a function.

$Y_1 = 2(x + 3)$

$Y_2 = -3x - 4$

When we graph these in the "standard viewing" window, we see the following:

NOTE

This window typically shows x-values from -10 to 10 and y-values from -10 to 10.

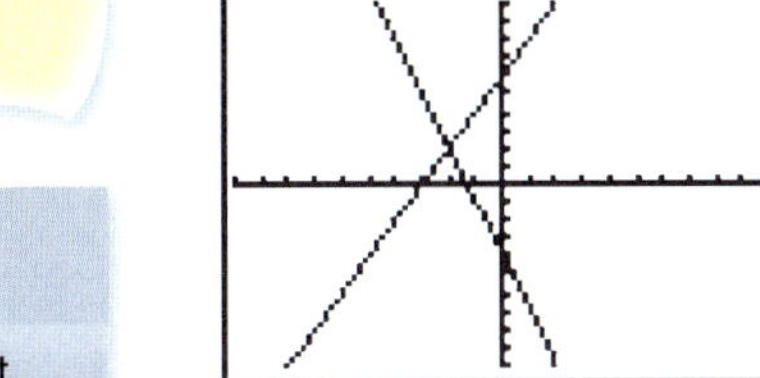

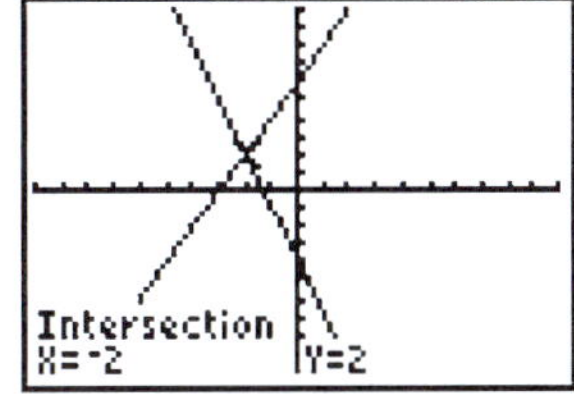

RECALL

We introduced the **intersect** utility in the Graphing Calculator Option segment at the end of Section 4.1.

Using the INTERSECT utility, we then see the view shown to the right.

Note that the calculator reports the intersection point as $(-2, 2)$. Since we are interested only in the x-value, the solution is $x = -2$. The solution set is $\{-2\}$.

Check Yourself 4

Use a graphing calculator to solve the equation.

$$-2x - 5 = x - 2$$

The graphing calculator is particularly effective when we are solving equations with "messy" coefficients. With technology, we can obtain a solution to any desired level of accuracy. Consider Example 5.

Example 5 Solving a Linear Equation with a Graphing Calculator

Use a graphing calculator to solve the equation. Give the solution accurate to the nearest hundredth.

$$2.05(x - 4.83) = -3.17(x + 0.29)$$

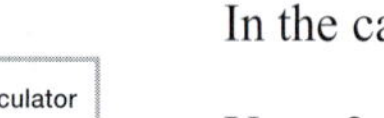

In the calculator we define

$$Y_1 = 2.05(x - 4.83)$$
$$Y_2 = -3.17(x + 0.29)$$

In the standard viewing window, we see this:

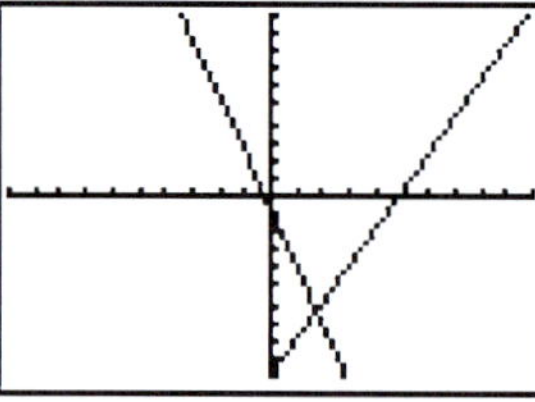

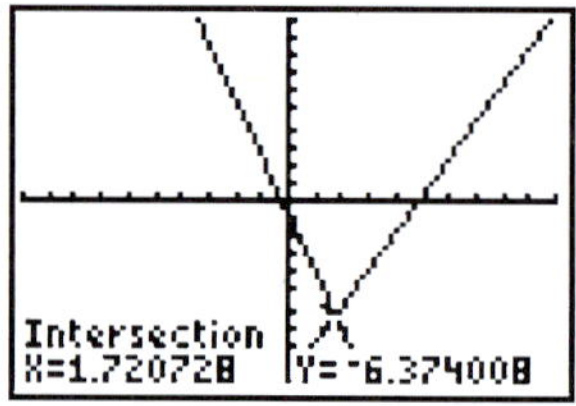

With the INTERSECT utility, we find the intersection point to be (1.720728, −6.374008). Because we want only the x-value, the solution (rounded to the nearest hundredth) is 1.72. The solution set is $\{1.72\}$.

Check Yourself 5

Solve the equation, using a graphing calculator. Give the solution accurate to the nearest hundredth.

$$-0.87x + 1.14 = -2.69(x + 4.05)$$

In Example 6, we turn to a business application.

Example 6 A Business and Finance Application

A manufacturer can produce and sell x items per week at a cost, in dollars, given by

$$C(x) = 30x + 800$$

The revenue from selling those items is given by

$$R(x) = 110x$$

Use a graphical approach to find the break-even point, which is the number of units at which the revenue equals the cost. That is, we wish to graphically solve the equation

$$110x = 30x + 800$$

Graphing the two functions, we have

NOTE

Try graphing these functions on your graphing calculator.

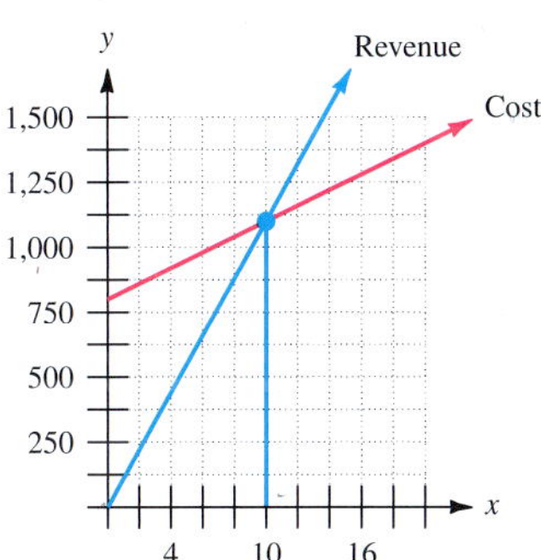

Drawing vertically from the intersection point to the x-axis, we see that the desired x-value (the break-even point) is 10 items per week. Note that if the company sells more than 10 units, it makes a profit since the revenue exceeds the cost.

Check Yourself 6

A manufacturer can produce and sell x items per week at a cost of

$$C(x) = 30x + 1{,}800$$

The revenue from selling those items is given by

$$R(x) = 120x$$

Use a graphical approach to find the break-even point.

Check Yourself ANSWERS

1. $f(x) = -3x + 6$
   $g(x) = 0$

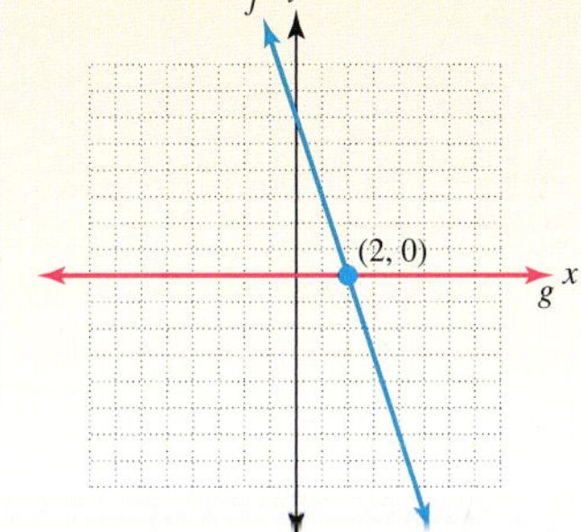

Solution set: $\{2\}$

2. $f(x) = -2x - 5$
   $g(x) = x - 2$

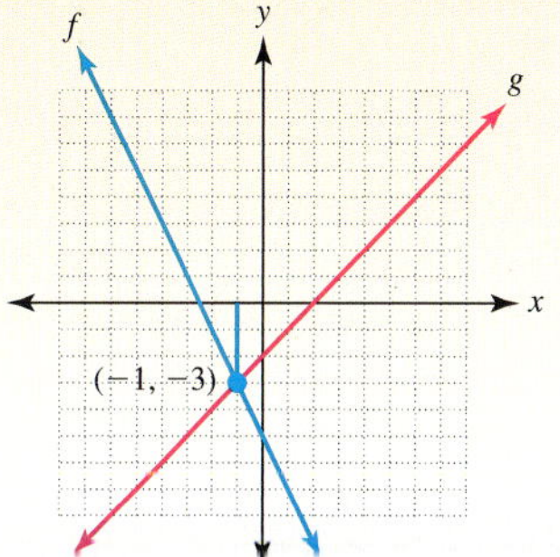

Solution set: $\{-1\}$

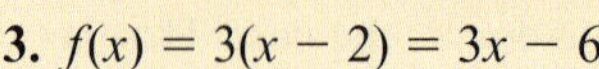

3. $f(x) = 3(x - 2) = 3x - 6$

$g(x) = -4x + 1$

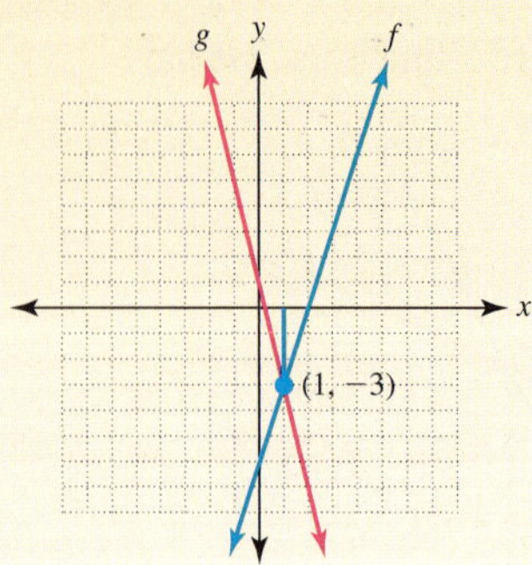

Solution set: $\{1\}$

4. $Y_1 = -2x - 5$

$Y_2 = x - 2$

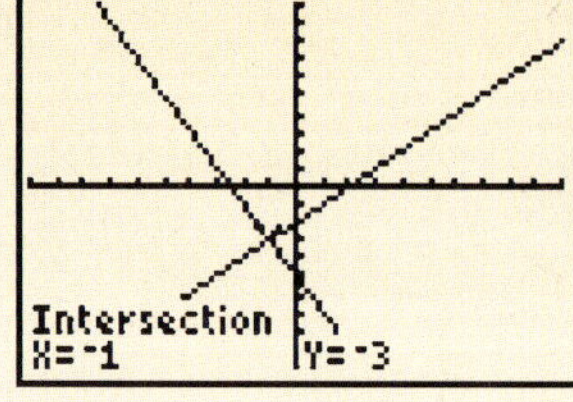

Solution set: $\{-1\}$

5. $Y_1 = -0.87x + 1.14$

$Y_2 = -2.69(x + 4.05)$

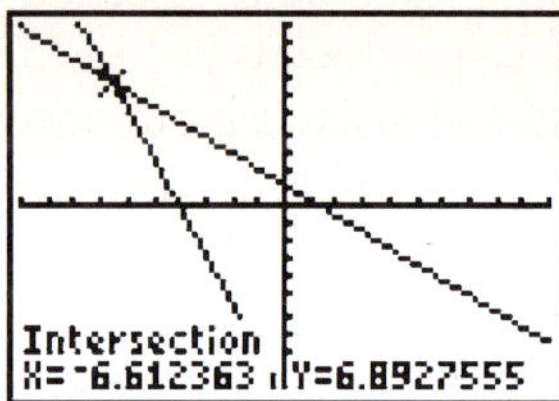

Solution set: $\{-6.61\}$

6. 20 items

Reading Your Text

SECTION 4.2

(a) When taking a graphical approach to solving a linear equation in one variable, the x-value at the point of intersection gives the ______________ to the equation.

(b) It is especially important to ______________ your work when solving an equation by graphing.

(c) You can use the ______________ utility of a graphing calculator to find the point where two curves intersect.

(d) Always use the ______________ equation to check a solution.

Basic Skills | Challenge Yourself | Calculator/Computer | Career Applications | Above and Beyond

4.2 exercises

< Objectives 1–3 >

Solve each equation graphically. Do not use a calculator.

1. $2x - 8 = 0$

2. $4x + 12 = 0$ > Videos

3. $7x - 7 = 0$

4. $2x - 6 = 0$

5. $5x - 8 = 2$

6. $4x + 5 = -3$

7. $2x - 3 = 7$

8. $5x + 9 = 4$

9. $4x - 2 = 3x + 1$ 

10. $6x + 1 = x + 6$

11. $\frac{7}{5}x - 3 = \frac{3}{10}x + \frac{5}{2}$

12. $2x - 3 = 3x - 2$

Name ______________________

Section __________ Date __________

Answers

1. ______________________
2. ______________________
3. ______________________
4. ______________________
5. ______________________
6. ______________________
7. ______________________
8. ______________________
9. ______________________
10. ______________________
11. ______________________
12. ______________________

Answers

13. ______

14. ______

15. ______

16. ______

17. ______

18. ______

19. ______

20. ______

21. ______

22. ______

13. $3(x - 1) = 4x - 5$

14. $2(x + 1) = 5x - 7$

15. $7\left(\frac{1}{5}x - \frac{1}{7}\right) = x + 1$

16. $2(3x - 1) = 12x + 4$

Basic Skills | **Challenge Yourself** | Calculator/Computer | Career Applications | Above and Beyond

Determine whether each statement is **true** *or* **false.**

17. When we solve an equation graphically, we let each side of the equation define a function.

18. After locating the point of intersection, we draw a line directly to the y-axis.

Complete each statement with **never, sometimes,** *or* **always.**

19. If each side of an equation is used to define a linear function, there will ________ be exactly two solutions to the equation.

20. If we have a zero on one side of an equation and an expression defining a linear function (with nonzero slope) on the other, the solution for the equation will ________ be the x-value where the linear graph crosses the x-axis.

21. **Business and Finance** A firm producing flashlights finds that its fixed cost is \$2,400 per week, and its marginal cost is \$4.50 per flashlight. The revenue is \$7.50 per flashlight, so the cost and revenue equations are, respectively,

$C(x) = 4.50x + 2,400$ and $R(x) = 7.50x$

Note that x represents the number of flashlights produced in the first equation and the number sold in the second. Find the break-even point for the firm (the point at which the revenue equals the cost). Use a graphical approach. > Videos

22. **Business and Finance** A company that produces portable television sets determines that its fixed cost is \$8,750 per month. The marginal cost is \$70 per set, and the revenue is \$105 per set. The cost and revenue equations, respectively, are

$C(x) = 70x + 8,750$ and $R(x) = 105x$

Note that x represents the number of TVs produced in the first equation and the number sold in the second. Find the number of sets the company must produce and sell in order to break even. Use a graphical approach.

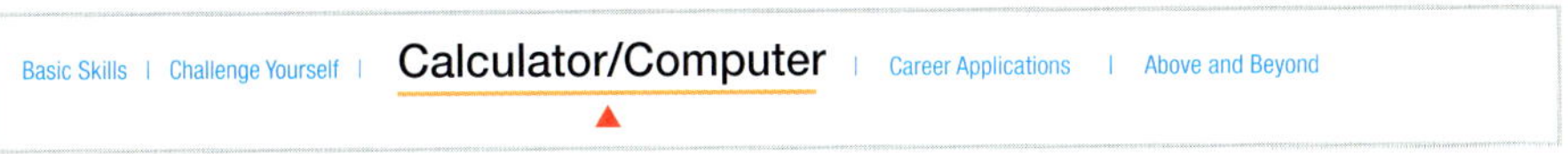

Solve each equation with a graphing calculator. Round your results to the nearest hundredth.

23. $4.17(x + 3.56) = 2.89(x + 0.35)$

24. $3.10(x - 2.57) = -4.15(x + 0.28)$ 

25. $3.61(x + 4.13) = 2.31(x - 2.59)$

26. $5.67(x - 2.13) = 1.14(x - 1.23)$

Basic Skills | Challenge Yourself | Calculator/Computer | **Career Applications** | Above and Beyond

ELECTRONICS TECHNOLOGY *Temperature sensors output voltage at a certain temperature. The output voltage varies with respect to temperature. For a particular sensor, the output voltage V for a given Celsius temperature C is given by*

$$V = 0.28C + 2.2$$

Use this information to complete exercises 27 and 28.

27. Determine the temperature (to the nearest tenth) if the sensor outputs 12.5 V.

> Videos

28. Determine the output voltage at 37°C.

ALLIED HEALTH *Yohimbine is used to reverse the effects of xylazine in deer. The recommended dose is 0.125 mg per kilogram of a deer's weight. We model the recommended dosage in terms of a deer's weight with the equation $d = 0.125w$. Use this information to complete exercises 29 and 30.*

29. What size fawn requires a 2.4-mg dose?

30. How much yohimbine should be administered to a 60-kg buck?

Answers

23. ______

24. ______

25. ______

26. ______

27. ______

28. ______

29. ______

30. ______

Basic Skills | Challenge Yourself | Calculator/Computer | Career Applications | **Above and Beyond**

Answers

31. ______

32. ______

33. ______

31. The graph below represents the rates that two different car rental agencies charge. The x-axis represents the number of miles driven (in hundreds of miles), and the y-axis represents the total charge. How would you use this graph to decide which agency to use?

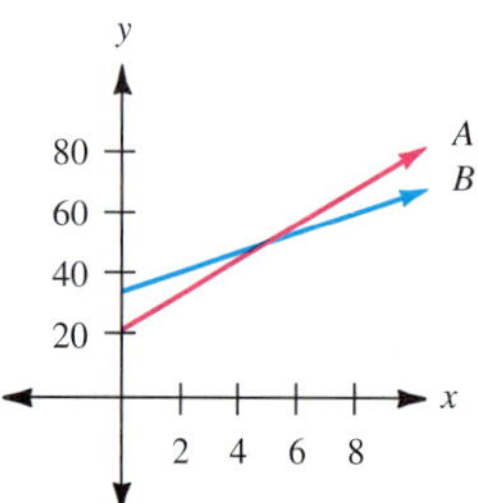

32. Graphs can be used to solve distance, time, and rate problems because graphs make pictures of the action.

(a) Consider this exercise: "Robert left on a trip, traveling at 45 mi/h. One-half hour later, Laura discovered that Robert forgot his luggage, and so she left along the same route, traveling at 54 mi/h, to catch up with him. When did Laura catch up with Robert?" How could drawing a graph help solve this problem? If you graph Robert's distance as a function of time and Laura's distance as a function of time, what does the slope of each line correspond to in the problem?

(b) Use a graph to solve this problem: Marybeth and Sam left her mother's house to drive home to Minneapolis along the interstate. They drove an average of 60 mi/h. After they had been gone for $\frac{1}{2}$ h, Marybeth's mother realized they had left their laptop computer. She grabbed it, jumped into her car, and pursued the two at 70 mi/h. Marybeth and Sam also noticed the missing computer, but not until 1 h after they had left. When they noticed that it was missing, they slowed to 45 mi/h while they considered what to do. After driving for another $\frac{1}{2}$ h, they turned around and drove back toward the home of Marybeth's mother at 65 mi/h. Where did they pass each other? How long had Marybeth's mother been driving when they met?

(c) Now that you have become an expert at this, try solving this problem by drawing a graph. It will require that you think about the slope and perhaps make several guesses when drawing the graphs. If you ride your new bicycle to class, it takes you 1.2 h. If you drive, it takes you 40 min. If you drive in traffic an average of 15 mi/h faster than you can bike, how far away from school do you live? Write an explanation of how you solved this problem by using a graph.

33. **Business and Finance** The family next door to you is trying to decide which health maintenance organization (HMO) to join. One parent has a job with health benefits for the employee only, but the rest of the family can be covered if the employee agrees to a payroll deduction. The choice is between The Empire Group, which would cost the family \$185 per month for coverage and \$25.50 for each office visit, and Group Vitality, which costs \$235 per month and \$4.00 for each office visit.

(a) Write an equation showing total yearly costs for each HMO. Graph the cost per year as a function of the number of visits, and put both graphs on the same axes.

(b) Write a note to the family explaining when The Empire Group would be better and when Group Vitality would be better. Explain how they can use your data and graph to help make a good decision. What other issues might be of concern to them?

Answers

1. $\{4\}$

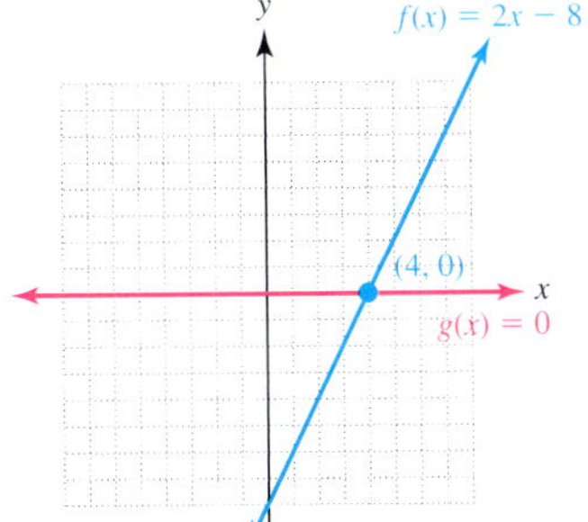

3. $\{1\}$

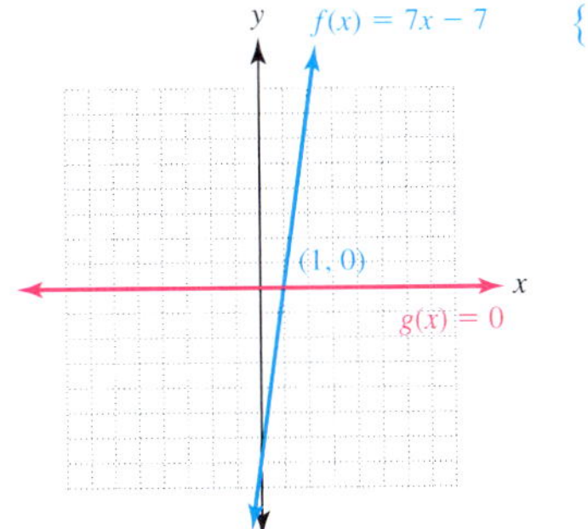

5. $\{2\}$

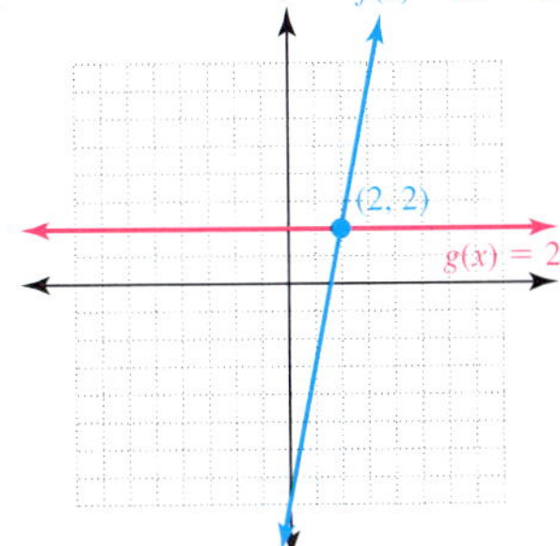

7. $\{5\}$

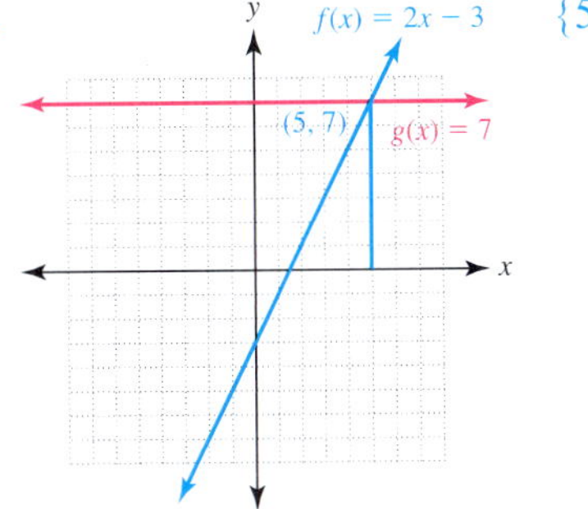

9. $\{3\}$

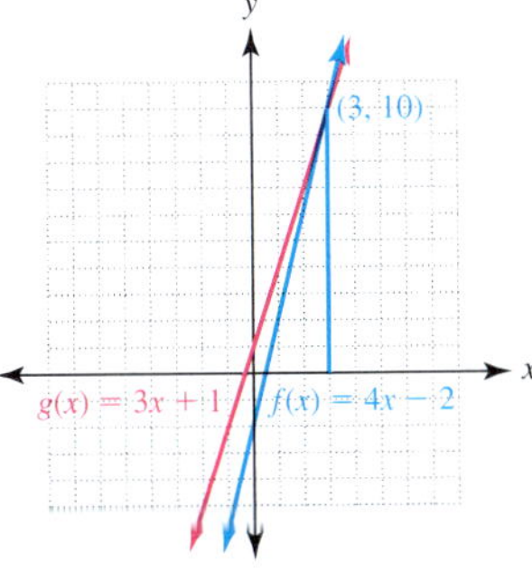

11. $\{5\}$

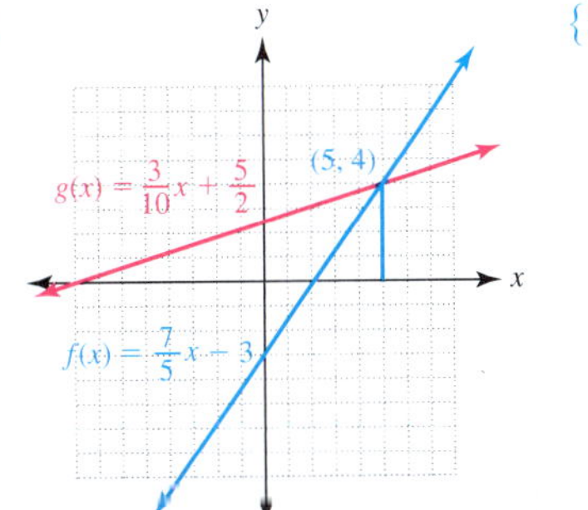

13. $\{2\}$

y, x, $(2, 3)$, $g(x) = 4x - 5$, $f(x) = 3(x - 1)$

15. $\{5\}$

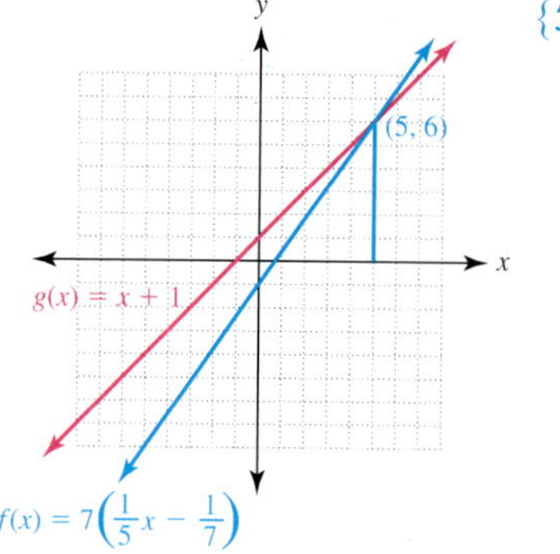

17. True **19.** never **21.** 800 flashlights

23. $\{-10.81\}$

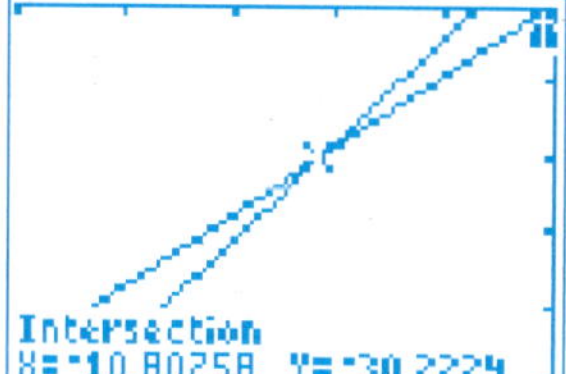
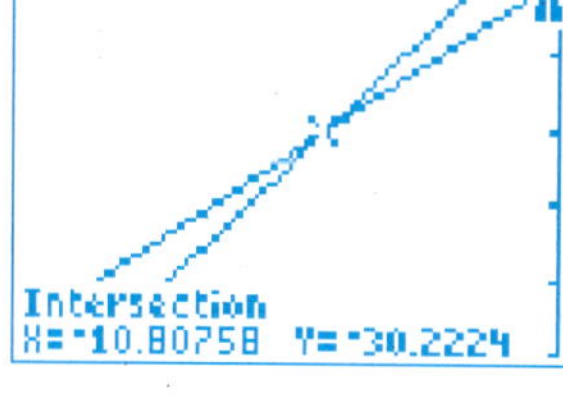

25. $\{-16.07\}$

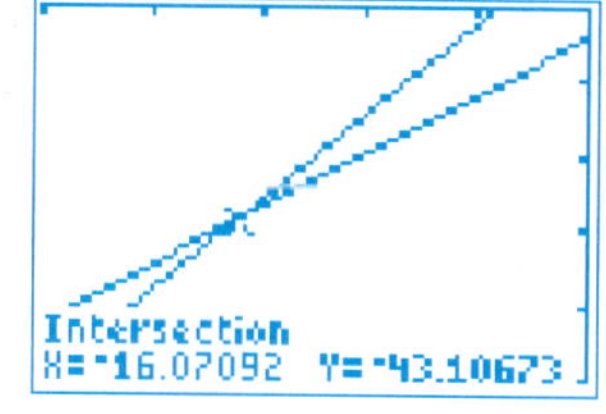

27. 36.8°C

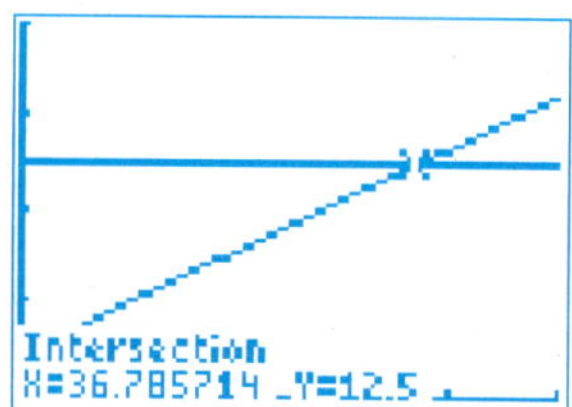

29. 19.2 kg

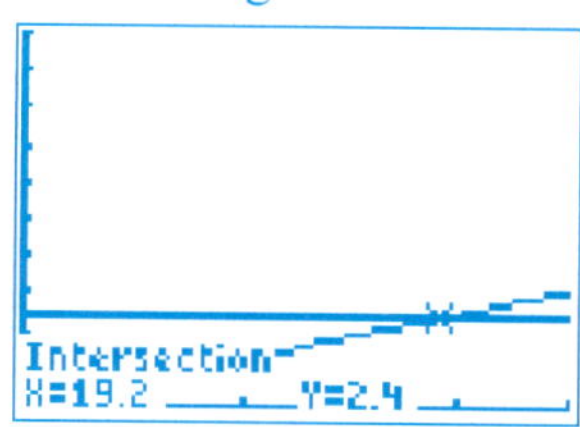

31. For a given number of miles, the lower graph gives the cheaper cost.

33. Above and Beyond

4.3 Systems of Equations in Two Variables with Applications

< 4.3 Objectives >

1 > Solve a system by the addition method

2 > Solve a system by the substitution method

3 > Use a system of equations to solve an application

Graphical solutions to linear systems are excellent for seeing and estimating solutions. The drawback comes in precision. No matter how carefully one graphs the lines, the displayed solution rarely leads to an exact solution. This problem is exaggerated when the solution includes fractions. In this section, we look at some methods that result in exact solutions.

One algebraic approach to solving a system of linear equations in two variables is the **addition method.** The basic idea to the addition method is to add the equations together so that one variable is eliminated.

NOTE

The *addition method* is also called the **elimination method.**

In Chapter 1, you learned that we can multiply both sides of an equation by some nonzero number and the result is an equivalent equation. You may need to do this to one or both equations before adding them in order to actually eliminate a variable.

The extra step is not necessary in Example 1. We will illustrate this step beginning with Example 2.

Example 1 Using the Addition Method to Solve a System

< Objective 1 >

Use the addition method to solve the system.

$$5x - 2y = 12$$
$$3x + 2y = 12$$

In this case, adding the equations eliminates the y-variable.

$8x = 24$ Remember to add the right sides of the equations together, as well.

Now, solve this last equation for x by dividing both sides by 8.

$$\frac{8x}{8} = \frac{24}{8}$$
$$x = 3$$

This last equation gives the x-value of the solution to the system of equations. We can take this value and substitute it into either of the original equations to find the y-value of the system's solution.

We substitute $x = 3$ into the first of the original equations in the system.

$5x - 2y = 12$ The first equation in the original system

$5(3) - 2y = 12$ Substitute $x = 3$ to find the y-value.

$15 - 2y = 12$

$-2y = -3$ Subtract 15 from both sides.

$y = \frac{3}{2}$ Divide both sides by -2: $\frac{-3}{-2} = \frac{3}{2}$.

NOTE

Using the other equation instead gives the same result.

$$3x + 2y = 12$$
$$3(3) + 2y = 12$$
$$9 + 2y = 12$$
$$2y = 3$$
$$y = \frac{3}{2}$$

Therefore, $\left(3, \frac{3}{2}\right)$ is the solution to the system of equations.

Check Yourself 1

Use the addition method to solve the system.

$$4x - 3y = 19$$
$$-4x + 5y = -25$$

RECALL

Multiplying both sides of an equation by the same nonzero number results in an equivalent equation.

Example 1 and the accompanying Check Yourself exercise were straightforward, in that adding the equations together eliminated one of the variables.

As we stated earlier in this section, we may need to multiply both sides of an equation by some nonzero number in order to eliminate a variable when we add the equations together. In fact, we may need to multiply both equations by (different) numbers to eliminate a variable. We see this in Example 2.

Example 2 Using the Addition Method to Solve a System

Use the addition method to solve the given system.

$$3x - 5y = 19$$
$$5x + 2y = 11$$

It should be clear that adding the two equations does not eliminate either variable. In this case, we decide which variable to eliminate and form an equivalent system by multiplying each equation by a constant.

We choose to eliminate the y-variable because the y-terms have different signs in the given system. The least common multiple of 5 and 2 is 10, so we multiply the first equation by 2 and the second equation by 5.

NOTE

We could multiply the first equation by 5 and the second equation by -3 to eliminate the x-variable.

$$3x - 5y = 19 \xrightarrow{\times 2} 6x - 10y = 38$$
$$5x + 2y = 11 \xrightarrow{\times 5} 25x + 10y = 55$$

Remember to multiply both sides in the equations.

This gives an equivalent system of equations. We can now eliminate a variable by adding the equations together.

$$\begin{aligned} 6x - 10y &= 38 \\ 25x + 10y &= 55 \\ \hline 31x \qquad &= 93 \end{aligned}$$

Divide both sides of this last equation by 31 to find the x-value of the solution.

$$\frac{31x}{31} = \frac{93}{31}$$
$$x = 3$$

Next, we substitute $x = 3$ into either of the original equations to find the y-value of the solution. We choose to substitute into the first equation.

$3x - 5y = 19$ Use an equation from the *original* system.

$3(3) - 5y = 19$ Substitute $x = 3$ into this equation.

$9 - 5y = 19$ Solve for y.

$-5y = 10$

$y = -2$

$\{(3, -2)\}$ is the solution set for the system of equations.

NOTE

The solution is unique. Because the lines have different slopes, there is a single point of intersection.

You should recall from Section 4.1 that we can check our solution by showing that it is a solution to each equation in the system.

Check

$$3x - 5y = 19$$
$$3(3) - 5(-2) \stackrel{?}{=} 19$$
$$9 + 10 \stackrel{?}{=} 19$$
$$19 = 19 \quad \text{True}$$

$$5x + 2y = 11$$
$$5(3) + 2(-2) \stackrel{?}{=} 11$$
$$15 - 4 \stackrel{?}{=} 11$$
$$11 = 11 \quad \text{True}$$

Check Yourself 2

Use the addition method to solve the system.

$$2x + 3y = -18$$
$$3x - 5y = 11$$

The following algorithm summarizes the addition method of solving linear systems of two equations in two variables.

Step by Step

Solving by the Addition Method

Step 1 If necessary, multiply one or both of the equations by a constant so that one of the variables can be eliminated by addition.

Step 2 Add the equations of the equivalent system formed in step 1.

Step 3 Solve the equation found in step 2.

Step 4 Substitute the value found in step 3 into either of the equations of the original system to find the corresponding value of the remaining variable.

Step 5 The ordered pair found in step 4 is the solution to the system. Check the solution by substituting the pair of values found in step 4 into both of the original equations.

Example 3 illustrates two special situations you may encounter while applying the addition method.

Example 3 Using the Addition Method to Solve a System

Use the addition method to solve each system.

(a) $4x + 5y = 20$
$8x + 10y = 19$

Multiply the first equation by -2. Then

$$\begin{aligned} -8x - 10y &= -40 \\ 8x + 10y &= 19 \\ \hline 0 &= -21 \end{aligned}$$

We add the two left sides to get 0 and the two right sides to get -21.

NOTE

The graph of this system is a pair of parallel lines.

The result $0 = -21$ is a *false* statement, which means that there is no point of intersection. Therefore, the system is inconsistent, and there is no solution.

(b) $5x - 7y = 9$
$15x - 21y = 27$

Multiply the first equation by -3. We then have

$$\begin{aligned} -15x + 21y &= -27 \\ 15x - 21y &= 27 \\ \hline 0 &= 0 \end{aligned}$$

We add the two equations.

NOTE

The solution set can be written $\{(x, y) \mid 5x - 7y = 9\}$. This means the set of all ordered pairs (x, y) that make $5x - 7y = 9$ a true statement.

Both variables have been eliminated, and the result is a *true* statement. If the two original equations were graphed, we would see that the two lines coincide. Thus, there are an infinite number of solutions, one for each point on that line. Recall that this is a *dependent system.*

Check Yourself 3

Use the addition method to solve each system.

(a) $3x + 2y = 8$
$9x + 6y = 11$

(b) $x - 2y = 8$
$3x - 6y = 24$

We summarize the results from Example 3.

Property

Inconsistent and Dependent Systems

When a system of two linear equations is solved:

1. If a false statement such as $3 = 4$ is obtained, then the system is inconsistent and has no solution.
2. If a true statement such as $8 = 8$ is obtained, then the system is dependent and has an infinite number of solutions.

A third method for finding the solutions of linear systems in two variables is called the **substitution method.** You may very well find the substitution method more difficult to apply in solving certain systems than the addition method, particularly when the equations involved in the substitution lead to fractions. However, the substitution method does have important extensions to systems involving higher-degree equations, as you will see in later mathematics classes.

To outline the technique, we solve one of the equations from the original system for one of the variables. That expression is then substituted into the *other* equation of the system to provide an equation in a single variable. That equation is solved, and the corresponding value for the other variable is found as before, as Example 4 illustrates.

Example 4 Using the Substitution Method to Solve a System

< Objective 2 >

(a) Use the substitution method to solve the system.

$$2x - 3y = -3$$
$$y = 2x - 1$$

Since the second equation is already solved for y, we substitute $2x - 1$ for y into the first equation.

NOTE

We now have an equation in the single variable x.

$$2x - 3(2x - 1) = -3$$

Solving for x gives

$$2x - 6x + 3 = -3$$
$$-4x = -6$$
$$x = \frac{3}{2}$$

We now substitute $x = \frac{3}{2}$ into the equation that was solved for y.

$$y = 2\left(\frac{3}{2}\right) - 1$$
$$= 3 - 1 = 2$$

The solution set for our system is $\left\{\left(\frac{3}{2}, 2\right)\right\}$.

Again, we check our proposed solution by showing that it is a solution to each equation in the system.

Check

$$2x - 3y = -3$$
$$2\left(\frac{3}{2}\right) - 3(2) \stackrel{?}{=} -3$$
$$3 - 6 \stackrel{?}{=} -3$$
$$-3 = -3 \quad \text{True}$$

$$y = 2x - 1$$
$$(2) \stackrel{?}{=} 2\left(\frac{3}{2}\right) - 1$$
$$2 \stackrel{?}{=} 3 - 1$$
$$2 = 2 \quad \text{True}$$

NOTE

Why did we choose to solve the second equation for y? We could have solved for x, so that

$x = \frac{y + 2}{3}$

We simply chose the easier case to avoid fractions.

(b) Use the substitution method to solve the system.

$$2x + 3y = 16$$
$$3x - y = 2$$

We start by solving the second equation for y.

$$3x - y = 2$$
$$-y = -3x + 2$$
$$y = 3x - 2$$

Substituting into the other equation yields

$$2x + 3(3x - 2) = 16$$
$$2x + 9x - 6 = 16$$
$$11x = 22$$
$$x = 2$$

NOTE

The solution should be checked in *both* equations of the original system.

We now substitute $x = 2$ into the equation that we solved for y.

$$y = 3(2) - 2$$
$$= 6 - 2 = 4$$

The solution set for the system is $\{(2, 4)\}$. We leave the check of this result to you.

Check Yourself 4

Use the substitution method to solve each system.

(a) $2x + 3y = 6$
$x = 3y + 6$

(b) $3x + 4y = -3$
$x + 4y = 1$

The following algorithm summarizes the substitution method for solving linear systems of two equations in two variables.

Step by Step

Solving by the Substitution Method

Step 1 If necessary, solve one of the equations of the original system for one of the variables.

Step 2 Substitute the expression obtained in step 1 into the *other* equation of the system to write an equation in a single variable.

Step 3 Solve the equation found in step 2.

Step 4 Substitute the value found in step 3 into the equation found in step 1 to find the corresponding value of the remaining variable.

Step 5 The ordered pair found in step 4 is the solution to the system of equations. Check the solution by substituting the pair of values found in step 4 into *both* equations of the original system.

As with the addition method, if both variables are eliminated after simplifying in step 3 and a true statement is formed, then we have a dependent system. If a false statement occurs, then the system is inconsistent.

A natural question at this point is, "How do you decide which solution method to use?" First, the graphical method can generally provide only approximate solutions. When exact solutions are necessary, one of the algebraic methods must be applied. Which method to use depends totally on the given system.

If you can easily solve for a variable in one of the equations, the substitution method should work well. However, if solving for a variable in either equation of the system leads to fractions, you may find the addition approach more efficient.

We are now ready to apply our equation-solving skills to solving applications and word problems. Being able to extend these skills to problem solving is an important goal, and the procedures developed here are used throughout the rest of the book.

Although we consider applications from a variety of areas in this section, all are approached with the same five-step strategy presented here.

Step by Step

Solving Applications

Step 1 Read the problem carefully to determine the unknown quantities.

Step 2 Choose variables to represent the unknown quantities.

Step 3 Translate the problem to the language of algebra to form a system of equations.

Step 4 Solve the system of equations.

Step 5 Answer the question from the original problem and verify your solution by returning to the original problem.

Example 5 **Solving a Mixture Problem**

< Objective 3 >

NOTES

Because we use *two* variables, we must form *two* equations.

The total value of the mixture comes from 100(9.50) = 950.

A coffee merchant has two types of coffee beans, one selling for \$8 per pound (lb) and the other for \$10 per pound. The beans are to be mixed to provide 100 lb of a mixture selling for \$9.50 per pound. How much of each type of coffee bean should be used to form 100 lb of the mixture?

Step 1 The unknowns are the amounts of the two types of beans.

Step 2 We use two variables to represent the two unknowns. Let x be the amount of the \$8 beans and y the amount of the \$10 beans.

Step 3 We now want to establish a system of two equations. One equation will be based on the *total amount* of the mixture, the other on the mixture's *value*.

$$x + y = 100$$ The mixture must weigh 100 lb.

$$8x + 10y = 950$$

Value of \$8 beans — Value of \$10 beans — Total value

Step 4 An easy approach to the solution of the system is to multiply the first equation by -8 and add the equations to eliminate x.

$$\begin{aligned} -8x - 8y &= -800 \\ 8x + 10y &= 950 \\ \hline 2y &= 150 \\ y &= 75 \text{ lb} \end{aligned}$$

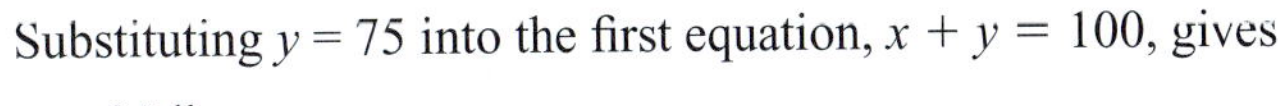

NOTE

$8(25) + 10(75) = 200 + 750$
$= 950$

$9.50(100) = 950$

Substituting $y = 75$ into the first equation, $x + y = 100$, gives

$x = 25$ lb

Step 5 We should use 25 lb of \$8 beans and 75 lb of \$10 beans.

To check the result, show that the value of the \$8 beans added to the value of the \$10 beans equals the desired value of the mixture.

Check Yourself 5

Peanuts, which sell for \$2.40 per pound, and cashews, which sell for \$6 per pound, are to be mixed to form a 60-lb mixture selling for \$3 per pound. How much of each type of nut should be used?

A related problem is illustrated in Example 6.

Example 6 Solving a Mixture Problem

A chemist has a 25% and a 50% acid solution. How much of each solution should be used to form 200 milliliters (mL) of a 35% acid solution?

Step 1 The unknowns in this case are the amounts of the 25% and 50% solutions to be used in forming the mixture.

Step 2 Again we use two variables to represent the two unknowns. Let x be the amount of the 25% solution and y the amount of the 50% solution. Let's draw a picture before proceeding to form a system of equations.

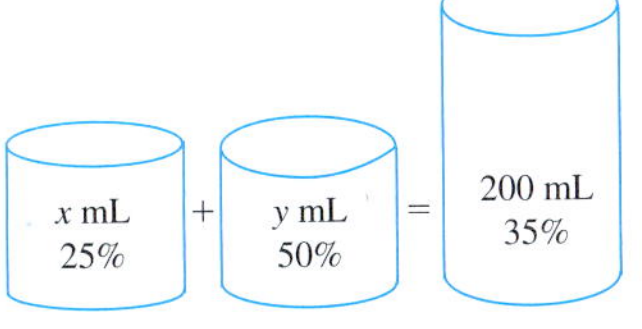

Drawing a sketch of a problem is often a valuable part of the problem-solving strategy.

Step 3 Now, to form our two equations, we want to consider two relationships: the *total amounts* combined and the *amounts of acid* combined.

$$x + y = 200 \quad \text{Total amounts combined}$$
$$0.25x + 0.50y = 0.35(200) \quad \text{Amounts of acid combined}$$

Step 4 Clear the second equation of decimals by multiplying it by 100. The solution then proceeds as before, with the result

$x = 120$ mL (25% solution)

$y = 80$ mL (50% solution)

Step 5 We need 120 mL of the 25% solution and 80 mL of the 50% solution.

To check, show that the amount of acid in the 25% solution, $(0.25)(120)$, added to the amount in the 50% solution, $(0.50)(80)$, equals the correct amount in the mixture, $(0.35)(200)$. We leave that to you.

Check Yourself 6

A pharmacist wants to prepare 300 mL of a 20% alcohol solution. How much of a 30% solution and a 15% solution should be used to form the desired mixture?

Applications that involve a constant rate of travel, or speed, require the use of the distance formula

$d = rt$

where

d = distance traveled
r = rate or speed
t = time

Example 7 Solving a Distance-Rate-Time Problem

A boat can travel 36 mi downstream in 2 h. Coming back upstream, the boat takes 3 h. What is the rate of the boat in still water? What is the rate of the current?

> **NOTE**
>
> Downstream the rate is then
>
> $x + y$
>
> Upstream, the rate is
>
> $x - y$

Step 1 We want to find the two rates.

Step 2 Let x be the rate of the boat in still water and y the rate of the current.

Step 3 To form a system, think about the following. Downstream, the rate of the boat is *increased* by the effect of the current. Upstream, the rate is *decreased.*

In many applications, it helps to lay out the information in tabular form. We will try that strategy here.

	d	r	t
Downstream	36	$x + y$	2
Upstream	36	$x - y$	3

Since $d = rt$, from the table we can easily form two equations:

$$36 = (x + y)(2)$$

$$36 = (x - y)(3)$$

Step 4 We clear the equations of parentheses and simplify, to write the equivalent system.

$$x + y = 18$$

$$x - y = 12$$

Solving, we have

$$x = 15 \text{ mi/h}$$

$$y = 3 \text{ mi/h}$$

Step 5 The rate of the current is 3 mi/h, and the rate of the boat in still water is 15 mi/h.

To check, verify the $d = rt$ equation in *both* the upstream and the downstream cases. We leave that to you.

Check Yourself 7

A plane flies 480 mi in an easterly direction, with the wind, in 4 h. Returning westerly along the same route, against the wind, the plane takes 6 h. What is the rate of the plane in still air? What is the rate of the wind?

Systems of equations have many applications in business settings. Example 8 illustrates one such application.

Example 8 A Business and Finance Application

A manufacturer produces a standard model and a deluxe model of a 32-inch (in.) television set. The standard model requires 12 h of labor to produce, while the deluxe model requires 18 h. The company has 360 h of labor available per week. The plant's capacity is a total of 25 sets per week. If all the available time and capacity are to be used, how many of each type of set should be produced?

Step 1 The unknowns in this case are the number of standard and deluxe models that can be produced.

Step 2 Let x be the number of standard models and y the number of deluxe models.

Step 3 Our system comes from the two given conditions that fix the total number of sets that can be produced and the total labor hours available.

NOTE

The choices for x and y could be reversed.

$$x + y = 25 \quad \leftarrow \text{Total number of sets}$$
$$12x + 18y = 360 \quad \leftarrow \text{Total labor hours available}$$

Labor hours, standard sets (12x) — Labor hours, deluxe sets (18y)

Step 4 Solving the system in step 3, we have

$x = 15$ and $y = 10$

Step 5 This tells us that to use all the available capacity, the plant should produce 15 standard sets and 10 deluxe sets per week.

We leave the check of this result to the reader.

Check Yourself 8

A manufacturer produces standard cassette players and compact disk players. Each cassette player requires 2 h of electronic assembly, and each CD player requires 3 h. The cassette players require 4 h of case assembly and the CD players 2 h. The company has 120 h of electronic assembly time available per week and 160 h of case assembly time. How many of each type of unit can be produced each week if all available assembly time is to be used?

Here is another application that leads to a system of two equations.

Example 9 A Business and Finance Application

Two car rental agencies have the following rate structures for a subcompact car. Company A charges \$20 per day plus 15¢ per mile. Company B charges \$18 per day plus 16¢ per mile. If you rent a car for 1 day, for what number of miles will the two companies have the same total charge?

Letting c represent the total a company will charge and m the number of miles driven, we calculate the rates.

For company A:

$$c = 20 + 0.15m$$

For company B:

$$c = 18 + 0.16m$$

The system can be solved most easily by substitution. Substituting $18 + 0.16m$ for c in the first equation gives

$$18 + 0.16m = 20 + 0.15m$$
$$0.01m = 2$$
$$m = 200 \text{ mi}$$

The graph of the system is shown below.

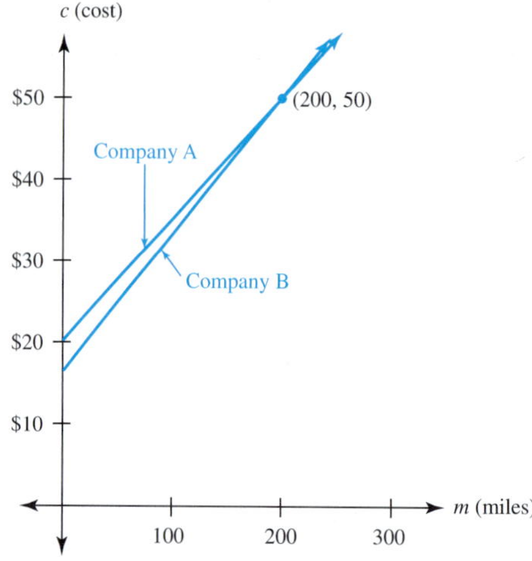

From the graph, how would you make a decision about which agency to use?

Check Yourself 9

For a compact car, the same two companies charge \$27 per day plus 20¢ per mile and \$24 per day plus 22¢ per mile. For a 2-day rental, for what mileage will the charges be the same? What is the total charge?

Check Yourself ANSWERS

1. $\left\{\left(\frac{5}{2}, -3\right)\right\}$ **2.** $\{(-3, -4)\}$
3. **(a)** Inconsistent system, no solution; **(b)** dependent system, an infinite number of solutions **4.** **(a)** $\left\{\left(4, -\frac{2}{3}\right)\right\}$; **(b)** $\left\{\left(-2, \frac{3}{4}\right)\right\}$
5. 50 lb of peanuts and 10 lb of cashews **6.** 100 mL of the 30% and 200 mL of the 15% **7.** Plane: 100 mi/h, wind: 20 mi/h
8. 30 cassette players and 20 CD players **9.** At 300 mi, \$114 charge

Reading Your Text

SECTION 4.3

(a) Multiplying both sides of an equation by the same nonzero number results in an ______________ equation.

(b) The solution to a system can be approximated by graphing the lines and locating the ______________ point.

(c) When solving a system results in a false statement such as $3 = 4$, then the system has ______________ solutions.

(d) The distance formula states that distance is equal to rate times ________.

Basic Skills | Challenge Yourself | Calculator/Computer | Career Applications | Above and Beyond

4.3 exercises

< Objective 1 >

Use the addition method to solve each system. If a unique solution does not exist, state whether the system is inconsistent or dependent.

1. $2x - y = 1$
$-2x + 3y = 5$

2. $x + 3y = 12$
$2x - 3y = 6$

3. $4x + 3y = -5$
$5x + 3y = -8$

4. $2x + 3y = 1$
$5x + 3y = 16$

5. $x + y = 3$
$3x - 2y = 4$

> Videos

6. $x - y = -2$
$2x + 3y = 21$

7. $2x + y = 8$
$-4x - 2y = -16$

8. $2x + 3y = 11$
$x - y = 3$

9. $5x - 2y = 31$
$4x + 3y = 11$

10. $2x - y = 4$
$6x - 3y = 10$

> Videos

11. $3x - 2y = 7$
$-6x + 4y = -15$

12. $3x + 4y = 0$
$5x - 3y = -29$

13. $-2x + 7y = 2$
$3x - 5y = -14$

14. $5x - 2y = 3$
$10x - 4y = 6$

< Objective 2 >

Use the substitution method to solve each system. If a unique solution does not exist, state whether the system is inconsistent or dependent.

15. $x - y = 7$
$y = 2x - 12$

16. $x - y = 4$
$x = 2y - 2$

17. $3x + 2y = -18$
$x = 3y + 5$

18. $3x - 18y = 4$
$x = 6y + 2$

19. $10x - 2y = 4$
$y = 5x - 2$

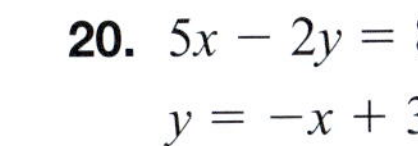

20. $5x - 2y = 8$
$y = -x + 3$

21. $3x + 4y = 9$
$y = 3x + 1$

22. $6x - 5y = 27$
$x = 5y + 2$

23. $x - 7y = 3$
$2x - 5y = 15$

24. $4x + 3y = -11$
$5x + y = -11$

Name ______________

Section ________ Date ________

Answers

1. ________ 2. ________

3. ________ 4. ________

5. ________ 6. ________

7. ________

8. ________ 9. ________

10. ________

11. ________

12. ________ 13. ________

14. ________

15. ________ 16. ________

17. ________

18. ________

19. ________

20. ________

21. ________ 22. ________

23. ________ 24. ________

Answers

25. ______
26. ______
27. ______
28. ______
29. ______
30. ______
31. ______
32. ______
33. ______
34. ______
35. ______
36. ______
37. ______
38. ______

25. $3x - 9y = 7$
$-x + 3y = 2$

26. $5x - 6y = 21$
$x - 2y = 5$

Use any method to solve each system.
Hint: *You can use multiplication to clear equations of fractions.*

27. $2x - 3y = 4$
$x = 3y + 6$

28. $5x + y = 2$
$5x - 3y = 6$

29. $4x - 3y = 0$
$5x + 2y = 23$

30. $7x - 2y = -17$
$x + 4y = 4$

31. $3x - y = 17$
$5x + 3y = 5$

32. $7x + 3y = -51$
$y = 2x + 9$

33. $\frac{1}{4}x - \frac{1}{6}y = 2$
$\frac{3}{4}x + y = 3$

34. $\frac{1}{5}x - \frac{1}{2}y = 0$
$x - \frac{3}{2}y = 4$

35. $\frac{2}{3}x + \frac{3}{5}y = -3$
$\frac{1}{3}x + \frac{2}{5}y = -3$

36. $\frac{3}{8}x - \frac{1}{2}y = -5$
$\frac{1}{4}x + \frac{3}{2}y = 4$

< Objective 3 >

In exercises 37 to 44, each application can be solved with a system of linear equations. Match the application with the appropriate system below.

(a) $12x + 5y = 116$
$8x + 12y = 112$

(b) $x + y = 4{,}000$
$0.03x + 0.05y = 180$

(c) $x + y = 200$
$0.20x + 0.60y = 90$

(d) $x + y = 36$
$y = 3x - 4$

(e) $2(x + y) = 36$
$3(x - y) = 36$

(f) $x + y = 200$
$6.50x + 4.50y = 980$

(g) $L = 2W + 3$
$2L + 2W = 36$

(h) $x + y = 120$
$2.20x + 5.40y = 360$

37. Number Problem One number is 4 less than 3 times another. If the sum of the numbers is 36, what are the two numbers?

38. Business and Finance Suppose a movie theater sold 200 adult and student tickets for a showing with a revenue of $980. If the adult tickets cost $6.50 and the student tickets cost $4.50, how many of each type of ticket were sold?

39. **Geometry** The length of a rectangle is 3 cm more than twice its width. If the perimeter of the rectangle is 36 cm, find the dimensions of the rectangle.

40. **Business and Finance** An order of 12 dozen roller-ball pens and 5 dozen ballpoint pens costs \$116. A later order for 8 dozen roller-ball pens and 12 dozen ballpoint pens costs \$112. What was the cost of 1 dozen of each type of pen?

41. **Business and Finance** A candy merchant wants to mix peanuts selling at \$2.20 per pound with cashews selling at \$5.40 per pound to form 120 lb of a mixed-nut blend that will sell for \$3 per pound. What amount of each type of nut should be used?

42. **Business and Finance** Donald has investments totaling \$4,000 in two accounts—one a savings account paying 3% interest and the other a bond paying 5%. If the annual interest from the two investments was \$180, how much did he have invested at each rate?

43. **Science and Medicine** A chemist wants to combine a 20% alcohol solution with a 60% solution to form 200 mL of a 45% solution. How much of each solution should be used to form the mixture?

44. **Science and Medicine** Xian was able to make a downstream trip of 36 mi in 2 h. Returning upstream, he took 3 h to make the trip. How fast can his boat travel in still water? What was the rate of the river's current?

Solve each application with a system of equations.

45. **Business and Finance** Suppose 750 tickets were sold for a concert with a total revenue of \$5,300. If adult tickets were \$8 and students tickets were \$4.50, how many of each type of ticket were sold?

46. **Business and Finance** Theater tickets sold for \$7.50 on the main floor and \$5 in the balcony. The total revenue was \$3,250, and there were 100 more main-floor tickets sold than balcony tickets. Find the number of each type of ticket sold.

47. **Geometry** The length of a rectangle is 3 in. less than twice its width. If the perimeter of the rectangle is 84 in., find the dimensions of the rectangle.

48. **Geometry** The length of a rectangle is 5 cm more than 3 times its width. If the perimeter of the rectangle is 74 cm, find the dimensions of the rectangle.

49. **Business and Finance** A garden store sold 8 bags of mulch and 3 bags of fertilizer for \$24. The next purchase was for 5 bags of mulch and 5 bags of fertilizer. The cost of that purchase was \$25. Find the cost of a single bag of mulch and a single bag of fertilizer.

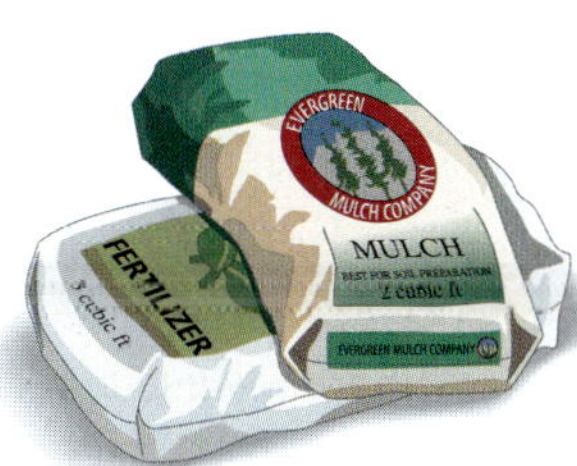

Answers

39. ____________

40. ____________

41. ____________

42. ____________

43. ____________

44. ____________

45. ____________

46. ____________

47. ____________

48. ____________

49. ____________

Answers

50.

51.

52.

53.

54.

55.

56.

57.

58.

59.

60.

50. **Business and Finance** The cost of an order for 10 computer disks and 3 packages of paper was \$22.50. The next order was for 30 disks and 5 packages of paper, and its cost was \$53.50. Find the price of a single disk and a single package of paper.

> Videos

51. **Business and Finance** A coffee retailer has two grades of decaffeinated beans, one selling for \$4 per pound and the other for \$6.50 per pound. She wishes to blend the beans to form a 150-lb mixture that will sell for \$4.75 per pound. How many pounds of each grade of bean should be used in the mixture?

52. **Business and Finance** A candy merchant sells jelly beans at \$3.50 per pound and gumdrops at \$4.70 per pound. To form a 200-lb mixture that will sell for \$4.40 per pound, how many pounds of each type of candy should be used?

53. **Business and Finance** Cheryl decided to divide \$12,000 into two investments—one a time deposit that pays 4% annual interest and the other a bond that pays 6%. If her annual interest was \$640, how much did she invest at each rate?

54. **Business and Finance** Miguel has \$1,500 more invested in a mutual fund paying 5% interest than in a savings account paying 3%. If he received \$155 in interest for 1 year, how much did he have invested in the two accounts?

55. **Science and Medicine** A chemist mixes a 10% acid solution with a 50% acid solution to form 400 mL of a 40% solution. How much of each solution should be used in the mixture?

56. **Science and Medicine** A laboratory technician wishes to mix a 70% saline solution and a 20% saline solution to prepare 500 mL of a 40% solution. What amount of each solution should be used?

57. **Science and Medicine** A boat traveled 36 mi up a river in 3 h. Returning downstream, the boat took 2 h. What is the boat's rate in still water, and what is the rate of the river's current?

58. **Science and Medicine** A jet flew east a distance of 1,800 mi with the jetstream in 3 h. Returning west, against the jetstream, the jet took 4 h. Find the jet's speed in still air and the rate of the jetstream.

59. **Number Problem** The sum of the digits of a two-digit number is 8. If the digits are reversed, the new number is 36 more than the original number. Find the original number. (*Hint:* If u represents the units digit of the number and t the tens digit, the original number can be represented by $10t + u$.)

60. **Number Problem** The sum of the digits of a two-digit number is 10. If the digits are reversed, the new number is 54 less than the original number. What was the original number?

61. Business and Finance A manufacturer produces a battery-powered calculator and a solar model. The battery-powered model requires 10 min of electronic assembly and the solar model 15 min. There is 450 min of assembly time available per day. Both models require 8 min for packaging, and 280 min of packaging time is available per day. If the manufacturer wants to use all the available time, how many of each unit should be produced per day?

62. Business and Finance A small tool manufacturer produces a standard model and a cordless model power drill. The standard model takes 2 h of labor to assemble and the cordless model 3 h. There is 72 h of labor available per week for the drills. Material costs for the standard drill are \$10, and for the cordless drill they are \$20. The company wishes to limit material costs to \$420 per week. How many of each model drill should be produced in order to use all the available resources?

chapter 4 > Make the Connection

63. Business and Finance In economics, a demand equation gives the quantity D that will be demanded by consumers at a given price p, in dollars. Suppose that $D = 210 - 4p$ for a particular product.

A supply equation gives the supply S that will be available from producers at price p. Suppose also that for the same product $S = 10p$.

The equilibrium point is that point where the supply equals the demand (here, where $S = D$). Use the given equations to find the equilibrium point.

64. Business and Finance Suppose the demand equation for a product is $D = 150 - 3p$ and the supply equation is $S = 12p$. Find the equilibrium point for the product.

chapter 4 > Make the Connection

65. Business and Finance Two car rental agencies have different rate structures for compact cars.

Company A: \$30/day and 22¢/mi

Company B: \$28/day and 26¢/mi

For a 2-day rental, at what number of miles will the charges be the same?

66. Construction Two construction companies submit these bids.

Company A: \$5,000 plus \$15 per square foot of building

Company B: \$7,000 plus \$12.50 per square foot of building

For what number of square feet of building will the bids of the two companies be the same?

Answers

61. ______

62. ______

63. ______

64. ______

65. ______

66. ______

Answers

67. ______

68. ______

69. ______

70. ______

71. ______

72. ______

73. ______

74. ______

75. ______

76. ______

Basic Skills | **Challenge Yourself** | Calculator/Computer | Career Applications | Above and Beyond

Complete each statement with **never, sometimes,** *or* **always.**

67. Both variables are ______ eliminated when the equations of a linear system are added.

68. A system is ______ both inconsistent and dependent.

69. It is ______ possible to use the addition method to solve a linear system.

70. The substitution method is ______ easier to use than the addition method.

For exercises 71 to 74, adjust the viewing window on your calculator so that you can see the point of intersection for the two lines representing the equations in the system. Then approximate the solution, expressing each coordinate to the nearest tenth.

71. $y = 2x - 3$
$2x + 3y = 1$

72. $3x - 4y = -7$
$2x + 3y = -1$

73. $5x - 12y = 8$
$7x + 2y = 44$

74. $9x - 3y = 10$
$x + 5y = 58$

Basic Skills | Challenge Yourself | Calculator/Computer | **Career Applications** | Above and Beyond

75. Construction Technology The beam shown in the figure is 24 feet long and has the load indicated on each end. > Videos

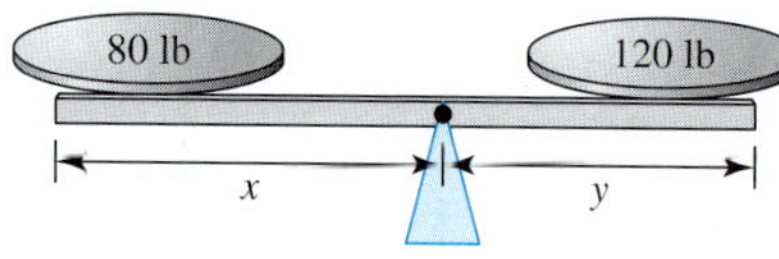

(a) Construct a system of equations in order to determine the point at which the beam balances. (*Hint:* The product of the length of a side and the load on that side must equal the product of the length of the other side and the load on that other side.)

(b) Find the necessary lengths, x and y, in order to balance the beam. Report your results to the nearest hundredth foot.

76. Manufacturing Technology Production for one week is up 2,600 units over the previous week. A total of 27,200 units were produced during the 2-week span.

(a) Construct a system of equations to determine the production for each of the 2 weeks.

(b) Determine the number of units produced each week.

77. **Allied Health** A medical lab technician needs to determine how much 9% sulfuric acid (H_2SO_4) solution to mix with 2% H_2SO_4 in order to produce 75 mL of 4% solution.

(a) Construct a system of equations to solve this application using x for the quantity of 9% solution needed.

(b) Determine the amount of each solution needed. Report your results to the nearest tenth mL.

78. **Allied Health** A medical lab technician needs to determine how much 20% saline solution to mix with 5% saline in order to produce 100 mL of 12% solution.

(a) Construct a system of equations to solve this application using x for the quantity of 20% solution needed.

(b) Determine the amount of each solution needed. Report your results to the nearest tenth mL.

Answers

77. ____

78. ____

79. ____

80. ____

81. ____

82. ____

83. ____

84. ____

Certain systems that are not linear can be solved with the methods of this section if we first substitute to change variables. For instance, the system

$$\frac{1}{x} + \frac{1}{y} = 4$$
$$\frac{1}{x} - \frac{3}{y} = -6$$

can be solved by the substitutions $u = \frac{1}{x}$ *and* $v = \frac{1}{y}$. *That gives the system* $u + v = 4$ *and* $u - 3v = -6$. *The system is then solved for u and v, and the corresponding values for x and y are found. Use this method to solve the systems in exercises 79 to 82.*

79. $\frac{1}{x} + \frac{1}{y} = 4$
$\frac{1}{x} - \frac{3}{y} = -6$

80. $\frac{1}{x} + \frac{3}{y} = 1$
$\frac{4}{x} + \frac{3}{y} = 3$

81. $\frac{2}{x} + \frac{3}{y} = 4$
$\frac{2}{x} - \frac{6}{y} = 10$

82. $\frac{4}{x} - \frac{3}{y} = -1$
$\frac{12}{x} - \frac{1}{y} = 1$

Here is another method for writing the equation of a line through two points. Given the coordinates of two points, substitute each pair of values into the equation $y = mx + b$. *This gives a system of two equations in variables m and b, which can be solved as before.*

In exercises 83 and 84, write the equation of the line through each pair of points, using the method outlined above.

83. (2, 1) and (4, 4)

84. (−3, 7) and (6, 1)

Answers

85. ______

86. ______

87. ______

85. We have discussed three different methods of solving a system of two linear equations in two unknowns: the graphical method, the addition method, and the substitution method. Discuss the strengths and weaknesses of each method.

86. Determine a system of two linear equations for which the solution is (3, 4). Are there other systems that have the same solution? If so, determine at least one more and explain why this can be true.

87. Suppose we have the linear system:

$$Ax + By = C$$
$$Dx + Ey = F$$

(a) Multiply the first equation by $-D$, multiply the second equation by A, and add. This will allow you to eliminate x. Solve for y and indicate what must be true about the coefficients in order for a unique value for y to exist.

(b) Now return to the original system and eliminate y instead of x. [*Hint*: Try multiplying the first equation by E and the second equation by $-B$.] Solve for x and again indicate what must be true about the coefficients for a unique value for x to exist.

Answers

1. $\{(2, 3)\}$ **3.** $\left\{\left(-3, \frac{7}{3}\right)\right\}$ **5.** $\{(2, 1)\}$ **7.** Infinite number of solutions, dependent system **9.** $\{(5, -3)\}$ **11.** No solutions, inconsistent system **13.** $\{(-8, -2)\}$ **15.** $\{(5, -2)\}$ **17.** $\{(-4, -3)\}$ **19.** Infinite number of solutions, dependent system **21.** $\left\{\left(\frac{1}{3}, 2\right)\right\}$ **23.** $\{(10, 1)\}$ **25.** No solutions, inconsistent system **27.** $\left\{\left(-2, -\frac{8}{3}\right)\right\}$ **29.** $\{(3, 4)\}$ **31.** $\{(4, -5)\}$ **33.** $\left\{\left(\frac{20}{3}, -2\right)\right\}$ **35.** $\{(9, -15)\}$ **37.** (d) **39.** (g) **41.** (h) **43.** (c) **45.** 550 adult tickets; 200 student tickets **47.** 27 in. by 15 in. **49.** Mulch \$1.80; fertilizer \$3.20 **51.** 105 lb of \$4 beans; 45 lb of \$6.50 beans **53.** \$8,000 bond; \$4,000 time deposit **55.** 100 mL of 10%; 300 mL of 50% **57.** 15 mi/h boat; 3 mi/h current **59.** 26 **61.** 15 battery-powered calculators; 20 solar models **63.** (15, 150) **65.** 100 mi **67.** sometimes **69.** always **71.** $\{(1.3, -0.5)\}$ **73.** $\{(5.8, 1.7)\}$ **75.** **(a)** $x + y = 24$; $90x = 280y$; **(b)** $x = 18.16$ ft, $y = 5.84$ ft **77.** **(a)** $x + y = 75$; $0.09x + 0.02y = 3$; **(b)** 9%: 21.4 mL; 2%: 53.6 mL **79.** $\left\{\left(\frac{2}{3}, \frac{2}{5}\right)\right\}$ **81.** $\left\{\left(\frac{1}{3}, -\frac{3}{2}\right)\right\}$ **83.** $y = \frac{3}{2}x - 2$ **85.** Above and Beyond **87.** **(a)** $y = \frac{AF - CD}{AE - BD}$, $AE - BD \neq 0$; **(b)** $x = \frac{CE - BF}{AE - BD}$, $AE - BD \neq 0$

4.4 Systems of Linear Equations in Three Variables

< 4.4 Objectives >

1 > Use the addition method to solve a system of linear equations in three variables

2 > Solve an application involving a system with three variables

Suppose an application involves three quantities that we want to label x, y, and z. A typical equation used for the solution might be

$$2x + 4y - z = 8$$

This is called a **linear equation in three variables.** A solution for such an equation is an **ordered triple** (x, y, z) of real numbers that satisfies the equation. For example, the ordered triple $(2, 1, 0)$ is a solution for the equation above since substituting 2 for x, 1 for y, and 0 for z results in a true statement.

$$\begin{aligned} 2x + 4y - z &= 8 \\ 2(2) + 4(1) - (0) &\stackrel{?}{=} 8 \\ 4 + 4 &\stackrel{?}{=} 8 \\ 8 &= 8 \qquad \text{True} \end{aligned}$$

NOTE

For a unique solution to exist, when *three variables* are involved, we must have at least *three equations.*

Of course, other solutions, in fact infinitely many, exist. You might want to verify that $(1, 1, -2)$ and $(3, 1, 2)$ are also solutions. To extend the concepts of Section 4.3, we want to consider systems of three linear equations in three variables, such as

$$\begin{aligned} x + y + z &= 5 \\ 2x - y + z &= 9 \\ x - 2y + 3z &= 16 \end{aligned}$$

NOTE

The choice of which variable to eliminate is yours.

Generally, you should pick the variable that allows the easiest computation.

The solution for such a system is the set of all ordered triples that satisfy each equation of the system. In this case, you should verify that $(2, -1, 4)$ is a solution for the system since that ordered triple makes each equation a true statement.

In this section, we consider the **addition method.** We then apply what we learn to solving applications.

The central idea is to choose *two pairs* of equations from the system and, by the addition method, to eliminate the *same variable* from each of those pairs. The method is best illustrated by example. We now proceed to see how we found the solution to the previous system.

Example 1 Solving a Linear System in Three Variables

< Objective 1 >

Solve the system.

$$\begin{aligned} x + y + z &= 5 \\ 2x - y + z &= 9 \\ x - 2y + 3z &= 16 \end{aligned}$$

First, we choose a variable to eliminate. Eliminating the y-variable seems reasonably convenient. Then we choose a pair of the equations and use the addition method to eliminate our chosen variable.

We choose to eliminate the y-variable by adding the first two equations together.

$$\begin{array}{r}x + y + z = 5\\ 2x - y + z = 9\\ \hline 3x \qquad + 2z = 14\end{array}$$

Next we choose a different pair of equations and eliminate the *same* variable. This time, we take the first and third equations. We multiply the first equation by 2 so that we eliminate the y-variable by adding the equations together.

$$x + y + z = 5 \xrightarrow{\times 2} 2x + 2y + 2z = 10$$

$$\begin{array}{r}2x + 2y + 2z = 10\\ x - 2y + 3z = 16\\ \hline 3x \qquad + 5z = 26\end{array}$$

We now have a pair of equations in x and z.

$$3x + 2z = 14$$
$$3x + 5z = 26$$

We can solve this system using any of the methods we studied earlier.

We choose to multiply the first equation by -1 and add the result to the second equation.

$$3x + 2z = 14 \xrightarrow{\times(-1)} -3x - 2z = -14$$

$$\begin{array}{r}-3x - 2z = -14\\ 3x + 5z = 26\\ \hline 3z = 12\end{array}$$

Divide by 3 to produce $z = 4$.

Substitute this result into one of the equations in the two-variable system to find the x-value of this solution.

$3x + 2z = 14$ Use one of the two-variable equations.

$3x + 2(4) = 14$ Substitute $z = 4$ and solve for x.

$3x + 8 = 14$ Subtract 8 from both sides.

$3x = 6$ Divide both sides by 3.

$x = 2$

Finally, substituting $x = 2$ and $z = 4$ into any of the equations in the *original* three-variable system enables us to find the appropriate y-value.

$$x + y + z = 5$$
$$(2) + y + (4) = 5$$
$$y + 6 = 5$$
$$y = -1$$

Therefore, $(2, -1, 4)$ is the solution to the given system of equations.

NOTE

We can use any of the original three-variable equations to find y.

You can check this result by substituting $(2, -1, 4)$ into each of the original equations and see that it is a solution in every case.

Check Yourself 1

Solve the system.

$$x - 2y + z = 0$$
$$2x + 3y - z = 16$$
$$3x - y - 3z = 23$$

One or more of the equations of a system may already have a missing variable. The elimination process is simplified in that case, as Example 2 illustrates.

Example 2 Solving a Linear System in Three Variables

Solve the system.

$$\begin{aligned} 2x + y - z &= -3 \\ y + z &= 2 \\ 4x - y + z &= 12 \end{aligned}$$

Since the middle equation involves only y and z, we can simply eliminate x from the pair of other equations. Multiply the first equation by -2 and add the result to the third equation to eliminate x.

$$\begin{aligned} -4x - 2y + 2z &= 6 \\ 4x - y + z &= 12 \\ \hline -3y + 3z &= 18 \\ y - z &= -6 \end{aligned}$$

NOTE

We now have a *second* equation in y and z.

We now form a system consisting of the pair of two-variable equations and solve as before.

$$\begin{aligned} y + z &= 2 \\ y - z &= -6 \\ \hline 2y &= -4 \\ y &= -2 \end{aligned}$$

Adding eliminates z.

Using the equation $y + z = 2$, if $y = -2$, we have

$$\begin{aligned} (-2) + z &= 2 \\ z &= 4 \end{aligned}$$

and from the first equation, if $y = -2$ and $z = 4$,

$$\begin{aligned} 2x + y - z &= -3 \\ 2x + (-2) - (4) &= -3 \\ 2x &= 3 \\ x &= \frac{3}{2} \end{aligned}$$

The solution set for the system is

$$\left\{\left(\frac{3}{2}, -2, 4\right)\right\}$$

We check our proposed solution by substituting $x = \frac{3}{2}$, $y = -2$, and $z = 4$ into each equation in the original system and see if true statements are formed.

Check

$$\begin{aligned} 2x + y - z &= -3 \\ 2\left(\frac{3}{2}\right) + (-2) - (4) &\stackrel{?}{=} -3 \\ 3 - 2 - 4 &\stackrel{?}{=} -3 \\ -3 &= -3 \end{aligned}$$

$$\begin{aligned} y + z &= 2 \\ (-2) + (4) &\stackrel{?}{=} 2 \\ 2 &= 2 \end{aligned}$$

$$\begin{aligned} 4x - y + z &= 12 \\ 4\left(\frac{3}{2}\right) - (-2) + (4) &\stackrel{?}{=} 12 \\ 6 + 2 + 4 &\stackrel{?}{=} 12 \\ 12 &= 12 \end{aligned}$$

Check Yourself 2

Solve the system.

$$\begin{aligned} x + 2y - z &= -3 \\ x - y + z &= 2 \\ x - z &= 3 \end{aligned}$$

The following algorithm summarizes the procedure for finding the solutions for a linear system of three equations in three variables.

Step by Step

Solving a System of Three Equations in Three Unknowns

Step 1 Choose a pair of equations from the system and use the addition method to eliminate one of the variables.

Step 2 Choose a *different* pair of equations and eliminate the *same* variable.

Step 3 Solve the system of two equations in two variables determined in steps 1 and 2.

Step 4 Substitute the values found above into one of the original equations and solve for the remaining variable.

Step 5 The solution is the ordered triple of values found in steps 3 and 4. It can be checked by substituting into each of the equations of the original system.

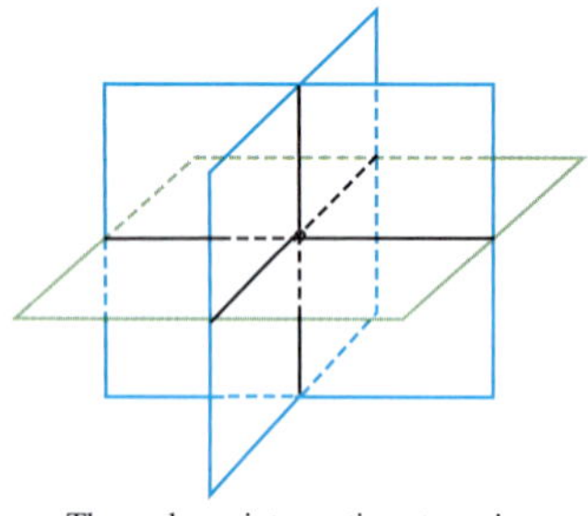

Three planes intersecting at a point

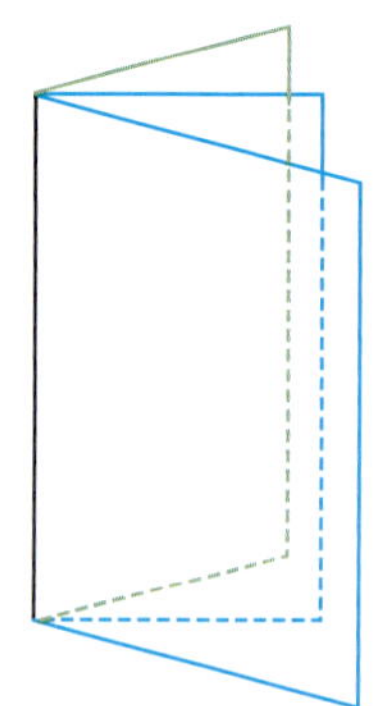

Three planes intersecting in a line

Systems of three equations in three variables may have (1) exactly one solution, (2) infinitely many solutions, or (3) no solution. Before we look at an algebraic approach in the second and third cases, you should understand the geometry involved.

The graph of a linear equation in three variables is a plane (a flat surface) in three dimensions. Two distinct planes are either parallel or they intersect in a line.

If three distinct planes intersect, that intersection will be either a single point (as in our first example) or a line (think of three pages in an open book—they intersect along the binding of the book).

Here are examples involving dependent systems with infinitely many solutions and inconsistent systems with no solutions.

Example 3 Solving a Dependent Linear System in Three Variables

Solve the system.

$$\begin{aligned} x + 2y - z &= 5 \\ x - y + z &= -2 \\ -5x - 4y + z &= -11 \end{aligned}$$

We begin as before by choosing two pairs of equations from the system and eliminating the same variable from each of the pairs. Adding the first two equations gives

$$2x + y = 3$$

Adding the first and third equations gives

$$-4x - 2y = -6$$

Now consider the system formed by these last two equations. We multiply the first equation in this new system by 2 and add again:

$$\begin{aligned} 4x + 2y &= 6 \\ -4x - 2y &= -6 \\ \hline 0 &= 0 \end{aligned}$$

This true statement tells us that the system has an infinite number of solutions (lying along a straight line). Again, such a system is dependent.

NOTE

There are ways of representing the solutions to such a system, as you will see in later courses.

Check Yourself 3

Solve the system.

$$\begin{aligned} 2x - y + 3z &= 3 \\ -x + y - 2z &= 1 \\ y - z &= 5 \end{aligned}$$

We have seen that a linear system in three variables can have exactly one solution (a consistent system) or, as in Example 3, infinitely many solutions (a dependent system). Consider now a third possibility: no solutions (an inconsistent system). It is illustrated in Example 4.

Example 4 Solving an Inconsistent Linear System in Three Variables

Solve the system.

$$\begin{aligned} 3x + y - 3z &= 1 \\ -2x - y + 2z &= 1 \\ -x - y + z &= 2 \end{aligned}$$

This time we eliminate variable y. Adding the first two equations gives

$$x - z = 2$$

Adding the first and third equations gives

$$2x - 2z = 3$$

Now, multiply the first equation in this new system by -2 and add the result to the second two-variable equation.

$$\begin{aligned} -2x + 2z &= -4 \\ 2x - 2z &= 3 \\ \hline 0 &= -1 \end{aligned}$$

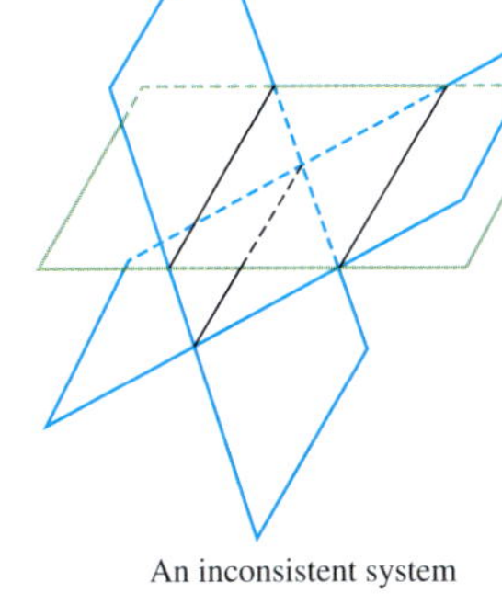

An inconsistent system

All the variables have been eliminated, and we have arrived at a contradiction, $0 = -1$. This means that the system is *inconsistent* and has no solutions. There is *no* point common to all three planes. The solution set is the empty set $\varnothing$.

Check Yourself 4

Solve the system.

$$\begin{aligned} x - y - z &= 0 \\ -3x + 2y + z &= 1 \\ 3x - y + z &= -1 \end{aligned}$$

We have by no means illustrated all possible types of inconsistent and dependent systems. Other possibilities involve either distinct parallel planes or planes that coincide. The solution techniques in these additional cases are similar to those illustrated in Examples 3 and 4.

In many instances, if an application involves three unknown quantities, you will find it useful to assign three variables to those quantities and then build a system of three equations from the given relationships in the problem. The extension of our problem-solving strategy is natural, as Example 5 illustrates.

Example 5 Solving a Number Problem

< Objective 2 >

NOTE

Sometimes it helps to choose variable letters that relate to the words, as is done here.

The sum of the digits of a three-digit number is 12. The tens digit is 2 less than the hundreds digit, and the units digit is 4 less than the sum of the other two digits. What is the number?

Step 1 The three unknowns are, of course, the three digits of the number.

Step 2 We now want to assign variables to each of three digits. Let u be the units digit, t the tens digit, and h the hundreds digit.

Step 3 There are three conditions given in the problem that allow us to write the necessary three equations. From those conditions

$$h + t + u = 12$$
$$t = h - 2$$
$$u = h + t - 4$$

NOTE

Take a moment now to go back to the original problem and pick out those conditions. That skill is a crucial part of the problem-solving strategy.

Step 4 There are various ways to approach the solution. To use addition, write the system in the equivalent form

$$h + t + u = 12$$
$$-h + t \qquad = -2$$
$$-h - t + u = -4$$

and solve by our earlier methods.

Step 5 The solution, which you can verify, is $h = 5$, $t = 3$, and $u = 4$. The desired number is 534.

To check, you should show that the digits of 534 meet each of the conditions of the original problem.

Check Yourself 5

The sum of the measures of the angles of a triangle is 180°. In a given triangle, the measure of the second angle is twice the measure of the first. The measure of the third angle is 30° less than the sum of the measures of the first two. Find the measure of each angle.

Let's continue with a slightly different application that will lead to a system of three equations.

Example 6 Solving an Investment Application

Monica decided to divide a total of \$42,000 into three investments: a savings account paying 5% interest, a time deposit paying 7%, and a bond paying 9%. Her total annual interest from the three investments was \$2,600, and the interest from the savings account was \$200 less than the total interest from the other two investments. How much did she invest at each rate?

NOTE

For 1 year, the interest formula is

$I = Pr$

(interest equals principal times rate).

Step 1 The three amounts are the unknowns.

Step 2 We let s be the amount invested at 5%, t the amount at 7%, and b the amount at 9%. Note that the interest from the savings account is then $0.05s$, and so on.

A table helps with the next step.

	5%	**7%**	**9%**
Principal	s	t	b
Interest	$0.05s$	$0.07t$	$0.09b$

Step 3 Again there are three conditions in the given problem. By using the table above, they lead to these equations.

$$s + t + b = 42{,}000 \quad \text{Total invested}$$

$$0.05s + 0.07t + 0.09b = 2{,}600 \quad \text{Total interest}$$

$$0.05s = 0.07t + 0.09b - 200$$

The savings interest was \$200 *less than* that from the other two investments.

NOTE

Find the interest earned from each investment, and verify that the conditions of the problem are satisfied.

Step 4 We clear the equation of decimals and solve as before, with the result

Step 5 $s = \$24{,}000 \qquad t = \$11{,}000 \qquad b = \$7{,}000$

We leave the check of this solution to you.

Check Yourself 6

Glenn has a total of \$11,600 invested in three accounts: a savings account paying 6% interest, a stock paying 8%, and a mutual fund paying 10%. The annual interest from the stock and mutual fund is twice that from the savings account, and the mutual fund returned \$120 more than the stock. How much did Glenn invest in each account?

Check Yourself ANSWERS

1. $\{(5, 1, -3)\}$ **2.** $\{(1, -3, -2)\}$

3. The system is dependent (there are an infinite number of solutions).

4. The system is inconsistent (there are no solutions).

5. The three angles are 35°, 70°, and 75°.

6. \$5,000 in savings, \$3,000 in stocks, and \$3,600 in the mutual fund

Reading Your Text

SECTION 4.4

(a) The solution to a linear equation in ____________ variables is an ordered triple.

(b) For a unique solution to exist to a system of equations when three variables are involved, we must have ____________ equations.

(c) If solving a system of equations results in a true statement such as $0 = 0$, the system has an ____________ number of solutions.

(d) If a system of equations has only one solution we call it a ____________ system.

4.4 exercises

Basic Skills | Challenge Yourself | Calculator/Computer | Career Applications | Above and Beyond

Name ____________

Section ________ Date ________

Answers

1. ____
2. ____
3. ____
4. ____
5. ____
6. ____
7. ____
8. ____
9. ____
10. ____
11. ____
12. ____
13. ____
14. ____
15. ____
16. ____

< Objective 1 >

Solve each system of equations. If a unique solution does not exist, state whether the system is inconsistent or dependent.

1. $\begin{aligned} x - y + z &= 3 \\ 2x + y + z &= 8 \\ 3x + y - z &= 1 \end{aligned}$

> Videos

2. $\begin{aligned} x - y - z &= 2 \\ 2x + y + z &= 8 \\ x + y + z &= 6 \end{aligned}$

3. $\begin{aligned} x + y + z &= 1 \\ 2x - y + 2z &= -1 \\ -x - 3y + z &= 1 \end{aligned}$

4. $\begin{aligned} x - y - z &= 6 \\ -x + 3y + 2z &= -11 \\ 3x + 2y + z &= 1 \end{aligned}$

5. $\begin{aligned} x + y + z &= 1 \\ -2x + 2y + 3z &= 20 \\ 2x - 2y - z &= -16 \end{aligned}$

6. $\begin{aligned} x + y + z &= -3 \\ 3x + y - z &= 13 \\ 3x + y - 2z &= 18 \end{aligned}$

7. $\begin{aligned} 2x + y - z &= 2 \\ -x - 3y + z &= -1 \\ -4x + 3y + z &= -4 \end{aligned}$

8. $\begin{aligned} x + 4y - 6z &= 8 \\ 2x - y + 3z &= -10 \\ 3x - 2y + 3z &= -18 \end{aligned}$

9. $\begin{aligned} 3x - y + z &= 5 \\ x + 3y + 3z &= -6 \\ x + 4y - 2z &= 12 \end{aligned}$

10. $\begin{aligned} 2x - y + 3z &= 2 \\ x - 2y + 3z &= 1 \\ 4x - y + 5z &= 5 \end{aligned}$

11. $\begin{aligned} x + 2y + z &= 2 \\ 2x + 3y + 3z &= -3 \\ 2x + 3y + 2z &= 2 \end{aligned}$

12. $\begin{aligned} x - 4y - z &= -3 \\ x + 2y + z &= 5 \\ 3x - 7y - 2z &= -6 \end{aligned}$

> Videos

13. $\begin{aligned} x + 3y - 2z &= 8 \\ 3x + 2y - 3z &= 15 \\ 4x + 2y + 3z &= -1 \end{aligned}$

14. $\begin{aligned} x + y - z &= 2 \\ 3x + 5y - 2z &= -5 \\ 5x + 4y - 7z &= -7 \end{aligned}$

15. $\begin{aligned} x + y - z &= 2 \\ x \qquad - 2z &= 1 \\ 2x - 3y - z &= 8 \end{aligned}$

16. $\begin{aligned} x + y + z &= 6 \\ x - 2y \qquad &= -7 \\ 4x + 3y + z &= 7 \end{aligned}$

17.
$$\begin{aligned} x - 3y + 2z &= 1 \\ 16y - 9z &= 5 \\ 4x + 4y - z &= 8 \end{aligned}$$

> Videos

18.
$$\begin{aligned} x - 4y + 4z &= -1 \\ y - 3z &= 5 \\ 3x - 4y + 6z &= 1 \end{aligned}$$

19.
$$\begin{aligned} x + 2y - 4z &= 13 \\ 3x + 4y - 2z &= 19 \\ 3x + 2z &= 3 \end{aligned}$$

20.
$$\begin{aligned} x + 2y - z &= 6 \\ -3x - 2y + 5z &= -12 \\ x - 2z &= 3 \end{aligned}$$

> Videos

Basic Skills | **Challenge Yourself** | Calculator/Computer | Career Applications | Above and Beyond

< Objective 2 >

Solve exercises 21 to 32 by choosing a variable to represent each unknown quantity and writing a system of equations.

21. Number Problem The sum of three numbers is 16. The largest number is equal to the sum of the other two, and 3 times the smallest number is 1 more than the largest. Find the three numbers.

22. Number Problem The sum of three numbers is 24. Twice the smallest number is 2 less than the largest number, and the largest number is equal to the sum of the other two. What are the three numbers?

23. Problem Solving A cashier has 25 coins consisting of nickels, dimes, and quarters with a value of \$4.90. If the number of dimes is 1 less than twice the number of nickels, how many of each type of coin does she have?

24. Business and Finance A theater has tickets at \$6 for adults, \$3.50 for students, and \$2.50 for children under 12 years old. A total of 278 tickets were sold for one showing with a total revenue of \$1,300. If the number of adult tickets sold was 10 less than twice the number of student tickets, how many of each type of ticket were sold for the showing?

> Videos

25. Geometry The perimeter of a triangle is 19 cm. If the length of the longest side is twice that of the shortest side and 3 cm less than the sum of the lengths of the other two sides, find the lengths of the three sides.

Answers

17. ______
18. ______
19. ______
20. ______
21. ______
22. ______
23. ______
24. ______
25. ______

Answers

26. ____

27. ____

28. ____

29. ____

30. ____

31. ____

32. ____

33. ____

34. ____

26. **Geometry** The measure of the largest angle of a triangle is 10° more than the sum of the measures of the other two angles and 10° less than 3 times the measure of the smallest angle. Find the measures of the three angles of the triangle.

27. **Business and Finance** Jovita divides \$12,000 into three investments: a savings account paying 4% annual interest, a bond paying 6%, and a money market fund paying 9%. The annual interest from the three accounts is \$860, and she has 2 times as much invested in the bond as in the savings account. What amount does she have invested in each account?

28. **Business and Finance** Adrienne has \$10,000 invested in a savings account paying 5%, a time deposit paying 7%, and a bond paying 10%. She has \$1,000 less invested in the bond than in her savings account, and she earned \$700 in annual interest. What has she invested in each account?

29. **Number Problem** The sum of the digits of a three-digit number is 9, and the tens digit of the number is twice the hundreds digit. If the digits are reversed in order, the new number is 99 more than the original number. What is the original number?

30. **Number Problem** The sum of the digits of a three-digit number is 9. The tens digit is 3 times the hundreds digit. If the digits are reversed in order, the new number is 99 less than the original number. Find the original three-digit number.

31. **Science and Medicine** Roy, Sally, and Jeff drive a total of 50 mi to work each day. Sally drives twice as far as Roy, and Jeff drives 10 mi farther than Sally. Use a system of three equations in three unknowns to find how far each person drives each day. > Videos

32. **Statistics** A parking lot has spaces reserved for motorcycles, cars, and vans. There are 5 more spaces reserved for vans than for motorcycles. There are 3 times as many car spaces as van and motorcycle spaces combined. If the parking lot has 180 total reserved spaces, how many of each type are there?

Determine whether each statement is **true** *or* **false.**

33. A system of three linear equations in three variables can have one unique solution, an infinite number of solutions, or no solution.

34. When solving a system, obtaining a statement such as $0 = 0$ means that the system is inconsistent.

Basic Skills | Challenge Yourself | Calculator/Computer | Career Applications | **Above and Beyond**

The solution process illustrated in this section can be extended to solving systems of more than three variables in a natural fashion. For instance, if four variables are involved, eliminate one variable in the system and then solve the resulting system in three variables as before. Substituting those three values into one of the original equations will provide the value for the remaining variable and the solution for the system.

In exercises 35 and 36, use this procedure to solve the system.

35.
$$\begin{aligned} x + 2y + 3z + w &= 0 \\ -x - y - 3z + w &= -2 \\ x - 3y + 2z + 2w &= -11 \\ -x + y - 2z + w &= 1 \end{aligned}$$

36.
$$\begin{aligned} x + y - 2z - w &= 4 \\ x - y + z + 2w &= 3 \\ 2x + y - z - w &= 7 \\ x - y + 2z + w &= 2 \end{aligned}$$

In some systems of equations there are more equations than variables. We can illustrate this situation with a system of three equations in two variables. To solve this type of system, pick any two of the equations and solve this system. Then substitute the solution obtained into the third equation. If a true statement results, the solution used is the solution to the entire system. If a false statement occurs, the system has no solution.

In exercises 37 and 38, use this procedure to solve each system.

37.
$$\begin{aligned} x - y &= 5 \\ 2x + 3y &= 20 \\ 4x + 5y &= 38 \end{aligned}$$

38.
$$\begin{aligned} 3x + 2y &= 6 \\ 5x + 7y &= 35 \\ 7x + 9y &= 8 \end{aligned}$$

39. Experiments have shown that cars, trucks, and buses emit different amounts of air pollutants. In one such experiment, a truck emitted 1.5 pounds (lb) of carbon dioxide (CO_2) per passenger-mile and 2 grams (g) of nitrogen oxide (NO) per passenger-mile. A car emitted 1.1 lb of CO_2 per passenger-mile and 1.5 g of NO per passenger-mile. A bus emitted 0.4 lb of CO_2 per passenger-mile and 1.8 g of NO per passenger-mile. A total of 85 passenger-miles was driven by the three vehicles, and 73.5 lb of CO_2 and 149.5 g of NO were collected. In the system of equations, T, C, and B represent the number of passenger-miles driven in a truck, car, and bus, respectively. Determine the passenger-miles driven by each vehicle.

$$\begin{aligned} T + C + B &= 85.0 \\ 1.5T + 1.1C + 0.4B &= 73.5 \\ 2T + 1.5C + 1.8B &= 149.5 \end{aligned}$$

40. Experiments have shown that cars, trucks, and trains emit different amounts of air pollutants. In one such experiment, a truck emitted 0.8 lb of carbon dioxide per passenger-mile and 1 g of nitrogen oxide per passenger-mile. A car emitted 0.7 lb of CO_2 per passenger-mile and 0.9 g of NO per passenger-mile. A train emitted 0.5 lb of CO_2 per passenger-mile and 4 g of NO per

Answers

35. ______

36. ______

37. ______

38. ______

39. ______

40. ______

Answers

41.

passenger-mile. A total of 141 passenger-miles was driven by the three vehicles, and 82.7 lb of CO_2 and 424.4 g of NO were collected. In the system of equations, T, C, and R represent the number of passenger-miles driven in a truck, car, and train, respectively. Determine the passenger-miles driven by each vehicle.

$$\begin{aligned} T + C + R &= 141.0 \\ 0.8T + 0.7C + 0.5R &= 82.7 \\ T + 0.9C + 4R &= 424.4 \end{aligned}$$

41. In Chapter 8 you will learn about quadratic functions and their graphs. A quadratic function has the form $y = ax^2 + bx + c$, where a, b, and c are specific numbers and $a \neq 0$. Three distinct points on the graph are enough to determine the equation.

(a) Suppose that (1, 5), (2, 10), and (3, 19) are on the graph of $y = ax^2 + bx + c$. Substituting the pair (1, 5) into this equation (that is, let $x = 1$ and $y = 5$) yields $5 = a + b + c$. Substituting each of the other ordered pairs yields $10 = 4a + 2b + c$ and $19 = 9a + 3b + c$. Solve the resulting system of equations to determine the values of a, b, and c. Then write the equation of the function.

(b) Repeat the work of part (a), using the three points (1, 2), (2, 9), and (3, 22).

Answers

1. $\{(1, 2, 4)\}$ **3.** $\{(-2, 1, 2)\}$ **5.** $\{(-4, 3, 2)\}$ **7.** Infinite number of solutions, dependent system **9.** $\left\{\left(3, \frac{1}{2}, -\frac{7}{2}\right)\right\}$ **11.** $\{(3, 2, -5)\}$ **13.** $\{(2, 0, -3)\}$ **15.** $\left\{\left(4, -\frac{1}{2}, \frac{3}{2}\right)\right\}$ **17.** No solutions, inconsistent system **19.** $\left\{\left(2, \frac{5}{2}, -\frac{3}{2}\right)\right\}$ **21.** 3, 5, 8 **23.** 3 nickels; 5 dimes; 17 quarters **25.** 4 cm, 7 cm, 8 cm **27.** \$2,000 savings; \$4,000 bond; \$6,000 money market **29.** 243 **31.** Roy 8 mi; Sally 16 mi; Jeff 26 mi **33.** True **35.** $\{(1, 2, -1, -2)\}$ **37.** $\{(7, 2)\}$ **39.** Truck: 20 passenger-miles; car: 25 passenger-miles; bus: 40 passenger-miles **41.** **(a)** $y = 2x^2 - x + 4$; **(b)** $y = 3x^2 - 2x + 1$

4.5 Systems of Linear Inequalities in Two Variables

< 4.5 Objectives >

1 > Graph systems of linear inequalities

2 > Solve an application involving a system of linear inequalities

Our previous work in this chapter dealt with finding the solution set of a system of linear equations. That solution set represented the points of intersection of the graphs of the equations in the system. In this section, we extend that idea to include systems of linear inequalities.

In this case, the solution set is the set of all ordered pairs that satisfy both inequalities. **The graph of the solution set of a system of linear inequalities** is then the intersection of the graphs of the individual inequalities. Here is an example.

Example 1 Solving a System of Linear Inequalities

< Objective 1 >

Solve the system of linear inequalities by graphing.

$$x + y > 4$$
$$x - y < 2$$

We start by graphing each inequality separately. We draw the boundary line and, using (0, 0) as a test point, we see that we should shade the half-plane above the line in both graphs.

NOTE

The boundary line is dashed to indicate that points on the line do *not* represent solutions.

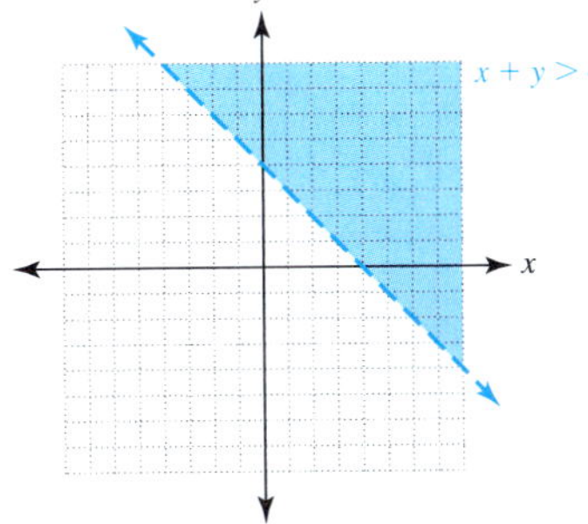

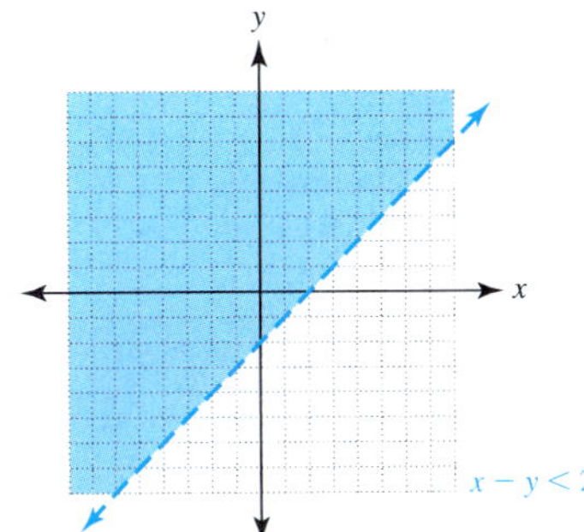

In practice, the graphs of the two inequalities are combined on the same set of axes, as shown below. The graph of the solution set of the original system is the intersection of the graphs drawn previously.

NOTES

Points on the lines are not included in the solution.

We want to show all ordered pairs that satisfy *both* statements.

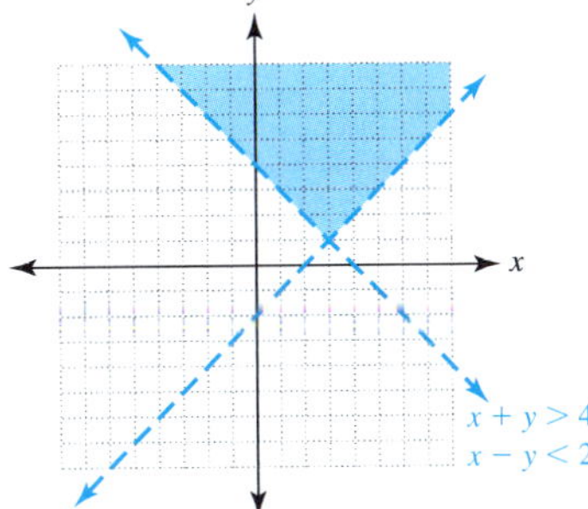

Check Yourself 1

Solve the system of linear inequalities by graphing.

$$2x - y < 4$$
$$x + y < 3$$

Most applications of systems of linear inequalities lead to **bounded regions.** This requires a system of three or more inequalities, as shown in Example 2.

Example 2 Solving a System of Linear Inequalities

Solve the system of linear inequalities by graphing.

$$x + 2y \le 6$$
$$x + y \le 5$$
$$x \ge 2$$
$$y \ge 0$$

On the same set of axes, we graph the boundary line of each of the inequalities. We then choose the appropriate half-planes (indicated by the arrow that is perpendicular to the line) in each case, and we locate the intersection of those regions for our graph.

NOTE

The vertices of the shaded region are given because they have particular significance in later applications of this concept. Can you see how the coordinates of the vertices were determined?

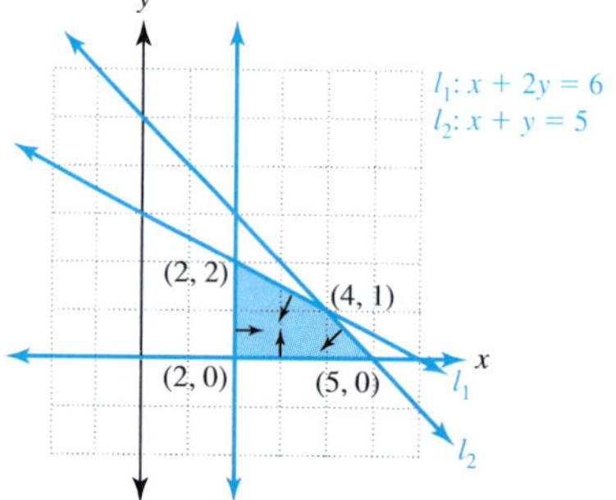

Check Yourself 2

Solve the system of linear inequalities by graphing.

$$2x - y \le 8$$
$$x + y \le 7$$
$$x \ge 0$$
$$y \ge 0$$

Let's expand on Example 8 in Section 4.3 to see an application of our work with systems of linear inequalities. Consider Example 3.

Example 3 A Business and Finance Application

< Objective 2 >

A manufacturer produces a standard model and a deluxe model of a 32-in. television set. The standard model requires 12 h of labor to produce, and the deluxe model requires 18 h. The labor available is limited to 360 h per week. Also, the plant capacity is limited to producing a total of 25 sets per week. Draw a graph of the region representing the number of sets that can be produced, given these conditions.

As suggested earlier, we let x represent the number of standard model sets produced and y the number of deluxe model sets. Since the labor is limited to 360 h, we have

$$12x + 18y \le 360$$

($12x$: 12 h per standard set; $18y$: 18 h per deluxe set)

NOTE

The total labor is limited to (or less than or equal to) 360 h.

The total production, here $x + y$ sets, is limited to 25, so we can write

$$x + y \le 25$$

For convenience in graphing, we divide both members of the first inequality by 6, to write the equivalent system

$$\begin{aligned} 2x + 3y &\le 60 \\ x + y &\le 25 \\ x &\ge 0 \\ y &\ge 0 \end{aligned}$$

NOTE

We have $x \ge 0$ and $y \ge 0$ because the number of sets produced cannot be negative.

We now graph the system of inequalities as before. The shaded area represents all possibilities in terms of the number of sets that can be produced. Only points with integer coordinates represent realistic solutions in the context of this application.

NOTE

The shaded area is called the **feasible region.** All points in the region meet the given conditions of the problem and represent possible production options.

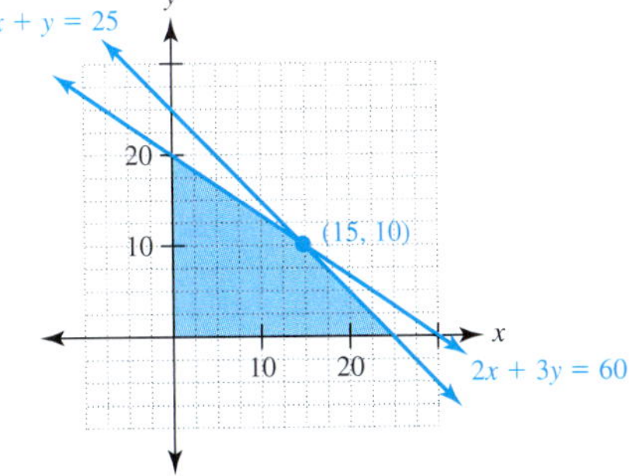

Check Yourself 3

A manufacturer produces TVs and CD players. The TVs require 10 h of labor to produce and the CD players require 20 h. The labor hours available are limited to 300 h per week. Existing orders require that at least 10 TVs and at least 5 CD players be produced per week. Draw a graph of the region representing the possible production options.

Check Yourself ANSWERS

1. $2x - y < 4$
$x + y < 3$

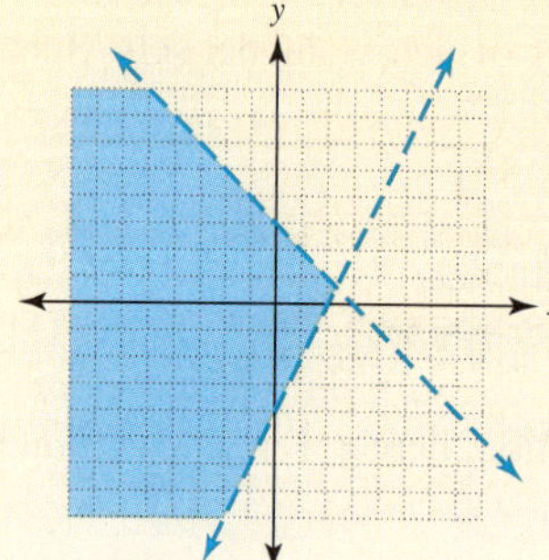

2. $2x - y \le 8$
$x + y \le 7$
$x \ge 0$
$y \ge 0$

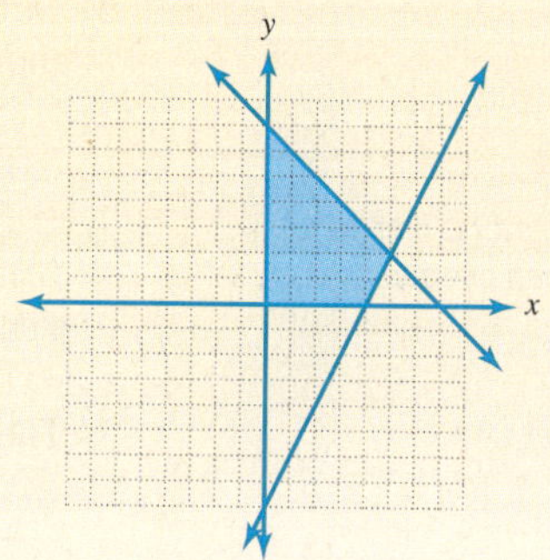

3. Let x be the number of TVs and y be the number of CD players. The system is

$$10x + 20y \le 300$$
$$x \ge 10$$
$$y \ge 5$$

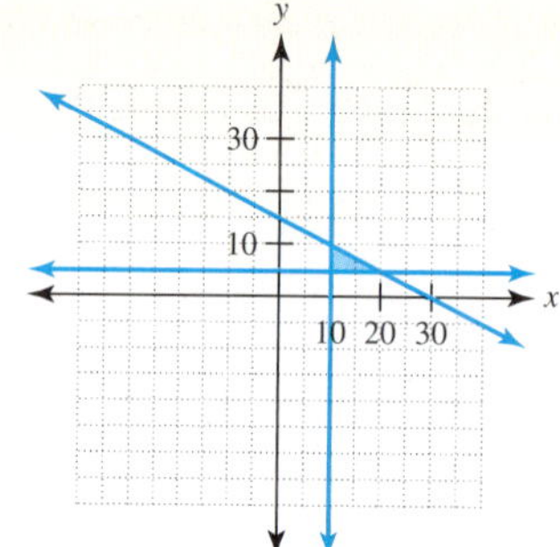

Reading Your Text 

SECTION 4.5

(a) The solution set of a linear system of inequalities is the set of all ordered pairs that ______________ both inequalities.

(b) Graphically, the solution set to a system of linear ______________ is given by the region shaded by every inequality in the system.

(c) When a ______________ line is dashed, it indicates that points on the line do not represent solutions.

(d) Most applications of systems of linear inequalities lead to ______________ regions.

Basic Skills | Challenge Yourself | Calculator/Computer | Career Applications | Above and Beyond

4.5 exercises

Boost *your* GRADE at ALEKS.com!

ALEKS

- Practice Problems
- Self-Tests
- NetTutor
- e-Professors
- Videos

Name ______________

Section ________ Date ________

< Objective 1 >

Solve each system of linear inequalities graphically.

1. $x + 2y \le 4$
$x - y \ge 1$

2. $3x - y > 6$
$x + y < 6$

3. $3x + y < 6$
$x + y > 4$

4. $2x + y \ge 8$
$x + y \ge 4$

5. $x + 3y \le 12$
$2x - 3y \le 6$

6. $x - 2y > 8$
$3x - 2y > 12$

7. $3x + 2y \le 12$
$x \ge 2$

8. $2x + y \le 6$
$y \ge 1$

9. $2x + y \le 8$
$x > 1$
$y > 2$

10. $3x - y \le 6$
$x \ge 1$
$y \le 3$

11. $x + 2y \le 8$
$2 \le x \le 6$
$y \ge 0$

12. $x + y < 6$
$0 \le y \le 3$
$x \ge 1$

Answers

1. ______
2. ______
3. ______
4. ______
5. ______
6. ______
7. ______
8. ______
9. ______
10. ______
11. ______
12. ______

Answers

13.

14.

15.

16.

17.

18.

19.

13. $3x + y \leq 6$
$x + y \leq 4$
$x \geq 0$
$y \geq 0$

14. $x - 2y \geq -2$
$x + 2y \leq 6$
$x \geq 0$
$y \geq 0$

15. $4x + 3y \leq 12$
$x + 4y \leq 8$
$x \geq 0$
$y \geq 0$

16. $2x + y \leq 8$
$x + y \geq 3$
$x \geq 0$
$y \geq 0$

17. $x - 4y \leq -4$
$x + 2y \leq 8$
$x \geq 2$

18. $x - 3y \geq -6$
$x + 2y \geq 4$
$x \leq 4$

Basic Skills | **Challenge Yourself** | Calculator/Computer | Career Applications | Above and Beyond

< Objective 2 >

In exercises 19 and 20, draw the appropriate graph.

19. **Business and Finance** A manufacturer produces both two-slice and four-slice toasters. The two-slice toaster takes 6 h of labor to produce, and the four-slice toaster takes 10 h. The labor available is limited to 300 h per week, and the total production capacity is 40 toasters per week. Draw a graph of the feasible region, given these conditions, where x is the number of two-slice toasters and y is the number of four-slice toasters.

20. **Business and Finance** A small firm produces both AM and AM/FM car radios. The AM radios take 15 h to produce, and the AM/FM radios take 20 h. The number of production hours is limited to 300 h per week. The plant's capacity is limited to a total of 18 radios per week, and existing orders require that at least 4 AM radios and at least 3 AM/FM radios be produced per week. Draw a graph of the feasible region, given these conditions, where x is the number of AM radios and y the number of AM/FM radios.

Determine whether each statement is **true** *or* **false.**

21. The feasible region in an application shows all the points that meet all the conditions of the problem.

22. The graph of the solution set of a system of three linear inequalities can be unbounded.

Complete each statement with **never, sometimes,** *or* **always.**

23. The graph of the solution set of a system of two linear inequalities __________ includes the origin.

24. The graph of the solution set of a system of two linear inequalities is __________ bounded.

Basic Skills | Challenge Yourself | Calculator/Computer | Career Applications | **Above and Beyond**

25. When you solve a system of linear inequalities, it is often easier to shade the region that is not part of the solution, rather than the region that is. Try this method, then describe its benefits.

26. Describe a system of linear inequalities for which there is no solution.

27. Write the system of inequalities whose graph is the shaded region.

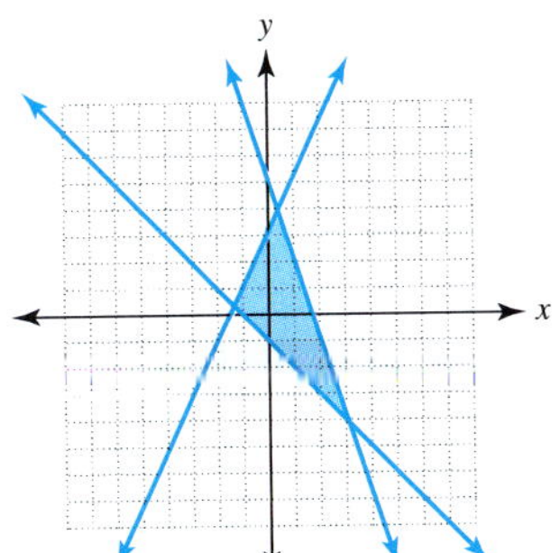

Answers

20. __________

21. __________

22. __________

23. __________

24. __________

25. __________

26. __________

27. __________

Answer

28. ______________________

28. Write the system of inequalities whose graph is the shaded region.

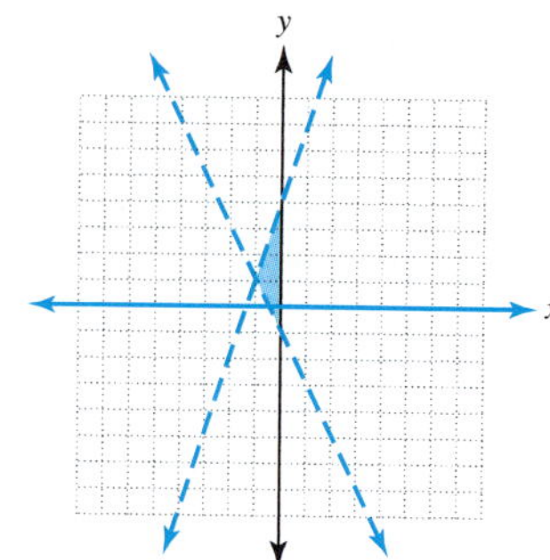

Answers

1.

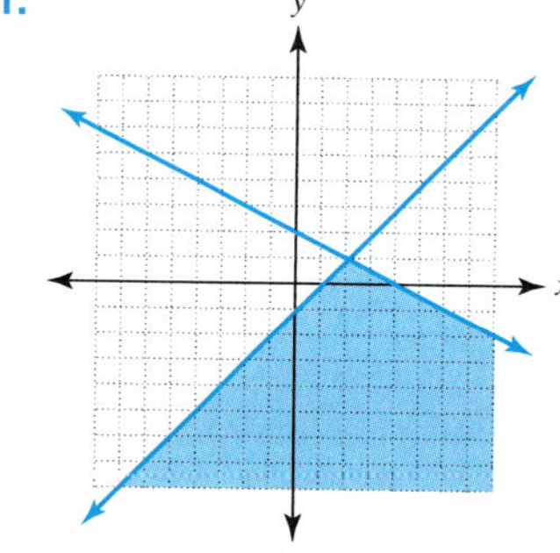

3.

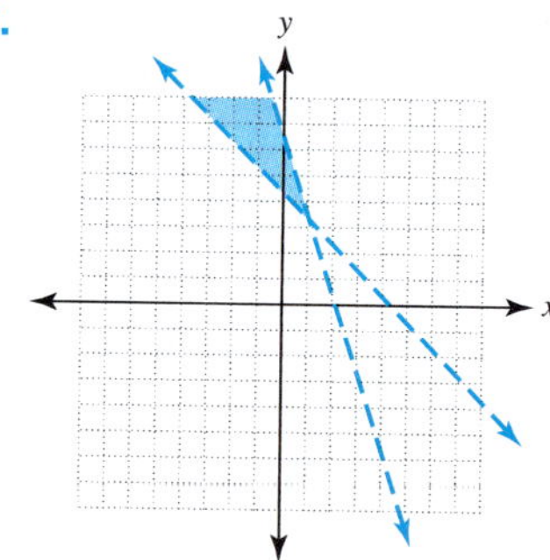

5.

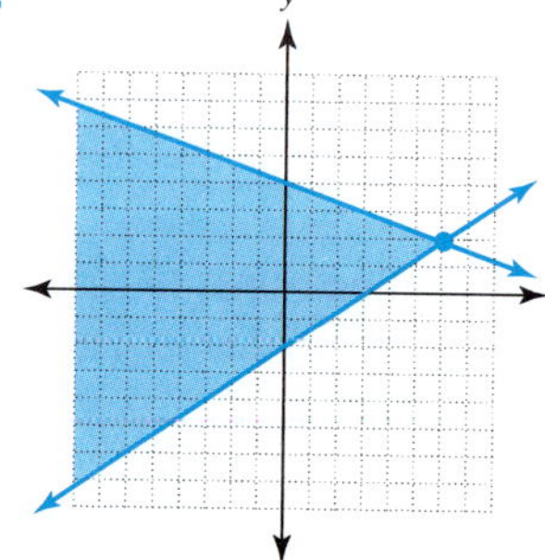

7.

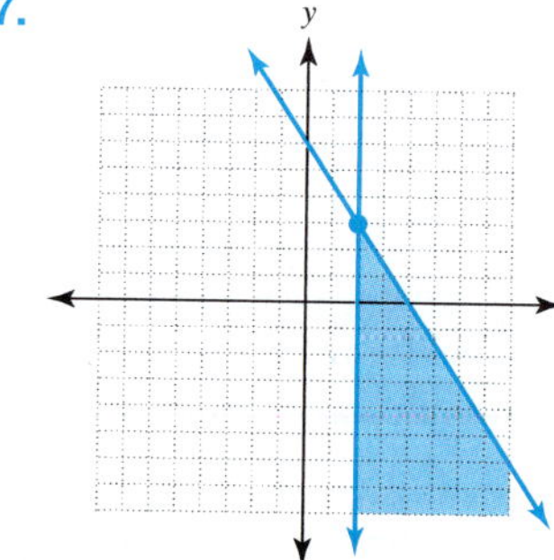

9.

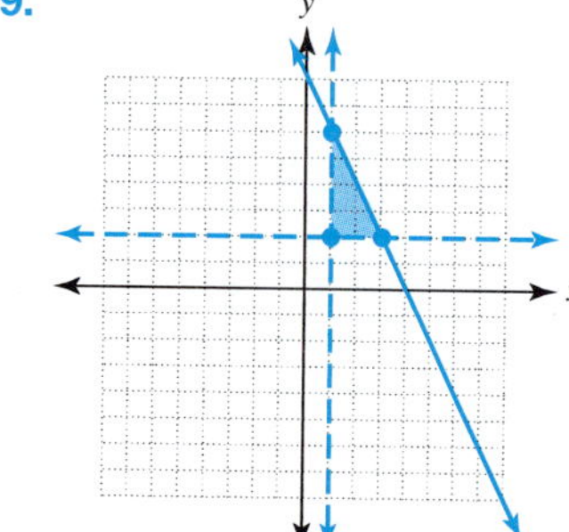

11.

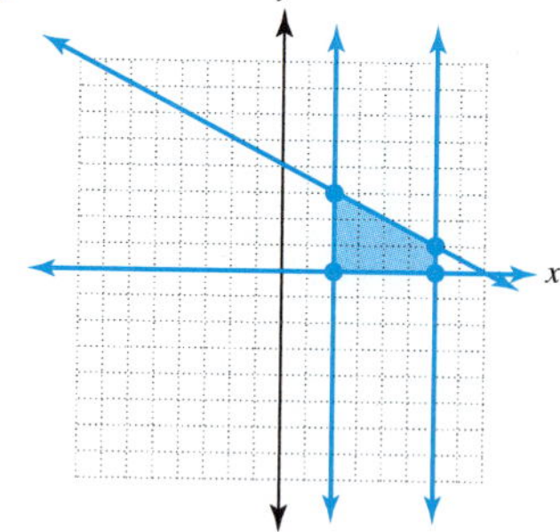

13.

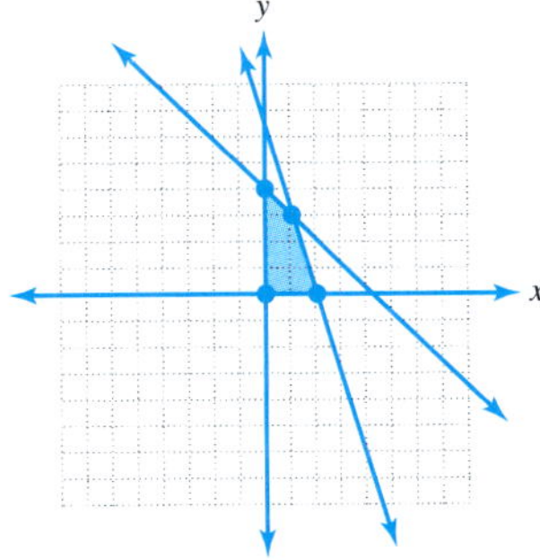

15.

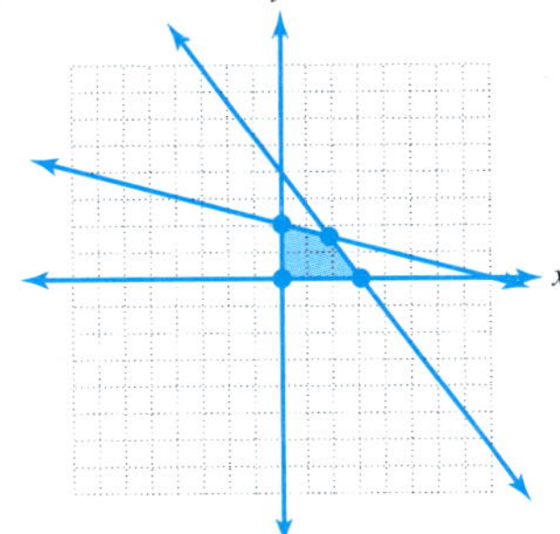

17.

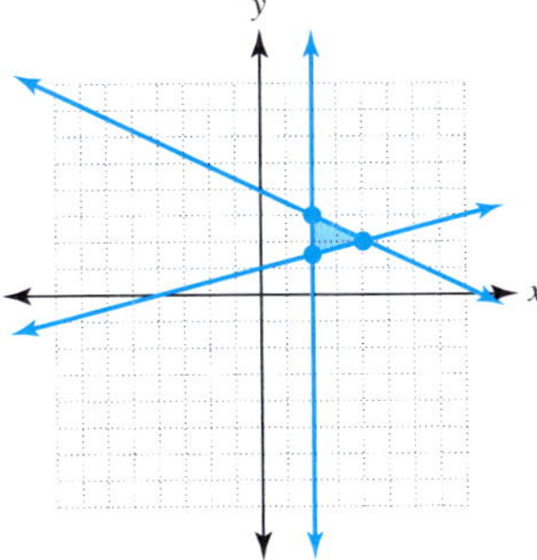

19.

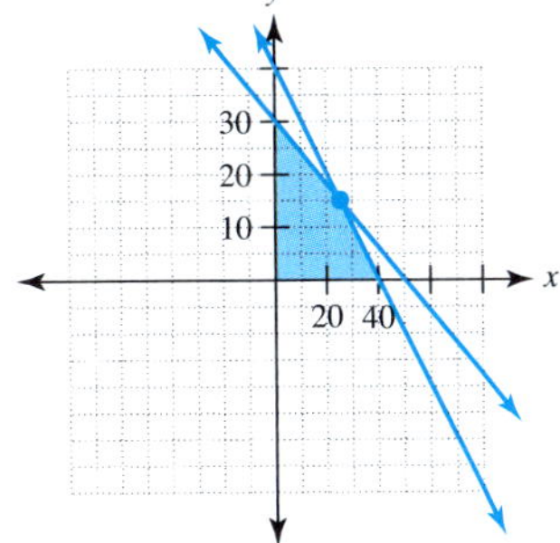

21. True **23.** sometimes **25.** Above and Beyond

27. $y \le 2x + 3$
$y \le -3x + 5$
$y \ge -x - 1$

Definition/Procedure	Example	Reference
Graphing Systems of Linear Equations		Section 4.1
A **system of linear equations** is two or more linear equations considered together. A solution for a linear system in two variables is an ordered pair of real numbers (x, y) that satisfies both equations in the system. There are three solution techniques: the graphing method, the addition method, and the substitution method.	The solution for the system $2x - y = 7$ $x + y = 2$ is $(3, -1)$. It is the only ordered pair that will satisfy each equation.	p. 398
Solving by the Graphing Method Graph each equation of the system on the same set of coordinate axes. If a solution exists, it will correspond to the point of intersection of the two lines. Such a system is called a **consistent system.** If a solution does not exist, there is no point at which the two lines intersect. Such lines are parallel, and the system is called an **inconsistent system.** If there are an infinite number of solutions, the lines coincide. Such a system is called a **dependent system.** You may or may not be able to determine exact solutions for the system of equations with this method.	To solve the system $2x - y = 7$ $x + y = 2$ by graphing:	p. 401

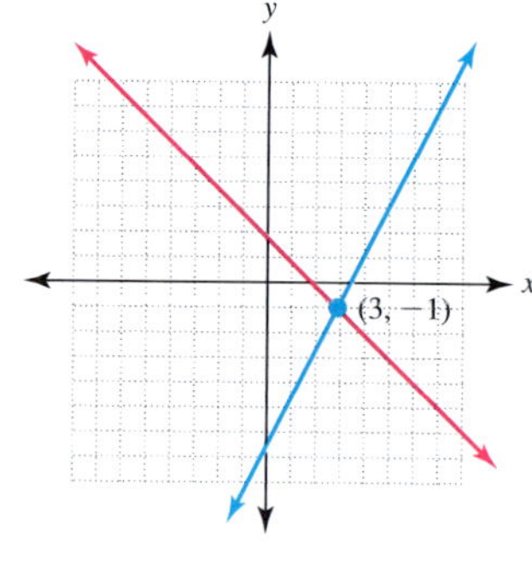

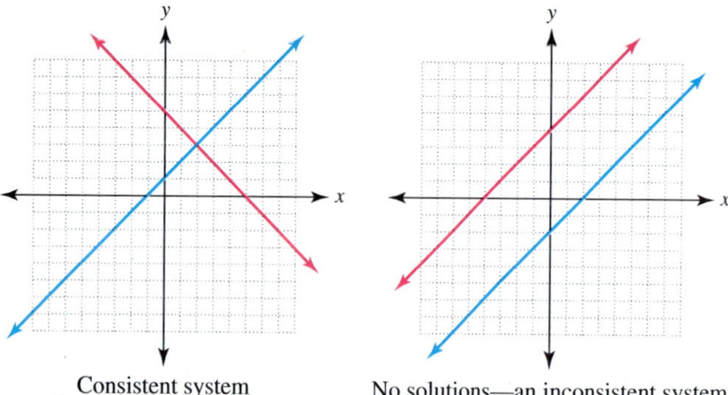

Consistent system

No solutions—an inconsistent system

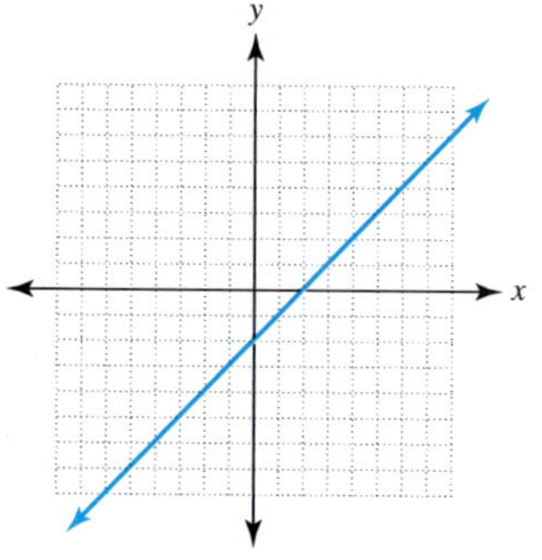

An infinite number of solutions—a dependent system

Definition/Procedure	Example	Reference
Solving Equations in One Variable Graphically		Section 4.2
Finding a Graphical Solution for an Equation **Step 1** Let each side of the equation represent a function of x. **Step 2** Graph the two functions on the same set of axes. **Step 3** Find the point of intersection of the two graphs. Draw a vertical line from the point of intersection to the x-axis, marking a point there. The x-value at this indicated point represents the solution to the original equation.	To solve $2x - 6 = 8x$, let $f(x) = 2x - 6$ $g(x) = 8x$ then graph both lines. 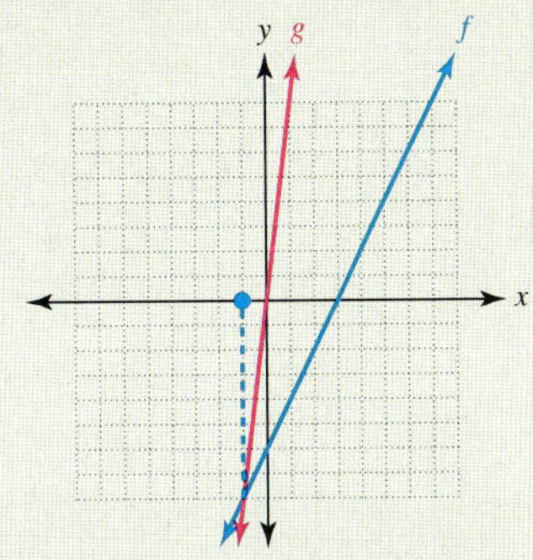 The intersection occurs when $x = -1$. The solution set is $\{-1\}$.	*p.* 416
Systems of Equations in Two Variables with Applications		Section 4.3
Solving by the Addition Method **Step 1** If necessary, multiply one or both of the equations by a constant so that one of the variables can be eliminated by addition. **Step 2** Add the equations of the equivalent system formed in step 1. **Step 3** Solve the equation found in step 2. **Step 4** Substitute the value found in step 3 into either of the equations of the original system to find the corresponding value of the remaining variable. **Step 5** The ordered pair found in step 4 is the solution to the system. Check the solution by substituting the pair of values found in step 4 into the equations of the original system.	To solve $5x - 2y = 11$ $2x + 3y = 12$ multiply the first equation by 3 and the second equation by 2. Then add to eliminate y. $19x = 57$ $x = 3$ Substituting 3 for x in the first equation gives $15 - 2y = 11$ $y = 2$ So $\{(3, 2)\}$ is the solution set.	*p.* 431

Continued

Definition/Procedure	Example	Reference
Solving by the Substitution Method **Step 1** If necessary, solve one of the equations of the original system for one of the variables. **Step 2** Substitute the expression obtained in step 1 into the other equation of the system to write an equation in a single variable. **Step 3** Solve the equation found in step 2. **Step 4** Substitute the value found in step 3 into the equation found in step 1 to find the corresponding value of the remaining variable. **Step 5** The ordered pair found in step 4 is the solution to the system of equations. Check the solution by substituting the pair of values found in step 4 into the equations of the original system.	To solve $3x - 2y = 6$ $6x + y = 2$ by substitution, solve the second equation for y. $y = -6x + 2$ Substituting into the first equation gives $3x - 2(-6x + 2) = 6$ and $x = \frac{2}{3}$ Substituting $\frac{2}{3}$ for x in the equation that we solved for y gives $y = (-6)\left(\frac{2}{3}\right) + 2$ $= -4 + 2 = -2$ The solution set is $\left\{\left(\frac{2}{3}, -2\right)\right\}$	*p.* 433
Applications of Systems of Linear Equations **Step 1** Read the problem carefully to determine the unknown quantities. **Step 2** Choose a variable to represent any unknown. **Step 3** Translate the problem to the language of algebra to form a system of equations. **Step 4** Solve the system of equations by any of the methods discussed. **Step 5** Answer the question in the original problem and verify your solution by returning to the original problem.	Also determine the condition that relates the unknown quantities. Use a different letter for each variable. A table or a sketch often helps in writing the equations of the system.	*p.* 434

Definition/Procedure	Example	Reference

Systems of Linear Equations in Three Variables

Section 4.4

A solution for a linear system of three equations in three variables is an ordered triple of numbers (x, y, z) that satisfies each equation in the system.

Solving a System of Three Equations in Three Unknowns

Step 1 Choose a pair of equations from the system and use the addition method to eliminate one of the variables.

Step 2 Choose a different pair of equations and eliminate the same variable.

Step 3 Solve the system of two equations in two variables determined in steps 1 and 2.

Step 4 Substitute the values found above into one of the original equations and solve for the remaining variable.

Step 5 The solution is the ordered triple of values found in steps 3 and 4. It can be checked by substituting into the other equations of the original system.

Solve.

$$\begin{aligned} x + y - z &= 6 \\ 2x - 3y + z &= -9 \\ 3x + y + 2z &= 2 \end{aligned}$$

Adding the first two equations gives

$3x - 2y = -3$

Multiplying the first equation by 2 and adding the result to the third equation gives

$5x + 3y = 14$

We solve the system consisting of the pair of two-variable equations as before to obtain

$x = 1 \qquad y = 3$

Substituting these values into one of the original equations gives

$z = -2$

The solution set is $\{(1, 3, -2)\}$.

p. 450

Systems of Linear Inequalities in Two Variables

Section 4.5

A **system of linear inequalities** is two or more linear inequalities considered together. The **graph of the solution set** of a system of linear inequalities is the intersection of the graphs of the individual inequalities.

Solving Systems of Linear Inequalities by Graphing

Step 1 Graph each inequality, shading the appropriate half-plane on the same set of coordinate axes.

Step 2 The graph of the system is the intersection of the regions shaded in step 1.

To solve the system

$$\begin{aligned} x + 2y &\le 8 \\ x + y &\le 6 \\ x &\ge 0 \\ y &\ge 0 \end{aligned}$$

by graphing:

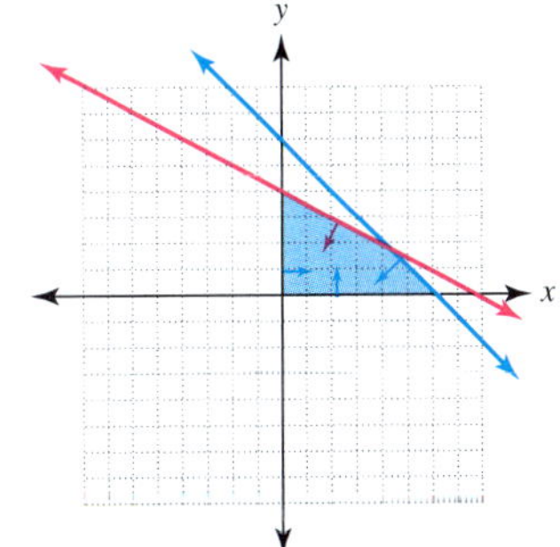

p. 459

summary exercises :: chapter 4

This summary exercise set is provided to give you practice with each of the objectives of this chapter. Each exercise is keyed to the appropriate chapter section. When you are finished, you can check your answers to the odd-numbered exercises in the back of the text. If you have difficulty with any of these questions, go back and reread the examples from that section. The answers to the even-numbered exercises appear in the *Instructor's Solutions Manual.* Your instructor will give you guidelines on how best to use these exercises in your instructional setting.

4.1 *Graph each system of equations and then solve the system.*

1. $x + y = 8$
$x - y = 4$

2. $x + 2y = 8$
$x - y = 5$

3. $2x + 3y = 12$
$2x + y = 8$

4. $x + 4y = 8$
$y = 1$

Use a graphing calculator to solve each system. Estimate your answer to the nearest hundredth. You may need to adjust the viewing window to see the point of intersection.

5. $44x + 35y = 1{,}115$
$11x - 27y = 850$

6. $15x - 48y = 935$
$25x + 51y = 1{,}051$

4.2 *Solve each equation graphically. Do not use a calculator.*

7. $3x - 6 = 0$

8. $4x + 3 = 7$

9. $3x + 5 = x + 7$

10. $4x - 3 = x - 6$

11. $\dfrac{6x - 1}{2} = 2(x - 1)$

12. $3x + 2 = 2x - 1$

13. $3(x - 2) = 2(x - 1)$

14. $3(x + 1) + 3 = -7(x + 2)$

4.3 *Use the addition method to solve each system. If a unique solution does not exist, state whether the given system is inconsistent or dependent.*

15. $x + 2y = 7$
$x - y = 1$

16. $x + 3y = 14$
$4x + 3y = 29$

17. $3x - 5y = 5$
$-x + y = -1$

18. $x - 4y = 12$
$2x - 8y = 24$

19. $6x + 5y = -9$
$-5x + 4y = 32$

20. $3x + y = 8$
$-6x - 2y = -10$

21. $5x - y = -17$
$4x + 3y = -6$

22. $4x - 3y = 1$
$6x + 5y = 30$

23. $x - \frac{1}{2}y = 8$
$\frac{2}{3}x + \frac{3}{2}y = -2$

24. $\frac{1}{5}x - 2y = 4$
$\frac{3}{5}x + \frac{2}{3}y = -8$

Use the substitution method to solve each system. If a unique solution does not exist, state whether the given system is inconsistent or dependent.

25. $2x + y = 23$
$x = y + 4$

26. $x - 5y = 26$
$y = x - 10$

27. $3x + y = 7$
$y = -3x + 5$

28. $2x - 3y = 13$
$x = 3y + 9$

29. $5x - 3y = 13$
$x - y = 3$

30. $4x - 3y = 6$
$x + y = 12$

31. $3x - 2y = -12$
$6x + y = 1$

32. $x - 4y = 8$
$-2x + 8y = -16$

Solve each problem by choosing a variable to represent each unknown quantity. Then write a system of equations that will allow you to solve for each variable.

33. Number Problem One number is 2 more than 3 times another. If the sum of the two numbers is 30, find the two numbers.

34. Problem Solving Suppose that a cashier has 78 \$5 and \$10 bills with a value of \$640. How many of each type of bill does she have?

35. **Business and Finance** Tickets for a basketball game sold at \$7 for an adult ticket and \$4.50 for a student ticket. If the revenue from 1,200 tickets was \$7,400, how many of each type of ticket were sold?

36. **Business and Finance** A purchase of 8 blank CDs and 4 blank DVDs costs \$36. A second purchase of 4 CDs and 5 DVDs costs \$30. What is the price of a single CD and of a single DVD?

37. **Geometry** The length of a rectangle is 4 cm less than twice its width. If the perimeter of the rectangle is 64 cm, find the dimensions of the rectangle.

38. **Business and Finance** A grocer in charge of bulk foods wishes to combine peanuts selling for \$2.25 per pound and cashews selling for \$6 per pound. What amount of each nut should be used to form a 120-lb mixture selling for \$3 per pound?

39. **Business and Finance** Reggie has two investments totaling \$17,000—one a savings account paying 6%, the other a time deposit paying 8%. If his annual interest is \$1,200, what does he have invested in each account?

40. **Science and Medicine** A pharmacist mixes a 20% alcohol solution and a 50% alcohol solution to form 600 mL of a 40% solution. How much of each solution should he use in forming the mixture?

41. **Science and Medicine** A jet flying east, with the wind, makes a trip of 2,200 mi in 4 h. Returning against the wind, the jet can travel only 1,800 mi in 4 h. What is the plane's rate in still air? What is the rate of the wind?

42. **Business and Finance** A manufacturer produces zip drives and flash drives. The zip drives require 20 min of component assembly time; the flash drives, 25 min. The manufacturer has 500 min of component assembly time available per day. Each drive requires 30 min for packaging and testing, and 690 min of that time is available per day. How many of each of the drives should be produced daily to use all the available time?

chapter 4 > Make the Connection

43. **Business and Finance** If the demand equation for a product is $D = 270 - 5p$ and the supply equation is $S = 13p$, find the equilibrium point.

chapter 4 > Make the Connection

44. **Business and Finance** Two car rental agencies have different rates for the rental of a compact automobile:

Company A: \$18 per day plus 12¢ per mile

Company B: \$20 per day plus 10¢ per mile

For a 3-day rental, at what number of miles will the charges from the two companies be the same?

4.4 *Use the addition method to solve each system. If a unique solution does not exist, state whether the given system is inconsistent or dependent.*

45. $x - y + z = 0$
$x + 4y - z = 14$
$x + y - z = 6$

46. $x - y + z = 3$
$3x + y + 2z = 15$
$2x - y + 2z = 7$

47. $x - y - z = 2$
$-2x + 2y + z = -5$
$-3x + 3y + z = -10$

48. $x - y = 3$
$2y + z = 5$
$x + 2z = 7$

49. $x + y - z = -1$
$x - y - 2z = 2$
$-5x - y - z = -1$

50. $2x + 3y + z = 7$
$-2x - 9y + 2z = 1$
$4x - 6y + 3z = 10$

Solve each problem by choosing a variable to represent each unknown quantity.

51. **Number Problem** The sum of three numbers is 15. The largest number is 4 times the smallest number, and it is also 1 more than the sum of the other two numbers. Find the three numbers.

52. **Number Problem** The sum of the digits of a three-digit number is 16. The tens digit is 3 times the hundreds digit, and the units digit is 1 more than the hundreds digit. What is the number?

53. **Business and Finance** A theater has orchestra tickets at \$10, box seat tickets at \$7, and balcony tickets at \$5. For one performance, a total of 360 tickets was sold, and the total revenue was \$3,040. If the number of orchestra tickets sold was 40 more than that of the other two types combined, how many of each type of ticket were sold for the performance?

54. **Geometry** The measure of the largest angle of a triangle is 15° less than 4 times the measure of the smallest angle and 30° more than the sum of the measures of the other two angles. Find the measures of the three angles of the triangle.

55. **Business and Finance** Rachel divided \$12,000 into three investments: a savings account paying 5%, a stock paying 7%, and a mutual fund paying 9%. Her annual interest from the investments was \$800, and the amount that she had invested at 5% was equal to the sum of the amounts invested in the other accounts. How much did she have invested in each type of account?

4.5 *Solve each system of linear inequalities.*

56. $\begin{aligned} x - y &< 7 \\ x + y &> 3 \end{aligned}$

57. $\begin{aligned} x - 2y &\le -2 \\ x + 2y &\le 6 \end{aligned}$

58. $\begin{aligned} x - 6y &< 6 \\ -x + y &< 4 \end{aligned}$

59. $\begin{aligned} 2x + y &\le 8 \\ x &\ge 1 \\ y &\ge 0 \end{aligned}$

60. $\begin{aligned} 2x + y &\le 6 \\ x &\ge 1 \\ y &\ge 0 \end{aligned}$

61. $\begin{aligned} 4x + y &\le 8 \\ x &\ge 0 \\ y &\ge 2 \end{aligned}$

62. $\begin{aligned} 4x + 2y &\le 8 \\ x + y &\le 3 \\ x &\ge 0 \\ y &\ge 0 \end{aligned}$

63. $\begin{aligned} 3x + y &\le 6 \\ x + y &\le 4 \\ x &\ge 0 \\ y &\ge 0 \end{aligned}$

self-test 4

Name ______

Section ______ Date ______

The purpose of this self-test is to help you assess your progress so that you can find concepts that you need to review before the next exam. Allow yourself about an hour to take this test. At the end of that hour, check your answers against those given in the back of this text. If you miss any, go back to the appropriate section to reread the examples until you have mastered that particular concept.

Solve each system. If a unique solution does not exist, state whether the given system is inconsistent or dependent.

1. $3x + y = -5$
$5x - 2y = -23$

2. $4x - 2y = -10$
$y = 2x + 5$

3. $9x - 3y = 4$
$-3x + y = -1$

4. $5x - 3y = 5$
$3x + 2y = -16$

5. $x - 2y = 5$
$2x + 5y = 10$

6. $5x - 3y = 20$
$4x + 9y = -3$

Solve each system.

7. $x - y + z = 1$
$-2x + y + z = 8$
$x + 5z = 19$

8. $x + 3y - 2z = -6$
$3x - y + 2z = 8$
$-2x + 3y - 4z = -11$

Solve each system of linear inequalities.

9. $x - 2y < 6$
$x + y < 3$

10. $3x + 4y \geq 12$
$x \geq 1$

11. $x + 2y \leq 8$
$x + y \leq 6$
$x \geq 0$
$y \geq 0$

Solve each equation graphically.

12. $4x - 7 = 5$

13. $6 - x = 4(x - 1)$

14. $-8x + 11 = 2x - 9$

15. $6(x - 1) = -3(x - 4)$

Solve each application by choosing a variable to represent each unknown quantity. Then write a system of equations that will allow you to solve for each variable.

16. An order for 30 computer disks and 12 printer ribbons totaled \$147. A second order for 12 more disks and 6 additional ribbons cost \$66. What was the cost per individual disk and ribbon?

17. A candy dealer wants to combine jawbreakers selling for \$2.40 per pound and licorice selling for \$3.90 per pound to form a 100-lb mixture that will sell for \$3 per pound. What amount of each type of candy should be used?

18. A small electronics firm assembles 5-in. portable television sets and 12-in. models. The 5-in. set requires 9 h of assembly time; the 12-in. set, 6 h. Each unit requires 5 h for packaging and testing. If 72 h of assembly time and 50 h of packaging and testing time are available per week, how many of each type of set should be finished if the firm wishes to use all its available capacity?

19. Hans decided to divide \$16,000 into three investments: a savings account paying 3% annual interest, a bond paying 5%, and a mutual fund paying 7%. His annual interest from the three investments was \$900, and he has as much in the mutual fund as in the savings account and bond combined. What amount did he invest in each type?

20. The fence around a rectangular yard requires 260 ft of fencing. The length is 20 ft less than twice the width. Find the dimensions of the yard.

Answers

1. ______
2. ______
3. ______
4. ______
5. ______
6. ______
7. ______
8. ______
9. ______
10. ______
11. ______
12. ______
13. ______
14. ______
15. ______
16. ______
17. ______
18. ______
19. ______
20. ______

cumulative review chapters 0-4

Name ____________________

Section ________ Date ________

Answers

1. ____________________
2. ____________________
3. ____________________
4. ____________________
5. ____________________
6. ____________________
7. ____________________
8. ____________________
9. ____________________
10. ____________________
11. ____________________
12. ____________________
13. ____________________
14. ____________________
15. ____________________

The following exercises are presented to help you review concepts from earlier chapters that you may have forgotten. This is meant as review material and not as a comprehensive exam. The answers are presented in the back of the text. If you have difficulty with any of these exercises, be certain to at least read through the summary related to that section.

Solve.

1. $3x - 2(x + 5) = 12 - 3x$

2. $2x - 7 < 3x - 5$

3. $x + 8 \geq 4x - 3$

4. $2x + 3(x - 2) = -4(x + 1) + 16$

Graph.

5. $5x + 7y = 35$

6. $2x + 3y < 6$

7. Solve the equation $P = P_0 + IRT$ for R.

8. Find the slope of the line connecting (4, 6) and (3, −1).

9. Write an equation of the line that passes through the points (−1, 4) and (5, −2).

10. Write an equation of the line passing through the point (3, 2) and parallel to the line $4x - 5y = 20$.

11. Find $f(-5)$ if $f(x) = 3x^2 - 4x - 5$.

Solve each system of equations.

12.
$$\begin{aligned} 2x + 3y &= 6 \\ 5x + 3y &= -24 \end{aligned}$$

13.
$$\begin{aligned} x + y + z &= 3 \\ 2x - y + 2z &= 0 \\ -x - 3y + z &= -9 \end{aligned}$$

Solve each application.

14. The length of a rectangle is 3 cm more than twice its width. If the perimeter of the rectangle is 54 cm, find the dimensions of the rectangle.

15. The sum of the digits of a two-digit number is 10. If the digits are reversed, the number is 36 less than the original number. What was the original number?

CHAPTER

5

Exponents and Polynomials

INTRODUCTION

People in business, finance, industry, and many academic disciplines use polynomial and exponential equations to solve problems and make predictions. We emphasize investment and business applications in this chapter's exercises and activity.

We evaluate investment strategies by estimating the future value of different choices. Although the future value of any investment is subject to many variables, mathematical models allow investors to compare different investment options.

You will be able to explore powerful ideas such as compound interest and savings by gaining experience working with polynomials and exponents.

CHAPTER 5 OUTLINE

5.1

Positive Integer Exponents

< 5.1 Objectives >

1 > Use exponential notation

2 > Simplify expressions with positive integer exponents

Doubling

Take a sheet of paper, $8\frac{1}{2}$ by 11 inches (in.), and cut it in half. Stack the pieces together and then cut this stack again in half. You should now have four pieces. If you continue this process, stacking and cutting, you double the number of pieces with each cut.

Cut Number	Number of Pieces
1	2
2	4
3	8
4	16
5	32
6	64
7	128
8	256

We can compute the height of the stack, for a given number of cuts, if we know the thickness of the original paper. Assuming the thickness to be $\frac{1}{500}$ in. = 0.002 in., the height after 8 cuts is

$$(256)(0.002) = 0.512$$

or a bit more than one-half inch.

If you have actually tried this, you know that it becomes *very* difficult to make even 8 cuts, and that the pieces become *very* small. If we continue to make cuts, we get a surprising result. How high do you think the stack would be if we could make 16 cuts? The number of pieces would be

$$\underbrace{2 \cdot 2 \cdot 2 \cdots 2}_{\text{16 times}}$$

which you can verify, with a calculator, to be 65,536 pieces. Then the height would be

$$(65{,}536)(0.002) \approx 131 \text{ in.}$$

which is almost 11 feet (ft) high!

The calculations done in the paper-cutting exercise are best represented with **exponents.** The way in which the number of pieces (and the height of the stack) grows is often called **exponential growth.**

We first presented exponential notation in Section 0.5, but we give a brief review here. Exponents are a shorthand form for writing repeated multiplication. Instead of writing

$2 \cdot 2 \cdot 2 \cdot 2 \cdot 2 \cdot 2 \cdot 2$

we write

2^7

RECALL

We call *a* the **base** of the expression and 5 the **exponent**, or **power**.

Instead of writing

$a \cdot a \cdot a \cdot a \cdot a$

we write

a^5

which we read as "*a* to the fifth power."

Definition

Exponential Expressions

For any real number *a* and any natural number *n*,

$$a^n = \underbrace{a \cdot a \cdots a}_{n \text{ factors}}$$

An expression of this type is said to be in **exponential form.**

Example 1 Using Exponential Notation

< Objective 1 >

Write each expression using exponential notation.

(a) $w \cdot w \cdot w \cdot w = w^4$

(b) $5y \cdot 5y \cdot 5y = (5y)^3$

Check Yourself 1

Write each expression using exponential notation.

(a) $3z \cdot 3z \cdot 3z \cdot 3z$ **(b)** $x \cdot x \cdot x \cdot x \cdot x \cdot x$

NOTE

We expand the expressions and then remove the parentheses.

Now consider what happens when we multiply two expressions in exponential form with the same base.

$$a^4 \cdot a^5 = \underbrace{(a \cdot a \cdot a \cdot a)}_{4 \text{ factors}}\underbrace{(a \cdot a \cdot a \cdot a \cdot a)}_{5 \text{ factors}}$$

$$= \underbrace{a \cdot a \cdot a \cdot a \cdot a \cdot a \cdot a \cdot a \cdot a}_{9 \text{ factors}}$$

$$= a^9$$

The product is simply the original base taken to the power that is the sum of the two original exponents. This leads to our first property of exponents.

Property

Product Rule for Exponents

For any real number a and positive integers m and n,

$$a^m \cdot a^n = \underbrace{(a \cdot a \cdots a)}_{m \text{ factors}}\underbrace{(a \cdot a \cdots a)}_{n \text{ factors}}$$

$$= \underbrace{a \cdot a \cdots a}_{m + n \text{ factors}}$$

$$= a^{m+n}$$

NOTE

Our first property of exponents:

$a^m \cdot a^n = a^{m+n}$

Example 2 illustrates the product rule for exponents.

Example 2 Using the Product Rule

< Objective 2 >

Simplify each expression.

(a) $b^4 \cdot b^6 = b^{4+6} = b^{10}$

(b) $(2a)^3 \cdot (2a)^4 = (2a)^{3+4} = (2a)^7$

(c) $(-2)^5(-2)^4 = (-2)^{5+4} = (-2)^9$

(d) $(10^7)(10^{11}) = 10^{7+11} = 10^{18}$

NOTE

In every case, the base stays the same.

Check Yourself 2

Simplify each expression.

(a) $(5b)^6(5b)^5$ **(b)** $(-3)^4(-3)^3$ **(c)** $10^8 \cdot 10^{12}$ **(d)** $(xy)^2(xy)^3$

Applying the commutative and associative properties of multiplication, we know that a product such as

$2x^3 \cdot 3x^2$

can be rewritten as

$(2 \cdot 3)(x^3 \cdot x^2)$

or as

$6x^5$

We expand on these ideas in Example 3.

Example 3 Using Properties of Exponents

RECALL

Multiply the coefficients and add the exponents by the product rule. With practice, you will not need to write the regrouping step.

Using the product rule for exponents together with the commutative and associative properties, simplify each expression.

(a) $(x^4)(x^2)(x^3)(x) = x^{10}$

(b) $(3x^4)(5x^2) = (3 \cdot 5)(x^4 \cdot x^2) = 15x^6$

(c) $(2x^5y)(9x^3y^4) = (2 \cdot 9)(x^5 \cdot x^3)(y \cdot y^4) = 18x^8y^5$

(d) $(-3x^2y^2)(-2x^4y^3) = (-3)(-2)(x^2 \cdot x^4)(y^2 \cdot y^3) = 6x^6y^5$

Check Yourself 3

Simplify each expression.

(a) $(x)(x^5)(x^3)$ **(b)** $(7x^5)(2x^2)$

(c) $(-2x^3y)(x^2y^2)$ **(d)** $(-5x^3y^2)(-x^2y^3)$

Now consider the quotient

$$\frac{a^6}{a^4}$$

If we write this in expanded form, we have

NOTE

Divide the numerator and denominator by the four common factors of a.

$$\frac{\overbrace{a \cdot a \cdot a \cdot a \cdot a \cdot a}^{\text{6 factors}}}{\underbrace{a \cdot a \cdot a \cdot a}_{\text{4 factors}}}$$

This can be simplified to

$$\frac{\overset{1}{\cancel{a}} \cdot \overset{1}{\cancel{a}} \cdot \overset{1}{\cancel{a}} \cdot \overset{1}{\cancel{a}} \cdot a \cdot a}{\underset{1}{\cancel{a}} \cdot \underset{1}{\cancel{a}} \cdot \underset{1}{\cancel{a}} \cdot \underset{1}{\cancel{a}}} \quad \text{or} \quad a^2$$

RECALL

$\frac{a}{a} = 1, a \neq 0.$

This means that

$$\frac{a^6}{a^4} = a^2$$

This leads to our second property of exponents.

Property

Quotient Rule for Exponents

For any nonzero real number a and positive integers $m > n$,

$$\frac{a^m}{a^n} = a^{m-n}$$

Example 4 illustrates this rule.

Example 4 Using Properties of Exponents

Simplify each expression.

RECALL

$a^1 = a$; there is no need to write the exponent 1 because it is understood.

(a) $\frac{x^{10}}{x^4} = x^{10-4} = x^6$ Subtract the exponents, applying the quotient rule.

(b) $\frac{a^8}{a^7} = a^{8-7} = a$

(c) $\frac{63w^8}{7w^5} = 9w^{8-5} = 9w^3$ *Divide* the coefficients and subtract the exponents.

(d) $\frac{-32a^4b^5}{8a^2b} = -4a^{4-2}b^{5-1} = -4a^2b^4$ Divide the coefficients and subtract the exponents for *each* variable.

(e) $\frac{10^{16}}{10^6} = 10^{16-6} = 10^{10}$

Check Yourself 4

Simplify each expression.

(a) $\frac{y^{12}}{y^5}$ (b) $\frac{x^9}{x^8}$ (c) $\frac{45r^8}{-9r^6}$ (d) $\frac{49a^6b^7}{7ab^3}$ (e) $\frac{10^{13}}{10^5}$

What happens when a product such as xy is raised to a power? Consider, for example, $(xy)^4$:

$$(xy)^4 = (xy)(xy)(xy)(xy)$$

$$= (x \cdot x \cdot x \cdot x)(y \cdot y \cdot y \cdot y)$$ We use the commutative and associative properties.

$$= x^4y^4$$

This is expressed in the product-power rule.

Property

Product-Power Rule for Exponents

For any real numbers a and b and positive integer n,

$(ab)^n = a^nb^n$

Example 5 illustrates this rule.

Example 5 Using the Product-Power Rule

Simplify each expression.

(a) $(2x)^3 = 2^3x^3 = 8x^3$

(b) $(-3x)^4 = (-3)^4x^4 = 81x^4$

Check Yourself 5

Simplify each expression.

(a) $(3x)^3$ **(b)** $(-2x)^4$

Now, consider the expression

$(3^2)^3$

This can be expanded to

$(3^2)(3^2)(3^2)$

and then expanded again to

$(3)(3)(3)(3)(3)(3) = 3^6$

This leads to the power rule for exponents.

Property

Power Rule for Exponents

For any real number a and positive integers m and n,

$(a^m)^n = a^{mn}$

Example 6 Using the Power Rule for Exponents

Simplify each expression.

(a) $(2^5)^3 = 2^{15}$

(b) $(x^2)^4 = x^8$

(c) $(2x^3)^3 = 2^3(x^3)^3 = 2^3x^9 = 8x^9$

NOTE

Part (c) also uses the product-power rule.

Check Yourself 6

Simplify each expression.

(a) $(3^4)^3$ **(b)** $(x^2)^6$ **(c)** $(3x^3)^4$

We have one final exponent property to develop. Suppose we have a quotient raised to a power. Consider the following:

$$\left(\frac{x}{2}\right)^3 = \frac{x}{2} \cdot \frac{x}{2} \cdot \frac{x}{2} = \frac{x \cdot x \cdot x}{2 \cdot 2 \cdot 2} = \frac{x^3}{2^3}$$

Note that the power, here 3, has been applied to the numerator x and to the denominator 2. This gives our fifth property of exponents.

Property

Quotient-Power Rule for Exponents

For any real numbers *a* and *b*, where *b* is not equal to 0, and positive integer *m*,

$$\left(\frac{a}{b}\right)^m = \frac{a^m}{b^m}$$

In words, to raise a quotient to a power, raise the numerator and denominator to that same power.

Example 7 illustrates the use of this property. Again, we may also have to apply the other properties when simplifying an expression.

Example 7 Using the Quotient-Power Rule for Exponents

Simplify each expression.

(a) $\left(\frac{3}{4}\right)^3 = \frac{3^3}{4^3} = \frac{27}{64}$

(b) $\left(\frac{x^3}{y^2}\right)^4 = \frac{(x^3)^4}{(y^2)^4} = \frac{x^{12}}{y^8}$

(c) $\left(\frac{r^2s^3}{t^4}\right)^2 = \frac{(r^2s^3)^2}{(t^4)^2} = \frac{(r^2)^2(s^3)^2}{(t^4)^2} = \frac{r^4s^6}{t^8}$

Check Yourself 7

Simplify each expression.

(a) $\left(\frac{2}{3}\right)^4$ (b) $\left(\frac{m^3}{n^4}\right)^5$ (c) $\left(\frac{a^2b^3}{c^5}\right)^2$

This table summarizes the five properties of exponents that we discussed in this section.

Property	General Form	Example
Product Rule	$a^m a^n = a^{m+n}$	$x^2 \cdot x^3 = x^5$
Quotient Rule	$\frac{a^m}{a^n} = a^{m-n}$	$\frac{5^7}{5^3} = 5^4$
Power Rule	$(a^m)^n = a^{mn}$	$(z^5)^4 = z^{20}$
Product-Power Rule	$(ab)^m = a^m b^m$	$(4x)^3 = 4^3x^3 = 64x^3$
Quotient-Power Rule	$\left(\frac{a}{b}\right)^m = \frac{a^m}{b^m}$	$\left(\frac{2}{3}\right)^6 = \frac{2^6}{3^6} = \frac{64}{729}$

Check Yourself ANSWERS

1. **(a)** $(3z)^4$; **(b)** x^6 **2.** **(a)** $(5b)^{11}$; **(b)** $(-3)^7$; **(c)** 10^{20}; **(d)** $(xy)^5$
3. **(a)** x^9; **(b)** $14x^7$; **(c)** $-2x^5y^3$; **(d)** $5x^5y^5$
4. **(a)** y^7; **(b)** x; **(c)** $-5r^2$; **(d)** $7a^5b^4$; **(e)** 10^8 **5.** **(a)** $27x^3$; **(b)** $16x^4$
6. **(a)** 3^{12}; **(b)** x^{12}; **(c)** $81x^{12}$ **7.** **(a)** $\frac{16}{81}$; **(b)** $\frac{m^{15}}{n^{20}}$; **(c)** $\frac{a^4b^6}{c^{10}}$

Reading Your Text

SECTION 5.1

(a) Exponents are a shorthand form for writing repeated _______________.

(b) An expression of the type a^n is said to be in _______________ form.

(c) When using the product rule for exponents, we multiply the coefficients and _______________ the exponents.

(d) To raise a quotient to a power, raise the numerator and _______________ to that same power.

5.1 exercises

Basic Skills | Challenge Yourself | Calculator/Computer | Career Applications | Above and Beyond

Name ____________________

Section ________ Date ________

Answers

1.	2.
3.	4.
5.	6.
7.	8.
9.	10.
11.	12.
13.	14.
15.	16.
17.	18.
19.	20.
21.	22.
23.	24.
25.	26.
27.	28.

< Objective 1 >

Write each expression in simplest exponential form.

1. $x^4 \cdot x^5$

2. $x^7 \cdot x^9$

3. $x^5 \cdot x^3 \cdot x^2$

4. $x^8 \cdot x^4 \cdot x^7$

5. $3^5 \cdot 3^2$

6. $(-3)^4(-3)^6$

7. $(-2)^3(-2)^5$

8. $4^3 \cdot 4^4$

9. $4 \cdot x^2 \cdot x^4 \cdot x^7$

10. $3 \cdot x^3 \cdot x^5 \cdot x^8$

11. $\left(\frac{1}{2}\right)^2\left(\frac{1}{2}\right)^3\left(\frac{1}{2}\right)$

12. $\left(-\frac{1}{3}\right)^4\left(-\frac{1}{3}\right)\left(-\frac{1}{3}\right)^5$

13. $(-2)^2(-2)^3(x^4)(x^5)$

14. $(-3)^4(-3)^2(x)^2(x)^6$

15. $(2x)^2(2x)^3(2x)^4$

16. $(-3x)^3(-3x)^5(-3x)^7$

< Objective 2 >

Use the product rule of exponents together with the commutative and associative properties to simplify the products.

17. $(x^2y^3)(x^4y^2)$

18. $(x^4y)(x^2y^3)$

19. $(x^3y^2)(x^4y^2)(x^2y^3)$

20. $(x^2y^3)(x^3y)(x^4y^2)$

21. $(2x^4)(3x^3)(-4x^3)$

22. $(2x^3)(-3x)(-4x^4)$

23. $(5x^2)(3x^3)(x)(-2x^3)$

24. $(4x^2)(2x)(x^2)(2x^3)$

25. $(5xy^3)(2x^2y)(3xy)$

26. $(-3xy)(5x^2y)(-2x^3y^2)$

27. $(x^2yz)(x^3y^5z)(x^4yz)$

28. $(xyz)(x^8y^3z^6)(x^2yz)(xyz^4)$

Use the quotient rule of exponents to simplify each expression.

29. $\frac{x^{10}}{x^7}$

30. $\frac{b^{23}}{b^{18}}$

31. $\frac{x^7y^{11}}{x^4y^3}$

32. $\frac{x^5y^9}{xy^4}$

33. $\frac{x^5y^4z^2}{xy^2z}$

34. $\frac{x^8y^6z^4}{x^3yz^3}$

35. $\frac{21x^4y^5}{7xy^2}$

36. $\frac{48x^6y^6}{12x^3y}$ 

Basic Skills | **Challenge Yourself** | Calculator/Computer | Career Applications | Above and Beyond

Simplify each expression.

37. $(-3x)(5x^5)$

38. $(-5x^2)(-2x^2)$

39. $(2x)^3$

40. $(-3x)^3$

41. $(x^3)^7$

42. $(-x^3)^5$

43. $(3x)(-2x)^3$

44. $(2x)(-3x)^3$

45. $(2x^3)^5$

46. $(-3x^2)^3$

47. $(-2x^2)^3(3x^2)^3$ 

48. $(-3x^2)^2(5x^2)^2$

49. $(3x^3)^2(x^2)^4$

50. $(2x^3)^4(3x^4)^2$

51. $\left(\frac{3}{4}\right)^2$

52. $\left(\frac{2}{3}\right)^2$

53. $\left(\frac{x}{5}\right)^3$

54. $\left(\frac{a}{2}\right)^4$

55. $\left(\frac{m^3}{n^2}\right)^3$

56. $\left(\frac{a^4}{b^3}\right)^4$

57. $\left(\frac{a^3b^2}{c^4}\right)^2$

58. $\left(\frac{x^5y^2}{z^4}\right)^3$

Answers

29. ______

30. ______

31. ______

32. ______

33. ______

34. ______

35. ______

36. ______

37. ______

38. ______

39. ______ 40. ______

41. ______ 42. ______

43. ______ 44. ______

45. ______ 46. ______

47. ______ 48. ______

49. ______ 50. ______

51. ______ 52. ______

53. ______ 54. ______

55. ______ 56. ______

57. ______ 58. ______

Answers

59. ____
60. ____
61. ____
62. ____
63. ____
64. ____
65. ____
66. ____
67. ____
68. ____
69. ____
70. ____
71. ____
72. ____
73. ____
74. ____
75. ____
76. ____
77. ____
78. ____
79. ____
80. ____
81. ____

59. $\left(\dfrac{2x^5}{y^3}\right)^2$

60. $\left(\dfrac{2x^5}{3x^3}\right)^3$

61. $(-8x^2y)(-3x^4y^5)^4$

62. $(5x^5y)^2(-3x^3y^4)^3$

63. $\left(\dfrac{3x^4y^9}{2x^2y^7}\right)\left(\dfrac{x^6y^3}{x^3y^2}\right)^2$

64. $\left(\dfrac{6x^5y^4}{5xy}\right)\left(\dfrac{x^3y^5}{xy^3}\right)^3$

Determine whether each statement is **true** *or* **false.**

65. Exponents are a shorthand form for writing repeated addition.

66. If we multiply x^m by x^n, we can write x raised to the $m + n$ power.

67. When we divide x^m by x^n, we can simply divide the exponents.

68. If we raise x^m to the t power, we can write x raised to the mt power.

Use your calculator to evaluate each expression.

69. 4^3

70. 5^7

71. $(-3)^4$

72. $(-4)^5$

73. $2^3 \cdot 2^5$

74. $3^4 \cdot 3^6$

75. $(3x^2)(2x^4)$, if $x = 2$

76. $(4x^3)(5x^4)$, if $x = 3$

77. $(2x^4)(4x^2)$, if $x = -2$

78. $(3x^5)(2x^3)$, if $x = -3$

79. $(-2x^3)(-3x^5)$, if $x = 2$

80. $(-3x^2)(-4x^4)$, if $x = 4$

81. **Business and Finance** The value A of a savings account that compounds interest annually is given by the formula

chapter 5 > Make the Connection

$A = P(1 + r)^t$

where

P = original amount (principal)
r = interest rate in decimal form
t = time in years

Find the amount of money in the account after 8 years if $2,000 was invested initially at 5% compounded annually.

82. **Business and Finance** Using the formula for compound interest in exercise 81, determine the amount of money in the account if the original investment is doubled. chapter 5 > Make the Connection

83. **Mechanical Engineering** The kinetic energy (in joules) of a falling object is given by

$$KE = \frac{1}{2}mv^2$$

in which m is the mass of the object and v is its velocity.

The velocity (in m/s) of a falling object is given by $v = 4.9t^2$, in which t represents the time since the object was dropped.

(a) Write an equation for the kinetic energy of a 12-kg object in terms of the time since it was dropped.

(b) What is the kinetic energy of a 12-kg object 4 s after it is dropped?

84. **Construction Technology** The load on a post is given by $P = 3L^2$, in which L is the length of the post, in inches. The change in the length (in inches) of the post when loaded is given by

$$\text{Contraction} = \frac{PL}{28{,}000{,}000}$$

(a) Express the contraction of a post in terms of its length.

(b) Report, to the nearest thousandth, the contraction of a post if its length is 120 in.

85. **Mechanical Engineering** The required depth of a beam is equal to four times the cube of its length. The moment of inertia of a four-inch wide beam is equal to $\frac{1}{9}$ of the cube of the depth of the beam. Express the moment of inertia of the beam in terms of its length.

86. **Mechanical Engineering** In a cantilevered beam, the specifications call for the span L of the beam to equal the square of the length of the cantilever c. The bending moment in the beam can be expressed as > Videos

$$M = \frac{wL^2}{8}$$

Express the bending moment M in terms of w and c.

Basic Skills | Challenge Yourself | Calculator/Computer | Career Applications | **Above and Beyond**

87. You have learned rules for working with exponents when multiplying, dividing, and raising an expression to a power.

(a) Explain each rule in your own words. Give numerical examples.

(b) Is there a rule for raising a *sum* to a power? That is, does $(a + b)^n = a^n + b^n$? Use numerical examples to explain why this is true in general or why it is not. Is it always true or always false?

Answers

82. ______

83. ______

84. ______

85. ______

86. ______

87. ______

Answers

88. ______

89. ______

90. ______

91. ______

92. ______

93. ______

94. ______

95. ______

96. ______

88. Work with another student to investigate the rate of inflation. The annual rate of inflation was about 3% from 1990 to 2004. This means that the value of the goods that you could buy for \$1 in 1990 would cost 3% more in 1991, 3% more than that in 1992, etc. If a movie ticket cost \$5.50 in 1990, what would it cost today if movie tickets just kept up with inflation? Construct a table to solve the problem.

Solve each problem.

89. Write x^{12} as a power of x^2.

90. Write y^{15} as a power of y^3. > Videos

91. Write a^{16} as a power of a^2.

92. Write m^{20} as a power of m^5.

93. Write each expression as a power of 8 (remember that $8 = 2^3$): 2^{12}, 2^{18}, $(2^5)^3$, $(2^7)^6$.

94. Write each expression as a power of 9: 3^8, 3^{14}, $(3^5)^8$, $(3^4)^7$.

95. What expression, raised to the third power, is $-8x^6y^9z^{15}$?

96. What expression, raised to the fourth power, is $81x^{12}y^8z^{16}$?

Answers

1. x^9 **3.** x^{10} **5.** 3^7 **7.** $(-2)^8$ **9.** $4x^{13}$ **11.** $\left(\frac{1}{2}\right)^6$
13. $(-2)^5x^9$ **15.** $(2x)^9$ **17.** x^6y^5 **19.** x^9y^7 **21.** $-24x^{10}$
23. $-30x^9$ **25.** $30x^4y^5$ **27.** $x^9y^7z^3$ **29.** x^3 **31.** x^3y^8 **33.** x^4y^2z
35. $3x^3y^3$ **37.** $-15x^6$ **39.** $8x^3$ **41.** x^{21} **43.** $-24x^4$ **45.** $32x^{15}$
47. $-216x^{12}$ **49.** $9x^{14}$ **51.** $\frac{9}{16}$ **53.** $\frac{x^3}{125}$ **55.** $\frac{m^9}{n^6}$ **57.** $\frac{a^6b^4}{c^8}$
59. $\frac{4x^{10}}{y^6}$ **61.** $-648x^{18}y^{21}$ **63.** $\frac{3x^8y^4}{2}$ **65.** False **67.** False
69. 64 **71.** 81 **73.** 256 **75.** 384 **77.** 512 **79.** 1,536
81. \$2,954.91 **83.** (a) $KE = 144.06t^4$; (b) 36,879.36 joules
85. $M = \frac{64}{9}L^9$ **87.** (a) Above and Beyond; (b) Above and Beyond
89. $(x^2)^6$ **91.** $(a^2)^8$ **93.** 8^4; 8^6; 8^5; 8^{14} **95.** $-2x^2y^3z^5$

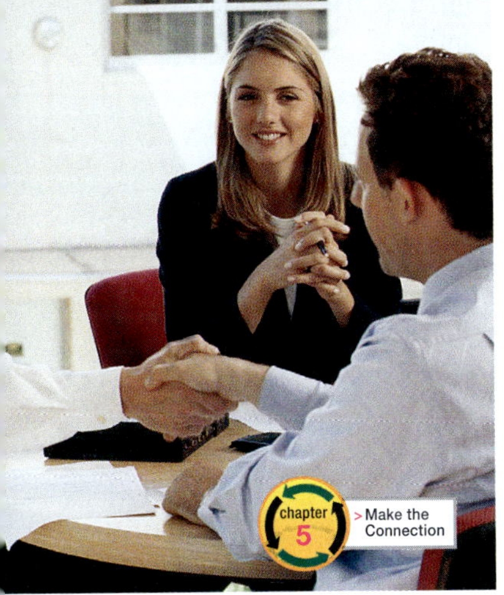

Activity 5 ::
Wealth and Compound Interest

Suppose that when you were born, an uncle put \$500 in the bank for you. He never deposited money again, but the bank paid 5% interest on the money every year on your birthday. How much money was in the bank after 1 year? After 2 years? After 1 year, the amount is $\$500 + 500(0.05)$, which can be written as $\$500(1 + 0.05)$ because of the distributive property. Because $1 + 0.05 = 1.05$, the amount in the bank after 1 year was $500(1.05)$. After 2 years, this amount was again multiplied by 1.05. How much is in the bank today? Complete the chart.

Birthday	Computation	Amount
0 (Day of birth)		\$500
1	\$500(1.05)	
2	\$500(1.05)(1.05)	
3	\$500(1.05)(1.05)(1.05)	
4	$\$500(1.05)^4$	
5	$\$500(1.05)^5$	
6		
7		
8		

(a) Write a formula for the amount in the bank on your *n*th birthday. About how many years does it take for the money to double? How many years does it take for it to double again? Can you see any connection between this and the rules for exponents? Explain why you think there may or may not be a connection.

(b) If the account earned 6% each year, how much more would it accumulate by the end of year 8? Year 21?

(c) Imagine that you start an Individual Retirement Account (IRA) at age 20, contributing \$2,500 each year for 5 years (total \$12,500) to an account that produces a return of 8% every year. You stop contributing and let the account grow. Using the information from the previous example, calculate the value of the account at age 65.

(d) Imagine that you don't start the IRA until you are 30. In an attempt to catch up, you invest \$2,500 into the same account, 8% annual return, each year for 10 years. You then stop contributing and let the account grow. What will its value be at age 65?

(e) What have you discovered as a result of these computations?

5.2 Zero and Negative Exponents and Scientific Notation

< 5.2 Objectives >

1 > Define a zero exponent

2 > Simplify expressions with negative exponents

3 > Write a number in scientific notation

4 > Solve an application involving scientific notation

In Section 5.1, we reviewed some properties of exponents, but all of the exponents were positive integers. In this section, we look at zero and negative exponents. First, we extend the quotient rule so that we can define a zero exponent.

Recall that, in the quotient rule, to divide two expressions that have the same base, we keep the base and subtract the exponents.

$$\frac{a^m}{a^n} = a^{m-n} \qquad \text{if } a \neq 0$$

NOTE

We must have $a \neq 0$. The form 0^0 is called **indeterminate** and is considered in later mathematics classes.

Now, suppose that we allow *m to equal n*. We then have

$$\frac{a^m}{a^m} = a^{m-m} = a^0 \qquad \text{if } a \neq 0$$

But we know that it is also true that

$$\frac{a^m}{a^m} = 1 \qquad \text{if } a \neq 0$$

Comparing these two equations, we see that the next definition is reasonable.

Definition

Zero Exponent	For any nonzero real number a, $a^0 = 1$

Example 1 The Zero Exponent

< Objective 1 >

Use the definition of the zero exponent to simplify each expression.

(a) $17^0 = 1$

(b) $(a^3b^2)^0 = 1$

(c) $6x^0 = 6 \cdot 1 = 6$

(d) $-3y^0 = -3$

NOTE

In $6x^0$, the exponent 0 is applied *only* to x.

> **NOTE**
>
> John Wallis (1616–1702), an English mathematician, was the first to fully discuss the meaning of 0, negative, and rational exponents. You will learn about rational exponents in Chapter 7.

Check Yourself 1

Simplify each expression.

(a) 25^0 **(b)** $(m^4n^2)^0$ **(c)** $8s^0$ **(d)** $-7t^0$

> **RECALL**
>
> If the product of two numbers is 1, they are reciprocals of each other.

When multiplying exponential expressions with the same base, the product rule says to keep the base and add the exponents.

$$a^m \cdot a^n = a^{m+n}$$

Now, what if we allow one of the exponents to be negative and apply the product rule? Suppose, for instance, that $m = 3$ and $n = -3$. Then

$$a^m \cdot a^n = a^3 \cdot a^{-3} = a^{3+(-3)}$$
$$= a^0 = 1$$

so $a^3 \cdot a^{-3} = 1$

If we divide both sides by a^3, we have

$$a^{-3} = \frac{1}{a^3}$$

This is the basis for this definition.

Definition

Negative Integer Exponents

For any nonzero real number a,

$$a^{-n} = \frac{1}{a^n}$$

and a^{-n} is the **multiplicative inverse** of a^n.

We can also say that a^{-n} is the reciprocal of a^n. Example 2 illustrates this definition.

Example 2 Using Properties of Exponents

< Objective 2 >

Simplify each expression.

(a) $y^{-5} = \frac{1}{y^5}$

(b) $4^{-2} = \frac{1}{4^2} = \frac{1}{16}$

(c) $(-3)^{-3} = \frac{1}{(-3)^3} = \frac{1}{-27} = -\frac{1}{27}$

> **NOTE**
>
> From this point on, to *simplify* means to write the expression with *positive exponents only.*
>
> We restrict all variables so that they represent nonzero real numbers.

Check Yourself 2

Simplify each expression.

(a) a^{-10} **(b)** 2^{-4} **(c)** $(-4)^{-2}$

Example 3 illustrates the case where coefficients are involved in an expression with negative exponents. As will be clear, some caution must be used. We must determine exactly what is included in the base of the exponent.

Example 3 Using Properties of Exponents

Simplify each expression.

(a) $2x^{-3} = 2 \cdot \frac{1}{x^3} = \frac{2}{x^3}$

The exponent -3 applies only to the variable x, and *not* to the coefficient 2.

(b) $4w^{-2} = 4 \cdot \frac{1}{w^2} = \frac{4}{w^2}$

(c) $(4w)^{-2} = \frac{1}{(4w)^2} = \frac{1}{16w^2}$

> CAUTION

The expressions $4w^{-2}$ and $(4w)^{-2}$ are *not* the same. Do you see why?

Check Yourself 3

Simplify each expression.

(a) $3w^{-4}$ **(b)** $10x^{-5}$ **(c)** $(2y)^{-4}$ **(d)** $-5t^{-2}$

Suppose that a variable with a negative exponent appears in the denominator of an expression. For instance, if we wish to simplify

$$\frac{1}{a^{-2}}$$

we can multiply numerator and denominator by a^2.

$$\frac{1}{a^{-2}} = \frac{1 \cdot a^2}{a^{-2} \cdot a^2} = \frac{a^2}{a^0} = \frac{a^2}{1} = a^2$$

Negative exponent in denominator

Positive exponent in numerator

So

$$\frac{1}{a^{-2}} = a^2$$

This leads to a property for negative exponents.

NOTE

$a^{-2} = \frac{1}{a^2}$

so

$\frac{1}{a^{-2}} = \frac{1}{\frac{1}{a^2}}$

We invert and multiply:

$\frac{1}{\frac{1}{a^2}} = 1 \div \frac{1}{a^2} = 1 \cdot \frac{a^2}{1} = a^2$

Property

Negative Exponents

For any nonzero real number a and integer n,

$$\frac{1}{a^{-n}} = a^n$$

Example 4 Using Properties of Exponents

Simplify each expression.

(a) $\dfrac{1}{y^{-3}} = y^3$

(b) $\dfrac{1}{2^{-5}} = 2^5 = 32$

(c) $\dfrac{3}{4x^{-2}} = \dfrac{3x^2}{4}$ The exponent -2 applies only to x, not to 4.

(d) $\dfrac{a^{-3}}{b^{-4}} = \dfrac{b^4}{a^3}$

Check Yourself 4

Simplify each expression.

(a) $\dfrac{1}{x^{-4}}$ (b) $\dfrac{1}{3^{-3}}$ (c) $\dfrac{2}{3a^{-2}}$ (d) $\dfrac{c^{-5}}{d^{-7}}$

RECALL

You can review these properties in Section 5.1.

The product and quotient rules for exponents apply to expressions that involve any integer exponent—positive, negative, or 0. Example 5 illustrates this concept.

Example 5 Using Properties of Exponents

Simplify each expression. Use only positive exponents to express the result.

(a) $x^3 \cdot x^{-7} = x^{3+(-7)}$ Add the exponents by the product rule.

$= x^{-4} = \dfrac{1}{x^4}$

(b) $\dfrac{m^{-5}}{m^{-3}} = m^{-5-(-3)} = m^{-5+3}$ Subtract the exponents by the quotient rule.

$= m^{-2} = \dfrac{1}{m^2}$

(c) $\dfrac{x^5x^{-3}}{x^{-7}} = \dfrac{x^{5+(-3)}}{x^{-7}} = \dfrac{x^2}{x^{-7}} = x^{2-(-7)} = x^9$ We apply first the product rule and then the quotient rule.

Check Yourself 5

Simplify each expression.

(a) $x^9 \cdot x^{-5}$ (b) $\dfrac{y^{-7}}{y^{-3}}$ (c) $\dfrac{a^{-3}a^2}{a^{-5}}$

How do we simplify a rational expression raised to a negative power? As we have seen, the properties of exponents can be extended to include negative exponents.

Suppose we wish to simplify $\left(\frac{x}{y}\right)^{-2}$.

$$\left(\frac{x}{y}\right)^{-2} = \frac{x^{-2}}{y^{-2}}$$ Use the quotient-power rule.

$$= \frac{y^2}{x^2} = \left(\frac{y}{x}\right)^2$$ Use the quotient-power rule again.

Property

Quotient Raised to a Negative Power

For any nonzero real numbers a and b,

$$\left(\frac{a}{b}\right)^{-n} = \left(\frac{b}{a}\right)^n$$

Example 6 Extending the Properties of Exponents

Simplify each expression.

(a) $\left(\frac{s^3}{t^2}\right)^{-2} = \left(\frac{t^2}{s^3}\right)^2 = \frac{t^4}{s^6}$

(b) $\left(\frac{m^2}{n^{-2}}\right)^{-3} = \left(\frac{n^{-2}}{m^2}\right)^3$

$= \frac{n^{-6}}{m^6} = \frac{1}{m^6n^6}$

Check Yourself 6

Simplify each expression.

(a) $\left(\frac{3t^2}{s^3}\right)^{-3}$ **(b)** $\left(\frac{x^5}{y^{-2}}\right)^{-3}$

Example 7 Using Properties of Exponents

Simplify each expression.

(a) $\left(\frac{3}{q^5}\right)^{-2} = \left(\frac{q^5}{3}\right)^2 = \frac{q^{10}}{9}$

(b) $\left(\frac{x^3}{y^4}\right)^{-3} = \left(\frac{y^4}{x^3}\right)^3 = \frac{(y^4)^3}{(x^3)^3} = \frac{y^{12}}{x^9}$

Check Yourself 7

Simplify each expression.

(a) $\left(\frac{r^4}{5}\right)^{-2}$ **(b)** $\left(\frac{a^4}{b^3}\right)^{-3}$

As you might expect, simplifying more complicated expressions often requires the use of more than one of the properties. Example 8 illustrates such cases.

Example 8 Using Properties of Exponents

Simplify each expression.

(a) $$\frac{(a^2)^{-3}(a^3)^4}{(a^{-3})^3} = \frac{a^{-6} \cdot a^{12}}{a^{-9}}$$ Apply the power rule to each factor.

$$= \frac{a^{-6+12}}{a^{-9}} = \frac{a^6}{a^{-9}}$$ Apply the product rule.

$$= a^{6-(-9)} = a^{6+9} = a^{15}$$ Apply the quotient rule.

(b) $$\frac{8x^{-2}y^{-5}}{12x^{-4}y^3} = \frac{8}{12} \cdot \frac{x^{-2}}{x^{-4}} \cdot \frac{y^{-5}}{y^3}$$ It helps to separate this expression into three fractions.

$$= \frac{2}{3} \cdot x^{-2-(-4)} \cdot y^{-5-3}$$

$$= \frac{2}{3} \cdot x^2 \cdot y^{-8} = \frac{2x^2}{3y^8}$$

(c) $$\left(\frac{pr^3s^{-5}}{p^3r^{-3}s^{-2}}\right)^{-2} = [(p^{1-3}r^{3-(-3)}s^{-5-(-2)})]^{-2}$$

$$= (p^{-2}r^6s^{-3})^{-2}$$ Apply the quotient rule inside the parentheses.

$$= (p^{-2})^{-2}(r^6)^{-2}(s^{-3})^{-2}$$ Apply the rule for a product to a power.

$$= p^4r^{-12}s^6 = \frac{p^4s^6}{r^{12}}$$ Apply the power rule.

> CAUTION

Another possible first step (and generally an efficient one) is to rewrite an expression by using our earlier properties.

$a^{-n} = \frac{1}{a^n}$ and $\frac{1}{a^{-n}} = a^n$

> CAUTION

A *common error* is to write

$$\frac{8x^{-2}y^{-5}}{12x^{-4}y^3} = \frac{12x^4}{8x^2y^3y^5}$$

This is *not* correct.

In Example 8(b), we could *correctly* begin

$$\frac{8x^{-2}y^{-5}}{12x^{-4}y^3} = \frac{8x^4}{12x^2y^3y^5}$$

The coefficients should not be moved along with the variables. Keep in mind that the exponents apply *only* to the variables in this expression. The coefficients remain *where they were* in the original expression when the expression is rewritten using this approach.

Check Yourself 8

Simplify each expression.

(a) $\frac{(x^5)^{-2}(x^2)^3}{(x^{-4})^3}$ (b) $\frac{12a^{-3}b^{-2}}{16a^{-2}b^3}$ (c) $\left(\frac{xy^{-3}z^{-5}}{x^{-4}y^{-2}z^3}\right)^{-3}$

> Calculator

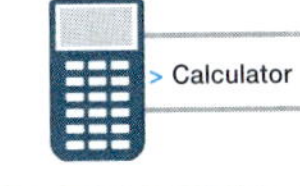

```
2300*1000
          2300000
Ans*1000
       2300000000
Ans*1000
           2.3E12
```

Let us now take a look at an important use of exponents, scientific notation.

We begin the discussion with a calculator exercise. On most scientific calculators, if you multiply 2.3 times 1,000, the display reads

2300

Multiply by 1,000 a second time. Now you should see

2300000

NOTE

Consider the table:

$2.3 = 2.3 \times 10^0$
$23 = 2.3 \times 10^1$
$230 = 2.3 \times 10^2$
$2{,}300 = 2.3 \times 10^3$
$23{,}000 = 2.3 \times 10^4$
$230{,}000 = 2.3 \times 10^5$

Multiplying by 1,000 a third time results in the display

2.3 09 or 2.3 E09 or 2300000000

And multiplying by 1,000 again yields

2.3 12 or 2.3 E12

Can you see what is happening? This is the way calculators display very large numbers. The number on the left is always between 1 and 10, and the number on the right indicates the number of places the decimal point must be moved to the right to put the answer in standard (or decimal) form.

This notation is used frequently in science. It is not uncommon in scientific applications of algebra to find yourself working with very large or very small numbers. Even in the time of Archimedes (287–212 B.C.E.), the study of such numbers was not unusual. Archimedes estimated that the universe was 23,000,000,000,000,000 m in diameter, which is the approximate distance light travels in $2\frac{1}{2}$ years. By comparison, Polaris (the North Star) is 680 light-years from the earth. We discuss light years in Example 10.

NOTE

Scientific notation is one of the few places where we still use the multiplication symbol $\times$.

In scientific notation, Archimedes' estimate for the diameter of the universe is

$$2.3 \times 10^{16} \text{ m}$$

We define scientific notation as follows.

Definition

Scientific Notation

Any positive number written in the form

$$a \times 10^n$$

in which $1 \le a < 10$ and n is an integer, is written in **scientific notation.**

RECALL

$10^0 = 1$, so
$2.3 \times 10^0 = 2.3 \times 1$
$= 2.3$

When a number is written in scientific notation, we look at the sign of the exponent to determine if the number is large or small. If the exponent is not negative, then the number is greater than or equal to one. If the exponent is negative, then the number is less than one.

$2.3 \times 10^2 = 230$ $\quad$ $2.3 \times 10^{-2} = 0.023$
$2.3 \times 10^1 = 23$ $\quad$ $2.3 \times 10^{-1} = 0.23$
$2.3 \times 10^0 = 2.3$

Example 9 Using Scientific Notation

< Objective 3 >

Write each number in scientific notation.

(a) $120{,}000. = 1.2 \times 10^5$
5 places — The power is 5.

(b) $88{,}000{,}000. = 8.8 \times 10^7$
7 places — The power is 7.

(c) $520{,}000{,}000. = 5.2 \times 10^8$
8 places

NOTES

Study the pattern for writing a number in scientific notation.

The exponent shows the *number of places* we move the decimal point so that the multiplier is a number between 1 and 10.

To convert back to standard or decimal form, we simply reverse the process.

(d) $4{,}000{,}000{,}000. = 4 \times 10^9$

9 places

(e) $0.0005 = 5 \times 10^{-4}$

4 places

(f) $0.0000000081 = 8.1 \times 10^{-9}$

9 places

Check Yourself 9

Write in scientific notation.

(a) 212,000,000,000,000,000
(b) 0.00079
(c) 5,600,000
(d) 0.0000007

Example 10 An Application of Scientific Notation

< Objective 4 >

(a) Light travels at an approximate speed of 3.05×10^8 meters per second (m/s). There are about 3.15×10^7 s in a year. How far does light travel in a year?

We multiply the distance traveled in 1 s by the number of seconds in a year. This yields

$$(3.05 \times 10^8)(3.15 \times 10^7) = (3.05 \cdot 3.15)(10^8 \cdot 10^7)$$
$$= 9.6075 \times 10^{15}$$

Multiply the coefficients, and add the exponents.

NOTE

$9.6075 \times 10^{15} \approx 10 \times 10^{15} = 10^{16}$

For our purposes we round the distance light travels in 1 year to 10^{16} m. This unit is called a **light-year,** and it is used to measure astronomical distances.

(b) The distance from Earth to the star Spica (in Virgo) is 2.2×10^{18} m. How many light-years is Spica from Earth?

NOTE

We divide the distance (in meters) by the number of meters in 1 light-year.

$$\frac{2.2 \times 10^{18}}{10^{16}} = 2.2 \times 10^{18-16}$$
$$= 2.2 \times 10^2 = 220 \text{ light-years}$$

Check Yourself 10

The farthest object that can be seen with the unaided eye is the Andromeda galaxy. This galaxy is 2.3×10^{22} m from earth. What is this distance in light-years?

Check Yourself ANSWERS

1. **(a)** 1; **(b)** 1; **(c)** 8; **(d)** -7 **2.** **(a)** $\frac{1}{a^{10}}$; **(b)** $\frac{1}{16}$; **(c)** $\frac{1}{16}$

3. **(a)** $\frac{3}{w^4}$; **(b)** $\frac{10}{x^5}$; **(c)** $\frac{1}{16y^4}$; **(d)** $-\frac{5}{t^2}$ **4.** **(a)** x^4; **(b)** 27; **(c)** $\frac{2a^2}{3}$; **(d)** $\frac{d^7}{c^5}$

5. **(a)** x^4; **(b)** $\frac{1}{y^4}$; **(c)** a^4 **6.** **(a)** $\frac{s^9}{27t^6}$; **(b)** $\frac{1}{x^{15}y^6}$

7. **(a)** $\frac{25}{r^8}$; **(b)** $\frac{b^9}{a^{12}}$ **8.** **(a)** x^8; **(b)** $\frac{3}{4ab^5}$; **(c)** $\frac{y^3z^{24}}{x^{15}}$

9. **(a)** 2.12×10^{17}; **(b)** 7.9×10^{-4}; **(c)** 5.6×10^6; **(d)** 7×10^{-7}

10. 2,300,000 light-years

Reading Your Text

SECTION 5.2

(a) When multiplying exponential expressions with the same base, the product rule says to keep the base and ______________ the exponents.

(b) a^{-n} is the multiplicative inverse, or ______________, of a^n.

(c) When a number is written in scientific notation and the exponent is not negative, then the number is greater than or equal to ______________.

(d) Light travels at an approximate speed of 3.05×10^8 ______________ per second.

Basic Skills | Challenge Yourself | Calculator/Computer | Career Applications | Above and Beyond

5.2 exercises

Boost *your* GRADE at ALEKS.com!

Name ______________

Section ________ Date ________

< Objectives 1 and 2 >

Simplify each expression.

1. x^{-5}
2. 3^{-3}
3. 5^{-2}
4. x^{-8}
5. $(-5)^{-2}$
6. $(-3)^{-3}$
7. $(-2)^{-3}$
8. $(-2)^{-4}$
9. $\left(\frac{2}{3}\right)^{-3}$
10. $\left(\frac{3}{4}\right)^{-2}$
11. $3x^{-2}$
12. $4x^{-3}$
13. $-5x^{-4}$
14. $(-2x)^{-4}$
15. $(-3x)^{-2}$
16. $-5x^{-2}$
17. $\frac{1}{x^{-3}}$
18. $\frac{1}{x^{-5}}$
19. $\frac{2}{5x^{-3}}$
20. $\frac{3}{4x^{-4}}$
21. $\frac{x^{-3}}{y^{-4}}$
22. $\frac{x^{-5}}{y^{-3}}$
23. $x^5 \cdot x^{-3}$
24. $y^{-4} \cdot y^5$
25. $a^{-9} \cdot a^6$
26. $w^{-5} \cdot w^3$
27. $z^{-2} \cdot z^{-8}$
28. $b^{-7} \cdot b^{-1}$

Answers

1.	2.
3.	4.
5.	6.
7.	8.
9.	10.
11.	12.
13.	14.
15.	16.
17.	18.
19.	20.
21.	22.
23.	24.
25.	26.
27.	28.

Answers

29. ______
30. ______
31. ______
32. ______
33. ______
34. ______
35. ______
36. ______
37. ______
38. ______
39. ______
40. ______
41. ______ 42. ______
43. ______ 44. ______
45. ______ 46. ______
47. ______ 48. ______
49. ______ 50. ______
51. ______ 52. ______
53. ______ 54. ______
55. ______ 56. ______
57. ______ 58. ______

29. $a^{-5} \cdot a^5$

30. $x^{-4} \cdot x^4$

31. $\dfrac{x^{-5}}{x^{-2}}$ > Videos

32. $\dfrac{x^{-3}}{x^{-6}}$

33. $(x^5)^3$

34. $(w^4)^6$

35. $(2x^{-3})(x^2)^4$

36. $(p^4)(3p^3)^2$

37. $(3a^{-4})(a^3)(a^2)$

38. $(5y^{-2})(2y)(y^5)$

39. $(x^4y)(x^2)^3(y^3)^0$

40. $(r^4)^2(r^2s)(s^3)^2$

41. $(ab^2c)(a^4)^4(b^2)^3(c^3)^4$

42. $(p^2qr^2)(p^2)(q^3)^2(r^2)^0$

43. $(x^5)^{-3}$

44. $(x^{-2})^{-3}$

45. $(b^{-4})^{-2}$

46. $(a^0b^{-4})^3$

47. $(x^5y^{-3})^2$

48. $(p^{-3}q^2)^{-2}$

49. $(x^{-4}y^{-2})^{-3}$

50. $(3x^{-2}y^{-2})^3$

51. $(2x^{-3}y^0)^{-5}$

52. $\dfrac{a^{-6}}{b^{-4}}$

53. $\dfrac{x^{-2}}{y^{-4}}$

54. $\left(\dfrac{x^{-3}}{y^2}\right)^{-3}$ > Videos

55. $\dfrac{x^{-4}}{y^{-2}}$

56. $\dfrac{(3x^{-4})^2(2x^2)}{x^6}$

57. $(4x^{-2})^2(3x^{-4})$

58. $(5x^{-4})^{-4}(2x^3)^{-5}$

< Objective 3 >

Express each number in scientific notation.

59. The distance from the earth to the sun: 93,000,000 mi.

60. The diameter of a grain of sand: 0.000021 m.

61. The diameter of the sun: 130,000,000,000 cm.

62. The number of oxygen atoms in 16 grams of oxygen gas: 602,000,000,000,000,000,000,000 (Avogadro's number).

63. The mass of the sun is approximately 1.98×10^{30} kg. If this were written in standard or decimal form, how many 0's would follow the digit 8?

64. Archimedes estimated the universe to be 2.3×10^{19} millimeters (mm) in diameter. If this number were written in standard or decimal form, how many 0's would follow the digit 3?

Write each expression in standard notation.

65. 8×10^{-3}

66. 7.5×10^{-6}

67. 2.8×10^{-5}

68. 5.21×10^{-4}

Write each number in scientific notation.

69. 0.0005

70. 0.000003

71. 0.00037

72. 0.000051

< Objective 4 >

73. Megrez, the nearest of the Big Dipper stars, is 6.6×10^{17} m from Earth. Approximately how long does it take light, traveling at 10^{16} m/yr, to travel from Megrez to Earth?

74. Alkaid, the most distant star in the Big Dipper, is 2.1×10^{18} m from Earth. Approximately how long does it take light to travel from Alkaid to Earth?

Answers

59. ______

60. ______

61. ______

62. ______

63. ______

64. ______

65. ______

66. ______

67. ______

68. ______

69. ______

70. ______

71. ______

72. ______

73. ______

74. ______

Basic Skills | **Challenge Yourself** | Calculator/Computer | Career Applications | Above and Beyond

Answers

75. ______
76. ______
77. ______
78. ______
79. ______
80. ______
81. ______
82. ______
83. ______
84. ______
85. ______
86. ______
87. ______
88. ______
89. ______
90. ______
91. ______ 92. ______
93. ______ 94. ______
95. ______ 96. ______
97. ______ 98. ______
99. ______ 100. ______

Determine whether each statement is **true** *or* **false.**

75. Any number raised to the zero power is zero.

76. 5^{-3} represents the multiplicative inverse of 5^3.

77. A fraction raised to the power $-n$ can be rewritten as the reciprocal fraction raised to the power n.

78. If we rewrite $3x^{-4}$ as a fraction, both the 3 and the x appear in the denominator.

Simplify each expression.

79. $(2x^5)^4(x^3)^2$

80. $(3x^2)^3(x^2)^4(x^2)$

81. $(2x^{-3})^3(3x^3)^2$

82. $(x^2y^3)^4(xy^3)^0$

83. $(xy^5z)^4(xyz^2)^8(x^6yz)^5$

84. $(x^2y^2z^2)^0(xy^2z)^2(x^3yz^2)$

85. $(3x^{-2})(5x^2)^2$

86. $(2a^3)^2(a^0)^5$

87. $(2w^3)^4(3w^{-5})^2$

88. $(3x^3)^2(2x^4)^5$

89. $\frac{3x^6}{2y^9} \cdot \frac{y^5}{x^3}$

90. $\frac{x^8}{y^6} \cdot \frac{2y^9}{x^3}$

91. $(-7x^2y)(-3x^5y^6)^4$

92. $\left(\frac{2w^5z^3}{3x^3y^9}\right)\left(\frac{x^5y^4}{w^4z^0}\right)^2$

93. $(2x^2y^{-3})(3x^{-4}y^{-2})$

94. $(-5a^{-2}b^{-4})(2a^5b^0)$

95. $\frac{(x^{-3})(y^2)}{y^{-3}}$

96. $\frac{6x^3y^{-4}}{24x^{-2}y^{-2}}$

97. $\frac{15x^{-3}y^2z^{-4}}{20x^{-4}y^{-3}z^2}$

98. $\frac{24x^{-5}y^{-3}z^2}{36x^{-2}y^3z^{-2}}$

99. $\frac{x^{-5}y^{-7}}{x^0y^{-4}}$

100. $\left(\frac{xy^3z^{-4}}{x^{-3}y^{-2}z^2}\right)^{-2}$

101. $\dfrac{x^{-2}y^{2}}{x^{3}y^{-2}} \cdot \dfrac{x^{-4}y^{2}}{x^{-2}y^{-2}}$ Videos

102. $\left(\dfrac{x^{-3}y^{3}}{x^{-4}y^{2}}\right)^{3} \cdot \left(\dfrac{x^{-2}y^{-2}}{xy^{4}}\right)^{-1}$

103. $x^{2n} \cdot x^{3n}$

104. $x^{n+1} \cdot x^{3n}$

105. $\dfrac{x^{n+3}}{x^{n+1}}$

106. $\dfrac{x^{n-4}}{x^{n-1}}$

107. $(y^{n})^{3n}$

108. $(x^{n+1})^{n}$

109. $\dfrac{x^{2n} \cdot x^{n+2}}{x^{3n}}$

110. $\dfrac{x^{n} \cdot x^{3n+5}}{x^{4n}}$

Evaluate each expression and write your result in scientific notation.

111. $(2 \times 10^{5})(4 \times 10^{4})$

112. $(2.5 \times 10^{7})(3 \times 10^{5})$

113. $\dfrac{6 \times 10^{9}}{3 \times 10^{7}}$

114. $\dfrac{4.5 \times 10^{12}}{1.5 \times 10^{7}}$

115. $\dfrac{(3.3 \times 10^{15})(6 \times 10^{15})}{(1.1 \times 10^{8})(3 \times 10^{6})}$

116. $\dfrac{(6 \times 10^{12})(3.2 \times 10^{8})}{(1.6 \times 10^{7})(3 \times 10^{2})}$

117. $(4 \times 10^{-3})(2 \times 10^{-5})$

118. $(1.5 \times 10^{-6})(4 \times 10^{2})$

119. $\dfrac{9 \times 10^{3}}{3 \times 10^{-2}}$

120. $\dfrac{7.5 \times 10^{-4}}{1.5 \times 10^{2}}$

121. The approximate amount of water on the Earth is 15,500 followed by 19 zeros, in liters. Write this number in scientific notation.

122. Use your result in exercise 121 to find the amount of water per person on the Earth if there were approximately 7.1 billion living people.

123. If there are 7.1×10^{9} people on the Earth and there is enough freshwater to provide each person with 6.57×10^{5} L, how much freshwater is on the Earth?

124. The United States uses an average of 2.6×10^{6} L of water per person each year. If there are 3.1×10^{8} people in the United States, how many liters of water does the United States use each year?

Answers

101. ________
102. ________
103. ________
104. ________
105. ________
106. ________
107. ________
108. ________
109. ________
110. ________
111. ________
112. ________
113. ________
114. ________
115. ________
116. ________
117. ________
118. ________
119. ________
120. ________
121. ________
122. ________
123. ________
124. ________

Answers

125. ______

126. ______

127. ______

128. ______

129. ______

130. ______

131. ______

Basic Skills | Challenge Yourself | Calculator/Computer | **Career Applications** | Above and Beyond

125. ELECTRONICS The resistance in a circuit is measured to be $12 \times 10^4\ \Omega$.

(a) Express the resistance in standard notation.

(b) Express the resistance in scientific notation.

ALLIED HEALTH Medical lab technicians can determine the concentration c of a solution, in moles per liter (mol/L), based on the absorbance A, which is a measure of the amount of light that passes through the solution, the molar absorption coefficient ε, and the length of the light path d, in cm, in the colorimeter using the formula

$$c = \frac{A}{\varepsilon d}$$

Use this information to complete exercises 126 and 127.

126. Determine the concentration of a solution with an absorbance of 0.254 mol/L, a molar absorption coefficient of 3.6×10^3, and a 1.4-cm light path. Use scientific notation to express your result.

127. Determine the concentration of a solution with an absorbance of 0.315 mol/L, a molar absorption coefficient of 8.2×10^6, and an 18-mm light path. Use scientific notation to express your result.

128. INFORMATION TECHNOLOGY The distance from the ground to a satellite in geostationary orbit is 42,245 km. How long will it take a light beam to reach the satellite (use scientific notation to express your result)? *Recall:* Light travels at an approximate speed of 3.05×10^8 m/s.

Basic Skills | Challenge Yourself | Calculator/Computer | Career Applications | **Above and Beyond**

129. Recall the paper "cut and stack" experiment described at the beginning of this chapter. If you had a piece of paper large enough to complete the task (and assuming that you *could* complete the task!), determine the height of the stack, given that the thickness of the paper is 0.002 in.

(a) After 30 cuts **(b)** After 50 cuts

130. Can $(a + b)^{-1}$ be written as $\frac{1}{a} + \frac{1}{b}$ by using the properties of exponents? If not, why not? Explain.

131. Write a short description of the difference between $(-4)^{-3}$, -4^{-3}, $(-4)^3$, and -4^3. Are any of these equal?

132. If $n > 0$, which expressions are negative?

$-n^{-3}$ $\quad n^{-3}$ $\quad (-n)^{-3}$ $\quad (-n)^3$ $\quad -n^3$

If $n < 0$, which of these expressions are negative? Explain what effect a negative in the exponent has on the sign of the result when an exponential expression is simplified.

133. You are offered a 28-day job in which you have a choice of two different pay arrangements. Plan 1 offers a flat $4,000,000 at the end of the 28th day on the job. Plan 2 offers 1¢ the first day, 2¢ the second day, 4¢ the third day, and so on, with the amount doubling each day. Make a table to decide which offer is better. Write a formula for the amount you make on the nth day and a formula for the total after n days. Which pay arrangement should you take? Why?

Answers

132. ______

133. ______

Answers

1. $\frac{1}{x^5}$ **3.** $\frac{1}{25}$ **5.** $\frac{1}{25}$ **7.** $-\frac{1}{8}$ **9.** $\frac{27}{8}$ **11.** $\frac{3}{x^2}$ **13.** $-\frac{5}{x^4}$ **15.** $\frac{1}{9x^2}$ **17.** x^3 **19.** $\frac{2x^3}{5}$ **21.** $\frac{y^4}{x^3}$ **23.** x^2 **25.** $\frac{1}{a^3}$ **27.** $\frac{1}{z^{10}}$ **29.** 1 **31.** $\frac{1}{x^3}$ **33.** x^{15} **35.** $2x^5$ **37.** $3a$ **39.** $x^{10}y$ **41.** $a^{17}b^8c^{13}$ **43.** $\frac{1}{x^{15}}$ **45.** b^8 **47.** $\frac{x^{10}}{y^6}$ **49.** $x^{12}y^6$ **51.** $\frac{x^{15}}{32}$ **53.** $\frac{y^4}{x^2}$ **55.** $\frac{y^2}{x^4}$ **57.** $\frac{48}{x^8}$ **59.** 9.3×10^7 **61.** 1.3×10^{11} **63.** 28 **65.** 0.008 **67.** 0.000028 **69.** 5×10^{-4} **71.** 3.7×10^{-4} **73.** 66 years **75.** False **77.** True **79.** $16x^{26}$ **81.** $\frac{72}{x^3}$ **83.** $x^{42}y^{33}z^{25}$ **85.** $75x^2$ **87.** $144w^2$ **89.** $\frac{3x^3}{2y^4}$ **91.** $-567x^{22}y^{25}$ **93.** $\frac{6}{x^2y^5}$ **95.** $\frac{y^5}{x^3}$ **97.** $\frac{3xy^5}{4z^6}$ **99.** $\frac{1}{x^5y^3}$ **101.** $\frac{y^8}{x^7}$ **103.** x^{5n} **105.** x^2 **107.** y^{3n^2} **109.** x^2 **111.** 8×10^9 **113.** 2×10^2 **115.** 6×10^{16} **117.** 8×10^{-8} **119.** 3×10^5 **121.** 1.55×10^{23} **123.** 4.66×10^{15} L **125. (a)** 120,000 Ω; **(b)** 1.2×10^5 Ω **127.** $c \approx 2.13 \times 10^{-8}$ **129. (a)** About 33.9 mi; **(b)** about 35.5 million mi **131.** Above and Beyond **133.** $p = \frac{2^{n-1}}{100}; t = \frac{2^n}{100} - 0.01$

5.3

Introduction to Polynomials

< 5.3 Objectives >

1 > Identify types of polynomials

2 > Find the degree of a polynomial

3 > Write polynomials in descending order

RECALL

We defined **term** in Section 1.3.

Our work in this section deals with the most common kind of algebraic expression, a *polynomial.* To define a polynomial, first recall the definition of the word *term.*

Definition

Term — A **term** can be written as a number or the product of a number and one or more variables and their exponents.

NOTE

In a polynomial, terms are separated by + and − signs.

For example, x^5, $3x$, $-4xy^2$, 8, $\frac{5}{x}$, and $-14\sqrt{x}$ are terms. A **polynomial** consists of one or more terms in which the only allowable exponents are the whole numbers, 0, 1, 2, 3, and so on. These terms are connected by addition or subtraction signs. The variable is never used as a divisor and never appears under a radical sign in a term of a polynomial.

Definition

Numerical Coefficient — In each term of a polynomial, the number factor is called the **numerical coefficient,** or more simply the **coefficient,** of that term.

Example 1 — Identifying Polynomials

NOTE

Each sign (+ or −) is attached to the term that *follows* that sign.

(a) $x + 3$ is a polynomial. The terms are x and 3. The coefficients are 1 and 3.

(b) $3x^2 - 2x + 5$ is also a polynomial. Its terms are $3x^2$, $-2x$, and 5. The coefficients are 3, −2, and 5.

(c) $5x^3 + 2 - \frac{3}{x}$ is *not* a polynomial because of the division by x in the third term.

NOTE

The prefix *mono* means 1. The prefix *bi* means 2. The prefix *tri* means 3. There are no special names for polynomials with more than three terms.

Check Yourself 1

Which are polynomials?

(a) $5x^2$ (b) $3y^3 - 2y + \frac{5}{y}$ (c) $4x^2 - 2x + 3$

Certain polynomials are given special names because of the number of terms that they have.

Definition

Monomial, Binomial, and Trinomial

A polynomial with exactly one term is called a **monomial.**

A polynomial with exactly two terms is called a **binomial.**

A polynomial with exactly three terms is called a **trinomial.**

Example 2 Identifying Types of Polynomials

< Objective 1 >

(a) $3x^2y$ is a monomial. It has exactly one term.

(b) $2x^3 + 5x$ is a binomial. It has exactly two terms, $2x^3$ and $5x$.

(c) $5x^2 - 4x + 3$ is a trinomial. Its three terms are $5x^2$, $-4x$, and 3.

RECALL

In a polynomial, the allowable exponents are the whole numbers 0, 1, 2, 3, and so on. The degree is a whole number.

Check Yourself 2

Classify each as a monomial, binomial, or trinomial.

(a) $5x^4 - 2x^3$ (b) $4x^7$ (c) $2x^2 + 5x - 3$

We also classify polynomials by their *degree.* The **degree** of a polynomial that has only one variable is the highest power of that variable appearing in any one term.

Example 3 Classifying Polynomials by Their Degree

< Objective 2 >

RECALL

In Section 5.2, you learned that $x^0 = 1$.

The highest power

(a) $5x^3 - 3x^2 + 4x$ has degree 3.

The highest power

(b) $4x - 5x^4 + 3x^3 + 2$ has degree 4.

(c) $8x$ has degree 1 because $8x = 8x^1$.

(d) 7 has degree 0 because $7 = 7 \cdot 1 = 7x^0$.

Note: Polynomials can have more than one variable, such as $4x^2y^3 + 5xy^2$. The degree is then the largest sum of the powers in any single term (here 2 + 3, or 5). In general, we work with polynomials in a single variable, such as x.

Check Yourself 3

Find the degree of each polynomial.

(a) $6x^5 - 3x^3 - 2$ (b) $5x$ (c) $3x^3 + 2x^6 - 1$ (d) 9

Polynomials are much easier if you get used to writing them in **descending order** (sometimes called *descending-exponent form*). When a polynomial has only one variable, this means that the term with the highest exponent is written first, then the term with the next-highest exponent is written, and so on.

We call the first term of a polynomial written in descending order the *leading term*. As stated above, this is the term that has the largest exponent of the variable. The power of the variable of the leading term is the same as the degree of the polynomial.

The numerical coefficient of the leading term is called the *leading coefficient*.

Example 4 Writing Polynomials in Descending Order

< Objective 3 >

The exponents get smaller from left to right.

(a) $5x^7 - 3x^4 + 2x^2$ is in descending order. The leading coefficient is 5.

(b) $4x^4 + 5x^6 - 3x^5$ is *not* in descending order. In descending order, the polynomial is written as

$5x^6 - 3x^5 + 4x^4$

The leading coefficient is 5.

You should see that the degree of the polynomial is 6, which is given by the exponent of the variable in the leading term.

Check Yourself 4

Write each polynomial in descending order and identify the leading coefficient.

(a) $5x^4 - 4x^5 + 7$ (b) $4x^3 + 9x^4 + 6x^8$

A polynomial can represent any number. Its value depends on the value given to the variable.

Example 5 Evaluating Polynomials

Given the polynomial

$3x^3 - 2x^2 - 4x + 1$

(a) Find the value of the polynomial when $x = 2$.

RECALL

We apply the order of operations rules. See Section 0.5 for a review.

Substituting 2 for x, we have

$3(2)^3 - 2(2)^2 - 4(2) + 1$
$= 3(8) - 2(4) - 4(2) + 1$
$= 24 - 8 - 8 + 1$
$= 9$

(b) Find the value of the polynomial when $x = -2$.

Now we substitute -2 for x.

$3(-2)^3 - 2(-2)^2 - 4(-2) + 1$
$= 3(-8) - 2(4) - 4(-2) + 1$
$= -24 - 8 + 8 + 1$
$= -23$

> **CAUTION**

Be particularly careful when dealing with powers of negative numbers!

Check Yourself 5

Find the value of the polynomial

$4x^3 - 3x^2 + 2x - 1$

when

(a) $x = 3$ **(b)** $x = -3$

Polynomials are used in almost every professional field. Many applications are related to predictions and forecasts. In allied health, polynomials can be used to calculate the concentration of a medication in the bloodstream after a given amount of time. The next example demonstrates just such an application.

Example 6 An Allied Health Application

The concentration of digoxin, a medication prescribed for congestive heart failure, in a patient's bloodstream t hours after injection is given by the polynomial $-0.0015t^2 + 0.0845t + 0.7170$, where concentration is measured in nanograms per milliliter (ng/mL). Determine the concentration of digoxin in a patient's bloodstream 19 hours after injection.

We are asked to evaluate the polynomial for the variable value $t = 19$.

$-0.0015t^2 + 0.0845t + 0.7170$

We substitute 19 for t in the polynomial.

$-0.0015(19)^2 + 0.0845(19) + 0.7170$
$= -0.0015(361) + 1.6055 + 0.7170$
$= -0.5415 + 1.6055 + 0.7170$
$= 1.781$

The concentration is 1.781 nanograms per milliliter.

Check Yourself 6

The concentration of a sedative, in micrograms per milliliter (mcg/mL), in a patient's bloodstream t hours after injection is given by the polynomial $-1.35t^2 + 10.01t + 7.38$. Determine the concentration of the sedative in the patient's bloodstream 3.5 hours after injection.

Check Yourself ANSWERS

1. **(a)** and **(c)** are polynomials. **2.** **(a)** binomial; **(b)** monomial; **(c)** trinomial **3.** **(a)** 5; **(b)** 1; **(c)** 6; **(d)** 0 **4.** **(a)** $-4x^5 + 5x^4 + 7, -4$; **(b)** $6x^8 + 9x^4 + 4x^3, 6$ **5.** **(a)** 86; **(b)** -142 **6.** 28.7 mcg/mL

Reading Your Text

SECTION 5.3

(a) A ______________ can be written as a number or the product of a number and one or more variables and their exponents.

(b) In each term of a polynomial, the number factor is called the numerical ______________.

(c) A polynomial with exactly two terms is called a ______________.

(d) The prefix ______________ means 3.

Basic Skills | Challenge Yourself | Calculator/Computer | Career Applications | Above and Beyond

5.3 exercises

< Objective 1 >

Which expressions are polynomials?

1. $7x^3$
2. $4x^3 - \frac{2}{x}$
3. $2x^5y^3 - 4x^2y^4$
4. 7
5. -7
6. $4x^3 + x$
7. $\frac{3 + x}{x^2}$
8. $5a^2 - 2a + 7$

For each polynomial, list the terms and the coefficients.

9. $2x^2 - 3x$
10. $5x^3 + x$
11. $4x^3 - 3x + 2$ 

12. $7x^2$

Classify each polynomial as a monomial, binomial, or trinomial where possible.

13. $4x^3 - 2x^2$
14. $4x^7$
15. $7y^2 + 4y + 5$
16. $3x^3$
17. $2x^4 - 3x^2 + 5x - 2$
18. $x^4 + \frac{5}{x} + 7$
19. $7x^{10}$
20. $4x^4 - 2x^2 + 5x - 7$
21. $x^5 - \frac{3}{x^2}$
22. $25x^2 - 16$

Name ______________

Section ________ Date ________

Answers

1. ______
2. ______
3. ______
4. ______
5. ______
6. ______
7. ______
8. ______
9. ______
10. ______
11. ______
12. ______ 13. ______
14. ______ 15. ______
16. ______
17. ______
18. ______
19. ______
20. ______
21. ______
22. ______

Answers

23. ______
24. ______
25. ______
26. ______
27. ______
28. ______
29. ______
30. ______
31. ______
32. ______
33. ______
34. ______
35. ______
36. ______
37. ______
38. ______
39. ______
40. ______
41. ______
42. ______
43. ______
44. ______
45. ______
46. ______
47. ______

< Objectives 2 and 3 >

Arrange in descending order, give the degree of each polynomial, and identify its leading coefficient.

23. $4x^5 - 3x^2$

24. $5x^2 + 3x^3 + 4$

25. $7x^7 - 5x^9 + 4x^3$

26. $4 - x + x^2$

27. $-9x$

28. $x^{17} - 3x^4$

29. $5x^2 - 3x^5 + x^6 - 7$ 

30. 15

Evaluate each polynomial for the given values of the variable.

31. $9x - 2$; $x = 1$ and $x = -1$

32. $5x - 5$; $x = 2$ and $x = -2$

33. $x^3 - 2x$; $x = 2$ and $x = -2$

34. $3x^2 + 7$; $x = 3$ and $x = -3$

35. $3x^2 + 4x - 2$; $x = 4$ and $x = -4$ 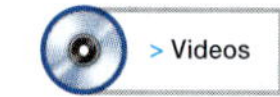

36. $2x^2 - 5x + 1$; $x = 2$ and $x = -2$

37. $-x^2 - x + 12$; $x = 3$ and $x = -4$

38. $-x^2 - 5x - 6$; $x = -3$ and $x = -2$

Basic Skills | **Challenge Yourself** | Calculator/Computer | Career Applications | Above and Beyond

Complete each statement with **never, sometimes,** *or* **always.**

39. A monomial is ______ a polynomial.

40. A binomial is ______ a trinomial. 

41. The degree of a trinomial is ______ 3.

42. A trinomial ______ has three terms.

43. A polynomial ______ has four or more terms.

44. A binomial ______ has two coefficients.

45. If x equals 0, the value of a polynomial in x ______ equals 0.

46. The coefficient of the leading term in a polynomial is ______ the largest coefficient of the polynomial.

47. **Business and Finance** The cost, in dollars, of typing a term paper is given as 3 times the number of pages plus \$20. Use x as the number of pages to be typed and write a polynomial to describe this cost. Find the cost of typing a 50-page paper. chapter 5 > Make the Connection

48. **Business and Finance** The cost, in dollars, of making suits is described as 20 times the number of suits plus \$150. Use x as the number of suits and write a polynomial to describe this cost. Find the cost of making seven suits. chapter 5 > Make the Connection

49. **Business and Finance** The revenue, in dollars, when x pairs of shoes are sold is given by $3x^2 - 95$. Find the revenue when 12 pairs of shoes are sold.

50. **Business and Finance** The cost, in dollars, of manufacturing x wing nuts is given by $0.07x + 13.3$. Find the cost when 375 wing nuts are made.

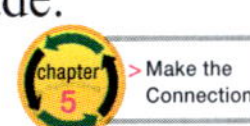

Basic Skills | Challenge Yourself | Calculator/Computer | **Career Applications** | Above and Beyond

Electronics Many devices in our homes consume energy, even when the device is not considered "on." For example, a remote-controlled TV has to power circuitry inside the device to recognize when the user turns the TV "on" via the remote control. Similarly, many microwave ovens have integrated clocks that require power to keep and display the time.

Assume that the total energy (in watt-hours, Wh) used by a certain television per day can be described by the expression $58t + 144$, in which t is the number of hours the television is actually "on" in a day. Use this information to complete exercises 51 and 52.

51. If the TV is not turned on in an entire day, how many watt-hours are consumed in the 24-hour period of non-operation?

52. If the TV is on for a total of 4.2 h in one day, how many watt-hours are consumed that day?

53. **Allied Health** One diabetic patient's morning blood glucose level in terms of the number of days t since the patient was diagnosed can be approximated by the polynomial

$0.472t^3 - 5.298t^2 + 11.802t + 93.143$

Estimate the patient's morning blood glucose level on the 5th day after being diagnosed.

54. **Mechanical Engineering** The deflection of a beam is given by

$$D = \frac{x^3}{3.8 \times 10^6}$$

Find the deflection for an 18-ft-long beam (that is, $x = 18$). Use scientific notation to express your result.

Answers

48. ______
49. ______
50. ______
51. ______
52. ______
53. ______
54. ______

Answers

1. Polynomial **3.** Polynomial **5.** Polynomial **7.** Not a polynomial **9.** $2x^2, -3x; 2, -3$ **11.** $4x^3, -3x, 2; 4, -3, 2$ **13.** Binomial **15.** Trinomial **17.** Not classified **19.** Monomial **21.** Not a polynomial **23.** $4x^5 - 3x^2, 5, 4$ **25.** $-5x^9 + 7x^7 + 4x^3, 9, -5$ **27.** $-9x, 1, -9$ **29.** $x^6 - 3x^5 + 5x^2 - 7, 6, 1$ **31.** $7, -11$ **33.** $4, -4$ **35.** 62, 30 **37.** 0, 0 **39.** always **41.** sometimes **43.** sometimes **45.** sometimes **47.** $3x + 20$; \$170 **49.** \$337 **51.** 144 Wh **53.** 78.703

5.4 Adding and Subtracting Polynomials

< 5.4 Objectives >

1 > Add two polynomials

2 > Subtract two polynomials

RECALL

The plus sign between the parentheses indicates addition.

Addition is always a matter of combining like quantities (two apples plus three apples, four books plus five books, and so on). If you keep that basic idea in mind, adding polynomials is just a matter of combining like terms. Suppose that you want to add

$5x^2 + 3x + 4$ and $4x^2 + 5x - 6$

Parentheses are sometimes used in adding, so for the sum of these polynomials, we can write

$(5x^2 + 3x + 4) + (4x^2 + 5x - 6)$

Now what about the parentheses? You can use the following rule.

Property

Removing Parentheses

When adding or subtracting polynomials:

If a plus sign (+) or nothing at all appears in front of parentheses, just remove the parentheses. No other changes are necessary.

If a minus sign (−) appears in front of a set of parentheses, the subtraction can be changed to addition by changing the sign of each term inside the parentheses.

NOTES

Just remove the parentheses. No other changes are necessary.

Use the associative and commutative properties when reordering and regrouping.

Now let's return to the addition.

$(5x^2 + 3x + 4) + (4x^2 + 5x - 6)$

$= 5x^2 + 3x + 4 + 4x^2 + 5x - 6$

Like terms

Like terms

Like terms

Collect like terms. (*Remember:* Like terms have the same variables raised to the same power.)

RECALL

We use the distributive property. For example,

$5x^2 + 4x^2$

$= (5 + 4)x^2$

$= 9x^2$

$= (5x^2 + 4x^2) + (3x + 5x) + (4 - 6)$

Finally, we combine like terms.

$= 9x^2 + 8x - 2$

As should be clear, much of this work can be done mentally. You can then write the sum directly by locating like terms and combining.

Example 1 Combining Like Terms

< Objective 1 >

Add $3x - 5$ and $2x + 3$.

Write the sum.

$$(3x - 5) + (2x + 3)$$

$$= 3x - 5 + 2x + 3 = 5x - 2$$

Like terms Like terms

Check Yourself 1

Add $6x^2 + 2x$ and $4x^2 - 7x$.

The same technique is used to find the sum of two trinomials.

Example 2 Adding Polynomials

Add $4a^2 - 7a + 5$ and $3a^2 + 3a - 4$.

Write the sum.

RECALL

Only the like terms are combined in the sum.

$$(4a^2 - 7a + 5) + (3a^2 + 3a - 4)$$

$$= 4a^2 - 7a + 5 + 3a^2 + 3a - 4 = 7a^2 - 4a + 1$$

Like terms

Like terms

Like terms

Check Yourself 2

Add $5y^2 - 3y + 7$ and $3y^2 - 5y - 7$.

Example 3 Adding Polynomials

Add $2x^2 + 7x$ and $4x - 6$.

Write the sum.

$$(2x^2 + 7x) + (4x - 6)$$

$$= 2x^2 + 7x + 4x - 6$$

These are the only like terms; $2x^2$ and -6 cannot be combined.

$$= 2x^2 + 11x - 6$$

Check Yourself 3

Add $5m^2 + 8$ and $8m^2 - 3m$.

As we mentioned in Section 5.3, writing polynomials in descending order usually makes the work easier. Look at Example 4.

Example 4 Adding Polynomials

Add $3x - 2x^2 + 7$ and $5 + 4x^2 - 3x$.

Write the polynomials in descending order; then add.

$$(-2x^2 + 3x + 7) + (4x^2 - 3x + 5)$$
$$= 2x^2 + 12$$

Check Yourself 4

Add $8 - 5x^2 + 4x$ and $7x - 8 + 8x^2$.

RECALL

This process is the same as simplifying an expression. You learned this in Section 1.3.

Subtracting polynomials requires an extra step since the associative and commutative properties do not apply to subtraction. Recall that we view subtraction as adding the opposite, so

$$a - b = a + (-b)$$

The opposite of a quantity with more than one term requires that we take the opposite of each term. For example,

$$-(a + b) = -a - b \quad \text{and} \quad -(a - b) = -a + b$$

Alternatively, the negative in front of a quantity can be understood as -1. Applying the distributive property, we get the same results.

$$-(a + b) = -1(a + b) = -a - b \quad \text{and} \quad -(a - b) = -1(a - b) = -a + b$$

We can now go on to subtracting polynomials.

Example 5 Removing Parentheses

Remove the parentheses.

(a) $-(2x + 3y) = -2x - 3y$ Change each sign to remove the parentheses.

(b) $m - (5n - 3p) = m - 5n + 3p$
Sign changes

(c) $2x - (-3y + z) = 2x + 3y - z$
Sign changes

NOTE

We use the distributive property.

$$-(2x + 3y) = (-1)(2x + 3y)$$
$$= -2x - 3y$$

Check Yourself 5

Remove the parentheses.

(a) $-(3m + 5n)$ **(b)** $-(5w - 7z)$
(c) $3r - (2s - 5t)$ **(d)** $5a - (-3b - 2c)$

Subtracting polynomials is just a matter of using the previous rule to remove the parentheses and then combining like terms. Consider Example 6.

Example 6 Subtracting Polynomials

< Objective 2 >

RECALL

The expression following *from* is written first in the problem.

(a) Subtract $5x - 3$ from $8x + 2$.

Write

$$(8x + 2) - (5x - 3)$$
$$= 8x + 2 - 5x + 3$$

Sign changes

$$= 3x + 5$$

(b) Subtract $4x^2 - 8x + 3$ from $8x^2 + 5x - 3$.

Write

$$(8x^2 + 5x - 3) - (4x^2 - 8x + 3)$$
$$= 8x^2 + 5x - 3 - 4x^2 + 8x - 3$$

Sign changes

$$= 4x^2 + 13x - 6$$

Check Yourself 6

(a) Subtract $7x + 3$ from $10x - 7$.
(b) Subtract $5x^2 - 3x + 2$ from $8x^2 - 3x - 6$.

Writing all polynomials in descending order makes locating and combining like terms much easier. Look at Example 7.

Example 7 Subtracting Polynomials

(a) Subtract $4x^2 - 3x^3 + 5x$ from $8x^3 - 7x + 2x^2$.

Write

$$(8x^3 + 2x^2 - 7x) - (-3x^3 + 4x^2 + 5x)$$
$$= 8x^3 + 2x^2 - 7x + 3x^3 - 4x^2 - 5x$$

Sign changes

$$= 11x^3 - 2x^2 - 12x$$

(b) Subtract $8x - 5$ from $-5x + 3x^2$.

Write

$$(3x^2 - 5x) - (8x - 5)$$
$$= 3x^2 - 5x - 8x + 5$$

Only the like terms can be combined.

$$= 3x^2 - 13x + 5$$

Check Yourself 7

(a) Subtract $7x - 3x^2 + 5$ from $5 - 3x + 4x^2$.
(b) Subtract $3a - 2$ from $5a + 4a^2$.

If you think back to addition and subtraction in arithmetic, you may remember that we arranged the work vertically. That is, we placed the numbers being added or subtracted under one another so that each column represented the same place value. This meant that in adding or subtracting columns, you always dealt with "like quantities."

It is also possible to use a vertical method for adding or subtracting polynomials. First rewrite the polynomials in descending order, then arrange them one under another, so that each column contains like terms. Finally, add or subtract in each column.

Example 8 Adding Using the Vertical Method

(a) Add $3x - 5$ and $x^2 + 2x + 4$.

Like terms

$$\begin{array}{r} 3x - 5 \\ x^2 + 2x + 4 \\ \hline x^2 + 5x - 1 \end{array}$$

(b) Add $2x^2 - 5x$, $3x^2 + 2$, and $6x - 3$.

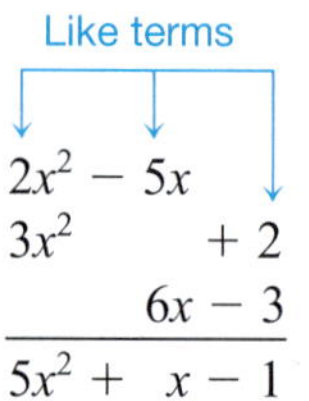

$$\begin{array}{lcr} 2x^2 & - 5x & \\ 3x^2 & & + 2 \\ & 6x & - 3 \\ \hline 5x^2 & + \ x & - 1 \end{array}$$

Check Yourself 8

Add $3x^2 + 5$, $x^2 - 4x$, and $6x + 7$.

Example 9 illustrates subtraction by the vertical method.

Example 9 Subtracting Using the Vertical Method

(a) Subtract $5x - 3$ from $8x - 7$.

Write

$$\begin{array}{r} 8x - 7 \\ -(5x - 3) \\ \hline \end{array}$$

To subtract, change each sign of $5x - 3$ to get $-5x + 3$, then add.

$$\begin{array}{r} 8x - 7 \\ -5x + 3 \\ \hline 3x - 4 \end{array}$$

NOTE

Since we are subtracting the entire quantity $5x - 3$, we place parentheses around $5x - 3$. Then we remove them according to the rule given on page 518.

(b) Subtract $5x^2 - 3x + 4$ from $8x^2 + 5x - 3$.

Write

$$\begin{array}{r} 8x^2 + 5x - 3 \\ -(5x^2 - 3x + 4) \\ \hline \end{array}$$

To subtract, change each sign of $5x^2 - 3x + 4$ to get $-5x^2 + 3x - 4$, then add.

$$\begin{array}{r} 8x^2 + 5x - 3 \\ -5x^2 + 3x - 4 \\ \hline 3x^2 + 8x - 7 \end{array}$$

Subtraction using the vertical method takes some practice. Take time to study the method carefully. We use it in long division in Section 5.6.

Check Yourself 9

Subtract, using the vertical method.

(a) $4x^2 - 3x$ from $8x^2 + 2x$

(b) $8x^2 + 4x - 3$ from $9x^2 - 5x + 7$

Check Yourself ANSWERS

1. $10x^2 - 5x$ **2.** $8y^2 - 8y$ **3.** $13m^2 - 3m + 8$ **4.** $3x^2 + 11x$
5. **(a)** $-3m - 5n$; **(b)** $-5w + 7z$; **(c)** $3r - 2s + 5t$; **(d)** $5a + 3b + 2c$
6. **(a)** $3x - 10$; **(b)** $3x^2 - 8$ **7.** **(a)** $7x^2 - 10x$; **(b)** $4a^2 + 2a + 2$
8. $4x^2 + 2x + 12$ **9.** **(a)** $4x^2 + 5x$; **(b)** $x^2 - 9x + 10$

Reading Your Text

SECTION 5.4

(a) If a ______________ sign appears in front of parentheses, simply remove the parentheses.

(b) If a minus sign appears in front of parenthesis, the subtraction can be changed to addition by changing the ______________ in front of each term inside the parentheses.

(c) When we change each sign inside parentheses, to subtract, we are using the ______________ property.

(d) When subtracting polynomials, the expression following the word *from* is written ______________ when writing the problem.

5.4 exercises

Basic Skills | Challenge Yourself | Calculator/Computer | Career Applications | Above and Beyond

Boost *your* GRADE at ALEKS.com!

ALEKS®

- Practice Problems
- Self-Tests
- NetTutor
- e-Professors
- Videos

Name ______________

Section ________ Date ________

Answers

1. ______ 2. ______
3. ______ 4. ______
5. ______ 6. ______
7. ______
8. ______
9. ______
10. ______
11. ______
12. ______ 13. ______
14. ______ 15. ______
16. ______
17. ______ 18. ______
19. ______
20. ______
21. ______ 22. ______
23. ______ 24. ______

< Objective 1 >

Add.

1. $5a - 7$ and $4a + 11$

2. $9x + 3$ and $3x - 4$

3. $8b^2 - 11b$ and $5b^2 - 7b$

4. $2m^2 + 3m$ and $6m^2 - 8m$

5. $3x^2 - 2x$ and $-5x^2 + 2x$

6. $9p^2 + 3p$ and $-13p^2 - 3p$

7. $2x^2 + 5x - 3$ and $3x^2 - 7x + 4$ 

8. $4d^2 - 8d + 7$ and $5d^2 - 6d - 9$

9. $3b^2 - 7$ and $2b - 7$

10. $4x - 3$ and $3x^2 - 9x$

11. $8y^3 - 5y^2$ and $5y^2 - 2y$

12. $9x^4 - 2x^2$ and $2x^2 + 3$

13. $2a^2 - 4a^3$ and $3a^3 + 2a^2$

14. $9m^3 - 2m$ and $-6m - 4m^3$

15. $7x^2 - 5 + 4x$ and $8 - 5x - 9x^2$

16. $5b^3 - 8b + 2b^2$ and $3b^2 - 7b^3 + 5b$

Remove the parentheses in each expression and simplify if possible.

17. $-(4a + 5b)$

18. $-(7x - 4y)$

19. $5a - (2b - 3c)$

20. $8x - (2y - 5z)$

21. $9r - (3r + 5s)$

22. $10m - (3m - 2n)$

23. $5p - (-3p + 2q)$ 

24. $8d - (-7c - 2d)$

< Objective 2 >

Subtract.

25. $x + 4$ from $2x - 3$

26. $5x + 1$ from $7x + 8$

27. $3m^2 - 2m$ from $4m^2 - 5m$

28. $9a^2 - 5a$ from $11a^2 - 10a$

29. $6y^2 + 5y$ from $4y^2 + 5y$

30. $9n^2 - 4n$ from $7n^2 - 4n$

31. $x^2 - 4x - 3$ from $3x^2 - 5x - 2$

32. $3x^2 - 2x + 4$ from $5x^2 - 8x - 3$

33. $3a + 7$ from $8a^2 - 9a$

34. $3x^3 + x^2$ from $4x^3 - 5x$

35. $4b^2 - 3b$ from $5b - 2b^2$

36. $7y - 3y^2$ from $3y^2 - 2y$

37. $5x^2 + 19 - 11x$ from $7x^2 - 11x + 31$

38. $4x - 2x^2 + 4x^3$ from $4x^3 + x - 3x^2$ 

Basic Skills | **Challenge Yourself** | Calculator/Computer | Career Applications | Above and Beyond

Complete each statement with **never, sometimes,** *or* **always.**

39. The sum of two trinomials is ________ another trinomial.

40. When subtracting one polynomial from a second polynomial, we ________ write the second polynomial first.

Perform the indicated operations.

41. Subtract $2b + 5$ from the sum of $5b - 4$ and $3b + 8$.

42. Subtract $5m - 7$ from the sum of $2m - 8$ and $9m - 2$.

43. Subtract $3x^2 + 2x - 1$ from the sum of $x^2 + 5x - 2$ and $2x^2 + 7x - 8$.

44. Subtract $4x^2 - 5x - 3$ from the sum of $x^2 - 3x - 7$ and $2x^2 - 2x + 9$.

45. Subtract $2x^2 - 3x$ from the sum of $4x^2 - 5$ and $2x - 7$.

Answers

25. ________
26. ________
27. ________
28. ________
29. ________
30. ________
31. ________
32. ________
33. ________
34. ________
35. ________
36. ________
37. ________
38. ________
39. ________
40. ________
41. ________
42. ________
43. ________
44. ________
45. ________

Answers

46. ____

47. ____

48. ____

49. ____

50. ____

51. ____

52. ____

53. ____

54. ____

55. ____

56. ____

57. ____

58. ____

59. ____

60. ____

61. ____

62. ____

46. Subtract $7a^2 + 8a - 15$ from the sum of $14a - 5$ and $7a^2 - 8$.

47. Subtract the sum of $3y^2 - 3y$ and $5y^2 + 3y$ from $2y^2 - 8y$. > Videos

48. Subtract the sum of $7r^3 - 4r^2$ and $-3r^3 + 4r^2$ from $2r^3 + 3r^2$.

Add, using the vertical method.

49. $4w^2 + 11$, $7w - 9$, and $2w^2 - 9w$

50. $3x^2 - 4x - 2$, $6x - 3$, and $2x^2 + 8$

51. $3x^2 + 3x - 4$, $4x^2 - 3x - 3$, and $2x^2 - x + 7$

52. $5x^2 + 2x - 4$, $x^2 - 2x - 3$, and $2x^2 - 4x - 3$

Subtract, using the vertical method.

53. $7a^2 - 9a$ from $9a^2 - 4a$

54. $6r^3 + 4r^2$ from $4r^3 - 2r^2$

55. $5x^2 - 6x + 7$ from $8x^2 - 5x + 7$

56. $8x^2 - 4x + 2$ from $9x^2 - 8x + 6$

57. $5x^2 - 3x$ from $8x^2 - 9$

58. $-13x^2 + 6x$ from $-11x^2 - 3$

Perform the indicated operations.

59. $[(9x^2 - 3x + 5) - (3x^2 + 2x - 1)] - (x^2 - 2x - 3)$

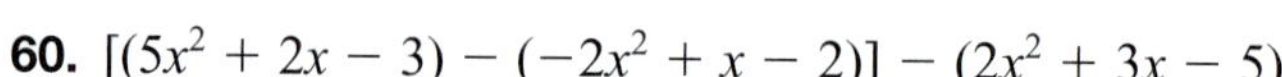

60. $[(5x^2 + 2x - 3) - (-2x^2 + x - 2)] - (2x^2 + 3x - 5)$

61. **Geometry** A rectangle has sides of $8x + 9$ and $6x - 7$. Find the polynomial that represents its perimeter.

62. **Geometry** A triangle has sides $3x + 7$, $4x - 9$, and $5x + 6$. Find the polynomial that represents its perimeter.

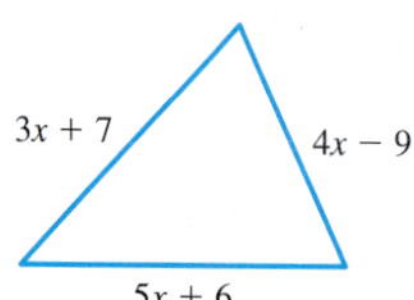

63. **Business and Finance** The cost of producing x units of an item is $150 + 25x$. The revenue for selling x units is $90x - x^2$. The profit is given by the revenue minus the cost. Find the polynomial that represents the profit. chapter 5 > Make the Connection

64. **Business and Finance** The revenue for selling y units is $3y^2 - 2y + 5$, and the cost of producing y units is $y^2 + y - 3$. Find the polynomial that represents profit. chapter 5 > Make the Connection

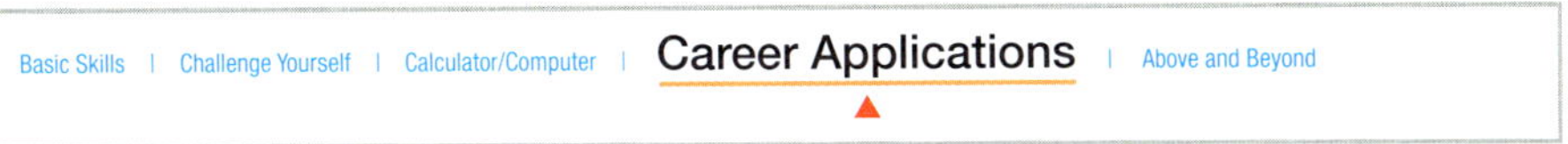

65. **Manufacturing Technology** The shear polynomial for a polymer is

$$0.4x^2 - 144x + 308$$

After vulcanization, the shear polynomial is increased by $0.2x^2 - 14x + 144$. Find the shear polynomial for the polymer after vulcanization.

66. **Allied Health** A diabetic patient's morning m and evening n blood glucose levels are given in terms of the number of days t since the patient was diagnosed and can be approximated by

$$m = 0.472t^3 - 5.298t^2 + 11.802t + 94.143$$
$$n = -1.083t^3 + 11.464t^2 - 29.524t + 117.429$$

Give the difference d between the morning and evening blood glucose levels in terms of the number of days since diagnosis.

Electrical Engineering The resistance (in Ω) of conductors, such as metals, varies in a relatively linear fashion, based on temperature. The common equation for the relationship between conductors at temperatures between 0°C and 100°C is

$$R_t = R_0(1 + \alpha t)$$

in which

R_t = Resistance of the conductor at temperature t (in °C)
R_0 = Resistance of the conductor at 0°C
α = Temperature coefficient of the conductor (a constant)
t = Temperature, in °C

Use this information in exercises 67 and 68.

67. Assuming $\alpha = 3.9 \times 10^{-3}$, express R_t in terms of t, if a certain piece of a copper conductor has a total resistance of $1.72 \times 10^{-8}\,\Omega$ at 0°C.

68. What is the resistance of the conductor described in exercise 67 if the temperature is 20°C?

Answers

63. ______

64. ______

65. ______

66. ______

67. ______

68. ______

Answers

69. ______

70. ______

Basic Skills | Challenge Yourself | Calculator/Computer | Career Applications | **Above and Beyond**

Find values for a, b, c, and d so that each equation is true.

69. $3ax^4 - 5x^3 + x^2 - cx + 2 = 9x^4 - bx^3 + x^2 - 2d$

70. $(4ax^3 - 3bx^2 - 10) - 3(x^3 + 4x^2 - cx - d) = x^3 - 6x + 8$

Answers

1. $9a + 4$ **3.** $13b^2 - 18b$ **5.** $-2x^2$ **7.** $5x^2 - 2x + 1$
9. $3b^2 + 2b - 14$ **11.** $8y^3 - 2y$ **13.** $-a^3 + 4a^2$ **15.** $-2x^2 - x + 3$
17. $-4a - 5b$ **19.** $5a - 2b + 3c$ **21.** $6r - 5s$ **23.** $8p - 2q$
25. $x - 7$ **27.** $m^2 - 3m$ **29.** $-2y^2$ **31.** $2x^2 - x + 1$
33. $8a^2 - 12a - 7$ **35.** $-6b^2 + 8b$ **37.** $2x^2 + 12$ **39.** sometimes
41. $6b - 1$ **43.** $10x - 9$ **45.** $2x^2 + 5x - 12$ **47.** $-6y^2 - 8y$
49. $6w^2 - 2w + 2$ **51.** $9x^2 - x$ **53.** $2a^2 + 5a$ **55.** $3x^2 + x$
57. $3x^2 + 3x - 9$ **59.** $5x^2 - 3x + 9$ **61.** $28x + 4$
63. $-x^2 + 65x - 150$ **65.** $0.6x^2 - 158x + 452$
67. $R_t = 1.72 \times 10^{-8} + 6.708 \times 10^{-11}t$
69. $a = 3; b = 5; c = 0; d = -1$

5.5 Multiplying Polynomials and Special Products

< 5.5 Objectives >

1 > Find the product of a monomial and a polynomial

2 > Find the product of two binomials

3 > Square a binomial

4 > Find the product of two binomials that differ only in sign

You have already had some experience multiplying polynomials. In Section 5.1 we stated the product rule for exponents and used that rule to find the product of two monomials. Let's review briefly.

Step by Step

To Multiply Monomials

Step 1 Multiply the coefficients.

Step 2 Use the product rule for exponents to combine the variables:

$ax^m \cdot bx^n = abx^{m+n}$

Here is an example in which we multiply two monomials.

Example 1 Multiplying Monomials

Multiply $3x^2y$ and $2x^3y^5$.

Write

$(3x^2y)(2x^3y^5)$

$= (3 \cdot 2)(x^2 \cdot x^3)(y \cdot y^5)$

Multiply the coefficients. Add the exponents.

$= 6x^5y^6$

NOTE

Once again we use the commutative and associative properties to rewrite the problem.

Check Yourself 1

Multiply.

(a) $(5a^2b)(3a^2b^4)$ (b) $(-3xy)(4x^3y^5)$

RECALL

You might want to review Section 0.4 before continuing.

Our next task is to find the product of a monomial and a polynomial. Here we use the distributive property, which we introduced in Section 0.4. That property leads to a rule for multiplication.

Property

To Multiply a Polynomial by a Monomial

Use the distributive property to multiply each term of the polynomial by the monomial and simplify the result.

Example 2 Multiplying a Monomial and a Binomial

< Objective 1 >

RECALL

Distributive property:

$a(b + c) = ab + ac$

NOTE

With practice you will do this step mentally.

(a) Multiply $2x + 3$ by x.

Write

$x(2x + 3)$

$= x \cdot 2x + x \cdot 3$

$= 2x^2 + 3x$

Multiply x by $2x$ and then by 3, the terms of the polynomial. That is, "distribute" the multiplication over the sum.

(b) Multiply $2a^3 + 4a$ by $3a^2$.

Write

$3a^2(2a^3 + 4a)$

$= 3a^2 \cdot 2a^3 + 3a^2 \cdot 4a = 6a^5 + 12a^3$

Check Yourself 2

Multiply.

(a) $2y(y^2 + 3y)$ (b) $3w^2(2w^3 + 5w)$

The patterns of Example 2 extend to *any* number of terms.

Example 3 Multiplying a Monomial and a Polynomial

Multiply.

(a) $3x(4x^3 + 5x^2 + 2)$

$= 3x \cdot 4x^3 + 3x \cdot 5x^2 + 3x \cdot 2 = 12x^4 + 15x^3 + 6x$

(b) $-5c(4c^2 - 8c)$

$= (-5c)(4c^2) - (-5c)(8c) = -20c^3 + 40c^2$

(c) $3c^2d^2(7cd^2 - 5c^2d^3)$

$= (3c^2d^2)(7cd^2) - (3c^2d^2)(5c^2d^3) = 21c^3d^4 - 15c^4d^5$

NOTE

We have shown all the steps of the process. With practice, you can write the product directly, and should try to do so.

Check Yourself 3

Multiply.

(a) $3(5a^2 + 2a + 7)$ (b) $4x^2(8x^3 - 6)$

(c) $-5m(8m^2 - 5m)$ (d) $9a^2b(3a^3b - 6a^2b^4)$

We apply the distributive property twice when multiplying a polynomial by a binomial. That is, every term in the binomial must multiply each term in the polynomial. We usually write the distributive property as

$$a(b + c) = ab + ac$$

We use this to multiply two binomials by treating the first binomial as a and the second binomial as $(b + c)$.

$$\underbrace{(x + y)}_{a}\underbrace{(w + z)}_{b+c} = \underbrace{(x + y)}_{a}\underbrace{w}_{b} + \underbrace{(x + y)}_{a}\underbrace{z}_{c}$$

NOTE

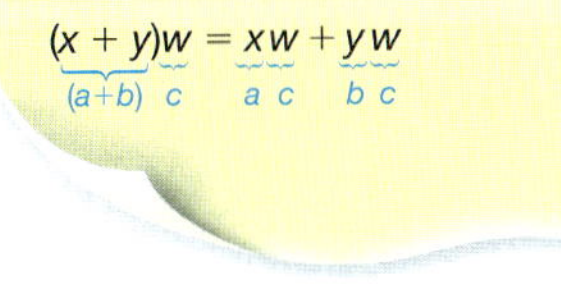

$$\underbrace{(x + y)}_{(a+b)}\underbrace{w}_{c} = \underbrace{x}_{a}\underbrace{w}_{c} + \underbrace{y}_{b}\underbrace{w}_{c}$$

We sometimes write the distributive property as

$$(a + b)c = ac + bc$$

We use this second form of the distributive property twice in this expansion.

$$(x + y)w + (x + y)z = xw + yw + xz + yz$$

The final step is to simplify the resulting polynomial, if possible.

We use this approach to multiply binomials in Example 4, though the pattern holds true when multiplying polynomials with more than two terms, as well.

Example 4 Multiplying Binomials

< Objective 2 >

(a) Multiply $x + 2$ by $x + 3$.

We can think of $x + 2$ as a single quantity and apply the distributive property.

$(x + 2)(x + 3)$ Multiply $x + 2$ by x and then by 3.

$= (x + 2)x + (x + 2)3$

$= x \cdot x + 2 \cdot x + x \cdot 3 + 2 \cdot 3$

$= x^2 + 2x + 3x + 6$

$= x^2 + 5x + 6$

NOTE

This ensures that each term, x and 2, of the first binomial is multiplied by each term, x and 3, of the second binomial.

(b) Multiply $a - 3$ by $a - 4$.

$(a - 3)(a - 4)$ Think of $a - 3$ as a single quantity and distribute.

$= (a - 3)a - (a - 3)(4)$

$= a \cdot a - 3 \cdot a - [(a \cdot 4) - (3 \cdot 4)]$

$= a^2 - 3a - (4a - 12)$ The parentheses are needed here because a *minus sign* precedes the binomial.

$= a^2 - 3a - 4a + 12$

$= a^2 - 7a + 12$

Check Yourself 4

Multiply.

(a) $(x + 4)(x + 5)$ **(b)** $(y + 5)(y - 6)$

Fortunately, there is a pattern to this kind of multiplication that allows you to write the product of the two binomials directly without going through all these steps. We call it the **FOIL method.** The reason for this name will be clear as we look at the process in greater detail.

To multiply $(x + 2)(x + 3)$:

Remember this by F.

1. $(x + 2)(x + 3)$ — $x \cdot x$ — Find the product of the *first* terms of the factors.

Remember this by O.

2. $(x + 2)(x + 3)$ — $x \cdot 3$ — Find the product of the *outer* terms.

Remember this by I.

3. $(x + 2)(x + 3)$ — $2 \cdot x$ — Find the product of the *inner* terms.

Remember this by L.

4. $(x + 2)(x + 3)$ — $2 \cdot 3$ — Find the product of the *last* terms.

NOTE

FOIL gives you an easy way of remembering the steps: *F*irst, *O*uter, *I*nner, and *L*ast.

Combining the four steps, we have

$(x + 2)(x + 3)$
$= x^2 + 3x + 2x + 6$
$= x^2 + 5x + 6$

With practice, the FOIL method lets you write the products quickly and easily. Consider Example 5, which illustrates this approach.

Example 5 Using the FOIL Method

Find each product, using the FOIL method.

(a) $(x + 4)(x + 5)$

F: $x \cdot x$; O: $5 \cdot x$; I: $4 \cdot x$; L: $4 \cdot 5$

$= x^2 + 5x + 4x + 20$
F O I L
$= x^2 + 9x + 20$

NOTE

When possible, you should combine the outer and inner products mentally and write just the final product.

(b) $(x - 7)(x + 3)$ Combine the outer and inner products as $-4x$.

F: $x \cdot x$; O: $3 \cdot x$; I: $-7 \cdot x$; L: $(-7)(3)$

$= x^2 - 4x - 21$

Check Yourself 5

Multiply.

(a) $(x + 6)(x + 7)$ (b) $(x + 3)(x - 5)$ (c) $(x - 2)(x - 8)$

You can also find the product of binomials with leading coefficients other than 1 or with more than one variable with the FOIL method.

Example 6 Using the FOIL Method

Find each product, using the FOIL method.

(a) $(4x - 3)(3x + 2)$

F: $(4x)(3x)$
O: $(4x)(2)$
I: $(-3)(3x)$
L: $(-3)(2)$

Combine: $-9x + 8x = -x$

$= 12x^2 - x - 6$

(b) $(3x - 5y)(2x - 7y)$

F: $(3x)(2x)$
O: $(3x)(-7y)$
I: $(-5y)(2x)$
L: $(-5y)(-7y)$

Combine: $-10xy + (-21xy) = -31xy$

$= 6x^2 - 31xy + 35y^2$

Here is a summary of our work in multiplying binomials.

Step by Step

To Multiply Two Binomials

Step 1 Multiply the first terms of the binomials (F).

Step 2 Then multiply the first term of the first binomial by the second term of the second binomial (O).

Step 3 Next multiply the second term of the first binomial by the first term of the second binomial (I).

Step 4 Finally, multiply the second terms of the binomials (L).

Step 5 Form the sum of the four terms found above, combining any like terms.

Check Yourself 6

Multiply.

(a) $(5x + 2)(3x - 7)$ **(b)** $(4a - 3b)(5a - 4b)$
(c) $(3m + 5n)(2m + 3n)$

The FOIL method works well for multiplying any two binomials. But what if one of the factors has three or more terms? The vertical format, shown in Example 7, works for factors with any number of terms.

Example 7 Using the Vertical Method

Multiply $x^2 - 5x + 8$ by $x + 3$.

Step 1

$$\begin{array}{r} x^2 - 5x + 8 \\ x + 3 \\ \hline 3x^2 - 15x + 24 \end{array}$$

Multiply each term of $x^2 - 5x + 8$ by 3.

Step 2

$$\begin{array}{r} x^2 - 5x + 8 \\ x + 3 \\ \hline 3x^2 - 15x + 24 \\ x^3 - 5x^2 + 8x \quad\;\; \\ \hline \end{array}$$

Now multiply each term by x.

Note that this line is shifted over so that like terms are in the same columns.

> **NOTE**
>
> Using this vertical method ensures that each term of one factor multiplies each term of the other. That's why it works!

Step 3

$$\begin{array}{r} x^2 - 5x + 8 \\ x + 3 \\ \hline 3x^2 - 15x + 24 \\ x^3 - 5x^2 + 8x \quad\;\; \\ \hline x^3 - 2x^2 - 7x + 24 \end{array}$$

Now add to combine like terms to write the product.

Check Yourself 7

Multiply $2x^2 - 5x + 3$ by $3x + 4$.

The vertical method works equally well when multiplying two trinomials, or even polynomials with more than three terms. The vertical method helps to ensure that we do not miss any of the necessary products.

Example 8 Using the Vertical Method to Multiply Trinomials

Multiply.

$(2x^2 + 3x - 5)(x^2 - 4x + 1)$

Step 1

$$\begin{array}{r} 2x^2 + 3x - 5 \\ x^2 - 4x + 1 \\ \hline 2x^2 + 3x - 5 \end{array}$$

Multiply each term of $2x^2 + 3x - 5$ by 1.

Step 2

$$\begin{array}{r} 2x^2 + 3x - 5 \\ x^2 - 4x + 1 \\ \hline 2x^2 + 3x - 5 \\ -8x^3 - 12x^2 + 20x \quad\;\; \end{array}$$

Now multiply each term by $-4x$.

As before, we align the resulting polynomials so that terms with the same degree line up.

Step 3

$$\begin{array}{r} 2x^2 + 3x - 5 \\ x^2 - 4x + 1 \\ \hline 2x^2 + 3x - 5 \\ -8x^3 - 12x^2 + 20x \quad\;\; \\ 2x^4 + 3x^3 - 5x^2 \qquad\qquad\quad \end{array}$$

Finally, multiply each term by x^2.

Step 4 Add the three polynomials by combining like terms.

$$\begin{array}{r} 2x^2 + 3x - 5 \\ x^2 - 4x + 1 \\ \hline 2x^2 + 3x - 5 \\ -8x^3 - 12x^2 + 20x \quad\;\; \\ 2x^4 + 3x^3 - 5x^2 \qquad\qquad\quad \\ \hline 2x^4 - 5x^3 - 15x^2 + 23x - 5 \end{array}$$

Check Yourself 8

Multiply $(-2x^2 + 3x - 7)(3x^2 - 5x + 8)$.

NOTE

Squaring a binomial **always** results in three terms.

Certain products occur frequently enough in algebra that it is worth learning special formulas for dealing with them. First, let's look at the **square of a binomial,** which is the product of two equal binomial factors.

$$(x + y)^2 = (x + y)(x + y) = x^2 + xy + xy + y^2$$
$$= x^2 + 2xy + y^2$$
$$(x - y)^2 = (x - y)(x - y) = x^2 - xy - xy + y^2$$
$$= x^2 - 2xy + y^2$$

The patterns above lead to another rule.

Step by Step

To Square a Binomial

Step 1 Find the first term of the square by squaring the first term of the binomial.

Step 2 Find the middle term of the square as twice the product of the two terms of the binomial.

Step 3 Find the last term of the square by squaring the last term of the binomial.

Example 9 Squaring a Binomial

< Objective 3 >

Multiply.

(a) $(x + 3)^2 = x^2 + 2 \cdot x \cdot 3 + 3^2$

Square of first term; Twice the product of the two terms; Square of the last term

$$= x^2 + 6x + 9$$

> **CAUTION**

A very common mistake in squaring binomials is to forget the middle term.

(b) $(3a + 4b)^2 = (3a)^2 + 2(3a)(4b) + (4b)^2$
$$= 9a^2 + 24ab + 16b^2$$

(c) $(y - 5)^2 = y^2 + 2 \cdot y \cdot (-5) + (-5)^2$
$$= y^2 - 10y + 25$$

(d) $(5c - 3d)^2 = (5c)^2 + 2(5c)(-3d) + (-3d)^2$
$$= 25c^2 - 30cd + 9d^2$$

Check Yourself 9

Multiply.

(a) $(2x + 1)^2$ **(b)** $(4x - 3y)^2$

Example 10 Squaring a Binomial

NOTE

You should see that $(2 + 3)^2 \neq 2^2 + 3^2$ because $5^2 \neq 4 + 9$.

Find $(y + 4)^2$.

$(y + 4)^2$ is *not* equal to $y^2 + 4^2$ or $y^2 + 16$

The correct square is

$(y + 4)^2 = y^2 + 8y + 16$

The middle term is twice the product of y and 4.

Check Yourself 10

Multiply.

(a) $(x + 5)^2$ (b) $(3a + 2)^2$

(c) $(y - 7)^2$ (d) $(5x - 2y)^2$

A second special product will be very important in Chapter 6. Suppose the form of a product is

$$(x + y)(x - y)$$

The two factors differ only in sign.

Let's see what happens when we multiply.

$$(x + y)(x - y)$$
$$= x^2 \underbrace{- xy + xy}_{= 0} - y^2$$
$$= x^2 - y^2$$

Property

Special Product

The product of two binomials that differ only in the sign between the terms is the square of the first term minus the square of the second term.

$$(a + b)(a - b) = a^2 - b^2$$

Let's look at the application of this rule in Example 11.

Example 11 **Multiplying Polynomials**

< Objective 4 >

Multiply each pair of factors.

(a) $(x + 5)(x - 5) = x^2 - 5^2$

Square of the first term; Square of the second term

$= x^2 - 25$

NOTE

$(2y)^2 = (2y)(2y)$
$= 4y^2$

(b) $(x + 2y)(x - 2y) = x^2 - (2y)^2$

Square of the first term; Square of the second term

$= x^2 - 4y^2$

(c) $(3m + n)(3m - n) = 9m^2 - n^2$

(d) $(4a - 3b)(4a + 3b) = 16a^2 - 9b^2$

Check Yourself 11

Find the products.

(a) $(a - 6)(a + 6)$ **(b)** $(x - 3y)(x + 3y)$
(c) $(5n + 2p)(5n - 2p)$ **(d)** $(7b - 3c)(7b + 3c)$

When you are finding the product of three or more factors, it is useful to first look for the pattern in which two binomials differ only in their sign. If you see it, finding this product first makes it easier to find the product of all the factors.

Example 12 Multiplying Polynomials

(a) $x(x - 3)(x + 3)$ — These binomials differ only in sign.

$= x(x^2 - 9)$

$= x^3 - 9x$

(b) $(x + 1)(x - 5)(x + 5)$ — These binomials differ only in sign.

$= (x + 1)(x^2 - 25)$ — With two binomials, use the FOIL method.

$= x^3 + x^2 - 25x - 25$

(c) $(2x - 1)(x + 3)(2x + 1)$

$(2x - 1)\ (x + 3)\ (2x + 1)$

$= (x + 3)(2x - 1)(2x + 1)$

$= (x + 3)(4x^2 - 1)$

$= 4x^3 + 12x^2 - x - 3$

These two binomials differ only in the sign of the second term. We can use the commutative property to rearrange the terms.

Check Yourself 12

Multiply.

(a) $3x(x - 5)(x + 5)$ **(b)** $(x - 4)(2x + 3)(2x - 3)$
(c) $(x - 7)(3x - 1)(x + 7)$

Check Yourself ANSWERS

1. (a) $15a^4b^5$; **(b)** $-12x^4y^6$ **2. (a)** $2y^3 + 6y^2$; **(b)** $6w^5 + 15w^3$

3. (a) $15a^2 + 6a + 21$; **(b)** $32x^5 - 24x^2$; **(c)** $-40m^3 + 25m^2$; **(d)** $27a^5b^2 - 54a^4b^5$

4. (a) $x^2 + 9x + 20$; **(b)** $y^2 - y - 30$

5. (a) $x^2 + 13x + 42$; **(b)** $x^2 - 2x - 15$; **(c)** $x^2 - 10x + 16$

6. (a) $15x^2 - 29x - 14$; **(b)** $20a^2 - 31ab + 12b^2$; **(c)** $6m^2 + 19mn + 15n^2$

7. $6x^3 - 7x^2 - 11x + 12$ **8.** $-6x^4 + 19x^3 - 52x^2 + 59x - 56$

9. (a) $4x^2 + 4x + 1$; **(b)** $16x^2 - 24xy + 9y^2$

10. (a) $x^2 + 10x + 25$; **(b)** $9a^2 + 12a + 4$; **(c)** $y^2 - 14y + 49$; **(d)** $25x^2 - 20xy + 4y^2$

11. (a) $a^2 - 36$; **(b)** $x^2 - 9y^2$; **(c)** $25n^2 - 4p^2$; **(d)** $49b^2 - 9c^2$

12. (a) $3x^3 - 75x$; **(b)** $4x^3 - 16x^2 - 9x + 36$; **(c)** $3x^3 - x^2 - 147x + 49$

Reading Your Text

SECTION 5.5

(a) When multiplying two monomials, we first multiply the ____________.

(b) When multiplying a polynomial by a monomial, use the ____________ property to multiply each term of the polynomial by the monomial.

(c) The FOIL method can only be used when multiplying two ____________.

(d) Squaring a binomial always results in ____________ terms.

Basic Skills | Challenge Yourself | Calculator/Computer | Career Applications | Above and Beyond

5.5 exercises

Boost *your* GRADE at ALEKS.com!

ALEKS

- Practice Problems
- Self-Tests
- NetTutor
- e-Professors
- Videos

Name ______________

Section ________ Date ________

< Objectives 1 and 2 >

Multiply.

1. $(5x^2)(3x^3)$
2. $(14a^4)(2a^7)$
3. $(-2b^2)(14b^8)$
4. $(14y^4)(-4y^6)$
5. $(-5p^7)(-8p^6)$
6. $(-6m^8)(9m^7)$
7. $(4m^5)(-3m)$
8. $(-5r^7)(-3r)$
9. $(4x^3y^2)(8x^2y)$
10. $(-3r^4s^2)(-7r^2s^5)$
11. $(-3m^5n^2)(2m^4n)$
12. $(3a^4b^2)(-14a^3b^4)$
13. $10(x + 3)$
14. $4(7b - 5)$
15. $3a(4a + 5)$
16. $5x(2x - 7)$
17. $3s^2(4s^2 - 7s)$
18. $3a^3(9a^2 + 15)$
19. $3x(5x^2 - 3x - 1)$
20. $5m(4m^3 - 3m^2 + 2)$
21. $3xy(2x^2y + xy^2 + 5xy)$
22. $5ab^2(ab - 3a + 5b)$
23. $6m^2n(3m^2n - 2mn + mn^2)$ 

24. $8pq^2(2pq - 3p + 5q)$

Answers

1. ______ 2. ______
3. ______ 4. ______
5. ______ 6. ______
7. ______ 8. ______
9. ______ 10. ______
11. ______ 12. ______
13. ______ 14. ______
15. ______ 16. ______
17. ______
18. ______
19. ______
20. ______
21. ______
22. ______
23. ______
24. ______

Answers

25. ______
26. ______
27. ______
28. ______
29. ______
30. ______
31. ______
32. ______
33. ______
34. ______
35. ______
36. ______
37. ______
38. ______
39. ______
40. ______
41. ______
42. ______
43. ______
44. ______
45. ______
46. ______
47. ______
48. ______

25. $(x + 3)(x + 2)$

26. $(a - 3)(a - 7)$

27. $(m - 5)(m - 9)$

28. $(b + 7)(b + 5)$

29. $(p - 8)(p + 7)$

30. $(x - 10)(x + 9)$

31. $(w - 7)(w + 6)$

32. $(s - 12)(s - 8)$

33. $(3x - 5)(x - 8)$

34. $(w + 5)(4w - 7)$

35. $(2x - 3)(3x + 4)$

36. $(7a - 2)(2a - 3)$

37. $(3a - b)(4a - 9b)$

38. $(7s - 3t)(3s + 8t)$

39. $(3p - 4q)(7p + 5q)$

40. $(5x - 4y)(2x - y)$

41. $(2x^2 - x + 8)(x^2 + 4x + 6)$

42. $(x^3 - 3x^2 + 5x + 1)(2x^2 + 4x - 1)$

< Objective 3 >

Find each square.

43. $(x + 3)^2$

44. $(y + 9)^2$

45. $(w - 6)^2$

46. $(a - 8)^2$

47. $(z + 12)^2$

48. $(p - 11)^2$

49. $(2a - 1)^2$

50. $(3x - 2)^2$

51. $(6m + 1)^2$

52. $(7b - 2)^2$

53. $(3x - y)^2$

54. $(5m + n)^2$

55. $(2r + 5s)^2$

56. $(3a - 4b)^2$

57. $(6a - 5b)^2$

58. $(7p + 6q)^2$

59. $\left(x + \frac{1}{2}\right)^2$ 

60. $\left(w - \frac{1}{4}\right)^2$

< Objective 4 >

Find each product.

61. $(x - 6)(x + 6)$

62. $(y + 8)(y - 8)$

63. $(m + 7)(m - 7)$

64. $(w - 10)(w + 10)$

65. $\left(x - \frac{1}{2}\right)\left(x + \frac{1}{2}\right)$

66. $\left(x + \frac{2}{3}\right)\left(x - \frac{2}{3}\right)$

67. $(p - 0.4)(p + 0.4)$

68. $(m - 1.1)(m + 1.1)$

69. $(a - 3b)(a + 3b)$

70. $(p + 4q)(p - 4q)$

71. $(3x + 2y)(3x - 2y)$

72. $(7x - y)(7x + y)$

73. $(8w + 5z)(8w - 5z)$

74. $(8c + 3d)(8c - 3d)$

75. $(5x - 9y)(5x + 9y)$ 

76. $(6s - 5t)(6s + 5t)$

Answers

49. ______
50. ______
51. ______
52. ______
53. ______
54. ______
55. ______
56. ______
57. ______
58. ______
59. ______
60. ______
61. ______ 62. ______
63. ______ 64. ______
65. ______ 66. ______
67. ______ 68. ______
69. ______ 70. ______
71. ______ 72. ______
73. ______
74. ______
75. ______
76. ______

Basic Skills | **Challenge Yourself** | Calculator/Computer | Career Applications | Above and Beyond

Answers

77. ______
78. ______
79. ______
80. ______
81. ______
82. ______
83. ______
84. ______
85. ______
86. ______
87. ______
88. ______
89. ______
90. ______
91. ______
92. ______
93. ______
94. ______
95. ______
96. ______
97. ______
98. ______

Multiply.

77. $2x(3x - 2)(4x + 1)$

78. $3x(2x + 1)(2x - 1)$

79. $5a(4a - 3)(4a + 3)$

80. $6m(3m - 2)(3m - 7)$

81. $3s(5s - 2)(4s - 1)$

82. $7w(2w - 3)(2w + 3)$

83. $(x - 2)(x + 1)(x - 3)$

84. $(y + 3)(y - 2)(y - 4)$

85. $(a - 1)^3$

86. $(x + 1)^3$

87. $\left(\frac{x}{2} + \frac{2}{3}\right)\left(\frac{2x}{3} - \frac{2}{5}\right)$

88. $\left(\frac{x}{3} + \frac{3}{4}\right)\left(\frac{3x}{4} - \frac{3}{5}\right)$

89. $[x + (y - 2)][x - (y - 2)]$

90. $[x + (3 - y)][x - (3 - y)]$

Determine whether each statement is **true** *or* **false.**

91. $(x + y)^2 = x^2 + y^2$

92. $(x - y)^2 = x^2 - y^2$

93. $(x + y)^2 = x^2 + 2xy + y^2$

94. $(x - y)^2 = x^2 - 2xy + y^2$

Complete each application.

95. GEOMETRY The length of a rectangle is given by $(3x + 5)$ centimeters (cm), and the width is given by $(2x - 7)$ cm. Express the area of the rectangle in terms of x.

96. GEOMETRY The base of a triangle measures $(3y + 7)$ in., and the height is $(2y - 3)$ in. Express the area of the triangle in terms of y.

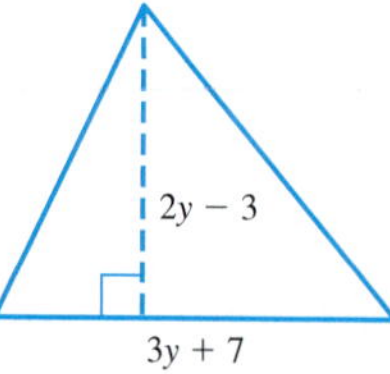

97. BUSINESS AND FINANCE The price of an item is given by $p = 12 - 0.05x$, where x is the number of items sold. If the revenue generated is found by multiplying the number of items sold by the price of an item, find the polynomial that represents the revenue. chapter 5 > Make the Connection

98. BUSINESS AND FINANCE The price of an item is given by $p = 1{,}000 - 0.02x^2$, where x is the number of items sold. Find the polynomial that represents the revenue generated from the sale of x items. chapter 5 > Make the Connection

99. **Business and Finance** Suppose an orchard is planted with trees in straight rows. If there are $5x - 4$ rows with $5x - 4$ trees in each row, find a polynomial that represents the number of trees in the orchard.

100. **Geometry** A square has sides of length $(3x - 2)$ cm. Express the area of the square as a polynomial.

101. **Geometry** The length and width of a rectangle are given by two consecutive odd integers. Write an expression for the area of the rectangle.

102. **Geometry** The length of a rectangle is 6 less than 3 times the width. Write an expression for the area of the rectangle.

Let x represent the unknown number and write an expression for each product.

103. The product of 6 more than a number and 6 less than that number

104. The square of 5 more than a number

105. The square of 4 less than a number

106. The product of 5 less than a number and 5 more than that number

Basic Skills | Challenge Yourself | Calculator/Computer | **Career Applications** | Above and Beyond

107. **Mechanical Engineering** The bending moment of a dual-span beam with cantilever is given by

$$M = \frac{wc^2}{8}$$

To account for a load on the cantilever, this expression needs to be multiplied by $\frac{c^2}{2}$. Create a single expression for the new bending moment.

108. **Mechanical Engineering** The maximum stress for a given allowable strain (deformation) for a certain material is given by the polynomial

$$\text{Stress} = 82.6x - 0.4x^2 + 322$$

in which x is the weight percent of nickel.

After heat-treating, the stress polynomial is multiplied by $0.08x + 0.9$. Find the stress (strength) after heat-treating.

Answers

99. ______

100. ______

101. ______

102. ______

103. ______

104. ______

105. ______

106. ______

107. ______

108. ______

Answers

109. ______

110. ______

111. ______

109. Mechanical Engineering The maximum stress for a given allowable strain (deformation) for a certain material is given by the polynomial

$$\text{Stress} = 86.2x - 0.6x^2 + 258$$

in which x is the allowable strain, in micrometers.

When an alloying substance is added to the material, the strength is increased by multiplying the stress polynomial by $\frac{p}{8.6}$, in which p is the percent of the alloying material.

Find a polynomial (in two variables) that describes the allowable stress (strength) of the material after alloying (accurate to two decimal places).

110. Construction Technology The shrinkage of a post of length L under a load is given by the product of $\frac{L}{28{,}000{,}000}$ and the load. If the load is equal to 3 times the square of the length, find an expression for the shrinkage of the post.

Basic Skills | Challenge Yourself | Calculator/Computer | Career Applications | **Above and Beyond**

111. Work with another student to complete this table and write the polynomial that represents the volume of the box. A paper box is to be made from a piece of cardboard 20 in. wide and 30 in. long. The box will be formed by cutting squares out of each of the four corners and folding up the sides to make a box.

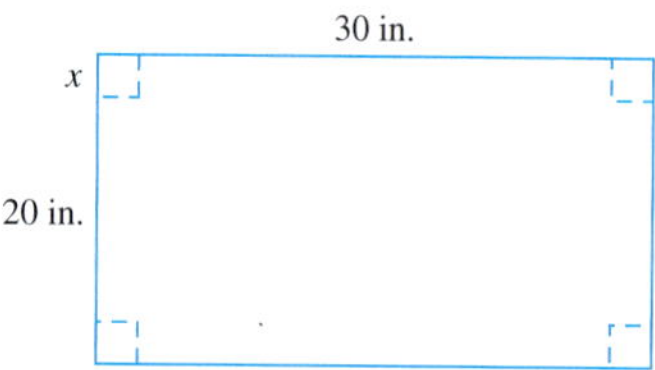

If x is the dimension of the side of the square cut out of the corner, when the sides are folded up the box will be x in. tall. You should use a piece of paper to try this to see how the box will be made. Complete the chart.

Length of Side of Corner Square	Length of Box	Width of Box	Depth of Box	Volume of Box
1 in.				
2 in.				
3 in.				
n in.				

Write general formulas for the width, length, and height of the box and a general formula for the *volume* of the box, and simplify it by multiplying. The variable will be the height, the side of the square cut out of the corners. What is the highest power of the variable in the polynomial you have written for the volume? Extend the table to decide what the dimensions are for a box with maximum volume. Draw a sketch of this box and write in the dimensions.

112. Complete the statement: $(a + b)^2$ is not equal to $a^2 + b^2$ because. . . . But wait! Isn't $(a + b)^2$ *sometimes* equal to $a^2 + b^2$? What do you think?

113. Is $(a + b)^3$ ever equal to $a^3 + b^3$? Explain.

114. In each figure, identify the length and the width of the outside square:

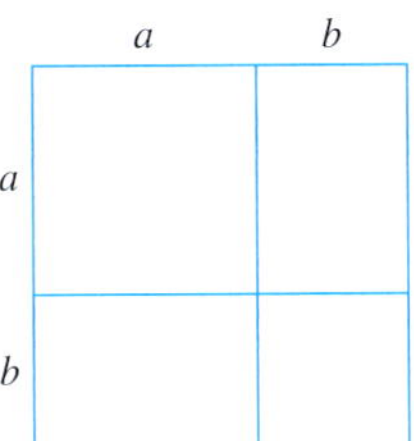

Length = __________

Width = __________

Area = __________

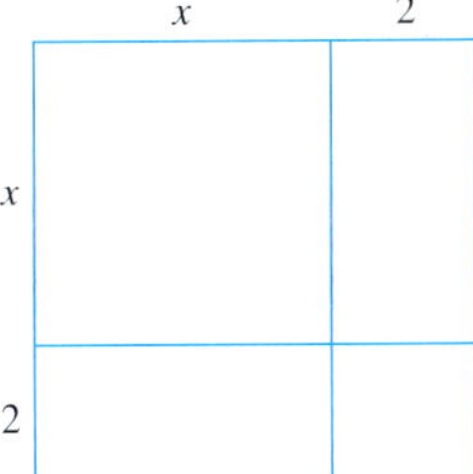

Length = __________

Width = __________

Area = __________

115. The square shown is x units on a side. The area is __________.

Draw a picture of what happens when the sides are doubled. The area is __________.

Continue the picture to show what happens when the sides are tripled. The area is __________.

If the sides are quadrupled, the area is __________.

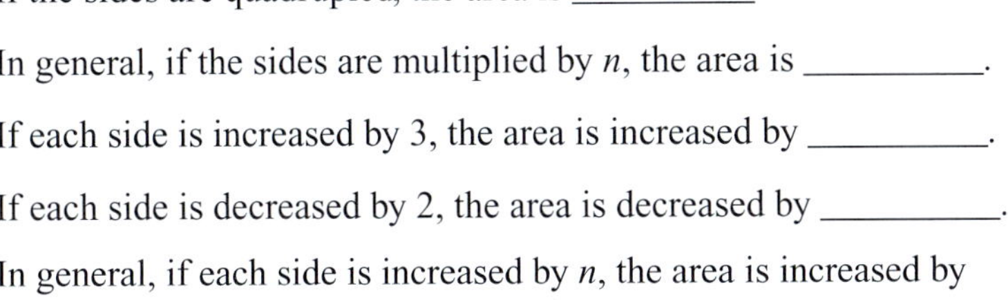

In general, if the sides are multiplied by n, the area is __________.

If each side is increased by 3, the area is increased by __________.

If each side is decreased by 2, the area is decreased by __________.

In general, if each side is increased by n, the area is increased by __________; and if each side is decreased by n, the area is decreased by __________.

x

x

Answers

112. __________

113. __________

114. __________

115. __________

Answers

116. ______

117. ______

118. ______

119. ______

120. ______

121. ______

Note that (28)(32) = (30 − 2)(30 + 2) = 900 − 4 = 896. Use this pattern to find each product.

116. (49)(51)

117. (27)(33)

118. (34)(26)

119. (98)(102)

120. (55)(65)

121. (64)(56) > Videos

Answers

1. $15x^5$ **3.** $-28b^{10}$ **5.** $40p^{13}$ **7.** $-12m^6$ **9.** $32x^5y^3$
11. $-6m^9n^3$ **13.** $10x + 30$ **15.** $12a^2 + 15a$ **17.** $12s^4 - 21s^3$
19. $15x^3 - 9x^2 - 3x$ **21.** $6x^3y^2 + 3x^2y^3 + 15x^2y^2$
23. $18m^4n^2 - 12m^3n^2 + 6m^3n^3$ **25.** $x^2 + 5x + 6$ **27.** $m^2 - 14m + 45$
29. $p^2 - p - 56$ **31.** $w^2 - w - 42$ **33.** $3x^2 - 29x + 40$
35. $6x^2 - x - 12$ **37.** $12a^2 - 31ab + 9b^2$ **39.** $21p^2 - 13pq - 20q^2$
41. $2x^4 + 7x^3 + 16x^2 + 26x + 48$ **43.** $x^2 + 6x + 9$ **45.** $w^2 - 12w + 36$
47. $z^2 + 24z + 144$ **49.** $4a^2 - 4a + 1$ **51.** $36m^2 + 12m + 1$
53. $9x^2 - 6xy + y^2$ **55.** $4r^2 + 20rs + 25s^2$ **57.** $36a^2 - 60ab + 25b^2$
59. $x^2 + x + \frac{1}{4}$ **61.** $x^2 - 36$ **63.** $m^2 - 49$ **65.** $x^2 - \frac{1}{4}$
67. $p^2 - 0.16$ **69.** $a^2 - 9b^2$ **71.** $9x^2 - 4y^2$ **73.** $64w^2 - 25z^2$
75. $25x^2 - 81y^2$ **77.** $24x^3 - 10x^2 - 4x$ **79.** $80a^3 - 45a$
81. $60s^3 - 39s^2 + 6s$ **83.** $x^3 - 4x^2 + x + 6$ **85.** $a^3 - 3a^2 + 3a - 1$
87. $\frac{x^2}{3} + \frac{11x}{45} - \frac{4}{15}$ **89.** $x^2 - y^2 + 4y - 4$ **91.** False **93.** True
95. $(6x^2 - 11x - 35)$ cm^2 **97.** $12x - 0.05x^2$ **99.** $25x^2 - 40x + 16$
101. $x(x + 2)$ or $x^2 + 2x$ **103.** $x^2 - 36$ **105.** $x^2 - 8x + 16$
107. $\frac{wc^4}{16}$ **109.** Stress $= -0.07x^2p + 10.02xp + 30p$
111. Above and Beyond **113.** Above and Beyond
115. Above and Beyond **117.** 891 **119.** 9,996 **121.** 3,584

5.6 Dividing Polynomials

< 5.6 Objectives >

1 > Find the quotient when a polynomial is divided by a monomial

2 > Find the quotient of two polynomials

In Section 5.1, we introduced the quotient rule for exponents to divide one monomial by another monomial. Let's review that process.

Step by Step

To Divide a Monomial by a Monomial

Step 1 Divide the coefficients.

Step 2 Use the quotient rule for exponents to combine the variables.

Example 1 Dividing Monomials

RECALL

The quotient rule says: If x is not zero,

$$\frac{x^m}{x^n} = x^{m-n}$$

Divide: $\frac{8}{2} = 4$

(a) $\frac{8x^4}{2x^2} = 4x^{4-2}$ Subtract the exponents.

$= 4x^2$

(b) $\frac{45a^5b^3}{9a^2b} = 5a^3b^2$

Check Yourself 1

Divide.

(a) $\frac{16a^5}{8a^3}$ (b) $\frac{28m^4n^3}{7m^3n}$

Now let's look at how this can be extended to divide any polynomial by a monomial. For example, to divide $12a^3 + 8a^2$ by $4a$, proceed as follows:

NOTE

Technically, this step depends on the distributive property and the definition of division.

$$\frac{12a^3 + 8a^2}{4a} = \frac{12a^3}{4a} + \frac{8a^2}{4a}$$

Divide each term in the numerator by the denominator $4a$.

Now do each division.

$= 3a^2 + 2a$

The preceding work leads us to the following rule.

Property

To Divide a Polynomial by a Monomial — Divide each term of the polynomial by the monomial. Then simplify the results.

Example 2 Dividing by Monomials

< Objective 1 >

Divide each term by 2.

(a) $\dfrac{4a^2 + 8}{2} = \dfrac{4a^2}{2} + \dfrac{8}{2}$

$= 2a^2 + 4$

Divide each term by 6y.

(b) $\dfrac{24y^3 - 18y^2}{6y} = \dfrac{24y^3}{6y} - \dfrac{18y^2}{6y}$

$= 4y^2 - 3y$

(c) $\dfrac{15x^2 + 10x}{-5x} = \dfrac{15x^2}{-5x} + \dfrac{10x}{-5x}$

Remember the rules for signs in division.

$= -3x - 2$

NOTE

With practice, you can just write the quotient.

(d) $\dfrac{14x^4 + 28x^3 - 21x^2}{7x^2} = \dfrac{14x^4}{7x^2} + \dfrac{28x^3}{7x^2} - \dfrac{21x^2}{7x^2}$

$= 2x^2 + 4x - 3$

(e) $\dfrac{9a^3b^4 - 6a^2b^3 + 12ab^4}{3ab} = \dfrac{9a^3b^4}{3ab} - \dfrac{6a^2b^3}{3ab} + \dfrac{12ab^4}{3ab}$

$= 3a^2b^3 - 2ab^2 + 4b^3$

Check Yourself 2

Divide.

(a) $\dfrac{20y^3 - 15y^2}{5y}$

(b) $\dfrac{8a^3 - 12a^2 + 4a}{-4a}$

(c) $\dfrac{16m^4n^3 - 12m^3n^2 + 8mn}{4mn}$

We are now ready to look at dividing one polynomial by another polynomial (with more than one term). The process is very much like long division in arithmetic, as Example 3 illustrates.

Example 3 Dividing by Binomials

< Objective 2 >

NOTE

The first term in the dividend, x^2, is divided by the first term in the divisor, x.

Divide $x^2 + 7x + 10$ by $x + 2$.

Step 1 $x + 2\overline{)x^2 + 7x + 10}$ with quotient x — Divide x^2 by x to get x.

Step 2

$$\begin{array}{r} x \\ x + 2\overline{)x^2 + 7x + 10} \\ \underline{x^2 + 2x} \end{array}$$

Multiply the divisor $x + 2$ by x.

Step 3

$$\begin{array}{r} x \\ x + 2\overline{)x^2 + 7x + 10} \\ \underline{x^2 + 2x} \\ 5x + 10 \end{array}$$

Subtract and bring down 10.

RECALL

To subtract $x^2 + 2x$, mentally change each sign to $-x^2 - 2x$ and then add. Take your time and be careful here. This is where most errors are made.

Step 4

$$\begin{array}{r} x + 5 \\ x + 2\overline{)x^2 + 7x + 10} \\ \underline{x^2 + 2x} \\ 5x + 10 \end{array}$$

Divide $5x$ by x to get 5.

Step 5

$$\begin{array}{r} x + 5 \\ x + 2\overline{)x^2 + 7x + 10} \\ \underline{x^2 + 2x} \\ 5x + 10 \\ \underline{5x + 10} \\ 0 \end{array}$$

Multiply $x + 2$ by 5 and then subtract.

NOTE

We repeat the process until the degree of the remainder is less than that of the divisor or until there is no remainder.

The quotient is $x + 5$.

Check Yourself 3

Divide $x^2 + 9x + 20$ by $x + 4$.

In Example 3, we showed all the steps separately to help you see the process. In practice, the work can be shortened.

Example 4 Dividing by Binomials

Divide $x^2 + x - 12$ by $x - 3$.

$$\begin{array}{r} x + 4 \\ x - 3\overline{)x^2 + x - 12} \\ \underline{x^2 - 3x} \\ 4x - 12 \\ \underline{4x - 12} \\ 0 \end{array}$$

The Steps

1. Divide x^2 by x to get x, the first term of the quotient.
2. Multiply $x - 3$ by x.
3. Subtract and bring down -12. Remember to mentally change the signs to $-x^2 + 3x$ and add.
4. Divide $4x$ by x to get 4, the second term of the quotient.
5. Multiply $x - 3$ by 4 and subtract.

NOTE

You might want to write out a problem like $408 \div 17$, to compare the steps.

The quotient is $x + 4$.

Check Yourself 4

Divide.

$(x^2 + 2x - 24) \div (x - 4)$

You may have a remainder in algebraic long division just as in arithmetic. Consider Example 5.

Example 5 Dividing by Binomials

Divide $4x^2 - 8x + 11$ by $2x - 3$.

$$
\begin{array}{r|l}
 & \underbrace{2x - 1}_{\text{Quotient}} \\
\underbrace{2x - 3}_{\text{Divisor}} & 4x^2 - 8x + 11 \\
 & 4x^2 - 6x \\
\hline
 & - 2x + 11 \\
 & - 2x + 3 \\
\hline
 & 8 \leftarrow \text{Remainder}
\end{array}
$$

This result can be written as

$$\frac{4x^2 - 8x + 11}{2x - 3}$$

$$= \underbrace{2x - 1}_{\text{Quotient}} + \frac{8 \leftarrow \text{Remainder}}{2x - 3 \leftarrow \text{Divisor}}$$

Check Yourself 5

Divide.

$(6x^2 - 7x + 15) \div (3x - 5)$

The division process shown in Examples 1 to 5 can be extended to dividends of a higher degree. The steps involved in the division process are exactly the same, as Example 6 illustrates.

Example 6 Dividing by Binomials

Divide $6x^3 + x^2 - 4x - 5$ by $3x - 1$.

$$
\begin{array}{r|l}
 & 2x^2 + x - 1 \\
3x - 1 & 6x^3 + x^2 - 4x - 5 \\
 & 6x^3 - 2x^2 \\
\hline
 & 3x^2 - 4x \\
 & 3x^2 - x \\
\hline
 & - 3x - 5 \\
 & - 3x + 1 \\
\hline
 & - 6
\end{array}
$$

This result can be written as

$$\frac{6x^3 + x^2 - 4x - 5}{3x - 1} = 2x^2 + x - 1 + \frac{-6}{3x - 1}$$

Check Yourself 6

Divide $4x^3 - 2x^2 + 2x + 15$ by $2x + 3$.

Suppose that the dividend is "missing" a term in some power of the variable. You can use 0 as the coefficient for the missing term. Consider Example 7.

Example 7 Dividing by Binomials

NOTE

Think of $0x$ as a placeholder. Writing it helps to align like terms.

Divide $x^3 - 2x^2 + 5$ by $x + 3$.

$$\begin{array}{r|l}
 & x^2 - 5x + 15 \\
\hline
x + 3 & x^3 - 2x^2 + 0x + 5 \\
 & \underline{x^3 + 3x^2} \\
 & \quad -5x^2 + 0x \\
 & \quad \underline{-5x^2 - 15x} \\
 & \qquad\quad 15x + 5 \\
 & \qquad\quad \underline{15x + 45} \\
 & \qquad\qquad\quad -40
\end{array}$$

Write $0x$ for the "missing" term in x.

This result can be written as

$$\frac{x^3 - 2x^2 + 5}{x + 3} = x^2 - 5x + 15 + \frac{-40}{x + 3}$$

Check Yourself 7

Divide.

$(4x^3 + x + 10) \div (2x - 1)$

You should always arrange the terms of the divisor and the dividend in descending order before starting the long division process, as illustrated in Example 8.

Example 8 Dividing by Binomials

Divide $5x^2 - x + x^3 - 5$ by $-1 + x^2$.

Write the divisor as $x^2 - 1$ and the dividend as $x^3 + 5x^2 - x - 5$.

$$\begin{array}{r|l}
 & x + 5 \\
\hline
x^2 - 1 & x^3 + 5x^2 - x - 5 \\
 & \underline{x^3 \qquad\quad - x} \\
 & \qquad 5x^2 \qquad - 5 \\
 & \qquad \underline{5x^2 \qquad - 5} \\
 & \qquad\qquad\qquad 0
\end{array}$$

Write $x^3 - x$, the product of x and $x^2 - 1$, so that like terms fall in the same columns.

Check Yourself 8

Divide.

$(5x^2 + 10 + 2x^3 + 4x) \div (2 + x^2)$

Check Yourself ANSWERS

1. **(a)** $2a^2$; **(b)** $4mn^2$ **2.** **(a)** $4y^2 - 3y$; **(b)** $-2a^2 + 3a - 1$; **(c)** $4m^3n^2 - 3m^2n + 2$ **3.** $x + 5$ **4.** $x + 6$ **5.** $2x + 1 + \dfrac{20}{3x - 5}$ **6.** $2x^2 - 4x + 7 + \dfrac{-6}{2x + 3}$ **7.** $2x^2 + x + 1 + \dfrac{11}{2x - 1}$ **8.** $2x + 5$

Reading Your Text

SECTION 5.6

(a) When dividing two monomials, we first divide the ____________.

(b) When dividing a polynomial by a monomial, divide each ____________ of the polynomial by the monomial.

(c) When doing long division of polynomials, we continue until the ____________ of the remainder is less than that of the divisor.

(d) When dividing polynomials, if the dividend is missing a term in some power of the variable we use ____________ as the coefficient for that term.

5.6 exercises

Boost *your* GRADE at ALEKS.com!

ALEKS®

- Practice Problems
- Self-Tests
- NetTutor
- e-Professors
- Videos

Name ______________________

Section __________ Date __________

< Objective 1 >

Divide.

1. $\dfrac{22x^9}{11x^5}$

2. $\dfrac{20a^7}{5a^5}$

3. $\dfrac{35m^3n^2}{7mn^2}$

4. $\dfrac{42x^5y^2}{6x^3y}$

5. $\dfrac{3a+6}{3}$

6. $\dfrac{3x-6}{3}$

7. $\dfrac{9b^2-12}{3}$

8. $\dfrac{10m^2+5m}{5}$

9. $\dfrac{28a^3-42a^2}{7a}$

10. $\dfrac{9x^3+12x^2}{3x}$

11. $\dfrac{12m^2+6m}{-3m}$

12. $\dfrac{20b^3-25b^2}{-5b}$

13. $\dfrac{18a^4-45a^3+63a^2}{9a^2}$

14. $\dfrac{21x^5-28x^4+14x^3}{7x}$

15. $\dfrac{20x^4y^2-15x^2y^3+10x^3y}{5x^2y}$

16. $\dfrac{16m^3n^3+24m^2n^2-40mn^3}{8mn^2}$

< Objective 2 >

17. $\dfrac{x^2-x-12}{x-4}$

18. $\dfrac{x^2+8x+15}{x+3}$

19. $\dfrac{x^2-x-20}{x+4}$ > Videos

20. $\dfrac{x^2-2x-35}{x+5}$

21. $\dfrac{x^2+x-30}{x-5}$

22. $\dfrac{3x^2+20x-32}{3x-4}$

Answers

1. ________ 2. ________

3. ________ 4. ________

5. ________ 6. ________

7. ________ 8. ________

9. ________ 10. ________

11. ____________________

12. ____________________

13. ____________________

14. ____________________

15. ____________________

16. ____________________

17. ________ 18. ________

19. ________ 20. ________

21. ________ 22. ________

Answers

23. ______
24. ______
25. ______
26. ______
27. ______
28. ______
29. ______ 30. ______
31. ______
32. ______
33. ______
34. ______
35. ______
36. ______
37. ______ 38. ______
39. ______ 40. ______
41. ______
42. ______
43. ______
44. ______
45. ______ 46. ______
47. ______
48. ______

23. $\dfrac{2x^2 - 3x - 5}{x - 3}$

24. $\dfrac{3x^2 + 17x - 12}{x + 6}$

25. $\dfrac{4x^2 - 18x - 15}{x - 5}$

26. $\dfrac{4x^2 - 11x - 24}{x - 5}$

27. $\dfrac{6x^2 - x - 10}{3x - 5}$

28. $\dfrac{4x^2 + 6x - 25}{2x + 7}$

29. $\dfrac{x^3 + x^2 - 4x - 4}{x + 2}$

30. $\dfrac{x^3 - 2x^2 + 4x - 21}{x - 3}$

31. $\dfrac{4x^3 + 7x^2 + 10x + 5}{4x - 1}$ > Videos

32. $\dfrac{2x^3 - 3x^2 + 4x + 4}{2x + 1}$

33. $\dfrac{x^3 - x^2 + 5}{x - 2}$

34. $\dfrac{x^3 + 4x - 3}{x + 3}$

35. $\dfrac{49x^3 - 2x}{7x - 3}$

36. $\dfrac{8x^3 - 6x^2 + 2x}{4x + 1}$

Basic Skills | **Challenge Yourself** | Calculator/Computer | Career Applications | Above and Beyond

Complete each statement with **never, sometimes,** *or* **always.**

37. When a trinomial is divided by a binomial, there is ______ a remainder.

38. A binomial divided by a binomial is ______ a binomial.

39. If a monomial exactly divides a trinomial, the result is ______ a trinomial.

40. For any positive integer n, if $x^n - 1$ is divided by $x - 1$, there is ______ a remainder.

41. $\dfrac{2x^2 - 8 - 3x + x^3}{x - 2}$

42. $\dfrac{x^2 - 18x + 2x^3 + 32}{x + 4}$

43. $\dfrac{x^4 - 1}{x - 1}$ > Videos

44. $\dfrac{x^4 + x^2 - 16}{x + 2}$

45. $\dfrac{x^3 - 3x^2 - x + 3}{x^2 - 1}$

46. $\dfrac{x^3 + 2x^2 + 3x + 6}{x^2 + 3}$

47. $\dfrac{x^4 + 2x^2 - 2}{x^2 + 3}$ > Videos

48. $\dfrac{x^4 + x^2 - 5}{x^2 - 2}$

49. $\dfrac{y^3 - 1}{y - 1}$

50. $\dfrac{y^3 - 8}{y - 2}$

51. $\dfrac{x^4 - 1}{x^2 - 1}$

52. $\dfrac{x^6 - 1}{x^3 - 1}$

Basic Skills | Challenge Yourself | Calculator/Computer | Career Applications | **Above and Beyond**

53. Find the value of c so that $\dfrac{y^2 - y + c}{y + 1} = y - 2$.

54. Find the value of c so that $\dfrac{x^3 + x^2 + x + c}{x^2 + 1} = x + 1$.

55. Write a summary of your work with polynomials. Explain how a polynomial is recognized, and explain the rules for the arithmetic of polynomials—how to add, subtract, multiply, and divide. What parts of this chapter do you feel you understand very well, and what part(s) do you still have questions about, or feel unsure of? Exchange papers with another student and compare your answers.

56. An interesting (and useful) thing about division of polynomials: To find out about this interesting thing, do this division. Compare your answer with that of another student.

$(x - 2)\overline{)2x^2 + 3x - 5}$ Is there a remainder?

Now, evaluate the polynomial $2x^2 + 3x - 5$ when $x = 2$. Is this value the same as the remainder?

Try $(x + 3)\overline{)5x^2 - 2x + 1}$. Is there a remainder?

Evaluate the polynomial $5x^2 - 2x + 1$ when $x = -3$. Is this value the same as the remainder?

What happens when there is no remainder?

Try $(x - 6)\overline{)3x^3 - 14x^2 - 23x - 6}$. Is the remainder zero?

Evaluate the polynomial $3x^3 - 14x^2 - 23x - 6$ when $x = 6$. Is this value zero? Write a description of the patterns you see. Make up several more examples and test your conjecture.

Answers

1. $2x^4$ 3. $5m^2$ 5. $a + 2$ 7. $3b^2 - 4$ 9. $4a^2 - 6a$ 11. $-4m - 2$ 13. $2a^2 - 5a + 7$ 15. $4x^2y - 3y^2 + 2x$ 17. $x + 3$ 19. $x - 5$ 21. $x + 6$ 23. $2x + 3 + \dfrac{4}{x - 3}$ 25. $4x + 2 + \dfrac{-5}{x - 5}$ 27. $2x + 3 + \dfrac{5}{3x - 5}$ 29. $x^2 - x - 2$ 31. $x^2 + 2x + 3 + \dfrac{8}{4x - 1}$ 33. $x^2 + x + 2 + \dfrac{9}{x - 2}$ 35. $7x^2 + 3x + 1 + \dfrac{3}{7x - 3}$ 37. sometimes 39. always 41. $x^2 + 4x + 5 + \dfrac{2}{x - 2}$ 43. $x^3 + x^2 + x + 1$ 45. $x - 3$ 47. $x^2 - 1 + \dfrac{1}{x^2 + 3}$ 49. $y^2 + y + 1$ 51. $x^2 + 1$ 53. $c = -2$ 55. Above and Beyond

Answers

49. ______

50. ______

51. ______

52. ______

53. ______

54. ______

55. ______

56. ______

summary :: chapter 5

Definition/Procedure	Example	Reference
Positive Integer Exponents		Section 5.1
Properties of Exponents For any nonzero real numbers a and b and integers m and n:		
Product Rule $a^m \cdot a^n = a^{m+n}$	$x^5 \cdot x^7 = x^{5+7} = x^{12}$	p. 482
Quotient Rule $\frac{a^m}{a^n} = a^{m-n}$ where $m > n$	$\frac{x^7}{x^5} = x^{7-5} = x^2$	p. 483
Product-Power Rule $(ab)^n = a^n b^n$	$(2y)^3 = 2^3y^3 = 8y^3$	p. 484
Power Rule $(a^m)^n = a^{mn}$	$(2^3)^4 = 2^{12}$	p. 485
Quotient-Power Rule $\left(\frac{a}{b}\right)^m = \frac{a^m}{b^m}$	$\left(\frac{2}{3}\right)^2 = \frac{2^2}{3^2} = \frac{4}{9}$	p. 486
Zero and Negative Exponents and Scientific Notation		Section 5.2
Zero Exponent For any real number a where $a \neq 0$, $a^0 = 1$	$5x^0 = 5 \cdot 1 = 5$	p. 494
Negative Integer Exponents For any nonzero real number a and whole number n, $a^{-n} = \frac{1}{a^n}$ and a^{-n} is the multiplicative inverse of a^n.	$x^{-3} = \frac{1}{x^3}$ $2y^{-5} = \frac{2}{y^5}$	p. 495
Quotient Raised to a Negative Power For nonzero numbers a and b and a whole number n, $\left(\frac{a}{b}\right)^{-n} = \left(\frac{b}{a}\right)^n$	$\left(\frac{x}{2}\right)^{-4} = \left(\frac{2}{x}\right)^4 = \frac{16}{x^4}$	p. 498

Definition/Procedure	Example	Reference
Scientific Notation Scientific notation is a useful way of expressing very large or very small numbers through the use of powers of 10. Any number written in the form $a \times 10^n$ in which $1 \le a < 10$ and n is an integer, is said to be written in scientific notation.	$38{,}000{,}000 = 3.8 \times 10^7$ 7 places	*p.* 500
Introduction to Polynomials		**Section 5.3**
Polynomial An algebraic expression made up of terms in which the exponents are whole numbers. These terms are connected by plus or minus signs. Each sign (+ or −) is attached to the term following that sign.	$4x^3 - 3x^2 + 5x$ is a polynomial.	*p.* 510
Term A number or the product of a number and variables and their exponents.	The terms of $4x^3 - 3x^2 + 5x$ are $4x^3$, $-3x^2$, and $5x$.	*p.* 510
Coefficient In each term of a polynomial, the number that is multiplied by the variable(s) is called the *numerical coefficient* or, more simply, the *coefficient* of that term.	The coefficients of $4x^3 - 3x^2 + 5x$ are 4, −3 and 5.	*p.* 510
Types of Polynomials A polynomial can be classified according to the number of terms it has. A *mono*mial has exactly one term. A *bi*nomial has exactly two terms. A *tri*nomial has exactly three terms.	$2x^3$ is a monomial. $3x^2 - 7x$ is a binomial. $5x^5 - 5x^3 + 2$ is a trinomial.	*p.* 511
Degree of a Polynomial with Only One Variable The highest power of the variable appearing in any one term.	The degree of $4x^5 - 5x^3 + 3x$ is 5.	*p.* 511
Descending Order The form of a polynomial when it is written with the highest-degree term first, the next-highest-degree term second, and so on. **Leading Term** The first term of a polynomial written in descending order. This is the term in which the variable has the largest exponent. The power of the variable of the leading term is the same as the degree of the polynomial. **Leading Coefficient** The numerical coefficient of the leading term.	$4x^5 - 5x^3 + 3x$ is written in descending order. The leading term is $4x^5$, so the leading coefficient is 4.	*p.* 512
Adding and Subtracting Polynomials		**Section 5.4**
Removing Signs of Grouping **1.** If a plus sign (+) or no sign at all appears in front of parentheses, just remove the parentheses. No other changes are necessary. **2.** If a minus sign (−) appears in front of parentheses, the parentheses can be removed by changing the sign of each term inside the parentheses.	$+(3x - 5) = +3x - 5$ $-(3x - 5) = -3x + 5$	*p.* 518

Continued

Definition/Procedure	Example	Reference
Adding Polynomials Remove the signs of grouping. Then collect and combine any like terms.	$(2x + 3) + (3x - 5)$ $= 2x + 3 + 3x - 5$ $= 5x - 2$	p. 519
Subtracting Polynomials Remove the signs of grouping by changing the sign of each term in the polynomial being subtracted. Then combine any like terms.	$(3x^2 + 2x) - (2x^2 + 3x - 1)$ $= 3x^2 + 2x - 2x^2 - 3x + 1$ Sign changes $= 3x^2 - 2x^2 + 2x - 3x + 1$ $= x^2 - x + 1$	p. 521
Multiplying Polynomials and Special Products		Section 5.5
To Multiply a Monomial by a Monomial Multiply the coefficients, and use the product rule for exponents to combine the variables: $ax^m \cdot bx^n = abx^{m+n}$	$(-2x^2y)(3x^3y)$ $= (-2)(3)(x^2x^3)(yy)$ $= -6x^5y^2$	p. 529
To Multiply a Polynomial by a Monomial Multiply each term of the polynomial by the monomial, and simplify the results.	$2x(x^2 + 4)$ $= 2x^3 + 8x$	p. 530
To Multiply a Binomial by a Binomial Use the FOIL method: F O I L $(a + b)(c + d) = a \cdot c + a \cdot d + b \cdot c + b \cdot d$	$(2x - 3)(3x + 5)$ $= 6x^2 + 10x - 9x - 15$ F O I L $= 6x^2 + x - 15$	p. 531
To Multiply a Polynomial by a Polynomial Arrange the polynomials vertically. Multiply each term of the upper polynomial by each term of the lower polynomial, and combine like terms.	$x^2 - 3x + 5$ $2x - 3$ $-3x^2 + 9x - 15$ $2x^3 - 6x^2 + 10x$ $2x^3 - 9x^2 + 19x - 15$	p. 533
The Square of a Binomial $(a + b)^2 = a^2 + 2ab + b^2$ **1.** The first term of the square is the square of the first term of the binomial. **2.** The middle term is twice the product of the two terms of the binomial. **3.** The last term is the square of the last term of the binomial.	$(2x - 5)^2$ $= 4x^2 + 2 \cdot 2x \cdot (-5) + 25$ $= 4x^2 - 20x + 25$	p. 535

Definition/Procedure	Example	Reference
The Product of Binomials That Differ Only in Sign Subtract the square of the second term from the square of the first term. $(a + b)(a - b) = a^2 - b^2$	$(2x - 5y)(2x + 5y)$ $= (2x)^2 - (5y)^2$ $= 4x^2 - 25y^2$	p. 536
Dividing Polynomials		Section 5.6
To Divide a Monomial by a Monomial Divide the coefficients, and use the quotient rule for exponents to combine the variables.	$\frac{28m^4n^3}{7m^3n} = \left(\frac{28}{7}\right)m^{4-3}n^{3-1}$ $= 4mn^2$	p. 547
To Divide a Polynomial by a Monomial Divide each term of the polynomial by the monomial.	$\frac{9x^4 + 6x^3 - 15x^2}{3x}$ $= 3x^3 + 2x^2 - 5x$	p. 548
To Divide a Polynomial by a Polynomial Use the long division method.	$x - 3\overline{)x^2 + 2x - 7}$ quotient $x + 5$; $x^2 - 3x$; $5x - 7$; $5x - 15$; 8. The result is $x + 5 + \frac{8}{x - 3}$	p. 549

summary exercises :: chapter 5

This summary exercise set is provided to give you practice with each of the objectives of this chapter. Each exercise is keyed to the appropriate chapter section. When you are finished, you can check your answers to the odd-numbered exercises in the back of the text. If you have difficulty with any of these questions, go back and reread the examples from that section. The answers to the even-numbered exercises appear in the *Instructor's Solutions Manual.* Your instructor will give you guidelines on how best to use these exercises in your instructional setting.

5.1–5.2 *Simplify each expression, using the properties of exponents.*

1. r^4r^9

2. $4x^{-5}$

3. $(2w)^{-3}$

4. $\dfrac{3}{m^{-4}}$

5. $y^{-5}y^2$

6. $\dfrac{w^{-7}}{w^{-3}}$

7. $\dfrac{x^{12}}{x^{15}}$

8. $(6c^0d^4)(-3c^2d^2)$

9. $(5a^2b^3)(2a^{-2}b^{-6})$

10. $\left(\dfrac{3m^2n^3}{p^4}\right)^3$

11. $\left(\dfrac{m^{-3}n^{-3}}{m^{-4}n^4}\right)^3$

12. $\left(\dfrac{r^{-5}}{s^4}\right)^{-2}$

13. $\left(\dfrac{x^3y^4}{x^6y^2}\right)^3$

14. $\left(\dfrac{a^8}{b^4}\right)\left(\dfrac{b^2}{2a^2}\right)^3$

15. $(2a^3)^0(-3a^4)^2$

16. $\left(\dfrac{2x^{-3}y^{-2}}{4x^{-5}y^{-4}}\right)^2\left(\dfrac{8x^5y^6}{4x^7y^4}\right)^3$

17. Write 0.0000425 in scientific notation.

18. Write 3.1×10^{-4} in standard notation.

5.3 *Classify each polynomial as a monomial, binomial, or trinomial, if possible.*

19. $6x^4 - 3x$

20. $7x^5$

21. $4x^5 - 8x^3 + 5$

22. $x^3 + 2x^2 - 5x + 3$

23. $-7a^4 - 9a^3$

Arrange in descending order and give the degree of each polynomial.

24. $5x^5 + 3x^2$

25. $13x^2$

26. $6x^2 + 4x^4 + 6$

27. $5 + x$

28. -8

29. $9x^4 - 3x + 7x^6$

5.4 *Add or subtract as indicated.*

30. Add $7a^2 + 3a$ and $14a^2 - 5a$

31. Add $5x^2 + 3x - 5$ and $4x^2 - 6x - 2$

32. Add $5y^3 - 3y^2$ and $4y + 3y^2$

33. Subtract $7x^2 - 23x$ from $11x^2 - 15x$

34. Subtract $2x^2 - 5x - 7$ from $7x^2 - 2x + 3$

35. Subtract $5x^2 + 3$ from $9x^2 - 4x$

Perform the indicated operations.

36. Subtract $5x - 3$ from the sum of $9x + 2$ and $-3x - 7$.

37. Subtract $5a^2 - 3a$ from the sum of $5a^2 + 2$ and $7a - 7$.

38. Subtract the sum of $16w^2 - 3w$ and $8w + 2$ from $7w^2 - 5w + 2$.

Add, using the vertical method.

39. $x^2 + 5x - 3$ and $2x^2 + 4x - 3$

40. $9b^2 - 7$ and $8b + 5$

41. $x^2 + 7$, $3x - 2$, and $4x^2 - 8x$

Subtract, using the vertical method.

42. $5x^2 - 3x + 2$ from $7x^2 - 5x - 7$

43. $8m - 7$ from $9m^2 - 7$

5.5 *Multiply.*

44. $(3a^4)(2a^3)$

45. $(2x^2)(3x^5)$

46. $(-9p^3)(-6p^2)$

47. $(3a^2b^3)(-7a^3b^4)$

48. $5(3x - 8)$

49. $2a(5a - 3)$

50. $(-5rs)(2r^2s - 5rs)$

51. $7mn(3m^2n - 2mn^2 + 5mn)$

52. $(x + 5)(x + 4)$

53. $(w - 9)(w - 10)$

54. $(a - 9b)(a + 9b)$

55. $(p - 3q)^2$

56. $(a + 4b)(a + 3b)$

57. $(b - 8)(2b + 3)$

58. $(3x - 5y)(2x - 3y)$

59. $(5r + 7s)(3r - 9s)$

60. $(y + 2)(y^2 - 2y + 3)$

61. $(b + 3)(b^2 - 5b - 7)$

62. $(x - 2)(x^2 + 2x + 4)$

63. $(m^2 - 3)(m^2 + 7)$

64. $2x(x + 5)(x - 6)$

65. $a(2a - 5b)(2a - 7b)$

Find each product.

66. $(x + 7)^2$

67. $(a - 7)^2$

68. $(2w - 5)^2$

69. $(3p + 4)^2$

70. $(a + 7b)^2$

71. $(8x - 3y)^2$

72. $(x - 5)(x + 5)$

73. $(y + 9)(y - 9)$

74. $(2m + 3)(2m - 3)$

75. $(4r - 5)(4r + 5)$

76. $(5r - 2s)(5r + 2s)$

77. $(7a + 3b)(7a - 3b)$

78. $3x(x - 4)^2$

79. $3c(c + 5d)(c - 5d)$

80. $(y - 4)(y + 5)(y + 4)$

5.6 *Divide.*

81. $\dfrac{9a^5}{3a^2}$

82. $\dfrac{24m^4n^2}{6m^2n}$

83. $\dfrac{15a - 10}{5}$

84. $\dfrac{32a^3 + 24a}{8a}$

85. $\dfrac{9r^2s^3 - 18r^3s^2}{-3rs^2}$

86. $\dfrac{35x^3y^2 - 21x^2y^3 + 14x^3y}{7x^2y}$

Perform the indicated long division.

87. $\dfrac{x^2 - 2x - 15}{x + 3}$

88. $\dfrac{2x^2 + 9x - 35}{2x - 5}$

89. $\dfrac{x^2 - 8x + 17}{x - 5}$

90. $\dfrac{6x^2 - x - 10}{3x + 4}$

91. $\dfrac{6x^3 + 14x^2 - 2x - 6}{6x + 2}$

92. $\dfrac{4x^3 + x + 3}{2x - 1}$

93. $\dfrac{3x^2 + x^3 + 5 + 4x}{x + 2}$

94. $\dfrac{2x^4 - 2x^2 - 10}{x^2 - 3}$

self-test 5

Name ________________

Section ________ Date ________

The purpose of this self-test is to help you assess your progress so that you can find concepts that you need to review before the next exam. Allow yourself about an hour to take this test. At the end of that hour, check your answers against those given in the back of this text. If you miss any, go back to the appropriate section to reread the examples until you have mastered that particular concept.

Use the properties of exponents to simplify each expression.

1. $(3x^2y)(-2xy^3)$

2. $\left(\dfrac{8m^2n^5}{2p^3}\right)^2$

3. $(x^4y^5)^2$

4. $\dfrac{9c^{-5}d^3}{18c^{-7}d^4}$

5. $(3x^2y)^3(-2xy^2)^2$

6. $(-3x^{-3}y)^2(4x^{-5}y^{-2})^{-1}$

7. $\dfrac{3x^0}{(2y)^0}$

Add.

8. $3x^2 - 7x + 2$ and $7x^2 - 5x - 9$

9. $7a^2 - 3a$ and $7a^3 + 4a^2$

Subtract.

10. $5x^2 - 2x + 5$ from $8x^2 + 9x - 7$

11. $5a^2 + a$ from the sum of $3a^2 - 5a$ and $9a^2 - 4a$

Multiply.

12. $5ab(3a^2b - 2ab + 4ab^2)$

13. $(x + 3y)(4x - 5y)$

14. $(3m + 2n)^2$

Divide.

15. $\dfrac{4x^3 - 5x^2 + 7x - 9}{x - 2}$

Arrange the polynomial in descending order. Give the coefficient and degree of each term. Then, give the degree of the polynomial.

16. $-3x^2 + 8x^4 - 7$

Classify each polynomial as a monomial, binomial, or trinomial.

17. $6x^2 + 7x$

18. $5x^2 + 8x - 8$

Use the vertical method to add or subtract.

19. Add $x^2 + 3, 5x - 9$, and $3x^2$.

20. Subtract $3x^2 - 5$ from $5x^2 - 7x$.

Answers

1. ________________
2. ________________
3. ________________
4. ________________
5. ________________
6. ________________
7. ________________
8. ________________
9. ________________
10. ________________
11. ________________
12. ________________
13. ________________
14. ________________
15. ________________
16. ________________
17. ________________
18. ________________
19. ________________
20. ________________

cumulative review chapters 0-5

Name ____________________

Section ________ Date ________

Answers

1. ____________________
2. ____________________
3. ____________________
4. ____________________
5. ____________________
6. ____________________
7. ____________________
8. ____________________
9. ____________________
10. ____________________
11. ____________________
12. ____________________
13. ____________________
14. ____________________
15. ____________________
16. ____________________
17. ____________________
18. ____________________

We offer the following exercises to help you review concepts from earlier chapters. This is meant as review material and not as a comprehensive exam. The answers are presented in the back of the text. If you have difficulty with any of these exercises, be certain to at least read through the summary related to that section.

Evaluate each expression if $x = 3$, $y = -2$, and $z = 4$.

1. $-2x^2 - 3y^2 + 5z$

2. $\dfrac{-4y + 3x^2}{5z + x - y}$

Solve each equation.

3. $11x - 7 = 10x$

4. $-\dfrac{2}{3}x = 24$

5. $7x - 5 = 3x + 11$

6. $7 - 3x = 5 - 6x$

7. $\dfrac{3}{5}x - 8 = 15 - \dfrac{2}{5}x$

8. $2(x - 3) + 5 = 2x - 1$

9. $4x - (2 - x) = 5x + 3$

10. $\dfrac{x + 1}{5} - \dfrac{2x - 3}{2} = 3$

Solve each inequality.

11. $7x - 5 > 8x + 10$

12. $6x - 9 < 3x + 6$

13. If $f(x) = 2x^3 - x^2 + 7$, evaluate $f(-2)$.

14. If $g(x) = 3x + 11$, solve $g(x) = 3$.

15. A hardware store sells a certain type of desk lamp for $39.95. Write a function modeling the revenue if it sells x lamps.

16. How much revenue does the hardware store earn by selling 17 of these lamps in one week?

17. Find the slope and y-intercept of the line represented by the equation $4x + y = 9$.

18. Find the slope of the line perpendicular to the line represented by the equation $3x + 9y = 10$.

Write the equation of the line that satisfies the given conditions.

19. L has slope -2 and y-intercept of $(0, 4)$.

20. L passes through the point $(3, 2)$ and is parallel to the line $4x - 5y = 20$.

21. L passes through the points $(1, -3)$ and $(3, 5)$.

22. L is perpendicular to the line $x - 2y = 3$ and passes through the point $(1, -2)$.

Perform the indicated operations.

23. $(x^2 - 3x + 5) + (2x^2 + 5x - 9)$

24. $(3x^2 - 8x - 7) - (2x^2 - 5x + 11)$

25. $4x(3x - 5)$

26. $(2x - 5)(3x + 8)$

27. $(x + 2)(x^2 - 3x + 5)$

28. $(2x + 7)(2x - 7)$

29. $(3x - 5)^2$

30. $5x(2x - 5)^2$

31. $\dfrac{32x^2y^3 - 16x^4y^2 + 8xy^2}{8xy^2}$

32. $\dfrac{2x^3 - 15x - 7}{x - 3}$

Answers

19. ____________

20. ____________

21. ____________

22. ____________

23. ____________

24. ____________

25. ____________

26. ____________

27. ____________

28. ____________

29. ____________

30. ____________

31. ____________

32. ____________

Answers

33. ____

34. ____

35. ____

36. ____

37. ____

38. ____

39. ____

40. ____

Use the properties of exponents to simplify each expression.

33. $(3x^2)^2(2x^3)$

34. $(x^4y^{-3})^4$

35. $(2x^0)^3(-3x^2y)^2$

36. $\dfrac{6x^3y^{-5}}{3x^{-2}y^6}$

37. Calculate. Write your answer in scientific notation.

$$\frac{(4.2 \times 10^7)(6.0 \times 10^{-3})}{1.2 \times 10^{-5}}$$

Solve each problem.

38. If 7 times a number decreased by 9 is 47, find the number.

39. The sum of two consecutive odd integers is 132. What are the two integers?

40. The length of a rectangle is 4 centimeters (cm) more than 5 times the width. If the perimeter is 56 cm, what are the dimensions of the rectangle?

CHAPTER

R

A Review of Elementary Algebra

INTRODUCTION

The first half of this text covers topics that are commonly grouped under the title *Elementary Algebra*. This transitional chapter provides an opportunity to prepare for the topics of the second half of the text. These topics are commonly referred to as *Intermediate Algebra*.

There are two elements to this transition. This chapter provides a thorough review of Chapters 1–5. Each of the five sections in this chapter covers the material from one of those five chapters. This review is useful both for students who have recently completed these chapters and for students using this text to cover the intermediate algebra topics.

The second element of the transition is a practice final. After taking that practice final, you may refer to either the original chapter or this abbreviated review to clarify those topics you found difficult.

In any case, it is important that you demonstrate mastery of the material in the first part of this text before moving on to the later topics.

CHAPTER R OUTLINE

R.1 From Arithmetic to Algebra

In Chapter 1, we worked with algebraic expressions. An *expression* is a meaningful collection of numbers, variables, and symbols of operation.

The collection

$3x - 2y + 5$

is an example of an expression. The collection

$2 + -3x \div \cdot \div 2x$

is not an expression because the symbols are not presented in a meaningful way. In Example 1, we translate expressions from words to symbols.

Example 1 Translating Expressions into Mathematical Symbols

Write each expression in mathematical symbols.

(a) The sum of a number and 5, divided by the number, is written

$$\frac{x + 5}{x}$$

(b) One-half of the base times the height is written

$$\frac{1}{2} \cdot b \cdot h \quad \text{or} \quad \frac{1}{2}bh$$

(c) A number times the quantity 5 less than the number is written

$x(x - 5)$

(d) Pi (π) times the radius squared is written

$\pi \cdot r^2$ or πr^2

(e) Two times length plus 2 times width is written

$2L + 2W$

In our next example, we evaluate an algebraic expression.

Step by Step

To Evaluate an Algebraic Expression

Step 1 Replace each variable with the assigned value.

Step 2 Do the arithmetic operations, following the rules for order of operations.

Example 2 Evaluating an Algebraic Expression

Use $x = 3$ and $y = -2$ to evaluate each expression.

(a) $2xy = 2(3)(-2)$
$= -12$

(b) $3x - 2y + 3 = 3(3) - 2(-2) + 3$
$= 9 + 4 + 3$
$= 16$

(c) $\frac{2x}{3y} = \frac{2(3)}{3(-2)}$
$= \frac{6}{-6}$
$= -1$

(d) $3x^2 - 5y^2 = 3(3)^2 - 5(-2)^2$
$= 27 - 20$
$= 7$

A *term* can be written as a number, or the product of a number and one or more variables and their exponents. If terms contain exactly the same variables raised to the same powers, they are called *like terms.*

Like terms in an expression can always be combined into a single term.

Step by Step

Combining Like Terms

To combine like terms, use the following steps.

Step 1 Add or subtract the numerical coefficients.

Step 2 Attach the common variable(s).

Example 3 Combining Like Terms

Simplify each expression by combining like terms.

(a) $5m + 3n + 2m + 6n = 7m + 9n$

(b) $3x^2y + 5xy^2 + x^2y + 4xy^2 = 4x^2y + 9xy^2$

(c) $3a - 2b - a + 4b = 2a + 2b$

Step by Step

Removing Parentheses

Case 1 If a plus sign (+) or nothing at all appears in front of parentheses, just remove the parentheses. No other changes are necessary.

Case 2 If a minus sign (−) appears in front of a set of parentheses, the parentheses can be removed by changing the sign in front of each term inside the parentheses. Each addition becomes subtraction, and each subtraction becomes addition.

Example 4 Removing Parentheses

Remove the parentheses and then simplify each expression.

(a) $(3x - 5) + (2x + 3) = 3x - 5 + 2x + 3$
$= 5x - 2$

(b) $(3x^2 + 2x - 1) + (x^2 - 5x + 3) = 3x^2 + 2x - 1 + x^2 - 5x + 3$
$= 4x^2 - 3x + 2$

(c) $(3x - 5) - (2x + 3) = 3x - 5 - 2x - 3$
$= x - 8$

(d) $(3x^2 + 2x - 1) - (x^2 - 5x + 3) = 3x^2 + 2x - 1 - x^2 + 5x - 3$
$= 2x^2 + 7x - 4$

An *equation* is a mathematical statement that two expressions are equal. The statements

$3x - 5 = 2x - 4 \qquad y = 2x + 1 \qquad \text{and} \qquad x + 3 = 0$

are examples of equations.

A *solution* for an equation is any value for the variable that makes the equation a true statement.

Given the equation $3x - 5 = 2x - 4$, 1 is a solution, because substituting 1 for x results in the true statement $-2 = -2$.

If we place every solution to a particular equation in set braces, in this case we have only $\{1\}$, we refer to it as the *solution set.*

Equations that have the same solution set are called *equivalent equations.* We can obtain equivalent equations by using the *addition property of equality.*

Property

The Addition Property of Equality

If $\quad a = b \quad$ then $\quad a + c = b + c$

In words, adding the same quantity to both sides of an equation gives an equivalent equation.

In Example 5, we use the addition property of equality to solve several equations.

Example 5 Solving Equations

RECALL

Subtracting 3 is the same as adding −3.

Use the addition property to solve each equation.

(a) $x + 3 = 1$ — Subtract 3 from each side of the equation.
$x = -2$ — The equation is solved. We could write the solution set as $\{-2\}$.

(b) $3x - 5 = 2x - 4$ — Add 5 to each side of the equation.
$3x = 2x + 1$ — Subtract $2x$ from each side of the equation.
$x = 1$ — The equation is solved. We can write the solution set as $\{1\}$.

(c) $4(5x - 2) = 19x + 4$ — Use the distributive property to remove parentheses.
$20x - 8 = 19x + 4$ — Add 8 to each side.
$20x = 19x + 12$ — Subtract $19x$ from each side.
$x = 12$ — The equation is solved. The solution set is $\{12\}$.

Check

$4(5(12) - 2) \stackrel{?}{=} 19(12) + 4$ Substitute 12 for *x*.

$4(60 - 2) \stackrel{?}{=} 228 + 4$ Use order of operations.

$4(58) \stackrel{?}{=} 232$

$232 = 232$ True

Perhaps the most important reason for learning mathematics is so that it can be applied. Such applications are called *word problems*. The following five-step approach helps organize the solution to a word problem.

Step by Step

Solving Word Problems

Step 1 Read the problem carefully. Then reread it to determine what you are asked to find.

Step 2 Choose a letter to represent one of the unknowns in the problem. Other unknowns should be written as expressions using that same variable.

Step 3 Translate the problem to the language of algebra to form an equation.

Step 4 Solve the equation.

Step 5 Answer the question and check your solution by returning to the original problem.

In Example 6, we solve a word problem. Although it is a problem that you can solve easily through trial and error, we present it to model the steps.

Example 6 Solving a Word Problem

The sum of a number and 13 is 29. Find the number.

Step 1 We are looking for a number.

Step 2 We will call the number *x*.

Step 3 The equation is $x + 13 = 29$.

Step 4 To solve the equation, we subtract 13 from each side of the equation, so $x = 16$.

Step 5 The number is 16. Check to make sure that $16 + 13 = 29$.

Given an equation such as $4x = 24$, the addition property is not enough to find a solution. We need a second property.

Property

The Multiplication Property of Equality

If $a = b$, then $ac = bc$, if $c \neq 0$.

In words, multiplying both sides of an equation by the same nonzero number yields an equivalent equation.

To solve an equation of the form

$ax = b$

we multiply both sides by the reciprocal of a, or $1/a$.

Example 7 illustrates this process.

Example 7 Using the Multiplication Property

RECALL

Multiplying by $\frac{1}{5}$ is the same as dividing by 5.

Solve each equation.

(a) $5m = 15$ Multiply each side of the equation by $\frac{1}{5}$.

$m = 3$

(b) $-\frac{1}{3}x = 2$ Multiply each side of the equation by -3.

$x = -6$

In most cases, we must combine the addition and multiplication rules to solve an equation. In such cases, we use the following steps.

Step by Step

Solving Linear Equations in One Variable

Step 1 Remove any grouping symbols by applying the distributive property.

Step 2 Multiply both sides by the least common multiple (LCM) of any denominators.

Step 3 Combine like terms on each side of the equation.

Step 4 Apply the addition property of equality.

Step 5 Apply the multiplication property of equality.

Step 6 Check the solution.

Example 8 Combining the Properties to Solve an Equation

Solve each equation.

(a) $2(3x - 5) = 8$

Step 1 $6x - 10 = 8$

Step 2 No denominators

Step 3 No like terms

Step 4 $6x = 18$

Step 5 $x = 3$

Step 6 Check to see that $2[3(3) - 5] = 8$ is a true statement.

(b) $\frac{3x + 1}{2} = \frac{2}{5}$

Step 1 No parentheses

Step 2 Multiply through by the LCM of 2 and 5, which is 10.

$15x + 5 = 4$

Step 3 No like terms

Step 4 $15x = -1$

Step 5 $x = -\frac{1}{15}$

Step 6 Check to see that

$$\frac{3\left(-\frac{1}{15}\right) + 1}{2} = \frac{2}{5}$$

Every equation can be classified as one of the following types:

Conditional equations are true for some variable values and not true for others.

Identities are true for every possible value of the variable.

Contradictions are never true.

Example 9 Classifying Equations

Decide whether each equation is a conditional equation, an identity, or a contradiction.

(a) $2x = x + x$

This statement is always true. The equation is an identity.

(b) $2x = 2x + 3$

This statement is never true. The equation is a contradiction.

(c) $2x + 1 = 3x - 4$

This equation is true only when $x = 5$. It is a conditional equation.

Inequalities are solved much as equations are. There are two important properties associated with solving inequalities. First, we have the addition property.

Property

The Addition Property of Inequalities

If $a < b$ then $a + c < b + c$

In words, adding the same quantity to both sides of an inequality gives an equivalent inequality.

Example 10 Using the Addition Property to Solve an Inequality

Solve and graph the solution set for $7x - 8 \le 6x + 2$.

$7x - 8 \le 6x + 2$ Add 8 to each side.

$7x \le 6x + 10$ Subtract $6x$ from each side.

$x \le 10$

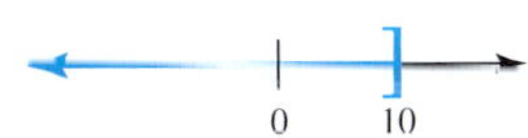

The graph indicates the set of real values less than or equal to 10.

Given an inequality such as $4x < 24$, the addition property is not enough to find a solution. We need a second property.

Property

The Multiplication Property of Inequalities

If $a < b$,

then $ac < bc$ when $c > 0$

and $ac > bc$ when $c < 0$

In words, multiplying both sides of an inequality by the same positive number yields an equivalent inequality. Multiplying by a negative number requires us to reverse the direction of the inequality to yield an equivalent inequality.

Example 11 Solving and Graphing Inequalities

Solve each inequality and then graph the solution set.

(a) $4x + 9 \geq x$ Subtract 9 from each side.

$4x \geq x - 9$ Subtract x from each side.

$3x \geq -9$ Multiply by $\frac{1}{3}$ on each side.

$x \geq -3$ Graph the solution set.

(b) $5 - 6x < 41$ Subtract 5 from each side.

$-6x < 36$ Divide by -6. Reverse the direction of the inequality.

$x > -6$ Graph the solution set.

R.1 exercises

Name ____________

Section ________ Date ________

Write each phrase, using symbols.

1. 5 less than a number x
2. A number x increased by 10
3. Twice a number x, decreased by 5
4. The quantity $x + y$ times the quantity x minus y
5. The product of p and the quantity 6 more than p
6. The sum of m and 5, divided by 3 less than n
7. $\frac{4}{3}$ times π times the cube of the radius r
8. $\frac{1}{3}$ times the product of the base b and the height h
9. One-half the product of the height h and the sum of two unequal sides b_1 and b_2
10. Twice the sum of x and y, minus 3 times the product of x and y

Evaluate each expression if $x = -4$ and $y = 2$.

11. $3x - 2y + 20$
12. $\frac{7x + 5y}{9y + 3x}$
13. $\frac{8y}{-2x}$
14. $(x - y)^2$
15. $x^2 - y^2$
16. $3xy - 5x$

Simplify each expression.

17. $11x^2 + 7x^2$
18. $-4p + 7q + 11p - 15q$
19. $4xy + 13y - 7xy - 5y$
20. $8r^3s^2 - 7r^2s^3 + 5r^3s^2 - 3r^2s^3$
21. $5x - (3x + 8)$
22. $9x + 11 - (-5x - 6)$

Answers

1. ________ 2. ________
3. ________ 4. ________
5. ________ 6. ________
7. ________ 8. ________
9. ________
10. ________
11. ________ 12. ________
13. ________ 14. ________
15. ________ 16. ________
17. ________ 18. ________
19. ________
20. ________
21. ________
22. ________

Is the number shown in parentheses a solution for the given equation?

23. $5x + 3 = -16 \quad (-3)$

24. $-7x + 9 = 30 \quad (-3)$

25. $-5(x - 7) = 6(5 - x) \quad (-5)$

26. $\frac{5}{6}x = 25 \quad (6)$

Solve each equation.

27. $7x - 9 = 15 + 6x$

28. $5x + 3 = 2(3x - 4)$

29. $3x + 2(x - 5) = 11 - (x + 3)$

30. $4 - \frac{x + 2}{3} = 1$

31. $5 - (2x - 7) = (9 - 4x) + 6$

32. $3x - 4 = 3(x + 2)$

Solve and graph the solution set for each inequality.

33. $4x + 5 < 3x - 7$

34. $9 - 7x > 13 - 8x$

35. $5x - 1 > 3(x + 1)$

36. $-7x + 3 \leq 2x + 21$

37. $2(x + 3) > 7(x + 2) + 2$

38. $7x + 5 < 2(x - 1) - 23$

39. $\frac{2}{3}\left(x + \frac{3}{4}\right) < \frac{5}{6}$

40. $4x - 3 \leq 9(x + 3)$

Solve each problem.

41. The sum of a number and 14 is 33. Find the number.

42. The sum of two consecutive odd integers is 20. Find the integers.

43. One number is 8 more than a second number. The sum of the two numbers is 22. Find the numbers.

44. Michelle earns $90 more per week than Dan. If their weekly salaries total $950, how much does Dan earn?

Answers

23. ____

24. ____

25. ____

26. ____

27. ____

28. ____

29. ____

30. ____

31. ____

32. ____

33. ____

34. ____

35. ____

36. ____

37. ____

38. ____

39. ____

40. ____

41. ____

42. ____

43. ____

44. ____

45. A customer purchases a radio for \$306.80 including a 4% sales tax. Find the price of the radio and the sales tax.

46. Woodville has a population 10% greater than that of Hightown. The total population of the two towns is 49,245. Find the population of each.

47. The length of a rectangle is 3 centimeters (cm) more than 4 times its width. If the perimeter of the rectangle is 46 cm, find the dimensions of the rectangle.

48. The Amazon River is 3 times as long as the Ohio-Allegheny river. Find the length of each if the difference in their lengths is 2,610 mi.

Answers

1. $x - 5$ **3.** $2x - 5$ **5.** $p(6 + p)$ **7.** $\frac{4}{3}\pi r^3$ **9.** $\frac{1}{2}h(b_1 + b_2)$

11. 4 **13.** 2 **15.** 12 **17.** $18x^2$ **19.** $-3xy + 8y$ **21.** $2x - 8$

23. No **25.** Yes **27.** $\{24\}$ **29.** $\{3\}$ **31.** $\left\{\frac{3}{2}\right\}$

33. $x < -12$ 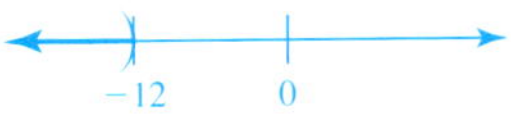

35. $x > 2$ 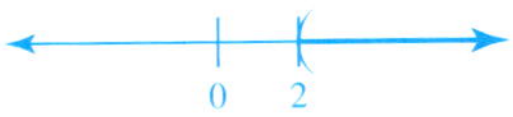

37. $x < -2$ 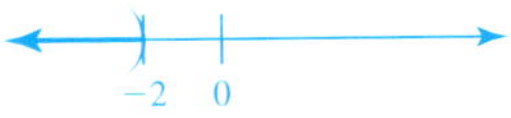

39. $x < \frac{1}{2}$ 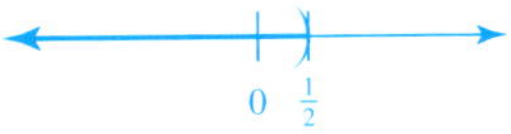

41. 19 **43.** 7, 15 **45.** Price: \$295; tax: \$11.80 **47.** 4 cm by 19 cm

Answers

45. ____________

46. ____________

47. ____________

48. ____________

R.2 Functions and Graphs

A *set* is a collcction of objects. Objects that belong to a set are called the *elements* of the set. Sets for which the elements are listed are said to be in *roster form.*

Example 1 Listing the Elements of a Set

Write each set in roster form.

(a) The set of all factors of 18

$\{1, 2, 3, 6, 9, 18\}$

(b) The set of all integers with an absolute value that is equal to 5

$\{-5, 5\}$

(c) The set of all even integers

$\{\ldots, -4, -2, 0, 2, 4, \ldots\}$

Not all sets can be described by using roster form. For example, the set of real numbers between 0 and 1 cannot be put in a list. Instead, we use *set-builder notation.* We write the set of real numbers greater than 0 but less than or equal to 1 as

$\{x \mid 0 < x \leq 1\}$

We can also describe a set by using *interval notation.* We write the above set as

$(0, 1]$

The parenthesis indicates that 0 is not included in the set, and the square bracket indicates that 1 is included.

Such sets can also be plotted on a number line. Our next example combines sets and their graphs.

Example 2 Plotting the Elements of a Set on a Number Line

Plot the elements of each set on the number line.

(a) $\{-2, 1, 5\}$

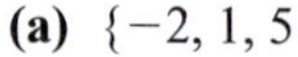

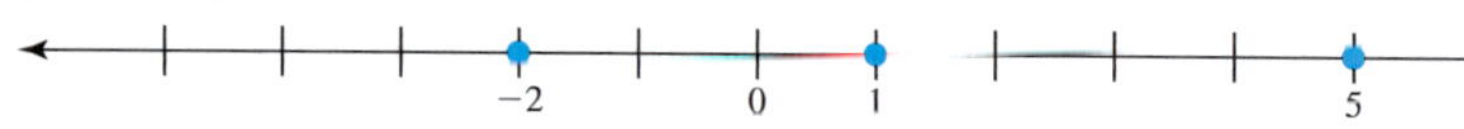

(b) $\{x \mid -3 < x < -1\}$

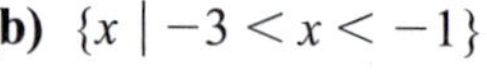

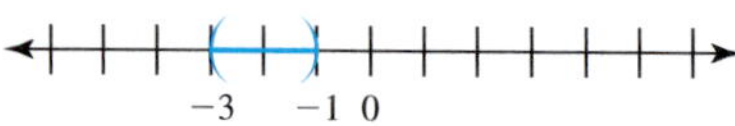

(c) $\{x \mid x \leq 5\}$

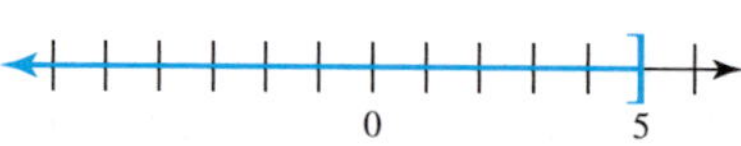

We can combine sets using the *union* and *intersection* operations. The union of two sets contains all the elements in one or both of the sets. The intersection of two sets contains the elements that are common to both sets. Our next example illustrates these operations.

Example 3 Finding Union and Intersection

Let $A = \{2, 4, 6, 8, 9\}$ and $B = \{1, 3, 4, 6, 7, 9\}$.

$A \cup B = \{1, 2, 3, 4, 6, 7, 8, 9\}$ and $A \cap B = \{4, 6, 9\}$

A linear equation in two variables is said to be in *standard form* if it is written as

$Ax + By = C$ where A and B are not both 0

A *solution* for an equation in two variables is an *ordered pair,* which, when substituted into the equation, results in a true statement.

Example 4 Finding Solutions for a Two-Variable Equation

Given the equation $3x - y = 5$:

(a) Determine whether the ordered pair $(-2, -11)$ is a solution.

Substituting -2 for x and -11 for y gives

$$3(-2) - (-11) \stackrel{?}{=} 5$$
$$-6 + 11 = 5$$

Since this is a true statement, $(-2, -11)$ is a solution.

(b) Find y if $x = 3$.

Substitute 3 for x.

$$3(3) - y = 5$$
$$9 - y = 5$$
$$-y = -4$$
$$y = 4$$

So, if $x = 3$, then $y = 4$.

(c) Complete the ordered pair (, 7) so that it is a solution.

Substituting 7 for y gives

$$3x - (7) = 5$$
$$3x = 12$$
$$x = 4$$

So (4, 7) is a solution.

The *rectangular,* or *Cartesian, coordinate system* consists of two number lines, one drawn horizontally (called the *x-axis*) and one drawn vertically (called the

y-*axis*). Their point of intersection is called the *origin,* and the axes divide the plane into four *quadrants.*

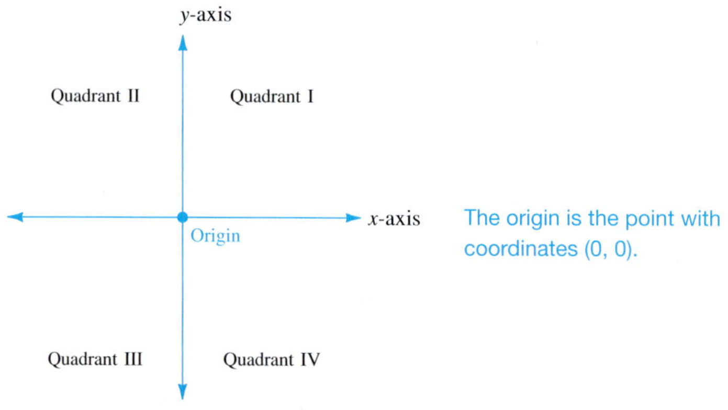

The origin is the point with coordinates (0, 0).

The two-dimensional picture shown here is referred to as the *plane.* There is a correspondence between *ordered pairs* of real numbers and *points in the plane.*

Example 5 Working with Ordered Pairs and Points in the Plane

(a)

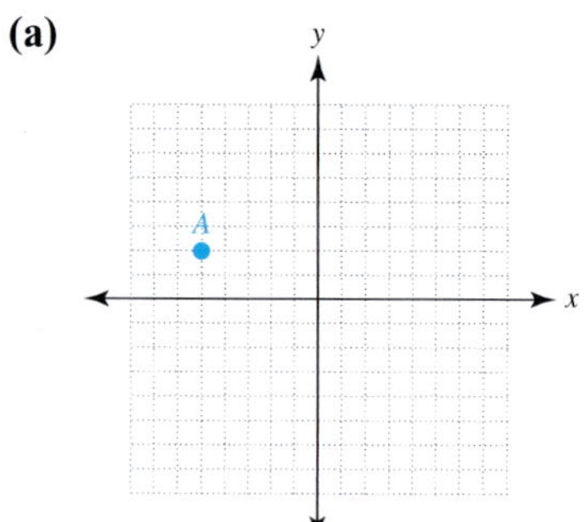

Name the ordered pair that corresponds to the given point A.

We see that A is 5 units to the *left* of the y-axis and 2 units *above* the x-axis. So point A has coordinates $(-5, 2)$.

(b)

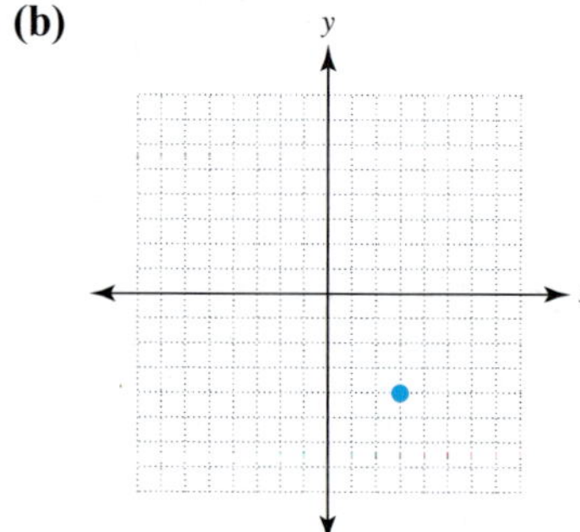

Graph the point corresponding to the ordered pair $(3, -4)$.

From the origin, move 3 units to the right, and then move 4 units down.

A *relation* is any set of ordered pairs. The set of first elements of the ordered pairs is called the *domain* of the relation, and the set of second elements of the ordered pairs is called the *range* of the relation.

The ordered pairs of a relation may be presented in a variety of forms. Among these are complete listings of the ordered pairs in set form or in table form.

Example 6 Finding the Domain and Range of a Relation

For each relation, find the domain and the range.

(a) $A = \{(-3, 5), (-1, 4), (2, 5)\}$

The domain is $D = \{-3, -1, 2\}$.

The range is $R = \{4, 5\}$.

(b) Suppose B contains the ordered pairs presented in the table.

x	y
-2	3
0	2
4	-1
4	-2

The domain is $D = \{-2, 0, 4\}$.

The range is $R = \{-2, -1, 2, 3\}$.

A *function* is a set of ordered pairs (a relation) in which no two first elements are equal.

Example 7 Identifying a Function

For each relation given in Example 6, determine whether the relation is a function.

(a) A is a function since no two first coordinates are equal.

(b) Since $(4, -1)$ and $(4, -2)$ are both ordered pairs of B, we see that B is *not* a function.

Another way of expressing a function is through the use of *$f(x)$ notation.*

Example 8 Evaluating a Function

Given $f(x) = x^2 - 3x + 2$, evaluate as indicated.

(a) $f(-5)$

Substituting -5 for x, we have

$$f(x) = x^2 - 3x + 2$$
$$f(-5) = (-5)^2 - 3(-5) + 2$$
$$= 25 + 15 + 2$$
$$= 42$$

(b) $f(2)$

$$f(2) = (2)^2 - 3(2) + 2$$
$$= 4 - 6 + 2$$
$$= 0$$

(c) $f(5h)$

$$f(5h) = (5h)^2 - 3(5h) + 2$$
$$= 25h^2 - 15h + 2$$

A set of ordered pairs can also be specified by means of a graph. In addition to finding the domain and range, we can determine from a graph whether a relation is a function. For this last purpose, we need the *vertical line test.*

Property

Vertical Line Test A relation is a function if no vertical line passes through two or more points on its graph.

Example 9 Identifying Functions, Domain, and Range

Determine whether the given graph is the graph of a function. Also provide the domain and range.

(a)

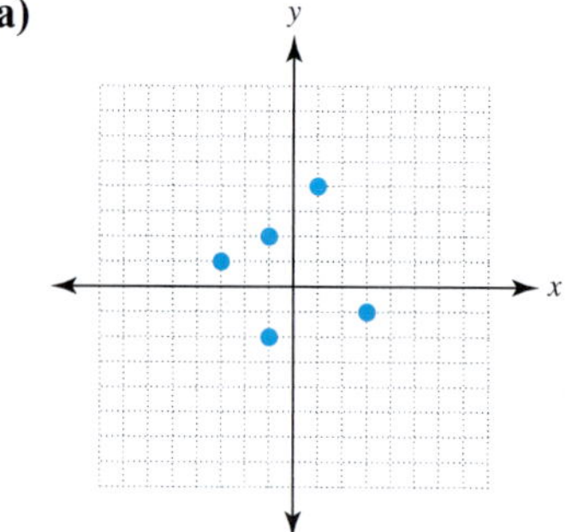

This is not a function. A vertical line at $x = -1$ passes through the two points $(-1, 2)$ and $(-1, -2)$.

The domain is $\{-3, -1, 1, 3\}$.

The range is $\{-2, -1, 1, 2, 4\}$.

(b)

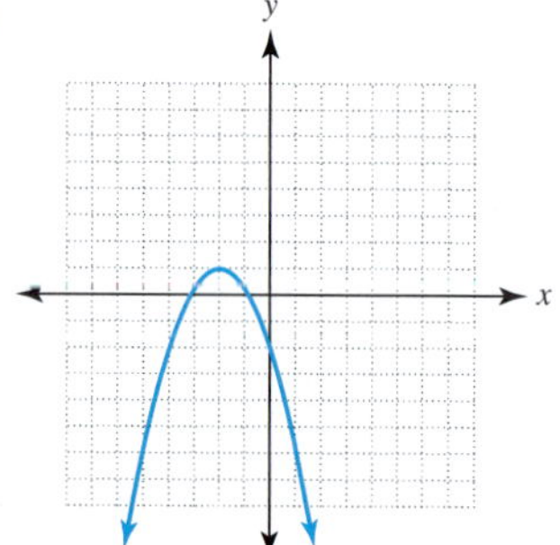

This is a function. The graph represents an infinite collection of ordered pairs, but since no vertical line passes through more than one point, it passes the vertical line test.

The x-values that are used in this graph consist of all real numbers, so

$$D = \{x \mid x \text{ is a real number}\} \qquad \text{or simply} \qquad D = \mathbb{R}$$

The y-values never go higher than 1, so the range is the set of all real numbers less than or equal to 1.

$$R = \{y \mid y \leq 1\}$$

It is frequently necessary to read function values from a graph. This generally involves one of two exercises: given x, find $f(x)$; or given $f(x)$, find x.

Example 10 Reading Values from a Graph

Given the graph of f, find the desired values.

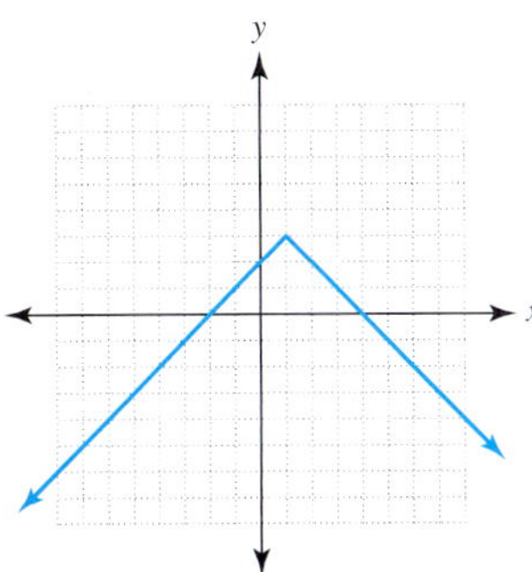

(a) Find $f(1)$.

Since $x = 1$, we move to 1 on the x-axis. Searching vertically, we find the point $(1, 3)$. So $f(1) = 3$.

(b) Find all x such that $f(x) = 1$.

We are given that the output, or y-value, is 1. So we move to 1 on the y-axis and search horizontally. There are two points with a y-value of 1: $(-1, 1)$ and $(3, 1)$. Since we want x-values, we report $x = -1$ and 3.

R.2 exercises

Boost *your* GRADE at ALEKS.com!

ALEKS®

- Practice Problems
- Self-Tests
- NetTutor
- e-Professors
- Videos

Name ____________________

Section ________ Date ________

Answers

1. ____________________
2. ____________________
3. ____________________
4. ____________________
5. ____________________
6. ____________________
7. ____________________
8. ____________________
9. ____________________
10. ____________________
11. ____________________
12. ____________________
13. ____________________
14. ____________________
15. ____________________
16. ____________________
17. ____________________
18. ____________________

Write each set in roster form.

1. The set of all factors of 24

2. The set of all odd whole numbers

3. The set of even integers less than 6

4. The set of integers between -4 and 3, inclusive

Use set-builder and interval notation to represent each set.

5. The set of all real numbers between -2 and 5, inclusive

6. The set of all real numbers less than or equal to -4

7. The set of all real numbers greater than -4 and less than 4

8. The set of all real numbers greater than -7

9. -3 0 4

10. 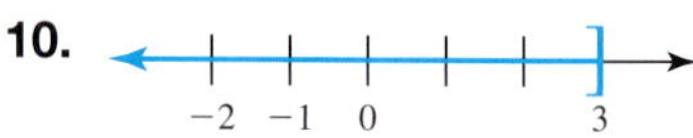

11. 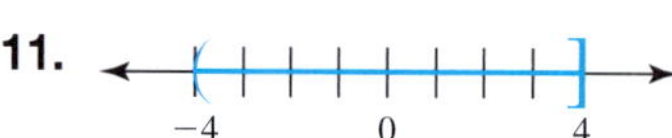

12. 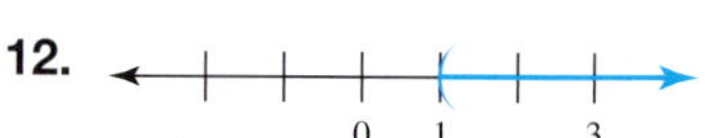

Graph each set on a number line.

13. $\{-5, -1, 2, 3, 5\}$

14. $\{x \mid x \le 6\}$

15. $\{x \mid -5 \le x \le 2\}$

16. $\{x \mid -3 \le x < 4\}$

Find $A \cup B$ and $A \cap B$.

17. $A = \{1, 4, 7, 10\}; B = \{2, 3, 7, 10\}$

18. $A = \{3, 5, 7, 9, 11\}; B = \{1, 3, 7, 8, 12\}$

Determine which of the ordered pairs are solutions for the given equation.

19. $4x - 2y = 8$ $(0, 4), (1, -2), (-2, 0), (2, 0)$

20. $3x - y = 5$ $(0, 5), (2, 0), (0, -5), (2, 1)$

21. $x = 4$ $(4, 0), (0, 4), (4, -1), (2, 3)$

22. $y = 5$ $(5, 0), (0, 5), (1, 5), (-2, 4)$

Complete the ordered pairs so that each is a solution for the given equation.

23. $x + y = 9$ $(4, \ ), (0, \ ), (\ , 0), (\ , -1)$

24. $5x - 3y = 15$ $(6, \ ), (0, \ ), (\ , 0), (\ , 10)$

Find four solutions for each equation. Note: *Your answers may vary from those shown in the answer section.*

25. $2x - y = 8$

26. $9x + 3y = 27$

Give the coordinates of the points shown on the graph.

27. A

28. B

29. C

30. D

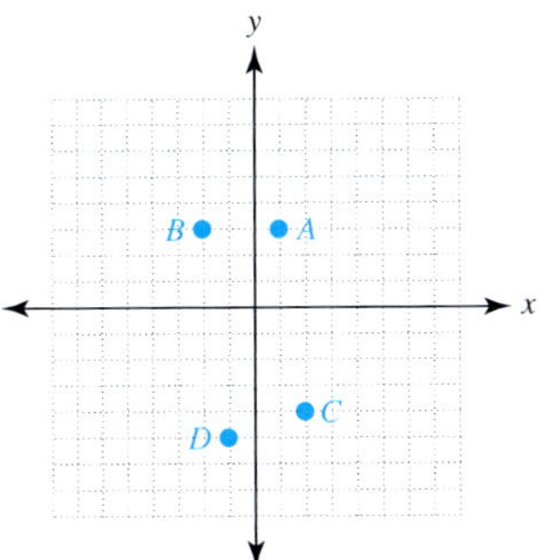

Plot each point on the rectangular coordinate system.

31. $A(2, -1)$

32. $B(-3, -4)$

33. $C(-2, 5)$

34. $D(4, 3)$

Find the domain and range of each relation.

35. $\{(-1, 3), (0, 5), (2, 4), (5, 7)\}$

36. $\{(-2, 4), (1, 0), (2, 5), (3, 7)\}$

37. $\{(-1, 3), (0, 3), (2, 3), (4, 3)\}$

38. $\{(1, -2), (1, 1), (1, 2), (1, 5)\}$

Answers

19. ______

20. ______

21. ______

22. ______

23. ______

24. ______

25. ______

26. ______

27. ______

28. ______

29. ______

30. ______

31. ______

32. ______

33. ______

34. ______

35. ______

36. ______

37. ______

38. ______

Answers

39. ____
40. ____
41. ____
42. ____
43. ____
44. ____
45. ____
46. ____
47. ____
48. ____
49. ____
50. ____
51. ____
52. ____
53. ____
54. ____

39.

x	y
-3	1
0	5
1	3
2	-4

40.

x	y
0	0
1	3
2	0
4	5

Determine which relations are also functions.

41. $\{(-2, 1), (0, 3), (1, 4), (2, 5)\}$

42. $\{(-3, -2), (-1, 0), (2, -1), (3, 0)\}$

43. $\{(-1, 2), (-1, -2), (2, 2), (4, 2)\}$

44. $\{(-2, 1), (-2, 3), (4, 5), (5, 6)\}$

45.

x	y
-3	0
-2	1
0	1
2	4

46.

x	y
-2	-2
0	5
-2	3
4	1

Evaluate each function for the value specified.

47. $f(x) = x^2 + 3x - 1 \quad f(-1)$

48. $f(x) = -2x^2 - 5x + 7 \quad f(2)$

49. $f(x) = -2x^2 + 5x + 15 \quad f(-3)$

50. $f(x) = 7x - 2 \quad f(x + h)$

Determine whether the given graph is the graph of a function. Provide the domain and range.

51.

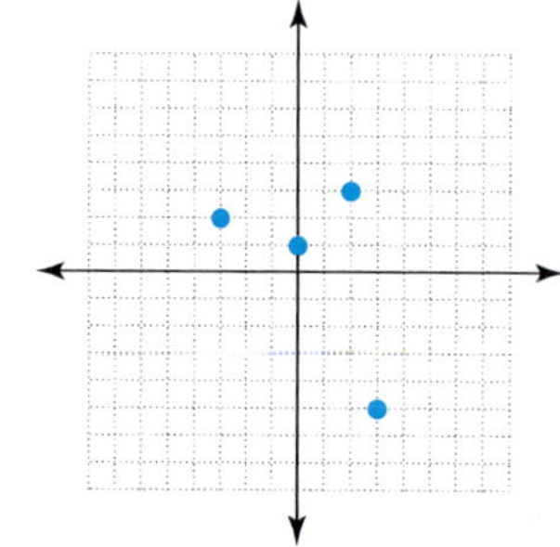

52.

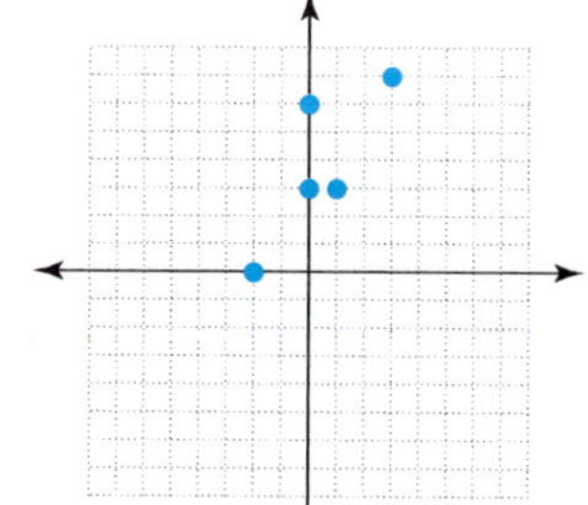

53.

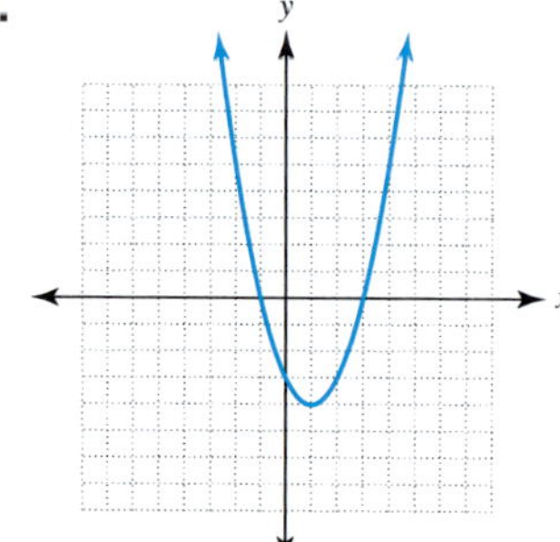

54.

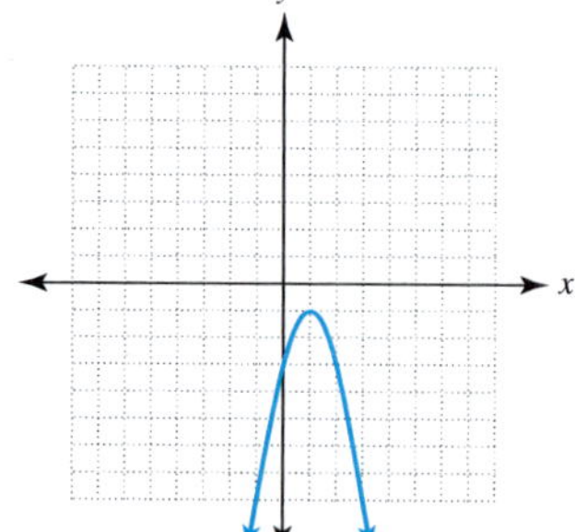

The graph of a function is shown. Find the indicated values.

55.

(a) $f(-1)$
(b) $f(5)$
(c) All x such that $f(x) = 3$
(d) All x such that $f(x) = 0$

56.

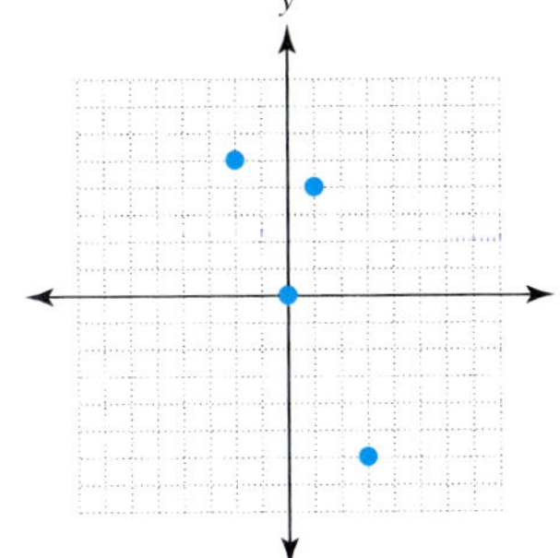

(a) $f(0)$
(b) $f(3)$
(c) All x such that $f(x) = 4$
(d) All x such that $f(x) = 5$

57.

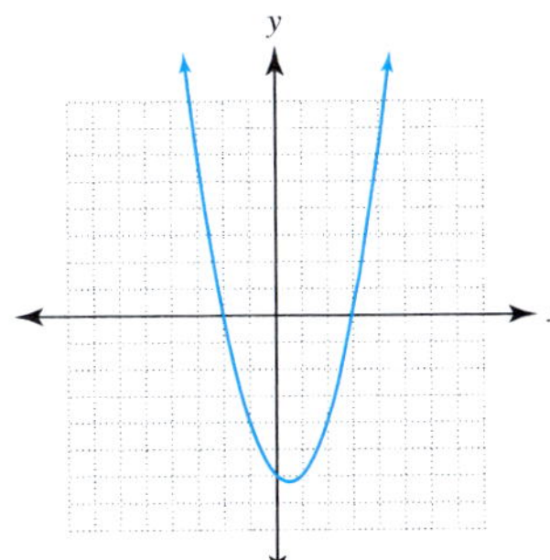

(a) $f(1)$
(b) $f(-2)$
(c) All x such that $f(x) = 0$
(d) All x such that $f(x) = -4$

58.

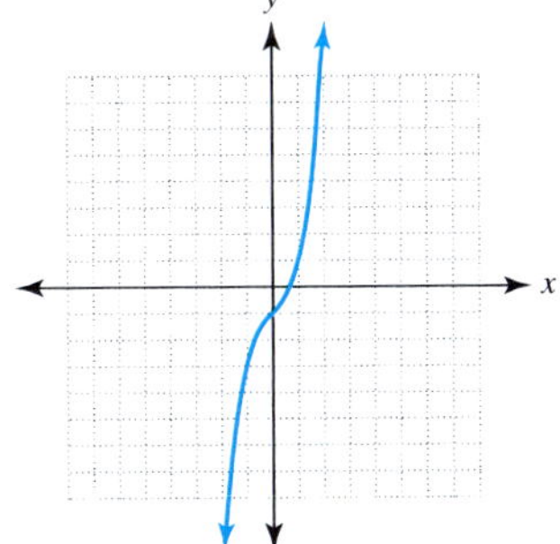

(a) $f(1)$
(b) $f(-1)$
(c) All x such that $f(x) = 8$
(d) All x such that $f(x) = -2$

Answers

55. ____________

56. ____________

57. ____________

58. ____________

Answers

1. $\{1, 2, 3, 4, 6, 8, 12, 24\}$ **3.** $\{\ldots, -4, -2, 0, 2, 4\}$
5. $\{x \mid -2 \le x \le 5\}$; $[-2, 5]$ **7.** $\{x \mid -4 < x < 4\}$; $(-4, 4)$
9. $\{x \mid -3 \le x \le 4]$; $[-3, 4]$ **11.** $\{x \mid -4 < x \le 4\}$; $(-4, 4]$
13. −5 −1 0 2 3 5 **15.** −5 0 2
17. $A \cup B = \{1, 2, 3, 4, 7, 10\}$; $A \cap B = \{7, 10\}$
19. $(1, -2), (2, 0)$ **21.** $(4, 0), (4, -1)$ **23.** $(4, 5), (0, 9), (9, 0), (10, -1)$
25. $(0, -8), (1, -6), (4, 0), (3, -2)$ **27.** $A(1, 3)$ **29.** $C(2, -4)$
31, 33.

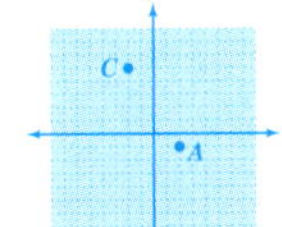

35. D: $\{-1, 0, 2, 5\}$; R: $\{3, 5, 4, 7\}$
37. D: $\{-1, 0, 2, 4\}$; R: $\{3\}$ **39.** D: $\{-3, 0, 1, 2\}$; R: $\{1, 5, 3, -4\}$
41. Function **43.** Not a function **45.** Function **47.** -3 **49.** -18
51. Function; D: $\{-3, 0, 2, 3\}$; R: $\{2, 1, 3, -5\}$
53. Function; D: $\mathbb{R}$; R: $\{y \mid y \ge -4\}$ **55. (a)** 3; **(b)** -3; **(c)** -1; **(d)** 3
57. (a) -6; **(b)** 0; **(c)** $-2, 3$; **(d)** $-1, 2$

R.3 Graphing Linear Functions

We show *all* the solutions of a linear equation in two variables by graphing points corresponding to solutions of the equation. The set of all ordered pairs that satisfy

$Ax + By = C$ where A and B are not both 0

produces a *straight-line graph* when the points corresponding to solutions are plotted. One method of creating such a graph is to solve for y and make a table.

Example 1 Graphing a Linear Equation

Draw the graph of $x - 2y = 8$.

Solve for y:

$$\begin{aligned} x - 2y &= 8 \\ -2y &= -x + 8 \\ y &= \frac{1}{2}x - 4 \end{aligned}$$

Make a table of values, where the x-values are multiples of 2.

x	y
-2	-5
0	-4
2	-3

Plot these points and draw a line through them.

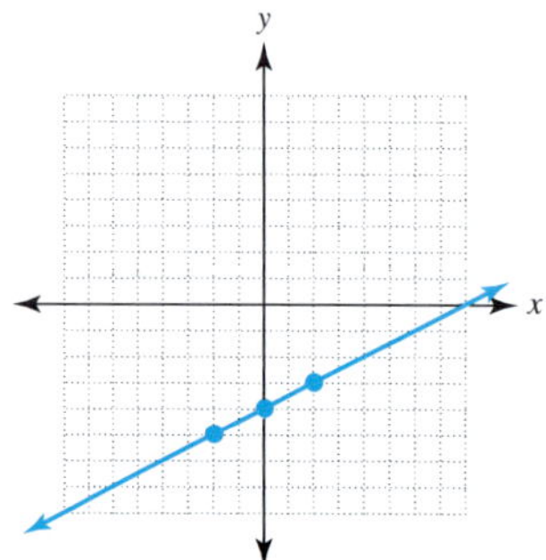

Another useful method for graphing linear equations is the *intercept method.*

Step by Step

Graphing a Line by the Intercept Method

- **Step 1** To find the x-intercept: Let $y = 0$ and solve for x.
- **Step 2** To find the y-intercept: Let $x = 0$ and solve for y.
- **Step 3** Graph the x- and y-intercepts.
- **Step 4** Draw a straight line through the intercepts.

Example 2 Using the Intercept Method to Graph a Line

Draw the graph of $3x - 4y = 12$.

Find the x-intercept. Let $y = 0$:

$$3x - 4(0) = 12$$
$$3x = 12$$
$$x = 4$$

So the x-intercept is $(4, 0)$.

Find the y-intercept. Let $x = 0$:

$$3(0) - 4y = 12$$
$$-4y = 12$$
$$y = -3$$

So the y-intercept is $(0, -3)$.

Plot the intercepts and draw a line through them.

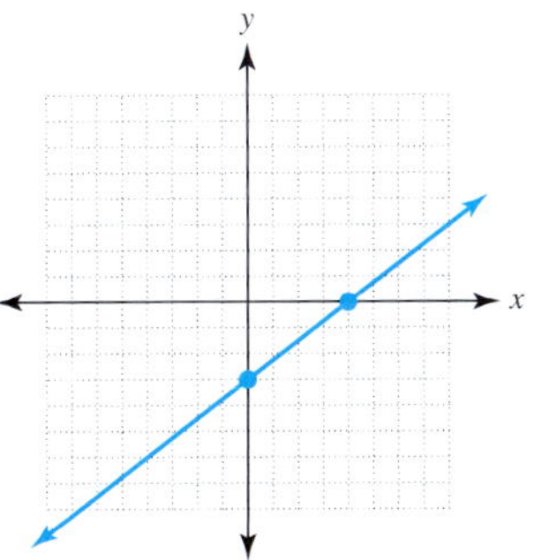

It is possible for the graph of a linear equation to be horizontal or vertical.

Property

Vertical and Horizontal Lines

1. The graph of $x = a$ is a *vertical line* crossing the x-axis at $(a, 0)$.
2. The graph of $y = b$ is a *horizontal line* crossing the y-axis at $(0, b)$.

Example 3 Creating Horizontal and Vertical Graphs

(a) Draw the graph of $x = -4$.

The line is vertical and crosses the x-axis at $(-4, 0)$.

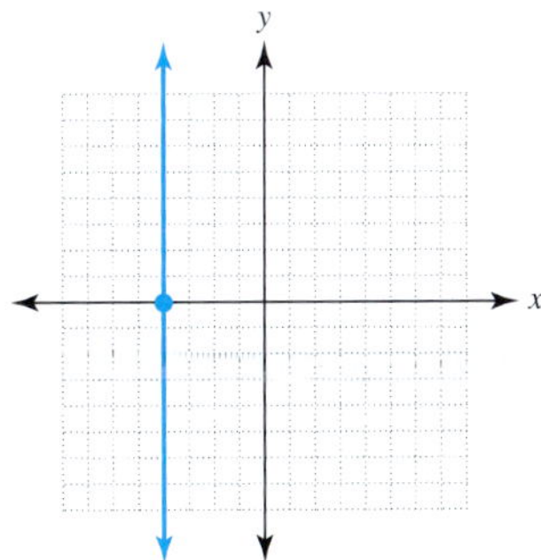

(b) Draw the graph of $y = 2$.

The line is horizontal and crosses the y-axis at (0, 2).

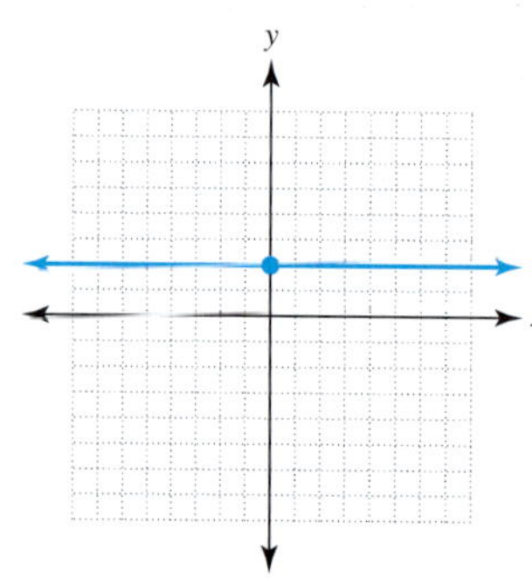

The *slope* of a line through two points $P(x_1, y_1)$ and $Q(x_2, y_2)$ is given by

$$m = \frac{\text{change in } y}{\text{change in } x}$$

$$= \frac{y_2 - y_1}{x_2 - x_1} \quad \text{if} \quad x_1 \neq x_2$$

Example 4 Finding Slope

Find the slope of the line that passes through the given points.

(a) $(-3, 2)$ and $(4, -1)$

Letting $(-3, 2) = (x_1, y_1)$ and $(4, -1) = (x_2, y_2)$, we have

$$m = \frac{y_2 - y_1}{x_2 - x_1}$$

$$= \frac{-1 - 2}{4 - (-3)}$$

$$= \frac{-3}{7}$$

$$= -\frac{3}{7}$$

(b) $(-4, 5)$ and $(2, 5)$

$$m = \frac{5 - 5}{2 - (-4)}$$

$$= \frac{0}{6}$$

$$= 0$$

(c) $(1, 6)$ and $(1, -3)$

$$m = \frac{6 - (-3)}{1 - 1}$$

$$= \frac{9}{0} \quad \text{which is undefined}$$

The slope and y-intercept of a line are very useful for graphing linear equations.

Definition

The Slope-Intercept Form for a Line

A linear equation is said to be in *slope-intercept form* when it is written as $y = mx + b$. When written in this form, m is the slope and the y-intercept is $(0, b)$.

Example 5 Using Slope-Intercept Form to Graph an Equation

Given the equation $4x - 3y = 6$:

(a) Find the slope and y-intercept. Solve for y:

$$4x - 3y = 6$$
$$-3y = -4x + 6$$
$$y = \frac{4}{3}x - 2$$

Since $m = \frac{4}{3}$, the slope is $\frac{4}{3}$.

Since $b = -2$, the y-intercept is $(0, -2)$.

(b) Draw the graph. Plot the y-intercept $(0, -2)$. Then, using the slope of $\frac{4}{3}$, move from the y-intercept with a rise of 4 and a run of 3 to plot a new point at $(3, 2)$. Draw a line through the plotted points.

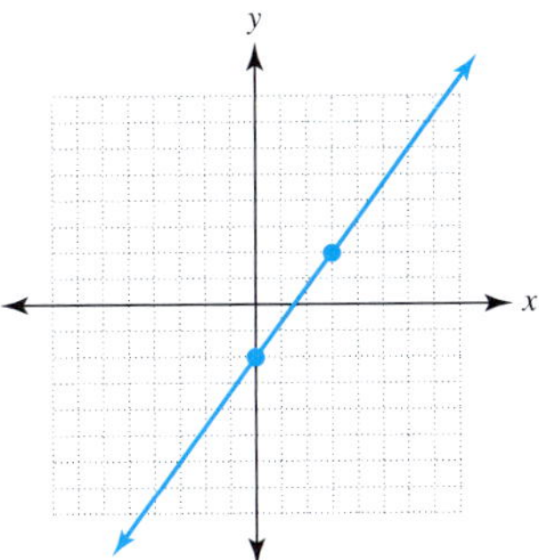

We can always write a linear equation as a function, provided that the equation does not represent a vertical line.

Step by Step

Writing a Linear Equation as a Function

Step 1 Write the equation in slope-intercept form, $y = mx + b$.

Step 2 Replace y with $f(x)$, so that $f(x) = mx + b$.

Example 6 Writing a Linear Equation as a Function

Use $f(x)$ notation to express the equation $2x + 5y = 10$ as a linear function.

Solve for y.

$$2x + 5y = 10$$
$$5y = -2x + 10$$
$$y = -\frac{2}{5}x + 2$$

Replace y with $f(x)$.

$$f(x) = -\frac{2}{5}x + 2$$

Definition

Parallel and Perpendicular Lines

Two lines are *parallel* if their slopes are equal: $m_1 = m_2$

Two lines are *perpendicular* if their slopes are negative reciprocals: $m_1 = -\frac{1}{m_2}$

In this case $m_1 \cdot m_2 = -1$; the product of their slopes is -1.

There is another useful form for a linear equation: the *point-slope form.*

Definition

Point-Slope Form for a Linear Equation

Given that a line has slope m and passes through a point (x_1, y_1), an equation of the line is

$$y - y_1 = m(x - x_1)$$

Example 7 Finding the Equation of a Line

Write the equation of the line that passes through $(-3, 5)$ and $(6, 2)$.

First, find the slope.

$$m = \frac{y_2 - y_1}{x_2 - x_1} = \frac{5 - 2}{-3 - 6} = \frac{3}{-9} = -\frac{1}{3}$$

Then, choosing a given point, say $(6, 2)$, use the point-slope form to write

$$y - y_1 = m(x - x_1)$$

$$y - 2 = -\frac{1}{3}(x - 6)$$

$$y - 2 = -\frac{1}{3}x + 2$$

$$y = -\frac{1}{3}x + 4$$

If we have two data points for a linear function, we can write its equation.

Example 8 Writing a Linear Function Using Two Data Points

Write the equation for a linear function f, given that $f(-1) = -3$ and $f(5) = 1$.

$f(-1) = -3$ indicates that $(-1, -3)$ is a point on the graph of f.
$f(5) = 1$ says that $(5, 1)$ is also a point.
Find the slope.

$$m = \frac{y_2 - y_1}{x_2 - x_1} = \frac{1 - (-3)}{5 - (-1)} = \frac{4}{6} = \frac{2}{3}$$

Using the point-slope form with $m = \frac{2}{3}$ and the point (5, 1), we have

$$y - 1 = \frac{2}{3}(x - 5)$$
$$y - 1 = \frac{2}{3}x - \frac{10}{3}$$
$$y = \frac{2}{3}x - \frac{10}{3} + 1$$
$$y = \frac{2}{3}x - \frac{7}{3}$$

We replace y with $f(x)$ to write the linear function.

$$f(x) = \frac{2}{3}x - \frac{7}{3}$$

Example 9 Interpreting the Slope of a Linear Function

A lemonade stand finds that the number of glasses G sold in a day can be predicted by the daily high temperature t (°F) by the linear function

$$G(t) = 1.5t - 74$$

Interpret the slope.

Because the units of $G(t)$ are glasses, and the units of t are degrees, the unit attached to the slope is $\frac{\text{glasses}}{\text{degree}}$. The slope is 1.5, so we can say that when the temperature goes up 1 degree, an additional 1.5 glasses of lemonade are sold.

Example 10 Finding the Equation of a Line

Write the equation of the line that passes through $(-1, -5)$ and is perpendicular to the line whose equation is $x + 2y = 6$.

First, find the slope of the line with equation $x + 2y = 6$:

$$x + 2y = 6$$
$$2y = -x + 6$$
$$y = -\frac{1}{2}x + 3$$

The slope of this line is $-\frac{1}{2}$, so the slope of our desired line must be 2.

Using $m = 2$ and the given point $(-1, -5)$, we have

$$y - (-5) = 2[x - (-1)]$$
$$y + 5 = 2(x + 1)$$
$$y + 5 = 2x + 2$$
$$y = 2x - 3$$

- To graph the linear inequality $Ax + By > C$, graph the equation $Ax + By = C$ with a dashed line and shade the appropriate half-plane (using a test point).
- To graph the linear inequality $Ax + By \geq C$, graph the equation $Ax + By = C$ with a solid line and shade the appropriate half-plane (using a test point).

Example 11 Graphing Linear Inequalities

(a) Graph the linear inequality $x - 2y < 4$.

We begin by solving for y.

$$y > \frac{1}{2}x - 2$$

We then use a dashed line to graph $y = \frac{1}{2}x - 2$.

Finally, we shade the region above the line to represent that y "is greater than" $\frac{1}{2}x - 2$.

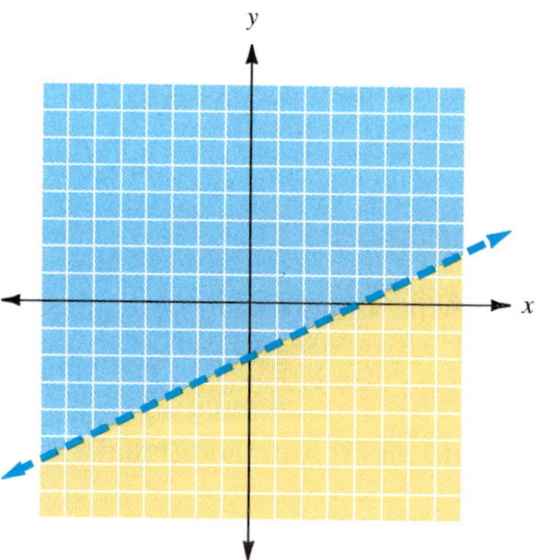

The shaded region represents the solution set for the inequality. Points on the border are not part of the solution set because we have strict inequality.

(b) Graph the linear inequality $3x + 2y \leq 5$.

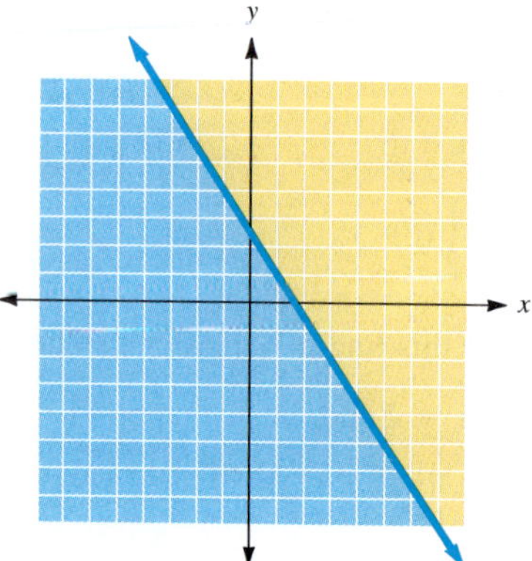

We use a solid line here because the inequality is "less than or equal to." Therefore, points on the border are part of the solution set. We shade the region below the line because the test point (0, 0) is part of the solution set.

R.3 exercises

Boost *your* GRADE at ALEKS.com!

ALEKS®

- Practice Problems
- Self-Tests
- NetTutor
- e-Professors
- Videos

Name ______________________

Section ________ Date ________

Graph the equation.

1. $x + y = 5$

2. $2x + y = 6$

3. $3x - y = 5$

4. $-4x - y = 7$

5. $y = 3x$

6. $y = -4x$

7. $2x + 5y = 10$

8. $3x - 5y = 15$

9. $x = 4$

10. $y = -2$

Find the slope of the line passing through each pair of points.

11. $(4, -2)$ and $(-1, 3)$

12. $(2, 7)$ and $(-1, -5)$

13. $(-3, 2)$ and $(7, 2)$

14. $(2, 5)$ and $(2, -3)$

Find the slope and y-intercept of the line represented by the given equation. Graph each equation.

15. $y = 3x + 2$

16. $3y + 7x = 21$

Answers

1. ______
2. ______
3. ______
4. ______
5. ______
6. ______
7. ______
8. ______
9. ______
10. ______
11. ______
12. ______
13. ______
14. ______
15. ______
16. ______

Answers

17. ______

18. ______

19. ______

20. ______

21. ______

22. ______

23. ______

24. ______

25. ______

26. ______

27. ______

28. ______

29. ______

30. ______

31. ______

32. ______

33. ______

17. $y + 5x = 3$

18. $4x - 3y = 12$

Write each linear equation as a function.

19. $y = 3x + 2$

20. $3y + 7x = 21$

21. $y + 5x = 3$

22. $4x - 3y = 12$

Write the equation for a linear function f using the two data points.

23. $f(-2) = -8$ and $f(3) = 2$

24. $f(-1) = 5$ and $f(2) = -4$

25. $f(-6) = 1$ and $f(4) = 5$

26. $f(1) = 6$ and $f(7) = -2$

Determine whether the lines are parallel, perpendicular, or neither.

27. L_1 through (3, 4) and (5, −6)
L_2 through (5, 4) and (10, 5)

28. L_1 with equation $5x - 4y = 20$
L_2 passes through (4, 7) and (8, 12)

Write an equation of the line satisfying the given geometric conditions.

29. L passes through the points (−6, 2) and (4, −3).

30. L passes through (3, 4) and is perpendicular to the line with equation $y = -\frac{3}{5}x + 2$.

31. L has y-intercept (0, 6) and is parallel to the line with equation $y - 3x = 5$.

32. L has x-intercept of (−2, 0) and y-intercept of (0, 8).

Interpret the slope.

33. A company that manufactures umbrellas finds that its weekly cost C can be expressed by the linear function

$C(x) = 8x + 350$

where x is the number of umbrellas produced in a week.

34. A company that sells flashlights finds that the number of flashlights N sold in a month can be expressed by the linear function n

$N(p) = 260 - 5p$

where p is the price, in dollars, of a flashlight.

Graph each linear inequality.

35. $x + y > 4$

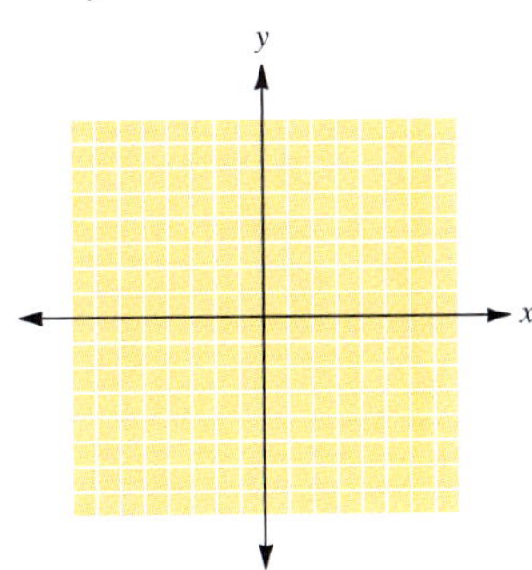

36. $x - 4y > -8$

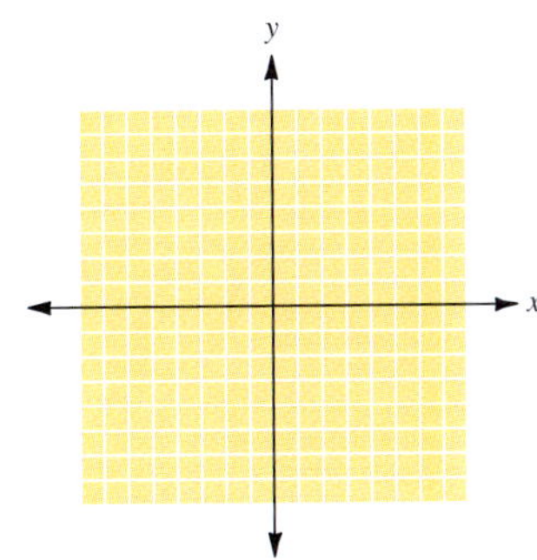

37. $2x + 3y \le 12$

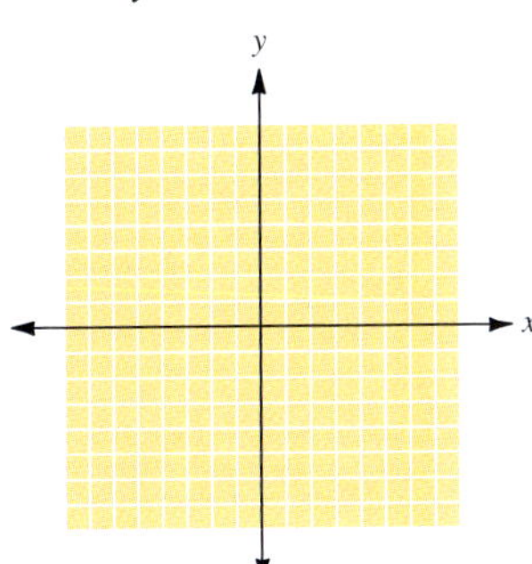

38. $3x - 5y \le 18$

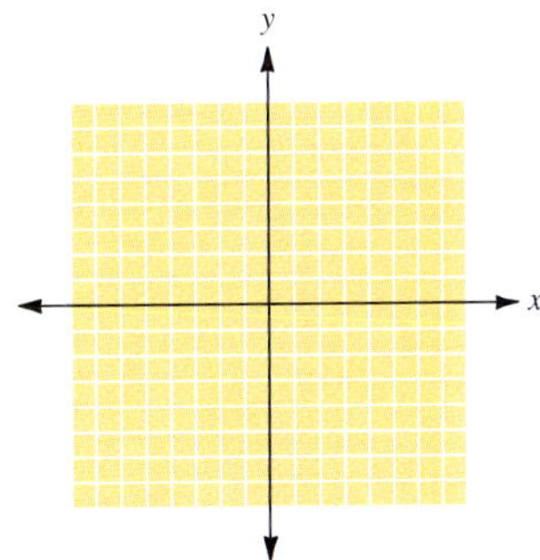

Answers

34. ______

35. ______

36. ______

37. ______

38. ______

Answers

1.

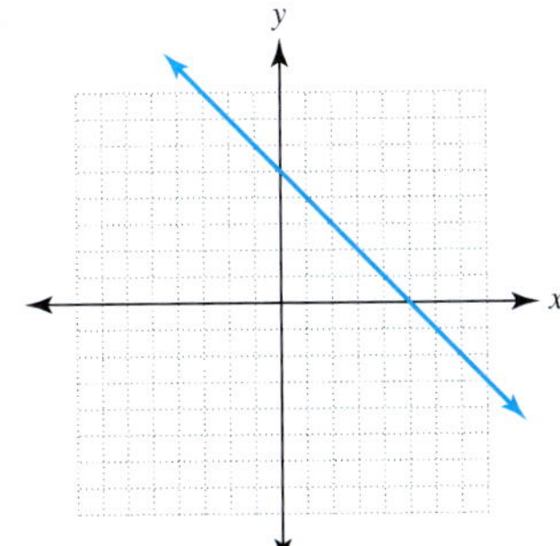

3.

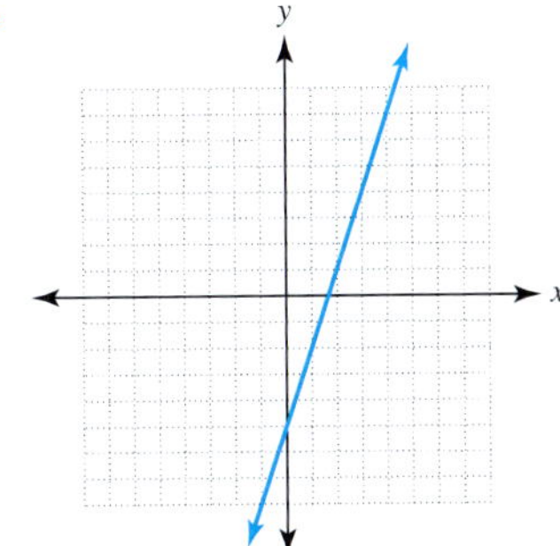

5.

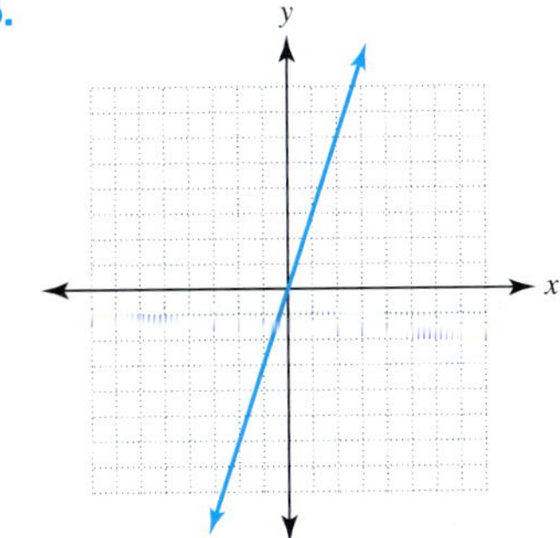

7.

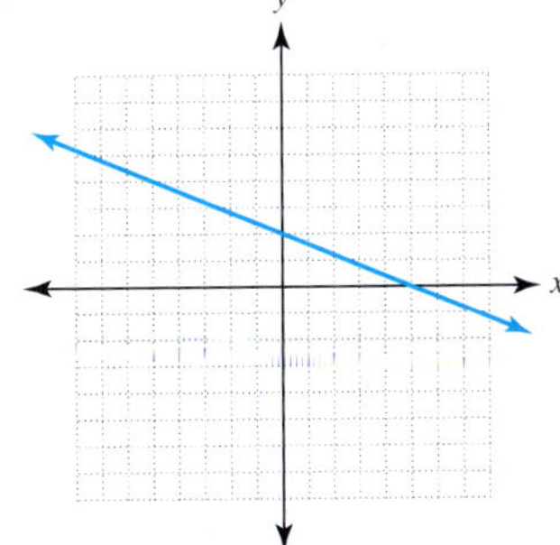

9.

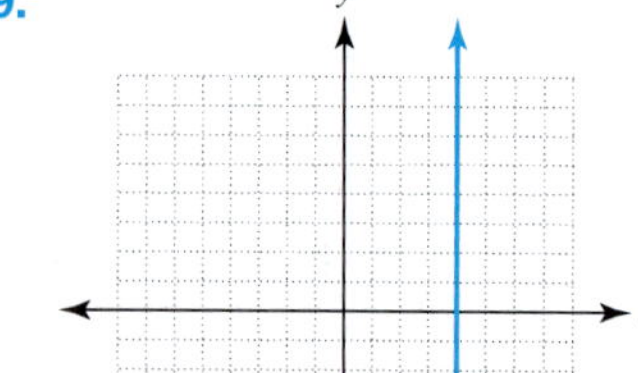

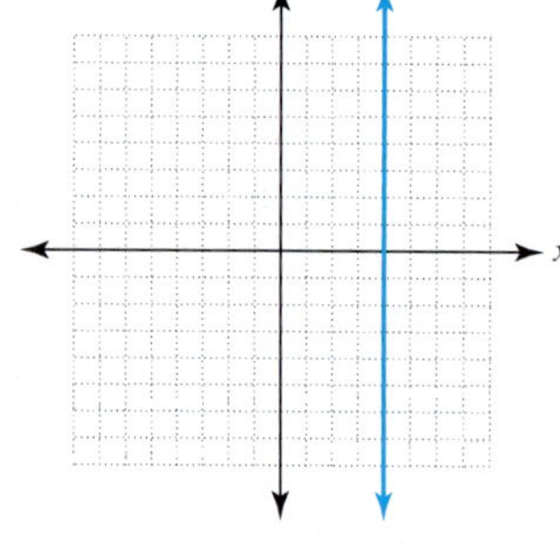

11. -1 **13.** 0

15. Slope 3, y-intercept $(0, 2)$

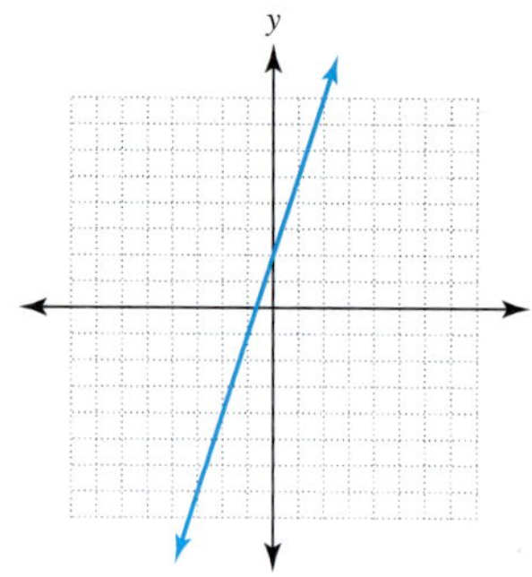

17. Slope -5, y-intercept $(0, 3)$

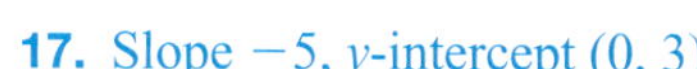

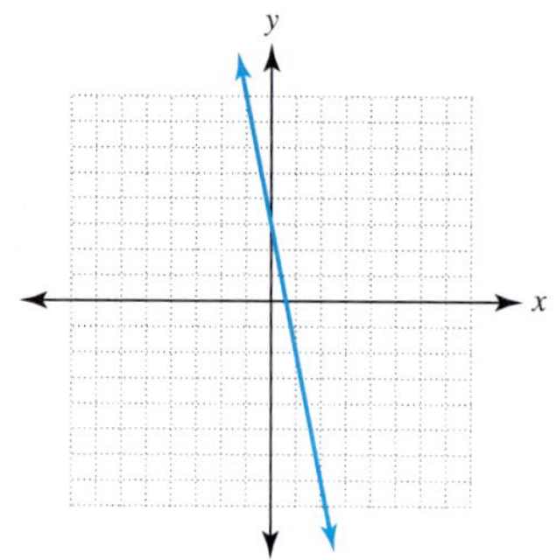

19. $f(x) = 3x + 2$ **21.** $f(x) = -5x + 3$ **23.** $f(x) = 2x - 4$

25. $f(x) = \frac{2}{5}x + \frac{17}{5}$ **27.** Perpendicular **29.** $y = -\frac{1}{2}x - 1$

31. $y = 3x + 6$ **33.** The slope is $8 \frac{\text{dollars}}{\text{umbrella}}$. We can say that for each additional umbrella produced, the cost increases by 8 dollars.

35. $x + y > 4$

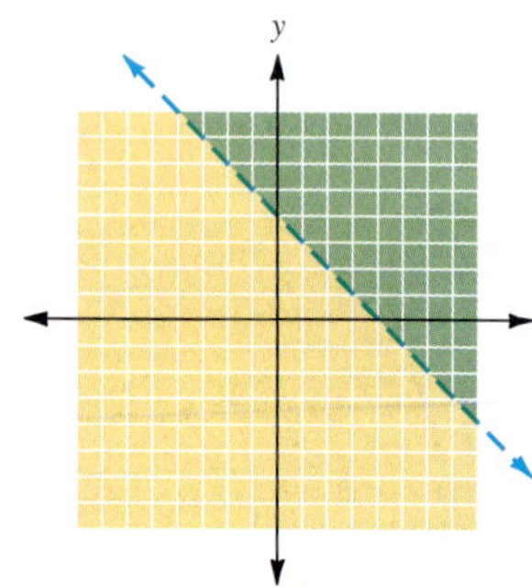

37. $2x + 3y \leq 12$

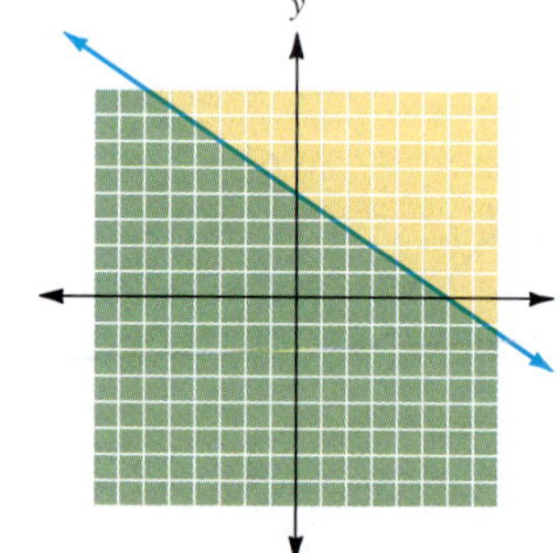

R.4 Systems of Linear Equations

We learned to solve systems of linear equations and inequalities in Chapter 4.

- A solution to a system of two linear equations in two variables is an ordered pair (x, y), which is a solution to both equations individually.
- Graphically, a solution to a system of equations is a point that is on the graphs of both equations.
- We can solve a system of equations graphically, by the addition (elimination) method, or by substitution.
- A system of two linear equations in two variables can have no solutions (inconsistent), one solution (consistent), or an infinite number of solutions (dependent).

Example 1 Solving a System by Graphing

Solve each system by graphing.

(a) $2x + y = 4$

$x - y = 5$

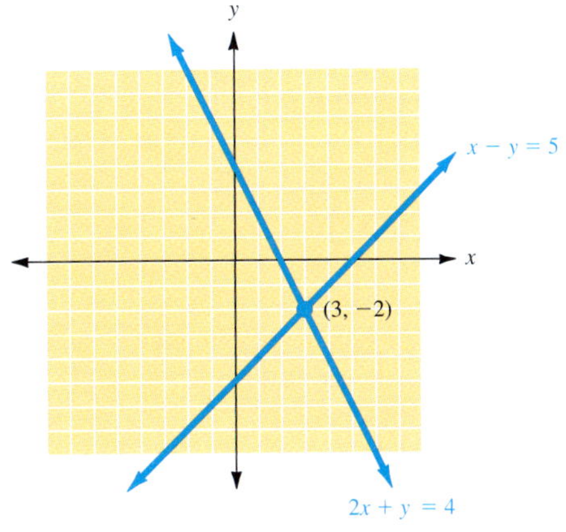

The solution set is $\{(3, -2)\}$.

(b) $2x - y = 4$

$6x - 3y = 18$

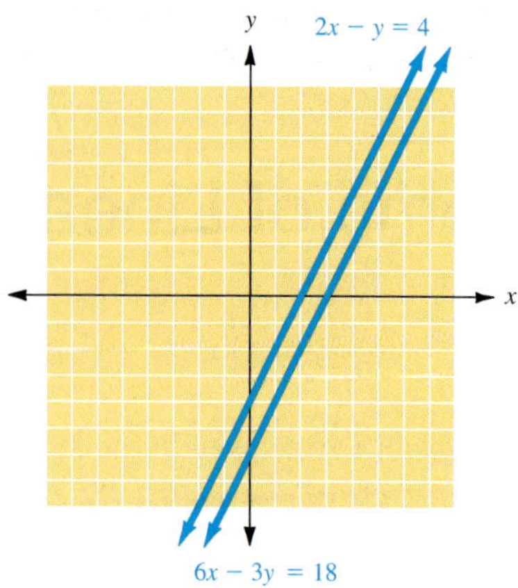

The system has no solutions (it is inconsistent).

Example 2 Using the Substitution Method to Solve a System

Solve each system by substitution.

(a) $2x - 3y = -3$

$y = 2x - 1$

Use the second equation to substitute for y in the first equation.

$$2x - 3(2x - 1) = -3$$
$$2x - 6x + 3 = -3$$
$$-4x = -6$$
$$x = \frac{3}{2}$$

Substitute this value for x into the second equation.

$$y = 2\left(\frac{3}{2}\right) - 1$$
$$y = 2$$

The solution set is $\left\{\left(\frac{3}{2}, 2\right)\right\}$.

(b) $2x - y = 2$

$4x - 2y = 4$

Solve the first equation for y.

$$y = 2x - 2$$

Substitute for y in the second equation.

$$4x - 2(2x - 2) = 4$$
$$4x - 4x + 4 = 4$$
$$4 = 4 \quad \text{True!}$$

Because this last statement is true, we have a dependent system. The solution set may be given as $\{(x, y) \mid y = 2x - 2\}$. There are an infinite number of solutions.

Example 3 Using the Addition Method to Solve a System

Solve each system with the addition method.

(a) $5x - 2y = 12$

$3x + 2y = 12$

Add the two equations together.

$8x = 24$

$x = 3$

Substitute this value for x into the first equation.

$$5(3) - 2y = 12$$
$$15 - 2y = 12$$
$$-2y = -3$$
$$y = \frac{-3}{-2}$$
$$y = \frac{3}{2}$$

Therefore, the solution set to our system is $\left\{\left(3, \frac{3}{2}\right)\right\}$.

(b) $2x + 3y = 5$

$4x - 2y = 7$

We choose to multiply the first equation by -2.

$$\begin{aligned} -4x - 6y &= -10 \\ 4x - 2y &= 7 \\ \hline -8y &= -3 \end{aligned}$$

$$y = \frac{3}{8}$$
$$2x + 3\left(\frac{3}{8}\right) = 5$$
$$2x + \frac{9}{8} = 5$$
$$2x = \frac{31}{8}$$
$$x = \frac{31}{16}$$

Therefore, the solution set is $\left\{\left(\frac{31}{16}, \frac{3}{8}\right)\right\}$.

A solution for a linear system of three variables is an ordered triple of numbers (x, y, z) that satisfies each equation in the system.

Step by Step

Solving a System of Three Equations in Three Unknowns

Step 1 Choose a pair of equations from the system and use the addition method to eliminate one of the variables.

Step 2 Choose a different pair of equations and eliminate the same variable.

Step 3 Solve the system of two equations in two variables determined in steps 1 and 2.

Step 4 Substitute the values found above into one of the original equations and solve for the remaining variable.

Step 5 The solution is the ordered triple of values found in steps 3 and 4. It can be checked by substituting into each of the equations of the original system.

Example 4 Solving a Linear System in Three Variables

Solve.

$$x + y - z = 6$$
$$2x - 3y + z = -9$$
$$3x + y + 2z = 2$$

Adding the first two equations gives

$$3x - 2y = -3$$

Multiply the first equation by 2 and add the result to the third equation.

$$5x + 3y = 14$$

The system consisting of these last two equations can be solved as before giving

$$x = 1 \qquad y = 3$$

Substituting these values into any of the original equations gives

$$z = -2$$

The solution set is $\{(1, 3, -2)\}$.

To solve a system of linear inequalities, recall these concepts.

- A solution to a system of inequalities is a point that satisfies all of the inequalities.
- The solution set of a system of inequalities is usually shown graphically, as a region.

Example 5 Solving a System of Linear Inequalities

Sketch the solution region for each system of inequalities.

(a) $y \geq 5 - 2x$

$y \leq 2x + 1$

We graph $y = 5 - 2x$ with a solid line and shade the region above the line to account for the "greater than or equal to" symbol. Similarly, we graph $y = 2x + 1$ and shade the region below the line because we have a "less than or equal to" symbol.

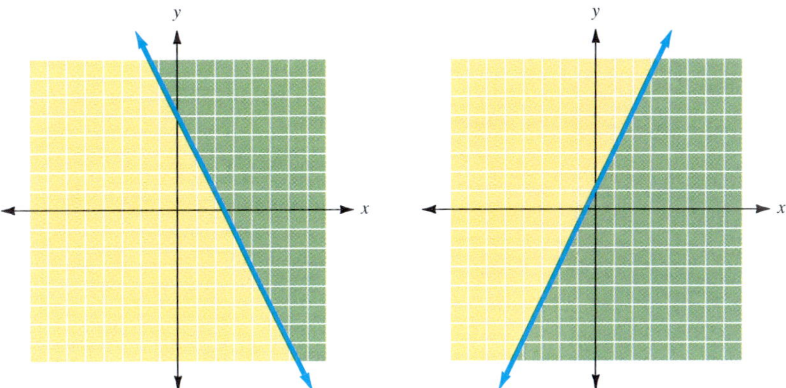

We combine these two into a single graph.

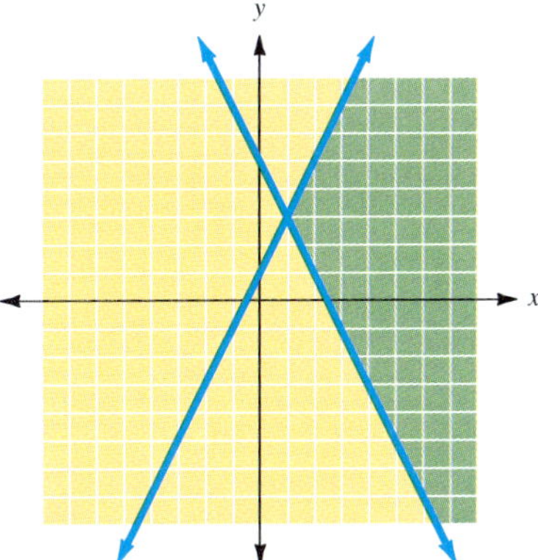

The darkened region represents the overlap of the two regions. That is, points in this region are solutions for both inequalities. Hence, points in this darkened region are solutions for the system. Points on the border of this region are also solutions because both inequalities include "or equal to."

(b) $y < -\frac{1}{2}x - 2$

$y \le \frac{1}{2}x + 1$

As in part (a), we sketch both inequalities. The overlap region (darkened) represents solutions to the system.

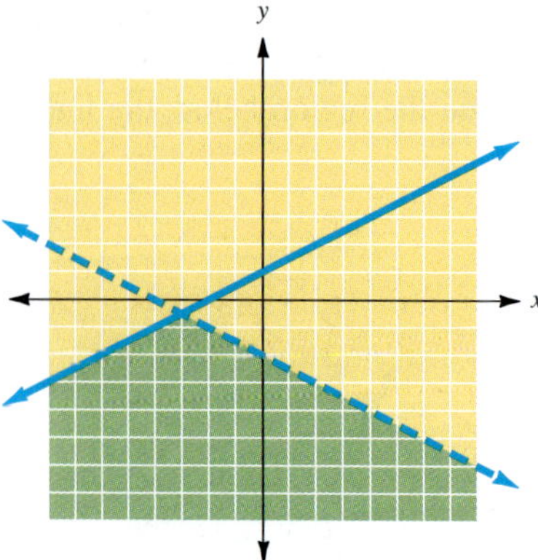

Only the solid line portion of the border of the darkened region is included in the solution set. Points on the dashed line do not represent solutions to the first (strict) inequality; therefore, these points are not solutions for the system of inequalities.

R.4 exercises

Solve each system graphically.

1. $x + y = 6$
$x - y = 4$

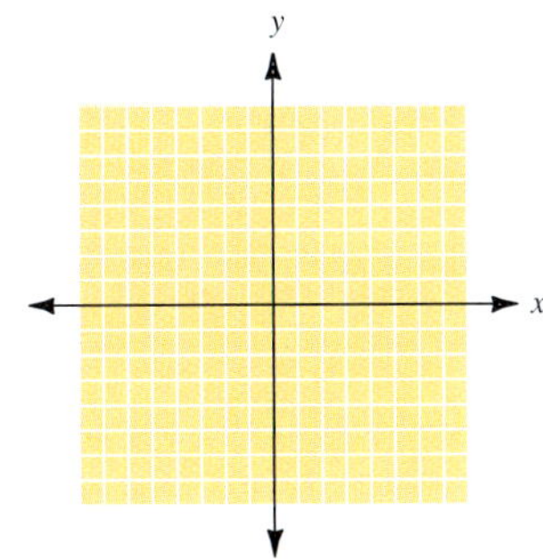

2. $3x + 2y = 12$
$y = 3$

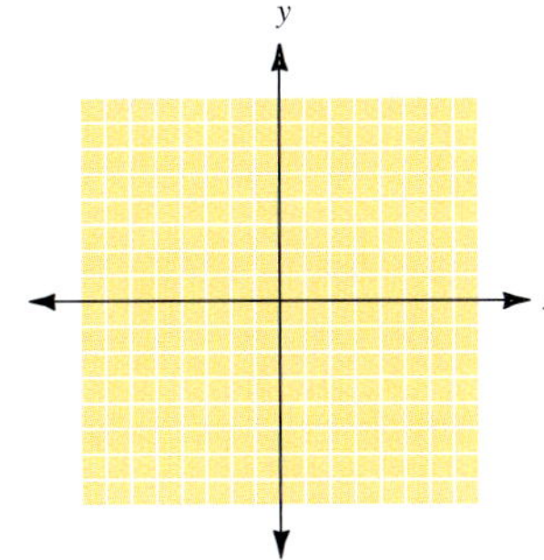

3. $3x - y = 3$
$3x - y = 6$

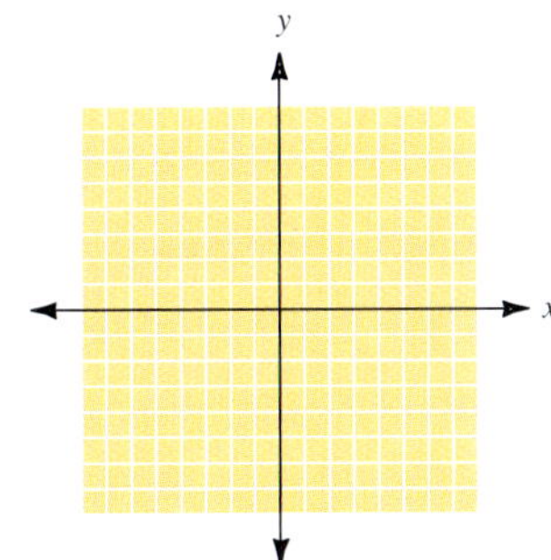

4. $-2x + 6y = -10$
$x - 3y = 5$

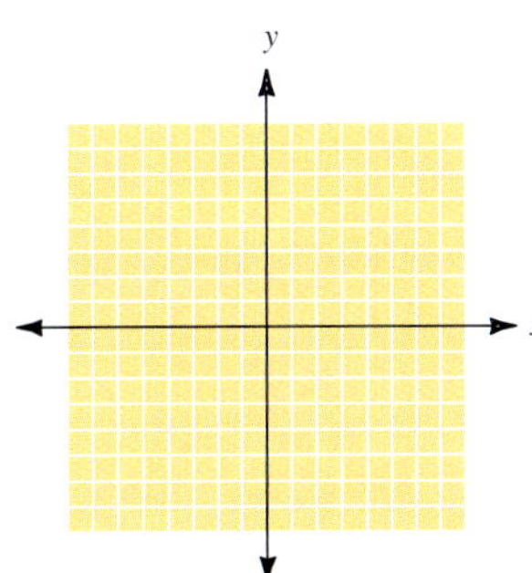

Use the substitution method to solve each system.

5. $x - y = 7$
$y = 2x - 12$

6. $x - y = 4$
$x = 2y - 2$

7. $4x + 3y = -11$
$5x + y = -11$

8. $8x - 4y = 16$
$-2x + y = -4$

Use the addition method to solve each system.

9. $-2x + 3y = 5$
$2x - y = 1$

10. $2x - y = 4$
$6x - 3y = 10$

11. $2x + 3y = 21$
$x - y = -2$

12. $5x + 4y = 5$
$7x - 6y = 36$

Name ______________________

Section ________ Date ________

Answers

1. ______________________
2. ______________________
3. ______________________
4. ______________________
5. ______________________
6. ______________________
7. ______________________
8. ______________________
9. ______________________
10. ______________________
11. ______________________
12. ______________________

Answers

13. ______

14. ______

15. ______

16. ______

17. ______

18. ______

19. ______

20. ______

13.
$$\begin{aligned} x - y + z &= 0 \\ x + 4y - z &= 14 \\ x + y - z &= 6 \end{aligned}$$

14.
$$\begin{aligned} x - y + z &= 3 \\ 3x + y + 2z &= 15 \\ 2x - y + 2z &= 7 \end{aligned}$$

15.
$$\begin{aligned} x - y - z &= 2 \\ -2x + 2y + z &= -5 \\ -3x + 3y + z &= -10 \end{aligned}$$

16.
$$\begin{aligned} x - y \quad &= 3 \\ 2y + z &= 5 \\ x \quad + 2z &= 7 \end{aligned}$$

Sketch a graph showing the solution set for each system of inequalities.

17. $y \geq 3x - 1$

$y \geq \frac{3}{4}x + 1$

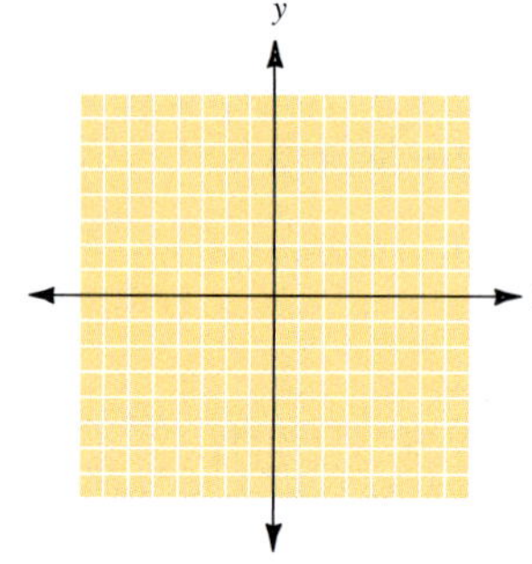

18. $y \leq 1 - x$

$y > \frac{x}{2} - 2$

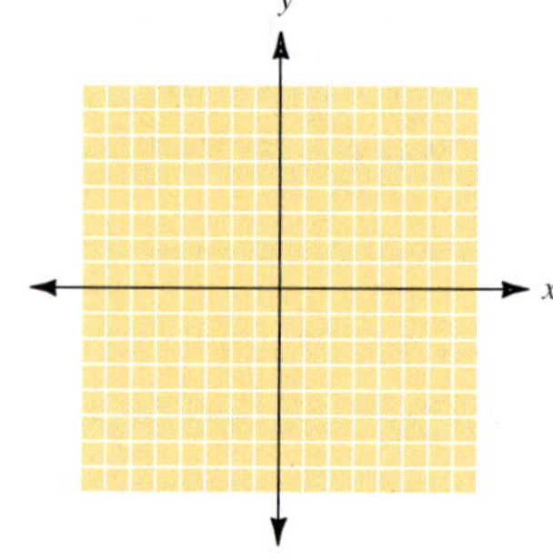

19. $y < x$

$y > x - 3$

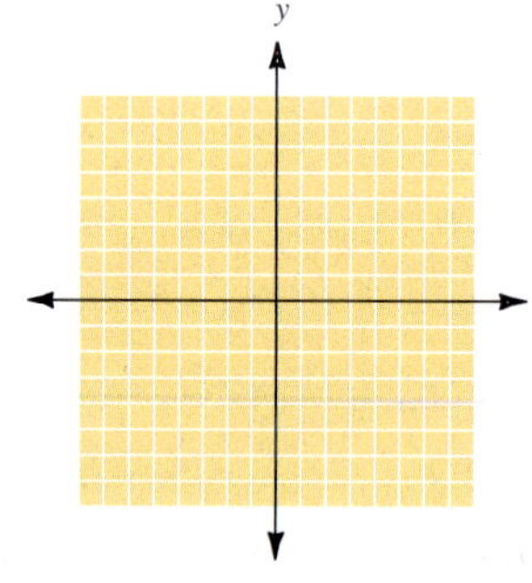

20. $y \geq \frac{7 - 3x}{2}$

$y < \frac{-3}{2}x$

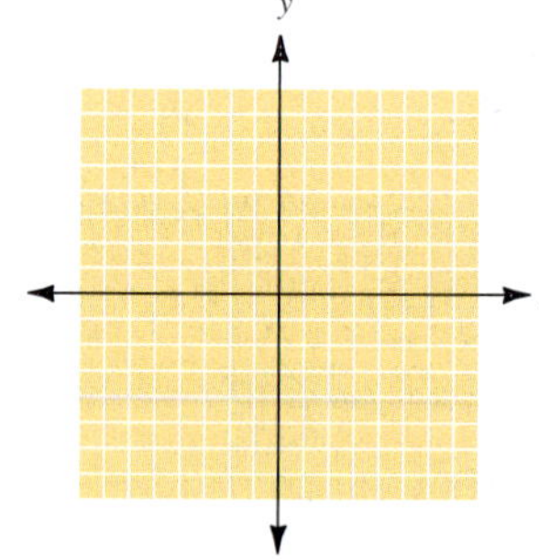

R.5 Exponents and Polynomials

Chapter 5 looked at the most common algebraic expressions, polynomials. To develop the definition of a polynomial, we first discussed exponents. Exponents are a shorthand for repeated multiplication. Instead of writing

$a \cdot a \cdot a \cdot a \cdot a \cdot a \cdot a$ we write a^7

which we read "a to the seventh power."

An expression of this type is said to be in *exponential form.* We call a the *base* of the expression and 7 the *exponent* or *power.*

Property

Product Rule for Exponents

$a^m \cdot a^n = a^{m+n}$

In words, when we multiply two expressions that have the same base, the resulting expression has the same base raised to the sum of the two exponents.

Example 1 **Using the Product Rule**

Simplify each product, first as a base to a single power and then, if possible, as a number.

(a) $a^5 \cdot a = a^6$ **(b)** $2^3 \cdot 2^5 = 2^8 = 256$

We use the *quotient rule for exponents* in Example 2.

Property

Quotient Rule for Exponents

$$\frac{a^m}{a^n} = a^{m-n}$$

In words, when we divide two expressions that have the same base, the resulting expression has the same base raised to the difference of the two exponents.

Example 2 **Using the Quotient Rule**

Simplify each quotient, first as a base to a single power and then, if possible, as a number.

(a) $\dfrac{a^9}{a^3} = a^6$ **(b)** $\dfrac{2^{12}}{2^3} = 2^9 = 512$

What does the quotient rule yield when m is equal to n?

$$\frac{a^m}{a^m} = a^{m-m} = a^0$$

But we know that

$$\frac{a^m}{a^m} = 1$$

This implies that $a^0 = 1$. In fact, any base (other than zero) raised to the 0 power equals 1.

Definition

The Zero Exponent

For any nonzero real number,

$a^0 = 1$

What if we allow one of the exponents to be negative and apply the product rule? Suppose, for instance, that $m = 3$ and $n = -3$. Then

$$a^m \cdot a^n = a^3 \cdot a^{-3} = a^{3+(-3)} = a^0 = 1$$

so

$$a^3 \cdot a^{-3} = 1$$

Dividing both sides by a^3, we get

$$a^{-3} = \frac{1}{a^3}$$

This leads to a definition.

Definition

Negative Integer Exponents

For any nonzero real number a and whole number n,

$$a^{-n} = \frac{1}{a^n}$$

Let's look at some examples.

Example 3 Working with Integer Exponents

Simplify each expression. Answers should not include negative exponents.

(a) $y^{-3} = \frac{1}{y^3}$

(b) $2^{-5} = \frac{1}{2^5} = \frac{1}{32}$ (or, in decimal form, 0.03125)

(c) $2(ab)^0 = 2(1) = 2$

When working with exponents, you must also keep three other properties in mind.

Property

Power Rule

For any nonzero real number a and integers m and n,

$(a^m)^n = a^{mn}$

Property

Product-Power Rule

For any nonzero real numbers a and b and integer n,

$(ab)^n = a^n b^n$

Property

Quotient-Power Rule For any nonzero real numbers a and b and integer n,

$$\left(\frac{a}{b}\right)^n = \frac{a^n}{b^n}$$

Example 4 Using the Properties of Exponents

Simplify each expression.

(a) $(a^3)^5 = a^{15}$

(b) $(2^4)^5 = 2^{20} = 1{,}048{,}576$

(c) $(ab)^5 = a^5b^5$

(d) $(2a)^5 = 2^5a^5 = 32a^5$

(e) $\left(\frac{2}{a}\right)^5 = \frac{2^5}{a^5} = \frac{32}{a^5}$

Scientific notation is a useful way of expressing very large or very small numbers using powers of 10. Any number written in the form

$a \times 10^n$

in which $1 \le a < 10$ and n is an integer, is said to be written in scientific notation.

Example 5 Using Scientific Notation

Write each number in scientific notation.

(a) $903{,}000{,}000{,}000. = 9.03 \times 10^{11}$

11 places

(b) $0.000892 = 8.92 \times 10^{-4}$

4 places

A *term* can be written as a number, or the product of a number and one or more variables and their exponents. A polynomial consists of one or more terms in which the only allowable exponents are whole numbers. In each term of a polynomial, the number factor is called the *coefficient.*

A polynomial with exactly one term is called a *monomial.* A polynomial with exactly two terms is called a *binomial.* A polynomial with exactly three terms is called a *trinomial.*

Polynomials are also classified by their *degree.* The *degree* of a polynomial that has only one variable is the highest power of that variable appearing in any one term.

The value of a polynomial depends on the value given to its variable(s).

Example 6 Identifying and Evaluating Polynomials

Classify each polynomial and give its degree. Evaluate it for the given value of the variable.

(a) $3x^2 + 5x - 4$ where $x = -2$

This is a trinomial. Its degree is 2. At $x = -2$, it has a value of

$3(-2)^2 + 5(-2) - 4$

$= 12 - 10 - 4$

$= -2$

(b) $-5x^4 + 7x$ where $x = 3$

This is a binomial. Its degree is 4. At $x = 3$, it has a value of

$-5(3)^4 + 7(3)$

$= -5(81) + 21$

$= -405 + 21 = -384$

Adding polynomials is simply a matter of adding like terms.

Example 7 Adding Polynomials

Add the two polynomials.

$(4a^2 + 3a - 5) + (3a^2 - 5a + 3)$

To add two polynomials, remove the parentheses and combine like terms.

$4a^2 + 3a - 5 + 3a^2 - 5a + 3$

$= 7a^2 - 2a - 2$

When subtracting a polynomial, take care to distribute the subtraction sign to every term in the second polynomial.

Example 8 Subtracting Polynomials

Subtract the two polynomials.

$(5a^2 + 6a - 5) - (3a^2 - 3a + 1)$

Note that distributing the subtraction sign changes addition to subtraction and subtraction to addition.

$= 5a^2 + 6a - 5 - 3a^2 + 3a - 1$

$= 2a^2 + 9a - 6$

When multiplying a polynomial by a monomial, distribute the monomial to each term of the polynomial and simplify the result.

Example 9 Multiplying a Polynomial by a Monomial

Multiply $2x^2 + 3x - 1$ by $2x^2$.

$$2x^2(2x^2 + 3x - 1) = 2x^2(2x^2) + 2x^2(3x) - 2x^2(1)$$
$$= 4x^4 + 6x^3 - 2x^2$$

We can use the FOIL method to multiply two binomials.

Step by Step

To Multiply Two Binomials

Step 1 Multiply the first terms of the binomials (F).

Step 2 Multiply the first term of the first binomial by the second term of the second binomial (O).

Step 3 Multiply the second term of the first binomial by the first term of the second binomial (I).

Step 4 Multiply the second terms of the binomials (L).

Step 5 Form the sum of the four terms found above, combining any like terms.

Example 10 Multiplying Two Binomials

Multiply $3x - 1$ by $2x + 3$.

$(3x - 1)(2x + 3)$

First: $(3x)(2x) = 6x^2$
Outer: $(3x)(3) = 9x$
Inner: $(-1)(2x) = -2x$
Last: $(-1)(3) = -3$

$= \underset{F}{6x^2} + \underset{O}{9x} - \underset{I}{2x} - \underset{L}{3}$ Now combine like terms.

$= 6x^2 + 7x - 3$

Step by Step

To Square a Binomial

Step 1 Find the first term of the square by squaring the first term of the binomial.

Step 2 Find the middle term of the square as twice the product of the two terms of the binomial.

Step 3 Find the last term of the square by squaring the last term of the binomial.

In symbols:

$(a + b)^2 = a^2 + 2ab + b^2$

Example 11 Squaring a Binomial

Square each binomial.

(a) $(x + 5)^2$

Step 1 Square x, which gives us: x^2

Step 2 Twice the product of the terms: $2 \cdot 5 \cdot x = 10x$

Step 3 Square the final term: $5^2 = 25$

We have

$(x + 5)^2 = x^2 + 10x + 25$

(b) $(2x - 3y)^2$

Step 1 Square $2x$, which gives us: $4x^2$

Step 2 Twice the product of the terms: $2 \cdot 2x \cdot (-3y) = -12xy$

Step 3 Square the final term: $(-3y)^2 = 9y^2$

We have

$(2x - 3y)^2 = 4x^2 - 12xy + 9y^2$

The product of two binomials that differ only in the sign between the terms is the square of the first term minus the square of the second term.

$(a - b)(a + b) = a^2 - b^2$

This pattern emerges because when we combine like terms, the sum of the inner and outer products is 0.

Example 12 Special Patterns in Multiplication

Multiply.

$(2x + 3y)(2x - 3y)$

$= (2x)^2 - (3y)^2$

$= 4x^2 - 9y^2$

To divide a monomial by a monomial, divide the coefficients and use the quotient rule for exponents to combine the variables.

To divide a polynomial by a monomial, divide each term of the polynomial by the monomial.

Example 13 Dividing Polynomials

Divide.

(a) $\dfrac{36x^5y^2}{9x^2y} = \left(\dfrac{36}{9}\right)x^{5-2}\,y^{2-1} = 4x^3y$

(b) $\dfrac{24x^4 - 4x^3 + 8x^2}{4x^2} = \dfrac{24x^4}{4x^2} - \dfrac{4x^3}{4x^2} + \dfrac{8x^2}{4x^2}$

$= 6x^2 - x + 2$

We can use long division to find the quotient when one polynomial is divided by another.

Example 14 Dividing a Polynomial by a Binomial

Divide $2x^2 + x - 6$ by $x + 1$.

$$\begin{array}{r} 2x \phantom{{}+x-6} \\ x+1\overline{)2x^2 + x - 6} \\ \underline{2x^2 + 2x} \phantom{{}-6} \end{array}$$

Start by dividing the lead term of the binomial (x) into the lead term of the trinomial ($2x^2$) and multiply the result ($2x$) by the binomial ($x + 1$).

$$\begin{array}{r} 2x - 1 \\ x+1\overline{)2x^2 + x - 6} \\ \underline{2x^2 + 2x} \phantom{{}-6} \\ -x - 6 \\ \underline{-x - 1} \\ -5 \end{array}$$

Subtract the product ($2x^2 + 2x$) to get $-x - 6$. Then divide again by $x + 1$.

The remainder is -5.

RECALL

We can also write the quotient and remainder as

$2x - 1 + \dfrac{-5}{x+1}$

Rewriting the result, we have

$(2x^2 + x - 6) \div (x + 1) = 2x - 1 - \dfrac{5}{x + 1}$

R.5 exercises

Boost *your* GRADE at ALEKS.com!

ALEKS

- Practice Problems
- Self-Tests
- NetTutor
- e-Professors
- Videos

Name ______

Section ______ Date ______

Simplify each expression.

1. x^8x^{10}

2. $\frac{x^{-2}x^{-6}}{x^7}$

3. $\frac{x^2y^3}{x^3y}$

4. $(3xy^2)^3$

5. $\frac{35x^{-4}y^{-5}z^3}{49x^{-6}y^{-8}z^4}$

6. $\left(\frac{x^{-1}y^{-2}}{y^{-3}}\right)^{-5}$

7. $(2x^{-2}y^3)^{-3}(-2x^{-3}y^{-2})^2$

8. $(3x^2y^{-5})^0(2x^{-3}y^{-2})^{-2}$

9. $(x^2y^3)^3(2xy^{-3})^2$

10. $(-2^0x^4)^{-1}(3x^{-3})^{-2}$

Write each number in scientific notation.

11. 0.00507

12. 80,630,000,000

Write each number in decimal notation.

13. 4.28×10^7

14. 5.6×10^{-4}

Classify each polynomial and then evaluate it for the given value of the variable.

15. $-3x^4$ where $x = -1$

16. $4x^2 - 7x + 9$ where $x = -3$

17. $-2x^2 + 7x$ where $x = 2$

18. $4x^3 - 3x - 5$ where $x = 3$

Perform the indicated operations.

19. $(5x^2 + 3) + (x^2 - 6)$

20. $(-3w^2 + 15w) + (-4 + w^2) + (-11w + 5)$

21. $(14n^2 - 13n - 8) + (-19n + 7n^2 - 3) + (6n^2 + 5)$

22. $(23r^2 - 6r + 5) - (-9r^2 + 7r - 8)$

Answers

1. ______ 2. ______
3. ______ 4. ______
5. ______ 6. ______
7. ______ 8. ______
9. ______ 10. ______
11. ______
12. ______
13. ______
14. ______
15. ______
16. ______
17. ______
18. ______
19. ______
20. ______
21. ______
22. ______

Answers

23. ______
24. ______
25. ______
26. ______
27. ______
28. ______
29. ______
30. ______
31. ______
32. ______
33. ______
34. ______
35. ______
36. ______
37. ______
38. ______
39. ______
40. ______
41. ______
42. ______
43. ______
44. ______

23. $(-6x^2 - 2x - 32) - (9x^2 + 13x - 19)$

24. $(2p^2 + 4p - 8) - (7p^2 - 8p + 5) + (14p^2 + 16p - 3)$

25. $-12y(-7y^2 - 9y)$

26. $2x(3x^2 + 4x - 3)$

27. $(7x + 2)(3x + 2)$

28. $(5t + 4)(7t - 1)$

29. $(3r - 2t)(4r - t)$

30. $(2m - 5)(2m + 5)$

31. $(3x - 4y)(3x + 4y)$

32. $(x + 8)^2$

33. $(x - 3)^2$

34. $(6x - 2y)^2$

35. $3x(x + 2)(x - 4)$

36. $-2y(3y - 1)(2y + 3)$

37. $x(5x - 2y)(5x + 2y)$

38. $-4x(5 - 3x) + 6x(3 - x)$

39. $(-2x^2)^3(x + 3)(2x - 1)$

40. $\dfrac{18x^2y^2 - 6xy^3}{2xy}$

41. $\dfrac{15n^4 - 5n^3 + 20n^2}{5n^2}$

42. $\dfrac{x^2 + 3x + 2}{x + 1}$

43. $\dfrac{y^2 - 5y - 6}{y + 2}$

44. $\dfrac{x^2 - 13x - 14}{x - 1}$

Answers

1. x^{18} **3.** $\dfrac{y^2}{x}$ **5.** $\dfrac{5x^2y^3}{7z}$ **7.** $\dfrac{1}{2y^{13}}$ **9.** $4x^8y^3$ **11.** 5.07×10^{-3} **13.** 42,800,000 **15.** Monomial, -3 **17.** Binomial, 6 **19.** $6x^2 - 3$ **21.** $27n^2 - 32n - 6$ **23.** $-15x^2 - 15x - 13$ **25.** $84y^3 + 108y^2$ **27.** $21x^2 + 20x + 4$ **29.** $12r^2 - 11rt + 2t^2$ **31.** $9x^2 - 16y^2$ **33.** $x^2 - 6x + 9$ **35.** $3x^3 - 6x^2 - 24x$ **37.** $25x^3 - 4xy^2$ **39.** $-16x^8 - 40x^7 + 24x^6$ **41.** $3n^2 - n + 4$ **43.** $y - 7 + \dfrac{8}{y + 2}$

final exam chapters 0-5

Evaluate each expression.

1. $4 - 5 + 3 \cdot 2^2 - 9$

2. $16 \div 8 + 4 \cdot 2$

3. $\frac{7}{8} + \frac{5}{12} - \frac{2}{3}$

4. $\frac{3}{5} \cdot \frac{6}{7} \div \frac{9}{10}$

Evaluate each expressions if $x = 3$ and $y = -5$.

5. $-12xy$

6. $x^5 + y^3$

7. $(x + y)(x - y)$

8. $3x^2y - 2xy^2$

9. Plot the elements of the set $\{x \mid -2 \le x \le 5\}$.

10. Use set-builder notation to describe this set.

Simplify each expression.

11. $7x^2y + 3x - 5x^2y + 2xy$

12. $(5x^2 + 4x - 3) - (4x^2 - 5x - 3)$

Solve each equation and check your result.

13. $7a - 3 = 6a + 8$

14. $\frac{2}{3}x = -22$

15. $12 - 5x = 4$

16. $5x - 2(x - 3) = 9$

17. $x + 4.8 = 1.2x + 1.1$

18. $5 + 2(x - 8) = 5x + 7$

19. $\frac{x + 2}{3} - \frac{2x + 5}{4} = 2$

Solve each inequality.

20. $2x - 7 < 5 + 4x$

21. $-8 + 3x < 5x + 4$

22. $-3 < 2x + 5 < 9$

23. $2x + 1 < -5$ or $2x + 1 > 5$

Combine like terms.

24. $2a - 6b - 5a + b$

25. $5mn^2 - 2m^2n + 3mn^2 - 7m^2n$

Name ______________________

Section __________ Date __________

Answers

1. ________ 2. ________

3. ________ 4. ________

5. ________ 6. ________

7. ________ 8. ________

9. ________

10. ________

11. ________

12. ________

13. ________

14. ________

15. ________

16. ________

17. ________

18. ________

19. ________

20. ________

21. ________

22. ________

23. ________

24. ________

25. ________

Answers

26. ______
27. ______
28. ______
29. ______
30. ______
31. ______
32. ______
33. ______
34. ______
35. ______
36. ______
37. ______
38. ______
39. ______
40. ______
41. ______
42. ______

Graph each function.

26. $f(x) = 3x$

27. $f(x) = \frac{4}{5}x + 4$

28. $f(x) = -3x - 3$

29. $f(x) = 4x + \frac{9}{2}$

30. Find the slope of the line through the points $(3, -4)$ and $(-1, 4)$.

31. Find the slope and y-intercept of the line represented by the equation $4x + y = 9$.

Find an equation of the line L that satisfies the given set of conditions.

32. L passes through $(-2, 5)$ and is parallel to the line with equation $y + 3x = 5$.

33. L passes through the points $(2, 3)$ and $(4, 7)$.

In exercises 34 to 37, determine whether each relation represents a function.

34. $\{(1, 1), (2, 1), (3, 1), (4, 1)\}$

35. $\{(1, 1), (2, 2), (1, -1)\}$

36.

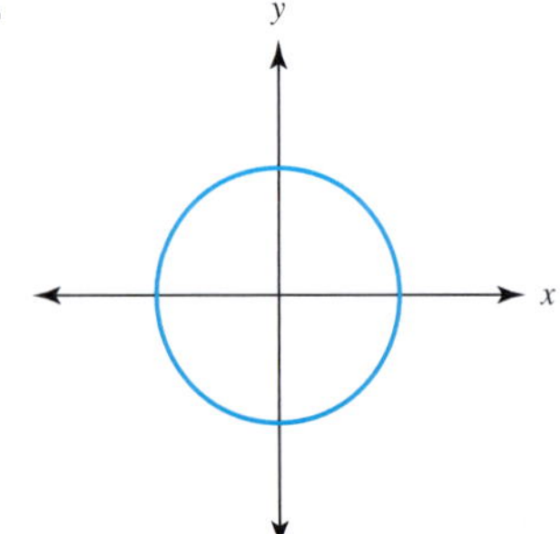

37.

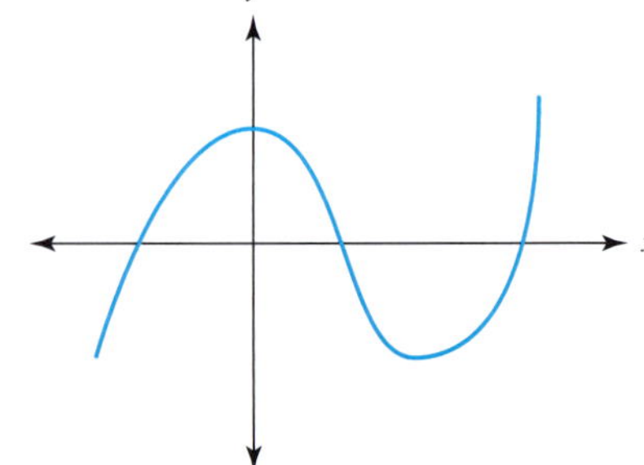

38. A company that makes and sells watches finds that its fixed costs are \$7,200 per week and its marginal cost is \$24 per watch. If the company sells the watches for \$69 each, find the number of watches that must be made and sold each week to break even.

39. The sum of three consecutive odd integers is 117. Find the three integers.

40. Solve for h: $S = 2\pi r^2 + 2\pi rh$

41. Simplify: $\frac{(2m^2n^3)^3}{4m^3n}$

42. Write in simplest form, using positive exponents: $\frac{x^4y^{-2}}{x^{-3}y^4}$

Determine which ordered pairs are solutions for the equation.

43. $3x + y = 6$ $(2, 0), (-2, 6), (3, -3), (-3, 15)$

Multiply or divide as indicated.

44. $(a - 3b)(a + 3b)$

45. $(x - 2y)^2$

46. $(x - 2)(x + 5)$

47. $(a - 3)(a + 4)$

48. $(9x^2 + 12x + 4) \div (3x + 2)$

49. $(3x^2 - 2) \div (x - 1)$

Solve each system by the subsitution method. If a unique solution does not exist, state whether the system is inconsistent or dependent.

50. $y = 2x + 11$
$x - 3y = -13$

51. $x = 7 - 2y$
$4y = 14 - 2x$

52. $y = -5x$
$6x - 3y = -7$

Solve each system by the addition method. If a unique solution does not exist, state whether the system is inconsistent or dependent.

53. $2x - 3y = 16$
$x + 3y = -1$

54. $4x + 3y = -2$
$-8x + 6y = 8$

55. $9x - 6y = 14$
$-6x + 4y = 7$

Solve each system. If a unique solution does not exist, state whether the system is inconsistent or dependent.

56. $3x - 2y - z = 9$
$-2x + y + z = -3$
$x - 3y + z = 19$

57. $4x - 3y = 3$
$2x - z = -2$
$2x + 6y + z = 2$

58. $x + 5y - 2z = 3$
$5x + 9y - 8z = 4$
$-3x + y + 4z = 2$

Answers

43. ____________________

44. ____________________

45. ____________________

46. ____________________

47. ____________________

48. ____________________

49. ____________________

50. ____________________

51. ____________________

52. ____________________

53. ____________________

54. ____________________

55. ____________________

56. ____________________

57. ____________________

58. ____________________

Answers

59. ______

60. ______

61. ______

62. ______

63. ______

64. ______

65. ______

66. ______

67. ______

Solve each system.

59. $y < \frac{1}{2}x - 1$

$x + y > -4$

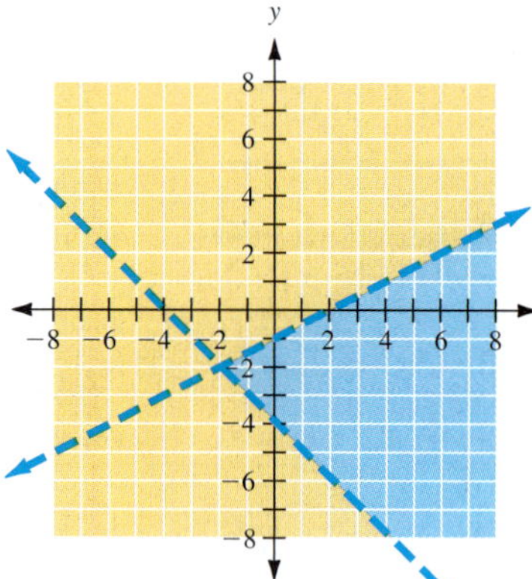

60. $2x + y \leq 4$

$x \geq 1$

$y \geq -2$

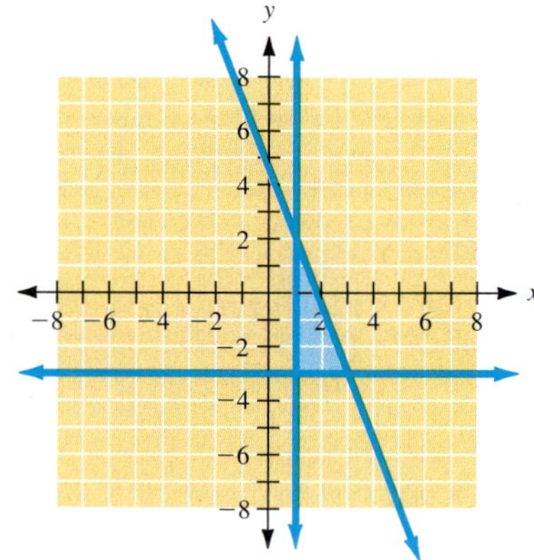

Solve each application.

61. The perimeter of a rectangle is 124 in. If the length of the rectangle is 1 in. more than its width, what are the dimensions of the rectangle?

62. Juan's biology text cost \$5 more than his mathematics text. Together they cost \$141. Find the cost of each text.

63. Boats can be rented for \$34.50 for the first 3 h and \$4.50 for each additional hour. How much will it cost to rent a boat for 5 h?

64. Eileen has been asked to prepare 900 mL of 29% acid solution. She finds containers marked 15% acid solution and 60% acid solution. How much of each should she use?

65. One car rental agency charges \$32 per day and 28¢ per mile. A second agency charges \$39 per day and 23¢ per mile. For a 6-day rental, at what number of miles will the charges from the two agencies be the same?

66. In a triangle, the sum of the smallest angle and the largest angle is twice the measure of the other angle. The largest angle is 24° less than twice the smallest. Find the measures of the three angles.

67. Tomas invested \$9,000 in three accounts. Part of the money is in a bond paying 11% interest, part is in a time deposit paying 7.5%, and the rest is in a savings account paying 5.5%. He invested \$1,400 more in the time deposit than in the other two accounts combined. The interest for 1 year from all the accounts was \$780.50. How much did Tomas invest in each account?

Factoring Polynomials

INTRODUCTION

Developing security codes and software is big business. Corporations all over the world sell encryption systems designed to keep data secure and safe.

In 1977, three professors from the Massachusetts Institute of Technology developed the RSA encryption system. They offered $100 to anyone who could break their security code, which was based on a number that has *129 digits.* They called the code RSA-129. For the code to be broken, the 129-digit number had to be factored into two prime numbers; that is, two prime numbers had to be found that when multiplied together give the 129-digit number. The three professors predicted that it would take *40 quadrillion* years to find the two numbers.

In April 1994, a research scientist, three computer hobbyists, and more than 600 volunteers from the Internet, using 1,600 computers, found the two numbers after 8 months of work and won the $100.

Software companies are waging a legal battle against the U.S. government because the government does not permit codes for which it does not have the key. The software firms claim that this prohibition costs them about $60 billion in lost sales because companies will not buy an encryption system knowing it can be monitored by the U.S. government.

CHAPTER 6 OUTLINE

6.1 An Introduction to Factoring

< 6.1 Objectives >

1 > Factor out the greatest common factor (GCF)

2 > Factor by grouping

In Chapter 5 you were given factors and asked to find a product. We are now going to reverse this process. You will be given a polynomial and asked to find its factors. This is called **factoring.**

We start with an example from arithmetic. To *multiply* $5 \cdot 7$, you write

$$5 \cdot 7 = 35$$

To *factor* 35, you write

$$35 = 5 \cdot 7$$

Factoring is the *reverse* of multiplication.

Now we look at factoring in algebra. You used the distributive property as

$$a(b + c) = ab + ac$$

For instance,

$$3(x + 5) = 3x + 15$$

NOTE

3 and $x + 5$ are the factors of $3x + 15$.

To use the distributive property in factoring, we apply that property in reverse.

$$ab + ac = a(b + c)$$

The distributive property lets us remove the common monomial factor a from the terms of $ab + ac$. To use this in factoring, the first step is to see whether each term of the polynomial has a common monomial factor. In our earlier example,

$$3x + 15 = 3 \cdot x + 3 \cdot 5$$

Common factor

So, by the distributive property,

$$3x + 15 = 3(x + 5)$$

The original terms are each divided by the greatest common factor to determine the expression in parentheses.

NOTES

Again, factoring is the reverse of multiplication.

Here is a diagram relating multiplication and factorization.

To check this, multiply $3(x + 5)$.

Multiplying

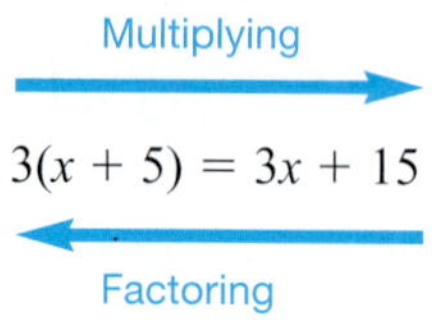

$$3(x + 5) = 3x + 15$$

Factoring

The first step in factoring a polynomial is to identify the *greatest common factor* (GCF) of a set of terms. This is the monomial with the largest common numerical coefficient and the largest power common to both variables.

Definition

Greatest Common Factor (GCF)

The **greatest common factor (GCF)** of a polynomial is the factor (usually monomial) with the highest degree and the largest numerical coefficient that is a factor of each term of the polynomial.

Example 1 Finding the GCF

< Objective 1 >

NOTE

Factoring out the GCF is the *first* step in any factoring problem.

Find the GCF for each list of terms.

(a) 9 and 12

The largest number that is a factor of both is 3.

(b) 10, 25, 150

The GCF is 5.

(c) x^4 and x^7

The largest power common to the two variables is x^4.

(d) $12a^3$ and $18a^2$

The GCF is $6a^2$.

Check Yourself 1

Find the GCF for each list of terms.

(a) 14, 24 **(b)** 9, 27, 81 **(c)** a^9, a^5 **(d)** $10x^5, 35x^4$

Step by Step

To Factor a Monomial from a Polynomial

Step 1 Find the *greatest common factor* for all the terms.
Step 2 Factor the GCF from each term; then apply the distributive property.
Step 3 Mentally check your factoring by multiplication.

Example 2 Factoring the GCF from a Binomial

NOTE

You should always check your result by multiplying to make sure that you get the original polynomial. Try it here. Multiply $4x$ by $2x + 3$.

(a) Factor $8x^2 + 12x$.

The largest common numerical factor of 8 and 12 is 4, and x is the variable factor with the largest common power. So $4x$ is the GCF. Write

$$8x^2 + 12x = 4x \cdot 2x + 4x \cdot 3$$

(GCF: the two factors $4x$)

Now, by the distributive property, we have

$$8x^2 + 12x = 4x(2x + 3)$$

NOTE

It is also true that

$6a^4 - 18a^2 = 3a(2a^3 - 6a)$

However, this is *not completely factored.* Do you see why? You want to find the common monomial factor with the *largest possible* coefficient and the *largest* exponent, in this case $6a^2$.

(b) Factor $6a^4 - 18a^2$.

The GCF in this case is $6a^2$. Write

$$6a^4 - 18a^2 = 6a^2 \cdot a^2 - 6a^2 \cdot 3$$

GCF

Again, using the distributive property yields

$$6a^4 - 18a^2 = 6a^2(a^2 - 3)$$

You should check this by multiplying.

Check Yourself 2

Factor each polynomial.

(a) $5x + 20$ **(b)** $6x^2 - 24x$ **(c)** $10a^3 - 15a^2$

The process is exactly the same for polynomials with more than two terms. Consider Example 3.

Example 3 Factoring the GCF from a Polynomial

NOTE

The GCF is 5.

(a) Factor $5x^2 - 10x + 15$.

$$5x^2 - 10x + 15 = 5 \cdot x^2 - 5 \cdot 2x + 5 \cdot 3$$

GCF

$$= 5(x^2 - 2x + 3)$$

NOTE

The GCF is $3a$.

(b) Factor $6ab + 9ab^2 - 15a^2$.

$$6ab + 9ab^2 - 15a^2 = 3a \cdot 2b + 3a \cdot 3b^2 - 3a \cdot 5a$$

GCF

$$= 3a(2b + 3b^2 - 5a)$$

NOTE

The GCF is $4a^2$.

(c) Factor $4a^4 + 12a^3 - 20a^2$.

$$4a^4 + 12a^3 - 20a^2 = 4a^2 \cdot a^2 + 4a^2 \cdot 3a - 4a^2 \cdot 5$$

GCF

$$= 4a^2(a^2 + 3a - 5)$$

RECALL

In each of these examples, you should check the result by multiplying the factors.

(d) Factor $6a^2b + 9ab^2 + 3ab$.

Mentally note that 3, a, and b are factors of each term, so

$$6a^2b + 9ab^2 + 3ab = 3ab(2a + 3b + 1)$$

Check Yourself 3

Factor each polynomial.

(a) $8b^2 + 16b - 32$
(b) $4xy - 8x^2y + 12x^3$
(c) $7x^4 - 14x^3 + 21x^2$
(d) $5x^2y^2 - 10xy^2 + 15x^2y$

If the leading coefficient of a polynomial is negative, we usually choose to factor out a GCF that has a negative coefficient. We must remember to be careful with the signs of the terms involved. Consider Example 4.

Example 4 Factoring out the GCF With a Negative Coefficient

NOTE

This will be useful in an upcoming section.

In each case, factor out the GCF with a negative coefficient.

(a) $-x^2 - 5x + 7$

Here, we factor out -1:

$$-x^2 - 5x + 7 = (-1)(x^2) + (-1)(5x) + (-1)(-7)$$
$$= -1[x^2 + 5x + (-7)]$$
$$= -1(x^2 + 5x - 7)$$

(b) $-10x^2y + 5xy^2 - 20xy$

We factor out the GCF, $-5xy$:

$$-10x^2y + 5xy^2 - 20xy = (-5xy)(2x) + (-5xy)(-y) + (-5xy)(4)$$
$$= -5xy(2x + (-y) + 4)$$
$$= -5xy(2x - y + 4)$$

Check Yourself 4

In each case, factor out the GCF with a negative coefficient.

(a) $-a^2 + 3a - 9$
(b) $-6a^3b^2 - 3a^2b + 12ab$

In some cases, we can factor a common binomial from an expression.

Example 5 Finding a Common Binomial Factor

(a) Factor $3x(x + y) + 2(x + y)$.

We see that the *binomial* $x + y$ is a common factor and can therefore be factored out.

$$3x(x + y) + 2(x + y)$$
$$= (x + y)(3x + 2)$$

(b) Factor $3x^2(x - y) + 6x(x - y) + 9(x - y)$.

We note that here the GCF is $3(x - y)$. Factoring as before, we have

$$3(x - y)(x^2 + 2x + 3)$$

Check Yourself 5

Completely factor each polynomial.

(a) $7a(a - 2b) + 3(a - 2b)$ **(b)** $4x^2(x + y) - 8x(x + y) - 16(x + y)$

If the terms of a polynomial have no common factor (other than 1), **factoring by grouping** is the preferred method, as illustrated in Example 6.

Example 6 Factoring by Grouping

< Objective 2 >

NOTE

Our example has *four* terms. That is the clue to try the factoring by grouping method.

Suppose we want to factor the polynomial

$ax - ay + bx - by$

As you can see, the polynomial has no common factors. However, look at what happens if we separate the polynomial into *two groups* of *two terms.*

$ax - ay + bx - by$

$= \underbrace{ax - ay} + \underbrace{bx - by}$

Now *each* group has a common factor, and we can write the polynomial as

$a(x - y) + b(x - y)$

In this form, we can see that $x - y$ is the GCF. Factoring out $x - y$, we get

$a(x - y) + b(x - y) = (x - y)(a + b)$

Check Yourself 6

Use the factoring by grouping method.

$x^2 - 2xy + 3x - 6y$

Be particularly careful of your treatment of addition and subtraction operations when you factor by grouping. Consider Example 7.

Example 7 Factoring by Grouping

Factor $2x^3 - 3x^2 - 6x + 9$.

We group the terms of the polynomial as follows:

$\underbrace{2x^3 - 3x^2} - \underbrace{6x + 9}$ Factor out the common factor of -3 from the second two terms.

RECALL

$9 = (-3)(-3)$

$= x^2(2x - 3) - 3(2x - 3)$

$= (2x - 3)(x^2 - 3)$

Check Yourself 7

Factor by grouping.

$3y^3 + 2y^2 - 6y - 4$

It may also be necessary to change the order of the terms as they are grouped. Look at Example 8.

Example 8 Factoring by Grouping

Factor $x^2 - 6yz + 2xy - 3xz$.

Grouping the terms as before, we have

$$\underbrace{x^2 - 6yz} + \underbrace{2xy - 3xz}$$

Do you see that we have accomplished nothing because there are no common factors in the first group?

We can, however, rearrange the terms to write the original polynomial as

$$\underbrace{x^2 + 2xy} - \underbrace{3xz - 6yz}$$

$$= x(x + 2y) - 3z(x + 2y)$$

We can now factor out the common factor of $x + 2y$ in both groups.

$$= (x + 2y)(x - 3z)$$

It is often true that the grouping can be done in more than one way. The factored form is always the same, no matter the method used.

Remember that you can always check your factoring by multiplying. Here we check by multiplying out $(x + 2y)(x - 3z)$:

$$(x + 2y)(x - 3z) = (x)(x) + (x)(-3z) + (2y)(x) + (2y)(-3z)$$
$$= x^2 - 3xz + 2xy - 6yz$$

By rearranging terms, we see that this is equal to the original expression.

Check Yourself 8

We can write the polynomial of Example 8 as

$x^2 - 3xz + 2xy - 6yz$

Factor, and verify that the factored form is the same in either case.

It might happen that a polynomial *cannot* be factored. We call such a polynomial a **prime polynomial.**

Check Yourself ANSWERS

1. **(a)** 2; **(b)** 9; **(c)** a^5; **(d)** $5x^4$ **2.** **(a)** $5(x + 4)$; **(b)** $6x(x - 4)$; **(c)** $5a^2(2a - 3)$
3. **(a)** $8(b^2 + 2b - 4)$; **(b)** $4x(y - 2xy + 3x^2)$; **(c)** $7x^2(x^2 - 2x + 3)$;
(d) $5xy(xy - 2y + 3x)$ **4.** **(a)** $-1(a^2 - 3a + 9)$; **(b)** $-3ab(2a^2b + a - 4)$
5. **(a)** $(a - 2b)(7a + 3)$; **(b)** $4(x + y)(x^2 - 2x - 4)$
6. $(x - 2y)(x + 3)$ **7.** $(3y + 2)(y^2 - 2)$ **8.** $(x - 3z)(x + 2y)$

Reading Your Text

SECTION 6.1

(a) The ______________ property lets us remove the common monomial factor a from the expression $ab + ac$.

(b) The first step in factoring is to identify the greatest ______________ factor.

(c) After factoring, it is always a good idea to check your result by ______________.

(d) If a polynomial has four terms, try the factoring by ______________ method.

Basic Skills | Challenge Yourself | Calculator/Computer | Career Applications | Above and Beyond

6.1 exercises

Boost *your* GRADE at ALEKS.com!

ALEKS

- Practice Problems
- Self-Tests
- NetTutor
- e-Professors
- Videos

Name ______________

Section ________ Date ________

< Objective 1 >

Find the greatest common factor for each list of terms.

1. 20, 22

2. 15, 35

3. 16, 32, 88

4. 44, 66, 143

5. x^2, x^5

6. y^7, y^9

7. a^3, a^6, a^9

8. b^4, b^6, b^8

9. $5x^4, 10x^5$

10. $8y^9, 24y^3$

11. $4a^4, 10a^7, 12a^{14}$

12. $9b^3, 6b^5, 12b^4$

13. $9x^2y, 12xy^2, 15x^2y^2$

14. $12a^3b^2, 18a^2b^3, 6a^4b^4$

15. $15ab^3, 10a^2bc, 25b^2c^3$

16. $12x^3, 15x^2y^2, 21y^5$

17. $15a^2bc^2, 9ab^2c^2, 6a^2b^2c^2$

> Videos

18. $18x^3y^2z^3, 27x^4y^2z^3, 81xy^2z$

19. $(x + y)^2, (x + y)^5$

20. $12(a + b)^4, 4(a + b)^3$

Factor each polynomial.

21. $10x + 5$

22. $5x - 15$

23. $24m - 32n$

24. $7p - 21q$

25. $12m^2 + 8m$

26. $30n^2 - 35n$

27. $10s^2 + 5s$

28. $12y^2 - 6y$

29. $24x^2 - 60x$

30. $14b^2 - 28b$

Answers

1.	2.
3.	4.
5.	6.
7.	8.
9.	10.
11.	12.
13.	14.
15.	16.
17.	18.
19.	20.
21.	22.
23.	24.
25.	26.
27.	28.
29.	30.

Answers

31. ______
32. ______
33. ______
34. ______
35. ______
36. ______
37. ______
38. ______
39. ______
40. ______
41. ______
42. ______
43. ______
44. ______
45. ______
46. ______
47. ______
48. ______
49. ______
50. ______
51. ______
52. ______
53. ______
54. ______
55. ______
56. ______

31. $15a^3 - 25a^2$

32. $36b^4 + 24b^2$

33. $6pq + 18p^2q$

34. $9xy - 18xy^2$

35. $7m^3n - 21mn^3$

36. $36p^2q^2 - 9pq$

37. $6x^2 - 18x + 30$

38. $7a^2 + 21a - 42$

39. $5a^3 - 15a^2 + 25a$

40. $5x^3 - 15x^2 + 25x$

41. $12x + 8xy - 28xy^2$

42. $4s + 6st - 14st^2$

43. $10x^2y + 15xy - 5xy^2$

44. $3ab^2 + 6ab - 15a^2b$

45. $10r^3s^2 + 25r^2s^2 - 15r^2s^3$

46. $28x^2y^3 - 35x^2y^2 + 42x^3y$

47. $9a^5 - 15a^4 + 21a^3 - 27a$

48. $8p^6 - 40p^4 + 24p^3 + 16p^2$

49. $15m^3n^2 - 20m^2n + 35mn^3 - 10mn$

50. $14ab^4 + 21a^2b^3 - 35a^3b^2 + 28ab^2$

51. $x(x - 9) + 5(x - 9)$

52. $y(y + 5) - 3(y + 5)$

53. $p(p - 2q) - q(p - 2q)$

54. $x(3x - 4y) - y(3x - 4y)$

55. $x(y - z) + 3(y - z)$

56. $2a(c - d) - b(c - d)$

Basic Skills | **Challenge Yourself** | Calculator/Computer | Career Applications | Above and Beyond

Factor out the GCF including a negative coefficient.

57. $-t^2 + 6t + 10$

58. $-u^2 - 4u + 9$

59. $-4m^2n^3 - 6mn^3 - 10n^2$

60. $-8a^4b^2 + 4a^2b^3 + 12ab^3$

< Objective 2 >

Factor each polynomial by grouping. (Hint: *You may have to rearrange terms.*)

61. $ab - ac + b^2 - bc$ 

62. $ax + 2a + bx + 2b$

63. $6r^2 + 12rs - r - 2s$

64. $2mn - 4m^2 + 3n - 6m$

65. $ab^2 - 2b^2 + 3a - 6$

66. $r^2s^2 - 3s^2 - 2r^2 + 6$

67. $x^2 + 3x - 4xy - 12y$

68. $a^2 - 12b + 3ab - 4a$

69. $m^2 - 6n^3 + 2mn^2 - 3mn$

70. $r^2 - 3rs^2 - 12s^3 + 4rs$

Determine whether each statement is **true** *or* **false.**

71. Factoring is the reverse of addition.

72. The key property used in factoring out the GCF is the distributive property.

Complete each statement with **never, sometimes,** *or* **always.**

73. If the GCF is factored out of a trinomial, the result is ______ the GCF times a trinomial.

74. If a four-term polynomial has no common factor (other than 1), factoring by grouping is ______ successful.

Determine whether each factoring is correct.

75. $x^2 - x - 6 = (x - 3)(x + 2)$

76. $x^2 - x - 12 = (x - 4)(x + 3)$

77. $x^2 + x - 12 = (x + 6)(x - 2)$

78. $x^2 + 2x - 8 = (x + 8)(x - 1)$

79. $2x^2 - 5x - 3 = (2x + 1)(x - 3)$

80. $6x^2 - 13x + 6 = (3x - 2)(2x - 3)$

Basic Skills | Challenge Yourself | Calculator/Computer | **Career Applications** | Above and Beyond

81. **Allied Health** A patient's protein secretion amount, in milligrams per day, is recorded over several days. Based on these observations, lab technicians determine that the polynomial $-t^3 - 6t^2 + 11t + 66$ provides a good approximation of the patient's protein secretion amount t days after testing begins. Factor this polynomial.

Answers

57. ______

58. ______

59. ______

60. ______

61. ______

62. ______

63. ______

64. ______

65. ______

66. ______

67. ______

68. ______

69. ______

70. ______

71. ______

72. ______

73. ______

74. ______

75. ______

76. ______

77. ______

78. ______

79. ______

80. ______

81. ______

Answers

82. ______

83. ______

84. ______

85. ______

86. ______

87. ______

88. ______

89. ______

90. ______

91. ______

92. ______

93. ______

82. **Allied Health** The concentration, in micrograms per milliliter (μg/mL), of the antibiotic chloramphenicol is given by $8t^2 - 2t^3$, in which t is the number of hours after the drug is taken. Factor this polynomial. > Videos

83. **Manufacturing Technology** Polymer pellets need to be as perfectly round as possible. In order to avoid the formation of flat spots during the hardening process, the pellets are kept off a surface by blasts of air. The height of a pellet above the surface t seconds after a blast is given by $v_0t - 4.9t^2$. Factor this expression.

84. **Information Technology** The total time to transmit a packet is given by the expression $T = d + 2p$, in which d is the quotient of the distance and the propagation velocity, and p is the quotient of the size of the packet and the information transfer rate. How long will it take to transmit a 1,500-byte packet 10 meters on an Ethernet if the information transfer rate is 100 MB per second and the propagation velocity is 2×10^8 m/s?
(*Hint:* Use 1 MB = 10^6 bytes.)

Basic Skills | Challenge Yourself | Calculator/Computer | Career Applications | **Above and Beyond**

85. The GCF of $2x - 6$ is 2. The GCF of $5x + 10$ is 5. Find the greatest common factor of the product $(2x - 6)(5x + 10)$.

86. The GCF of $3z + 12$ is 3. The GCF of $4z + 8$ is 4. Find the GCF of the product $(3z + 12)(4z + 8)$.

87. The GCF of $2x^3 - 4x$ is $2x$. The GCF of $3x + 6$ is 3. Find the GCF of the product $(2x^3 - 4x)(3x + 6)$.

88. State, in a sentence, the rule that exercises 85 to 87 illustrate.

89. For the monomials x^4y^2, x^8y^6, and x^9y^4, explain how you can determine the GCF by inspecting exponents.

90. It is not possible to use the grouping method to factor $2x^3 + 6x^2 + 8x + 4$. Is it correct to conclude that the polynomial is prime? Justify your answer.

91. **Geometry** The area of a rectangle with width t is given by $33t - t^2$. Factor the expression and determine the length of the rectangle in terms of t.

92. **Geometry** The area of a rectangle of length x is given by $3x^2 + 5x$. Find the width of the rectangle.

93. For centuries, mathematicians have found factoring numbers into prime factors a fascinating subject. A prime number is a number that cannot be written as a product of any whole numbers but 1 and itself. The list of primes begins with 2 because 1 is not considered a prime number and then goes on: 3, 5, 7, 11, ... What are the first 10 primes? What are the primes less than 100? If you list the numbers from 1 to 100 and then cross out all numbers that are multiples of 2, 3,

5, and 7, what is left? Are all the numbers not crossed out prime? Write a paragraph to explain why this might be so. You might want to investigate the Sieve of Eratosthenes, a system from 230 B.C.E. for finding prime numbers.

94. If we could make a list of all the prime numbers, what number would be at the end of the list? Because there are an infinite number of prime numbers, there is no "largest prime number." But is there some formula that will give us all the primes? Here are some formulas proposed over the centuries:

$n^2 + n + 17$ $\quad 2n^2 + 29$ $\quad n^2 - n + 11$

In all these expressions, $n = 1, 2, 3, 4, \ldots$, that is, a positive integer beginning with 1. Investigate these expressions with a partner. Do the expressions give prime numbers when they are evaluated for these values of n? Do the expressions give *every* prime in the range of resulting numbers? Can you put in *any* positive number for n?

95. How are primes used in coding messages and for security? Work together to decode the messages. The messages are coded using this code: After the numbers are factored into prime factors, the power of 2 gives the number of the letter in the alphabet. This code would be easy for a code breaker to figure out, but you might make up a code that would be more difficult to break.

(a) 1310720, 229376, 1572864, 1760, 460, 2097152, 336

(b) 786432, 142, 4608, 278528, 1344, 98304, 1835008, 352, 4718592, 5242880

(c) Code a message, using this rule. Exchange your message with a partner to decode it.

96. One concept used in computer encryption involves factoring a large number that is the product of two prime numbers. If the original number is very large, it is extremely difficult and time-consuming to find the prime factors. Try to factor each number below. (These are not considered large!)

(a) 1,739 **(b)** 5,429 **(c)** 19,177 **(d)** 163,747

Answers

94. ________

95. ________

96. ________

Answers

1. 2 **3.** 8 **5.** x^2 **7.** a^3 **9.** $5x^4$ **11.** $2a^4$ **13.** $3xy$ **15.** $5b$ **17.** $3abc^2$ **19.** $(x + y)^2$ **21.** $5(2x + 1)$ **23.** $8(3m - 4n)$ **25.** $4m(3m + 2)$ **27.** $5s(2s + 1)$ **29.** $12x(2x - 5)$ **31.** $5a^2(3a - 5)$ **33.** $6pq(1 + 3p)$ **35.** $7mn(m^2 - 3n^2)$ **37.** $6(x^2 - 3x + 5)$ **39.** $5a(a^2 - 3a + 5)$ **41.** $4x(3 + 2y - 7y^2)$ **43.** $5xy(2x + 3 - y)$ **45.** $5r^2s^2(2r + 5 - 3s)$ **47.** $3a(3a^4 - 5a^3 + 7a^2 - 9)$ **49.** $5mn(3m^2n - 4m + 7n^2 - 2)$ **51.** $(x - 9)(x + 5)$ **53.** $(p - 2q)(p - q)$ **55.** $(y - z)(x + 3)$ **57.** $-1(t^2 - 6t - 10)$ **59.** $-2n^2(2m^2n + 3mn + 5)$ **61.** $(b - c)(a + b)$ **63.** $(r + 2s)(6r - 1)$ **65.** $(a - 2)(b^2 + 3)$ **67.** $(x + 3)(x - 4y)$ **69.** $(m - 3n)(m + 2n^2)$ **71.** False **73.** always **75.** Correct **77.** Incorrect **79.** Correct **81.** $(t + 6)(-t^2 + 11)$ **83.** $t(v_0 - 4.9t)$ **85.** 10 **87.** $6x$ **89.** Above and Beyond **91.** $33 - t$ **93.** Above and Beyond **95.** Above and Beyond

Activity 6 :: ISBNs and the Check Digit

Each activity in this text is designed to either enhance your understanding of the topics of the chapter, provide you with a mathematical extension of those topics, or both. The activities can be undertaken by one student, but they are better suited for a small group project. Occasionally it is only through discussion that different facets of the activity become apparent.

If you look at the back of your textbook, you should see a long number and a bar code. The number is called the International Standard Book Number or ISBN.

The ISBN system was first developed in 1966 by Gordon Foster at Trinity College in Dublin, Ireland. When first developed, ISBNs were 9 digits long, but by 1970, an international agreement extended them to 10 digits.

In 2007, 13-digits became the standard for ISBN numbers. This is the number on the back of your text. Each ISBN has five blocks of numbers. A common form is XXX-X-XX-XXXXXX-X, though it can vary.

- The first block or set of digits is either 978 or 979. This set was added in 2007 to increase the number of ISBNs available for new books.
- The second set of digits represents the language of the book. Zero represents English.
- The third set represents the publisher. This block is usually two or three digits long.
- The fourth set is the book code and is assigned by the publisher. This block is usually five or six digits long.
- The fifth and final block is a one-digit *check digit*.

Consider the ISBN assigned to this text: 978-0-07-338419-1 (*Elementary and Intermediate Algebra, 4/e,* by Baratto, Bergman, and Hutchison). The check digit in this ISBN is the final digit, 1. It ensures that the book has a valid ISBN.

To use the check digit, we use the algorithm that follows.

Step by Step: Validating an ISBN

Step 1 Identify the first 12 digits of the ISBN (omit the check digit).

Step 2 Multiply the first digit by one, the second by 3, the third by 1, the fourth by 3, and continue alternating until each of the first twelve digits has been multiplied.

Step 3 Add all 12 of these products together.

Step 4 Take only the units digit of this sum and subtract it from 10.

Step 5 If the difference found in step 4 is the same as the check digit, then the ISBN is valid.

We can use the ISBN from this text, 978-0-07-338419, to see how this works.

To do so, we multiply the first digit by one, the second by three, the third by one, the fourth by 3, again, and so on. Then we add these products together. We call this a *weighted sum*.

$$9 \cdot 1 + 7 \cdot 3 + 8 \cdot 1 + 0 \cdot 3 + 0 \cdot 1 + 7 \cdot 3 + 3 \cdot 1 + 3 \cdot 3 + 8 \cdot 1 + 4 \cdot 3 + 1 \cdot 1 + 9 \cdot 3$$
$$= 9 + 21 + 8 + 0 + 0 + 21 + 3 + 9 + 8 + 12 + 1 + 27$$
$$= 119$$

The units digit is 9. We subtract this from 10.

$$10 - 9 = 1$$

The last digit in the ISBN 978-0-07-338419-1 is 1. This matches the difference above and so this text has a valid ISBN number.

Determine whether each set of numbers represents a valid ISBN.

1. 978-0-07-038023-6
2. 978-0-07-327374-7
3. 978-0-553-34948-1
4. 978-0-07-000317-3
5. 978-0-14-200066-3

For each valid ISBN, go online and find the book associated with that ISBN.

6.2

Factoring Special Polynomials

< 6.2 Objectives >

1 > Factor the difference of squares

2 > Factor the sum and difference of cubes

3 > Factor a perfect square trinomial

In Section 5.5, we introduced some special products. Recall the formula for the product of a sum and difference of two terms.

$(a + b)(a - b) = a^2 - b^2$

This also means that a binomial of the form $a^2 - b^2$, called the **difference of squares,** has as its factors $a + b$ and $a - b$.

To use this idea for factoring, we can write

Property

Factoring a Difference of Squares $a^2 - b^2 = (a + b)(a - b)$

NOTE

The exponent must be even.

To apply this pattern, look for **perfect squares.** Perfect square numbers are 1, 4, 9, 16, 25, 36, and so on, because $1^2 = 1$, $2^2 = 4$, $3^2 = 9$, $4^2 = 16$, and so on. Since $(x^1)^2 = x^2$, $(x^2)^2 = x^4$, $(x^3)^2 = x^6$, $(x^4)^2 = x^8$, $(x^5)^2 = x^{10}$, and so on, variables that have even exponents are perfect squares.

Example 1 Factoring the Difference of Squares

< Objective 1 >

Factor $x^2 - 16$.

Think $x^2 - 4^2$.

Because $x^2 - 16$ is a difference of squares, we have

$x^2 - 16 = (x + 4)(x - 4)$

NOTE

You could also write $(x - 4)(x + 4)$. The order does not matter because multiplication is commutative.

Check Yourself 1

Factor $m^2 - 49$.

Anytime an expression is a difference of squares, it can be factored.

Example 2 Factoring the Difference of Squares

Factor $4a^2 - 9$.

Think $(2a)^2 - 3^2$.

So $$4a^2 - 9 = (2a)^2 - (3)^2$$
$$= (2a + 3)(2a - 3)$$

Check Yourself 2

Factor $9b^2 - 25$.

The process for factoring a difference of squares does not change when more than one variable is involved.

Example 3 Factoring the Difference of Squares

NOTE

Think $(5a)^2 - (4b^2)^2$.

Factor $25a^2 - 16b^4$.

$$25a^2 - 16b^4 = (5a)^2 - (4b^2)^2 = (5a + 4b^2)(5a - 4b^2)$$

Check Yourself 3

Factor $49c^4 - 9d^2$.

We now consider an example that combines common-term factoring with difference-of-squares factoring. Note that the common factor is always factored out as the *first step*.

Example 4 Removing the GCF First

NOTE

Step 1
Factor out the GCF.
Step 2
Factor the remaining binomial.

Factor $32x^2y - 18y^3$.

Note that $2y$ is a common factor, so

$$32x^2y - 18y^3 = 2y\underbrace{(16x^2 - 9y^2)}_{\text{Difference of squares}}$$
$$= 2y(4x + 3y)(4x - 3y)$$

Check Yourself 4

Factor $50a^3 - 8ab^2$.

You may have to apply the difference of two squares method *more than once* to completely factor a polynomial.

Example 5 Factoring the Difference of Two Squares

er binomial $m^2 + 9n^2$, is a *sum of* ...s, and cannot be ...ed.

Factor $m^4 - 81n^4$.

$$m^4 - 81n^4 = (m^2 + 9n^2)(m^2 - 9n^2)$$

Do you see that we are not done? Since $m^2 - 9n^2$ is still factorable, we can continue to factor as shown.

$$m^4 - 81n^4 = (m^2 + 9n^2)(m + 3n)(m - 3n)$$

Check Yourself 5

Factor $x^4 - 16y^4$.

Two additional factoring patterns are the **sum and difference of cubes.** Unlike the sum of squares, we *can* factor the sum of cubes.

Property

The Sum or Difference of Two Cubes

$$a^3 + b^3 = (a + b)(a^2 - ab + b^2)$$

$$a^3 - b^3 = (a - b)(a^2 + ab + b^2)$$

NOTE

The exponent must be a multiple of 3.

We are now looking for **perfect cubes.** Perfect cube numbers are 1, 8, 27, 64, and so on. Since $(x^1)^3 = x^3$, $(x^2)^3 = x^6$, $(x^3)^3 = x^9$, $(x^4)^3 = x^{12}$, $(x^5)^3 = x^{15}$, and so on, variables that have exponents that are multiples of 3 are perfect cubes.

Example 6 Factoring the Sum or Difference of Two Cubes

< Objective 2 >

NOTE

We are now looking for perfect cubes—the exponents must be multiples of 3 and the coefficients perfect cubes—1, 8, 27, 64, and so on.

(a) Factor $x^3 + 27$.

The first term is the cube of x, and the second is the cube of 3, so we can apply the $a^3 + b^3$ equation. Letting $a = x$ and $b = 3$, we have

$$x^3 + 27 = (x)^3 + (3)^3 = (x + 3)(x^2 - 3x + 9)$$

(b) Factor $8w^3 - 27z^3$.

This is a difference of cubes, so use the $a^3 - b^3$ equation.

$$8w^3 - 27z^3 = (2w)^3 - (3z)^3 = (2w - 3z)[(2w)^2 + (2w)(3z) + (3z)^2]$$

$$= (2w - 3z)(4w^2 + 6wz + 9z^2)$$

(c) Factor $5a^3b - 40b^4$.

RECALL

Looking for a *common factor* should be your first step. Remember to write the GCF as a part of the final factored form.

First note the common factor of $5b$.

$$5a^3b - 40b^4 = 5b(a^3 - 8b^3)$$

The binomial is the difference of cubes, so

$$= 5b[(a)^3 - (2b)^3]$$

$$= 5b(a - 2b)(a^2 + 2ab + 4b^2)$$

Check Yourself 6

Factor completely.

(a) $27x^3 + 8y^3$ **(b)** $3a^4 - 24ab^3$

NOTE

We also have
$(a - b)^2 = a^2 - 2ab + b^2$

In Section 5.5, we presented a pattern for squaring a binomial. The result is always a trinomial. In general,

$$(a + b)^2 = a^2 + 2ab + b^2$$

The expression on the right, $a^2 + 2ab + b^2$, is called a **perfect square trinomial**, and if we see it, we know that it can be factored as $(a + b)(a + b)$, or $(a + b)^2$.

Property

Factoring a Perfect Square Trinomial

$$a^2 + 2ab + b^2 = (a + b)^2$$

and

$$a^2 - 2ab + b^2 = (a - b)^2$$

Example 7 Factoring a Perfect Square Trinomial

< Objective 3 >

Factor $x^2 + 10x + 25$.

To determine that this is a perfect square trinomial, we observe

$x^2 = (x)^2$	The first term must be a perfect square.
$25 = (5)^2$	The third term must be a perfect square.
$10x = 2 \cdot x \cdot 5$	The middle term must be 2 times the product of x and 5.

Since these three conditions are met, we can factor the expression.

$$x^2 + 10x + 25 = (x + 5)^2$$

NOTE

Be sure to expand $(x + 5)^2$ to check your work.

We check this result.

$$(x + 5)^2 = x^2 + 2 \cdot x \cdot 5 + 5^2$$
$$= x^2 + 10x + 25$$

Check Yourself 7

Factor $y^2 + 14y + 49$.

If the middle term of a trinomial is negative, it may still be a perfect square trinomial.

Example 8 Factoring a Perfect Square Trinomial

> CAUTION

In a perfect square trinomial, any constant term must follow a "plus" sign.

Factor $x^2 - 6x + 9$.

The first term is the square of x, and the third term is the square of 3.

Noting that 2 times the product of x and 3 is $6x$, we can factor

$$x^2 - 6x + 9 = (x - 3)^2$$

Check Yourself 8

Factor $t^2 - 16t + 64$

As before, the process does not change when more than one variable is involved.

Example 9 Factoring a Perfect Square Trinomial

Factor $9x^2 + 30xy + 25y^2$.

We note $9x^2 = (3x)^2$

$25y^2 = (5y)^2$

And, noting that 2 times the product of $3x$ and $5y$ is

$2(3x)(5y) = 30xy$

We can factor this as

$9x^2 + 30xy + 25y^2 = (3x + 5y)^2$

NOTE

Check this result by expanding $(3x + 5y)^2$.

Check Yourself 9

Factor $4v^2 - 28vw + 49w^2$.

We conclude with a table summarizing the special product factoring that we have seen.

Factoring Special Polynomials	
Difference of squares	$a^2 - b^2 = (a + b)(a - b)$
Sum of cubes	$a^3 + b^3 = (a + b)(a^2 - ab + b^2)$
Difference of cubes	$a^3 - b^3 = (a - b)(a^2 + ab + b^2)$
Perfect square trinomial	$a^2 + 2ab + b^2 = (a + b)^2$
	$a^2 - 2ab + b^2 = (a - b)^2$

Check Yourself ANSWERS

1. $(m + 7)(m - 7)$ **2.** $(3b + 5)(3b - 5)$ **3.** $(7c^2 + 3d)(7c^2 - 3d)$
4. $2a(5a + 2b)(5a - 2b)$ **5.** $(x^2 + 4y^2)(x + 2y)(x - 2y)$
6. **(a)** $(3x + 2y)(9x^2 - 6xy + 4y^2)$; **(b)** $3a(a - 2b)(a^2 + 2ab + 4b^2)$
7. $(y + 7)^2$ **8.** $(t - 8)^2$ **9.** $(2v - 7w)^2$

Reading Your Text

SECTION 6.2

(a) Numbers such as 1, 4, 9, 16, and 25 are called __________ squares.

(b) Anytime an expression is the __________ of squares, it can be factored.

(c) The __________ factor is always factored out as the first step.

(d) Numbers such as 1, 8, 27, and 64 are called perfect __________.

Basic Skills | Challenge Yourself | Calculator/Computer | Career Applications | Above and Beyond

6.2 exercises

Boost *your* GRADE at ALEKS.com!

ALEKS

- Practice Problems
- Self-Tests
- NetTutor
- e-Professors
- Videos

Name ______

Section ______ Date ______

State whether each binomial is a difference of squares.

1. $25x^2 + 9y^2$

2. $5x^2 - 7y^2$

3. $16a^2 - 25b^2$

4. $9n^2 - 16m^2$

5. $16r^2 + 4$

6. $9p^2 - 54$

7. $16a^2 - 12b^3$

8. $9a^2b^2 - 16c^2d^2$

9. $a^2b^2 - 25$ 

10. $8x^6 - 27y^3$

< Objectives 1 and 2 >

Factor each binomial completely.

11. $m^2 - n^2$

12. $r^2 - 9$

13. $x^2 - 169$

14. $c^2 - d^2$

15. $49 - y^2$

16. $196 - y^2$

17. $9b^2 - 16$

18. $36 - x^2$

19. $16w^2 - 49$

20. $4x^2 - 25$

21. $4s^2 - 9r^2$

22. $64y^2 - x^2$

Answers

1. ______ 2. ______
3. ______ 4. ______
5. ______ 6. ______
7. ______ 8. ______
9. ______ 10. ______
11. ______
12. ______
13. ______
14. ______
15. ______
16. ______
17. ______
18. ______
19. ______
20. ______
21. ______
22. ______

Answers

23. ______
24. ______
25. ______
26. ______
27. ______
28. ______
29. ______
30. ______
31. ______
32. ______
33. ______
34. ______
35. ______
36. ______
37. ______
38. ______
39. ______
40. ______
41. ______
42. ______
43. ______
44. ______
45. ______
46. ______

23. $9w^2 - 49z^2$ 

24. $25x^2 - 81y^2$

25. $49a^2 - 9b^2$

26. $64m^2 - 9n^2$

27. $x^4 - 36$

28. $y^6 - 49$

29. $x^2y^2 - 16$

30. $m^2n^2 - 64$

31. $25 - a^2b^2$

32. $49 - w^2z^2$

33. $r^4 - 4s^2$

34. $p^2 - 9q^4$

35. $81a^2 - 100b^6$

36. $64x^4 - 25y^4$

37. $18x^3 - 2xy^2$ 

38. $50a^2b - 2b^3$

39. $12m^3n - 75mn^3$

40. $63p^4 - 7p^2q^2$

41. $16a^4 - 81b^4$

42. $81x^4 - y^4$

43. $y^3 + 125$

44. $y^3 - 8$

45. $m^3 - 125$

46. $b^3 + 27$

47. $a^3b^3 - 27$

48. $p^3q^3 - 64$

49. $8w^3 + z^3$

50. $c^3 - 27d^3$

51. $r^3 - 64s^3$ > Videos

52. $125x^3 + y^3$

53. $8x^3 - 27y^3$

54. $64m^3 + 27n^3$

55. $3a^3 + 81b^3$

56. $4x^3 - 32y^3$

< Objective 3 >

State whether each trinomial is a perfect square trinomial.

57. $m^2 + 8m + 16$

58. $n^2 - 12n + 36$

59. $x^2 + 14x - 49$

60. $4u^2 + 20uv + 25v^2$

61. $9m^2 - 12mn + 4n^2$

62. $6y^2 + 30y + 25$

Factor each trinomial completely, or write "not factorable."

63. $x^2 - 4x + 4$

64. $u^2 + 18u + 81$

65. $y^2 + 4y + 8$

66. $t^2 - 6t + 36$

67. $16a^2 + 24a + 9$

68. $25a^2 - 20ab + 4b^2$

Basic Skills | **Challenge Yourself** | Calculator/Computer | Career Applications | Above and Beyond

Complete each statement with **never, sometimes,** *or* **always.**

69. A "difference-of-squares" binomial __________ factors.

70. A "sum-of-squares" binomial of degree 2 __________ factors.

71. A "sum-of-cubes" binomial __________ factors.

72. In attempting to factor a binomial, we should __________ factor out a GCF, first, if possible.

Answers

47. ____________________

48. ____________________

49. ____________________

50. ____________________

51. ____________________

52. ____________________

53. ____________________

54. ____________________

55. ____________________

56. ____________________

57. __________ 58. __________

59. __________ 60. __________

61. __________ 62. __________

63. __________ 64. __________

65. ____________________

66. ____________________

67. __________ 68. __________

69. __________ 70. __________

71. __________ 72. __________

Answers

73. ______

74. ______

75. ______

76. ______

77. ______

78. ______

79. ______

80. ______

81. ______

82. ______

83. ______

84. ______

85. ______

86. ______

Factor each expression.

73. $x^2(x + y) - y^2(x + y)$ 

74. $a^2(b - c) - 16b^2(b - c)$

75. $2m^2(m - 2n) - 18n^2(m - 2n)$

76. $3a^3(2a + b) - 27ab^2(2a + b)$

Basic Skills | Challenge Yourself | Calculator/Computer | **Career Applications** | Above and Beyond

77. Manufacturing Technology The difference d in the calculated maximum deflection between two similar cantilevered beams is given by the formula

$$d = \left(\frac{w}{8EI}\right)(l_1^2 - l_2^2)(l_1^2 + l_2^2)$$

Rewrite the formula in its completely factored form.

78. Manufacturing Technology The work W done by a steam turbine is given by the formula

$$W = \frac{1}{2}m(v_1^2 - v_2^2)$$

Factor the right-hand side of this equation.

79. Allied Health A toxic chemical is introduced into a protozoan culture. The number of deaths per hour is given by the polynomial $338 - 2t^2$, in which t is the number of hours after the chemical is introduced. Factor this expression.

80. Allied Health Radiation therapy is one technique used to control cancer. After treatment, the total number of cancerous cells, in thousands, can be estimated by $144 - 4t^2$, in which t is the number of days of treatment. Factor this expression.

Basic Skills | Challenge Yourself | Calculator/Computer | Career Applications | **Above and Beyond**

81. Find the value for k so that $kx^2 - 25$ has the factors $2x + 5$ and $2x - 5$.

82. Find the value for k so that $9m^2 - kn^2$ has the factors $3m + 7n$ and $3m - 7n$.

83. Find the value for k so that $2x^3 - kxy^2$ has the factors $2x$, $x - 3y$, and $x + 3y$.

84. Find the value for k so that $20a^3b - kab^3$ has the factors $5ab$, $2a - 3b$, and $2a + 3b$.

85. Complete this statement: "To factor a number, you. . . ."

86. Complete this statement: "To factor an algebraic expression into prime factors means. . . ."

87. What binomial multiplied by $25x^2 - 15xy + 9y^2$ gives the sum of two cubes? What is the result of the multiplication?

88. What binomial when multiplied by $9x^2 + 6xy + 4y^2$ gives the difference of two cubes? What is the result of the multiplication?

89. What are the characteristics of a perfect cube monomial?

90. Suppose you factored the polynomial $4x^2 - 16$ as

$$4x^2 - 16 = (2x + 4)(2x - 4)$$

Is this completely factored? If not, what is the final form?

Answers

87. ______

88. ______

89. ______

90. ______

Answers

1. No **3.** Yes **5.** No **7.** No **9.** Yes **11.** $(m + n)(m - n)$
13. $(x + 13)(x - 13)$ **15.** $(7 + y)(7 - y)$ **17.** $(3b + 4)(3b - 4)$
19. $(4w + 7)(4w - 7)$ **21.** $(2s + 3r)(2s - 3r)$ **23.** $(3w + 7z)(3w - 7z)$
25. $(7a + 3b)(7a - 3b)$ **27.** $(x^2 + 6)(x^2 - 6)$ **29.** $(xy + 4)(xy - 4)$
31. $(5 + ab)(5 - ab)$ **33.** $(r^2 + 2s)(r^2 - 2s)$ **35.** $(9a + 10b^3)(9a - 10b^3)$
37. $2x(3x + y)(3x - y)$ **39.** $3mn(2m + 5n)(2m - 5n)$
41. $(4a^2 + 9b^2)(2a + 3b)(2a - 3b)$ **43.** $(y + 5)(y^2 - 5y + 25)$
45. $(m - 5)(m^2 + 5m + 25)$ **47.** $(ab - 3)(a^2b^2 + 3ab + 9)$
49. $(2w + z)(4w^2 - 2wz + z^2)$ **51.** $(r - 4s)(r^2 + 4rs + 16s^2)$
53. $(2x - 3y)(4x^2 + 6xy + 9y^2)$ **55.** $3(a + 3b)(a^2 - 3ab + 9b^2)$
57. yes **59.** no **61.** yes **63.** $(x - 2)^2$ **65.** not factorable
67. $(4a + 3)^2$ **69.** always **71.** always **73.** $(x + y)^2(x - y)$
75. $2(m - 2n)(m + 3n)(m - 3n)$ **77.** $d = \left(\frac{w}{8EI}\right)(l_1 + l_2)(l_1 - l_2)(l_1^2 + l_2^2)$
79. $2(13 - t)(13 + t)$ **81.** 4 **83.** 18 **85.** Above and Beyond
87. $5x + 3y$; $125x^3 + 27y^3$ **89.** Above and Beyond

6.3 Factoring Trinomials: Trial and Error

< 6.3 Objectives >

1 > Factor a trinomial of the form $x^2 + bx + c$

2 > Factor a trinomial of the form $ax^2 + bx + c$

3 > Completely factor a trinomial

4 > Factor a trinomial that is quadratic in form

Recall that the product of two binomials may be a trinomial of the form

$$ax^2 + bx + c$$

This suggests that some trinomials may be factored as the product of two binomials. In fact, factoring trinomials in this way is probably the most common type of factoring that you will encounter in algebra. One process for factoring a trinomial into a product of two binomials is called *trial and error*.

As before, we introduce the factoring technique with a multiplication example.

$$(x + 3)(x + 4) = x^2 + 4x + 3x + 12$$
$$= x^2 + 7x + 12$$

Product of first terms, x and x

Sum of inner and outer products, $3x$ and $4x$

Product of last terms, 3 and 4

To reverse the multiplication process, we see that the product of the *first* terms of the binomial factors is the *first* term of the given trinomial, the product of the *last* terms of the binomial factors is the *last* term of the trinomial, and the *middle* term of the trinomial must equal the sum of the *outer* and *inner* products. This leads to some sign patterns in factoring a trinomial.

Property

Factoring Trinomials

		Factoring Sign Pattern
$x^2 + bx + c$	Both signs are positive.	$(x + \quad)(x + \quad)$
$x^2 - bx + c$	The constant is positive, and the x coefficient is negative.	$(x - \quad)(x - \quad)$
$x^2 + bx - c$ or $x^2 - bx - c$	The constant is negative.	$(x + \quad)(x - \quad)$

In the examples, the coefficients and factors involve only integers.

Example 1 Factoring Trinomials of the Form $x^2 + bx + c$

< Objective 1 >

Factor $x^2 + 7x + 10$.

Both signs are positive, so the desired sign pattern is

$$(x + \_\_)(x + \_\_)$$

We want two positive integers whose product is the constant term $c = 10$. Our choices are 1 and 10 or 2 and 5.

Because the coefficient of the middle term is $b = 7$, we need the factor pair whose sum is 7.

NOTE

With practice, you can do much of this work mentally. We show the factors and their sums here, and in later examples, to emphasize the process.

Factors of 10	Sum
1, 10	11
2, 5	7

The correct factorization is

$$x^2 + 7x + 10 = (x + 2)(x + 5)$$

We multiply the factors to check our answer.

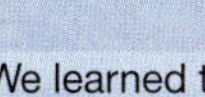

RECALL

We learned to multiply binomials in Section 5.5.

$$(x + 2)(x + 5) = x^2 + 5x + 2x + 10$$
$$= x^2 + 7x + 10 \quad \text{The original polynomial}$$

Check Yourself 1

Factor $x^2 + 8x + 15$.

Example 2 Factoring Trinomials when $a = 1$

Factor $x^2 - 9x + 14$. Do you see that the sign pattern must be as follows?

$$(x - \_\_)(x - \_\_)$$

We then want two factors of 14 whose sum is -9.

Factors of 14	Sum
−1, −14	−15
−2, −7	−9

NOTE

We use two negative factors of 14 since the coefficient of the x-term is negative while the constant is positive.

Because the desired middle term is $-9x$, the correct factors are

$$x^2 - 9x + 14 = (x - 2)(x - 7)$$

We leave it to you to check the answer.

Check Yourself 2

Factor $x^2 - 12x + 32$.

In Example 3, we look at applying our factoring technique to a trinomial whose constant term is negative.

Example 3 Factoring Trinomials when $c < 0$

NOTE

Because the constant is negative, the signs in the binomial factors must be *opposite*.

Factor

$x^2 + 4x - 12$

In this case, the sign pattern is

$(x - \_\_)(x + \_\_)$

Here we want two integers whose product is -12 and whose sum is 4. Again let's look at the possible factors:

Factors of −12	Sum
1, −12	−11
−1, 12	11
3, −4	−1
−3, 4	1
2, −6	−4
−2, 6	4

From the information in the table, we see that the correct factors are

$x^2 + 4x - 12 = (x - 2)(x + 6)$

Check Yourself 3

Factor $x^2 - 7x - 18$.

So far we have considered only trinomials of the form $x^2 + bx + c$. Suppose that the leading coefficient is *not* 1. In general, to factor the trinomial $ax^2 + bx + c$ (with $a \neq 1$), we must consider binomial factors of the form

$(\_\_\ x + \_\_)(\_\_\ x + \_\_)$

where one or both of the coefficients of x in the binomial factors are greater than 1. We look at a multiplication example for some clues to the technique. Consider

$$(2x + 3)(3x + 5) = 6x^2 + 19x + 15$$

Product of $2x$ and $3x$

Sum of outer and inner products, $10x$ and $9x$

Product of 3 and 5

We show one way to reverse the process and factor in Example 4.

Example 4 Factoring Trinomials when $a \neq 1$

< Objective 2 >

To factor $5x^2 + 9x + 4$, we must have the sign pattern

$$(\_x + \_)(\_x + \_)$$

This product must be 4.

This product must be 5.

Factors of 5	Factors of 4
1, 5	1, 4
	4, 1
	2, 2

NOTE

The leading coefficient is no longer 1, so we must be prepared to try both 1, 4 and 4, 1.

Therefore, the possible binomial factors are

$(x + 1)(5x + 4)$

$(x + 4)(5x + 1)$

$(x + 2)(5x + 2)$

Checking the middle term of each product, we see that the proper factorization is

$$5x^2 + 9x + 4 = (x + 1)(5x + 4)$$

Check Yourself 4

Factor $6x^2 - 17x + 7$.

The sign patterns discussed earlier remain the same when the leading coefficient is not 1. Look at Example 5 involving a trinomial with a negative constant.

Example 5 Factoring Trinomials when $a \neq 1$

(a) Factor $6x^2 + 7x - 3$. The sign pattern is

$$(\_x + \_)(\_x - \_)$$

Factors of 6	Factors of −3
1, 6	1, −3
2, 3	−1, 3

There are eight possible binomial factors:

$(x + 1)(6x - 3)$	$(2x + 1)(3x - 3)$
$(x - 1)(6x + 3)$	$(2x - 1)(3x + 3)$
$(x + 3)(6x - 1)$	$(3x + 1)(2x - 3)$
$(x - 3)(6x + 1)$	$(3x - 1)(2x + 3)$

NOTE

Can we simplify this search? One clue: If the trinomial has no common factors (other than 1), then a binomial factor can have no common factor. This means that we do not need to consider $6x - 3$, $6x + 3$, $3x - 3$, and $3x + 3$.

Again, checking the middle terms, we find the correct factors

$$6x^2 + 7x - 3 = (3x - 1)(2x + 3)$$

Factoring certain trinomials in more than one variable involves similar techniques, as illustrated below.

(b) Factor

$$4x^2 - 16xy + 7y^2$$

From the first term of the trinomial, we see that possible first terms for the binomial factors are $4x$ and x or $2x$ and $2x$. The last term of the trinomial tells us that the only choices for the last terms of the binomial factors are y and $7y$. So given the sign of the middle and last terms, the only possible factors are

$$(4x - 7y)(x - y)$$
$$(4x - y)(x - 7y)$$
$$(2x - 7y)(2x - y)$$

From the middle term of the original trinomial we see that $2x - 7y$ and $2x - y$ are the proper factors.

Check Yourself 5

Factor $6a^2 + 11ab - 10b^2$.

Recall that the *first step* in any factoring process is to remove any existing common factors. As before, it may be necessary to combine common-term factoring with other methods (such as factoring a trinomial into a product of binomials) to completely factor a polynomial. Look at Example 6.

Example 6 Factoring Trinomials with a Common Factor

< Objective 3 >

RECALL

Factor out the common factor of 2.

(a) Factor

$$2x^2 - 16x + 30$$

First note the common factor of 2. So we can write

$$2x^2 - 16x + 30 = 2(x^2 - 8x + 15)$$

Now, as the second step, examine the trinomial factor. By the trial-and-error method we know that

$$x^2 - 8x + 15 = (x - 3)(x - 5)$$

and we have

$$2x^2 - 16x + 30 = 2(x - 3)(x - 5)$$

in completely factored form.

(b) Factor

$$6x^3 + 15x^2y - 9xy^2$$

There is a common factor of $3x$ in each term of the trinomial. Factoring out that common factor, we have

$$6x^3 + 15x^2y - 9xy^2 = 3x(2x^2 + 5xy - 3y^2)$$

Again, considering the trinomial factor, we see that $2x^2 + 5xy - 3y^2$ has factors of $2x - y$ and $x + 3y$. And the original trinomial becomes

$$3x(2x - y)(x + 3y)$$

in completely factored form.

Check Yourself 6

Factor.

(a) $9x^2 - 39x + 36$ (b) $24a^3 + 4a^2b - 8ab^2$

Occasionally, we need to factor an expression that is not quadratic, but is **quadratic in form.** Consider the expression $x^4 + 5x^2 + 6$.

Example 7 Factoring an Expression That Is Quadratic in Form

< Objective 4 >

Factor $x^4 + 5x^2 + 6$.

Observing that x^4 is the square of x^2, we factor

$$(x^2 + 2)(x^2 + 3)$$

Multiplying these binomials shows that we factored correctly.

$$(x^2 + 2)(x^2 + 3) = x^4 + 3x^2 + 2x^2 + 6 = x^4 + 5x^2 + 6$$

Since the binomials $x^2 + 2$ and $x^2 + 3$ cannot be further factored, we are done.

$$x^4 + 5x^2 + 6 = (x^2 + 2)(x^2 + 3)$$

Check Yourself 7

Factor $y^4 + 9y^2 + 8$.

When factoring such an expression, it is important to check that it is in fact completely factored.

Example 8 Factoring an Expression That Is Quadratic in Form

Factor $x^4 - 13x^2 + 36$.

Since x^4 is the square of x^2, we begin with

$$(x^2 - \quad)(x^2 - \quad)$$

Looking for two integers whose product is 36 but whose sum is 13, we choose 4 and 9.

$$(x^2 - 4)(x^2 - 9)$$

Now we note that each binomial is a difference of squares.

$$x^4 - 13x^2 + 36 = (x + 2)(x - 2)(x + 3)(x - 3)$$

Check Yourself 8

Factor $t^4 - 17t^2 + 16$.

One final note: When factoring, we require that all coefficients be integers. Given this restriction, not all polynomials are factorable over the integers.

To factor $x^2 - 9x + 12$, we know that the only possible binomial factors (using integers as coefficients) are

$(x - 1)(x - 12)$

$(x - 2)(x - 6)$

$(x - 3)(x - 4)$

You can verify that *none* of these pairs gives the correct middle term of $-9x$. We then say that the original trinomial is not factorable using integers as coefficients.

RECALL

We refer to such polynomials as **prime**.

Check Yourself ANSWERS

1. $(x + 3)(x + 5)$ **2.** $(x - 4)(x - 8)$ **3.** $(x - 9)(x + 2)$
4. $(2x - 1)(3x - 7)$ **5.** $(3a - 2b)(2a + 5b)$
6. **(a)** $3(x - 3)(3x - 4)$; **(b)** $4a(3a + 2b)(2a - b)$
7. $(y^2 + 8)(y^2 + 1)$ **8.** $(t + 1)(t - 1)(t + 4)(t - 4)$

Reading Your Text

SECTION 6.3

(a) Some trinomials may be factored as the product of two ____________.

(b) The product of the first terms of ____________ factors is the first term of the given trinomial.

(c) When the constant in a trinomial is negative, the signs of the binomial factors must be ____________.

(d) The first step in any factoring process is to remove any existing ____________ factors.

Basic Skills | Challenge Yourself | Calculator/Computer | Career Applications | Above and Beyond

6.3 exercises

Boost *your* GRADE at ALEKS.com!

ALEKS

- Practice Problems
- Self-Tests
- NetTutor
- e-Professors
- Videos

Name ____________

Section ______ Date ______

Determine whether each statement is **true** *or* **false.**

1. $x^2 + 2x - 3 = (x + 3)(x - 1)$

2. $y^2 + 3y + 18 = (y - 6)(y + 3)$

3. $x^2 - 10x - 24 = (x - 6)(x + 4)$

4. $a^2 + 9a - 36 = (a - 12)(a + 4)$

5. $x^2 - 16x + 64 = (x - 8)(x - 8)$

6. $w^2 - 12w - 45 = (w + 9)(w - 5)$

7. $25y^2 - 10y + 1 = (5y - 1)(5y + 1)$ 

8. $6x^2 + 5xy - 4y^2 = (6x - 2y)(x + 2y)$

9. $10p^2 - pq - 3q^2 = (5p - 3q)(2p + q)$

10. $6a^2 + 13a + 6 = (2a - 3)(3a - 2)$

For each trinomial, label a, b, and c.

11. $x^2 + 7x - 5$

12. $x^2 + 5x + 11$

13. $x^2 - 3x + 8$

14. $x^2 + 7x - 15$

15. $3x^2 + 5x - 8$

16. $3x^2 - 5x + 7$

17. $4x^2 + 8x + 11$

18. $5x^2 + 7x - 9$

19. $-7x^2 - 3x + 2$

20. $7x^2 + 9x - 18$

Answers

1. ______ 2. ______
3. ______ 4. ______
5. ______ 6. ______
7. ______ 8. ______
9. ______ 10. ______
11. ______
12. ______
13. ______
14. ______
15. ______
16. ______
17. ______
18. ______
19. ______
20. ______

< Objectives 1 and 2 >

Factor completely.

21. $x^2 + 10x + 24$

22. $x^2 + 3x - 10$

23. $x^2 - 9x + 20$

24. $x^2 - 8x + 15$

25. $x^2 - 2x - 63$

26. $x^2 + 6x - 55$

27. $x^2 + 8x + 14$

28. $x^2 - 11x + 24$

29. $x^2 - 11x + 28$

30. $y^2 - y - 21$

31. $-s^2 - 13s - 30$

32. $-b^2 - 11b - 28$

33. $a^2 - 2a - 48$

34. $x^2 - 17x + 60$

35. $x^2 - 8x + 7$

36. $x^2 + 7x - 18$

37. $x^2 + 11x + 24$

38. $x^2 - 11x + 10$

39. $-x^2 + 14x - 49$

40. $-s^2 + 4s + 32$

41. $p^2 - 10p - 28$

42. $x^2 - 11x - 60$

43. $x^2 + 5x - 66$

44. $a^2 - 5a - 24$

Answers

21. ______
22. ______
23. ______
24. ______
25. ______
26. ______
27. ______
28. ______
29. ______
30. ______
31. ______
32. ______
33. ______
34. ______
35. ______
36. ______
37. ______
38. ______
39. ______
40. ______
41. ______
42. ______
43. ______
44. ______

45. $c^2 + 19c + 60$

46. $t^2 - 4t - 60$

47. $n^2 + 5n - 50$

48. $x^2 - 16x + 65$

49. $x^2 + 7xy + 10y^2$

50. $x^2 - 8xy + 12y^2$

51. $x^2 + xy - 12y^2$

52. $m^2 - 8mn + 16n^2$

53. $x^2 - 13xy + 40y^2$ > Videos

54. $r^2 - 9rs - 36s^2$

55. $6x^2 + 19x + 10$

56. $6x^2 - 7x - 3$

< Objective 3 >

57. $9x^2 - 12x + 4$

58. $20x^2 - 23x + 6$

59. $12x^2 - 8x - 15$

60. $16a^2 + 40a + 25$

61. $3y^2 + 7y - 6$

62. $12x^2 + 11x - 15$

63. $8x^2 - 27x - 20$ > Videos

64. $24v^2 + 5v - 36$

65. $2x^2 + 3xy + y^2$

66. $3x^2 - 5xy + 2y^2$

67. $5a^2 - 8ab - 4b^2$

68. $5x^2 + 7xy - 6y^2$

69. $9x^2 + 4xy - 5y^2$

70. $16x^2 + 32xy + 15y^2$

71. $6m^2 - 17mn + 12n^2$

72. $15x^2 - xy - 6y^2$

Answers

45. __________
46. __________
47. __________
48. __________
49. __________
50. __________
51. __________
52. __________
53. __________
54. __________
55. __________
56. __________
57. __________
58. __________
59. __________
60. __________
61. __________
62. __________
63. __________
64. __________
65. __________
66. __________
67. __________
68. __________
69. __________
70. __________
71. __________
72. __________

Answers

73. ________

74. ________

75. ________

76. ________

77. ________

78. ________

79. ________

80. ________

81. ________

82. ________

83. ________

84. ________

85. ________

86. ________

87. ________

88. ________

89. ________

90. ________

91. ________

92. ________

93. ________

94. ________

95. ________

96. ________

73. $36a^2 - 3ab - 5b^2$

74. $10q^2 + 14qr - 12r^2$

75. $x^2 + 4xy + 4y^2$

76. $25b^2 - 80bc + 64c^2$

77. $20x^2 - 20x - 15$

78. $24x^2 - 18x - 6$

79. $8m^2 + 12m + 4$

80. $14x^2 - 20x + 6$

81. $15r^2 - 21rs + 6s^2$

82. $10x^2 + 5xy - 30y^2$

83. $2x^3 - 2x^2 - 4x$

84. $2y^3 + y^2 - 3y$

85. $2y^4 + 5y^3 + 3y^2$ > Videos

86. $4z^3 - 18z^2 - 10z$

87. $36a^3 - 66a^2 + 18a$

88. $20n^4 - 22n^3 - 12n^2$

89. $9p^2 + 30pq + 21q^2$

90. $12x^2 + 2xy - 24y^2$

< Objective 4 >

91. $u^4 - 5u^2 + 4$

92. $y^4 - 29y^2 + 100$

93. $w^4 - 5w^2 - 36$

94. $t^4 - 15t^2 - 16$

95. $2y^4 - 12y^2 - 54$ (*Hint:* Remember to look for a GCF.)

96. $3x^4 - 24x^2 + 48$ (*Hint:* Remember to look for a GCF.)

Basic Skills | **Challenge Yourself** | Calculator/Computer | Career Applications | Above and Beyond

Complete each statement with **never, sometimes,** *or* **always.**

97. In factoring $x^2 + bx + c$, if c is a prime number then the trinomial is _________ factorable.

98. If a GCF has already been factored out of a trinomial, we will _________ find a common factor in one (or both) of the binomial factors.

99. In factoring $x^2 + bx + c$, if c is negative then the signs in the binomial factors are _________ opposites.

100. In factoring $x^2 + bx + c$, if c is positive then the signs in the binomial factors are _________ both negative.

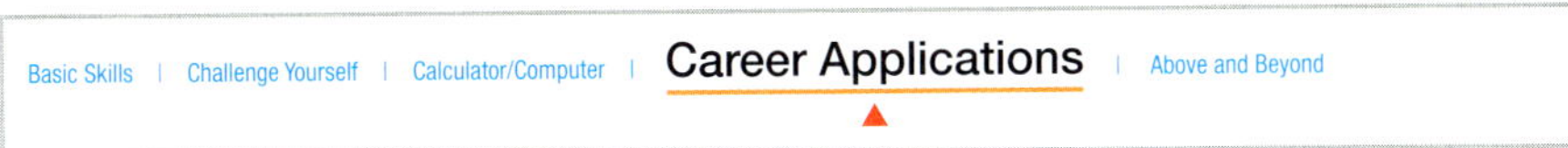

101. Mechanical Engineering The bending stress on an overhanging beam is given by the expression $310(x^2 - 36x + 128)$. Factor this expression.

102. Automotive Technology The acceleration curve for low gear in a car is described by the equation $a = \frac{1}{20}(x^2 - 16x - 80)$. Rewrite this equation by factoring the right-hand side.

103. Construction Technology The stress-strain curve of a weld is given by the formula $s = 325 + 60l - l^2$. Rewrite this formula by factoring the right-hand side.

104. Manufacturing Technology The maximum stress for a given allowable strain (deformation) of a certain material is given by the polynomial equation

$$\text{Stress} = 85.8x - 0.6x^2 - 1{,}537.2$$

in which x is the allowable strain, in micrometers. Factor the right-hand side of this equation. *Hint:* Factor out -0.6 first, and then rearrange the polynomial. > Videos

Find positive integer values for k so that each polynomial can be factored.

105. $x^2 + kx + 8$

106. $x^2 + kx + 9$

107. $x^2 - kx + 16$

108. $x^2 - kx + 17$

Answers

97. _________

98. _________

99. _________

100. _________

101. _________

102. _________

103. _________

104. _________

105. _________

106. _________

107. _________

108. _________

Answers

109. ____

110. ____

111. ____

112. ____

113. ____

114. ____

109. $x^2 - kx - 5$

110. $x^2 - kx - 7$

111. $x^2 + 3x + k$

112. $x^2 + 5x + k$

113. $x^2 + 2x - k$

114. $x^2 + x - k$

Answers

1. True **3.** False **5.** True **7.** False **9.** False **11.** $a = 1; b = 7; c = -5$ **13.** $a = 1; b = -3; c = 8$ **15.** $a = 3; b = 5; c = -8$ **17.** $a = 4; b = 8; c = 11$ **19.** $a = -7; b = -5; c = 2$ **21.** $(x + 4)(x + 6)$ **23.** $(x - 5)(x - 4)$ **25.** $(x - 9)(x + 7)$ **27.** Not factorable **29.** $(x - 4)(x - 7)$ **31.** $-1(s + 10)(s + 3)$ **33.** $(a - 8)(a + 6)$ **35.** $(x - 1)(x - 7)$ **37.** $(x + 3)(x + 8)$ **39.** $-1(x - 7)(x - 7)$ **41.** Not factorable **43.** $(x + 11)(x - 6)$ **45.** $(c + 4)(c + 15)$ **47.** $(n + 10)(n - 5)$ **49.** $(x + 2y)(x + 5y)$ **51.** $(x - 3y)(x + 4y)$ **53.** $(x - 5y)(x - 8y)$ **55.** $(3x + 2)(2x + 5)$ **57.** $(3x - 2)(3x - 2)$ **59.** $(6x + 5)(2x - 3)$ **61.** $(3y - 2)(y + 3)$ **63.** $(8x + 5)(x - 4)$ **65.** $(2x + y)(x + y)$ **67.** $(5a + 2b)(a - 2b)$ **69.** $(9x - 5y)(x + y)$ **71.** $(3m - 4n)(2m - 3n)$ **73.** $(12a - 5b)(3a + b)$ **75.** $(x + 2y)^2$ **77.** $5(2x - 3)(2x + 1)$ **79.** $4(2m + 1)(m + 1)$ **81.** $3(5r - 2s)(r - s)$ **83.** $2x(x - 2)(x + 1)$ **85.** $y^2(2y + 3)(y + 1)$ **87.** $6a(3a - 1)(2a - 3)$ **89.** $3(p + q)(3p + 7q)$ **91.** $(u + 2)(u - 2)(u + 1)(u - 1)$ **93.** $(w + 3)(w - 3)(w^2 + 4)$ **95.** $2(y + 3)(y - 3)(y^2 + 3)$ **97.** sometimes **99.** always **101.** $310(x - 4)(x - 32)$ **103.** $s = (5 + l)(65 - l)$ **105.** 6 or 9 **107.** 8 or 10 or 17 **109.** 4 **111.** 2 **113.** 3, 8, 15, 24, . . .

6.4 Factoring Trinomials: The *ac* Method

< 6.4 Objectives >

1 > Use the *ac* test to determine factorability

2 > Factor a trinomial using the *ac* method

3 > Completely factor a trinomial

Factoring trinomials is more time-consuming when the coefficient of the first term is not 1. Consider the product

$$(5x + 2)(2x + 3) = 10x^2 + 19x + 6$$

Factors of $10x^2$ — Factors of 6

Do you see the additional difficulty? In order to factor the polynomial on the right, we need to consider all possible factors of the first coefficient (10 in the example) as well as those of the third term (6 in our example).

In the previous section, we used the trial-and-error method to factor trinomials. We also learned that not all trinomials can be factored. In this section, we look at trinomials again, but with a slightly different approach. We first learn to determine whether a trinomial is factorable. We then use the result of that analysis to factor the trinomial, without guessing.

Some students prefer the trial-and-error method for factoring because it is generally faster and more intuitive. Other students prefer the method of this section (called the ***ac* method**) because it yields the answer in a systematic way. It does not matter which method you choose. Either method works to factor a trinomial. We are introducing you to both so you can determine which method you prefer.

To introduce the *ac* method, we first factor trinomials of the form $x^2 + bx + c$. Then we will apply the *ac* method to factor trinomials whose leading coefficient is not 1 (usually written as $ax^2 + bx + c$).

First, we consider some trinomials that are already factored.

Example 1 Matching Trinomials and Their Factors

Determine which of the following are true statements.

(a) $x^2 - 2x - 8 = (x - 4)(x + 2)$

This is a true statement. Using the FOIL method, we see that

$$(x - 4)(x + 2) = x^2 + 2x - 4x - 8 = x^2 - 2x - 8$$

(b) $x^2 - 6x + 5 = (x - 2)(x - 3)$

This is not a true statement.

$$(x - 2)(x - 3) = x^2 - 3x - 2x + 6 = x^2 - 5x + 6$$

(c) $x^2 + 5x - 14 = (x - 2)(x + 7)$

This is true.

$(x - 2)(x + 7) = x^2 + 7x - 2x - 14 = x^2 + 5x - 14$

(d) $x^2 - 8x - 15 = (x - 5)(x - 3)$

This is false.

$(x - 5)(x - 3) = x^2 - 3x - 5x + 15 = x^2 - 8x + 15$

Check Yourself 1

Determine which of the following are true statements.

(a) $2x^2 - 2x - 3 = (2x - 3)(x + 1)$
(b) $3x^2 + 11x - 4 = (3x - 1)(x + 4)$
(c) $2x^2 - 7x + 3 = (x - 3)(2x - 1)$

The first step in learning to factor a trinomial by the *ac* method is to identify its coefficients. So that we are consistent, we write the trinomial in standard $ax^2 + bx + c$ form, then label the three coefficients as a, b, and c.

Example 2 Identifying the Coefficients of $ax^2 + bx + c$

If necessary, rewrite the trinomial in $ax^2 + bx + c$ form. Then give the values for a, b, and c, where a is the coefficient of the x^2-term, b is the coefficient of the x-term, and c is the constant.

(a) $x^2 - 3x - 18$

$a = 1 \quad b = -3 \quad c = -18$

RECALL

The minus sign is attached to the coefficient.

(b) $x^2 - 24x + 23$

$a = 1 \quad b = -24 \quad c = 23$

(c) $x^2 + 8 - 11x$

First rewrite the trinomial in descending order.

$x^2 - 11x + 8$

$a = 1 \quad b = -11 \quad c = 8$

Check Yourself 2

If necessary, rewrite the trinomials in $ax^2 + bx + c$ form. Then label a, b, and c, where a is the coefficient of the x^2-term, b is the coefficient of the x-term, and c is the constant.

(a) $x^2 + 5x - 14$ **(b)** $x^2 - 18x + 17$ **(c)** $x - 6 + 2x^2$

Not all trinomials can be factored. To discover whether a trinomial is factorable, we try the ***ac* test.**

Property

The ac Test

A trinomial of the form $ax^2 + bx + c$ is factorable if (and only if) there are two integers m and n such that

$ac = mn$ and $b = m + n$

In other words, we are looking for two integers whose product is the same as $a \cdot c$ and whose sum is b.

In Example 3 we look for m and n to determine whether each trinomial is factorable.

Example 3 Using the ac Test

< Objective 1 >

Use the ac test to determine which trinomials can be factored. Find the values of m and n for each trinomial that can be factored.

(a) $x^2 - 3x - 18$

First, we find the values of a, b, and c, so that we can find ac.

$a = 1 \qquad b = -3 \qquad c = -18$

$ac = 1(-18) = -18$ and $b = -3$

Then we look for two integers m and n such that $mn = ac$ and $m + n = b$. In this case, that means

$mn = -18$ and $m + n = -3$

We now look at all pairs of integers with a product of -18. We then look at the sum of each pair of integers.

mn	$m + n$
$1(-18) = -18$	$1 + (-18) = -17$
$2(-9) = -18$	$2 + (-9) = -7$
$3(-6) = -18$	$3 + (-6) = -3$
$6(-3) = -18$	
$9(-2) = -18$	
$18(-1) = -18$	

We need look no further than 3 and -6.

NOTE

We could have chosen $m = -6$ and $n = 3$ as well.

The two integers with a product of ac and a sum of b are 3 and -6. We can say that

$m = 3$ and $n = -6$

Because we found values for m and n, we know that $x^2 - 3x - 18$ is factorable.

(b) $x^2 - 24x + 23$

We find that

$a = 1 \qquad b = -24 \qquad c = 23$

$ac = 1(23) = 23$ and $b = -24$

So $mn = 23$ and $m + n = -24$

We now calculate integer pairs, looking for two numbers with a product of 23 and a sum of -24.

mn	$m + n$
$1(23) = 23$	$1 + 23 = 24$
$-1(-23) = 23$	$-1 + (-23) = -24$

$m = -1$ and $n = -23$

So $x^2 - 24x + 23$ is factorable.

(c) $x^2 - 11x + 8$

We find that $a = 1$, $b = -11$, and $c = 8$. Therefore, $ac = 8$ and $b = -11$. Thus, $mn = 8$ and $m + n = -11$. We calculate integer pairs.

mn	$m + n$
$1(8) = 8$	$1 + 8 = 9$
$2(4) = 8$	$2 + 4 = 6$
$-1(-8) = 8$	$-1 + (-8) = -9$
$-2(-4) = 8$	$-2 + (-4) = -6$

There are no other pairs of integers with a product of 8, and none of these pairs has a sum of -11. The trinomial $x^2 - 11x + 8$ is not factorable.

(d) $2x^2 + 7x - 15$

We find that $a = 2$, $b = 7$, and $c = -15$. Therefore, $ac = 2(-15) = -30$ and $b = 7$. Thus, $mn = -30$ and $m + n = 7$. We calculate integer pairs.

mn	$m + n$
$1(-30) = -30$	$1 + (-30) = -29$
$2(-15) = -30$	$2 + (-15) = -13$
$3(-10) = -30$	$3 + (-10) = -7$
$5(-6) = -30$	$5 + (-6) = -1$
$6(-5) = -30$	$6 + (-5) = 1$
$10(-3) = -30$	$10 + (-3) = 7$

There is no need to go any further. We see that 10 and -3 have a product of -30 and a sum of 7, so

$m = 10$ and $n = -3$

Therefore, $2x^2 + 7x - 15$ is factorable.

Check Yourself 3

Use the *ac* test to determine which trinomials can be factored. Find the values of *m* and *n* for each trinomial that can be factored.

(a) $x^2 - 7x + 12$
(b) $x^2 + 5x - 14$
(c) $3x^2 - 6x + 7$
(d) $2x^2 + x - 6$

So far we have used the results of the *ac* test only to determine whether a trinomial is factorable. The results can also be used to help factor the trinomial. Now we factor the trinomials from the previous example, using the results of the *ac* test.

Example 4 Using the Results of the *ac* Test to Factor

< Objective 2 >

Rewrite the middle term as the sum of two terms, then factor by grouping.

(a) $x^2 - 3x - 18$

We find that $a = 1$, $b = -3$, and $c = -18$, so $ac = -18$ and $b = -3$. We are looking for two numbers m and n where $mn = -18$ and $m + n = -3$.

In Example 3(a), we looked at every pair of integers whose product (mn) was -18, to find a pair that had a sum $(m + n)$ of -3. We found the two integers to be 3 and -6, because $3(-6) = -18$ and $3 + (-6) = -3$, so $m = 3$ and $n = -6$.

We use that result to rewrite the middle term as the sum of $3x$ and $-6x$.

$$x^2 + 3x - 6x - 18$$

We factor this by grouping.

$$\begin{aligned} x^2 + 3x - 6x - 18 &= x(x + 3) - 6(x + 3) \\ &= (x + 3)(x - 6) \end{aligned}$$

RECALL

After factoring, immediately check your work. Here, multiply $(x + 3)(x - 6)$ and confirm that the product is $x^2 - 3x - 18$.

(b) $x^2 - 24x + 23$

We use the results from Example 3(b), in which we found $m = -1$ and $n = -23$, to rewrite the middle term of the equation.

$$x^2 - 24x + 23 = x^2 - x - 23x + 23$$

Then we factor by grouping.

$$\begin{aligned} x^2 - x - 23x + 23 &= (x^2 - x) - (23x - 23) \\ &= x(x - 1) - 23(x - 1) \\ &= (x - 1)(x - 23) \end{aligned}$$

RECALL

Again, check by multiplying.

(c) $2x^2 + 7x - 15$

From Example 3(d), we know that this trinomial is factorable, and $m = 10$ and $n = -3$. We use that result to rewrite the middle term of the trinomial.

$$\begin{aligned} 2x^2 + 7x - 15 &= 2x^2 + 10x - 3x - 15 \\ &= (2x^2 + 10x) - (3x + 15) \\ &= 2x(x + 5) - 3(x + 5) \\ &= (x + 5)(2x - 3) \end{aligned}$$

RECALL

Check!

Careful readers will note that we did not ask you to factor Example 3(c), $x^2 - 11x + 8$. Recall that, by the *ac* method, we determined that this trinomial was not factorable.

Check Yourself 4

Use the results of Check Yourself 3 to rewrite the middle term as the sum of two terms, then factor by grouping.

(a) $x^2 - 7x + 12$ **(b)** $x^2 + 5x - 14$ **(c)** $2x^2 + x - 6$

Now look at some examples that require us to first find m and n and then factor the trinomial.

Example 5 Rewriting Middle Terms to Factor

Rewrite the middle term as the sum of two terms, and then factor by grouping.

(a) $2x^2 - 13x - 7$

We find that $a = 2$, $b = -13$, and $c = -7$, so $mn = ac = -14$ and $m + n = b = -13$. Therefore,

mn	$m + n$
$1(-14) = -14$	$1 + (-14) = -13$

RECALL

Check that

$(2x + 1)(x - 7)$
$= 2x^2 - 13x - 7$

So $m = 1$ and $n = -14$. We rewrite the middle term of the trinomial.

$$\begin{aligned} 2x^2 - 13x - 7 &= 2x^2 + x - 14x - 7 \\ &= (2x^2 + x) - (14x + 7) \\ &= x(2x + 1) - 7(2x + 1) \\ &= (2x + 1)(x - 7) \end{aligned}$$

(b) $6x^2 - 5x - 6$

We find that $a = 6, b = -5$, and $c = -6$, so $mn = ac = -36$ and $m + n = b = -5$.

mn	$m + n$
$1(-36) = -36$	$1 + (-36) = -35$
$2(-18) = -36$	$2 + (-18) = -16$
$3(-12) = -36$	$3 + (-12) = -9$
$4(-9) = -36$	$4 + (-9) = -5$

RECALL

Multiply to check.

So $m = 4$ and $n = -9$. We rewrite the middle term of the trinomial.

$$\begin{aligned} 6x^2 - 5x - 6 &= 6x^2 + 4x - 9x - 6 \\ &= (6x^2 + 4x) - (9x + 6) \\ &= 2x(3x + 2) - 3(3x + 2) \\ &= (3x + 2)(2x - 3) \end{aligned}$$

RECALL

Our first step is always to try factoring out the GCF. To make certain that you have not missed the GCF, check the factors of your answer to be certain that each is factored completely.

Check Yourself 5

Rewrite the middle term as the sum of two terms and then factor by grouping.

(a) $2x^2 - 7x - 15$ **(b)** $6x^2 - 5x - 4$

Be certain to check trinomials and binomial factors for any common monomial factor. (There is no common factor in the binomial unless it is also a common factor in the original trinomial.) Example 6 shows the factoring out of monomial factors.

Example 6 Factoring Out Common Factors

< Objective 3 >

> **NOTE**
>
> If we had not removed the GCF in the first step, we would have gotten either $(x - 1)(3x + 15)$ or $(3x - 3)(x + 5)$ after factoring. Neither of these is factored completely.

Completely factor the trinomial

$$3x^2 + 12x - 15$$

We first factor out the common factor of 3.

$$3x^2 + 12x - 15 = 3(x^2 + 4x - 5)$$

Finding m and n for the trinomial $x^2 + 4x - 5$ yields $mn = -5$ and $m + n = 4$.

mn	$m + n$
$1(-5) = -5$	$1 + (-5) = -4$
$5(-1) = -5$	$5 + (-1) = 4$

So $m = 5$ and $n = -1$. This gives us

$$\begin{aligned} 3x^2 + 12x - 15 &= 3(x^2 + 4x - 5) \\ &= 3(x^2 + 5x - x - 5) \\ &= 3[(x^2 + 5x) - (x + 5)] \\ &= 3[x(x + 5) - (x + 5)] \\ &= 3[(x + 5)(x - 1)] \\ &= 3(x + 5)(x - 1) \end{aligned}$$

> **RECALL**
>
> Again, multiply this result to check.

Check Yourself 6

Completely factor the trinomial.

$6x^3 + 3x^2 - 18x$

Not all possible product pairs need to be tried to find m and n. A look at the sign pattern of the trinomial eliminates many of the possibilities. Assuming the leading coefficient is positive, there are four possible sign patterns. If the leading coefficient is negative, factor out -1 and then consider the remaining polynomial, whose leading coefficient is now positive.

Pattern	Example	Conclusion
1. b and c are both positive.	$2x^2 + 13x + 15$	m and n must both be positive.
2. b is negative and c is positive.	$x^2 - 7x + 12$	m and n must both be negative.
3. b is positive and c is negative.	$x^2 + 3x - 10$	m and n are of opposite signs. (The value with the larger absolute value is positive.)
4. b and c are both negative.	$x^2 - 3x - 10$	m and n are of opposite signs. (The value with the larger absolute value is negative.)

Check Yourself ANSWERS

1. **(a)** False; **(b)** true; **(c)** true
2. **(a)** $a = 1, b = 5, c = -14$; **(b)** $a = 1, b = -18, c = 17$; **(c)** $a = 2, b = 1, c = -6$ **3.** **(a)** Factorable, $m = -3, n = -4$; **(b)** factorable, $m = 7, n = -2$; **(c)** not factorable; **(d)** factorable, $m = 4, n = -3$
4. **(a)** $x^2 - 3x - 4x + 12 = (x - 3)(x - 4)$; **(b)** $x^2 + 7x - 2x - 14 = (x + 7)(x - 2)$; **(c)** $2x^2 + 4x - 3x - 6 = (x + 2)(2x - 3)$
5. **(a)** $2x^2 - 10x + 3x - 15 = (x - 5)(2x + 3)$; **(b)** $6x^2 - 8x + 3x - 4 = (2x + 1)(3x - 4)$
6. $3x(x + 2)(2x - 3)$

Reading Your Text

SECTION 6.4

(a) The first step in learning to factor a trinomial by the *ac* method is to identify its ______________.

(b) If the leading coefficient of a trinomial is positive, there are ______________ possible sign patterns.

(c) To discover whether a trinomial is ______________, we try the *ac* test.

(d) Our first step is always to try factoring out the ______________.

Basic Skills | Challenge Yourself | Calculator/Computer | Career Applications | Above and Beyond

6.4 exercises

State whether each statement is **true** *or* **false.**

1. $x^2 + 2x - 3 = (x + 3)(x - 1)$

2. $y^2 + 3y + 18 = (y - 6)(y + 3)$

3. $x^2 - 10x - 24 = (x - 6)(x + 4)$

4. $a^2 + 9a - 36 = (a - 12)(a + 4)$

5. $x^2 - 16x + 64 = (x - 8)(x - 8)$

6. $w^2 - 12w - 45 = (w + 9)(w - 5)$

7. $25y^2 - 10y + 1 = (5y - 1)(5y + 1)$ 

8. $6x^2 + 5xy - 4y^2 = (6x - 2y)(x + 2y)$

9. $10p^2 - pq - 3q^2 = (5p - 3q)(2p + q)$

10. $6a^2 + 13a + 6 = (2a - 3)(3a - 2)$

For each trinomial, label a, b, and c.

11. $x^2 + 7x - 5$

12. $x^2 + 5x + 11$

13. $x^2 - 3x + 8$

14. $x^2 + 7x - 15$

15. $3x^2 + 5x - 8$

16. $3x^2 - 5x + 7$

17. $4x^2 + 8x + 11$

18. $5x^2 + 7x - 9$

19. $-7x^2 - 5x + 2$

20. $-7x^2 + 9x - 18$

Name ______________________

Section __________ Date __________

Answers

1. ________ 2. ________
3. ________ 4. ________
5. ________ 6. ________
7. ________ 8. ________
9. ________ 10. ________
11. ________
12. ________
13. ________
14. ________
15. ________
16. ________
17. ________
18. ________
19. ________
20. ________

Answers

21. ____
22. ____
23. ____
24. ____
25. ____
26. ____
27. ____
28. ____
29. ____
30. ____
31. ____
32. ____
33. ____
34. ____
35. ____
36. ____
37. ____
38. ____
39. ____
40. ____
41. ____
42. ____
43. ____
44. ____

< Objective 1 >

Use the ac test to determine which of the trinomials can be factored. Find the values of m and n for each trinomial that can be factored.

21. $x^2 + x - 6$

22. $x^2 - x - 6$

23. $x^2 + 3x - 1$

24. $x^2 - 3x + 7$

25. $x^2 - 5x + 6$

26. $x^2 - x + 2$

27. $2x^2 + 5x - 3$

28. $3x^2 - 14x - 5$

29. $6x^2 - 19x + 10$ 

30. $4x^2 + 5x + 6$

< Objectives 2–3 >

Rewrite the middle term as the sum of two terms and then factor by grouping.

31. $x^2 + 6x + 8$

32. $x^2 + 3x - 10$

33. $x^2 - 9x + 20$

34. $x^2 - 8x + 15$

35. $x^2 - 2x - 63$

36. $x^2 + 6x - 55$

Rewrite the middle term as the sum of two terms and then factor completely.

37. $x^2 + 10x + 24$

38. $x^2 - 11x + 24$

39. $x^2 - 11x + 28$

40. $y^2 - y - 20$

41. $s^2 + 13s + 30$

42. $b^2 + 11b + 28$

43. $a^2 - 2a - 48$

44. $x^2 - 17x + 60$

45. $x^2 - 8x + 7$

46. $x^2 + 7x - 18$

47. $x^2 - 7x - 18$

48. $x^2 - 11x + 10$

49. $x^2 - 14x + 49$

50. $s^2 - 4s - 32$

51. $-p^2 + 10p + 24$

52. $-x^2 + 11x + 60$

53. $x^2 + 5x - 66$

54. $a^2 - 5a - 24$

55. $c^2 + 19c + 60$

56. $t^2 - 4t - 60$

57. $-n^2 - 5n + 50$

58. $-x^2 + 16x - 63$

59. $x^2 + 7xy + 10y^2$

60. $x^2 - 8xy + 12y^2$

61. $x^2 + xy - 12y^2$

62. $m^2 - 8mn + 16n^2$

63. $x^2 - 13xy + 40y^2$

64. $r^2 - 9rs - 36s^2$

65. $-6x^2 - 19x - 10$

66. $-6x^2 + 7x + 3$

67. $15x^2 + x - 6$

68. $12w^2 + 19w + 4$

Answers

45. ____________

46. ____________

47. ____________

48. ____________

49. ____________

50. ____________

51. ____________

52. ____________

53. ____________

54. ____________

55. ____________

56. ____________

57. ____________

58. ____________

59. ____________

60. ____________

61. ____________

62. ____________

63. ____________

64. ____________

65. ____________

66. ____________

67. ____________

68. ____________

69. $6m^2 + 25m - 25$

70. $12x^2 + x - 20$

71. $9x^2 - 12x + 4$

72. $20x^2 - 23x + 6$

73. $12x^2 - 8x - 15$

74. $16a^2 + 40a + 25$

75. $3y^2 + 7y - 6$

76. $12x^2 + 11x - 15$

77. $8x^2 - 27x - 20$ > Videos

78. $24v^2 + 5v - 36$

79. $2x^2 + 3xy + y^2$

80. $3x^2 - 5xy + 2y^2$

81. $5a^2 - 8ab - 4b^2$

82. $5x^2 + 7xy - 6y^2$

83. $9x^2 + 4xy - 5y^2$

84. $16x^2 + 32xy + 15y^2$

Basic Skills | **Challenge Yourself** | Calculator/Computer | Career Applications | Above and Beyond

Determine whether each statement is **true** *or* **false.**

85. A trinomial can always be factored into the product of two binomials.

86. Using the *ac* method requires the use of factoring by grouping.

Complete each statement with **never, sometimes,** *or* **always.**

87. In factoring $x^2 + bx + c$, if c is negative then the signs in the binomial factors are __________ opposite signs.

88. In factoring $x^2 + bx + c$, if c is positive then the signs in the binomial factors are __________ both negative.

Factor completely.

89. $x^2 + 4xy + 4y^2$

90. $25b^2 - 80bc + 64c^2$

91. $20x^2 - 20x - 15$

92. $24x^2 - 18x - 6$

93. $8m^2 + 12m + 4$

94. $14x^2 - 20x + 6$

Answers

69. ______

70. ______

71. ______

72. ______

73. ______

74. ______

75. ______

76. ______

77. ______

78. ______

79. ______

80. ______

81. ______

82. ______

83. ______

84. ______

85. ______

86. ______

87. ______

88. ______

89. ______

90. ______

91. ______

92. ______

93. ______

94. ______

95. $15r^2 - 21rs + 6s^2$

96. $10x^2 + 5xy - 30y^2$

97. $2x^3 - 2x^2 - 4x$

98. $2y^3 + y^2 - 3y$

99. $2y^4 + 5y^3 + 3y^2$ > Videos

100. $4z^3 - 18z^2 - 10z$

101. $36a^3 - 66a^2 + 18a$

102. $20n^4 - 22n^3 - 12n^2$

103. $9p^2 + 30pq + 21q^2$

104. $12x^2 + 2xy - 24y^2$

Each trinomial is "quadratic in form." For a brief discussion of factoring such expressions, see page 649. Factor each polynomial completely.

105. $u^4 - 5u^2 + 4$

106. $y^4 - 29y^2 + 100$

107. $w^4 - 5w^2 - 36$

108. $t^4 - 15t^2 - 16$

109. $2y^4 - 12y^2 - 54$ (*Hint:* Remember to look for a GCF.)

110. $3x^4 - 24x^2 + 48$ (*Hint:* Remember to look for a GCF.)

Basic Skills | Challenge Yourself | Calculator/Computer | **Career Applications** | Above and Beyond

111. **Agricultural Technology** The yield Y of a crop is given by the equation

$$Y = -0.05x^2 + 1.5x + 140$$

Rewrite this equation by factoring the right-hand side.
(*Hint:* Begin by factoring out -0.05.)

112. **Construction Technology** The profit curve P for a welding shop is given by the equation

$$P = 2x^2 - 143x - 1{,}360$$

Rewrite this equation by factoring the right-hand side.

113. **Allied Health** The number N of people who are sick t days after the outbreak of a flu epidemic is given by the equation

$$N = 50 + 25t - 3t^2$$

Rewrite this equation by factoring the right-hand side.

Answers

95. ______
96. ______
97. ______
98. ______
99. ______
100. ______
101. ______
102. ______
103. ______
104. ______
105. ______
106. ______
107. ______
108. ______
109. ______
110. ______
111. ______
112. ______
113. ______

Answers

114. ______

115. ______

116. ______

117. ______

118. ______

119. ______

120. ______

121. ______

122. ______

123. ______

124. ______

114. Mechanical Engineering The flow rate through a hydraulic hose can be found using the equation

$$2Q^2 + Q - 21 = 0$$

Rewrite this equation by factoring the left-hand side.

Basic Skills | Challenge Yourself | Calculator/Computer | Career Applications | **Above and Beyond**

Find positive integer values for k for which each polynomial can be factored.

115. $x^2 + kx + 8$

116. $x^2 + kx + 9$

117. $x^2 - kx + 16$ 

118. $x^2 - kx + 17$

119. $x^2 - kx - 5$

120. $x^2 - kx - 7$

121. $x^2 + 3x + k$

122. $x^2 + 5x + k$

123. $x^2 + 2x - k$

124. $x^2 + x - k$

Answers

1. True **3.** False **5.** True **7.** False **9.** True
11. $a = 1; b = 7; c = -5$ **13.** $a = 1; b = -3; c = 8$
15. $a = 3; b = 5; c = -8$ **17.** $a = 4; b = 8; c = 11$
19. $a = -7; b = -5; c = 2$ **21.** Factorable; $3, -2$ **23.** Not factorable
25. Factorable; $-3, -2$ **27.** Factorable; $6, -1$ **29.** Factorable; $-15, -4$
31. $2x + 4x; (x + 2)(x + 4)$ **33.** $-5x - 4x; (x - 5)(x - 4)$
35. $-9x + 7x; (x - 9)(x + 7)$ **37.** $(x + 4)(x + 6)$ **39.** $(x - 4)(x - 7)$
41. $(s + 10)(s + 3)$ **43.** $(a - 8)(a + 6)$ **45.** $(x - 1)(x - 7)$
47. $(x - 9)(x + 2)$ **49.** $(x - 7)(x - 7)$ **51.** $-1(p - 12)(p + 2)$
53. $(x + 11)(x - 6)$ **55.** $(c + 4)(c + 15)$ **57.** $-1(n + 10)(n - 5)$
59. $(x + 2y)(x + 5y)$ **61.** $(x - 3y)(x + 4y)$ **63.** $(x - 5y)(x - 8y)$
65. $-1(3x + 2)(2x + 5)$ **67.** $(5x - 3)(3x + 2)$ **69.** $(6m - 5)(m + 5)$
71. $(3x - 2)(3x - 2)$ **73.** $(6x + 5)(2x - 3)$ **75.** $(3y - 2)(y + 3)$
77. $(8x + 5)(x - 4)$ **79.** $(2x + y)(x + y)$ **81.** $(5a + 2b)(a - 2b)$
83. $(9x - 5y)(x + y)$ **85.** False **87.** always **89.** $(x + 2y)^2$
91. $5(2x - 3)(2x + 1)$ **93.** $4(2m + 1)(m + 1)$ **95.** $3(5r - 2s)(r - s)$
97. $2x(x - 2)(x + 1)$ **99.** $y^2(2y + 3)(y + 1)$ **101.** $6a(3a - 1)(2a - 3)$
103. $3(p + q)(3p + 7q)$ **105.** $(u + 2)(u - 2)(u + 1)(u - 1)$
107. $(w + 3)(w - 3)(w^2 + 4)$ **109.** $2(y + 3)(y - 3)(y^2 + 3)$
111. $Y = -0.05\,(x + 40)(x - 70)$ **113.** $N = -(3t + 5)(t - 10)$
115. 6 or 9 **117.** 8 or 10 or 17 **119.** 4 **121.** 2
123. 3, 8, 15, 24, . . .

6.5 Strategies in Factoring

< 6.5 Objectives >

1 > Recognize factoring patterns

2 > Apply appropriate factoring strategies

You have seen a variety of techniques for factoring polynomials in this chapter. This section reviews those techniques and presents some guidelines for choosing an appropriate strategy or combination of strategies.

1. Always look for a greatest common factor. If you find a GCF (other than 1), factor out the GCF as your first step. If the leading coefficient is negative, factor out the GCF including a negative coefficient.

To factor $5x^2y - 10xy + 25xy^2$, the GCF is $5xy$, so

$$5x^2y - 10xy + 25xy^2 = 5xy(x - 2 + 5y)$$

2. Now look at the number of terms in the polynomial you are trying to factor.

(a) If the polynomial is a *binomial,* consider the special binomial formulas.

(i) To factor $x^2 - 49y^2$, recognize the difference of squares, so

$$x^2 - 49y^2 = (x + 7y)(x - 7y)$$

(ii) The binomial

$$x^2 + 121$$

is the sum of squares and cannot be further factored.

(iii) To factor $t^3 - 64$, recognize the difference of cubes, so

$$t^3 - 64 = (t - 4)(t^2 + 4t + 16)$$

(iv) The binomial $z^3 + 1$ is the sum of cubes, so

$$z^3 + 1 = (z + 1)(z^2 - z + 1)$$

(b) If the polynomial is a *trinomial,* try to factor it as a product of two binomials. You can use either the trial-and-error method or the *ac* method.

To factor $2x^2 - x - 6$, a consideration of possible factors leads to

$$2x^2 - x - 6 = (2x + 3)(x - 2)$$

(c) If the polynomial has *more than three terms,* try to factor by grouping.

To factor $2x^2 - 3xy + 10x - 15y$, group the first two terms, and then the last two, and factor out common factors.

$$2x^2 - 3xy + 10x - 15y = x(2x - 3y) + 5(2x - 3y)$$

Now factor out the common binomial factor $(2x - 3y)$.

$$2x^2 - 3xy + 10x - 15y = (2x - 3y)(x + 5)$$

3. Always factor the polynomial completely. After you apply one of the techniques given in part 2, another one may be necessary.

RECALL

$a^2 - b^2 = (a + b)(a - b)$

The sum of squares $a^2 + b^2$ cannot be factored.

$a^3 - b^3 = (a - b)(a^2 + ab + b^2)$

$a^3 + b^3 = (a + b)(a^2 - ab + b^2)$

(a) To factor $6x^3 + 22x^2 - 40x$, first factor out the common factor of $2x$. So

$$6x^3 + 22x^2 - 40x = 2x(3x^2 + 11x - 20)$$

Now continue to factor the trinomial as before and

$$6x^3 + 22x^2 - 40x = 2x(3x - 4)(x + 5)$$

(b) To factor $x^3 - x^2y - 4x + 4y$, first we proceed by grouping.

$$x^3 - x^2y - 4x + 4y = x^2(x - y) - 4(x - y)$$
$$= (x - y)(x^2 - 4)$$

Now because $x^2 - 4$ is a difference of two squares, we continue to factor and obtain

$$x^3 - x^2y - 4x + 4y = (x - y)(x + 2)(x - 2)$$

4. Always check your answer by multiplying.

Example 1 Recognizing Factoring Patterns

< Objective 1 >

For each expression, state the appropriate first step for factoring the polynomial.

(a) $9x^2 - 18x - 72$

Find the GCF.

(b) $x^2 - 3x + 2xy - 6y$

Group the terms.

(c) $x^4 - 81y^4$

Factor the difference of squares.

(d) $3x^2 + 7x + 2$

Use the *ac* method (or trial and error).

Check Yourself 1

For each expression, state the appropriate first step for factoring the polynomial.

(a) $5x^2 + 2x - 3$ **(b)** $a^4b^4 - 16$

(c) $3x^2 + 3x - 60$ **(d)** $2a^2 - 5a + 4ab - 10b$

Remember that some polynomials are simply not factorable! If we try all steps in the above plan and are unable to break down the polynomial, answer "not factorable."

Example 2 Factoring Polynomials

< Objective 2 >

Factor $2xy + 10x + 6y + 30$.

$2xy + 10x + 6y + 30$ We note a GCF of 2, and factor.

$= 2(xy + 5x + 3y + 15)$

RECALL

Check by multiplying the factors together. You should get the original polynomial.

Because the polynomial has four terms, we move to step 3 and try grouping.

$= 2[(xy + 5x) + (3y + 15)]$

$= 2[x(y + 5) + 3(y + 5)]$

$= 2[(y + 5)(x + 3)]$

$= 2(y + 5)(x + 3)$

The binomial factors are first degree with 1 as their leading coefficient, so they cannot be further factored. We finish by checking our work.

$$\begin{aligned} 2(y + 5)(x + 3) &= 2(y \cdot x + y \cdot 3 + 5 \cdot x + 5 \cdot 3) \\ &= 2(xy + 3y + 5x + 15) \\ &= 2xy + 6y + 10x + 30 \\ &= 2xy + 10x + 6y + 30 \end{aligned}$$ The original polynomial

Check Yourself 2

Factor $4mn - 12m - 20n + 60$.

Be sure to keep your eyes open for factors that can be further factored. This is illustrated in Example 3.

Example 3 Factoring Polynomials

Factor $3mn^4 - 48m$.

$3mn^4 - 48m$ — Factor out the GCF.

$= 3m(n^4 - 16)$ — Note the difference-of-squares binomial.

$= 3m(n^2 - 4)(n^2 + 4)$ — Note another difference of squares.

$= 3m(n - 2)(n + 2)(n^2 + 4)$

The only binomial that could possibly factor is $n^2 + 4$, but since this is a sum of squares, it does not. We are done.

Check Yourself 3

Factor $3x^2y - 75y$.

Remember to always start with step 1: Factor out the GCF.

Example 4 Factoring Polynomials

Factor $-6x^2y + 18xy + 60y$.

$-6x^2y + 18xy + 60y$ — We find a GCF of $-6y$.

$= -6y(x^2 - 3x - 10)$ — We factor the trinomial using trial and error or the *ac* method.

$= -6y(x - 5)(x + 2)$

Check Yourself 4

Factor $-5xy^2 - 15xy + 90x$.

Do not become frustrated if your factoring attempts do not seem to produce results. You may have a polynomial that does not factor.

Example 5 Factoring Polynomials

Factor $9m^2 - 8$.

We cannot find a GCF greater than 1, so we proceed to step 2. We do have a binomial, but it does not fit any of our special patterns: $9m^2$ is a perfect square, but 8 is not. And 8 is a perfect cube, but $9m^2$ is not. So we conclude that the given binomial is not factorable.

Check Yourself 5

Factor $2m^3 - 16$.

Example 6 Factoring Polynomials

Factor $x^2y^2 - 9x^2 + y^2 - 9$.

There is no GCF greater than 1. Since we have four terms, we try grouping:

$$x^2y^2 - 9x^2 + y^2 - 9$$
$$= x^2(y^2 - 9) + (y^2 - 9)$$
$$= (x^2 + 1)(y^2 - 9)$$ We note the difference of squares.
$$= (x^2 + 1)(y - 3)(y + 3)$$

Since $x^2 + 1$ is a sum of squares, we are done.

Check Yourself 6

Factor $x^2y^2 - 3x^2 - 4y^2 + 12$.

Check Yourself ANSWERS

1. **(a)** *ac* method (or trial and error); **(b)** factor the difference of squares; **(c)** find the GCF; **(d)** group the terms
2. $4(m - 5)(n - 3)$ **3.** $3y(x - 5)(x + 5)$ **4.** $-5x(y + 6)(y - 3)$
5. $2(m - 2)(m^2 + 2m + 4)$ **6.** $(x - 2)(x + 2)(y^2 - 3)$

Reading Your Text

SECTION 6.5

(a) The _______ of squares is not factorable.

(b) If a polynomial consists of four terms, try to factor by _______.

(c) When we multiply two binomial factors, we get the original _______.

(d) We can factor a trinomial using trial and error or the _______ method.

Basic Skills | Challenge Yourself | Calculator/Computer | Career Applications | Above and Beyond

6.5 exercises

< Objectives 1 and 2 >

Completely factor each polynomial.

1. $x^2 - 5x - 14$

2. $y^2 - 3y - 40$

3. $a^2 + 10a + 24$

4. $n^2 + 11n + 18$

5. $y^2 - 10y + 21$

6. $z^2 - 12z + 20$

7. $w^3 - 125$

8. $1 - z^3$

9. $2t^2 + 9t - 5$

10. $3t^2 - 11t - 4$

11. $4y^2 - 81$

12. $9m^2 - 25n^2$

13. $3a^2 + 6a - 21$

14. $5b^2 - 15b + 30$

15. $8a^3 + 27$

16. $y^3 - 64$

17. $4t^3 - 4$

18. $4m^4 - 4$

Name ______________

Section ________ Date ________

Answers

1. ______________
2. ______________
3. ______________
4. ______________
5. ______________
6. ______________
7. ______________
8. ______________
9. ______________
10. ______________
11. ______________
12. ______________
13. ______________
14. ______________
15. ______________
16. ______________
17. ______________
18. ______________

6.5 exercises

Answers

19. ______

20. ______

21. ______

22. ______

23. ______

24. ______

25. ______

26. ______

27. ______

28. ______

29. ______

30. ______

31. ______

32. ______

33. ______

34. ______

35. ______

36. ______

37. ______

38. ______

39. ______

40. ______

19. $3x^2 + 30x + 75$

20. $4x^2 - 24x + 36$

21. $xy - 2x + 6y - 12$

22. $yz + 5y - 3z - 15$

23. $2x^2 + 5x + 4$

24. $4x^2 + 12x - 72$

25. $a^2 - 10a + 25$

26. $x^2 + 6xy + 9y^2$

27. $2abx - 14ax - 8bx + 56x$

28. $3ax + 12x + 3a + 12$

29. $3x^2 + x - 10$

30. $3x^2 - 2x + 4$

31. $5n^2 + 20n + 30$

32. $3ay^2 - 48a$

33. $3y^3 - 81$

34. $6t^2 - 12t + 24$

35. $3x^2y + 9xy + 9y$

36. $18mn^2 - 15mn - 12m$

37. $9x^3 \quad 4x$

38. $5uv^2 - 20uv + 30u$

39. $2n^4 - 16n$

40. $8x^3 - 2x$

41. $-3x^2 + 12x + 15$

42. $-5y^2 - 30y - 40$

43. $x^2y^2 - 4x^2 - 9y^2 + 36$

44. $40x^2 + 12x - 72$

45. $8x^2 + 8x + 2$

46. $m^2n^2 - 4n^2 - 25m^2 + 100$

47. $x^3 + 5x^2 - 4x - 20$

48. $x^3 - 2x^2 - 9x + 18$

49. $x^4 + 3x^2 - 10$

50. $y^4 - 2y^2 - 24$

51. $t^4 + t^2 - 20$

52. $w^4 - 10w^2 + 9$

Basic Skills | **Challenge Yourself** | Calculator/Computer | Career Applications | Above and Beyond

Determine whether each statement is **true** *or* **false.**

53. Factoring a polynomial may require more than one technique.

54. Every polynomial can be factored.

55. If we are trying to factor a four-term polynomial, we should try one of the special patterns.

56. No matter what type of polynomial we are trying to factor, we should always check for a GCF first.

Answers

1. $(x - 7)(x + 2)$ **3.** $(a + 6)(a + 4)$ **5.** $(y - 7)(y - 3)$
7. $(w - 5)(w^2 + 5w + 25)$ **9.** $(2t - 1)(t + 5)$ **11.** $(2y + 9)(2y - 9)$
13. $3(a^2 + 2a - 7)$ **15.** $(2a + 3)(4a^2 - 6a + 9)$
17. $4(t - 1)(t^2 + t + 1)$ **19.** $3(x + 5)^2$ **21.** $(x + 6)(y - 2)$
23. Not factorable **25.** $(a - 5)^2$ **27.** $2x(a - 4)(b - 7)$
29. $(3x - 5)(x + 2)$ **31.** $5(n^2 + 4n + 6)$ **33.** $3(y - 3)(y^2 + 3y + 9)$
35. $3y(x^2 + 3x + 3)$ **37.** $x(3x - 2)(3x + 2)$
39. $2n(n - 2)(n^2 + 2n + 4)$ **41.** $-3(x + 1)(x - 5)$
43. $(x - 3)(x + 3)(y - 2)(y + 2)$ **45.** $2(2x + 1)^2$
47. $(x + 5)(x - 2)(x + 2)$ **49.** $(x^2 - 2)(x^2 + 5)$
51. $(t + 2)(t - 2)(t^2 + 5)$ **53.** True **55.** False

Answers

41. ________

42. ________

43. ________

44. ________

45. ________

46. ________

47. ________

48. ________

49. ________

50. ________

51. ________

52. ________

53. ________

54. ________

55. ________

56. ________

6.6 Solving Quadratic Equations by Factoring

< 6.6 Objectives >

1 > Solve quadratic equations by factoring

2 > Find the zeros of a quadratic function

The factoring techniques you learned provide you with the tools to solve equations that can be written in the form

$$ax^2 + bx + c = 0 \qquad a \neq 0$$

This is a quadratic equation in one variable, here x. You can recognize such a quadratic equation by the fact that the highest power of the variable x is the second power.

where a, b, and c are constants.

An equation written in the form $ax^2 + bx + c = 0$ is called a **quadratic equation in standard form.** Using factoring to solve quadratic equations requires the **zero-product principle,** which says that if the product of two factors is 0, then one or both of the factors must be equal to 0. In symbols:

Property

Zero-Product Principle If $a \cdot b = 0$, then $a = 0$ or $b = 0$ or $a = b = 0$.

We apply this principle to solving quadratic equations in Example 1.

Example 1 Solving Equations by Factoring

< Objective 1 >

NOTE

To use the zero-product principle, 0 must be on one side of the equation.

RECALL

We first studied functions in Chapter 2.

Solve

$$x^2 - 3x - 18 = 0$$

Factoring on the left, we have

$$(x - 6)(x + 3) = 0$$

By the zero-product principle, we know that one or both of the factors must be zero. We can then write

$$x - 6 = 0 \qquad \text{or} \qquad x + 3 = 0$$

Solving each equation gives

$$x = 6 \qquad \text{or} \qquad x = -3$$

The two solutions are 6 and -3.

The solutions are sometimes called the **roots** of the equation. These roots have an important connection to the graph of the function $f(x) = x^2 - 3x - 18$, also written $y = x^2 - 3x - 18$. The graph of this function forms a curve, which we will study

NOTE

Graph the equation
$y = x^2 - 3x - 18$
on your graphing calculator.
Use the ZERO utility to show

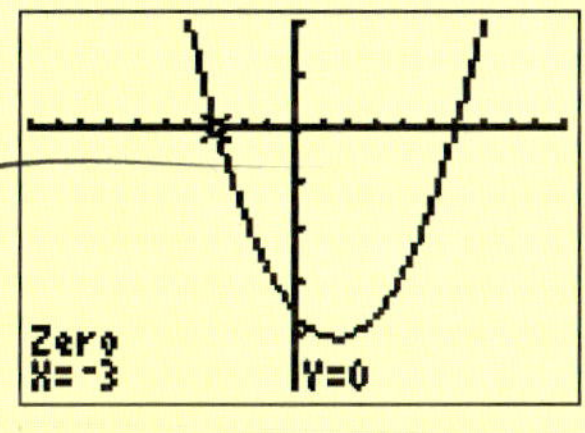

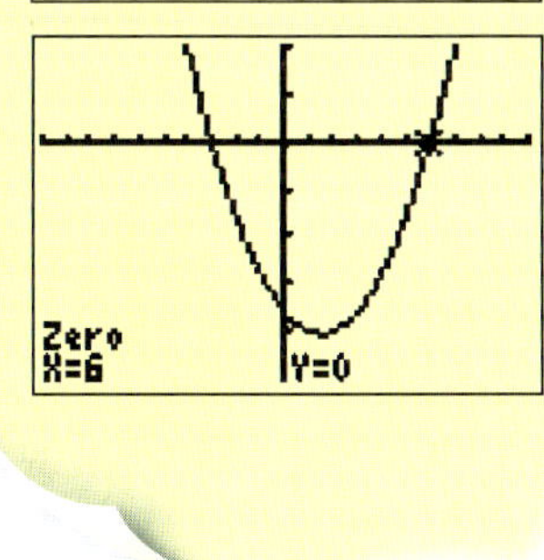

in Chapter 8. This particular curve crosses the x-axis at two points, $(-3, 0)$ and $(6, 0)$. -3 and 6 are called the **zeros** of this function, since $f(-3) = 0$ and $f(6) = 0$. So, the solutions of the equation $x^2 - 3x - 18 = 0$, -3 and 6, are zeros of the function $f(x) = x^2 - 3x - 18$, and they tell us the points where the graph of f crosses the x-axis, $(-3, 0)$ and $(6, 0)$.

Quadratic equations can be checked in the same way as linear equations were checked: by substitution. For instance, if $x = 6$, we have

$$(6)^2 - 3 \cdot (6) - 18 \stackrel{?}{=} 0$$
$$36 - 18 - 18 \stackrel{?}{=} 0$$
$$0 = 0$$

which is a true statement. We leave it to you to check the solution -3.

Check Yourself 1

Solve $x^2 - 9x + 20 = 0$.

Other factoring techniques may also be used when solving quadratic equations. Example 2 illustrates this concept.

Example 2 Solving Equations by Factoring

> **CAUTION**
>
> A *common mistake* is to forget the statement $x = 0$ when solving equations of this type. Be sure to include the *two* values of x that satisfy the equation: $x = 0$ and $x = 5$.

(a) Solve $x^2 - 5x = 0$.

Again, factor the left side of the equation and apply the zero-product principle.

$x(x - 5) = 0$

Now

$x = 0$ or $x - 5 = 0$

$x = 5$

The two solutions are 0 and 5. The solution set is $\{0, 5\}$.

(b) Solve $x^2 - 9 = 0$.

Factoring yields

$(x + 3)(x - 3) = 0$

$x + 3 = 0$ or $x - 3 = 0$

$x = -3$ $\quad x = 3$

The solution set is $\{-3, 3\}$, which may be written as $\{\pm 3\}$.

NOTE

The symbol $\pm$ is read "plus or minus."

Check Yourself 2

Solve by factoring.

(a) $x^2 + 8x = 0$ (b) $x^2 - 16 = 0$

Example 3 illustrates a crucial point. Our solving technique depends on the zero-product principle, which means that the product of factors *must be equal to 0.*

Example 3 Solving Equations by Factoring

> **CAUTION**
>
> Consider the equation
>
> $x(2x - 1) = 3$
>
> Students are sometimes tempted to write
>
> $x = 3 \quad \text{or} \quad 2x - 1 = 3$
>
> This is *not correct.*
>
> If $a \cdot b = 0$, then a or b **must** be 0. But if $a \cdot b = 3$, we do not know that a or b is 3, for example, it could be that $a = 6$ and $b = 1/2$ (or many other possibilities).

Solve $2x^2 - x = 3$.

The first step in solving is to write the equation in standard form (that is, with 0 on one side of the equation). So start by subtracting 3 from both sides of the equation.

$2x^2 - x - 3 = 0$ Make sure all nonzero terms are on one side of the equation. The other side must be 0.

You can now factor and solve using the zero-product principle.

$(2x - 3)(x + 1) = 0$

$2x - 3 = 0 \quad \text{or} \quad x + 1 = 0$

$2x = 3 \qquad x = -1$

$x = \dfrac{3}{2}$

The solution set is $\left\{\dfrac{3}{2}, -1\right\}$.

Check Yourself 3

Solve $3x^2 = 5x + 2$.

In all of the previous examples, the quadratic equations had two distinct real-number solutions. That may not always be the case.

Example 4 Solving Equations by Factoring

Solve $x^2 - 6x + 9 = 0$.

Factoring gives

$(x - 3)(x - 3) = 0$

and

$x - 3 = 0 \quad \text{or} \quad x - 3 = 0$

$x = 3 \qquad x = 3$

The solution set is $\{3\}$.

A quadratic (or second-degree) equation always has *two* solutions. When an equation such as this one has two solutions that are the same number, we call 3 the **repeated** (or **double**) **solution** of the equation.

Even though a quadratic equation always has two solutions, they may not always be real numbers. You will learn more about this in Chapters 7 and 8.

Check Yourself 4

Solve $x^2 + 6x + 9 = 0$.

Always examine the quadratic expression of an equation for common factors. Finding one makes your work easier, as Example 5 illustrates.

Example 5 Solving Equations by Factoring

Solve $3x^2 - 3x - 60 = 0$.

First, note the common factor 3 in the quadratic expression of the equation. Factoring out the 3, we have

$3(x^2 - x - 20) = 0$

Now divide both sides of the equation by 3.

$$\frac{3(x^2 - x - 20)}{3} = \frac{0}{3}$$

or $x^2 - x - 20 = 0$

We can now factor and solve as before.

$(x - 5)(x + 4) = 0$

$x - 5 = 0 \quad \text{or} \quad x + 4 = 0$

$x = 5 \qquad x = -4 \quad \text{or} \quad \{-4, 5\}$

NOTE

The advantage of dividing both sides by 3 is that the coefficients in the quadratic expression become smaller making the expression easier to factor.

Check Yourself 5

Solve $2x^2 - 10x - 48 = 0$.

Fractions may seem to complicate matters, but there is a nice way to eliminate them. Recall from Section 1.5 that we can choose to multiply both sides of an equation by a nonzero constant without affecting the solution set.

Example 6 Clearing Fractions from a Quadratic Equation

Solve $\frac{x^2}{5} + \frac{x}{10} - 1 = 0$.

Noting denominators of 5, 10, and 1, we choose the least common multiple of these, namely 10, and multiply.

$$10\left(\frac{x^2}{5}\right) + 10\left(\frac{x}{10}\right) - 10(1) = 10(0)$$

$2x^2 + x - 10 = 0$ Fractions have been "cleared."

$(2x + 5)(x - 2) = 0$ We factor as usual, and solve.

$2x + 5 = 0 \quad \text{or} \quad x - 2 = 0$

$x = -\frac{5}{2} \qquad x = 2 \quad \text{or} \quad \left\{-\frac{5}{2}, 2\right\}$

NOTE

The is often called "clearing fractions."

Check Yourself 6

Solve $x^2 - \frac{x}{6} = \frac{1}{3}$.

Here is a summary of the steps to follow when solving a quadratic equation by factoring.

Step by Step

Solving Quadratic Equations by Factoring

Step 1 Add or subtract the necessary terms on both sides of the equation so that the equation is in standard form (set equal to 0).

Step 2 Factor the quadratic expression.

Step 3 Set each factor equal to 0.

Step 4 Solve the resulting equations to find the solutions.

Step 5 Check each solution by substituting in the original equation.

We will have many occasions to work with quadratic functions. The standard form of such a function is $f(x) = ax^2 + bx + c$, where a is not 0. When working with this type of function, we often need to find the zeros. These are input values that result in an output value of 0. As mentioned earlier, the zeros are also the solutions to the equation $ax^2 + bx + c = 0$. Graphically, these values are the x-coordinates of the x-intercepts of the graph of f.

Definition

Zeros of a Quadratic Function

The zeros of the function $f(x) = ax^2 + bx + c$, where $a \neq 0$, are the solutions to the equation $ax^2 + bx + c = 0$. These zeros give the x-intercepts of the graph of the function.

Example 7 Finding the Zeros of a Quadratic Function

< Objective 2 >

NOTE

The x-intercepts of the graph of the function $f(x) = x^2 - x - 2$ are $(-1, 0)$ and $(2, 0)$, so -1 and 2 are the zeros of the function.

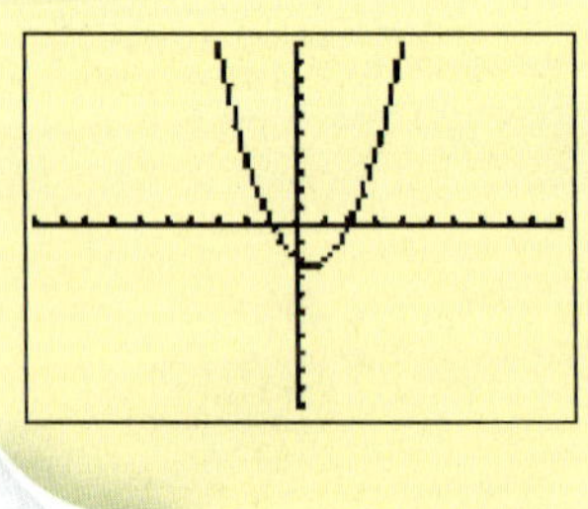

Find the zeros of the function

$$f(x) = x^2 - x - 2$$

To find the zeros of the function, set $f(x) = 0$, and solve.

$$x^2 - x - 2 = 0$$

$$(x - 2)(x + 1) = 0$$

$$x - 2 = 0 \quad \text{or} \quad x + 1 = 0$$

$$x = 2 \qquad\qquad x = -1$$

The zeros are -1 and 2.

Check Yourself 7

Find the zeros of $f(x) = 2x^2 - x - 3$.

Check Yourself ANSWERS

1. $\{4, 5\}$ **2. (a)** $\{-8, 0\}$; **(b)** $\{-4, 4\}$ **3.** $\left\{-\frac{1}{3}, 2\right\}$ **4.** $\{-3\}$

5. $\{-3, 8\}$ **6.** $\left\{-\frac{1}{2}, \frac{2}{3}\right\}$ **7.** -1 and $\frac{3}{2}$

Reading Your Text

SECTION 6.6

(a) An equation of the form $ax^2 + bx + c = 0$ is called a quadratic equation in _______________ form.

(b) Using factoring to solve a quadratic equation requires the _______________ principle.

(c) Solutions are sometimes called _______________ of an equation.

(d) The zeros of a function tell us the points where the graph of f crosses the _______________.

6.6 exercises

Basic Skills | Challenge Yourself | Calculator/Computer | Career Applications | Above and Beyond

Name ______________

Section ________ Date ________

Answers

1. ________ 2. ________
3. ________ 4. ________
5. ________ 6. ________
7. ________ 8. ________
9. ________ 10. ________
11. ________ 12. ________
13. ________ 14. ________
15. ________ 16. ________
17. ________ 18. ________
19. ________ 20. ________
21. ________
22. ________
23. ________
24. ________

< Objectives 1 and 2 >

Solve each quadratic equation by factoring.

1. $x^2 - 3x - 10 = 0$

2. $x^2 - 5x + 4 = 0$

3. $x^2 - 2x - 15 = 0$

4. $x^2 + 4x - 32 = 0$

5. $x^2 - 11x + 30 = 0$

6. $x^2 + 13x + 36 = 0$

7. $x^2 - 4x - 21 = 0$

8. $x^2 + 5x - 36 = 0$

9. $x^2 - 5x = 50$

10. $x^2 + 14x = -33$ 

11. $x^2 = 5x + 84$

12. $x^2 = 6x + 27$

13. $x^2 - 8x = 0$

14. $x^2 + 7x = 0$

15. $x^2 + 10x = 0$

16. $x^2 - 9x = 0$

17. $x^2 = 5x$

18. $x^2 = 11x$

19. $x^2 - 25 = 0$

20. $x^2 - 49 = 0$

21. $9x^2 = 25$

22. $x^2 = 169$

23. $4x^2 + 12x + 9 = 0$

24. $9x^2 - 30x + 25 = 0$

25. $2x^2 - 17x + 36 = 0$

26. $5x^2 + 17x - 12 = 0$

27. $5x^2 + 9x = 18$

28. $12x^2 = 25x - 12$

29. $6x^2 = 7x - 2$

30. $4x^2 - 3 = x$

31. $2m^2 = 12m + 54$

32. $5x^2 - 55x = 60$

33. $7x^2 - 63x = 0$

34. $6x^2 - 9x = 0$

35. $5x^2 = 15x$

36. $7x^2 = -49x$ 

37. $\frac{x^2}{8} - \frac{x}{4} - 1 = 0$

38. $\frac{x^2}{15} + \frac{x}{5} - \frac{2}{3} = 0$

39. $\frac{x^2}{6} + \frac{3x}{2} + 3 = 0$

40. $\frac{x^2}{10} - \frac{x}{2} = \frac{3}{5}$

41. $\frac{2x^2}{3} + \frac{x}{15} = \frac{1}{5}$

42. $\frac{4x^2}{5} - x - \frac{3}{10} = 0$

43. $x(x + 2) = 15$

44. $x(x + 3) = 28$

45. $x(2x - 3) = 9$

46. $x(3x + 1) = 52$

47. $2x(3x + 1) = 28$ > Videos

48. $3x(2x - 1) = 30$

49. $(x - 3)(x - 1) = 15$

50. $(x + 3)(x - 2) = 14$

51. $(x - 5)(x + 2) = 18$

52. $(3x - 5)(x + 2) = 14$

Answers

25. ______ 26. ______

27. ______ 28. ______

29. ______ 30. ______

31. ______ 32. ______

33. ______

34. ______

35. ______ 36. ______

37. ______ 38. ______

39. ______ 40. ______

41. ______ 42. ______

43. ______

44. ______

45. ______

46. ______

47. ______

48. ______

49. ______

50. ______

51. ______

52. ______

Answers

53. ______

54. ______

55. ______

56. ______

57. ______

58. ______

59. ______

60. ______

61. ______

62. ______

63. ______

64. ______

65. ______

66. ______

67. ______

68. ______

69. ______

70. ______

71. ______

Find the zeros of each function.

53. $f(x) = x^2 - 6x + 5$

54. $f(x) = x^2 + 2x - 8$

55. $f(x) = x^2 - 9x$

56. $f(x) = 6x^2 - x - 2$

Complete each statement with **never, sometimes,** *or* **always.**

57. To solve a quadratic equation by factoring, we ______ work to place zero on one side of the equation.

58. A quadratic equation in standard form can ______ have three solutions.

Write an equation that has the given solution. (Hint: *Write the binomial factors and then find their product.*)

59. $\{-4, 5\}$

60. $\{0, 5\}$

61. $\{2, 6\}$ 

62. $\{-4, 4\}$

The zero-product principle can be extended to three or more factors. If $a \cdot b \cdot c = 0$, then at least one of these factors is 0. Use this information to solve each equation.

63. $x^3 - 3x^2 - 10x = 0$

64. $x^3 + 8x^2 + 15x = 0$

65. $x^3 - 9x = 0$

66. $x^3 = 16x$

Extend the ideas in the previous exercises to find solutions for each equation. (Hint: *Apply factoring by grouping in exercises 67 and 68.*)

67. $x^3 + x^2 - 4x - 4 = 0$

68. $x^3 - 5x^2 - x + 5 = 0$

69. $x^4 - 10x^2 + 9 = 0$

70. $x^4 - 5x^2 + 4 = 0$

Basic Skills | Challenge Yourself | Calculator/Computer | **Career Applications** | Above and Beyond

71. **Agricultural Technology** The height h (in feet) of a drop of water above an irrigation nozzle, in terms of the time t (in seconds) since the drop left the nozzle, is given by the formula

$$h = v_0t - 16t^2$$

If the initial velocity is $v_0 = 80$ ft/s, how many seconds need to pass for a drop to be 75 ft high?

72. **Manufacturing Technology** A piece of stainless steel warps due to the heat created during welding. The shape of the warping is approximated by the curve

$$w = \frac{a^2 - 16}{64}$$

At what value of a is $w = 0$?

73. **Allied Health** The number N of people who are sick t days after the outbreak of a flu epidemic is given by the equation

> Videos

$$N = 50 + 25t - 3t^2$$

How many days will it take until no one is infected?

74. **Mechanical Engineering** The flow rate through a hydraulic hose can be found using the equation

$$2Q^2 + Q - 21 = 0$$

Find the flow rate by solving the equation (you need to consider only positive solutions).

Basic Skills | Challenge Yourself | Calculator/Computer | Career Applications | **Above and Beyond**

The net productivity of a forested wetland is related to the amount of water moving through the wetland and can be modeled by a quadratic equation. In exercises 75 to 78, y represents the amount of wood produced and x represents the amount of water present, in cubic centimeters (cm³). Determine where the productivity is zero in each wetland represented by the equations.

75. $y = -3x^2 + 300x$

76. $y = -4x^2 + 500x$

77. $y = -6x^2 + 792x$

78. $y = -7x^2 + 1{,}022x$

79. **Business and Finance** The manager of a bicycle shop knows that the cost of selling x bicycles is $C = 20x + 60$ and the revenue from selling x bicycles is $R = x^2 - 8x$. Find the break-even value of x. (Recall that break-even occurs when cost equals revenue.)

80. **Business and Finance** A company that produces computer games has found that its daily operating cost in dollars is $C = 40x + 150$ and its daily revenue in dollars is $R = 65x - x^2$. For what value(s) of x will the company break even?

Answers

72. ______

73. ______

74. ______

75. ______

76. ______

77. ______

78. ______

79. ______

80. ______

Answers

1. $\{-2, 5\}$ **3.** $\{-3, 5\}$ **5.** $\{5, 6\}$ **7.** $\{-3, 7\}$ **9.** $\{-5, 10\}$
11. $\{-7, 12\}$ **13.** $\{0, 8\}$ **15.** $\{-10, 0\}$ **17.** $\{0, 5\}$ **19.** $\{-5, 5\}$
21. $\left\{-\frac{5}{3}, \frac{5}{3}\right\}$ **23.** $\left\{-\frac{3}{2}\right\}$ **25.** $\left\{4, \frac{9}{2}\right\}$ **27.** $\left\{-3, \frac{6}{5}\right\}$ **29.** $\left\{\frac{1}{2}, \frac{2}{3}\right\}$
31. $\{-3, 9\}$ **33.** $\{0, 9\}$ **35.** $\{0, 3\}$ **37.** $\{-2, 4\}$ **39.** $\{-3, -6\}$
41. $\left\{-\frac{3}{5}, \frac{1}{2}\right\}$ **43.** $\{-5, 3\}$ **45.** $\left\{-\frac{3}{2}, 3\right\}$ **47.** $\left\{-\frac{7}{3}, 2\right\}$
49. $\{-2, 6\}$ **51.** $\{-4, 7\}$ **53.** 1 and 5 **55.** 0 and 9 **57.** always
59. $x^2 - x - 20 = 0$ **61.** $x^2 - 8x + 12 = 0$ **63.** $\{-2, 0, 5\}$
65. $\{-3, 0, 3\}$ **67.** $\{-2, -1, 2\}$ **69.** $\{-3, -1, 1, 3\}$
71. 1.25 s and 3.75 s **73.** 10 days **75.** 0 cm^3, 100 cm^3
77. 0 cm^3, 132 cm^3 **79.** 30 bicycles

6.7 Problem Solving with Factoring

< 6.7 Objective >

1 > Use factoring to solve applications

With the techniques introduced in this chapter for solving equations by factoring, we can now examine a new group of applications. Recall that the key to problem solving lies in a step-by-step, organized approach to the process. You might want to take time now to review the five-step process introduced in Section 1.4. All the examples in this section make use of that model.

You will find that these applications typically lead to quadratic equations that can be solved by factoring. Remember that quadratic equations can have two, one, or no distinct real-number solutions. Step 5, which includes the process of verifying or checking your solutions, is particularly important here. By checking solutions, you may find that both, only one, or none of the derived solutions satisfy the physical conditions stated in the original problem.

We begin with a numerical application.

Example 1 Solving a Number Application

< Objective 1 >

One integer is 3 less than twice another. If their product is 35, find the two integers.

Step 1 The unknowns are the two integers.

Step 2 Let x represent the first integer. Then

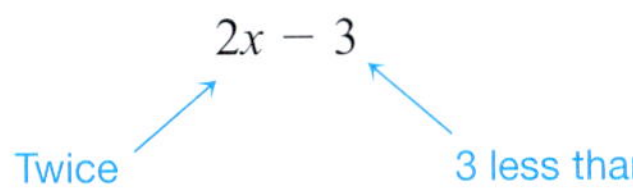

represents the second.

Step 3 Form an equation.

$$\underbrace{x(2x - 3)}_{\text{Product of the two integers}} = 35$$

Step 4 Remove the parentheses and solve.

$$2x^2 - 3x = 35$$

$$2x^2 - 3x - 35 = 0$$

Factor on the left.

$$(2x + 7)(x - 5) = 0$$

$$2x + 7 = 0 \quad \text{or} \quad x - 5 = 0$$

$$2x = -7 \qquad\qquad x = 5$$

$$x = -\frac{7}{2}$$

> **NOTE**
>
> These represent solutions to the equation, but not the answer to the original problem.

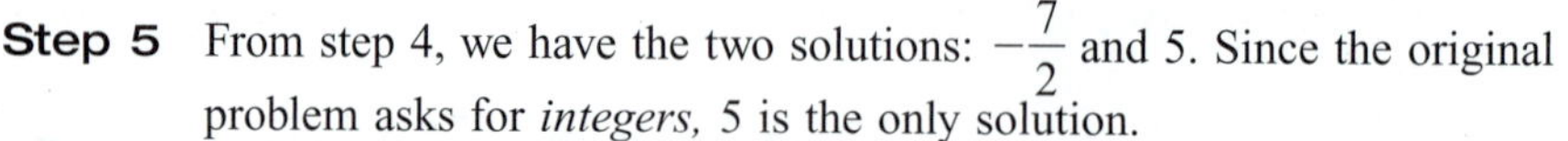

Step 5 From step 4, we have the two solutions: $-\frac{7}{2}$ and 5. Since the original problem asks for *integers,* 5 is the only solution.

Using 5 for x, we see that the other integer is $2x - 3 = 2(5) - 3 = 7$. So the desired integers are 5 and 7.

To verify, note that (a) 7 is 3 less than 2 times 5, and (b) the product of 5 and 7 is 35.

Check Yourself 1

One integer is 2 more than 3 times another. If their product is 56, what are the two integers?

Problems involving consecutive integers may also lead to quadratic equations. Recall that consecutive integers can be represented by x, $x + 1$, $x + 2$, and so on. Consecutive even (or odd) integers are represented by x, $x + 2$, $x + 4$, and so on.

Example 2 Solving a Number Application

The sum of the squares of two consecutive integers is 85. What are the two integers?

Step 1 The unknowns are the two consecutive integers.

Step 2 Let x be the first integer and $x + 1$ the second integer.

Step 3 Form the equation.

$$\underbrace{x^2 + (x + 1)^2}_{\text{Sum of squares}} = 85$$

Step 4 Solve.

$$x^2 + (x + 1)^2 = 85 \quad \text{Remove parentheses.}$$
$$x^2 + x^2 + 2x + 1 = 85$$
$$2x^2 + 2x - 84 = 0 \quad \text{Note the common factor 2, and divide both sides of the equation by 2.}$$
$$x^2 + x - 42 = 0$$
$$(x + 7)(x - 6) = 0$$
$$x + 7 = 0 \quad \text{or} \quad x - 6 = 0$$
$$x = -7 \qquad\qquad x = 6$$

Step 5 The work in step 4 leads to two possibilities, -7 or 6. Since *both numbers are integers,* both meet the conditions of the original problem. There are then two pairs of consecutive integers that work:

-7 and -6 (where $x = -7$ and $x + 1 = -6$)

or, 6 and 7 (where $x = 6$ and $x + 1 = 7$)

To check: $(-7)^2 + (-6)^2 = 49 + 36 = 85$ ✓

and: $(6)^2 + (7)^2 = 36 + 49 = 85$ ✓

Check Yourself 2

The sum of the squares of two consecutive even integers is 100. Find the two integers.

We proceed to applications involving geometry.

Example 3 Solving a Geometric Application

The length of a rectangle is 3 cm greater than its width. If the area of the rectangle is 108 cm^2, what are the dimensions of the rectangle?

Step 1 We are asked to find the dimensions (the length and the width) of the rectangle.

Step 2 Whenever geometric figures are involved in an application, start by drawing, and *then labeling,* a sketch of the problem. Let x represent the width and $x + 3$ the length.

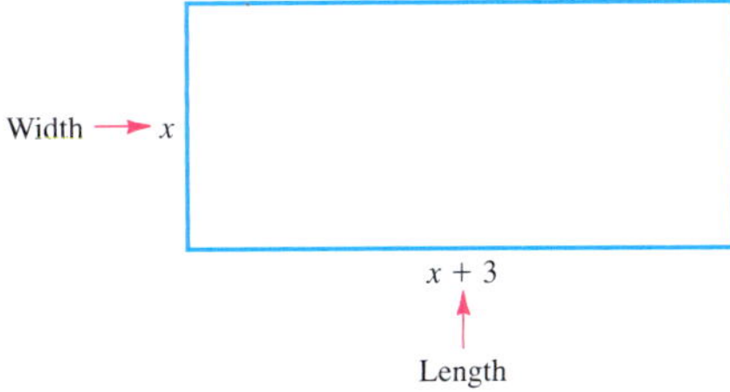

Step 3 Once the drawing is correctly labeled, the next step should be easy. The area of a rectangle is the product of its length and width, so

$$x(x + 3) = 108$$

Step 4 Solve the equation.

$$x(x + 3) = 108$$

$$x^2 + 3x - 108 = 0 \quad \text{Multiply and write in standard form.}$$

$$(x + 12)(x - 9) = 0 \quad \text{Factor and solve as before.}$$

$$x + 12 = 0 \quad \text{or} \quad x - 9 = 0$$

$$x = -12 \qquad x = 9$$

Step 5 We reject -12 cm as a solution. A length cannot be negative, and so we only consider 9 cm in finding the required dimensions.

The width x is 9 cm, and the length $x + 3$ is 12 cm. Since this gives a rectangle of area 108 cm^2, the solution is verified.

Check Yourself 3

In a triangle, the base is 4 in. less than its height. If its area is 30 $in.^2$, find the length of the base and the height of the triangle. (*Note:* The formula for the area of a triangle is $A = \frac{1}{2}bh$.)

We look at another geometric application.

Example 4 Solving a Rectangular Box Application

An open box is formed from a rectangular piece of cardboard, whose length is 2 in. more than its width, by cutting 2-in. squares from each corner and folding up the sides. If the volume of the box is to be 96 $in.^3$, what must be the dimensions of the original piece of cardboard?

Step 1 We are asked for the dimensions of the sheet of cardboard.

Step 2 Again, sketch the problem.

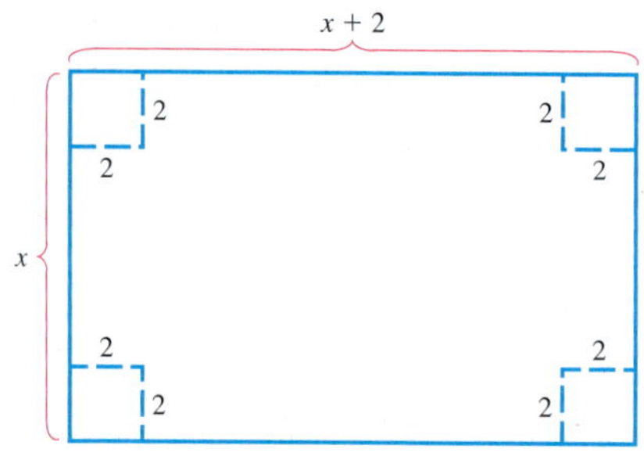

NOTE

The original width of the cardboard was x. Removing two 2-in. squares leaves $x - 4$ for the width of the box. Similarly, the length of the box is $x - 2$. Do you see why?

Step 3 To form an equation for volume, we sketch the completed box.

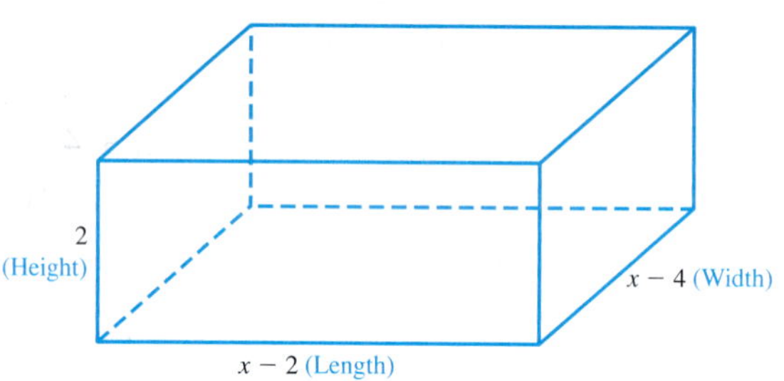

Since volume is the product of height, length, and width,

$2(x - 2)(x - 4) = 96$

Step 4

$$2(x - 2)(x - 4) = 96 \quad \text{Divide both sides by 2.}$$
$$(x - 2)(x - 4) = 48 \quad \text{Multiply on the left.}$$
$$x^2 - 6x + 8 = 48 \quad \text{Write in standard form.}$$
$$x^2 - 6x - 40 = 0 \quad \text{Solve as before.}$$
$$(x - 10)(x + 4) = 0$$
$$x = 10 \text{ in.} \quad \text{or} \quad x = -4 \text{ in.}$$

Step 5 Again, we only consider the positive solution. The width x of the original piece of cardboard is 10 in., and its length $x + 2$ is 12 in. The dimensions of the completed box is 6 in. by 8 in. by 2 in., which gives the required volume of 96 in.^{3}.

Check Yourself 4

A similar box is to be made by cutting 3-in. squares from a piece of cardboard that is 4 in. longer than it is wide. If the required volume is 180 in.^{3}, find the dimensions of the original piece of cardboard.

We now turn to another field for an application that leads to solving a quadratic equation. Many equations of motion in physics involve quadratic equations.

Example 5 Solving a Thrown Ball Application

Suppose that a person throws a ball directly upward, releasing the ball at a height of 5 ft above the ground. If the ball is thrown with an initial velocity of 80 ft/s, the height of the ball, in feet, above the ground after t seconds is given by

$h = -16t^2 + 80t + 5$

When is the ball at a height of 69 ft?

Step 1 The height h of the ball and the time t since the ball was released are the unknowns.

Step 2 We want to know the value(s) of t for which $h = 69$.

Step 3 Write $69 = -16t^2 + 80t + 5$.

Step 4 Solve for t.

$$69 = -16t^2 + 80t + 5$$
$$0 = -16t^2 + 80t - 64 \qquad \text{We need 0 on one side.}$$
$$0 = t^2 - 5t + 4 \qquad \text{Divide both sides by } -16.$$
$$0 = (t - 1)(t - 4) \qquad \text{Factor.}$$
$$t - 1 = 0 \quad \text{or} \quad t - 4 = 0$$
$$t = 1 \qquad\qquad t = 4$$

Step 5 By substituting 1 for t, we confirm that $h = 69$.

$$h = -16(1)^2 + 80(1) + 5 = -16 + 80 + 5 = 69$$

You should also check that $h = 69$ when $t = 4$.

The ball is at a height of 69 ft at $t = 1$ second (on the way up) and at $t = 4$ seconds (on the way down).

Check Yourself 5

A ball is thrown vertically upward from the top of a building 100 m high with an initial velocity of 25 m/s. After t seconds, the height h, in meters, is given by

$h = -5t^2 + 25t + 100$

When is the ball at a height of 130 m?

In Example 6, we must pay particular attention to the solutions that satisfy the given physical conditions.

Example 6 Solving a Thrown Ball Application

A ball is thrown vertically upward from the top of a building 60 m high with an initial velocity of 20 m/s. After t seconds, the height h is given by

$h = -5t^2 + 20t + 60$

(a) When is the ball at a height of 35 m?

Steps 1 and 2 We want to know the value(s) of t for which $h = 35$.

Step 3 Write $35 = -5t^2 + 20t + 60$.

Step 4 Solve for t.

$$35 = -5t^2 + 20t + 60 \quad \text{We need a 0 on one side.}$$
$$0 = -5t^2 + 20t + 25 \quad \text{Divide through by } -5.$$
$$0 = t^2 - 4t - 5 \quad \text{Factor.}$$
$$0 = (t - 5)(t + 1)$$
$$t - 5 = 0 \quad \text{or} \quad t + 1 = 0$$
$$t = 5 \qquad\qquad t = -1$$

NOTE

When $t = 5$,
$h = -5(5)^2 + 20(5) + 60$
$= -125 + 100 + 60$
$= 35$

Step 5 Note that $t = -1$ does not make sense in the context of this problem. (It represents 1 s *before* the ball is released!) We can easily check that when $t = 5$, $h = 35$. The ball is at a height of 35 m at 5 s after release.

(b) When does the ball hit the ground?

Steps 1 and 2 When the ball hits the ground, its height is 0 m.

Step 3 Write $0 = -5t^2 + 20t + 60$.

Step 4 Solve for t.

$$0 = -5t^2 + 20t + 60 \quad \text{Divide through by } -5.$$
$$0 = t^2 - 4t - 12 \quad \text{Factor.}$$
$$0 = (t - 6)(t + 2)$$
$$t - 6 = 0 \quad \text{or} \quad t + 2 = 0$$
$$t = 6 \qquad\qquad t = -2$$

Step 5 As before, we reject the negative solution $t = -2$. The ball hits the ground after 6 s. You should verify that when $t = 6$, $h = 0$.

Check Yourself 6

A ball is thrown vertically upward from the top of a building 150 ft high with an initial velocity of 64 ft/s. After t seconds, the height h is given by

$$h = -16t^2 + 64t + 150$$

When is the ball at a height of 70 ft?

An important application that leads to a quadratic equation involves the stopping distance of a car and its relation to the speed of the car. This is illustrated in Example 7.

Example 7 Solving a Stopping Distance Application

The stopping distance d, in feet, of a car that is traveling at x mi/h on a particular surface is approximated by the equation

$$d = \frac{x^2}{20} + x$$

If the stopping distance of this car is 240 ft, what is its speed?

Step 1 The stopping distance d, in feet, and the car's speed x, in miles per hour, are the unknowns.

Step 2 We want to know the value of x such that $d = 240$.

Step 3 Write $240 = \frac{x^2}{20} + x$.

Step 4 Solve for x.

$$240 = \frac{x^2}{20} + x$$

$$4{,}800 = x^2 + 20x \quad \text{Multiply through by 20.}$$

$$0 = x^2 + 20x - 4{,}800 \quad \text{We need a 0 on one side.}$$

$$0 = (x + 80)(x - 60) \quad \text{Factor.}$$

$$x + 80 = 0 \quad \text{or} \quad x - 60 = 0$$

$$x = -80 \qquad x = 60$$

Step 5 Because x represents the speed of the car, we reject the value $x = -80$. We verify that if $x = 60$, $d = 240$.

$$d = \frac{60^2}{20} + 60 = \frac{3{,}600}{20} + 60$$

$$= 180 + 60 = 240$$

The speed of the car is 60 mi/h.

Check Yourself 7

If the stopping distance of this car is 175 ft, what is its speed? Use the equation

$$d = \frac{x^2}{20} + x$$

Check Yourself ANSWERS

1. 4, 14 **2.** −8, −6 or 6, 8 **3.** Base 6 in.; height 10 in.
4. 12 in. by 16 in. **5.** 2 s or 3 s **6.** 5 s **7.** 50 mi/h

Reading Your Text

SECTION 6.7

(a) Many applications lead to quadratic equations that can be solved by ____________.

(b) A quadratic equation has two, one, or ____________ distinct real-number solutions.

(c) ____________ integers can be represented by x, $x + 1$, $x + 2$, and so on.

(d) We always reject a ____________ solution when we are solving for a variable representing a distance measurement.

6.7 exercises

Boost *your* GRADE at ALEKS.com!

ALEKS

- Practice Problems
- Self-Tests
- NetTutor
- e-Professors
- Videos

Name ______________

Section ________ Date ________

Answers

1. ______________
2. ______________
3. ______________
4. ______________
5. ______________
6. ______________
7. ______________
8. ______________
9. ______________
10. ______________
11. ______________
12. ______________
13. ______________
14. ______________
15. ______________
16. ______________
17. ______________

Basic Skills | Challenge Yourself | Calculator/Computer | Career Applications | Above and Beyond

< Objective 1 >

Solve each application.

1. **Number Problem** One integer is 3 more than twice another. If the product of those integers is 65, find the two integers.

2. **Number Problem** One positive integer is 5 less than 3 times another, and their product is 78. What are the two integers?

3. **Number Problem** The sum of two integers is 10, and their product is 24. Find the two integers.

4. **Number Problem** The sum of two integers is 12. If the product of the two integers is 27, what are the two integers?

5. **Number Problem** The product of two consecutive integers is 72. What are the two integers?

6. **Number Problem** If the product of two consecutive odd integers is 63, find the two integers.

7. **Number Problem** The sum of the squares of two consecutive whole numbers is 61. Find the two whole numbers.

8. **Number Problem** If the sum of the squares of two consecutive even integers is 100, what are the two integers? > Videos

9. **Number Problem** The sum of two integers is 9, and the sum of the squares of those two integers is 41. Find the two integers.

10. **Number Problem** The sum of two whole numbers is 12. If the sum of the squares of those numbers is 74, what are the two numbers?

11. **Number Problem** The sum of the squares of three consecutive integers is 50. Find the three integers.

12. **Number Problem** If the sum of the squares of three consecutive odd positive integers is 83, what are the three integers?

13. **Number Problem** Twice the square of a positive integer is 12 more than 5 times that integer. What is the integer?

14. **Number Problem** Find an integer such that if 10 is added to the integer's square, the result is 40 more than that integer.

15. **Geometry** The width of a rectangle is 3 ft less than its length. If the area of the rectangle is 70 ft^2, what are the dimensions of the rectangle? > Videos

16. **Geometry** The length of a rectangle is 5 cm more than its width. If the area of the rectangle is 84 cm^2, find the dimensions of the rectangle.

17. **Geometry** The length of a rectangle is 2 cm more than 3 times its width. If the area of the rectangle is 85 cm^2, find the dimensions of the rectangle.

18. **Geometry** If the length of a rectangle is 3 ft less than twice its width and the area of the rectangle is 54 ft^2, what are the dimensions of the rectangle?

19. **Geometry** The length of a rectangle is 1 cm more than its width. If the length of the rectangle is doubled, the area of the rectangle is increased by 30 cm^2. What were the dimensions of the original rectangle?

20. **Geometry** The height of a triangle is 2 in. more than the length of the base. If the base is tripled in length, the area of the new triangle is 48 in.^{2} more than the original. Find the height and base of the original triangle.

21. **Geometry** A box is to be made from a rectangular piece of tin that is twice as long as it is wide. To accomplish this, a 10-cm square is cut from each corner, and the sides are folded up. The volume of the finished box is to be 4,000 cm^3. Find the dimensions of the original piece of tin.

Hint: To solve this equation, use the given sketch of the piece of tin. Note that the original dimensions are represented by x and $2x$. Do you see why? Also recall that the volume of the resulting box is the product of the length, width, and height.

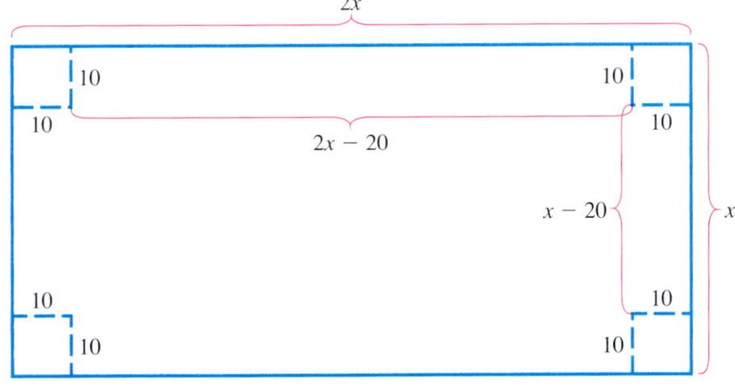

22. **Geometry** An open box is formed from a square piece of material by cutting 2-in. squares from each corner of the material and folding up the sides. If the volume of the box that is formed is to be 72 in.^{3}, what was the size of the original piece of material? > Videos

23. **Geometry** An open carton is formed from a rectangular piece of cardboard that is 4 ft longer than it is wide, by removing 1-ft squares from each corner and folding up the sides. If the volume of the carton is then 32 ft^3, what were the dimensions of the original piece of cardboard?

24. **Geometry** A box that has a volume of 1,936 in.^{3} was made from a square piece of tin. The square piece cut from each corner had sides of length 4 in. What were the original dimensions of the square?

25. **Geometry** A square piece of cardboard is to be formed into a box. After 5-cm squares are cut from each corner and the sides are folded up, the resulting box will have a volume of 4,500 cm^3. Find the length of a side of the original piece of cardboard.

26. **Geometry** A rectangular piece of cardboard has a length that is 2 cm longer than twice its width. If 2-cm squares are cut from each of its corners, it can

Answers

18. ______
19. ______
20. ______
21. ______
22. ______
23. ______
24. ______
25. ______
26. ______

Answers

27. ______

28. ______

29. ______

30. ______

31. ______

32. ______

33. ______

34. ______

be folded into a box that has a volume of 280 cm^3. What were the original dimensions of the piece of cardboard?

27. **Science and Medicine** If a ball is thrown vertically upward from the ground, with an initial velocity of 64 ft/s, its height h after t seconds is given by

$h = -16t^2 + 64t$

How long does it take the ball to return to the ground?

28. **Science and Medicine** If a ball is thrown vertically upward from the ground, with an initial velocity of 64 ft/s, its height h after t seconds is given by

$h = -16t^2 + 64t$

How long does it take the ball to reach a height of 48 ft on the way up?

29. **Science and Medicine** If a ball is thrown vertically upward from the ground, with an initial velocity of 96 ft/s, its height h after t seconds is given by

$h = -16t^2 + 96t$

How long does it take the ball to return to the ground?

30. **Science and Medicine** If a ball is thrown vertically upward from the ground, with an initial velocity of 96 ft/s, its height h after t seconds is given by

$h = -16t^2 + 96t$

How long does it take the ball to pass through a height of 128 ft on the way back down to the ground?

31. **Science and Medicine** If a ball is thrown vertically upward from the roof of a building 192 ft high with an initial velocity of 64 ft/s, its approximate height h after t seconds is given by > Videos

$h = -16t^2 + 64t + 192$

How long does it take the ball to fall back to the ground?

32. **Science and Medicine** If a ball is thrown vertically upward from the roof of a building 192 ft high with an initial velocity of 64 ft/s, its approximate height h after t seconds is given by

$h = -16t^2 + 64t + 192$

When will the ball reach a height of 240 ft?

33. **Science and Medicine** If a ball is thrown vertically upward from the roof of a building 192 ft high with an initial velocity of 96 ft/s, its approximate height h after t seconds is given by

$h = -16t^2 + 96t + 192$

How long does it take the ball to return to the thrower?

34. **Science and Medicine** If a ball is thrown vertically upward from the roof of a building 192 ft high with an initial velocity of 96 ft/s, its approximate height h after t seconds is given by

$h = -16t^2 + 96t + 192$

When will the ball reach a height of 272 ft?

35. **SCIENCE AND MEDICINE** If a ball is thrown upward from the roof of a 105-m building with an initial velocity of 20 m/s, its approximate height h after t seconds is given by

$$h = -5t^2 + 20t + 105$$

How long will it take the ball to fall to the ground?

36. **SCIENCE AND MEDICINE** If a ball is thrown upward from the roof of a 105-m building with an initial velocity of 20 m/s, its approximate height h after t seconds is given by

$$h = -5t^2 + 20t + 105$$

When will the ball reach a height of 80 m?

Determine whether each statement is **true** *or* **false.**

37. To find the area of a rectangle, we add the length and the width.

38. If a ball is thrown upward, the ball might reach a specified height two times.

Complete each statement with **never, sometimes,** *or* **always.**

39. The product of two consecutive integers is ______________ even.

40. The sum of two consecutive integers is ______________ even.

SCIENCE AND MEDICINE *Use the equation* $d = \frac{x^2}{20} + x$, *which relates the speed of a car x, in miles per hour, to the stopping distance d, in feet, on a particular road surface.*

41. If the stopping distance of a car is 120 ft, how fast is the car traveling?

42. How fast is a car going if it requires 75 ft to stop?

43. Marcus is driving at high speed on the freeway when he spots a vehicle stopped ahead of him. If Marcus's car is 315 ft from the vehicle, what is the maximum speed at which Marcus can be traveling and still stop in time?

44. Juliana is driving at high speed on a country highway when she sees a deer in the road ahead. If Juliana's car is 400 ft from the deer, what is the maximum speed at which Juliana can be traveling and still stop in time?

45. **BUSINESS AND FINANCE** Suppose that the cost C, in dollars, of producing x chairs is given by

$$C = 2x^2 - 40x + 2{,}400$$

How many chairs can be produced for $5,400?

Answers

35. ______________

36. ______________

37. ______________

38. ______________

39. ______________

40. ______________

41. ______________

42. ______________

43. ______________

44. ______________

45. ______________

Answers

46. ____________

47. ____________

48. ____________

49. ____________

50. ____________

51. ____________

52. ____________

46. **Business and Finance** Suppose that the profit P, in dollars, of producing and selling x appliances is given by

$$P = -3x^2 + 240x - 1{,}800$$

How many appliances must be produced and sold to achieve a profit of $3,000?

47. **Business and Finance** The relationship between the number x of calculators that a company can sell per month and the price of each calculator p is given by $x = 1{,}700 - 100p$. Find the price at which a calculator should be sold to produce a monthly revenue of $7,000. (*Hint:* Revenue $= xp$.)

48. **Business and Finance** A small manufacturer's weekly profit in dollars is given by

$$P = -3x^2 + 270x$$

Find the number of items x that must be produced to realize a profit of $4,200.

49. **Business and Finance** Suppose that a manufacturer's weekly profit in dollars is given by

$$P = -2x^2 + 240x$$

How many items x must be produced to realize a profit of $5,400?

Basic Skills | Challenge Yourself | Calculator/Computer | **Career Applications** | Above and Beyond

50. **Allied Health** A patient's body temperature T (°F) can be approximated by the formula

> Videos

$$T = 0.4t^2 - 2.6t + 103$$

in which t is the number of hours since the patient took the analgesic acetaminophen.

Determine the amount of time before the patient's temperature rises back up to 100°F.

51. **Allied Health** A healthy person's blood glucose level g (in mg per 100 mL) can be approximated by the formula

$$g = -480t^2 + 400t + 80$$

in which t is the number of hours since the person ate a meal. How long after eating will the person's blood glucose level return to 80 mg per 100 mL?

52. **Mechanical Engineering** The rotational moment M in a shaft is given by the formula

$$M = -30x + 2x^2$$

At what x-value is the moment equal to 152?

Answers

1. 5, 13 **3.** 4, 6 **5.** $-9, -8$ or 8, 9 **7.** 5, 6 **9.** 4, 5 **11.** $-5, -4, -3$ or 3, 4, 5 **13.** 4 **15.** 7 ft by 10 ft **17.** 5 cm by 17 cm **19.** 5 cm by 6 cm **21.** 30 cm by 60 cm **23.** 6 ft by 10 ft **25.** 40 cm **27.** 4 s **29.** 6 s **31.** 6 s **33.** 6 s **35.** 7 s **37.** False **39.** always **41.** 40 mi/h **43.** 70 mi/h **45.** 50 chairs **47.** $7 or $10 **49.** 30 or 90 items **51.** $\frac{5}{6}$ h or 50 min

Definition/Procedure	Example	Reference
An Introduction to Factoring		**Section 6.1**
Common Monomial Factor A single term that is a factor of every term of the polynomial. The greatest common factor (GCF) is the common monomial factor that has the largest possible numerical coefficient and the largest possible exponents.	$4x^2$ is the greatest common monomial factor of $8x^4 - 12x^3 + 16x^2$.	p. 621
Factoring a Monomial from a Polynomial 1. Determine the greatest common factor. 2. Apply the distributive property in the form $ab + ac = a(b + c)$ ↑ The greatest common factor	$8x^4 - 12x^3 + 16x^2$ $= 4x^2(2x^2 - 3x + 4)$	p. 621
Factoring by Grouping When there are four terms of a polynomial, factor the first pair and factor the last pair. If these two pairs have a common binomial factor, factor that out. The result will be the product of two binomials.	$4x^2 - 6x + 10x - 15$ $= 2x(2x - 3) + 5(2x - 3)$ $= (2x - 3)(2x + 5)$	p. 623
Factoring Special Polynomials		**Section 6.2**
Factoring a Difference of Squares Use the form $a^2 - b^2 = (a + b)(a - b)$	To factor $16x^2 - 25y^2$, ↓ ↓ think $(4x)^2 - (5y)^2$ so, $16x^2 - 25y^2$ $= (4x + 5y)(4x - 5y)$	p. 634
Factoring a Difference of Cubes Use the form $a^3 - b^3 = (a - b)(a^2 + ab + b^2)$	Factor $x^3 - 64$. $x^3 - 64 = x^3 - 4^3$ $= (x - 4)(x^2 + 4x + 16)$	p. 636
Factoring a Sum of Cubes Use the form $a^3 + b^3 = (a + b)(a^2 - ab + b^2)$	Factor $x^3 + 8y^3$. $x^3 + 8y^3 = x^3 + (2y)^3$ $= (x + 2y)(x^2 - 2xy + 4y^2)$	p. 636
Factoring a Perfect Square Trinomial Use one of the forms $a^2 + 2ab + b^2 = (a + b)^2$ $a^2 - 2ab + b^2 = (a - b)^2$	Factor $25x^2 + 40xy + 16y^2$. $25x^2 + 40xy + 16y^2$ $= (5x)^2 + 2(5x \cdot 4y) + (4y)^2$ $= (5x + 4y)^2$	p. 637

Continued

Definition/Procedure	Example	Reference

Factoring Trinomials: Trial-and-Error

Section 6.3

p. 644

	Factoring Sign Pattern	Example
Both signs are positive	$(x + \quad)(x + \quad)$	$x^2 + 5x + 6$
The constant is positive and the x coefficient is negative	$(x - \quad)(x - \quad)$	$x^2 - 1x + 6$
The constant is negative	$(x + \quad)(x - \quad)$	$x^2 - 1x - 12$ or $x^2 + 4x - 12$

Factoring Trinomials

Sections 6.3–6.4

p. 658

The *ac* Test A trinomial of the form $ax^2 + bx + c$ is factorable if (and only if) there are two integers m and n such that

$$ac = mn \qquad b = m + n$$

To Factor a Trinomial In general, you can apply these steps.

Step 1 Write the trinomial in standard $ax^2 + bx + c$ form.

Step 2 Label the three coefficients a, b, and c.

Step 3 Find two integers m and n such that

$$ac = mn \qquad \text{and} \qquad b = m + n$$

Step 4 Rewrite the trinomial as

$$ax^2 + mx + nx + c$$

Step 5 Factor by grouping.

Not all trinomials are factorable. To discover whether a trinomial is factorable, try the ***ac* test.**

Example:

Given $2x^2 - 5x - 3$

$a = 2 \quad b = -5 \quad c = -3$

$ac = -6 \quad b = -5$

$m = -6 \quad n = 1$

$mn = -6 \quad m + n = -5$

$2x^2 - 6x + x - 3$

$2x(x - 3) + 1(x - 3)$

$(2x + 1)(x - 3)$

Strategies in Factoring

Section 6.5

p. 671

1. **Always look for a greatest common factor.** If you find a GCF (other than 1), factor out the GCF as your first step. If the leading coefficient is negative, factor out the GCF with a negative coefficient.
2. Now look at the **number of terms** in the polynomial you are trying to factor.
   - **(a)** If the polynomial is a *binomial,* consider the special binomial formulas.
   - **(b)** If the polynomial is a *trinomial,* try to factor it as a product of two binomials. You can use either the trial-and-error method or the *ac* method.
   - **(c)** If the polynomial has *more than three terms,* try factoring by grouping.
3. You should always **factor the polynomial completely.** So after you apply one of the techniques given in part 2, another one may be necessary.
4. You can always **check your answer by multiplying.**

Example:

Factor completely.

$-2x^4 + 32$

$= -2(x^4 - 16)$

$= -2(x^2 + 4)(x^2 - 4)$

$= -2(x^2 + 4)(x + 2)(x - 2)$

To check, we multiply.

$-2(x^2 + 4)(x + 2)(x - 2)$

$= -2(x^2 + 4)(x^2 - 2x + 2x - 4)$

$= -2(x^2 + 4)(x^2 - 4)$

$= -2(x^4 - 4x^2 + 4x^2 - 16)$

$= -2(x^4 - 16)$

$= -2x^4 + 32$

Definition/Procedure	Example	Reference
Solving Quadratic Equations by Factoring		**Section 6.6**
1. Add or subtract the necessary terms on both sides of the equation so that the equation is in standard form (set equal to 0). **2.** Factor the quadratic expression. **3.** Set each factor equal to 0. **4.** Solve the resulting equations to find the solutions. **5.** Check each solution by substituting in the original equation.	To solve: $x^2 + 7x = 30$ $x^2 + 7x - 30 = 0$ $(x + 10)(x - 3) = 0$ $x + 10 = 0 \quad \text{or} \quad x - 3 = 0$ $x = -10 \quad \text{or} \quad x = 3$ *Check* $(-10)^2 + 7(-10) \stackrel{?}{=} 30$ $100 + (-70) = 30$ ✓ $(3)^2 + 7(3) \stackrel{?}{=} 30$ $9 + 21 = 30$ ✓ Solution set: $\{-10, 3\}$	*p.* 682

summary exercises :: chapter 6

This summary exercise set is provided to give you practice with each of the objectives of this chapter. Each exercise is keyed to the appropriate chapter section. When you are finished, you can check your answers to the odd-numbered exercises in the back of the text. If you have difficulty with any of these questions, go back and reread the examples from that section. The answers to the even-numbered exercises appear in the *Instructor's Solutions Manual.* Your instructor will give you guidelines on how best to use these exercises in your instructional setting.

6.1 *Completely factor each polynomial.*

1. $14a - 35$

2. $9m^2 - 21m$

3. $24s^2t - 16s^2$

4. $18a^2b + 36ab^2$

5. $27s^4 + 18s^3$

6. $3x^3 - 6x^2 + 15x$

7. $18m^2n^2 - 27m^2n + 36m^2n^3$

8. $81x^7y^6 + 63x^4y^6$

9. $8a^2b + 24ab - 16ab^2$

10. $3x^2y - 6xy^3 + 9x^3y - 12xy^2$

11. $2x(3x + 4y) - y(3x + 4y)$

12. $5(w - 3z) - w(w - 3z)$

6.2

13. $p^2 - 49$

14. $25a^2 - 16$

15. $9n^2 - 25m^2$

16. $16r^2 - 49s^2$

17. $25 - z^2$

18. $a^4 - 16b^2$

19. $25a^2 - 36b^2$

20. $x^{10} - 9y^2$

21. $3w^3 - 12wz^2$

22. $16a^4 - 49b^2$

23. $2m^2 - 72n^4$

24. $3w^3z - 12wz^3$

25. $x^2 - 4x + 5x - 20$

26. $x^2 + 7x - 2x - 14$

27. $8x^2 + 6x - 20x - 15$

28. $12x^2 - 9x - 28x + 21$

29. $6x^3 + 9x^2 - 4x^2 - 6x$

30. $3x^4 + 6x^3 + 5x^3 + 10x^2$

31. $y^2 - 49$

32. $9x^2 - 64$

33. $4x^2 - 1$

34. $3n - 75n^3$

35. $x^2 - 18x + 81$

36. $x^2 + 12x + 36$

37. $16m^2 - 49n^2$

38. $50m^3 - 18mn^2$

39. $a^4 - 16b^4$

40. $m^3 - 64$

41. $8x^3 + 1$

42. $8c^3 - 27d^3$

43. $125m^3 + 64n^3$

44. $2x^4 + 54x$

6.3–6.5

45. $x^2 + 12x + 20$

46. $a^2 + a - 2$

47. $w^2 - 15w + 54$

48. $r^2 - 9r - 36$

49. $x^2 - 8xy - 48y^2$

50. $a^2 + 17ab + 30b^2$

51. $14x^2 + 43x - 21$

52. $2a^2 + 3a - 35$

53. $-x^2 - 9x - 20$

54. $x^2 - 10x + 24$

55. $a^2 - 7a + 12$

56. $w^2 + 13w + 40$

57. $x^2 + 16x + 64$

58. $-r^2 + 15r - 36$

59. $b^2 - 4bc - 21c^2$

60. $m^2n + 4mn - 32n$

61. $m^3 + 2m^2 - 35m$

62. $2x^2 - 2x - 40$

63. $3y^3 - 48y^2 + 189y$

64. $3b^3 - 15b^2 - 42b$

65. $3x^2 + 8x + 5$

66. $5w^2 + 13w - 6$

67. $2b^2 - 9b + 9$

68. $8x^2 + 2x - 3$

69. $10x^2 - 11x + 3$

70. $4a^2 + 7a - 15$

71. $16y^2 - 8xy - 15x^2$

72. $8x^2 + 14xy - 15y^2$

73. $-8x^3 + 36x^2 + 20x$

74. $9x^2 - 15x - 6$

75. $6x^3 - 3x^2 - 9x$

76. $-3x^2 + 3xy + 18y^2$

6.4 *Use the ac test to determine which trinomials can be factored. Find the values of m and n for each trinomial that can be factored.*

77. $x^2 - x - 30$

78. $x^2 + 3x + 2$

79. $2x^2 - 11x + 12$

80. $4x^2 - 23x + 15$

6.5 *Factor each polynomial completely.*

81. $5x^4 - 5x^3 - 210x^2$

82. $81n^4 - 3n$

83. $72y - 2y^3$

84. $xy - 8x - 5y + 40$

85. $10x^5 + 80x^4 + 160x^3$

86. $3x^2 + 8x - 15$

87. $x^3 + 2x^2 - 25x - 50$

88. $54m + 16m^4$

6.6 *Solve each equation by factoring.*

89. $x^2 + 5x - 6 = 0$

90. $x^2 + 2x - 8 = 0$

91. $x^2 + 7x = 30$

92. $x^2 - 6x = 40$

93. $x^2 + x = 20$

94. $x^2 = 28 - 3x$

95. $x^2 - 10x = 0$

96. $x^2 = 12x$

97. $x^2 - 25 = 0$

98. $x^2 = 225$

99. $2x^2 - x - 3 = 0$

100. $3x^2 - 4x = 15$

101. $3x^2 + 9x - 30 = 0$

102. $4x^2 + 24x = -32$

103. $x(x - 5) = 36$

104. $(x - 2)(2x + 1) = 33$

105. $x^3 - 2x^2 - 15x = 0$

106. $x^3 + x^2 - 4x - 4 = 0$

6.7 *Solve each application.*

107. NUMBER PROBLEM One integer is 15 more than another. If the product of the integers is -54, find the two integers.

108. NUMBER PROBLEM The sum of the squares of two consecutive odd integers is 130. What are the two integers?

109. GEOMETRY The length of a rectangle is 6 ft more than its width. If the area of the rectangle is 216 ft^2, find the dimensions of the rectangle.

110. GEOMETRY An open carton is formed from a rectangular piece of cardboard that is 5 in. longer than it is wide, by removing 4-in. squares from each corner and folding up the sides. If the volume of the carton is then 200 $in.^3$, what were the dimensions of the original piece of cardboard?

111. SCIENCE AND MEDICINE If a ball is thrown vertically upward from the roof of a building 20 ft high with an initial velocity of 80 ft/s, its approximate height h after t seconds is given by

$$h = -16t^2 + 80t + 20$$

How long does it take the ball to pass through a height of 84 ft on the way back down to the ground?

112. CONSTRUCTION A rectangular garden has dimensions 15 ft by 20 ft. The garden is to be enlarged with a strip of equal width surrounding it. If the resulting garden is to be 294 ft^2 larger than before, how wide should the strip be?

113. BUSINESS AND FINANCE Suppose that the cost, in dollars, of producing x stereo systems is given by

$$C(x) = 3{,}000 - 60x + 3x^2$$

How many systems can be produced for $7,500?

114. BUSINESS AND FINANCE The demand equation for a certain type of computer paper when sold at price p (in dollars) is predicted to be

$$D = -3p + 69$$

The supply equation is predicted to be

$$S = -p^2 + 24p - 3$$

Find the equilibrium price.

Hint: The equilibrium price occurs when demand is equal to supply.

self-test 6

CHAPTER 6

Name ____________

Section ________ Date ________

Answers

1. ____________
2. ____________
3. ____________
4. ____________
5. ____________
6. ____________
7. ____________
8. ____________
9. ____________
10. ____________
11. ____________
12. ____________
13. ____________
14. ____________
15. ____________
16. ____________
17. ____________
18. ____________
19. ____________
20. ____________

The purpose of this self-test is to help you assess your progress so that you can find concepts that you need to review before the next exam. Allow yourself about an hour to take this test. At the end of that hour, check your answers against those given in the back of this text. If you miss any, go back to the appropriate section to reread the examples until you have mastered that particular concept.

Factor each polynomial completely.

1. $32a^2b - 50b^3$

2. $x^2 + 2x - 5x - 10$

3. $7b + 42$

4. $x^4 - 81$

5. $9x^2 - 12xy + 4y^2$

6. $5x^2 - 10x + 20$

7. $16y^2 - 49x^2$

8. $8x^2 - 2xy - 3y^2$

9. $27y^3 - 8x^3$

10. $3w^2 + 10w + 7$

11. $6x^2 + 4x - 15x - 10$

12. $a^2 - 5a - 14$

13. $-6x^3 - 3x^2 + 30x$

14. $y^2 + 12yz + 20z^2$

Solve each equation.

15. $x^2 - 11x = -30$

16. $2x^2 + 16x + 30 = 0$

17. $x^2 - 2x - 3 = 0$

18. $6x^2 - 7x = 3$

Solve each application.

19. GEOMETRY The length of a rectangle is 4 cm less than twice the width. If the area is 240 cm^2, what is the length of the rectangle?

20. SCIENCE AND MEDICINE If a ball is thrown upward from the roof of an 18-meter building with an initial velocity of 20 m/s, its approximate height h after t seconds is given by

$$h = -5t^2 + 20t + 18$$

When is the ball at a height of 38 m?

cumulative review chapters 0-6

Name ____________

Section ________ Date ________

We offer the following exercises to help you review concepts from earlier chapters. This is meant as review material and not as a comprehensive exam. The answers are presented in the back of the text. If you have difficulty with any of these exercises, be certain to at least read through the summary related to that section.

1. If $x = -3$, $y = 2$, and $z = -4$, find the value of the expression $\dfrac{3x^2 - 2z + 1}{5y + 2x}$.

Solve each equation.

2. $7a - 3 = 6a + 8$

3. $\frac{2}{3}x = -22$

Solve for the indicated variable.

4. $A = P + Prt$ for r

5. $P = 2L + 2W$ for W

Solve each inequality.

6. $2x - 7 < 5 + 4x$

7. $2x - 9 \le 7x - 3(x - 1)$

Solve each system.

8. $3x - 5y = 5$
$-x + y = -1$

9. $2x - 3y = 13$
$x = 3y + 9$

10. Find the slope and y-intercept of the line represented by the equation $2x + 5y = 10$.

Write the equation of the line that satisfies the given conditions.

11. L passes through the points $(-2, 1)$ and $(1, 7)$.

12. L has y-intercept $(0, -2)$ and is parallel to the line with equation $3x + y = 5$.

Perform the indicated operations. Simplify your results.

13. $7x^2y + 3xy - 5x^2y + 2xy$

14. $(-3x^2 + 5x - 6) + (2x^2 - 3x + 5)$

15. $(5x^2 + 4x - 3) - (4x^2 - 5x - 1)$

16. $3x(x^2 - 2x + 2)$

17. $(2x - 5)(x + 1)$

18. $(x + 3)(x^2 - 3x + 9)$

Answers

1. ____________
2. ____________
3. ____________
4. ____________
5. ____________
6. ____________
7. ____________
8. ____________
9. ____________
10. ____________
11. ____________
12. ____________
13. ____________
14. ____________
15. ____________
16. ____________
17. ____________
18. ____________

Answers

19. ______
20. ______
21. ______
22. ______
23. ______
24. ______
25. ______
26. ______
27. ______
28. ______
29. ______
30. ______
31. ______
32. ______
33. ______
34. ______
35. ______
36. ______
37. ______
38. ______

19. $(4x - 3)(2x + 5)$

20. $(a - 3b)(a + 3b)$

21. $(x - 2y)^2$

22. $2x(5x + 3y)(5x - 3y)$

Use the properties of exponents to simplify each expression.

23. $(-2x^{-2}y^3)^3$

24. $\dfrac{27a^5b^7}{9ab}$

25. $(2x^3y^{-2})^{-2}(x^{-3}y^{-1})$

26. $\dfrac{16x^2y^{-3}z^{-4}}{8x^{-1}y^2z^{-5}}$

Factor each expression.

27. $12x + 20$

28. $25x^2 - 49y^2$

29. $12x^2 - 15x + 8x - 10$

30. $2x^2 - 13x + 15$

Solve each quadratic equation.

31. $x^2 + 2x = 15$

32. $x^2 - 9 = 0$

Perform the indicated division.

33. $\dfrac{24x^2y^5 - 15x^4y^2 + 9xy}{3xy}$

34. $\dfrac{x^2 + 3x + 2}{x - 1}$

Solve each application.

35. Number Problem Three times a number decreased by 5 is 46. Find the number.

36. Business and Finance Juan's biology text cost \$5 more than his mathematics text. Together they cost \$81. Find the cost of the biology text.

37. Science and Medicine The equation $d = \dfrac{x^2}{20} + x$ relates the speed of a car x, in miles per hour, to the stopping distance d, in feet. How fast is a car going if it requires 240 ft to stop?

38. Business and Finance Suppose that a manufacturer's weekly profit in dollars is given by

$P = -5x^2 + 300x$

How many items x must be produced to realize a profit of \$2,500?

CHAPTER

7

Radicals and Exponents

INTRODUCTION

Applications of mathematics arise anytime we take a measurement. Some examples of measurable quantities include time, temperature, and distance. Often, we take measurements when conducting an experiment. Sometimes, experiments lead us to new discoveries in science and math and provide us with some of the most interesting applications.

The 17th century Italian scientist Galileo (1564–1642) conducted some important experiments related to motion. In particular, Galileo worked with pendulums (he called them *pulsilogia*) to develop his theories. You will conduct experiments similar to Galileo's in this chapter's activity.

CHAPTER 7 OUTLINE

7.1 Roots and Radicals

< 7.1 Objectives >

1 > Evaluate expressions containing radicals

2 > Use a calculator to estimate or evaluate radical expressions

3 > Simplify expressions that contain radicals

4 > Apply the Pythagorean theorem

5 > Use the distance formula

6 > Write the equation of a circle and sketch its graph

In Chapter 5 we reviewed the properties of integer exponents. In this chapter, we work to extend those properties. To achieve that objective, we must develop a notation that "reverses" the power process.

The statement

$$x^2 = 9$$

is read as "x squared equals 9."

NOTE

A negative number has *no* real square roots.

In this section we are concerned with the relationship between the base x and the number 9. Equivalently, we can say that "x is a square root of 9."

We know from experience that x must be 3 (because $3^2 = 9$) or -3 (because $(-3)^2 = 9$). We see that 9 has the two square roots, 3 and -3. In fact, every positive number has *two* square roots, one positive and one negative. In general,

If $x^2 = a$, we say x is a *square root* of a.

We also know that

$$3^3 = 27$$

and similarly we call 3 a *cube root* of 27. Here 3 is the *only* real number with that property. Every real number (positive or negative) has exactly *one* real cube root.

Definition

Roots

In general, we state that if

$$x^n = a$$

then x is an *nth root* of a.

NOTE

The symbol $\sqrt{\ }$ first appeared in print in 1525. In Latin, "radix" means root, and this was contracted to a small *r*. The present symbol may have been used because it resembled the manuscript form of that small *r*.

We are now ready for new notation. The symbol $\sqrt{\ }$ is called a *radical sign*. We saw earlier that 3 is the positive square root of 9.

We call 3 the *principal square root* of 9, and we write

$$\sqrt{9} = 3$$

Every positive number has two square roots, one positive and one negative. The principal square root is always the positive one.

In some cases we want to indicate the negative square root; to do so, we write

$$-\sqrt{9} = -3$$

to indicate the negative root.

If both square roots need to be indicated, we can write

$$\pm\sqrt{9} = \pm 3$$

NOTE

The index of 2 for square roots is generally not written. We understand that

$\sqrt{a}$

is the principal square root of *a*.

Every radical expression contains three parts, as shown below. The principal *n*th root of *a* is written as

$$\sqrt[n]{a}$$

Example 1 Evaluating Radical Expressions

< Objective 1 >

Evaluate, if possible.

(a) $\sqrt{49} = 7$

(b) $-\sqrt{49} = -7$

(c) $\pm\sqrt{49} = \pm 7$

(d) $\sqrt{-49}$ is not a real number.

Let's examine part (d) more carefully. Suppose that for some real number x,

$$x = \sqrt{-49}$$

By our earlier definition, this means that

$$x^2 = -49$$

NOTE

Neither 7^2 nor $(-7)^2$ equals -49.
We consider imaginary numbers in Section 7.6.

which is impossible. There is no real square root for -49. We call $\sqrt{-49}$ an *imaginary number.*

Check Yourself 1

Evaluate, if possible.

(a) $\sqrt{64}$ **(b)** $-\sqrt{64}$ **(c)** $\pm\sqrt{64}$ **(d)** $\sqrt{-64}$

NOTE

Indices is the plural of index.

Our next example considers cube roots and radicals with higher indices.

Example 2 Evaluating Radical Expressions

Evaluate, if possible.

(a) $\sqrt[3]{64} = 4$ because $4^3 = 64$

> **NOTE**
>
> The cube root of a *negative* number is *negative.*

> **NOTE**
>
> In general, an *even* root of a *negative* number is *not real;* it is *imaginary.*

(b) $-\sqrt[3]{64} = -4$

(c) $\sqrt[3]{-64} = -4$ because $(-4)^3 = -64$

(d) $\sqrt[4]{81} = 3$ because $3^4 = 81$

(e) $\sqrt[4]{-81}$ is not a real number.

(f) $\sqrt[5]{32} = 2$ because $2^5 = 32$

(g) $\sqrt[5]{-32} = -2$ because $(-2)^5 = -32$

Check Yourself 2

Evaluate.

(a) $\sqrt[3]{125}$ **(b)** $-\sqrt[3]{125}$ **(c)** $\sqrt[3]{-125}$ **(d)** $\sqrt[4]{16}$

(e) $\sqrt[5]{243}$ **(f)** $\sqrt[4]{-16}$ **(g)** $\sqrt[5]{-243}$

All of the numbers in our previous examples and exercises were chosen so that the results would be *rational numbers.* That is, our radicands were

Perfect squares: 1, 4, 9, 16, 25, . . .

Perfect cubes: 1, 8, 27, 64, 125, . . .

and so on.

The square root of a number that is *not* a perfect square (or the cube root of a number that is *not* a perfect cube) is not a rational number.

Expressions such as $\sqrt{2}$, $\sqrt{3}$, and $\sqrt{5}$ are *irrational numbers.* A calculator with a square root key [√] gives decimal approximations for such numbers.

Example 3 Estimating Radical Expressions

< Objective 2 >

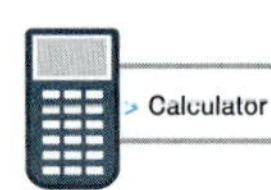


```
√(17)
        4.123105626
√(28)
        5.291502622
```

Use a calculator to find decimal approximations for each number. Round all answers to three decimal places.

(a) $\sqrt{17}$

On many calculators, the square root is shown as the "2nd function" or "inverse" of x^2. Press [2nd] [√] and type 17. Then type a closing parenthesis [)] and press [ENTER]. The display should read 4.123105626. Rounded to three decimal places, the result is 4.123.

(b) $\sqrt{28}$

The display should read 5.291502622. Rounded to three decimal places, the result is 5.292.

```
ERR:NONREAL ANS
1:Quit
2:Goto
```

(c) $\sqrt{-11}$

Press [2nd] [√], and then type [(−)] 11 [)], and press [ENTER]. The display will say NONREAL ANS (or something similar). This indicates that −11 does not have a real square root.

Check Yourself 3

Use a calculator to find decimal approximations for each number. Round each answer to three decimal places.

(a) $\sqrt{13}$ **(b)** $\sqrt{38}$ **(c)** $\sqrt{-21}$

To evaluate roots other than square roots using a graphing calculator, we may utilize the menu found by pressing the [MATH] key, and choosing the symbol $\sqrt[x]{\ }$.

Example 4 Estimating Radical Expressions

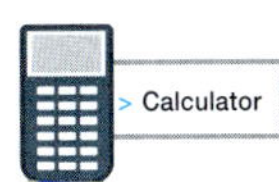

Use a calculator to find decimal approximations of each number. Round each answer to three decimal places.

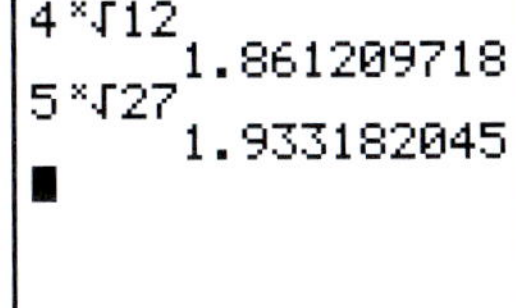

(a) $\sqrt[4]{12}$

Begin by typing the index 4. Then press the [MATH] key, and choose the symbol $\sqrt[x]{\ }$. Now type 12, and press [ENTER]. The display should read 1.861209718. Rounded to three decimal places, the result is 1.861.

(b) $\sqrt[5]{27}$

Type the index 5, get the symbol $\sqrt[x]{\ }$ as before, and then type 27. Pressing [ENTER] should show 1.933182045. Rounded to three decimal places, the result is 1.933.

Check Yourself 4

Use a calculator to find decimal approximations of each number. Round each answer to three decimal places.

(a) $\sqrt[4]{35}$ **(b)** $\sqrt[5]{29}$

A certain amount of caution should be exercised in dealing with principal even roots. For example, consider the statement

$$\sqrt{x^2} = x$$

First, let $x = 2$: $\sqrt{2^2} = \sqrt{4} = 2$

Now, let $x = -2$: $\sqrt{(-2)^2} = \sqrt{4} = 2$

So the statement $\sqrt{x^2} = x$ is true when x is positive, but $\sqrt{x^2} = x$ is not true when x is negative.

In fact, if x is negative, we have

$$\sqrt{x^2} = -x$$

Putting these ideas together gives

$$\sqrt{x^2} = \begin{cases} x & \text{when } x \geq 0 \\ -x & \text{when } x < 0 \end{cases}$$

Earlier, you studied absolute value, and we make a connection here.

$$|2| = 2 \quad \text{and} \quad |-2| = 2$$

NOTE

We can extend this last statement to

$\sqrt[n]{x^n} = |x|$

when n is *even*.

So, $|x| = x$ when x is positive, and $|x| = -x$ when x is negative.

Putting these ideas together, gives

$$|x| = \begin{cases} x & \text{when } x \geq 0 \\ -x & \text{when } x < 0 \end{cases}$$

We can summarize the discussion by writing

$$\sqrt{x^2} = |x|$$

Example 5 Evaluating Radical Expressions

Evaluate.

NOTE

Alternatively, we could write

$\sqrt{(-4)^2} = \sqrt{16} = 4$

(a) $\sqrt{5^2} = 5$

(b) $\sqrt{(-4)^2} = |-4| = 4$

(c) $\sqrt[4]{2^4} = 2$

(d) $\sqrt[4]{(-3)^4} = |-3| = 3$

Check Yourself 5

Evaluate.

(a) $\sqrt{6^2}$ (b) $\sqrt{(-6)^2}$ (c) $\sqrt[4]{3^4}$ (d) $\sqrt[4]{(-3)^4}$

Roots with indices that are odd do *not* require absolute values. For instance,

$$\sqrt[3]{3^3} = \sqrt[3]{27} = 3$$
$$\sqrt[3]{(-3)^3} = \sqrt[3]{-27} = -3$$

and we see that

$$\sqrt[n]{x^n} = x \qquad \text{when } n \text{ is odd}$$

To summarize, we can write

$$\sqrt[n]{x^n} = \begin{cases} |x| & \text{when } n \text{ is even} \\ x & \text{when } n \text{ is odd} \end{cases}$$

We turn now to an example in which variables are involved in the radicand.

Example 6 Simplifying Radical Expressions

< Objective 3 >

NOTE

We can determine the power of the variable in the root by dividing the power in the radicand by the index. In Example 6(d), 8 ÷ 4 = 2.

Simplify each expression.

(a) $\sqrt[3]{a^3} = a$

(b) $\sqrt{16m^2} = 4|m|$

(c) $\sqrt[5]{32x^5} = 2x$

(d) $\sqrt[4]{x^8} = x^2$ because $(x^2)^4 = x^8$

(e) $\sqrt[3]{27y^6} = 3y^2$ Do you see why?

Check Yourself 6

Simplify.

(a) $\sqrt[4]{x^4}$ (b) $\sqrt{49w^2}$ (c) $\sqrt[5]{a^{10}}$ (d) $\sqrt[3]{8y^9}$

NOTE

"Distance" or "length" is always a nonnegative number.

An important result that involves radicals is the **Pythagorean theorem.** You may recall from earlier math courses that this theorem gives a relationship between the lengths of the sides of a right triangle (a triangle with a 90° angle).

Property

The Pythagorean Theorem

In any right triangle, the square of the longest side (the hypotenuse) is equal to the sum of the squares of the two shorter sides (the legs).

$c^2 = a^2 + b^2$

NOTE

This is only true in the case of right triangles.

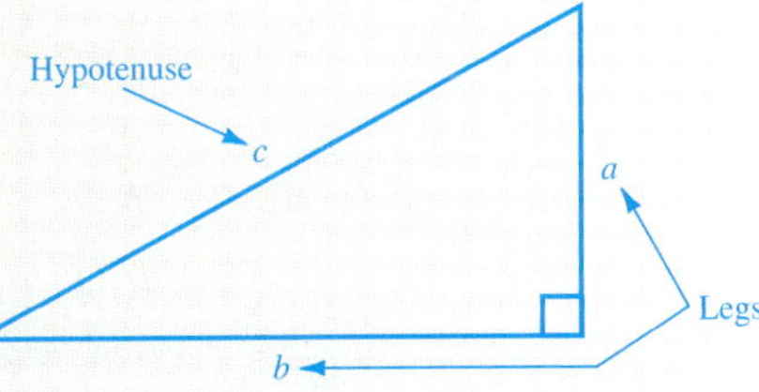

The conclusion of this theorem says that $a^2 + b^2 = c^2$. This means that $\sqrt{a^2 + b^2} = \sqrt{c^2}$. Since the lengths a, b, and c are all positive numbers, $\sqrt{c^2} = c$. So we may write: $c = \sqrt{a^2 + b^2}$ in a right triangle.

Example 7 Using the Pythagorean Theorem

< Objective 4 >

Suppose the two legs of a right triangle are 7 and 12. How long is the hypotenuse? We should always draw a sketch when solving a geometric problem.

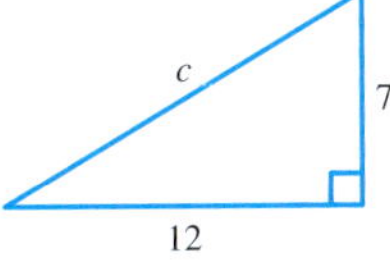

The Pythagorean theorem tells us that $c = \sqrt{(12)^2 + (7)^2} = \sqrt{144 + 49} = \sqrt{193}$.

NOTE

$13 < \sqrt{193} < 14$ because $13^2 = 169$ and $14^2 = 196$.

Check Yourself 7

Suppose the two legs of a right triangle are 11 and 14. How long is the hypotenuse?

As long as two sides of a right triangle have known lengths, we can determine the length of the third side. In the next example, the hypotenuse and one leg have known lengths.

Example 8 Using the Pythagorean Theorem

The hypotenuse of a right triangle is 23 and one leg is 16. How long is the other leg? As in Example 7, drawing a sketch helps.

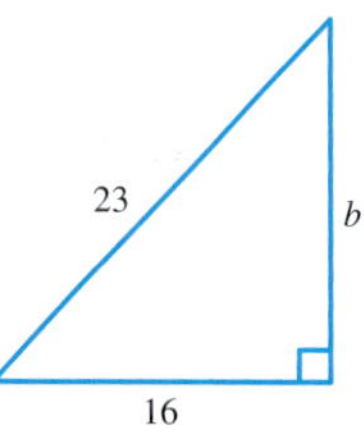

Let $c = 23$ and $a = 16$. Since $a^2 + b^2 = c^2$, we write

$$b^2 = c^2 - a^2$$
$$b = \sqrt{c^2 - a^2}$$

So, $b = \sqrt{(23)^2 - (16)^2} = \sqrt{529 - 256} = \sqrt{273}$

The length of the other leg is $\sqrt{273}$.

NOTE

$16 < \sqrt{273} < 17$

Check Yourself 8

The hypotenuse of a right triangle is 19 and one leg is 12. How long is the other leg?

Now suppose that we want to find the distance d between two points in the coordinate plane, say (3, 1) and (8, 7).

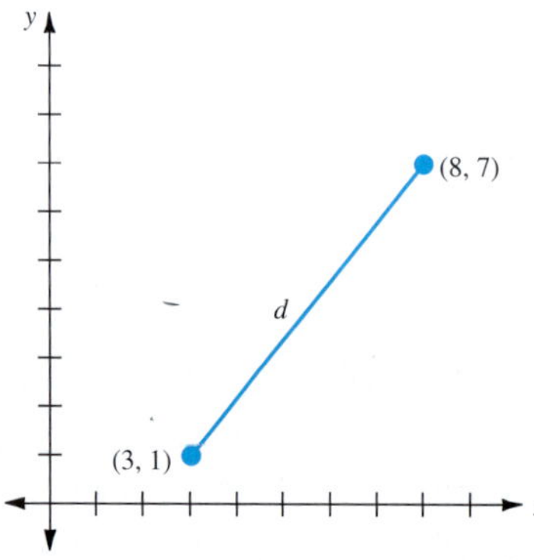

We can make a right triangle here.

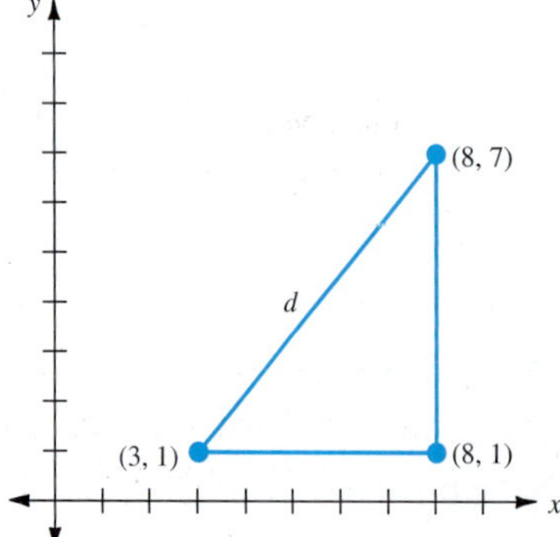

The length of the horizontal leg of the triangle is $8 - 3 = 5$ units, while the length of the vertical leg of this triangle is $7 - 1 = 6$ units. Using the Pythagorean theorem, we write

$$d = \sqrt{(8 - 3)^2 + (7 - 1)^2} = \sqrt{5^2 + 6^2} = \sqrt{25 + 36} = \sqrt{61}$$

We generalize this to construct the distance formula.

Property

The Distance Formula

The distance, d, between two points (x_1, y_1) and (x_2, y_2) can be found using the formula

$$d = \sqrt{(x_2 - x_1)^2 + (y_2 - y_1)^2}$$

Example 9 demonstrates the use of this formula.

Example 9 Finding the Distance Between Two Points

< Objective 5 >

Find the distance between each pair of points.

(a) (3, −5) and (−5, −5)

Let $(x_1, y_1) = (3, -5)$ and $(x_2, y_2) = (-5, -5)$. Plugging those values into the distance formula gives

$$d = \sqrt{(-5 - 3)^2 + [-5 - (-5)]^2} = \sqrt{(-8)^2 + 0^2} = \sqrt{64} = 8$$

The distance between the two points is 8 units.

(b) (−4, 7) and (1, 5)

Let $(x_1, y_1) = (-4, 7)$ and $(x_2, y_2) = (1, 5)$. Plugging those values into the distance formula gives

$$d = \sqrt{[1 - (-4)]^2 + (5 - 7)^2} = \sqrt{5^2 + (-2)^2} = \sqrt{29}$$

The distance between the two points is $\sqrt{29}$ units.

Check Yourself 9

Find the distance between each pair of points.

(a) (−2, 7) and (−5, 7) **(b)** (3, −5) and (7, −4)

The distance formula gives us a method for describing the equation of a circle in the coordinate plane. Consider the definition of a circle.

Definition

Circle

A **circle** is the set of all points in the plane equidistant from a fixed point, called the **center** of the circle. The distance between the center of the circle and any point on the circle is called the **radius** of the circle.

Suppose a circle has its center at a point with coordinates (h, k) and radius r. If (x, y) represents any point on the circle, then, by its definition, the distance from (h, k) to (x, y) is r. Applying the distance formula, we have

$$r = \sqrt{(x - h)^2 + (y - k)^2}$$

Squaring both sides gives an equation of a circle.

$$r^2 = (x - h)^2 + (y - k)^2$$

Property

Equation of a Circle

The equation of a circle with center (h, k) and radius r is

$$(x - h)^2 + (y - k)^2 = r^2$$

A special case is the circle centered at the origin with radius r. Then $(h, k) = (0, 0)$, and its equation is

$$x^2 + y^2 = r^2$$

The circle equation can be used in two ways. Given the center and radius of the circle, we can write its equation; or given its equation, we can find the center and radius of a circle.

Example 10 Finding the Equation of a Circle

< Objective 6 >

Find the equation of a circle with center at $(2, -1)$ and radius 3. Sketch the circle.

Let $(h, k) = (2, -1)$ and $r = 3$. Applying the circle equation yields

$$(x - 2)^2 + [y - (-1)]^2 = 3^2$$
$$(x - 2)^2 + (y + 1)^2 = 9$$

To sketch the circle, we first locate its center. Then we determine four points 3 units to the right and left and up and down from the center of the circle. Drawing a smooth curve through those four points completes the graph.

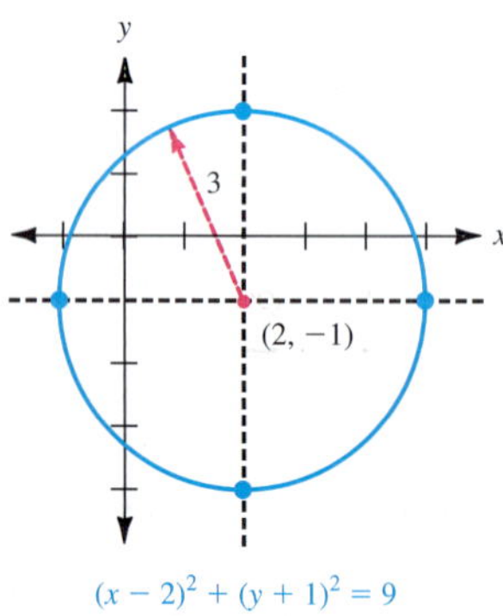

$(x - 2)^2 + (y + 1)^2 = 9$

Check Yourself 10

Find the equation of the circle with center at $(-2, 1)$ and radius 5. Sketch the circle.

Now, given an equation for a circle, we can also find the radius and center and then sketch the circle.

Example 11 Finding the Center and Radius of a Circle

Find the center and radius of the circle with equation

$$(x - 5)^2 + (y + 2)^2 = 16$$

Remember, the general form is

$$(x - h)^2 + (y - k)^2 = r^2$$

Our equation "fits" this form when it is written as

Note: $y + 2 = y - (-2)$

$$(x - 5)^2 + [y - (-2)]^2 = 4^2$$

So the center is at $(5, -2)$, and the radius is 4. The graph is shown.

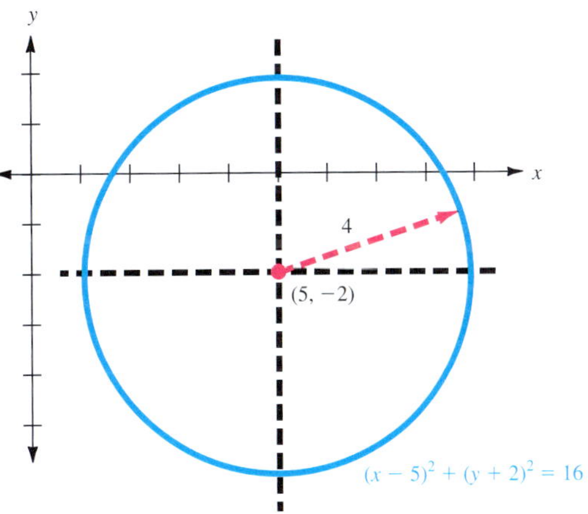

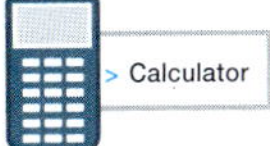

NOTE

A circle can be graphed on the calculator by solving for y, then graphing both the upper half and lower half of the circle. For example, consider the circle with equation $(x - 1)^2 + (y + 2)^2 = 9$.

$$(x - 1)^2 + (y + 2)^2 = 9$$

$$(y + 2)^2 = 9 - (x - 1)^2$$

$$y + 2 = \pm\sqrt{9 - (x - 1)^2}$$

$$y = -2 \pm \sqrt{9 - (x - 1)^2}$$

Now graph the two functions

$$y = -2 + \sqrt{9 - (x - 1)^2}$$

and

$$y = -2 - \sqrt{9 - (x - 1)^2}$$

on your calculator. (The display screen may need to be squared to obtain the shape of a circle.)

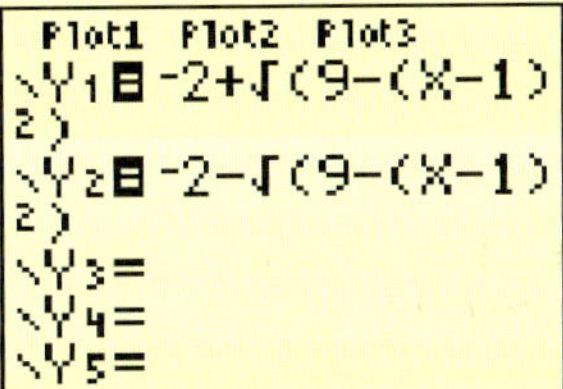

WINDOW
Xmin=-9.096774...
Xmax=9.0967741...
Xscl=1
Ymin=-6
Ymax=6
Yscl=1
Xres=1

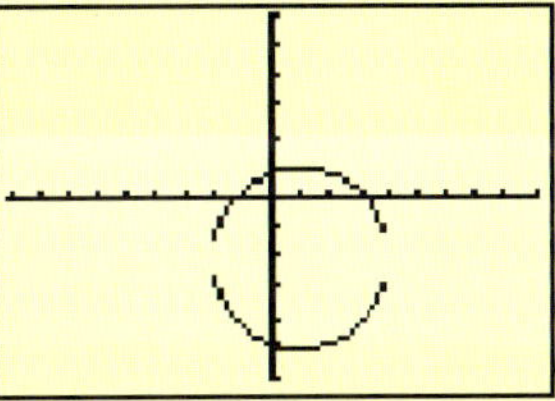

The "gaps" in the graph are a limitation of graphing calculators technology. When sketching these graphs, be sure to fill in the gaps to form a full circle.

Check Yourself 11

Find the center and radius of the circle with equation

$$(x + 3)^2 + (y - 2)^2 = 25$$

Sketch the circle.

Check Yourself ANSWERS

1. **(a)** 8; **(b)** -8; **(c)** ± 8; **(d)** not a real number **2.** **(a)** 5; **(b)** -5; **(c)** -5; **(d)** 2; **(e)** 3; **(f)** not a real number; **(g)** -3
3. **(a)** 3.606; **(b)** 6.164; **(c)** not a real number **4.** **(a)** 2.432; **(b)** 1.961
5. **(a)** 6; **(b)** 6; **(c)** 3; **(d)** 3 **6.** **(a)** $|x|$; **(b)** $7|w|$; **(c)** a^2; **(d)** $2y^3$
7. $\sqrt{317}$ **8.** $\sqrt{217}$ **9.** **(a)** 3 units; **(b)** $\sqrt{17}$ units
10. $(x + 2)^2 + (y - 1)^2 = 25$ **11.** Center $(-3, 2)$; radius 5

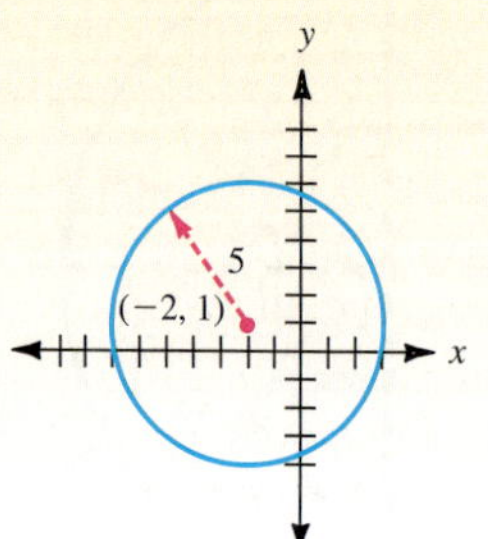

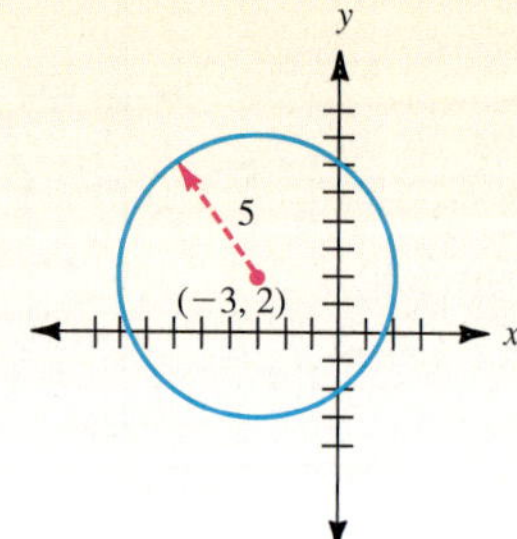

Reading Your Text

SECTION 7.1

(a) If $x^2 = a$, we say x is a ______________ of a.

(b) Every positive number has ______________ square roots.

(c) The cube root of a negative number is ______________.

(d) The distance between the center of a circle and any point on the circle is called the ______________ of the circle.

Basic Skills | Challenge Yourself | Calculator/Computer | Career Applications | Above and Beyond

7.1 exercises

Boost *your* GRADE at ALEKS.com!

ALEKS

- Practice Problems
- Self-Tests
- NetTutor
- e-Professors
- Videos

Name ______________

Section ________ Date ________

< Objective 1 >

Evaluate each root, if possible.

1. $\sqrt{49}$

2. $\sqrt{36}$

3. $-\sqrt{36}$

4. $-\sqrt{81}$

5. $\pm\sqrt{81}$ > Videos

6. $\pm\sqrt{49}$

7. $\sqrt{-49}$

8. $\sqrt{-25}$

9. $\sqrt[3]{27}$

10. $\sqrt[3]{64}$

11. $\sqrt[3]{-64}$

12. $-\sqrt[3]{125}$ > Videos

13. $-\sqrt[3]{216}$

14. $\sqrt[3]{-27}$

15. $\sqrt[4]{81}$

16. $\sqrt[5]{32}$

17. $\sqrt[5]{-32}$

18. $\sqrt[4]{-81}$

19. $-\sqrt[4]{16}$

20. $\sqrt[5]{-243}$

21. $\sqrt[4]{-16}$

22. $-\sqrt[5]{32}$

23. $-\sqrt[5]{243}$

24. $-\sqrt[4]{625}$

Answers

1. ________ 2. ________

3. ________ 4. ________

5. ________ 6. ________

7. ________

8. ________

9. ________ 10. ________

11. ________ 12. ________

13. ________ 14. ________

15. ________ 16. ________

17. ________

18. ________

19. ________ 20. ________

21. ________

22. ________ 23. ________

24. ________

Answers

25. ____
26. ____
27. ____
28. ____
29. ____
30. ____
31. ____
32. ____
33. ____
34. ____
35. ____ 36. ____
37. ____ 38. ____
39. ____ 40. ____
41. ____ 42. ____
43. ____ 44. ____
45. ____ 46. ____
47. ____ 48. ____
49. ____ 50. ____

25. $\sqrt{\dfrac{4}{9}}$

26. $\sqrt{\dfrac{9}{25}}$ > Videos

27. $\sqrt[3]{\dfrac{8}{27}}$

28. $\sqrt[3]{-\dfrac{27}{64}}$

29. $\sqrt{6^2}$

30. $\sqrt{9^2}$

31. $\sqrt{(-3)^2}$

32. $\sqrt{(-5)^2}$

33. $\sqrt[3]{4^3}$

34. $\sqrt[3]{(-5)^3}$

35. $\sqrt[4]{3^4}$

36. $\sqrt[4]{(-2)^4}$

< Objective 3 >

Simplify each root.

37. $\sqrt{x^2}$

38. $\sqrt[3]{w^3}$

39. $\sqrt[5]{y^5}$

40. $\sqrt[7]{z^7}$

41. $\sqrt{9x^2}$

42. $\sqrt{81y^2}$

43. $\sqrt{a^4b^6}$

44. $\sqrt{w^6z^{10}}$

45. $\sqrt{16x^4}$ 

46. $\sqrt{49y^6}$

< Objective 4 >

Find the missing length in each triangle. Express your answer in radical form where appropriate.

47.

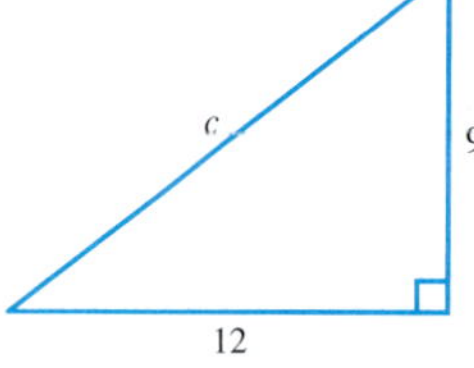

48.

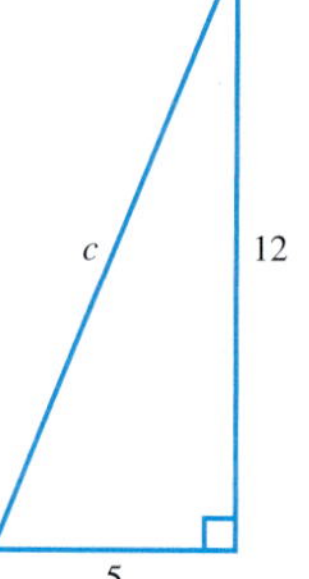

49.

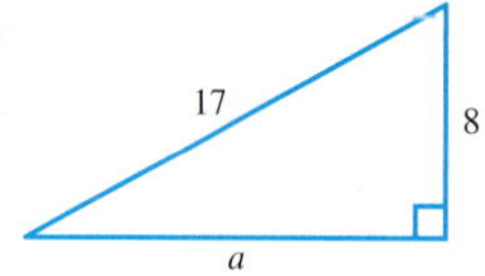

50.

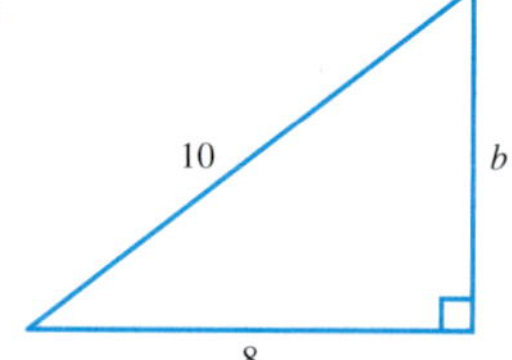

51.

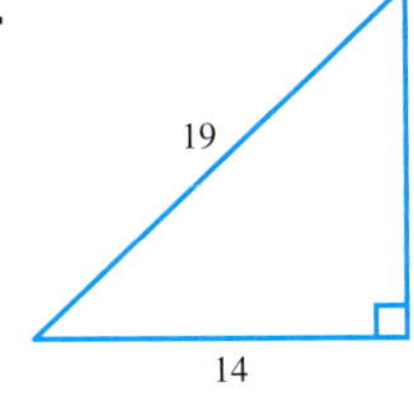

52.

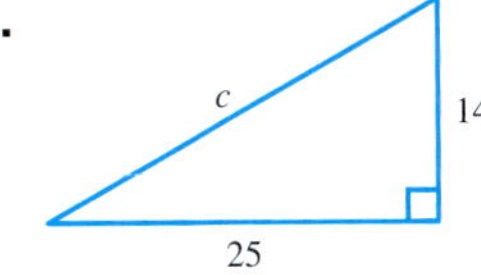

< Objective 5 >

Use the distance formula to find the distance between each pair of points.

53. $(2, -6)$ and $(2, -9)$

54. $(-3, 7)$ and $(4, 7)$

55. $(-7, 1)$ and $(-4, 0)$

56. $(-17, -5)$ and $(-12, -3)$

< Objective 6 >

Find the center and radius of each circle. Then graph it.

57. $x^2 + y^2 = 4$

58. $x^2 + y^2 = 25$

59. $(x - 1)^2 + y^2 = 9$

60. $x^2 + (y + 2)^2 = 16$

61. $(x - 4)^2 + (y + 1)^2 = 16$

> Videos

62. $(x + 3)^2 + (y + 2)^2 = 25$

Write the equation of each circle pictured.

63.

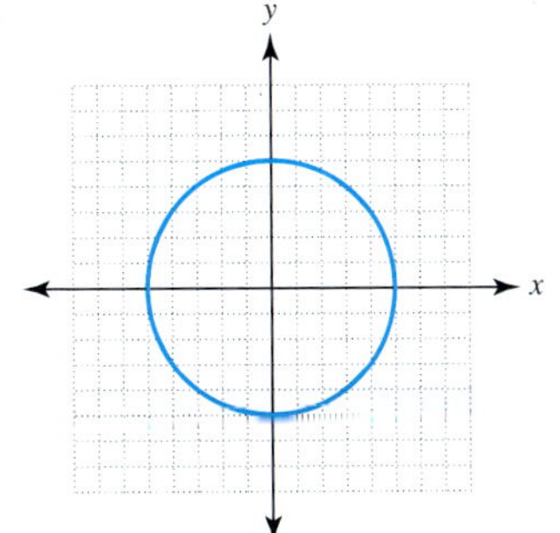

64.

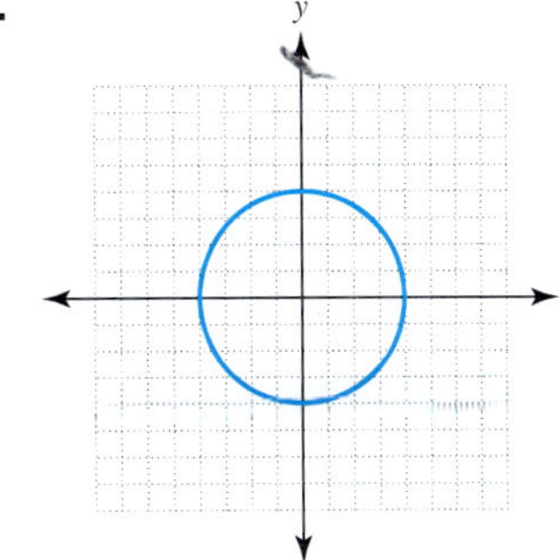

Answers

51. ______

52. ______

53. ______

54. ______

55. ______

56. ______

57. ______

58. ______

59. ______

60. ______

61. ______

62. ______

63. ______

64. ______

Answers

65. __________
66. __________
67. __________
68. __________
69. __________
70. __________
71. __________
72. __________
73. __________
74. __________
75. _____ 76. _____
77. _____ 78. _____
79. __________
80. __________
81. __________
82. __________
83. __________
84. __________
85. __________
86. __________

65. 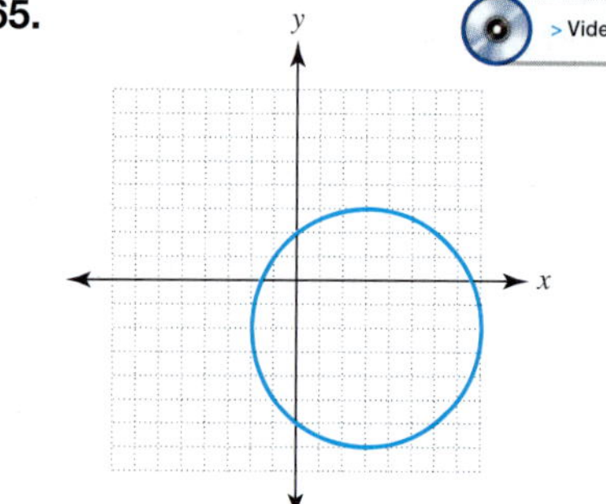

66. 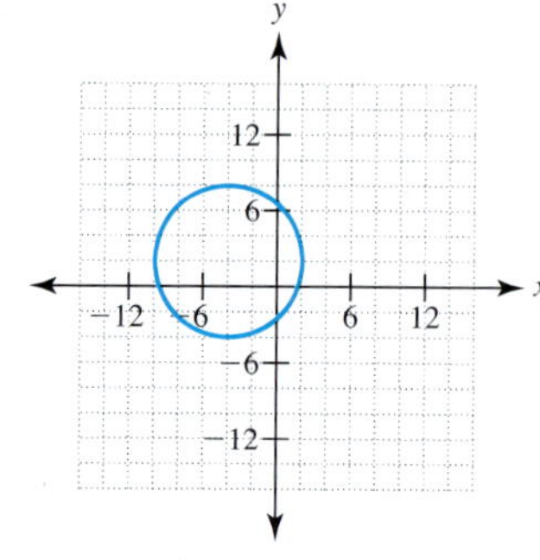

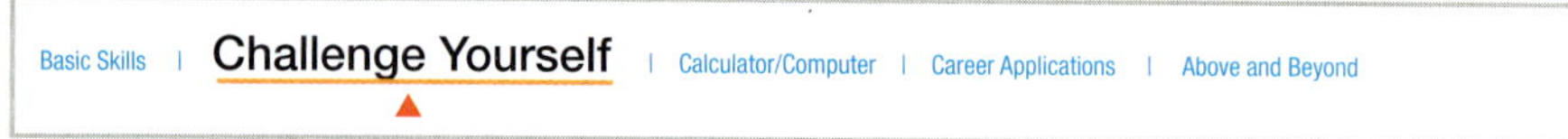

Complete each statement with **never, sometimes,** *or* **always.**

67. The principal square root of a number is __________ negative.

68. The cube root of a number is __________ a real number.

Determine whether each statement is **true** *or* **false.**

69. Given the equation $x^2 + y^2 = 9$, the graph is a circle with center at (0, 0).

70. Given the equation $x^2 + y^2 = 9$, the graph is a circle with radius 9.

Each equation defines a relation. Write the domain and the range of each relation.

71. $(x + 3)^2 + (y - 2)^2 = 16$

72. $(x - 1)^2 + (y - 5)^2 = 9$

73. $x^2 + (y - 3)^2 = 25$ 

74. $(x + 2)^2 + y^2 = 36$

Basic Skills | Challenge Yourself | **Calculator/Computer** | Career Applications | Above and Beyond

< Objective 2 >

Use a calculator to evaluate each root. Round each answer to three decimal places.

75. $\sqrt{15}$

76. $\sqrt{29}$

77. $\sqrt{156}$

78. $\sqrt{213}$

79. $\sqrt{-15}$

80. $\sqrt{-79}$

81. $\sqrt[3]{83}$

82. $\sqrt[3]{97}$

83. $\sqrt[5]{123}$

84. $\sqrt[5]{283}$

85. $\sqrt[3]{-15}$

86. $\sqrt[5]{-29}$

87. **Mechanical Engineering** The time, in seconds, that it takes for an object to fall, from rest, is given by $t = \frac{1}{4}\sqrt{d}$, in which d is the distance fallen (ft). Find the time required for an object to fall to the ground from a building that is 800 ft high. Report your result to the nearest hundredth second.

88. **Mechanical Engineering** Use the information in exercise 87 to find the time required for an object to fall to the ground from a 1,400-ft high building. Report your result to the nearest hundredth second.

Basic Skills | Challenge Yourself | Calculator/Computer | Career Applications | **Above and Beyond**

89. Is there any prime number whose square root is an integer? Explain your answer.

90. Find two consecutive integers whose square roots are also consecutive integers.

91. Use a calculator to complete each exercise.

(a) Choose a number greater than 1 and find its square root. Then find the square root of the result and continue in this manner, observing the successive square roots. Do these numbers seem to be approaching a certain value? If so, what?

(b) Choose a number greater than 0 but less than 1 and find its square root. Then find the square root of the result, and continue in this manner, observing successive square roots. Do these numbers seem to be approaching a certain value? If so, what?

92. **(a)** Can a number be equal to its own square root?

(b) Other than the number(s) found in part (a), is a number always greater than its square root? Investigate.

93. Let a and b be positive numbers. If a is greater than b, is it always true that the square root of a is greater than the square root of b? Investigate.

94. Suppose that a weight is attached to a string of length L, and the other end of the string is held fixed. If we pull the weight and then release it, allowing the weight to swing back and forth, we can observe the behavior of a simple pendulum. The period T is the time required for the weight to complete a full cycle, swinging forward and then back. The formula below describes the relationship between T and L. chapter 7 > Make the Connection

$$T = 2\pi\sqrt{\frac{L}{g}}$$

If L is expressed in centimeters, then $g = 980 \text{ cm/s}^2$. For each string length, calculate the corresponding period. Round to the nearest tenth of a second.

(a) 30 cm **(b)** 50 cm **(c)** 70 cm **(d)** 90 cm **(e)** 110 cm

Answers

87. ______

88. ______

89. ______

90. ______

91. ______

92. ______

93. ______

94. ______

Answers

95. ______

96. ______

95. In parts (a) through (f), evaluate when possible.

(a) $\sqrt{4 \cdot 9}$ **(b)** $\sqrt{4} \cdot \sqrt{9}$ **(c)** $\sqrt{9 \cdot 16}$

(d) $\sqrt{9} \cdot \sqrt{16}$ **(e)** $\sqrt{(-4)(-25)}$ **(f)** $\sqrt{-4} \cdot \sqrt{-25}$

(g) Based on parts (a) through (f), make a general conjecture concerning $\sqrt{ab}$. Be careful to specify any restrictions on possible values for a and b.

96. In parts (a) through (d), evaluate when possible.

(a) $\sqrt{9 + 16}$ **(b)** $\sqrt{9} + \sqrt{16}$ **(c)** $\sqrt{36 + 64}$ **(d)** $\sqrt{36} + \sqrt{64}$

(e) Based on parts (a) through (d), what can you say about $\sqrt{a + b}$ and $\sqrt{a} + \sqrt{b}$?

Answers

1. 7 **3.** -6 **5.** ± 9 **7.** Not a real number **9.** 3 **11.** -4 **13.** -6 **15.** 3 **17.** -2 **19.** -2 **21.** Not a real number **23.** -3 **25.** $\frac{2}{3}$ **27.** $\frac{2}{3}$ **29.** 6 **31.** 3 **33.** 4 **35.** 3 **37.** $|x|$ **39.** y **41.** $3|x|$ **43.** $|a^2b^3|$ **45.** $4x^2$ **47.** 15 **49.** 15 **51.** $\sqrt{165}$ **53.** 3 units **55.** $\sqrt{10}$ units

57.

$x^2 + y^2 = 4$
Center: (0, 0);
radius: 2

59.

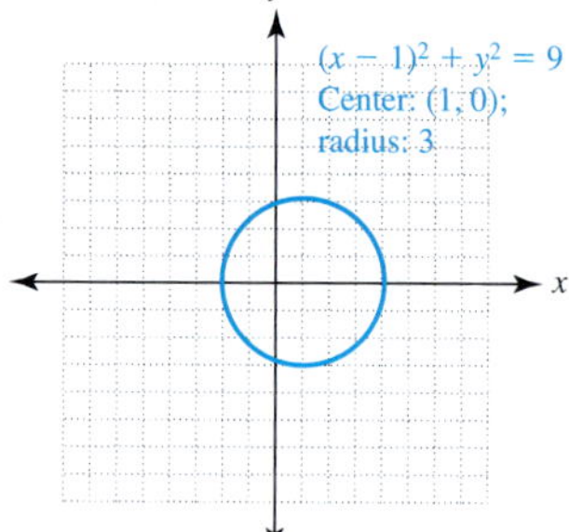

61.

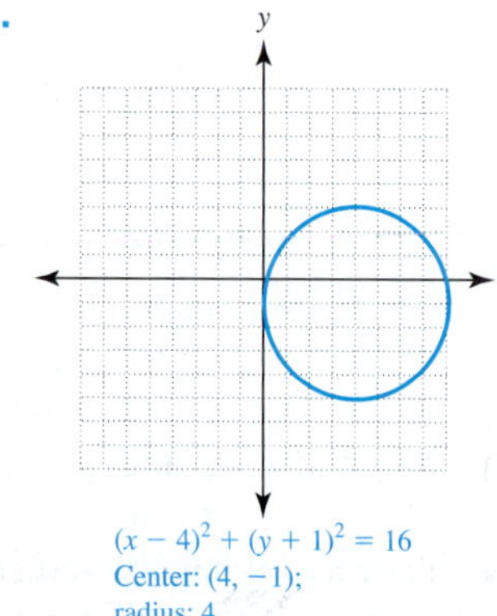

63. $x^2 + y^2 = 25$ **65.** $(x - 3)^2 + (y + 2)^2 = 25$ **67.** never **69.** True **71.** Domain $\{x \mid -7 \le x \le 1\}$, range $\{y \mid -2 \le y \le 6\}$ **73.** Domain $\{x \mid -5 \le x \le 5\}$, range $\{y \mid -2 \le y \le 8\}$ **75.** 3.873 **77.** 12.490 **79.** Not a real number **81.** 4.362 **83.** 2.618 **85.** -2.466 **87.** 7.07 s **89.** No **91.** Above and Beyond **93.** Above and Beyond **95.** **(a)** 6; **(b)** 6; **(c)** 12; **(d)** 12; **(e)** 10; **(f)** not possible; **(g)** Above and Beyond

Activity 7 :: The Swing of a Pendulum

The action of a pendulum seems simple. Scientists have studied the characteristics of a swinging pendulum and found them to be quite useful. In 1851 in Paris, Jean Foucault (pronounced "Foo-koh") used a pendulum to clearly demonstrate the rotation of Earth about its own axis.

A pendulum can be as simple as a string or cord with a weight fastened to one end. The other end is fixed, and the weight is allowed to swing. We define the **period** of a pendulum to be the amount of time required for the pendulum to make one complete swing (back and forth). The question we pose is: How does the *period* of a pendulum relate to the *length* of the pendulum?

For this activity, you need a piece of string that is approximately 1 m long. Fasten a weight (such as a small hexagonal nut) to one end, and then place clear marks on the string every 10 cm up to 70 cm, measured from the center of the weight.

1. Working with one or two partners, hold the string at the mark that is 10 cm from the weight. Pull the weight to the side with your other hand and let it swing freely. To estimate the period, let the weight swing through 30 periods, record the time in the given table, and then divide by 30. Round your result to the nearest hundredth of a second and record it. (*Note:* If you are unable to perform the experiment and collect your own data, you can use the sample data collected in this manner and presented at the end of this activity.)

   Repeat the described procedure for each length indicated in the given table.

Length of string, cm	10	20	30	40	50	60	70
Time for 30 periods, s							
Time for 1 period, s							

2. Let L represent the length of the pendulum and T represent the time period that results from swinging that pendulum. Fill out the table:

L	10	20	30	40	50	60	70
T							

3. Which variable, L or T, is viewed here as the independent variable?

4. On graph paper, draw horizontal and vertical axes, but plan to graph the data points in the first quadrant only. Explain why this is reasonable.

5. With the independent variable marked on the horizontal axis, scale the axes appropriately, keeping an eye on your data.

6. Plot your data points. Should you connect them with a smooth curve?

7. What period T would correspond to a string length of 0? Include this point on your graph.

8. Use your graph to predict the period for a string length of 80 cm.

9. Verify your prediction by measuring the period when the string is held at 80 cm (as described in step 1). How close did your experimental estimate come to the prediction made in step 8?

You have created a graph showing T as a function of L. The shape of the graph may not be familiar to you yet. In fact, the shape of your pendulum graph fits that of a square-root function.

Sample Data

Length of string, cm	10	20	30	40	50	60	70
Time for 30 periods, s	19	27	33	38	42	46	49

7.2 Simplifying Radical Expressions

< 7.2 Objectives >

1 > Simplify a radical expression by using the product property

2 > Simplify a radical expression by using the quotient property

NOTE

A precise set of conditions for a radical to be in simplified form follows.

In the previous section, we introduced radical notation. For some applications, we want all radical expressions written in *simplified form*. To accomplish this objective, we need two basic properties. In stating these properties, and in our subsequent examples, we assume that all variables represent positive real numbers whenever the index of a radical is even.

To develop our first property, consider an expression such as

$$\sqrt{25 \cdot 4}$$

One approach to simplify the expression would be

$$\sqrt{25 \cdot 4} = \sqrt{100} = 10$$

Now what happens if we separate the original radical as follows?

$$\sqrt{25 \cdot 4} = \sqrt{25} \cdot \sqrt{4}$$
$$= 5 \cdot 2 = 10$$

The result is the same, and this suggests our first property for radicals.

Property

Product Property for Radicals

$$\sqrt[n]{ab} = \sqrt[n]{a} \cdot \sqrt[n]{b}$$

In words, the radical of a product is equal to the product of the radicals. Both a and b are assumed to be positive real numbers when n is an even integer.

The second property we need is similar.

Property

Quotient Property for Radicals

$$\sqrt[n]{\frac{a}{b}} = \frac{\sqrt[n]{a}}{\sqrt[n]{b}}$$

In words, the radical of a quotient is the quotient of the radicals.

An example of the property above is

$$\sqrt{\frac{100}{4}} = \frac{\sqrt{100}}{\sqrt{4}}$$

Check to see that each side is equal to 5.

With these two properties, we are ready to define the simplified form for a radical expression. A radical is in simplified form if the following three conditions are satisfied.

Definition

Simplified Form for a Radical Expression

1. The radicand has no factor raised to a power greater than or equal to the index.
2. No fraction appears in the radical.
3. No radical appears in a denominator.

> **CAUTION**

Be careful! Students sometimes assume that because

$\sqrt{ab} = \sqrt{a} \cdot \sqrt{b}$

it should also be true that

$\sqrt{a + b} = \sqrt{a} + \sqrt{b}$ This is *not* true.

You can easily see that this is *not* true. Let $a = 9$ and $b = 16$ in the statement.

Our initial example deals with satisfying the first of the above conditions. We want to find the largest perfect-square factor (in the case of a square root) in the radicand and then apply the product property to simplify the expression.

Example 1 Simplifying Radical Expressions

< Objective 1 >

Simplify each expression.

(a) $\sqrt{18} = \sqrt{9 \cdot 2}$

$= \sqrt{9} \cdot \sqrt{2}$ Apply the product property.

$= 3\sqrt{2}$

(b) $\sqrt{75} = \sqrt{25 \cdot 3}$

$= \sqrt{25} \cdot \sqrt{3}$

$= 5\sqrt{3}$

(c) $\sqrt{27x^3} = \sqrt{9x^2 \cdot 3x}$

$= \sqrt{9x^2} \cdot \sqrt{3x} = 3x\sqrt{3x}$

(d) $\sqrt{72a^3b^4} = \sqrt{36a^2b^4 \cdot 2a}$

$= \sqrt{36a^2b^4} \cdot \sqrt{2a}$

$= 6ab^2\sqrt{2a}$

NOTES

The largest perfect-square factor of 18 is 9.

The largest perfect-square factor of 75 is 25.

The largest perfect-square factor of $27x^3$ is $9x^2$. Note that the exponent must be *even* in a perfect square.

RECALL

We assume that all variables represent positive real numbers when the index of a radical is even.

Check Yourself 1

Simplify each expression.

(a) $\sqrt{45}$ (b) $\sqrt{200}$ (c) $\sqrt{75p^5}$ (d) $\sqrt{98m^3n^4}$

Writing a cube root in simplest form involves finding factors of the radicand that are perfect cubes, as illustrated in Example 2. The process illustrated in this example can be extended in an identical fashion to simplify radical expressions with any index.

Example 2 Simplifying Radical Expressions

Simplify each expression.

NOTE

In a perfect cube, the exponent must be a *multiple of* 3.

(a) $\sqrt[3]{48} = \sqrt[3]{8 \cdot 6}$
$= \sqrt[3]{8} \cdot \sqrt[3]{6} = 2\sqrt[3]{6}$

(b) $\sqrt[3]{24x^4} = \sqrt[3]{8x^3 \cdot 3x}$
$= \sqrt[3]{8x^3} \cdot \sqrt[3]{3x} = 2x\sqrt[3]{3x}$

(c) $\sqrt[3]{54a^7b^4} = \sqrt[3]{27a^6b^3 \cdot 2ab}$
$= \sqrt[3]{27a^6b^3} \cdot \sqrt[3]{2ab} = 3a^2b\sqrt[3]{2ab}$

Check Yourself 2

Simplify each expression.

(a) $\sqrt[3]{128w^4}$ (b) $\sqrt[3]{40x^5y^7}$ (c) $\sqrt[4]{48a^8b^5}$

Satisfying our second condition for a radical to be in simplified form (no fractions should appear inside the radical) requires the second property for radicals. Consider the next example.

Example 3 Simplifying Radical Expressions

< Objective 2 >

Write each expression in simplified form.

NOTE

Apply the quotient property.

(a) $\sqrt{\frac{5}{9}} = \frac{\sqrt{5}}{\sqrt{9}}$
$= \frac{\sqrt{5}}{3}$

(b) $\sqrt{\frac{a^4}{25}} = \frac{\sqrt{a^4}}{\sqrt{25}} = \frac{a^2}{5}$

(c) $\sqrt[3]{\frac{5x^2}{8}} = \frac{\sqrt[3]{5x^2}}{\sqrt[3]{8}} = \frac{\sqrt[3]{5x^2}}{2}$

Check Yourself 3

Write each expression in simplified form.

(a) $\sqrt{\frac{7}{16}}$ (b) $\sqrt{\frac{3}{25a^2}}$ (c) $\sqrt[3]{\frac{5x}{27}}$

We begin our next example by applying the quotient property for radicals. However, an additional step is required because the third condition (that no radical appears in a denominator) must also be satisfied during the process.

Example 4 Rationalizing the Denominator

Write $\sqrt{\frac{3}{5}}$ in simplified form.

$$\sqrt{\frac{3}{5}} = \frac{\sqrt{3}}{\sqrt{5}}$$

The application of the quotient property satisfies the second condition—there are now no fractions *inside* a radical. However, we have a radical in the denominator, violating the third condition. The expression is not simplified until that radical is removed.

To remove the radical in the denominator, we multiply the numerator and denominator by the *same* expression, here $\sqrt{5}$. This is called *rationalizing the denominator.*

NOTE

The point here is to arrive at a perfect square inside the radical in the denominator. This is done by multiplying the numerator and denominator by $\sqrt{5}$ because

$\sqrt{5} \cdot \sqrt{5} = \sqrt{5^2} = \sqrt{25} = 5$

$$\frac{\sqrt{3}}{\sqrt{5}} = \frac{\sqrt{3} \cdot \sqrt{5}}{\sqrt{5} \cdot \sqrt{5}} \qquad \frac{\sqrt{5}}{\sqrt{5}} = 1$$

$$= \frac{\sqrt{15}}{\sqrt{25}} = \frac{\sqrt{15}}{5}$$

Check Yourself 4

Simplify $\sqrt{\frac{3}{7}}$.

We now look at some further examples that involve rationalizing the denominator of an expression.

Example 5 Rationalizing the Denominator

Write each expression in simplified form.

(a) $\frac{3}{\sqrt{8}} = \frac{3 \cdot \sqrt{2}}{\sqrt{8} \cdot \sqrt{2}}$ Multiply numerator and denominator by $\sqrt{2}$.

$$= \frac{3\sqrt{2}}{\sqrt{16}} = \frac{3\sqrt{2}}{4}$$

(b) $\sqrt[3]{\frac{5}{4}} = \frac{\sqrt[3]{5}}{\sqrt[3]{4}}$

Now note that

$$\sqrt[3]{4} \cdot \sqrt[3]{2} = \sqrt[3]{8} = 2$$

so multiplying the numerator and denominator by $\sqrt[3]{2}$ produces a perfect cube inside the radical in the denominator. Continuing, we have

NOTE

Why did we use $\sqrt[3]{2}$?

$\sqrt[3]{4} \cdot \sqrt[3]{2} = \sqrt[3]{2^2} \cdot \sqrt[3]{2}$
$= \sqrt[3]{2^3}$

and the exponent is a multiple of 3.

$$\frac{\sqrt[3]{5}}{\sqrt[3]{4}} = \frac{\sqrt[3]{5} \cdot \sqrt[3]{2}}{\sqrt[3]{4} \cdot \sqrt[3]{2}}$$

$$= \frac{\sqrt[3]{10}}{\sqrt[3]{8}} = \frac{\sqrt[3]{10}}{2}$$

Check Yourself 5

Simplify each expression.

(a) $\frac{5}{\sqrt{12}}$ (b) $\sqrt[3]{\frac{2}{9}}$

Our next example illustrates the process of rationalizing a denominator when variables are involved in a rational expression.

Example 6 Rationalizing Variable Denominators

Simplify each expression.

(a) $\sqrt{\frac{8x^3}{3y}}$

By the quotient property, we have

$$\sqrt{\frac{8x^3}{3y}} = \frac{\sqrt{8x^3}}{\sqrt{3y}}$$

Because the numerator can be simplified in this case, we start with that procedure.

$$\frac{\sqrt{8x^3}}{\sqrt{3y}} = \frac{\sqrt{4x^2}\cdot\sqrt{2x}}{\sqrt{3y}} = \frac{2x\sqrt{2x}}{\sqrt{3y}}$$

Multiplying the numerator and denominator by $\sqrt{3y}$ rationalizes the denominator.

$$\frac{2x\sqrt{2x}\cdot\sqrt{3y}}{\sqrt{3y}\cdot\sqrt{3y}} = \frac{2x\sqrt{6xy}}{\sqrt{9y^2}} = \frac{2x\sqrt{6xy}}{3y}$$

(b) $\frac{2}{\sqrt[3]{3x}}$

To satisfy the third condition, we must remove the radical from the denominator. For this we need a perfect cube inside the radical in the denominator. Multiplying the numerator and denominator by $\sqrt[3]{9x^2}$ provides the perfect cube.

$$\frac{2\sqrt[3]{9x^2}}{\sqrt[3]{3x}\cdot\sqrt[3]{9x^2}} = \frac{2\sqrt[3]{9x^2}}{\sqrt[3]{27x^3}}$$

$$= \frac{2\sqrt[3]{9x^2}}{3x}$$

NOTE

$\sqrt[3]{9x^2} = \sqrt[3]{3^2x^2}$

so

$\sqrt[3]{3x}\cdot\sqrt[3]{9x^2} = \sqrt[3]{3^3x^3}$

and each exponent is a multiple of 3.

Check Yourself 6

Simplify each expression.

(a) $\sqrt{\frac{12a^3}{5b}}$ (b) $\frac{3}{\sqrt[3]{2w^2}}$

We summarize our work to this point in simplifying radical expressions.

Step by Step

Simplifying Radical Expressions

NOTE

In the case of a cube root, steps 1 and 2 refer to perfect cubes, etc.

Step 1 To satisfy the first condition, determine the largest perfect-square factor of the radicand. Apply the product property to "remove" that factor from inside the radical.

Step 2 To satisfy the second condition, use the quotient property to write the expression in the form

$$\frac{\sqrt{a}}{\sqrt{b}}$$

If b is a perfect square, simplify the radical in the denominator. If not, proceed to step 3.

Step 3 Multiply the numerator and denominator of the radical expression by an appropriate radical to simplify and remove the radical in the denominator. Simplify the resulting expression when necessary.

Check Yourself ANSWERS

1. (a) $3\sqrt{5}$; **(b)** $10\sqrt{2}$; **(c)** $5p^2\sqrt{3p}$; **(d)** $7mn^2\sqrt{2m}$

2. (a) $4w\sqrt[3]{2w}$; **(b)** $2xy^2\sqrt[3]{5x^2y}$; **(c)** $2a^2b\sqrt[4]{3b}$

3. (a) $\frac{\sqrt{7}}{4}$; **(b)** $\frac{\sqrt{3}}{5a}$; **(c)** $\frac{\sqrt[3]{5x}}{3}$ **4.** $\frac{\sqrt{21}}{7}$ **5. (a)** $\frac{5\sqrt{3}}{6}$; **(b)** $\frac{\sqrt[3]{6}}{3}$

6. (a) $\frac{2a\sqrt{15ab}}{5b}$; **(b)** $\frac{3\sqrt[3]{4w}}{2w}$

Reading Your Text

SECTION 7.2

(a) The radical of a product is equal to the ______________ of the radicals.

(b) In the simplified form for a radical, no fraction appears in the ______________.

(c) In the simplified form for a radical, no radical appears in a ______________.

(d) To simplify a square-root expression, we first determine the largest ______________ factor of the radicand.

Basic Skills | Challenge Yourself | Calculator/Computer | Career Applications | Above and Beyond

7.2 exercises

< Objective 1 >

Simplify each expression. Assume all variables represent positive real numbers.

1. $\sqrt{12}$

2. $\sqrt{24}$

3. $\sqrt{50}$

4. $\sqrt{28}$

5. $-\sqrt{108}$ > Videos

6. $\sqrt{32}$

7. $\sqrt{52}$

8. $-\sqrt{96}$

9. $\sqrt{60}$

10. $\sqrt{150}$

11. $-\sqrt{125}$

12. $\sqrt{128}$

13. $\sqrt[3]{16}$

14. $\sqrt[3]{-54}$

15. $\sqrt[3]{-48}$

16. $\sqrt[3]{250}$

17. $\sqrt[3]{135}$

18. $\sqrt[3]{-160}$

19. $\sqrt[4]{32}$

20. $\sqrt[4]{96}$

21. $\sqrt{18z^2}$

22. $\sqrt{45a^2}$

23. $\sqrt{63x^4}$

24. $\sqrt{54w^4}$

25. $\sqrt{98m^3}$

26. $\sqrt{75a^5}$ > Videos

27. $\sqrt{80x^2y^3}$

28. $\sqrt{108p^5q^2}$

29. $\sqrt[3]{40b^3}$

30. $\sqrt[3]{16x^3}$

Name ____________

Section ______ Date ______

Answers

1. ______ 2. ______

3. ______ 4. ______

5. ______ 6. ______

7. ______ 8. ______

9. ______ 10. ______

11. ______ 12. ______

13. ______ 14. ______

15. ______ 16. ______

17. ______ 18. ______

19. ______ 20. ______

21. ______ 22. ______

23. ______ 24. ______

25. ______ 26. ______

27. ______ 28. ______

29. ______ 30. ______

Answers

31. ______
32. ______
33. ______
34. ______
35. ______
36. ______
37. ______
38. ______
39. ______
40. ______
41. ______ 42. ______
43. ______ 44. ______
45. ______ 46. ______
47. ______
48. ______
49. ______
50. ______
51. ______ 52. ______
53. ______ 54. ______
55. ______ 56. ______

31. $\sqrt[3]{48p^9}$

32. $\sqrt[3]{-80a^6}$

33. $\sqrt[3]{54m^7}$

34. $\sqrt[3]{250x^{13}}$

35. $\sqrt[3]{56x^6y^5z^4}$

36. $-\sqrt[3]{250a^4b^{15}c^9}$

37. $\sqrt[4]{32x^8}$

38. $\sqrt[4]{96w^5z^{13}}$

39. $\sqrt[4]{128a^{12}b^{17}}$

40. $\sqrt[5]{64w^{10}}$

< Objective 2 >

41. $\sqrt{\frac{5}{16}}$

42. $\sqrt{\frac{a^6}{49}}$

43. $\sqrt{\frac{5}{9y^4}}$

44. $\sqrt{\frac{7}{25x^2}}$

45. $\sqrt[3]{\frac{5}{8}}$

46. $\sqrt[3]{\frac{4x^2}{27}}$ > Videos

Use the distance formula (Section 7.1) to find the distance between each pair of points.

47. $(-7, 1)$ and $(0, 0)$

48. $(-18, -5)$ and $(-12, -3)$

49. $(22, -13)$ and $(18, -9)$ > Videos

50. $(-12, -17)$ and $(-9, -11)$

Basic Skills | **Challenge Yourself** | Calculator/Computer | Career Applications | Above and Beyond

Write each expression in simplified form. Assume all variables represent positive real numbers.

51. $\frac{5}{\sqrt{7}}$

52. $\sqrt{\frac{5}{8}}$

53. $\frac{7}{\sqrt{12}}$

54. $\frac{\sqrt{5}}{\sqrt{11}}$

55. $\frac{2\sqrt{3}}{\sqrt{10}}$

56. $\frac{3\sqrt{5}}{\sqrt{3}}$

57. $\sqrt[3]{\dfrac{7}{4}}$

58. $\sqrt[3]{\dfrac{5}{9}}$

59. $\dfrac{5}{\sqrt[3]{16}}$

60. $\sqrt{\dfrac{3}{x}}$ > Videos

61. $\sqrt{\dfrac{12}{w}}$

62. $\dfrac{\sqrt{18}}{\sqrt{a}}$

63. $\dfrac{\sqrt{8m^3}}{\sqrt{5n}}$

64. $\sqrt{\dfrac{24x^5}{7y}}$

65. $\sqrt[3]{\dfrac{5}{y}}$

66. $\sqrt[3]{\dfrac{7}{x^2}}$

67. $\dfrac{3}{\sqrt[3]{2x}}$

68. $\dfrac{5}{\sqrt[3]{3a}}$

69. $\sqrt[3]{\dfrac{2}{5x^2}}$

70. $\sqrt[3]{\dfrac{5}{7w^2}}$

71. $\dfrac{\sqrt[3]{5}}{\sqrt[3]{4a^2}}$

72. $\dfrac{\sqrt[3]{2}}{\sqrt[3]{9m^2}}$

73. $\sqrt[3]{\dfrac{a^5}{b^7}}$

74. $\sqrt[3]{\dfrac{w^7}{z^{10}}}$

Determine whether each statement is **true** *or* **false.**

75. $\sqrt{16x^{16}} = 4x^8$

76. $\sqrt{x^2 + y^2} = x + y$

77. $\dfrac{\sqrt{x^2 - 25}}{\sqrt{x - 5}} = \sqrt{x + 5}$

78. $\sqrt[3]{x^6} \cdot \sqrt[3]{x^3 - 1} = x^2\sqrt[3]{x - 1}$

79. $\sqrt[3]{(8b^6)^2} = \left(\sqrt[3]{8b^6}\right)^2$

80. $\dfrac{\sqrt[3]{8x^3}}{\sqrt[3]{2x}} = \sqrt[3]{4x^2}$

81. For nonnegative numbers a and b, $\sqrt{ab} = \sqrt{a}\sqrt{b}$.

82. For nonnegative numbers a and b, $\sqrt{a + b} = \sqrt{a} + \sqrt{b}$.

Answers

57. ________ 58. ________

59. ________ 60. ________

61. ________ 62. ________

63. ________

64. ________

65. ________ 66. ________

67. ________ 68. ________

69. ________ 70. ________

71. ________ 72. ________

73. ________

74. ________

75. ________

76. ________

77. ________

78. ________

79. ________

80. ________

81. ________

82. ________

Answers

83. ______

84. ______

85. ______

86. ______

87. ______

88. ______

83. For positive numbers a and b, $\sqrt{\dfrac{a}{b}} = \dfrac{\sqrt{a}}{\sqrt{b}}$.

84. For nonnegative numbers a and b, $\sqrt{a - b} = \sqrt{a} - \sqrt{b}$.

Basic Skills | Challenge Yourself | Calculator/Computer | **Career Applications** | Above and Beyond

Mechanical Engineering *In the Chapter 7 Activity, "The Swing of a Pendulum," you worked with the relationship between the period and length of a pendulum. The general model for this relationship is given by*

chapter 7 > Make the Connection

$$T = k\sqrt{L}$$

in which k is a gravitational constant.

Use this information to complete exercises 85 to 87.

85. Compute the value of k for each given T-value. Report your results to the nearest thousandth.

L	10	20	30	40	50	60	70
T	0.633	0.9	1.1	1.267	1.4	1.533	1.633
k							

86. Find the average (mean) of the values you found for k in exercise 85.

87. Fill in the row for time T in the table below using your own results from the Chapter 7 Activity. Then use your results for T to complete the row for k and find the mean k-value.

L	10	20	30	40	50	60	70
T							
k							

88. Physicists have found that the time of a pendulum's period as a function of its length can be modeled with

$$T = 2\pi\sqrt{\frac{L}{g}}$$

chapter 7 > Make the Connection

in which g is the gravitational constant $g = 980\,\dfrac{\text{cm}}{\text{s}^2}$, L is measured in cm, and T is in seconds.

(a) Use the properties of radicals to simplify the radical and rewrite it in the form

$$T = k\sqrt{L}$$

Round k to the nearest thousandth.

(b) How does the theoretical k found in part (a) compare with the experimental period found in exercise 86? Exercise 87?

Basic Skills | Challenge Yourself | Calculator/Computer | Career Applications | **Above and Beyond**

Simplify.

89. $\dfrac{7\sqrt{x^2y^4}\cdot\sqrt{36xy}}{6\sqrt{x^{-6}y^{-2}}\cdot\sqrt{49x^{-1}y^{-3}}}$

90. $\dfrac{3\sqrt[3]{32c^{12}d^2}\cdot\sqrt[3]{2c^5d^4}}{4\sqrt[3]{9c^8d^{-2}}\cdot\sqrt[3]{3c^{-3}d^{-4}}}$

91. Explain the difference between a pair of binomials in which the middle sign is changed and the opposite of a binomial. To illustrate, use $4 - \sqrt{7}$.

92. Find the missing binomial.

$(\sqrt{3} - 2)(\qquad) = -1$

93. Use a calculator to evaluate the expression in parts (a) through (d). Round your answers to the nearest hundredth.

(a) $3\sqrt{5} + 4\sqrt{5}$ **(b)** $7\sqrt{5}$ **(c)** $2\sqrt{6} + 3\sqrt{6}$ **(d)** $5\sqrt{6}$

(e) Based on parts (a) through (d), make a conjecture concerning $a\sqrt{m} + b\sqrt{m}$. Check your conjecture on an example of your own that is similar to parts (a) through (d).

Answers

89. ______

90. ______

91. ______

92. ______

93. ______

Answers

1. $2\sqrt{3}$ **3.** $5\sqrt{2}$ **5.** $-6\sqrt{3}$ **7.** $2\sqrt{13}$ **9.** $2\sqrt{15}$ **11.** $-5\sqrt{5}$
13. $2\sqrt[3]{2}$ **15.** $-2\sqrt[3]{6}$ **17.** $3\sqrt[3]{5}$ **19.** $2\sqrt[4]{2}$ **21.** $3z\sqrt{2}$
23. $3x^2\sqrt{7}$ **25.** $7m\sqrt{2m}$ **27.** $4xy\sqrt{5y}$ **29.** $2b\sqrt[3]{5}$
31. $2p^3\sqrt[3]{6}$ **33.** $3m^2\sqrt[3]{2m}$ **35.** $2x^2yz\sqrt[3]{7y^2z}$ **37.** $2x^2\sqrt[4]{2}$
39. $2a^3b^4\sqrt[4]{8b}$ **41.** $\dfrac{\sqrt{5}}{4}$ **43.** $\dfrac{\sqrt{5}}{3y^2}$ **45.** $\dfrac{\sqrt[3]{5}}{2}$ **47.** $5\sqrt{2}$ units
49. $4\sqrt{2}$ units **51.** $\dfrac{5\sqrt{7}}{7}$ **53.** $\dfrac{7\sqrt{3}}{6}$ **55.** $\dfrac{\sqrt{30}}{5}$ **57.** $\dfrac{\sqrt[3]{14}}{2}$
59. $\dfrac{5\sqrt[3]{4}}{4}$ **61.** $\dfrac{2\sqrt{3w}}{w}$ **63.** $\dfrac{2m\sqrt{10mn}}{5n}$ **65.** $\dfrac{\sqrt[3]{5y^2}}{y}$
67. $\dfrac{3\sqrt[3]{4x^2}}{2x}$ **69.** $\dfrac{\sqrt[3]{50x}}{5x}$ **71.** $\dfrac{\sqrt[3]{10a}}{2a}$ **73.** $\dfrac{a\sqrt[3]{a^2b^2}}{b^3}$ **75.** True
77. True **79.** True **81.** True **83.** True

85.

L	10	20	30	40	50	60	70
T	0.633	0.9	1.1	1.267	1.4	1.533	1.633
k	0.2	0.201	0.201	0.2	0.198	0.198	0.195

87. Answers will vary **89.** x^5y^5 **91.** Above and Beyond
93. **(a)** 15.65; **(b)** 15.65; **(c)** 12.25; **(d)** 12.25; **(e)** Above and Beyond

7.3 Operations on Radical Expressions

< 7.3 Objectives >

1 > Add and subtract radical expressions

2 > Multiply radical expressions

3 > Divide radical expressions

Adding and subtracting radical expressions exactly parallels our earlier work with polynomials containing like terms.

To add $3x^2 + 4x^2$, we have

$$3x^2 + 4x^2 = (3 + 4)x^2$$
$$= 7x^2$$

RECALL

This uses the distributive property.

Keep in mind that we are able to simplify or combine the above expressions because of like terms in x^2. (Recall that like terms have the same variable factor raised to the same power.)

We *cannot* combine terms such as

$$4a^3 + 3a^2 \qquad \text{or} \qquad 3x - 5y$$

By extending these ideas, we conclude that radical expressions can be combined *only* if they are *similar*, that is, if the expressions contain the same radicand with the same index. This is illustrated in Example 1.

Example 1 Adding and Subtracting Radical Expressions

< Objective 1 >

Add or subtract as indicated.

(a) $3\sqrt{7} + 2\sqrt{7} = (3 + 2)\sqrt{7}$
$= 5\sqrt{7}$

(b) $7\sqrt{3} - 4\sqrt{3} = (7 - 4)\sqrt{3} = 3\sqrt{3}$

(c) $5\sqrt{10} - 3\sqrt{10} + 2\sqrt{10} = (5 - 3 + 2)\sqrt{10}$
$= 4\sqrt{10}$

(d) $2\sqrt{5} + 3\sqrt{3}$ cannot be combined or simplified.

(e) $\sqrt{7} + \sqrt[3]{7}$ cannot be simplified.

(f) $5\sqrt{x} + 2\sqrt{x} = (5 + 2)\sqrt{x}$
$= 7\sqrt{x}$

(g) $5\sqrt{3ab} - 2\sqrt{3ab} + 3\sqrt{3ab} = (5 - 2 + 3)\sqrt{3ab} = 6\sqrt{3ab}$

(h) $\sqrt[3]{3x^2} + \sqrt[3]{3x}$ cannot be simplified. The radicands are *not* the same.

NOTES

Apply the distributive property.

In (d), the expressions have different radicands, 5 and 3.

In (e), the expressions have different indices, 2 and 3.

Check Yourself 1

Add or subtract as indicated.

(a) $5\sqrt{3} + 2\sqrt{3}$

(b) $7\sqrt{5} - 2\sqrt{5} + 3\sqrt{5}$

(c) $2\sqrt{3} + 3\sqrt{2}$

(d) $\sqrt{2y} + 5\sqrt{2y} - 3\sqrt{2y}$

(e) $2\sqrt[3]{3m} - 5\sqrt[3]{3m}$

(f) $\sqrt{5x} - \sqrt[3]{5x}$

Often it is necessary to simplify radical expressions using the methods of Section 7.2 before they can be combined. Example 2 illustrates how the product property is applied.

Example 2 Adding and Subtracting Radical Expressions

Add or subtract as indicated.

(a) $\sqrt{48} + 2\sqrt{3}$

In this form, the radicals cannot be combined. However, the first radical can be simplified by our earlier methods because 48 has the perfect-square factor 16.

$$\sqrt{48} = \sqrt{16 \cdot 3} = 4\sqrt{3}$$

With this result we can proceed as before.

$$\sqrt{48} + 2\sqrt{3} = 4\sqrt{3} + 2\sqrt{3}$$
$$= (4 + 2)\sqrt{3} = 6\sqrt{3}$$

(b) $\sqrt{50} - \sqrt{32} + \sqrt{98} = 5\sqrt{2} - 4\sqrt{2} + 7\sqrt{2}$

$$= (5 - 4 + 7)\sqrt{2} = 8\sqrt{2}$$

NOTE

$\sqrt{50} = \sqrt{25 \cdot 2}$

$\sqrt{32} = \sqrt{16 \cdot 2}$

$\sqrt{98} = \sqrt{49 \cdot 2}$

(c) $x\sqrt{2x} + 3\sqrt{8x^3}$

Note that

$$3\sqrt{8x^3} = 3\sqrt{4x^2 \cdot 2x}$$
$$= 3\sqrt{4x^2} \cdot \sqrt{2x}$$
$$= 3 \cdot 2x\sqrt{2x} = 6x\sqrt{2x}$$

So

$$x\sqrt{2x} + 3\sqrt{8x^3} = x\sqrt{2x} + 6x\sqrt{2x}$$
$$= (x + 6x)\sqrt{2x} = 7x\sqrt{2x}$$

(d) $\sqrt[3]{2a} - \sqrt[3]{16a} + \sqrt[3]{54a} = \sqrt[3]{2a} - 2\sqrt[3]{2a} + 3\sqrt[3]{2a}$

$$= 2\sqrt[3]{2a}$$

NOTE

$\sqrt[3]{16a} = \sqrt[3]{8 \cdot 2a}$

$\sqrt[3]{54a} = \sqrt[3]{27 \cdot 2a}$

Check Yourself 2

Add or subtract as indicated.

(a) $\sqrt{125} + 3\sqrt{5}$

(b) $\sqrt{75} - \sqrt{27} + \sqrt{48}$

(c) $5\sqrt{24y^3} - y\sqrt{6y}$

(d) $\sqrt[3]{81x} - \sqrt[3]{3x} + \sqrt[3]{24x}$

It may be necessary to apply the quotient property before combining rational expressions as shown in Example 3.

Example 3 Adding and Subtracting Radical Expressions

Add or subtract as indicated.

(a) $2\sqrt{6} + \sqrt{\dfrac{2}{3}}$

NOTE

Multiply by $\dfrac{\sqrt{3}}{\sqrt{3}}$.

We apply the quotient property to the *second term* and rationalize the denominator.

$$\sqrt{\frac{2}{3}} = \frac{\sqrt{2}}{\sqrt{3}} = \frac{\sqrt{2}\cdot\sqrt{3}}{\sqrt{3}\cdot\sqrt{3}} = \frac{\sqrt{6}}{3} \qquad \frac{\sqrt{3}}{\sqrt{3}} = 1$$

So

$$2\sqrt{6} + \sqrt{\frac{2}{3}}$$

$$= 2\sqrt{6} + \frac{\sqrt{6}}{3} \qquad \text{Factor out the } \sqrt{6} \text{ from these two terms.}$$

$$= \left(2 + \frac{1}{3}\right)\sqrt{6} = \frac{7}{3}\sqrt{6}$$

NOTE

Note that $\dfrac{\sqrt{6}}{3}$ and $\dfrac{1}{3}\sqrt{6}$ are equivalent.

(b) $\sqrt{20x} - \sqrt{\dfrac{x}{5}}$

Again we first simplify the two expressions. So

$$\sqrt{20x} - \sqrt{\frac{x}{5}} = 2\sqrt{5x} - \frac{\sqrt{x}\cdot\sqrt{5}}{\sqrt{5}\cdot\sqrt{5}}$$

$$= 2\sqrt{5x} - \frac{\sqrt{5x}}{5}$$

$$= 2\sqrt{5x} - \frac{1}{5}\sqrt{5x}$$

$$= \left(2 - \frac{1}{5}\right)\sqrt{5x} = \frac{9}{5}\sqrt{5x}$$

NOTE

$$\sqrt{20x} = \sqrt{4\cdot 5x} = \sqrt{4}\sqrt{5x} = 2\sqrt{5x}$$

Check Yourself 3

Add or subtract as indicated.

(a) $3\sqrt{7} + \sqrt{\dfrac{1}{7}}$ **(b)** $\sqrt{40x} - \sqrt{\dfrac{2x}{5}}$

The next example illustrates how addition of fractions may be applied when working with radical expressions.

Example 4 Adding Radical Expressions

Add $\dfrac{\sqrt{5}}{3} + \dfrac{2}{\sqrt{5}}$.

Our first step is to rationalize the denominator of the second fraction, to write the sum as

$$\frac{\sqrt{5}}{3} + \frac{2\sqrt{5}}{\sqrt{5}\cdot\sqrt{5}}$$

or $\dfrac{\sqrt{5}}{3} + \dfrac{2\sqrt{5}}{5}$

The LCD of the fractions is 15, and rewriting each fraction with that denominator, we have

$$\frac{\sqrt{5} \cdot 5}{3 \cdot 5} + \frac{2\sqrt{5} \cdot 3}{5 \cdot 3} = \frac{5\sqrt{5} + 6\sqrt{5}}{15}$$
$$= \frac{11\sqrt{5}}{15}$$

Check Yourself 4

Subtract $\frac{3}{\sqrt{10}} - \frac{\sqrt{10}}{5}$.

In Section 7.2 we introduced the product and quotient properties for radical expressions. At that time they were used to simplify radicals.

If we turn those properties around, we have ways to multiply and divide radical expressions.

Property

Multiplying Radical Expressions

$\sqrt[n]{a} \cdot \sqrt[n]{b} = \sqrt[n]{a \cdot b}$

In words, the product of two roots is the root of the product of the radicands.

The use of this multiplication property is illustrated in Example 5. We assume that all variables represent positive real numbers.

Example 5 Multiplying Radical Expressions

< Objective 2 >

NOTE

Just multiply the radicands.

Multiply.

(a) $\sqrt{7} \cdot \sqrt{5} = \sqrt{7 \cdot 5} = \sqrt{35}$

(b) $\sqrt{3x} \cdot \sqrt{10y} = \sqrt{3x \cdot 10y}$
$= \sqrt{30xy}$

(c) $\sqrt[3]{4x} \cdot \sqrt[3]{7x} = \sqrt[3]{4x \cdot 7x}$
$= \sqrt[3]{28x^2}$

Check Yourself 5

Multiply.

(a) $\sqrt{6} \cdot \sqrt{7}$ **(b)** $\sqrt{5a} \cdot \sqrt{11b}$ **(c)** $\sqrt[3]{3y} \cdot \sqrt[3]{5y}$

Keep in mind that all radical expressions should be written in simplified form. Often we have to apply the methods of Section 7.2 to simplify a product once it has been formed.

Example 6 Multiplying Radical Expressions

NOTE

$\sqrt{18}$ is *not* in simplified form. 9 is a perfect-square factor of 18.

Multiply and simplify.

(a) $\sqrt{3} \cdot \sqrt{6} = \sqrt{18}$

$= \sqrt{9 \cdot 2} = \sqrt{9}\sqrt{2}$

$= 3\sqrt{2}$

(b) $\sqrt{5x} \cdot \sqrt{15x} = \sqrt{75x^2}$

$= \sqrt{25x^2 \cdot 3} = \sqrt{25x^2} \cdot \sqrt{3}$

$= 5x\sqrt{3}$

NOTE

Now we want a factor that is a *perfect cube.*

(c) $\sqrt[3]{4a^2b} \cdot \sqrt[3]{10a^2b^2} = \sqrt[3]{40a^4b^3} = \sqrt[3]{8a^3b^3 \cdot 5a}$

$= \sqrt[3]{8a^3b^3} \cdot \sqrt[3]{5a} = 2ab\sqrt[3]{5a}$

Check Yourself 6

Multiply and simplify.

(a) $\sqrt{10} \cdot \sqrt{20}$ **(b)** $\sqrt{6x} \cdot \sqrt{15x}$ **(c)** $\sqrt[3]{9p^2q^2} \cdot \sqrt[3]{6pq^2}$

We are now ready to combine multiplication with the techniques for adding and subtracting radicals. This allows us to multiply radical expressions with more than one term. Consider these examples.

Example 7 Using the Distributive Property

Multiply and simplify.

NOTES

We distribute $\sqrt{2}$ over the sum $\sqrt{5} + \sqrt{7}$ to multiply.

Distribute $\sqrt{3}$.
$\sqrt{18} = \sqrt{9 \cdot 2} = 3\sqrt{2}$
$\sqrt{45} = \sqrt{9 \cdot 5} = 3\sqrt{5}$

(a) $\sqrt{2}(\sqrt{5} + \sqrt{7})$

Distributing $\sqrt{2}$, we have

$\sqrt{2} \cdot \sqrt{5} + \sqrt{2} \cdot \sqrt{7} = \sqrt{10} + \sqrt{14}$

The expression cannot be simplified further.

(b) $\sqrt{3}(\sqrt{6} + 2\sqrt{15}) = \sqrt{3} \cdot \sqrt{6} + \sqrt{3} \cdot 2\sqrt{15}$

$= \sqrt{18} + 2\sqrt{45}$

$= 3\sqrt{2} + 6\sqrt{5}$

(c) $\sqrt{x}(\sqrt{2x} + \sqrt{8x}) = \sqrt{x} \cdot \sqrt{2x} + \sqrt{x} \cdot \sqrt{8x}$

$= \sqrt{2x^2} + \sqrt{8x^2}$

$= x\sqrt{2} + 2x\sqrt{2} = 3x\sqrt{2}$

NOTE

Alternatively, we could choose to simplify $\sqrt{8x}$ in the original expression as our first step. We leave it to the reader to verify that the result is the same.

Check Yourself 7

Multiply and simplify.

(a) $\sqrt{3}(\sqrt{10} + \sqrt{2})$ **(b)** $\sqrt{2}(3 + 2\sqrt{6})$ **(c)** $\sqrt{a}(\sqrt{3a} + \sqrt{12a})$

If both of the radical expressions involved in a multiplication have two terms, we must apply the patterns for multiplying polynomials developed in Chapter 5. Here is an example to illustrate.

Example 8 Multiplying Radical Binomials

Multiply and simplify.

(a) $(\sqrt{3} + 1)(\sqrt{3} + 5)$

To write the desired product, we use the FOIL pattern for multiplying binomials.

$(\sqrt{3} + 1)(\sqrt{3} + 5)$

$$= \underbrace{\sqrt{3} \cdot \sqrt{3}}_{\text{First}} + \underbrace{5 \cdot \sqrt{3}}_{\text{Outer}} + \underbrace{1 \cdot \sqrt{3}}_{\text{Inner}} + \underbrace{1 \cdot 5}_{\text{Last}}$$

$$= 3 + 6\sqrt{3} + 5$$

$$= 8 + 6\sqrt{3}$$

NOTE

Combine the outer and inner products.

(b) $(\sqrt{6} + \sqrt{2})(\sqrt{6} - \sqrt{2})$

Multiplying as before, we have

$$\sqrt{6} \cdot \sqrt{6} - \sqrt{6} \cdot \sqrt{2} + \sqrt{6} \cdot \sqrt{2} - \sqrt{2} \cdot \sqrt{2} = 6 + 0 - 2 = 4$$

NOTE

The form of the product

$(a + b)(a - b)$

gives $a^2 - b^2$

When a and b are square roots, the product is rational.

Two binomial radical expressions that differ *only* in the sign of the second term are called *conjugates* of each other. So

$\sqrt{6} + \sqrt{2}$ and $\sqrt{6} - \sqrt{2}$

are conjugates, and their product does *not* contain a radical—the product is a rational number. That is always the case with two conjugates and has particular significance later in this section.

(c) $(\sqrt{2} + \sqrt{5})^2 = (\sqrt{2} + \sqrt{5})(\sqrt{2} + \sqrt{5})$

Multiplying as before, we have

$$(\sqrt{2} + \sqrt{5})^2 = \sqrt{2} \cdot \sqrt{2} + \sqrt{2} \cdot \sqrt{5} + \sqrt{2} \cdot \sqrt{5} + \sqrt{5} \cdot \sqrt{5}$$

$$= 2 + \sqrt{10} + \sqrt{10} + 5$$

$$= 7 + 2\sqrt{10}$$

NOTE

We apply the multiplication pattern for binomials.

$(\sqrt{2} + \sqrt{5})^2$ can also be handled using our earlier formula for the square of a binomial

$$(a + b)^2 = a^2 + 2ab + b^2$$

in which $a = \sqrt{2}$ and $b = \sqrt{5}$.

Check Yourself 8

Multiply and simplify.

(a) $(\sqrt{2} + 3)(\sqrt{2} + 5)$ **(b)** $(\sqrt{5} - \sqrt{3})(\sqrt{5} + \sqrt{3})$ **(c)** $(\sqrt{7} - \sqrt{3})^2$

We are now ready to state our basic property for dividing radical expressions. Again, it is simply a restatement of our earlier quotient property.

Property

Dividing Radical Expressions

$$\frac{\sqrt[n]{a}}{\sqrt[n]{b}} = \sqrt[n]{\frac{a}{b}}$$

In words, the quotient of two roots is the root of the quotient of the radicands.

Although we illustrate this property in one of the examples that follow, dividing rational expressions is most often carried out by rationalizing the denominator. This process can be separated into two types of problems: those with a monomial divisor and those with binomial divisors. The next series of examples illustrates.

Example 9 Dividing Radical Expressions

< Objective 3 >

Simplify each expression. Assume that all variables represent positive real numbers.

(a) $\dfrac{3}{\sqrt{5}} = \dfrac{3 \cdot \sqrt{5}}{\sqrt{5} \cdot \sqrt{5}} = \dfrac{3\sqrt{5}}{5}$

We multiply the numerator and denominator by $\sqrt{5}$ to rationalize the denominator.

(b) $\dfrac{\sqrt{7x}}{\sqrt{10y}} = \dfrac{\sqrt{7x} \cdot \sqrt{10y}}{\sqrt{10y} \cdot \sqrt{10y}}$

$= \dfrac{\sqrt{70xy}}{10y}$

NOTE

$\sqrt[3]{2} \cdot \sqrt[3]{4} = \sqrt[3]{2} \cdot \sqrt[3]{2^2}$

$= \sqrt[3]{2^3}$

$= 2$

(c) $\dfrac{3}{\sqrt[3]{2}} = \dfrac{3\sqrt[3]{4}}{\sqrt[3]{2} \cdot \sqrt[3]{4}}$

In this case we want a perfect cube in the denominator, so we multiply the numerator and denominator by $\sqrt[3]{4}$.

$= \dfrac{3\sqrt[3]{4}}{2}$

These division problems are similar to those we saw in Section 7.2 when we simplified radical expressions. They are shown here to illustrate this case of division with radicals.

Check Yourself 9

Simplify each expression.

(a) $\dfrac{5}{\sqrt{7}}$ (b) $\dfrac{\sqrt{3a}}{\sqrt{5b}}$ (c) $\dfrac{5}{\sqrt[3]{9}}$

Our division property is particularly useful when the radicands in the numerator and denominator have common factors. Consider Example 10.

Example 10 Dividing Radical Expressions

Simplify

$$\frac{\sqrt{10}}{\sqrt{15a}}$$

NOTE

5 is a common factor of the radicands in the numerator and denominator.

We apply the division property so that the radicand can be reduced as a fraction.

$$\frac{\sqrt{10}}{\sqrt{15a}} = \sqrt{\frac{10}{15a}} = \sqrt{\frac{2}{3a}}$$

In the radicand, divide numerator and denominator by 5.

Now we use the quotient property and rationalize the denominator.

$$\sqrt{\frac{2}{3a}} = \frac{\sqrt{2}}{\sqrt{3a}} = \frac{\sqrt{2}\cdot\sqrt{3a}}{\sqrt{3a}\cdot\sqrt{3a}}$$ Multiply numerator and denominator by $\sqrt{3a}$.

$$= \frac{\sqrt{6a}}{3a}$$ Use the multiplication property in the numerator and in the denominator.

Check Yourself 10

Simplify $\dfrac{\sqrt{15}}{\sqrt{18x}}$.

We now turn our attention to a second type of division problem involving radical expressions. Here the divisors (the denominators) are binomials. This uses the idea of conjugates that we saw in Example 8.

Example 11 Rationalizing Radical Denominators

Rationalize each denominator.

(a) $\dfrac{6}{\sqrt{6}+\sqrt{2}}$

Recall that $\sqrt{6}-\sqrt{2}$ is the conjugate of $\sqrt{6}+\sqrt{2}$, and the product of conjugates is *always a rational number*. Therefore, to rationalize the denominator, we multiply by $\sqrt{6}-\sqrt{2}$.

NOTES

If a radical expression has a sum or difference in the denominator, multiply the numerator and denominator by the *conjugate* of the denominator to rationalize.

See Example 8(b) for the details of the multiplication in the denominator.

$$\frac{6}{\sqrt{6}+\sqrt{2}} = \frac{6(\sqrt{6}-\sqrt{2})}{(\sqrt{6}+\sqrt{2})(\sqrt{6}-\sqrt{2})}$$

$$= \frac{6(\sqrt{6}-\sqrt{2})}{4}$$

$$= \frac{3(\sqrt{6}-\sqrt{2})}{2}$$

(b) $\dfrac{\sqrt{5}+\sqrt{3}}{\sqrt{5}-\sqrt{3}}$

Multiply numerator and denominator by $\sqrt{5}+\sqrt{3}$.

NOTE

Combine like terms, factor, and divide the numerator and denominator by 2 to simplify.

$$\frac{(\sqrt{5}+\sqrt{3})(\sqrt{5}+\sqrt{3})}{(\sqrt{5}-\sqrt{3})(\sqrt{5}+\sqrt{3})} = \frac{5+\sqrt{15}+\sqrt{15}+3}{5-3}$$

$$= \frac{8+2\sqrt{15}}{2} = \frac{2(4+\sqrt{15})}{2}$$

$$= 4+\sqrt{15}$$

Check Yourself 11

Rationalize the denominator.

(a) $\dfrac{4}{\sqrt{3}-\sqrt{2}}$ (b) $\dfrac{\sqrt{6}+\sqrt{3}}{\sqrt{6}-\sqrt{3}}$

Check Yourself ANSWERS

1. **(a)** $7\sqrt{3}$; **(b)** $8\sqrt{5}$; **(c)** cannot be simplified; **(d)** $3\sqrt{2y}$; **(e)** $-3\sqrt[3]{3m}$; **(f)** cannot be simplified **2.** **(a)** $8\sqrt{5}$; **(b)** $6\sqrt{3}$; **(c)** $9y\sqrt{6y}$; **(d)** $4\sqrt[3]{3x}$
3. **(a)** $\frac{22}{7}\sqrt{7}$; **(b)** $\frac{9}{5}\sqrt{10x}$ **4.** $\frac{\sqrt{10}}{10}$ **5.** **(a)** $\sqrt{42}$; **(b)** $\sqrt{55ab}$; **(c)** $\sqrt[3]{15y^2}$
6. **(a)** $10\sqrt{2}$; **(b)** $3x\sqrt{10}$; **(c)** $3pq\sqrt[3]{2q}$
7. **(a)** $\sqrt{30} + \sqrt{6}$; **(b)** $3\sqrt{2} + 4\sqrt{3}$; **(c)** $3a\sqrt{3}$
8. **(a)** $17 + 8\sqrt{2}$; **(b)** 2; **(c)** $10 - 2\sqrt{21}$ **9.** **(a)** $\frac{5\sqrt{7}}{7}$; **(b)** $\frac{\sqrt{15ab}}{5b}$; **(c)** $\frac{5\sqrt[3]{3}}{3}$
10. $\frac{\sqrt{30x}}{6x}$ **11.** **(a)** $4(\sqrt{3} + \sqrt{2})$; **(b)** $3 + 2\sqrt{2}$

Reading Your Text

SECTION 7.3

(a) Adding and subtracting radical expressions parallels our earlier work with ____________ containing like terms.

(b) The product of two roots is the root of the product of the ____________.

(c) Dividing radical expressions is most often carried out by ____________ the denominator.

(d) If a radical expression has a sum or difference in the denominator, multiply the numerator and denominator by the ____________ of the denominator to rationalize.

Basic Skills | Challenge Yourself | Calculator/Computer | Career Applications | Above and Beyond

7.3 exercises

< Objective 1 >

Add or subtract as indicated. Assume that all variables represent positive real numbers.

1. $3\sqrt{5} + 4\sqrt{5}$
2. $5\sqrt{6} + 3\sqrt{6}$
3. $11\sqrt{3a} - 8\sqrt{3a}$
4. $2\sqrt{5w} + 3\sqrt{5w}$
5. $7\sqrt{m} + 6\sqrt{n}$
6. $8\sqrt{a} - 6\sqrt{b}$
7. $2\sqrt[3]{2} + 7\sqrt[3]{2}$
8. $5\sqrt[4]{3} - 2\sqrt[4]{3}$
9. $8\sqrt{6} - 2\sqrt{6} + 3\sqrt{6}$ 

10. $8\sqrt{3} + 2\sqrt{3} - 7\sqrt{3}$

Simplify the radical expressions when necessary. Then add or subtract as indicated. Assume that all variables represent positive real numbers.

11. $\sqrt{20} + \sqrt{5}$
12. $\sqrt{27} + \sqrt{3}$
13. $4\sqrt{28} - \sqrt{63}$
14. $2\sqrt{40} + \sqrt{90}$
15. $\sqrt{98} - \sqrt{18} + \sqrt{8}$
16. $\sqrt{108} - \sqrt{27} + \sqrt{75}$
17. $\sqrt[3]{81} + \sqrt[3]{3}$
18. $\sqrt[3]{16} - \sqrt[3]{2}$
19. $\sqrt{54w} - \sqrt{24w}$
20. $\sqrt{27p} + \sqrt{75p}$
21. $\sqrt{18x^3} + \sqrt{8x^3}$ > Videos
22. $\sqrt{125y^3} - \sqrt{20y^3}$
23. $\sqrt[3]{16w^5} + 2w\sqrt[3]{2w^2} - \sqrt[3]{2w^5}$
24. $\sqrt[4]{2z^7} - z\sqrt[4]{32z^3} + \sqrt[4]{162z^7}$
25. $\sqrt{3} + \sqrt{\dfrac{1}{3}}$
26. $\sqrt{6} - \sqrt{\dfrac{1}{6}}$
27. $\sqrt[3]{48} - \sqrt[3]{\dfrac{3}{4}}$
28. $\sqrt[3]{96} + \sqrt[3]{\dfrac{4}{9}}$

Boost your GRADE at ALEKS.com!

ALEKS

- Practice Problems
- Self-Tests
- NetTutor
- e-Professors
- Videos

Name ____________

Section ________ Date ________

Answers

1. ______ 2. ______
3. ______ 4. ______
5. ______
6. ______
7. ______ 8. ______
9. ______ 10. ______
11. ______ 12. ______
13. ______ 14. ______
15. ______ 16. ______
17. ______ 18. ______
19. ______ 20. ______
21. ______ 22. ______
23. ______ 24. ______
25. ______ 26. ______
27. ______ 28. ______

Answers

29. ______ 30. ______
31. ______ 32. ______
33. ______ 34. ______
35. ______ 36. ______
37. ______ 38. ______
39. ______ 40. ______
41. ______ 42. ______
43. ______ 44. ______
45. ______
46. ______
47. ______ 48. ______
49. ______ 50. ______
51. ______ 52. ______
53. ______ 54. ______
55. ______
56. ______
57. ______
58. ______
59. ______
60. ______
61. ______
62. ______

29. $\dfrac{\sqrt{6}}{2} + \dfrac{1}{\sqrt{6}}$

30. $\dfrac{\sqrt{10}}{2} - \dfrac{1}{\sqrt{10}}$

31. $\dfrac{\sqrt{12}}{3} - \dfrac{1}{\sqrt{3}}$ > Videos

32. $\dfrac{\sqrt{20}}{5} + \dfrac{2}{\sqrt{5}}$

< Objective 2 >

Multiply each expression and simplify.

33. $\sqrt{a} \cdot \sqrt{11}$

34. $\sqrt{10} \cdot \sqrt{w}$

35. $\sqrt{3} \cdot \sqrt{7} \cdot \sqrt{2}$

36. $\sqrt{5} \cdot \sqrt{7} \cdot \sqrt{3}$

37. $\sqrt[3]{4} \cdot \sqrt[3]{9}$

38. $\sqrt[3]{5} \cdot \sqrt[3]{7}$

39. $\sqrt{3} \cdot \sqrt{12}$

40. $\sqrt{5} \cdot \sqrt{20}$

41. $\sqrt[3]{9p^2} \cdot \sqrt[3]{6p}$

42. $\sqrt[3]{25x^2} \cdot \sqrt[3]{10x^2}$

43. $\sqrt[3]{4x^2y} \cdot \sqrt[3]{10xy^3}$ 

44. $\sqrt[3]{18r^2s^2} \cdot \sqrt[3]{9r^2s}$

45. $\sqrt{2}(\sqrt{3} + 5)$

46. $\sqrt{3}(\sqrt{5} - 7)$

47. $\sqrt{3}(5\sqrt{2} - \sqrt{18})$

48. $\sqrt{2}(2\sqrt{10} + \sqrt{40})$

49. $\sqrt{x}(\sqrt{3x} + \sqrt{27x})$

50. $\sqrt{y}(\sqrt{8y} - \sqrt{2y})$

51. $\sqrt[3]{4}(\sqrt[3]{4} + \sqrt[3]{32})$

52. $\sqrt[3]{6}(\sqrt[3]{32} - \sqrt[3]{4})$

53. $(\sqrt{2} + 3)(\sqrt{2} - 4)$

54. $(\sqrt{3} - 1)(\sqrt{3} + 5)$

55. $(\sqrt{2} + 3\sqrt{5})(\sqrt{2} - 2\sqrt{5})$

56. $(\sqrt{6} - 2\sqrt{3})(\sqrt{6} - 3\sqrt{3})$

57. $(\sqrt{5} + 3)(\sqrt{5} - 3)$

58. $(\sqrt{10} + 2)(\sqrt{10} - 2)$

59. $(\sqrt{a} + \sqrt{3})(\sqrt{a} - \sqrt{3})$

60. $(\sqrt{m} + \sqrt{7})(\sqrt{m} - \sqrt{7})$

61. $(\sqrt{3} - 5)^2$

62. $(\sqrt{5} + \sqrt{2})^2$

63. $(\sqrt{a} + 3)^2$

64. $(\sqrt{x} - 4)^2$

65. $(\sqrt{x} + \sqrt{y})^2$

66. $(\sqrt{r} - \sqrt{s})^2$

Basic Skills | **Challenge Yourself** | Calculator/Computer | Career Applications | Above and Beyond

< Objective 3 >

Rationalize the denominator in each expression. Simplify when necessary.

67. $\dfrac{\sqrt{3}}{\sqrt{7}}$

68. $\dfrac{\sqrt{5}}{\sqrt{3}}$

69. $\dfrac{\sqrt{2a}}{\sqrt{3b}}$

70. $\dfrac{\sqrt{5x}}{\sqrt{6y}}$

71. $\dfrac{3}{\sqrt[3]{4}}$

72. $\dfrac{2}{\sqrt[3]{9}}$

73. $\dfrac{1}{2 + \sqrt{3}}$

74. $\dfrac{2}{3 - \sqrt{2}}$

75. $\dfrac{8}{3 - \sqrt{5}}$

76. $\dfrac{20}{4 + \sqrt{6}}$

77. $\dfrac{\sqrt{6} + \sqrt{3}}{\sqrt{6} - \sqrt{3}}$

78. $\dfrac{\sqrt{7} - \sqrt{5}}{\sqrt{7} + \sqrt{5}}$

79. $\dfrac{\sqrt{w} + 3}{\sqrt{w} - 3}$

80. $\dfrac{\sqrt{x} - 5}{\sqrt{x} + 5}$

81. $\dfrac{\sqrt{x} - \sqrt{y}}{\sqrt{x} + \sqrt{y}}$

82. $\dfrac{\sqrt{m} + \sqrt{n}}{\sqrt{m} - \sqrt{n}}$

Simplify each radical expression.

83. $x\sqrt[3]{8x^4} + 4\sqrt[3]{27x^7}$

84. $\sqrt[3]{8x^2} - \sqrt[3]{27x^2}$

85. $\dfrac{\sqrt{2x^2 + 3x}}{\sqrt{x}}$

86. $\dfrac{\sqrt{x^2 - 9}}{\sqrt{x + 3}}$

Answers

63. ______

64. ______

65. ______

66. ______

67. ______ 68. ______

69. ______ 70. ______

71. ______ 72. ______

73. ______ 74. ______

75. ______

76. ______

77. ______

78. ______

79. ______

80. ______

81. ______

82. ______

83. ______

84. ______

85. ______

86. ______

Answers

87. ________

88. ________

89. ________

90. ________

91. ________

92. ________

93. ________

Determine whether each statement is **true** *or* **false.**

87. Two radical expressions can be added only if they have the same radicand with the same index.

88. Two radical expressions can be multiplied only if they have the same radicand with the same index.

Complete each statement with **never, sometimes,** *or* **always.**

89. Conjugate square-root expressions ________ have a product that contains no radical sign.

90. When we multiply two radicals, the product is ________ rational.

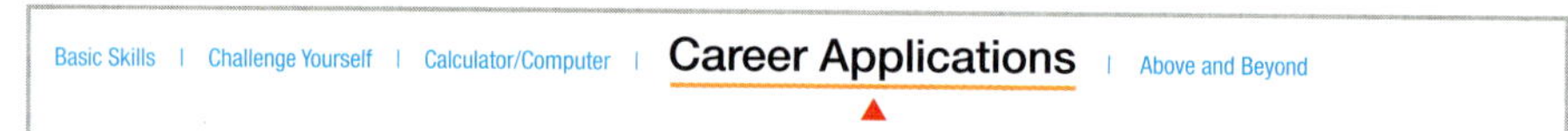

Mechanical Engineering *In the Chapter 7 Activity, "The Swing of a Pendulum," you worked with the relationship between the period and length of a pendulum. The general model for this relationship is given by*

chapter 7 > Make the Connection

$$T = 2\pi\sqrt{\frac{L}{g}}$$

in which g is a gravitational constant.

Use this information to complete exercises 91 and 92.

91. If we measure T in seconds and L in feet, then $g = 32\,\frac{\text{ft}}{\text{s}^2}$. Use this value for g and simplify the pendulum period function (rationalize the denominator).

92. If L is measured in inches and T in seconds, then g must be expressed differently. Use this new value for g and simplify the pendulum period function.

93. **Construction Technology** Plans call for the dimensions of a rectangular room as shown (in feet).

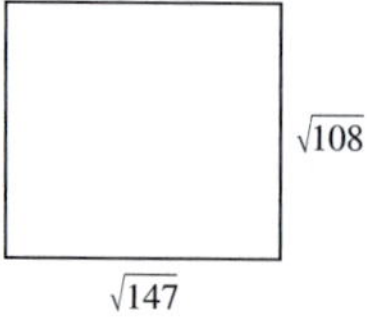

(a) Find the perimeter of the room (give a simplified exact-value result).

(b) Find the area of the room (give a simplified exact-value result).

94. **Construction Technology** Plans call for the dimensions of a rectangular room to be given in terms of an unknown x, as shown in the figure.

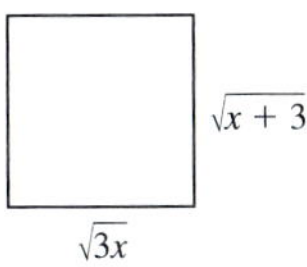

(a) Find the perimeter of the room (give a simplified exact-value result).

(b) Find the area of the room (give a simplified exact-value result).

Answer

94. ____________________

Answers

1. $7\sqrt{5}$ **3.** $3\sqrt{3a}$ **5.** Cannot be simplified **7.** $9\sqrt[3]{2}$ **9.** $9\sqrt{6}$ **11.** $3\sqrt{5}$ **13.** $5\sqrt{7}$ **15.** $6\sqrt{2}$ **17.** $4\sqrt[3]{3}$ **19.** $\sqrt{6w}$ **21.** $5x\sqrt{2x}$ **23.** $3w\sqrt[3]{2w^2}$ **25.** $\frac{4}{3}\sqrt{3}$ **27.** $\frac{3}{2}\sqrt[3]{6}$ **29.** $\frac{2}{3}\sqrt{6}$ **31.** $\frac{1}{3}\sqrt{3}$ **33.** $\sqrt{11a}$ **35.** $\sqrt{42}$ **37.** $\sqrt[3]{36}$ **39.** 6 **41.** $3p\sqrt[3]{2}$ **43.** $2xy\sqrt[3]{5y}$ **45.** $\sqrt{6} + 5\sqrt{2}$ **47.** $2\sqrt{6}$ **49.** $4x\sqrt{3}$ **51.** $6\sqrt[3]{2}$ **53.** $-10 - \sqrt{2}$ **55.** $-28 + \sqrt{10}$ **57.** -4 **59.** $a - 3$ **61.** $28 - 10\sqrt{3}$ **63.** $a + 6\sqrt{a} + 9$ **65.** $x + 2\sqrt{xy} + y$ **67.** $\frac{\sqrt{21}}{7}$ **69.** $\frac{\sqrt{6ab}}{3b}$ **71.** $\frac{3\sqrt[3]{2}}{2}$ **73.** $2 - \sqrt{3}$ **75.** $6 + 2\sqrt{5}$ **77.** $3 + 2\sqrt{2}$ **79.** $\frac{w + 6\sqrt{w} + 9}{w - 9}$ **81.** $\frac{x - 2\sqrt{xy} + y}{x - y}$ **83.** $14x^2\sqrt[3]{x}$ **85.** $\sqrt{2x + 3}$ **87.** True **89.** always **91.** $T = \frac{\pi}{4}\sqrt{2L}$ **93.** **(a)** $26\sqrt{3}$ ft; **(b)** 126 ft^2

7.4 Solving Radical Equations

< 7.4 Objectives >

1 > Solve an equation containing a radical expression

2 > Solve an equation containing two radical expressions

3 > Solve an equation containing a cube root

In this section, we establish procedures for solving equations involving radicals. The basic technique we use involves raising both sides of an equation to some power. However, doing so requires some caution.

For example, suppose we begin with the equation $x = 1$. Squaring both sides gives us

$$x^2 = 1$$
$$x^2 - 1 = 0$$
$$(x + 1)(x - 1) = 0$$

so the solutions appear to be 1 and -1.

Clearly -1 is not a solution to the original equation, $x = 1$. We refer to -1 as an *extraneous solution.*

We must be aware of the possibility of extraneous solutions anytime we raise both sides of an equation to any *even power.* Having said that, we are now prepared to introduce the power property of equality.

> CAUTION

Always check your answers when applying the power property to solve an equation that contains radical expressions.

Property

The Power Property of Equality

Given any two expressions a and b and any positive integer n,

if $a = b$ then $a^n = b^n$

While you never lose a solution by applying the power property, you often gain an extraneous one. Because of this, it is very important that you *check all solutions* when you use the power property to solve an equation.

Example 1 Solving a Radical Equation

< Objective 1 >

Solve $\sqrt{x + 2} = 3$.

Squaring each side, we have

$$(\sqrt{x + 2})^2 = 3^2$$
$$x + 2 = 9$$
$$x = 7$$

Substituting 7 into the original equation, we find

$$\sqrt{7 + 2} \stackrel{?}{=} 3$$
$$\sqrt{9} \stackrel{?}{=} 3$$
$$3 = 3$$

Because this is a true statement, 7 is the solution for the equation.

NOTE

$(\sqrt{x + 2})^2 = x + 2$

This is why squaring both sides of the equation removes the radical.

Check Yourself 1

Solve the equation $\sqrt{x - 5} = 4$.

Example 2 Solving a Radical Equation

NOTE

Applying the power property only removes the radical if that radical is isolated on one side of the equation.

Solve $\sqrt{4x + 5} + 1 = 0$.

We must *first isolate the radical* on the left side.

$$\sqrt{4x + 5} = -1$$

At this point, we should recognize that there can be no solution to this equation. The radical sign means "the positive square root of," so its value cannot be -1. If we fail to notice this, we would continue as follows.

Squaring both sides, we have

$$(\sqrt{4x + 5})^2 = (-1)^2$$
$$4x + 5 = 1$$

and solving for x, we find that

$$x = -1$$

Now we check the solution by substituting -1 for x in the original equation.

NOTE

This is clearly a false statement, so -1 is *not* a solution for the original equation.

$$\sqrt{4(-1) + 5} + 1 \stackrel{?}{=} 0$$
$$\sqrt{1} + 1 \stackrel{?}{=} 0$$

and $2 \neq 0$

Because -1 is an extraneous solution, there are *no solutions* to the original equation.

Check Yourself 2

Solve $\sqrt{3x - 2} + 2 = 0$.

We now consider an example that involves squaring a binomial.

Example 3 Solving a Radical Equation

NOTE

These problems can also be solved graphically. With a graphing utility, plot the two graphs $Y_1 = \sqrt{x + 3}$ and $Y_2 = x + 1$. Note that the graphs have one point of intersection, where $x = 1$.

Solve $\sqrt{x + 3} = x + 1$.

We can square each side, as before.

$$(\sqrt{x + 3})^2 = (x + 1)^2$$
$$x + 3 = x^2 + 2x + 1$$

Simplifying this gives us the quadratic equation

$$x^2 + x - 2 = 0$$

Factoring, we have

$(x - 1)(x + 2) = 0$

which gives us the possible solutions

$x = 1 \qquad \text{or} \qquad x = -2$

Now we check for extraneous solutions and find that $x = 1$ is a valid solution, but that $x = -2$ does not yield a true statement. Therefore, 1 is the only solution.

NOTES

We solved similar equations in Section 6.6.

Verify this for yourself by substituting 1 and then −2 for x in the original equation.

Check Yourself 3

Solve $\sqrt{x - 5} = x - 7$.

> CAUTION

Be careful! Sometimes (as in Example 3), one side of the equation contains a binomial. In that case, we must remember the middle term when we square the binomial. The square of a binomial is *always a trinomial.*

It is not always the case that one of the solutions is extraneous. We may have zero, one, or two valid solutions when we generate a quadratic from a radical equation.

In Example 4 we see a case in which both of the derived solutions satisfy the equation.

Example 4 Solving a Radical Equation

Solve $\sqrt{7x + 1} - 1 = 2x$.

First, *we isolate the term involving the radical.*

$\sqrt{7x + 1} = 2x + 1$

We can now square both sides of the equation.

$7x + 1 = 4x^2 + 4x + 1$

Write the quadratic equation in standard form.

$4x^2 - 3x = 0$

Factoring gives

$x(4x - 3) = 0$

which yields two possible solutions

$x = 0 \qquad \text{or} \qquad x = \frac{3}{4}$

Checking the solutions by substitution, we find that both values for x give true statements.

Letting $x = 0$, we have

$$\sqrt{7(0) + 1} - 1 \stackrel{?}{=} 2(0)$$
$$\sqrt{1} - 1 \stackrel{?}{=} 0$$
$$0 = 0 \qquad \text{True!}$$

NOTE

With a graphing utility plot $Y_1 = \sqrt{7x + 1} - 1$ and $Y_2 = 2x$. Where do they intersect?

Letting $x = \frac{3}{4}$, we have

$$\sqrt{7\left(\frac{3}{4}\right) + 1} - 1 \stackrel{?}{=} 2\left(\frac{3}{4}\right)$$

$$\sqrt{\frac{25}{4}} - 1 \stackrel{?}{=} \frac{3}{2} \qquad 7\left(\frac{3}{4}\right) + 1 = \frac{21}{4} + \frac{4}{4} = \frac{25}{4}$$

$$\frac{5}{2} - 1 \stackrel{?}{=} \frac{3}{2}$$

$$\frac{3}{2} = \frac{3}{2} \qquad \text{True!}$$

Check Yourself 4

Solve $\sqrt{5x + 1} - 1 = 3x$.

Sometimes when an equation involves more than one radical, we must apply the power property more than once. In such a case, it is generally best to avoid having to work with two radicals on the same side of the equation. Example 5 illustrates one approach to solving such equations.

Example 5 Solving an Equation Containing Two Radicals

< Objective 2 >

> **NOTE**
>
> $1 + \sqrt{2x - 6}$ is a binomial of the form $a + b$, in which $a = 1$ and $b = \sqrt{2x - 6}$. The square on the right then has the form $a^2 + 2ab + b^2$.

Solve $\sqrt{x - 2} - \sqrt{2x - 6} = 1$.

First we isolate $\sqrt{x - 2}$ by adding $\sqrt{2x - 6}$ to both sides of the equation. This gives

$$\sqrt{x - 2} = 1 + \sqrt{2x - 6}$$

Then squaring each side, we have

$$(\sqrt{x - 2})^2 = (1 + \sqrt{2x - 6})^2$$

$$x - 2 = 1 + 2\sqrt{2x - 6} + 2x - 6$$

Now we isolate the radical on the right side.

$$x - 2x - 2 - 1 + 6 = 2\sqrt{2x - 6}$$

$$-x + 3 = 2\sqrt{2x - 6}$$

We must square again to remove that radical.

$$(-x + 3)^2 = (2\sqrt{2x - 6})^2 \qquad \text{Square both the 2 and the } \sqrt{2x - 6}.$$

$$x^2 - 6x + 9 = 4(2x - 6)$$

$$x^2 - 6x + 9 = 8x - 24 \qquad \text{Solve the quadratic equation that results.}$$

$$x^2 - 14x + 33 = 0$$

$$(x - 3)(x - 11) = 0$$

So

$$x = 3 \quad \text{or} \quad x = 11$$

are the possible solutions. Checking the possible solutions, we find that $x = 3$ yields the only valid solution. You should verify that for yourself.

Check Yourself 5

Solve $\sqrt{x + 3} - \sqrt{2x + 4} + 1 = 0$.

Earlier in this section, we noted that extraneous roots are possible whenever we raise both sides of the equation to an *even power*. In Example 6, we raise both sides of the equation to an odd power. We will still check the solutions, but in this case it is simply a check of our work and not a search for extraneous solutions.

Example 6 **Solving a Radical Equation**

< Objective 3 >

> **NOTE**
> Because a *cube root* is involved, we *cube* both sides to remove the radical.

Solve $\sqrt[3]{x^2 + 23} = 3$.

Cubing each side gives

$x^2 + 23 = 27$

which results in the quadratic equation

$x^2 - 4 = 0$

> **NOTE**
> Raising both sides to an odd-numbered power does not introduce extraneous solutions.

This has two solutions

$x = 2 \qquad \text{or} \qquad x = -2$

Checking the solutions, we find that both result in true statements. Again you should verify this result.

Check Yourself 6

Solve $\sqrt[3]{x^2 - 8} - 2 = 0$.

We summarize our work in this section with an algorithm for solving equations involving radicals.

Step by Step

Solving Equations Involving Radicals

Step 1 Isolate a radical term on one side of the equation.

Step 2 Raise each side of the equation to the smallest power that eliminates the isolated radical.

Step 3 If any radicals remain in the equation derived in step 2, return to step 1 and continue the solution process.

Step 4 Solve the resulting equation to determine any possible solutions.

Step 5 Check all solutions to determine whether extraneous solutions may have resulted from step 2.

Check Yourself ANSWERS

1. {21} **2.** No solutions **3.** {9} **4.** $\left\{0, -\frac{1}{9}\right\}$ **5.** {6} **6.** {4, −4}

Reading Your Text

SECTION 7.4

(a) We must look for ______________ solutions anytime we raise both sides of an equation to an even power.

(b) Given any two expressions a and b and any positive ______________ n, if $a = b$ then $a^n = b^n$.

(c) Applying the power property only removes a radical if that radical term is ______________ on one side of the equation.

(d) We may have zero, one, or two valid ______________ when we generate a quadratic equation from a radical equation.

7.4 exercises

Boost *your* GRADE at ALEKS.com!

ALEKS®

- Practice Problems
- Self-Tests
- NetTutor
- e-Professors
- Videos

Name ____________

Section ________ Date ________

Answers

1. ________ 2. ________
3. ________ 4. ________
5. ________ 6. ________
7. ________ 8. ________
9. ________
10. ________
11. ________
12. ________
13. ________ 14. ________
15. ________ 16. ________
17. ________ 18. ________
19. ________ 20. ________
21. ________ 22. ________
23. ________ 24. ________
25. ________ 26. ________

Basic Skills | Challenge Yourself | Calculator/Computer | Career Applications | Above and Beyond

< Objectives 1–3 >

Solve each equation. Be sure to check your solutions.

1. $\sqrt{x} = 2$

2. $\sqrt{x} - 3 = 0$

3. $2\sqrt{y} - 1 = 0$

4. $3\sqrt{2z} = 9$

5. $\sqrt{m + 5} = 3$ > Videos

6. $\sqrt{y + 7} = 5$

7. $\sqrt{2x + 4} - 4 = 0$

8. $\sqrt{3x + 3} - 6 = 0$

9. $\sqrt{3x - 2} + 2 = 0$

10. $\sqrt{4x + 1} + 3 = 0$

11. $\sqrt{x - 1} = \sqrt{1 - x}$

12. $\sqrt{x + 1} = \sqrt{1 + x}$
(*Hint:* Both radicands must be nonnegative.)

13. $\sqrt{w + 3} = \sqrt{3 + w}$
(*Hint:* Both radicands must be nonnegative.)

14. $\sqrt{w - 3} = \sqrt{3 - w}$

15. $\sqrt{2x - 3} + 1 = 3$

16. $\sqrt{3x + 1} - 2 = -1$

17. $2\sqrt{3z + 2} - 1 = 5$

18. $3\sqrt{4q - 1} - 2 = 7$

19. $\sqrt{15 - 2x} = x$

20. $\sqrt{48 - 2y} = y$

21. $\sqrt{x + 5} = x - 1$

22. $\sqrt{2x - 1} = x - 8$

23. $\sqrt{3m - 2} + m = 10$

24. $\sqrt{2x + 1} + x = 7$

25. $\sqrt{t + 9} + 3 = t$ 

26. $\sqrt{2y + 7} + 4 = y$

27. $\sqrt{6x+1} - 1 = 2x$

28. $\sqrt{7x+1} - 1 = 3x$

29. $\sqrt[3]{x-5} = 3$

30. $\sqrt[3]{x+6} = 2$

31. $\sqrt[3]{x^2-1} = 2$

32. $\sqrt[3]{x^2+11} = 3$

Basic Skills | **Challenge Yourself** | Calculator/Computer | Career Applications | Above and Beyond

Solve each equation. Be sure to check your solutions.

33. $\sqrt{2x} = \sqrt{x+1}$

34. $\sqrt{3x} = \sqrt{5x-1}$

35. $2\sqrt{3r} = \sqrt{r+11}$

36. $5\sqrt{2q-7} = \sqrt{15q}$

37. $\sqrt{x+2} + 1 = \sqrt{x+4}$ > Videos

38. $\sqrt{x+5} - 1 = \sqrt{x+3}$

39. $\sqrt{4m-3} - 2 = \sqrt{2m-5}$

40. $\sqrt{2c-1} = \sqrt{3c+1} - 1$

41. $\sqrt{x+1} + \sqrt{x} = 1$

42. $\sqrt{z-1} - \sqrt{6-z} = 1$

43. $\sqrt{5x+6} - \sqrt{x+3} = 3$

44. $\sqrt{5y+6} - \sqrt{3y+4} = 2$

45. $\sqrt{y^2+12y} - 3\sqrt{5} = 0$

46. $\sqrt{x^2+2x} - 2\sqrt{6} = 0$

47. $\sqrt{\dfrac{x-3}{x+2}} = \dfrac{2}{3}$

48. $\dfrac{\sqrt{x-2}}{x-2} = \dfrac{x-5}{\sqrt{x-2}}$

49. $\sqrt{\sqrt{t}+5} = 3$

50. $\sqrt{\sqrt{s}-1} = \sqrt{s-7}$

Answers

27. ______

28. ______

29. ______

30. ______

31. ______

32. ______

33. ______

34. ______

35. ______

36. ______

37. ______

38. ______

39. ______

40. ______

41. ______

42. ______

43. ______

44. ______

45. ______ 46. ______

47. ______ 48. ______

49. ______ 50. ______

Answers

51. ______

52. ______

53. ______

54. ______

55. ______

56. ______

57. ______

58. ______

59. ______

60. ______

61. ______

62. ______

63. ______

64. ______

65. ______

66. ______

67. ______

68. ______

Complete each statement with **never, sometimes,** *or* **always.**

51. When we raise both sides of an equation to an even power, we ______ obtain extraneous solutions.

52. Before applying the power property, we ______ try to isolate a radical term on one side of the equation.

53. To remove square roots from an equation, it is ______ necessary to square both sides twice.

54. If we raise both sides of an equation to an odd power, we ______ obtain extraneous solutions.

Solve each application.

55. The sum of an integer and its square root is 12. Find the integer.

56. The difference between an integer and its square root is 12. What is the integer?

57. The sum of an integer and twice its square root is 24. What is the integer?

58. The sum of an integer and 3 times its square root is 40. Find the integer.

The function $d = \sqrt{2h}$ *can be used to estimate the distance d to the horizon (in miles) from a given height (in feet).*

59. If a plane flies at 30,000 ft, how far away is the horizon?

60. Janine was looking out across the ocean from her hotel room on the beach. Her eyes were 250 ft above the ground. She saw a ship on the horizon. Approximately how far was the ship from her?

61. Given a distance d to the horizon, give the height in terms of d that would allow you to see that far.

62. Use the result of exercise 61 to estimate the height required to see 100 miles to the horizon.

When a car comes to a sudden stop, you can determine the skidding distance (in feet) for a given speed (in miles per hour) by using the formula $s = 2\sqrt{5x}$, *in which s is skidding distance and x is speed. Calculate the skidding distance for each speed.*

63. 55 mi/h

64. 65 mi/h

65. 75 mi/h

66. 40 mi/h

67. Given the skidding distance s, what formula would allow you to calculate the speed in miles per hour?

68. Use the formula obtained in exercise 67 to determine the speed of a car in miles per hour if the skid marks were 35 ft long.

Use a graphing calculator to solve each equation. Express solutions to the nearest hundredth. (Hint: *Define* Y_1 *by the expression on the left side of the equation, and define* Y_2 *by the expression on the right side. Graph these functions and locate any intersection points. For each such point, the x-value represents a solution.*)

69. $\sqrt{x+4} = x - 3$

70. $\sqrt{2-x} = x + 4$

71. $3 - 2\sqrt{x+4} = 2x - 5$

72. $5 - 3\sqrt{2-x} = 3 - 4x$

Basic Skills | Challenge Yourself | Calculator/Computer | **Career Applications** | Above and Beyond

Mechanical Engineering *In the Chapter 7 Activity, "The Swing of a Pendulum," you worked with the relationship between the period and length of a pendulum. The general model for this relationship is given by*

$$T = 2\pi\sqrt{\frac{L}{g}}$$

in which g is a gravitational constant.

Use this information to complete exercises 73 and 74.

73. Express the length of a pendulum as a function of the time of its period (that is, solve the above model for L).

74. The Foucault (pronounced "foo-koh") pendulum in the Smithsonian Institution (Washington, D.C.) had an 8-s period. Use $g = 980\ \frac{\text{cm}}{\text{s}^2}$ to determine the pendulum's length, to the nearest cm.

75. **Construction Technology** Plans for the dimensions of a rectangular room are shown here (in feet). Find the length of the room's diagonal (give a simplified exact-value result).

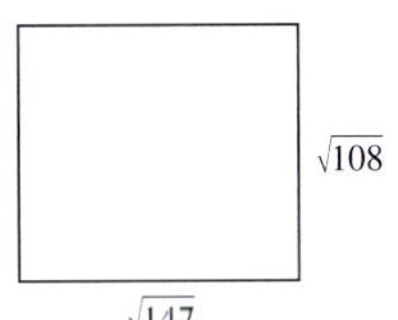

76. **Construction Technology** Plans call for the dimensions of a rectangular room to be given in terms of an unknown x, as shown here. Find the length of the room's diagonal.

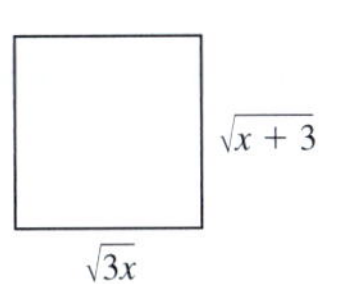

Answers

69. ______

70. ______

71. ______

72. ______

73. ______

74. ______

75. ______

76. ______

Basic Skills | Challenge Yourself | Calculator/Computer | Career Applications | **Above and Beyond**

Answers

77.

78.

79.

80.

81.

82.

83.

84.

85.

86.

87.

88.

89.

90.

91.

92.

77. For what values of x is $\sqrt{(x-1)^2} = x - 1$ a true statement?

78. For what values of x is $\sqrt[3]{(x-1)^3} = x - 1$ a true statement?

Solve for the indicated variable.

79. $h = \sqrt{pq}$ for q

80. $c = \sqrt{a^2 + b^2}$ for a

81. $v = \sqrt{2gR}$ for R

82. $v = \sqrt{2gR}$ for g

83. $r = \sqrt{\dfrac{S}{2\pi}}$ for S

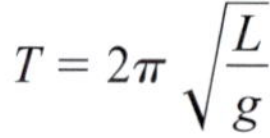

84. $r = \sqrt{\dfrac{3V}{4\pi}}$ for V

85. $r = \sqrt{\dfrac{2V}{\pi h}}$ for V

86. $r = \sqrt{\dfrac{2V}{\pi h}}$ for h

87. $d = \sqrt{(x-1)^2 + (y-2)^2}$ for x

88. $d = \sqrt{(x-1)^2 + (y-2)^2}$ for y

A weight suspended on the end of a string is a pendulum. *The most common example of a pendulum (this side of Edgar Allan Poe) is the kind found in many clocks.*

The regular back-and-forth motion of the pendulum is periodic, *and one such cycle of motion is called a* period. *The time, in seconds, that it takes for one period is given by the radical equation*

$$T = 2\pi\sqrt{\frac{L}{g}}$$

in which g is the force of gravity ($10\ m/s^2$) and L is the length of the pendulum, in meters.

89. Find the period (to the nearest hundredth of a second) if the pendulum is 0.9 m long.

90. Find the period if the pendulum is 0.049 m long.

91. Solve the equation for length L.

92. How long is a pendulum if its period is 1 s?

Answers

1. $\{4\}$ **3.** $\left\{\frac{1}{4}\right\}$ **5.** $\{4\}$ **7.** $\{6\}$ **9.** No solutions **11.** $\{1\}$
13. $\{w \mid w \geq -3\}$ **15.** $\left\{\frac{7}{2}\right\}$ **17.** $\left\{\frac{7}{3}\right\}$ **19.** $\{3\}$ **21.** $\{4\}$
23. $\{6\}$ **25.** $\{7\}$ **27.** $\left\{0, \frac{1}{2}\right\}$ **29.** $\{32\}$ **31.** $\{3, -3\}$
33. $\{1\}$ **35.** $\{1\}$ **37.** $\left\{-\frac{7}{4}\right\}$ **39.** $\{3, 7\}$ **41.** $\{0\}$ **43.** $\{6\}$
45. $\{-15, 3\}$ **47.** $\{7\}$ **49.** $\{16\}$ **51.** sometimes **53.** sometimes
55. 9 **57.** 16 **59.** $\approx$245 mi **61.** $\frac{d^2}{2}$ **63.** 33 ft **65.** 39 ft
67. $x = \frac{s^2}{20}$ **69.** $\{6.19\}$ **71.** $\{1.63\}$ **73.** $L = \frac{g}{4\pi^2}T^2$
75. $\sqrt{255}$ ft **77.** $\{x \mid x \geq 1\}$ **79.** $q = \frac{h^2}{p}$ **81.** $R = \frac{v^2}{2g}$
83. $S = 2\pi r^2$ **85.** $V = \frac{\pi h r^2}{2}$ **87.** $x = 1 \pm \sqrt{d^2 - (y-2)^2}$
89. 1.88 s **91.** $L = \frac{T^2 g}{4\pi^2}$

7.5 Rational Exponents

< 7.5 Objectives >

1 > Simplify expressions containing rational exponents

2 > Use a calculator to estimate the value of an expression containing rational exponents

3 > Write an expression in radical or exponential form

In Section 7.1, we discussed roots and radical notation. In this section, we develop notation using exponents to provide an alternate way of writing roots.

This notation involves **rational numbers as exponents.** To start the development, we extend all the previous properties of exponents to include rational exponents.

Given that extension, suppose that

$$a = 4^{1/2}$$

Squaring both sides of the equation yields

$$a^2 = (4^{1/2})^2$$

or

$$a^2 = 4^1$$
$$a^2 = 4$$

NOTE

We will see later in this section that the property $(x^m)^n = x^{mn}$ holds for rational numbers m and n.

From this last equation we see that a is the number whose square is 4; that is, a is the principal square root of 4. Using our earlier notation, we can write

$$a = \sqrt{4}$$

But we began with

$$a = 4^{1/2}$$

and to be consistent, we must have

$$4^{1/2} = \sqrt{4}$$

NOTE

$4^{1/2}$ indicates the *principal square root* of 4.

This argument can be repeated for any exponent of the form $\frac{1}{n}$, so it seems reasonable to make the following definition.

Definition

Rational Exponents

If a is any real number and n is a positive integer ($n > 1$), then

$$a^{1/n} = \sqrt[n]{a}$$

We restrict a so that a is nonnegative when n is even. In words, $a^{1/n}$ indicates the principal nth root of a.

Example 1 illustrates the use of rational exponents to represent roots.

Example 1 Writing Expressions in Radical Form

< Objective 1 >

NOTES

$27^{1/3}$ is the *cube root* of 27.

$32^{1/5}$ is the *fifth root* of 32.

Write each expression in radical form and then simplify.

(a) $25^{1/2} = \sqrt{25} = 5$

(b) $27^{1/3} = \sqrt[3]{27} = 3$

(c) $-36^{1/2} = -\sqrt{36} = -6$

(d) $(-36)^{1/2} = \sqrt{-36}$ is not a real number.

(e) $32^{1/5} = \sqrt[5]{32} = 2$

Check Yourself 1

Write each expression in radical form and simplify.

(a) $8^{1/3}$ **(b)** $-64^{1/2}$ **(c)** $81^{1/4}$

NOTE

The two radical forms for $a^{m/n}$ are equivalent, and the choice of which form to use generally depends on whether we are evaluating numerical expressions or rewriting expressions containing variables in radical form.

We are now ready to extend our exponent notation to allow *any* rational exponent, again assuming that our previous exponent properties must still be valid. Note that

$$a^{m/n} = (a^{1/n})^m = (a^m)^{1/n} \qquad \text{because} \qquad \frac{m}{n} = (m)\left(\frac{1}{n}\right) = \left(\frac{1}{n}\right)(m)$$

From our earlier work, we know that $a^{1/n} = \sqrt[n]{a}$, and combining this with the above observation, we offer the following definition for $a^{m/n}$.

Definition

Rational Exponents

For any real number a and positive integers m and n with $n > 1$,

$$a^{m/n} = (\sqrt[n]{a})^m = \sqrt[n]{a^m}$$

We apply this extension of our rational-exponent notation in Example 2.

Example 2 Simplifying Expressions with Rational Exponents

Simplify each expression.

(a) $9^{3/2} = (9^{1/2})^3 = (\sqrt{9})^3$

$= 3^3 = 27$

(b) $\left(\dfrac{16}{81}\right)^{3/4} = \left[\left(\dfrac{16}{81}\right)^{1/4}\right]^3 = \left(\sqrt[4]{\dfrac{16}{81}}\right)^3$

$= \left(\dfrac{2}{3}\right)^3 = \dfrac{8}{27}$

(c) $(-8)^{2/3} = [(-8)^{1/3}]^2 = (\sqrt[3]{-8})^2$

$= (-2)^2 = 4$

NOTE

This illustrates why we use $(\sqrt[n]{a})^m$ for $a^{m/n}$ when evaluating numerical expressions. The numbers involved are smaller and easier to work with.

In (a) we could also have evaluated the expression as

$$9^{3/2} = \sqrt{9^3} = \sqrt{729}$$
$$= 27$$

Check Yourself 2

Simplify each expression.

(a) $16^{3/4}$ (b) $\left(\frac{8}{27}\right)^{2/3}$ (c) $(-32)^{3/5}$

Now we want to extend our rational exponent notation. Using the definition of negative exponents, we can write

$$a^{-m/n} = \frac{1}{a^{m/n}}$$

Example 3 illustrates the use of negative rational exponents.

Example 3 Simplifying Expressions with Rational Exponents

Simplify each expression.

(a) $16^{-1/2} = \frac{1}{16^{1/2}} = \frac{1}{4}$ $16^{1/2} = \sqrt{16} = 4$

(b) $27^{-2/3} = \frac{1}{27^{2/3}} = \frac{1}{(\sqrt[3]{27})^2} = \frac{1}{3^2} = \frac{1}{9}$

Check Yourself 3

Simplify each expression.

(a) $16^{-1/4}$ (b) $81^{-3/4}$

You can use a graphing calculator to evaluate expressions containing rational exponents by using the [∧] and parentheses keys.

Example 4 Using a Calculator to Estimate Powers

< Objective 2 >

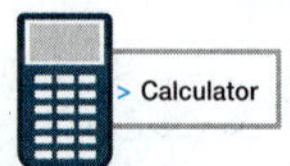

RECALL

Fractions indicate division.

Use a graphing calculator to evaluate each expression. Round all answers to three decimal places.

(a) $45^{2/5}$

Enter 45 and press the [∧] key. Then use the keystrokes

[(] 2 [÷] 5 [)] You *must* use parentheses for the entire exponent.

Press [ENTER], and the display will read 4.584426407. Rounded to three decimal places, the result is 4.584.

NOTE

```
45^(2/5)
         4.584426407
38^(-2/3)
          .088473037
```

(b) $38^{-2/3}$

Enter 38 and press the $\boxed{\wedge}$ key. Then use the keystrokes

$\boxed{(}$ $\boxed{(-)}$ 2 $\boxed{\div}$ 3 $\boxed{)}$

Press $\boxed{\text{ENTER}}$, and the display will read 0.088473037. Rounded to three decimal places, the result is 0.088.

Check Yourself 4

Use a calculator to evaluate each expression. Round each answer to three decimal places.

(a) $23^{3/5}$ **(b)** $18^{-4/7}$

As we mentioned earlier in this section, we assume that all our previous exponent properties continue to hold for rational exponents. Those properties are restated here.

Property

Properties of Exponents

For any nonzero real numbers *a* and *b* and rational numbers *m* and *n*,

1. Product rule $a^m \cdot a^n = a^{m+n}$
2. Quotient rule $\dfrac{a^m}{a^n} = a^{m-n}$
3. Power rule $(a^m)^n = a^{mn}$
4. Product-power rule $(ab)^m = a^m b^m$
5. Quotient-power rule $\left(\dfrac{a}{b}\right)^m = \dfrac{a^m}{b^m}$

We restrict *a* and *b* to being nonnegative real numbers when *m* or *n* indicates an even root.

Example 5 illustrates the use of our extended properties to simplify expressions involving rational exponents. Here, we assume that all variables represent positive real numbers.

Example 5 Simplifying Expressions

Simplify each expression.

NOTES

Product rule—add the exponents.

Quotient rule—subtract the exponents.

(a) $x^{2/3} \cdot x^{1/2} = x^{2/3+1/2}$

$= x^{4/6+3/6} = x^{7/6}$

(b) $\dfrac{w^{3/4}}{w^{1/2}} = w^{3/4-1/2}$

$= w^{3/4-2/4} = w^{1/4}$

NOTE

Power rule—multiply the exponents.

(c) $(a^{2/3})^{3/4} = a^{(2/3)(3/4)}$ $\quad \frac{2}{\not 3} \cdot \frac{\not 3}{4} = \frac{2}{4} = \frac{1}{2}$

$= a^{1/2}$

Check Yourself 5

Simplify each expression.

(a) $z^{3/4} \cdot z^{1/2}$ (b) $\dfrac{x^{5/6}}{x^{1/3}}$ (c) $(b^{5/6})^{2/5}$

As you would expect from your previous experience with exponents, simplifying expressions often involves using several exponent properties.

Example 6 Simplifying Expressions

Simplify each expression.

(a) $(x^{2/3} \cdot y^{5/6})^{3/2}$

$= (x^{2/3})^{3/2} \cdot (y^{5/6})^{3/2}$ Product-power rule

$= x^{(2/3)(3/2)} \cdot y^{(5/6)(3/2)} = xy^{5/4}$ Power rule

(b) $\left(\dfrac{r^{-1/2}}{s^{1/3}}\right)^6 = \dfrac{(r^{-1/2})^6}{(s^{1/3})^6}$ Quotient-power rule

$= \dfrac{r^{-3}}{s^2} = \dfrac{1}{r^3s^2}$ Power rule

(c) $\left(\dfrac{4a^{-2/3} \cdot b^2}{a^{1/3} \cdot b^{-4}}\right)^{1/2} = \left(\dfrac{4b^2 \cdot b^4}{a^{1/3} \cdot a^{2/3}}\right)^{1/2} = \left(\dfrac{4b^6}{a}\right)^{1/2}$ Simplify inside the parentheses first.

$= \dfrac{(4b^6)^{1/2}}{a^{1/2}} = \dfrac{4^{1/2}(b^6)^{1/2}}{a^{1/2}}$

$= \dfrac{2b^3}{a^{1/2}}$

Check Yourself 6

Simplify each expression.

(a) $(a^{3/4} \cdot b^{1/2})^{2/3}$ (b) $\left(\dfrac{w^{1/2}}{z^{-1/4}}\right)^4$ (c) $\left(\dfrac{8x^{-3/4}y}{x^{1/4} \cdot y^{-5}}\right)^{1/3}$

We can also use the relationships between rational exponents and radicals to write expressions involving rational exponents as radicals and vice versa.

Example 7 Writing Expressions in Radical Form

< Objective 3 >

Write each expression in radical form.

(a) $a^{3/5} = \sqrt[5]{a^3}$ $\quad a^{m/n} = \sqrt[n]{a^m}$

(b) $(mn)^{3/4} = \sqrt[4]{(mn)^3}$

$= \sqrt[4]{m^3n^3}$

(c) $2y^{5/6} = 2\sqrt[6]{y^5}$ The exponent applies *only* to the variable y.

(d) $(2y)^{5/6} = \sqrt[6]{(2y)^5}$ Now the exponent applies to $2y$ because of the parentheses.

$$= \sqrt[6]{32y^5}$$

Check Yourself 7

Write each expression in radical form.

(a) $(ab)^{2/3}$ (b) $3x^{3/4}$ (c) $(3x)^{3/4}$

Example 8 Writing Expressions in Exponential Form

Use rational exponents to write each expression and simplify.

(a) $\sqrt[3]{5x} = (5x)^{1/3}$

(b) $\sqrt{9a^2b^4} = (9a^2b^4)^{1/2}$

$$= 9^{1/2}(a^2)^{1/2}(b^4)^{1/2} = 3ab^2$$

(c) $\sqrt[4]{16w^{12}z^8} = (16w^{12}z^8)^{1/4}$

$$= 16^{1/4}(w^{12})^{1/4}(z^8)^{1/4} = 2w^3z^2$$

Check Yourself 8

Use rational exponents to write each expression and simplify.

(a) $\sqrt{7a}$ (b) $\sqrt[3]{27p^6q^9}$ (c) $\sqrt[4]{81x^8y^{16}}$

Our final example applies various multiplication patterns to terms involving rational exponents.

Example 9 Multiplying Terms That Involve Rational Exponents

Use the appropriate property to find each product.

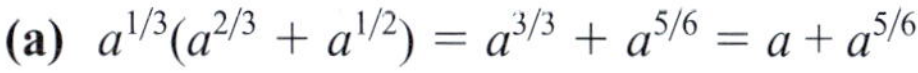

(a) $a^{1/3}(a^{2/3} + a^{1/2}) = a^{3/3} + a^{5/6} = a + a^{5/6}$ We use the distributive property.

(b) $(a^{2/3} - a^{1/2})(a^{2/3} + a^{1/2}) = a^{4/3} - a^{2/2} = a^{4/3} - a$ This is the difference of squares.

NOTES

The expression in (a) cannot be simplified further.

The pattern in (b) results in the difference of two squares.

Check Yourself 9

Use the appropriate properties to find each product.

(a) $b^{1/3}(b^{2/3} + b^{1/3})$ (b) $(b^{2/3} - b^{1/2})^2$

Check Yourself ANSWERS

1. **(a)** $\sqrt[3]{8} = 2$; **(b)** $-\sqrt{64} = -8$; **(c)** $\sqrt[4]{81} = 3$ **2.** **(a)** 8; **(b)** $\frac{4}{9}$; **(c)** -8
3. **(a)** $\frac{1}{2}$; **(b)** $\frac{1}{27}$ **4.** **(a)** 6.562; **(b)** 0.192 **5.** **(a)** $z^{5/4}$; **(b)** $x^{1/2}$; **(c)** $b^{1/3}$
6. **(a)** $a^{1/2}b^{1/3}$; **(b)** w^2z; **(c)** $\frac{2y^2}{x^{1/3}}$ **7.** **(a)** $\sqrt[3]{a^2b^2}$; **(b)** $3\sqrt[4]{x^3}$; **(c)** $\sqrt[4]{27x^3}$
8. **(a)** $(7a)^{1/2}$; **(b)** $(27p^6q^9)^{1/3} = 3p^2q^3$; **(c)** $(81x^8y^{16})^{1/4} = 3x^2y^4$
9. **(a)** $b + b^{2/3}$; **(b)** $b^{4/3} - 2b^{7/6} + b$

Reading Your Text

SECTION 10.5

(a) If a is any ____________ number and n is a positive integer greater than one, then $a^{1/n} = \sqrt[n]{a}$, where a is nonnegative when n is even.

(b) $a^{1/n}$ indicates the ____________ nth root of a.

(c) We can write expressions involving rational exponents as ____________.

(d) In the expression $2y^{1/3}$, the ____________ applies only to the variable y.

Basic Skills | Challenge Yourself | Calculator/Computer | Career Applications | Above and Beyond

< Objective 1 >

Evaluate each expression.

1. $49^{1/2}$

2. $100^{1/2}$

3. $-25^{1/2}$

4. $(-81)^{1/2}$

5. $(-49)^{1/2}$

6. $-49^{1/2}$

7. $27^{1/3}$

8. $(-64)^{1/3}$ 

9. $81^{1/4}$

10. $-81^{1/4}$

11. $\left(\frac{4}{9}\right)^{1/2}$

12. $\left(\frac{27}{8}\right)^{1/3}$

13. $27^{2/3}$

14. $16^{3/2}$

15. $(-8)^{4/3}$

16. $64^{2/3}$

17. $32^{2/5}$

18. $-81^{3/4}$

19. $81^{3/2}$

20. $(-243)^{3/5}$

21. $\left(\frac{8}{27}\right)^{2/3}$

22. $\left(\frac{16}{9}\right)^{3/2}$

< Objective 2 >

Evaluate each expression and use a calculator to check each answer.

23. $49^{-1/2}$

24. $27^{-1/3}$

25. $81^{-1/4}$

26. $121^{-1/2}$

27. $9^{-3/2}$ 

28. $16^{-3/4}$

29. $64^{-5/6}$

30. $16^{-3/2}$

31. $\left(\frac{4}{25}\right)^{-1/2}$

32. $\left(\frac{27}{8}\right)^{-2/3}$

Name ________________

Section ________ Date ________

Answers

1. ________ 2. ________

3. ________ 4. ________

5. ________ 6. ________

7. ________ 8. ________

9. ________ 10. ________

11. ________ 12. ________

13. ________ 14. ________

15. ________ 16. ________

17. ________ 18. ________

19. ________ 20. ________

21. ________ 22. ________

23. ________ 24. ________

25. ________ 26. ________

27. ________ 28. ________

29. ________ 30. ________

31. ________ 32. ________

7.5 exercises

Simplify each expression. Assume all variables represent positive real numbers.

Answers

33. ______
34. ______
35. ______
36. ______
37. ______
38. ______
39. ______
40. ______
41. ______ 42. ______
43. ______ 44. ______
45. ______ 46. ______
47. ______ 48. ______
49. ______ 50. ______
51. ______ 52. ______
53. ______ 54. ______
55. ______ 56. ______
57. ______ 58. ______
59. ______ 60. ______
61. ______ 62. ______

33. $x^{1/2} \cdot x^{1/2}$

34. $a^{2/3} \cdot a^{1/3}$

35. $y^{5/7} \cdot y^{1/7}$

36. $m^{1/4} \cdot m^{5/4}$

37. $b^{2/3} \cdot b^{3/2}$

38. $p^{5/6} \cdot p^{2/3}$

39. $\frac{x^{2/3}}{x^{1/3}}$

40. $\frac{a^{5/6}}{a^{1/6}}$

41. $\frac{s^{7/5}}{s^{2/5}}$

42. $\frac{z^{9/2}}{z^{3/2}}$

43. $\frac{w^{7/6}}{w^{1/2}}$

44. $\frac{b^{7/6}}{b^{2/3}}$

45. $(x^{3/4})^{4/3}$

46. $(y^{4/3})^{3/4}$

47. $(a^{2/5})^{3/2}$

48. $(p^{3/4})^{2/3}$

49. $(y^{-2/3})^{6}$

50. $(w^{-2/3})^{6}$

51. $(a^{2/3} \cdot b^{3/2})^{6}$ > Videos

52. $(p^{3/4} \cdot q^{5/2})^{4}$

53. $(x^{2/7} \cdot y^{3/7})^{7}$

54. $(3m^{3/4} \cdot n^{5/4})^{4}$

55. $(s^{3/4} \cdot t^{1/4})^{4/3}$

56. $(x^{5/2} \cdot y^{5/7})^{2/5}$

57. $(8p^{3/2} \cdot q^{5/2})^{2/3}$

58. $(16a^{1/3} \cdot b^{2/3})^{3/4}$

59. $(x^{3/5} \cdot y^{3/4} \cdot z^{3/2})^{2/3}$

60. $(p^{5/6} \cdot q^{2/3} \cdot r^{5/3})^{3/5}$

61. $\frac{a^{5/6} \cdot b^{3/4}}{a^{1/3} \cdot b^{1/2}}$

62. $\frac{x^{2/3} \cdot y^{3/4}}{x^{1/2} \cdot y^{1/2}}$

63. $\dfrac{(r^{-1} \cdot s^{1/2})^3}{r \cdot s^{-1/2}}$

64. $\dfrac{(w^{-2} \cdot z^{-1/4})^6}{w^{-8}z^{1/2}}$

65. $\left(\dfrac{x^{12}}{y^8}\right)^{1/4}$

66. $\left(\dfrac{q^{12}}{p^{28}}\right)^{1/4}$

67. $\left(\dfrac{m^{-1/4}}{n^{1/2}}\right)^4$

68. $\left(\dfrac{r^{1/5}}{s^{-1/2}}\right)^{10}$

69. $\left(\dfrac{r^{-1/2} \cdot s^{3/4}}{t^{1/4}}\right)^4$

70. $\left(\dfrac{a^{1/3} \cdot b^{-1/6}}{c^{-1/6}}\right)^6$

71. $\left(\dfrac{8x^3 \cdot y^{-6}}{z^{-9}}\right)^{1/3}$

72. $\left(\dfrac{16p^{-4} \cdot q^6}{r^2}\right)^{-1/2}$

73. $\left(\dfrac{16m^{-3/5} \cdot n^2}{m^{1/5} \cdot n^{-2}}\right)^{1/4}$ 

74. $\left(\dfrac{27x^{5/6} \cdot y^{-4/3}}{x^{-7/6} \cdot y^{5/3}}\right)^{1/3}$

75. $\left(\dfrac{x^{3/2} \cdot y^{1/2}}{z^2}\right)^{1/2}\left(\dfrac{x^{3/4} \cdot y^{3/2}}{z^{-3}}\right)^{1/3}$

76. $\left(\dfrac{p^{1/2} \cdot q^{4/3}}{r^{-4}}\right)^{3/4}\left(\dfrac{p^{15/8} \cdot q^{-3}}{r^6}\right)^{1/3}$

< Objective 3 >

Write each expression in radical form. Do not simplify.

77. $a^{3/4}$

78. $m^{5/6}$

79. $2x^{2/3}$

80. $3m^{-2/5}$

81. $3x^{2/5}$

82. $2y^{-3/4}$

83. $(3x)^{2/5}$

84. $(2y)^{-3/4}$

Write each expression using rational exponents, and simplify where necessary.

85. $\sqrt{7a}$

86. $\sqrt{25w^4}$

87. $\sqrt[3]{8m^6n^9}$

88. $\sqrt[5]{32r^{10}s^{15}}$

Answers

63. ______ 64. ______

65. ______ 66. ______

67. ______ 68. ______

69. ______ 70. ______

71. ______ 72. ______

73. ______

74. ______

75. ______

76. ______

77. ______

78. ______

79. ______

80. ______

81. ______

82. ______

83. ______

84. ______

85. ______

86. ______

87. ______

88. ______

Basic Skills | **Challenge Yourself** | Calculator/Computer | Career Applications | Above and Beyond

Answers

89. ______
90. ______
91. ______
92. ______
93. ______
94. ______
95. ______
96. ______
97. ______
98. ______
99. ______
100. ______
101. ______
102. ______
103. ______
104. ______
105. ______
106. ______
107. ______
108. ______
109. ______
110. ______

Determine whether each statement is **true** *or* **false.**

89. $a^{m/n}$ means the same thing as $a^{n/m}$.

90. If $a > 0$, then $(a^m)^n$ is equal to $(a^n)^m$.

Complete each statement with **never, sometimes,** *or* **always.**

91. A negative number raised to a rational number power is ________ a real number.

92. An expression with a rational number exponent can ________ be rewritten as a radical.

Simplify each expression and write your answer in scientific notation.

93. $(4 \times 10^8)^{1/2}$

94. $(8 \times 10^6)^{1/3}$

95. $(16 \times 10^{-12})^{1/4}$

96. $(9 \times 10^{-4})^{1/2}$

97. $(16 \times 10^{-8})^{1/2}$

98. $(16 \times 10^{-8})^{3/4}$

99. $(27 \times 10^6)^{1/3}$

100. $(64 \times 10^6)^{-1/3}$

Apply the appropriate multiplication patterns and simplify your result.

101. $a^{1/2}(a^{3/2} + a^{3/4})$

102. $2x^{1/4}(3x^{3/4} - 5x^{-1/4})$

103. $(a^{1/2} + 2)(a^{1/2} - 2)$

104. $(w^{1/3} - 3)(w^{1/3} + 3)$

105. $(m^{1/2} + n^{1/2})(m^{1/2} - n^{1/2})$

> Videos

106. $(x^{1/3} + y^{1/3})(x^{1/3} - y^{1/3})$

107. $(x^{1/2} + 2)^2$

108. $(a^{1/3} - 3)^2$

109. $(r^{1/2} + s^{1/2})^2$

110. $(p^{1/2} - q^{1/2})^2$

As is suggested by several of the preceding exercises, certain expressions containing rational exponents are factorable. For instance, to factor $x^{2/3} - x^{1/3} - 6$, let $u = x^{1/3}$. Note that $x^{2/3} = (x^{1/3})^2 = u^2$.

Substituting, we have $u^2 - u - 6$, and factoring yields $(u - 3)(u + 2)$ or $(x^{1/3} - 3)(x^{1/3} + 2)$.

Use this technique to factor each expression.

111. $x^{2/3} + 4x^{1/3} + 3$

112. $y^{2/5} - 2y^{1/5} - 8$

113. $a^{4/5} - 7a^{2/5} + 12$

114. $w^{4/3} + 3w^{2/3} - 10$

115. $x^{4/3} - 4$

116. $x^{2/5} - 16$

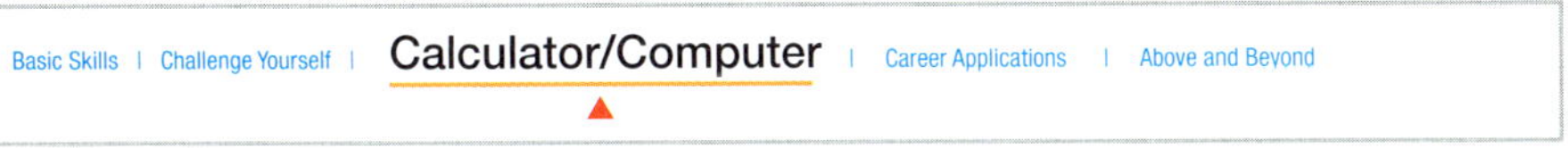

Use a calculator to evaluate each expression. Round each answer to three decimal places.

117. $46^{3/5}$

118. $23^{2/7}$

119. $12^{-2/5}$

120. $36^{-3/4}$

Basic Skills | Challenge Yourself | Calculator/Computer | **Career Applications** | Above and Beyond

AGRICULTURAL TECHNOLOGY *One formula that researchers encounter when investigating rainfall runoff in regions of semiarid farmland is*

$$t = C\left(\frac{L}{xy^2}\right)^{1/3}$$

Use this information to complete exercises 121 and 122.

121. Evaluate t when $C = 20$, $L = 600$, $x = 3$, and $y = 5$.

122. Solve the given formula for L.

AGRICULTURAL TECHNOLOGY *The average velocity of water in an open irrigation ditch is given by the formula*

$$V = \frac{1.5x^{2/3}y^{1/2}}{z}$$

Use this information to complete exercises 123 and 124.

123. Find the average velocity when $x = 27$, $y = 16$, and $z = 12$.

124. Rewrite the average velocity formula using radicals in place of rational exponents and fractions in place of decimals.

Answers

111. ______

112. ______

113. ______

114. ______

115. ______

116. ______

117. ______

118. ______

119. ______

120. ______

121. ______

122. ______

123. ______

124. ______

Basic Skills | Challenge Yourself | Calculator/Computer | Career Applications | **Above and Beyond**

Answers

125. ______
126. ______
127. ______
128. ______
129. ______
130. ______
131. ______
132. ______
133. ______
134. ______
135. ______
136. ______
137. ______
138. ______
139. ______
140. ______

Perform the indicated operations. Assume that n represents a positive integer and that the denominators are not zero.

125. $x^{3n} \cdot x^{2n}$

126. $p^{1-n} \cdot p^{n+3}$

127. $(y^2)^{2n}$

128. $(a^{3n})^3$

129. $\dfrac{r^{n+2}}{r^n}$

130. $\dfrac{w^n}{w^{n-3}}$

131. $(a^3 \cdot b^2)^{2n}$

132. $(c^4 \cdot d^2)^{3m}$

133. $\left(\dfrac{x^{n+2}}{x^n}\right)^{1/2}$

134. $\left(\dfrac{b^n}{b^{n-3}}\right)^{1/3}$

Write each expression in exponential form, simplify, and give the result as a single radical.

135. $\sqrt{\sqrt{x}}$

136. $\sqrt[3]{\sqrt{a}}$

137. $\sqrt[4]{\sqrt{y}}$

138. $\sqrt{\sqrt[3]{w}}$

139. The geometric mean is used to measure average inflation rates or interest rates. If prices increased by 15% over 5 years, then the average *annual* rate of inflation is obtained by taking the fifth root of 1.15:

$(1.15)^{1/5} = 1.0283$ or $\approx 2.8\%$

The 1 is added to 0.15 because we are taking the original price and adding 15% of that price. We could write that as

$P + 0.15P$

Factoring, we get

$$P + 0.15P = P(1 + 0.15)$$
$$= P(1.15)$$

From December 1990 through February 1997, the Bureau of Labor Statistics computed an inflation rate of 16.2%, which is equivalent to an annual growth rate of 2.46%. From December 1990 through February 1997 is 75 months. To what exponent was 1.162 raised to obtain this average annual growth rate?

140. On your calculator, try evaluating $(-9)^{4/2}$ two ways:

(a) $[(-9)^4]^{1/2}$ **(b)** $[(-9)^{1/2}]^4$

Discuss the results.

141. Describe the difference between x^{-2} and $x^{1/2}$.

142. Some rational exponents, such as $\frac{1}{2}$, can easily be rewritten as terminating decimals (0.5). Others, such as $\frac{1}{3}$, cannot. What is it that determines which rational numbers can be rewritten as terminating decimals?

143. Use the properties of exponents to decide what x should be to make each statement true. Explain your choices regarding which properties of exponents you decide to use.

(a) $(a^{2/3})^x = a$ (b) $(a^{5/6})^x = \frac{1}{a}$

(c) $a^{2x} \cdot a^{3/2} = 1$ (d) $\left(\sqrt{a^{2/3}}\right)^x = a$

Answers

141. ______________

142. ______________

143. ______________

Answers

1. 7 **3.** -5 **5.** Not a real number **7.** 3 **9.** 3 **11.** $\frac{2}{3}$ **13.** 9 **15.** 16 **17.** 4 **19.** 729 **21.** $\frac{4}{9}$ **23.** $\frac{1}{7}$ **25.** $\frac{1}{3}$ **27.** $\frac{1}{27}$ **29.** $\frac{1}{32}$ **31.** $\frac{5}{2}$ **33.** x **35.** $y^{6/7}$ **37.** $b^{13/6}$ **39.** $x^{1/3}$ **41.** s **43.** $w^{2/3}$ **45.** x **47.** $a^{3/5}$ **49.** $\frac{1}{y^4}$ **51.** a^4b^9 **53.** x^2y^3 **55.** $st^{1/3}$ **57.** $4pq^{5/3}$ **59.** $x^{2/5}y^{1/2}z$ **61.** $a^{1/2}b^{1/4}$ **63.** $\frac{s^2}{r^4}$ **65.** $\frac{x^3}{y^2}$ **67.** $\frac{1}{mn^2}$ **69.** $\frac{s^3}{r^2t}$ **71.** $\frac{2xz^3}{y^2}$ **73.** $\frac{2n}{m^{1/5}}$ **75.** $xy^{3/4}$ **77.** $\sqrt[4]{a^3}$ **79.** $2\sqrt[3]{x^2}$ **81.** $3\sqrt[5]{x^2}$ **83.** $\sqrt[5]{9x^2}$ **85.** $(7a)^{1/2}$ **87.** $2m^2n^3$ **89.** False **91.** sometimes **93.** 2×10^4 **95.** 2×10^{-3} **97.** 4×10^{-4} **99.** 3×10^2 **101.** $a^2 + a^{5/4}$ **103.** $a - 4$ **105.** $m - n$ **107.** $x + 4x^{1/2} + 4$ **109.** $r + 2r^{1/2}s^{1/2} + s$ **111.** $(x^{1/3} + 1)(x^{1/3} + 3)$ **113.** $(a^{2/5} - 3)(a^{2/5} - 4)$ **115.** $(x^{2/3} - 2)(x^{2/3} + 2)$ **117.** 9.946 **119.** 0.370 **121.** 40 **123.** 4.5 **125.** x^{5n} **127.** y^{4n} **129.** r^2 **131.** $a^{6n}b^{4n}$ **133.** x **135.** $\sqrt[4]{x}$ **137.** $\sqrt[8]{y}$ **139.** $\frac{4}{25}$ **141.** Above and Beyond **143.** (a) $\frac{3}{2}$; (b) $-\frac{6}{5}$; (c) $-\frac{3}{4}$; (d) 3

7.6 Complex Numbers

< 7.6 Objectives >

1 > Use the imaginary number i

2 > Add and subtract complex numbers

3 > Multiply and divide complex numbers

Radicals such as

$$\sqrt{-4} \quad \text{and} \quad \sqrt{-49}$$

are *not* real numbers because no real number squared produces a negative number. Our work in this section extends our number system to include these **imaginary numbers** which allows us to consider radicals such as $\sqrt{-4}$.

First we offer a definition.

Definition

The Imaginary Number i

The number i is defined as

$$i = \sqrt{-1}$$

so that

$$i^2 = -1$$

This definition of the number i gives us an alternate means of indicating the square root of a negative number.

Property

Writing an Imaginary Number

When a is a positive real number,

$$\sqrt{-a} = \sqrt{a}i \quad \text{or} \quad i\sqrt{a}$$

Example 1 Using the Number i

< Objective 1 >

NOTE

Some courses do not cover complex numbers at this level. The use of complex numbers in ensuing sections will be indicated with the symbol $\mathbb{C}$ so that they may be skipped if desired.

Write each expression as a multiple of i.

(a) $\sqrt{-4} = \sqrt{4}i = 2i$

(b) $-\sqrt{-9} = -\sqrt{9}i = -3i$

(c) $\sqrt{-8} = \sqrt{8}i = 2\sqrt{2}i$ or $2i\sqrt{2}$

(d) $\sqrt{-7} = \sqrt{7}i$ or $i\sqrt{7}$

NOTE

We simplify $\sqrt{8}$ as $2\sqrt{2}$. We write the *i in front of* the radical to make it clear that *i* is *not part of* the radicand.

Check Yourself 1

Write each radical as a multiple of *i*.

(a) $\sqrt{-25}$ (b) $\sqrt{-24}$

We are now ready to define complex numbers in terms of the number *i*.

Definition

Complex Number

A **complex number** is any number that can be written in the form

$a + bi$

in which *a* and *b* are real numbers and

$i = \sqrt{-1}$ so that $i^2 = -1$

NOTE

The term *imaginary number* was introduced by René Descartes in 1637. Euler used *i* to indicate $\sqrt{-1}$ in 1748, but it was not until 1832 that Gauss used the term *complex number*.

NOTES

The first application of these numbers was made by Charles Steinmetz (1865–1923) in explaining the behavior of electric circuits.

$5i$ is also called a **pure imaginary** number.

The real numbers can be considered a subset of the set of complex numbers.

The form $a + bi$ is called the **standard form** of a complex number. We call *a* the **real part** of the complex number and *b* the **imaginary part.** Some examples follow.

$3 + 7i$ is an example of a complex number with real part 3 and imaginary part 7.

$5i$ is also a complex number because it can be written as $0 + 5i$.

-3 is a complex number because it can be written as $-3 + 0i$.

The basic operations of addition and subtraction on complex numbers are defined here.

Property

Adding and Subtracting Complex Numbers

For the complex numbers $a + bi$ and $c + di$,

$(a + bi) + (c + di) = (a + c) + (b + d)i$

$(a + bi) - (c + di) = (a - c) + (b - d)i$

In words, we add or subtract the real parts and the imaginary parts of the complex numbers separately.

Example 2 illustrates these properties.

Example 2 Adding and Subtracting Complex Numbers

< Objective 2 >

Perform the indicated operations.

NOTE

Regrouping is essentially a matter of combining like terms.

(a) $(5 + 3i) + (6 - 7i) = (5 + 6) + (3 - 7)i$

$= 11 - 4i$

(b) $5 + (7 - 5i) = (5 + 7) + (-5i)$

$= 12 - 5i$

(c) $(8 - 2i) - (3 - 4i) = (8 - 3) + [-2 - (-4)]i$ Distribute the − sign.

$= 5 + 2i$

Check Yourself 2

Perform the indicated operations.

(a) $(4 - 7i) + (3 - 2i)$ **(b)** $-7 + (-2 + 3i)$ **(c)** $(-4 + 3i) - (-2 - i)$

Because complex numbers are binomials, the product of two complex numbers is found by applying our earlier multiplication pattern for binomials, as Example 3 illustrates.

Example 3 Multiplying Complex Numbers

< Objective 3 >

Multiply.

NOTE

We use the definition of i to replace i^2 with −1 and then simplify.

(a) $(2 + 3i)(3 - 4i)$

$= 2 \cdot 3 + 2(-4i) + (3i)3 + (3i)(-4i)$

$= 6 + (-8i) + 9i + (-12i^2)$

$= 6 - 8i + 9i + (-12)(-1)$

$= 6 + i + 12$

$= 18 + i$

(b) $(1 - 2i)(3 - 4i)$

$= 1 \cdot 3 + 1(-4i) + (-2i)3 + (-2i)(-4i)$

$= 3 + (-4i) + (-6i) + 8i^2$

$= 3 - 10i + 8(-1)$

$= 3 - 10i - 8$

$= -5 - 10i$

Check Yourself 3

Multiply $(2 - 5i)(3 - 2i)$.

Example 3 suggests a pattern when multiplying complex numbers.

Property

Multiplying Complex Numbers

For the complex numbers $a + bi$ and $c + di$,

$$\begin{aligned}(a + bi)(c + di) &= ac + adi + bci + bdi^2 \\ &= ac + adi + bci - bd \\ &= (ac - bd) + (ad + bc)i\end{aligned}$$

This formula for the general product of two complex numbers can be memorized. However, you will find it much easier to get used to the multiplication pattern as it is applied to complex numbers than to memorize this formula.

There is one particular product form that will seem very familiar. We call $a + bi$ and $a - bi$ **complex conjugates.** For instance,

$3 + 2i$ and $3 - 2i$

are complex conjugates.

Consider the product

$$\begin{aligned}(3 + 2i)(3 - 2i) &= 3^2 - (2i)^2 \\ &= 9 - 4i^2 = 9 - 4(-1) \\ &= 9 + 4 = 13\end{aligned}$$

The product of $3 + 2i$ and $3 - 2i$ is a real number. In general, we can write the product of two complex conjugates as

$$(a + bi)(a - bi) = a^2 + b^2$$

The fact that this product is always a real number is very useful when dividing complex numbers.

Example 4 Multiplying Complex Numbers

NOTE

We get the same result when we apply the formula above with $a = 7$ and $b = 4$.

Multiply.

$$\begin{aligned}(7 - 4i)(7 + 4i) &= 7^2 - (4i)^2 \\ &= 7^2 - 4^2(-1) \\ &= 7^2 + 4^2 \\ &= 49 + 16 = 65\end{aligned}$$

Check Yourself 4

Multiply $(5 + 3i)(5 - 3i)$.

We are now ready to divide complex numbers. Generally, we find the quotient by multiplying the numerator and denominator by the conjugate of the denominator, as Example 5 illustrates.

Example 5 Dividing Complex Numbers

NOTES

Think of $3i$ as $0 + 3i$ and of its conjugate as $0 - 3i$, or $-3i$.

Multiplying the numerator and denominator in the original expression by i yields the same result. Try it yourself.

We multiply by $\frac{3-2i}{3-2i}$, which equals 1.

To write a complex number in standard form, we separate the real and imaginary parts.

Divide.

(a) $\frac{6+9i}{3i}$

$$\frac{6+9i}{3i} = \frac{(6+9i)(-3i)}{(3i)(-3i)}$$

The conjugate of $3i$ is $-3i$, so we multiply the numerator and denominator by $-3i$.

$$= \frac{-18i - 27i^2}{-9i^2}$$

$$= \frac{-18i - 27(-1)}{(-9)(-1)}$$

$$= \frac{27 - 18i}{9} = \frac{27}{9} - \frac{18i}{9}$$

$$= 3 - 2i$$

(b) $$\frac{3-i}{3+2i} = \frac{(3-i)(3-2i)}{(3+2i)(3-2i)}$$

$$= \frac{9 - 6i - 3i + 2i^2}{9 - 4i^2}$$

$$= \frac{9 - 9i - 2}{9 + 4}$$

$$= \frac{7 - 9i}{13} = \frac{7}{13} - \frac{9}{13}i$$

(c) $$\frac{2+i}{4-5i} = \frac{(2+i)(4+5i)}{(4-5i)(4+5i)}$$

$$= \frac{8 + 10i + 4i + 5i^2}{16 - 25i^2}$$

$$= \frac{8 + 14i - 5}{16 + 25}$$

$$= \frac{3 + 14i}{41} = \frac{3}{41} + \frac{14}{41}i$$

Check Yourself 5

Divide.

(a) $\frac{5+i}{5-3i}$ **(b)** $\frac{4+10i}{2i}$

We have seen that $i = \sqrt{-1}$ and $i^2 = -1$. We can use these two values to develop a table for the powers of i.

$i = i$

$i^2 = -1$

$i^3 = i^2 \cdot i = -1 \cdot i = -i$

$i^4 = i^2 \cdot i^2 = -1 \cdot (-1) = 1$

$i^5 = i^4 \cdot i = 1 \cdot i = i$

$i^6 = i^4 \cdot i^2 = 1 \cdot (-1) = -1$

$i^7 = i^4 \cdot i^3 = 1 \cdot (-i) = -i$

$i^8 = i^4 \cdot i^4 = 1 \cdot 1 = 1$

NOTE

The symbol $\mathbb{C}$ is often used to denote the set of complex numbers.

This pattern i, -1, $-i$, 1 repeats forever. You will see it again in the exercise set (and then in many subsequent math classes!).

We conclude this section with the following diagram summarizing the structure of the system of complex numbers.

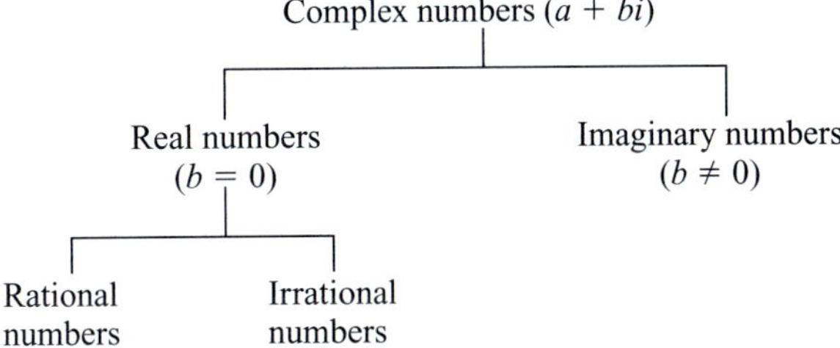

Check Yourself ANSWERS

1. **(a)** $5i$; **(b)** $2i\sqrt{6}$ **2.** **(a)** $7 - 9i$; **(b)** $-9 + 3i$; **(c)** $-2 + 4i$

3. $-4 - 19i$ **4.** 34 **5.** **(a)** $\frac{11}{17} + \frac{10}{17}i$; **(b)** $5 - 2i$

Reading Your Text

SECTION 7.6

(a) The ______________ number i is defined as $i = \sqrt{-1}$ so that $i^2 = -1$.

(b) A ______________ number is any number that can be written in the form $a + bi$.

(c) To add two complex numbers, add the real parts and add the ______________ parts.

(d) $a + bi$ and $a - bi$ are called complex ______________.

7.6 exercises

Basic Skills | Challenge Yourself | Calculator/Computer | Career Applications | Above and Beyond

Name ____________

Section ________ Date ________

Answers

1. ________ 2. ________

3. ________ 4. ________

5. ________ 6. ________

7. ________ 8. ________

9. ________ 10. ________

11. ________ 12. ________

13. ________ 14. ________

15. ________ 16. ________

17. ________ 18. ________

19. ________ 20. ________

21. ________

22. ________

23. ________

< Objective 1 >

Write each expression as a multiple of i. Simplify your results where possible.

1. $\sqrt{-16}$

2. $\sqrt{-36}$

3. $-\sqrt{-121}$

4. $-\sqrt{-25}$

5. $\sqrt{-21}$

6. $\sqrt{-23}$

7. $\sqrt{-12}$

8. $\sqrt{-24}$

9. $-\sqrt{-108}$ 

10. $-\sqrt{-192}$

< Objective 2 >

Perform the indicated operations.

11. $(5 + 4i) - (6 + 5i)$

12. $(2 + 3i) + (4 + 5i)$

13. $(-3 - 2i) + (2 + 3i)$

14. $(-5 - 3i) + (-2 + 7i)$

15. $(5 + 4i) - (3 + 2i)$

16. $(7 + 6i) - (3 + 5i)$

17. $(8 - 5i) - (3 + 2i)$

18. $(7 - 3i) - (-2 - 5i)$

19. $(3 + i) + (4 + 5i) - 7i$

20. $(3 - 2i) + (2 + 3i) + 7i$

21. $(2 + 3i) - (3 - 5i) + (4 + 3i)$ 

22. $(5 - 7i) + (7 + 3i) - (2 - 7i)$

23. $(7 + 3i) - [(3 + i) - (2 - 5i)]$

24. $(8 - 2i) - [(4 + 3i) - (-2 + i)]$

25. $(5 + 3i) + (-5 - 3i)$

26. $(9 - 11i) + (-9 + 11i)$

< Objective 3 >

Find each product and write your answer in standard form.

27. $3i(3 + 5i)$

28. $2i(7 + 3i)$

29. $6i(2 - 5i)$

30. $2i(6 + 3i)$

31. $-2i(4 - 3i)$

32. $-5i(2 - 7i)$

33. $6i\left(\frac{2}{3} + \frac{5}{6}i\right)$

34. $4i\left(\frac{1}{2} + \frac{3}{4}i\right)$

35. $(4 + 3i)(4 - 3i)$

36. $(5 - 2i)(3 - i)$

37. $(4 - 3i)(2 + 5i)$

38. $(7 + 2i)(3 - 2i)$

39. $(-2 - 3i)(-3 + 4i)$ 

40. $(-5 - i)(-3 - 4i)$

41. $(7 - 3i)^2$

42. $(3 + 7i)^2$

Write the conjugate of each complex number. Then find the product of the given number and its conjugate.

43. $3 - 2i$

44. $5 + 2i$

45. $3 + 2i$

46. $7 - i$

47. $-3 - 2i$

48. $-5 - 7i$

49. $5i$

50. $-3i$

Answers

24. ______

25. ______

26. ______

27. ______ 28. ______

29. ______ 30. ______

31. ______ 32. ______

33. ______ 34. ______

35. ______ 36. ______

37. ______ 38. ______

39. ______

40. ______

41. ______

42. ______

43. ______

44. ______

45. ______

46. ______

47. ______

48. ______

49. ______

50. ______

Answers

51. ________
52. ________
53. ________
54. ________
55. ________
56. ________
57. ________
58. ________
59. ________
60. ________
61. ________
62. ________
63. ________
64. ________
65. ________
66. ________
67. ________
68. ________

Find each quotient, and write your answer in standard form.

51. $\dfrac{3 + 2i}{i}$

52. $\dfrac{5 - 3i}{-i}$

53. $\dfrac{5 - 2i}{3i}$

54. $\dfrac{8 + 12i}{-4i}$

55. $\dfrac{3}{2 + 5i}$

56. $\dfrac{5}{2 - 3i}$

57. $\dfrac{13}{2 + 3i}$

58. $\dfrac{-17}{3 + 5i}$

59. $\dfrac{2 + 3i}{4 + 3i}$

60. $\dfrac{4 - 2i}{5 - 3i}$

61. $\dfrac{3 - 4i}{3 + 4i}$

62. $\dfrac{7 + 2i}{7 - 2i}$

Basic Skills | **Challenge Yourself** | Calculator/Computer | Career Applications | Above and Beyond

Complete each statement with **never, sometimes,** *or* **always.**

63. The product of two complex numbers is ________ a real number.

64. If $a \neq 0$ and $b \neq 0$, the square of a complex number $a + bi$ is ________ a real number.

65. The product of a complex number and its conjugate is ________ a real number.

66. A real number can ________ be viewed as a complex number in $a + bi$ form.

Basic Skills | Challenge Yourself | Calculator/Computer | Career Applications | **Above and Beyond**

67. The first application of complex numbers was suggested by the Norwegian surveyor Caspar Wessel in 1797. He found that complex numbers could be used to represent distance and direction on a two-dimensional grid. Why would a surveyor care about such a thing?

68. To what sets of numbers does 1 belong?

We defined $\sqrt{-4} = \sqrt{4}i = 2i$ *in the process of expressing the square root of a negative number as a multiple of i.*

Particular care must be taken with products where two negative radicands are involved. For instance,

$$\sqrt{-3} \cdot \sqrt{-12} = (i\sqrt{3})(i\sqrt{12})$$
$$= i^2\sqrt{36} = (-1)\sqrt{36} = -6$$

is correct. However, if we try to apply the product property for radicals, we have

$$\sqrt{-3} \cdot \sqrt{-12} \stackrel{?}{=} \sqrt{(-3)(-12)} = \sqrt{36} = 6$$

which is not *correct. The property* $\sqrt{a} \cdot \sqrt{b} = \sqrt{ab}$ *is not applicable in the case where a and b are both negative. Radicals such as* $\sqrt{-a}$ *must be written in the standard form* $i\sqrt{a}$ before *multiplying to use the rules for real-valued radicals.*

Find each product.

69. $\sqrt{-5} \cdot \sqrt{-7}$

70. $\sqrt{-3} \cdot \sqrt{-10}$

71. $\sqrt{-2} \cdot \sqrt{-18}$

72. $\sqrt{-4} \cdot \sqrt{-25}$

73. $\sqrt{-6} \cdot \sqrt{-15}$

74. $\sqrt{-5} \cdot \sqrt{-30}$

75. $\sqrt{-10} \cdot \sqrt{-10}$

76. $\sqrt{-11} \cdot \sqrt{-11}$

Simplify each power of i.

77. i^{10}

78. i^{9}

79. i^{20}

80. i^{15}

81. i^{38}

82. i^{40}

83. i^{51}

84. i^{61}

85. Show that a square root of i is $\frac{\sqrt{2}}{2} + \frac{\sqrt{2}}{2}i$. That is, $\left(\frac{\sqrt{2}}{2} + \frac{\sqrt{2}}{2}i\right)^2 = i$.

86. You know that 2 is a cube root of 8, but did you know that there are two more cube roots of 8? Now that you have studied multiplication of complex numbers, show that $-1 + i\sqrt{3}$ and $-1 - i\sqrt{3}$ are also cube roots of 8.

Answers

69. ____
70. ____
71. ____
72. ____
73. ____
74. ____
75. ____
76. ____
77. ____
78. ____
79. ____
80. ____
81. ____
82. ____
83. ____
84. ____
85. ____
86. ____

Answers

1. $4i$ **3.** $-11i$ **5.** $i\sqrt{21}$ **7.** $2i\sqrt{3}$ **9.** $-6i\sqrt{3}$ **11.** $-1 - i$
13. $-1 + i$ **15.** $2 + 2i$ **17.** $5 - 7i$ **19.** $7 - i$ **21.** $3 + 11i$
23. $6 - 3i$ **25.** $0 + 0i$ or 0 **27.** $-15 + 9i$ **29.** $30 + 12i$
31. $-6 - 8i$ **33.** $-5 + 4i$ **35.** 25 **37.** $23 + 14i$ **39.** $18 + i$
41. $40 - 42i$ **43.** $3 + 2i$; 13 **45.** $3 - 2i$; 13 **47.** $-3 + 2i$; 13
49. $-5i$; 25 **51.** $2 - 3i$ **53.** $-\frac{2}{3} - \frac{5}{3}i$ **55.** $\frac{6}{29} - \frac{15}{29}i$
57. $2 - 3i$ **59.** $\frac{17}{25} + \frac{6}{25}i$ **61.** $-\frac{7}{25} - \frac{24}{25}i$ **63.** sometimes
65. always **67.** Above and Beyond **69.** $-\sqrt{35}$ **71.** -6
73. $-3\sqrt{10}$ **75.** -10 **77.** -1 **79.** 1 **81.** -1 **83.** $-i$
85. Above and Beyond

Definition/Procedure	Example	Reference

Roots and Radicals

Section 7.1

Square Roots Every positive number has two square roots. The positive or principal square root of a number a is denoted

$$\sqrt{a}$$

The negative square root is written as

$$-\sqrt{a}$$

$\sqrt{25} = 5$

5 is the principal square root of 25 because $5^2 = 25$.

$-\sqrt{49} = -7$

p. 712

Higher Roots Cube roots, fourth roots, and so on are denoted by using an index and a radical. The principal nth root of a is written as

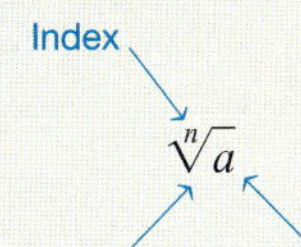

$\sqrt[3]{27} = 3$

$\sqrt[3]{-64} = -4$

$\sqrt[4]{81} = 3$

$\sqrt{(-5)^2} = 5$

$\sqrt[3]{(-3)^3} = -3$

$\sqrt{m^2} = |m|$

$\sqrt[3]{27x^3} = 3x$

p. 713

Radicals Containing Variables In general,

$$\sqrt[n]{x^n} = \begin{cases} |x| & \text{if } n \text{ is even} \\ x & \text{if } n \text{ is odd} \end{cases}$$

p. 716

Pythagorean Theorem
In a right triangle, $c^2 = a^2 + b^2$.

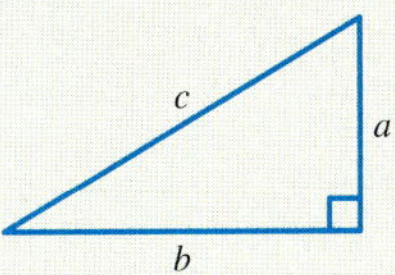

If $c = 17$ and $b = 14$, to find a:

$c^2 = a^2 + b^2$

$17^2 = a^2 + 14^2$

$17^2 - 14^2 = a^2$

$93 = a^2$

$a = \sqrt{93}$

p. 717

Distance Formula
The distance d between two points (x_1, y_1) and (x_2, y_2) is

$$d = \sqrt{(x_2 - x_1)^2 + (y_2 - y_1)^2}$$

Given $(-5, -2)$ and $(3, 9)$,

$$d = \sqrt{(-5 - 3)^2 + (-2 - 9)^2}$$
$$= \sqrt{(-8)^2 + (-11)^2}$$
$$= \sqrt{64 + 121} = \sqrt{185}$$

p. 719

Circles The standard form for the circle with center (h, k) and radius r is

$$(x - h)^2 + (y - k)^2 = r^2$$

Determining the center and radius of the circle from its equation allows us to easily graph the circle.

Given the equation

$(x - 2)^2 + (y + 3)^2 = 4$

we see that the center is at $(2, -3)$ and the radius is 2.

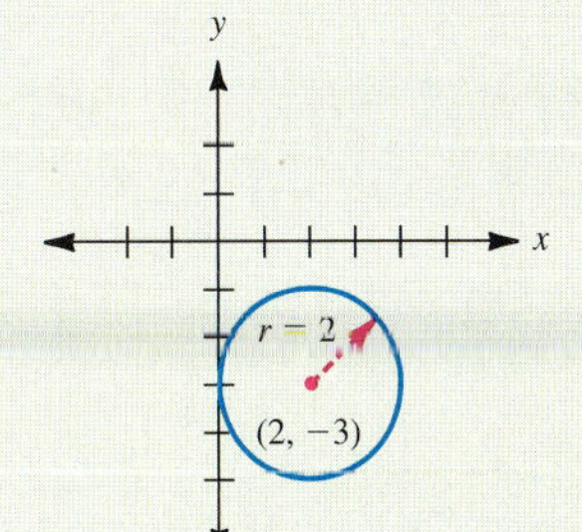

p. 719

Continued

Definition/Procedure	Example	Reference
Simplifying Radical Expressions		**Section 7.2**
Simplifying radical expressions entails applying two properties for radicals.		
Product Property $\sqrt[n]{ab} = \sqrt[n]{a} \cdot \sqrt[n]{b}$ (*a* and *b* nonnegative)	$\sqrt{35} = \sqrt{5 \cdot 7}$ $= \sqrt{5} \cdot \sqrt{7}$	*p.* 731
Quotient Property $\sqrt[n]{\dfrac{a}{b}} = \dfrac{\sqrt[n]{a}}{\sqrt[n]{b}}$ $b \neq 0$ (*a* and *b* nonnegative)	$\sqrt{\dfrac{2}{5}} = \dfrac{\sqrt{2}}{\sqrt{5}}$	*p.* 731
Simplified Form for Radicals A radical is in *simplified form* if the following three conditions are satisfied. **1.** The radicand has no factor raised to a power greater than or equal to the index. **2.** No fraction appears in the radical. **3.** No radical appears in a denominator. **Note:** Satisfying the third condition may require *rationalizing the denominator.*	$\sqrt{18x^3} = \sqrt{9x^2 \cdot 2x}$ $= \sqrt{9x^2} \cdot \sqrt{2x}$ $= 3x\sqrt{2x}$ $\sqrt{\dfrac{5}{9}} = \dfrac{\sqrt{5}}{\sqrt{9}} = \dfrac{\sqrt{5}}{3}$ $\sqrt{\dfrac{3}{7x}} = \dfrac{\sqrt{3}}{\sqrt{7x}} = \dfrac{\sqrt{3} \cdot \sqrt{7x}}{\sqrt{7x} \cdot \sqrt{7x}}$ $= \dfrac{\sqrt{21x}}{\sqrt{49x^2}} = \dfrac{\sqrt{21x}}{7x}$	*p.* 732
Operations on Radical Expressions		**Section 7.3**
Radical expressions may be combined by using addition or subtraction only if they are *similar,* that is, if they have the same radicand with the same index. Similar radicals are combined by application of the distributive property.	$8\sqrt{5} + 3\sqrt{5} = (8 + 3)\sqrt{5}$ $= 11\sqrt{5}$ $2\sqrt{18} - 4\sqrt{2}$ $= 2\sqrt{9 \cdot 2} - 4\sqrt{2}$ $= 2\sqrt{9} \cdot \sqrt{2} - 4\sqrt{2}$ $= 2 \cdot 3\sqrt{2} - 4\sqrt{2}$ $= 6\sqrt{2} - 4\sqrt{2} = (6 - 4)\sqrt{2}$ $= 2\sqrt{2}$	*p.* 742
Multiplication To multiply two radical expressions, we use $\sqrt[n]{a} \cdot \sqrt[n]{b} = \sqrt[n]{ab}$ and simplify the product. If binomial expressions are involved, we use the distributive property or the FOIL method.	$\sqrt{3x} \cdot \sqrt{6x^2} = \sqrt{18x^3}$ $= \sqrt{9x^2 \cdot 2x}$ $= \sqrt{9x^2} \cdot \sqrt{2x}$ $= 3x\sqrt{2x}$ $\sqrt{2}(5 + \sqrt{8}) = \sqrt{2} \cdot 5 + \sqrt{2} \cdot \sqrt{8}$ $= 5\sqrt{2} + 4$ $(3 + \sqrt{2})(5 - \sqrt{2})$ $= 15 - 3\sqrt{2} + 5\sqrt{2} - 2$ $= 13 + 2\sqrt{2}$	*p.* 745

Definition/Procedure	Example	Reference
Division To divide two radical expressions, rationalize the denominator by multiplying the numerator and denominator by the appropriate radical. If the divisor (the denominator) is a binomial, multiply the numerator and denominator by the conjugate of the denominator.	$\frac{5}{\sqrt{8}} = \frac{5 \cdot \sqrt{2}}{\sqrt{8} \cdot \sqrt{2}} = \frac{5\sqrt{2}}{\sqrt{16}} = \frac{5\sqrt{2}}{4}$ **Note:** $3 + \sqrt{5}$ is the conjugate of $3 - \sqrt{5}$. $\frac{2}{3 - \sqrt{5}} = \frac{2(3 + \sqrt{5})}{(3 - \sqrt{5})(3 + \sqrt{5})} = \frac{2(3 + \sqrt{5})}{4} = \frac{3 + \sqrt{5}}{2}$	*p.* 748
Solving Radical Equations		**Section 7.4**
Power Property of Equality If $a = b$ then $a^n = b^n$	If $\sqrt{x+1} = 5$ then $(\sqrt{x+1})^2 = 5^2$; $x + 1 = 25$; $x = 24$	*p.* 756
Solving Equations Involving Radicals **Step 1** Isolate a radical term on one side of the equation. **Step 2** Raise each side of the equation to the smallest power that will eliminate the isolated radical. **Step 3** If any radicals remain in the equation derived in step 2, return to step 1 and continue the solution process. **Step 4** Solve the resulting equation to determine any possible solutions. **Step 5** Check all solutions to determine whether extraneous solutions may have resulted from step 2.	Given $\sqrt{x} + \sqrt{x+7} = 7$; $\sqrt{x} = 7 - \sqrt{x+7}$; $x = 49 - 14\sqrt{x+7} + (x + 7)$; $x = 56 + x - 14\sqrt{x+7}$; $-56 = -14\sqrt{x+7}$; $4 = \sqrt{x+7}$; $16 = x + 7$; $x = 9$. Check: $\sqrt{9} + \sqrt{9+7} \stackrel{?}{=} 7$; $3 + 4 = 7$ ✓ Solution set: $\{9\}$	*p.* 760
Rational Exponents Rational exponents are an alternate way of indicating roots. We use the following definition. If a is any real number and n is a positive integer ($n > 1$), $a^{1/n} = \sqrt[n]{a}$ We restrict a so that a is nonnegative when n is even. We also define the following. For any real number a and positive integers m and n, with $n > 1$, then $a^{m/n} = (\sqrt[n]{a})^m = \sqrt[n]{a^m}$	$36^{1/2} = \sqrt{36} = 6$; $-27^{1/3} = -\sqrt[3]{27} = -3$; $243^{1/5} = \sqrt[5]{243} = 3$; $25^{-1/2} = \frac{1}{\sqrt{25}} = \frac{1}{5}$; $27^{2/3} = (\sqrt[3]{27})^2 = 3^2 = 9$; $(a^4b^8)^{3/4} = \sqrt[4]{(a^4b^8)^3} = \sqrt[4]{a^{12}b^{24}} = a^3b^6$	**Section 7.5** *p.* 768

Continued

Definition/Procedure	Example	Reference
Properties of Exponents The following five properties for exponents continue to hold for rational exponents. **Product Rule** $a^m \cdot a^n = a^{m+n}$	$x^{1/2} \cdot x^{1/3} = x^{1/2+1/3} = x^{5/6}$	p. 771
Quotient Rule $\dfrac{a^m}{a^n} = a^{m-n}$	$\dfrac{x^{3/2}}{x^{1/2}} = x^{3/2-1/2} = x^{2/2} = x$	p. 771
Power Rule $(a^m)^n = a^{m \cdot n}$	$(x^{1/3})^5 = x^{1/3 \cdot 5} = x^{5/3}$	p. 771
Product-Power Rule $(ab)^m = a^m b^m$	$(2xy)^{1/2} = 2^{1/2}x^{1/2}y^{1/2}$	p. 771
Quotient-Power Rule $\left(\dfrac{a}{b}\right)^m = \dfrac{a^m}{b^m}$	$\left(\dfrac{x^{1/3}}{3}\right)^2 = \dfrac{(x^{1/3})^2}{3^2}$ $= \dfrac{x^{2/3}}{9}$	p. 771
Complex Numbers		**Section 7.6**
The number i is defined as $i = \sqrt{-1}$ so that $i^2 = -1$ A **complex number** is any number that can be written in the form $a + bi$ in which a and b are real numbers.	$\sqrt{-16} = 4i$ $\sqrt{-8} = 2i\sqrt{2}$	p. 782
Addition and Subtraction For the complex numbers $a + bi$ and $c + di$, $(a + bi) + (c + di) = (a + c) + (b + d)i$ and $(a + bi) - (c + di) = (a - c) + (b - d)i$	$(2 + 3i) + (-3 - 5i)$ $= (2 - 3) + (3 - 5)i$ $= -1 - 2i$ $(5 - 2i) - (3 - 4i)$ $= (5 - 3) + [-2 - (-4)]i$ $= 2 + 2i$	p. 783

Definition/Procedure	Example	Reference
Multiplication For the complex numbers $a + bi$ and $c + di$, $(a + bi)(c + di) = (ac - bd) + (ad + bc)i$ **Note:** It is generally easier to use the FOIL multiplication pattern and the definition of i than to apply the above formula.	$(2 + 5i)(3 - 4i)$ $= 6 - 8i + 15i - 20i^2$ $= 6 + 7i - 20(-1)$ $= 26 + 7i$	p. 785
Division To divide two complex numbers, we multiply the numerator and denominator by the complex conjugate of the denominator and write the result in standard form.	$\frac{3 + 2i}{3 - 2i} = \frac{(3 + 2i)(3 + 2i)}{(3 - 2i)(3 + 2i)}$ $= \frac{9 + 6i + 6i + 4i^2}{9 - 4i^2}$ $= \frac{9 + 12i + 4(-1)}{9 - 4(-1)}$ $= \frac{5 + 12i}{13}$ $= \frac{5}{13} + \frac{12}{13}i$	p. 786

summary exercises :: chapter 7

This summary exercise set is provided to give you practice with each of the objectives of this chapter. Each exercise is keyed to the appropriate chapter section. When you are finished, you can check your answers to the odd-numbered exercises in the back of the text. If you have difficulty with any of these questions, go back and reread the examples from that section. The answers to the even-numbered exercises appear in the *Instructor's Solutions Manual.* Your instructor will give you guidelines on how best to use these exercises in your instructional setting.

7.1 *Evaluate each root over the set of real numbers.*

1. $\sqrt{121}$

2. $-\sqrt{64}$

3. $\sqrt{-81}$

4. $\sqrt[3]{64}$

5. $\sqrt[3]{-64}$

6. $\sqrt[4]{81}$

7. $\sqrt{\frac{9}{16}}$

8. $\sqrt[3]{-\frac{8}{27}}$

9. $\sqrt{8^2}$

Simplify each expression. Assume that all variables represent positive real numbers.

10. $\sqrt{4x^2}$

11. $\sqrt{a^4}$

12. $\sqrt{36y^2}$

13. $\sqrt{49w^4z^6}$

14. $\sqrt[3]{x^9}$

15. $\sqrt[3]{-27b^6}$

16. $\sqrt[3]{8r^3s^9}$

17. $\sqrt[4]{16x^4y^8}$

18. $\sqrt[5]{32p^5q^{15}}$

Find the distance between each pair of points.

19. $(-4, 3)$ and $(-1, 1)$

20. $(-1, -2)$ and $(1, 3)$

21. $(-2, -5)$ and $(3, -1)$

Find the center and radius of the circle graphed by each equation.

22. $x^2 + y^2 = 81$

23. $(x - 3)^2 + y^2 = 36$

24. $(x + 2)^2 + (y - 1)^2 = 25$

Graph each equation.

25. $x^2 + y^2 = 9$

26. $(x - 2)^2 + y^2 = 9$

27. $(x + 3)^2 + (y + 3)^2 = 25$

7.2 *Use the product property to write each expression in simplified form.*

28. $\sqrt{45}$

29. $-\sqrt{75}$

30. $\sqrt{60x^2}$

31. $\sqrt{108a^3}$

32. $\sqrt[3]{32}$

33. $\sqrt[3]{-80w^4z^3}$

Use the quotient property to write each expression in simplified form.

34. $\sqrt{\dfrac{9}{16}}$

35. $\sqrt{\dfrac{7}{36}}$

36. $\sqrt{\dfrac{y^4}{49}}$

37. $\sqrt{\dfrac{2x}{9}}$

38. $\sqrt{\dfrac{5}{16x^2}}$

39. $\sqrt[3]{\dfrac{5a^2}{27}}$

7.3 *Simplify each expression if necessary. Then add or subtract as indicated.*

40. $7\sqrt{10} + 4\sqrt{10}$

41. $5\sqrt{3x} - 2\sqrt{3x}$

42. $7\sqrt[3]{2x} + 3\sqrt[3]{2x}$

43. $8\sqrt{10} - 3\sqrt{10} + 2\sqrt{10}$

44. $\sqrt{72} + \sqrt{50}$

45. $\sqrt{54} - \sqrt{24}$

46. $9\sqrt{7} - 2\sqrt{63}$

47. $\sqrt{20} - \sqrt{45} + 2\sqrt{125}$

48. $2\sqrt[3]{16} + 3\sqrt[3]{54}$

49. $\sqrt{27w^3} - w\sqrt{12w}$

50. $\sqrt[3]{128a^5} + 6a\sqrt[3]{2a^2}$

51. $\sqrt{20} + \dfrac{3}{\sqrt{5}}$

52. $\sqrt{72x} - \sqrt{\dfrac{x}{2}}$

53. $\sqrt[3]{81a^4} - a\sqrt[3]{\dfrac{a}{9}}$

54. $\dfrac{\sqrt{15}}{3} - \dfrac{1}{\sqrt{15}}$

Multiply and simplify each expression.

55. $\sqrt{3x} \cdot \sqrt{7y}$

56. $\sqrt{6x^2} \cdot \sqrt{18}$

57. $\sqrt[3]{4a^2b} \cdot \sqrt[3]{ab^2}$

58. $\sqrt{5}\,(\sqrt{3} + 2)$

59. $\sqrt{6}\,(\sqrt{8} - \sqrt{2})$

60. $\sqrt{a}\,(\sqrt{5a} + \sqrt{125a})$

61. $(\sqrt{3} + 5)(\sqrt{3} - 7)$

62. $(\sqrt{7} - \sqrt{2})(\sqrt{7} + \sqrt{3})$

63. $(\sqrt{5} - 2)(\sqrt{5} + 2)$

64. $(\sqrt{7} - \sqrt{3})(\sqrt{7} + \sqrt{3})$

65. $(2 + \sqrt{3})^2$

66. $(\sqrt{5} - \sqrt{2})^2$

Rationalize the denominator, and simplify each expression.

67. $\sqrt{\dfrac{3}{7}}$

68. $\dfrac{\sqrt{12}}{\sqrt{x}}$

69. $\dfrac{\sqrt{10a}}{\sqrt{5b}}$

70. $\sqrt[3]{\dfrac{3}{a^2}}$

71. $\dfrac{2}{\sqrt[3]{3x}}$

72. $\dfrac{\sqrt[3]{x^2}}{\sqrt[3]{y^5}}$

Divide and simplify each expression.

73. $\dfrac{1}{3 + \sqrt{2}}$

74. $\dfrac{11}{5 - \sqrt{3}}$

75. $\dfrac{\sqrt{5} - 2}{\sqrt{5} + 2}$

76. $\dfrac{\sqrt{x} - 3}{\sqrt{x} + 3}$

7.4 *Solve each equation. Be sure to check your solutions.*

77. $\sqrt{x - 5} = 4$

78. $\sqrt{3x - 2} + 2 = 5$

79. $\sqrt{y + 7} = y - 5$

80. $\sqrt{2x - 1} + x = 8$

81. $\sqrt[3]{5x + 2} = 3$

82. $\sqrt[3]{x^2 + 2} - 3 = 0$

83. $\sqrt{z + 7} = 1 + \sqrt{z}$

84. $\sqrt{4x + 5} - \sqrt{x - 1} = 3$

7.5 *Evaluate each expression.*

85. $49^{1/2}$

86. $-100^{1/2}$

87. $(-27)^{1/3}$

88. $16^{1/4}$

89. $64^{2/3}$

90. $25^{3/2}$

91. $\left(\frac{4}{9}\right)^{3/2}$

92. $49^{-1/2}$

93. $81^{-3/4}$

Use the properties of exponents to simplify each expression.

94. $x^{3/2} \cdot x^{5/2}$

95. $b^{2/3} \cdot b^{3/2}$

96. $\frac{r^{8/5}}{r^{3/5}}$

97. $\frac{a^{5/4}}{a^{1/2}}$

98. $(x^{3/5})^{2/3}$

99. $(y^{-4/3})^6$

100. $(x^{4/5}y^{3/2})^{10}$

101. $(16x^{1/3} \cdot y^{2/3})^{3/4}$

102. $\left(\frac{x^{-2}y^{-1/6}}{x^{-4}y}\right)^3$

103. $\left(\frac{27y^3z^{-6}}{x^{-3}}\right)^{1/3}$

Write each expression in radical form.

104. $x^{3/4}$

105. $(w^2z)^{2/5}$

106. $3a^{2/3}$

107. $(3a)^{2/3}$

Write each expression using rational exponents, and simplify when necessary.

108. $\sqrt[5]{7x}$

109. $\sqrt{16w^4}$

110. $\sqrt[3]{27p^3q^9}$

111. $\sqrt[4]{16a^8b^{16}}$

7.6 *Write each root as a multiple of i. Simplify your result.*

112. $\sqrt{-49}$

113. $\sqrt{-13}$

114. $-\sqrt{-60}$

Perform the indicated operations.

115. $(2 + 3i) + (3 - 5i)$

116. $(7 - 3i) + (-3 - 2i)$

117. $(5 - 3i) - (2 + 5i)$

118. $(-4 + 2i) - (-1 - 3i)$

Find each product.

119. $4i(7 - 2i)$

120. $(5 - 2i)(3 + 4i)$

121. $(3 - 4i)^2$

122. $(2 - 3i)(2 + 3i)$

Find each quotient, and write your answer in standard form.

123. $\dfrac{5 - 15i}{5i}$

124. $\dfrac{10}{3 - 4i}$

125. $\dfrac{3 - 2i}{3 + 2i}$

126. $\dfrac{5 + 10i}{2 + i}$

self-test 7

Name ______________

Section ________ Date ________

The purpose of this self-test is to help you assess your progress so that you can find concepts that you need to review before the next exam. Allow yourself about an hour to take this test. At the end of that hour, check your answers against those given in the back of this text. If you miss any, go back to the appropriate section to reread the examples until you have mastered that particular concept.

Simplify each expression. Assume that all variables represent positive real numbers in all subsequent problems.

1. $\sqrt[3]{9p^7q^5}$

2. $\sqrt{\dfrac{5x}{8y}}$

3. $\sqrt{6x}\,(\sqrt{18x} - \sqrt{2x})$

4. $\sqrt{49a^4}$

5. $\dfrac{7x}{\sqrt{64y^2}}$

6. $\dfrac{\sqrt{6} - \sqrt{3}}{\sqrt{6} + \sqrt{3}}$

7. $(16x^4)^{3/2}$

8. $\left(\dfrac{4x^{1/5}y^{1/5}}{x^{-7/5}y^{3/5}}\right)^{5/2}$

9. $\sqrt[3]{-27w^6z^9}$

10. $\sqrt[3]{4x^5}\,\sqrt[3]{8x^6}$

Use a calculator to evaluate each root. Round your answers to the nearest tenth.

11. $\sqrt{43}$

12. $\sqrt[3]{\dfrac{73}{27}}$

Solve each equation and check your solutions.

13. $\sqrt{x - 7} - 2 = 0$

14. $\sqrt{3w + 4} + w = 8$

Simplify each expression.

15. $\dfrac{3}{\sqrt[3]{9x}}$

16. $\sqrt{7x^3}\,\sqrt{2x^4}$

17. $\sqrt[3]{54m^4} + m\sqrt[3]{16m}$

18. $\sqrt{3x^3} + x\sqrt{75x} - \sqrt{27x^3}$

Write the expression in radical form and simplify.

19. $(a^7b^3)^{2/5}$

Answers

1. ______________
2. ______________
3. ______________
4. ______________
5. ______________
6. ______________
7. ______________
8. ______________
9. ______________
10. ______________
11. ______________
12. ______________
13. ______________
14. ______________
15. ______________
16. ______________
17. ______________
18. ______________
19. ______________

Answers

20. ______

21. ______

22. ______

23. ______

24. ______

25. ______

26. ______

27. ______

28. ______

29. ______

30. ______

Write each expression using rational exponents. Then simplify.

20. $\sqrt[3]{125p^9q^6}$

Simplify each expression.

21. $5\sqrt{20} + 4\sqrt{45}$

22. $\sqrt{32x^5y^7z^6}$

23. $(27m^{3/2}n^{-6})^{2/3}$

24. $\left(\dfrac{16r^{-1/3}s^{5/3}}{rs^{-7/3}}\right)^{3/4}$

25. $(\sqrt{11} - \sqrt{5})^2$

Find the distance between each pair of points. Express your answer in radical form and as a decimal, rounded to the nearest thousandth.

26. $(-2, -3)$ and $(5, 2)$

27. $(-5, 8)$ and $(6, 0)$

Give the center and radius of the circle whose equation is given.

28. $(x - 4)^2 + (y + 1)^2 = 49$

Simplify each expression and write your answer in standard form.

29. $(-5 + 3i)(7 - 6i)$

30. $\dfrac{4 + 6i}{2 - 3i}$

Name ______________

Section ________ Date ________

We offer the following exercises to help you review concepts from earlier chapters. This is meant as review material and not as a comprehensive exam. The answers are presented in the back of the text. If you have difficulty with any of these exercises, be certain to at least read through the summary related to that section.

1. Solve the equation $7x - 6(x - 1) = 2(5 + x) + 11$.

2. If $f(x) = 3x^6 - 4x^3 + 9x^2 - 11$, find $f(-1)$.

3. Find the equation of the line that has a y-intercept of $(0, -6)$ and is parallel to the line given by the equation $6x - 4y = 18$.

4. The sum of three consecutive integers is 54. Find the three integers.

Simplify each expression.

5. $5x^2 - 8x + 11 - (-3x^2 - 2x + 8) - (-2x^2 - 4x + 3)$

6. $(5x + 3)(2x - 9)$

Factor each expression completely.

7. $2x^3 + x^2 - 3x$

8. $9x^4 - 36y^4$

9. $4x^2 + 8xy - 5x - 10y$

10. $x^4 - 13x^2 - 48$

11. Find the slope of the line whose equation is $4x + 3y = 7$.

12. Write the equation of the line that passes through the point $(-4, 2)$ and is perpendicular to the line with equation $y = 4x + 5$.

13. A company that produces computer games has found that its daily operating cost in dollars is $C = 30x + 500$ and its daily revenue in dollars is $R = 75x - x^2$. For what value(s) of x will the company break even?

Answers

1. ______________

2. ______________

3. ______________

4. ______________

5. ______________

6. ______________

7. ______________

8. ______________

9. ______________

10. ______________

11. ______________

12. ______________

13. ______________

Answers

14. ______
15. ______
16. ______
17. ______
18. ______
19. ______
20. ______
21. ______
22. ______
23. ______
24. ______
25. ______
26. ______
27. ______

Simplify each radical expression.

14. $\sqrt{3x^3y}\,\sqrt{4x^5y^6}$

15. $(\sqrt{3} - 5)(\sqrt{2} + 3)$

Graph each equation.

16. $y = 3x - 5$

17. $x = -5$

18. $2x - 3y = 12$

Solve each system of equations.

19. $4x - 3y = 15$
$x + y = 2$

20. $6x - 5y = 27$
$x = 5y + 2$

Simplify.

21. $\left(\dfrac{x^2y^{-3}}{x^5y^4}\right)^{-2}$

22. $(-12x^3y^2)(-18xy^3)$

23. $\left(\dfrac{64x^3y^9}{27}\right)^{1/3}$

Solve.

24. $\sqrt{x + 3} = 2$

25. Solve the inequality.

$5x - (2 - 3x) \geq 6 + 10x$

26. Find the zeros of the function $f(x) = 2x^2 + 9x - 5$.

27. The length of a rectangle is 3 inches less than twice its width. If the perimeter of the rectangle is 96 inches, find the dimensions of the rectangle.

CHAPTER

8

Quadratic Functions

INTRODUCTION

Perhaps no field is more strongly related to mathematics than engineering. Nearly every aspect of engineering can be traced to the algebra that you are learning from this text. The activity presented in this chapter introduces a very important engineering application—that of measuring stress and strain. Whether designing a bridge, a dam, or an airplane wing, determining stress and strain loads is one of the most important steps in the process.

CHAPTER 8 OUTLINE

8.1 Solving Quadratic Equations

< 8.1 Objectives >

1 > Use factoring to solve quadratic equations

2 > Use the square-root method to solve quadratic equations

3 > Solve quadratic equations by completing the square

Recall that a quadratic equation is an equation of the form $ax^2 + bx + c = 0$, in which a is not equal to zero.

In Section 6.6, we factored quadratic expressions and then used the zero-product principle to solve such equations. Of course, this only works when the quadratic expression is factorable. In this chapter, we learn other techniques for solving quadratic equations. One such technique is called the **square-root method.** After reviewing the factoring approach, we will introduce you to the square-root method.

Example 1 Solving Equations by Factoring

< Objective 1 >

Solve the quadratic equation $2x^2 + x - 10 = 0$ by factoring.

Write the equation in factored form.

$(x - 2)(2x + 5) = 0$

RECALL

The zero-product principle states that if $ab = 0$, then $a = 0$ or $b = 0$ or both.

Use the zero-product principle to set each factor equal to 0 and solve both equations.

$x - 2 = 0$ and $2x + 5 = 0$

Solving, we have

$x = 2$ and $x = -\frac{5}{2}$ or $\left\{-\frac{5}{2}, 2\right\}$

Check Yourself 1

Solve each equation by factoring.

(a) $x^2 + x - 12 = 0$ **(b)** $3x^2 - x - 10 = 0$

Example 2 Solving Equations by Factoring

Solve the quadratic equation $x^2 = 16$ by factoring.

Write the equation in standard form.

$x^2 - 16 = 0$

Factoring gives

$(x + 4)(x - 4) = 0$

NOTE

Here, we factor the quadratic expression as a difference of squares.

Finally, the solutions are

$x = -4$ and $x = 4$ or $\{\pm 4\}$

Check Yourself 2

Solve each quadratic equation.

(a) $x^2 = 25$ **(b)** $5x^2 = 180$

NOTE

Be sure to include *both* the positive and the negative square roots when you use the square-root method.

The equation in Example 2 can be solved in an alternative fashion. We can use what is called the **square-root method.** Take another look at the equation

$$x^2 = 16$$

In Chapter 7, we learned that if $x^2 = 16$, then we can say that "x is a square root of 16." We know from experience that x must be 4 or -4.

In symbols, we write

$$x = \sqrt{16} \quad \text{or} \quad x = -\sqrt{16}$$

which we simplify to

$$x = 4 \quad \text{or} \quad x = -4$$

The solution set is $\{-4, 4\}$ or, in more compact form, $\{\pm 4\}$.

This discussion leads us to a general result.

Property

Square-Root Property

If $x^2 = k$, then

$$x = \sqrt{k} \quad \text{or} \quad x = -\sqrt{k}$$

Example 3 further illustrates this property.

Example 3 Using the Square-Root Method

< Objective 2 >

Use the square-root method to solve each equation.

(a) $x^2 = 9$

By the square-root property,

$$x = \sqrt{9} \quad \text{or} \quad x = -\sqrt{9}$$
$$= 3 \qquad\qquad = -3 \quad \text{or} \quad \{\pm 3\}$$

(b) $x^2 - 17 = 0$

NOTE

With a calculator, we can approximate $\sqrt{17} \approx 4.123$ (rounded to three decimal places).

Add 17 to both sides of the equation.

$$x^2 = 17$$
$$x = \sqrt{17} \quad \text{or} \quad x = -\sqrt{17} \quad \text{or} \quad \{\pm\sqrt{17}\} \quad \text{or} \quad \{-\sqrt{17}, \sqrt{17}\}$$

(c) $4x^2 - 3 = 0$

$$4x^2 = 3$$
$$x^2 = \frac{3}{4}$$
$$x = \pm\sqrt{\frac{3}{4}} = \pm\frac{\sqrt{3}}{\sqrt{4}}$$
$$x = \pm\frac{\sqrt{3}}{2} \quad \text{or} \quad \left\{\pm\frac{\sqrt{3}}{2}\right\}$$

NOTE

In Example 3(d), we see that complex-number solutions may result.

(d) $x^2 + 1 = 0 \quad (\mathbb{C})$

$$x^2 = -1$$

$$x = \pm\sqrt{-1}$$

$$x = \pm i \quad \text{or} \quad \{\pm i\} \quad \text{(nonreal)}$$

Check Yourself 3

Solve each equation.

(a) $x^2 = 5$ **(b)** $x^2 - 2 = 0$ **(c)** $9x^2 - 8 = 0$ **(d)** $x^2 + 9 = 0$ $(\mathbb{C})$

We can also use the approach in Example 3 to solve an equation like

$$(x + 3)^2 = 16$$

We can say that the quantity inside the parentheses, $x + 3$, must be equal to 4 or -4.

In symbols,

$$x + 3 = \sqrt{16} \quad \text{or} \quad x + 3 = -\sqrt{16}$$

$$x + 3 = 4 \quad \text{or} \quad x + 3 = -4$$

Solving for x yields

$$x = -3 + 4 \quad \text{or} \quad x = -3 - 4$$
$$= 1 \qquad\qquad = -7$$

The solution set is $\{-7, 1\}$.

To check, substitute into the original equation:

$$[(-7) + 3]^2 \stackrel{?}{=} 16$$

$$(-4)^2 \stackrel{?}{=} 16$$

$$16 = 16 \quad \text{True}$$

and

$$[(1) + 3]^2 \stackrel{?}{=} 16$$

$$(4)^2 \stackrel{?}{=} 16$$

$$16 = 16 \quad \text{True}$$

Example 4 Using the Square-Root Method

NOTE

The two solutions $5 + \sqrt{5}$ and $5 - \sqrt{5}$ are abbreviated as $5 \pm \sqrt{5}$.

Use the square-root method to solve each equation.

(a) $(x - 5)^2 - 5 = 0$

$(x - 5)^2 = 5$ Add 5 to both sides.

$x - 5 = \pm\sqrt{5}$ Use the square-root property.

$x = 5 \pm \sqrt{5}$ or $\{5 \pm \sqrt{5}\}$ Add 5 to both sides.

(b) $9(y + 1)^2 - 2 = 0$

$9(y + 1)^2 = 2$ Add 2 to both sides.

$(y + 1)^2 = \dfrac{2}{9}$ Divide both sides by 9.

$y + 1 = \pm\sqrt{\dfrac{2}{9}}$ Use the square-root property.

$$y + 1 = \pm\frac{\sqrt{2}}{3}$$ Use the quotient property for radicals.

$$y = -1 \pm \frac{\sqrt{2}}{3}$$ Add −1 to both sides.

$$= \frac{-3}{3} \pm \frac{\sqrt{2}}{3}$$ Use a common denominator of 3.

$$= \frac{-3 \pm \sqrt{2}}{3}$$ Combine the fractions.

The solution set is $\left\{\frac{-3 \pm \sqrt{2}}{3}\right\}$.

Check Yourself 4

Use the square-root method to solve each equation.

(a) $(x - 2)^2 - 3 = 0$ **(b)** $4(x - 1)^2 = 3$

Graphing Calculator Option

Using the Memory Feature to Check Solutions

In Section 1.2, you learned how to use the memory features of a graphing calculator to evaluate expressions. We can use that same approach to check complicated solutions of equations.

In Example 4, we solved the equation

$$9(y + 1)^2 - 2 = 0$$

and obtained the solution set $\left\{\frac{-3 \pm \sqrt{2}}{3}\right\}$.

To check, store the value $\frac{-3 + \sqrt{2}}{3}$ in memory location **Y**:

```
(-3+√(2))/3→Y
     -.5285954792
█
```

Then evaluate the expression on the left side of the original equation, entering memory cell **Y** where you see y in the equation. We expect the result to be 0.

```
(-3+√(2))/3→Y
     -.5285954792
9(Y+1)²-2
                0
```

The solution $\frac{-3 + \sqrt{2}}{3}$ checks.

Graphing Calculator Check

Check the other solution, $\frac{-3 - \sqrt{2}}{3}$, for this same equation.

ANSWER

It checks. When $\frac{-3 - \sqrt{2}}{3}$ is substituted into $9(y + 1)^2 - 2$, the output is 0.

NOTE

If $(x + h)^2 = k$, then
$x + h = \pm\sqrt{k}$
and
$x = -h \pm \sqrt{k}$

Not all quadratic equations can be solved directly by factoring or using the square-root method. We must extend our techniques.

The square-root method is useful in this process because any quadratic equation can be written in the form

$(x + h)^2 = k$

which yields the solutions

$x = -h \pm \sqrt{k}$

The process of changing an equation in standard form

$ax^2 + bx + c = 0$

to the form

$(x + h)^2 = k$

is called the method of **completing the square,** and it is based on the relationship between the middle term and the last term of any perfect-square trinomial.

We look at three perfect-square trinomials to see whether we can detect a pattern:

$$x^2 + 4x + 4 = (x + 2)^2$$
$$x^2 - 6x + 9 = (x - 3)^2$$
$$x^2 + 8x + 16 = (x + 4)^2$$

NOTE

This relationship is true *only* if the leading, or x^2, coefficient is 1.

Note that in each case the last (or constant) term is the square of one-half of the coefficient of x in the middle (or linear) term. For example, in the second equation,

$$x^2 \underbrace{- 6}x + 9 = (x - 3)^2$$

$\frac{1}{2}$ of this coefficient is -3, and $(-3)^2 = 9$, the constant.

Use the third equation, $x^2 + 8x + 16 = (x + 4)^2$, to verify this relationship for yourself. To summarize, in perfect-square trinomials, the constant is always the square of one-half the coefficient of x.

In the next example, we use **completing the square** to produce perfect-square trinomials.

Example 5 Completing the Square

Determine the constant that must be added to the given expression to produce a perfect-square trinomial.

(a) $x^2 + 10x$

The coefficient of x is 10. One-half the coefficient of x is 5.

We square the number 5: $5^2 = 25$. So, 25 is the constant that must be added.

Also, $x^2 + 10x + 25$ is a perfect-square trinomial, since $x^2 + 10x + 25 = (x + 5)^2$.

(b) $x^2 - 12x$

The coefficient of x is -12, and one-half of that is -6. Squaring -6 gives 36, so 36 is the desired constant.

And, $x^2 - 12x + 36$ is a perfect-square trinomial, since $x^2 - 12x + 36 = (x - 6)^2$.

(c) $x^2 + 7x$

This one is a bit messier. One-half the coefficient of x is $\frac{7}{2}$. Squaring this gives $\frac{49}{4}$.

This is the constant that we need.

Note that $x^2 + 7x + \frac{49}{4} = \left(x + \frac{7}{2}\right)^2$.

NOTE

In each part of Example 5, we wrote the resulting trinomial as a binomial squared.

Check Yourself 5

In each case, determine the constant needed to produce a perfect-square trinomial. Then write the trinomial as a binomial squared.

(a) $x^2 - 16x$ (b) $x^2 + 3x$

We are now ready to use **completing the square** to solve quadratic equations.

Example 6 Completing the Square to Solve an Equation

< Objective 3 >

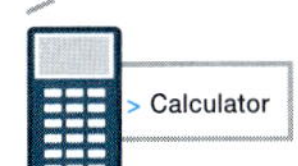

Solve $x^2 + 8x - 7 = 0$ by completing the square.

First, we rewrite the equation with the constant on the *right-hand side.*

$$x^2 + 8x = 7$$

Our objective is to have a perfect-square trinomial on the left-hand side. We know that we must add the square of one-half of the x coefficient to complete the square. In this case, that value is 16, so now we add 16 to each side of the equation.

$$x^2 + 8x + 16 = 7 + 16 \qquad \frac{1}{2} \cdot 8 = 4 \text{ and } 4^2 = 16$$

Factor the perfect-square trinomial on the left, and add on the right.

$$(x + 4)^2 = 23$$

Now use the square-root property.

$$x + 4 = \pm\sqrt{23}$$

Subtracting 4 from both sides of the equation gives

$$x = -4 \pm \sqrt{23} \quad \text{or} \quad \{-4 \pm \sqrt{23}\}$$

NOTE

If we graph the related function $y = x^2 + 8x - 7$ in the standard viewing window, we see that the x-intercepts are just to the right of -9 and just to the left of 1.

Be certain that you see how these points relate to the exact solutions $-4 + \sqrt{23}$ and $-4 - \sqrt{23}$

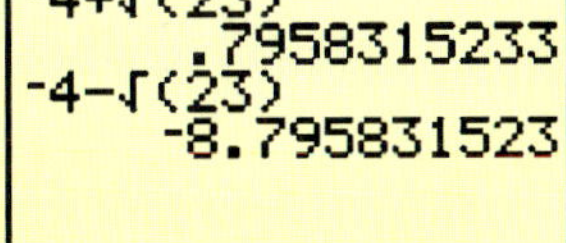

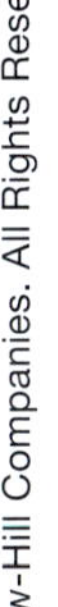

Check Yourself 6

Solve $x^2 - 6x - 2 = 0$ by completing the square.

Step by Step

Completing the Square

Step 1 Isolate the constant on the right side of the equation.

Step 2 Divide both sides of the equation by the coefficient of the x^2-term if that coefficient is not equal to 1.

Step 3 Add the square of one-half of the coefficient of the linear term to both sides of the equation. This gives a perfect-square trinomial on the left side of the equation.

Step 4 Write the left side of the equation as the square of a binomial, and simplify on the right side.

Step 5 Use the square-root property, and then solve the resulting linear equations.

Example 7 Completing the Square to Solve an Equation

Solve $x^2 + 5x - 3 = 0$ by completing the square.

$$x^2 + 5x - 3 = 0$$

$$x^2 + 5x = 3 \qquad \text{Add 3 to both sides.}$$

$$x^2 + 5x + \left(\frac{5}{2}\right)^2 = 3 + \left(\frac{5}{2}\right)^2 \qquad \text{Make the left-hand side a perfect square.}$$

$$\left(x + \frac{5}{2}\right)^2 = \frac{37}{4}$$

$$x + \frac{5}{2} = \pm\frac{\sqrt{37}}{2} \qquad \text{Use the square-root property.}$$

$$x = \frac{-5 \pm \sqrt{37}}{2} \quad \text{or} \quad \left\{\frac{-5 \pm \sqrt{37}}{2}\right\}$$

NOTES

Add the square of one-half of the x coefficient to both sides of the equation.

$\frac{1}{2} \cdot 5 = \frac{5}{2}$

$3 + \left(\frac{5}{2}\right)^2 = 3 + \frac{25}{4}$

$= \frac{12}{4} + \frac{25}{4} = \frac{37}{4}$

Check Yourself 7

Solve $x^2 + 3x - 7 = 0$ by completing the square.

Some equations have nonreal or complex solutions, as Example 8 illustrates.

Example 8 Completing the Square to Solve an Equation

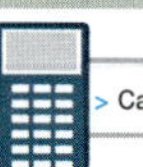

Solve $x^2 + 4x + 13 = 0$ by completing the square. (C)

$$x^2 + 4x + 13 = 0$$

$$x^2 + 4x = -13 \qquad \text{Subtract 13 from both sides.}$$

$$x^2 + 4x + 4 = -13 + 4 \qquad \text{Add } \left[\frac{1}{2}(4)\right]^2 \text{ to both sides.}$$

$$(x + 2)^2 = -9 \qquad \text{Factor the left-hand side.}$$

$$x + 2 = \pm\sqrt{-9} \qquad \text{Use the square-root property.}$$

$$x + 2 = \pm i\sqrt{9} \qquad \text{Simplify the radical.}$$

$$x + 2 = \pm 3i$$

$$x = -2 \pm 3i \quad \text{or} \quad \{-2 \pm 3i\} \qquad \text{nonreal}$$

NOTE

The graph of $y = x^2 + 4x + 13$ does not intersect the x-axis.

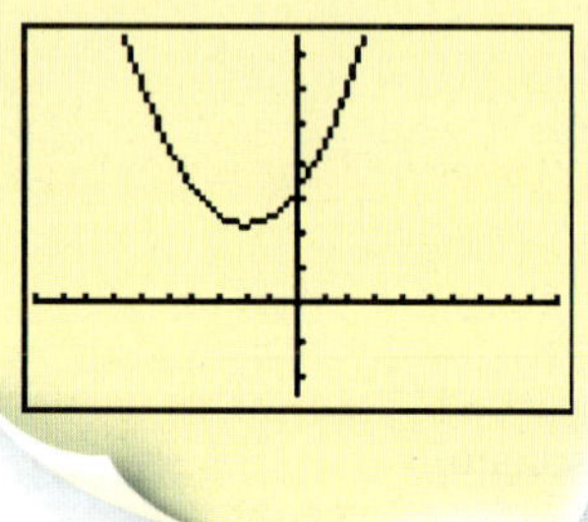

Check Yourself 8

Solve $x^2 + 10x + 41 = 0$. (ℂ)

Example 9 illustrates a situation in which the leading coefficient of the quadratic expression is not equal to 1. An extra step is required in such cases.

Example 9 Completing the Square to Solve an Equation

> CAUTION

Before you can complete the square on the left, the coefficient of x^2 must be equal to 1. If it is not, we must *divide* both sides of the equation by that coefficient.

Solve $4x^2 + 8x - 7 = 0$ by completing the square.

$$4x^2 + 8x - 7 = 0$$

$$4x^2 + 8x = 7 \qquad \text{Add 7 to both sides.}$$

$$x^2 + 2x = \frac{7}{4} \qquad \text{Divide both sides by 4.}$$

$$x^2 + 2x + 1 = \frac{7}{4} + 1 \qquad \text{Complete the square on the left.}$$

$$(x + 1)^2 = \frac{11}{4}$$

$$x + 1 = \pm\sqrt{\frac{11}{4}} \qquad \text{Use the square-root property.}$$

$$x = -1 \pm \sqrt{\frac{11}{4}}$$

$$= -1 \pm \frac{\sqrt{11}}{2}$$

$$= \frac{-2 \pm \sqrt{11}}{2} \quad \text{or} \quad \left\{\frac{-2 \pm \sqrt{11}}{2}\right\}$$

Check Yourself 9

Solve $4x^2 - 8x + 3 = 0$ by completing the square.

Check Yourself ANSWERS

1. **(a)** $\{-4, 3\}$; **(b)** $\left\{-\frac{5}{3}, 2\right\}$ **2.** **(a)** $\{-5, 5\}$; **(b)** $\{-6, 6\}$

3. **(a)** $\{\sqrt{5}, -\sqrt{5}\}$; **(b)** $\{\sqrt{2}, -\sqrt{2}\}$; **(c)** $\left\{\frac{2\sqrt{2}}{3}, -\frac{2\sqrt{2}}{3}\right\}$; **(d)** $\{3i, -3i\}$ nonreal

4. **(a)** $\{2 \pm \sqrt{3}\}$; **(b)** $\left\{\frac{2 \pm \sqrt{3}}{2}\right\}$

5. **(a)** constant: 64; $x^2 - 16x + 64 = (x - 8)^2$;

(b) constant: $\frac{9}{4}$; $x^2 + 3x + \frac{9}{4} = \left(x + \frac{3}{2}\right)^2$

6. $\{3 \pm \sqrt{11}\}$ **7.** $\left\{\frac{-3 \pm \sqrt{37}}{2}\right\}$ **8.** $\{-5 \pm 4i\}$ nonreal **9.** $\left\{\frac{1}{2}, \frac{3}{2}\right\}$

Reading Your Text

SECTION 8.1

(a) A ____________ equation is an equation of the form $ax^2 + bx + c = 0$, in which a is not equal to zero.

(b) The ____________ property states that, if $x^2 = k$, then $x = \sqrt{k}$ or $x = -\sqrt{k}$.

(c) The process of changing an equation in standard form to the form $(x + h)^2 = k$ is called ____________ the square.

(d) When completing the square, we first isolate the ____________ on the right side of the equation.

Basic Skills | Challenge Yourself | Calculator/Computer | Career Applications | Above and Beyond

8.1 exercises

Boost *your* GRADE at ALEKS.com!

ALEKS

- Practice Problems
- Self-Tests
- NetTutor
- e-Professors
- Videos

Name ______________

Section ________ Date ________

< Objective 1 >

Solve each equation by factoring.

1. $x^2 + 9x + 14 = 0$

2. $x^2 + 5x + 6 = 0$

3. $z^2 - 2z - 35 = 0$

4. $q^2 - 5q - 24 = 0$

5. $2x^2 - 5x - 3 = 0$

6. $3x^2 + 10x - 8 = 0$

7. $6y^2 - y - 2 = 0$

8. $21z^2 + z - 2 = 0$

< Objective 2 >

Use the square-root method to solve each equation.

9. $x^2 = 121$

10. $x^2 = 144$

11. $y^2 = 7$

12. $p^2 = 18$

13. $2x^2 - 12 = 0$

14. $5x^2 = 65$

15. $2t^2 + 12 = 4$ $(\mathbb{C})$

16. $3u^2 - 5 = -32$ $(\mathbb{C})$

17. $(x + 1)^2 = 12$ 

18. $(2x - 3)^2 = 5$

19. $(3z + 1)^2 - 5 = 0$

20. $(3p - 4)^2 + 9 = 0$ $(\mathbb{C})$

Find the constant that must be added to each binomial expression to form a perfect-square trinomial.

21. $x^2 + 18x$

22. $r^2 - 14r$

23. $y^2 - 8y$

24. $w^2 + 16w$

Answers

1. ______ 2. ______
3. ______ 4. ______
5. ______ 6. ______
7. ______ 8. ______
9. ______ 10. ______
11. ______
12. ______
13. ______
14. ______
15. ______
16. ______
17. ______
18. ______
19. ______
20. ______
21. ______ 22. ______
23. ______ 24. ______

Answers

25. ______ 26. ______

27. ______ 28. ______

29. ______ 30. ______

31. ______ 32. ______

33. ______

34. ______

35. ______

36. ______

37. ______ 38. ______

39. ______

40. ______

41. ______

42. ______

43. ______

44. ______

45. ______

46. ______

47. ______

48. ______

49. ______

50. ______

25. $x^2 - 3x$ 

26. $z^2 + 7z$

27. $n^2 + n$

28. $x^2 - x$

29. $x^2 + \frac{1}{5}x$

30. $x^2 - \frac{1}{3}x$

31. $x^2 - \frac{1}{6}x$

32. $y^2 - \frac{1}{4}y$

< Objective 3 >

Solve each equation by completing the square.

33. $x^2 + 10x = 4$

34. $x^2 - 14x - 7 = 0$

35. $y^2 - 2y = 8$

36. $z^2 + 4z - 72 = 0$

37. $x^2 - 2x - 5 = 0$ 

38. $x^2 - 3x = 10$

39. $x^2 + 10x + 13 = 0$

40. $x^2 + 3x - 17 = 0$

41. $z^2 - 5z - 7 = 0$

42. $q^2 - 8q + 20 = 0$ ($\mathbb{C}$)

43. $m^2 - 3m - 5 = 0$

44. $y^2 + y - 5 = 0$

Basic Skills | **Challenge Yourself** | Calculator/Computer | Career Applications | Above and Beyond

Solve each equation.

45. $x^2 + \frac{1}{2}x = 1$

46. $x^2 - \frac{1}{3}x = 2$

47. $2x^2 + 2x - 1 = 0$

48. $5x^2 - 6x = 3$

49. $3x^2 - 8x = 2$

50. $4x^2 + 8x - 1 = 0$

51. $3x^2 - 2x + 12 = 0$ (ℂ)

52. $7y^2 - 2y + 3 = 0$ (ℂ)

53. $x^2 + 10x + 28 = 0$ (ℂ)

54. $x^2 - 2x + 10 = 0$ (ℂ)

Complete each statement with **never, sometimes,** *or* **always.**

55. An equation of the form $(x - h)^2 = k$, where k is positive, ________ has two distinct solutions.

56. An equation of the form $x^2 = k$ ________ has real-number solutions.

57. When completing the square for $x^2 + bx$, the number added is ________ negative.

58. A quadratic equation can ________ be solved by completing the square.

59. Consider this representation of "completing the square": Suppose we wish to complete the square for $x^2 + 10x$. A square with dimensions x by x has area equal to x^2.

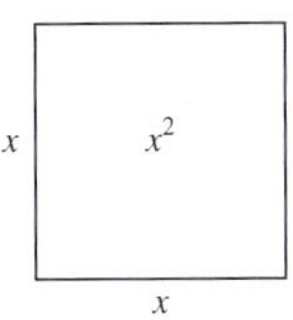

We divide the quantity $10x$ by 2 and get $5x$. If we extend the base x by 5 units and draw the rectangle attached to the square, the rectangle's dimensions are 5 by x with an area of $5x$.

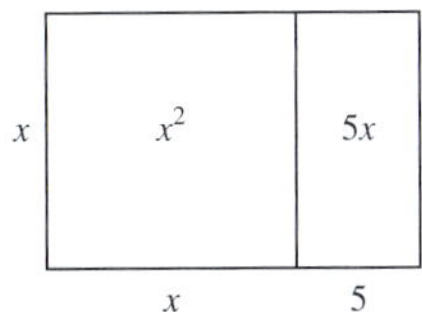

Now we extend the height by 5 units, and we draw another rectangle whose area is $5x$.

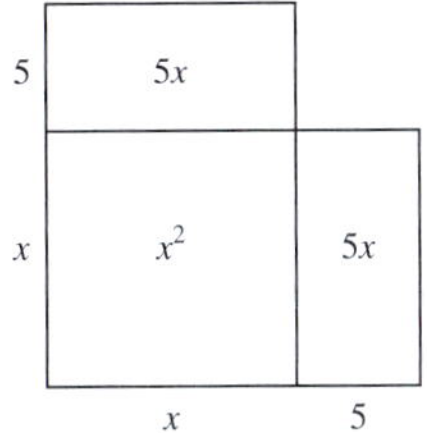

(a) What is the total area represented in the figure so far?

(b) How much area must be added to the figure to "complete the square"?

(c) Write the area of the completed square as a binomial squared.

60. Repeat the process described in exercise 59 with $x^2 + 16x$.

Answers

51. ____________________

52. ____________________

53. ____________________

54. ____________________

55. ____________________

56. ____________________

57. ____________________

58. ____________________

59. ____________________

60. ____________________

Answers

61. ______

62. ______

63. ______

64. ______

65. ______

66. ______

67. ______

68. ______

Basic Skills | Challenge Yourself | **Calculator/Computer** | Career Applications | Above and Beyond

Use your graphing calculator to find the graph. Approximate the x-intercepts for each graph. (You may have to adjust the viewing window to see both intercepts.) Round your answers to the nearest tenth.

61. $y = x^2 + 12x - 2$

62. $y = x^2 - 14x - 7$

63. $y = x^2 - 2x - 8$

64. $y = x^2 + 4x - 72$

65. On your graphing calculator, view the graph of $f(x) = x^2 + 1$.

(a) What can you say about the x-intercepts of the graph?

(b) Determine the zeros of the function, using the square-root method. (ℂ)

(c) How does your answer to part (a) relate to your answer to part (b)?

66. On your graphing calculator, view the graph of $f(x) = x^2 + 4$.

(a) What can you say about the x-intercepts of the graph?

(b) Determine the zeros of the function, using the square-root method. (ℂ)

(c) How does your answer to part (a) relate to your answer to part (b)?

Basic Skills | Challenge Yourself | Calculator/Computer | **Career Applications** | Above and Beyond

67. **Mechanical Engineering** The rotational moment in a shaft is given by the formula

$$M = 2x^2 - 30x$$

Find the value of x when the moment is 152.

68. **Mechanical Engineering** The deflection d of a beam loaded with a single, concentrated load is described by the equation

$$d = \frac{x^2 - 64}{200}$$

Find the location x if $d = 0.085$ in.

69. **Information Technology** The demand equation for a certain computer chip is given by

> Videos

$$D = -4p + 50$$

The supply equation for the same chip is predicted to be

$$S = -p^2 + 20p - 6$$

Find the equilibrium price.
Hint: The *equilibrium price* occurs when demand equals supply.

70. **Allied Health** A toxic chemical is introduced into a protozoan culture. The number of deaths per hour N is given by the equation

$$N = 363 - 3t^2$$

in which t is the number of hours after the chemical's introduction. How long will it take before the protozoa stop dying?

Basic Skills | Challenge Yourself | Calculator/Computer | Career Applications | Above and Beyond

71. Why must the leading coefficient of the quadratic expression be set equal to 1 before you can use the technique of completing the square?

72. What relationship exists between the solution(s) of a quadratic equation and the graph of a quadratic function?

Find the constant that must be added to each binomial to form a perfect-square trinomial. Let x be the variable; other letters represent constants.

73. $x^2 + 2ax$

74. $x^2 + 2abx$

75. $x^2 + 3ax$

76. $x^2 + abx$

77. $a^2x^2 + 2ax$

78. $a^2x^2 + 4abx$

Solve each equation by completing the square.

79. $x^2 + 2ax = 4$

80. $x^2 + 2ax - 8 = 0$

Answers

69. ______
70. ______
71. ______
72. ______
73. ______
74. ______
75. ______
76. ______
77. ______
78. ______
79. ______
80. ______

Answers

1. $\{-7, -2\}$ **3.** $\{-5, 7\}$ **5.** $\left\{-\frac{1}{2}, 3\right\}$ **7.** $\left\{-\frac{1}{2}, \frac{2}{3}\right\}$ **9.** $\{-11, 11\}$
11. $\{-\sqrt{7}, \sqrt{7}\}$ **13.** $\{-\sqrt{6}, \sqrt{6}\}$ **15.** $\{\pm 2i\}$ nonreal
17. $\{-1 \pm 2\sqrt{3}\}$ **19.** $\left\{\frac{-1 \pm \sqrt{5}}{3}\right\}$ **21.** 81 **23.** 16 **25.** $\frac{9}{4}$
27. $\frac{1}{4}$ **29.** $\frac{1}{100}$ **31.** $\frac{1}{144}$ **33.** $\{-5 \pm \sqrt{29}\}$ **35.** $\{-2, 4\}$
37. $\{1 \pm \sqrt{6}\}$ **39.** $\{-5 \pm 2\sqrt{3}\}$ **41.** $\left\{\frac{5 \pm \sqrt{53}}{2}\right\}$ **43.** $\left\{\frac{3 \pm \sqrt{29}}{2}\right\}$
45. $\left\{\frac{-1 \pm \sqrt{17}}{4}\right\}$ **47.** $\left\{\frac{-1 \pm \sqrt{3}}{2}\right\}$ **49.** $\left\{\frac{4 \pm \sqrt{22}}{3}\right\}$
51. $\left\{\frac{1 \pm i\sqrt{35}}{3}\right\}$ nonreal **53.** $\{-5 \pm i\sqrt{3}\}$ nonreal **55.** always
57. never **59.** **(a)** $x^2 + 5x + 5x$; **(b)** 25; **(c)** $x^2 + 10x + 25 = (x + 5)^2$

61.

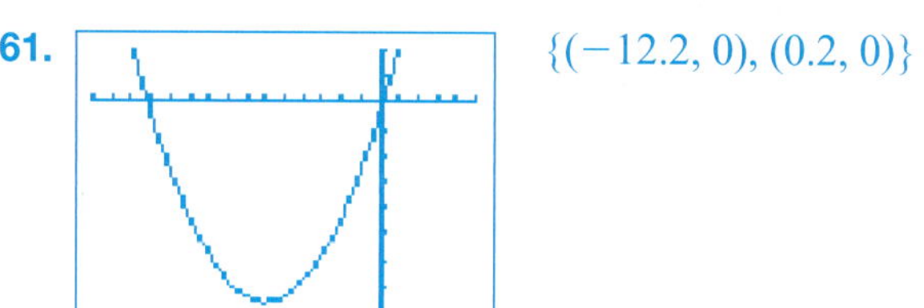

$\{(-12.2, 0), (0.2, 0)\}$

63.

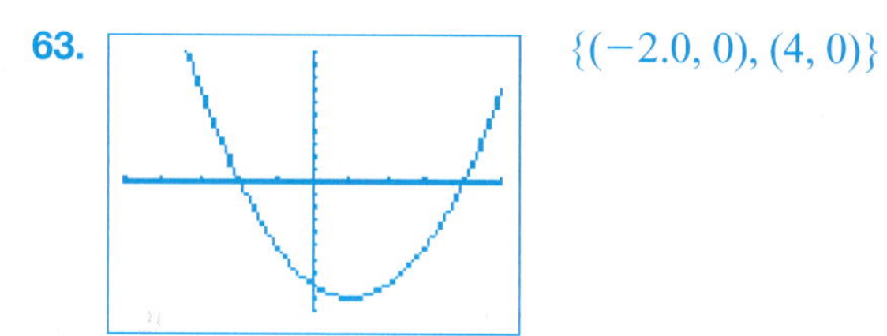

$\{(-2.0, 0), (4, 0)\}$

65. **(a)** There are none; **(b)** $x = \pm i$ nonreal; **(c)** If the graph of $f(x)$ has no x-intercepts, the zeros of the function are not real.

67. 19 **69.** \$2.62 **71.** Above and Beyond **73.** a^2 **75.** $\frac{9}{4}a^2$
77. 1 **79.** $\{-a \pm \sqrt{4 + a^2}\}$

Activity 8 ::
Stress-Strain Curves

A stress-strain curve describes how a material reacts to an applied force (that is, its strength). A typical curve is shown here.

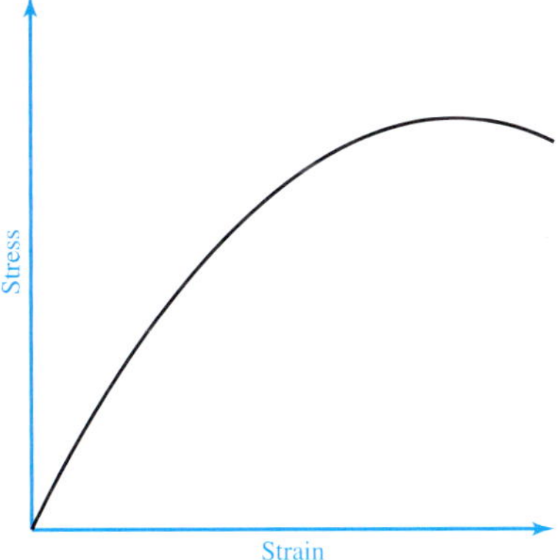

The stress describes the force applied to the material, and the strain describes the amount that the material deforms. When the curve shifts upward, the material is stronger. There are several methods of strengthening a material, causing an upward shift in the curve.

For a given material, the curve can be approximated by the equation

Stress $= 97x - 0.4x^2 + 495$

1. Find the values for the strain x so that the stress is zero.
2. Find the midpoint between the two values you found in exercise 1.
3. Compute the stress at the point found in exercise 2.
4. Describe the significance of the point found in exercise 3 in the context of this application.
5. A material cannot safely be used at the maximum stress. The safe, allowable, stress is usually set at 70% of the maximum stress. Find the safe, allowable stress for this material.
6. Find the strain at the stress point found in exercise 5.

8.2 The Quadratic Formula

< 8.2 Objectives >

1 > Use the quadratic formula to solve quadratic equations

2 > Use the discriminant to determine the nature of the solutions of a quadratic equation

3 > Use the Pythagorean theorem to solve a geometric application

Every quadratic equation can be solved by using the quadratic formula. In this section, we describe how the quadratic formula is derived, and then use it to solve equations. Recall that a quadratic equation is any equation that can be written in the form

$$ax^2 + bx + c = 0 \qquad \text{in which } a \neq 0$$

Step by Step

Deriving the Quadratic Formula

Step 1 Isolate the constant on the right side of the equation.

$$ax^2 + bx = -c$$

Step 2 Divide both sides by the coefficient of the x^2-term.

$$x^2 + \frac{b}{a}x = -\frac{c}{a}$$

Step 3 Add the square of one-half the x-coefficient to both sides.

$$x^2 + \frac{b}{a}x + \frac{b^2}{4a^2} = -\frac{c}{a} + \frac{b^2}{4a^2}$$

Step 4 Factor the left side to write it as the square of a binomial. Then apply the square-root property.

$$\left(x + \frac{b}{2a}\right)^2 = \frac{-4ac + b^2}{4a^2}$$

$$x + \frac{b}{2a} = \pm\sqrt{\frac{b^2 - 4ac}{4a^2}}$$

Step 5 Solve the resulting linear equations.

$$x = -\frac{b}{2a} \pm \frac{\sqrt{b^2 - 4ac}}{2a}$$

Step 6 Simplify.

$$= \frac{-b \pm \sqrt{b^2 - 4ac}}{2a}$$

We use the result derived above to state the **quadratic formula,** a formula that allows us to find the solutions for any quadratic equation.

Property

The Quadratic Formula

Given any quadratic equation in the form

$$ax^2 + bx + c = 0 \qquad \text{in which } a \neq 0$$

the two solutions to the equation are found by using the formula

$$x = \frac{-b \pm \sqrt{b^2 - 4ac}}{2a}$$

Our first example uses an equation in standard form.

Example 1 Using the Quadratic Formula

< Objective 1 >

Use the quadratic formula to solve.

$6x^2 - 7x - 3 = 0$

First, we determine the values for a, b, and c. Here,

$a = 6 \qquad b = -7 \qquad c = -3$

Substituting those values into the quadratic formula gives

NOTE

Because $b^2 - 4ac = 121$ is a perfect square, the two solutions are rational numbers.

$$x = \frac{-(-7) \pm \sqrt{(-7)^2 - 4(6)(-3)}}{2(6)}$$

Simplifying inside the radical gives us

$$x = \frac{7 \pm \sqrt{121}}{12}$$

$$= \frac{7 \pm 11}{12}$$

$$x = \frac{7 + 11}{12} \quad \text{or} \quad x = \frac{7 - 11}{12}$$

$$x = \frac{18}{12} \quad \text{or} \quad x = \frac{-4}{12}$$

NOTE

Compare these solutions to the x-intercepts of the graph of

$y = 6x^2 - 7x - 3$

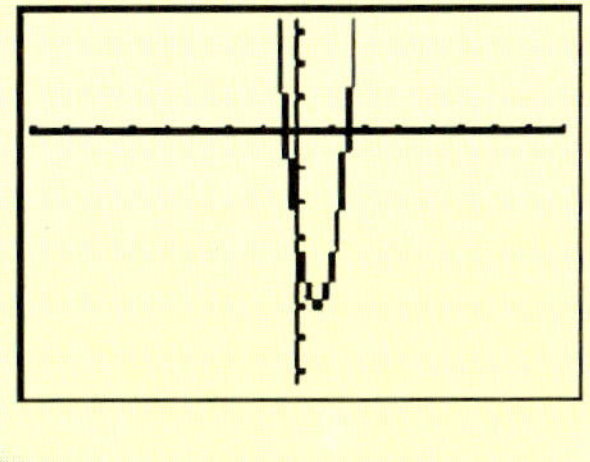

This gives us the solutions

$$x = \frac{3}{2} \quad \text{or} \quad x = -\frac{1}{3} \quad \text{or} \quad \left\{\frac{3}{2}, -\frac{1}{3}\right\}$$

Since the solutions for the equation of this example are rational, the original equation could have been solved by factoring.

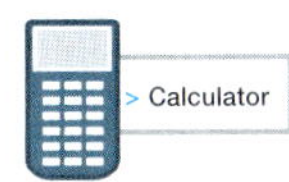

Check Yourself 1

Use the quadratic formula to solve.

$3x^2 + 2x - 8 = 0$

To use the quadratic formula, we must write the equation in standard form. Example 2 illustrates this.

Example 2 Using the Quadratic Formula

Use the quadratic formula to solve.

$9x^2 = 12x - 4$

NOTE

The equation *must be in standard form* to determine a, b, and c.

First, the equation must be written in standard form.

$9x^2 - 12x + 4 = 0$

Second, we find the values of a, b, and c.

$a = 9 \qquad b = -12 \qquad c = 4$

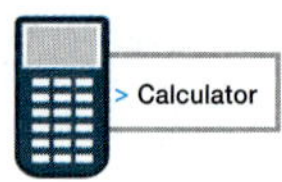

NOTE

The graph of

$y = 9x^2 - 12x + 4$

intersects the x-axis only at the point $\left(\frac{2}{3}, 0\right)$.

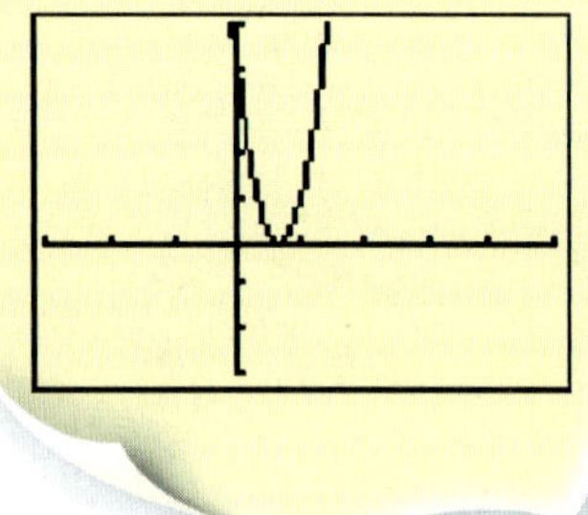

Substitute these values into the quadratic formula.

$$x = \frac{-(-12) \pm \sqrt{(-12)^2 - 4(9)(4)}}{2(9)}$$

$$= \frac{12 \pm \sqrt{0}}{18}$$

and simplifying yields

$$x = \frac{2}{3} \quad \text{or} \quad \left\{\frac{2}{3}\right\}$$

Look again at the original equation, and now try the factoring method.

$$9x^2 = 12x - 4$$

$$9x^2 - 12x + 4 = 0 \qquad \text{The trinomial on the left is a perfect-square trinomial.}$$

$$(3x - 2)(3x - 2) = 0$$

At this point, it is clear that there is exactly one solution, $x = \frac{2}{3}$.

Since the factor $3x - 2$ is repeated, and since solutions are often called *roots,* we say that $x = \frac{2}{3}$ is a **repeated root** of the equation.

We always find a *repeated root* if the quadratic equation, placed in standard form, has a perfect-square trinomial on one side. If we use the quadratic formula to solve such an equation, the value of the radicand $b^2 - 4ac$ is 0.

Check Yourself 2

Use the quadratic formula to solve the equation

$4x^2 - 4x = -1$

So far, all of our solutions have been rational numbers. That is not always the case, as Example 3 illustrates.

Example 3 Using the Quadratic Formula

Use the quadratic formula to solve

$$x^2 - 3x = 5$$

NOTE

Decimal approximations for the solutions can be found with a calculator. To the nearest hundredth, we have −1.19 and 4.19.

```
(3-√(29))/2
       -1.192582404
(3+√(29))/2
        4.192582404
```

Once again, to use the quadratic formula, we write the equation in standard form.

$$x^2 - 3x - 5 = 0$$

We now determine values for a, b, and c and substitute.

$$x = \frac{-(-3) \pm \sqrt{(-3)^2 - 4(1)(-5)}}{2(1)}$$

Simplifying as before, we have

$$x = \frac{3 \pm \sqrt{29}}{2} \quad \text{or} \quad \left\{\frac{3 \pm \sqrt{29}}{2}\right\}$$

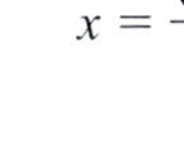

Check Yourself 3

Use the quadratic formula to solve $2x^2 = x + 7$.

Example 4 requires some special care in simplifying the solution.

Example 4 Using the Quadratic Formula

Using the quadratic formula, solve

$$3x^2 - 6x + 2 = 0$$

Here, we have $a = 3$, $b = -6$, and $c = 2$. Substituting gives

$$x = \frac{-(-6) \pm \sqrt{(-6)^2 - 4(3)(2)}}{2(3)}$$

$$= \frac{6 \pm \sqrt{12}}{6}$$

We now look for the largest perfect-square factor of 12, the radicand.

Simplifying, we note that $\sqrt{12}$ is equal to $\sqrt{4 \cdot 3}$, or $2\sqrt{3}$. We can then write the solutions as

$$x = \frac{6 \pm 2\sqrt{3}}{6} = \frac{2(3 \pm \sqrt{3})}{6} = \frac{3 \pm \sqrt{3}}{3}$$

> CAUTION

Students are sometimes tempted to simplify this result to

$$\frac{6 \pm 2\sqrt{3}}{6} \stackrel{?}{=} 1 \pm 2\sqrt{3}$$

This is *not a valid step.* We must divide *each term* in the numerator by 2 when simplifying the expression.

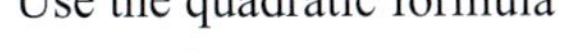

Check Yourself 4

Use the quadratic formula to solve

$$x^2 - 4x = 6$$

Next, we examine a case in which the solutions are nonreal or complex numbers.

Example 5 Using the Quadratic Formula (ℂ)

NOTES

The solutions are nonreal anytime $b^2 - 4ac$ is negative.

The graph of $y = x^2 - 2x + 2$ does not intersect the x-axis, so there are no real solutions.

Use the quadratic formula to solve

$$x^2 - 2x + 2 = 0$$

Labeling the coefficients, we find that

$$a = 1 \qquad b = -2 \qquad c = 2$$

Applying the quadratic formula, we have

$$x = \frac{2 \pm \sqrt{-4}}{2}$$

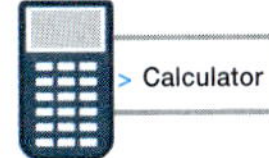

> Calculator

and noting that $\sqrt{-4}$ is $2i$, we can simplify to

$$x = 1 \pm i \qquad \text{or} \qquad \{1 \pm i\} \quad \text{(nonreal)}$$

Check Yourself 5

Solve by using the quadratic formula.

$$x^2 - 4x + 6 = 0 \quad (ℂ)$$

In attempting to solve a quadratic equation, you should first try the factoring method. If this method does not work, you can apply the quadratic formula or the square-root method to find the solution. This algorithm outlines the steps.

Step by Step

Solving a Quadratic Equation Using the Quadratic Formula

Step 1 Write the equation in standard form (one side is equal to 0).

$$ax^2 + bx + c = 0$$

Step 2 Determine the values for a, b, and c.

Step 3 Substitute those values into the quadratic formula.

$$x = \frac{-b \pm \sqrt{b^2 - 4ac}}{2a}$$

Step 4 Simplify.

NOTE

Graphically, we can see the number of real solutions as the number of times the related quadratic function intersects the x-axis.

Given a quadratic equation, the radicand $b^2 - 4ac$ determines the number of real solutions. For example, if $b^2 - 4ac$ is a negative number, we would be taking the square root of that negative number, and therefore we would obtain two nonreal solutions. Because of the information that this quantity $b^2 - 4ac$ gives us, it has a name: it is called the **discriminant.**

Definition

The Discriminant

Given the equation $ax^2 + bx + c = 0$, the quantity $b^2 - 4ac$ is called the **discriminant.**

If $b^2 - 4ac \begin{cases} < 0 & \text{there are } \textit{no real solutions,} \text{ but two nonreal solutions} \\ = 0 & \text{there is } \textit{one real solution} \text{ (a double solution)} \\ > 0 & \text{there are } \textit{two distinct real solutions} \end{cases}$

NOTE

Although the solutions are not necessarily distinct or real, every second-degree equation has two solutions.

Example 6 Analyzing the Discriminant

< Objective 2 >

How many real solutions are there for each quadratic equation?

(a) $x^2 + 7x - 15 = 0$

The discriminant $(7)^2 - 4(1)(-15)$ is 109. This indicates that there are two real solutions.

(b) $3x^2 - 5x + 7 = 0$

The discriminant $b^2 - 4ac = -59$ is negative. There are no real solutions.

(c) $9x^2 - 12x + 4 = 0$

The discriminant is 0. There is exactly one real solution (a double solution).

RECALL

We can use the quadratic formula to find any solutions.

Check Yourself 6

How many real solutions are there for each quadratic equation?

(a) $2x^2 - 3x + 2 = 0$ **(b)** $3x^2 + x - 11 = 0$
(c) $4x^2 - 4x + 1 = 0$ **(d)** $x^2 = -5x - 7$

Frequently, as in Examples 3 and 4, the solutions of a quadratic equation involve square roots. When we are solving algebraic equations, it is generally best to leave solutions in this form. However, if an equation results from an application, we often estimate the root and sometimes accept only positive solutions. Consider these applications involving thrown balls that can be solved with the quadratic formula.

Example 7 Solving a Thrown-Ball Application

NOTE

Here, h measures the height above the ground, in feet, t seconds (s) after the ball is thrown upward.

If a ball is launched from the ground with an initial velocity of 80 ft/s the equation to find the height h of the ball after t seconds is

$$h = 80t - 16t^2$$

Find the time it takes the ball to reach a height of 48 ft.

We substitute 48 for h, and then we rewrite the equation in standard form.

$$(48) = 80t - 16t^2$$

$$16t^2 - 80t + 48 = 0$$

To simplify the computation, we divide both sides of the equation by the common factor 16.

$$t^2 - 5t + 3 = 0$$

NOTE

There are two solutions because the ball reaches the height *twice*, once on the way up and once on the way down.

We use the quadratic formula to solve for t.

$$t = \frac{5 \pm \sqrt{13}}{2}$$

This gives us two solutions, $\frac{5 + \sqrt{13}}{2}$ and $\frac{5 - \sqrt{13}}{2}$. But, because we have specified units of time, we generally estimate the answer to the nearest tenth or hundredth of a second.

In this case, estimating to the nearest tenth of a second gives solutions of 0.7 and 4.3 s.

Check Yourself 7

The equation to find the height h of a ball thrown with an initial velocity of 64 ft/s is

$$h(t) = 64t - 16t^2$$

Find the time it takes the ball to reach a height of 32 ft (to the nearest tenth of a second).

Example 8 Solving a Thrown-Ball Application

The height h of a ball thrown downward from the top of a 240-ft building with an initial velocity of 64 ft/s is given by

$$h(t) = 240 - 64t - 16t^2$$

When will the ball reach a height of 176 ft?
Let $h(t) = 176$, and write the equation in standard form.

$$(176) = 240 - 64t - 16t^2$$
$$0 = 64 - 64t - 16t^2$$
$$16t^2 + 64t - 64 = 0$$

Divide both sides of the equation by 16 to simplify the computation.

$$t^2 + 4t - 4 = 0$$

Applying the quadratic formula with $a = 1$, $b = 4$, and $c = -4$ yields

$$t = -2 \pm 2\sqrt{2}$$

Estimating these solutions, we have $t = -4.8$ and $t = 0.8$ s, but of these two values only the *positive value* makes any sense. (To accept the negative solution would be to say that the ball reached the specified height before it was thrown.)

NOTE

The graph of

$h(t) = 240 - 64t - 16t^2$

shows the height h at any time t.

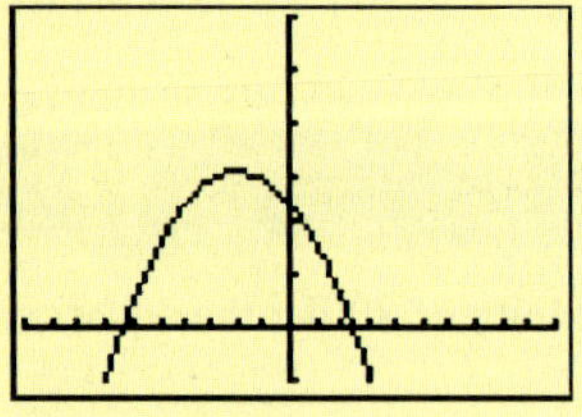

The ball has a height of 176 ft at approximately 0.8 s.

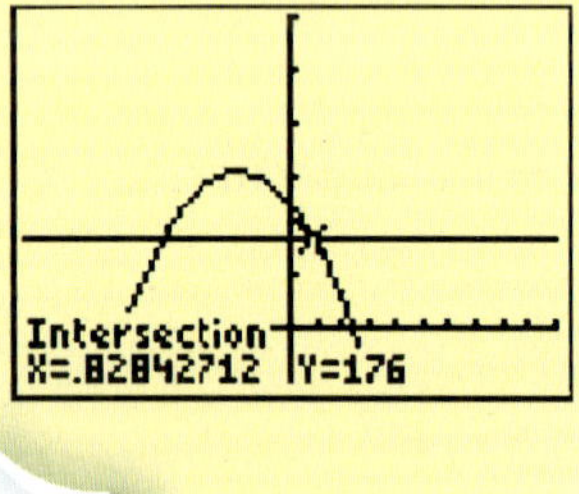

Check Yourself 8

The height h of a ball thrown upward from the top of a 96-ft building with an initial velocity of 16 ft/s is given by

$$h(t) = 96 + 16t - 16t^2$$

When will the height of the ball be 32 ft? (Estimate your answer to the nearest tenth of a second.)

Recall from Section 7.1 that the **Pythagorean theorem** gives an important relationship between the lengths of the sides of a right triangle (a triangle with a 90° angle).

We restate this important theorem here.

Definition

The Pythagorean Theorem

In any right triangle, the square of the longest side (the hypotenuse) is equal to the sum of the squares of the two shorter sides (the legs).

$$c^2 = a^2 + b^2$$

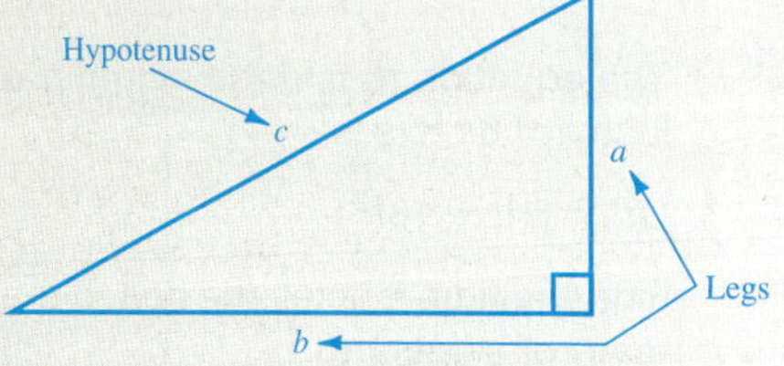

In Example 9, the solution of the quadratic equation contains a radical. When working with applications, we are usually interested in decimal approximations rather than exact radicals. Checking the "reasonableness" of our answer is very important, but we should not expect a decimal approximation to check *exactly*.

Example 9 A Triangular Application

< Objective 3 >

RECALL

The sum of the squares of the legs of the triangle is equal to the square of the hypotenuse.

One leg of a right triangle is 4 cm longer than the other leg. The length of the hypotenuse of the triangle is 12 cm. Find the length of the two legs, accurate to the nearest hundredth of a centimeter.

As in any geometric problem, a sketch of the information helps us visualize the problem.

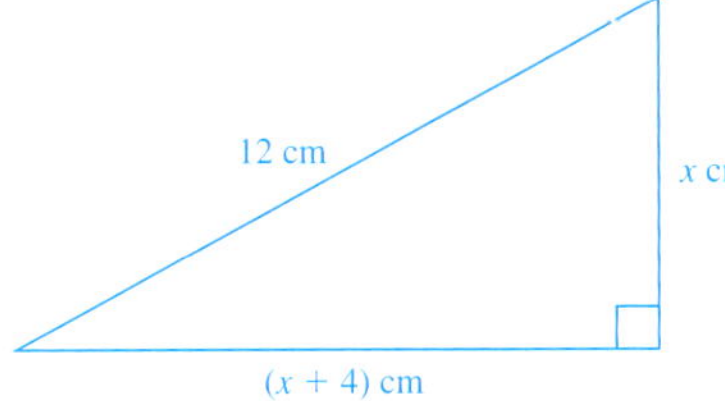

We assign the variable x to the shorter leg and $x + 4$ to the other leg.

NOTES

Dividing both sides of a quadratic equation by a common factor is always a prudent step. It simplifies your work with the quadratic formula.

```
(-4+√(272))/2
          6.246211251
(-4-√(272))/2
         -10.24621125
```

Now we apply the Pythagorean theorem to write an equation for the solution.

$$x^2 + (x + 4)^2 = (12)^2$$

$$x^2 + x^2 + 8x + 16 = 144$$

or $\quad 2x^2 + 8x - 128 = 0$

Dividing both sides by 2 gives

$$x^2 + 4x - 64 = 0$$

Using the quadratic formula, we get

$$x = \frac{-4 \pm \sqrt{272}}{2}$$

We are interested in lengths rounded to the nearest hundredth, so there is no need to simplify the radical expression. Using a calculator, we can reject one solution (do you see why?) and find $x = 6.25$ cm for the other.

This means that the two legs have lengths of 6.25 and 10.25 cm. How do we check? The sides must "reasonably" satisfy the Pythagorean theorem:

$$6.25^2 + 10.25^2 \approx 12^2$$

$$144.125 \approx 144$$

This indicates that our answer is reasonable.

NOTE

Do you see why we should not expect an "exact" check here?

Check Yourself 9

One leg of a right triangle is 2 cm longer than the other. The hypotenuse is 1 cm less than twice the length of the shorter leg. Find the length of each side of the triangle, accurate to the nearest tenth.

Check Yourself ANSWERS

1. $\left\{-2, \frac{4}{3}\right\}$ **2.** $\left\{\frac{1}{2}\right\}$ **3.** $\left\{\frac{1 \pm \sqrt{57}}{4}\right\}$ **4.** $\{2 \pm \sqrt{10}\}$
5. $\{2 \pm i\sqrt{2}\}$ nonreal **6.** **(a)** None; **(b)** two; **(c)** one; **(d)** none
7. 0.6 and 3.4 s **8.** 2.6 s **9.** Approximately 4.3, 6.3, and 7.7 cm

Reading Your Text

SECTION 8.2

(a) Every quadratic equation can be solved by using the quadratic ______________.

(b) If the solutions to a quadratic equation are rational, the original equation can be solved by the method of ______________.

(c) Given a quadratic equation in standard form, $b^2 - 4ac$ is called the ______________.

(d) Given a quadratic equation in standard form, if $b^2 - 4ac < 0$ there are ______________ real solutions.

Basic Skills | Challenge Yourself | Calculator/Computer | Career Applications | Above and Beyond

8.2 exercises

< Objective 1 >

Solve each quadratic equation first by factoring and then with the quadratic formula.

1. $x^2 - 5x - 14 = 0$

2. $x^2 - 2x - 35 = 0$

3. $t^2 + 8t - 65 = 0$

4. $q^2 + 3q - 130 = 0$

5. $3x^2 + x - 10 = 0$

6. $3x^2 + 2x - 1 = 0$

7. $16t^2 - 24t + 9 = 0$

8. $6m^2 - 23m + 10 = 0$

Solve each quadratic equation **(a)** *by completing the square and* **(b)** *with the quadratic formula.*

9. $x^2 - 4x - 7 = 0$

10. $x^2 + 6x - 1 = 0$

11. $x^2 + 3x - 27 = 0$

12. $t^2 + 4t - 7 = 0$

13. $3x^2 - 5x + 1 = 0$

14. $2x^2 - 6x + 1 = 0$

15. $2q^2 - 2q - 1 = 0$

16. $3r^2 - 2r + 4 = 0$ $(\mathbb{C})$

17. $3x^2 - x - 2 = 0$

18. $2x^2 - 5x + 3 = 0$

19. $2y^2 - y - 5 = 0$

20. $3m^2 + 2m - 1 = 0$

Use the quadratic formula to solve each equation.

21. $x^2 - 4x + 3 = 0$ 

22. $x^2 - 7x + 3 = 0$

23. $p^2 - 8p + 16 = 0$

24. $u^2 + 7u - 30 = 0$

Name ______________________

Section ________ Date ________

Answers

1. ________ 2. ________

3. ________ 4. ________

5. ________ 6. ________

7. ________ 8. ________

9. ________

10. ________

11. ________

12. ________

13. ________ 14. ________

15. ________

16. ________

17. ________ 18. ________

19. ________ 20. ________

21. ________

22. ________

23. ________ 24. ________

Answers

25. ______ 26. ______
27. ______
28. ______
29. ______
30. ______
31. ______
32. ______
33. ______
34. ______
35. ______ 36. ______
37. ______
38. ______
39. ______ 40. ______
41. ______ 42. ______
43. ______ 44. ______
45. ______ 46. ______
47. ______ 48. ______
49. ______ 50. ______
51. ______ 52. ______
53. ______
54. ______
55. ______ 56. ______
57. ______
58. ______

25. $-x^2 + 3x + 5 = 0$ 

26. $2x^2 - 3x - 7 = 0$

27. $-3s^2 + 2s - 1 = 0$ (ℂ)

28. $5t^2 - 2t - 2 = 0$

(Hint: *Clear the equations of fractions or remove grouping symbols, as needed.*)

29. $2x^2 - \frac{1}{2}x - 5 = 0$

30. $3x^2 + \frac{1}{3}x - 3 = 0$

31. $5t^2 - 2t - \frac{2}{3} = 0$

32. $3y^2 + 2y + \frac{3}{4} = 0$ (ℂ)

33. $(x - 2)(x + 3) = 4$ > Videos

34. $(x + 1)(x - 8) = 3$

35. $(t + 1)(2t - 4) - 7 = 0$

36. $(2w + 1)(3w - 2) = 1$

37. $\frac{2x^2}{3} - \frac{7x}{3} = 1$

38. $\frac{x^2}{3} + x = \frac{1}{3}$

39. $t^2 - \frac{3}{2} = \frac{3t}{2}$

40. $p^2 - \frac{1}{4} = \frac{3p}{2}$

41. $5 + 2y = y^2$

42. $6 - 2x = x^2$

< Objective 2 >

For each quadratic equation, find the value of the discriminant and give the number of real solutions.

43. $2x^2 - 5x = 0$

44. $3x^2 + 8x = 0$

45. $m^2 - 18m + 81 = 0$

46. $4p^2 + 12p + 9 = 0$

47. $3x^2 - 7x + 1 = 0$ > Videos

48. $2x^2 - x + 5 = 0$

49. $2w^2 - 5w + 11 = 0$

50. $6q^2 - 5q + 2 = 0$

Solve each quadratic equation. Use any applicable method.

51. $x^2 - 8x + 16 = 0$

52. $4x^2 + 12x + 9 = 0$

53. $3t^2 - 7t + 1 = 0$

54. $2z^2 - z + 5 = 0$ (ℂ)

55. $5y^2 - 2y = 0$

56. $7z^2 - 6z - 2 = 0$

57. $(x - 1)(2x + 7) = -6$

58. $4x^2 - 3 = 0$

59. $x^2 + 9 = 0$ (ℂ)

60. $(4x - 5)(x + 2) = 1$

61. $x^2 - 5x = 10 - 2x$

62. $x^2 + x = -2 - x$ (ℂ)

Basic Skills | **Challenge Yourself** | Calculator/Computer | Career Applications | Above and Beyond

63. SCIENCE AND MEDICINE The equation

$$h(t) = 112t - 16t^2$$

is the equation for the height of an arrow, shot upward from the ground with an initial velocity of 112 ft/s, in which t is the time, in seconds, after the arrow leaves the ground.

(a) Find the time it takes for the arrow to reach a height of 112 ft.

(b) Find the time it takes for the arrow to reach a height of 144 ft.

Express your answers to the nearest tenth of a second.

64. SCIENCE AND MEDICINE The equation

$$h(t) = 320 - 32t - 16t^2$$

is the equation for the height of a ball, thrown downward from the top of a 320-ft building with an initial velocity of 32 ft/s, in which t is the time after the ball is thrown down from the top of the building.

(a) Find the time it takes for the ball to reach a height of 240 ft.

(b) Find the time it takes for the ball to reach a height of 96 ft.

Express your answers to the nearest tenth of a second.

65. NUMBER PROBLEM The product of two consecutive integers is 72. What are the two integers?

66. NUMBER PROBLEM The sum of the squares of two consecutive whole numbers is 61. Find the two whole numbers.

67. GEOMETRY The width of a rectangle is 3 ft less than its length. If the area of the rectangle is 70 ft², what are the dimensions of the rectangle?

68. GEOMETRY The length of a rectangle is 5 cm more than its width. If the area of the rectangle is 84 cm², find the dimensions.

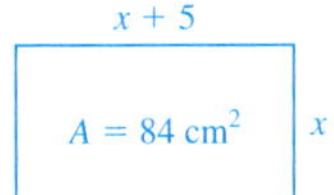

< Objective 3 >

69. GEOMETRY One leg of a right triangle is twice the length of the other. The hypotenuse is 6 m long. Find the length of each leg. Round to the nearest tenth of a meter.

> Videos

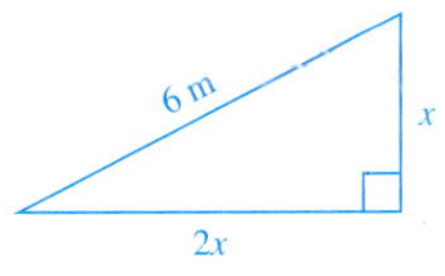

Answers

59. ____

60. ____

61. ____

62. ____

63. ____

64. ____

65. ____

66. ____

67. ____

68. ____

69. ____

Answers

70. ________

71. ________

72. ________

73. ________

74. ________

75. ________

76. ________

77. ________

70. **GEOMETRY** One leg of a right triangle is 2 ft longer than the shorter side. If the length of the hypotenuse is 14 ft, how long is each leg (nearest tenth of a ft)?

71. **SCIENCE AND MEDICINE** If a ball is thrown vertically upward from the ground, with an initial velocity of 64 ft/s, its height h after t seconds is given by $h(t) = 64t - 16t^2$.

    (a) How long does it take the ball to return to the ground? [*Hint:* Let $h(t) = 0$.]

    (b) How long does it take the ball to reach a height of 48 ft on the way up?

72. **SCIENCE AND MEDICINE** If a ball is thrown vertically upward from the ground, with an initial velocity of 96 ft/s, its height h after t seconds is given by $h(t) = 96t - 16t^2$.

    (a) How long does it take the ball to return to the ground?

    (b) How long does it take the ball to pass through a height of 128 ft on the way back down to the ground?

73. **BUSINESS AND FINANCE** Suppose that the cost $C(x)$, in dollars, of producing x chairs is given by

    $C(x) = 2{,}400 - 40x + 2x^2$

    How many chairs can be produced for \$5,400?

74. **BUSINESS AND FINANCE** Suppose that the profit $T(x)$, in dollars, of producing and selling x microwave ovens is given by

    $T(x) = -3x^2 + 240x - 1{,}800$

    How many microwaves must be produced and sold to achieve a profit of \$3,000?

75. **GEOMETRY** One leg of a right triangle is 1 in. shorter than the other leg. The hypotenuse is 3 in. longer than the shorter side. Find the length of each side, to the nearest tenth of an inch.

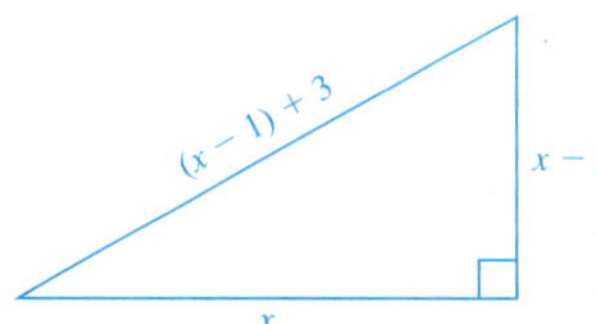

76. **GEOMETRY** The hypotenuse of a given right triangle is 5 cm longer than the shorter leg. The length of the shorter leg is 2 cm less than the length of the longer leg. Find the lengths of the three sides, to the nearest tenth of a centimeter.

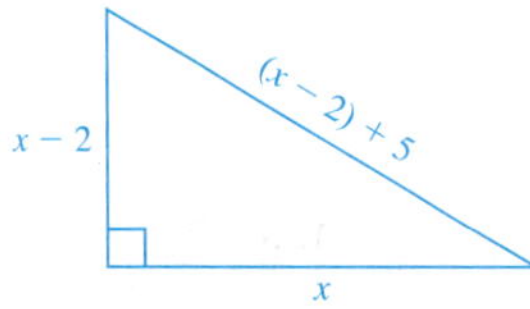

77. **BUSINESS AND FINANCE** A small manufacturer's weekly profit, in dollars, is given by

    $P(x) = -3x^2 + 270x$

    Find the number of DVD players x that must be produced to realize a profit of \$5,100.

78. **Business and Finance** Suppose the profit, in dollars, is given by

$$P(x) = -2x^2 + 240x$$

Now how many DVD players must be sold to realize a profit of $5,100?

79. **Business and Finance** The demand equation for a certain computer chip is given by

$$D = -2p + 14$$

The supply equation is predicted to be

$$S = -p^2 + 16p - 2$$

Find the equilibrium price.

80. **Business and Finance** The demand equation for a certain type of printer is predicted to be

$$D = -200p + 36{,}000$$

The supply equation is predicted to be

$$S = -p^2 + 400p - 24{,}000$$

Find the equilibrium price.

81. **Science and Medicine** If a ball is thrown upward from the roof of a building 70 m tall with an initial velocity of 15 m/s, its approximate height h after t seconds is given by

$$h(t) = 70 + 15t - 5t^2$$

Note: The difference between this equation and the one we used in Example 8 has to do with the units used. When we used feet, the t^2-coefficient was -16 (because the acceleration due to gravity is approximately 32 ft/s^2). When we use meters as the height, the t^2-coefficient is -5 because that same acceleration becomes approximately 10 m/s^2. Use this information to complete each exercise.

(a) How long does it take the ball to fall back to the ground?

(b) When will the ball reach a height of 80 m?

Express your answers to the nearest tenth of a second.

82. **Science and Medicine** Changing the initial velocity to 25 m/s only changes the t-coefficient. Our new equation becomes

$$h(t) = 70 + 25t - 5t^2$$

(a) How long will it take the ball to return to the thrower?

(b) When will the ball reach a height of 85 m?

Express your answers to the nearest tenth of a second.

Answers

78. ____________

79. ____________

80. ____________

81. ____________

82. ____________

Answers

83. ______

84. ______

85. ______

86. ______

87. ______

88. ______

89. ______

90. ______

The only part of the height equation that we have not discussed is the constant. You have probably noticed that the constant is always equal to the initial height of the ball (70 m in our previous exercises). Now, we ask *you* to develop an equation.

83. Science and Medicine A ball is thrown upward from the roof of a 100-m building with an initial velocity of 20 m/s. Use this information to complete each exercise.

(a) Find the equation for the height h of the ball after t seconds.

(b) How long will it take the ball to fall back to the ground?

(c) When will the ball reach a height of 75 m?

(d) Will the ball ever reach a height of 125 m? (*Hint:* Check the discriminant.)

Express your answers to the nearest tenth of a meter.

84. Science and Medicine A ball is thrown upward from the roof of a 100-ft building with an initial velocity of 20 ft/s. Use this information to complete each exercise.

(a) Find the height h of the ball after t seconds.

(b) How long will it take the ball to fall back to the ground?

(c) When will the ball reach a height of 80 ft?

(d) Will the ball ever reach a height of 120 ft? Explain.

Express your answers to the nearest tenth of a foot.

Complete each statement with **never, sometimes,** *or* **always.**

85. The quadratic formula can ________ be used to solve a quadratic equation.

86. If the value of $b^2 - 4ac$ is negative, the equation $ax^2 + bx + c = 0$ ________ has real-number solutions.

87. The solutions of a quadratic equation are ________ irrational.

88. To apply the quadratic formula, a quadratic equation must ________ be written in standard form.

89. (a) Use the quadratic formula to solve $x^2 - 3x - 5 = 0$. For each solution give a decimal approximation to the nearest tenth.

(b) Graph the function $f(x) = x^2 - 3x - 5$ on your graphing calculator. Use a zoom utility and estimate the x-intercepts to the nearest tenth.

(c) Describe the connection between parts (a) and (b).

90. (a) Solve the equation using any appropriate method.

$$x^2 - 2x = 3$$

(b) Graph the functions on your graphing calculator.

$$f(x) = x^2 - 2x \qquad \text{and} \qquad g(x) = 3$$

Estimate the points of intersection of the graphs of f and g. In particular, note the x-coordinates of these points.

(c) Describe the connection between parts (a) and (b).

Basic Skills | Challenge Yourself | Calculator/Computer | **Career Applications** | Above and Beyond

91. AGRICULTURAL TECHNOLOGY The Scribner log rule is used to calculate the volume, in cubic feet, of a 16-ft log given the diameter (inside the bark) of the smaller end, in inches. The formula for the log rule is

$$V = 0.79D^2 - 2D - 4$$

Find the diameter of the small end of a log if its volume is 272 ft^3.

92. AGRICULTURAL TECHNOLOGY A walkway of constant width is to be installed around a rectangular garden. The garden measures 5 m by 8 m; the total area (garden and walkway, combined) is limited to 100 m^2. What is the maximum width of the walkway? Report your results with two decimal places of precision.

93. ALLIED HEALTH Radiation therapy is one technique used to control cancer. After such a treatment, the number of cancerous cells N, in thousands, that remain in a particular patient can be estimated by the formula

> Videos

$$N = -3t^2 - 6t + 140$$

in which t is the number of days of treatment. According to the model, how many days of treatment are required to kill all of a patient's cancer cells?

94. ALLIED HEALTH An experimental drug is being tested on a bacteria colony. It is found that t days after the colony is treated, the number of bacteria N per cubic centimeter is given by the formula

$$N = -20t^2 - 120t + 1{,}000$$

In how many days will the colony be reduced to 200 bacteria per cubic centimeter?

Basic Skills | Challenge Yourself | Calculator/Computer | Career Applications | **Above and Beyond**

95. Can the solution of a quadratic equation with integer coefficients include one real and one imaginary number? Justify your answer.

96. Explain how the discriminant is used to predict the nature of the solutions of a quadratic equation.

Solve each equation for x.

97. $x^2 + y^2 = z^2$

98. $2x^2y^2z^2 = 1$

99. $x^2 - 36a^2 = 0$

100. $ax^2 - 9b^2 = 0$

101. $2x^2 + 5ax - 3a^2 = 0$

102. $3x^2 - 16bx + 5b^2 = 0$

103. $2x^2 + ax - 2a^2 = 0$

104. $3x^2 - 2bx - 2b^2 = 0$

Answers

91. ______

92. ______

93. ______

94. ______

95. ______

96. ______

97. ______

98. ______

99. ______

100. ______

101. ______

102. ______

103. ______

104. ______

Answers

105. ______

106. ______

107. ______

108. ______

109. ______

110. ______

105. Given that the polynomial $x^3 - 3x^2 - 15x + 25 = 0$ has as one of its solutions $x = 5$, find the other two solutions. (*Hint:* If you divide the given polynomial by $x - 5$, the quotient will be a quadratic expression. The remaining solutions will be the solutions for the resulting equation.)

106. Given that $2x^3 + 2x^2 - 5x - 2 = 0$ has as one of its solutions $x = -2$, find the other two solutions. (*Hint:* In this case, divide the original polynomial by $x + 2$.)

107. Find all the zeros of the function $f(x) = x^3 + 1$. ($\mathbb{C}$)

108. Find the zeros of the function $f(x) = x^2 + x + 1$. ($\mathbb{C}$)

109. Find all six solutions to the equation $x^6 - 1 = 0$. (*Hint:* Factor the left-hand side of the equation first as the difference of squares, then as the sum and difference of cubes.) ($\mathbb{C}$)

110. Find all six solutions to $x^6 = 64$. ($\mathbb{C}$)

Answers

1. $\{-2, 7\}$ **3.** $\{-13, 5\}$ **5.** $\left\{-2, \frac{5}{3}\right\}$ **7.** $\left\{\frac{3}{4}\right\}$ **9.** $\{2 \pm \sqrt{11}\}$
11. $\left\{\frac{-3 \pm 3\sqrt{13}}{2}\right\}$ **13.** $\left\{\frac{5 \pm \sqrt{13}}{6}\right\}$ **15.** $\left\{\frac{1 \pm \sqrt{3}}{2}\right\}$ **17.** $\left\{-\frac{2}{3}, 1\right\}$
19. $\left\{\frac{1 \pm \sqrt{41}}{4}\right\}$ **21.** $\{1, 3\}$ **23.** $\{4\}$ **25.** $\left\{\frac{3 \pm \sqrt{29}}{2}\right\}$
27. $\left\{\frac{1 \pm i\sqrt{2}}{3}\right\}$ nonreal **29.** $\left\{\frac{1 \pm \sqrt{161}}{8}\right\}$ **31.** $\left\{\frac{3 \pm \sqrt{39}}{15}\right\}$
33. $\left\{\frac{-1 \pm \sqrt{41}}{2}\right\}$ **35.** $\left\{\frac{1 \pm \sqrt{23}}{2}\right\}$ **37.** $\left\{\frac{7 \pm \sqrt{73}}{4}\right\}$
39. $\left\{\frac{3 \pm \sqrt{33}}{4}\right\}$ **41.** $\{1 \pm \sqrt{6}\}$ **43.** 25, two **45.** 0, one
47. 37, two **49.** -63, none **51.** $\{4\}$ **53.** $\left\{\frac{7 \pm \sqrt{37}}{6}\right\}$ **55.** $\left\{0, \frac{2}{5}\right\}$
57. $\left\{\frac{-5 \pm \sqrt{33}}{4}\right\}$ **59.** $\{-3i, 3i\}$ nonreal **61.** $\{-2, 5\}$
63. **(a)** 1.2 or 5.8 s; **(b)** 1.7 or 5.3 s **65.** $-9, -8$ or 8, 9 **67.** 7 ft by 10 ft
69. 2.7 cm and 5.4 cm **71.** **(a)** 4 s; **(b)** 1 s **73.** 50 chairs
75. 5.5 in., 6.5 in., 8.5 in. **77.** 63 or 27 DVD players
79. \$0.94 or \$17.06 **81.** **(a)** 5.5 s; **(b)** 1 s, 2 s
83. **(a)** $h(t) = 100 + 20t - 5t^2$; **(b)** 6.9 s; **(c)** 5 s; **(d)** no **85.** always
87. sometimes **89.** **(a)** $\{-1.2, 4.2\}$; **(b)** $-1.2, 4.2$; **(c)** The solutions to the quadratic equation are the x-intercepts of the graph. **91.** 20 in. **93.** 6 days
95. Above and Beyond **97.** $\{\pm\sqrt{z^2 - y^2}\}$ **99.** $\{-6a, 6a\}$
101. $\left\{-3a, \frac{a}{2}\right\}$ **103.** $\left\{\frac{-a \pm |a|\sqrt{17}}{4}\right\}$ **105.** $\{-1 \pm \sqrt{6}\}$
107. $\left\{-1, \frac{1 \pm i\sqrt{3}}{2}\right\}$ two nonreal zeros
109. $\left\{-1, 1, \frac{1 \pm i\sqrt{3}}{2}, \frac{-1 \pm i\sqrt{3}}{2}\right\}$ four nonreal solutions

8.3 An Introduction to Parabolas

< 8.3 Objectives >

1 > Find the axis of symmetry and vertex of a parabola

2 > Graph a parabola

In Section 3.1, you learned to graph a linear equation. We discovered that the graph of every linear equation in two variables is a straight line. In this section, we consider the graph of a quadratic equation in two variables.

Consider a quadratic equation,

$$y = ax^2 + bx + c \qquad a \neq 0$$

This equation defines a quadratic function with x as the input variable and y as the output variable. The graph of a quadratic function is a curve called a **parabola.**

Property

Shape of a Parabola

For a function

$$f(x) = ax^2 + bx + c \qquad a \neq 0$$

the parabola opens upward or downward, as follows:

1. If $a > 0$, the parabola opens *upward*.
2. If $a < 0$, the parabola opens *downward*.

$$f(x) = ax^2 + bx + c$$

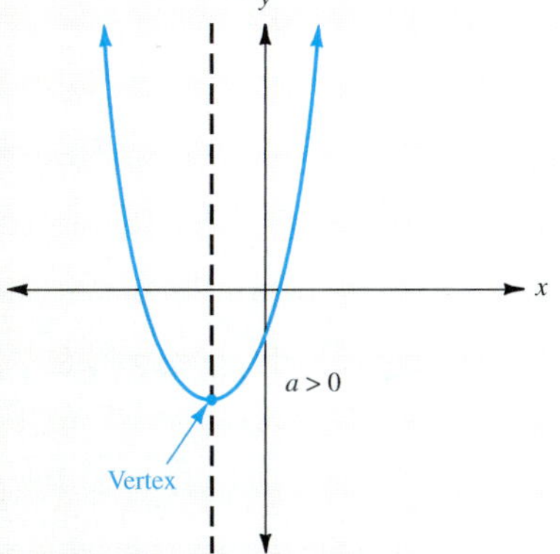

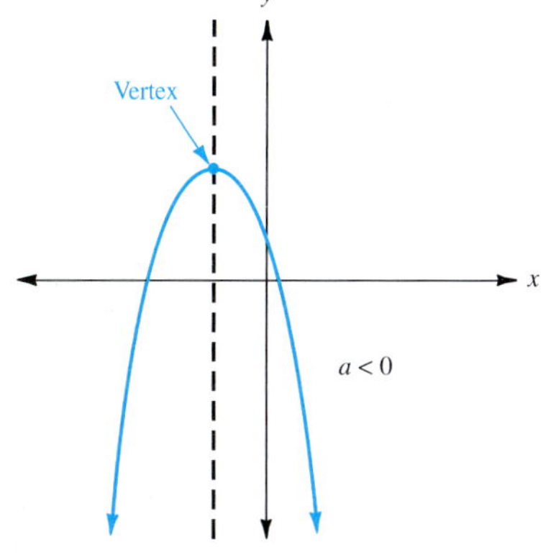

NOTE

You can think of the vertex as the **turning point** of the parabola.

Two concepts regarding parabolas can be made by observation. Consider the previous illustrations.

1. There is always a **minimum** (or lowest) point on a parabola that opens upward. There is always a **maximum** (or highest) point on a parabola that opens downward. In either case, that maximum or minimum value occurs at the **vertex** of the parabola.
2. Every parabola has an **axis of symmetry.** In the case of parabolas that open upward or downward, the axis of symmetry is a vertical line that splits the graph

into two pieces, each a mirror image of the other. The axis of symmetry always passes through the vertex.

This figure summarizes these observations.

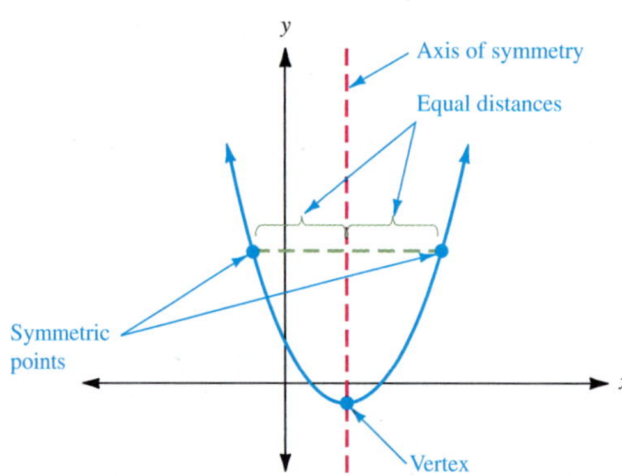

Our objective is to be able to quickly sketch a parabola. This can be done with *as few as three points* if those points are carefully chosen. For this purpose we want to find the vertex and two symmetric points.

First, we see how the coordinates of the vertex can be determined from the standard equation

$$y = ax^2 + bx + c$$

If $x = 0$, then $y = c$, and so $(0, c)$ gives the point where the parabola intersects the y-axis (the y-intercept).

Look at the sketch. To determine the coordinates of the symmetric point (x_1, c), note that it lies along the horizontal line $y = c$.

NOTE

For this discussion, we assume that the vertex is not on the *y*-axis.
If the vertex *is* on the *y*-axis, the axis of symmetry is the *y*-axis itself, and its equation is $x = 0$.

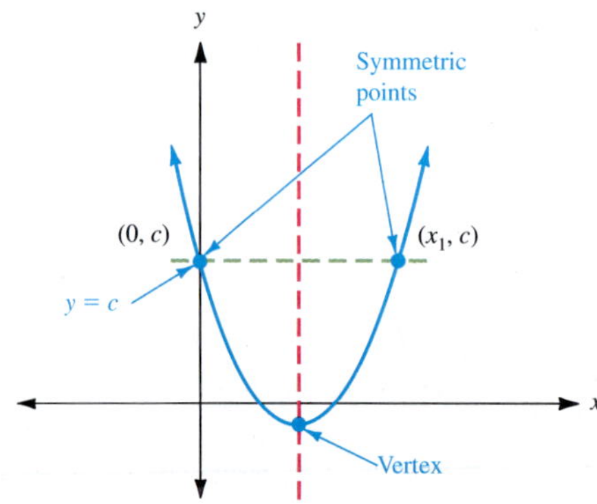

Therefore, let $y = c$ in the quadratic equation.

$c = ax^2 + bx + c$

$0 = ax^2 + bx$ Subtract c from both sides.

$0 = x(ax + b)$ Factor and solve.

and

$x = 0$ or $x = -\dfrac{b}{a}$

We now know that

$(0, c)$ and $\left(-\dfrac{b}{a}, c\right)$

> **NOTE**
>
> To ensure that we remember the negative, we sometimes write
>
> $x = -\frac{b}{2a}$

are the coordinates of the symmetric points shown. Since the axis of symmetry must be midway between these points, the x-value along that axis is given by

$$x = \frac{0 + (-b/a)}{2} = \frac{-b}{2a}$$

Since the vertex for any parabola lies on the axis of symmetry, we know that the x-coordinate of the vertex is $-\frac{b}{2a}$, and the corresponding y-coordinate can be found by evaluating the function at $x = -\frac{b}{2a}$:

Property

Axis of Symmetry and Vertex of a Parabola

If $f(x) = ax^2 + bx + c \quad a \neq 0$

then the equation of the axis of symmetry is

$$x = -\frac{b}{2a}$$

and the coordinates of the vertex of the graph of f are

$$\left(-\frac{b}{2a}, f\left(-\frac{b}{2a}\right)\right)$$

We now know how to find the vertex of a parabola, and if two symmetric points can be determined, we are well on our way to the desired graph. Perhaps the simplest case occurs when the quadratic expression of the given equation is factorable. In such cases, the two x-intercepts (determined by the factors) give two symmetric points that are easily found. Example 1 illustrates such a case.

Example 1 Graphing a Parabola

< Objectives 1 and 2>

Draw the graph of the function

$f(x) = x^2 + 2x - 8$

First, find the axis of symmetry. In this equation, $a = 1$, $b = 2$, and $c = -8$ so we have

$$x = -\frac{b}{2a} = -\frac{(2)}{2 \cdot (1)} = -\frac{2}{2} = -1$$

Thus, $x = -1$ is the axis of symmetry.

Second, find the vertex. Since the vertex of the parabola lies on the axis of symmetry, let $x = -1$ in the original equation. If $x = -1$,

$$f(-1) = (-1)^2 + 2(-1) - 8 = -9$$

and $(-1, -9)$ is the vertex of the parabola.

Third, find two symmetric points. Note that the quadratic expression in this case is factorable, and so setting $f(x) = 0$ in the original equation quickly gives two symmetric points (the x-intercepts).

$$0 = x^2 + 2x - 8$$
$$= (x + 4)(x - 2)$$

So when $f(x) = 0$,

$x + 4 = 0$ or $x - 2 = 0$

$x = -4$ $\quad x = 2$

and the x-intercepts are $(-4, 0)$ and $(2, 0)$.

> **NOTES**
>
> Sketch the information to help solve the problem. Begin by drawing—as a dashed line—the axis of symmetry.

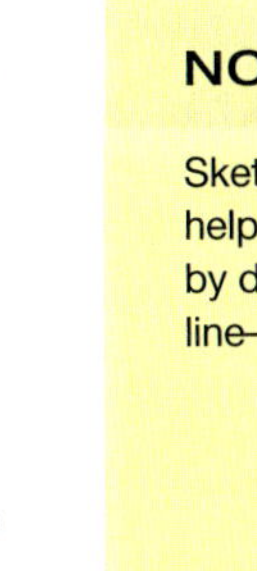

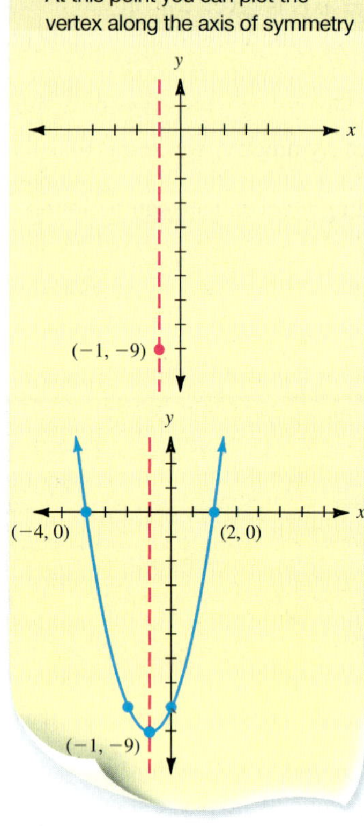

Fourth, draw a smooth curve connecting the points found above, to form the parabola.

You could find additional pairs of symmetric points at this time if necessary. For instance, the symmetric points $(0, -8)$ and $(-2, -8)$ are easily located.

Check Yourself 1

Graph the function

$$f(x) = -x^2 - 2x + 3$$

(*Hint:* The parabola opens downward because the coefficient of x^2 is negative.)

A similar process works if the quadratic expression is *not* factorable. In that case, one of two things happens:

1. The x-intercepts are irrational and therefore not particularly helpful in the graphing process.
2. The x-intercepts do not exist.

Consider Example 2.

Example 2 Graphing a Parabola

Graph the function

$$f(x) = x^2 - 6x + 3$$

First, find the axis of symmetry. Here $a = 1$, $b = -6$, and $c = 3$. So

$$x = \frac{-b}{2a} = \frac{-(-6)}{2(1)} = \frac{6}{2} = 3$$

Thus, $x = 3$ is the axis of symmetry.

Second, find the vertex. If $x = 3$,

$$f(3) = (3)^2 - 6 \cdot (3) + 3 = -6$$

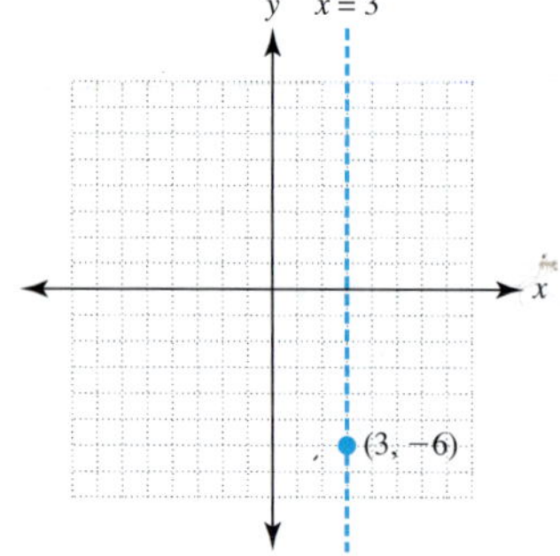

and $(3, -6)$ is the vertex of the desired parabola.

Third, find two symmetric points. Here the quadratic expression is not factorable, so we need to find another pair of symmetric points.

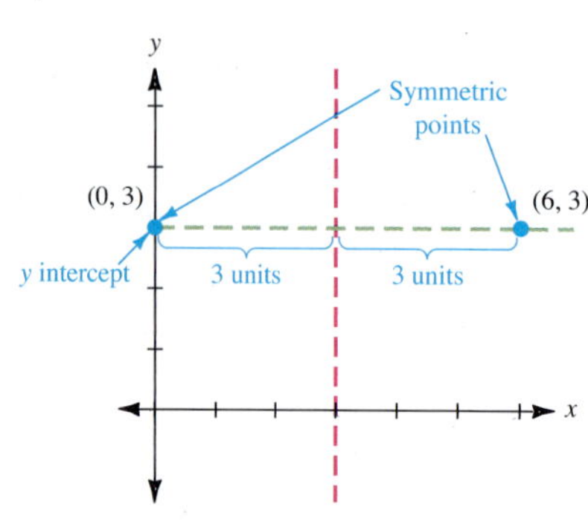

Note that (0, 3) is the y-intercept of the parabola. We found the axis of symmetry at $x = 3$ in step 1. The point symmetric to (0, 3) lies along the horizontal line through the y-intercept at the same distance (3 units) from the axis of symmetry. Hence, (6, 3) is our symmetric point.

Fourth, draw a smooth curve connecting the points found above to form the parabola.

An alternate method is available in step 3. Observing that (0, 3) is the y-intercept and that the symmetric point lies along the line $y = 3$, set $f(x) = 3$ in the original equation.

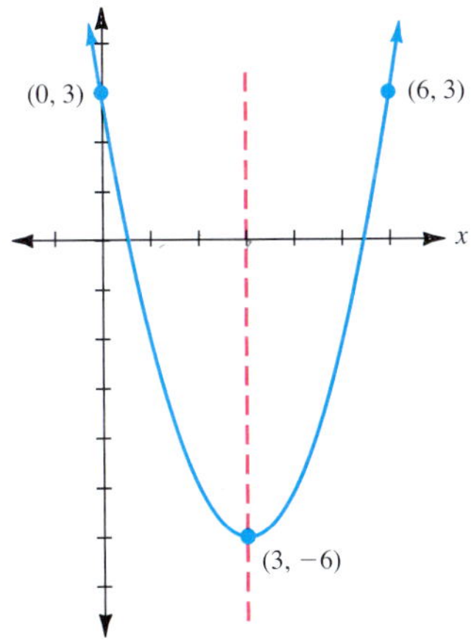

$$3 = x^2 - 6x + 3$$
$$0 = x^2 - 6x$$
$$0 = x(x - 6)$$

so

$$x = 0 \qquad \text{or} \qquad x - 6 = 0$$
$$x = 6$$

and (0, 3) and (6, 3) are the desired symmetric points.

Check Yourself 2

Graph the function.

$f(x) = x^2 + 4x + 5$

The axis of symmetry can also be found directly from the two x-intercepts. To do this we must first introduce a new idea, that of the **midpoint.**

Every pair of points has a *midpoint*. The midpoint of A (x_1, y_1) and B (x_2, y_2) is the point on line AB that is an equal distance from A and B. This formula can be used to find the midpoint M.

$$M = \left(\frac{x_1 + x_2}{2}, \frac{y_1 + y_2}{2}\right)$$

In words, the x-coordinate of the midpoint is the average (mean) of the two x-values, and the y-coordinate of the midpoint is the average of the two y-values.

Example 3 Finding the Midpoint

(a) Find the midpoint of (2, 0) and (10, 0).

$$M = \left(\frac{2 + 10}{2}, \frac{0 + 0}{2}\right) = \left(\frac{12}{2}, \frac{0}{2}\right) = (6, 0)$$

(b) Find the midpoint of (5, 7) and (−1, −3).

$$M = \left(\frac{5 + (-1)}{2}, \frac{7 + (-3)}{2}\right) = \left(\frac{4}{2}, \frac{4}{2}\right) = (2, 2)$$

(c) Find the midpoint of (−7, 3) and (8, 6).

$$M = \left(\frac{-7 + 8}{2}, \frac{3 + 6}{2}\right) = \left(\frac{1}{2}, \frac{9}{2}\right)$$

Check Yourself 3

Find the midpoint of each pair of points.

(a) (2, 7) and (−4, −1) **(b)** (−3, −6) and (5, −2)

(c) (−2, 3) and (3, 8)

We can use the midpoint to find the axis of symmetry.

Example 4 Graphing a Parabola

Graph the function $f(x) = 2x^2 - x - 6$.

$f(x) = 2x^2 - x - 6$

$f(x) = (2x + 3)(x - 2)$ Factor the expression.

$0 = (2x + 3)(x - 2)$ Find the x-intercepts.

$x = -\frac{3}{2}$ or $x = 2$

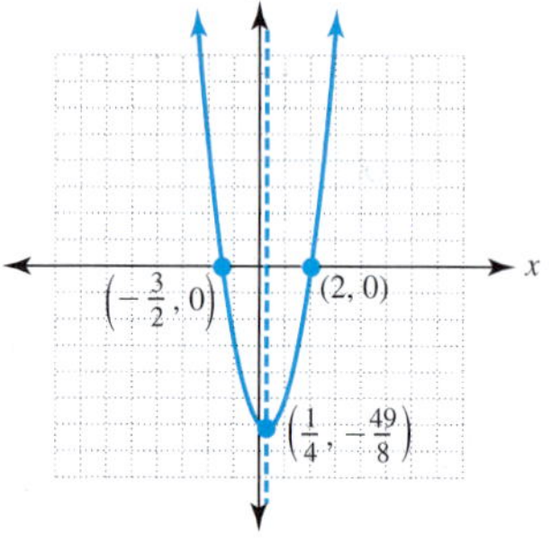

The x-intercepts are $\left(-\frac{3}{2}, 0\right)$ and (2, 0).

The midpoint is

$$\left(\frac{-\frac{3}{2} + 2}{2}, \frac{0 + 0}{2}\right) = \left(\frac{-\frac{3}{2} + \frac{4}{2}}{2}, 0\right) = \left(\frac{1}{4}, 0\right)$$

The axis of symmetry is $x = \frac{1}{4}$. The vertex is $\left(\frac{1}{4}, -\frac{49}{8}\right)$.

Check Yourself 4

Graph the function

$f(x) = 2x^2 + 5x - 3$

It is not typical for quadratic expressions to be factorable. As a result, we generally use the formula $x = \frac{-b}{2a}$ to find the axis of symmetry.

Example 5 Graphing a Parabola

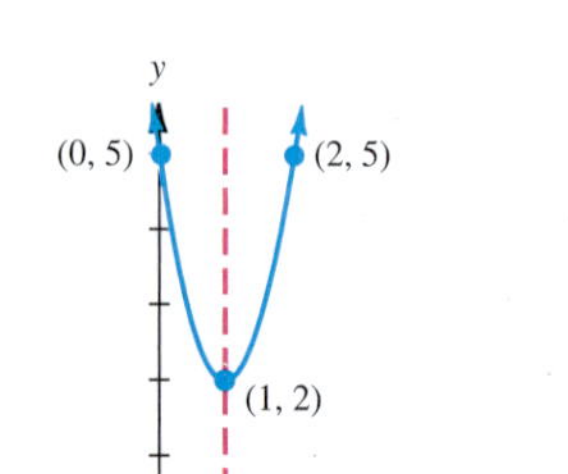

Graph the function $f(x) = 3x^2 - 6x + 5$.

First, find the axis of symmetry.

$$x = \frac{-b}{2a} = \frac{-(-6)}{2(3)} = \frac{6}{6} = 1$$

Second, find the vertex. If $x = 1$,

$f(1) = 3(1)^2 - 6 \cdot 1 + 5 = 2$

So (1, 2) is the vertex.

Third, find symmetric points. Again the quadratic expression is not factorable, so we use the y-intercept (0, 5) and its symmetric point (2, 5).

Fourth, connect the points with a smooth curve to form the parabola. Compare this curve to those in previous examples. Note that the parabola is "tighter" about the axis of symmetry. That is because the x^2 coefficient is larger.

Check Yourself 5

Graph the function.

$$f(x) = \frac{1}{2}x^2 - 3x - 1$$

This algorithm summarizes our work.

Step by Step

Graphing a Parabola

Step 1 Find the axis of symmetry.

Step 2 Find the vertex.

Step 3 Determine two symmetric points.

Note: Use the x-intercepts if the quadratic expression is factorable. If the vertex is not on the y-axis, you can use the y-intercept and its symmetric point. Another option is to simply choose an x-value that does not match the axis of symmetry, compute the corresponding y-value, and then locate its symmetric point.

Step 4 Draw a smooth curve connecting the points found above to form the parabola. You may choose to find additional pairs of symmetric points.

Check Yourself ANSWERS

1.

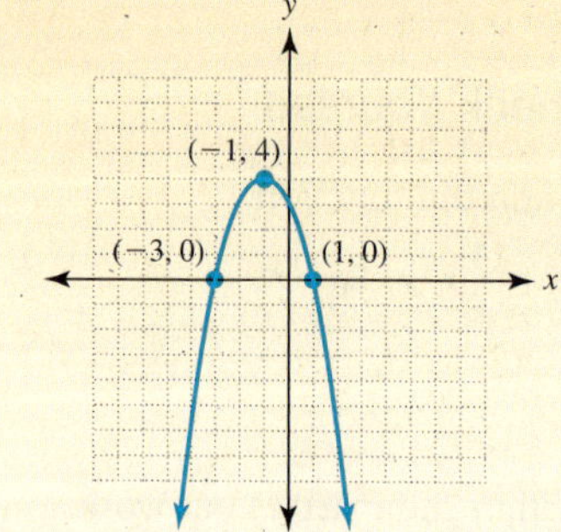

2.

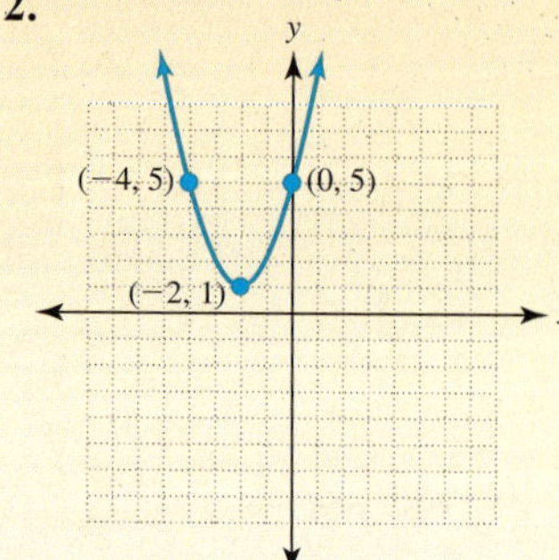

3. **(a)** $(-1, 3)$; **(b)** $(1, -4)$; **(c)** $\left(\frac{1}{2}, \frac{11}{2}\right)$

4.

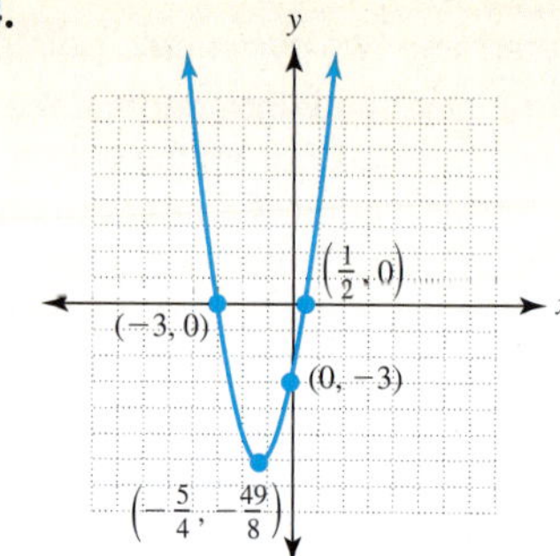

5.

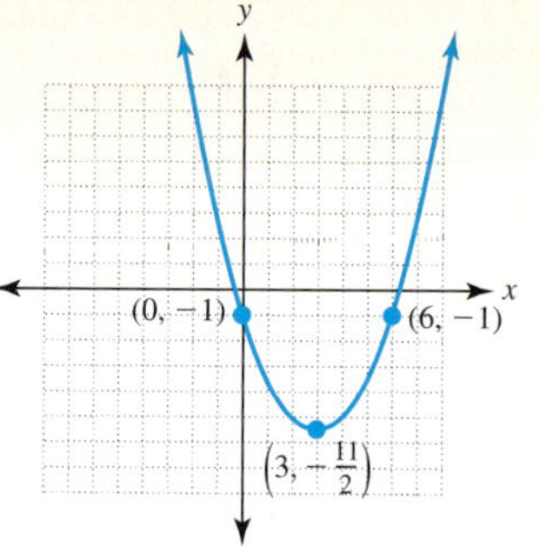

Reading Your Text

SECTION 8.3

(a) In an equation of the form $y = ax^2 + bx + c$, the parabola opens ______________ if $a < 0$.

(b) There is always a ______________ point on a parabola if it opens upward.

(c) The vertex of a parabola lies on the ______________ of symmetry.

(d) The axis of symmetry is a ______________ line that splits the graph into two pieces, each a mirror image of the other.

Basic Skills | Challenge Yourself | Calculator/Computer | Career Applications | Above and Beyond

8.3 exercises

Boost *your* GRADE at ALEKS.com!

ALEKS®

- Practice Problems
- Self-Tests
- NetTutor
- e-Professors
- Videos

Name ______________________

Section ________ Date ________

Match each graph with one of the equations.

(a) $y = x^2 + 2$

(b) $y = 2x^2 - 1$

(c) $y = 2x + 1$

(d) $y = x^2 - 3x$

(e) $y = -x^2 - 4x$

(f) $y = -2x + 1$

(g) $y = x^2 + 2x - 3$

(h) $y = -x^2 + 6x - 8$

1.

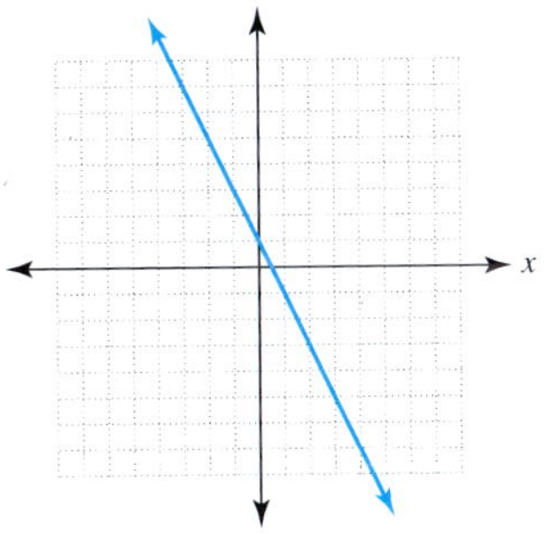

2.

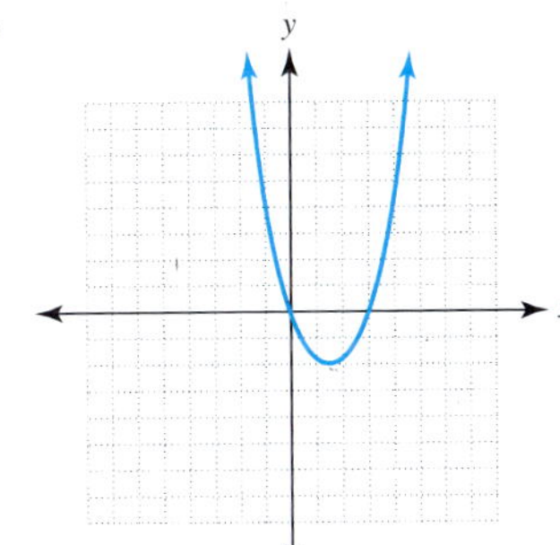

3.

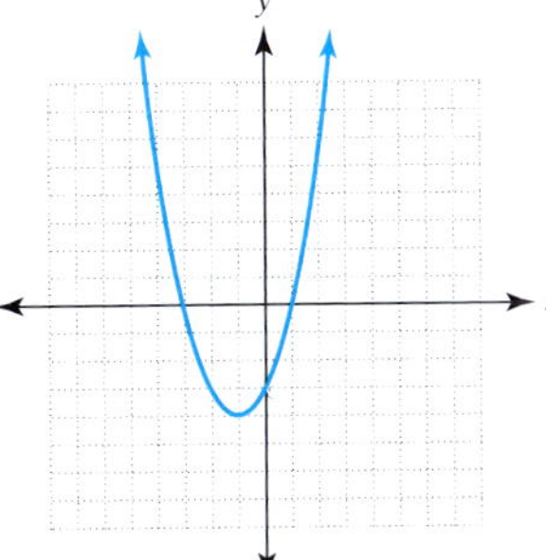

4.

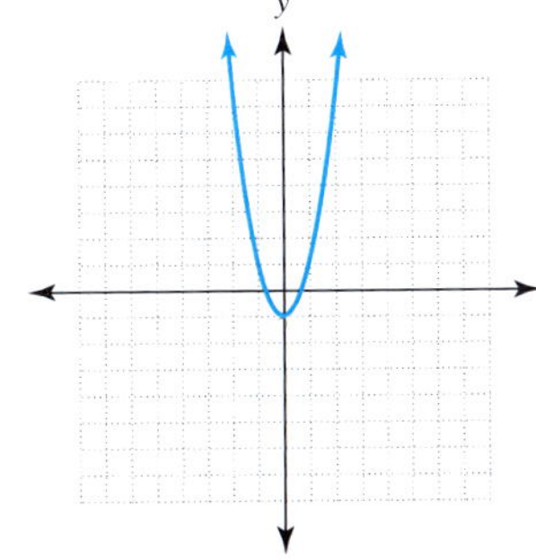

5.

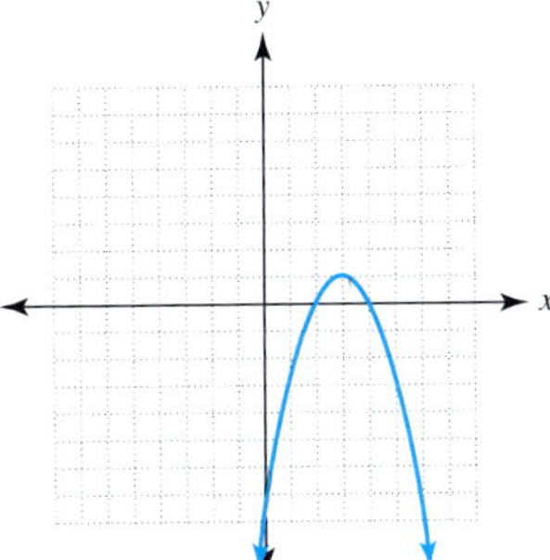

6.

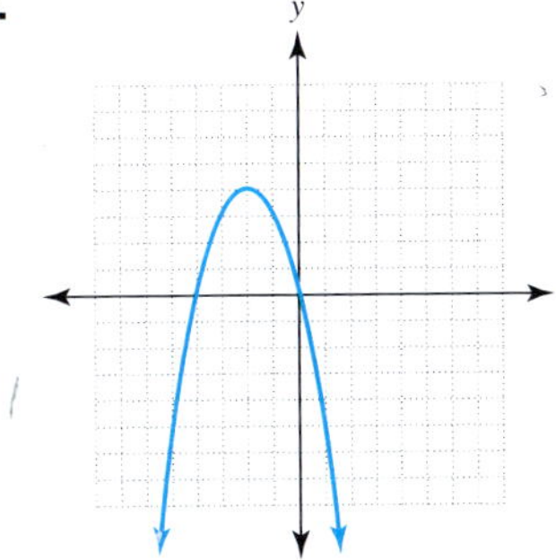

7.

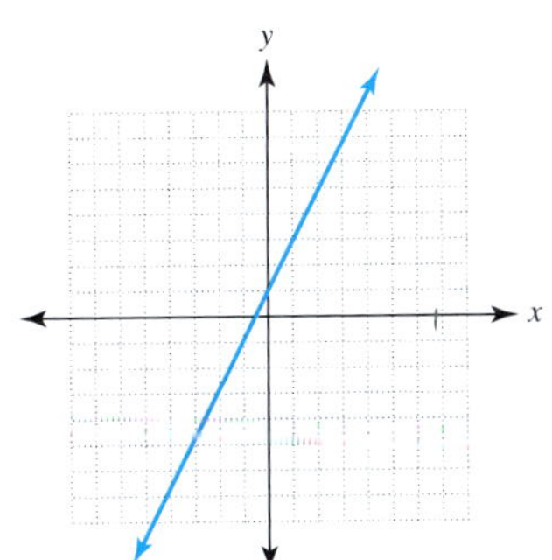

8.

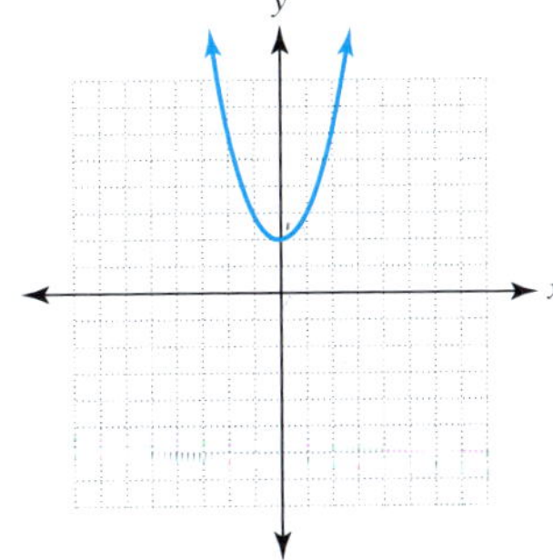

Answers

1. ______________________

2. ______________________

3. ______________________

4. ______________________

5. ______________________

6. ______________________

7. ______________________

8. ______________________

Answers

9.

10.

11.

12.

13.

14.

15.

16.

17.

18.

19.

20.

21.

22.

23.

24.

25.

26.

27.

28.

Which of the given conditions apply to the graphs of each equation? Note that more than one condition may apply.

(a) The parabola opens upward.
(b) The parabola opens downward.
(c) The parabola has two x-intercepts.
(d) The parabola has one x-intercept.
(e) The parabola has no x-intercept.

9. $y = x^2 - 3$

10. $y = -x^2 + 4x$

11. $y = x^2 - 3x - 4$

12. $y = x^2 - 2x + 2$

13. $y = -x^2 - 3x + 10$

14. $y = x^2 - 8x + 16$

Find the midpoint of each pair of points.

15. $(-2, 6)$ and $(1, 7)$

16. $\left(\frac{1}{2}, -2\right)$ and $\left(-3, \frac{1}{2}\right)$

In exercises 17 to 20, you are given a point on a parabola and the axis of symmetry. Locate the point symmetric to the given one.

17. $(6, 5)$; axis of symmetry: $x = 4$

18. $(-5, 2)$; axis of symmetry: $x = -2$

19. $(-1, -7)$; axis of symmetry: $x = 3$

20. $(4, -2)$; axis of symmetry: $x = -1$

In exercises 21 to 24, find the equation of the axis of symmetry, given two points on a parabola.

21. $(-1, 5)$ and $(5, 5)$

22. $(-3, 0)$ and $(6, 0)$

23. $(-6, 0)$ and $(-1, 0)$

24. $(-4, -6)$ and $(10, -6)$

< Objectives 1–2 >

In exercises 25 to 36, find the equation of the axis of symmetry, the coordinates of the vertex, and the x-intercepts. Sketch the graph of each function.

25. $f(x) = -x^2 + 4$ 

26. $f(x) = x^2 + 2x$

27. $f(x) = -x^2 - 2x$

28. $f(x) = -x^2 - 3x$

29. $f(x) = x^2 - 6x + 5$

30. $f(x) = x^2 + x - 6$

31. $f(x) = x^2 - 5x + 6$ > Videos

32. $f(x) = x^2 + 6x + 5$

33. $f(x) = x^2 - 6x + 8$

34. $f(x) = -x^2 - 3x + 4$

35. $f(x) = -x^2 - 6x - 5$

36. $f(x) = -x^2 + 6x - 8$

In exercises 37 to 48, find the equation of the axis of symmetry, the coordinates of the vertex, and at least two symmetric points. Sketch the graph of each function. (Note: *A sample answer is provided for the two symmetric points.*)

37. $f(x) = x^2 - 2x - 1$

38. $f(x) = x^2 + 4x + 6$

39. $f(x) = x^2 - 4x - 1$

40. $f(x) = -x^2 + 6x - 5$

Answers

29. ______
30. ______
31. ______
32. ______
33. ______
34. ______
35. ______
36. ______
37. ______
38. ______
39. ______
40. ______

Answers

41. ____________

42. ____________

43. ____________

44. ____________

45. ____________

46. ____________

47. ____________

48. ____________

49. ____________

50. ____________

51. ____________

52. ____________

53. ____________

54. ____________

41. $f(x) = -x^2 + 3x - 3$

42. $f(x) = x^2 + 5x + 3$

43. $f(x) = 2x^2 + 4x - 1$ 

44. $f(x) = \frac{1}{2}x^2 - x - 1$

45. $f(x) = -\frac{1}{3}x^2 + x - 3$

46. $f(x) = -2x^2 - 4x - 1$

47. $f(x) = 3x^2 + 12x + 5$

48. $f(x) = -3x^2 + 6x + 1$

Basic Skills | **Challenge Yourself** | Calculator/Computer | Career Applications | Above and Beyond

Complete each statement with **never, sometimes,** *or* **always.**

49. The vertex of a parabola is ______ located on its axis of symmetry.

50. The vertex of a parabola is ______ the highest point on the graph.

51. The graph of $y = ax^2 + bx + c$ ______ intersects the x-axis.

52. The graph of $y = ax^2 + bx + c$ ______ has more than two x-intercepts.

53. The graph of $y = ax^2 + bx + c$ ______ intersects the y-axis.

54. The graph of $y = ax^2 + bx + c$ ______ intersects the y-axis more than once.

Answers

1. (f) **3.** (g) **5.** (h) **7.** (c) **9.** (a), (c) **11.** (a), (c)

13. (b), (c) **15.** $\left(-\frac{1}{2}, \frac{13}{2}\right)$ **17.** (2, 5) **19.** (7, −7) **21.** $x = 2$

23. $x = -\frac{7}{2}$

25. $x = 0$; vertex (0, 4); (−2, 0) and (2, 0)

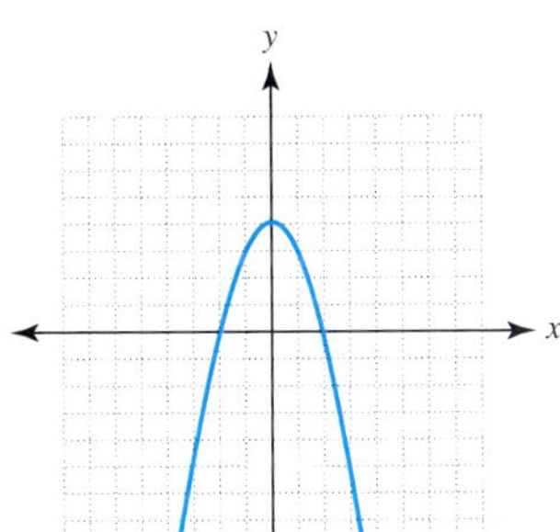

27. $x = -1$; vertex (−1, 1); (−2, 0) and (0, 0)

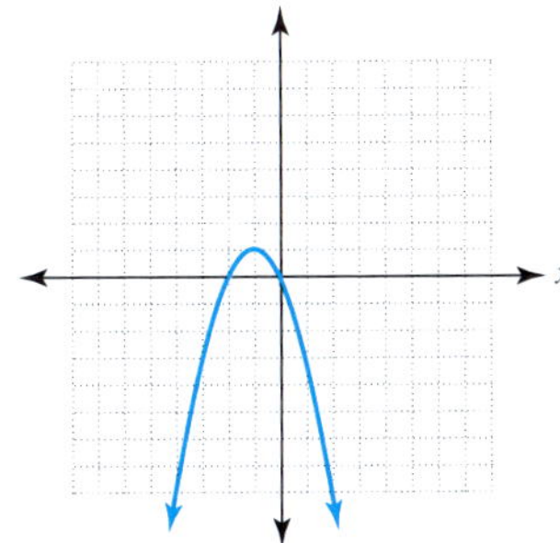

29. $x = 3$; vertex (3, −4); (1, 0) and (5, 0)

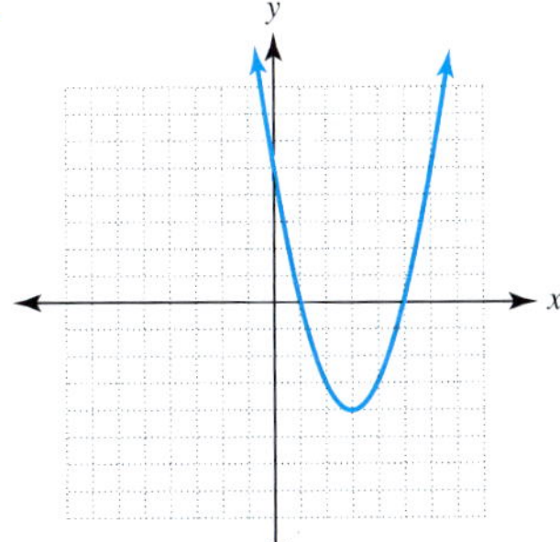

31. $x = \frac{5}{2}$; vertex $\left(\frac{5}{2}, -\frac{1}{4}\right)$; (3, 0) and (2, 0)

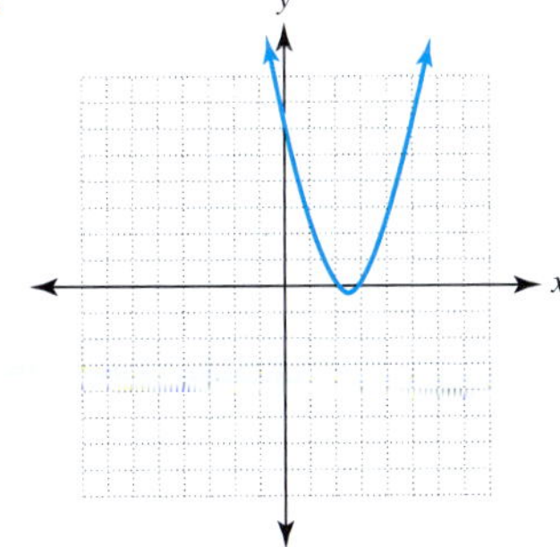

33. 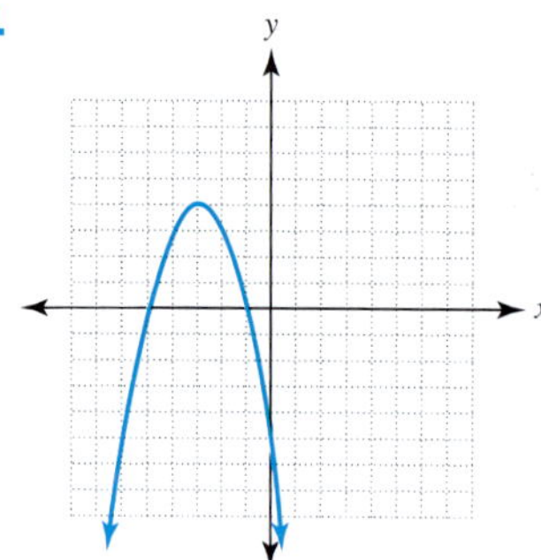

$x = 3$; vertex $(3, -1)$; $(2, 0)$ and $(4, 0)$

35. 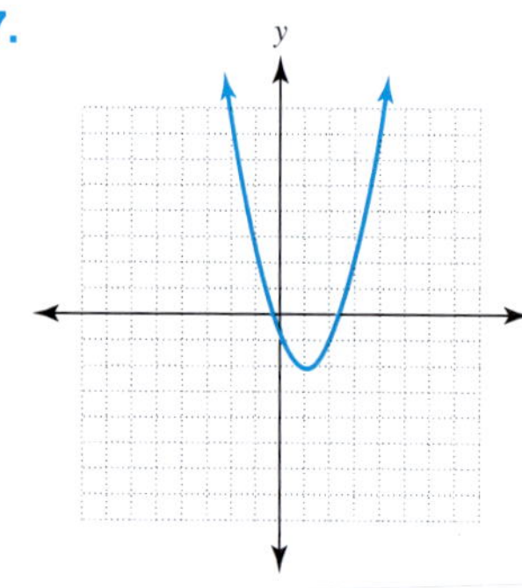

$x = -3$; vertex $(-3, 4)$; $(-5, 0)$ and $(-1, 0)$

37. 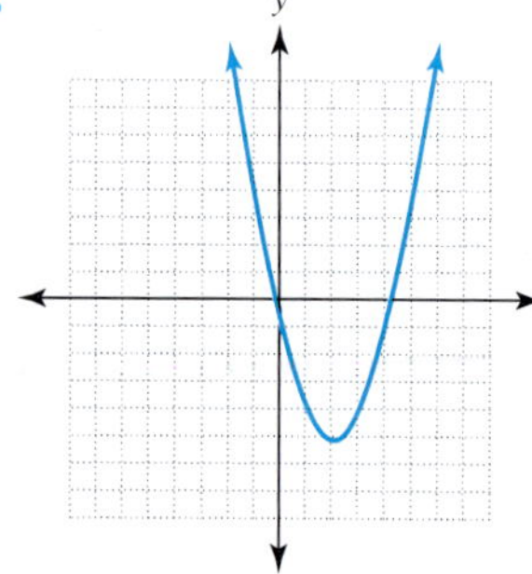

$x = 1$; vertex $(1, -2)$; $(0, -1)$ and $(2, -1)$

39. 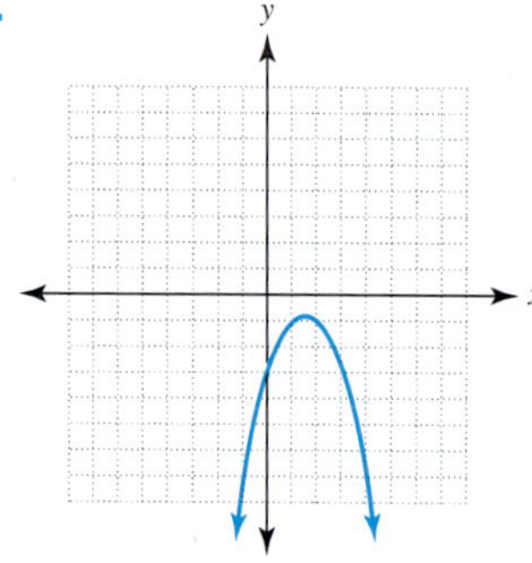

$x = 2$; vertex $(2, -5)$; $(0, -1)$ and $(4, -1)$

41.

$x = \frac{3}{2}$; vertex $\left(\frac{3}{2}, -\frac{3}{4}\right)$; $(0, -3)$ and $(3, -3)$

43.

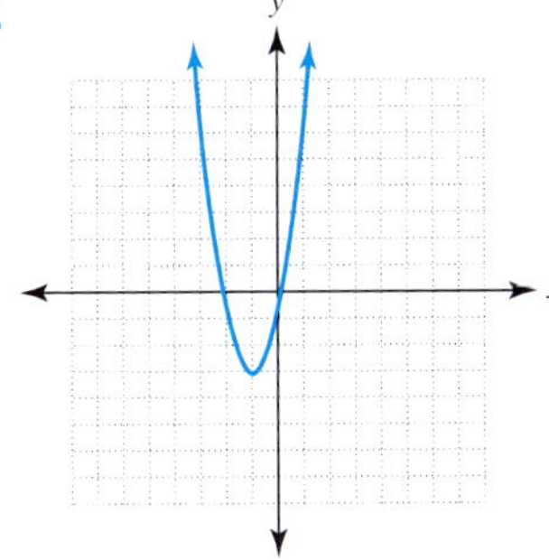

$x = -1$; vertex $(-1, -3)$; $(0, -1)$ and $(-2, -1)$

45.

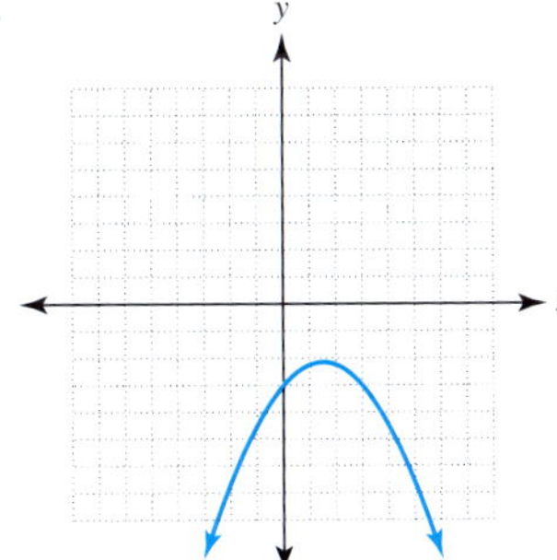

$x = \frac{3}{2}$; vertex $\left(\frac{3}{2}, -\frac{9}{4}\right)$; $(0, -3)$ and $(3, -3)$

47.

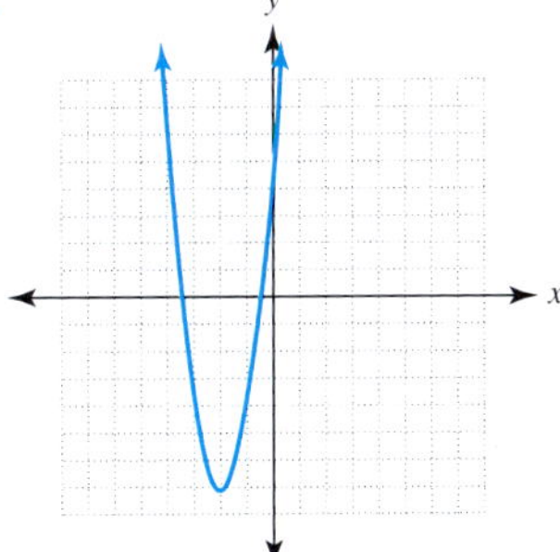

$x = -2$; vertex $(-2, -7)$; $(0, 5)$ and $(-4, 5)$

49. always **51.** sometimes **53.** always

8.4 Problem Solving with Quadratics

< 8.4 Objectives >

1 > Solve a quadratic equation by graphing

2 > Solve an application involving a quadratic equation

3 > Solve an equation that is quadratic in form

We have seen that quadratic equations can be solved in three different ways: by factoring (Section 6.6), by completing the square (Section 8.1), or by using the quadratic formula (Section 8.2). Having studied the graphs of quadratic functions (Section 8.3), we now look at a fourth technique for solving quadratic equations, a graphical method. Unlike the other methods, the graphical technique may yield only an approximation of the solution(s). This, however, may be perfectly adequate in applications; in such situations, we are generally more interested in the decimal form of a number than in the "exact radical" form. And, using current technology, we are able to approximate solutions with great precision.

To graphically solve the equation

$$ax^2 + bx + c = 0$$

we define functions f and g as

$$f(x) = ax^2 + bx + c$$
$$g(x) = 0$$

and we ask, for what values of x do the two graphs intersect? Now, the graph of f is a parabola, and the graph of g is simply the x-axis. So solutions for the original equation are simply the x-values where the parabola intersects the x-axis. We need only look at the x-intercepts!

Example 1 Solving a Quadratic Equation Graphically

< Objective 1 >

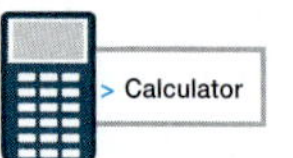

Use a graphing calculator to solve the equation. Give solutions to the nearest thousandth.

$$0.4x^2 - x - 2.5 = 0$$

In the calculator, we define $Y_1 = 0.4x^2 - x - 2.5$, and we view the graph in the standard viewing window:

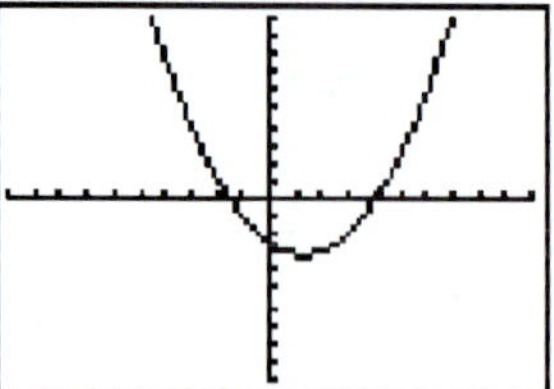

We are interested in the x-intercepts. Your calculator has a "ZERO" or "ROOT" utility that allows you to locate the x-intercepts. Using this, we find

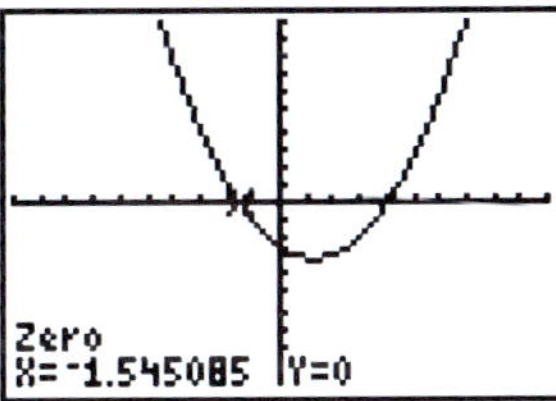

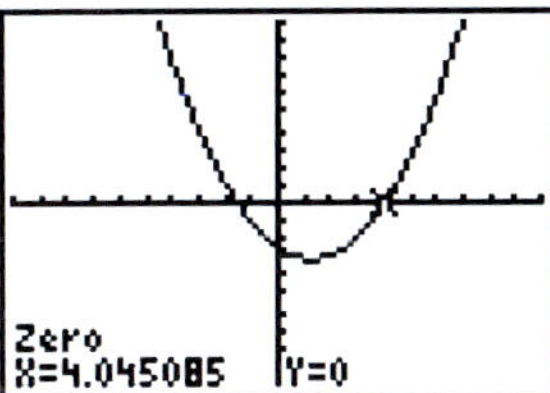

So, to the nearest thousandth, the solution set is $\{-1.545, 4.045\}$.

Check Yourself 1

Use a graphing calculator to solve the equation. Give solutions accurate to the nearest thousandth.

$-0.3x^2 - 0.4x + 2.75 = 0$

We often need to apply our knowledge of parabolas when solving graphically on a calculator. Consider the next example.

Example 2 Solving a Quadratic Equation Graphically

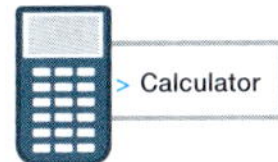

Use a graphing calculator to solve the equation. Give solutions accurate to the nearest thousandth.

$-x^2 + 16x + 160 = 0$

When we define $Y_1 = -x^2 + 16x + 160$ and view the graph in the standard viewing window, we see

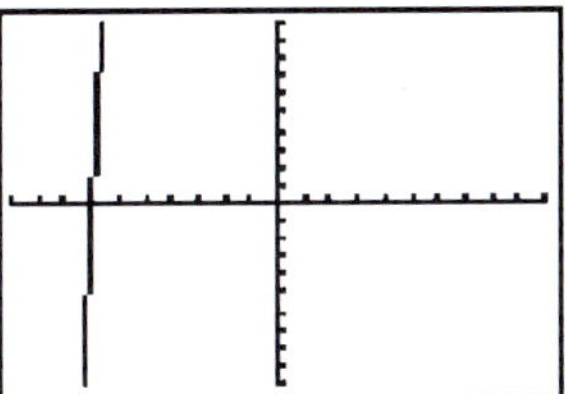

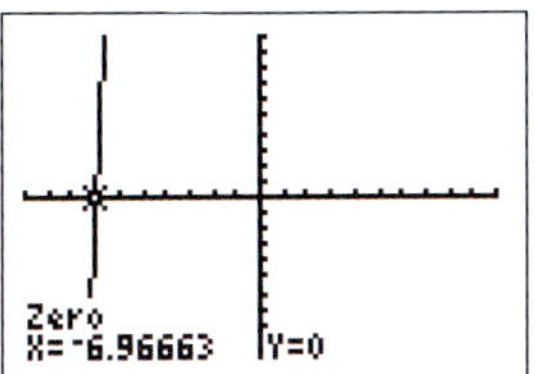

We expect to see a parabola, and we expect that it opens down. (Do you see why?) Therefore, we must be looking at the left portion of the parabola. Further, we know that the graph climbs for a while, turns, and comes back down to cross the x-axis somewhere to the right. While we have this view, we find the x-intercept (to the nearest thousandth) to be $(-6.967, 0)$.

Using the TABLE utility, we try values to the right of 10 because the parabola must intersect the x-axis somewhere to the right.

X	Y1
10	220
20	80
30	-260

X=

Notice:

when $x = 20$, the y-value is 80;

when $x = 30$, the y-value is -260.

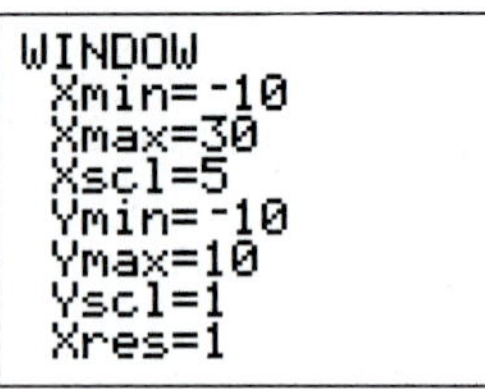

We conclude that the graph must cross the x-axis somewhere between $x = 20$ and $x = 30$. If we simply change the window to be $-10 \le x \le 30$, with x-scale of 5, we see

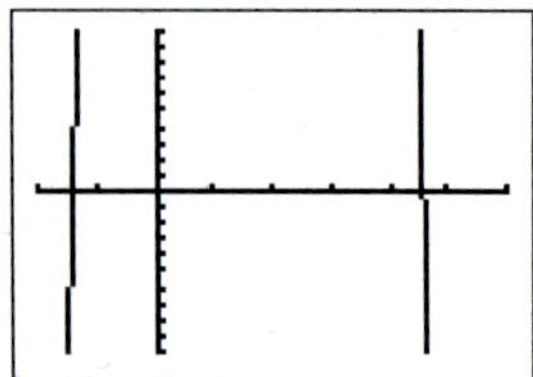

Using the ZERO utility, we find the right-hand x-intercept to be (22.967, 0). To the nearest thousandth, the solution set is $\{-6.967, 22.967\}$.

Check Yourself 2

Use a graphing calculator to solve the equation. Give solutions to the nearest thousandth.

$x^2 + 24x - 265 = 0$

From graphs of equations of the form $y = ax^2 + bx + c$, we know that if $a > 0$, then the vertex is the lowest point on the graph (the minimum value). Also, if $a < 0$, then the vertex is the highest point on the graph (the maximum value). We use this to solve a variety of problems in which we want to find the maximum or minimum value of a variable. Here are just two of many typical examples.

Example 3 An Application Involving a Quadratic Function

< Objective 2 >

A software company sells a word-processing program for personal computers. It has found that the monthly profit P, in dollars, from selling x copies of the program is approximated by

$$P = -0.3x^2 + 90x - 1{,}500$$

We have a quadratic function where x is the input variable and P is the output variable.

$$P(x) = -0.3x^2 + 90x - 1{,}500$$

Find the number of copies of the program that should be sold in order to maximize the profit, and find the maximum profit.

Since the profit function is quadratic, the graph must be a parabola. Also, since the coefficient of x^2 is negative, the parabola must open downward, and thus the vertex will give the maximum value for the profit P. To find the vertex,

$$x = \frac{-b}{2a} = \frac{-(90)}{2(-0.3)} = \frac{-90}{-0.6} = 150$$

The maximum profit must occur when $x = 150$, so we substitute that value into the original equation.

$$\begin{aligned} P(150) &= -0.3(150)^2 + (90)(150) - 1{,}500 \\ &= \$5{,}250 \end{aligned}$$

The maximum profit occurs when 150 copies are sold in a month. That profit would be $5,250.

NOTE

View a graph of $P(x)$ using this window: $0 \le x \le 300$ and $-1{,}500 \le y \le 6{,}000$. Then use the MAXIMUM utility in your calculator to find the vertex.

Check Yourself 3

A company that sells portable radios finds that its weekly profit P, in dollars, and the number of radios sold x are related by

$$P(x) = -0.2x^2 + 40x - 100$$

Find the number of radios that should be sold to have the largest weekly profit and find the amount of that profit.

Example 4 An Application Involving a Quadratic Function

A farmer has 3,600 ft of fence to enclose a rectangular area of a lot. Find the largest possible area that can be enclosed.

As usual, when dealing with geometric figures, we start by drawing a sketch of the problem.

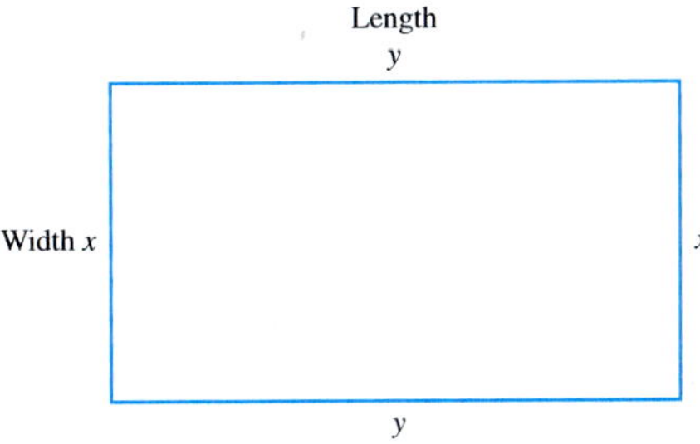

RECALL

Area = length × width

The perimeter of the region is $2x + 2y$

First, we can write the area A as

$$A = xy$$

Since 3,600 ft of fence is to be used, we know that

$$2x + 2y = 3{,}600$$
$$2y = 3{,}600 - 2x$$
$$y = 1{,}800 - x$$

Substituting for y in the area formula, we have

$$A = xy$$
$$A = x(1{,}800 - x)$$
$$A = 1{,}800x - x^2$$
$$A = -x^2 + 1{,}800x$$

Here we have a quadratic function where x is the input variable and A is the output variable.

$$A(x) = -x^2 + 1{,}800x$$

Again, the graph of a is a parabola opening downward, and the largest possible area occurs at the vertex. As before, to find the vertex

$$x = \frac{-1{,}800}{2(-1)} = \frac{-1{,}800}{-2} = 900$$

and the largest possible area is

$$A(900) = -(900)^2 + 1{,}800(900) = 810{,}000 \text{ ft}^2$$

NOTE

The width x is 900 ft, so we have

$$y = 1{,}800 - 900$$
$$= 900 \text{ ft}$$

Therefore, the length is also 900 ft. The desired region is a square.

Check Yourself 4

We want to enclose three sides of the largest possible rectangular area by using 900 ft of fence. Assume that an existing wall makes the fourth side. What are the dimensions of the rectangle?

We turn our attention now to solving equations that are not quadratic, but are **quadratic in form.**

Recall that in Section 6.3 we factored expressions that are quadratic in form. Now suppose that we need to solve the equation

$$x^4 - 5x^2 + 6 = 0$$

NOTE

This is an example of a fourth-degree polynomial equation. You will study these in a later math course. Expect to find four solutions.

The key to observing a "quadratic in form" situation is to note that x^4 is the square of x^2.

We will use the technique of substitution. We choose another variable, say u, and let $u = x^2$. This implies that $u^2 = x^4$. Changing to an equation that involves u gives

$$u^2 - 5u + 6 = 0$$

This certainly looks like a quadratic equation! Consider Example 5.

Example 5 Solving an Equation That Is Quadratic in Form

< Objective 3 >

Solve

$$x^4 - 5x^2 + 6 = 0$$

As explained above, we let $u = x^2$, which implies that $u^2 = x^4$. Rewriting the equation in terms of u gives

$$u^2 - 5u + 6 = 0$$

We can solve for u using factoring.

$$(u - 2)(u - 3) = 0$$

So $u = 2$ or $u = 3$.

But we want to solve for x (in our original equation), so we now return to equations involving x.

NOTE

This is often called "back-substituting."

$x^2 = 2$ or $x^2 = 3$ Remember that $u = x^2$.

Solving these by the square-root method, we have

$$x = \pm\sqrt{2} \quad \text{or} \quad x = \pm\sqrt{3}$$

Each of these four values can be checked in the original equation. They all work! The solution set is $\{\pm\sqrt{2}, \pm\sqrt{3}\}$.

Check Yourself 5

Solve $x^4 - 6x^2 + 8 = 0$.

It is quite possible that such an equation has nonreal solutions as well as real-number solutions.

Example 6 Solving an Equation That Is Quadratic in Form (ℂ)

Solve $x^4 - 4x^2 - 12 = 0$.

This equation is quadratic in form (x^4 is the square of x^2).

Let $u = x^2$. Then $u^2 = x^4$, and

$$u^2 - 4u - 12 = 0$$
$$(u - 6)(u + 2) = 0$$
$$u = 6 \quad \text{or} \quad u = -2$$

Back-substituting,

$$x^2 = 6 \quad \text{or} \quad x^2 = -2$$

Solving these equations yields two real solutions and two nonreal solutions

$$x = \pm\sqrt{6} \quad \text{or} \quad x = \pm\sqrt{-2} = \pm i\sqrt{2}$$

The solution set is $\{\sqrt{6}, -\sqrt{6}, i\sqrt{2}, -i\sqrt{2}\}$.

NOTE

Again we find four solutions for a fourth-degree polynomial equation.

Check Yourself 6 (ℂ)

Solve $x^4 + 5x^2 - 14 = 0$.

In our next example, we show a radical equation that could be solved by the methods studied in Section 7.4. However, we can also solve it using a "quadratic in form" approach.

Example 7 Solving an Equation That Is Quadratic in Form

Solve $x - 3\sqrt{x} - 10 = 0$.

Note that in $x - 3\sqrt{x} - 10$, x is the square of $\sqrt{x}$.

So, we let $u = \sqrt{x}$, which means $u^2 = x$.

$$u^2 - 3u - 10 = 0$$
$$(u - 5)(u + 2) = 0$$
$$u = 5 \quad \text{or} \quad u = -2$$

Back-substituting,

$$\sqrt{x} = 5 \quad \text{or} \quad \sqrt{x} = -2$$

Noting that $\sqrt{x}$ cannot be negative, we only need to solve the first of these.

If $\sqrt{x} = 5$, then $x = 25$.

Checking this value in the original equation, we see

$$(25) - 3\sqrt{(25)} - 10 \stackrel{?}{=} 0$$
$$25 - 3(5) - 10 \stackrel{?}{=} 0$$
$$25 - 15 - 10 = 0 \quad \text{True}$$

The solution set is $\{25\}$.

If we had not noticed that the statement $\sqrt{x} = -2$ has no solution, we would have proceeded as shown.

$$\sqrt{x} = -2$$

$$(\sqrt{x})^2 = (-2)^2 \quad \text{We square both sides.}$$

$$x = 4$$

Checking this value in the original equation, we see

$$(4) - 3\sqrt{(4)} - 10 \stackrel{?}{=} 0$$

$$4 - 3(2) - 10 \stackrel{?}{=} 0$$

$$4 - 6 - 10 \stackrel{?}{=} 0$$

$$-12 = 0 \quad \text{False}$$

This means that $x = 4$ is an extraneous solution, and we conclude that the solution set is $\{25\}$.

Check Yourself 7

Solve $x - 11\sqrt{x} + 24 = 0$.

If an equation is quadratic in form, but is not factorable, we use the quadratic formula to find solutions. This is likely to yield some "messy" solutions in radical form, so we emphasize decimal approximations here. A nice method for checking such messy solutions employs the graphing calculator.

Example 8 Solving an Equation That Is Quadratic in Form

Solve $2x^4 - 3x^2 - 8 = 0$, finding only real number solutions to the nearest thousandth.
We let $u = x^2$, so that $u^2 = x^4$.

$$2u^2 - 3u - 8 = 0$$

The quadratic expression on the left does not factor, so

$$u = \frac{-(-3) \pm \sqrt{(-3)^2 - 4(2)(-8)}}{2(2)} = \frac{3 \pm \sqrt{9 + 64}}{4} = \frac{3 \pm \sqrt{73}}{4}$$

$$u = \frac{3 + \sqrt{73}}{4} \quad \text{or} \quad u = \frac{3 - \sqrt{73}}{4}$$

Back-substituting,

$$x^2 = \frac{3 + \sqrt{73}}{4} \quad \text{or} \quad x^2 = \frac{3 - \sqrt{73}}{4}$$

With your calculator, note that $\frac{3 - \sqrt{73}}{4}$ is negative, so we only obtain real-number solutions from the first equation, $x^2 = \frac{3 + \sqrt{73}}{4}$.

```
√((3+√(73))/4)
        1.698823398
```

Thus, $x = \pm\sqrt{\frac{3 + \sqrt{73}}{4}}$. This is what we mean by messy!

Using your calculator, you should find that, to the nearest thousandth, $x = \pm 1.699$.

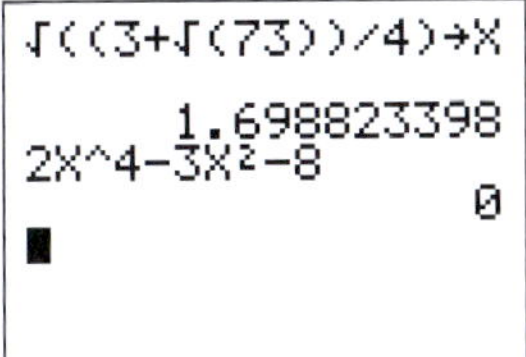

A nice way to check these values using your graphing calculator involves the use of the "store" key STO➡.

First, compute the value of $\sqrt{\dfrac{3+\sqrt{73}}{4}}$, and store it in a memory location of your choosing, say X. Then type the expression $2X^4 - 3X^2 - 8$ and the result should be 0 (or very nearly 0). Check the value $-\sqrt{\dfrac{3+\sqrt{73}}{4}}$ in the same manner.

Check Yourself 8

Solve $3x^4 - 5x^2 - 6 = 0$, finding only real number solutions accurate to the nearest thousandth.

When using your calculator to check in this manner, be aware that the calculator's result may not appear to be exactly what you expect. For example, consider this screenshot.

```
√((5+√(97))/6)→X
          1.573152769
3X^4-5X²-6
               -1E-12
```

The expected result was 0, but we see $-1E - 12$. This is calculator notation for -1×10^{-12}, which, as a decimal, is -0.000000000001, which is extremely close to 0. Remember that an irrational number like $\sqrt{97}$ has a decimal representation that never ends and never repeats. A calculator can carry only a finite number of decimal places (such as 16). Therefore, the value for $\sqrt{97}$ in a calculator is approximate, not exact. When it is then used in further calculations, **approximation errors** occur. To use a calculator wisely, we must recognize these situations, and realize that, in the above example, the value entered for X does check!

Graphing Calculator Option

Applying Quadratic Regression

Suppose we collect some data in the form of ordered pairs, and, when plotted, the resulting scatterplot indicates that a parabola might fit the data pretty well. You can use the quadratic regression utility in your graphing calculator to find the equation for such a quadratic function.

Consider how the number of Blackberry subscribers has increased through several years.

Fiscal year	2000	2001	2002	2003	2004	2005
Number of subscribers (in thousands)	25	165	321	534	1,070	2,510

```
L1     L2     L3     1
0      25     ------
1      165
2      321
3      534
4      1070
5      2510
       ------
L1(7)=
```

We clear data lists [L1] and [L2]: STAT 4:ClrList 2nd [L1] , 2nd [L2] ENTER.

Then we enter the data into [L1] and [L2]: STAT 1:Edit, and type in the numbers.

Now exit the data editor: 2nd QUIT.

```
Plot1 Plot2 Plot3
On Off
Type:
Xlist:L1
Ylist:L2
Mark: ▪ + ·
```

To make and view a scatterplot: 2nd [STAT PLOT] ENTER; press "On"; for "Type" select the first icon; "Xlist" should say [L1] and "Ylist" should say [L2]; for "Mark"

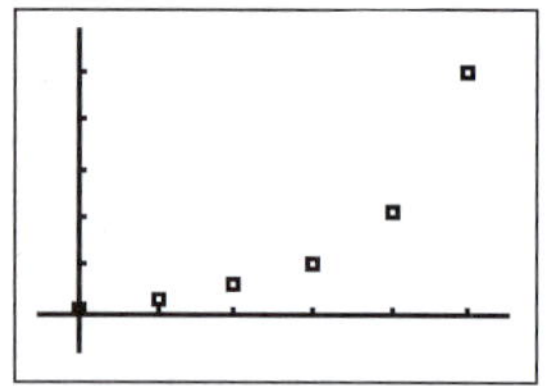

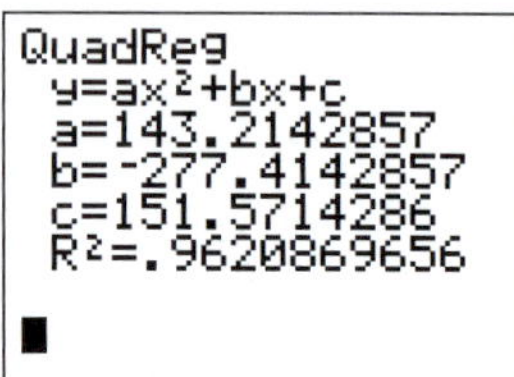

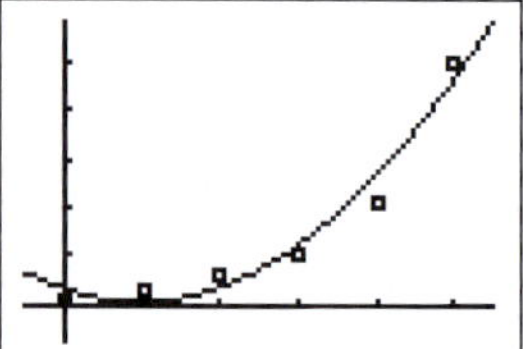

choose the first symbol; press [Y=] and delete (or turn off) any existing equations; press [ZOOM] 9:ZoomStat. (To improve the scaling, go to [WINDOW] and choose appropriate numbers for Xscl and Yscl. Then [GRAPH].)

To find the "best fitting" quadratic function: [STAT] CALC 5:QuadReg [2nd] [L1] [,] [2nd] [L2] [ENTER]. To four decimal places, we have

$$y = 143.2143x^2 - 277.4143x + 151.5714$$

To view the graph of this function on the scatterplot, enter its equation on the [Y=] screen and press [GRAPH].

Of course, this function may be used to predict the number of subscribers in the year 2006 (which you could check!). We must warn, however, that it is risky to predict beyond the scope of the data. You may wish to review the discussion of **extrapolation** in the Graphing Calculator Option in Chapter 3.

Graphing Calculator Check

The table below shows the number of retail prescription drug sales (in millions) in the United States for several years. Using your graphing calculator (let 0 represent 1997), apply quadratic regression to fit a quadratic function to these data. Round coefficients to four decimal place precision.

Year	1997	1998	1999	2000	2001	2002	2003	2004
Number of prescriptions (in millions)	2,316	2,481	2,707	2,865	3,009	3,139	3,215	3,274

ANSWER

$y = -12.3929x^2 + 227.4167x + 2296.6667$

Check Yourself ANSWERS

1. $\{-3.767, 2.434\}$ **2.** $\{-32.224, 8.224\}$ **3.** 100 radios, $1,900
4. Width 225 ft, length 450 ft **5.** $\{-\sqrt{2}, \sqrt{2}, -2, 2\}$
6. $\{-\sqrt{2}, \sqrt{2}, -i\sqrt{7}, i\sqrt{7}\}$ two nonreal solutions
7. $\{9, 64\}$ **8.** $\{-1.573, 1.573\}$

Reading Your Text

SECTION 8.4

(a) The graphical technique for solving equations may yield only ____________ solutions.

(b) From graphs of equations of the form $y = ax^2 + bx + c$, we know that if $a > 0$, then the vertex is the __________ point on the graph.

(c) With a fourth-degree polynomial equation, expect to find ___________ solutions.

(d) If an equation is quadratic in form, but is not factorable, we can use the ____________.

Basic Skills | Challenge Yourself | Calculator/Computer | Career Applications | Above and Beyond

8.4 exercises

Boost *your* GRADE at ALEKS.com!

ALEKS

- Practice Problems
- Self-Tests
- NetTutor
- e-Professors
- Videos

Name ______

Section ______ Date ______

< Objective 2 >

Solve each application.

1. **Business and Finance** A company's weekly profit P is related to the number of items sold by $P(x) = -0.3x^2 + 60x - 400$. Find the number of items that should be sold each week in order to maximize the profit. Then find the amount of that weekly profit.

2. **Business and Finance** A company's monthly profit P is related to the number of items sold by $P(x) = -0.2x^2 + 50x - 800$. How many items should be sold each month to obtain the largest possible profit? What is the amount of that profit?

3. **Construction** A builder wants to enclose the largest possible rectangular area with 2,000 ft of fencing. What should be the dimensions of the rectangle, and what is the area of that rectangle?

4. **Construction** A farmer wants to enclose a rectangular area along a river on three sides. If 1,600 ft of fencing is to be used, what dimensions give the maximum enclosed area? Find that maximum area.

5. **Science and Medicine** A ball is thrown upward into the air with an initial velocity of 96 ft/s. If h gives the height of the ball at time t, then the equation relating h and t is

$$h = -16t^2 + 96t$$

Find the maximum height the ball will attain.

6. **Science and Medicine** A ball is thrown upward into the air with an initial velocity of 64 ft/s. If h gives the height of the ball at time t, then the equation relating h and t is

$$h = -16t^2 + 64t$$

Find the maximum height the ball will attain.

Basic Skills | Challenge Yourself | Calculator/Computer | Career Applications | Above and Beyond

<Objective 3>

Solve. Express solutions in simplified form.

7. $x^4 - 14x^2 + 45 = 0$

8. $x^4 - 18x^2 + 32 = 0$

9. $6x^4 - 7x^2 + 2 = 0$

10. $12x^4 - 7x^2 + 1 = 0$

11. $x^4 + x^2 - 20 = 0$ (C)

12. $x^4 - 6x^2 - 27 = 0$ (C)

Answers

1. ______
2. ______
3. ______
4. ______
5. ______
6. ______
7. ______
8. ______
9. ______
10. ______
11. ______
12. ______

Answers

13. ______
14. ______
15. ______
16. ______
17. ______
18. ______
19. ______
20. ______
21. ______
22. ______
23. ______
24. ______
25. ______
26. ______
27. ______
28. ______
29. ______
30. ______
31. ______
32. ______
33. ______
34. ______
35. ______
36. ______
37. ______

13. $x^4 + 11x^2 + 28 = 0$ (ℂ)

14. $x^4 + 11x^2 + 18 = 0$ (ℂ)

15. $x - 8\sqrt{x} + 15 = 0$

16. $x - 10\sqrt{x} + 24 = 0$

17. $x - 4\sqrt{x} - 21 = 0$

18. $x - 6\sqrt{x} - 16 = 0$

Find real number solutions. Round your results to the nearest thousandth.

19. $x^4 - 7x^2 - 4 = 0$

20. $x^4 - 5x^2 + 3 = 0$

21. $2x^4 - 7x^2 + 4 = 0$

22. $2x^4 - 9x^2 - 3 = 0$

Basic Skills | Challenge Yourself | **Calculator/Computer** | Career Applications | Above and Beyond

< Objective 1 >

Use the graph of the related parabola to estimate the solutions to each equation. Round answers to the nearest thousandth.

23. $0 = x^2 + x - 12$

24. $0 = x^2 + 3x + 2$

25. $0 = 6x^2 - 19x$

26. $0 = 7x^2 - 15x$

27. $0 = 9x^2 + 12x - 7$

28. $0 = 3x^2 + 9x + 5$

29. $0 = x^2 + 2x - 7$

30. $0 = x^2 - 8x + 11$

For each quadratic function, use your graphing calculator to determine **(a)** *the vertex of the parabola and* **(b)** *the range of the function.*

31. $f(x) = 2(x - 3)^2 + 1$

32. $g(x) = 3(x + 4)^2 + 2$

33. $f(x) = -(x - 1)^2 + 2$

34. $g(x) = -(x + 2)^2 - 1$

35. $f(x) = 3(x + 1)^2 - 2$

36. $g(x) = -2(x - 4)^2$

Each table shows a relationship between speed (miles per hour) and gasoline consumption (miles per gallon, MPG) for a vehicle. In each case, use quadratic regression to find a quadratic function that best fits the data. Round coefficients to the nearest thousandth.

37. Oldsmobile

Speed	5	10	15	20	25	30	35
MPG	5.1	7.9	11.4	12.5	15.6	19.0	21.2

Speed	40	45	50	55	60	65	70	75
MPG	23.0	23.0	27.3	29.1	28.2	25.0	22.9	21.6

38. Chevrolet

Speed	5	10	15	20	25	30	35
MPG	7.9	18.0	16.3	19.9	22.7	26.3	24.3

Speed	40	45	50	55	60	65	70	75
MPG	26.7	27.3	26.3	25.1	22.6	21.8	20.1	18.1

39. Jeep

Speed	5	10	15	20	25	30	35
MPG	8.2	11.2	17.5	24.7	21.8	21.6	25.0

Speed	40	45	50	55	60	65	70	75
MPG	25.5	25.4	24.8	24.0	23.2	21.3	20.0	19.1

40. Honda

Speed	5	10	15	20	25	30	35
MPG	11.2	16.1	21.4	25.1	27.3	28.0	28.7

Speed	40	45	50	55	60	65	70	75
MPG	29.5	30.1	30.2	29.9	28.3	27.1	23.8	23.1

Basic Skills | Challenge Yourself | Calculator/Computer | **Career Applications** | Above and Beyond

ALLIED HEALTH *The number of people infected t days after the outbreak of a flu epidemic is modeled by the equation*

$P = -t^2 + 120t + 20$

Use this model to complete exercises 41 and 42.

41. How many days after the outbreak will the maximum number of people be sick?

42. What is the maximum number of people that will be infected at one time?

ALLIED HEALTH *A patient's body temperature (T°F) t hours after taking the analgesic acetaminophen can be approximated by the formula*

$T = 0.4t^2 - 2.6t + 103$

Use this model to complete exercises 43 and 44.

43. When will the patient's temperature reach its minimum?

Answers

38. ______

39. ______

40. ______

41. ______

42. ______

43. ______

Answers

44. ________

45. ________

46. ________

47. ________

48. ________

49. ________

44. What will the patient's minimum temperature be? (Round your answer to the nearest tenth.)

Basic Skills | Challenge Yourself | Calculator/Computer | Career Applications | **Above and Beyond**

Describe a viewing window that includes the vertex and all intercepts for the graph of each function.

45. $f(x) = 3x^2 - 25$

46. $f(x) = 9x^2 - 5x - 7$

47. $f(x) = -2x^2 + 5x - 7$

48. $f(x) = -5x^2 + 2x + 7$

49. Explain how to determine the domain and range of the function $f(x) = a(x - h)^2 + k$.

Answers

1. 100 items, \$2,600 **3.** 500 ft by 500 ft; 250,000 ft^2 **5.** 144 ft
7. $\{\pm 3, \pm\sqrt{5}\}$ **9.** $\left\{\pm\frac{\sqrt{2}}{2}, \pm\frac{\sqrt{6}}{3}\right\}$
11. $\{\pm 2, \pm i\sqrt{5}\}$ two nonreal solutions
13. $\{\pm 2i, \pm i\sqrt{7}\}$ four nonreal solutions **15.** $\{9, 25\}$ **17.** $\{49\}$
19. $\{\pm 2.744\}$ **21.** $\{\pm 1.668, \pm 0.848\}$ **23.** $\{-4, 3\}$ **25.** $\{0, 3.167\}$
27. $\{-1.772, 0.439\}$ **29.** $\{-3.828, 1.828\}$ **31.** **(a)** $(3, 1)$; **(b)** $y \geq 1$
33. **(a)** $(1, 2)$; **(b)** $y \leq 2$ **35.** **(a)** $(-1, -2)$; **(b)** $y \geq -2$
37. $y = -0.008x^2 + 0.946x - 1.174$ **39.** $y = -0.010x^2 + 0.899x + 5.424$
41. 60 days **43.** 3.25 h
45. $-3 \leq x \leq 3$; $-25 \leq y \leq 0$ (this is a sample answer—yours may be different)
47. $-2 \leq x \leq 4$; $-10 \leq y \leq 0$ (this is a sample answer—yours may be different)
49. Above and Beyond

Definition/Procedure	Example	Reference
Solving Quadratic Equations		Section 8.1
Square-Root Property If $x^2 = k$, when k is any real number, then $x = \sqrt{k}$ or $x = -\sqrt{k}$.	To solve: $(x - 3)^2 = 5$ $x - 3 = \pm\sqrt{5}$ $x = 3 \pm \sqrt{5}$	*p.* 809
Completing the Square **Step 1** Isolate the constant on the right side of the equation. **Step 2** Divide both sides of the equation by the coefficient of the x^2-term if that coefficient is not equal to 1. **Step 3** Add the square of one-half of the coefficient of the linear term to both sides of the equation. This gives a perfect-square trinomial on the left side of the equation. **Step 4** Write the left side of the equation as the square of a binomial, and simplify the right side. **Step 5** Use the square-root property, and then solve the resulting linear equations.	To solve: $x^2 + x = \frac{1}{2}$ $x^2 + x + \left(\frac{1}{2}\right)^2 = \frac{1}{2} + \left(\frac{1}{2}\right)^2$ $\left(x + \frac{1}{2}\right)^2 = \frac{3}{4}$ $x + \frac{1}{2} = \pm\sqrt{\frac{3}{4}}$ $x = \frac{-1 \pm \sqrt{3}}{2}$	*p.* 814
The Quadratic Formula		Section 8.2
Any quadratic equation can be solved by using this algorithm. **Step 1** Write the equation in standard form (set it equal to 0). $ax^2 + bx + c = 0$ **Step 2** Determine the values for a, b, and c. **Step 3** Substitute those values into the quadratic formula $x = \frac{-b \pm \sqrt{b^2 - 4ac}}{2a}$ **Step 4** Write the solutions in simplest form.	To solve: $x^2 - 2x = 4$ write the equation as $x^2 - 2x - 4 = 0$ $a = 1 \quad b = -2 \quad c = -4$ $x = \frac{-(-2) \pm \sqrt{(-2)^2 - 4(1)(-4)}}{2 \cdot (1)}$ $= \frac{2 \pm \sqrt{20}}{2}$ $= \frac{2 \pm 2\sqrt{5}}{2}$ $= 1 \pm \sqrt{5}$	*p.* 828
The Discriminant The expression $b^2 - 4ac$ is called the **discriminant** for a quadratic equation. There are three possibilities: **1.** If $b^2 - 4ac < 0$, there are no real solutions (but two imaginary solutions). **2.** If $b^2 - 4ac = 0$, there is one real solution (a double solution). **3.** If $b^2 - 4ac > 0$, there are two distinct real solutions.	Given $2x^2 - 5x + 3 = 0$ $a = 2 \quad b = -5 \quad c = 3$ $b^2 - 4ac = 25 - 4(2)(3)$ $= 25 - 24$ $= 1$ There are two distinct solutions.	*p.* 828

continued

Definition/Procedure	Example	Reference

An Introduction to Parabolas

Section 8.3

Axis of Symmetry The axis of symmetry is a vertical line midway between any pair of symmetric points on a parabola. The axis of symmetry passes through the vertex of the parabola.

p. 841

Vertex of a Parabola If

$$f(x) = ax^2 + bx + c \qquad a \neq 0$$

then the coordinates of the vertex of the graph of f are

$$\left(\frac{-b}{2a}, f\left(\frac{-b}{2a}\right)\right)$$

p. 843

To Graph a Parabola:

Step 1 Find the axis of symmetry.

Step 2 Find the vertex.

Step 3 Determine two symmetric points.
Note: Use the x-intercepts if the quadratic expression is factorable. If the vertex is not on the y-axis, you can use the y-intercept and its symmetric point. Another option is to simply choose an x-value that does not match the axis of symmetry, compute the corresponding y-value, and then locate its symmetric point.

Step 4 Draw a smooth curve connecting the points found in step 3 to form the parabola. You may choose to find additional pairs of symmetric points.

p. 847

Graph the function

$$f(x) = x^2 - 4x - 12$$

1. Find the axis of symmetry.

$$x = \frac{-b}{2a} = \frac{-(-4)}{2(1)}$$
$$= \frac{4}{2} = 2$$

so $x = 2$ is the axis of symmetry.

2. Find the vertex. Let $x = 2$ in the original equation.

$$f(2) = (2)^2 - 4(2) - 12$$
$$f(2) = 4 - 8 - 12$$
$$f(2) = -16$$

The vertex is $(2, -16)$.

3. Find two symmetric points.

$$0 = x^2 - 4x - 12$$
$$0 = (x - 6)(x + 2)$$
$$x - 6 = 0 \qquad x + 2 = 0$$
$$x = 6 \qquad x = -2$$

Two symmetric points are $(6, 0)$ and $(-2, 0)$.

4. Draw a smooth curve connecting the points found.

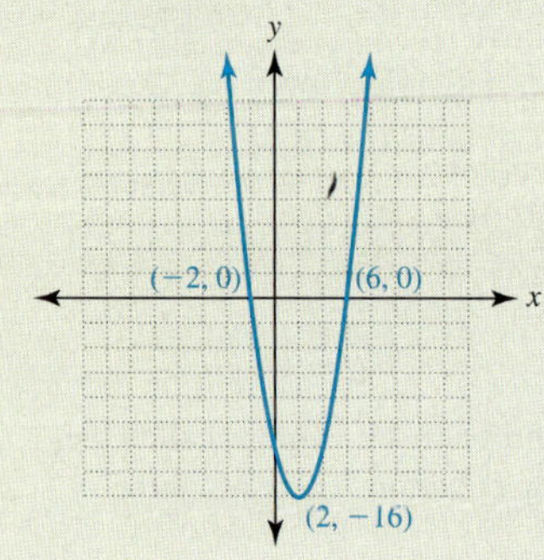

Definition/Procedure	Example	Reference
Problem Solving with Quadratics		**Section 8.4**
Solving Quadratic Equations Graphically		
To solve the equation $ax^2 + bx + c = 0$ **1.** Graph the function $Y = ax^2 + bx + c$ **2.** Use the ZERO or ROOT utility to determine the x-intercepts of the graph. These values are the solutions to the original equation.	Solve the equation graphically. $0.5x^2 + 3x - 2 = 0$ 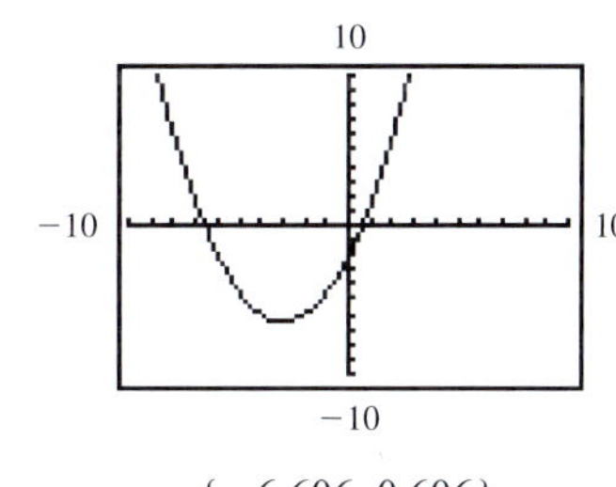 $\{-6.606, 0.606\}$	*p.* 856
Solving Equations Quadratic in Form Given an equation with a trinomial on one side and 0 on the other, if the variable part of the first term is the square of the variable part of the second term, then the trinomial is "quadratic in form," and may be solved by a substitution technique.	Solve: $x^4 - 14x^2 + 45 = 0$ Let $u = x^2$. Then $u^2 = x^4$. Rewrite in terms of u. $u^2 - 14u + 45 = 0$ Solve for u. $(u - 5)(u - 9) = 0$ $u = 5$ or $u = 9$ Back-substitute. $x^2 = 5$ or $x^2 = 9$ Solve for x. $x = \pm\sqrt{5}$ or $x = \pm 3$ Solution set: $\{\pm\sqrt{5}, \pm 3\}$.	*p.* 860

summary exercises :: chapter 8

This summary exercise set is provided to give you practice with each of the objectives of this chapter. Each exercise is keyed to the appropriate chapter section. When you are finished, you can check your answers to the odd-numbered exercises in the back of the text. If you have difficulty with any of these questions, go back and reread the examples from that section. The answers to the even-numbered exercises appear in the *Instructor's Solutions Manual.* Your instructor will give you guidelines on how best to use these exercises in your instructional setting.

8.1 *Use the square-root method to solve each equation.*

1. $x^2 - 12 = 0$

2. $3y^2 - 15 = 0$

3. $(x - 2)^2 = 20$

4. $(3x - 2)^2 - 15 = 0$

Find the constant that must be added to each binomial to form a perfect-square trinomial.

5. $x^2 - 14x$

6. $y^2 + 3y$

Solve each equation by completing the square.

7. $x^2 - 4x - 5 = 0$

8. $x^2 - 8x - 9 = 0$

9. $w^2 - 10w - 3 = 0$

10. $y^2 + 3y - 1 = 0$

11. $2x^2 - 6x + 1 = 0$

12. $3x^2 + 4x - 1 = 0$

8.2 *Solve each equation by using the quadratic formula.*

13. $x^2 - 5x - 24 = 0$

14. $w^2 + 10w + 25 = 0$

15. $x^2 = 5x - 2$

16. $2y^2 - 5y + 2 = 0$

17. $3y^2 + 4y = 1$

18. $3y^2 + 4y + 7 = 0$ $(\mathbb{C})$

19. $(x - 5)(x + 3) = 13$

20. $\frac{1}{x^2} - \frac{4}{x} + 1 = 0$

21. $3x^2 + 2x + 5 = 0$ $(\mathbb{C})$

22. $(x - 1)(2x + 3) = -5$ $(\mathbb{C})$

For each quadratic equation, use the discriminant to determine the number of real solutions.

23. $x^2 - 3x + 3 = 0$

24. $x^2 + 4x = 2$

25. $4x^2 - 12x + 9 = 0$

26. $2x^2 + 3 = 3x$

27. **Number Problem** The sum of two integers is 12, and their product is 32. Find the two integers.

28. **Number Problem** The product of two consecutive, positive, even integers is 80. What are the two integers?

29. **Number Problem** Twice the square of a positive integer is 10 more than 8 times that integer. Find the integer.

30. **GEOMETRY** The length of a rectangle is 2 ft more than its width. If the area of the rectangle is 80 ft^2, what are the dimensions of the rectangle?

31. **GEOMETRY** The length of a rectangle is 3 cm less than twice its width. The area of the rectangle is 35 cm^2. Find the length and width of the rectangle.

32. **GEOMETRY** An open box is formed by cutting 3-in. squares from each corner of a rectangular piece of cardboard that is 3 in. longer than it is wide. If the box is to have a volume of 120 $in.^3$, what must be the size of the original piece of cardboard?

33. **BUSINESS AND FINANCE** Suppose that a manufacturer's weekly profit P is given by

$$P = -3x^2 + 240x$$

where x is the number of patio chairs manufactured and sold. Find the number of patio chairs that must be manufactured and sold if the profit is to be at least $4,500.

34. **SCIENCE AND MEDICINE** If a ball is thrown vertically upward from the ground with an initial velocity of 64 ft/s, its approximate height is given by

$$h(t) = -16t^2 + 64t$$

When will the ball's height be at least 48 ft?

35. **GEOMETRY** The length of a rectangle is 1 cm more than twice its width. If the length is doubled, the area of the new rectangle is 36 cm^2 more than that of the old. Find the dimensions of the original rectangle.

36. **GEOMETRY** One leg of a right triangle is 4 in. longer than the other. The hypotenuse of the triangle is 8 in. longer than the shorter leg. What are the lengths of the three sides of the triangle?

37. **GEOMETRY** The diagonal of a rectangle is 9 ft longer than the width of the rectangle, and the length is 7 ft more than its width. Find the dimensions of the rectangle.

38. **SCIENCE AND MEDICINE** If a ball is thrown vertically upward from the ground, the height h after t seconds is given by

$$h = 128t - 16t^2$$

(a) How long does it take the ball to return to the ground?

(b) How long does it take the ball to reach a height of 240 ft on the way up?

39. **GEOMETRY** One leg of a right triangle is 2 m longer than the other. If the length of the hypotenuse is 8 m, find the length of the other two legs.

40. **SCIENCE AND MEDICINE** Suppose that the height (in meters) of a golf ball, hit off a raised tee, is approximated by

$$h(t) = -5t^2 + 10t + 10$$

t seconds after the ball is hit. When will the ball hit the ground?

Find the real zeros of each function.

41. $f(x) = x^2 - x - 2$

42. $f(x) = 6x^2 + 7x + 2$

43. $f(x) = -2x^2 - 7x - 6$

44. $f(x) = -x^2 - 1$

8.3 Find the equation of the axis of symmetry and the coordinates for the vertex of each quadratic function.

45. $f(x) = x^2$

46. $f(x) = x^2 + 2$

47. $f(x) = x^2 - 5$

48. $f(x) = (x - 3)^2$

49. $f(x) = (x + 2)^2$

50. $f(x) = -(x - 3)^2$

51. $f(x) = (x + 3)^2 + 1$

52. $f(x) = -(x + 2)^2 - 3$

53. $f(x) = -(x - 5)^2 - 2$

54. $f(x) = 2(x - 2)^2 - 5$

55. $f(x) = -x^2 + 2x$

56. $f(x) = x^2 - 4x + 3$

57. $f(x) = -x^2 - x + 6$

58. $f(x) = x^2 + 4x + 5$

59. $f(x) = -x^2 - 6x + 4$

8.4 *Use a graphing calculator to estimate the solutions to each equation. Round answers to the nearest thousandth.*

60. $0 = x^2 - 2x - 5$

61. $0 = 2x^2 + 5x - 9$

62. $0 = x^2 - 5x + 5$

63. $0 = x^2 - 7x + 5$

64. $0 = 2x^2 - 4x - 5$

Graph each function.

65. $f(x) = x^2$

66. $f(x) = x^2 + 2$

67. $f(x) = x^2 - 5$

68. $f(x) = (x - 3)^2$

69. $f(x) = (x + 2)^2$

70. $f(x) = -(x - 3)^2$

71. $f(x) = (x + 3)^2 + 1$

72. $f(x) = -(x + 2)^2 - 3$

73. $f(x) = x^2 - 4x$

74. $f(x) = -x^2 + 2x$

75. $f(x) = x^2 + 2x - 3$

76. $f(x) = x^2 - 4x + 3$

77. $f(x) = -x^2 - x + 6$

78. $f(x) = -x^2 + 3x + 4$

79. $f(x) = x^2 + 4x + 5$

80. $f(x) = x^2 - 6x + 4$

81. $f(x) = x^2 - 2x + 4$

82. $f(x) = -x^2 + 2x - 2$

83. $f(x) = 2x^2 - 4x + 1$

84. $f(x) = \frac{1}{2}x^2 - 4x$

Solve. Express solutions in simplified form.

85. $x^4 - 13x^2 + 40 = 0$

86. $x^4 - 13x^2 + 12 = 0$

87. $x^4 - 5x^2 - 36 = 0$ $(\mathbb{C})$

88. $x^4 + 2x^2 - 63 = 0$ $(\mathbb{C})$

89. $x - 7\sqrt{x} + 12 = 0$

90. $x - 6\sqrt{x} - 27 = 0$

Solve. Find real number solutions to the nearest thousandth.

91. $x^4 - 6x^2 - 10 = 0$

92. $x^4 - 9x^2 + 5 = 0$

self-test 8

CHAPTER 8

Name ______________

Section ________ Date ________

Answers

1. ______________
2. ______________
3. ______________
4. ______________
5. ______________
6. ______________
7. ______________
8. ______________
9. ______________
10. ______________
11. ______________
12. ______________
13. ______________
14. ______________
15. ______________
16. ______________
17. ______________
18. ______________
19. ______________

The purpose of this self-test is to help you assess your progress so that you can find concepts that you need to review before the next exam. Allow yourself about an hour to take this test. At the end of that hour, check your answers against those given in the back of this text. If you miss any, go back to the appropriate section to reread the examples until you have mastered that particular concept.

Find the equation of the axis of symmetry and the coordinates of the vertex of each equation.

1. $y = -3(x + 2)^2 + 1$

2. $y = x^2 - 4x - 5$

3. $y = -2x^2 + 6x - 3$

4. $y = (x - 3)^2 - 2$

5. $y = x^2 - 6x + 2$

Solve each equation by completing the square.

6. $m^2 + 3m - 1 = 0$

7. $2x^2 - 10x + 3 = 0$

8. Find the zeros of the function $f(x) = 3x^2 - 10x - 8$.

Use a graphing calculator to estimate the solutions to each equation. Round your answers to the nearest thousandth.

9. $0 = x^2 + 3x - 7$

10. $0 = 4x^2 + 2x - 5$

Graph each function.

11. $f(x) = (x - 5)^2$

12. $f(x) = (x + 2)^2 - 3$

13. $f(x) = -2(x - 3)^2 - 1$

14. $f(x) = 3x^2 + 9x + 2$

Solve each equation by factoring.

15. $2x^2 + 7x + 3 = 0$

16. $6x^2 = 10 - 11x$

17. $4x^3 - 9x = 0$

Solve.

18. The product of two consecutive, positive, odd integers is 63. Find the two integers.

19. Suppose that the height (in feet) of a ball thrown upward from a raised platform is approximated by

$h(t) = -16t^2 + 32t + 32$

t seconds after the ball is released. How long will it take the ball to hit the ground?

Use the quadratic formula to solve each equation.

20. $x^2 - 5x - 3 = 0$

21. $x^2 + 4x = 7$

22. $15x^2 = 2x + 8$

23. $2x^2 + 2x + 5 = 0$ (ℂ)

Solve. Express solutions in simplified form.

24. $x^4 - 15x^2 + 36 = 0$

25. $x^4 - 4x^2 - 32 = 0$ (ℂ)

26. $x - 11\sqrt{x} + 30 = 0$

Use the square-root method to solve each equation.

27. $4w^2 - 20 = 0$

28. $(x - 1)^2 = 10$

29. $4(x - 1)^2 = 23$

Solve. Find real-number solutions to the nearest thousandth.

30. $2x^4 - 7x^2 - 1 = 0$

Answers

20. ____________

21. ____________

22. ____________

23. ____________

24. ____________

25. ____________

26. ____________

27. ____________

28. ____________

29. ____________

30. ____________

cumulative review chapters 0-8

Name ____________

Section ________ Date ________

We offer the following exercises to help you review concepts from earlier chapters. This is meant as review material and not as a comprehensive exam. The answers are presented in the back of the text. If you have difficulty with any of these exercises, be certain to at least read through the summary related to that section.

Answers

1. ____________
2. ____________
3. ____________
4. ______ 5. ______
6. ____________
7. ____________
8. ____________
9. ____________
10. ____________
11. ______ 12. ______
13. ______ 14. ______
15. ______ 16. ______
17. ______ 18. ______
19. ______ 20. ______
21. ______ 22. ______
23. ____________
24. ____________
25. ____________

Graph each equation.

1. $2x - 3y = 6$

2. $y = -\frac{1}{3}x - 2$

3. $y = 4$

Find the slope of the line determined by each set of points.

4. $(-4, 7)$ and $(-3, 4)$

5. $(-2, 3)$ and $(-5, -1)$

6. Let $f(x) = 6x^2 - 5x + 1$. Evaluate $f(-2)$.

7. Simplify the function $f(x) = (x^2 - 1)(x + 3)$.

8. Completely factor the expression $x^3 + x^2 - 6x$.

9. Simplify the expression $\sqrt{\frac{2}{3}} + 7\sqrt{6}$.

10. Simplify the expression $\sqrt{72x^3y^5}$.

Solve each equation.

11. $2x - 7 = 0$

12. $3x - 5 = 5x + 3$

13. $0 = (x - 3)(x + 5)$

14. $x^2 - 3x + 2 = 0$

15. $x^2 + 7x - 30 = 0$

16. $x^2 - 3x - 3 = 0$

17. $(x - 3)^2 = 5$

18. $x^3 - 2x^2 = 15x$

19. $\frac{x}{3} - \frac{4}{9} = \frac{5}{18}$

20. $3 - \sqrt{2x + 2} = x$

Solve the inequality.

21. $x - 2 \leq 7$

22. Find the distance between $(-1, -4)$ and $(6, 1)$.

Solve each word problem. Show the equation used for the solution.

23. Five times a number decreased by 7 is -72. Find the number.

24. One leg of a right triangle is 4 ft longer than the shorter leg. If the hypotenuse is 28 ft, how long is each leg?

25. Suppose that a manufacturer's weekly profit P is given by

$P = -4x^2 + 320x$

where x is the number of receivers manufactured and sold. Find the number of receivers that must be manufactured and sold to guarantee a profit of \$4,956.

CHAPTER 9

Rational Expressions

INTRODUCTION

As a college student, you will find that most of what you are taught can be thought of as some combination of communicating and problem solving. Too often in mathematics, students learn to be problem solvers without learning how to communicate a solution. In the world of business and industry, the difficulty one encounters first is usually describing or understanding the problem that needs to be solved. Once the problem is understood and described, it is frequently easy to solve.

The activity in this chapter is designed to help you learn and practice the art of communicating mathematical information.

CHAPTER 9 OUTLINE

9.1 Simplifying Rational Expressions

< 9.1 Objectives >

1 > Evaluate rational expressions
2 > Avoid division by zero
3 > Simplify rational expressions
4 > Identify rational functions
5 > Write a rational function in simplified form

Our work in this chapter focuses on **rational expressions.** What is a rational expression? Roughly speaking, it is a fraction that may have variables. (We give a more precise definition below.) All of your experience working with fractions will help you deal with the rational expressions in this chapter.

Some examples of rational expressions are

$$\frac{8}{x} \qquad \frac{2x-3}{x+5} \qquad \frac{x+6}{x^2-2x-15}$$

Recall that a **rational number** is a number of the form $\frac{a}{b}$, where a and b are integers and b is not 0. Just as a rational number can be thought of as $\frac{\text{integer}}{\text{integer}}$, a rational expression can be thought of as $\frac{\text{polynomial}}{\text{polynomial}}$.

Definition

Rational Expression

A **rational expression** is an expression of the form $\frac{P}{Q}$, where P and Q are polynomials and Q cannot be 0.

We often need to find the value of a rational expression for a given value of the variable. Consider Example 1.

Example 1 Evaluating a Rational Expression

< Objective 1 >

Evaluate each expression for the given value of the variable.

(a) $\frac{3x}{2x-5}$ for $x = -3$

$$\frac{3(-3)}{2(-3)-5} = \frac{-9}{-6-5} = \frac{-9}{-11} = \frac{9}{11}$$

Substitute −3 for x.

(b) $\frac{2x+7}{x^2-x-6}$ for $x = -2$

$$\frac{2(-2)+7}{(-2)^2-(-2)-6} = \frac{-4+7}{4+2-6} = \frac{3}{0}$$

Undefined

This expression is undefined for $x = -2$.

Check Yourself 1

Evaluate each expression for the given value of the variable.

(a) $\dfrac{5x}{3x - 2}$ for $x = 4$ **(b)** $\dfrac{x^2 - 9}{x^2 - 1}$ for $x = -5$

In part (b) of Example 1, we saw that the given expression is undefined when $x = -2$. This is so because the denominator polynomial, $x^2 - x - 6$, has a value of 0 when $x = -2$. We cannot divide by 0.

You have probably noticed the emphasis placed on the idea that the denominator cannot be 0, whether we are speaking of rational numbers or of rational expressions. Undoubtedly you have met this idea many times.

To review why division by 0 is undefined, think of division using a "fits into" concept. For example,

$\dfrac{8}{2} = 4$ How many times does 2 "fit into" 8? 4 times.

$\dfrac{8}{1} = 8$ How many times does 1 "fit into" 8? 8 times.

$\dfrac{8}{\frac{1}{2}} = 16$ How many times does $\frac{1}{2}$ "fit into" 8? 16 times.

$\dfrac{8}{0.1} = 80$ How many times does 0.1 "fit into" 8? 80 times.

$\dfrac{8}{0.01} = 800$ How many times does 0.01 "fit into" 8? 800 times.

Note that as the denominator becomes smaller, approaching 0, the quotient gets larger. Ask yourself: How many times does 0 "fit into" 8? Your answer would have to be an *infinitely large* number! This is one reason why $\frac{8}{0}$ is undefined.

Because of this, when we work with rational expressions we must take care to avoid division by 0. We ask the question "For what values of the variable is the denominator polynomial equal to 0?" These are values that cause the value of the rational expression to be undefined.

Example 2 Avoiding Division by Zero

< Objective 2 >

> **NOTE**
>
> A fraction is undefined when its denominator is equal to 0.
>
> When $x = 5$, $\dfrac{x}{x - 5}$ becomes $\dfrac{(5)}{(5) - 5}$, or $\dfrac{5}{0}$.

For what values of x are the expressions undefined?

(a) $\dfrac{x}{x - 5}$

To answer this question, we must find where the denominator is 0.

$x - 5 = 0$

$x = 5$

The expression $\dfrac{x}{x - 5}$ is undefined for $x = 5$.

(b) $\dfrac{3}{x + 5}$

Again, set the denominator equal to 0.

$x + 5 = 0$

$x = -5$

The expression $\dfrac{3}{x + 5}$ is undefined for $x = -5$.

Check Yourself 2

For what values of the variable are the expressions undefined?

(a) $\dfrac{1}{r+7}$ **(b)** $\dfrac{5}{2x-9}$

It may be necessary to factor the denominator to determine the values of x for which the expression is undefined.

Example 3 Avoiding Division by Zero

For each rational expression, find the values such that the expression is undefined.

(a) $\dfrac{x+6}{x^2+2x-15}$

$= \dfrac{x+6}{(x+5)(x-3)}$ Factor the denominator.

Set the denominator equal to 0 to find the "problem" values, and solve.

$(x+5)(x-3) = 0$

$x + 5 = 0$ or $x - 3 = 0$

$x = -5$ or $x = 3$

We find that when $x = -5$ or $x = 3$, the expression is undefined.

(b) $\dfrac{x^2+x-12}{3x^2+x-2}$

$= \dfrac{x^2+x-12}{(3x-2)(x+1)}$ We factor the denominator to find the "problem" values.

$(3x-2)(x+1) = 0$ Set the denominator equal to 0 and solve.

$3x - 2 = 0$ or $x + 1 = 0$

$x = \dfrac{2}{3}$ or $x = -1$

The expression is undefined when $x = \dfrac{2}{3}$ or $x = -1$.

Check Yourself 3

Find the values for which the expression is undefined.

$\dfrac{x^2-2x-3}{2x^2-3x-20}$

Generally, we want to write rational expressions in the simplest possible form. Your past experience with fractions will help you here. Recall that

$$\frac{3}{5} = \frac{3\cdot 2}{5\cdot 2} = \frac{6}{10}$$

so $\dfrac{3}{5}$ and $\dfrac{6}{10}$

name equivalent fractions. Similarly,

$$\frac{10}{15} = \frac{5 \cdot 2}{5 \cdot 3} = \frac{2}{3}$$

$$\text{so} \quad \frac{10}{15} \quad \text{and} \quad \frac{2}{3}$$

name equivalent fractions.

We can always multiply or divide the numerator and denominator of a fraction by the same nonzero number. The same pattern is true in algebra.

Property

Fundamental Principle of Rational Expressions

For polynomials P, Q, and R,

$$\frac{P}{Q} = \frac{PR}{QR} \quad \text{where } Q \neq 0 \text{ and } R \neq 0$$

NOTE

In fact, most of the methods in this chapter depend on factoring polynomials.

This property can be used in two ways. We can multiply or divide the numerator and denominator of a rational expression by the same nonzero polynomial. The result is always equivalent to the original expression.

In simplifying arithmetic fractions, we used this principle to divide the numerator and denominator by all common factors. With arithmetic fractions, those common factors are generally easy to recognize. Given rational expressions where the numerator and denominator are polynomials, we must determine those factors as our first step. The most important tools for simplifying expressions are the factoring techniques in Chapter 6.

Example 4 Simplifying Rational Expressions

< Objective 3 >

Simplify each rational expression. Assume the denominators are not 0.

NOTE

We find the common factors 4, x, and y in the numerator and denominator. We divide the numerator and denominator by the common factor $4xy$. Note that

$$\frac{4xy}{4xy} = 1$$

(a) $$\frac{4x^2y}{12xy^2} = \frac{4xy \cdot x}{4xy \cdot 3y}$$

$$= \frac{x}{3y}$$

(b) $$\frac{3x - 6}{x^2 - 4} = \frac{3(x - 2)}{(x + 2)(x - 2)}$$ Factor the numerator and the denominator.

We can now divide the numerator and denominator by the common factor $x - 2$.

$$\frac{3\cancel{(x - 2)}}{(x + 2)\cancel{(x - 2)}} = \frac{3}{x + 2}$$

and the rational expression is in simplest form.

Be careful! Given the expression

$$\frac{x + 2}{x + 3}$$

students are sometimes tempted to "cancel" the variable x, as in

$$\frac{x + 2}{x + 3} \stackrel{?}{=} \frac{2}{3}$$ This is wrong for every nonzero x.

> CAUTION

Pick any value other than 0 for the variable x and substitute. You will quickly see that

$$\frac{x + 2}{x + 3} \neq \frac{2}{3}$$

This is not a valid operation. We can only divide by common *factors,* and in this expression the variable x is a *term* in both the numerator and the denominator. In this expression, x is *not* a factor of the numerator, nor is x a factor of the denominator. The numerator and denominator of a rational expression must be factored *before* common factors are divided out. Therefore,

$$\frac{x+2}{x+3}$$

is in simplest possible form.

Check Yourself 4

Simplify each expression.

(a) $\frac{36a^3b}{9ab^2}$ (b) $\frac{x^2-25}{4x+20}$

We use the same techniques when trinomials need to be factored.

Example 5 Simplifying Rational Expressions

Simplify each rational expression.

(a) $\frac{5x^2-5}{x^2-4x-5}$

$= \frac{5(x^2-1)}{x^2-4x-5}$ Completely factor the expressions in both the numerator and denominator.

$= \frac{5(x-1)(x+1)}{(x-5)(x+1)}$ Divide by the common factor.

$= \frac{5(x-1)}{x-5}$

NOTE

Divide by the common factor $x + 1$, using the fact that

$$\frac{x+1}{x+1} = 1$$

if $x \neq -1$.

(b) $\frac{2x^2+x-6}{2x^2-x-3}$

$= \frac{(x+2)(2x-3)}{(x+1)(2x-3)}$

$= \frac{x+2}{x+1}$

(c) $\frac{x^3+2x^2-3x-6}{x^3+8}$

$= \frac{x^2(x+2)-3(x+2)}{(x+2)(x^2-2x+4)}$

$= \frac{(x+2)(x^2-3)}{(x+2)(x^2-2x+4)}$

$= \frac{x^2-3}{x^2-2x+4}$

NOTE

In part (c) we factor the numerator by grouping and use the sum of cubes in the denominator.

Check Yourself 5

Simplify each rational expression.

(a) $\dfrac{x^2 - 5x + 6}{3x^2 - 6x}$ (b) $\dfrac{3x^2 + 14x - 5}{3x^2 + 2x - 1}$

RECALL

$\dfrac{a-b}{a-b} = 1$

but

$\dfrac{a-b}{b-a} = -1$

Simplifying certain algebraic expressions involves recognizing a particular pattern. Verify for yourself that

$3 - 9 = -(9 - 3)$

In general, it is true that

$a - b = -(-a + b) = -(b - a) = -1(b - a)$

Dividing the above equation by $b - a$, gives us the result shown below.

Property

Polynomial Opposites $\dfrac{a-b}{b-a} = \dfrac{-(b-a)}{b-a} = -1$ if $a \neq b$

We use this property to complete Example 6.

Example 6 Simplifying Rational Expressions

Simplify each rational expression.

NOTE

$\dfrac{x-2}{2-x} = -1$

(a) $\dfrac{2x-4}{4-x^2} = \dfrac{2\overset{-1}{\cancel{(x-2)}}}{(2+x)\underset{1}{\cancel{(2-x)}}}$

$= \dfrac{2(-1)}{2+x} = \dfrac{-2}{2+x}$ or $-\dfrac{2}{x+2}$

(b) $\dfrac{9-x^2}{x^2+2x-15} = \dfrac{(3+x)\overset{-1}{\cancel{(3-x)}}}{(x+5)\underset{1}{\cancel{(x-3)}}}$

$= \dfrac{(3+x)(-1)}{x+5} = \dfrac{-x-3}{x+5}$ or $-\dfrac{x+3}{x+5}$

Step by Step

Simplifying Rational Expressions

Step 1 Completely factor both the numerator and the denominator of the expression.

Step 2 Divide the numerator and denominator by *all* common factors.

Step 3 The resulting expression will be in simplest form (or in lowest terms).

Check Yourself 6

Simplify each rational expression.

(a) $\dfrac{5x - 20}{16 - x^2}$ (b) $\dfrac{x^2 - 6x - 27}{81 - x^2}$

Definition

Rational Function

A **rational function** is a function that is defined by a rational expression. It can be written as

$$f(x) = \frac{P}{Q}$$

where P and Q are polynomials. The function is *not* defined for any value of x for which $Q = 0$.

Example 7 Identifying Rational Functions

< Objective 4 >

Identify the rational functions?

(a) $f(x) = 3x^3 - 2x + 5$ This is a rational function; it can be written as $\frac{3x^3 - 2x + 5}{1}$.

(b) $f(x) = \frac{3x^2 - 5x + 2}{2x - 1}$ This is a rational function; it is the ratio of two polynomials.

(c) $f(x) = 3x^3 + 3\sqrt{x}$ This is not a rational function. Since $3\sqrt{x} = 3x^{1/2}$, as seen in Section 7.5, $3\sqrt{x}$ cannot be a term of a polynomial.

RECALL

In Chapter 5, you learned that the exponents in a polynomial must always be whole numbers.

Check Yourself 7

Identify the rational functions?

(a) $f(x) = x^5 - 2x^4 - 1$ **(b)** $f(x) = \frac{x^2 - x + 7}{\sqrt{x} - 1}$

(c) $f(x) = \frac{3x^3 + 3x}{2x + 1}$

If we determine the values of x for which a rational function is undefined, then we can describe the domain of f as the set of all real numbers *except* those identified "problem" values.

Example 8 Simplifying a Rational Function

< Objective 5 >

(a) Determine the values of x for which $f(x) = \frac{x^2 - 2x - 24}{2x^2 + 7x - 4}$ is undefined, and write the domain of f.

Factoring, we have

$$f(x) = \frac{x^2 - 2x - 24}{2x^2 + 7x - 4} = \frac{(x - 6)(x + 4)}{(2x - 1)(x + 4)}$$

Focusing on the denominator, the values of x for which f is undefined are $x = \frac{1}{2}$ and $x = -4$.

The domain of f is $\left\{x \,\middle|\, x \neq \frac{1}{2} \text{ or } -4\right\}$.

(b) Write the function in simplified form, including the domain.
Dividing by the common factor $x + 4$, we have

$$f(x) = \frac{x - 6}{2x - 1}, \text{ where } x \neq \frac{1}{2} \text{ or } -4$$

Check Yourself 8

Given $f(x) = \dfrac{x^2 + 7x + 10}{x^2 - 5x - 14}$, write the simplified form of f, including the appropriate domain.

Check Yourself ANSWERS

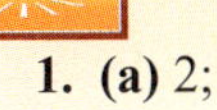

1. **(a)** 2; **(b)** $\frac{2}{3}$ **2.** **(a)** $r = -7$; **(b)** $x = \frac{9}{2}$ **3.** $x = -\frac{5}{2}$ or 4

4. **(a)** $\frac{4a^2}{b}$; **(b)** $\frac{x - 5}{4}$ **5.** **(a)** $\frac{x - 3}{3x}$; **(b)** $\frac{x + 5}{x + 1}$

6. **(a)** $\frac{-5}{x + 4}$ or $-\frac{5}{x + 4}$; **(b)** $\frac{-x - 3}{x + 9}$ or $-\frac{x + 3}{x + 9}$

7. **(a)** A rational function; **(b)** not a rational function; **(c)** a rational function

8. $f(x) = \frac{x + 5}{x - 7}, x \neq -2$ or 7

Reading Your Text

SECTION 9.1

(a) A rational expression is the ratio of two ____________.

(b) A fraction is undefined when its ____________ is equal to zero.

(c) A rational number is the ratio of two ____________.

(d) When simplifying a fraction, we divide by common ____________.

9.1 exercises

Basic Skills | Challenge Yourself | Calculator/Computer | Career Applications | Above and Beyond

Name ______________

Section ________ Date ________

Answers

1. ______ 2. ______
3. ______ 4. ______
5. ______ 6. ______
7. ______ 8. ______
9. ______ 10. ______
11. ______
12. ______
13. ______ 14. ______
15. ______ 16. ______
17. ______ 18. ______
19. ______ 20. ______
21. ______ 22. ______
23. ______ 24. ______
25. ______ 26. ______
27. ______ 28. ______
29. ______ 30. ______

< Objective 1 >

Evaluate each expression for the given value of the variable.

1. $\dfrac{3x}{2x-1}$ for $x = 5$

2. $\dfrac{4x}{5x-6}$ for $x = 2$

3. $\dfrac{3x+10}{x+2}$ for $x = -4$

4. $\dfrac{4x-7}{2x-1}$ for $x = -2$

5. $\dfrac{x^2+x}{x^2+2x}$ for $x = -2$

6. $\dfrac{4x-5}{2x^2-x+3}$ for $x = -1$

7. $\dfrac{2-3x}{x^2-4}$ for $x = -1$

8. $\dfrac{3x+1}{x^2-5x+6}$ for $x = 3$

< Objective 2 >

For what values of the variable is each rational expression undefined?

9. $\dfrac{x}{x-3}$

10. $\dfrac{y}{y+7}$

11. $\dfrac{x+5}{3}$

12. $\dfrac{x-6}{4}$

13. $\dfrac{2x-3}{2x-1}$

14. $\dfrac{4x-5}{5x+2}$

15. $\dfrac{2x+5}{x}$

16. $\dfrac{3x-7}{x}$

17. $\dfrac{x(x+1)}{x+2}$

18. $\dfrac{x+2}{3x-7}$

19. $\dfrac{5-3x}{2x}$

20. $\dfrac{2x+7}{3x+\frac{1}{3}}$

< Objective 3 >

Simplify each expression. Assume the denominators are not 0.

21. $\dfrac{14}{21}$

22. $\dfrac{45}{75}$

23. $\dfrac{4x^5}{6x^2}$

24. $\dfrac{30x^8}{25x^3}$

25. $\dfrac{10x^2y^5}{25xy^2}$

26. $\dfrac{18a^2b^3}{24a^4b^3}$

27. $\dfrac{-36x^5y^3}{21x^2y^5}$

28. $\dfrac{-15x^3y^3}{-20xy^2}$

29. $\dfrac{28a^5b^3c^2}{84a^2bc^4}$

30. $\dfrac{-52p^5q^3r^2}{39p^3q^5r^2}$

31. $\dfrac{6x - 24}{x^2 - 16}$

32. $\dfrac{x^2 - 25}{3x - 15}$

33. $\dfrac{x^2 + 2x + 1}{6x + 6}$ Videos

34. $\dfrac{5y^2 - 10y}{y^2 + y - 6}$

35. $\dfrac{x^2 - 13x + 36}{x^2 - 81}$

36. $\dfrac{2m^2 + 11m - 21}{4m^2 - 9}$

37. $\dfrac{3b^2 - 7b - 6}{b - 3}$

38. $\dfrac{a^2 - 9b^2}{a^2 + 8ab + 15b^2}$

39. $\dfrac{2y^2 + 3yz - 5z^2}{2y^2 + 11yz + 15z^2}$

40. $\dfrac{6x^2 - x - 2}{3x^2 - 5x + 2}$

41. $\dfrac{x^3 - 64}{x^2 - 16}$

42. $\dfrac{r^2 - rs - 6s^2}{r^3 + 8s^3}$

43. $\dfrac{a^4 - 81}{a^2 + 5a + 6}$

44. $\dfrac{x^4 - 625}{x^2 - 2x - 15}$

45. $\dfrac{xy - 2x + 3y - 6}{x^2 + 8x + 15}$

46. $\dfrac{cd - 3c + 5d - 15}{d^2 - 7d + 12}$

47. $\dfrac{x^2 + 3x - 18}{x^3 - 3x^2 - 2x + 6}$

48. $\dfrac{y^2 + 2y - 35}{y^2 - 8y + 15}$

49. $\dfrac{2m - 10}{25 - m^2}$ 

50. $\dfrac{5x - 20}{16 - x^2}$

51. $\dfrac{121 - x^2}{2x^2 - 21x - 11}$

52. $\dfrac{2x^2 - 7x + 3}{9 - x^2}$

< Objective 4 >

Identify the rational functions.

53. $f(x) = -7x^2 + 2x - 5$

54. $f(x) = \dfrac{x^3 - 2x^2 + 7}{\sqrt{x} + 2}$

55. $f(x) = \dfrac{x^2 - x - 1}{x + 2}$

56. $f(x) = \dfrac{\sqrt{x} - x + 3}{x - 2}$

57. $f(x) = 5x^2 - \sqrt[3]{x}$

58. $f(x) = \dfrac{x^2 - x + 5}{x}$

Answers

31. ______ 32. ______

33. ______ 34. ______

35. ______ 36. ______

37. ______ 38. ______

39. ______ 40. ______

41. ______

42. ______

43. ______

44. ______

45. ______

46. ______

47. ______

48. ______

49. ______

50. ______

51. ______

52. ______

53. ______

54. ______

55. ______

56. ______

57. ______

58. ______

Answers

59. ______
60. ______
61. ______
62. ______
63. ______
64. ______
65. ______
66. ______
67. ______
68. ______
69. ______
70. ______
71. ______ 72. ______
73. ______ 74. ______
75. ______ 76. ______
77. ______ 78. ______
79. ______
80. ______

< Objective 5 >

Rewrite each function in simplified form, including the appropriate domain.

59. $f(x) = \dfrac{x^2 - x - 2}{x + 1}$

60. $f(x) = \dfrac{x^2 + x - 12}{x + 4}$

61. $f(x) = \dfrac{3x^2 + 5x - 2}{x + 2}$

62. $f(x) = \dfrac{2x^2 - 7x + 5}{2x - 5}$

63. $f(x) = \dfrac{x^2 + 4x + 4}{5(x + 2)}$

64. $f(x) = \dfrac{x^2 - 6x + 9}{7(x - 3)}$

65. $f(x) = \dfrac{x^2 - 2x - 8}{x^2 - x - 6}$

66. $f(x) = \dfrac{x^2 + 4x - 5}{x^2 + 9x + 20}$

67. $f(x) = \dfrac{x^2 + 4x + 3}{x^2 + 7x + 6}$

68. $f(x) = \dfrac{x^2 + 7x + 10}{x^2 - 6x - 16}$

69. $f(x) = \dfrac{x^2 - 4x + 3}{x^2 - 1}$

70. $f(x) = \dfrac{x^2 - 6x + 8}{x^2 - 16}$

Basic Skills | **Challenge Yourself** | Calculator/Computer | Career Applications | Above and Beyond

Determine whether each statement is **true** *or* **false.**

71. If we multiply both numerator and denominator by the same nonzero expression, we obtain an equivalent rational expression.

72. If we add the same nonzero expression to both numerator and denominator, we obtain an equivalent rational expression.

Complete each statement with **never, sometimes,** *or* **always.**

73. A rational expression is ______ the ratio of two polynomials.

74. A value of x that causes the denominator to be zero can ______ be used as a value for the variable in a rational expression.

Simplify.

75. $\dfrac{2(x + h) - 2x}{(x + h) - x}$

76. $\dfrac{-3(x + h) - (-3x)}{(x - h) - x}$

77. $\dfrac{3(x + h) - 3 - (3x - 3)}{(x + h) - x}$

78. $\dfrac{2(x + h) + 5 - (2x + 5)}{(x + h) - x}$

79. $\dfrac{(x + h)^2 - x^2}{(x + h) - x}$

80. $\dfrac{(x + h)^3 - x^3}{(x + h) - x}$

81. **Business and Finance** A company has a setup cost of \$3,500 for the production of a new product. The cost to produce a single unit is \$8.75.

(a) Write a rational function that gives the average cost per unit when x units are produced.

(b) Find the average cost when 50 units are produced. 

82. **Business and Finance** The total revenue from the sale of a popular video is approximated by the rational function

$$R(x) = \frac{300x^2}{x^2 + 9}$$

where x is the number of months since the video has been released and $R(x)$ gives the total revenue in hundreds of dollars.

(a) Find the total revenue generated by the end of the first month.

(b) Find the total revenue generated by the end of the second month.

(c) Find the total revenue generated by the end of the third month.

(d) Find the revenue in the second month only.

83. If we view the graph of a rational function on a graphing calculator, we often see "unusual" behavior near x-values for which the function is undefined. Consider the rational function

$$f(x) = \frac{1}{x - 3}$$

(a) For what value(s) of x is the function undefined?

(b) Complete the table.

x	$f(x)$
4	
3.1	
3.01	
3.001	
3.0001	

(c) What do you observe concerning $f(x)$ as x is chosen close to 3, but slightly larger than 3?

(d) Complete the table.

x	$f(x)$
2	
2.9	
2.99	
2.999	
2.9999	

Answers

81. ______

82. ______

83. ______

Answers

84. ______

(e) What do you observe concerning $f(x)$ as x is chosen close to 3, but slightly smaller than 3?

(f) Graph the function on your graphing calculator. Describe the behavior of the graph of f near $x = 3$.

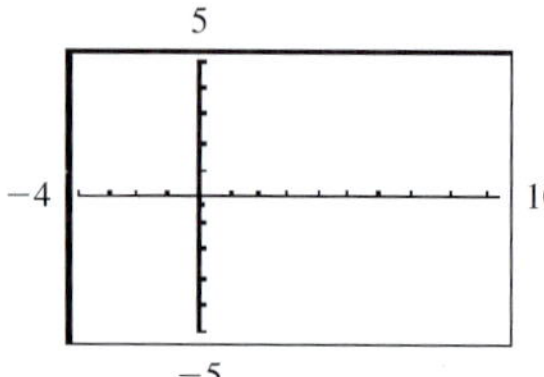

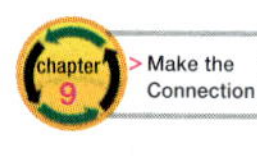

84. If we view the graph of a rational function on a graphing calculator, we often see "unusual" behavior near x-values for which the function is undefined. Consider the rational function

$$f(x) = \frac{1}{x + 2}$$

(a) For what value(s) of x is the function undefined?

(b) Complete the table.

x	$f(x)$
−1	
−1.9	
−1.99	
−1.999	
−1.9999	

(c) What do you observe concerning $f(x)$ as x is chosen close to −2, but slightly larger than −2?

(d) Complete the table.

x	$f(x)$
−3	
−2.1	
−2.01	
−2.001	
−2.0001	

(e) What do you observe concerning $f(x)$ as x is chosen close to −2, but slightly smaller than −2?

(f) Graph the function on your graphing calculator. Describe the behavior of the graph of f near $x = -2$.

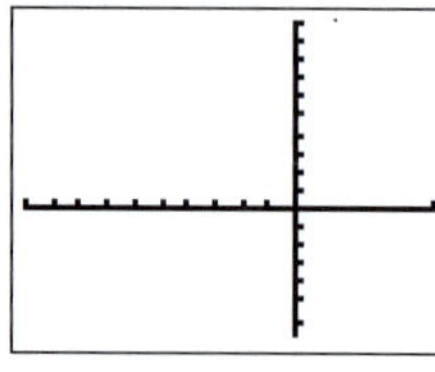

85. **Manufacturing Technology** The safe load of a drop-hammer-style pile driver is given from the formula

$$p = \frac{6whs + 6wh}{3s^2 + 6s + 3}$$

Simplify the rational expression.

86. **Mechanical Engineering** The shape of a beam loaded with a single concentrated load is described by the expression

$$\frac{x^2 - 64}{200}$$

Factor the numerator of this expression.

87. **Allied Health** A 4-year old child is upset because his 9-year-old sister tells him that he will never catch up to her in age. Write an expression for the ratio of the younger child's age x to the older child's age.

88. **Allied Health** Use the expression constructed in exercise 87 to argue that the significance of the difference in their ages reduces with time.

89. Explain why this statement is false.

$$\frac{6m^2 + 2m}{2m} = 6m^2 + 1$$

90. State and explain the fundamental principle of rational expressions.

91. The rational expression $\frac{x^2 - 4}{x + 2}$ can be simplified to $x - 2$. Is this reduction true for all values of x? Explain.

92. What is meant by a rational expression in lowest terms?

Answers

85. ____
86. ____
87. ____
88. ____
89. ____
90. ____
91. ____
92. ____

Answers

1. $\frac{5}{3}$ **3.** 1 **5.** undefined **7.** $-\frac{5}{3}$ **9.** 3 **11.** Never undefined
13. $\frac{1}{2}$ **15.** 0 **17.** -2 **19.** 0 **21.** $\frac{2}{3}$ **23.** $\frac{2x^3}{3}$ **25.** $\frac{2xy^3}{5}$
27. $\frac{-12x^3}{7y^2}$ **29.** $\frac{a^3b^2}{3c^2}$ **31.** $\frac{6}{x+4}$ **33.** $\frac{x+1}{6}$ **35.** $\frac{x-4}{x+9}$
37. $3b+2$ **39.** $\frac{y-z}{y+3z}$ **41.** $\frac{x^2+4x+16}{x+4}$ **43.** $\frac{(a^2+9)(a-3)}{a+2}$
45. $\frac{y-2}{x+5}$ **47.** $\frac{x+6}{x^2-2}$ **49.** $\frac{-2}{m+5}$ **51.** $\frac{-11-x}{2x+1} = -\frac{x+11}{2x+1}$
53. Rational **55.** Rational **57.** Not rational
59. **(a)** $f(x) = x-2; x \neq -1$; **(b)** $(-1, -3)$ **61.** $f(x) = 3x-1; x \neq -2$
63. $f(x) = \frac{x+2}{5}; x \neq -2$ **65.** $f(x) = \frac{x-4}{x-3}, x \neq -2, 3$
67. $f(x) = \frac{x+3}{x+6}, x \neq -6, -1$ **69.** $f(x) = \frac{x-3}{x+1}, x \neq -1, 1$
71. True **73.** always **75.** 2 **77.** 3 **79.** $2x+h$
81. **(a)** $R(x) = \frac{3{,}500 + 8.75x}{x}$; **(b)** \$78.75
83. **(a)** 3; **(b)**

x	$f(x)$
4	1
3.1	10
3.01	100
3.001	1,000
3.0001	10,000

(d)

x	$f(x)$
2	−1
2.9	−10
2.99	−100
2.999	−1,000
2.9999	−10,000

(f)

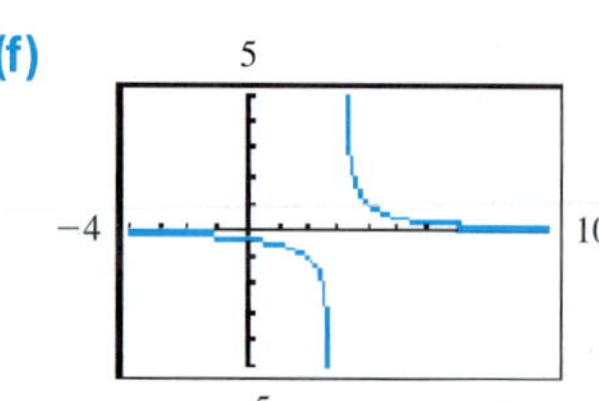

85. $\frac{2wh}{s+1}$ **87.** $\frac{x}{x+5}$ **89.** Above and Beyond
91. Above and Beyond

9.2 Multiplying and Dividing Rational Expressions

< 9.2 Objectives >

1 > Multiply and divide rational expressions

2 > Multiply and divide rational functions

Once again, we turn to an example from arithmetic to begin our discussion of multiplying rational expressions. Recall that to multiply two fractions, we multiply the numerators and multiply the denominators. For instance,

$$\frac{2}{5} \cdot \frac{3}{7} = \frac{2 \cdot 3}{5 \cdot 7} = \frac{6}{35}$$

In algebra, the pattern is exactly the same.

Property

Multiplying Rational Expressions

For polynomials P, Q, R, and S,

$$\frac{P}{Q} \cdot \frac{R}{S} = \frac{PR}{QS} \qquad \text{where } Q \neq 0 \text{ and } S \neq 0$$

Example 1 **Multiplying Rational Expressions**

< Objective 1 >

NOTE

For all problems with rational expressions, assume the denominators are not 0.

Multiply.

$$\frac{2x^3}{5y^2} \cdot \frac{10y}{3x^2} = \frac{20x^3y}{15x^2y^2}$$

$$= \frac{5x^2y \cdot 4x}{5x^2y \cdot 3y} \qquad \text{Divide by the common factor } 5x^2y \text{ to simplify.}$$

$$= \frac{4x}{3y}$$

Check Yourself 1

Multiply.

$$\frac{9a^2b^3}{5ab^4} \cdot \frac{20ab^2}{27ab^3}$$

We find it best to divide by any common factors before multiplying, as Example 2 illustrates.

Example 2 Multiplying Rational Expressions

NOTE

We use the factoring methods in Chapter 6 to simplify rational expressions.

Multiply and simplify.

(a) $\frac{x}{x^2 - 3x} \cdot \frac{6x - 18}{9x}$ Factor.

$= \frac{\overset{1}{\cancel{x}}}{\underset{1}{\cancel{x}}\underset{1}{\cancel{(x-3)}}} \cdot \frac{\overset{2}{\cancel{6}}\overset{1}{\cancel{(x-3)}}}{\underset{3}{\cancel{9}}x}$ Divide by the common factors of 3, x, and $x - 3$.

$= \frac{2}{3x}$

(b) $\frac{x^2 - y^2}{5x^2 - 5xy} \cdot \frac{10xy}{x^2 + 2xy + y^2}$ Factor and divide by the common factors of 5, x, $x - y$, and $x + y$.

$= \frac{\overset{1}{\cancel{(x+y)}}\overset{1}{\cancel{(x-y)}}}{\underset{1}{\cancel{5}}\underset{1}{\cancel{x}}\underset{1}{\cancel{(x-y)}}} \cdot \frac{\overset{2}{\cancel{10}}\overset{1}{\cancel{x}}y}{\underset{1}{\cancel{(x+y)}}(x+y)}$

$= \frac{2y}{x + y}$

RECALL

$\frac{2 - x}{x - 2} = -1$

(c) $\frac{4}{x^2 - 2x} \cdot \frac{10x - 5x^2}{8x + 24}$

$= \frac{\overset{1}{\cancel{4}}}{\underset{1}{\cancel{x}}\underset{1}{\cancel{(x-2)}}} \cdot \frac{5\overset{1}{\cancel{x}}\overset{-1}{\cancel{(2-x)}}}{\underset{2}{\cancel{8}}(x+3)}$

$= \frac{-5}{2(x + 3)}$

Check Yourself 2

Multiply and simplify.

(a) $\frac{x^2 - 5x - 14}{4x^2} \cdot \frac{8x + 56}{x^2 - 49}$ (b) $\frac{x}{2x - 6} \cdot \frac{3x - x^2}{2}$

This algorithm summarizes our work in multiplying rational expressions.

Step by Step

Multiplying Rational Expressions

Step 1 Write each numerator and denominator in completely factored form.

Step 2 Divide by any common factors appearing in both the numerator and the denominator.

Step 3 Multiply as needed to form the product.

RECALL

To divide fractions, we multiply by the reciprocal of the divisor. That is, invert the *divisor* (the second fraction) and multiply.

To divide rational expressions, you can again use your experience from arithmetic. Recall that

$$\frac{3}{5} \div \frac{2}{3} = \frac{3}{5} \cdot \frac{3}{2} = \frac{9}{10}$$

Once more, the pattern in algebra is identical.

Property

Dividing Rational Expressions

For polynomials P, Q, R, and S,

$$\frac{P}{Q} \div \frac{R}{S} = \frac{P}{Q} \cdot \frac{S}{R} = \frac{PS}{QR}$$

where $Q \neq 0$, $R \neq 0$, and $S \neq 0$.

Example 3 Dividing Rational Expressions

Divide and simplify.

NOTE

Invert the divisor and multiply.

(a) $\frac{3x^2}{8x^3y} \div \frac{9x^2y^2}{4y^4} = \frac{3x^2}{8x^3y} \cdot \frac{4y^4}{9x^2y^2} = \frac{y}{6x^3}$

(b) $\frac{2x^2 + 4xy}{9x - 18y} \div \frac{4x + 8y}{3x - 6y} = \frac{2x^2 + 4xy}{9x - 18y} \cdot \frac{3x - 6y}{4x + 8y}$

$$= \frac{2x(x + 2y)}{9(x - 2y)} \cdot \frac{3(x - 2y)}{4(x + 2y)} = \frac{x}{6}$$

> CAUTION

Invert the divisor, then factor.

(c) $\frac{2x^2 - x - 6}{4x^2 + 6x} \div \frac{x^2 - 4}{4x} = \frac{2x^2 - x - 6}{4x^2 + 6x} \cdot \frac{4x}{x^2 - 4}$

$$= \frac{(2x + 3)(x - 2)}{2x(2x + 3)} \cdot \frac{4x}{(x + 2)(x - 2)}$$

$$= \frac{2}{x + 2}$$

Check Yourself 3

Divide and simplify.

(a) $\frac{5xy}{7x^3} \div \frac{10y^2}{14x^3}$

(b) $\frac{3x - 9y}{2x + 10y} \div \frac{x^2 - 3xy}{4x^2 + 20xy}$

(c) $\frac{x^2 - 9}{x^3 - 27} \div \frac{x^2 - 2x - 15}{2x^2 - 10x}$

Here is an algorithm summarizing our work in dividing rational expressions.

Step by Step

Dividing Rational Expressions

Step 1 Invert the divisor (the *second* rational expression) to write the problem as one of multiplication.

Step 2 Proceed with the algorithm for multiplying rational expressions.

The product of two rational functions is always a rational function. Given two rational functions $f(x)$ and $g(x)$, we can rename the product, so

$$h(x) = f(x) \cdot g(x)$$

This is always true for values of x for which both f and g are defined. So, for example, $h(1) = f(1) \cdot g(1)$ as long as both $f(1)$ and $g(1)$ exist. Example 4 illustrates this concept.

Example 4 Multiplying Rational Functions

< Objective 2 >

Consider the rational functions

$$f(x) = \frac{x^2 - 3x - 10}{x + 1} \quad \text{and} \quad g(x) = \frac{x^2 - 4x - 5}{x - 5}$$

NOTE

$$f(0) = \frac{(0)^2 - 3(0) - 10}{(0) + 1} = \frac{-10}{1} = -10$$

$$g(0) = \frac{(0)^2 - 4(0) - 5}{(0) - 5} = \frac{-5}{-5} = 1$$

(a) $f(0) \cdot g(0)$

Because $f(0) = -10$ and $g(0) = 1$, we have $f(0) \cdot g(0) = (-10)(1) = -10$.

(b) $f(5) \cdot g(5)$

Although we can find $f(5)$, $g(5)$ is undefined. The number 5 is excluded from the domain of the function. Therefore, $f(5) \cdot g(5)$ is undefined.

(c) $h(x) = f(x) \cdot g(x)$

$$= \frac{x^2 - 3x - 10}{x + 1} \cdot \frac{x^2 - 4x - 5}{x - 5}$$

$$= \frac{(x - 5)(x + 2)}{\cancel{(x + 1)}} \cdot \frac{\cancel{(x + 1)}\cancel{(x - 5)}}{\cancel{(x - 5)}}$$

Factor the numerators, and divide by the common factors.

$$= (x - 5)(x + 2) \qquad x \neq -1, x \neq 5$$

NOTE

$f(x)$ is undefined for $x = -1$, and $g(x)$ is undefined for $x = 5$. Therefore, $h(x)$ is undefined for both of these values.

(d) $h(0)$

$$h(0) = (0 - 5)(0 + 2) = -10$$

(e) $h(5)$

Although the temptation is to substitute 5 for x in part (c), notice that the function is undefined when x is -1 or 5. As was true in part (b), the function is undefined at that point.

Check Yourself 4

Consider

$$f(x) = \frac{x^2 - 2x - 8}{x + 2} \quad \text{and} \quad g(x) = \frac{x^2 - 3x - 10}{x - 4}$$

(a) $f(0) \cdot g(0)$ **(b)** $f(4) \cdot g(4)$ **(c)** $h(x) = f(x) \cdot g(x)$

(d) $h(0)$ **(e)** $h(4)$

When we divide two rational functions to create a third rational function, we must be certain to exclude values for which the polynomial in the denominator is equal to zero, as Example 5 illustrates.

Example 5 Dividing Rational Functions

Consider the rational functions

$$f(x) = \frac{x^3 - 2x^2}{x + 2} \quad \text{and} \quad g(x) = \frac{x^2 - 3x + 2}{x - 4}$$

(a) Find $\dfrac{f(0)}{g(0)}$.

Because $f(0) = 0$ and $g(0) = -\dfrac{1}{2}$, we have

$$\frac{f(0)}{g(0)} = \frac{0}{-\frac{1}{2}} = 0$$

(b) Find $\dfrac{f(1)}{g(1)}$.

Although we can find both $f(1)$ and $g(1)$, $g(1) = 0$, so division is undefined. The value 1 is excluded from the domain of the quotient.

(c) Find $h(x) = \dfrac{f(x)}{g(x)}$.

$$h(x) = \frac{f(x)}{g(x)}$$

Note that −2 is excluded from the domain of f and 4 is excluded from the domain of g.

$$= \frac{\frac{x^3 - 2x^2}{x + 2}}{\frac{x^2 - 3x + 2}{x - 4}}$$

Invert and multiply.

$$= \frac{x^3 - 2x^2}{x + 2} \cdot \frac{x - 4}{x^2 - 3x + 2}$$

$$= \frac{x^2\overset{1}{\cancel{(x - 2)}}}{x + 2} \cdot \frac{x - 4}{(x - 1)\underset{1}{\cancel{(x - 2)}}}$$

Because $(x - 1)(x - 2)$ is part of the denominator, 1 and 2 are excluded from the domain of h.

$$= \frac{x^2(x - 4)}{(x + 2)(x - 1)} \qquad x \neq -2, 1, 2, 4$$

(d) For which values of x is $h(x)$ undefined?

The function $h(x)$ will be undefined for any value of x that would cause division by zero. So $h(x)$ is undefined for the values −2, 1, 2, and 4.

Check Yourself 5

Given the rational functions

$$f(x) = \frac{x^2 - 2x + 1}{x + 3} \quad \text{and} \quad g(x) = \frac{x^2 - 5x + 4}{x - 2}$$

(a) Find $\dfrac{f(0)}{g(0)}$. **(b)** Find $\dfrac{f(1)}{g(1)}$. **(c)** Find $h(x) = \dfrac{f(x)}{g(x)}$.

(d) For which values of x is $h(x)$ undefined?

Check Yourself ANSWERS

1. $\frac{4a}{3b^2}$ **2. (a)** $\frac{2(x+2)}{x^2}$; **(b)** $\frac{-x^2}{4}$ **3. (a)** $\frac{x}{y}$; **(b)** 6; **(c)** $\frac{2x}{x^2+3x+9}$

4. (a) -10; **(b)** undefined; **(c)** $h(x) = (x-5)(x+2), x \neq -2, x \neq 4$; **(d)** -10; **(e)** undefined

5. (a) $-\frac{1}{6}$; **(b)** undefined; **(c)** $h(x) = \frac{(x-1)(x-2)}{(x+3)(x-4)}$; **(d)** $x \neq -3, 1, 2, 4$

Reading Your Text

SECTION 9.2

(a) To multiply two fractions, we multiply the numerators and ____________ the denominators.

(b) Before multiplying rational expressions, we write each numerator and denominator in completely ____________ form.

(c) To divide fractions, we multiply by the ____________ of the divisor.

(d) The product of two rational expressions is ____________ a rational expression.

Basic Skills | Challenge Yourself | Calculator/Computer | Career Applications | Above and Beyond

9.2 exercises

Boost *your* GRADE at ALEKS.com!

ALEKS®

- Practice Problems
- Self-Tests
- NetTutor
- e-Professors
- Videos

Name ____________

Section ________ Date ________

< Objective 1 >

Compute, as indicated. Express your result in simplest form.

1. $\frac{x^2}{3} \cdot \frac{6x}{x^4}$

2. $\frac{-y^3}{10} \cdot \frac{15y}{y^6}$

3. $\frac{x}{8x^4} \div \frac{x^5}{24}$

4. $\frac{p^5}{8} \div \frac{-p^2}{12p}$

5. $\frac{4xy^2}{15x^3} \cdot \frac{25xy}{16y^3}$

6. $\frac{3x^3y}{10xy^3} \cdot \frac{5xy^2}{-9xy^3}$

7. $\frac{8b^3}{15ab} \div \frac{2ab^2}{20ab^3}$

8. $\frac{5x^3y^3}{8x^5} \div \frac{15x^3y}{32x^3y^2}$

9. $\frac{m^3n}{2mn} \cdot \frac{6mn^2}{m^3n} \div \frac{3mn}{5m^2n}$

10. $\frac{4cd^2}{5cd} \cdot \frac{3c^3d}{2c^2d} \div \frac{9cd}{20cd^3}$

11. $\frac{6x + 18}{4x} \cdot \frac{16x^3}{3x + 9}$

12. $\frac{a^2 - 3a}{5a} \cdot \frac{20a^2}{3a - 9}$

13. $\frac{3b - 15}{6b} \div \frac{4b - 20}{9b^2}$

14. $\frac{7m^2 + 28m}{4m} \div \frac{5m + 20}{12m^2}$

15. $\frac{x^2 - 3x - 10}{5x} \cdot \frac{15x^2}{3x - 15}$

16. $\frac{y^2 - 8y}{4y} \cdot \frac{12y^2}{y^2 - 64}$

17. $\frac{c^2 + 2c - 8}{6c} \div \frac{5c + 20}{18c}$

18. $\frac{m^2 - 64}{6m^2} \div \frac{2m - 16}{24m^5}$

19. $\frac{x^2 - x - 12}{3x - 12} \cdot \frac{15x^3}{x^2 - 9}$

20. $\frac{y^2 + 7y + 10}{y^2 + 5y} \cdot \frac{2y}{y^2 - 4}$

21. $\frac{d^2 - 3d - 18}{16d - 96} \div \frac{d^2 - 9}{20d}$

22. $\frac{b^2 + 2b - 8}{b^2 - 2b} \div \frac{b^2 - 16}{4b}$

23. $\frac{2x^2 - x - 3}{3x^2 + 7x + 4} \cdot \frac{3x^2 - 11x - 20}{4x^2 - 9}$

Answers

1. ________ 2. ________
3. ________ 4. ________
5. ________ 6. ________
7. ________ 8. ________
9. ________ 10. ________
11. ________ 12. ________
13. ________ 14. ________
15. ________ 16. ________
17. ________
18. ________
19. ________
20. ________
21. ________
22. ________
23. ________

Answers

24. ______
25. ______
26. ______
27. ______
28. ______
29. ______
30. ______
31. ______
32. ______
33. ______
34. ______
35. ______
36. ______
37. ______
38. ______
39. ______
40. ______

24. $\dfrac{4p^2 - 1}{2p^2 - 9p - 5} \cdot \dfrac{3p^2 - 13p - 10}{9p^2 - 4}$

25. $\dfrac{a^2 - 9}{2a^2 - 6a} \div \dfrac{2a^2 + 5a - 3}{4a^2 - 1}$

26. $\dfrac{2x^2 - 5x - 7}{4x^2 - 9} \div \dfrac{5x^2 + 5x}{2x^2 + 3x}$

> Videos

27. $\dfrac{2w - 6}{w^2 + 2w} \cdot \dfrac{3w}{3 - w}$

28. $\dfrac{3y - 15}{y^2 + 3y} \cdot \dfrac{4y}{5 - y}$

29. $\dfrac{a - 7}{2a + 6} \div \dfrac{21 - 3a}{a^2 + 3a}$

30. $\dfrac{x - 5}{x^2 + 3x} \div \dfrac{25 - 5x}{2x + 6}$

> Videos

31. $\dfrac{x^2 - 9y^2}{2x^2 - xy - 15y^2} \cdot \dfrac{4x + 10y}{x^2 + 3xy}$

32. $\dfrac{2a^2 - 7ab - 15b^2}{2ab - 10b^2} \cdot \dfrac{2a^2 - 3ab}{4a^2 - 9b^2}$

33. $\dfrac{3m^2 - 5mn + 2n^2}{9m^2 - 4n^2} \div \dfrac{m^3 - m^2n}{9m^2 + 6mn}$

34. $\dfrac{2x^2y - 5xy^2}{4x^2 - 25y^2} \div \dfrac{4x^2 + 20xy}{2x^2 + 15xy + 25y^2}$

35. $\dfrac{x^3 + 8}{x^2 - 4} \cdot \dfrac{5x - 10}{x^3 - 2x^2 + 4x}$

36. $\dfrac{a^3 - 27}{a^2 - 9} \div \dfrac{a^3 + 3a^2 + 9a}{3a^3 + 9a^2}$

Basic Skills | **Challenge Yourself** | Calculator/Computer | Career Applications | Above and Beyond

< Objective 2 >

37. Let $f(x) = \dfrac{x^2 - 3x - 4}{x + 2}$ and $g(x) = \dfrac{x^2 - 2x - 8}{x - 4}$. Find **(a)** $f(0) \cdot g(0)$; **(b)** $f(4) \cdot g(4)$; **(c)** $h(x) = f(x) \cdot g(x)$; **(d)** $h(0)$; and **(e)** $h(4)$.

38. Let $f(x) = \dfrac{x^2 - 4x + 3}{x + 5}$ and $g(x) = \dfrac{x^2 + 7x + 10}{x - 3}$. Find **(a)** $f(1) \cdot g(1)$; **(b)** $f(3) \cdot g(3)$; **(c)** $h(x) = f(x) \cdot g(x)$; **(d)** $h(1)$; and **(e)** $h(3)$.

39. Let $f(x) = \dfrac{2x^2 - 3x - 5}{x + 2}$ and $g(x) = \dfrac{3x^2 + 5x - 2}{x + 1}$. Find **(a)** $f(1) \cdot g(1)$; **(b)** $f(-2) \cdot g(-2)$; **(c)** $h(x) = f(x) \cdot g(x)$; **(d)** $h(1)$; and **(e)** $h(-2)$.

40. Let $f(x) = \dfrac{x^2 - 1}{x - 3}$ and $g(x) = \dfrac{x^2 - 9}{x - 1}$. Find **(a)** $f(2) \cdot g(2)$; **(b)** $f(3) \cdot g(3)$; **(c)** $h(x) = f(x) \cdot g(x)$; **(d)** $h(2)$; and **(e)** $h(3)$.

41. Let $f(x) = \frac{3x^2 + x - 2}{x - 2}$ and $g(x) = \frac{x^2 - 4x - 5}{x + 4}$. Find **(a)** $\frac{f(0)}{g(0)}$; **(b)** $\frac{f(1)}{g(1)}$; **(c)** $h(x) = \frac{f(x)}{g(x)}$; and **(d)** the values of x for which $h(x)$ is undefined.

42. Let $f(x) = \frac{x^2 + x}{x - 5}$ and $g(x) = \frac{x^2 - x - 6}{x - 5}$. Find **(a)** $\frac{f(0)}{g(0)}$; **(b)** $\frac{f(2)}{g(2)}$; **(c)** $h(x) = \frac{f(x)}{g(x)}$; and **(d)** the values of x for which $h(x)$ is undefined.

Determine whether each statement is **true** *or* **false.**

43. When we multiply two rational expressions, we multiply the numerators together and we multiply the denominators together.

44. When we divide two rational expressions, we invert the second rational expression, and then we divide.

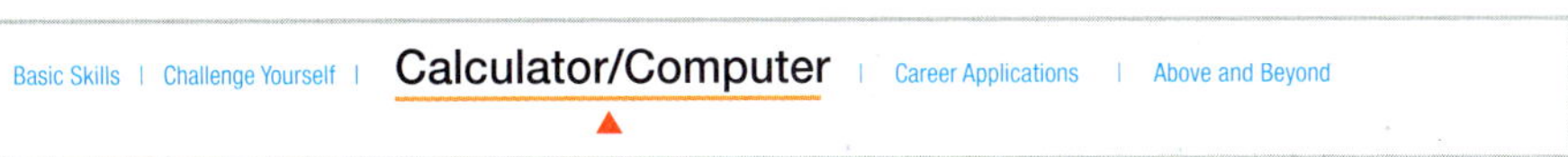

The results from multiplying and dividing rational expressions can be checked by using a graphing calculator. To do this, define one expression in Y_1 and the other in Y_2. Then define the operation in Y_3 as $Y_1 \cdot Y_2$ or $Y_1 \div Y_2$. Put your simplified result in Y_4 (sorry, you still must simplify algebraically). Deselect the graphs for Y_1 and Y_2. If you have correctly simplified the expression, the graphs of Y_3 and Y_4 will appear to be identical. Use this technique to check your multiplication and division in exercises 45 to 48.

45. $\frac{x^3 - 3x^2 + 2x - 6}{x^2 - 9} \cdot \frac{5x^2 + 15x}{20x}$

46. $\frac{3a^3 + a^2 - 9a - 3}{15a^2 + 5a} \cdot \frac{3a^2 + 9}{a^4 - 9}$

47. $\frac{x^4 - 16}{x^2 + x - 6} \div (x^3 + 4x)$

48. $\frac{w^3 + 27}{w^2 + 2w - 3} \div (w^3 - 3w^2 + 9w)$

Basic Skills | Challenge Yourself | Calculator/Computer | **Career Applications** | Above and Beyond

49. **Agricultural Technology** Herbicides constitute approximately $\frac{2}{3}$ of all pesticides used in the United States. Insecticides account for another $\frac{1}{4}$ of the pesticides used in the United States. Write a simplified expression for the ratio of herbicides to insecticides used in the United States.

50. **Agricultural Technology** Fungicides account for approximately $\frac{1}{10}$ of the pesticides used in the United States. Insecticides account for another $\frac{1}{4}$ of the pesticides used. Write a simplified expression for the ratio of fungicides to insecticides used in the United States.

Answers

41. ______
42. ______
43. ______
44. ______
45. ______
46. ______
47. ______
48. ______
49. ______
50. ______

Answers

51. ______________________

52. ______________________

51. Construction Technology Plans call for the dimensions of a rectangular room to be given in terms of an unknown x. Find the area of the room shown, in terms of x.

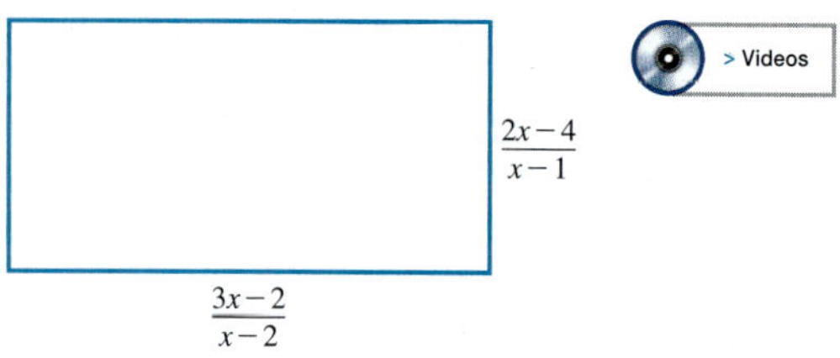

52. Construction Technology Plans call for the dimensions of a rectangular room to be given in terms of an unknown x. Find the area of the room shown, in terms of x.

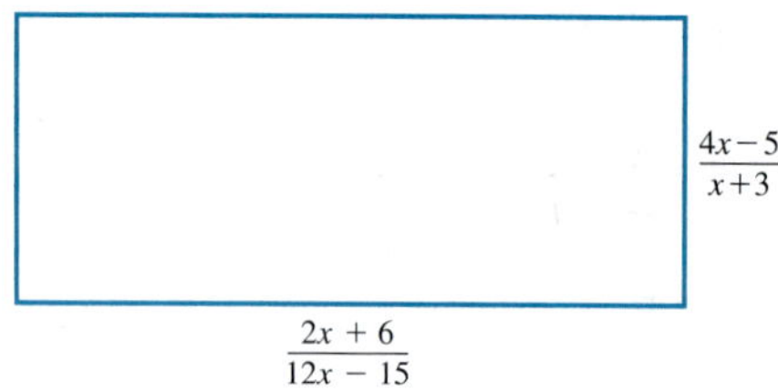

Answers

1. $\frac{2}{x}$ **3.** $\frac{3}{x^8}$ **5.** $\frac{5}{12x}$ **7.** $\frac{16b^3}{3a}$ **9.** $5mn$ **11.** $8x^2$ **13.** $\frac{9b}{8}$
15. $x^2 + 2x$ **17.** $\frac{3(c-2)}{5}$ **19.** $\frac{5x^3}{x-3}$ **21.** $\frac{5d}{4(d-3)}$ **23.** $\frac{x-5}{2x+3}$
25. $\frac{2a+1}{2a}$ **27.** $\frac{-6}{w+2}$ **29.** $\frac{-a}{6}$ **31.** $\frac{2}{x}$ **33.** $\frac{3}{m}$ **35.** $\frac{5}{x}$
37. **(a)** -4; **(b)** undefined; **(c)** $h(x) = (x+1)(x-4), x \neq -2, x \neq 4$; **(d)** -4; **(e)** undefined
39. **(a)** -6; **(b)** undefined; **(c)** $h(x) = (2x-5)(3x-1), x \neq -2, x \neq -1$; **(d)** -6; **(e)** undefined
41. **(a)** $-\frac{4}{5}$; **(b)** $\frac{5}{4}$; **(c)** $h(x) = \frac{(3x-2)(x+4)}{(x-2)(x-5)}$; **(d)** $-4, -1, 2, 5$
43. True **45.** $\frac{x^2+2}{4}$ **47.** $\frac{x+2}{x(x+3)}$ **49.** $\frac{8}{3}$ **51.** $\frac{2(3x-2)}{x-1}$

9.3 Adding and Subtracting Rational Expressions

< 9.3 Objectives >

1 > Add and subtract rational expressions

2 > Add and subtract rational functions

Recall that adding or subtracting two arithmetic fractions with the same denominator is straightforward. The same is true in algebra. To add or subtract two rational expressions with the same denominator, we add or subtract their numerators and then write that sum or difference over the common denominator.

Property

Adding or Subtracting Rational Expressions

$$\frac{P}{R} + \frac{Q}{R} = \frac{P + Q}{R}$$

and $$\frac{P}{R} - \frac{Q}{R} = \frac{P - Q}{R}$$

where $R \neq 0$.

Example 1 Adding and Subtracting Rational Expressions

< Objective 1 >

NOTE

Since we have common denominators, we simply perform the indicated operations on the numerators.

Perform the indicated operations.

$$\frac{3}{2a^2} - \frac{1}{2a^2} + \frac{5}{2a^2} = \frac{3 - 1 + 5}{2a^2}$$

$$= \frac{7}{2a^2}$$

Check Yourself 1

Perform the indicated operations.

$$\frac{5}{3y^2} + \frac{4}{3y^2} - \frac{7}{3y^2}$$

We always express the sum or difference of rational expressions in simplest form as shown in Example 2.

Example 2 Adding and Subtracting Rational Expressions

Add or subtract as indicated.

(a) $\frac{5x}{x^2 - 9} + \frac{15}{x^2 - 9}$

$= \frac{5x + 15}{x^2 - 9}$ Add the numerators.

$= \frac{5(x + 3)}{(x - 3)(x + 3)} = \frac{5}{x - 3}$ Factor and divide by the common factor.

(b) $\dfrac{3x + y}{2x} - \dfrac{x - 3y}{2x} = \dfrac{(3x + y) - (x - 3y)}{2x}$

Be sure to *enclose the second numerator* in parentheses.

$= \dfrac{3x + y - x + 3y}{2x}$

Remove the parentheses by *changing each sign.*

$= \dfrac{2x + 4y}{2x} = \dfrac{2(x + 2y)}{2x}$

Factor and divide by the common factor of 2.

$= \dfrac{x + 2y}{x}$

Check Yourself 2

Perform the indicated operations.

(a) $\dfrac{6a}{a^2 - 2a - 8} + \dfrac{12}{a^2 - 2a - 8}$ (b) $\dfrac{5x - y}{3y} - \dfrac{2x - 4y}{3y}$

NOTE

By **inspection,** we mean you look at the denominators and find the LCD.

Now, what if our rational expressions *do not* have common denominators? In that case, we find the least common denominator (LCD). The **least common denominator** is the simplest polynomial that is divisible by each of the individual denominators. Each expression in the desired sum or difference is then "built up" to an equivalent expression having that LCD as a denominator. We can then add or subtract as before.

Although in many cases we can find the LCD by inspection, we can state an algorithm for finding the LCD that is similar to the one used in arithmetic.

Step by Step

Finding the Least Common Denominator

Step 1 Write each of the denominators in completely factored form.

Step 2 Write the LCD as the product of each prime factor to the highest power to which it appears in the factored form of any of the individual denominators.

Example 3 illustrates the procedure.

Example 3 Finding the LCD for Two Rational Expressions

Find the LCD for each pair of rational expressions.

(a) $\dfrac{3}{4x^2}$ and $\dfrac{5}{6xy}$

Factor the denominators.

$4x^2 = 2^2 \cdot x^2$

$6xy = 2 \cdot 3 \cdot x \cdot y$

The LCD must have the factors

$2^2 \cdot 3 \cdot x^2 \cdot y$

so $12x^2y$ is the LCD.

NOTE

You may be able to find this LCD by inspecting the numerical coefficients and the variable factors.

NOTE

It is generally best to leave the LCD in factored form.

(b) $\frac{7}{x-3}$ and $\frac{2}{x+5}$

Here, neither denominator can be factored. The LCD must have the factors $x - 3$ and $x + 5$. So the LCD is

$$(x-3)(x+5)$$

Check Yourself 3

Find the LCD for each pair of rational expressions.

(a) $\frac{3}{8a^3}$ and $\frac{5}{6a^2}$ **(b)** $\frac{4}{x+7}$ and $\frac{3}{x-5}$

In Example 4, we see how factoring techniques are applied.

Example 4 Finding the LCD for Two Rational Expressions

Find the LCD for each pair of rational expressions.

(a) $\frac{2}{x^2-x-6}$ and $\frac{1}{x^2-9}$

Factoring the denominators gives

$$x^2 - x - 6 = (x+2)(x-3)$$

and

$$x^2 - 9 = (x+3)(x-3)$$

The LCD for the given denominators is then

$$(x+2)(x-3)(x+3)$$

NOTE

The LCD must contain *each* of the factors appearing in the original denominators.

(b) $\frac{5}{x^2-4x+4}$ and $\frac{3}{x^2+2x-8}$

Again, we factor.

$$x^2 - 4x + 4 = (x-2)^2$$
$$x^2 + 2x - 8 = (x-2)(x+4)$$

The LCD is then

$$(x-2)^2(x+4)$$

NOTE

The LCD must contain $(x-2)^2$ as a factor since $x - 2$ appears *twice* as a factor in the first denominator.

Check Yourself 4

Find the LCD for each pair of rational expressions.

(a) $\frac{3}{x^2-2x-15}$ and $\frac{5}{x^2-25}$ **(b)** $\frac{5}{y^2+6y+9}$ and $\frac{3}{y^2-y-12}$

In Example 5, the concept of the LCD is applied in adding and subtracting rational expressions.

Example 5 Adding and Subtracting Rational Expressions

Add or subtract as indicated.

NOTE

In each case, we are multiplying by 1 $\left(\frac{x}{x}\right.$ in the first fraction and $\frac{2y}{2y}$ in the second fraction$\left.\right)$, which is why the resulting fractions are equivalent to the original ones.

(a) $\frac{5}{4xy} + \frac{3}{2x^2}$

For the denominators $2x^2$ and $4xy$, the LCD is $4x^2y$. We rewrite each of the rational expressions with the LCD as a denominator.

$$\frac{5}{4xy} + \frac{3}{2x^2} = \frac{5 \cdot x}{4xy \cdot x} + \frac{3 \cdot 2y}{2x^2 \cdot 2y}$$

$$= \frac{5x}{4x^2y} + \frac{6y}{4x^2y} = \frac{5x + 6y}{4x^2y}$$

Multiply the first rational expression by $\frac{x}{x}$ and the second by $\frac{2y}{2y}$ to form the LCD of $4x^2y$.

(b) $\frac{3}{a - 3} - \frac{2}{a}$

For the denominators a and $a - 3$, the LCD is $a(a - 3)$. We rewrite each of the rational expressions with that LCD as a denominator.

$$\frac{3}{a - 3} - \frac{2}{a}$$

$$= \frac{3a}{a(a - 3)} - \frac{2(a - 3)}{a(a - 3)}$$

$$= \frac{3a - 2(a - 3)}{a(a - 3)}$$ Subtract the numerators.

$$= \frac{3a - 2a + 6}{a(a - 3)} = \frac{a + 6}{a(a - 3)}$$ Remove the parentheses in the numerator, and combine like terms.

Check Yourself 5

Perform the indicated operations.

(a) $\frac{3}{2ab} + \frac{4}{5b^2}$ **(b)** $\frac{5}{y + 2} - \frac{3}{y}$

We now proceed to Example 6, in which factoring will be required to form the LCD.

Example 6 Adding and Subtracting Rational Expressions

Add or subtract as indicated.

(a) $\frac{-5}{x^2 - 3x - 4} + \frac{8}{x^2 - 16}$

We first factor the two denominators.

$$x^2 - 3x - 4 = (x + 1)(x - 4)$$
$$x^2 - 16 = (x + 4)(x - 4)$$

We see that the LCD must be

$$(x + 1)(x + 4)(x - 4)$$

NOTE

We use the facts that

$$\frac{x+4}{x+4} = 1$$

and $\frac{x+1}{x+1} = 1$

Again, rewriting the original expressions with factored denominators gives

$$\frac{-5}{(x+1)(x-4)} + \frac{8}{(x-4)(x+4)}$$

$$= \frac{-5(x+4)}{(x+1)(x-4)(x+4)} + \frac{8(x+1)}{(x-4)(x+4)(x+1)}$$

$$= \frac{-5(x+4) + 8(x+1)}{(x+1)(x-4)(x+4)}$$ Add the numerators.

$$= \frac{-5x - 20 + 8x + 8}{(x+1)(x-4)(x+4)}$$

$$= \frac{3x - 12}{(x+1)(x-4)(x+4)}$$ Combine like terms in the numerator.

$$= \frac{3(x-4)}{(x+1)(x-4)(x+4)}$$ Factor.

$$= \frac{3}{(x+1)(x+4)}$$ Divide by the common factor $x - 4$.

(b) $\dfrac{5}{x^2 - 5x + 6} - \dfrac{3}{4x - 12}$

Again, factor the denominators.

$$x^2 - 5x + 6 = (x-2)(x-3)$$
$$4x - 12 = 4(x-3)$$

The LCD is $4(x-2)(x-3)$, and proceeding as before, we have

$$\frac{5}{(x-2)(x-3)} - \frac{3}{4(x-3)}$$

$$= \frac{5 \cdot 4}{4(x-2)(x-3)} - \frac{3(x-2)}{4(x-2)(x-3)}$$

$$= \frac{20 - 3(x-2)}{4(x-2)(x-3)}$$ Now subtract the numerators.

$$= \frac{20 - 3x + 6}{4(x-2)(x-3)} = \frac{-3x + 26}{4(x-2)(x-3)}$$ Simplify the numerator.

Check Yourself 6

Add or subtract as indicated.

(a) $\dfrac{-4}{x^2 - 4} + \dfrac{7}{x^2 - 3x - 10}$ **(b)** $\dfrac{5}{3x - 9} - \dfrac{2}{x^2 - 9}$

Example 7 looks slightly different from those you have seen thus far, but the reasoning involved in performing the subtraction is exactly the same.

Example 7 Subtracting Rational Expressions

Subtract.

$$3 - \frac{5}{2x - 1}$$

To perform the subtraction, remember that 3 is equivalent to the fraction $\frac{3}{1}$, so

$$3 - \frac{5}{2x - 1} = \frac{3}{1} - \frac{5}{2x - 1}$$

For the denominators 1 and $2x - 1$, the LCD is just $2x - 1$. We now rewrite the first expression with that denominator.

$$3 - \frac{5}{2x - 1} = \frac{3(2x - 1)}{2x - 1} - \frac{5}{2x - 1}$$

$$= \frac{3(2x - 1) - 5}{2x - 1}$$ Subtract the numerators.

$$= \frac{6x - 8}{2x - 1}$$ Simplify the numerator.

Check Yourself 7

Subtract.

$$\frac{4}{3x + 1} - 3$$

Example 8 uses an observation from Section 9.1. Recall that

$$a - b = -(b - a)$$
$$= -1(b - a)$$

Example 8 Adding and Subtracting Rational Expressions

NOTE

Use

$$\frac{-1}{-1} = 1$$

Note that

$(-1)(5 - x) = x - 5$

The fractions now have a common denominator, and we can add as before.

Add.

$$\frac{x^2}{x - 5} + \frac{3x + 10}{5 - x}$$

Your first thought might be to use a denominator of $(x - 5)(5 - x)$. However, we can simplify our work considerably if we multiply the numerator and denominator of the second fraction by -1 to find a common denominator.

$$\frac{x^2}{x - 5} + \frac{3x + 10}{5 - x}$$

$$= \frac{x^2}{x - 5} + \frac{(-1)(3x + 10)}{(-1)(5 - x)}$$

$$= \frac{x^2}{x - 5} + \frac{-3x - 10}{x - 5}$$ Add the numerators.

$$= \frac{x^2 - 3x - 10}{x - 5}$$ Factor the numerator.

$$= \frac{(x + 2)(x - 5)}{x - 5}$$ Simplify.

$$= x + 2$$

Check Yourself 8

Add.

$$\frac{x^2}{x-7} + \frac{10x-21}{7-x}$$

The sum of two rational functions is always a rational function. Given two rational functions $f(x)$ and $g(x)$, we can rename the sum, so $h(x) = f(x) + g(x)$. This is always true for values of x for which both f and g are defined. So, for example,

$$h(-2) = f(-2) + g(-2)$$

as long as both $f(-2)$ and $g(-2)$ exist.

Example 9 Adding Rational Functions

< Objective 2 >

Consider

$$f(x) = \frac{3x}{x+5} \qquad \text{and} \qquad g(x) = \frac{x}{x-4}$$

(a) Find $f(1) + g(1)$.

Because $f(1) = \frac{1}{2}$ and $g(1) = -\frac{1}{3}$, we have

$$f(1) + g(1) = \frac{1}{2} + \left(-\frac{1}{3}\right)$$

$$= \frac{3}{6} + \left(-\frac{2}{6}\right) = \frac{1}{6}$$

(b) Find $h(x) = f(x) + g(x)$.

$$h(x) = f(x) + g(x)$$

$$= \frac{3x}{x+5} + \frac{x}{x-4}$$

$$= \frac{3x(x-4) + x(x+5)}{(x+5)(x-4)} = \frac{3x^2 - 12x + x^2 + 5x}{(x+5)(x-4)}$$

$$= \frac{4x^2 - 7x}{(x+5)(x-4)} \qquad x \neq -5, 4$$

(c) Find the ordered pair $(1, h(1))$.

$$h(1) = \frac{-3}{-18} = \frac{1}{6}$$

The ordered pair is $\left(1, \frac{1}{6}\right)$.

Check Yourself 9

Given

$$f(x) = \frac{x}{2x-5} \qquad \text{and} \qquad g(x) = \frac{2x}{3x-1}$$

(a) Find $f(1) + g(1)$.

(b) Find $h(x) = f(x) + g(x)$.

(c) Find the ordered pair $(1, h(1))$.

When subtracting rational functions, we must be careful with the signs in the numerator of the expression being subtracted.

Example 10 Subtracting Rational Functions

Consider

$$f(x) = \frac{3x}{x+5} \quad \text{and} \quad g(x) = \frac{x-2}{x-4}$$

(a) Find $f(1) - g(1)$.

Because $f(1) = \frac{1}{2}$ and $g(1) = \frac{1}{3}$,

$$\begin{aligned} f(1) - g(1) &= \frac{1}{2} - \frac{1}{3} \\ &= \frac{3}{6} - \frac{2}{6} \\ &= \frac{3-2}{6} \\ &= \frac{1}{6} \end{aligned}$$

(b) Find $h(x) = f(x) - g(x)$.

$$\begin{aligned} h(x) &= \frac{3x}{x+5} - \frac{x-2}{x-4} \\ &= \frac{3x(x-4)}{(x+5)(x-4)} - \frac{(x-2)(x+5)}{(x-4)(x+5)} \\ &= \frac{3x(x-4) - (x-2)(x+5)}{(x+5)(x-4)} && \text{Subtract numerators.} \\ &= \frac{(3x^2 - 12x) - (x^2 + 3x - 10)}{(x+5)(x-4)} && \text{Combine like terms.} \\ &= \frac{2x^2 - 15x + 10}{(x+5)(x-4)} \qquad x \neq -5, 4 \end{aligned}$$

(c) Find the ordered pair $(1, h(1))$.

$$h(1) = \frac{-3}{-18} = \frac{1}{6}$$

The ordered pair is $\left(1, \frac{1}{6}\right)$.

Check Yourself 10

Given

$$f(x) = \frac{x}{2x-5} \quad \text{and} \quad g(x) = \frac{2x-1}{3x-1}$$

(a) Find $f(1) - g(1)$.

(b) Find $h(x) = f(x) - g(x)$.

(c) Find the ordered pair $(1, h(1))$.

Check Yourself ANSWERS

1. $\frac{2}{3y^2}$ **2. (a)** $\frac{6}{a-4}$; **(b)** $\frac{x+y}{y}$ **3. (a)** $24a^3$; **(b)** $(x+7)(x-5)$
4. (a) $(x-5)(x+5)(x+3)$; **(b)** $(y+3)^2(y-4)$
5. (a) $\frac{8a+15b}{10ab^2}$; **(b)** $\frac{2y-6}{y(y+2)}$
6. (a) $\frac{3}{(x-2)(x-5)}$; **(b)** $\frac{5x+9}{3(x+3)(x-3)}$ **7.** $\frac{-9x+1}{3x+1}$ **8.** $x-3$
9. (a) $\frac{2}{3}$; **(b)** $h(x) = \frac{7x^2-11x}{(2x-5)(3x-1)}, x \neq \frac{5}{2}, \frac{1}{3}$; **(c)** $\left(1, \frac{2}{3}\right)$
10. (a) $-\frac{5}{6}$; **(b)** $h(x) = \frac{-x^2+11x-5}{(2x-5)(3x-1)}, x \neq \frac{5}{2}, \frac{1}{3}$; **(c)** $\left(1, -\frac{5}{6}\right)$

Reading Your Text

SECTION 9.3

(a) The least common denominator (LCD) of two rational expressions is the simplest ______________ that is divisible by each of the denominators.

(b) To find the LCD we first write the denominators in completely ______________ form.

(c) An LCD must contain each of the factors appearing in the original ______________.

(d) Assuming that $h(x) = f(x) + g(x)$, $h(2) = f(2) + g(2)$ as long as both $f(2)$ and $g(2)$ ______________.

9.3 exercises

Basic Skills | Challenge Yourself | Calculator/Computer | Career Applications | Above and Beyond

Name ____________

Section ________ Date ________

Answers

1. ________ 2. ________
3. ________ 4. ________
5. ________ 6. ________
7. ________ 8. ________
9. ________ 10. ________
11. ________ 12. ________
13. ________ 14. ________
15. ________ 16. ________
17. ________ 18. ________
19. ________ 20. ________
21. ________ 22. ________
23. ________
24. ________

< Objective 1 >

Perform the indicated operations. Express your results in simplest form.

1. $\frac{9}{4x^3} + \frac{3}{4x^3}$

2. $\frac{11}{3b^3} - \frac{2}{3b^3}$

3. $\frac{5}{3a + 7} + \frac{2}{3a + 7}$

4. $\frac{6}{5x + 3} - \frac{3}{5x + 3}$

5. $\frac{2x}{x - 3} - \frac{6}{x - 3}$ 

6. $\frac{6w}{w + 4} + \frac{24}{w + 4}$

7. $\frac{y^2}{2y + 8} + \frac{3y - 4}{2y + 8}$ 

8. $\frac{x^2}{4x - 12} - \frac{9}{4x - 12}$

9. $\frac{5m - 2}{m - 6} - \frac{3m + 10}{m - 6}$

10. $\frac{3b - 8}{b - 6} + \frac{b - 16}{b - 6}$

11. $\frac{x - 7}{x^2 - x - 6} + \frac{2x - 2}{x^2 - x - 6}$

12. $\frac{5x - 12}{x^2 - 8x + 15} - \frac{3x - 2}{x^2 - 8x + 15}$

13. $\frac{3}{2x} + \frac{4}{5x}$

14. $\frac{4}{5w} - \frac{3}{4w}$

15. $\frac{6}{a} + \frac{3}{a^2}$

16. $\frac{3}{p} - \frac{7}{p^2}$

17. $\frac{2}{m} - \frac{2}{n}$

18. $\frac{5}{x} + \frac{10}{y}$

19. $\frac{3}{4b^2} - \frac{5}{3b^3}$

20. $\frac{4}{5x^3} - \frac{3}{2x^2}$

21. $\frac{3}{b} - \frac{1}{b - 3}$

22. $\frac{4}{c} + \frac{3}{c + 1}$

23. $\frac{2}{x + 1} + \frac{3}{x + 2}$ 

24. $\frac{4}{y - 1} + \frac{2}{y + 3}$

25. $\dfrac{5}{y-3} - \dfrac{1}{y+1}$

26. $\dfrac{4}{x+5} - \dfrac{3}{x-1}$

27. $\dfrac{3w}{w-6} + \dfrac{4w}{w-2}$

28. $\dfrac{3n}{n+5} + \dfrac{n}{n-4}$

29. $\dfrac{3x}{3x-2} - \dfrac{2x}{2x+1}$

30. $\dfrac{5c}{5c-1} + \dfrac{2c}{2c-3}$

31. $\dfrac{5}{x-9} + \dfrac{4}{9-x}$

32. $\dfrac{5}{a-5} - \dfrac{3}{5-a}$

33. $\dfrac{3}{x^2-16} + \dfrac{2}{x-4}$

34. $\dfrac{5}{y^2+5y+6} + \dfrac{2}{y+2}$

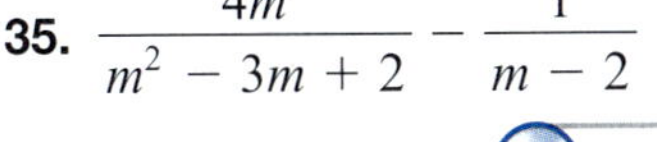

35. $\dfrac{4m}{m^2-3m+2} - \dfrac{1}{m-2}$

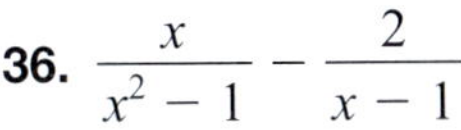

36. $\dfrac{x}{x^2-1} - \dfrac{2}{x-1}$

Basic Skills | **Challenge Yourself** | Calculator/Computer | Career Applications | Above and Beyond

Complete each statement with **never, sometimes,** *or* **always.**

37. When adding rational expressions, we ____________ have to ensure that the denominators are equal.

38. When multiplying rational expressions, we ____________ have to ensure that the denominators are equal.

39. The least common denominator for two rational expressions is ____________ created by multiplying together the two denominators.

40. When adding rational expressions, we can ____________ add the denominators together and then add the numerators together.

< Objective 2 >

Find **(a)** $f(1) + g(1)$; **(b)** $h(x) = f(x) + g(x)$; *and* **(c)** *the ordered pair* $(1, h(1))$.

41. $f(x) = \dfrac{3x}{x+1}$ and $g(x) = \dfrac{2x}{x-3}$

42. $f(x) = \dfrac{4x}{x-4}$ and $g(x) = \dfrac{x+4}{x+1}$

Answers

25. ____________

26. ____________

27. ____________

28. ____________

29. ____________

30. ____________

31. ____________

32. ____________

33. ____________

34. ____________

35. ____________

36. ____________

37. ____________

38. ____________

39. ____________

40. ____________

41. ____________

42. ____________

43. $f(x) = \frac{x}{x+1}$ and $g(x) = \frac{1}{x^2 + 2x + 1}$

44. $f(x) = \frac{x+2}{x-4}$ and $g(x) = \frac{x+3}{x+4}$

Find **(a)** $f(1) - g(1)$; **(b)** $h(x) = f(x) - g(x)$; *and* **(c)** *the ordered pair* $(1, h(1))$.

45. $f(x) = \frac{x+5}{x-5}$ and $g(x) = \frac{x-5}{x+5}$

46. $f(x) = \frac{2x}{x-4}$ and $g(x) = \frac{3x}{x+7}$

47. $f(x) = \frac{x+9}{4x-36}$ and $g(x) = \frac{x-9}{x^2 - 18x + 81}$

48. $f(x) = \frac{4x+1}{x+5}$ and $g(x) = -\frac{2}{x}$

Evaluate each expression at the given variable value(s).

49. $\frac{5x+5}{x^2+3x+2} - \frac{x-3}{x^2+5x-6}$, $x = -4$

50. $\frac{y-3}{y^2-6y+8} + \frac{2y-6}{y^2-4}$, $y = 3$

51. $\frac{2m+2n}{m^2-n^2} + \frac{m-2n}{m^2+2mn+n^2}$, $m = 3, n = 2$

52. $\frac{w-3z}{w^2-2wz+z^2} - \frac{w+2z}{w^2-z^2}$, $w = 2, z = 1$

53. $\frac{1}{a-3} - \frac{1}{a+3} + \frac{2a}{a^2-9}$, $a = 4$

54. $\frac{1}{m+1} + \frac{1}{m-3} - \frac{4}{m^2-2m-3}$, $m = -2$

55. $\frac{3w^2+16w-8}{w^2+2w-8} + \frac{w}{w+4} - \frac{w-1}{w-2}$, $w = 3$

56. $\frac{4x^2-7x-45}{x^2-6x+5} - \frac{x+2}{x-1} - \frac{x}{x-5}$, $x = -3$

Answers

43. ______

44. ______

45. ______

46. ______

47. ______

48. ______

49. ______

50. ______

51. ______

52. ______

53. ______

54. ______

55. ______

56. ______

57. $\dfrac{a^2 - 9}{2a^2 - 5a - 3} \cdot \left(\dfrac{1}{a - 2} + \dfrac{1}{a + 3}\right), \quad a = -3$

58. $\dfrac{m^2 - 2mn + n^2}{m^2 + 2mn - 3n^2} \cdot \left(\dfrac{2}{m - n} - \dfrac{1}{m + n}\right), \quad m = 4, n = -3$

Basic Skills | Challenge Yourself | **Calculator/Computer** | Career Applications | Above and Beyond

As we saw in Section 9.2 exercises, the graphing calculator can be used to check our work. In exercises 59 to 64, enter the first rational expression in Y_1 *and the second in* Y_2. *In* Y_3, *you will enter either* $Y_1 + Y_2$ *or* $Y_1 - Y_2$. *Enter your algebraically simplified rational expression in* Y_4. *The graphs of* Y_3 *and* Y_4 *will appear to be identical if you have correctly simplified the expression.*

59. $\dfrac{6y}{y^2 - 8y + 15} + \dfrac{9}{y - 3}$

60. $\dfrac{8a}{a^2 - 8a + 12} + \dfrac{4}{a - 2}$

61. $\dfrac{6x}{x^2 - 10x + 24} - \dfrac{18}{x - 6}$

62. $\dfrac{21p}{p^2 - 3p - 10} - \dfrac{15}{p - 5}$

63. $\dfrac{2}{z^2 - 4} + \dfrac{3}{z^2 + 2z - 8}$

64. $\dfrac{5}{x^2 - 3x - 10} + \dfrac{2}{x^2 - 25}$

Answers

57. ____________________

58. ____________________

59. ____________________

60. ____________________

61. ____________________

62. ____________________

63. ____________________

64. ____________________

Answers

1. $\frac{3}{x^3}$ **3.** $\frac{7}{3a+7}$ **5.** 2 **7.** $\frac{y-1}{2}$ **9.** 2 **11.** $\frac{3}{x+2}$
13. $\frac{23}{10x}$ **15.** $\frac{3(2a+1)}{a^2}$ **17.** $\frac{2(n-m)}{mn}$ **19.** $\frac{9b-20}{12b^3}$
21. $\frac{2b-9}{b(b-3)}$ **23.** $\frac{5x+7}{(x+1)(x+2)}$ **25.** $\frac{4(y+2)}{(y-3)(y+1)}$
27. $\frac{w(7w-30)}{(w-6)(w-2)}$ **29.** $\frac{7x}{(3x-2)(2x+1)}$ **31.** $\frac{1}{x-9}$
33. $\frac{2x+11}{(x+4)(x-4)}$ **35.** $\frac{3m+1}{(m-1)(m-2)}$ **37.** always
39. sometimes **41.** **(a)** $\frac{1}{2}$; **(b)** $h(x) = \frac{5x^2-7x}{(x+1)(x-3)}, x \neq -1, 3$; **(c)** $\left(1, \frac{1}{2}\right)$
43. **(a)** $\frac{3}{4}$; **(b)** $h(x) = \frac{x^2+x+1}{(x+1)^2}, x \neq -1$; **(c)** $\left(1, \frac{3}{4}\right)$
45. **(a)** $-\frac{5}{6}$; **(b)** $h(x) = \frac{20x}{(x-5)(x+5)}, x \neq 5, -5$; **(c)** $\left(1, -\frac{5}{6}\right)$
47. **(a)** $-\frac{3}{16}$; **(b)** $h(x) = \frac{(x+5)}{4(x-9)}, x \neq 9$; **(c)** $\left(1, -\frac{3}{16}\right)$
49. $-\frac{16}{5}$ **51.** $\frac{49}{25}$ **53.** 2 **55.** 8 **57.** Undefined
59. $\frac{15}{y-5}$ **61.** $-\frac{12}{x-4}$ **63.** $\frac{5z+14}{(z+2)(z-2)(z+4)}$

9.4 Complex Fractions

< 9.4 Objectives >

1 > Use the fundamental principle to simplify complex fractions

2 > Use division to simplify complex fractions

Our work in this section deals with two methods for simplifying complex fractions. We begin with a definition. A **complex fraction** is a fraction that has a fraction in its numerator or denominator (or both). Some examples are

$$\frac{\frac{5}{6}}{\frac{3}{4}} \qquad \frac{\frac{4}{x}}{\frac{3}{x+1}} \qquad \frac{1+\frac{1}{x}}{1-\frac{1}{x}}$$

RECALL

Fundamental principle:

$\frac{P}{Q} = \frac{PR}{QR}$

if $Q \neq 0$ and $R \neq 0$.

Two methods can be used to simplify complex fractions. Method 1 involves the fundamental principle, and Method 2 involves inverting and multiplying.

Recall that by the *fundamental principle* we can always multiply the numerator and denominator of a fraction by the same nonzero quantity. In simplifying a complex fraction, we multiply the numerator and denominator by the LCD of all fractions that appear within the complex fraction.

Here the denominators are 5 and 10, so we can write

$$\frac{\frac{3}{5}}{\frac{7}{10}} = \frac{\frac{3}{5} \cdot 10}{\frac{7}{10} \cdot 10} = \frac{6}{7}$$

NOTE

We are multiplying by $\frac{10}{10}$ or 1.

Our second approach interprets the complex fraction as division and applies our earlier work in dividing fractions in which we *invert and multiply.*

$$\frac{\frac{3}{5}}{\frac{7}{10}} = \frac{3}{5} \div \frac{7}{10} = \frac{3}{5} \cdot \frac{10}{7} = \frac{6}{7}$$

Invert and multiply.

Which method is better? The answer depends on the expression you are trying to simplify. Both approaches are effective, and you should be familiar with both. With practice you will be able to tell which method may be easier to use in a particular situation.

Let's look at the same two methods applied to the simplification of an algebraic complex fraction.

Example 1 Simplifying Complex Fractions

< Objective 1 >

Simplify.

$$\frac{1+\frac{2x}{y}}{2-\frac{x}{y}}$$

Method 1 The LCD of 1, $\frac{2x}{y}$, 2, and $\frac{x}{y}$ is y. So we multiply the numerator and denominator by y.

$$\frac{1 + \frac{2x}{y}}{2 - \frac{x}{y}} = \frac{\left(1 + \frac{2x}{y}\right) \cdot y}{\left(2 - \frac{x}{y}\right) \cdot y}$$

$$= \frac{1 \cdot y + \frac{2x}{y} \cdot y}{2 \cdot y - \frac{x}{y} \cdot y}$$ Distribute y over the numerator and denominator.

$$= \frac{y + 2x}{2y - x}$$ Simplify.

Method 2 In this approach, we must *first work separately* in the numerator and denominator to form single fractions.

NOTE

Make sure you understand the steps in forming a single fraction in the numerator and denominator.

$$\frac{1 + \frac{2x}{y}}{2 - \frac{x}{y}} = \frac{\frac{y}{y} + \frac{2x}{y}}{\frac{2y}{y} - \frac{x}{y}} = \frac{\frac{y + 2x}{y}}{\frac{2y - x}{y}}$$

$$= \frac{y + 2x}{y} \cdot \frac{y}{2y - x}$$ Invert the divisor and multiply.

$$= \frac{y + 2x}{2y - x}$$

Check Yourself 1

Simplify.

$$\frac{\frac{x}{y} - 1}{\frac{2x}{y} + 2}$$

Again, simplifying a complex fraction means writing an equivalent simple fraction in lowest terms, as Example 2 illustrates.

Example 2 Simplifying Complex Fractions

Simplify.

$$\frac{1 - \frac{2y}{x} + \frac{y^2}{x^2}}{1 - \frac{y^2}{x^2}}$$

We choose the first method of simplification in this case. The LCD of all the fractions that appear is x^2. So we multiply the numerator and denominator by x^2.

$$\frac{1 - \frac{2y}{x} + \frac{y^2}{x^2}}{1 - \frac{y^2}{x^2}} = \frac{\left(1 - \frac{2y}{x} + \frac{y^2}{x^2}\right) \cdot x^2}{\left(1 - \frac{y^2}{x^2}\right) \cdot x^2}$$

Distribute x^2 over the numerator and denominator, and simplify.

$$= \frac{x^2 - 2xy + y^2}{x^2 - y^2}$$

Factor the numerator and denominator.

$$= \frac{(x - y)(x - y)}{(x + y)(x - y)} = \frac{x - y}{x + y}$$

Divide by the common factor $x - y$.

Check Yourself 2

Simplify.

$$\frac{1 + \frac{5}{x} + \frac{6}{x^2}}{1 - \frac{9}{x^2}}$$

In Example 3, we illustrate the second method of simplification for purposes of comparison.

Example 3 Simplifying Complex Fractions

< Objective 2 >

NOTES

Again, take time to make sure you understand how the numerator and denominator are rewritten as single fractions.

Method 2 is probably the more efficient in this case. The LCD of the denominators would be $(x + 2)(x - 1)$, leading to a more complicated process if we use Method 1.

Simplify.

$$\frac{1 - \frac{1}{x + 2}}{x - \frac{2}{x - 1}}$$

$$\frac{1 - \frac{1}{x + 2}}{x - \frac{2}{x - 1}} = \frac{\frac{x + 2}{x + 2} - \frac{1}{x + 2}}{\frac{x(x - 1)}{x - 1} - \frac{2}{x - 1}} = \frac{\frac{x + 2 - 1}{x + 2}}{\frac{x(x - 1) - 2}{x - 1}} = \frac{\frac{x + 1}{x + 2}}{\frac{x^2 - x - 2}{x - 1}}$$

$$= \frac{x + 1}{x + 2} \cdot \frac{x - 1}{x^2 - x - 2} = \frac{\overset{1}{\cancel{x + 1}}}{x + 2} \cdot \frac{x - 1}{(x - 2)\underset{1}{\cancel{(x + 1)}}} = \frac{x - 1}{(x + 2)(x - 2)}$$

Check Yourself 3

Simplify.

$$\frac{2 + \frac{5}{x - 3}}{x - \frac{1}{2x + 1}}$$

Complex fractions can show up when we solve certain problems known as "work" problems.

Example 4 Simplifying a Complex Fraction

NOTE

We will study work problems in Section 9.6.

If one person can install a storm door in u hours, and a second person can do the same installation in v hours, then the time required to install a door working together is given by the expression

$$\frac{1}{\frac{1}{u}+\frac{1}{v}}$$

Simplify this expression.

Using Method 1, we multiply the numerator and denominator by uv:

$$\frac{1}{\frac{1}{u}+\frac{1}{v}} = \frac{1 \cdot uv}{\left(\frac{1}{u}+\frac{1}{v}\right) \cdot uv}$$

$$= \frac{uv}{\left(\frac{1}{u}\right)uv + \left(\frac{1}{v}\right)uv}$$ Distribute the multiplication of uv in the denominator.

$$= \frac{uv}{v+u}$$

Check Yourself 4

Simplify.

$$\frac{5}{\frac{1}{t}-\frac{1}{u}}$$

In a later math class, you will encounter an important expression dealing with functions: $\frac{f(x+h)-f(x)}{h}$. This expression involves a complex fraction if $f(x)$ is itself a rational function. Consider Example 5.

Example 5 Simplifying a Complex Fraction

Let $f(x) = \frac{2}{x}$. Simplify the expression $\frac{f(x+h)-f(x)}{h}$.

First we note that $f(x+h) = \frac{2}{x+h}$.

Now: $$\frac{f(x+h)-f(x)}{h} = \frac{\frac{2}{x+h}-\frac{2}{x}}{h}$$

Using Method 1, we multiply the numerator and denominator by $x(x + h)$.

$$\frac{\left(\frac{2}{x+h} - \frac{2}{x}\right) \cdot [x(x+h)]}{h \cdot [x(x+h)]}$$

$$= \frac{\left(\frac{2}{x+h}\right) \cdot [x(x+h)] - \left(\frac{2}{x}\right) \cdot [x(x+h)]}{h \cdot [x(x+h)]}$$

Distribute $x \cdot (x + h)$ in the numerator.

$$= \frac{\frac{(2)(x)(x+h)}{x+h} - \frac{(2)(x)(x+h)}{x}}{hx(x+h)}$$

Multiply in the numerator.

$$= \frac{2x - 2(x+h)}{hx(x+h)}$$

Cancel common factors in the numerator.

$$= \frac{2x - 2x - 2h}{hx(x+h)} = \frac{-2h}{hx(x+h)} = \frac{-2}{x(x+h)}$$

Simplify the numerator, and divide by the common factor h.

NOTE

Try using Method 2. You should get the same result.

Check Yourself 5

Let $f(x) = \frac{3}{x}$. Simplify the expression $\frac{f(x+h) - f(x)}{h}$.

This algorithm summarizes our work with complex fractions.

Step by Step

Simplifying Complex Fractions

Method 1

Step 1 Multiply the numerator and denominator of the complex fraction by the LCD of all the fractions that appear within the numerator and denominator.

Step 2 Simplify the resulting rational expression, writing the expression in lowest terms.

Method 2

Step 1 Write the numerator and denominator of the complex fraction as single fractions, if necessary.

Step 2 Invert the denominator and multiply as before, writing the result in lowest terms.

Check Yourself ANSWERS

1. $\dfrac{x - y}{2(x + y)}$ 2. $\dfrac{x + 2}{x - 3}$ 3. $\dfrac{2x + 1}{(x - 3)(x + 1)}$ 4. $\dfrac{5tu}{u - t}$
5. $\dfrac{-3}{x(x + h)}$

Reading Your Text

SECTION 9.4

(a) A ______________ fraction is a fraction that has a fraction in its numerator or denominator (or both).

(b) By the ______________ principle we can always multiply the numerator and denominator of a fraction by the same nonzero quantity.

(c) When dividing fractions, we ______________ the second fraction and multiply.

(d) No matter which method we use, the final step requires that we write the result in ______________ terms.

Basic Skills | Challenge Yourself | Calculator/Computer | Career Applications | Above and Beyond

9.4 exercises

< Objectives 1 and 2 >

Simplify each complex fraction.

1. $\dfrac{\frac{2}{3}}{\frac{6}{8}}$

2. $\dfrac{\frac{5}{6}}{\frac{10}{15}}$

3. $\dfrac{\frac{3}{4}+\frac{1}{3}}{\frac{1}{6}+\frac{3}{4}}$

4. $\dfrac{\frac{3}{4}+\frac{1}{2}}{\frac{7}{8}-\frac{1}{4}}$ > Videos

5. $\dfrac{2+\frac{1}{3}}{3-\frac{1}{5}}$

6. $\dfrac{1+\frac{3}{4}}{2-\frac{1}{8}}$

7. $\dfrac{\frac{x}{8}}{\frac{x^2}{4}}$

8. $\dfrac{\frac{x^2}{12}}{\frac{x^5}{18}}$

9. $\dfrac{\frac{3}{m}}{\frac{6}{m^2}}$

10. $\dfrac{\frac{15}{x^2}}{\frac{20}{x^3}}$

11. $\dfrac{\frac{y+1}{y}}{\frac{y-1}{2y}}$

12. $\dfrac{\frac{x+3}{4x}}{\frac{x-3}{2x}}$

13. $\dfrac{\frac{a+2b}{3a}}{\frac{a^2+2ab}{9b}}$

14. $\dfrac{\frac{m-3n}{4m}}{\frac{m^2-3mn}{8n}}$

15. $\dfrac{\frac{x-3}{x^2-25}}{\frac{x^2+x-12}{x^2+5x}}$

16. $\dfrac{\frac{x+5}{x^2-6x}}{\frac{x^2-25}{x^2-36}}$ > Videos

17. $\dfrac{2-\frac{1}{x}}{2+\frac{1}{x}}$

18. $\dfrac{3+\frac{1}{b}}{3-\frac{1}{b}}$

19. $\dfrac{\frac{1}{x}-\frac{1}{y}}{\frac{1}{xy}}$

20. $\dfrac{\frac{4}{xy}}{\frac{1}{y}-\frac{1}{x}}$

Name ____________

Section ________ Date ________

Answers

1. ________ 2. ________

3. ________ 4. ________

5. ________ 6. ________

7. ________ 8. ________

9. ________ 10. ________

11. ________ 12. ________

13. ________ 14. ________

15. ________

16. ________

17. ________

18. ________

19. ________

20. ________

Answers

21. ______
22. ______
23. ______
24. ______
25. ______
26. ______
27. ______
28. ______
29. ______
30. ______
31. ______
32. ______
33. ______
34. ______
35. ______
36. ______
37. ______
38. ______
39. ______

21. $\dfrac{\dfrac{x^2}{y^2} - 1}{\dfrac{x}{y} + 1}$

22. $\dfrac{\dfrac{m}{n} + 2}{\dfrac{m^2}{n^2} - 4}$

23. $\dfrac{1 + \dfrac{3}{a} - \dfrac{4}{a^2}}{1 + \dfrac{2}{a} - \dfrac{3}{a^2}}$

24. $\dfrac{1 - \dfrac{2}{x} - \dfrac{8}{x^2}}{1 - \dfrac{1}{x} - \dfrac{6}{x^2}}$

25. $\dfrac{\dfrac{x^2}{y} + 2x + y}{\dfrac{1}{y^2} - \dfrac{1}{x^2}}$

26. $\dfrac{\dfrac{a}{b} + 1 - \dfrac{2b}{a}}{\dfrac{1}{b^2} - \dfrac{4}{a^2}}$

27. $\dfrac{2 - \dfrac{2}{x + 1}}{2 + \dfrac{2}{x + 1}}$

28. $\dfrac{3 - \dfrac{4}{m + 2}}{3 + \dfrac{4}{m + 2}}$

29. $\dfrac{1 - \dfrac{1}{y - 1}}{y - \dfrac{8}{y + 2}}$

30. $\dfrac{1 + \dfrac{1}{x + 2}}{x - \dfrac{18}{x - 3}}$

31. $\dfrac{\dfrac{1}{x - 3} + \dfrac{1}{x + 3}}{\dfrac{1}{x - 3} - \dfrac{1}{x + 3}}$

32. $\dfrac{\dfrac{2}{m - 2} + \dfrac{1}{m - 3}}{\dfrac{2}{m - 2} - \dfrac{1}{m - 3}}$

33. $\dfrac{\dfrac{x}{x + 1} + \dfrac{1}{x - 1}}{\dfrac{x}{x - 1} - \dfrac{1}{x + 1}}$

34. $\dfrac{\dfrac{y}{y - 4} + \dfrac{1}{y + 2}}{\dfrac{4}{y - 4} - \dfrac{1}{y + 2}}$

35. $\dfrac{\dfrac{a + 1}{a - 1} - \dfrac{a - 1}{a + 1}}{\dfrac{a + 1}{a - 1} + \dfrac{a - 1}{a + 1}}$

36. $\dfrac{\dfrac{x + 2}{x - 2} - \dfrac{x - 2}{x + 2}}{\dfrac{x + 2}{x - 2} + \dfrac{x - 2}{x + 2}}$

Basic Skills | **Challenge Yourself** | Calculator/Computer | Career Applications | Above and Beyond

37. $1 + \dfrac{1}{1 + \dfrac{1}{x}}$

38. $1 + \dfrac{1}{1 - \dfrac{1}{y}}$

39. $1 + \dfrac{1}{1 + \dfrac{1}{1 + \dfrac{1}{x}}}$

40. Extend the "continued fraction" patterns in exercises 37 and 39 to write the next complex fraction.

41. Simplify the complex fraction in exercise 40.

Determine whether each statement is **true** *or* **false.**

42. We can always rewrite a complex fraction as a simple fraction.

43. The complex fraction $\dfrac{2}{\dfrac{3}{4}}$ is the same as the complex fraction $\dfrac{\dfrac{2}{3}}{4}$.

For each function, find and simplify the expression $\dfrac{f(x+h)-f(x)}{h}$.

44. $f(x) = \dfrac{5}{x}$

45. $f(x) = \dfrac{1}{2x}$

46. $f(x) = -\dfrac{1}{x}$

47. $f(x) = -\dfrac{3}{x}$

Basic Skills | Challenge Yourself | **Calculator/Computer** | Career Applications | Above and Beyond

Use the table utility on your graphing calculator to complete each table. Comment on the equivalence of the two expressions.

48.

x	-3	-2	-1	0	1	2	3
$\dfrac{1-\dfrac{2}{x}}{1-\dfrac{4}{x^2}}$							
$\dfrac{x}{x+2}$							

49.

x	-3	-2	-1	0	1	2	3
$\dfrac{-8+\dfrac{20}{x}}{4-\dfrac{25}{x^2}}$							
$\dfrac{-4x}{2x+5}$							

Answers

40. ______

41. ______

42. ______

43. ______

44. ______

45. ______

46. ______

47. ______

48. ______

49. ______

Answers

50. ____________

51. ____________

52. ____________

53. ____________

54. ____________

55. ____________

Basic Skills | Challenge Yourself | Calculator/Computer | **Career Applications** | Above and Beyond

Allied Health *Total compliance (C_T) for a patient is based on lung compliance (C_L) and chest-wall compliance (C_{CW}). It is measured in centimeters of water (cm H_2O) and computed using the formula*

$$C_T = \frac{1}{\frac{1}{C_L} + \frac{1}{C_{CW}}}$$

Use this formula to complete exercises 50 and 51.

50. Simplify the total compliance formula. 

51. Determine the total compliance for a patient whose lung compliance is 0.15 cm H_2O and whose chest-wall compliance is 0.20 cm H_2O. Report your result accurate to three decimal places.

Electrical Engineering *Generally, the wiring in buildings is arranged so all electric devices are in parallel. This way, if one device is disconnected, the current to the other devices is not interrupted. The equivalent single resistance (measured in ohms, Ω) for two devices connected in parallel (see figure) is given by the formula*

$$R_{eq} = (R_1^{-1} + R_2^{-1})^{-1}$$

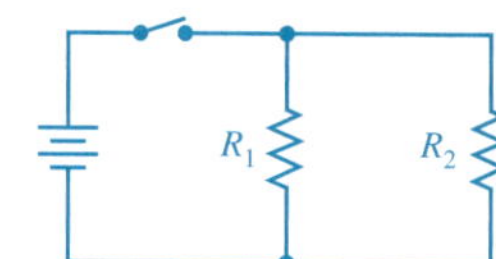

Use the formula to complete exercises 52 and 53.

52. Write the resistance formula without exponents. Simplify the resistance formula.

53. Use the resistance formula to determine the equivalent total resistance of the parallel circuit shown. Report your result to the nearest ohm.

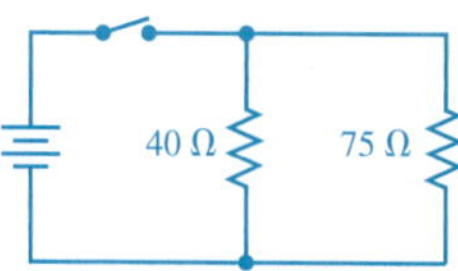

Basic Skills | Challenge Yourself | Calculator/Computer | Career Applications | **Above and Beyond**

54. Outline the two different methods used to simplify a complex fraction. What are the advantages of each method?

55. Can the expression $\frac{x^2 + y^2}{x + y}$ be written as $\frac{x^2}{x} + \frac{y^2}{y}$? If not, is there a correct simplified form?

56. Write and simplify a complex fraction that is the reciprocal of $x + \dfrac{6}{x - 1}$.

57. Let $f(x) = \dfrac{3}{x}$. Write and simplify a complex fraction whose numerator is $f(3 + h) - f(3)$ and whose denominator is h.

58. Write and simplify a complex fraction that is the arithmetic mean of $\dfrac{1}{x}$ and $\dfrac{1}{x - 1}$.

59. Use a reference work to find the first six Fibonnacci numbers. Compare this set of numbers to your results in exercises 37, 39, and 41.

Suppose you drive at 40 mi/h from city A to city B. You then return along the same route from city B to city A at 50 mi/h. What is your average rate for the round trip? Your obvious guess would be 45 mi/h, but you are in for a surprise.

Suppose that the cities are 200 mi apart. Your time from city A to city B is the distance divided by the rate, or

$$\frac{200 \text{ mi}}{40 \text{ mi/h}} = 5 \text{ h}$$

Similarly, your time from city B to city A is

$$\frac{200 \text{ mi}}{50 \text{ mi/h}} = 4 \text{ h}$$

The total time is then 9 h, and now using rate equals distance divided by time, we have

$$\frac{400 \text{ mi}}{9 \text{ h}} = \frac{400}{9} \text{ mi/h} = 44\frac{4}{9} \text{ mi/h}$$

Note that the rate for the round trip is independent of the distance involved. For instance, try the previous computations if cities A and B are 400 mi apart.

The answer to the problem is the complex fraction

$$R = \frac{2}{\dfrac{1}{R_1} + \dfrac{1}{R_2}}$$

where R_1 = rate going
R_2 = rate returning
R = rate for round trip

Use this information to solve exercises 60 to 63.

60. Verify that if $R_1 = 40$ mi/h and $R_2 = 50$ mi/h, then $R = 44\frac{4}{9}$ mi/h, by simplifying the complex fraction *after* substituting those values.

61. Simplify the given complex fraction first. *Then* substitute 40 for R_1 and 50 for R_2 to calculate R.

62. Repeat exercise 60, where $R_1 = 50$ mi/h and $R_2 = 60$ mi/h.

63. Use the procedure in exercise 61 with the above values for R_1 and R_2.

Answers

56. ______

57. ______

58. ______

59. ______

60. ______

61. ______

62. ______

63. ______

Answers

64. ______

65. ______

64. Mathematicians have shown that there are situations in which the method used to determine the number of U.S. representatives each state gets may not be fair, and a state may not get its basic quota of representatives. They give the table below of a hypothetical seven states and their populations as an example.

State	Population	Exact Quota	Rounded Quota	Actual Number of Reps.
A	325	1.625	2	2
B	788	3.940	4	4
C	548	2.740	3	3
D	562	2.810	3	3
E	4,263	21.315	21	21
F	3,219	16.095	16	15
G	295	1.475	1	2
Total	10,000	50		50

In this case, the total population of all states is 10,000, and there are 50 representatives in all, so there should be no more than 10,000/50, or 200, people per representative. The quotas are found by dividing the population by 200. Whether a state *A* should get an additional representative before another state *E* should get one is decided in this method by using the simplified inequality below. For each state, we find the ratio of the state's population to the square root of the product of the possible number of representatives and one more than this integer. We then compare these ratios for each two states. If

$$\frac{A}{\sqrt{a(a+1)}} > \frac{E}{\sqrt{e(e+1)}}$$

is true, then *A* gets an extra representative before *E* does.

(a) If you go through the process of comparing the inequality for each pair of states, state *F* loses a representative to state *G*. Do you see how this happens? Will state *F* complain?

chapter 9 > Make the Connection

(b) Alexander Hamilton, one of the signers of the Constitution, proposed that the extra representative positions be given one at a time to states with the largest remainder until all the "extra" positions were filled. How would this affect the table? Do you agree or disagree?

65. In Italy in the 1500s, Pietro Antonio Cataldi expressed square roots as infinite, continued fractions. It is not a difficult process to follow. For instance, if you want the square root of 5, then let

$$x + 1 = \sqrt{5}$$

Squaring both sides gives

$$(x + 1)^2 = 5 \qquad \text{or} \qquad x^2 + 2x + 1 = 5$$

which can be written

$$x(x + 2) = 4$$

$$x = \frac{4}{x + 2}$$

One can continue replacing x with $\frac{4}{2 + x}$:

$$x = \cfrac{4}{2 + \cfrac{4}{2 + \cfrac{4}{2 + \cfrac{4}{2 + \cdots}}}}$$

to obtain

$\sqrt{5} - 1$

(a) Evaluate the complex fraction above (ignore the three dots) and then add 1, and see how close it is to the square root of 5. What should you put where the ellipsis (. . .) is? Try a number you think is close to $\sqrt{5} - 1$. How far would you have to go to get the square root correct to the nearest hundredth?

(b) Develop an infinite complex fraction for $\sqrt{10} - 1$.

66. Here is yet another method for simplifying a complex fraction. Suppose we want to simplify

$$\frac{\frac{3}{5}}{\frac{7}{10}}$$

Multiply the numerator and denominator of the complex fraction by $\frac{10}{7}$.

(a) What principle allows you to do this?

(b) Why was $\frac{10}{7}$ chosen?

(c) When learning to divide fractions, you may have heard the saying "Yours is not to reason why . . . just invert and multiply." How does this method serve to explain the "reason why" we invert and multiply?

Answers

66. ______________________

Answers

1. $\frac{8}{9}$ 3. $\frac{13}{11}$ 5. $\frac{5}{6}$ 7. $\frac{1}{2x}$ 9. $\frac{m}{2}$ 11. $\frac{2(y+1)}{y-1}$ 13. $\frac{3b}{a^2}$

15. $\frac{x}{(x-5)(x+4)}$ 17. $\frac{2x-1}{2x+1}$ 19. $y-x$ 21. $\frac{x-y}{y}$ 23. $\frac{a+4}{a+3}$

25. $\frac{x^2y(x+y)}{(x-y)}$ 27. $\frac{x}{x+2}$ 29. $\frac{y+2}{(y-1)(y+4)}$ 31. $\frac{x}{3}$ 33. 1

35. $\frac{2a}{a^2+1}$ 37. $\frac{2x+1}{x+1}$ 39. $\frac{3x+2}{2x+1}$ 41. $\frac{5x+3}{3x+2}$ 43. False

45. $\frac{-1}{2x(x+h)}$ 47. $\frac{3}{x(x+h)}$

49.

x	−3	−2	−1	0	1	2	3
$\dfrac{1-\frac{2}{x}}{1-\frac{4}{x^2}}$	3	Error	−1	Error	0.3333	Error	0.6
$\dfrac{x}{x+2}$	3	Error	−1	0	0.3333	0.5	0.6

The expressions are equivalent.

51. 0.086 cm H_2O 53. 26 Ω 55. Above and Beyond

57. $\dfrac{\frac{3}{3+h}-1}{h}=\dfrac{-1}{3+h}$ 59. Above and Beyond 61. $44\frac{4}{9}$ mi/h

63. $54\frac{6}{11}$ mi/h 65. Above and Beyond

9.5 Introduction to Graphing Rational Functions

< 9.5 Objectives >

1 > Find the domain and intercepts for a rational function

2 > Draw the graph of a rational function

3 > Identify the asymptotes for a rational function

Earlier in this text, you learned about the graphs of linear and quadratic functions. These graphs are very predictable: lines and parabolas. Unfortunately, the story for rational functions is not so straightforward. The graphs of these functions can vary widely. If we confine our study to a simpler subgroup of rational functions, there are some consistent patterns and characteristics. In this introductory section, that is exactly what we do.

To begin, we focus on the domain of the function and the intercepts for the graph.

Example 1 Finding the Domain and Intercepts for a Rational Function

< Objective 1 >

For each function, give the domain of f, and find the intercepts for the graph of f.

RECALL

To find the domain of a rational function, focus on the denominator.

(a) $f(x) = \frac{8}{x}$

The domain of f is the set of all real numbers except 0; that is, $\{x|x \neq 0\}$.

Recall that to find the y-intercept, we input 0 for x and find the resulting output value, since we are searching for a point of the form (0, ?). But 0 is not in the domain so there is no y-intercept.

To find any x-intercepts, remember that we are searching for points of the form (?, 0). That is, we should let the output value be 0 and solve for x.

But, $0 = \frac{8}{x}$ has no solution. There are no x-intercepts.

(b) $f(x) = \frac{6}{x - 2}$

Focusing on the denominator, we see that if $x = 2$, the function is undefined. So, the domain of f is $\{x|x \neq 2\}$.

To find the y-intercept, let $x = 0$.

$$f(0) = \frac{6}{0 - 2} = -3$$

The y-intercept is $(0, -3)$.

To find x-intercepts, let $y = 0$.

$$0 = \frac{6}{x - 2}$$

This equation has no solution, since 6 divided by a number can never be 0. There are no x-intercepts.

Check Yourself 1

For each function, give the domain of f, and find the intercepts for the graph of f.

(a) $f(x) = \dfrac{5}{x}$ (b) $f(x) = \dfrac{-3}{x + 2}$

A simplified fraction equals 0 if, and only if, the numerator equals 0. Looking back at Example 1, notice that the numerator for each given rational expression cannot equal 0. When searching for x-intercepts, we need only determine when the numerator has the value 0.

Let us now consider a slightly more complicated function.

Example 2 Finding the Domain and Intercepts for a Rational Function

For the function $f(x) = \dfrac{x - 5}{x - 3}$, give the domain of f, and find the intercepts for the graph of f.

The domain of f is the set of all real numbers except 3. (Do you see why?) So, the domain of f is $\{x | x \neq 3\}$.

To find the y-intercept, set $x = 0$: $f(0) = \dfrac{0 - 5}{0 - 3} = \dfrac{5}{3}$

RECALL

To find the x-intercepts of a rational function, focus on the numerator.

So $\left(0, \dfrac{5}{3}\right)$ is the y-intercept.

To find x-intercepts, let the output be 0: $0 = \dfrac{x - 5}{x - 3}$

This is only true if the numerator has the value 0. So, set $x - 5$ equal to 0, and solve. This gives us $x = 5$. There is an x-intercept at (5, 0).

Check Yourself 2

For the function $f(x) = \dfrac{x + 4}{x - 2}$, give the domain of f, and find the intercepts for the graph of f.

We turn our attention now to the graph of a rational function.

Example 3 Drawing the Graph of a Rational Function

< Objective 2 >

Draw the graph of $f(x) = \dfrac{8}{x}$.

We know the function is undefined at $x = 0$, and that there are no intercepts for the graph of f. (See Example 1.)

Making a table of some easy-to-compute points, we have

x	$f(x)$
−8	−1
−4	−2
−2	−4
−1	−8
0	undef.
1	8
2	4
4	2
8	1

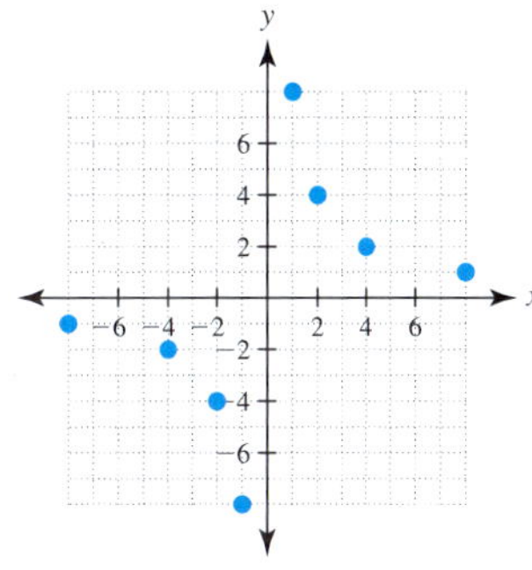

Connecting these with smooth curves, we have

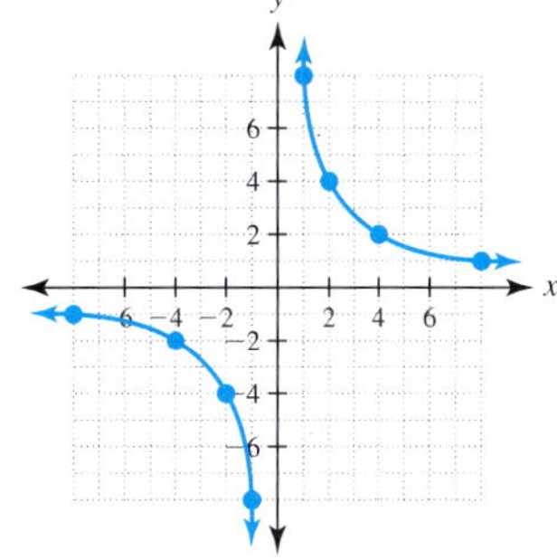

Check Yourself 3

Sketch the graph of $f(x) = \dfrac{-6}{x}$.

Looking at the graph sketched in Example 3, two questions arise: **(1)** What is the behavior of the graph near the undefined location (that is, near $x = 0$)? **(2)** What is the behavior of the graph out to the sides (that is, for large positive and negative values of x)?

1. As x-values are chosen close to 0, but slightly to the right of 0, we see that the output values grow to be huge positive numbers.

x	1	0.1	0.01	0.001	0.0001
$f(x)$	8	80	800	8,000	80,000

As x-values are chosen close to 0, but slightly to the left of 0, we see that the output values grow to be huge negative numbers.

x	−1	−0.1	−0.01	−0.001	−0.0001
$f(x)$	−8	−80	−800	−8,000	−80,000

The two pieces of the graph are approaching the y-axis (but never reaching it!). The y-axis is a vertical line whose equation is $x = 0$, and the graph of f approaches this without touching it. We call this vertical line a **vertical asymptote**, and, again, its equation is $x = 0$.

NOTE

Think of an asymptote as a "guideline" for the graph.

2. As large positive x-values are chosen, we see that the output values approach 0.

x	10	100	1,000	10,000
$f(x)$	0.8	0.08	0.008	0.0008

As large negative x-values are chosen, we see that the output values also approach 0, but from below the x-axis.

x	-10	-100	$-1,000$	$-10,000$
$f(x)$	-0.8	-0.08	-0.008	-0.0008

As we look out to the sides, then, the graph approaches the x-axis. The x-axis is a horizontal line whose equation is $y = 0$, and it serves as a **horizontal asymptote.**

This sort of **asymptotic** behavior is a typical characteristic of the graphs of many rational functions. To sketch the graph, we must learn to identify the asymptotes.

Example 4 Identifying the Asymptotes for a Rational Function

< Objective 3 >

Identify the asymptotes for the graph of each rational function.

(a) $f(x) = \dfrac{2}{x + 3}$

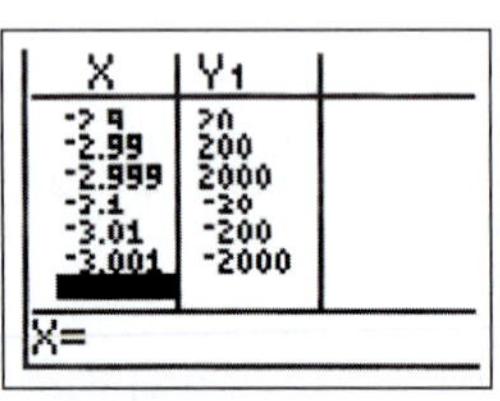

To search for vertical asymptotes, we focus on the denominator. Since f is undefined at $x = -3$, we suspect that there is a vertical asymptote at that location. Using a calculator, we see that as x-values are chosen close to $x = -3$, the outputs are growing huge (positively on one side of -3, and negatively on the other side of -3). The equation of the vertical asymptote is $x = -3$.

To find horizontal asymptotes, we choose large input values for x, and observe the resulting values for $f(x)$. In this case, we see that the outputs approach 0. So, there is a horizontal asymptote with equation $y = 0$.

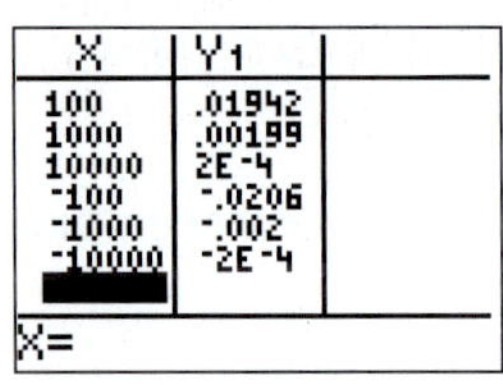

(b) $f(x) = \dfrac{x + 4}{x - 1}$

To find vertical asymptotes, look at the denominator. With some checking on a calculator, we find there is a vertical asymptote with equation $x = 1$.

To find horizontal asymptotes, let x get "huge." As x-values are chosen with large positive numbers (and with large negative numbers), the outputs approach 1. There is a horizontal asymptote at $y = 1$.

Check Yourself 4

Identify the asymptotes for the graph of each rational function.

(a) $f(x) = \dfrac{4}{x + 2}$ **(b)** $f(x) = \dfrac{x + 5}{x + 1}$

Property

Vertical and Horizontal Asymptotes

A **vertical asymptote** is a vertical guideline toward which the graph of a function approaches, but does not touch. Its equation is of the form

$x = c$

As x-values are chosen close to c, output values grow larger and larger (positively or negatively).

A **horizontal asymptote** is a horizontal guideline toward which the graph of a function approaches, for large positive or negative values of x. Its equation is of the form

$y = b$

As large positive or large negative x-values are chosen, output values become close to b.

We use these steps to put together a sketch of a rational function f.

Step by Step

Step 1 Determine the domain of f.
Step 2 Find the intercepts of the graph of f. Plot these.
Step 3 Locate vertical and horizontal asymptotes. Draw these as dotted lines.
Step 4 Choose a few easy-to-compute points to plot.
Step 5 Connect plotted points with a smooth curve, allowing the curve to approach the asymptotes.

Example 5 Drawing the Graph of a Rational Function

Draw the graph of each function.

(a) $f(x) = \dfrac{4}{x + 2}$

The domain of f is $\{x \mid x \neq -2\}$.

Since $f(0) = 2$, the y-intercept is $(0, 2)$. There are no x-intercepts.
There is a vertical asymptote with equation $x = -2$.
There is a horizontal asymptote with equation $y = 0$.

So far, we have

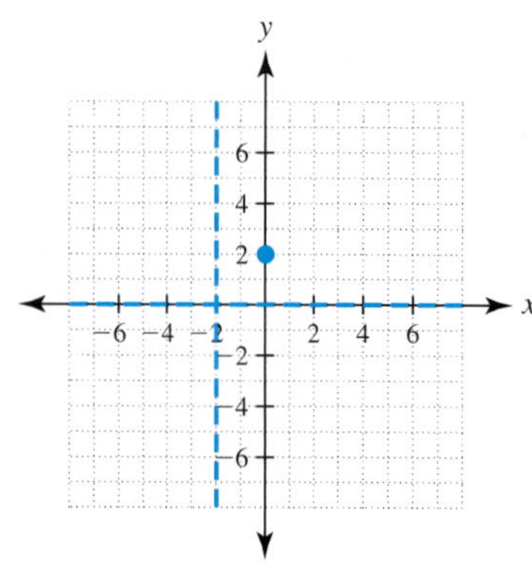

We choose convenient x-values and find some points.

x	−6	−4	−3	−1	2	6
$f(x)$	−1	−2	−4	4	1	0.5

Plotting these, and connecting with a smooth curve gives

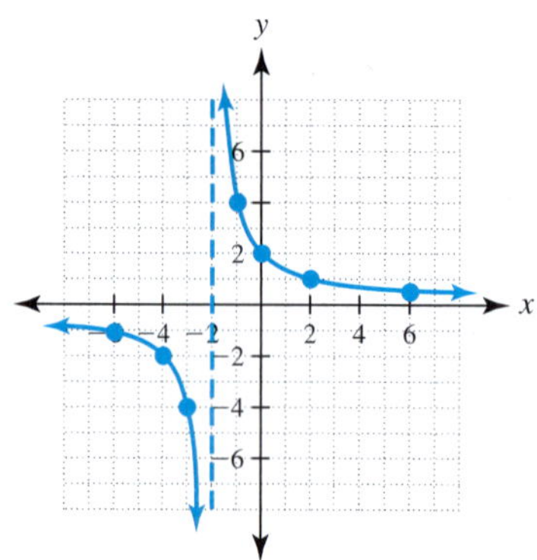

(b) $f(x) = \dfrac{x + 5}{x + 1}$

The domain of f is $\{x \mid x \neq -1\}$. Since $f(0) = 5$, the y-intercept is $(0, 5)$. The x-intercept is $(-5, 0)$. We have a vertical asymptote at $x = -1$. We have a horizontal asymptote at $y = 1$.

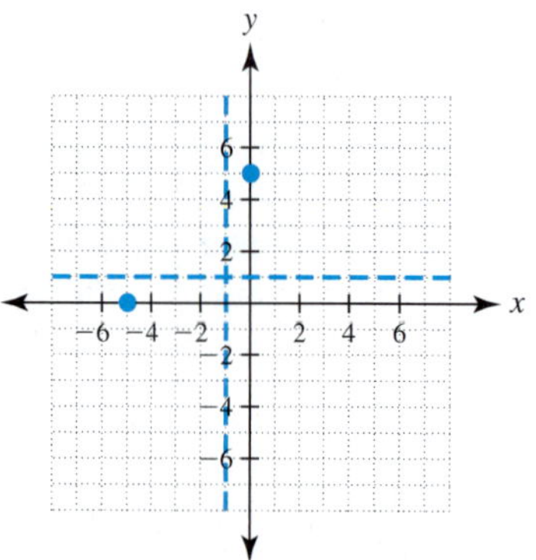

RECALL

You can use your graphing calculator's TABLE utility to find "nice" points.

Next choose convenient x-values.

x	-9	-3	-2	1	3	7
$f(x)$	0.5	-1	-3	3	2	1.5

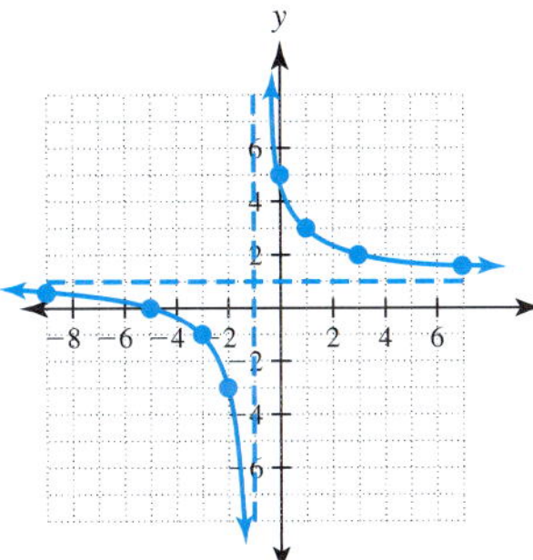

(c) $f(x) = \dfrac{2x - 6}{x + 2}$

The domain of f is $\{x \mid x \neq -2\}$. Since $f(0) = -3$, the y-intercept is $(0, -3)$. To find an x-intercept, set $2x - 6$ equal to 0 and solve.

$$2x - 6 = 0$$
$$x = 3$$

So $(3, 0)$ is an x-intercept.

There is a vertical asymptote with equation $x = -2$.

Letting x get "huge," we find that output values approach 2. So there is a horizontal asymptote with equation $y = 2$.

NOTE

You can use the TABLE utility in your graphing calculator to find points like these.

x	-7	-6	-4	-1	2	8
$f(x)$	4	4.5	7	-8	-0.5	1

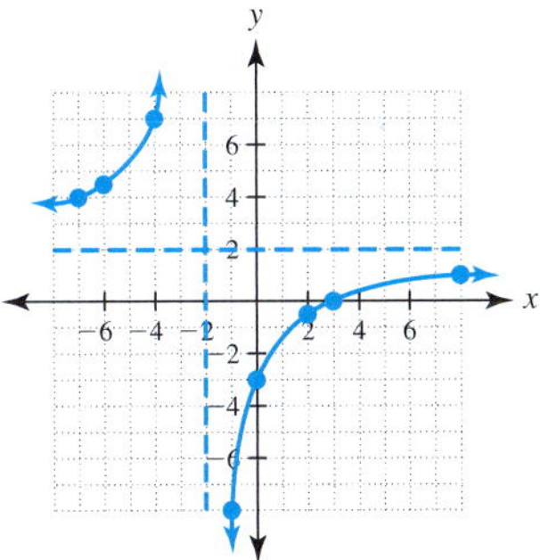

Check Yourself 5

Draw the graph of $f(x) = \dfrac{3 - x}{x - 2}$.

If you look back over the examples of this section, you see that each function has been one of two forms:

1. $f(x) = \dfrac{a}{x - c}$ (assuming $a \neq 0$)

2. $f(x) = \dfrac{a(x - b)}{x - c}$ (assuming $a \neq 0, b \neq c$)

Before studying the box below, try to establish on your own the intercepts and asymptotes for each of the two forms.

Property

Intercepts and Asymptotes for Rational Functions

For the rational function $f(x) = \frac{a}{x - c}$, where $a \neq 0$, there are no x-intercepts, and the y-intercept is $\left(0, -\frac{a}{c}\right)$ if $c \neq 0$.

There is a vertical asymptote with equation $x = c$.

There is a horizontal asymptote with equation $y = 0$.

For the rational function $f(x) = \frac{a(x - b)}{x - c}$, where $a \neq 0$ and $b \neq c$:

The y-intercept is $\left(0, \frac{ab}{c}\right)$ if $c \neq 0$.

The x-intercept is $(b, 0)$.

There is a vertical asymptote with equation $x = c$.

There is a horizontal asymptote with equation $y = a$.

In future courses, you will have the opportunity to study the graphs of rational functions to a greater extent. Those covered in this section should provide you with a good introduction.

Check Yourself ANSWERS

1. **(a)** Domain: $\{x|x \neq 0\}$; no intercepts; **(b)** domain: $\{x|x \neq -2\}$; y-intercept: $\left(0, -\frac{3}{2}\right)$; no x-intercept **2.** Domain: $\{x|x \neq 2\}$; y-intercept: $(0, -2)$; x-intercept: $(-4, 0)$ **3.**

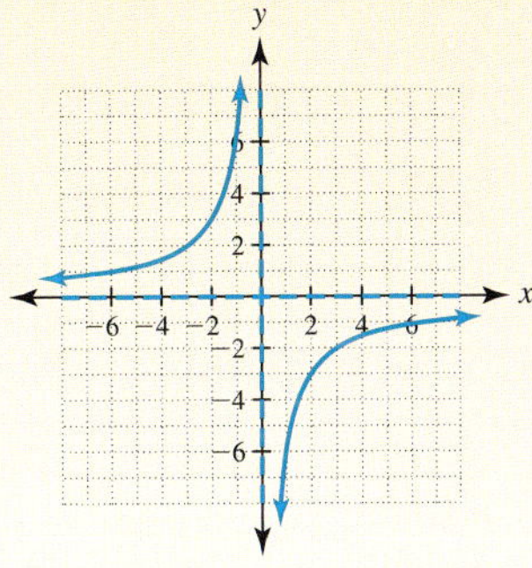

4. **(a)** Vertical asymptote: $x = -2$; horizontal asymptote: $y = 0$; **(b)** vertical asymptote: $x = -1$; horizontal asymptote: $y = 1$

5.

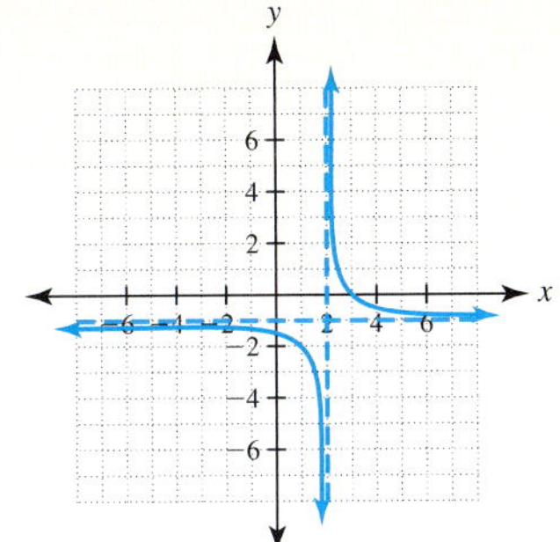

Reading Your Text

SECTION 9.5

(a) To find the ______________, we input 0 for x and find the resulting output value.

(b) To find the domain, focus on the ______________.

(c) To search for ______________ asymptotes, we focus on the denominator.

(d) To find ______________ asymptotes, we choose large input values of x.

9.5 exercises

Basic Skills | Challenge Yourself | Calculator/Computer | Career Applications | Above and Beyond

Name ____________

Section ________ Date ________

Answers

1. ____________
2. ____________
3. ____________
4. ____________
5. ____________
6. ____________
7. ____________
8. ____________
9. ____________
10. ____________
11. ____________
12. ____________
13. ____________
14. ____________

< Objective 1 >

For each function, determine the domain of f, and find all intercepts for the graph of f.

1. $f(x) = \frac{7}{x}$

2. $f(x) = \frac{-5}{x}$

3. $f(x) = \frac{3}{x - 9}$

4. $f(x) = \frac{12}{x - 3}$

5. $f(x) = \frac{8}{2 - x}$

6. $f(x) = \frac{6}{3 - x}$

7. $f(x) = \frac{x - 6}{x - 4}$

8. $f(x) = \frac{x + 9}{x + 3}$

9. $f(x) = \frac{2 - x}{x - 4}$

10. $f(x) = \frac{10 - x}{x + 2}$

11. $f(x) = \frac{2x - 10}{x + 4}$

12. $f(x) = \frac{3x + 9}{x - 1}$

13. $f(x) = \frac{2 - 4x}{x - 3}$

14. $f(x) = \frac{3 - 5x}{x + 1}$

< Objective 3 >

Identify the asymptotes for the graph of each function.

15. $f(x) = \frac{7}{x}$

16. $f(x) = \frac{-5}{x}$

17. $f(x) = \frac{3}{x-9}$

18. $f(x) = \frac{12}{x-3}$

19. $f(x) = \frac{8}{2-x}$

20. $f(x) = \frac{6}{3-x}$

21. $f(x) = \frac{x-6}{x-4}$

22. $f(x) = \frac{x+9}{x+3}$

23. $f(x) = \frac{2-x}{x-4}$

24. $f(x) = \frac{10-x}{x+2}$

25. $f(x) = \frac{2x-10}{x+4}$

26. $f(x) = \frac{3x+9}{x-1}$

27. $f(x) = \frac{2-4x}{x-3}$

28. $f(x) = \frac{3-5x}{x+1}$

< Objective 2 >

Draw the graph of each function. Use a dotted line for each asymptote and clearly plot several points.

29. $f(x) = \frac{6}{x}$

30. $f(x) = \frac{12}{x}$

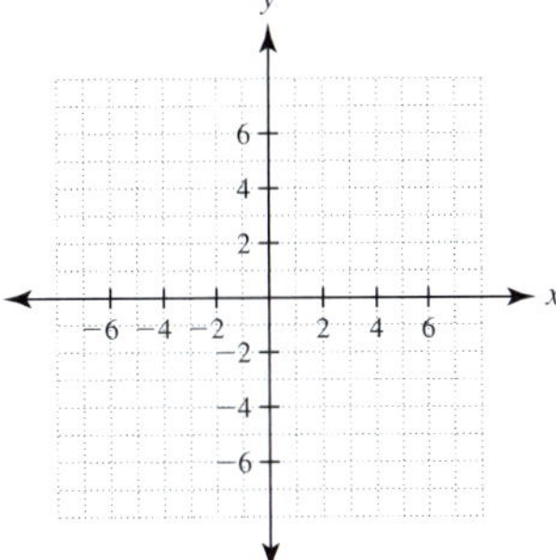

31. $f(x) = \frac{-8}{x}$

32. $f(x) = \frac{-10}{x}$

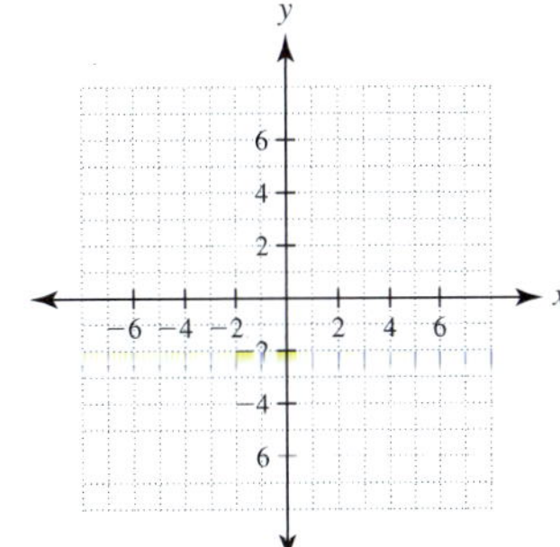

Answers

15. ______

16. ______

17. ______

18. ______

19. ______

20. ______

21. ______

22. ______

23. ______

24. ______

25. ______

26. ______

27. ______

28. ______

29. ______

30. ______

31. ______

32. ______

Answers

33. ______

34. ______

35. ______

36. ______

37. ______

38. ______

39. ______

40. ______

33. $f(x) = \dfrac{10}{x - 2}$

34. $f(x) = \dfrac{12}{x - 3}$

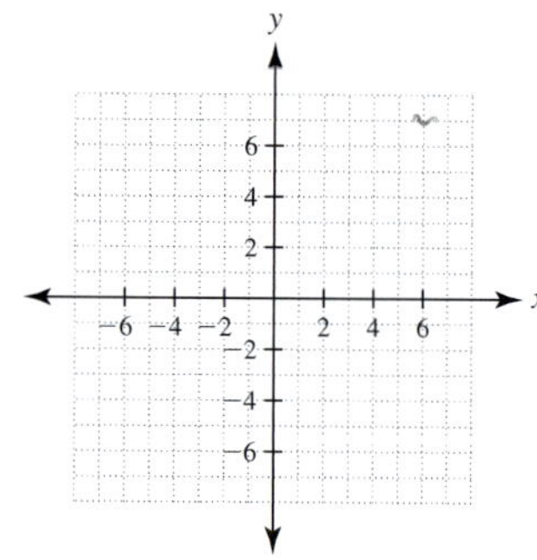

35. $f(x) = \dfrac{6}{3 - x}$

36. $f(x) = \dfrac{8}{2 - x}$

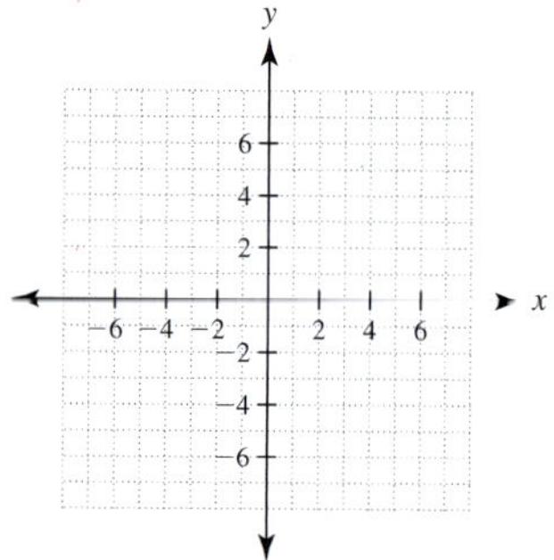

37. $f(x) = \dfrac{x + 5}{x - 2}$

38. $f(x) = \dfrac{x - 6}{x - 1}$

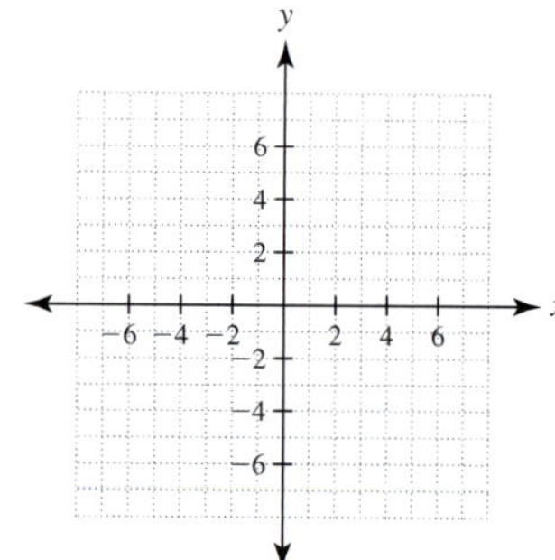

39. $f(x) = \dfrac{4 - x}{x + 2}$

40. $f(x) = \dfrac{5 - x}{x - 1}$

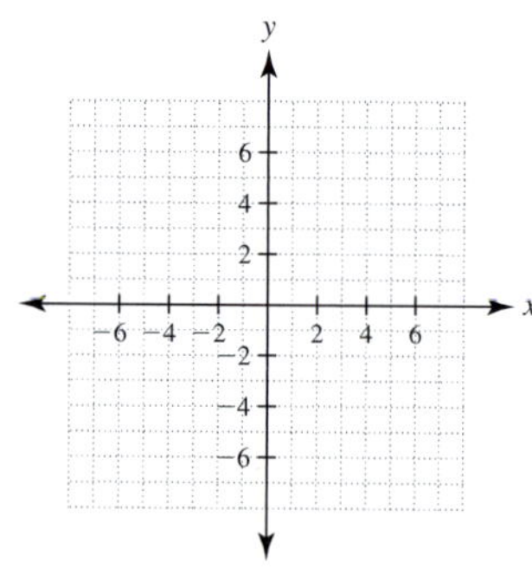

41. $f(x) = \dfrac{2x - 6}{x + 4}$

42. $f(x) = \dfrac{3x + 9}{x - 1}$

43. $f(x) = \dfrac{2 - 4x}{x - 3}$

44. $f(x) = \dfrac{3 - 5x}{x + 1}$

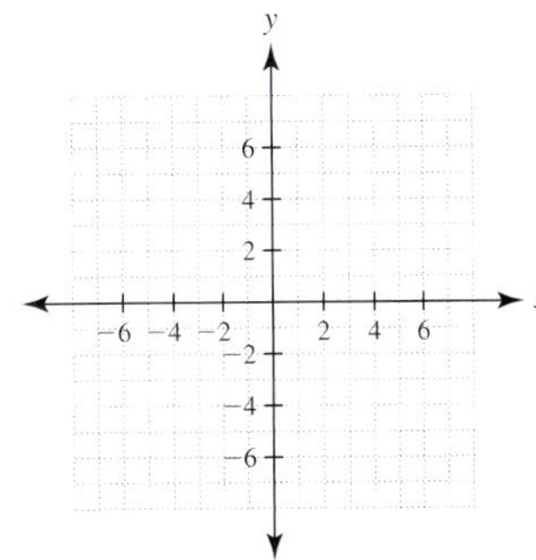

Basic Skills | **Challenge Yourself** | Calculator/Computer | Career Applications | Above and Beyond

Complete each statement with **never, sometimes,** or **always.**

45. The graph of $f(x) = \dfrac{a}{x}$, $a \neq 0$, ____________ has an x-intercept.

46. The graph of $f(x) = \dfrac{a}{x}$, $a \neq 0$, ____________ has a y-intercept.

47. The graph of $f(x) = \dfrac{a(x - b)}{x - c}$, $a \neq 0$, $c \neq 0$, and $b \neq c$, ____________ has an x-intercept.

48. The graph of $f(x) = \dfrac{a(x - b)}{x - c}$, $a \neq 0$, $c \neq 0$, and $b \neq c$, ____________ has a y-intercept.

49. The graph of $f(x) = \dfrac{a}{x - c}$, $a \neq 0$, ____________ has a vertical asymptote that lies exactly on the y-axis.

50. If $a > 0$, the graph of $f(x) = \dfrac{a}{x}$ ____________ has points in Quadrants II and IV.

Answers

41. ____________

42. ____________

43. ____________

44. ____________

45. ____________

46. ____________

47. ____________

48. ____________

49. ____________

50. ____________

Basic Skills | Challenge Yourself | **Calculator/Computer** | Career Applications | Above and Beyond

Answers

51. ____________

52. ____________

53. ____________

54. ____________

55. ____________

Each given function has a horizontal asymptote. Use your calculator to complete the table, rounding answers to four decimal places. Based on these values, give the equation of the horizontal asymptote.

51. $f(x) = \dfrac{7x - 5}{2x + 3}$

x	100	1,000	10,000	−100	−1,000	−10,000
$f(x)$						

52. $f(x) = \dfrac{8x + 6}{3x - 5}$

x	100	1,000	10,000	−100	−1,000	−10,000
$f(x)$						

53. $f(x) = \dfrac{3x + 7}{8 - 9x}$

x	100	1,000	10,000	−100	−1,000	−10,000
$f(x)$						

54. $f(x) = \dfrac{2x - 9}{5 - 4x}$

x	100	1,000	10,000	−100	−1,000	−10,000
$f(x)$						

55. Based on your answers to the previous exercises, give the equation of the horizontal asymptote for $f(x) = \dfrac{ax + b}{cx + d}$ (assuming there is no common factor in the numerator and denominator).

Basic Skills | Challenge Yourself | Calculator/Computer | Career Applications | **Above and Beyond**

We have not yet considered what happens to the graph of a rational function if there is a common factor in the numerator and denominator. In each exercise, be sure to begin by factoring the numerator and denominator. Discuss the behavior of each graph, starting with the domain.

56. $f(x) = \dfrac{x^2 - 2x - 8}{x + 2}$

Use your calculator to check output values close to $x = -2$. What happens to the graph at $x = -2$? View the graph on your calculator.

57. $f(x) = \dfrac{3x - 15}{x^2 - 9x + 20}$

Use your calculator to check output values close to $x = 5$. What happens to the graph at $x = 5$? View the graph on your calculator.

58. $f(x) = \dfrac{x^2 - 4x + 3}{x - 1}$

Use your calculator to check output values close to $x = 1$. What happens to the graph at $x = 1$? View the graph on your calculator.

59. $f(x) = \dfrac{5x + 10}{x^2 + 7x + 10}$

Use your calculator to check output values close to $x = -2$. What happens to the graph at $x = -2$? View the graph on your calculator.

Answers

56. ______

57. ______

58. ______

59. ______

Answers

1. $\{x | x \neq 0\}$; no intercepts **3.** $\{x | x \neq 9\}$; y-int: $\left(0, -\dfrac{1}{3}\right)$; no x-int

5. $\{x | x \neq 2\}$; y-int: $(0, 4)$; no x-int **7.** $\{x | x \neq 4\}$; y-int: $\left(0, \dfrac{3}{2}\right)$; x-int: $(6, 0)$

9. $\{x | x \neq 4\}$; y-int: $\left(0, -\dfrac{1}{2}\right)$; x-int: $(2, 0)$ **11.** $\{x | x \neq -4\}$; y-int: $\left(0, -\dfrac{5}{2}\right)$; x-int: $(5, 0)$ **13.** $\{x | x \neq 3\}$; y-int: $\left(0, -\dfrac{2}{3}\right)$; x-int: $\left(\dfrac{1}{2}, 0\right)$

15. vert: $x = 0$; horiz: $y = 0$ **17.** vert: $x = 9$; horiz: $y = 0$
19. vert: $x = 2$; horiz: $y = 0$ **21.** vert: $x = 4$; horiz: $y = 1$
23. vert: $x = 4$; horiz: $y = -1$ **25.** vert: $x = -4$; horiz: $y = 2$
27. vert: $x = 3$; horiz: $y = -4$

29.

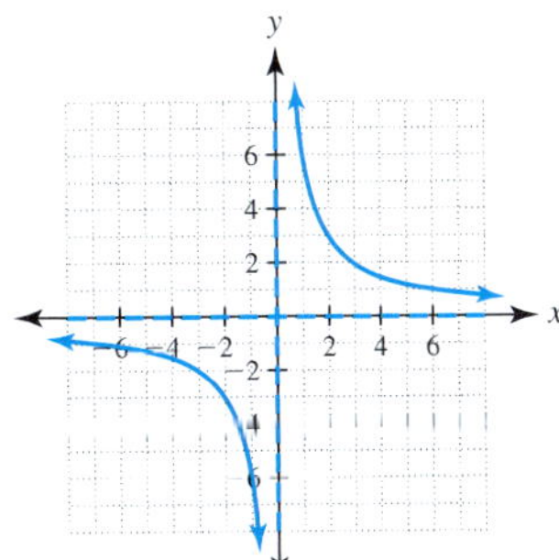

31.

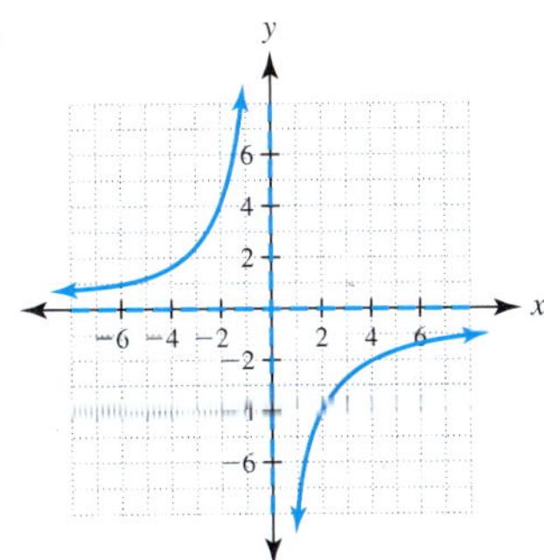

33.

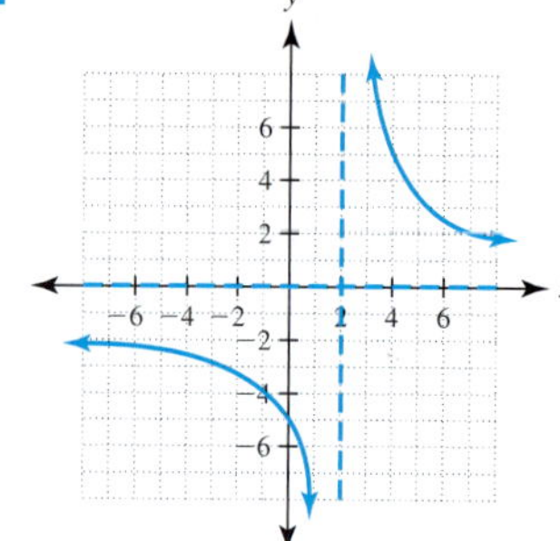

35.

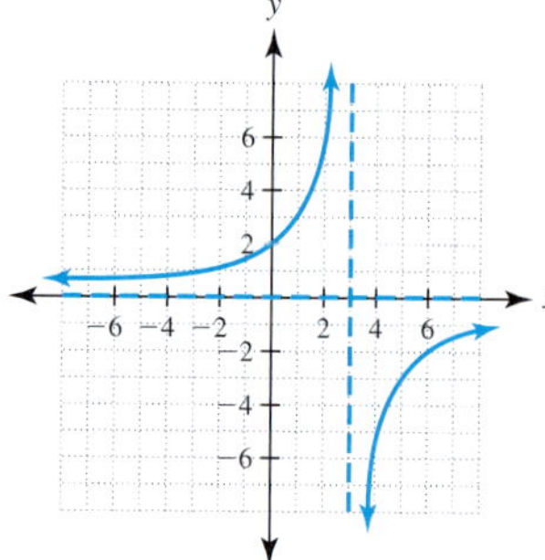

37.

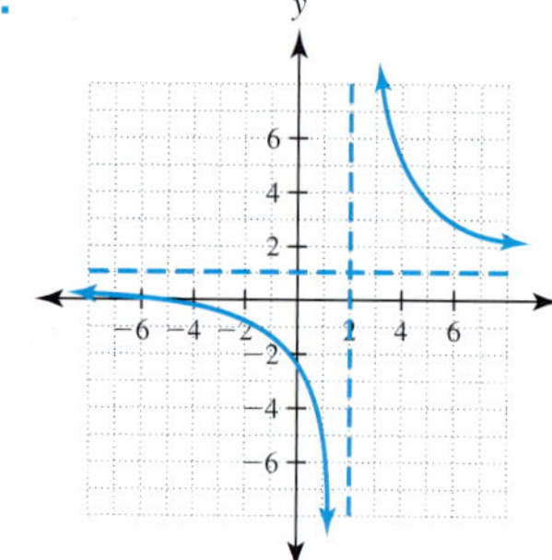

39.

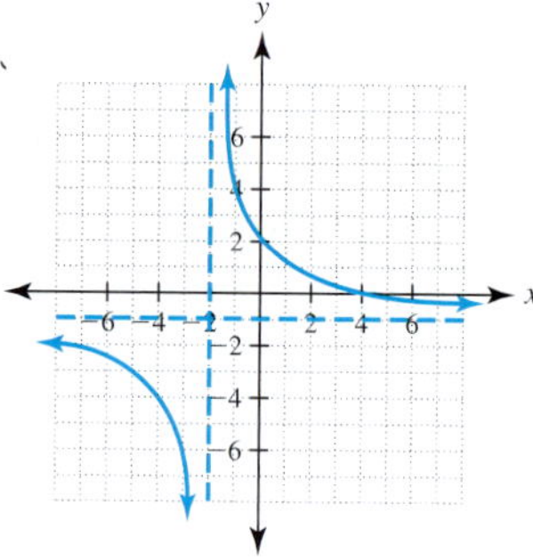

41.

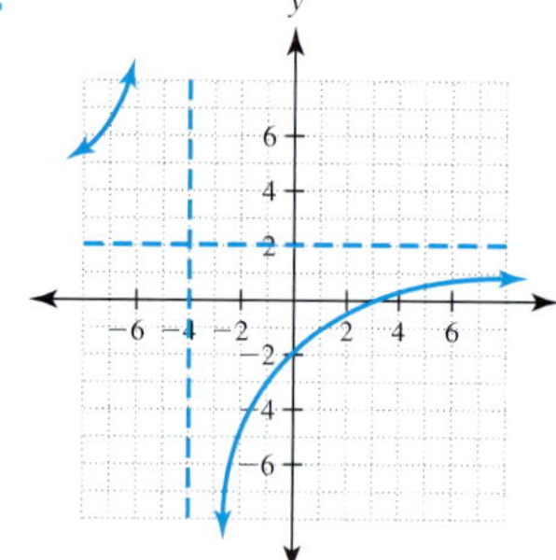

43.

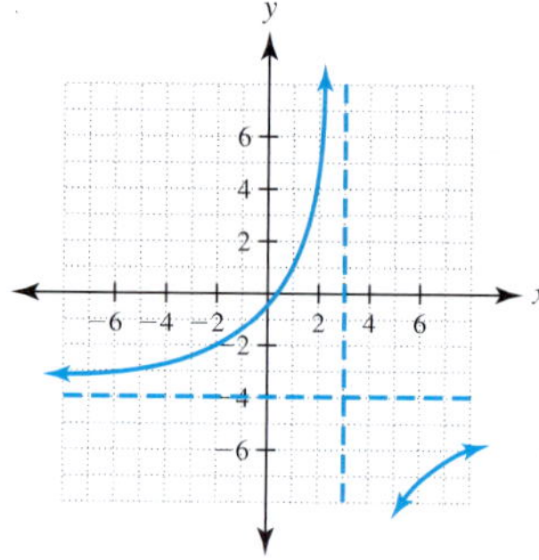

45. never 47. always 49. sometimes

51.

x	100	1,000	10,000	−100	−1,000	−10,000
$f(x)$	3.4236	3.4923	3.4992	3.5787	3.5078	3.5008

$y = \frac{7}{2}$

53.

x	100	1,000	10,000	−100	−1,000	−10,000
$f(x)$	−0.3442	−0.3344	−0.3334	−0.3227	−0.3323	−0.3332

$y = -\frac{1}{3}$

55. $y = \frac{a}{c}$ 57. Above and Beyond 59. Above and Beyond

Activity 9 :: Communicating Mathematical Ideas

Organizations concerned with mathematics education, such as NCTM (National Council of Teachers of Mathematics) and AMATYC (American Mathematical Association of Two-Year Colleges), have long recognized the importance of communication. Your drive to explore and investigate, your ability to solve problems, and your skill in presenting your findings are all key factors for success in today's world.

This activity is designed to help you practice and develop these skills. As you work through the problems below, focus on effectively communicating your work to others.

1. A rectangle with an area of 30 cm^2 has length l and width w.

(a) Use the area constraint to write the perimeter of the rectangle as a function of its width w.

(b) Based on the physical constraints of this application, what is the domain of the function?

(c) Find the minimum perimeter of the rectangle.

(d) Find the dimensions that yield this minimum perimeter.

2. In general, for a cylinder of radius r and height h, the volume and surface area are given by

$$V_{\text{Cyl}} = \pi r^2 h$$
$$S_{\text{Cyl}} = 2\pi rh + 2\pi r^2$$

We wish to manufacture a metal tank (cylinder) that holds 80 cu ft of fluid.

(a) Use the volume formula to express the height of the tank as a function of its radius.

(b) Use (a) to express the surface area as a function of the radius (substitute for the height).

(c) Provide a graph of the surface area function found in (b).

(d) Use your calculator to approximate (one decimal place) the radius that produces the minimum surface area (that is, uses the minimum amount of metal).

(e) What is the minimum surface area of a cylinder that holds 80 cu ft of fluid (use (b) or use the graph)?

(f) Find the height of this minimum surface area cylinder (use the formula from (a)).

3. Consider the rational function $f(x) = \dfrac{2x + 3}{x - 3}$.

(a) Give the domain of the function.

(b) Give any y-intercepts of the function (write any answers as ordered pairs).

(c) Give any x-intercepts of the function (write any answers as ordered pairs).

9.6 Solving Rational Equations

< 9.6 Objectives >

1 > Solve rational equations in one variable

2 > Solve literal equations involving rational expressions

3 > Solve applications involving rational expressions

Applications often result in equations involving rational expressions. Our objective in this section is to develop methods to find solutions for such equations.

The usual technique for solving such equations is to multiply both sides of the equation by the least common denominator (LCD) of all the rational expressions appearing in the equation. The resulting equation will be cleared of fractions, and we can then proceed to solve the equation with techniques that you have already learned. Example 1 illustrates the process.

Example 1 Clearing Equations of Fractions

Solve.

$$\frac{2x}{3} + \frac{x}{5} = 13$$

For the denominators 3 and 5, the LCD is 15. Multiplying both sides of the equation by 15 gives

$$15\left(\frac{2x}{3} + \frac{x}{5}\right) = 15 \cdot 13$$

$$15 \cdot \frac{2x}{3} + 15 \cdot \frac{x}{5} = 15 \cdot 13 \quad \text{Distribute 15 on the left.}$$

$$\frac{\overset{5}{\cancel{15}} \cdot 2x}{\underset{1}{\cancel{3}}} + \frac{\overset{3}{\cancel{15}} \cdot x}{\underset{1}{\cancel{5}}} = 195$$

$$10x + 3x = 195 \quad \text{Simplify. The equation is now cleared of fractions.}$$

$$13x = 195$$

$$x = 15$$

The solution set is $\{15\}$.

To check, substitute 15 in the original equation.

$$\frac{2x}{3} + \frac{x}{5} = 13$$

$$\frac{2(15)}{3} + \frac{(15)}{5} \stackrel{?}{=} 13$$

$$10 + 3 \stackrel{?}{=} 13$$

$$13 = 13 \quad \text{A true statement}$$

So 15 is the solution for the equation.

> **CAUTION**
>
> A common mistake is to confuse an *equation* such as
>
> $\frac{2x}{3} + \frac{x}{5} = 13$
>
> and an *expression* such as
>
> $\frac{2x}{3} + \frac{x}{5}$

Let's compare.

Equation: $\frac{2x}{3} + \frac{x}{5} = 13$

Here we want to *solve the equation for* x, as in Example 1. We multiply both sides by the LCD to clear fractions and proceed as before.

Expression: $\frac{2x}{3} + \frac{x}{5}$

Here we want to find *a third fraction* that is equivalent to the given expression. We write each fraction as an equivalent fraction with the LCD as a common denominator.

$$\frac{2x}{3} + \frac{x}{5} = \frac{2x \cdot 5}{3 \cdot 5} + \frac{x \cdot 3}{5 \cdot 3}$$
$$= \frac{10x}{15} + \frac{3x}{15} = \frac{10x + 3x}{15}$$
$$= \frac{13x}{15}$$

Check Yourself 1

Solve.

$$\frac{3x}{2} - \frac{x}{3} = 7$$

The process is similar when variables are in the denominators. Consider Example 2.

Example 2 Solving an Equation Involving Rational Expressions

< Objective 1 >

Solve.

$$\frac{7}{4x} - \frac{3}{x^2} = \frac{1}{2x^2}$$

> **NOTE**
>
> We assume that x cannot be 0. Do you see why?

For $4x$, x^2, and $2x^2$, the LCD is $4x^2$. So, multiplying both sides by $4x^2$, we have

$$4x^2\left(\frac{7}{4x} - \frac{3}{x^2}\right) = 4x^2 \cdot \frac{1}{2x^2} \quad \text{Distribute } 4x^2 \text{ on the left side.}$$

$$4x^2 \cdot \frac{7}{4x} - 4x^2 \cdot \frac{3}{x^2} = 4x^2 \cdot \frac{1}{2x^2} \quad \text{Simplify.}$$

$$\frac{\overset{x}{\cancel{4x^2}} \cdot 7}{\underset{1}{\cancel{4x}}} - \frac{\overset{4}{\cancel{4x^2}} \cdot 3}{\underset{1}{\cancel{x^2}}} = \frac{\overset{2}{\cancel{4x^2}} \cdot 1}{\underset{1}{\cancel{2x^2}}}$$

$$7x - 12 = 2$$
$$7x = 14$$
$$x = 2$$

The solution set is $\{2\}$.

We leave it to you to check the solution $x = 2$. Be sure to return to the original equation and substitute 2 for x.

Check Yourself 2

Solve.

$$\frac{5}{2x} - \frac{4}{x^2} = \frac{7}{2x^2}$$

Example 3 illustrates the same solution process when there are binomials in the denominators.

Example 3 Solving an Equation Involving Rational Expressions

NOTE

Here we assume that x *cannot* have the value -2 or 3.

Solve.

$$\frac{4}{x+2} + 3 = \frac{3x}{x-3}$$

The LCD is $(x + 2)(x - 3)$. Multiplying by that LCD gives

$$(x+2)(x-3)\left(\frac{4}{x+2}\right) + (x+2)(x-3)(3) = (x+2)(x-3)\left(\frac{3x}{x-3}\right)$$

Simplifying each term gives

$$4(x-3) + 3(x+2)(x-3) = 3x(x+2)$$

We now clear the parentheses and proceed as before.

$$\begin{aligned} 4x - 12 + 3x^2 - 3x - 18 &= 3x^2 + 6x \\ 3x^2 + x - 30 &= 3x^2 + 6x \\ x - 30 &= 6x \\ -5x &= 30 \\ x &= -6 \end{aligned}$$

The solution set is $\{-6\}$.

Check

$$\frac{4}{(-6)+2} + 3 \stackrel{?}{=} \frac{3(-6)}{(-6)-3}$$

$$\frac{-4}{-4} + 3 \stackrel{?}{=} \frac{-18}{-9}$$

$$-1 + 3 = 2$$ A true statement

Check Yourself 3

Solve.

$$\frac{5}{x-4} + 2 = \frac{2x}{x-3}$$

Factoring plays an important role in solving equations containing rational expressions.

Example 4 Solving an Equation Involving Rational Expressions

Solve.

$$\frac{3}{x-3} - \frac{7}{x+3} = \frac{2}{x^2-9}$$

In factored form, the denominator on the right side is $(x - 3)(x + 3)$, which forms the LCD, and we multiply each term by that LCD.

$$(x-3)(x+3)\left(\frac{3}{x-3}\right) - (x-3)(x+3)\left(\frac{7}{x+3}\right) = (x-3)(x+3)\left[\frac{2}{(x-3)(x+3)}\right]$$

Again, simplifying each term on the right and left sides, we have

$$3(x + 3) - 7(x - 3) = 2$$
$$3x + 9 - 7x + 21 = 2$$
$$-4x = -28$$
$$x = 7$$

The solution set is $\{7\}$.

Be sure to check this result by substitution in the original equation.

Check Yourself 4

Solve $\dfrac{4}{x - 4} - \dfrac{3}{x + 1} = \dfrac{5}{x^2 - 3x - 4}$.

Whenever we multiply both sides of an equation by an expression containing a variable, there is the possibility that a proposed solution may make that factor 0. As we pointed out earlier, multiplying by 0 does not give an equivalent equation, and therefore verifying solutions by substitution serves not only as a check of our work but also as a check for extraneous solutions. Consider Example 5.

Example 5 Solving an Equation Involving Rational Expressions

Solve.

$$\frac{x}{x - 2} - 7 = \frac{2}{x - 2}$$

NOTES

We must assume that $x \neq 2$.

Each of the three terms is multiplied by $x - 2$.

The LCD is $x - 2$, and multiplying, we have

$$\left(\frac{x}{x - 2}\right)(x - 2) - 7(x - 2) = \left(\frac{2}{x - 2}\right)(x - 2)$$

Simplifying yields

$$x - 7(x - 2) = 2$$
$$x - 7x + 14 = 2$$
$$-6x = -12$$
$$x = 2$$

To check this result, substitute 2 for x,

$$\frac{(2)}{(2) - 2} - 7 \stackrel{?}{=} \frac{2}{(2) - 2}$$
$$\frac{2}{0} - 7 \stackrel{?}{=} \frac{2}{0}$$

> **CAUTION**

Because division by 0 is undefined, we conclude that 2 is *not a solution* for the original equation. It is an extraneous solution. The original equation has no solutions.

The solution set is empty. The set is written $\{\ \}$ or $\varnothing$.

Check Yourself 5

Solve $\dfrac{x - 3}{x - 4} = 4 + \dfrac{1}{x - 4}$.

Equations involving rational expressions may also lead to quadratic equations, as illustrated in Example 6.

Example 6 Solving an Equation Involving Rational Expressions

NOTE

Assume $x \neq 3$ and $x \neq 4$.

Solve.

$$\frac{x}{x-4} = \frac{15}{x-3} - \frac{2x}{x^2 - 7x + 12}$$

After we factor the denominator on the right, the LCD for the denominators $x - 3$, $x - 4$, and $x^2 - 7x + 12$ is $(x - 3)(x - 4)$. Multiplying by that LCD, we have

$$(x-3)(x-4)\left(\frac{x}{x-4}\right) = (x-3)(x-4)\left(\frac{15}{x-3}\right) - (x-3)(x-4)\left[\frac{2x}{(x-3)(x-4)}\right]$$

Simplifying yields

$$x(x-3) = 15(x-4) - 2x \qquad \text{Remove the parentheses.}$$

$$x^2 - 3x = 15x - 60 - 2x \qquad \text{Write in standard form and factor.}$$

$$x^2 - 16x + 60 = 0$$

$$(x-6)(x-10) = 0$$

So $x = 6$ or $x = 10$

Verify that 6 and 10 are both solutions for the original equation. The solution set is $\{6, 10\}$.

Check Yourself 6

Solve $\dfrac{3x}{x+2} - \dfrac{2}{x+3} = \dfrac{36}{x^2 + 5x + 6}$.

This algorithm summarizes our work in solving equations containing rational expressions.

Step by Step

Solving Equations Containing Rational Expressions

Step 1 Clear the equation of fractions by multiplying both sides of the equation by the LCD of all the fractions that appear.

Step 2 Solve the equation resulting from step 1.

Step 3 Check all solutions by substitution in the original equation.

Rational equations are needed when we are asked to find the missing sides of a triangle. Two triangles are said to be **similar** if they have the same shape. They may or may not be the same size.

When two triangles are similar, the ratios of the related sides are equal. In the triangles shown here, note that side a is related to side d and side b is related to side e.

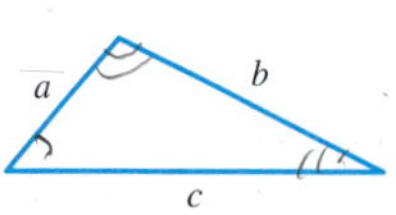

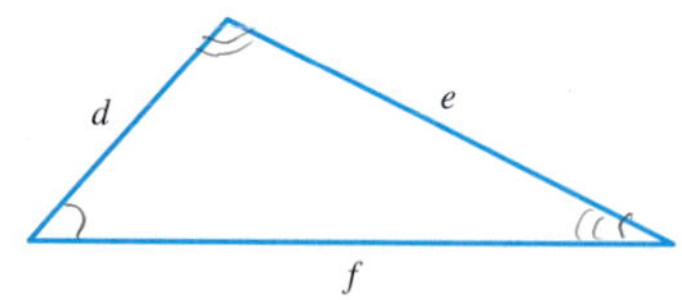

Because these are similar triangles, we can say that $\frac{a}{b} = \frac{d}{e}$. This idea is used in Example 7.

Example 7 **Finding the Lengths of the Sides of Similar Triangles**

Given that these two triangles are similar triangles, find the lengths of the indicated sides.

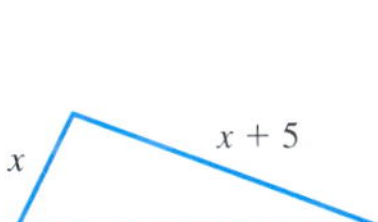

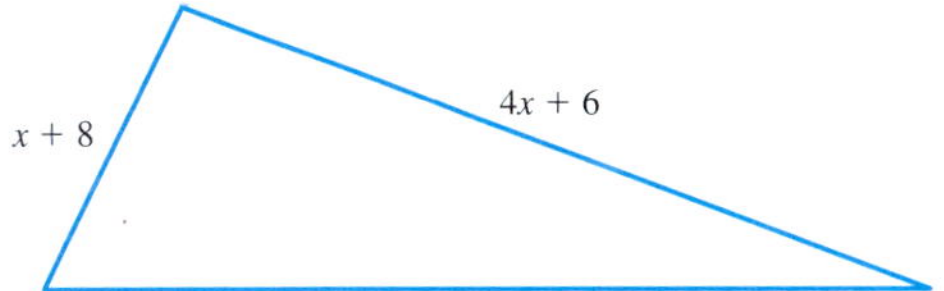

From the equation above, we know that $\frac{a}{b} = \frac{d}{e}$; in this case, we have

$$\frac{x}{x+5} = \frac{x+8}{4x+6}$$

Multiplying by the common denominator gives the equation

$$x(4x + 6) = (x + 8)(x + 5)$$
$$4x^2 + 6x = x^2 + 13x + 40$$
$$3x^2 - 7x - 40 = 0$$
$$(3x + 8)(x - 5) = 0$$
$$x = -\frac{8}{3} \quad \text{or} \quad x = 5$$

We know that x represents the length of one side of a triangle, so it must be a positive number. We can now find the four indicated lengths.

$$x = 5$$
$$x + 5 = 10$$
$$x + 8 = 13$$
$$4x + 6 = 26$$

Check Yourself 7

Given that the two triangles below are similar triangles, find the lengths of the indicated sides.

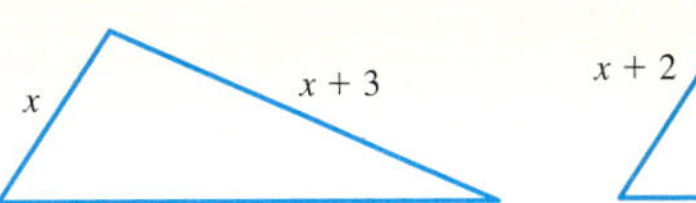

The method in this section may also be used to solve certain literal equations for a specified variable. Consider Example 8.

Example 8 Solving a Literal Equation

< Objective 2 >

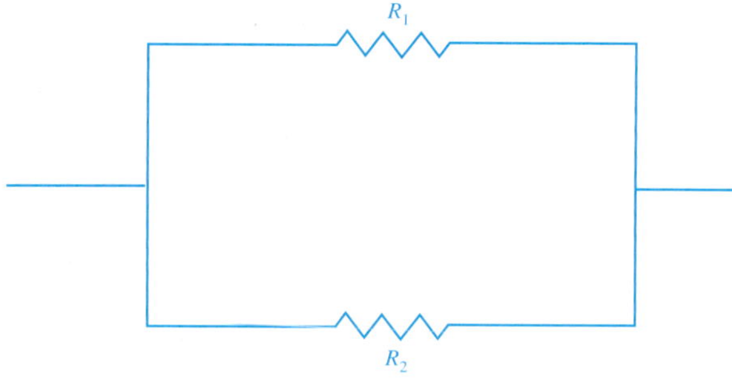

NOTE

This is a parallel electric circuit. The symbol for a resistor is

If two resistors with resistances R_1 and R_2 are connected in parallel, the combined resistance R can be found from

$$\frac{1}{R} = \frac{1}{R_1} + \frac{1}{R_2}$$

RECALL

The numbers 1 and 2 are *subscripts.* We read R_1 as "R sub 1" and R_2 as "R sub 2."

Solve the formula for R.

First, the LCD is RR_1R_2, and we multiply:

$$RR_1R_2 \cdot \frac{1}{R} = RR_1R_2 \cdot \frac{1}{R_1} + RR_1R_2 \cdot \frac{1}{R_2}$$

Simplifying yields

$$R_1R_2 = RR_2 + RR_1 \qquad \text{Factor out } R \text{ on the right.}$$

$$R_1R_2 = R(R_2 + R_1) \qquad \text{Divide by } R_2 + R_1 \text{ to isolate } R.$$

$$\frac{R_1R_2}{R_2 + R_1} = R \qquad \text{or} \qquad R = \frac{R_1R_2}{R_1 + R_2}$$

Check Yourself 8

Solve for D_1.

$$\frac{1}{F} = \frac{1}{D_1} + \frac{1}{D_2}$$

NOTE

This formula involves the focal length of a convex lens.

We have previously used the relationship among distance, rate, and time to solve certain motion problems.

Definition

The Distance Formula $d = r \cdot t$ Distance equals rate times time.

Sometimes this formula leads to a rational equation, as in Example 9.

Example 9 Solving a Motion Problem

< Objective 3 >

A boat can travel 16 mi/h in still water. If the boat can travel 5 mi downstream in the same time it takes to travel 3 mi upstream, what is the rate of the river's current?

Step 1 We want to find the rate of the current.

Step 2 Let c be the rate of the current. Then $16 + c$ is the rate of the boat going downstream and $16 - c$ is the rate going upstream.

Step 3 We use a chart to help us set up an appropriate equation.

> **RECALL**
>
> Because $d = rt$ we know that $t = \frac{d}{r}$.

	d	r	t
Downstream	5 mi	$16 + c$	$\frac{5}{16 + c}$
Upstream	3 mi	$16 - c$	$\frac{3}{16 - c}$

The key to finding our equation is noting that the time is the same upstream and downstream, so

$$\frac{5}{16 + c} = \frac{3}{16 - c}$$

Step 4 Multiplying both sides by the LCD $(16 + c)(16 - c)$ yields

$$5(16 - c) = 3(16 + c)$$
$$80 - 5c = 48 + 3c$$
$$32 = 8c$$
$$c = 4$$

Step 5 The current is moving at 4 mi/h.

To check, verify that $\frac{5}{16 + 4} = \frac{3}{16 - 4}$.

Check Yourself 9

A plane flew 540 mi into a steady 30 mi/h wind. The pilot then returned along the same route with a tailwind. If the entire trip took $7\frac{1}{2}$ h, what would his speed have been in still air?

Another application that frequently results in rational equations is the work problem. Example 10 illustrates.

Example 10 Solving a Work Problem

> **CAUTION**
>
> **Error 1:** Students sometimes try adding the two times. If we add 40 min and 80 min, we get 120 min. Is this a reasonable answer? Certainly not! If one printer does the job in 40 min, why would it take 120 min for two printers to do it? It wouldn't.

One computer printer can print a company's paychecks in 40 minutes (min). A second printer can print them in 80 min. If both printers are working, how long will it take to print the paychecks?

Before we learn how to solve work problems, we should look at a couple of common errors made in attempting to solve such problems.

Although the method in the margin on the next page labeled **Error 2** does not solve the problem, it does give us guidelines for a reasonable answer. When the printers work together, it will take them somewhere between 20 and 40 min to finish the job.

To tackle a work problem such as this, we use the concept of **work rate.** If the first printer can complete a task in 40 minutes, then it can complete $\frac{1}{40}$ of the task in a minute, and we say that its work rate is $\frac{1}{40}$ task per minute. The second printer's work rate, then, is $\frac{1}{80}$ task per minute.

The **work accomplished** is the amount of the task that is completed, and is found by multiplying the work rate by the time working. For example, if the first printer runs for 10 minutes, it will accomplish

$$\left(\frac{1 \text{ task}}{40 \text{ min}}\right) \cdot (10 \text{ min}) = \frac{1}{4} \text{ task}$$

That is, one-fourth of the task will be completed.

Property

Work Principle #1

Given an object A that completes a task in time a and works for t units of time, then A's **work rate** is $\frac{1}{a}$ and the work accomplished W is

$$W = \frac{1}{a} \cdot t = \frac{t}{a}$$

From this follows our second Work Principle.

Property

Work Principle #2

Given an object A that completes a task in time a and a second object B that completes the same task in time b, then the work W accomplished in t units of time, if A and B work together, is

$$W = \frac{1}{a} \cdot t + \frac{1}{b} \cdot t = \frac{t}{a} + \frac{t}{b}$$

NOTE

This second principle can easily be extended to three or more objects.

Now we can solve the problem.

Step 1 We are looking for the time it takes to print the paychecks.

Step 2 Let t be the time it takes for both printers, working together, to complete 1 task.

Step 3 Since W (the work accomplished) is 1 task, we have

$$\frac{t}{40} + \frac{t}{80} = 1$$

Step 4 Multiply by the LCD, 80.

$$2t + t = 80$$

$$3t = 80$$

$$t = \frac{80}{3} = 26\frac{2}{3}$$

Step 5 The time required is $26\frac{2}{3}$ min, or 26 min 40 s.

To verify this answer, we find what fraction of the job each printer does in this time. The first printer does $\frac{\frac{80}{3}}{40} = \frac{2}{3}$ of the job. The second printer does $\frac{\frac{80}{3}}{80} = \frac{1}{3}$ of the job. Together, they do the entire job $\left(\frac{2}{3} + \frac{1}{3} = 1\right)$.

> **CAUTION**

Error 2: A reasonable approach would be to give one-half of the job to each printer. The first printer would finish its half of the job in 20 min. The second would finish its half in 40 min. The first printer would be idle for the final 20 min, so we know the job could be finished faster.

Check Yourself 10

It would take Sasha 48 days to paint the house. Natasha could do it in 36 days. How long would it take them to paint the house if they worked together?

In the next example, our "objects" need to complete more than one task.

Example 11 Solving a Work Problem

Suppose that an air circulator can completely replace the air in a room in 3 hours, and a second circulator can do the same job in 5 hours. If the air must be completely replaced two times, how long will it take the two circulators working together?

Step 1 We want the time that the two machines will be working to finish 2 complete tasks.

Step 2 Again, let t be the time that the machines will be working.

Step 3 Our equation is now

$$\frac{t}{3} + \frac{t}{5} = 2$$

Step 4 Multiplying by 15 gives

$$5t + 3t = 30$$

$$8t = 30$$

$$t = \frac{30}{8} = \frac{15}{4} = 3\frac{3}{4}$$

Step 5 The time required is $3\frac{3}{4}$ hours, or 3 hr 45 min. To check, we see that the first circulator can do $\frac{1}{3}$ task per hour, so if it runs for $3\frac{3}{4}$ hours it will accomplish $\left(\frac{1}{3}\right)\left(\frac{15}{4}\right) = \frac{5}{4}$ of a task. The other circulator can do $\frac{1}{5}$ task per hour, so it will accomplish $\left(\frac{1}{5}\right)\left(\frac{15}{4}\right) = \frac{3}{4}$ of a task. Together, they will accomplish $\frac{5}{4} + \frac{3}{4} = 2$ complete tasks.

Check Yourself 11

If one pipe can fill a large tub in 12 minutes, and a second pipe can fill the same tub in 15 minutes, how long will it take the two pipes together to fill four tubs of the same size?

Example 12 Number Analysis

The numerator of a fraction exceeds the denominator by 5. When the numerator is increased by 4 and the denominator is decreased by 8, the fraction is equivalent to 2. Find the fraction.

Step 1 We want to find the original fraction.

Step 2 Let x represent the denominator of the original fraction.

The original numerator is $x + 5$.

The original fraction is $\frac{x + 5}{x}$.

The new numerator is $(x + 5) + 4$, or $x + 9$.

The new denominator is $x - 8$.

The new fraction is $\frac{x + 9}{x - 8}$.

Step 3 Since the new fraction must have the value of 2, the equation is

$$\frac{x + 9}{x - 8} = 2$$

Step 4 Multiplying both sides of the equation by the LCD of $x - 8$ yields

$$\begin{aligned} x + 9 &= 2(x - 8) \\ x + 9 &= 2x - 16 \\ 9 + 16 &= 2x - x \\ 25 &= x \end{aligned}$$

Step 5 The original denominator is 25.

The original numerator is $x + 5$, or 30.

The original fraction is $\frac{30}{25}$.

You should check by returning to the original statement of the problem to see that the answer gives a new fraction equivalent to 2.

Check Yourself 12

Five added to twice the numerator of a fraction results in the denominator of the fraction. When 1 is added to both the numerator and denominator, the resulting fraction is $\frac{1}{3}$. Find the original fraction.

Check Yourself ANSWERS

1. {6} **2.** {3} **3.** {9} **4.** {−11} **5.** { } or ∅ **6.** $\left\{-5, \frac{8}{3}\right\}$

7. The lengths are 6 and 9 on the first triangle and 8 and 12 on the second.

8. $D_1 = \frac{FD_2}{D_2 - F}$ **9.** 150 mi/h **10.** $20\frac{4}{7}$ days

11. $26\frac{2}{3}$ min, or 26 min 40 s **12.** $\frac{3}{11}$

Reading Your Text

SECTION 9.6

(a) Whenever we multiply both sides of an equation by an expression containing a ______________, there is the possibility that a proposed solution may make that factor equal to zero.

(b) Our final step in solving an equation is always to ______________ the answer by substituting it into the original equation.

(c) Two triangles are said to be ______________ if they have the same shape.

(d) The distance formula states that distance is equal to ______________ times time.

9.6 exercises

Basic Skills | Challenge Yourself | Calculator/Computer | Career Applications | Above and Beyond

Name ____________

Section ________ Date ________

Answers

1. ____________
2. ____________
3. ____________
4. ____________
5. ____________
6. ____________
7. ____________
8. ____________
9. ______ 10. ______
11. ______ 12. ______
13. ______ 14. ______
15. ______ 16. ______
17. ______ 18. ______

Decide whether each is an expression or an equation. If it is an equation, find a solution. If it is an expression, write it as a single fraction.

1. $\frac{x}{4} - \frac{x}{5} = 2$

2. $\frac{x}{4} - \frac{x}{7} = 3$

3. $\frac{x}{2} - \frac{x}{5}$

4. $\frac{x}{7} - \frac{x}{14}$

5. $\frac{3x+1}{4} = x - 1$

6. $\frac{3x-1}{2} - \frac{x}{5} - \frac{x+3}{4}$

7. $\frac{x}{4} = \frac{x}{12} + \frac{1}{2}$

8. $\frac{2x-1}{3} + \frac{x}{2}$

< Objective 1 >

Solve each equation.

9. $\frac{x}{3} + \frac{3}{2} = \frac{x}{6} + \frac{7}{3}$

10. $\frac{x}{12} - \frac{2}{3} = \frac{x}{6} + \frac{3}{4}$

11. $\frac{4}{x} + \frac{3}{4} = \frac{10}{x}$

12. $\frac{3}{x} = \frac{5}{3} - \frac{7}{x}$

13. $\frac{5}{4x} - \frac{1}{2} = \frac{1}{2x}$

14. $\frac{7}{6x} - \frac{1}{3} = \frac{1}{2x}$

15. $\frac{4}{x+5} = \frac{3}{x+3}$

16. $\frac{5}{x-2} = \frac{4}{x+1}$

17. $\frac{9}{x} + 2 = \frac{2x}{x+3}$ 

18. $\frac{6}{x} + 3 = \frac{3x}{x+1}$

9.6 exercises

19. $\frac{3}{x+2} - \frac{5}{x} = \frac{13}{x+2}$

20. $\frac{7}{x} - \frac{2}{x-3} = \frac{6}{x}$

21. $\frac{3}{2} + \frac{2}{2x-4} = \frac{1}{x-2}$

22. $\frac{10}{2x+6} + \frac{2}{x+3} = \frac{1}{2}$

23. $\frac{x}{3x+12} + \frac{x-1}{x+4} = \frac{5}{3}$

24. $\frac{x}{4x-12} - \frac{x-4}{x-3} = \frac{1}{8}$

25. $\frac{16-3x}{x+6} + \frac{x+3}{5} = \frac{3x-2}{15}$

26. $\frac{x+1}{x-2} - \frac{x+3}{x} = \frac{6}{x^2-2x}$

> Videos

27. $\frac{1}{x-2} - \frac{2}{x+2} = \frac{2}{x^2-4}$

28. $\frac{1}{x+4} + \frac{1}{x-4} = \frac{12}{x^2-16}$

29. $\frac{7}{x+5} - \frac{1}{x-5} = \frac{x}{x^2-25}$

30. $\frac{2}{x-2} = \frac{3}{x+2} + \frac{x}{x^2-4}$

31. $\frac{11}{x+2} - \frac{5}{x^2-x-6} = \frac{1}{x-3}$

32. $\frac{3}{x-4} - \frac{4}{x^2-3x-4} = \frac{1}{x+1}$

33. $\frac{5}{x-2} - \frac{3}{x+3} = \frac{24}{x^2+x-6}$

34. $\frac{3}{x+1} - \frac{5}{x+6} = \frac{2}{x^2+7x+6}$

35. $\frac{x}{x-3} - 2 = \frac{3}{x-3}$

36. $\frac{x}{x-5} + 2 = \frac{5}{x-5}$

37. $\frac{2}{x^2-3x} - \frac{1}{x^2+2x} = \frac{2}{x^2-x-6}$

38. $\frac{2}{x^2-x} - \frac{4}{x^2+5x-6} = \frac{3}{x^2+6x}$

39. $\frac{2}{x^2-4x+3} - \frac{3}{x^2-9} = \frac{2}{x^2+2x-3}$

40. $\frac{2}{x^2-4} - \frac{1}{x^2+x-2} = \frac{3}{x^2-3x+2}$

Answers

19. ______
20. ______
21. ______
22. ______
23. ______
24. ______
25. ______
26. ______
27. ______
28. ______
29. ______
30. ______
31. ______
32. ______
33. ______
34. ______
35. ______
36. ______
37. ______ 38. ______
39. ______ 40. ______

Answers

41. ______

42. ______

43. ______

44. ______

45. ______

46. ______

47. ______

48. ______

49. ______

50. ______

51. ______

52. ______

53. ______

54. ______

55. ______

56. ______

57. ______

58. ______

41. $\frac{7}{x-5} - \frac{3}{x+5} = \frac{40}{x^2-25}$

42. $\frac{3}{x-3} - \frac{18}{x^2-9} = \frac{5}{x+3}$

43. $\frac{2x}{x-3} + \frac{2}{x-5} = \frac{3x}{x^2-8x+15}$

44. $\frac{x}{x-4} = \frac{5x}{x^2-x-12} - \frac{3}{x+3}$

45. $\frac{2x}{x+2} = \frac{5}{x^2-x-6} - \frac{1}{x-3}$

46. $\frac{3x}{x-1} = \frac{2}{x-2} - \frac{2}{x^2-3x+2}$

47. $\frac{7}{x-2} + \frac{16}{x+3} = 3$

48. $\frac{5}{x-2} + \frac{6}{x+2} = 2$

49. $\frac{11}{x-3} - 1 = \frac{10}{x+3}$

50. $\frac{17}{x-4} - 2 = \frac{10}{x+2}$

Two similar triangles are given. Find the indicated sides.

51.

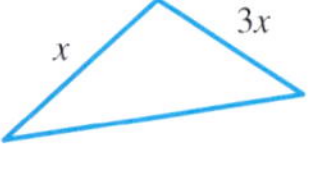

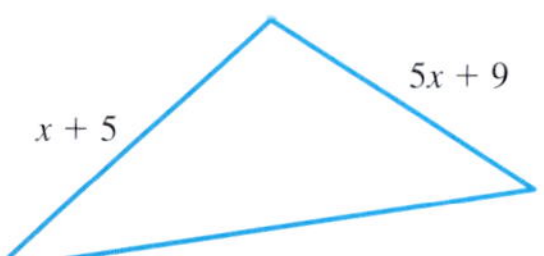

52.

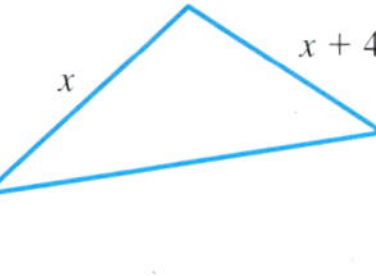

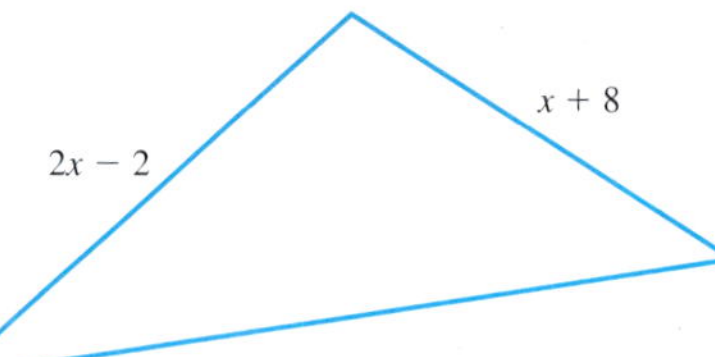

Basic Skills | **Challenge Yourself** | Calculator/Computer | Career Applications | Above and Beyond

< Objective 2 >

Solve each equation for the indicated variable.

53. $\frac{1}{x} = \frac{1}{a} - \frac{1}{b}$ for x

54. $\frac{1}{x} = \frac{1}{a} + \frac{1}{b}$ for a

55. $\frac{1}{R} = \frac{1}{R_1} + \frac{1}{R_2}$ for R_1

56. $\frac{1}{F} = \frac{1}{D_1} + \frac{1}{D_2}$ for D_2

57. $y = \frac{x+1}{x-1}$ for x

58. $y = \frac{x-3}{x-2}$ for x

Determine whether each statement is **true** *or* **false.**

59. When solving a rational equation, we need to multiply both numerator and denominator of a rational expression by the same nonzero quantity.

60. When solving a rational equation, we usually multiply each side of the equation by the same nonzero quantity, to clear it of fractions.

Complete each statement with **never, sometimes,** *or* **always.**

61. We must ________ check a possible solution to a rational equation, in case we have obtained an extraneous solution.

62. After clearing fractions in a rational equation, we ________ obtain a quadratic equation to solve.

< Objective 3 >

Solve each application.

63. Number Problem One number is 4 times another number. The sum of the reciprocals of the numbers is $\frac{5}{24}$. Find the two numbers.

64. Number Problem The sum of the reciprocals of two consecutive even integers is equal to 10 times the reciprocal of the product of those integers. Find the two integers.

65. Number Problem If the same number is subtracted from the numerator and denominator of $\frac{11}{15}$, the result is $\frac{1}{3}$. Find that number.

66. Number Problem If the numerator of $\frac{8}{9}$ is multiplied by a number and that same number is subtracted from the denominator, the result is 10. What is that number?

67. Science and Medicine A motorboat can travel 20 mi/h in still water. If the boat can travel 3 mi downstream on a river in the same time it takes to travel 2 mi upstream, what is the rate of the river's current?

Answers

59. ________

60. ________

61. ________

62. ________

63. ________

64. ________

65. ________

66. ________

67. ________

Answers

68.

69.

70.

71.

72.

73.

74.

75.

76.

68. **Science and Medicine** Janet and Michael took a canoe trip, traveling 6 mi upstream along a river, against a 2 mi/h current. They then returned downstream to the starting point of their trip. If their entire trip took 4 h, what was their rate in still water? > Videos

69. **Science and Medicine** A plane flew 720 mi with a steady 30 mi/h tailwind. The pilot then returned to the starting point, flying against the same wind. If the round-trip flight took 10 h, what was the plane's airspeed?

70. **Science and Medicine** A small jet has an airspeed (the rate in still air) of 300 mi/h. During one day's flights, the pilot noted that the plane could fly 85 mi with a tailwind in the same time it took to fly 65 mi against the same wind. What was the rate of the wind?

71. **Business and Finance** One computer printer can print a company's weekly payroll checks in 60 min. A second printer would take 90 min to complete the job. How long would it take the two printers, operating together, to print the checks?

72. **Business and Finance** An electrician can wire a house in 20 h. If she works with an apprentice, the same job can be completed in 12 h. How long would it take the apprentice, working alone, to wire the house?

73. **Science and Medicine** Po Ling can bicycle 75 mi in the same time it takes her to drive 165 mi. If her driving rate is 30 mi/h faster than her rate on the bicycle, find each rate.

74. **Science and Medicine** A passenger train can travel 275 mi in the same time a freight train takes to travel 225 mi. If the speed of the passenger train is 10 mi/h more than that of the freight train, find the speed of each train.

75. **Science and Medicine** A light plane took 1 h longer to fly 540 mi on the first portion of a trip than to fly 360 mi on the second. If the rate was the same for each portion, what was the flying time for each leg of the trip?

76. **Science and Medicine** Gilbert took 2 h longer to drive 240 mi on the first day of a business trip than to drive 144 mi on the second day. If his rate was the same both days, what was his driving time for each day?

77. Science and Medicine One road crew can pave a section of highway in 15 h. A second crew, working with newer equipment, can do the same job in 10 h. How long would it take to pave that same section of highway if both crews worked together?

78. Science and Medicine A landscaper can prepare and seed a new lawn in 12 h. If he works with an assistant, the job takes 8 h. How long would it take the assistant, working alone, to complete the job?

79. Science and Medicine An experienced roofer can work twice as fast as her helper. Working together, they can shingle a new section of roof in 4 h. How long would it take the experienced roofer, working alone, to complete the same job?

80. Science and Medicine Virginia can complete her company's monthly report in 5 h less time than Carl. If they work together, the report will take them 6 h to finish. How long would it take Virginia, working alone?

Basic Skills | Challenge Yourself | Calculator/Computer | **Career Applications** | Above and Beyond

81. Electronics Technology A 60-in. piece of wire is to be cut into two pieces whose lengths have the ratio 5 to 7. Find the length of each piece.

82. Construction Technology A 21-ft-long board is cut into two pieces so that the ratio of their lengths is 3 to 4. Find the lengths of the two pieces.

Answers

77. ____________

78. ____________

79. ____________

80. ____________

81. ____________

82. ____________

Answers

83. ______

84. ______

85. ______

Electrical Engineering *One formula for determining the equivalent resistance in a parallel circuit is*

$$\frac{1}{R_{eq}} = \frac{1}{R_1} + \frac{1}{R_2}$$

Use this formula to complete exercises 83 and 84.

83. Find the unknown resistance in the figure shown, to the nearest hundredth of an ohm, if the total (equivalent) resistance is 0.41 Ω.

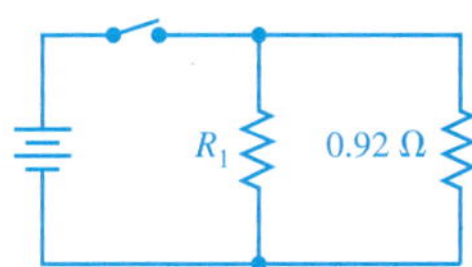

84. Find the unknown resistance in the figure shown, to the nearest ohm, if the total (equivalent) resistance is 26 Ω.

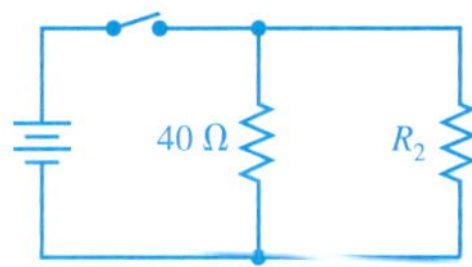

Basic Skills | Challenge Yourself | Calculator/Computer | Career Applications | **Above and Beyond**

85. What special considerations must be made when an equation contains rational expressions with variables in the denominator? chapter 9 > Make the Connection

Answers

1. Equation, $\{40\}$ **3.** Expression, $\frac{3x}{10}$ **5.** Equation, $\{5\}$
7. Equation, $\{3\}$ **9.** $\{5\}$ **11.** $\{8\}$ **13.** $\left\{\frac{3}{2}\right\}$ **15.** $\{3\}$
17. $\left\{-\frac{9}{5}\right\}$ **19.** $\left\{-\frac{2}{3}\right\}$ **21.** No solution or $\varnothing$ **23.** $\{-23\}$ **25.** $\{9\}$
27. $\{4\}$ **29.** $\{8\}$ **31.** $\{4\}$ **33.** $\left\{\frac{3}{2}\right\}$ **35.** No solution or $\varnothing$
37. $\{7\}$ **39.** $\{5\}$ **41.** $\left\{-\frac{5}{2}\right\}$ **43.** $\left\{-\frac{1}{2}, 6\right\}$ **45.** $\left\{-\frac{1}{2}\right\}$
47. $\left\{-\frac{1}{3}, 7\right\}$ **49.** $\{-8, 9\}$
51. Sides are 3 and 9 on the first triangle and 8 and 24 on the second.
53. $\frac{ab}{b-a}$ **55.** $\frac{RR_2}{R_2 - R}$ **57.** $\frac{y+1}{y-1}$ **59.** False **61.** always
63. 6, 24 **65.** 9 **67.** 4 mi/h **69.** 150 mi/h **71.** 36 min
73. Bicycling: 25 mi/h, driving: 55 mi/h **75.** First leg: 3 h, second leg: 2 h
77. 6 h **79.** 6 h **81.** 25 in., 35 in. **83.** 0.74 Ω
85. Above and Beyond

Definition/Procedure	Example	Reference
Simplifying Rational Expressions		Section 9.1
Rational expressions have the form $\frac{P}{Q}$ in which P and Q are polynomials and $Q \neq 0$.	$\frac{x^2 - 5x}{x - 3}$ is a rational expression. The variable x cannot have the value 3.	p. 880
Fundamental Principle of Rational Expressions For polynomials P, Q, and R, $\frac{P}{Q} = \frac{PR}{QR}$ when $Q \neq 0$ and $R \neq 0$. This principle can be used in two ways. We can multiply or divide the numerator and denominator of a rational expression by the same nonzero polynomial.	This uses the fact that $\frac{R}{R} = 1$ when $R \neq 0$.	p. 883
Simplifying Rational Expressions **Step 1** Completely factor both the numerator and the denominator of the expression. **Step 2** Divide the numerator and denominator by *all* common factors. **Step 3** The resulting expression is in simplest form (or in lowest terms).	$\frac{x^2 - 4}{x^2 - 2x - 8} = \frac{(x-2)(x+2)}{(x-4)(x+2)} = \frac{x-2}{x-4}$	p. 885
Identifying Rational Functions A rational function is a function that is defined by a rational expression. It can be written as $f(x) = \frac{P(x)}{Q(x)}$ in which $P(x)$ and $Q(x)$ are polynomial functions, $Q(x) \neq 0$.	$f(x) = \frac{2x^2 - 3x}{x + 1}$ and $g(x) = \frac{3}{x^2 - 3}$ are both rational functions	p. 886
Multiplying and Dividing Rational Expressions		Section 9.2
Multiplying Rational Expressions For polynomials P, Q, R, and S, $\frac{P}{Q} \cdot \frac{R}{S} = \frac{PR}{QS}$ when $Q \neq 0, S \neq 0$	$\frac{2x-6}{x^2-9} \cdot \frac{x^2+3x}{6x+24} = \frac{2(x-3)}{(x-3)(x+3)} \cdot \frac{x(x+3)}{6(x+4)} = \frac{x}{3(x+4)}$	p. 895
In practice, we apply this algorithm to multiply two rational expressions. **Step 1** Write each numerator and denominator in completely factored form. **Step 2** Divide by any common factors appearing in both the numerator and the denominator. **Step 3** Multiply as needed to form the product.		p. 896

Continued

Definition/Procedure	Example	Reference
Dividing Rational Expressions For polynomials P, Q, R, and S, $\frac{P}{Q} \div \frac{R}{S} = \frac{P}{Q} \cdot \frac{S}{R} = \frac{PS}{QR}$ when $Q \neq 0, R \neq 0, S \neq 0$ To divide two rational expressions, **Step 1** Invert the divisor (the *second* rational expression) to write the problem as one of multiplication. **Step 2** Proceed as in the algorithm for the multiplication of rational expressions.	$\frac{5y}{2y-8} \div \frac{10y^2}{y^2-y-12}$ $= \frac{5y}{2y-8} \cdot \frac{y^2-y-12}{10y^2}$ $= \frac{5y}{2(y-4)} \cdot \frac{(y-4)(y+3)}{10y^2}$ $= \frac{y+3}{4y}$	p. 897
Adding and Subtracting Rational Expressions		**Section 9.3**
Adding and Subtracting Rational Expressions To add or subtract rational expressions with the same denominator, add or subtract their numerators and then write that sum over the common denominator. The result should be written in lowest terms. In symbols, $\frac{P}{R} + \frac{Q}{R} = \frac{P+Q}{R}$ and $\frac{P}{R} - \frac{Q}{R} = \frac{P-Q}{R}$ when $R \neq 0$.	$\frac{5w}{w^2-16} - \frac{20}{w^2-16}$ $= \frac{5w-20}{w^2-16}$ $= \frac{5(w-4)}{(w+4)(w-4)}$ $= \frac{5}{w+4}$	p. 905
Least Common Denominator The least common denominator (LCD) of a group of rational expressions is the simplest polynomial that is divisible by each of the individual denominators of the rational expressions. To find the LCD, **Step 1** Write each of the denominators in completely factored form. **Step 2** Write the LCD as the product of each prime factor, to the highest power to which it appears in the factored form of any of the individual denominators.	To find the LCD for $\frac{2}{x^2+2x+1}$ and $\frac{3}{x^2+x}$ write $x^2+2x+1 = (x+1)(x+1)$ $x^2+x = x(x+1)$ The LCD is $x(x+1)(x+1)$	p. 906
To add or subtract rational expressions with different denominators, we first find the LCD by the procedure outlined above. We then rewrite each of the rational expressions with that LCD as a common denominator. Then we can add or subtract as before.	$\frac{2}{(x+1)^2} - \frac{3}{x(x+1)}$ $= \frac{2 \cdot x}{(x+1)^2 x} - \frac{3(x+1)}{x(x+1)(x+1)}$ $= \frac{2x - 3(x+1)}{x(x+1)(x+1)}$ $= \frac{-x-3}{x(x+1)(x+1)}$	p. 908

Definition/Procedure	Example	Reference

Complex Fractions

Section 9.4

p. 919

Complex fractions are fractions that have a fraction in their numerator or denominator (or both).

There are two commonly used methods for simplifying complex fractions.

Method 1

1. Multiply the numerator and denominator of the complex fraction by the LCD of all the fractions that appear within the numerator and denominator.
2. Simplify the resulting rational expression, writing the result in lowest terms.

Method 2

1. Write the numerator and denominator of the complex fraction as single fractions, if necessary.
2. Invert the denominator and multiply as before, writing the result in lowest terms.

Simplify $\dfrac{1 - \dfrac{2}{x}}{1 - \dfrac{4}{x^2}}$.

Method 1:

$$\frac{\left(1 - \dfrac{2}{x}\right)x^2}{\left(1 - \dfrac{4}{x^2}\right)x^2}$$

$$= \frac{x^2 - 2x}{x^2 - 4} = \frac{x(x - 2)}{(x + 2)(x - 2)}$$

$$= \frac{x}{x + 2}$$

Method 2:

$$\frac{\dfrac{x - 2}{x}}{\dfrac{x^2 - 4}{x^2}}$$

$$= \frac{x - 2}{x} \cdot \frac{x^2}{x^2 - 4}$$

$$= \frac{x - 2}{x} \cdot \frac{x^2}{(x + 2)(x - 2)}$$

$$= \frac{x}{x + 2}$$

Continued

Definition/Procedure	Example	Reference

Introduction to Graphing Rational Functions

Section 9.5

To draw the graph of a rational function:

Step 1 Determine the domain of f.

Step 2 Find the intercepts of the graph of f. Plot these.

Step 3 Locate vertical and horizontal asymptotes. Draw these as dotted lines.

Step 4 Choose a few easy-to-compute points to plot.

Step 5 Connect plotted points with a smooth curve, allowing the curve to approach the asymptotes.

To find a vertical asymptote: **(1)** locate any x-value for which f is undefined by setting the denominator equal to 0; **(2)** investigate values of $f(x)$ close to an undefined x-value.

To find a horizontal asymptote, investigate $f(x)$ values for large positive and large negative values of x.

Draw the graph of $f(x) = \dfrac{6 - 2x}{x + 3}$.

Domain: $\{x \mid x \neq -3\}$

y-intercept: $(0, 2)$

x-intercept: $(3, 0)$

Vertical asymptote: $x = -3$

Horizontal asymptote: $y = -2$

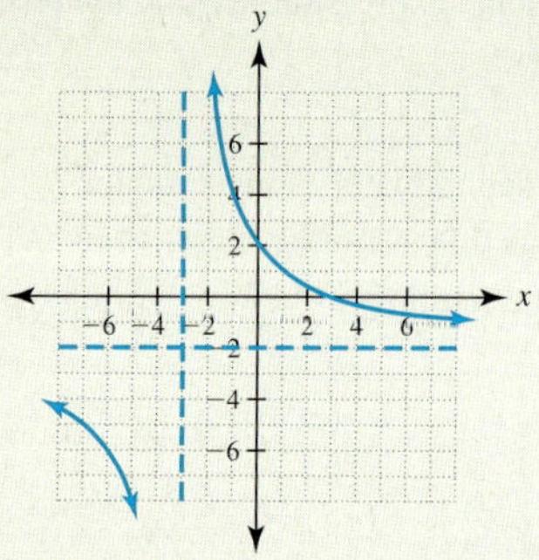

p. 937

Solving Rational Equations

Section 9.6

To solve an equation involving rational expressions,

Step 1 Clear the equation of fractions by multiplying both sides of the equation by the LCD of all the fractions that appear.

Step 2 Solve the equation resulting from step 1.

Step 3 Check all solutions by substitution in the original equation.

Solve.

$$\frac{3}{x - 3} - \frac{2}{x + 2} = \frac{19}{x^2 - x - 6}$$

Multiply by the LCD $(x - 3)(x + 2)$.

$$3(x + 2) - 2(x - 3) = 19$$
$$3x + 6 - 2x + 6 = 19$$
$$x = 7$$

Check:

$$\frac{3}{4} - \frac{2}{9} \stackrel{?}{=} \frac{19}{36}$$
$$\frac{19}{36} = \frac{19}{36} \quad \text{True}$$

p. 954

summary exercises :: chapter 9

This summary exercise set is provided to give you practice with each of the objectives of this chapter. Each exercise is keyed to the appropriate chapter section. When you are finished, you can check your answers to the odd-numbered exercises in the back of the text. If you have difficulty with any of these questions, go back and reread the examples from that section. The answers to the even-numbered exercises appear in the *Instructor's Solutions Manual.* Your instructor will give you guidelines on how best to use these exercises in your instructional setting.

9.1 *For what value(s) of the variable is each rational expression defined?*

1. $\frac{x}{2}$

2. $\frac{3}{y}$

3. $\frac{2}{x-5}$

4. $\frac{4x}{3x-2}$

Simplify each rational expression.

5. $\frac{18x^5}{24x^3}$

6. $\frac{15m^3n}{-5mn^2}$

7. $\frac{7y-49}{y-7}$

8. $\frac{5x-20}{x^2-16}$

9. $\frac{9-x^2}{x^2+2x-15}$

10. $\frac{3w^2+8w-35}{2w^2+13w+15}$

11. $\frac{6a^2-ab-b^2}{9a^2-b^2}$

12. $\frac{6w-3z}{8w^3-z^3}$

Simplify the given function. Indicate any value for x for which the function is undefined.

13. $f(x) = \frac{x^2-3x-4}{x+1}$

14. $f(x) = \frac{x^2+x-6}{x-2}$

9.2 *Multiply or divide as indicated. Express your results in simplest form.*

15. $\frac{x^7}{36} \cdot \frac{24}{x^4}$

16. $\frac{a^3b}{4ab^2} \div \frac{ab}{12ab^2}$

17. $\frac{6y-18}{9y} \cdot \frac{10}{5y-15}$

18. $\frac{m^2-3m}{m^2-5m+6} \cdot \frac{m^2-4}{m^2+7m+10}$

19. $\frac{a^2-2a}{a^2-4} \div \frac{2a^2}{3a+6}$

20. $\frac{r^2+2rs}{r^3-r^2s} \div \frac{5r+10s}{r^2-2rs+s^2}$

21. $\frac{x^2-2xy-3y^2}{x^2-xy-2y^2} \cdot \frac{x^2-4y^2}{x^2-8xy+15y^2}$

22. $\frac{w^3+3w^2+2w+6}{w^4-4} \div (w^3+27)$

23. Let $f(x) = \frac{x^2-16}{x-5}$ and $g(x) = \frac{x^2-25}{x+4}$. Find **(a)** $f(3) \cdot g(3)$; **(b)** $h(x) = f(x) \cdot g(x)$; and **(c)** $h(3)$.

24. Let $f(x) = \frac{2x^2-5x-3}{x-4}$ and $g(x) = \frac{x^2-3x-4}{2x^2+5x+2}$. Find **(a)** $f(3) \cdot g(3)$; **(b)** $h(x) = f(x) \cdot g(x)$; and **(c)** $h(3)$.

9.3 *Perform the indicated operations. Express your results in simplified form.*

25. $\dfrac{5x+7}{x+4} - \dfrac{2x-5}{x-4}$

26. $\dfrac{5}{6x^2} + \dfrac{3}{4x}$

27. $\dfrac{2}{x-5} - \dfrac{1}{x}$

28. $\dfrac{2}{y+5} + \dfrac{3}{y+4}$

29. $\dfrac{2}{3m-3} - \dfrac{5}{2m-2}$

30. $\dfrac{7}{x-3} - \dfrac{5}{3-x}$

31. $\dfrac{6}{3x+3} - \dfrac{6}{5x-5}$

32. $\dfrac{2a}{a^2-9a+20} + \dfrac{8}{a-4}$

33. $\dfrac{2}{s-1} - \dfrac{6s}{s^2+s-2}$

34. $\dfrac{4}{x^2-9} - \dfrac{3}{x^2-4x+3}$

35. $\dfrac{x^2-14x-8}{x^2-2x-8} + \dfrac{2x}{x-4} - \dfrac{3}{x+2}$

36. $\dfrac{w^2+2wz+z^2}{w^2-wz-2z^2} \cdot \left(\dfrac{3}{w+z} - \dfrac{1}{w-z}\right)$

37. Let $f(x) = \dfrac{2x}{x-2}$ and $g(x) = \dfrac{x}{x-3}$. Find **(a)** $f(4) + g(4)$; **(b)** $h(x) = f(x) + g(x)$; and **(c)** the ordered pair $(4, h(4))$.

38. Let $f(x) = \dfrac{x+2}{x-2}$ and $g(x) = \dfrac{x+1}{x-7}$. Find **(a)** $f(3) + g(3)$; **(b)** $h(x) = f(x) + g(x)$; and **(c)** the ordered pair $(3, h(3))$.

9.4 *Simplify each complex fraction.*

39. $\dfrac{\frac{x^3}{15}}{\frac{x^5}{10}}$

40. $\dfrac{\frac{y-1}{y^2-4}}{\frac{y^2-1}{y^2-y-2}}$

41. $\dfrac{1+\frac{a}{b}}{1-\frac{a}{b}}$

42. $\dfrac{2-\frac{x}{y}}{4-\frac{x^2}{y^2}}$

43. $\dfrac{\frac{1}{r^2}-\frac{1}{s^2}}{\frac{1}{r}-\frac{1}{s}}$

44. $\dfrac{1-\frac{1}{x+2}}{1+\frac{1}{x+2}}$

45. $\dfrac{1-\frac{2}{x-1}}{x+\frac{3}{x-4}}$

46. $\dfrac{\frac{w}{w+1}-\frac{1}{w-1}}{\frac{w}{w-1}+\frac{1}{w+1}}$

47. $\dfrac{1}{1-\dfrac{1}{1-\frac{1}{y-1}}}$

48. $1-\dfrac{1}{1+\dfrac{1}{1-\frac{1}{x}}}$

49. $\dfrac{1-\frac{1}{x-1}}{x-\frac{8}{x+2}}$

50. $\dfrac{1}{1-\dfrac{1}{1+\frac{1}{y+1}}}$

9.5 *Determine the domain of each function.*

51. $f(x) = \frac{x-3}{x+4}$

52. $f(x) = \frac{7}{5-x}$

53. $f(x) = \frac{8x}{x-2}$

54. $f(x) = \frac{x-6}{x}$

55. $f(x) = \frac{2x-6}{x+1}$

56. $f(x) = \frac{4-2x}{x-4}$

Find all intercepts for the graph of each function.

57. $f(x) = \frac{x-3}{x+4}$

58. $f(x) = \frac{7}{5-x}$

59. $f(x) = \frac{8x}{x-2}$

60. $f(x) = \frac{x-6}{x}$

61. $f(x) = \frac{2x-6}{x+1}$

62. $f(x) = \frac{4-2x}{x-4}$

Identify all asymptotes for each function.

63. $f(x) = \frac{x-3}{x+4}$

64. $f(x) = \frac{7}{5-x}$

65. $f(x) = \frac{8x}{x-2}$

66. $f(x) = \frac{x-6}{x}$

67. $f(x) = \frac{2x-6}{x+1}$

68. $f(x) = \frac{4-2x}{x-4}$

Sketch the graph of each function.

69. $f(x) = \frac{x-3}{x+4}$

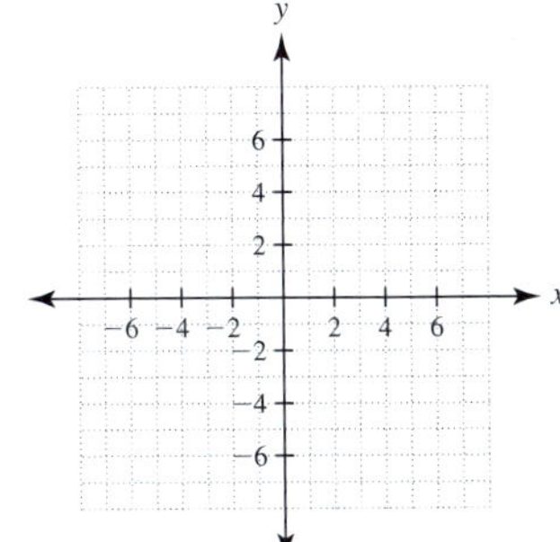

70. $f(x) = \frac{x-6}{x}$

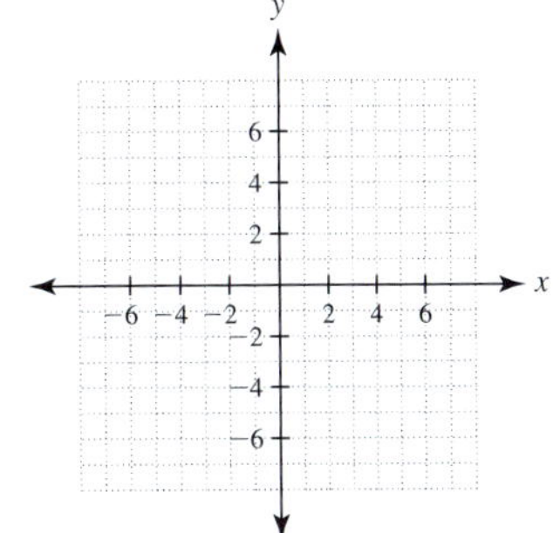

71. $f(x) = \frac{2x-6}{x+1}$

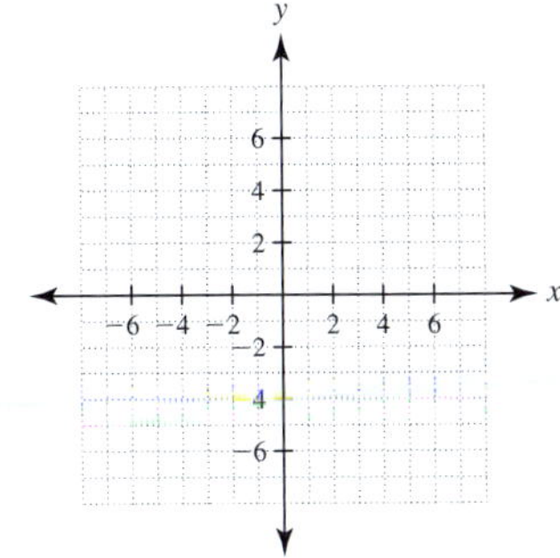

72. $f(x) = \frac{4-2x}{x-4}$

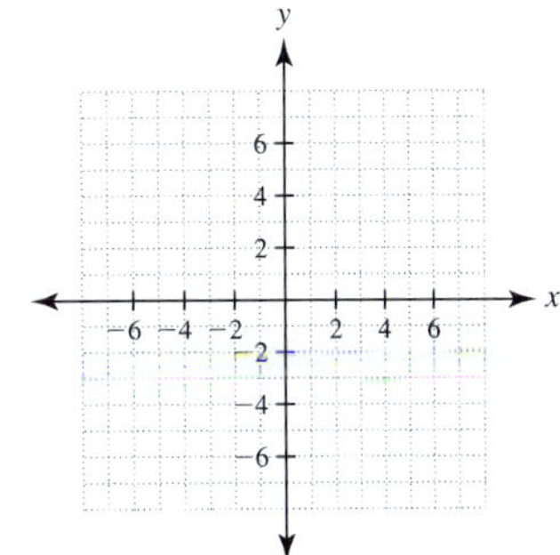

9.6 *Solve each equation.*

73. $\frac{1}{2x} + \frac{1}{3x} = \frac{1}{6}$

74. $\frac{5}{2x^2} - \frac{1}{4x} = \frac{1}{x}$

75. $\frac{x}{x-2} + 1 = \frac{x+4}{x-2}$

76. $\frac{2x-1}{x-3} - \frac{5}{x-3} = 1$

77. $\frac{2}{3x+1} = \frac{1}{x+2}$

78. $\frac{5}{x+1} + \frac{1}{x-2} = \frac{7}{x+1}$

79. $\frac{4}{x-1} - \frac{5}{3x-7} = \frac{3}{x-1}$

80. $\frac{7}{x} - \frac{1}{x-3} = \frac{9}{x^2-3x}$

81. $\frac{2}{x-3} - \frac{11}{x^2-9} = \frac{3}{x+3}$

82. $\frac{5}{x+3} + \frac{1}{x-5} = \frac{1}{x+3}$

83. $\frac{2}{x-4} = \frac{x}{x-2} - \frac{x+4}{x^2-6x+8}$

84. $\frac{x}{x-5} = \frac{3x}{x^2-7x+10} + \frac{8}{x-2}$

In exercises 85 and 86, two similar triangles are given. Find the lengths of the indicated sides.

85.

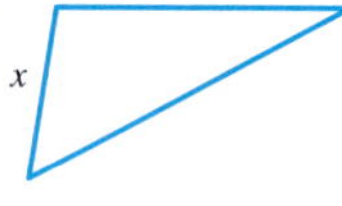

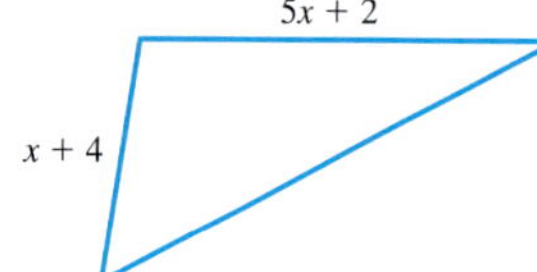

86.

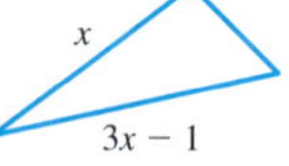

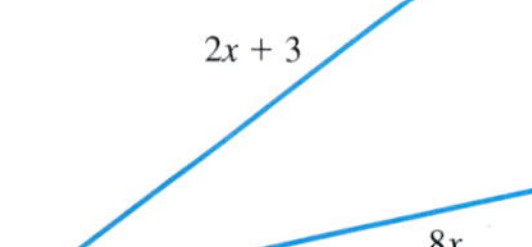

Solve.

87. Number Problem The sum of the reciprocals of two consecutive integers is equal to 11 times the reciprocal of the product of those two integers. What are the two integers?

88. Science and Medicine Karl drove 224 mi on the expressway for a business meeting. On his return, he decided to use a shorter route of 200 mi, but road construction slowed his speed by 6 mi/h. If the trip took the same time each way, what was his average speed in each direction?

89. Construction An electrician can wire a certain model home in 20 h while it would take her apprentice 30 h to wire the same model. How long would it take the two of them, working together, to wire the house?

90. Science and Medicine A light plane took 1 h longer to fly 540 mi on the first portion of a trip than to fly 360 mi on the second. If the rate was the same for each portion, what was the flying time for each leg of the trip?

self-test 9

Name ______________________

Section __________ Date __________

The purpose of this self-test is to help you assess your progress so that you can find concepts that you need to review before the next exam. Allow yourself about an hour to take this test. At the end of that hour, check your answers against those given in the back of this text. If you miss any, go back to the appropriate section to reread the examples until you have mastered that particular concept.

Perform the indicated operations, and simplify.

1. $\dfrac{m^2 - 3m}{m^2 - 9} \div \dfrac{4m}{m^2 - m - 12}$

2. $\dfrac{2}{x + 3} + \dfrac{12}{x^2 - 9}$

3. $\dfrac{3 - \dfrac{x}{y}}{9 - \dfrac{x^2}{y^2}}$

4. $\dfrac{3ab^2}{5ab^3} \cdot \dfrac{20a^2b}{21b}$

5. $\dfrac{3}{x^2 - 3x - 4} + \dfrac{5}{x^2 - 16}$

6. $\dfrac{x^2 - 3x}{5x^2} \cdot \dfrac{10x}{x^2 - 4x + 3}$

7. $\dfrac{9x^2 - 9x - 4}{6x^2 - 11x + 3} \cdot \dfrac{15 - 10x}{3x - 4}$

8. $\dfrac{1 - \dfrac{10}{z + 3}}{2 - \dfrac{12}{z - 1}}$

9. $\dfrac{x^2 + 3xy}{2x^3 - x^2y} \div \dfrac{x^2 + 6xy + 9y^2}{4x^2 - y^2}$

10. $\dfrac{6x}{x^2 - x - 2} - \dfrac{2}{x + 1}$

11. $\dfrac{5}{x - 2} - \dfrac{1}{x}$

Solve.

12. $\dfrac{5}{x} - \dfrac{x - 3}{x + 2} = \dfrac{22}{x^2 + 2x}$

Answers

1. ______________________
2. ______________________
3. ______________________
4. ______________________
5. ______________________
6. ______________________
7. ______________________
8. ______________________
9. ______________________
10. ______________________
11. ______________________
12. ______________________

Answers

13. ______

14. ______

15. ______

16. ______

17. ______

18. ______

19. ______

20. ______

Simplify.

13. $\dfrac{3w^2 + w - 2}{3w^2 - 8w + 4}$

14. $\dfrac{x^3 + 2x^2 - 3x}{x^3 - 3x^2 + 2x}$

15. $\dfrac{-21x^5y^3}{28xy^5}$

16. If the numerator of $\dfrac{4}{7}$ is multiplied by a number and that same number is added to the denominator, the result is $\dfrac{6}{5}$. What was that number?

Sketch the graph of each function.

17. $f(x) = \dfrac{3x}{x + 1}$

18. $f(x) = \dfrac{-10}{x}$

Simplify. Indicate any value of x for which the function is undefined.

19. $f(x) = \dfrac{x^2 - 5x + 4}{x - 4}$

20. **(a)** Find the intercepts for the graph of the function $f(x) = \dfrac{4x - 1}{3x + 2}$.

(b) Identify the asymptotes for the graph of f.

cumulative review chapters 0-9

Name ____________

Section ________ Date ________

We offer the following exercises to help you review concepts from earlier chapters. This is meant as review material and not as a comprehensive exam. The answers are presented in the back of the text. If you have difficulty with any of these exercises, be certain to at least read through the summary related to that section.

Answers

1. ____________
2. ____________
3. ____________
4. ____________
5. ____________
6. ____________
7. ____________
8. ____________
9. ____________
10. ____________
11. ____________
12. ____________
13. ____________

1. Solve the equation $5x - 3(2x + 6) = 4 - (3x - 2)$.

2. If $f(x) = 5x^4 - 3x^2 + 7x - 9$, find $f(-1)$.

3. Find the equation of the line that is parallel to the line $6x + 7y = 42$ and has a y-intercept of $(0, -3)$.

4. Find the x- and y-intercepts of the equation $7x - 6y = -42$.

Simplify each polynomial function.

5. $f(x) = 3x - 2[x - (3x - 1)] + 6x(x - 2)$

6. $f(x) = x(2x - 1)(x + 3)$

7. Find the domain of the function $f(x) = \dfrac{x^2 + x}{x}$.

8. Evaluate the expression $6^2 - (16 \div 8 \cdot 2) - 4^2$.

Factor each polynomial completely.

9. $6x^3 + 7x^2 - 3x$

10. $16x^{16} - 9y^8$

Simplify each rational expression.

11. $\dfrac{5}{x - 1} - \dfrac{2x + 6}{x^2 + 2x - 3}$

12. $\dfrac{x + 1}{x^2 - 5x - 6} \div \dfrac{x^2 - 1}{x - 6}$

13. $\dfrac{1 - \dfrac{3}{x + 3}}{\dfrac{1}{x^2 - 9}}$

Answers

14. ____

15. ____

16. ____

17. ____

18. ____

19. ____

20. ____

21. ____

22. ____

23. ____

24. ____

25. ____

14. The height of a ball thrown into the air from a platform can be determined by the function

$$h(t) = -16t^2 + 58t + 15$$

To the nearest hundredth of a second, when will the ball be at a height of 50 ft?

Solve each equation.

15. $7x + (x - 10) = -12(x - 5)$

16. $x^4 - 18x^2 + 32 = 0$

17. $-4(7x + 6) = 8(5x + 12)$

18. $6 - 2\sqrt{x - 3} = x$

19. $\dfrac{5}{x} = \dfrac{2}{x + 3}$

Solve the inequality.

20. $-4(-2x - 7) > -6x$

21. If the hypotenuse of a right triangle has length 22 cm, and one leg has length 15 cm, find the length of the other leg, to the nearest tenth of a centimeter.

22. Identify the asymptotes for the graph of $f(x) = \dfrac{5 - 2x}{x - 5}$.

23. Simplify the expression $\left(\dfrac{a^{-2}b}{a^3b^{-2}}\right)^2$.

Solve each application.

24. When each works alone, Barry can mow a lawn in 3 h less time than Don. When they work together, it takes 2 h. How long does it take each to do the job by himself?

25. The length of a rectangle is 2 cm less than twice the width. The area of the rectangle is 180 cm^2. Find the length and width of the rectangle.

Exponential and Logarithmic Functions

INTRODUCTION

There are many applications of mathematics in the field of chemistry. Some of the most important of these occur in pharmacology. Pharmacologists use exponential and logarithmic functions to model drug absorption and elimination. After a drug is taken, it is distributed throughout the body via the circulatory system. For a medicine to be effective, there must be enough of the substance in the body to achieve the desired effect but not enough to cause harm. The therapeutic level is maintained by taking the right dosage at timed intervals determined by the rate at which the body absorbs or eliminates the medication.

CHAPTER 10 OUTLINE

10.1 Algebra of Functions

< 10.1 Objectives >

1 > Find the sum or difference of two functions

2 > Find the product of two functions

3 > Find the quotient of two functions

4 > Find the domain of the sum, difference, product, or quotient of two functions

In this chapter, we begin with a deeper study of functions. The first two sections focus on ways that functions can be combined to form new functions, and the third section deals with the *inverse* of a function. The remainder of the chapter introduces two very important, and highly applied, groups of functions: exponential and logarithmic functions.

The profit that a company earns on an item is determined by subtracting the cost of making the item from the total revenue the company receives from selling the item. This is an example of **combining functions.** It can be written as

$$P(x) = R(x) - C(x)$$

Many applications of functions involve the combination of two or more component functions. In this section, we look at several properties that allow for the addition, subtraction, multiplication, and division of functions.

Definition

Sum of Two Functions

The **sum of two functions** f and g is written as $f + g$ and is defined as

$$(f + g)(x) = f(x) + g(x)$$

for every value of x that is in the domain of both functions f and g.

Definition

Difference of Two Functions

The **difference of two functions** f and g is written as $f - g$ and is defined as

$$(f - g)(x) = f(x) - g(x)$$

for every value of x that is in the domain of both functions f and g.

Example 1 Finding the Sum or Difference of Two Functions

< Objective 1 >

Suppose the functions f and g are defined by the tables.

x	$f(x)$
−4	−8
0	6
2	5
1	−2

x	$g(x)$
−4	3
0	−5
2	7
3	1

RECALL

From the first table, we know that $f(-4) = -8$. From the second table we see that $g(-4) = 3$

(a) Evaluate $(f + g)(-4)$.

$$(f + g)(-4) = f(-4) + g(-4)$$
$$= -8 + 3$$
$$= -5$$

(b) Evaluate $(f - g)(0)$.

$$(f - g)(0) = f(0) - g(0)$$
$$= 6 - (-5) \qquad f(0) = 6;\ g(0) = -5$$
$$= 11$$

(c) Evaluate $(f + g)(3)$.

$$(f + g)(3) = f(3) + g(3)$$

We can find $g(3)$, but not $f(3)$ since 3 is not in the domain of f. This means that 3 is not in the domain of $f + g$. So $(f + g)(3)$ does not exist.

(d) Find the domain of $f + g$.

We need to find all values of x that are in the domains of *both f and g*. Therefore,

$$D = \{-4, 0, 2\}$$

Check Yourself 1

Suppose the functions f and g are defined by the tables.

x	$f(x)$
−3	7
−1	0
5	−3
6	2

x	$g(x)$
−3	−4
2	8
5	−6
7	0

(a) Evaluate $(f + g)(5)$.

(b) Evaluate $(f - g)(-3)$.

(c) Evaluate $(f - g)(-1)$.

(d) Find the domain of $f - g$.

In Example 2, we look at functions that are defined by equations rather than tables.

Example 2 Finding the Sum or Difference of Two Functions

You are given the functions $f(x) = 2x - 1$ and $g(x) = -3x + 4$.

(a) Find $(f + g)(x)$.

$$(f + g)(x) = f(x) + g(x)$$
$$= (2x - 1) + (-3x + 4) = -x + 3$$

(b) Find $(f - g)(x)$.

$$(f - g)(x) = f(x) - g(x)$$
$$= (2x - 1) - (-3x + 4) = 5x - 5$$

(c) Evaluate $(f + g)(2)$.

If we use the definition of the sum of two functions, we find that

$$(f + g)(2) = f(2) + g(2) \qquad f(2) = 2(2) - 1 = 3$$
$$= 3 + (-2) = 1 \qquad g(2) = -3(2) + 4 = -2$$

As an alternative, we could use part (a) and say

$$(f + g)(x) = -x + 3$$

Therefore,

$$(f + g)(2) = -(2) + 3 = 1$$

Check Yourself 2

You are given the functions $f(x) = -2x - 3$ and $g(x) = 5x - 1$.

(a) Find $(f + g)(x)$. **(b)** Find $(f - g)(x)$. **(c)** Evaluate $(f + g)(2)$.

In the next example, we find the domain of the sum of two functions.

Example 3 Finding the Sum of Two Functions

< Objective 4 >

You are given the functions $f(x) = 2x - 4$ and $g(x) = \frac{1}{x}$.

(a) Find $(f + g)(x)$.

$$(f + g)(x) = (2x - 4) + \left(\frac{1}{x}\right) = 2x - 4 + \frac{1}{x}$$

(b) Find the domain of $f + g$.

The domain of $f + g$ is the set of all numbers in the domain of f and also in the domain of g. The domain of f consists of all real numbers. The domain of g consists of all real numbers except 0 because we cannot divide by 0. The domain of $f + g$ is the set of all real numbers except 0. We write $D = \{x \mid x \neq 0\}$.

Check Yourself 3

You are given $f(x) = -3x + 1$ and $g(x) = \frac{1}{x - 2}$.

(a) Find $(f + g)(x)$. **(b)** Find the domain of $f + g$.

Definition

Product of Two Functions

The **product of two functions** f and g is written as $f \cdot g$ and is defined as

$(f \cdot g)(x) = f(x) \cdot g(x)$

for every value of x that is in the domain of both functions f and g.

Definition

Quotient of Two Functions

The **quotient of two functions** f and g is written as $f \div g$ or $\frac{f}{g}$ and is defined as

$(f \div g)(x) = f(x) \div g(x)$

for every value of x that is in the domain of both functions f and g, such that $g(x) \neq 0$.

Example 4 Finding the Product or Quotient of Two Functions

< Objectives 2–3 >

Suppose we have the functions f and g defined by the tables.

x	$f(x)$
-3	7
-1	0
5	-3
7	2

x	$g(x)$
-3	-4
2	8
5	-6
7	0

(a) Evaluate $(f \cdot g)(-3)$.

$$\begin{aligned}(f \cdot g)(-3) &= f(-3) \cdot g(-3) \\ &= (7)(-4) \\ &= -28\end{aligned}$$

(b) Evaluate $(f \div g)(5)$.

$$\begin{aligned}(f \div g)(5) &= f(5) \div g(5) \\ &= (-3) \div (-6) \\ &= \frac{1}{2}\end{aligned}$$

(c) Evaluate $(f \div g)(7)$.

$(f \div g)(7) = f(7) \div g(7) = 2 \div 0$, which is undefined. Therefore, 7 is not in the domain of $f \div g$. So we answer that $(f \div g)(7)$ does not exist.

(d) Find the domain of $f \cdot g$.

We want all values of x that are in the domains *both* of f and of g. Then

$D = \{-3, 5, 7\}$

(e) Find the domain of $f \div g$.

Again we want all values of x that are in the domains of *both* f and g, but we must exclude any x-value such that $g(x) = 0$. Since $g(7) = 0$, 7 cannot be in the domain of $f \div g$. Thus,

$D = \{-3, 5\}$

Check Yourself 4

Suppose we have the functions *f* and *g* defined by the tables.

x	$f(x)$
-5	-1
-2	3
0	4
3	0

x	$g(x)$
-5	0
0	-6
1	7
3	2

(a) Find $(f \cdot g)(0)$.
(b) Find $(f \div g)(3)$.
(c) Find the domain of $f \cdot g$.
(d) Find the domain of $f \div g$.

We now consider functions defined by equations.

Example 5 Finding the Product of Two Functions

RECALL

From Section 5.5 you should recall that the product of two binomials $(x + a)(x + b)$ is $x^2 + bx + ax + ab$

You are given $f(x) = x - 1$ and $g(x) = x + 5$. Find $(f \cdot g)(x)$.

$(f \cdot g)(x) = f(x) \cdot g(x) = (x - 1)(x + 5) = x^2 + 5x - x - 5 = x^2 + 4x - 5$

Check Yourself 5

Given $f(x) = x - 3$ and $g(x) = x + 2$, find $(f \cdot g)(x)$.

Example 6 Finding the Quotient of Two Functions

You are given $f(x) = x - 1$ and $g(x) = x + 5$.

(a) Find $(f \div g)(x)$.

$$(f \div g)(x) = f(x) \div g(x) = (x - 1) \div (x + 5) = \frac{x - 1}{x + 5}$$

(b) Find the domain of $f \div g$.

The domain is the set of all real numbers except -5, because $g(-5) = 0$ and division by 0 is undefined. We write $D = \{x \mid x \neq -5\}$.

Check Yourself 6

You are given $f(x) = x - 3$ and $g(x) = x + 2$.

(a) Find $(f \div g)(x)$. **(b)** Find the domain of $f \div g$.

Check Yourself ANSWERS

1. **(a)** -9; **(b)** 11; **(c)** does not exist; **(d)** $D = \{-3, 5\}$

2. **(a)** $3x - 4$; **(b)** $-7x - 2$; **(c)** 2

3. **(a)** $-3x + 1 + \frac{1}{x - 2}$; **(b)** $D = \{x \mid x \neq 2\}$

4. **(a)** -24; **(b)** 0; **(c)** $D = \{-5, 0, 3\}$; **(d)** $D = \{0, 3\}$

5. $x^2 - x - 6$ **6.** **(a)** $\frac{x - 3}{x + 2}$; **(b)** $D = \{x \mid x \neq -2\}$

Reading Your Text

SECTION 10.1

(a) The sum of two functions can be defined for every value x that is in the ______________ of both functions.

(b) The ______________ of two functions f and g is written as $f \cdot g$.

(c) The ______________ of two functions f and g is written as $f \div g$.

(d) The product of two ______________ $(x + a)(x + b)$ is $x^2 + bx + ax + ab$.

10.1 exercises

Basic Skills | Challenge Yourself | Calculator/Computer | Career Applications | Above and Beyond

Boost your GRADE at ALEKS.com!

ALEKS

- Practice Problems
- Self-Tests
- NetTutor
- e-Professors
- Videos

Name ______________

Section ________ Date ________

Answers

1. ________ 2. ________
3. ________ 4. ________
5. ________
6. ________ 7. ________
8. ________ 9. ________
10. ________
11. ________
12. ________
13. ________
14. ________
15. ________
16. ________
17. ________
18. ________

< Objectives 1 and 4 >

Use the tables to find the desired values.

x	$f(x)$
−3	−5
0	7
2	3
5	−3

x	$g(x)$
0	8
1	−3
2	4
7	−1

x	$h(x)$
−4	−1
0	−7
2	0
5	6

x	$k(x)$
−5	4
0	7
3	0
7	−3

1. $(f + g)(2)$ 

2. $(f + h)(5)$

3. $(k - g)(7)$

4. $(h - f)(5)$

5. $(f + k)(2)$ 

6. $(g - f)(0)$

7. Find the domain of $f + g$.

8. Find the domain of $h + k$.

9. Find the domain of $g - h$.

10. Find the domain of $k - f$.

Find **(a)** $(f + g)(x)$; **(b)** $(f - g)(x)$; **(c)** $(f + g)(3)$; *and* **(d)** $(f - g)(2)$.

11. $f(x) = -4x + 5$; $g(x) = 7x - 4$

12. $f(x) = 9x - 3$; $g(x) = -3x + 5$

13. $f(x) = 8x - 2$; $g(x) = -5x + 6$

14. $f(x) = -7x + 9$; $g(x) = 2x - 1$

15. $f(x) = x^2 + x - 1$; $g(x) = -3x^2 - 2x + 5$

16. $f(x) = -3x^2 - 2x + 5$; $g(x) = 5x^2 + 3x - 6$

17. $f(x) = -x^3 - 5x + 8$; $g(x) = 2x^2 + 3x - 4$ 

18. $f(x) = 2x^3 + 3x^2 - 5$; $g(x) = -4x^2 + 5x - 7$

Find **(a)** $(f + g)(x)$ *and* **(b)** *the domain of* $f + g$.

19. $f(x) = -9x + 11;\ g(x) = 15x - 7$

20. $f(x) = -11x + 3;\ g(x) = 8x - 5$

21. $f(x) = 3x + 2;\ g(x) = \dfrac{1}{x - 2}$

22. $f(x) = -2x + 5;\ g(x) = \dfrac{3}{x + 1}$

23. $f(x) = x^2 + x - 5;\ g(x) = \dfrac{2}{3x + 1}$ > Videos

24. $f(x) = 3x^2 - 5x + 1;\ g(x) = \dfrac{-2}{2x - 3}$

< Objectives 2 and 3 >

Use the tables to find the desired values.

x	$f(x)$
-3	15
0	-4
2	5
4	9

x	$g(x)$
-4	18
0	12
2	-3
5	4

x	$h(x)$
-4	6
-3	-3
-2	-4
5	9

x	$k(x)$
-2	2
0	0
2	3
4	18

25. $(f \cdot g)(2)$

26. $(h \cdot k)(-2)$

27. $(f \cdot k)(0)$

28. $(g \cdot h)(5)$

29. $(f \div h)(-3)$

30. $(g \div h)(-4)$

31. $(g \div k)(0)$

32. $(f \div k)(4)$

33. Find the domain of $f \div g$.

34. Find the domain of $h \div k$.

Answers

19. ______

20. ______

21. ______

22. ______

23. ______

24. ______

25. ______

26. ______

27. ______

28. ______

29. ______

30. ______

31. ______

32. ______

33. ______

34. ______

Answers

35. ________

36. ________

37. ________

38. ________

39. ________

40. ________

41. ________

42. ________

43. ________

44. ________

45. ________

46. ________

47. ________

48. ________

Find **(a)** $(f \cdot g)(x)$; **(b)** $(f \div g)(x)$; *and* **(c)** *the domain of* $f \div g$.

35. $f(x) = 2x - 1$; $g(x) = x - 3$

36. $f(x) = -x + 3$; $g(x) = x + 4$

37. $f(x) = 3x + 2$; $g(x) = 2x - 1$

> Videos

38. $f(x) = -3x + 5$; $g(x) = -x + 2$

39. $f(x) = 2 - x$; $g(x) = 5 + 2x$

40. $f(x) = x + 5$; $g(x) = 1 - 3x$

Basic Skills | **Challenge Yourself** | Calculator/Computer | Career Applications | Above and Beyond

Complete each statement with **never, sometimes,** *or* **always.**

41. The domain of $f + g$ is ________ the same as the domain of f.

42. If $g(a) = 0$, then the domain of $f \div g$ will ________ contain a.

43. If f and g are linear functions that include the variable x, then the function $f \cdot g$ will ________ be a second-degree (quadratic) function.

44. If f and g are linear functions, then the function $f + g$ will ________ be a second-degree (quadratic) function.

In business, the profit $P(x)$ obtained from selling x units of a product is equal to the revenue $R(x)$ minus the cost $C(x)$. In exercises 45 and 46, find the profit $P(x)$ for selling x units.

45. $R(x) = 25x$; $C(x) = x^2 + 4x + 50$

46. $R(x) = 20x$; $C(x) = x^2 + 2x + 30$

Let $V(t)$ be the velocity of an object that has been thrown in the air. It can be shown that $V(t)$ is the combination of three functions: the initial velocity V_0 (this is a constant), the acceleration due to gravity g (this is also a constant), and the time that has elapsed t.

$V(t) = V_0 + g \cdot t$

Find the velocity as a function of time t.

47. $V_0 = 10$ m/s; $g = -9.8$ m/s^2

48. $V_0 = 64$ ft/s; $g = -32$ ft/s^2

The revenue produced from the sale of an item can be found by multiplying the price $p(x)$ by the quantity sold x. In exercises 49 and 50, find the revenue produced from selling x items.

49. $p(x) = 119 - 6x$

50. $p(x) = 1{,}190 - 36x$

Answers

49. ______

50. ______

Answers

1. 7 **3.** -2 **5.** Does not exist **7.** $\{0, 2\}$ **9.** $\{0, 2\}$
11. (a) $3x + 1$; **(b)** $-11x + 9$; **(c)** 10; **(d)** -13 **13. (a)** $3x + 4$; **(b)** $13x - 8$; **(c)** 13; **(d)** 18 **15. (a)** $-2x^2 - x + 4$; **(b)** $4x^2 + 3x - 6$; **(c)** -17; **(d)** 16
17. (a) $-x^3 + 2x^2 - 2x + 4$; **(b)** $-x^3 - 2x^2 - 8x + 12$; **(c)** -11; **(d)** -20
19. (a) $6x + 4$; **(b)** $\mathbb{R}$ **21. (a)** $3x + 2 + \frac{1}{x - 2}$; **(b)** $\{x \mid x \neq 2\}$
23. (a) $x^2 + x - 5 + \frac{2}{3x + 1}$; **(b)** $\left\{x \mid x \neq -\frac{1}{3}\right\}$ **25.** -15 **27.** 0
29. -5 **31.** Undefined **33.** $\{0, 2\}$
35. (a) $(2x - 1)(x - 3) = 2x^2 - 7x + 3$; **(b)** $\frac{2x - 1}{x - 3}$; **(c)** $\{x \mid x \neq 3\}$
37. (a) $6x^2 + x - 2$; **(b)** $\frac{3x + 2}{2x - 1}$; **(c)** $\left\{x \mid x \neq \frac{1}{2}\right\}$
39. (a) $-2x^2 - x + 10$; **(b)** $\frac{2 - x}{5 + 2x}$; **(c)** $\left\{x \mid x \neq -\frac{5}{2}\right\}$
41. sometimes **43.** always **45.** $P(x) = -x^2 + 21x - 50$
47. $V(t) = 10 - 9.8t$ **49.** $R(x) = 119x - 6x^2$

10.2 Composition of Functions

< 10.2 Objectives >

1 > Evaluate the composition of two functions given in table form

2 > Evaluate the composition of two functions given in equation form

3 > Write a function as a composition of two simpler functions

4 > Solve an application involving function composition

In Section 10.1, we learned that two functions can be combined by using any of the standard operations of arithmetic. There is still another way to combine two functions, called **composition.**

Definition

Composition of Functions

The **composition** of functions f and g is the function $f \circ g$, where

$(f \circ g)(x) = f(g(x))$

The domain of the composition is the set of all elements x in the domain of g for which $g(x)$ is in the domain of f.

Composition may be thought of as a chaining together of functions. To understand the meaning of $(f \circ g)(x)$, note that first the function g acts on x, producing $g(x)$, and then the function f acts on $g(x)$.

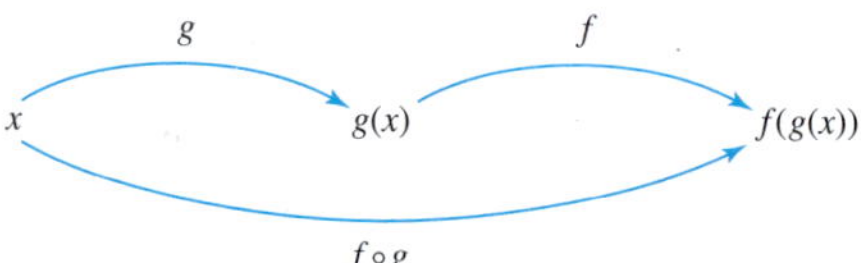

Consider Example 1, involving functions given in table form.

Example 1 Composing Two Functions

< Objective 1 >

Suppose the functions f and g are defined by the tables.

x	$g(x)$
−2	5
1	0
3	−4
8	2

x	$f(x)$
−4	8
0	6
2	5
1	−2

NOTE

Think, what does g do to 1? And then, what does f do to the result?

(a) Evaluate $(f \circ g)(1)$.

Since $(f \circ g)(1) = f(g(1))$, we first find $g(1)$. In the table for g we see that $g(1) = 0$. We then find $f(0)$. In the table for f we see that $f(0) = 6$. We then have

$$(f \circ g)(1) = f(g(1)) = f(0) = 6$$

or $\quad (f \circ g)(1) = 6$

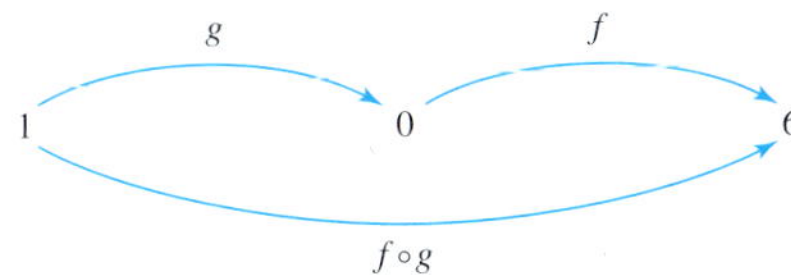

> CAUTION

When you are evaluating $(f \circ g)(x)$, the first function to act is g, not f.

(b) Evaluate $(f \circ g)(8)$.

Since $(f \circ g)(8) = f(g(8))$, we first note that $g(8) = 2$. Next we note that $f(2) = 5$. So

$$(f \circ g)(8) = f(g(8)) = f(2) = 5$$

or $\quad (f \circ g)(8) = 5$

NOTE

-2 is not in the domain of $f \circ g$.

(c) Evaluate $(f \circ g)(-2)$.

Since $(f \circ g)(-2) = f(g(-2))$, we see that $g(-2) = 5$. But we cannot compute $f(5)$ because 5 is not in the domain of f. So $(f \circ g)(-2)$ does not exist.

(d) Evaluate $(f \circ g)(3)$.

Since $(f \circ g)(3) = f(g(3))$, we first find $g(3) = -4$. We then find that $f(-4) = 8$. Altogether

$$(f \circ g)(3) = f(g(3)) = f(-4) = 8$$

or $\quad (f \circ g)(3) = 8$

Check Yourself 1

Using the functions given in Example 1, evaluate

(a) $(g \circ f)(-4)$ **(b)** $(g \circ f)(1)$ **(c)** $(g \circ f)(2)$

Typically, we encounter composition of functions in equation form. Example 2 demonstrates how f and g are chained together to form the new function $f \circ g$.

Example 2 Composing Two Functions

< Objective 2 >

NOTE

The function g turns 0 into 3. The function f then turns 3 into 7.

Suppose we have the functions $f(x) = x^2 - 2$ and $g(x) = x + 3$.

(a) Evaluate $(f \circ g)(0)$.

Since $(f \circ g)(0) = f(g(0))$, we first find $g(0)$.

$g(0) = 0 + 3 = 3$

Then

$f(3) = 3^2 - 2 = 7$

Altogether

$f(g(0)) = f(3) = 7$

So

$(f \circ g)(0) = 7$

(b) Evaluate $(f \circ g)(4)$.

Since $(f \circ g)(4) = f(g(4))$, we first find $g(4)$.

$g(4) = 4 + 3 = 7$

Then

$f(7) = 7^2 - 2 = 47$

So

$(f \circ g)(4) = f(g(4)) = f(7) = 47$ or $(f \circ g)(4) = 47$

(c) Find $(f \circ g)(x)$.

Since $(f \circ g)(x) = f(g(x))$, we first note that $g(x) = x + 3$.

Then

$$f(g(x)) = f(x + 3) = (x + 3)^2 - 2$$
$$= x^2 + 6x + 9 - 2 = x^2 + 6x + 7$$

So

$(f \circ g)(x) = x^2 + 6x + 7$

Check Yourself 2

Suppose that $f(x) = x^2 + x$ and $g(x) = x - 1$. Evaluate each composition.

(a) $(f \circ g)(0)$ **(b)** $(f \circ g)(-2)$ **(c)** $(f \circ g)(x)$

In our next example of composed functions, we need to pay attention to the domains of the functions involved.

Example 3 Composing Two Functions

Suppose we have the functions $f(x) = \sqrt{x}$ and $g(x) = 3 - x$. Evaluate each composition.

(a) $(f \circ g)(1) = f(g(1))$ $g(1) = 3 - (1) = 2$

$= f(2)$

$= \sqrt{2}$

(b) $(f \circ g)(-1) = f(g(-1))$ $g(-1) = 3 - (-1) = 4$

$= f(4)$

$= \sqrt{4}$

$= 2$

(c) $(f \circ g)(7) = f(g(7))$ $\quad g(7) = 3 - (7) = -4$

$= f(-4)$

$= \sqrt{-4}$

Since this is not a real number, we say that $(f \circ g)(7)$ does not exist. And 7 is not in the domain of $f \circ g$.

NOTE

Graph the function $Y = \sqrt{3 - x}$ in your graphing calculator. Note that the graph exists only for $x \le 3$.

(d) $(f \circ g)(x) = f(g(x))$

$= f(3 - x)$

$= \sqrt{3 - x}$

This function produces real-number values only if $3 - x \ge 0$, that is, only if $x \le 3$. This is the domain of $f \circ g$.

Check Yourself 3

Suppose that $f(x) = \dfrac{1}{x}$ and $g(x) = x^2 - 4$. Find each composition.

(a) $(f \circ g)(0)$ **(b)** $(f \circ g)(-2)$ **(c)** $(f \circ g)(x)$

The order of composition is important. In general, $(f \circ g)(x) \ne (g \circ f)(x)$. Using the functions given in Example 2, we saw that $(f \circ g)(x) = x^2 + 6x + 7$ whereas

$$(g \circ f)(x) = g(f(x)) = g(x^2 - 2) = x^2 - 2 + 3 = x^2 + 1$$

which is not the same as $(f \circ g)(x)$.

Often it is convenient to write a given function as the composition of two simpler functions. While the choice of simpler functions is not unique, a good choice makes many applications easier to solve.

Example 4 Writing a Function as the Composition of Two Functions

< Objective 3 >

Use the functions $f(x) = x + 3$ and $g(x) = x^2$ to express the given function as a composition of f and g.

NOTE

Can you see how we are using the order of operations here?

(a) $h(x) = (x + 3)^2$

When a value is substituted for x, the first action that occurs is that 3 is added to the input. The function f does exactly that. The second action that occurs is that of squaring. The function g does this. Since f acts first and then g (to carry out the total action of h),

$$h(x) = g(f(x)) = (g \circ f)(x)$$

It is easily checked that $(g \circ f)(x) = h(x)$.

$$(g \circ f)(x) = g(f(x)) = g(x + 3) = (x + 3)^2 = h(x)$$

(b) $k(x) = x^2 + 3$

Now when a value is substituted for x, the first action that occurs is a squaring action (which is what g does). The second action is that of adding 3 (which is what f does).

$$k(x) = f(g(x)) = (f \circ g)(x)$$

Check:

$$(f \circ g)(x) = f(g(x)) = f(x^2) = x^2 + 3 = k(x)$$

Check Yourself 4

Use the functions $f(x) = \sqrt{x}$ and $g(x) = x + 2$ to express the given functions as compositions of f and g.

(a) $h(x) = \sqrt{x + 2}$ **(b)** $k(x) = \sqrt{x} + 2$

There are many examples that involve the composition of functions, as illustrated in Example 5.

Example 5 Solving an Application Involving Function Composition

< Objective 4 >

At Kinky's Duplication Salon, customers pay \$2 plus 4¢ per page copied. Duplication consultant Vinny makes a commission of 5% of the bill for each job he sends to Kinky's.

(a) Express a customer's bill B as a function of the number of pages copied p.

$B(p) = 0.04p + 2$ The bill is \$0.04 times the number of pages, plus \$2.

(b) Express Vinny's commission V as a function of each bill B.

$V(B) = 0.05B$ Vinny's commission is 5% of the bill.

(c) Use function composition to express Vinny's commission V as a function of the number of pages a customer has copied p.

Since "commission" is a function of "bill," and "bill" is a function of "pages," the composition creates "commission" as a function of "pages."

$V(B) = V(B(p))$ Substitute $B(p)$ for B, creating a composition.

$= V(0.04p + 2)$ Since $B(p) = 0.04p + 2$

$= 0.05(0.04p + 2)$ Input the quantity $0.04p + 2$ into the function V.

$= 0.002p + 0.1$

So $V(p) = 0.002p + 0.1$.

(d) Use the function in part (c) to find Vinny's commission on a job consisting of 2,000 pages.

$V(p) = 0.002(2,000) + 0.1$

$= 4 + 0.1 = 4.1$

Therefore, Vinny's commission is \$4.10.

Check Yourself 5

On his regular route, Gonzalo averages 62 mi/h between Charlottesville and Lawrenceville. His van averages 24 mi/gal, and his gas tank holds 12 gal of fuel. Assume that his tank is full when he starts the trip from Charlottesville to Lawrenceville.

(a) Express the fuel left in the tank as a function of n, the number of gallons used.

(b) Express the number of gallons used as a function of m, the number of miles driven.

(c) Express the fuel left in the tank as a function of m, the number of miles driven.

Check Yourself ANSWERS

1. **(a)** 2; **(b)** 5; **(c)** does not exist **2.** **(a)** 0; **(b)** 6; **(c)** $x^2 - x$

3. **(a)** $-\frac{1}{4}$; **(b)** does not exist; **(c)** $\frac{1}{x^2 - 4}$ **4.** **(a)** $(f \circ g)(x)$; **(b)** $(g \circ f)(x)$

5. **(a)** $f(n) = 12 - n$; **(b)** $g(m) = \frac{m}{24}$; **(c)** $(f \circ g)(m) = 12 - \frac{m}{24}$

Reading Your Text

SECTION 10.2

(a) ______________ two functions can be thought of as a chaining together of the functions.

(b) When you are evaluating $(f \circ g)(x)$ the first function to act on x is ______________.

(c) Often it is convenient to write a given function as the composition of two ______________ functions.

(d) When evaluating a function, we use the order of ______________.

10.2 exercises

Basic Skills | Challenge Yourself | Calculator/Computer | Career Applications | Above and Beyond

Boost *your* GRADE at ALEKS.com!

ALEKS®

- Practice Problems
- Self-Tests
- NetTutor
- e-Professors
- Videos

Name ______________

Section ________ Date ________

Answers

1. ______________
2. ______________
3. ______________
4. ______________
5. ______________
6. ______________
7. ______________
8. ______________
9. ______________
10. ______________
11. ______________
12. ______________
13. ______________

< Objective 1 >

For exercises 1 to 12, use the tables to find the desired values.

x	$f(x)$
-3	-1
-1	7
2	6
3	-3

x	$g(x)$
-2	4
1	-2
4	3
6	2

x	$h(x)$
-2	5
0	0
1	-2
3	6

x	$k(x)$
-2	0
0	4
2	3
3	-4

1. $(f \circ g)(4)$

2. $(g \circ f)(2)$

3. $(h \circ g)(1)$ 

4. $(g \circ h)(1)$

5. $(g \circ h)(3)$

6. $(k \circ h)(0)$

7. $(h \circ k)(0)$

8. $(k \circ g)(1)$

9. $(f \circ h)(3)$

10. $(k \circ g)(4)$

11. $(k \circ k)(2)$

12. $(f \circ f)(-3)$

< Objective 2 >

Evaluate each composition.

13. $f(x) = x - 3$ and $g(x) = 2x + 1$

(a) $(f \circ g)(0)$ **(b)** $(f \circ g)(-2)$ **(c)** $(f \circ g)(3)$ **(d)** $(f \circ g)(x)$

14. $f(x) = x - 1$ and $g(x) = 3x + 4$

(a) $(f \circ g)(0)$ (b) $(f \circ g)(-2)$ (c) $(f \circ g)(3)$ (d) $(f \circ g)(x)$

15. $f(x) = 3x + 1$ and $g(x) = 4x - 3$

(a) $(f \circ g)(0)$ (b) $(f \circ g)(-2)$ (c) $(g \circ f)(3)$ (d) $(g \circ f)(x)$

16. $f(x) = 4x - 2$ and $g(x) = -2x + 5$

(a) $(f \circ g)(0)$ (b) $(f \circ g)(-2)$ (c) $(g \circ f)(3)$ (d) $(g \circ f)(x)$

17. $f(x) = x^2$ and $g(x) = x + 3$ 

(a) $(f \circ g)(0)$ (b) $(f \circ g)(-2)$ (c) $(g \circ f)(3)$ (d) $(g \circ f)(x)$

18. $f(x) = x^2 + 3$ and $g(x) = 3x$

(a) $(f \circ g)(0)$ (b) $(f \circ g)(-2)$ (c) $(g \circ f)(3)$ (d) $(g \circ f)(x)$

19. $f(x) = 2x^2 - 1$ and $g(x) = -2x$

(a) $(g \circ f)(0)$ (b) $(g \circ f)(-2)$ (c) $(f \circ g)(3)$ (d) $(f \circ g)(x)$

20. $f(x) = x^2 + 3$ and $g(x) = 3x$

(a) $(g \circ f)(0)$ (b) $(g \circ f)(-2)$ (c) $(f \circ g)(3)$ (d) $(f \circ g)(x)$

Basic Skills | **Challenge Yourself** | Calculator/Computer | Career Applications | Above and Beyond

Determine whether each statement is **true** *or* **false.**

21. $(f \circ g)(x)$ is the same as $f(x) \cdot g(x)$.

22. $(f \circ g)(x)$ is always the same as $(g \circ f)(x)$.

Complete each statement with **never, sometimes,** *or* **always.**

23. To compute $(f \circ g)(5)$, we must ___________ find $g(5)$ first.

24. To compute $(f \circ g)(5)$, $g(5)$ must ___________ be in the domain of f.

Answers

14. ___________

15. ___________

16. ___________

17. ___________

18. ___________

19. ___________

20. ___________

21. ___________

22. ___________

23. ___________

24. ___________

Answers

25.

26.

27.

28.

29.

30.

31.

32.

33.

34.

35.

< Objective 3 >

Rewrite the function h as a composite of functions f and g.

25. $f(x) = 3x$ $\quad g(x) = x + 2$ $\quad h(x) = 3x + 2$

26. $f(x) = x - 4$ $\quad g(x) = 7x$ $\quad h(x) = 7x - 4$

27. $f(x) = x + 5$ $\quad g(x) = \sqrt{x}$ $\quad h(x) = \sqrt{x + 5}$

28. $f(x) = x + 5$ $\quad g(x) = \sqrt{x}$ $\quad h(x) = \sqrt{x} + 5$

29. $f(x) = x^2$ $\quad g(x) = x - 5$ $\quad h(x) = x^2 - 5$

30. $f(x) = x^2$ $\quad g(x) = x - 5$ $\quad h(x) = (x - 5)^2$

31. $f(x) = x - 3$ $\quad g(x) = \dfrac{2}{x}$ $\quad h(x) = \dfrac{2}{x - 3}$

32. $f(x) = x - 3$ $\quad g(x) = \dfrac{2}{x}$ $\quad h(x) = \dfrac{2}{x} - 3$

33. $f(x) = x - 1$ $\quad g(x) = x^2 + 2$ $\quad h(x) = x^2 + 1$

> Videos

34. $f(x) = x - 1$ $\quad g(x) = x^2 + 2$ $\quad h(x) = x^2 - 2x + 3$

< Objective 4 >

35. Carine and Jacob are getting married at East Fork Estates. The wedding will cost \$1,000 plus \$40 per guest. They have read that typically 80% of the people invited actually attend a wedding.

(a) Write a function to represent the number of people N expected to attend if v are invited.

(b) Write a function to represent the cost C of the wedding for N guests.

(c) Write a function to represent the cost C of the wedding if v people are invited.

36. On his regular route, Gonzalo averages 62 mi/h between Charlottesville and Lawrenceville. His van averages 24 mi/gal, and his gas tank holds 12 gal of fuel. Assume that his tank is full when he starts the trip from Charlottesville to Lawrenceville.

(a) Express the number of miles driven as a function of t, the time on the road.

(b) Express the number of gallons used as a function of M, the miles driven.

(c) Express the number of gallons used as a function of t, the time on the road.

37. When she arrives in London, Bichvan receives an exchange rate of 0.69 British pound for each U.S. dollar. In Reykjavik, she receives an exchange rate of 146.41 Icelandic kronas for each British pound. When she returns to the United States, how much (in U.S. dollars) should she expect to receive in exchange for 12,000 Icelandic kronas? > Videos

38. If the exchange rate for Japanese yen is 127.3 and the exchange rate for Indian rupees is 48.37 (both from U.S. dollars), then what is the exchange rate from Japanese yen to Indian rupees?

Answers

36. __________

37. __________

38. __________

Answers

1. -3 **3.** 5 **5.** 2 **7.** Does not exist **9.** Does not exist **11.** -4 **13.** **(a)** -2; **(b)** -6; **(c)** 4; **(d)** $2x - 2$ **15.** **(a)** -8; **(b)** -32; **(c)** 37; **(d)** $12x + 1$ **17.** **(a)** 9; **(b)** 1; **(c)** 12; **(d)** $x^2 + 3$ **19.** **(a)** 2; **(b)** -14; **(c)** 71; **(d)** $8x^2 - 1$ **21.** False **23.** always **25.** $h(x) = (g \circ f)(x)$ **27.** $h(x) = (g \circ f)(x)$ **29.** $h(x) = (g \circ f)(x)$ **31.** $h(x) = (g \circ f)(x)$ **33.** $h(x) = (f \circ g)(x)$ **35.** **(a)** $N(v) = 0.8v$; **(b)** $C(N) = 40N + 1{,}000$; **(c)** $C(v) = 32v + 1{,}000$ **37.** \$118.78

10.3 Inverse Relations and Functions

< 10.3 Objectives >

1 > Find the inverse of a function given in equation form

2 > Find the inverse of a function given in table form

3 > Graph a relation and its inverse

4 > Identify a one-to-one function

5 > Determine whether the inverse of a function is also a function

6 > Restrict the domain of a function

Composition of functions leads us to the question: Can we chain together (i.e., compose) two functions in such a way that one function "undoes" the other?

Suppose, for example, that $f(x) = \frac{x-5}{3}$ and $g(x) = 3x + 5$. Let's pick a convenient x-value for f, say, $x = 8$. Now $f(8) = \frac{8-5}{3} = \frac{3}{3} = 1$. The function f turns 8 into 1. Now let g act on this result: $g(1) = 3(1) + 5 = 8$. The function g turns 1 back into 8.

When we view the composition of g and f, acting on 8, we see g "undoing" f's actions:

$(g \circ f)(8) = g(f(8)) = g(1) = 8$

In general, a function g that undoes the action of f is called the **inverse** of f.

Definition

Inverse Functions

Functions f and g are said to be **inverse functions** if

$(g \circ f)(x) = x$ for all x in the domain of f

and $(f \circ g)(x) = x$ for all x in the domain of g

NOTE

The notation f^{-1} has a different meaning from the negative exponent, as in x^{-1} or $\frac{1}{x}$.

If g is the inverse of f, we denote the function g as f^{-1}.

A natural question now is, given a function f, how do we find the inverse function f^{-1}? One way to find such a function f^{-1} is to analyze the actions of f, noting the order of operations involved, and then define f^{-1} by using the *opposite* operations *in the reverse order*.

Example 1 Finding the Inverse of a Function

< Objective 1 >

Given $f(x) = \frac{x-5}{3}$, find its inverse function f^{-1}.

When we substitute a value for x into f, two actions occur.

1. The number 5 is subtracted.
2. Division by 3 occurs.

To design an inverse function f^{-1}, we use the *opposite* operations *in the reverse order.*

1. Multiply by 3.

2. Add 5.

We conclude that

$f^{-1}(x) = 3x + 5$

To verify, we must check that $(f^{-1} \circ f)(x) = x$ and $(f \circ f^{-1})(x) = x$.

$$(f^{-1} \circ f)(x) = f^{-1}(f(x)) = f^{-1}\left(\frac{x-5}{3}\right) = 3\left(\frac{x-5}{3}\right) + 5 = x - 5 + 5 = x$$

$$(f \circ f^{-1})(x) = f(f^{-1}(x)) = f(3x+5) = \frac{(3x+5)-5}{3} = \frac{3x}{3} = x$$

Check Yourself 1

Given $f(x) = \dfrac{x+1}{4}$, find f^{-1}.

To develop another technique for finding inverses, let us revisit functions defined by tables.

Example 2 Finding the Inverse of a Function

< Objective 2 >

Find the inverse of the function f.

x	$f(x)$
−4	8
0	6
2	5
1	−2

The inverse of f is easily found by reversing the order of the input values and output values.

x	$f^{-1}(x)$
8	−4
6	0
5	2
−2	1

While f turns −4 into 8, for example, f^{-1} turns 8 back into −4.

Check Yourself 2

Find the inverse of the function f.

x	$f(x)$
-2	5
1	0
4	-4
8	2

In Example 2, if we write y in place of $f(x)$, we see that we are just interchanging the roles of x and y in order to create the inverse f^{-1}. This suggests the following technique for finding the inverse of a function that is given in equation form.

Step by Step

Finding the Inverse of a Function

- **Step 1** Given a function $f(x)$, write y in place of $f(x)$.
- **Step 2** Interchange the variables x and y.
- **Step 3** Solve for y.
- **Step 4** Write $f^{-1}(x)$ in place of y.
- **Step 5** Check that $f^{-1}(f(x)) = x$ and $f(f^{-1}(x)) = x$.

Example 3 Finding the Inverse of a Function

Find the inverse of $f(x) = 2x - 4$.

Begin by writing y in place of $f(x)$.

$$y = 2x - 4 \qquad \text{Interchange } x \text{ and } y.$$

$$x = 2y - 4 \qquad \text{Solve for } y. \text{ Add 4 to both sides.}$$

$$x + 4 = 2y \qquad \text{Divide both sides by 2.}$$

$$\frac{x+4}{2} = y \qquad \text{Write } f^{-1}(x) \text{ in place of } y.$$

So

$$f^{-1}(x) = \frac{x+4}{2} \qquad \text{or} \qquad f^{-1}(x) = \frac{1}{2}x + 2$$

RECALL

$$\frac{x+4}{2} = \frac{x}{2} + \frac{4}{2} = \frac{1}{2}x + 2$$

Check that $(f^{-1} \circ f)(x) = x$ and $(f \circ f^{-1})(x) = x$.

$$(f^{-1} \circ f)(x) = f^{-1}(f(x)) = f^{-1}(2x - 4) = \frac{(2x-4)+4}{2} = \frac{2x-4+4}{2} = \frac{2x}{2} = x$$

$$(f \circ f^{-1})(x) = f(f^{-1}(x)) = f\left(\frac{x+4}{2}\right) = 2\left(\frac{x+4}{2}\right) - 4 = x + 4 - 4 = x$$

Check Yourself 3

Find the inverse of $f(x) = \dfrac{x-7}{5}$.

The graphs of relations and their inverses are connected in an interesting way. First, note that the graphs of the ordered pairs (a, b) and (b, a) always have symmetry about the line $y = x$.

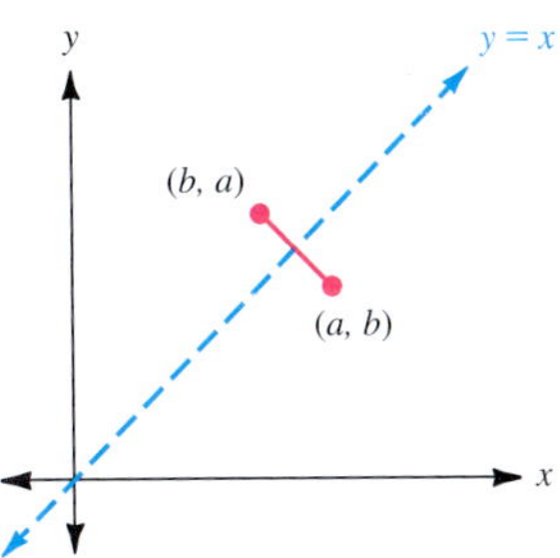

Now, with the above symmetry in mind, we consider Example 4.

Example 4 Graphing a Relation and Its Inverse

< Objective 3 >

Graph the relation f from Example 3 along with its inverse.

Recall that

$f(x) = 2x - 4$

and

$f^{-1}(x) = \frac{1}{2}x + 2$

The graphs of f and f^{-1} are shown here.

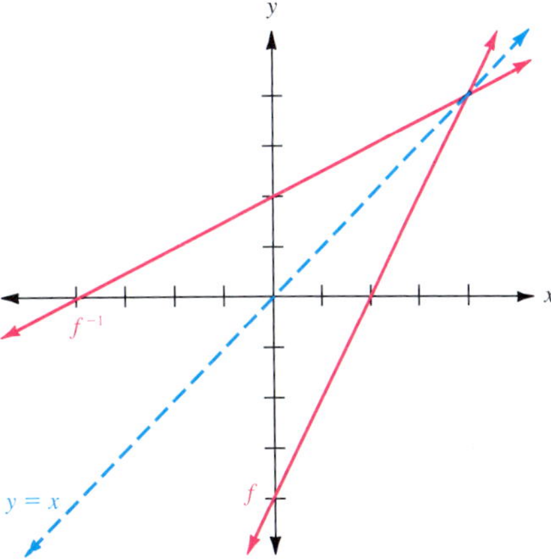

The graphs of f and f^{-1} are symmetric about the line $y = x$. That symmetry follows from our earlier observation about the pairs (a, b) and (b, a) because we simply reversed the roles of x and y in forming the inverse relation.

Check Yourself 4

Find the inverse of the function $f(x) = 3x + 6$ and graph both $f(x)$ and $f^{-1}(x)$.

In our work so far, we have seen techniques for finding the inverse of a function. However, it is quite possible that the inverse obtained may not be a function.

Example 5 Finding the Inverse of a Function

NOTE

Interchange the elements of the ordered pairs.

Find the inverse of each function.

(a) $f = \{(1, 3), (2, 4), (3, 9)\}$

Its inverse is

$\{(3, 1), (4, 2), (9, 3)\}$

which is also a function.

(b) $g = \{(1, 3), (2, 6), (3, 6)\}$

Its inverse is

$\{(3, 1), (6, 2), (6, 3)\}$

which is *not* a function.

NOTE

It is not a function because 6 maps to both 2 and 3.

Check Yourself 5

Write the inverse of each function. Which of the inverses are also functions?

(a) $\{(-1, 2), (0, 3), (1, 4)\}$ **(b)** $\{(2, 5), (3, 7), (4, 5)\}$

Can we predict in advance whether the inverse of a function is also a function? The answer is yes.

We already know that for a relation to be a function, no element in its domain can be associated with more than one element in its range. Since, in creating an inverse, the x-values and y-values are interchanged, the inverse of a function f will also be a function only if *no element in the range of f can be associated with more than one element in its domain*. That is, no two ordered pairs of f can have the same y-coordinate. This leads us to a definition.

Definition

One-to-One Function

A function f is **one-to-one** if no two distinct domain elements are paired with the same range element.

We then have this property.

Property

Inverse of a Function

The inverse of a function f is also a function if f is one-to-one.

Note that in Example 5(a)

$f = \{(1, 3), (2, 4), (3, 9)\}$

is a one-to-one function and its inverse is also a function. However, the function in Example 5(b)

$$g = \{(1, 3), (2, 6), (3, 6)\}$$

is *not* a one-to-one function, and its inverse is *not* a function.

Our result regarding a one-to-one function and its inverse also has a convenient graphical interpretation. Here we graph the function g from Example 5.

$$g = \{(1, 3), (2, 6), (3, 6)\}$$

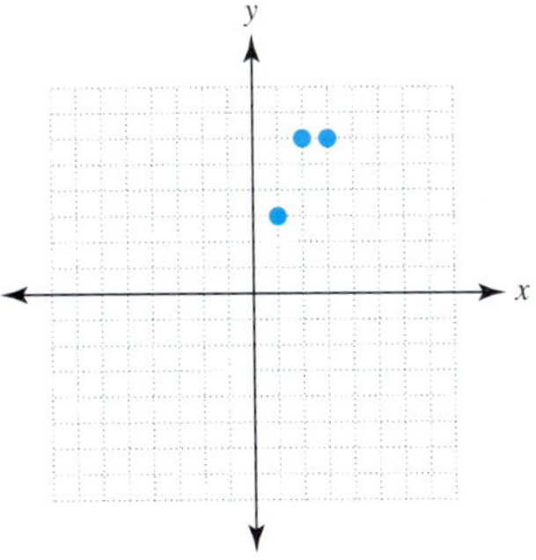

Again, g is *not* a one-to-one function, because two points, namely (2, 6) and (3, 6), have the same range element. As a result, a horizontal line may be drawn that passes through two points.

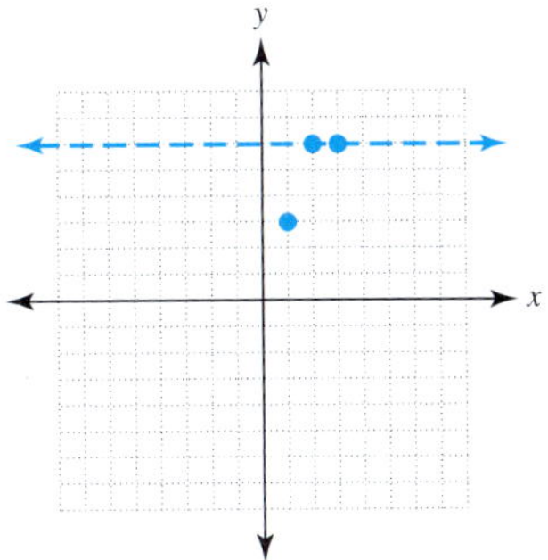

This means that when we form the inverse by reversing the coordinates, the resulting relation is *not* a function. Points (6, 2) and (6, 3) are part of the inverse, and the resulting graph fails the vertical line test.

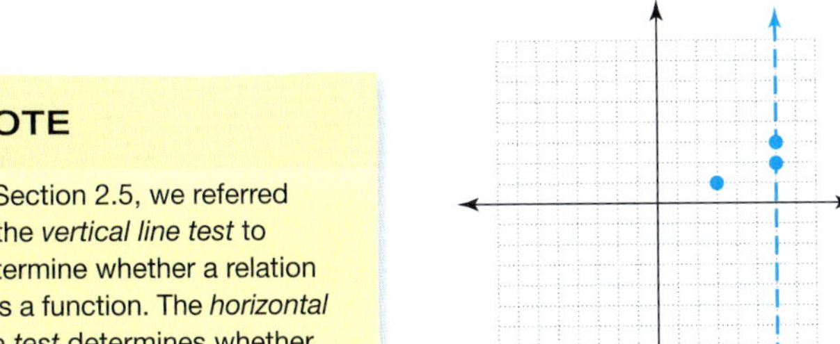

NOTE

In Section 2.5, we referred to the *vertical line test* to determine whether a relation was a function. The *horizontal line test* determines whether a function is one-to-one.

This leads to the **horizontal line test.**

Property

Horizontal Line Test A function is one-to-one if no horizontal line passes through two or more points on its graph.

Now we have a graphical way to determine whether the inverse of a function f is a function.

Property

Inverse of a Function The inverse of a function f is also a function if the graph of f passes the horizontal line test.

This is a very useful property, as Example 6 illustrates.

Example 6 Identifying One-to-One Functions

< Objectives 4 and 5 >

For each function, determine **(i)** whether the function is one-to-one and **(ii)** whether the inverse is also a function.

(a)

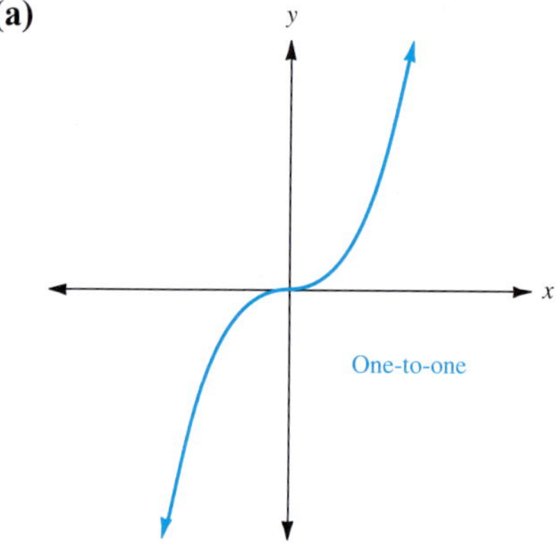

(i) Because no horizontal line passes through two or more points of the graph, f is one-to-one.

(ii) Because f is one-to-one, its inverse is also a function.

(b)

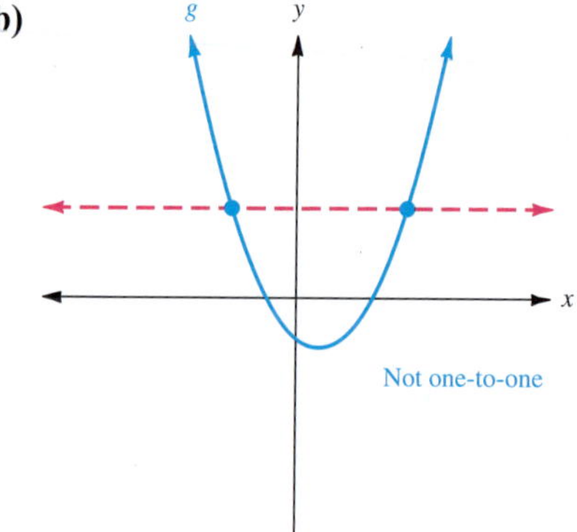

RECALL

The parabola given by g represents a function. It passes the vertical line test.

(i) Because a horizontal line can meet the graph of g at two points, g is *not* a one-to-one function.

(ii) Because g is not one-to-one, its inverse is not a function.

Check Yourself 6

For each function, determine (i) whether the function is one-to-one and (ii) whether the inverse is also a function.

(a)

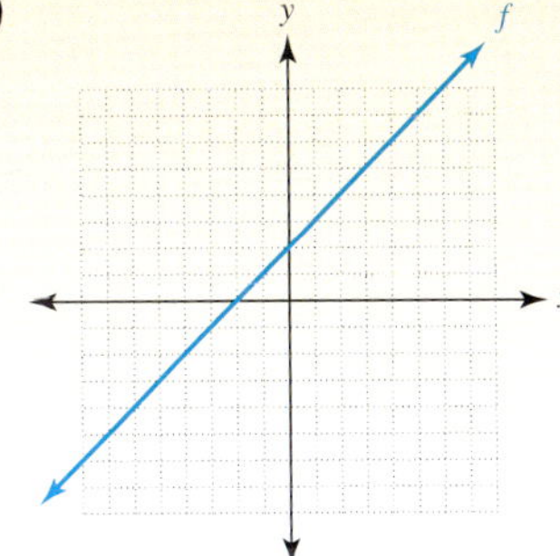

(b)

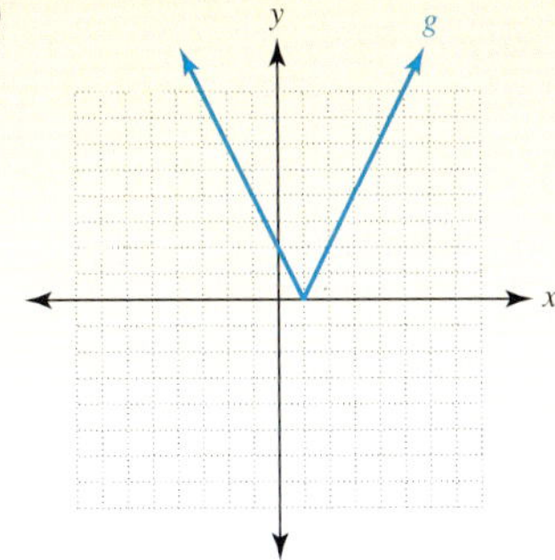

When a function is not one-to-one, we can restrict the domain of the function so that it is one-to-one (and, as a result, the inverse is a function).

Example 7 Restricting the Domain of a Function

< Objective 6 >

Consider the function $f(x) = x^2 + 3$.

(a) Restrict the domain of f so that f is one-to-one.

The graph of f is a parabola, shifted vertically upward 3 units (below left).

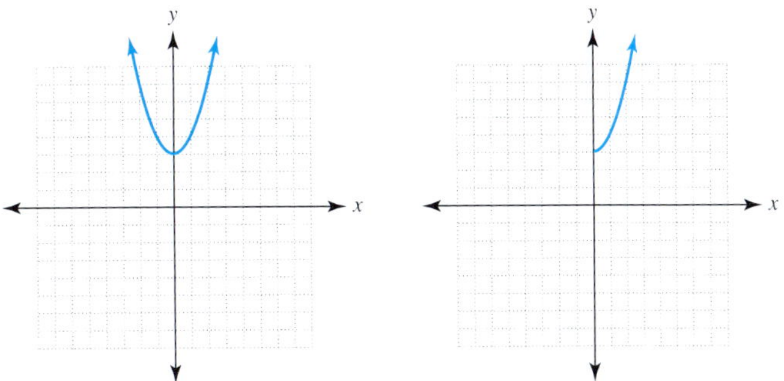

We restrict the domain of f to be $D = \{x \mid x \geq 0\}$ so the graph passes the horizontal line test. The restricted f is stated as

$$f(x) = x^2 + 3 \qquad x \geq 0$$

(b) Using the restricted f, find f^{-1}.

Letting y replace $f(x)$, we write

$y = x^2 + 3$ Interchange x and y.

$x = y^2 + 3$

This produces a parabola with horizontal axis, failing the vertical line test. Solve for y.

$$x - 3 = y^2$$

$$y = \pm\sqrt{x - 3}$$

We choose $y = \sqrt{x - 3}$ (do you see why?) and write

$$f^{-1}(x) = \sqrt{x - 3}$$

NOTE

We could also restrict the domain of f to be

$D = \{x \mid x \leq 0\}$

in order to pass the horizontal line test.

(c) Graph the restricted f and f^{-1} on the same axes.

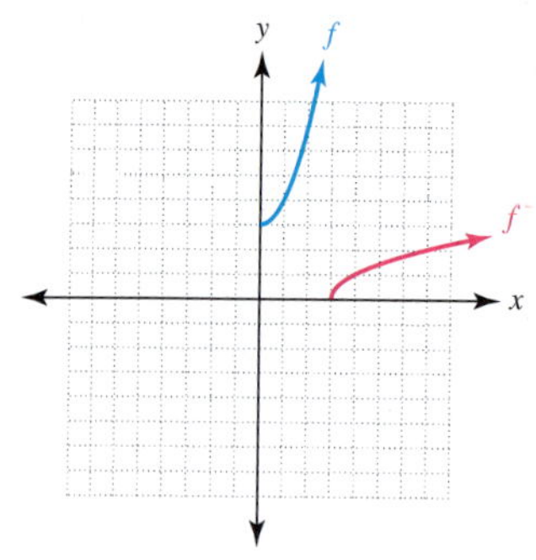

NOTE

Observe that the graphs of f (restricted) and f^{-1} are symmetric with respect to the $y = x$ line.

The domain of the restricted f (all real numbers greater than or equal to 0) is the same as the range of f^{-1}. Further, the range of the restricted f (all real numbers greater than or equal to 3) is the same as the domain of f^{-1}.

Check Yourself 7

Given the function $f(x) = x^2 - 2$,

(a) Restrict the domain of f so that f is one-to-one.
(b) Using the restricted f, find f^{-1}.
(c) Graph the restricted f and f^{-1} on the same axes.

Check Yourself ANSWERS

1. $f^{-1}(x) = 4x - 1$

2.

x	$f^{-1}(x)$
5	−2
0	1
−4	4
2	8

3. $f^{-1}(x) = 5x + 7$

4. $f^{-1}(x) = \dfrac{x - 6}{3}$ or $f^{-1}(x) = \dfrac{1}{3}x - 2$

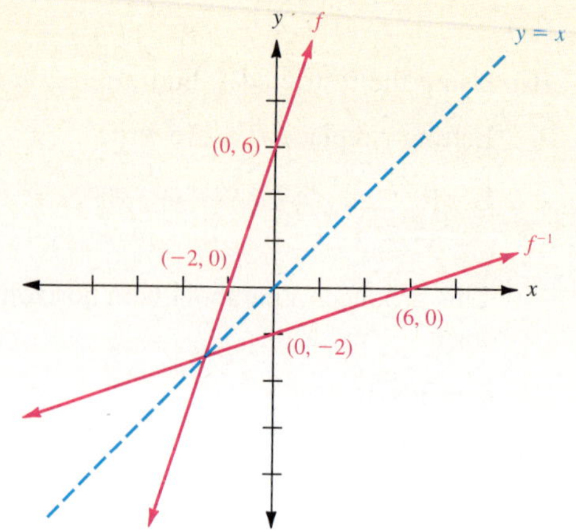

5. (a) $\{(2, -1), (3, 0), (4, 1)\}$; a function; **(b)** $\{(5, 2), (7, 3), (5, 4)\}$; not a function

6. (a) One-to-one; the inverse is a function; **(b)** Not one-to-one; the inverse is not a function

7. (a) $f(x) = x^2 - 2, x \geq 0$; **(b)** $f^{-1}(x) = \sqrt{x + 2}$;

(c)

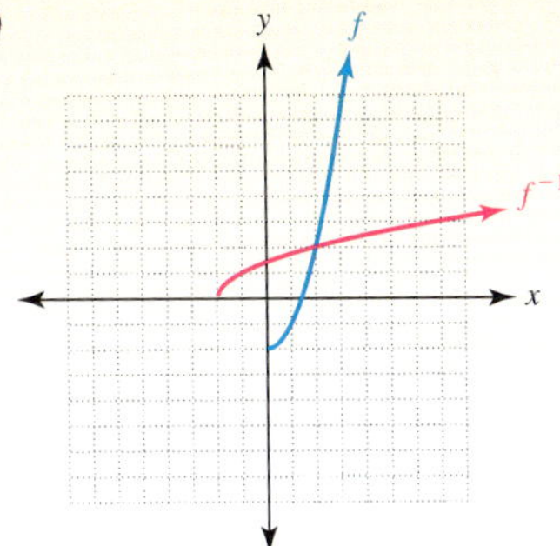

Reading Your Text

SECTION 10.3

(a) In general, a function g that undoes the action of f is called the ____________ of f.

(b) The graphs of f^{-1} and f are ____________ about the line $y = x$.

(c) The inverse of function f is also a function if f is ____________.

(d) A function is one-to-one if no ____________ line passes through two or more points on its graph.

10.3 exercises

Basic Skills | Challenge Yourself | Calculator/Computer | Career Applications | Above and Beyond

Name ______________

Section ________ Date ________

Answers

1. ______________
2. ______________
3. ______________
4. ______________
5. ______________
6. ______________
7. ______________
8. ______________
9. ______________
10. ______________
11. ______________
12. ______________
13. ______________
14. ______________
15. ______________
16. ______________

< Objective 1 >

Find the inverse function f^{-1} for the given function f.

1. $f(x) = 3x + 5$

2. $f(x) = -3x - 7$

3. $f(x) = \dfrac{x - 1}{2}$

4. $f(x) = \dfrac{x + 1}{3}$

5. $f(x) = 2x - 3$ 

6. $f(x) = -5x + 3$

7. $f(x) = \dfrac{x + 4}{3}$

8. $f(x) = \dfrac{x - 5}{7}$

9. $f(x) = \dfrac{x}{3} + 5$ 

10. $f(x) = \dfrac{2x}{5} - 7$

< Objective 2 >

Find the inverse of each function. In each case, determine whether the inverse is also a function.

11.

x	$f(x)$
−2	−5
2	6
3	4
4	3

12.

x	$f(x)$
−5	2
−3	3
1	3
5	2

13. > Videos

x	$f(x)$
−4	3
−2	7
3	5
7	−4

14.

x	$f(x)$
−5	2
−3	−4
0	2
6	−4

15. $f = \{(2, 3), (3, 4), (4, 5)\}$

16. $g = \{(1, 4), (2, 3), (3, 4)\}$

17. $f = \{(1, 5), (2, 5), (3, 5)\}$

18. $g = \{(4, 7), (2, 6), (6, 9)\}$

19. $f = \{(2, 3), (3, 5), (4, 7)\}$

20. $g = \{(-1, 0), (2, 0), (0, -1)\}$

Find the inverse function f^{-1} for the given function f. (Hint: *Use the method demonstrated in Example 3.*)

21. $f(x) = \dfrac{6 - 5x}{3}$

22. $f(x) = \dfrac{4x - 7}{3}$

23. $f(x) = \dfrac{11x}{5} + 2$

24. $f(x) = 5 - \dfrac{7x}{3}$

< Objective 3 >

For each function f, find its inverse f^{-1}. Then graph both on the same set of axes.

25. $f(x) = 3x - 6$

26. $f(x) = 4x + 8$

27. $f(x) = -2x + 6$ 

28. $f(x) = -3x - 6$

< Objectives 4 and 5 >

Determine whether the given function is one-to-one. In each case, decide whether the inverse is a function.

29. $f = \{(-3, 5), (-2, 3), (0, 2), (1, 4), (6, 5)\}$ > Videos

30. $g = \{(-3, 7), (0, 4), (2, 5), (4, 1)\}$

31.

x	$f(x)$
−3	4
0	−3
2	1
6	2
8	0

32.

x	$g(x)$
−2	−6
−1	2
3	0
4	2

Answers

17. ________

18. ________

19. ________

20. ________

21. ________

22. ________

23. ________

24. ________

25. ________

26. ________

27. ________

28. ________

29. ________

30. ________

31. ________

32. ________

Answers

33. ______

34. ______

35. ______

36. ______

37. ______

38. ______

39. ______

40. ______

41. ______

42. ______

33.

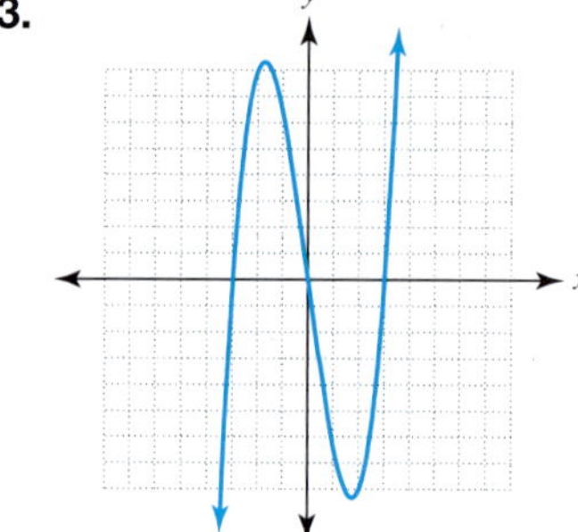

34.

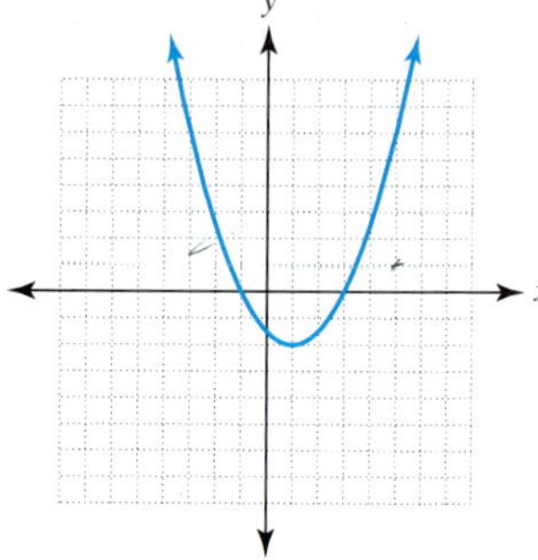

35.

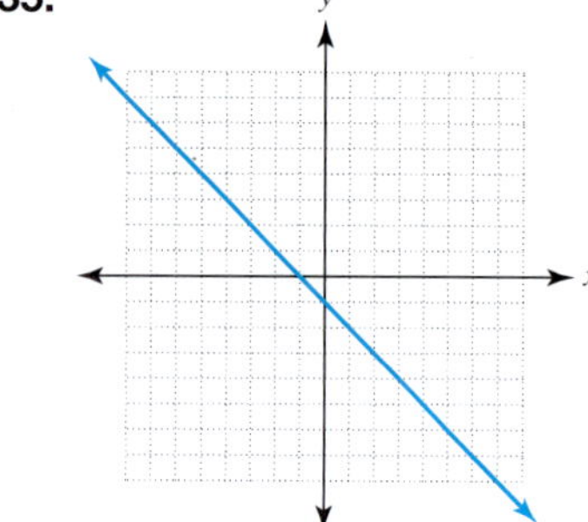

36.

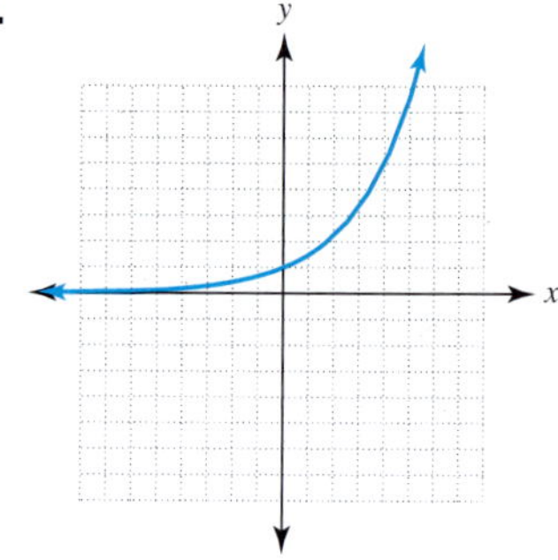

Basic Skills | **Challenge Yourself** | Calculator/Computer | Career Applications | Above and Beyond

Determine whether each statement is **true** *or* **false.**

37. The inverse of a linear function with nonzero slope is itself a function.

38. The inverse of a quadratic function is itself a function.

39. The graphs of a function and its inverse are symmetric to each other over the y-axis.

40. If f has an inverse function f^{-1}, and $f(a) = b$, then $f^{-1}(b) = a$.

Complete each statement with **never, sometimes,** *or* **always.**

41. The inverse of a function is __________ a function.

42. If the graph of a function passes the horizontal line test, then the graph of the inverse __________ passes the vertical line test.

If $f(x) = 3x - 6$, *then* $f^{-1}(x) = \frac{1}{3}x + 2$. *Evaluate as indicated.*

43. $f(6)$

44. $f^{-1}(6)$

45. $f(f^{-1}(6))$

46. $f^{-1}(f(6))$

47. $f(f^{-1}(x))$

48. $f^{-1}(f(x))$

If $g(x) = \frac{x+1}{2}$, *then* $g^{-1}(x) = 2x - 1$. *Evaluate as indicated.*

49. $g(3)$

50. $g^{-1}(3)$

51. $g(g^{-1}(3))$

52. $g^{-1}(g(3))$

53. $g(g^{-1}(x))$

54. $g^{-1}(g(x))$

Let $h(x) = 2x + 8$. *Evaluate as indicated.*

55. $h(4)$

56. $h^{-1}(4)$

57. $h(h^{-1}(4))$

58. $h^{-1}(h(4))$

59. $h(h^{-1}(x))$

60. $h^{-1}(h(x))$

Suppose that f and g are one-to-one functions.

61. If $f(5) = 7$, find $f^{-1}(7)$.

62. If $g^{-1}(4) = 9$, find $g(9)$.

Let f be a linear function; that is, let $f(x) = mx + b$.

63. Find $f^{-1}(x)$.

64. Based on exercise 63, if the slope of f is 3, what is the slope of f^{-1}?

65. Based on exercise 63, if the slope of f is $\frac{2}{5}$, what is the slope of f^{-1}?

Answers

43. ______
44. ______
45. ______
46. ______
47. ______
48. ______
49. ______
50. ______
51. ______
52. ______
53. ______
54. ______
55. ______
56. ______
57. ______
58. ______
59. ______
60. ______
61. ______
62. ______
63. ______
64. ______
65. ______

Answers

66. ______

67. ______

Basic Skills | Challenge Yourself | Calculator/Computer | Career Applications | **Above and Beyond**

66. An inverse process is an operation that undoes a procedure. If the procedure is wrapping a present, describe in detail the inverse process.

67. If the procedure is the series of steps that take you from home to your classroom, describe the inverse process.

Answers

1. $f^{-1}(x) = \dfrac{x-5}{3}$ **3.** $f^{-1}(x) = 2x + 1$ **5.** $f^{-1}(x) = \dfrac{x+3}{2}$
7. $f^{-1}(x) = 3x - 4$ **9.** $f^{-1}(x) = 3x - 15$
11. $\{(-5, -2), (6, 2), (4, 3), (3, 4)\}$; a function
13. $\{(3, -4), (7, -2), (5, 3), (-4, 7)\}$; a function
15. $\{(3, 2), (4, 3), (5, 4)\}$; a function **17.** $\{(5, 1), (5, 2), (5, 3)\}$; not a function
19. $\{(3, 2), (5, 3), (7, 4)\}$; a function **21.** $f^{-1}(x) = \dfrac{6-3x}{5}$
23. $f^{-1}(x) = \dfrac{5x-10}{11}$

25.

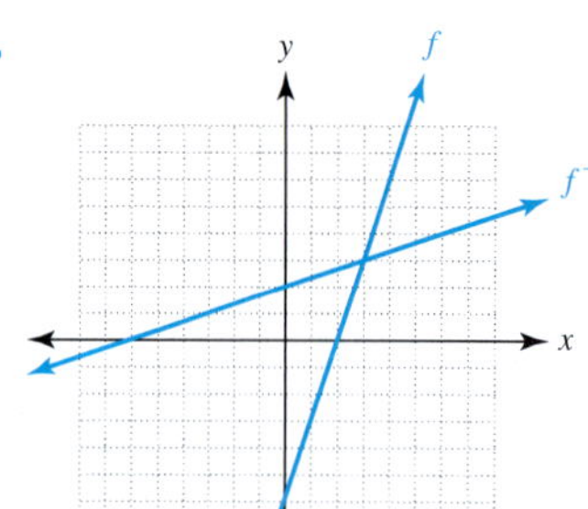

$f^{-1}(x) = \dfrac{x+6}{3} = \dfrac{1}{3}x + 2$

27.

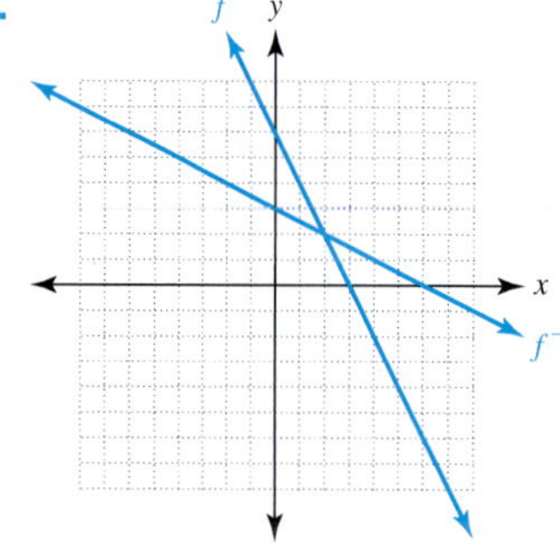

$f^{-1}(x) = \dfrac{6-x}{2} = -\dfrac{1}{2}x + 3$

29. Not one-to-one; inverse is not a function
31. One-to-one; inverse is a function
33. Not one-to-one; inverse is not a function
35. One-to-one; inverse is a function **37.** True **39.** False
41. sometimes **43.** 12 **45.** 6 **47.** x **49.** 2 **51.** 3 **53.** x
55. 16 **57.** 4 **59.** x **61.** 5 **63.** $f^{-1}(x) = \dfrac{x-b}{m}$ **65.** $\dfrac{5}{2}$
67. Above and Beyond

10.4 Exponential Functions

< 10.4 Objectives >

1 > Graph an exponential function

2 > Solve an application of exponential functions

3 > Solve an elementary exponential equation

Up to this point in the book, we have worked with polynomials and other functions in which the variable was used as a base. We now want to turn to a new class of functions, **exponential functions.**

Exponential functions involve the variable as an *exponent.* The introduction of these functions allows us to consider many further applications, including population growth, radioactive decay, and compound interest.

Definition

Exponential Functions

An **exponential function** is a function that can be expressed in the form

$f(x) = b^x$

in which $b > 0$ and $b \neq 1$. We call b the **base** of the exponential function.

Here are some examples of exponential functions.

$$f(x) = 2^x \qquad g(x) = 3^x \qquad h(x) = \left(\frac{1}{2}\right)^x$$

As we have done with other new functions, we begin by finding some function values. Then we use that information to graph the function.

Example 1 **Graphing an Exponential Function**

< Objective 1 >

Graph the exponential function

$f(x) = 2^x$

First, choose convenient values for x.

RECALL

$2^{-2} = \frac{1}{2^2} = \frac{1}{4}$

$$f(0) = 2^0 = 1 \qquad f(1) = 2^1 = 2 \qquad f(-1) = 2^{-1} = \frac{1}{2}$$

$$f(2) = 2^2 = 4 \qquad f(-2) = 2^{-2} = \frac{1}{4} \qquad f(3) = 2^3 = 8 \qquad f(-3) = 2^{-3} = \frac{1}{8}$$

Next, form a table from these values. Then plot the corresponding points and connect them with a smooth curve for the desired graph.

x	$f(x)$
−3	0.125
−2	0.25
−1	0.5
0	1
1	2
2	4
3	8

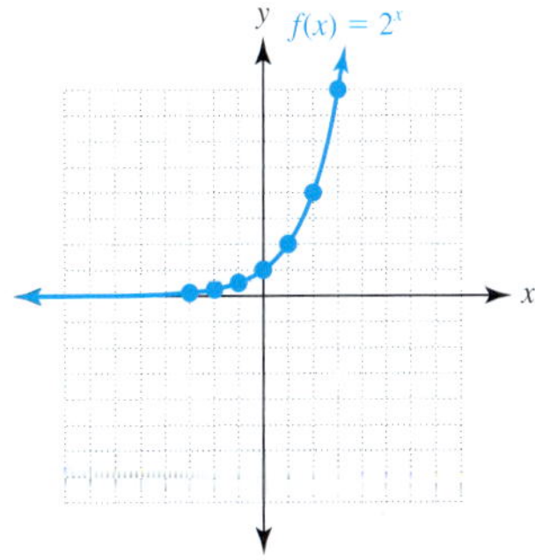

NOTES

There is no value for x such that

$2^x = 0$

so the graph never touches the x-axis.

We call $y = 0$ (or the x-axis) the **horizontal asymptote.**

Let's examine some characteristics of the graph of the exponential function. First, the vertical line test shows that this is indeed the graph of a function. Also note that the horizontal line test shows that the function is one-to-one.

The graph *approaches* the x-axis on the left, but it does *not intersect* the x-axis. The y-intercept is (0, 1) because $2^0 = 1$. To the right, the function values get larger. We say that the values *grow without bound.* This same language may be applied to linear or quadratic functions.

Check Yourself 1

Sketch the graph of the exponential function

$g(x) = 3^x$

We now look at an example in which the base of the function is less than 1.

Example 2 Graphing an Exponential Function

RECALL

$\left(\frac{1}{2}\right)^x = 2^{-x}$

Graph the exponential function

$$f(x) = \left(\frac{1}{2}\right)^x$$

First, choose convenient values for x.

$$f(0) = \left(\frac{1}{2}\right)^0 = 1 \qquad f(1) = \left(\frac{1}{2}\right)^1 = \frac{1}{2} \qquad f(-1) = \left(\frac{1}{2}\right)^{-1} = 2$$

$$f(2) = \left(\frac{1}{2}\right)^2 = \frac{1}{4} \qquad f(-2) = \left(\frac{1}{2}\right)^{-2} = 4$$

$$f(3) = \left(\frac{1}{2}\right)^3 = \frac{1}{8} \qquad f(-3) = \left(\frac{1}{2}\right)^{-3} = 8$$

Again, form a table of values and graph the desired function.

NOTE

By the vertical and horizontal line tests, this is the graph of a one-to-one function.

x	$f(x)$
−3	8
−2	4
−1	2
0	1
1	0.5
2	0.25
3	0.125

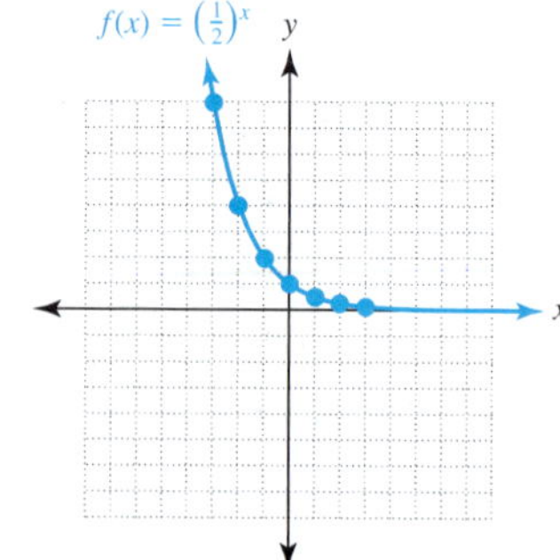

NOTE

The base of a *growth function* is *greater than* 1.

The base of a *decay function* is *less than* 1 but greater than 0.

Comparing this graph with that of Example 1, we see that this graph also represents a one-to-one function. As in Example 1, the graph does not intersect the x-axis but approaches that axis, here on the right. The values for the function again grow without bound, but this time on the left. The y-intercept for both graphs occurs at (0, 1).

The graph of Example 1 is *increasing* (going up) as we move from left to right. That function is an example of a **growth function.**

The graph of Example 2 is *decreasing* (going down) as we move from left to right. It is an example of a **decay function.**

Check Yourself 2

Sketch the graph of the exponential function

$$g(x) = \left(\frac{1}{3}\right)^x$$

This algorithm summarizes our work thus far in this section.

Step by Step

Graphing an Exponential Function

Step 1 Establish a table of values by considering the function in the form $y = b^x$.

Step 2 Plot points from that table of values and connect them with a smooth curve to form the graph.

NOTE

The use of the letter *e* as a base originated with Leonhard Euler (1707–1783), and *e* is sometimes called *Euler's number* for that reason.

> Calculator

NOTE

Graph $y = e^x$ on your calculator. You may find the $[e^x]$ key to be the second (or inverse) function to the LN key. Note that e^1 is approximately 2.71828.

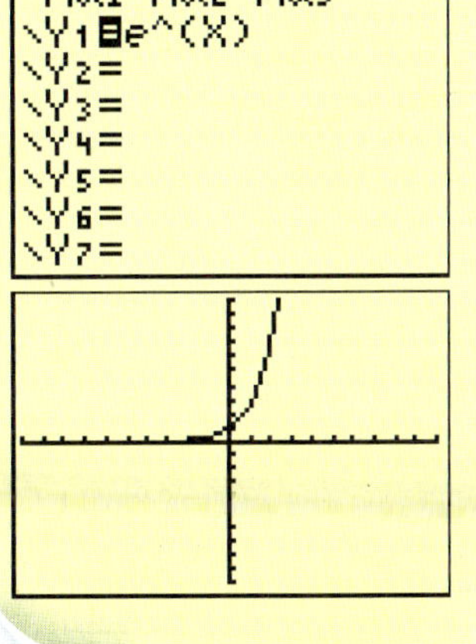

The graphs of exponential functions, $f(x) = b^x$, have these properties.

(a) The y-intercept is (0, 1).

(b) The graphs approach, but do not touch, the x-axis.

(c) The graphs represent one-to-one functions.

(d) If $b > 1$, the graph increases from left to right. If $0 < b < 1$, the graph decreases from left to right.

We used bases of 2 and $\frac{1}{2}$ for the exponential functions in our examples because they provided convenient computations. A far more important base for an exponential function is an irrational number named e. In fact, when e is used as a base, the function defined by

$$f(x) = e^x$$

is called *the* exponential function.

The significance of this number will be made clear in later courses, particularly calculus. For our purposes, e can be approximated as

$$e \approx 2.71828$$

The graph of $f(x) = e^x$ is shown below. Of course, it is very similar to the graphs seen earlier in this section.

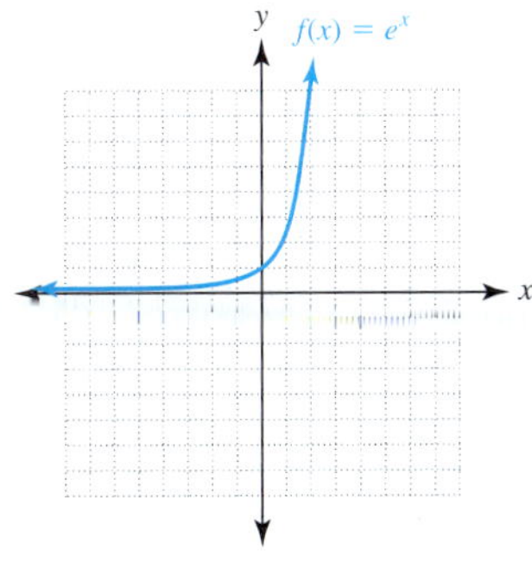

Exponential expressions involving the base e occur frequently in real-world applications. Example 3 illustrates two such applications.

Example 3 An Exponential Application

< Objective 2 >

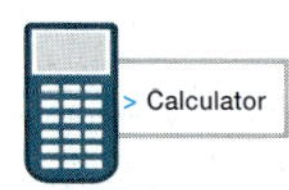

(a) Suppose that the population of a city is presently 20,000 and that the population is expected to have a continuous growth rate of 5% per year. The equation

$$P(t) = 20{,}000e^{0.05t}$$

gives the town's population after t years. Find the population in 5 years.

Let $t = 5$ in the original equation to obtain

$$P(5) = 20{,}000e^{(0.05)(5)} \approx 25{,}681$$

which is the expected population 5 years from now.

NOTE

Be certain that you enclose the entire exponent (0.05×5) in parentheses, or else the calculator will misinterpret your intended order of operations.

(b) Suppose \$1,000 is invested at an annual rate of 8%, compounded continuously. The equation

$$A(t) = 1{,}000e^{0.08t}$$

gives the amount in the account after t years. Find the amount after 9 years.

Let $t = 9$ in the original equation to obtain

$$A(9) = 1{,}000e^{(0.08)(9)} \approx 2{,}054$$

which is the amount in the account after 9 years.

In 9 years, the amount in the account is a little more than *double* the original principal.

Continuous compounding gives the highest accumulation of interest at any rate. However, daily compounding results in an amount of interest that is only slightly less.

Check Yourself 3

If \$1,000 is invested at an annual rate of 6%, compounded continuously, then the equation for the amount in the account after t years is

$$A(t) = 1{,}000e^{0.06t}$$

Use your calculator to find the amount in the account after 12 years.

In some applications, we see a variation of the basic exponential function $f(x) = b^x$. What happens to the graph of such a function if we add (or subtract) a constant? The graph of the new function $f(x) = b^x + k$ is a familiar graph translated vertically k units. Consider the next example.

Example 4 Graphing a Variation of an Exponential Function

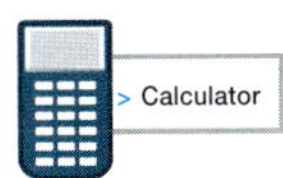

Graph the function $f(x) = 2^x - 3$.

Again, we create a table of values and connect points with a smooth curve.

x	$f(x)$
−3	−2.875
−2	−2.75
−1	−2.5
0	−2
1	−1
2	1
3	5

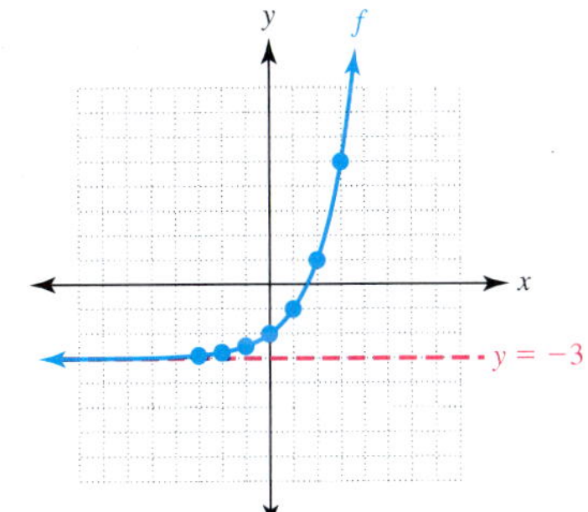

NOTE

Graph this function on your graphing calculator. Graph the line given by $y = -3$ on the same screen.

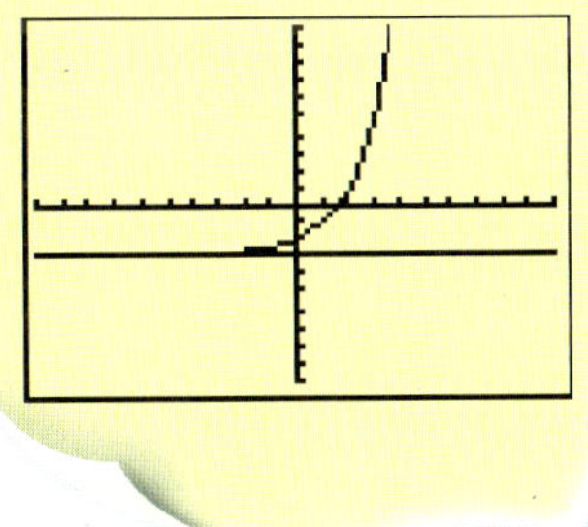

The graph of f appears to be the familiar graph of $g(x) = 2^x$ shifted down 3 units. Instead of a y-intercept of (0, 1), we see a y-intercept of (0, −2). Instead of a horizontal asymptote of $y = 0$, we see a horizontal asymptote of $y = -3$.

Check Yourself 4

Sketch the graph of the function $f(x) = \left(\frac{1}{2}\right)^x + 2$.

We generalize the previous result.

Property

Graphing a Function of the Form $y = b^x + k$

All such graphs have these properties.

1. The y-intercept is $(0, k + 1)$.
2. There is a horizontal asymptote with equation $y = k$.

As we observed, exponential functions are always one-to-one. This yields an important property that can be used to solve certain types of equations involving exponents.

Property

Property of Exponential Equations

If $b > 0$ and $b \neq 1$, then

$$b^m = b^n \quad \text{if and only if} \quad m = n$$

Example 5 Solving an Exponential Equation

< Objective 3 >

(a) Solve $2^x = 8$ for x.

We recognize that 8 is a power of 2, and we can write the equation as

$2^x = 2^3$ Write with equal bases.

Applying the property above, we have

$x = 3$ Set exponents equal.

and 3 is the solution. The solution set is $\{3\}$.

NOTE

The answer can easily be checked by substitution. Letting $x = 2$ gives

$3^{2(2)} = 3^4 = 81$

(b) Solve $3^{2x} = 81$ for x.

Since $81 = 3^4$, we can write

$$3^{2x} = 3^4$$
$$2x = 4$$
$$x = 2$$

We see that 2 is the solution for the equation.

NOTE

$\frac{1}{16} = \frac{1}{2^4} = 2^{-4}$

Verify the solution.

$2^{-5+1} \stackrel{?}{=} \frac{1}{16}$

$2^{-4} \stackrel{?}{=} \frac{1}{16}$

$\frac{1}{16} = \frac{1}{16}$ True

(c) Solve $2^{x+1} = \frac{1}{16}$ for x.

Again, we write $\frac{1}{16}$ as a power of 2, so that

$$2^{x+1} = 2^{-4}$$

Then $x + 1 = -4$

$$x = -5$$

The solution set is $\{-5\}$.

Check Yourself 5

Solve each equation.

(a) $2^x = 16$ **(b)** $4^{x+1} = 64$ **(c)** $3^{2x} = \frac{1}{81}$

Graphing Calculator Option

Applying Exponential Regression

We first looked at exponential functions of the form $f(x) = b^x$. We then extended that to consider the form $f(x) = b^x + k$. Another way to modify the basic form is to multiply by a constant a: $f(x) = ab^x$. This form is available as a regression model in your graphing calculator.

Suppose we collect some data that suggests a pattern of exponential growth. Finding an exponential function that best fits the data was once a tedious and time-consuming process. It is now quite simple and quick on a graphing calculator. Consider this data, representing the population, in millions, of California as it has grown through the years.

Year	1890	1910	1930	1950	1970	1990
Years since 1890	0	20	40	60	80	100
CA population (millions)	1.21	2.38	5.68	10.59	19.97	29.76

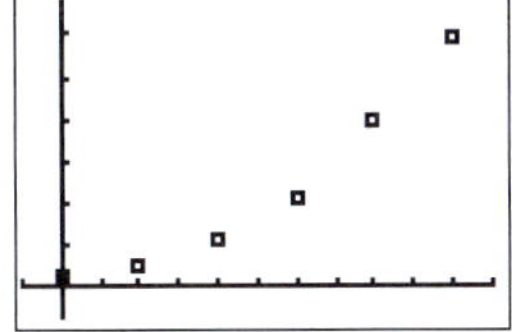

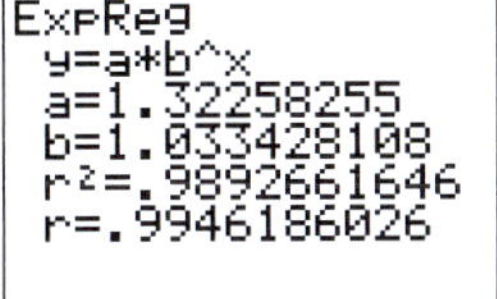

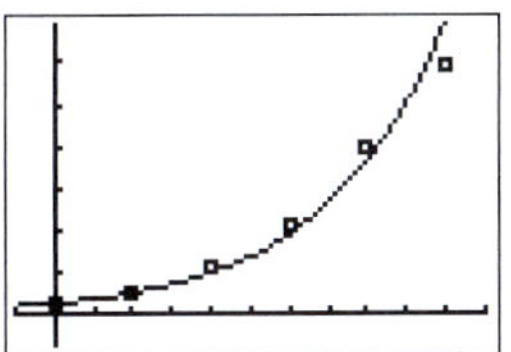

We begin by plotting the data. This gives us a chance to look for a pattern. We clear data lists [L1] and [L2]: STAT 4:ClrList 2nd [L1] , 2nd [L2] ENTER . Then we enter the data into [L1] and [L2]: STAT 1:Edit, and type in the numbers. Now exit the data editor: 2nd [QUIT].

To make and view a scatterplot: 2nd [STAT PLOT] ENTER ; press "On"; for "Type" select the first icon; "Xlist" should say [L1] and "Ylist" should say [L2]; for "Mark" choose the first symbol; press Y= and delete (or turn off) any existing equations; press ZOOM 9:ZoomStat. (To improve the scaling, go to WINDOW and choose appropriate numbers for Xscl and Yscl. Then GRAPH .)

We do see a pattern that suggests exponential growth. To find the "best fitting" exponential function: STAT CALC 0:ExpReg 2nd [L1] , 2nd [L2] ENTER . We have, accurate to four decimal places,

$$y = (1.3226)(1.0334)^x$$

To view the graph of this function on the scatterplot, enter its equation on the Y= screen and press GRAPH .

This function may be used to predict the population of California in the year 2010, for example. We must warn again that it is risky to predict too far beyond the scope of the data.

Graphing Calculator Check

The table below shows the U.S. public debt (in billions of dollars) since 1950. Using your graphing calculator (let 0 represent 1950), apply exponential regression to fit an exponential function to these data. Round coefficients accurate to four decimal places.

End of fiscal year	1950	1960	1970	1980	1990	2000	2005	2007
U.S. public debt (billions $)	257.4	290.2	389.2	930.2	3,233	5,674	7,933	9,008

ANSWERS

$y = (159.2319)(1.0722)^x$

Check Yourself ANSWERS

1. $y = g(x) = 3^x$

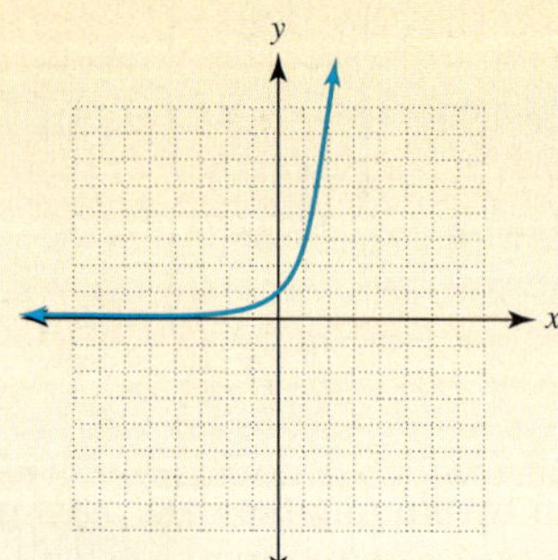

2. $y = \left(\frac{1}{3}\right)^x$

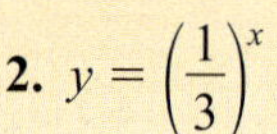

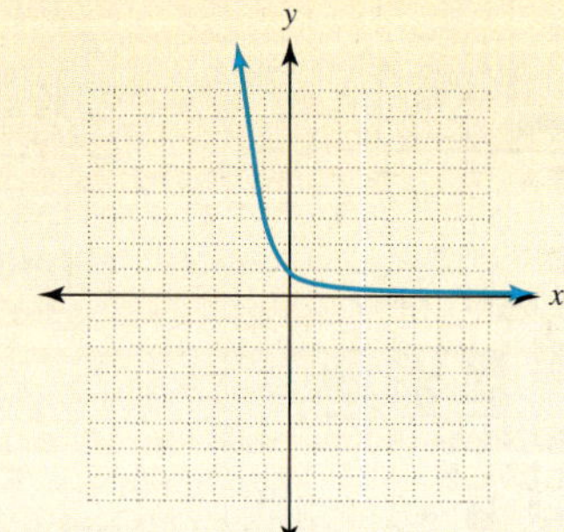

3. $2,054.43

4.

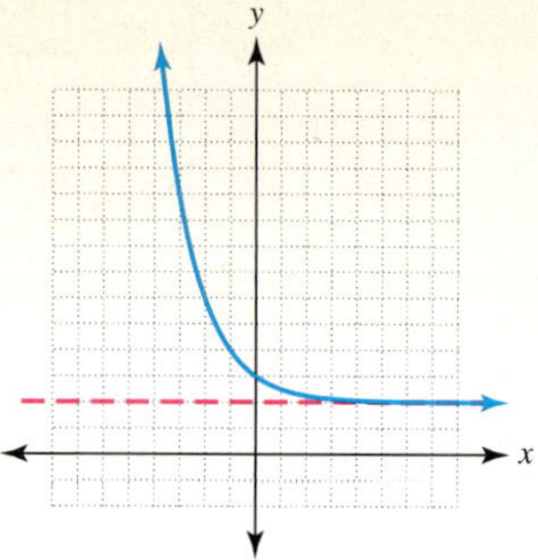

5. **(a)** $\{4\}$; **(b)** $\{2\}$; **(c)** $\{-2\}$

Reading Your Text

SECTION 10.4

(a) An ______________ function is a function that can be expressed in the form $f(x) = b^x$.

(b) Given the function $f(x) = b^x$, we call b the ______________ of the function.

(c) The base of a growth function is ______________ than one.

(d) When used as a mathematical constant, the letter e is sometimes called ______________ number.

Basic Skills | Challenge Yourself | Calculator/Computer | Career Applications | Above and Beyond

10.4 exercises

< Objective 1 >

Match the graphs in exercises 1 to 8 with the appropriate equation.

(a) $y = \left(\frac{1}{2}\right)^x$ (b) $y = 2x - 1$ (c) $y = 2^x$ (d) $y = x^2$

(e) $y = 1^x$ (f) $y = 5^x$ (g) $y = 1 - 2x$ (h) $y = 2^x - 1$

1.

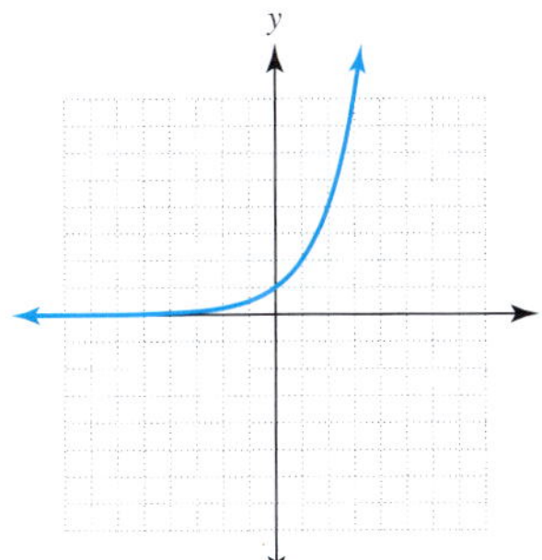

2.

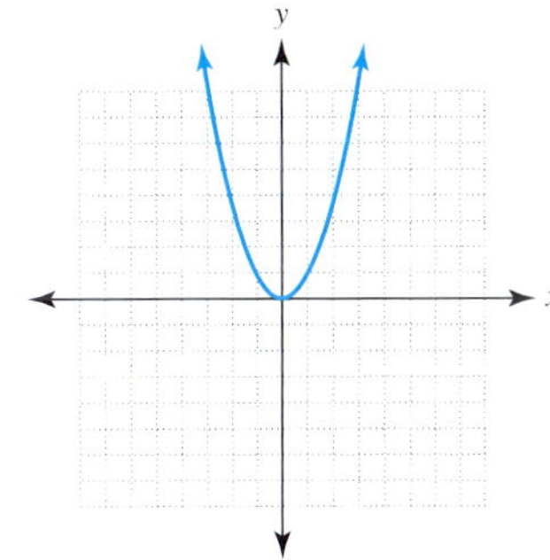

3.

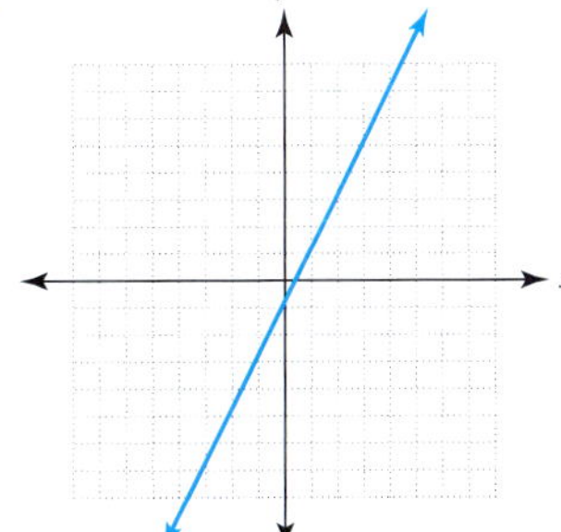

4.

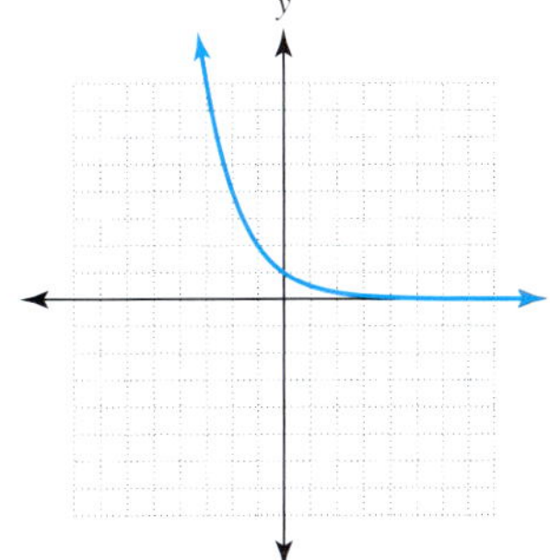

5.

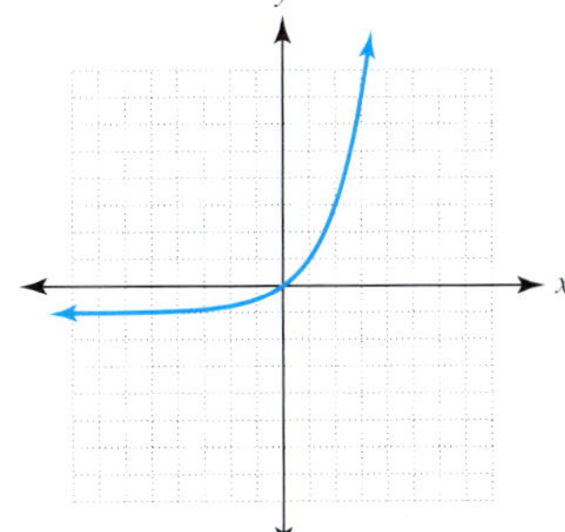

6.

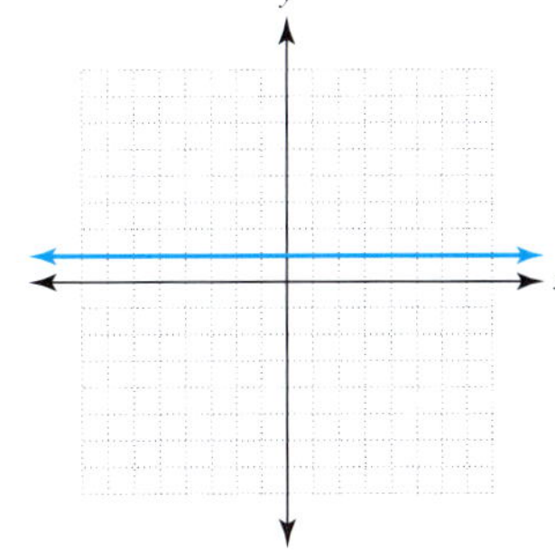

7.

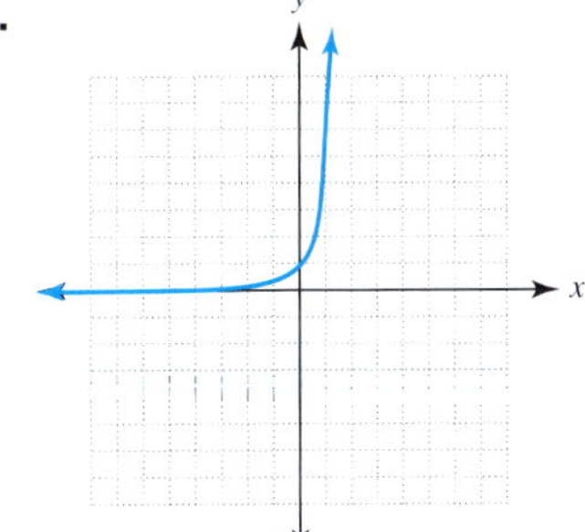

8.

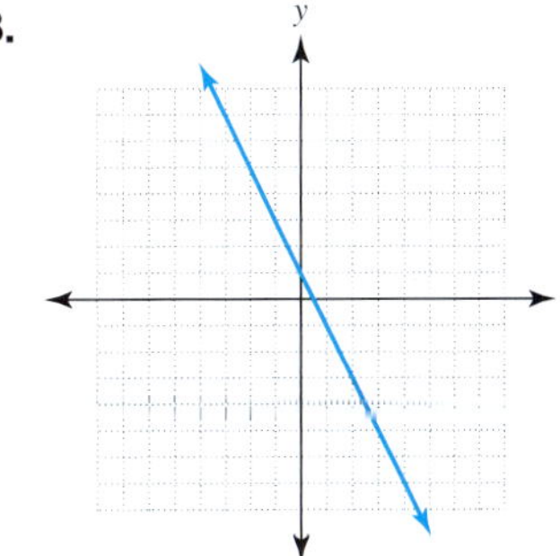

Name ____________

Section ________ Date ________

Answers

1. ____________
2. ____________
3. ____________
4. ____________
5. ____________
6. ____________
7. ____________
8. ____________

Answers

9. ______
10. ______
11. ______
12. ______
13. ______
14. ______
15. ______
16. ______
17. ______
18. ______
19. ______
20. ______
21. ______
22. ______
23. ______
24. ______
25. ______
26. ______
27. ______
28. ______
29. ______
30. ______

Let $f(x) = 4^x$ and evaluate.

9. $f(0)$

10. $f(4)$

11. $f(2)$

12. $f(-2)$

Let $g(x) = 4^{x+1}$ and evaluate.

13. $g(-1)$

14. $g(1)$

15. $g(2)$

16. $g(-2)$

Let $h(x) = 4^x + 1$ and evaluate.

17. $h(-1)$

18. $h(1)$

19. $h(2)$ > Videos

20. $h(-2)$

Let $f(x) = \left(\frac{1}{4}\right)^x$ and evaluate.

21. $f(1)$

22. $f(-1)$

23. $f(0)$

24. $f(2)$

Graph each exponential function.

25. $y = 4^x$

26. $y = \left(\frac{1}{4}\right)^x$

27. $y = \left(\frac{2}{3}\right)^x$

28. $y = \left(\frac{3}{2}\right)^x$

29. $y = 3 \cdot 2^x$

30. $y = 2 \cdot 3^x$

31. $y = 3^x$

32. $y = 2^{x-1}$

33. $y = 2^{2x}$

34. $y = \left(\frac{1}{2}\right)^{2x}$ 

35. $y = e^{-x}$

36. $y = e^{2x}$

Graph each function. Sketch each horizontal asymptote as a dotted line.

37. $y = 3^x + 2$

38. $y = \left(\frac{1}{3}\right)^x - 4$

< Objective 3 >

Solve each exponential equation.

39. $2^x = 128$

40. $4^x = 64$

41. $10^x = 10{,}000$

42. $5^x = 625$

43. $3^x = \frac{1}{9}$

44. $2^x = \frac{1}{16}$

45. $4^{2x} = 64$ > Videos

46. $3^{2x} = 81$

47. $3^{x-1} = 81$

48. $4^{x-1} = 16$

49. $3^{x-1} = \frac{1}{27}$

50. $2^{x-3} = \frac{1}{16}$

Answers

31. ____
32. ____
33. ____
34. ____
35. ____
36. ____
37. ____
38. ____
39. ____
40. ____
41. ____
42. ____
43. ____
44. ____
45. ____
46. ____
47. ____
48. ____
49. ____
50. ____

Basic Skills | Challenge Yourself | Calculator/Computer | Career Applications | Above and Beyond

Answers

51. ____________

52. ____________

53. ____________

54. ____________

55. ____________

56. ____________

57. ____________

58. ____________

59. ____________

60. ____________

61. ____________

62. ____________

Assume $b > 0$, $b \neq 1$. Complete each statement with **never, sometimes,** *or* **always.**

51. The function $g(x) = b^x$ is ____________ a decay function.

52. The graph of $f(x) = b^x$ ____________ intersects the y-axis.

53. The graph of $f(x) = b^x$ ____________ intersects the x-axis.

54. The function $g(x) = b^x$ is ____________ one-to-one.

< Objective 2 >

Science and Medicine *Suppose it takes 1 h for a certain bacterial culture to double by dividing in half. If there are 100 bacteria in the culture to start, then the number of bacteria in the culture after x hours is given by $N(x) = 100 \cdot 2^x$. Use this function to complete exercises 55 to 58.*

55. Find the number of bacteria in the culture after 2 h.

56. Find the number of bacteria in the culture after 3 h.

57. Find the number of bacteria in the culture after 5 h.

58. Graph the relationship between the number of bacteria in the culture and the number of hours. Be sure to choose an appropriate scale for the N-axis.

Science and Medicine *The half-life of radium is 1,690 years. That is, after a 1,690-year period, one-half of the original amount of radium will have decayed into another substance. If the original amount of radium was 64 grams (g), the formula relating the amount of radium left after time t is given by $R(t) = 64 \cdot 2^{-t/1{,}690}$. Use this formula to complete exercises 59 to 62.*

chapter 10 > Make the Connection

59. Find the amount of radium left after 1,690 years.

60. Find the amount of radium left after 3,380 years.

61. Find the amount of radium left after 5,070 years.

62. Graph the relationship between the amount of radium remaining and time. Be sure to use appropriate scales for the R- and t-axes.

Business and Finance *If \$1,000 is invested in a savings account with an interest rate of 8%, compounded annually, the amount in the account after t years is given by* $A(t) = 1{,}000(1 + 0.08)^t$. *Use a calculator to complete each exercise.*

63. Find the amount in the account after 2 years.

64. Find the amount in the account after 5 years.

65. Find the amount in the account after 9 years.

66. Graph the relationship between the amount in the account and time. Be sure to choose appropriate scales for the A- and t-axes.

Science and Medicine *The so-called learning curve in psychology applies to learning a skill, such as typing, in which the performance level progresses rapidly at first and then levels off with time. One can approximate N, the number of words per minute (wpm) that a person can type after t weeks of training, with the equation* $N = 80\,(1 - e^{-0.06t})$. *Use a calculator to complete exercises 67 and 68.*

67. **(a)** N after 10 weeks; **(b)** N after 20 weeks; **(c)** N after 30 weeks

68. Graph the relationship between the number of words per minute N and the number of weeks of training t.

69. Sales in the organic food business have grown steadily in recent years. The table below shows annual amounts, in billions of dollars, for such sales in the United States. Using your graphing calculator, apply exponential regression to fit an exponential function to these data. Round coefficients accurate to four decimal places. Let x be the number of years since 1997.

Year	1997	1998	1999	2000	2001	2002	2003	2004	2005
Sales	3.59	4.29	5.04	6.10	7.36	8.64	10.38	11.90	13.83

70. The table below shows the declining temperature of some coffee, in degrees Celsius (°C), as 50 minutes passed. Using your graphing calculator, apply exponential regression to fit an exponential function to these data. Round coefficients accurate to four decimal places.

Minutes	0	5	8	11	15	18	24
Temp (°C)	83	76.5	70.5	65	61	57.5	52.5

Minutes	25	30	34	38	42	45	50
Temp (°C)	51	47.5	45	43	41	39.5	38

Answers

63. ____________

64. ____________

65. ____________

66. ____________

67. ____________

68. ____________

69. ____________

70. ____________

Answers

71. ______

72. ______

73. ______

74. ______

71. The stopping distance for a car depends on (among other things) the speed that the car is traveling. The tables show the stopping distances of a certain car, for various speeds. Using your graphing calculator, apply exponential regression to fit an exponential function to these data. Round coefficients accurate to four decimal places.

Speed (m/hr)	5	15	25	35	45
Distance (ft)	7	24	48	72	90

Speed (m/hr)	55	65	75	85	95
Distance (ft)	146	207	336	603	840

72. After a drug is introduced into a patient's bloodstream, the concentration of the drug begins to drop as time passes. The table below shows concentration levels, in mg/ml, of a certain drug over an 8-hour period. Using your graphing calculator, apply exponential regression to fit an exponential function to these data. Round coefficients accurate to four decimal places.

Hour (h)	0	1	2	3	4	5	6	7	8
Concentration (mg/ml)	1.50	0.95	0.60	0.40	0.25	0.15	0.10	0.07	0.04

Basic Skills | Challenge Yourself | Calculator/Computer | **Career Applications** | Above and Beyond

73. **Agricultural Technology** The biomass per acre of a cornfield grows exponentially over time. The amount of biomass is given by the function

$B(t) = (1.132)^t$

in which B is the biomass per acre, in pounds, and t represents the growing time, in days.

(a) How much biomass is in a 1-acre field after 80 days of growth (nearest pound)?

(b) By how much will the biomass increase between day 80 and day 90 (nearest pound)?

74. **Mechanical Engineering** The intensity of light transmitted through a certain material is reduced by 6% per mm of thickness. The percentage of light transmitted is found using the function

$P(T) = (0.94)^T$

in which T represents the material's thickness, in mm.

(a) Find the percentage (to the nearest whole percent) of light transmitted if the material is 34 mm thick.

(b) How thick is the material (to the nearest tenth mm) if exactly half the light is transmitted?

75. Electronics Capacitors are used to store energy. When a capacitor reaches the point when it cannot store anymore energy, it is said to be *charged*. The charging process is not instantaneous. Instead, it can be modeled by the function

$$V_c(t) = V_{ps}(1 - e^{-t/\tau})$$

in which V_c is the stored voltage measured across the capacitor, V_{ps} is the voltage based on the power supply, t is the time the capacitor has been charging, in seconds, and τ is a (time) constant.

Find the stored capacitive voltage (to the nearest whole volt) after 150 s if the supply voltage in a circuit is 14 volts and has a time constant of 103 s.

76. Construction Technology The temperature of a piece of metal, as it cools, is given by the formula

> Videos

$$T(t) = T_r + (T_m - T_r)e^{-t/\tau}$$

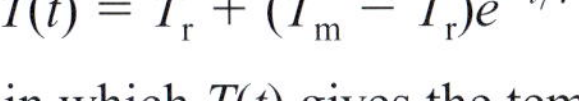

in which $T(t)$ gives the temperature of the metal t minutes after it begins to cool. T_r represents the ambient (room) temperature, T_m is the temperature that the metal was heated to, and τ is a (time) constant specific to a particular metal.

For a metal that has a time constant of $\tau = 3.1$, what will its temperature be, to the nearest hundredth degree F, 20 min after being heated if the room temperature is 72°F, and the metal is heated to 280°F?

77. Find two different calculators that have $\boxed{e^x}$ keys. Describe how to use the function on each of the calculators.

78. Are there any values of x for which e^x produces an exact answer on the calculator? Why are other answers not exact?

A possible calculator sequence for evaluating the expression

$$\left(1 + \frac{1}{n}\right)^n$$

where $n = 10$ *is*

[(] 1 [+] 1 [÷] 10 [)] [∧] 10 [ENTER]

which yields approximately 2.5937.

Find $\left(1 + \dfrac{1}{n}\right)^n$ *for each value of* n*. Round your answers to the nearest ten-thousandth.*

79. $n = 100$

80. $n = 1{,}000$

81. $n = 10{,}000$

82. $n = 100{,}000$

83. $n = 1{,}000{,}000$

84. What did you observe from the results of exercises 79 to 83?

Answers

75. ______

76. ______

77. ______

78. ______

79. ______

80. ______

81. ______

82. ______

83. ______

84. ______

85. Graph the exponential function defined by $y = 2^x$.

86. Graph the function defined by $x = 2^y$ on the same set of axes as the previous graph. What do you observe? (*Hint:* To graph $x = 2^y$, choose convenient values for y and then compute the corresponding values for x.)

87. Suppose you have a large piece of paper whose thickness is 0.003 in. If you tear the paper in half and stack the pieces, the height of the stack is (0.003)(2) in., or 0.006 in. If you now tear the stack in half again and then stack the pieces, the stack is $(0.003)(2)(2) = (0.003)(2^2)$ in., or 0.012 in. high.

(a) Define a function that gives the height h of the stack (in inches) after n tears.

(b) After which tear will the stack's height exceed 8 in.?

(c) Compute the height of the stack after the 15th tear. You will need to convert your answer to the appropriate units.

88. Use your graphing calculator to find all three points of intersection of the graphs of $f(x) = x^2$ and $g(x) = 2^x$. Give coordinates accurate to two decimal places.

Answers

85. ____

86. ____

87. ____

88. ____

Answers

1. (c) **3.** (b) **5.** (h) **7.** (f) **9.** 1 **11.** 16 **13.** 1 **15.** 64

17. $\frac{5}{4}$ **19.** 17 **21.** $\frac{1}{4}$ **23.** 1

25.

$y = 4^x$

27.

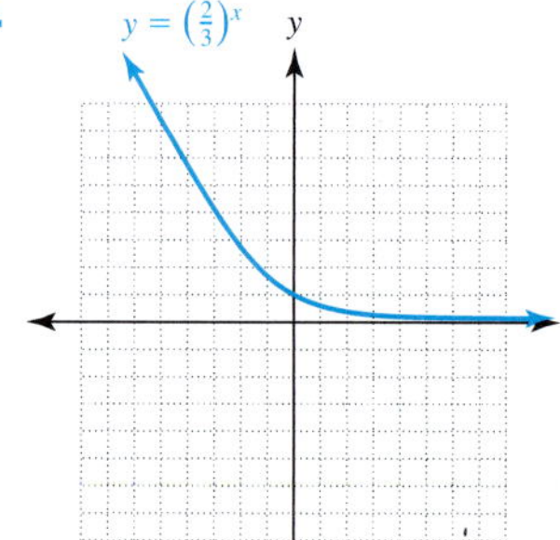

29.

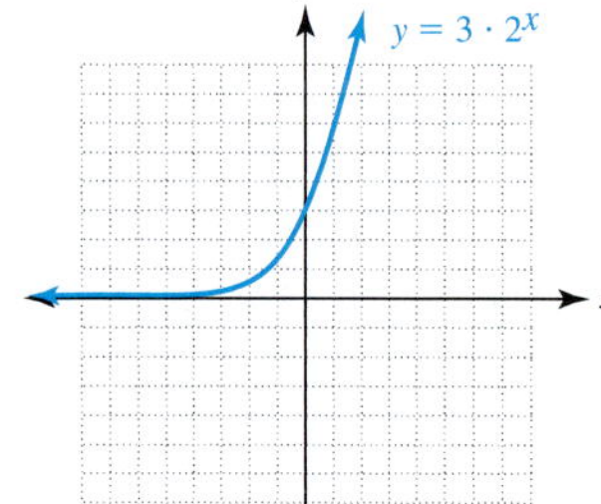

31.

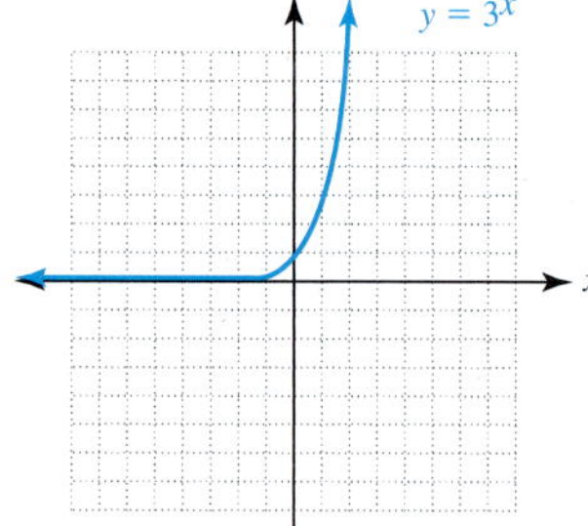

33.

35.

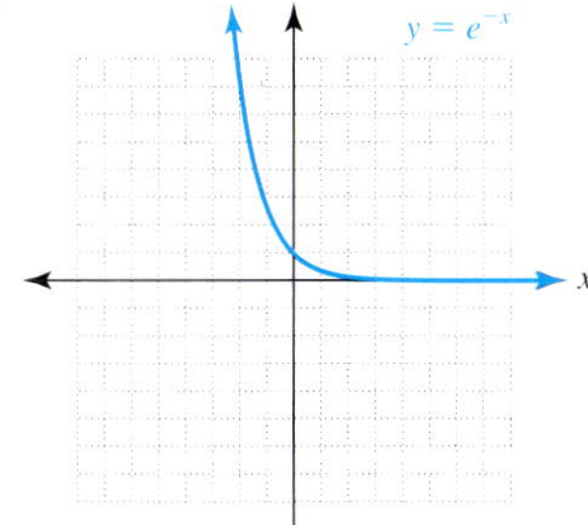

37.

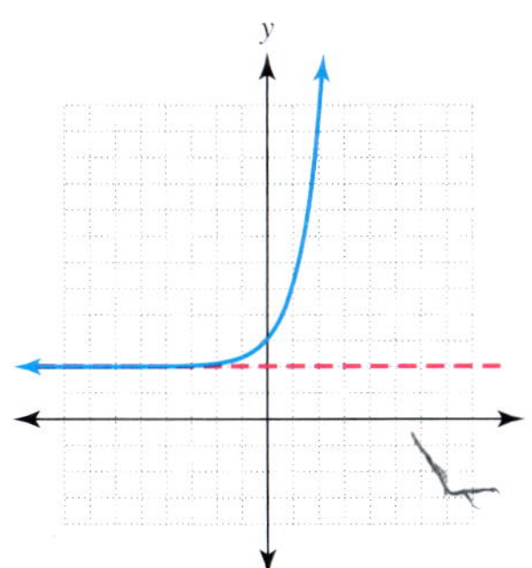

39. $\{7\}$ **41.** $\{4\}$ **43.** $\{-2\}$ **45.** $\left\{\frac{3}{2}\right\}$ **47.** $\{5\}$ **49.** $\{-2\}$
51. sometimes **53.** never **55.** 400 bacteria **57.** 3,200 bacteria
59. 32 g **61.** 8 g **63.** \$1,166.40 **65.** \$1,999
67. **(a)** 36; **(b)** 56; **(c)** 67 **69.** $y = (3.6315)(1.1863)^x$
71. $y = (10.0645)(1.0491)^x$ **73.** **(a)** 20,310 lb; **(b)** 49,864 lb **75.** 11 V
77. Above and Beyond **79.** 2.7048 **81.** 2.7181 **83.** 2.71828
85.

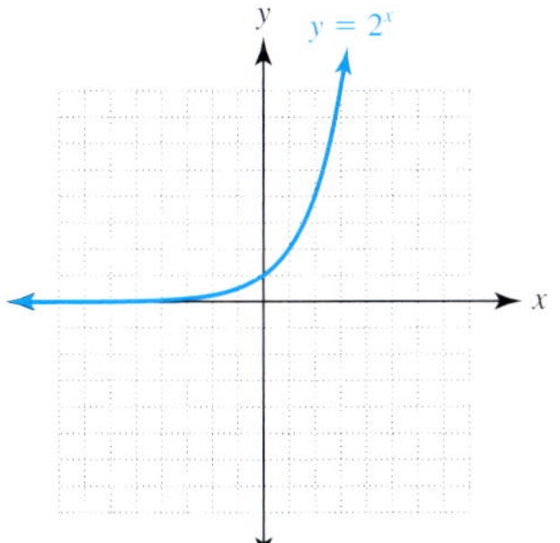

87. **(a)** $h = (0.003)(2^n)$; **(b)** 12; **(c)** 8.192 ft

Activity 10 :: Half-Life and Decay

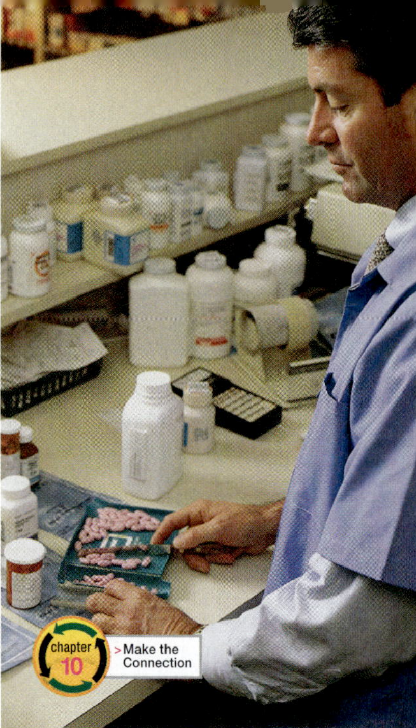

You may have encountered the idea of half-life in connection with radioactive substances. Given an initial amount of some radioactive material, half of that material will remain after an amount of time known as the **half-life** of the material. As the material continues to decay, the amount that remains may be modeled by an **exponential** function.

You can simulate the decay of radioactive material. Working with two or three partners, obtain approximately 20 wooden cubes, and place a marker on just one side of each cube. (Dice may be used: simply choose one number to be the marked side.)

1. Count the number of cubes you have. This is your *initial amount of radioactive material.* Record this number.
2. Roll the entire set of cubes. The cube(s) that show the marked side up have decayed. Remove these, and record the number that are still active.
3. Roll the remaining radioactive cubes, remove those that have decayed, and record the amount remaining.
4. Continue in this manner, filling out a table like that shown, until all cubes have decayed. Note that the variable x represents the number of rolls, and y represents the number of cubes still active.

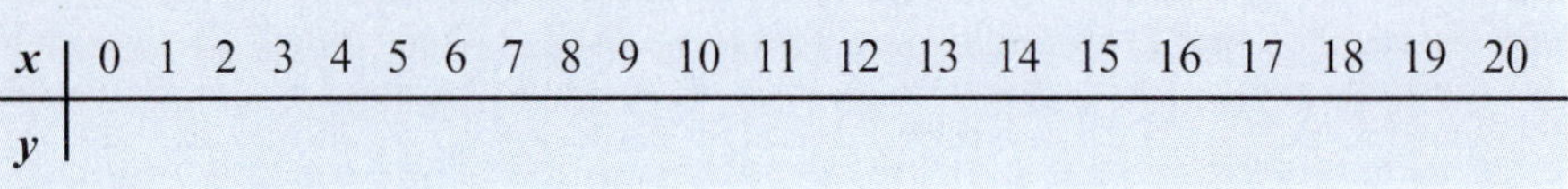

x	0	1	2	3	4	5	6	7	8	9	10	11	12	13	14	15	16	17	18	19	20
y																					

If dice are not available, use the sample data on the next page to complete exercises 5–9.

5. Draw a scatter plot of the ordered pairs (x, y) in your table.
6. Repeat the entire procedure (with the same set of cubes), completing a new table. Plot this set of ordered pairs on the *same* coordinate system you made in step 5.
7. Do this a third time, again adding the points to your scatter plot.
8. Now, draw a smooth curve that seems (to you) to fit the points best.
9. Use your graph to determine the approximate half-life for your cubes. For example, if you began with 22 cubes, see how many rolls it took for 11 to remain. Confirm your estimate of the half-life by checking elsewhere on the graph. For example, estimate the number of rolls corresponding to 16 cubes, and see about how many rolls it took, from that point, for 8 to remain.

Sample Data

x	0	1	2	3	4	5	6	7	8	9	10	11	12	13	14	15	16	17	18	19	20
y	22	19	16	14	10	8	6	5	4	4	4	3	3	3	3	3	3	3	2	2	0

x	0	1	2	3	4	5	6	7	8	9	10	11	12	13	14	15	16	17	18
y	22	20	15	13	10	9	9	7	5	1	1	1	1	1	1	0			

x	0	1	2	3	4	5	6	7	8	9	10	11	12	13	14	15	16	17	18
y	22	15	12	8	7	7	5	5	5	4	4	4	3	2	2	0			

10.5 Logarithmic Functions

< 10.5 Objectives >

1 > Graph a logarithmic function

2 > Convert between logarithmic and exponential equations

3 > Evaluate a logarithmic expression

4 > Solve an elementary logarithmic equation

Given our experience with exponential functions in Section 10.4 and inverse functions in Section 10.3, we can now introduce the logarithmic function.

The Scottish mathematician John Napier (1550–1617) is credited with the invention of logarithms. The development of the logarithm grew out of a desire to ease the work involved in numerical computations, particularly in the field of astronomy. Today the availability of calculators has made logarithms unnecessary as a computational tool.

NOTE

Napier coined the word *logarithm* from the Greek words "logos" (a ratio) and "arithmos" (a number).

However, the concept of the logarithm and the properties of logarithmic functions are still very important in the solutions of particular equations in calculus and in the applied sciences.

Again, the applications for this new function are numerous. The Richter scale for measuring the intensity of an earthquake and the decibel scale for measuring the intensity of sound both use logarithms.

To develop the idea of a logarithmic function, we return to the definition of an exponential function.

$f(x) = b^x \qquad b > 0, b \neq 1$

RECALL

If f is a one-to-one function, then its inverse is also a function.

Letting y replace $f(x)$ and interchanging the roles of x and y, we have the inverse function

$b^y = x$

Presently, we have no way to solve the equation $b^y = x$ for y. So, to write the inverse in a more useful form, we offer a definition.

Definition

Logarithm of x to base b

The **logarithm of x to base b** is denoted

$\log_b x$

and $y = \log_b x \qquad$ if and only if $\qquad b^y = x$

We can now write an inverse function, using this new notation, as

$f^{-1}(x) = \log_b x \qquad b > 0, b \neq 1$

NOTE

The restrictions on the base are the same as those for the exponential function.

In general, any function defined in this form is called a **logarithmic function.**

The logarithm of x to base b should really be thought of as a function, and, technically, we should write $y = \log_b(x)$. This emphasizes that x is the input variable for the function whose name is $\log_b$ and whose output variable is y. However, it is common practice to drop the parentheses, and simply write $y = \log_b x$.

The important meaning here is that y is the power (or exponent) we place on b to produce x. For example, $3 = \log_2 8$ means 3 is the exponent we place on 2 to produce 8. So, $y = \log_b x$ is equivalent to $b^y = x$.

Property

Logarithmic and Exponential Functions

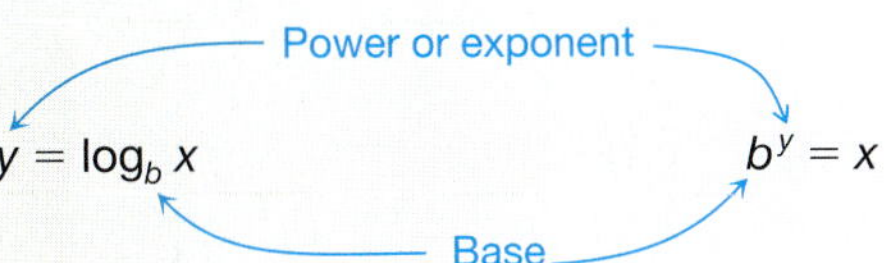

The logarithm y is the power to which we must raise b to get x. In other words, *a logarithm is simply a power or an exponent.*

We begin our work by graphing a typical logarithmic function.

Example 1 Graphing a Logarithmic Function

< Objective 1 >

NOTE

The base is 2, and the logarithm or power is y.

NOTE

What do the vertical and horizontal line tests tell you about this graph?

Graph the logarithmic function

$y = \log_2 x$

Since $y = \log_2 x$ is equivalent to the exponential form

$2^y = x$

we can find ordered pairs satisfying this equation by choosing convenient values for y and calculating the corresponding values for x.

Letting y take on values from -3 to 3 yields the table of values shown here. As before, we plot points from the ordered pairs and connect them with a smooth curve to form the graph of the function.

x	y
$\frac{1}{8}$	-3
$\frac{1}{4}$	-2
$\frac{1}{2}$	-1
1	0
2	1
4	2
8	3

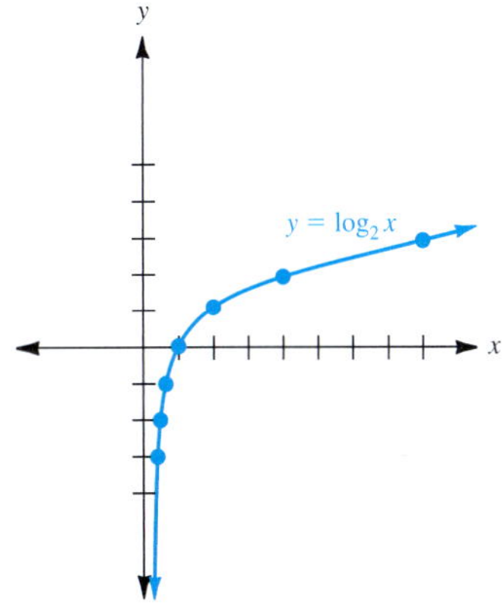

We observe that the graph represents a one-to-one function whose domain is $\{x \mid x > 0\}$ and whose range is the set of all real numbers.

For base 2 (or for any base greater than 1), the function increases over its domain.

Recall from Section 10.3 that the graphs of a function and its inverse are always reflections of each other about the line $y = x$. Since we have defined the logarithmic function as the inverse of an exponential function, we can anticipate the same relationship.

The graphs of

$f(x) = 2^x$ and $f^{-1}(x) = \log_2 x$

are shown in the figure.

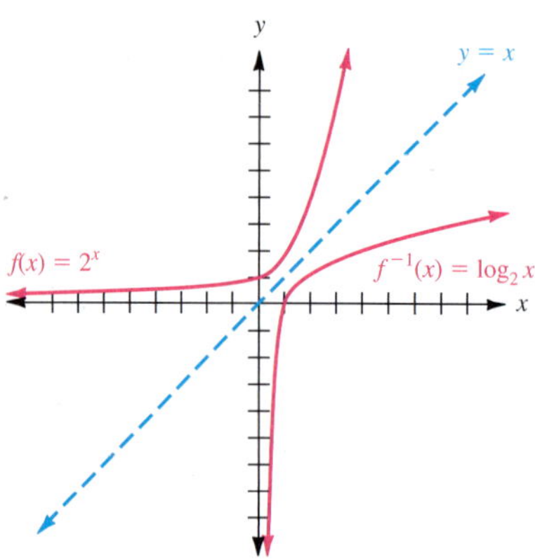

We see that the graphs of f and f^{-1} are indeed reflections of each other about the line $y = x$. In fact, this relationship provides an alternate method of sketching $y = \log_b x$. We can sketch the graph of $y = b^x$ and then reflect that graph about line $y = x$ to form the graph of the logarithmic function.

Check Yourself 1

Graph the logarithmic function defined by

$y = \log_3 x$

(*Hint:* Consider the equivalent form $3^y = x$.)

Let us summarize some facts regarding logarithmic and exponential functions.

Property

Inverse Functions

$y = b^x$ and $y = \log_b x$ are inverse functions.

1. Their graphs are symmetric with respect to the line $y = x$.
2. Because the point (0,1) is on the graph of $y = b^x$ (it is the y-intercept), the point (1, 0) is on the graph of $y = \log_b x$ (it is the x-intercept).
3. Because the line $y = 0$ is the asymptote for $y = b^x$ (it is horizontal), the line $x = 0$ is the asymptote for $y = \log_b x$ (it is vertical).
4. The domain of $y = b^x$ is the set of all real numbers, and the range is the set of all positive real numbers.
5. The domain of $y = \log_b x$ is the set of all positive real numbers, and the range is the set of all real numbers.

For our work in this chapter, it is necessary for us to convert between exponential and logarithmic forms. The conversion is straightforward. You need only keep in mind the basic relationship

$y = \log_b x$ means $b^y = x$

Example 2 **Writing Equations in Logarithmic Form**

< Objective 2 >

Convert to logarithmic form.

(a) $3^4 = 81$ is equivalent to $\log_3 81 = 4$.

(b) $10^3 = 1{,}000$ is equivalent to $\log_{10} 1{,}000 = 3$.

(c) $2^{-3} = \frac{1}{8}$ is equivalent to $\log_2 \frac{1}{8} = -3$.

(d) $9^{1/2} = 3$ is equivalent to $\log_9 3 = \frac{1}{2}$.

Check Yourself 2

Convert each statement to logarithmic form.

(a) $4^3 = 64$

(b) $10^{-2} = 0.01$

(c) $3^{-3} = \frac{1}{27}$

(d) $27^{1/3} = 3$

Example 3 shows how to write a logarithmic expression in exponential form.

Example 3 Writing Equations in Exponential Form

NOTE

In (a), the base is 2; the logarithm, which is the power, is 3.

Convert to exponential form.

(a) $\log_2 8 = 3$ is equivalent to $2^3 = 8$.

(b) $\log_{10} 100 = 2$ is equivalent to $10^2 = 100$.

(c) $\log_3 \frac{1}{9} = -2$ is equivalent to $3^{-2} = \frac{1}{9}$.

(d) $\log_{25} 5 = \frac{1}{2}$ is equivalent to $25^{1/2} = 5$.

Check Yourself 3

Convert to exponential form.

(a) $\log_2 32 = 5$

(b) $\log_{10} 1{,}000 = 3$

(c) $\log_4 \frac{1}{16} = -2$

(d) $\log_{27} 3 = \frac{1}{3}$

Certain logarithms can be directly calculated by changing an expression to the equivalent exponential form, as Example 4 illustrates.

Example 4 Evaluating Logarithmic Expressions

< Objective 3 >

RECALL

$b^m = b^n$ if and only if $m = n$.

(a) Evaluate $\log_3 27$.

If $x = \log_3 27$, in exponential form we have

$$3^x = 27$$
$$3^x = 3^3$$
$$x = 3$$

We then have $\log_3 27 = 3$.

NOTE

Rewrite each side as a power of the same base.

(b) Evaluate $\log_{10} \frac{1}{10}$.

If $x = \log_{10} \frac{1}{10}$, we can write

$$10^x = \frac{1}{10}$$
$$= 10^{-1}$$

We then have $x = -1$ and

$$\log_{10} \frac{1}{10} = -1$$

Check Yourself 4

Evaluate each logarithm.

(a) $\log_2 64$ **(b)** $\log_3 \frac{1}{27}$

The relationship between exponents and logarithms also allows us to solve certain equations involving logarithms where two of the quantities in the equation $y = \log_b x$ are known, as Example 5 illustrates.

Example 5 Solving Logarithmic Equations

< Objective 4 >

(a) Solve $\log_5 x = 3$ for x.

Since $\log_5 x = 3$, in exponential form we have

$$5^3 = x$$
$$x = 125$$

The solution set is $\{125\}$.

(b) Solve $y = \log_4 \frac{1}{16}$ for y.

The original equation is equivalent to

$$4^y = \frac{1}{16}$$
$$= 4^{-2}$$

We then have $y = -2$ as the solution.

(c) Solve $\log_b 81 = 4$ for b.

In exponential form the equation becomes

$$b^4 = 81$$
$$b = 3$$

The solution set is $\{3\}$.

NOTE

Keep in mind that the base must be *positive*, so we do not consider the possible solution $b = -3$.

Check Yourself 5

Solve each equation.

(a) $\log_4 x = 4$ **(b)** $\log_b \frac{1}{8} = -3$ **(c)** $y = \log_9 3$

We use the **decibel scale** to measure the loudness of various sounds. If I represents the intensity of a given sound and I_0 represents the intensity of a "threshold sound," then the decibel (dB) rating L of the given sound is given by

$$L = 10 \log_{10} \frac{I}{I_0}$$

where $I_0 = 10^{-16}$ watt per square centimeter (W/cm^2). Consider Example 6.

NOTES

Loudness can be measured in **bels (B),** a unit named for Alexander Graham Bell. This unit is rather large, so a more practical unit is the **decibel (dB),** a unit that is one-tenth as large.

The constant I_0 is the intensity of the minimum sound level detectable by the human ear.

Example 6 A Decibel Application

NOTE

To evaluate $\log_{10} 10^2$, think "To what power must we raise 10 to obtain 10^2?" Answer: 2.

(a) A whisper has intensity $I = 10^{-14}$. Its decibel rating is

$$L = 10 \log_{10} \frac{10^{-14}}{10^{-16}}$$
$$= 10 \log_{10} 10^2$$
$$= 10 \cdot 2$$
$$= 20$$

NOTE

Again, think "10 to what power produces 10^{12}?" Answer: 12.

(b) A rock concert has intensity $I = 10^{-4}$. Its decibel rating is

$$L = 10 \log_{10} \frac{10^{-4}}{10^{-16}}$$
$$= 10 \log_{10} 10^{12}$$
$$= 10 \cdot 12$$
$$= 120$$

Check Yourself 6

Ordinary conversation has intensity $I = 10^{-12}$. Find its rating on the decibel scale.

NOTES

The scale was named after Charles Richter, a U.S. geologist.

A *zero-level* earthquake is the quake of least intensity that is measurable by a seismograph.

Geologists use the **Richter scale** to convert seismographic readings, which give the intensity of the shock waves of an earthquake, to a measure of the magnitude of that earthquake.

The magnitude M of an earthquake is given by

$$M = \log_{10} \frac{a}{a_0}$$

where a is the intensity of its shock waves and a_0 is the intensity of the shock wave of a zero-level earthquake.

Example 7 A Richter Scale Application

How many times stronger is an earthquake measuring 5 on the Richter scale than one measuring 4 on the Richter scale?

RECALL

The ratio of a_1 to a_2 is $\frac{a_1}{a_2}$

Suppose a_1 is the intensity of the earthquake with magnitude 5 and a_2 is the intensity of the earthquake with magnitude 4. We want to find $\frac{a_1}{a_2}$. Then

$$5 = \log_{10} \frac{a_1}{a_0} \quad \text{and} \quad 4 = \log_{10} \frac{a_2}{a_0}$$

We convert these logarithmic expressions to exponential form.

$$10^5 = \frac{a_1}{a_0} \quad \text{and} \quad 10^4 = \frac{a_2}{a_0}$$

or

$$a_1 = a_0 \cdot 10^5 \qquad \text{and} \qquad a_2 = a_0 \cdot 10^4$$

We want the ratio of the intensities of the two earthquakes, so

$$\frac{a_1}{a_2} = \frac{a_0 \cdot 10^5}{a_0 \cdot 10^4} = 10^1 = 10$$

The earthquake of magnitude 5 is *10 times stronger* than the earthquake of magnitude 4.

NOTE

$\frac{a_1}{a_2} = 10$ is equivalent to $a_1 = 10a_2$, which says that a_1 is 10 times the size of a_2.

Check Yourself 7

How many times stronger is an earthquake of magnitude 6 than one of magnitude 4?

Check Yourself ANSWERS

1. $y = \log_3 x$

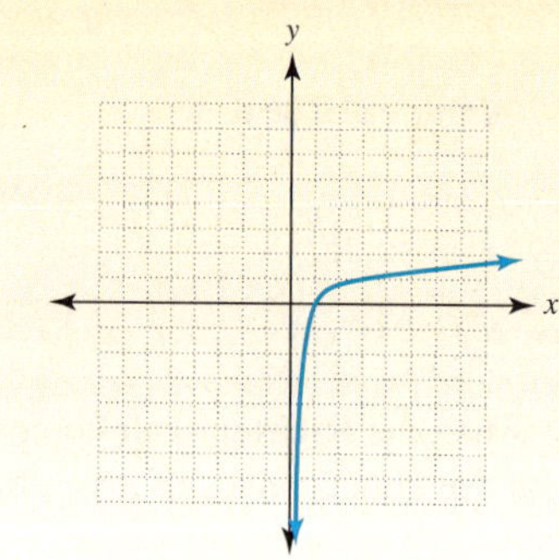

2. **(a)** $\log_4 64 = 3$; **(b)** $\log_{10} 0.01 = -2$; **(c)** $\log_3 \frac{1}{27} = -3$; **(d)** $\log_{27} 3 = \frac{1}{3}$

3. **(a)** $2^5 = 32$; **(b)** $10^3 = 1{,}000$; **(c)** $4^{-2} = \frac{1}{16}$; **(d)** $27^{1/3} = 3$

4. **(a)** $\log_2 64 = 6$; **(b)** $\log_3 \frac{1}{27} = -3$ **5.** **(a)** $\{256\}$; **(b)** $\{2\}$; **(c)** $\left\{\frac{1}{2}\right\}$

6. 40 dB **7.** 100 times

Reading Your Text

SECTION 10.5

(a) The ______________ of x to the base b is denoted $y = \log_b x$.

(b) Given $y = \log_b x$, the logarithm y is the ______________ to which we must raise b to get x.

(c) The point (0, 1) is on the graph of $y = b^x$. It is the ______________.

(d) ______________ can be measured in bels.

Basic Skills | Challenge Yourself | Calculator/Computer | Career Applications | Above and Beyond

10.5 exercises

Name ________________

Section ________ Date ________

< Objective 1 >

Sketch the graph of the function defined by each equation.

1. $y = \log_4 x$

2. $y = \log_{10} x$

3. $y = \log_2 (x - 1)$

4. $y = \log_3 (x + 1)$

5. $y = \log_8 x$

6. $y = \log_3 x + 1$

< Objective 2 >

Convert each statement to logarithmic form.

7. $2^5 = 32$

8. $3^5 = 243$

9. $10^2 = 100$

10. $5^3 = 125$

11. $3^0 = 1$

12. $10^0 = 1$

13. $4^{-2} = \frac{1}{16}$

14. $3^{-4} = \frac{1}{81}$

15. $10^{-2} = \frac{1}{100}$

16. $3^{-3} = \frac{1}{27}$

Answers

1. ________
2. ________
3. ________
4. ________
5. ________
6. ________
7. ________
8. ________
9. ________
10. ________
11. ________
12. ________
13. ________
14. ________
15. ________
16. ________

Answers

17. ______
18. ______
19. ______
20. ______
21. ______
22. ______
23. ______
24. ______
25. ______
26. ______
27. ______ 28. ______
29. ______ 30. ______
31. ______ 32. ______
33. ______ 34. ______
35. ______
36. ______
37. ______ 38. ______
39. ______ 40. ______
41. ______
42. ______

17. $16^{1/2} = 4$

18. $125^{1/3} = 5$

19. $64^{-1/3} = \frac{1}{4}$

20. $36^{-1/2} = \frac{1}{6}$

21. $27^{2/3} = 9$

22. $9^{3/2} = 27$

23. $27^{-2/3} = \frac{1}{9}$

24. $16^{-3/2} = \frac{1}{64}$

Convert each statement to exponential form.

25. $\log_2 16 = 4$

26. $\log_5 5 = 1$

27. $\log_5 1 = 0$

28. $\log_3 27 = 3$

29. $\log_{10} 10 = 1$ > Videos

30. $\log_2 32 = 5$

31. $\log_5 125 = 3$

32. $\log_{10} 1 = 0$

33. $\log_3 \frac{1}{27} = -3$

34. $\log_5 \frac{1}{25} = -2$

35. $\log_{10} 0.001 = -3$

36. $\log_{10} \frac{1}{1{,}000} = -3$

37. $\log_{16} 4 = \frac{1}{2}$

38. $\log_{125} 5 = \frac{1}{3}$

39. $\log_8 4 = \frac{2}{3}$

40. $\log_9 27 = \frac{3}{2}$

41. $\log_{25} \frac{1}{5} = -\frac{1}{2}$

42. $\log_{64} \frac{1}{16} = -\frac{2}{3}$

< Objective 3 >

Evaluate each logarithm.

43. $\log_2 64$

44. $\log_3 81$

45. $\log_4 64$ > Videos

46. $\log_{10} 10{,}000$

47. $\log_3 \frac{1}{81}$

48. $\log_4 \frac{1}{64}$

49. $\log_{10} \frac{1}{100}$

50. $\log_5 \frac{1}{25}$

51. $\log_{25} 5$

52. $\log_{27} 3$

< Objective 4 >

Solve each equation.

53. $y = \log_5 25$

54. $\log_2 x = 4$

55. $\log_b 256 = 4$

56. $y = \log_3 1$

57. $\log_{10} x = 2$

58. $\log_b 125 = 3$

59. $y = \log_5 5$

60. $y = \log_3 81$

61. $\log_{3/2} x = 3$

62. $\log_b \frac{4}{9} = 2$

63. $\log_b \frac{1}{25} = -2$ > Videos

64. $\log_3 x = -3$

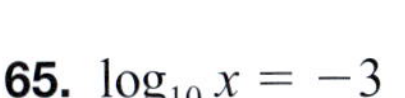

65. $\log_{10} x = -3$ > Videos

66. $y = \log_2 \frac{1}{16}$

67. $y = \log_8 \frac{1}{64}$

68. $\log_b \frac{1}{100} = -2$

69. $\log_{27} x = \frac{1}{3}$

70. $y = \log_{100} 10$

71. $\log_b 5 = \frac{1}{2}$

72. $\log_{64} x = \frac{2}{3}$

73. $y = \log_{27} \frac{1}{9}$

74. $\log_b \frac{1}{8} = -\frac{3}{4}$

Answers

43. ______ 44. ______

45. ______ 46. ______

47. ______ 48. ______

49. ______ 50. ______

51. ______ 52. ______

53. ______ 54. ______

55. ______ 56. ______

57. ______ 58. ______

59. ______ 60. ______

61. ______ 62. ______

63. ______ 64. ______

65. ______

66. ______

67. ______

68. ______

69. ______

70. ______

71. ______

72. ______

73. ______

74. ______

Basic Skills | **Challenge Yourself** | Calculator/Computer | Career Applications | Above and Beyond

Answers

75. __________
76. __________
77. __________
78. __________
79. __________
80. __________
81. __________
82. __________
83. __________

Determine whether each statement is **true** *or* **false.**

75. $m = \log_k n$ means the same as $k^n = m$.

76. The inverse of an exponential function is a logarithmic function.

Complete each statement with **never, sometimes,** *or* **always.**

77. The graph of $f(x) = \log_b x$ ________ has a vertical asymptote.

78. The graph of $f(x) = \log_b x$ ________ passes through the point (1, 0).

Use the decibel formula

$$L = 10 \log_{10} \frac{I}{I_0}$$

where $I_0 = 10^{-16}$ W/cm² *to solve each problem.*

79. Science and Medicine A television commercial has a volume with intensity $I = 10^{-11}$ W/cm². Find its rating in decibels.

80. Science and Medicine The sound of a jet plane on takeoff has an intensity $I = 10^{-2}$ W/cm². Find its rating in decibels.

81. Science and Medicine The sound of a computer printer has an intensity of $I = 10^{-9}$ W/cm². Find its rating in decibels.

82. Science and Medicine The sound of a busy street has an intensity of $I = 10^{-8}$ W/cm². Find its rating in decibels.

The formula for the decibel rating L can be solved for the intensity of the sound as $I = 10^{-16} \cdot 10^{L/10}$. *Use this formula in exercises 83 to 87.*

83. Science and Medicine Find the intensity of the sound in an airport waiting area if the decibel rating is 80.

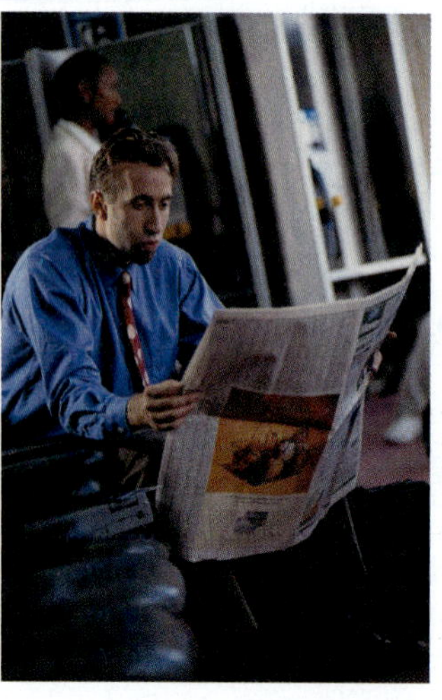

84. **Science and Medicine** Find the intensity of the sound of conversation in a crowded room if the decibel rating is 70.

85. **Science and Medicine** What is the ratio of intensity of a sound of 80 dB to that of 70 dB?

86. **Science and Medicine** What is the ratio of intensity of a sound of 60 dB to one measuring 40 dB?

87. **Science and Medicine** What is the ratio of intensity of a sound of 70 dB to one measuring 40 dB?

88. Derive the formula for intensity provided above. (*Hint*: First divide both sides of the decibel formula by 10. Then write the equation in exponential form.)

Solve exercises 89 to 92 by using the earthquake formula

$$M = \log_{10} \frac{a}{a_0}$$

89. **Technology** An earthquake has an intensity a of $10^6 \cdot a_0$, where a_0 is the intensity of the zero-level earthquake. What was its magnitude?

90. **Technology** The great San Francisco earthquake of 1906 had an intensity of $10^{8.3} \cdot a_0$. What was its magnitude?

91. **Technology** An earthquake can begin causing damage to buildings with a magnitude of 5 on the Richter scale. Find its intensity in terms of a_0.

92. **Technology** An earthquake may cause moderate building damage with a magnitude of 6 on the Richter scale. Find its intensity in terms of a_0.

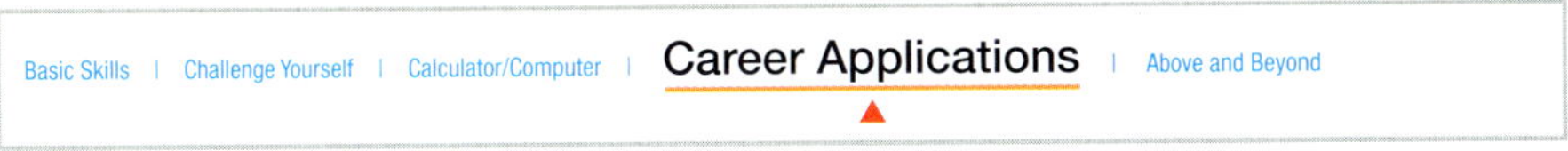

Allied Health *The molar concentration of hydrogen ions $[H^+]$ in an aqueous solution is equal to the product of the normality N and the percent ionization (in decimal form). The acidity level, or pH, of a solution is a function of the molar concentration and is given by the formula*

$$\text{pH} = -\log([H^+])$$

Use this information to complete exercises 93 and 94.

Hints: When "log" is written without a base, we always assume the base is 10. The LOG key on your graphing calculator is the log base 10 function. To find $\log_{10} 42$, for example, press LOG 42) ENTER. The result should be 1.623, to the nearest thousandth.

93. Determine the pH (to the nearest thousandth) of a 0.15-N acid solution that is 65% ionized.

Answers

84. ______

85. ______

86. ______

87. ______

88. ______

89. ______

90. ______

91. ______

92. ______

93. ______

Answers

94. ______

95. ______

96. ______

97. ______

98. ______

94. Find the pH (to the nearest tenth) of a chemical that contains 3.8×10^{-4} moles of H^+ per liter.

95. **Electrical Engineering** One formula for sound level, in decibels, is

$$\text{db} = 10 \log_{10}\left(\frac{I}{I_0}\right)$$

in which I is the sound intensity, in watts per sq m, and I_0 is the base sound level, 10^{-12} W/m^2 (the lowest sound discernible to most people).

Find the decibel level in a factory in which the sound intensity is 3.2×10^{-9} W/m^2. Report your result to the nearest decibel (dB).

96. **Construction Technology** The strength of concrete depends on its curing time. The longer it takes to cure, the stronger the concrete will be. If the desired strength of the concrete is known (to a maximum of 3,000 psi after 48 h), the required curing time is found using the formula

$$t = 48 \log_e\left(\frac{s}{3{,}000}\right) + 48$$

in which s is the desired strength and t is measured in hours.

Determine the curing time necessary to produce concrete with a strength of 2,241 psi.

Hint: When "log" is written with base e, this may be calculated by using the [LN] key on your graphing calculator. To find $\log_e 42$, for example, press [LN] 42 [ )] [ENTER]. The result should be 3.738, to the nearest thousandth.

Basic Skills | Challenge Yourself | Calculator/Computer | Career Applications | **Above and Beyond**

97. The **learning curve** describes the relationship between learning and time. Its graph is a logarithmic curve in the first quadrant. Describe that curve as it relates to learning.

98. In what other scientific fields would you expect to encounter a discussion of logarithms?

The half-life *of a radioactive substance is the time it takes for one-half of the original amount of the substance to decay to a nonradioactive element. The half-life of radioactive waste is very important in figuring how long the waste must be kept isolated from the environment in some sort of storage facility. Half-lives of various radioactive waste products vary from a few seconds to millions of years. It usually takes at least 10 half-lives for a radioactive waste product to be considered safe.*

The half-life of a radioactive substance can be determined by the formula

$$\log_e \frac{1}{2} = -\lambda x$$

where λ = radioactive decay constant

x = half-life

Approximate the half-lives of these important radioactive waste products, given the radioactive decay constant (RDC). Report your results to the nearest year. (To compute a logarithm base e on your calculator, see the hint in exercise 96.)

99. Plutonium-239, RDC = 0.000029

100. Strontium-90, RDC = 0.024755

101. Thorium-230, RDC = 0.000009

102. Cesium-135, RDC = 0.00000035

103. How many years will it be before each waste product is considered safe?

104. **(a)** Evaluate $\log_2 (4 \cdot 8)$. **(b)** Evaluate $\log_2 4$ and $\log_2 8$. **(c)** Write an equation that connects the result of part (a) with the results of part (b).

105. **(a)** Evaluate $\log_3 (9 \cdot 81)$. **(b)** Evaluate $\log_3 9$ and $\log_3 81$. **(c)** Write an equation that connects the result of part (a) with the results of part (b).

106. Based on exercises 104 and 105, propose a statement that connects $\log_a(mn)$ with $\log_a m$ and $\log_a n$.

Answers

99. ______

100. ______

101. ______

102. ______

103. ______

104. ______

105. ______

106. ______

Answers

1. 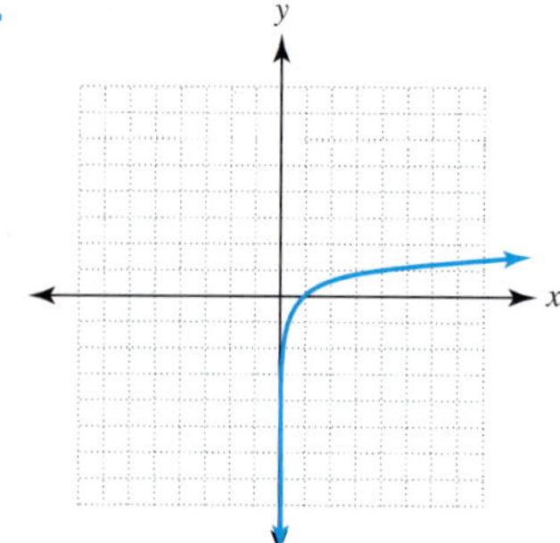

3. 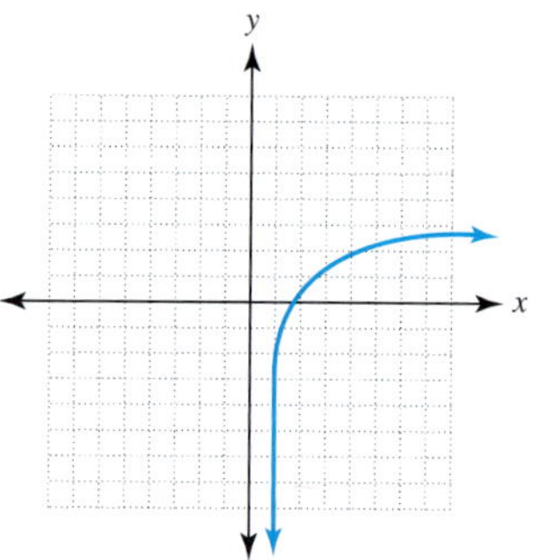

5. 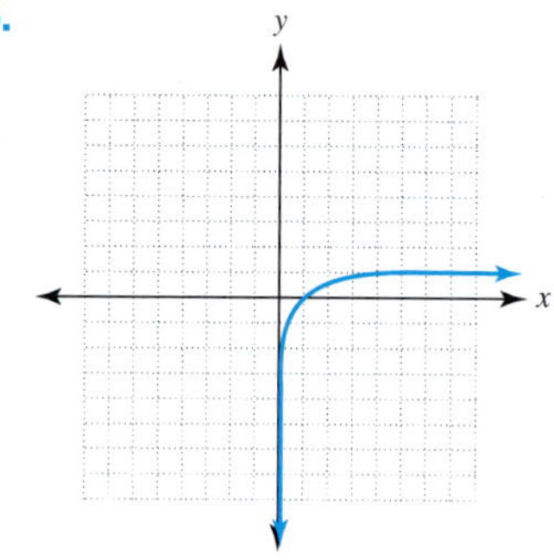

7. $\log_2 32 = 5$ **9.** $\log_{10} 100 = 2$ **11.** $\log_3 1 = 0$ **13.** $\log_4 \frac{1}{16} = -2$
15. $\log_{10} \frac{1}{100} = -2$ **17.** $\log_{16} 4 = \frac{1}{2}$ **19.** $\log_{64} \frac{1}{4} = -\frac{1}{3}$
21. $\log_{27} 9 = \frac{2}{3}$ **23.** $\log_{27} \frac{1}{9} = -\frac{2}{3}$ **25.** $2^4 = 16$ **27.** $5^0 = 1$
29. $10^1 = 10$ **31.** $5^3 = 125$ **33.** $3^{-3} = \frac{1}{27}$ **35.** $10^{-3} = 0.001$
37. $16^{1/2} = 4$ **39.** $8^{2/3} = 4$ **41.** $25^{-1/2} = \frac{1}{5}$ **43.** 6 **45.** 3
47. -4 **49.** -2 **51.** $\frac{1}{2}$ **53.** $\{2\}$ **55.** $\{4\}$ **57.** $\{100\}$
59. $\{1\}$ **61.** $\left\{\frac{27}{8}\right\}$ **63.** $\{5\}$ **65.** $\left\{\frac{1}{1{,}000}\right\}$ **67.** $\{-2\}$
69. $\{3\}$ **71.** $\{25\}$ **73.** $\left\{-\frac{2}{3}\right\}$ **75.** False **77.** always
79. 50 dB **81.** 70 dB **83.** 10^{-8} W/cm^2 **85.** 10 **87.** 1,000
89. 6 **91.** $10^5 \cdot a_0$ **93.** 1.011 **95.** 35 dB
97. Above and Beyond **99.** 23,902 yr **101.** 77,016 yr
103. Pu-239: 239,020 yr; Sr-90: 280 yr; Th-230: 770,160 yr; Cs-135: 19,804,210 yr
105. **(a)** 6; **(b)** 2, 4; **(c)** $\log_3 (9 \cdot 81) = \log_3 9 + \log_3 81$

10.6 Properties of Logarithms

< 10.6 Objectives >

1 > Apply the properties of logarithms

2 > Evaluate logarithmic expressions with any base

3 > Solve applications involving logarithms

4 > Estimate the value of an antilogarithm

In this section we develop and use the properties of logarithms. These properties are applied in a variety of areas that lead to exponential or logarithmic equations.

Since a logarithm is an exponent, it seems reasonable that our knowledge of the properties of exponents should lead to useful properties for logarithms. That is, in fact, the case.

We start with two basic facts that follow immediately from the definition of the logarithm.

Property

Properties of Logarithms

For $b > 0$ and $b \neq 1$,

1. $\log_b b = 1$ Since $b^1 = b$

2. $\log_b 1 = 0$ Since $b^0 = 1$

We know that the logarithmic function $y = \log_b x$ and the exponential function $y = b^x$ are inverses of each other. So, for $f(x) = b^x$, we have $f^{-1}(x) = \log_b x$.

For any one-to-one function f,

$$f^{-1}(f(x)) = x \quad \text{for any } x \text{ in domain of } f$$

and $$f(f^{-1}(x)) = x \quad \text{for any } x \text{ in domain of } f^{-1}$$

NOTE

The inverse "undoes" what f does to x.

Since $f(x) = b^x$ is a one-to-one function, we can apply these results to the case where

$$f(x) = b^x \quad \text{and} \quad f^{-1}(x) = \log_b x$$

to derive some additional properties.

Property

Properties of Logarithms

3. $\log_b b^x = x$

4. $b^{\log_b x} = x$ for $x > 0$

Since logarithms are exponents, we can again turn to the familiar exponent rules to derive some further properties of logarithms.

We know that

$$\log_b M = x \quad \text{if and only if} \quad b^x = M$$

and $$\log_b N = y \quad \text{if and only if} \quad b^y = N$$

Then $M \cdot N = b^x \cdot b^y = b^{x+y}$

From this last equation we see that $x + y$ is the power to which we must raise b to get the product MN. In logarithmic form, that becomes

$$\log_b MN = x + y$$

Now, since $x = \log_b M$ and $y = \log_b N$, we can substitute and write

$$\log_b MN = \log_b M + \log_b N$$

This is the first of the basic logarithmic properties presented here. The remaining properties may all be proved by arguments similar to those presented above.

Property

Properties of Logarithms

NOTE

In all cases, $M, N > 0$, $b > 0$, $b \neq 1$, and $p \neq 0$.

Product Property

$$\log_b MN = \log_b M + \log_b N$$

Quotient Property

$$\log_b \frac{M}{N} = \log_b M - \log_b N$$

Power Property

$$\log_b M^p = p \log_b M$$

Many applications of logarithms require using these properties to write a single logarithmic expression as the sum or difference of simpler expressions, as Example 1 illustrates.

Example 1 Using the Properties of Logarithms

< Objective 1 >

RECALL

$\sqrt{a} = a^{1/2}$

Use the properties of logarithms to expand each expression.

(a) $\log_b xy = \log_b x + \log_b y$ Product property

(b) $\log_b \frac{xy}{z} = \log_b xy - \log_b z$ Quotient property

$= \log_b x + \log_b y - \log_b z$ Product property

(c) $\log_{10} x^2y^3 = \log_{10} x^2 + \log_{10} y^3$ Product property

$= 2 \log_{10} x + 3 \log_{10} y$ Power property

(d) $\log_b \sqrt{\frac{x}{y}} = \log_b \left(\frac{x}{y}\right)^{1/2}$ Definition of exponent

$= \frac{1}{2} \log_b \frac{x}{y}$ Power property

$= \frac{1}{2} (\log_b x - \log_b y)$ Quotient property

Check Yourself 1

Expand each expression, using the properties of logarithms.

(a) $\log_b x^2y^3z$ **(b)** $\log_{10} \sqrt{\frac{xy}{z}}$

In some cases, we reverse the process and use the properties to write a single logarithm, given a sum or difference of logarithmic expressions.

Example 2 Rewriting Logarithmic Expressions

Write each expression as a single logarithm with coefficient 1.

(a) $2 \log_b x + 3 \log_b y$

$= \log_b x^2 + \log_b y^3$ Power property

$= \log_b x^2y^3$ Product property

(b) $5 \log_{10} x + 2 \log_{10} y - \log_{10} z$

$= \log_{10} x^5y^2 - \log_{10} z$

$= \log_{10} \dfrac{x^5y^2}{z}$ Quotient property

(c) $\dfrac{1}{2}(\log_2 x - \log_2 y)$

$= \dfrac{1}{2}\left(\log_2 \dfrac{x}{y}\right)$

$= \log_2 \left(\dfrac{x}{y}\right)^{1/2}$ Power property

$= \log_2 \sqrt{\dfrac{x}{y}}$

Check Yourself 2

Write each expression as a single logarithm with coefficient 1.

(a) $3 \log_b x + 2 \log_b y - 2 \log_b z$ **(b)** $\frac{1}{3}(2 \log_2 x - \log_2 y)$

Example 3 illustrates the basic concept of the use of logarithms as a computational aid.

Example 3 Approximating Logarithms Using Properties

< Objective 2 >

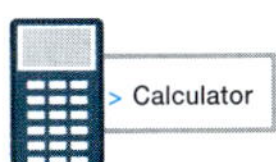

NOTES

We wrote the logarithms correct to three decimal places and will follow this practice throughout the remainder of this chapter.

Keep in mind, however, that this is an approximation and that $10^{0.301}$ only approximates 2. Verify this with your calculator.

Suppose $\log_{10} 2 = 0.301$ and $\log_{10} 3 = 0.477$. Evaluate, as indicated.

(a) $\log_{10} 6$

Since $6 = 2 \cdot 3$,

$\log_{10} 6 = \log_{10} (2 \cdot 3)$
$= \log_{10} 2 + \log_{10} 3$
$= 0.301 + 0.477$
$= 0.778$

(b) $\log_{10} 18$

Since $18 = 2 \cdot 3 \cdot 3$,

$\log_{10} 18 = \log_{10} (2 \cdot 3 \cdot 3)$
$= \log_{10} 2 + \log_{10} 3 + \log_{10} 3$
$= 1.255$

(c) $\log_{10} \frac{1}{9}$

Since $\frac{1}{9} = \frac{1}{3^2}$,

$$\log_{10} \frac{1}{9} = \log_{10} \frac{1}{3^2}$$

$$= \log_{10} 1 - \log_{10} 3^2$$ $\log_b 1 = 0$ for any base b.

$$= 0 - 2 \log_{10} 3$$

$$= -0.954$$

NOTE

Verify each answer with your calculator.

```
log(1/9)
     -.9542425094
log(16)
      1.204119983
log(√(3))
      .2385606274
```

(d) $\log_{10} 16$

Since $16 = 2^4$,

$$\log_{10} 16 = \log_{10} 2^4 = 4 \log_{10} 2$$

$$= 1.204$$

(e) $\log_{10} \sqrt{3}$

Since $\sqrt{3} = 3^{1/2}$,

$$\log_{10} \sqrt{3} = \log_{10} 3^{1/2} = \frac{1}{2} \log_{10} 3$$

$$= 0.239$$

Check Yourself 3

Given the values for $\log_{10} 2$ and $\log_{10} 3$, evaluate as indicated.

(a) $\log_{10} 12$ **(b)** $\log_{10} 27$ **(c)** $\log_{10} \sqrt[3]{2}$

When "log" is written without a base, we always assume the base is 10. The LOG key on your calculator is the log base 10 function. To find $\log_{10} 16$, for example, press LOG 16) ENTER. The result should be 1.204, to the nearest thousandth. There are in fact two logarithm functions built into your graphing calculator, both of which are frequently used in mathematics.

Logarithms to base 10

Logarithms to base e

Of course, logarithms to base 10 are convenient because our number system has base 10. We call logarithms to base 10 **common logarithms,** and it is customary to omit the base in writing a common (or base-10) logarithm. So

Definition

The Common Logarithm, log

$\log N$ means $\log_{10} N$

> **NOTE**
>
> When no base for a log is written, it is assumed to be 10.

The table shows the common logarithms for various powers of 10.

Exponential Form	Logarithmic Form
$10^3 = 1{,}000$	$\log 1{,}000 = 3$
$10^2 = 100$	$\log 100 = 2$
$10^1 = 10$	$\log 10 = 1$
$10^0 = 1$	$\log 1 = 0$
$10^{-1} = 0.1$	$\log 0.1 = -1$
$10^{-2} = 0.01$	$\log 0.01 = -2$
$10^{-3} = 0.001$	$\log 0.001 = -3$

Example 4 Approximating Logarithms with a Calculator

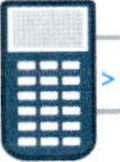

> **NOTE**
>
> The number 4.8 lies between 1 and 10, so log 4.8 lies between 0 and 1.

Verify each with a calculator.

(a) $\log 4.8 = 0.681$

(b) $\log 48 = 1.681$

(c) $\log 480 = 2.681$

(d) $\log 4{,}800 = 3.681$

(e) $\log 0.48 = -0.319$

```
log(4.8)
          .6812412374
log(48)
          1.681241237
log(480)
          2.681241237
```

```
log(4800)
          3.681241237
log(0.48)
          -.3187587626
```

> **NOTES**
>
> $480 = 4.8 \times 10^2$
>
> and
>
> $\log(4.8 \times 10^2)$
> $= \log 4.8 + \log 10^2$
> $= \log 4.8 + 2$
> $= 2 + \log 4.8$
>
> The value of log 0.48 is really $-1 + 0.681$. Your calculator combines the signed numbers.

Check Yourself 4

Use your calculator to evaluate each logarithm, rounded to three decimal places.

(a) log 2.3 **(b)** log 23 **(c)** log 230
(d) log 2,300 **(e)** log 0.23 **(f)** log 0.023

Now we look at an application of common logarithms from chemistry. Common logarithms are used to define the pH of a solution. This is a scale that measures whether a solution is acidic or basic.

The pH of a solution is defined as

$$\text{pH} = -\log\,[\text{H}^+]$$

where $[\text{H}^+]$ is the hydrogen ion concentration, in moles per liter (mol/L), in the solution.

Example 5 A Chemistry Application

< Objective 3 >

> **NOTES**
>
> A solution with pH = 7 is neutral. It is **acidic** if the pH is less than 7 and **basic** if the pH is greater than 7.
>
> In general, $\log_b b^x = x$, so $\log 10^{-7} = -7$.

Find the pH of each substance. Determine whether each is a base or an acid.

(a) Rainwater: $[\text{H}^+] = 1.6 \times 10^{-7}$

From the definition,

$$\begin{aligned}\text{pH} &= -\log\,[\text{H}^+] \\ &= -\log\,(1.6 \times 10^{-7}) && \text{Use the product rule.} \\ &= -(\log 1.6 + \log 10^{-7}) \\ &\approx -[0.204 + (-7)] \\ &\approx -(-6.796) = 6.796\end{aligned}$$

Rain is slightly acidic.

(b) Household ammonia: $[H^+] = 2.3 \times 10^{-8}$

$$\begin{aligned} pH &= -\log(2.3 \times 10^{-8}) \\ &= -(\log 2.3 + \log 10^{-8}) \\ &\approx -[0.362 + (-8)] \\ &\approx 7.638 \end{aligned}$$

Ammonia is slightly basic.

(c) Vinegar: $[H^+] = 2.9 \times 10^{-3}$

$$\begin{aligned} pH &= -\log(2.9 \times 10^{-3}) \\ &= -(\log 2.9 + \log 10^{-3}) \\ &\approx 2.538 \end{aligned}$$

Vinegar is very acidic.

Check Yourself 5

Find the pH for each solution. Are they acidic or basic?

(a) Orange juice: $[H^+] = 6.8 \times 10^{-5}$
(b) Drain cleaner: $[H^+] = 5.2 \times 10^{-13}$

Many applications require reversing the process. That is, given the logarithm of a number, we must be able to find that number. The process is straightforward.

Example 6 Solving a Logarithmic Equation

< Objective 4 >

> Calculator

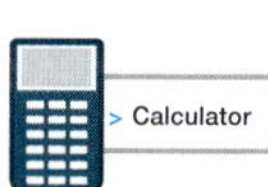

Suppose that $\log x = 2.1567$. We want to find a number x whose logarithm is 2.1567.

Rewriting in exponential form,

$\log_{10} x = 2.1567$ is equivalent to $10^{2.1567} = x$

On your graphing calculator, note that the inverse function for [LOG] is $[10^x]$. So, you can either press

[2nd] $[10^x]$ 2.1567 [)] [ENTER]

or, you can directly type

10 [^] 2.1567 [ENTER]

Both give the result 143.450, rounded to the nearest thousandth, sometimes called the **antilogarithm** of 2.1567.

It is important to keep in mind that $y = \log x$ and $y = 10^x$ are inverse functions.

Check Yourself 6

In each case, find x to the nearest thousandth.

(a) $\log x = 0.828$
(b) $\log x = 1.828$
(c) $\log x = 2.828$
(d) $\log x = -0.172$

Now we return to a chemistry application that requires us to find an antilogarithm.

Example 7 A Chemistry Application

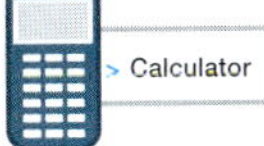
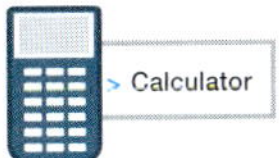

Calculator

```
10^(-6.2)
    6.309573445E-7
```

Suppose that the pH for tomato juice is 6.2. Find the hydrogen ion concentration $[H^+]$.

Recall from our earlier formula that

$$pH = -\log [H^+]$$

In this case, we have

$$6.2 = -\log [H^+]$$

or

$$\log [H^+] = -6.2$$

The desired value for $[H^+]$ is the antilogarithm of -6.2. To find $[H^+]$, type [2nd] [10^x] [(-)] 6.2 [)] [ENTER].

The result is 0.00000063, and we can write

$$[H^+] = 6.3 \times 10^{-7}$$

Check Yourself 7

The pH for eggs is 7.8. Find $[H^+]$ for eggs.

NOTE

Natural logarithms are also called **Napierian logarithms** after Napier. The importance of this system of logarithms was not fully understood until later developments in the calculus.

As we mentioned, there are two systems of logarithms in common use. The second type of logarithm uses the number e as a base, and we call logarithms to base e **natural logarithms.** As with common logarithms, a convenient notation has developed.

Definition

The Natural Logarithm, ln

The **natural logarithm** is a logarithm to base e, and it is denoted $\ln x$, where

$\ln x = \log_e x$ The restrictions on the domain of the natural logarithmic function are the same as before. The function is defined only if $x > 0$.

Since $y = \ln x$ means $y = \log_e x$, we can easily convert this to $e^y = x$, which leads us directly to these facts.

$\ln 1 = 0$ Because $e^0 = 1$

$\ln e = 1$ Because $e^1 = e$

$\left.\begin{aligned} \ln e^2 &= 2 \\ \ln e^3 &= 3 \\ \ln e^{-5} &= -5 \end{aligned}\right\}$ Because $\ln e^x = x$

We want to emphasize the inverse relationship that exists between logarithmic functions and exponential functions.

Property

Inverse Functions

For any base b,

$\log_b b^x = x$ (for all real x)

$b^{\log_b x} = x$ (for $x > 0$)

So, for common logarithms,

$\log 10^x = x$

$10^{\log x} = x$

And, for natural logarithms,

$\ln e^x = x$

$e^{\ln x} = x$

Example 8 Using the Property of Inverses

NOTE

Each of these can be easily confirmed on a calculator. But you should learn to quickly recognize these forms.

Simplify.

(a) $\log 10^8 = 8$

(b) $\ln e^{-6} = -6$

(c) $10^{\log 7} = 7$

(d) $e^{\ln 4} = 4$

Check Yourself 8

Simplify.

(a) $\ln e^{1.2}$ **(b)** $10^{\log 4.5}$ **(c)** $\log 10^{-5}$ **(d)** $e^{\ln 3.7}$

Example 9 Approximating Logarithms with a Calculator

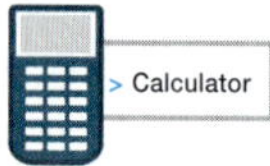

NOTE

```
ln(2)
      .6931471806
```

To evaluate natural logarithms, we use a calculator. To find the value of ln 2, use the sequence

[ln] 2 [)] [ENTER]

The result is 0.693 (to three decimal places).

Check Yourself 9

Use a calculator to evaluate each logarithm. Round to the nearest thousandth.

(a) ln 3 **(b)** ln 6 **(c)** ln 4 **(d)** $\ln \sqrt{3}$

Of course, the properties of logarithms are applied in the same way, no matter what the base.

Example 10 Approximating Logarithms Using Properties

RECALL

$\log_b MN = \log_b M + \log_b N$

$\log_b M^p = p \log_b M$

If ln 2 = 0.693 and ln 3 = 1.099, evaluate each logarithm.

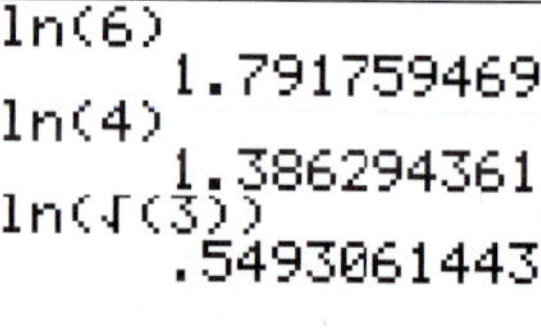

(a) $\ln 6 = \ln (2 \cdot 3) = \ln 2 + \ln 3 = 1.792$

(b) $\ln 4 = \ln 2^2 = 2 \ln 2 = 1.386$

(c) $\ln \sqrt{3} = \ln 3^{1/2} = \frac{1}{2} \ln 3 = 0.550$

Verify these results with your calculator.

Check Yourself 10

Use ln 2 = 0.693 and ln 3 = 1.099 to evaluate each logarithm.

(a) ln 12 **(b)** ln 27

It may also be necessary to find x, given $\ln x$. The key here is to remember that $y = \ln x$ and $y = e^x$ are inverse functions.

Example 11 Solving a Logarithmic Equation

Suppose that $\ln x = 4.1685$. We want to find a number x whose logarithm, base e, is 4.1685. Rewriting in exponential form,

$\ln x = 4.1685$ is equivalent to $e^{4.1685} = x$

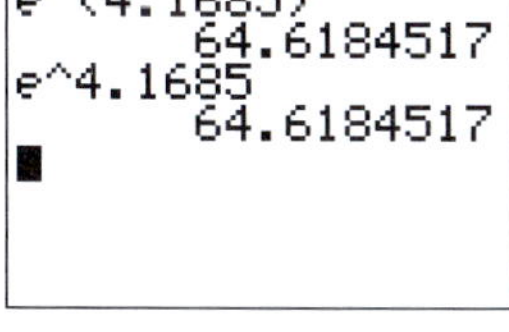

On your graphing calculator, note that the inverse function for [LN] is [e^x]. So, you can either press

[2nd] [e^x] 4.1685 [ )] [ENTER]

or, you can directly type

[2nd] [÷] [^] 4.1685 [ENTER]

Both give the result 64.618, rounded to the nearest thousandth.

Check Yourself 11

In each case, find x to the nearest thousandth.

(a) $\ln x = 2.065$ **(b)** $\ln x = -2.065$ **(c)** $\ln x = 7.293$

The natural logarithm function plays an important role in both theoretical and applied mathematics. Example 12 illustrates just one of the many applications of this function.

Example 12 A Learning Curve Application

RECALL

We read $S(t)$ as "S of t."

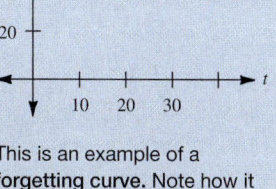

This is an example of a **forgetting curve.** Note how it drops more rapidly at first. Compare this curve to the learning curve drawn in Section 10.4, exercise 68.

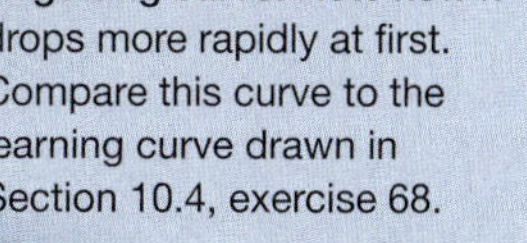

A class of students took a mathematics examination and received an average score of 76. In a psychological experiment, the students are retested at weekly intervals over the same material. If t is measured in weeks, then the new average score after t weeks is given by

$$S(t) = 76 - 5 \ln (t + 1)$$

(a) Find the score after 10 weeks.

$$S(10) = 76 - 5 \ln (10 + 1)$$
$$= 76 - 5 \ln 11 \approx 64$$

(b) Find the score after 20 weeks.

$$S(20) = 76 - 5 \ln (20 + 1) \approx 61$$

(c) Find the score after 30 weeks.

$$S(30) = 76 - 5 \ln (30 + 1) \approx 59$$

Check Yourself 12

The average score for a group of biology students, retested after time t (in months), is given by

$$S(t) = 83 - 9 \ln (t + 1)$$

Find the average score rounded to the nearest tenth after

(a) 3 months **(b)** 6 months

We conclude this section with one final property of logarithms. This property allows us to quickly find the logarithm of a number to any base. Although work with logarithms with bases other than 10 or e is relatively infrequent, the relationship between logarithms of different bases is interesting in itself.

Suppose we want to find $\log_2 5$. This means we want to find the power to which 2 should be raised to produce 5. Now, if there were a log base 2 function ($\log_2 x$) on the calculator, we could obtain this directly. But since there is not, we must take another approach. If we write

$$\log_2 5 = x$$

then we have $2^x = 5$.

Taking the logarithm to base 10 of both sides of the equation yields

$$\log 2^x = \log 5$$

or $\quad x \log 2 = \log 5 \qquad$ Use the power property of logarithms.

> CAUTION
>
> Do not cancel the logs.

Now, dividing both sides of this equation by log 2 gives

$$x = \frac{\log 5}{\log 2}$$

We can now find a value for x with the calculator. Dividing with the calculator log 5 by log 2, we get an approximate answer of 2.322.

> **NOTE**
>
> This says $2^{2.322} \approx 5$.

Since $x = \log_2 5$ and $x = \dfrac{\log 5}{\log 2}$, then

$$\log_2 5 = \frac{\log 5}{\log 2}$$

Before leaving this, note that when we took the logarithm (base 10) of both sides, we could also have taken the logarithm, base e, of both sides.

$$2^x = 5$$

$$\ln 2^x = \ln 5$$

$$x \ln 2 = \ln 5$$

$$x = \frac{\ln 5}{\ln 2} \approx 2.322$$

So, $\log_2 5 = \dfrac{\log_e 5}{\log_e 2} = \dfrac{\log_{10} 5}{\log_{10} 2}$.

Generalizing our result gives us the change-of-base formula.

Property

Change-of-Base Formula

For positive real numbers a and x,

$$\log_a x = \frac{\log_b x}{\log_b a}$$

The logarithm on the left side has base a while the logarithms on the right side have base b. This allows us to calculate the logarithm to base a of any positive number, using the corresponding logarithms to base b (or any other base), as Example 13 illustrates.

Example 13 Using the Change-of-Base Formula

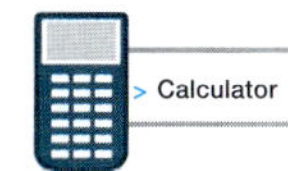

Find $\log_5 15$.

From the change-of-base formula with $a = 5$ and $b = 10$,

$$\log_5 15 = \frac{\log_{10} 15}{\log_{10} 5}$$

$$= 1.683$$

The graphing calculator sequence for the above computation is

[log] 15 [)] [÷] [log] 5 [)] [ENTER]

The result is 1.683, rounded to the nearest thousandth.

NOTES

We wrote $\log_{10} 15$ rather than log 15 to emphasize the change-of-base formula.

$\log_5 5 = 1$ and $\log_5 25 = 2$, so the result for $\log_5 15$ must be between 1 and 2.

We could choose base e so that $\log_5 15 = \frac{\ln 15}{\ln 5}$ instead.

```
log(15)/log(5)
      1.682606194
ln(15)/ln(5)
      1.682606194
```

Remember to close the parentheses in the numerator when entering these expressions into a calculator.

Check Yourself 13

Use the change-of-base formula to find $\log_8 32$.

> CAUTION

A couple of cautions are in order.

1. We cannot "cancel" logs. There is the temptation to write

$$\frac{\log 15}{\log 5} = \frac{15}{5} = 3$$

This is quite wrong!

2. There is also the temptation to write

$$\frac{\log 15}{\log 5} = \log 15 - \log 5$$

This is also quite wrong.

It is true that $\log\left(\frac{15}{5}\right) = \log 15 - \log 5$ (the Quotient Property), but this is very different from $\frac{\log 15}{\log 5}$. Be sure you note the difference.

Graphing Calculator Option

Applying Logarithmic Regression

A general form of logarithmic functions is available as a regression model in your graphing calculator: $y = a + b \ln x$.

Suppose we have collected some data that suggest a pattern of logarithmic growth. A sample scatterplot that exhibits this is:

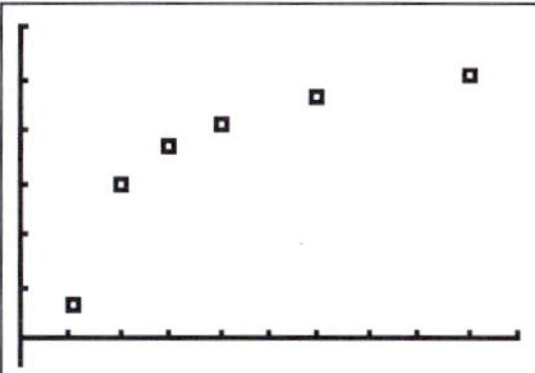

We notice relatively rapid growth for small values of x, followed by growth that seems to be slowing. Look at the data, which show how the time (in seconds) for a dropped tennis ball to complete its third bounce varies according to the height (in inches) of the drop.

Height of drop (in.)	40	45	50	55	60
Time to third bounce (s)	1.75	1.87	1.99	2.07	2.12

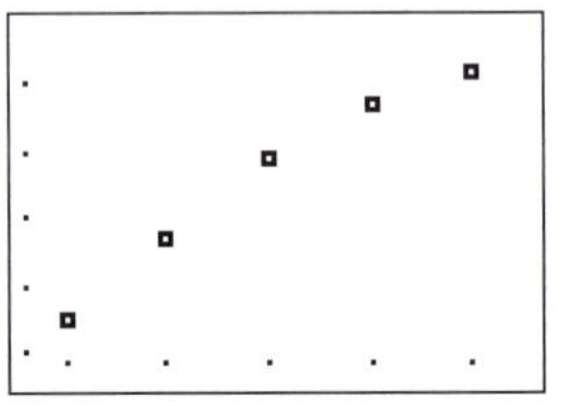

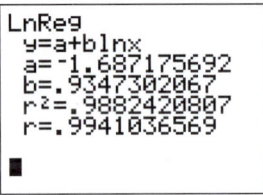

We plot the data, making a scatterplot:

Clear data lists [L1] and [L2]: STAT 4:ClrList 2nd [L1] , 2nd [L2] ENTER.

Enter the data into [L1] and [L2]: STAT 1:Edit, and type in the numbers. Exit the data editor: 2nd [QUIT].

Make the scatterplot: 2nd [STAT PLOT] ENTER; press "On"; for "Type" select the first icon; "Xlist" should say [L1] and "Ylist" should say [L2] ; for "Mark" choose the first symbol; press Y= and delete (or turn off) any existing equations; press ZOOM 9:ZoomStat. (To improve the scaling, go to WINDOW and choose appropriate numbers for Xscl and Yscl. Then GRAPH .)

```
LnReg
 y=a+blnx
 a=-1.687175692
 b=.9347302067
 r²=.9882420807
 r=.9941036569
```

To find the "best fitting" logarithmic function: STAT CALC 9:LnReg 2nd [L1] , 2nd [L2] ENTER. We have, accurate to four decimal places,

$$y = -1.6872 + 0.9347 \ln x$$

To view the graph of this function on the scatterplot, enter its equation on the Y= screen and press GRAPH .

Graphing Calculator Check

The table shows the systolic blood pressure p (in mm of Hg) for children of varying weights w (in pounds). Using your graphing calculator, apply logarithmic regression to fit a logarithmic function to these data. Round coefficients accurate to four decimal places.

Weight, w	44	61	81	113	131
Blood pressure, p	91	98	103	110	112

ANSWER

$p = 17.9243 + 19.3850 \ln w$

Check Yourself ANSWERS

1. **(a)** $2 \log_b x + 3 \log_b y + \log_b z$; **(b)** $\frac{1}{2}(\log_{10} x + \log_{10} y - \log_{10} z)$

2. **(a)** $\log_b \frac{x^3y^2}{z^2}$; **(b)** $\log_2 \sqrt[3]{\frac{x^2}{y}}$ **3.** **(a)** 1.079; **(b)** 1.431; **(c)** 0.100

4. **(a)** 0.362; **(b)** 1.362; **(c)** 2.362; **(d)** 3.362; **(e)** −0.638; **(f)** −1.638

5. **(a)** 4.167, acidic; **(b)** 12.284, basic **6.** **(a)** 6.730; **(b)** 67.298; **(c)** 672.977; **(d)** 0.673 **7.** $[H^+] = 1.6 \times 10^{-8}$ **8.** **(a)** 1.2; **(b)** 4.5; **(c)** −5; **(d)** 3.7

9. **(a)** 1.099; **(b)** 1.792; **(c)** 1.386; **(d)** 0.549 **10.** **(a)** 2.485; **(b)** 3.297

11. **(a)** 7.885; **(b)** 0.127; **(c)** 1,469.974 **12.** **(a)** 70.5; **(b)** 65.5

13. $\log_8 32 = \frac{\log 32}{\log 8} \approx 1.667$

Reading Your Text

SECTION 10.6

(a) By definition, a logarithm is an ______________.

(b) The logarithmic property $\log_b M^p = p \log_b M$ is called the ____________ property.

(c) We call logarithms to the base 10 ______________ logarithms.

(d) A solution whose pH = 7 is ______________.

10.6 exercises

Basic Skills | Challenge Yourself | Calculator/Computer | Career Applications | Above and Beyond

Boost *your* GRADE at ALEKS.com!

ALEKS

- Practice Problems
- Self-Tests
- NetTutor
- e-Professors
- Videos

Name ____________

Section ________ Date ________

Answers

1. ____________
2. ____________
3. ____________
4. ____________
5. ____________
6. ____________
7. ____________
8. ____________
9. ____________
10. ____________
11. ____________
12. ____________
13. ____________
14. ____________
15. ____________
16. ____________
17. ____________
18. ____________

< Objective 1 >

Use the properties of logarithms to expand each expression.

1. $\log_b 5x$

2. $\log_3 7x$

3. $\log_6 \dfrac{x}{7}$

4. $\log_b \dfrac{2}{y}$

5. $\log_3 a^2$

6. $\log_5 y^4$

7. $\log_5 \sqrt{x}$

8. $\log \sqrt[3]{z}$

9. $\log_b x^2y^4$

10. $\log_7 x^3z^2$

11. $\log_4 y^2\sqrt{x}$

12. $\log_b x^3\sqrt[3]{z}$

13. $\log_b \dfrac{x^2y}{z}$

14. $\log_5 \dfrac{3}{xy}$

15. $\log \dfrac{xy^2}{\sqrt{z}}$ 

16. $\log_4 \dfrac{x^3\sqrt{y}}{z^2}$

17. $\log_5 \sqrt[3]{\dfrac{xy}{z^2}}$

18. $\log_b \sqrt[4]{\dfrac{x^2y}{z^3}}$

Write each expression as a single logarithm.

19. $\log_b x + \log_b y$

20. $\log_5 x - \log_5 y$

21. $3 \log_5 x - 2 \log_5 y$

22. $3 \log_b x + \log_b z$

23. $\log_b x + \frac{1}{2} \log_b y$

24. $\frac{1}{2} \log_b x - 3 \log_b z$

25. $\log_b x - 2 \log_b y - \log_b z$

26. $2 \log_5 x - (3 \log_5 y + \log_5 z)$

27. $\frac{1}{2} \log_6 y - 3 \log_6 z$ > Videos

28. $\log_b x - \frac{1}{3} \log_b y - 4 \log_b z$

29. $\frac{1}{3}(2 \log_b x + \log_b y - \log_b z)$

30. $\frac{1}{5}(2 \log_4 x - \log_4 y + 3 \log_4 z)$

< Objective 2 >

Given that $\log 2 = 0.301$ *and* $\log 3 = 0.477$, *find each logarithm.*

31. $\log 24$

32. $\log 36$

33. $\log 8$

34. $\log 81$

35. $\log \sqrt{2}$ 

36. $\log \sqrt[3]{3}$

37. $\log \frac{1}{4}$

38. $\log \frac{1}{27}$

Basic Skills | **Challenge Yourself** | Calculator/Computer | Career Applications | Above and Beyond

Determine whether each statement is **true** *or* **false**.

39. $\log_m x^n = n \log_m x$

40. $(\log_m x) \cdot (\log_m y) = \log_m xy$

41. $\log_m m = 0$

42. $\log_m x - \log_m y = \log_m \frac{x}{y}$

In exercises 43 to 46, simplify.

43. $10^{\log 8.2}$

44. $\log 10^{-1.3}$

45. $\ln e^{5.8}$

46. $e^{\ln 2.6}$

Answers

19. ______
20. ______
21. ______
22. ______
23. ______
24. ______
25. ______
26. ______
27. ______
28. ______
29. ______
30. ______
31. ______ 32. ______
33. ______ 34. ______
35. ______ 36. ______
37. ______ 38. ______
39. ______ 40. ______
41. ______ 42. ______
43. ______ 44. ______
45. ______ 46. ______

10.6 exercises

Answers

47. ______
48. ______
49. ______
50. ______
51. ______
52. ______
53. ______ 54. ______
55. ______ 56. ______
57. ______ 58. ______
59. ______ 60. ______
61. ______ 62. ______
63. ______ 64. ______
65. ______ 66. ______
67. ______ 68. ______
69. ______ 70. ______
71. ______ 72. ______
73. ______ 74. ______

Estimate each logarithm by "trapping" it between consecutive integers. To estimate $\log_4 52$, *we note that* $4^2 = 16$ *and* $4^3 = 64$, *so* $\log_4 52$ *must lie between 2 and 3.*

47. $\log_3 25$

48. $\log_5 30$

49. $\log_2 70$

50. $\log_2 19$

51. $\log 680$

52. $\log 6{,}800$

Without a calculator, use the properties of logarithms to evaluate each expression.

53. $\log 5 + \log 2$

54. $\log 25 + \log 4$

55. $\log_3 45 - \log_3 5$

56. $10 \log_4 2$

Basic Skills | Challenge Yourself | **Calculator/Computer** | Career Applications | Above and Beyond

Use your calculator to find each logarithm.

57. $\log 7.3$

58. $\log 68$

59. $\log 680$

60. $\log 6{,}800$

61. $\log 0.72$

62. $\log 0.068$

63. $\ln 2$

64. $\ln 3$

65. $\ln 10$

66. $\ln 30$

< Objective 4 >

Solve for x. Round to the nearest thousandth.

67. $\log x = 0.749$

68. $\log x = 1.749$

69. $\log x = 3.749$

70. $\log x = -0.251$

71. $\ln x = 1.238$

72. $\ln x - 3.141$

73. $\ln x = -0.786$

74. $\ln x = -3.141$

< Objective 3 >

You are given the hydrogen ion concentration $[H^+]$ *for each solution. Use the formula* $pH = -\log[H^+]$ *to find each* pH. *Are the solutions acidic or basic?*

75. Blood: $[H^+] = 3.8 \times 10^{-8}$

76. Lemon juice: $[H^+] = 6.4 \times 10^{-3}$

Given the pH of the solutions, approximate the hydrogen ion concentration $[H^+]$.

77. Wine: pH = 4.7 

78. Household ammonia: pH = 7.8

The average score on a final examination for a group of psychology students, retested after time t (in weeks), is given by

$$S = 85 - 8\ln(t + 1)$$

Find the average score on the retests.

79. After 3 weeks

80. After 12 weeks

Use the change-of-base formula to approximate each logarithm.

81. $\log_3 25$

82. $\log_5 30$ 

83. The table shows measurements taken for several trees of the same species. The measurements were diameter at the base, in centimeters (cm), and height, in meters (m). Using your graphing calculator, apply logarithmic regression to fit a logarithmic function to these data. Round coefficients accurate to four decimal places.

x (diameter)	2.6	4.6	9.8	14.5	15.8	27.0
y (height)	1.93	4.15	11.50	11.85	13.25	15.80

84. The table below shows measurements taken for several trees of the same species. The measurements were diameter at the base, in centimeters (cm), and crown width, in meters (m). Using your graphing calculator, apply logarithmic regression to fit a logarithmic function to these data. Round coefficients accurate to four decimal places.

x (diameter)	2.6	4.6	9.8	14.5	15.8	27.0
y (crown width)	0.5	1.6	3.6	3.7	4.0	6.5

Answers

75. ______

76. ______

77. ______

78. ______

79. ______

80. ______

81. ______

82. ______

83. ______

84. ______

Basic Skills | Challenge Yourself | Calculator/Computer | Career Applications | **Above and Beyond**

Answers

85. ______

86. ______

87. ______

88. ______

89. ______

90. ______

91. ______

The amount of a radioactive substance remaining after time t is given by

$$A = e^{\lambda t + \ln A_0}$$

where A is the amount remaining after time t, A_0 is the original amount of the substance, and λ is the radioactive decay constant. Assume t is measured in years.

85. How much plutonium-239 will remain after 50,000 years if 24 kg was originally stored? Plutonium-239 has a radioactive decay constant of -0.000029.

86. How much plutonium-241 will remain after 100 years if 52 kg was originally stored? Plutonium-241 has a radioactive decay constant of -0.053319.

87. How much strontium-90 was originally stored if after 56 years it is discovered that 15 kg still remains? Strontium-90 has a radioactive decay constant of -0.024755.

chapter 10 > Make the Connection

88. How much cesium-137 was originally stored if after 90 years it is discovered that 20 kg still remains? Cesium-137 has a radioactive decay constant of -0.023105.

chapter 10 > Make the Connection

89. Which keys on your calculator are function keys and which are operation keys? What is the difference?

90. How is the pH factor relevant to your selection of a hair-care product?

91. **(a)** Use the change-of-base formula to write $\log_3 8$ in terms of base-10 logarithms. Then use your calculator to find $\log_3 8$ rounded to three decimal places.

(b) Use the change-of-base formula to write $\log_3 8$ in terms of base-e logarithms. Then use your calculator to find $\log_3 8$ rounded to three decimal places.

(c) Compare your answers to parts (a) and (b).

Answers

1. $\log_b 5 + \log_b x$ **3.** $\log_6 x - \log_6 7$ **5.** $2 \log_3 a$ **7.** $\frac{1}{2} \log_5 x$

9. $2 \log_b x + 4 \log_b y$ **11.** $2 \log_4 y + \frac{1}{2} \log_4 x$

13. $2 \log_b x + \log_b y - \log_b z$ **15.** $\log x + 2 \log y - \frac{1}{2} \log z$

17. $\frac{1}{3}(\log_5 x + \log_5 y - 2 \log_5 z)$ **19.** $\log_b xy$ **21.** $\log_5 \frac{x^3}{y^2}$

23. $\log_b x\sqrt{y}$ **25.** $\log_b \frac{x}{y^2 z}$ **27.** $\log_6 \frac{\sqrt{y}}{z^3}$ **29.** $\log_b \sqrt[3]{\frac{x^2 y}{z}}$

31. 1.380 **33.** 0.903 **35.** 0.151 **37.** -0.602 **39.** True

41. False **43.** 8.2 **45.** 5.8 **47.** Between 2 and 3

49. Between 6 and 7 **51.** Between 2 and 3 **53.** 1 **55.** 2

57. 0.863 **59.** 2.833 **61.** -0.143 **63.** 0.693 **65.** 2.303

67. 5.610 **69.** 5,610.480 **71.** 3.449 **73.** 0.456 **75.** 7.42, basic

77. 2×10^{-5} **79.** 74 **81.** 2.930 **83.** $y = -4.2465 + 6.2220 \ln x$

85. 5.6 kg **87.** 60 kg **89.** Above and Beyond

91. **(a)** $\frac{\log 8}{\log 3}$, 1.893; **(b)** $\frac{\ln 8}{\ln 3}$, 1.893; **(c)** same

10.7 Logarithmic and Exponential Equations

< 10.7 Objectives >

1 > Solve a logarithmic equation

2 > Solve an exponential equation

3 > Solve an application involving an exponential equation

The properties of logarithms developed in Section 10.6 are necessary for solving equations involving logarithms and exponents. Our work in this section considers techniques to solve both types of equations. We start with a definition.

Definition

Logarithmic Equations

A **logarithmic equation** is an equation in which a variable is contained in a logarithmic expression.

We solved some simple examples in Section 10.5. Recall that to solve $\log_3 x = 4$ for x, we simply convert the logarithmic equation to exponential form. Here,

$$3^4 = x$$

so $\quad x = 81$

and $\{81\}$ is the solution set for the given equation.

Now, what if the logarithmic equation involves more than one logarithmic term? Example 1 illustrates how the properties of logarithms must then be applied.

Example 1 **Solving a Logarithmic Equation**

< Objective 1 >

NOTE

We apply the product rule for logarithms:

$\log_b M + \log_b N = \log_b MN$

Solve each logarithmic equation.

(a) $\log_5 x + \log_5 3 = 2$

The original equation can be written as

$$\log_5 3x = 2$$

Now, since only a single logarithm is involved, we write the equation in the equivalent exponential form.

$$5^2 = 3x$$

$$3x = 25$$

$$x = \frac{25}{3}$$

To check this, we substitute $\frac{25}{3}$ into the original equation.

$$\log_5\left(\frac{25}{3}\right) + \log_5(3) \stackrel{?}{=} 2$$

To proceed, we either simplify the left side using logarithm properties, or we convert these logarithms to base 10 logs or base e logs and use a calculator. To illustrate the latter,

$$\frac{\log\left(\frac{25}{3}\right)}{\log(5)} + \frac{\log(3)}{\log(5)} \stackrel{?}{=} 2$$

```
log(25/3)/log(5)
+log(3)/log(5)
                2
```

So $\left\{\frac{25}{3}\right\}$ is the solution set.

RECALL

Since no base is written, it is assumed to be 10.

(b) $\log x + \log(x - 3) = 1$

Write the equation as

$$\log x(x - 3) = 1$$

or $\quad 10^1 = x(x - 3)$

We now have

$$x^2 - 3x = 10$$

$$x^2 - 3x - 10 = 0$$

$$(x - 5)(x + 2) = 0$$

Possible solutions are $x = 5$ or $x = -2$.

Note that substitution of -2 into the original equation gives

$$\log(-2) + \log(-5) = 1$$

Since logarithms of negative numbers are *not* defined, -2 is an extraneous solution and we must reject it. Substituting 5 gives

$$\log 5 + \log(5 - 3) \stackrel{?}{=} 1$$

$$\log 5 + \log 2 \stackrel{?}{=} 1$$

On a calculator, this checks.

```
log(5)+log(2)
                1
```

The only solution for the original equation is 5.

NOTES

Checking possible solutions is particularly important here.

To check $x = -2$

```
log(-2)+log(-5)
```

```
ERR:NONREAL ANS
1:Quit
2:Goto
```

Check Yourself 1

Solve $\log_2 x + \log_2 (x + 2) = 3$ for x.

The quotient property is used in a similar fashion for solving logarithmic equations. Consider Example 2.

Example 2 Solving a Logarithmic Equation

NOTE

We apply the quotient rule for logarithms:

$\log_b M - \log_b N = \log_b \frac{M}{N}$

Solve each equation.

(a) $\log_5 x - \log_5 2 = 2$

Rewrite the original equation as

$$\log_5 \frac{x}{2} = 2$$

Now, $5^2 = \frac{x}{2}$

$$\frac{x}{2} = 25$$

$$x = 50$$

Check: $\log_5 50 - \log_5 2 \stackrel{?}{=} 2$

Using change-of-base,

$$\frac{\log 50}{\log 5} - \frac{\log 2}{\log 5} \stackrel{?}{=} 2$$

```
log(50)/log(5)-l
og(2)/log(5)
               2
```

The solution set is $\{50\}$.

(b) $\log_3 (x + 1) - \log_3 x = 3$

$$\log_3 \left(\frac{x+1}{x}\right) = 3$$

$$3^3 = \frac{x+1}{x}$$

$$27x = x + 1$$

$$26x = 1$$

$$x = \frac{1}{26}$$

To check this, we substitute $\frac{25}{3}$ into the original equation.

$$\log_5\left(\frac{25}{3}\right) + \log_5(3) \stackrel{?}{=} 2$$

To proceed, we either simplify the left side using logarithm properties, or we convert these logarithms to base 10 logs or base e logs and use a calculator. To illustrate the latter,

$$\frac{\log\left(\frac{25}{3}\right)}{\log(5)} + \frac{\log(3)}{\log(5)} \stackrel{?}{=} 2$$

```
log(25/3)/log(5)
+log(3)/log(5)
                2
```

RECALL

Since no base is written, it is assumed to be 10.

So $\left\{\frac{25}{3}\right\}$ is the solution set.

(b) $\log x + \log(x - 3) = 1$

Write the equation as

$\log x(x - 3) = 1$

or $10^1 = x(x - 3)$

We now have

$$x^2 - 3x = 10$$
$$x^2 - 3x - 10 = 0$$
$$(x - 5)(x + 2) = 0$$

Possible solutions are $x = 5$ or $x = -2$.

Note that substitution of -2 into the original equation gives

$$\log(-2) + \log(-5) = 1$$

Since logarithms of negative numbers are *not* defined, -2 is an extraneous solution and we must reject it. Substituting 5 gives

$$\log 5 + \log(5 - 3) \stackrel{?}{=} 1$$
$$\log 5 + \log 2 \stackrel{?}{=} 1$$

On a calculator, this checks.

```
log(5)+log(2)
               1
```

NOTES

Checking possible solutions is particularly important here.

To check $x = -2$

```
log(-2)+log(-5)
```

```
ERR:NONREAL ANS
1:Quit
2:Goto
```

The only solution for the original equation is 5.

This algorithm summarizes our work in solving logarithmic equations.

Step by Step

Solving Logarithmic Equations

Step 1 Use the properties of logarithms to combine terms containing logarithmic expressions into a single term.

Step 2 Write the equation formed in step 1 in exponential form.

Step 3 Solve for the indicated variable.

Step 4 Check your solutions to make sure that possible solutions do not result in the logarithms of negative numbers.

Now we look at **exponential equations,** which are equations in which the variable appears as an exponent.

We solved some elementary exponential equations in Section 10.4. In solving an equation such as

$3^x = 81$

we wrote the right-hand member as a power of 3, so that

$3^x = 3^4$

or $x = 4$

NOTE

Again, we want to write both sides as a power of the same base, here 3.

The technique here works only when both sides of the equation can be conveniently expressed as powers of the same base. If that is not the case, we use logarithms to solve the equation, as illustrated in Example 4.

Example 4 Solving an Exponential Equation

< Objective 2 >

Solve $3^x = 5$, rounded to the nearest thousandth.

We begin by taking the common logarithm of both sides of the original equation.

RECALL

If $M = N$
then $\log_b M = \log_b N$

$\log 3^x = \log 5$

Now we apply the power property so that the variable becomes a coefficient on the left.

$x \log 3 = \log 5$

> CAUTION

This is *not* log 5 − log 3, a common error.

Dividing both sides of the equation by log 3 isolates x, and we have

$$x = \frac{\log 5}{\log 3}$$ Remember: "logs" *cannot* be canceled!

$$\approx 1.465$$ (to three decimal places)

NOTE

```
log(5)/log(3)
        1.464973521
3^1.465
        5.000145454
3^1.464973521
        5.000000002
```

The solution 1.465 is not exact. You can verify the approximate solution on a calculator. Raise 3 to power 1.465. You should see a result close to 5, but not exactly 5.

Check Yourself 4

Solve $2^x = 10$, rounded to the nearest thousandth.

Example 5 shows how to solve an equation with a more complicated exponent.

Example 5 Solving an Exponential Equation

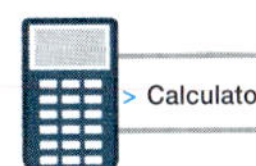

NOTES

On the left, we apply
$\log_b M^p = p \log_b M$

On a graphing calculator, the sequence is
(log 8) ÷ log 5) −
1) ÷ 2 ENTER

```
(log(8)/log(5)-1
)/2
        .1460148371
```

Solve $5^{2x+1} = 8$.

The process begins as in Example 4.

$$\log 5^{2x+1} = \log 8$$

$$(2x + 1) \log 5 = \log 8$$

$$2x + 1 = \frac{\log 8}{\log 5}$$

$$2x = \frac{\log 8}{\log 5} - 1$$

$$x = \frac{1}{2}\left(\frac{\log 8}{\log 5} - 1\right)$$

$$x \approx 0.146$$

The solution set is $\{0.146\}$.

Check Yourself 5

Solve $3^{2x-1} = 7$.

The procedure is similar if the variable appears as an exponent in more than one term of the equation.

Example 6 Solving an Exponential Equation

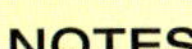

NOTES

Use the power property to write the variables as coefficients.

We now isolate x on the left.

To check the reasonableness of this result, use your calculator to verify that
$3^{1.710} \approx 2^{2.710}$

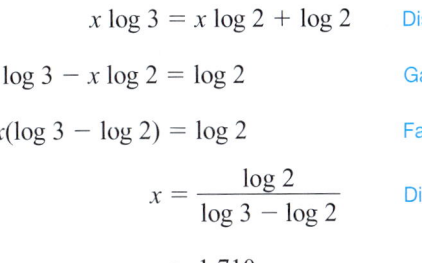

```
3^1.71
        6.544513163
2^2.71
        6.543216468
```

Solve $3^x = 2^{x+1}$.

$$\log 3^x = \log 2^{x+1}$$

$$x \log 3 = (x + 1) \log 2 \quad \text{Apply the power property.}$$

$$x \log 3 = x \log 2 + \log 2 \quad \text{Distribute } x + 1.$$

$$x \log 3 - x \log 2 = \log 2 \quad \text{Gather terms with } x \text{ on the left side.}$$

$$x(\log 3 - \log 2) = \log 2 \quad \text{Factor out } x.$$

$$x = \frac{\log 2}{\log 3 - \log 2} \quad \text{Divide by } (\log 3 - \log 2).$$

$$\approx 1.710$$

The solution set is $\{1.710\}$.

Check Yourself 6

Solve $5^{x+1} = 3^{x+2}$.

Here is an algorithm summarizing our work with solving exponential equations.

Step by Step

Solving Exponential Equations

Step 1 Try to write each side of the equation as a power of the same base. Then equate the exponents to form an equation.

Step 2 If the above procedure is not applicable, take the common logarithm of both sides of the original equation.

Step 3 Use the power rule for logarithms to write an equivalent equation with the variables as coefficients.

Step 4 Solve the resulting equation.

Step 5 Check the solutions.

There are many applications of our work with exponential equations. We look at a financial application in Example 7.

Example 7 An Interest Application

< Objective 3 >

If an investment of P dollars earns interest at an annual interest rate r and the interest is compounded n times per year, then the amount in the account after t years is given by

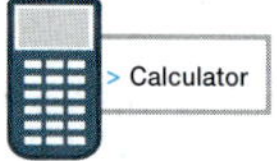

$$A = P\left(1 + \frac{r}{n}\right)^{nt}$$

If \$1,000 is placed in an account with an annual interest rate of 6%, find out how long it will take the money to double when interest is compounded annually and how long when compounded quarterly.

NOTE

Since the interest is compounded *once* per year, $n = 1$.

(a) Compound interest annually.

Using the formula with $A = 2{,}000$ (we want the original \$1,000 to double), $P = 1{,}000$, $r = 0.06$, and $n = 1$, we have

$$2{,}000 = 1{,}000(1 + 0.06)^t$$

Dividing both sides by 1,000 yields

$$2 = (1.06)^t$$

NOTE

From accounting, we have the **rule of 72,** which states that the doubling time is approximately 72 divided by the interest rate as a percentage. Here $\frac{72}{6} =$ 12 years.

We now have an exponential equation that we can solve.

$$\log 2 = \log (1.06)^t$$

$$= t \log 1.06$$

$$\text{or} \quad t = \frac{\log 2}{\log 1.06}$$

$$\approx 11.9 \text{ years}$$

It takes just a little less than 12 years for the money to double.

NOTE

Since the interest is compounded 4 times per year, $n = 4$.

(b) Compound interest quarterly.

Now $n = 4$ in the formula, so

$$2{,}000 = 1{,}000\left(1 + \frac{0.06}{4}\right)^{4t}$$

$$2 = (1.015)^{4t}$$

$$\log 2 = \log (1.015)^{4t}$$

$$\log 2 = 4t \log 1.015$$

$$\frac{\log 2}{4 \log 1.015} = t$$

$$t \approx 11.6 \text{ years}$$

The doubling time is reduced by approximately 3 months by the more frequent compounding.

Check Yourself 7

Find the doubling time in Example 7 if the interest is compounded monthly.

Problems involving rates of growth or decay can also be solved by using exponential equations.

Example 8 A Population Application

A town's population is presently 10,000. Given a projected continuous growth rate of 7% per year, t years from now the population P will be given by

$$P = 10{,}000e^{0.07t}$$

In how many years will the town's population double?

We want the time t when P will be 20,000 (doubled in size). So

$$20{,}000 = 10{,}000e^{0.07t}$$ Divide both sides by 10,000.

$$2 = e^{0.07t}$$

In this case, we take the *natural logarithm* of both sides of the equation. This is because e is involved in the equation.

$$\ln 2 = \ln e^{0.07t}$$

$$\ln 2 = 0.07t \ln e$$ Apply the power property.

$$\ln 2 = 0.07t$$

$$\frac{\ln 2}{0.07} = t$$

$$t \approx 9.9 \text{ years}$$

The population will double in approximately 9.9 years.

RECALL

$\ln e = 1$

To check, type: 10,000 [2nd] [e^x] .07 [×] 9.9 [)] [ENTER]

```
10000e^(.07*9.9)
        19997.05661
```

which is close to 20,000.

Check Yourself 8

If $1,000 is invested in an account with an annual interest rate of 6%, compounded continuously, the amount A in the account after t years is given by

$$A = 1{,}000e^{0.06t}$$

Find the time t that it will take for the amount to double ($A = 2{,}000$). Compare this time with the result of the Check Yourself 7 Exercise. Which is shorter? Why?

Check Yourself ANSWERS

1. $\{2\}$ **2.** $\left\{\frac{1}{8}\right\}$ **3.** $\left\{\frac{3}{2}\right\}$ **4.** $\{3.322\}$ **5.** $\{1.386\}$ **6.** $\{1.151\}$ **7.** 11.58 years **8.** 11.55 years. The doubling time is shorter, because interest is compounded more frequently.

Reading Your Text

SECTION 10.7

(a) A logarithmic ______________ is an equation that contains a logarithmic expression.

(b) If no base for a logarithm is written, it is assumed to be ______________.

(c) Equations in which the ______________ appears as an exponent are called exponential equations.

(d) The final step in solving logarithmic equations is to check for ______________ solutions.

Basic Skills | Challenge Yourself | Calculator/Computer | Career Applications | Above and Beyond

10.7 exercises

Boost *your* GRADE at ALEKS.com!

ALEKS®

- Practice Problems
- Self-Tests
- NetTutor
- e-Professors
- Videos

Name ______________

Section ________ Date ________

< Objective 1 >

Solve each logarithmic equation.

1. $\log_5 x = 3$

2. $\log_3 x = -2$

3. $\log (x + 1) = 2$

4. $\log_5 (3x + 2) = 3$

5. $\log_2 x + \log_2 8 = 6$

6. $\log 5 + \log x = 2$

7. $\log_3 x - \log_3 6 = 3$

8. $\log_4 x - \log_4 8 = 3$

9. $\log_2 x + \log_2 (x + 2) = 3$

10. $\log_3 x + \log_3 (2x + 3) = 2$

11. $\log_7 (x + 1) + \log_7 (x - 5) = 1$

12. $\log_2 (x + 2) + \log_2 (x - 5) = 3$

13. $\log x - \log (x - 2) = 1$

14. $\log_5 (x + 5) - \log_5 x = 2$

15. $\log_3 (x + 1) - \log_3 (x - 2) = 2$

> Videos

16. $\log (x + 2) - \log (2x - 1) = 1$

17. $\log (x + 5) - \log (x - 2) = \log 5$

18. $\log_3 (x + 12) - \log_3 (x - 3) = \log_3 6$

19. $\log_2 (x^2 - 1) - \log_2 (x - 2) = 3$

20. $\log (x^2 + 1) - \log (x - 2) = 1$

< Objective 2 >

Solve each exponential equation. If your solution is an approximation, round to three decimal places.

21. $6^x = 1{,}296$

22. $4^x = 64$

23. $2^{x+1} = \frac{1}{8}$

24. $9^x = 3$

Answers

1. ______ 2. ______
3. ______ 4. ______
5. ______ 6. ______
7. ______ 8. ______
9. ______ 10. ______
11. ______ 12. ______
13. ______ 14. ______
15. ______ 16. ______
17. ______ 18. ______
19. ______ 20. ______
21. ______ 22. ______
23. ______ 24. ______

Answers

25. ______
26. ______
27. ______
28. ______
29. ______
30. ______
31. ______
32. ______
33. ______
34. ______
35. ______
36. ______
37. ______
38. ______
39. ______
40. ______
41. ______
42. ______
43. ______
44. ______

25. $8^x = 2$

26. $3^{2x-1} = 27$

27. $3^x = 7$ > Videos

28. $5^x = 30$

29. $4^{x+1} = 12$

30. $3^{2x} = 5$

31. $7^{3x} = 50$

32. $6^{x-3} = 21$

33. $5^{3x-1} = 15$

34. $8^{2x+1} = 20$

35. $4^x = 3^{x+1}$ > Videos

36. $5^x = 2^{x+2}$

37. $2^{x+1} = 3^{x-1}$

38. $3^{2x+1} = 5^{x+1}$

Basic Skills | **Challenge Yourself** | Calculator/Computer | Career Applications | Above and Beyond

< Objective 3 >

Use the formula

$$A = P\left(1 + \frac{r}{n}\right)^{nt}$$

to complete exercises 39 to 42. Round your answers to two decimal places.

39. Business and Finance If \$5,000 is placed in an account with an annual interest rate of 9%, how long will it take the amount to double if the interest is compounded annually? > Videos

40. Repeat exercise 39 if the interest is compounded semiannually.

41. Repeat exercise 39 if the interest is compounded quarterly.

42. Repeat exercise 39 if the interest is compounded monthly.

Suppose the number of bacteria present in a culture after t hours is given by $N(t) = N_0 \cdot 2^{t/2}$, where N_0 is the initial number of bacteria. Use the formula to complete exercises 43 to 46.

43. How long will it take the bacteria to increase from 12,000 to 20,000?

44. How long will it take the bacteria to increase from 12,000 to 50,000?

45. How long will it take the bacteria to triple? (*Hint:* Let $N = 3N_0$.)

46. How long will it take the culture to increase to 5 times its original size? (*Hint:* Let $N = 5N_0$.)

Science and Medicine *The radioactive element strontium-90 has a half-life of approximately 28 years. That is, in a 28-year period, one-half of the initial amount will have decayed into another substance. If A_0 is the initial amount of the element, then the amount A remaining after t years is given by* chapter 10 > Make the Connection

$$A = A_0\left(\frac{1}{2}\right)^{t/28}$$

Use the formula to complete exercises 47 to 50.

47. If the initial amount of the element is 100 g, in how many years will 60 g remain? chapter 10 > Make the Connection

48. If the initial amount of the element is 100 g, in how many years will 20 g remain? chapter 10 > Make the Connection

49. In how many years will 75% of the original amount remain? (*Hint:* Let $A = 0.75A_0$.) chapter 10 > Make the Connection

50. In how many years will 10% of the original amount remain? (*Hint:* Let $A = 0.1A_0$.) chapter 10 > Make the Connection

Given projected growth, t years from now a city's population P can be approximated by $P(t) = 25{,}000e^{0.045t}$. Use the formula to complete exercises 51 and 52.

51. How long will it take the city's population to reach 35,000?

52. How long will it take the population to double?

The number of bacteria in a culture after t hours is given by $N(t) = N_0e^{0.03t}$, where N_0 is the initial number of bacteria in the culture. Use the formula to complete exercises 53 and 54.

53. In how many hours will the size of the culture double? 

54. In how many hours will the culture grow to 4 times its original population?

Answers

45. ______
46. ______
47. ______
48. ______
49. ______
50. ______
51. ______
52. ______
53. ______
54. ______

Answers

55. ______

56. ______

57. ______

58. ______

59. ______

60. ______

61. ______

The atmospheric pressure P, in inches of mercury (in. Hg), at an altitude h feet above sea level is approximated by $P(t) = 30e^{-0.00004h}$. *Use the formula to complete exercises 55 and 56.*

55. Find the altitude at which the pressure is 25 in. Hg.

56. Find the altitude at which the pressure is 20 in. Hg.

Carbon-14 dating is used to determine the age of specimens and is based on the radioactive decay of the element carbon-14. This decay begins once a plant or animal dies. If A_0 *is the initial amount of carbon-14, then the amount remaining after t years is* $A(t) = A_0e^{-0.000124t}$. *Use the formula to complete exercises 57 and 58.*

57. Estimate the age of a specimen if 70% of the original amount of carbon-14 remains. chapter 10 > Make the Connection

58. Estimate the age of a specimen if 20% of the original amount of carbon-14 remains. chapter 10 > Make the Connection

Basic Skills | Challenge Yourself | Calculator/Computer | **Career Applications** | Above and Beyond

59. **Allied Health** Chemists assign a pH-value (a measure of a solution's acidity) as a function of the concentration of hydrogen ions (H^+, measured in moles per liter M) according to Sorenson's 1909 model,

$$\text{pH} = -\log([H^+])$$

The most acidic rainfall ever measured occurred in Scotland in 1974. What was the hydrogen ion concentration given that its pH was 2.4? (Report your results with two decimal places in scientific notation.)

60. **Automotive Technology** One formula for noise level N (in dB) is

$$N = 10\log\left(\frac{I}{10^{-12}\text{ W/m}^2}\right)$$

One city ordinance requires that the maximum noise level for a car exhaust be 100 dB. What is the maximum sound intensity I (in W/m^2) allowed (to the nearest hundredth)?

61. **Information Technology** One problem associated with using the Internet to perform research is the broken or "dead link." This refers to Web pages that include links to other pages that yield an error message "file not found" or direct the user to a website that is different from the original one. Although links cited in research articles generally last longer than those on the Web, they still may not last very long.

Researchers at the University of Iowa examined links cited in articles accepted by the Communications-Technology division of the Association for Education in Journalism and Mass Communication. They found that such links have a half-life of approximately 1.25 yr. This means that 1.25 years (15 months) after the initial linkage, half the links are no longer valid.

Using this information, we can build a function to estimate the number of valid links t years after the initial citation.

$L(t) = a(0.57435)^t$

in which a represents the total number of links cited in an article, and $L(t)$ gives the number still valid after t years.

(a) If an article cites 30 Internet links, how many would you expect to be "live" after 6 months? 5 years?

(b) If you peruse an article that is 2 years old, and 12 links are still valid, how many links would you expect to be invalid?

Answers

62. ______

63. ______

64. ______

Basic Skills | Challenge Yourself | Calculator/Computer | Career Applications | **Above and Beyond**

62. Manufacturing Technology Aceto Balsamico Tradizionale (authentic, traditional balsamic vinegar) from the Modena and Reggio regions of Italy's Emilia-Romagna province sells for anywhere from \$50 to \$600 per ounce.

The producers of this vinegar typically fill 60-liter barrels to 75% capacity. After 10 years, only 60% of the original contents (by volume) are present in the barrel.

A bottle of this vinegar, for sale, is 100 mL. Aged 12 years (the minimum allowed), a bottle sells for roughly \$75. Aged 20 years, a bottle sells for roughly \$110. Aged 30 years, a bottle sells for roughly \$200. Further aging can bring the price of a bottle as high as \$600.

(a) How much vinegar is initially placed in a 60-liter barrel?

(b) How much vinegar is left in the barrel after 10 years? 20 years? (*Note:* In fact, barrels are changed each year with a mixing of newer and older varieties of vinegar.) Report your results to the nearest liter.

(c) Construct a function to model the amount of vinegar in a barrel t years after its initial fill.

(d) How many bottles does a barrel produce after 12 years? 20 years? 30 years?

(e) How much time needs to pass before the original barrel produces only one bottle of vinegar?

63. In some of the earlier exercises, we talked about bacteria cultures that double in size every few minutes. Can this go on forever? Explain.

64. The population of the United States has been doubling every 45 years. Is it reasonable to assume that this rate will continue? What factors will start to limit that growth?

Answers

65. ______

66. ______

67. ______

68. ______

69. ______

Use your calculator to describe the graph of each equation, then explain the result.

65. $y = \log 10^x$

66. $y = 10^{\log x}$

67. $y = \ln e^x$

68. $y = e^{\ln x}$

69. In this section, we solved the equation $3^x = 2^{x+1}$ by first applying the logarithmic function, base 10, to each side of the equation. Try this again, but this time apply the natural logarithm function to each side of the equation. Compare the solutions that result from the two approaches.

Answers

1. $\{125\}$ **3.** $\{99\}$ **5.** $\{8\}$ **7.** $\{162\}$ **9.** $\{2\}$ **11.** $\{6\}$
13. $\left\{\frac{20}{9}\right\}$ **15.** $\left\{\frac{19}{8}\right\}$ **17.** $\left\{\frac{15}{4}\right\}$ **19.** $\{3, 5\}$ **21.** $\{4\}$ **23.** $\{-4\}$
25. $\left\{\frac{1}{3}\right\}$ **27.** $\{1.771\}$ **29.** $\{0.792\}$ **31.** $\{0.670\}$ **33.** $\{0.894\}$
35. $\{3.819\}$ **37.** $\{4.419\}$ **39.** 8.04 yr **41.** 7.79 yr **43.** 1.47 h
45. 3.17 h **47.** 20.6 yr **49.** 11.6 yr **51.** 7.5 yr **53.** 23.1 h
55. 4,558 ft **57.** 2,876 yr **59.** 3.98×10^{-3} M **61. (a)** 23; 2; **(b)** 24
63. Above and Beyond
65. The graph is that of $y = x$. The two functions undo each other.
67. The graph is that of $y = x$. The two functions undo each other.
69. Above and Beyond

Definition/Procedure	Example	Reference
Algebra of Functions		**Section 10.1**
The **sum of two functions** f and g is written $f + g$. It is defined as $(f + g)(x) = f(x) + g(x)$	Let $f(x) = 2x + 1$ and $g(x) = 3x^2$. $(f + g)(x) = (2x + 1) + (3x^2)$ $= 3x^2 + 2x + 1$	p. 982
The **difference of two functions** f and g is written $f - g$. It is defined as $(f - g)(x) = f(x) - g(x)$	$(f - g)(x) = (2x + 1) - (3x^2)$ $= -3x^2 + 2x + 1$	p. 982
The **product of two functions** f and g is written $f \cdot g$. It is defined as $(f \cdot g)(x) = f(x) \cdot g(x)$	$(f \cdot g)(x) = (2x + 1)(3x^2)$ $= 6x^3 + 3x^2$	p. 985
The **quotient of two functions** f and g is written $f \div g$. It is defined as $(f \div g)(x) = f(x) \div g(x) \quad g(x) \neq 0$	$(f \div g)(x) = (2x + 1) \div (3x^2)$ $= \dfrac{2x + 1}{3x^2}$	p. 985
Composition of Functions		**Section 10.2**
The composition of two functions f and g is written $f \circ g$. It is defined as $(f \circ g)(x) = f(g(x))$	Let $f(x) = 2x + 1$ and $g(x) = 3x^2$ $(f \circ g)(x) = 2(3x^2) + 1$ $= 6x^2 + 1$	p. 992
Inverse Relations and Functions		**Section 10.3**
The **inverse** of a relation is formed by interchanging the components of each ordered pair in the given relation. If a relation (or function) is specified by an equation, interchange the roles of x and y in the defining equation to form the inverse.	The inverse of the relation $\{(1, 2), (2, 3), (4, 3)\}$ is $\{(2, 1), (3, 2), (3, 4)\}$ To find the inverse of $f(x) = 4x - 8$ $y = 4x - 8$ change y to x and x to y $x = 4y - 8$ so $4y = x + 8$ $y = \frac{1}{4}(x + 8)$ $y = \frac{1}{4}x + 2$ $f^{-1}(x) = \frac{1}{4}x + 2$	p. 1003

Continued

Definition/Procedure	Example	Reference

The inverse of a function f may or may not be a function. If the inverse *is* also a function, we denote that inverse as f^{-1}, read "the inverse of f."

A function f has an inverse f^{-1}, which is also a function, if and only if f is a **one-to-one** function. That is, no two ordered pairs in the function have the same second component.

The **horizontal line test** can be used to determine whether a function is one-to-one.

If $f(x) = 4x - 8$, then

$$f^{-1}(x) = \frac{1}{4}x + 2$$

p. 1006

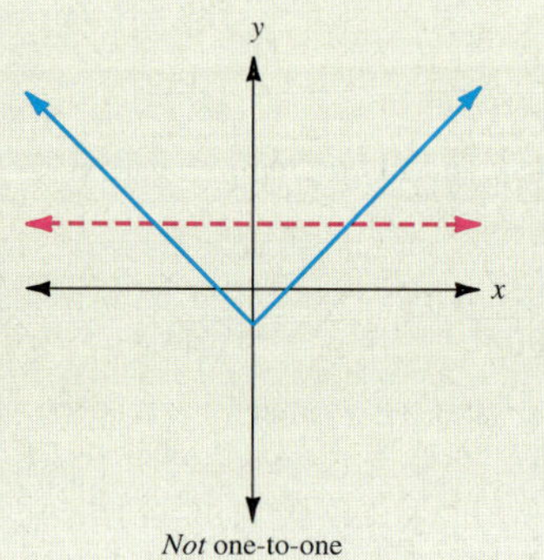

Not one-to-one

p. 1008

Finding Inverse Relations and Functions

1. Interchange the x- and y-components of the ordered pairs of the given relation or the roles of x and y in the defining equation.
2. If the relation was described in equation form, solve the defining equation of the inverse for y.
3. If desired, graph the relation and its inverse on the same set of axes. The two graphs will be symmetric about the line $y = x$.

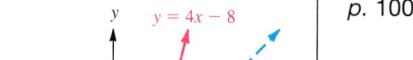

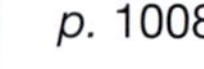

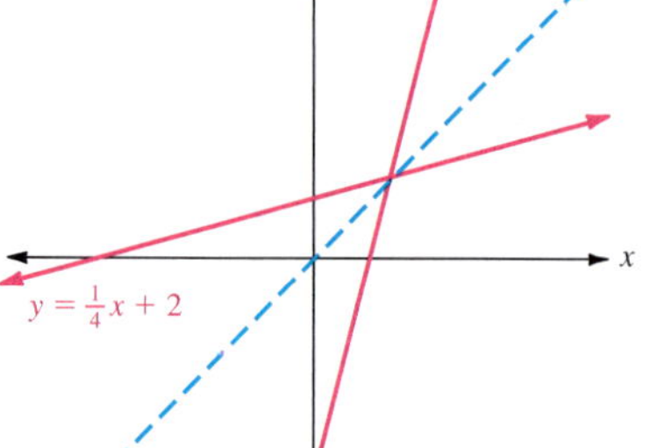

Exponential Functions

Section 10.4

An **exponential function** is any function defined by an equation of the form

$y = f(x) = b^x \qquad b > 0, b \neq 1$

If b is greater than 1, the function is always increasing (a **growth function**). If b is less than 1, the function is always decreasing (a **decay function**).

In both cases, the exponential function is one-to-one. The domain is the set of all real numbers, and the range is the set of positive real numbers.

The function defined by $f(x) = e^x$, in which e is an irrational number (approximately 2.71828), is called *the* exponential function.

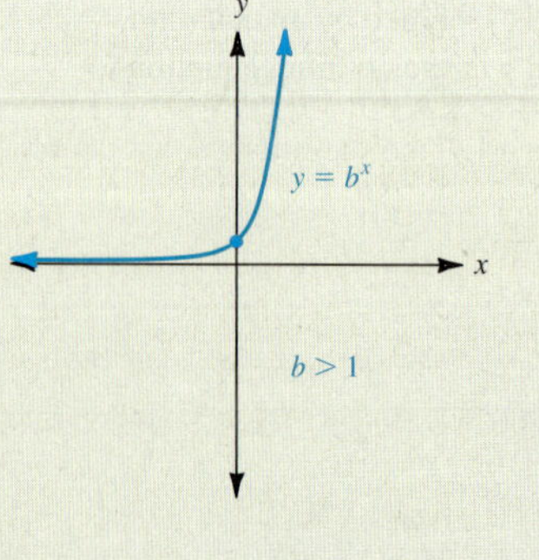

p. 1017

Definition/Procedure	Example	Reference

Graphing an Exponential Function

Step 1 Establish a table of values by considering the function in the form $y = b^x$.

Step 2 Plot points from that table of values and connect them with a smooth curve to form the graph.

Properties of Exponential Graphs

1. If $b > 1$, the graph increases from left to right. If $0 < b < 1$, the graph decreases from left to right.
2. All exponential graphs have these properties in common.
   (a) The y-intercept is $(0, 1)$.
   (b) The graph approaches, but does not touch, the x-axis.
   (c) The graphs represent one-to-one functions.

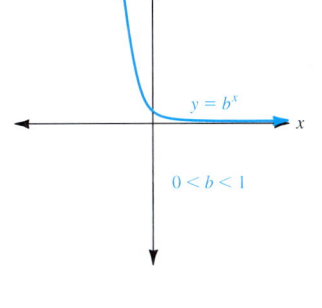

p. 1019

Logarithmic Functions

Section 10.5

In the expression

$y = \log_b x$

y is called the *logarithm of x to base b,* when $b > 0$ and $b \neq 1$.

An expression such as $y = \log_b x$ is said to be in **logarithmic form.**

An expression such as $x = b^y$ is said to be in **exponential form.**

$y = \log_b x$ means the same as $x = b^y$

A logarithm is an exponent or a power. The logarithm of x to base b is the power to which we must raise b to get x.

A **logarithmic function** is any function defined by an equation of the form

$f(x) = \log_b x \qquad b > 0, b \neq 1$

The logarithm function is the inverse of the corresponding exponential function. The function is one-to-one with domain $\{x \mid x > 0\}$ and range composed of the set of all real numbers.

$\log_3 9 = 2$ is in logarithmic form.
$3^2 = 9$ is the exponential form.
$\log_3 9 = 2$ is equivalent to $3^2 = 9$.
2 is the power to which we must raise 3 to get 9.

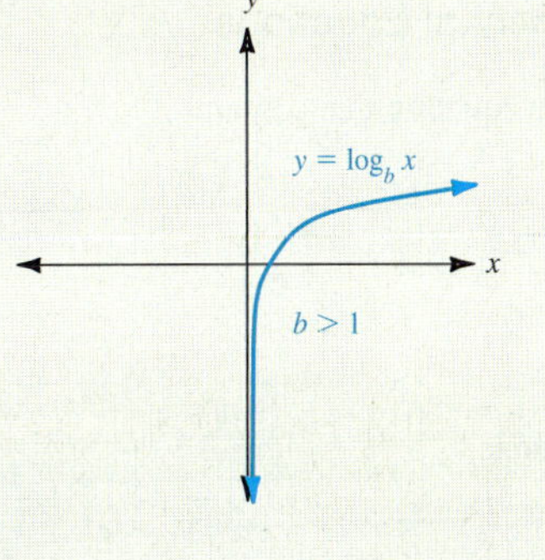

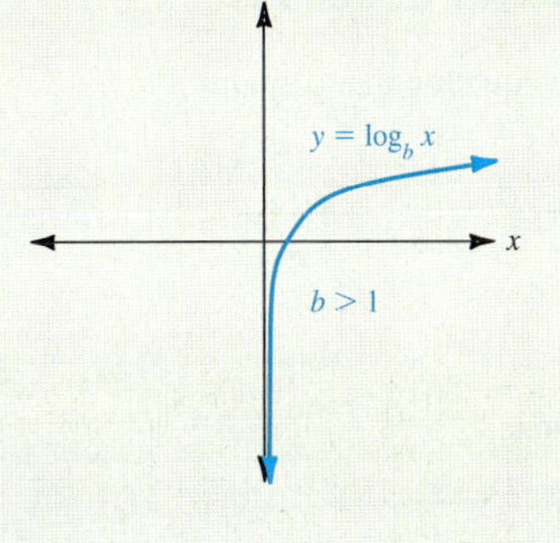

p. 1036

Properties of Logarithms

Section 10.6

If M, N, and b are positive real numbers with $b \neq 1$ and if p is any real number, then we can state these properties of logarithms.

1. $\log_b b = 1$
2. $\log_b 1 = 0$
3. $b^{\log_b x} = x$
4. $\log_b b^x = x$

$\log 10 = 1$
$\log_2 1 = 0$
$3^{\log_3 2} = 2$
$\log_5 5^x = x$

p. 1051

Continued

Definition/Procedure	Example	Reference
Product Property $\log_b MN = \log_b M + \log_b N$	$\log_3 x + \log_3 y = \log_3 xy$	p. 1052
Quotient Property $\log_b \frac{M}{N} = \log_b M - \log_b N$	$\log_5 8 - \log_5 3 = \log_5 \frac{8}{3}$	p. 1052
Power Property $\log_b M^p = p \log_b M$	$\log 3^2 = 2 \log 3$	p. 1052
Common logarithms are logarithms to base 10. For convenience, we omit the base in writing common logarithms: $\log M = \log_{10} M$	$\log_{10} 1{,}000 = \log 1{,}000$ $= \log 10^3 = 3$	p. 1054
Natural logarithms are logarithms to base e. By custom we also omit the base in writing natural logarithms: $\ln M = \log_e M$	$\ln 3 = \log_e 3$	p. 1057
Logarithmic and Exponential Equations		**Section 10.7**
A **logarithmic equation** is an equation that contains a logarithmic expression. $\log_2 x = 5$ is a logarithmic equation.	To solve $\log_2 x = 5$: Write the equation in the equivalent exponential form to solve $x = 2^5$ or $x = 32$	p. 1070
Solving Logarithmic Equations **Step 1** Use the properties of logarithms to combine terms containing logarithmic expressions into a single term. **Step 2** Write the equation formed in step 1 in exponential form. **Step 3** Solve for the indicated variable. **Step 4** Check your solutions to make sure that possible solutions do not result in the logarithms of negative numbers or zero.	To solve $\log_4 x + \log_4 (x - 6) = 2$ $\log_4 x(x - 6) = 2$ $x(x - 6) = 4^2$ $x^2 - 6x - 16 = 0$ $(x - 8)(x + 2) = 0$ $x = 8$ or $x = -2$ Because substituting -2 for x in the original equation results in the logarithm of a negative number, we reject that answer. The only solution is 8.	p. 1074

Definition/Procedure	Example	Reference
An **exponential equation** is an equation in which the variable appears as an exponent.	To solve $4^x = 64$: Because $64 = 4^3$, write $4^x = 4^3$ or $x = 3$	p. 1074
Solving Exponential Equations **Step 1** Try to write each side of the equation as a power of the same base. Then equate the exponents to form an equation. **Step 2** If the above procedure is not applicable, take the common logarithm of both sides of the original equation. **Step 3** Use the power rule for logarithms to write an equivalent equation with the variables as coefficients. **Step 4** Solve the resulting equation. **Step 5** Check the solutions.	$2^{x+3} = 5^x$ $\log 2^{x+3} = \log 5^x$ $(x + 3) \log 2 = x \log 5$ $x \log 2 + 3 \log 2 = x \log 5$ $x \log 2 - x \log 5 = -3 \log 2$ $x (\log 2 - \log 5) = -3 \log 2$ $x = \dfrac{-3 \log 2}{\log 2 - \log 5} \approx 2.269$	p. 1076

summary exercises :: chapter 10

This summary exercise set is provided to give you practice with each of the objectives of this chapter. Each exercise is keyed to the appropriate chapter section. When you are finished, you can check your answers to the odd-numbered exercises in the back of the text. If you have difficulty with any of these questions, go back and reread the examples from that section. The answers to the even-numbered exercises appear in the *Instructor's Solutions Manual.* Your instructor will give you guidelines on how best to use these exercises in your instructional setting.

10.1 *In exercises 1 to 8, use the tables to find the desired values.*

x	$f(x)$
-2	3
-1	8
3	0
7	-6
8	-4

x	$g(x)$
-4	5
-2	-1
3	2
5	-3
7	0

1. $(f + g)(-2)$

2. $(g - f)(7)$

3. $(f - g)(-1)$

4. $(f \cdot g)(3)$

5. $(f \div g)(7)$

6. $(g \div f)(-2)$

7. Find the domain of $g - f$.

8. Find the domain of $g \div f$.

In exercises 9 to 12, $f(x)$ and $g(x)$ are given. Find $(f + g)(x)$.

9. $f(x) = 4x^2 + 5x - 3$ and $g(x) = -2x^2 + x - 5$

10. $f(x) = -3x^3 + 2x^2 - 5$ and $g(x) = 4x^3 - 4x^2 + 5x + 6$

11. $f(x) = 2x^4 + 4x^2 + 5$ and $g(x) = x^3 - 5x^2 + 6x$

12. $f(x) = 3x^3 + 5x - 5$ and $g(x) = -2x^3 + 2x^2 + 5x$

In exercises 13 to 16, $f(x)$ and $g(x)$ are given. Find $(f - g)(x)$.

13. $f(x) = 7x^2 - 2x + 3$ and $g(x) = 2x^2 - 5x - 7$

14. $f(x) = 9x^2 - 4x$ and $g(x) = 5x^2 + 3$

15. $f(x) = 8x^2 + 5x$ and $g(x) = 4x^2 - 3x$

16. $f(x) = -2x^2 - 3x$ and $g(x) = -3x^2 + 4x - 5$

In exercises 17 to 20, find the product $(f \cdot g)(x)$.

17. $f(x) = 2x$ and $g(x) = 3x - 5$

18. $f(x) = x + 1$ and $g(x) = -3x$

19. $f(x) = 3x$ and $g(x) = x^2$

20. $f(x) = 2x$ and $g(x) = x^2 - 5$

In exercises 21 to 24, find the quotient $(f \div g)(x)$ and state the domain of the resulting function.

21. $f(x) = 2x$ and $g(x) = x - 3$

22. $f(x) = x + 1$ and $g(x) = 2x - 4$

23. $f(x) = 3x$ and $g(x) = x^2$

24. $f(x) = 2x^2$ and $g(x) = x - 5$

10.2 *In exercises 25 to 30, use the tables to find the desired values.*

x	$f(x)$
−2	3
−1	8
3	0
7	−6
8	−4

x	$g(x)$
−4	5
−2	−1
3	2
5	−3
7	0

25. $(f \circ g)(-2)$

26. $(g \circ f)(8)$

27. $(g \circ f)(-2)$

28. $(f \circ g)(-4)$

29. $(g \circ g)(-4)$

30. $(f \circ f)(-2)$

In exercises 31 to 34, evaluate the indicated composite functions in each part.

31. $f(x) = x - 3$ and $g(x) = 3x + 1$

(a) $(f \circ g)(0)$ **(b)** $(f \circ g)(-2)$ **(c)** $(f \circ g)(3)$ **(d)** $(f \circ g)(x)$

32. $f(x) = 5x - 1$ and $g(x) = -4x + 5$

(a) $(f \circ g)(0)$ **(b)** $(f \circ g)(-2)$ **(c)** $(g \circ f)(3)$ **(d)** $(g \circ f)(x)$

33. $f(x) = x^2$ and $g(x) = x - 5$

(a) $(f \circ g)(0)$ **(b)** $(f \circ g)(-2)$ **(c)** $(g \circ f)(3)$ **(d)** $(g \circ f)(x)$

34. $f(x) = x^2 + 3$ and $g(x) = -2x$

(a) $(g \circ f)(0)$ **(b)** $(g \circ f)(-2)$ **(c)** $(f \circ g)(3)$ **(d)** $(f \circ g)(x)$

In exercises 35 and 36, rewrite the function h as a composite of functions f and g.

35. $f(x) = -2x$, $g(x) = x + 2$, $h(x) = -2x - 4$

36. $f(x) = 6x$, $g(x) = x^2 + 5$, $h(x) = 36x^2 + 5$

10.3 *Find the inverse function f^{-1} for the given function.*

37. $f(x) = -2x + 3$

38. $f(x) = 4x + 5$

39. $f(x) = \dfrac{x - 3}{4}$

40. $f(x) = \dfrac{x}{4} - 3$

Find the inverse of each function. In each case, determine whether the inverse is also a function.

41.

x	$f(x)$
−3	1
−1	2
2	3
3	4

42.

x	$f(x)$
−2	5
0	4
3	5
5	4

43. $\{(1, 5), (2, 7), (3, 9)\}$

44. $\{(3, 1), (5, 1), (7, 1)\}$

45. $\{(2, 4), (4, 3), (6, 4)\}$

46. $\{(-2, 6), (0, 0), (3, 8)\}$

For each function f, find its inverse f^{-1}. Then graph both on the same set of axes.

47. $f(x) = 5x + 3$

48. $f(x) = -3x + 9$

49. $f(x) = \dfrac{x - 4}{5}$

50. $f(x) = \dfrac{3 - x}{2}$

Determine whether the given function is one-to-one. In each case determine whether the inverse is a function.

51. $f = \{(-1, 2), (1, 3), (2, 5), (4, 7)\}$

52. $f = \{(-3, 2), (0, 2), (1, 2), (3, 2)\}$

53.

x	$f(x)$
−1	2
3	4
4	5
5	6

54.

x	$f(x)$
1	3
2	4
3	5
4	3

55.

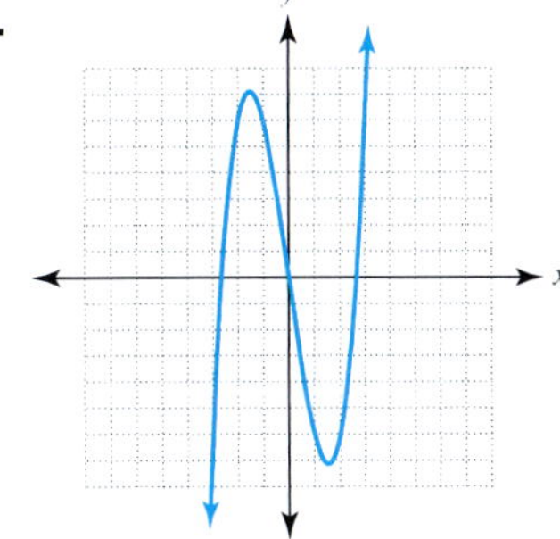

56.

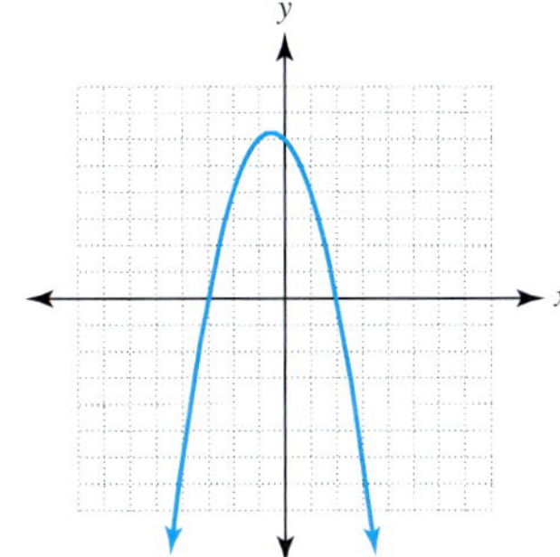

Let $f(x) = 4x + 12$ *and* $f^{-1}(x) = \frac{x}{4} - 3$ *and evaluate each expression.*

57. $f(-2)$

58. $f^{-1}(4)$

59. $f(f^{-1}(8))$

60. $f^{-1}(f(-4))$

61. $f(f^{-1}(x))$

62. $f^{-1}(f(x))$

10.4 *Graph the exponential functions defined by each equation.*

63. $y = 3^x$

64. $y = \left(\frac{3}{4}\right)^x$

Solve each equation.

65. $5^x = 125$

66. $2^{2x+1} = 32$

67. $3^{x-1} = \frac{1}{9}$

If it takes 2 h for the population of a certain bacterial culture to double (by dividing in half), then the number N of bacteria in the culture after t hours is given by $N = 1{,}000 \cdot 2^{t/2}$, *where the initial population of the culture was 1,000. Using this formula, find the number in the culture.*

68. After 4 h

69. After 12 h

70. After 15 h

10.5 *Graph each logarithmic function.*

71. $y = \log_3 x$

72. $y = \log_2 (x - 1)$

Convert each statement to logarithmic form.

73. $2^5 = 32$

74. $10^3 = 1{,}000$

75. $5^0 = 1$

76. $5^{-2} = \frac{1}{25}$

77. $25^{1/2} = 5$

78. $16^{3/4} = 8$

Convert each statement to exponential form.

79. $\log_4 64 = 3$

80. $\log 100 = 2$

81. $\log_{81} 9 = \frac{1}{2}$

82. $\log_5 25 = 2$

83. $\log 0.001 = -3$

84. $\log_{32} \frac{1}{2} = -\frac{1}{5}$

Solve each equation for the unknown variable.

85. $y = \log_5 125$

86. $\log_b \frac{1}{9} = -2$

87. $\log_7 x = 2$

88. $y = \log_5 1$

89. $\log_b 3 = \frac{1}{2}$

90. $y = \log_9 3$

91. $y = \log_8 2$

The decibel (dB) rating for the loudness of a sound is given by

$$L = 10 \log \frac{I}{I_0}$$

where I is the intensity of that sound in watts per square centimeter and I_0 is the intensity of the "threshold" sound $I_0 = 10^{-16}$ W/cm². Find the decibel rating of each of the given sounds.

92. A table saw in operation with intensity $I = 10^{-6}$ W/cm²

93. The sound of a passing car horn with intensity $I = 10^{-8}$ W/cm²

The formula for the decibel rating of a sound can be solved for the intensity of the sound as

$$I = I_0 \cdot 10^{L/10}$$

where L is the decibel rating of the given sound.

94. What is the ratio of intensity of a 60-dB sound to one of 50 dB?

95. What is the ratio of intensity of a 60-dB sound to one of 40 dB?

The magnitude of an earthquake on the Richter scale is given by

$$M = \log \frac{a}{a_0}$$

where a is the intensity of the shock wave of the given earthquake and a_0 is the intensity of the shock wave of a zero-level earthquake. Use that formula to complete exercises 96 and 97.

96. The Alaskan earthquake of 1964 had an intensity of $10^{8.4}a_0$. What was its magnitude on the Richter scale?

97. Find the ratio of intensity of an earthquake of magnitude 7 to an earthquake of magnitude 6.

10.6 *Use the properties of logarithms to expand each expression.*

98. $\log_b x^2y$

99. $\log_4 \frac{y^3}{5}$

100. $\log_5 \frac{x^2y}{z^3}$

101. $\log_5 x^3yz^2$

102. $\log \frac{xy}{\sqrt{z}}$

103. $\log_b \sqrt[3]{\frac{x^2y}{z}}$

Use the properties of logarithms to write each expression as a single logarithm.

104. $\log x + 2 \log y$

105. $3 \log_b x - 2 \log_b z$

106. $\log_b x + \log_b y - \log_b z$

107. $2 \log_5 x - 3 \log_5 y - \log_5 z$

108. $\log x - \frac{1}{2} \log y$

109. $\frac{1}{3}(\log_b x - 2 \log_b y)$

Given that $\log 2 = 0.301$ *and* $\log 3 = 0.477$, *find each logarithm. Verify your results with a calculator.*

110. $\log 18$

111. $\log 16$

112. $\log \frac{1}{8}$

113. $\log \sqrt{3}$

Use your calculator to find the pH of each solution, given the hydrogen ion concentration $[H^+]$ *for each solution, where*

$$\text{pH} = -\log [\text{H}^+]$$

Are the solutions acidic or basic?

114. Coffee: $[\text{H}^+] = 5 \times 10^{-6}$

115. Household detergent: $[\text{H}^+] = 3.2 \times 10^{-10}$

Given the pH of these solutions, find the hydrogen ion concentration $[H^+]$.

116. Lemonade: pH = 3.5

117. Ammonia: pH = 10.2

The average score on a final examination for a group of chemistry students, retested after time t (in weeks), is given by

$$S(t) = 81 - 6 \ln (t + 1)$$

Find the average score on the retests after the given times.

118. After 5 weeks

119. After 10 weeks

120. After 15 weeks

121. Graph these results.

The formula for converting from a logarithm with base b to a logarithm with base a is

$$\log_a x = \frac{\log_b x}{\log_b a}$$

Use that formula to find each logarithm.

122. $\log_4 20$

123. $\log_8 60$

10.7 *Solve each logarithmic equation.*

124. $\log_3 x + \log_3 5 = 3$

125. $\log_5 x - \log_5 10 = 2$

126. $\log_3 x + \log_3 (x + 6) = 3$

127. $\log_5 (x + 3) + \log_5 (x - 1) = 1$

128. $\log x - \log (x - 1) = 1$

129. $\log_2 (x + 3) - \log_2 (x - 1) = \log_2 3$

Solve each exponential equation. Give your results rounded to three decimal places.

130. $3^x = 243$

131. $5^x = \frac{1}{25}$

132. $5^x = 10$

133. $4^{x-1} = 8$

134. $6^x = 2^{2x+1}$

135. $2^{x+1} = 3^{x-1}$

If an investment of P dollars earns interest at an annual rate of 12% and the interest is compounded n times per year, then the amount A in the account after t years is

$$A(t) = P\left(1 + \frac{0.12}{n}\right)^{nt}$$

Use that formula to complete each exercise.

136. If \$1,000 is invested and the interest is compounded quarterly, how long will it take the amount in the account to double?

137. If \$3,000 is invested and the interest is compounded monthly, how long will it take the amount in the account to reach \$8,000?

A certain radioactive element has a half-life of 50 years. The amount A of the substance remaining after t years is given by

$$A(t) = A_0 \cdot 2^{-t/50}$$

where A_0 is the initial amount of the substance. Use this formula to complete each exercise.

138. If the initial amount of the substance is 100 milligrams (mg), after how long will 40 mg remain?

139. After how long will only 10% of the original amount of the substance remain?

A city's population is presently 50,000. Given the projected growth, t years from now the population P will be given by $P(t) = 50{,}000e^{0.08t}$. Use this formula to complete each exercise.

140. How long will it take the population to reach 70,000?

141. How long will it take the population to double?

The atmospheric pressure, in inches of mercury, at an altitude h miles above the surface of the earth, is approximated by $P(h) = 30e^{-0.021h}$. Use this formula to complete each exercise.

142. Find the altitude at the top of Mt. McKinley in Alaska if the pressure is 27.7 in. Hg.

143. Find the altitude of an airliner in flight if the pressure outside is 26.1 in. Hg.

self-test 10

Name ______________________

Section ________ Date ________

The purpose of this self-test is to help you assess your progress so that you can find concepts that you need to review before the next exam. Allow yourself about an hour to take this test. At the end of that hour, check your answers against those given in the back of this text. If you miss any, go back to the appropriate section to reread the examples until you have mastered that particular concept.

Convert each statement to logarithmic form.

1. $10^4 = 10{,}000$

2. $27^{2/3} = 9$

Given the two functions $f(x) = 5x - 7$ *and* $f^{-1}(x) = \dfrac{x + 7}{5}$, *evaluate each expression.*

3. $f(f^{-1}(x))$

4. $f^{-1}(f(x))$

Use the properties of logarithms to expand the expression.

5. $\log_5 \sqrt{\dfrac{xy^2}{z}}$

In exercises 6 to 9, use $f(x) = x^2 - 1$ *and* $g(x) = 3x - 2$.

6. Find $(f \cdot g)(x)$ and state the domain of the resulting function.

7. Find $(g \div f)(x)$ and state the domain of the resulting function.

8. Find $(f \circ g)(x)$.

9. Find $(g \circ f)(x)$.

Solve each exponential equation.

10. $5^x = \dfrac{1}{25}$

11. $3^{2x-1} = 81$

Convert each statement to exponential form.

12. $\log_5 125 = 3$

13. $\log 0.01 = -2$

In exercises 14 and 15, determine whether the given function is one-to-one. In each case, determine whether the inverse is a function.

14. $f = \{(3, 4), (5, -1), (6, 2), (7, -1)\}$

15.

x	$f(x)$
1	3
3	8
5	11
7	5

Use the properties of logarithms to write the expression as a single logarithm.

16. $\dfrac{1}{3}(\log_b x - 2\log_b z)$

Answers

1. ______________________
2. ______________________
3. ______________________
4. ______________________
5. ______________________
6. ______________________
7. ______________________
8. ______________________
9. ______________________
10. ______________________
11. ______________________
12. ______________________
13. ______________________
14. ______________________
15. ______________________
16. ______________________

Answers

17. ______

18. ______

19. ______

20. ______

21. ______

22. ______

23. ______

24. ______

25. ______

26. ______

27. ______

28. ______

29. ______

30. ______

Solve each exponential equation. Round results accurate to three decimal places.

17. $3^{x+1} = 4$

18. $5^x = 3^{x+1}$

For the given $f(x)$ and $g(x)$, find **(a)** $(f + g)(x)$ *and* **(b)** $(f - g)(x)$.

19. $f(x) = -3x^3 + 5x^2 - 2x - 7$ and $g(x) = -2x^2 + 7x - 2$

20. Graph the logarithmic function defined by the equation $y = \log_4 x$.

Solve each logarithmic equation.

21. $\log_6(x + 1) + \log_6(x - 4) = 2$

22. $\log(2x + 1) - \log(x - 1) = 1$

In exercises 23 to 25, find the inverse function f^{-1} for the given function.

23. $f(x) = \dfrac{x - 3}{5}$

24.

x	$f(x)$
−3	−2
1	2
4	5
5	6

25. $\{(-3, 1), (4, 2), (5, -2)\}$

Graph the exponential function defined by each equation.

26. $y = 4^x$

27. $y = \left(\dfrac{2}{3}\right)^x$

Solve each equation for the unknown variable.

28. $y = \log_2 64$

29. $\log_b \dfrac{1}{16} = -2$

30. $\log_{25} x = \dfrac{1}{2}$

cumulative review chapters 0-10

Name ____________

Section ______ Date ______

We offer the following exercises to help you review concepts from earlier chapters. This is meant as review material and not as a comprehensive exam. The answers are presented in the back of the text. If you have difficulty with any of these exercises, be certain to at least read through the summary related to that section.

Solve each equation.

1. $2x - 3(x + 2) = 4(5 - x) + 7$

2. $2^{3x} = 32$

3. $\log x - \log (x - 1) = 1$

Graph.

4. $5x - 3y = 15$

5. $-8(2 - x) \geq y$

6. Find an equation of the line that passes through the points $(2, -1)$ and $(-3, 5)$.

7. Solve the linear inequality

$$3x - 2(x - 5) \geq 20$$

Simplify each expression.

8. $4x^2 - 3x + 8 - 2(x^2 + 5) - 3(x - 1)$

9. $(3x + 1)(2x - 5)$

Completely factor each expression.

10. $2x^2 - x - 10$

11. $25x^3 - 16xy^2$

Perform the indicated operations.

12. $\frac{2}{x - 4} - \frac{3}{x - 5}$

13. $\frac{x^2 - x - 6}{x^2 + 2x - 15} \div \frac{x - 2}{x + 5}$

Simplify each radical expression.

14. $\sqrt{18} + \sqrt{50} - 3\sqrt{32}$

15. $(3\sqrt{2} + 2)(3\sqrt{2} + 2)$

16. $\frac{5}{\sqrt{5} - \sqrt{2}}$

17. Find three consecutive odd integers whose sum is 237.

Solve each equation.

18. $x^2 + x - 2 = 0$

19. $2x^2 - 6x - 5 = 0$

20. Solve the equation for R.

$$\frac{1}{R} = \frac{1}{R_1} + \frac{1}{R_2}$$

Answers

1. ____________
2. ____________
3. ____________
4. ____________
5. ____________
6. ____________
7. ____________
8. ____________
9. ____________
10. ____________
11. ____________
12. ____________
13. ____________
14. ____________
15. ____________
16. ____________
17. ____________
18. ____________
19. ____________
20. ____________

Answers

21. ______
22. ______
23. ______
24. ______
25. ______
26. ______
27. ______
28. ______
29. ______
30. ______
31. ______
32. ______
33. ______
34. ______
35. ______
36. ______
37. ______
38. ______
39. ______
40. ______

If $f(x) = 3x - 1$ *and* $g(x) = x^2 - x - 6$:

21. Find $(f \cdot g)(x)$ including the domain.

22. Find $(f \div g)(x)$ including the domain.

If $f(x) = 2x - 1$ *and* $g(x) = -3x + 2$:

23. Find **(a)** $(f \circ g)(x)$ and **(b)** $(f \circ g)(5)$.

24. Find **(a)** $(g \circ f)(x)$ and **(b)** $(g \circ f)(5)$.

Factor each polynomial completely.

25. $14a^2b^2 - 21a^2b + 35ab^2$

26. $x^2 - 3xy + 5x - 15y$

27. $25c^2 - 64d^2$

28. $27x^3 - 1$

29. $16a^4 + 2ab^3$

30. $x^2 - 2x - 48$

31. $10x^2 - 39x + 14$

32. $6x^3 + 3x^2 - 45x$

Simplify.

33. $\dfrac{x^3y}{4xy^2} \div \dfrac{xy}{12xy^2}$

34. $\dfrac{7}{3 - y} - \dfrac{5}{y - 3}$

35. $\dfrac{4}{m^2 - 9} - \dfrac{3}{m^2 - 4m + 3}$

36. $\dfrac{2x}{x^2 - 9x + 20} + \dfrac{8}{x - 4}$

37. The length of the longer leg of a right triangle is 4 cm more than twice the length of the shorter leg. The length of the longer leg is 2 cm shorter than the length of the hypotenuse. Find the lengths of all three sides.

Solve each equation.

38. $\log_4 x + \log_4 (x - 6) = 2$

39. $\sqrt{2x - 1} + x = 8$

40. Find the center and the radius of the circle whose equation is

$(x + 5)^2 + (y - 2)^2 = 16$

A Searching the Internet

< A Objectives >

1 > Search the Internet for math help

2 > Evaluate the results of an Internet search

There are many resources available to help you when you have difficulty with your math work. Your instructor can answer many of your questions, but there are other resources to help you learn, as well.

Studying with friends and classmates is a great way to learn math. Your school may have a "math lab" where instructors or peers provide tutoring services. And this text provides examples and exercises to help you learn and understand new concepts.

Another place to go for help is the **Internet.** There are many math tutorials on the Web. This activity is designed to introduce you to searching the Web and evaluating what you find there.

If you are new to computers or the Internet, your instructor or a classmate can help you get started. You will need to access the Internet through one of the many **Web browsers** such as Microsoft's Internet Explorer, Mozilla Firefox, Netscape Navigator, AOL's browser, or Opera.

First, you need to connect to the Internet. Then, you need to access a page containing a **search engine.** Many *default* home pages contain a *search* field. If yours does not, several of the more popular search engines are at the sites:

http://www.ask.com

http://www.dogpile.com

http://www.google.com

http://www.yahoo.com

Access one of these search engines or use one from another site as you work through this activity.

1. Type the word *integers* in the search field. You should see a (long) list of websites related to your search.

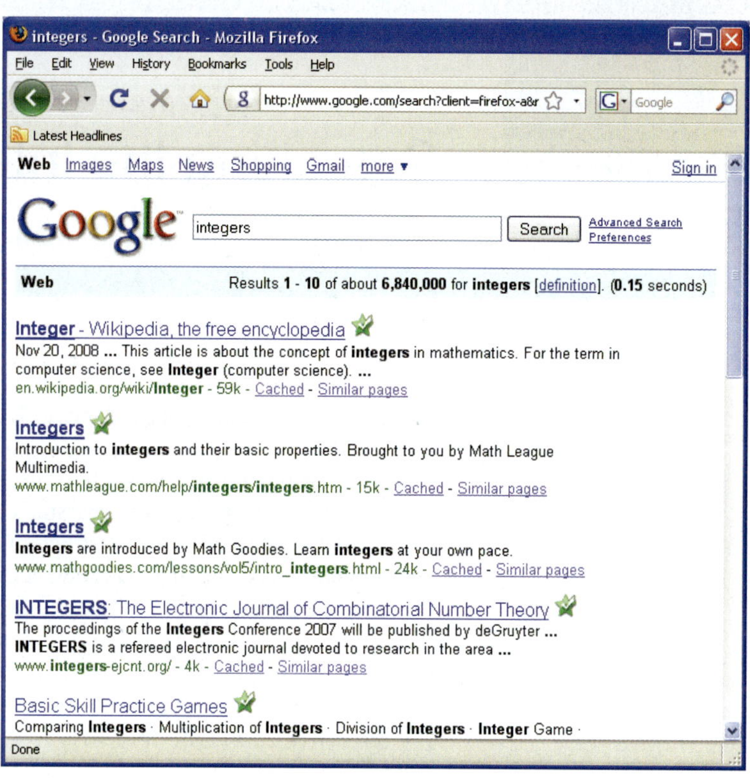

2. Look at the page titles and descriptions. Find a page that has an introduction to integers and click on that link.

3. Write two or three sentences describing the layout of the Web page. Is it "user friendly"? Are the topics presented in an easy-to-find and useful way? Are the colors and images helpful?

4. Choose a topic such as integer multiplication or even some math game. Describe the instruction that the website has for the topic. In what format is the information given? Is there an interactive component to the instruction?

5. Does the website offer free tutoring services? If so, try to get some help with a homework problem. Briefly evaluate the tutoring services.

6. Chapter 4 in our text introduces you to *systems of equations*. Are there activities or links on the website related to systems of equations? Do they appear to be helpful to a student having difficulty with this topic?

7. Return to your search engine. Find a second math Web page by typing "systems of equations" (including the quote marks) into the search field. Choose a page that offers instruction, tutoring, and activities related to systems of equations. Save the link for this page: this is called a bookmark, favorite, or preference, depending on your browser. If you find yourself struggling with systems of equations in Chapter 4, try using this page to get some additional help.

B.1 Solving Linear Inequalities in One Variable Graphically

< B.1 Objective >

1 > Solve linear inequalities in one variable graphically

In Section 4.2, we looked at the graphical approach to solving a linear equation. In this appendix, we use the graphs of linear functions to determine the solutions of a linear inequality.

Linear inequalities in one variable x are obtained from linear equations by replacing the symbol for equality $(=)$ with one of the inequality symbols $(<, >, \leq, \geq)$.

The general form for a linear inequality in one variable is

$$x < a$$

where the symbol $<$ can be replaced with $>$, $\leq$, or $\geq$. Examples of linear inequalities in one variable include

$$x \geq -3 \qquad 2x + 5 > 7 \qquad 2x - 3 \leq 5x + 6$$

Recall that the solution set for an equation is the set of all values for the variable (or ordered pair) that make the equation a true statement. Similarly, the solution set for an inequality is the set of all values that make the inequality a true statement. In Example 1, we look at a graphical approach to solving an inequality.

Example 1 **Solving a Linear Inequality Graphically**

< Objective 1 >

Use a graph to find the solution set to the inequality

$$2x + 5 > 7$$

First, rewrite the inequality as a comparison of two functions. Here $f(x) > g(x)$, in which $f(x) = 2x + 5$ and $g(x) = 7$.

Now graph the two functions on a single set of axes.

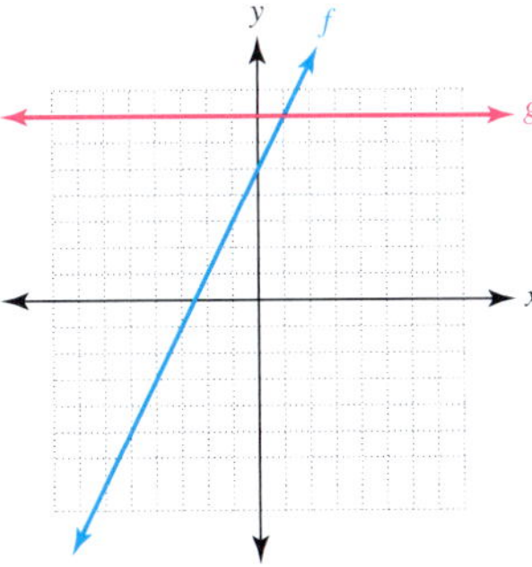

Here we ask the question, For what values of x is the graph of f above the graph of g?

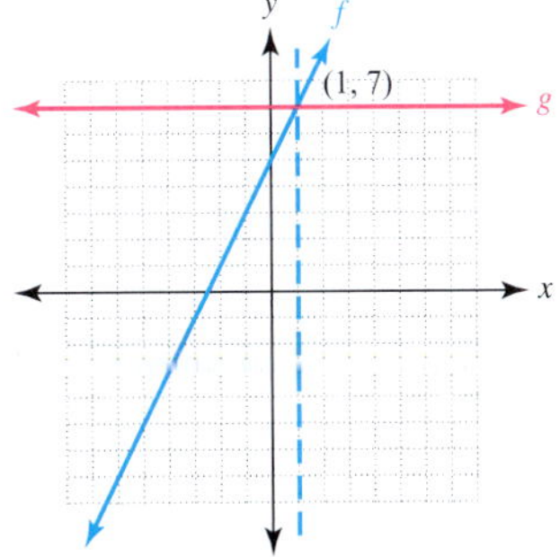

Next, draw a vertical dotted line through the point of intersection of the two functions. In this case, there will be a vertical line through the point (1, 7).

NOTE

A dotted line is used to indicate that the x-value of 1 is not included.

NOTE

The solution set contains the x-values that make the original statement $2x + 5 > 7$ true.

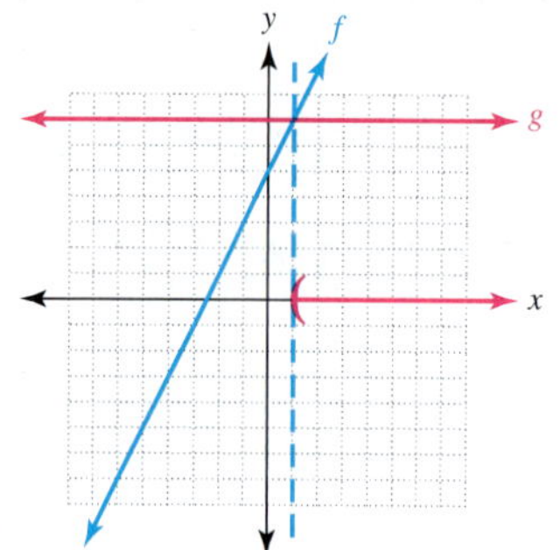

The solution set is every x-value that results in $f(x)$ being greater than $g(x)$, which is every x-value to the right of the dotted line.

We can express the solution set in set-builder notation as $\{x \mid x > 1\}$.

In Example 1, the function $g(x) = 7$ resulted in a horizontal line. In Example 2, we see that the same method works for comparing any two functions.

Example 2 Solving an Inequality Graphically

Solve the inequality graphically.

$2x - 3 \geq 5x$

First, rewrite the inequality as a comparison of two functions. Here, $f(x) \geq g(x)$, $f(x) = 2x - 3$, and $g(x) = 5x$. Now graph the two functions on a single set of axes.

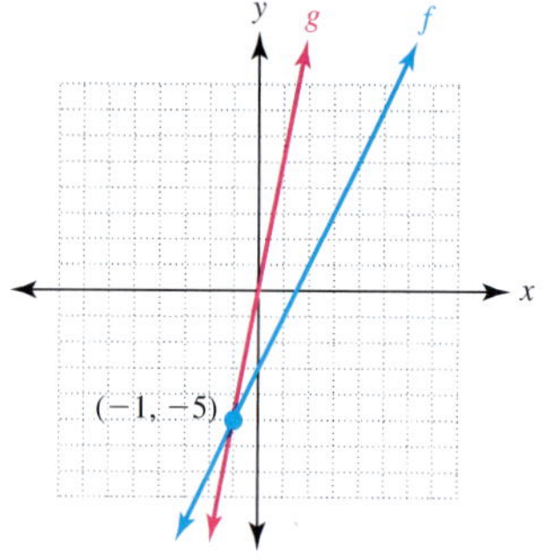

As in Example 1, draw a vertical line through the point of intersection of the two functions. The vertical line will go through the point $(-1, -5)$. In this case, the line is included (greater than or *equal to*), so the line is solid, not dotted.

Again, we need to mark every x-value that makes the statement true. In this case, that is every x for which the line representing $f(x)$ is above or intersects the line representing $g(x)$. That is the region in which $f(x)$ is greater than or equal to $g(x)$. We mark the x-values to the left of the line, but we also want to include the x-value on the line, so we make it a bracket rather than a parenthesis.

NOTE

A solid line is used to indicate that the x-value of -1 is included.

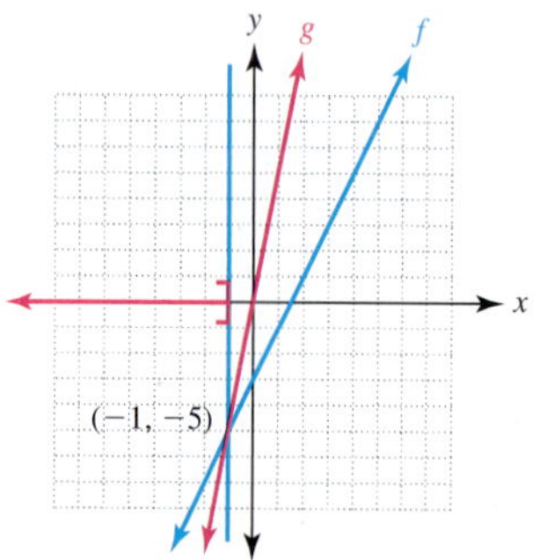

Finally, we express the solutions in set notation. We see that the solution set is every x-value less than or equal to -1, so we write

$\{x \mid x \leq -1\}$

The algorithm below summarizes our work in this section.

Step by Step

Solving an Inequality in One Variable Graphically

Step 1 Rewrite the inequality as a comparison of two functions.

$f(x) < g(x) \qquad f(x) > g(x) \qquad f(x) \leq g(x) \qquad f(x) \geq g(x)$

Step 2 Graph the two functions on a single set of axes.

Step 3 Draw a vertical line through the point of intersection of the two graphs. Use a dotted line if equality is not included ($<$ or $>$). Use a solid line if equality is included ($\leq$ or $\geq$).

Step 4 Mark the x-values that make the inequality a true statement.

Step 5 Write the solutions in set-builder notation.

It is possible for a linear inequality to have no solutions, or to be true for all real numbers. Using graphical methods makes these situations clear.

Example 3 Solving a Linear Inequality Graphically

Solve each inequality graphically.

(a) $5 + \frac{1}{2}x > \frac{x + 4}{2}$

Let

$$f(x) = 5 + \frac{1}{2}x = \frac{1}{2}x + 5$$

and
$$g(x) = \frac{x + 4}{2} = \frac{1}{2}x + 2$$

We note that the graphs of f and g have the same slope of $\frac{1}{2}$ and therefore are parallel. The graphs are shown below.

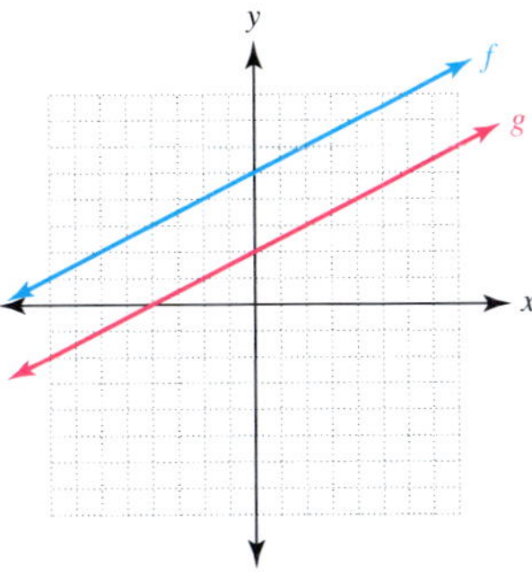

We ask, for what values of x is the graph of f above the graph of g? Clearly, f is *always* above g. So the original statement is true for *all* real numbers, and the solution set is $\mathbb{R}$.

(b) $\dfrac{9 - x}{3} \le \dfrac{-x - 6}{3}$

Let $\quad f(x) = \dfrac{9 - x}{3} = 3 - \dfrac{x}{3} = -\dfrac{1}{3}x + 3$

and $\quad g(x) = \dfrac{-x - 6}{3} = -\dfrac{1}{3}x - 2$

Again, we note that the graphs of f and g have the same slope.

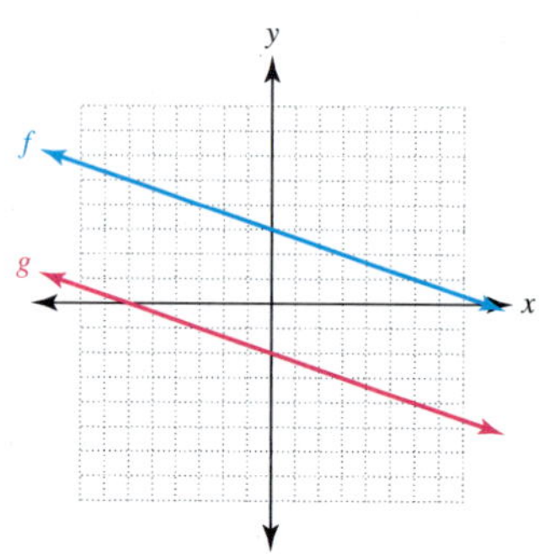

Now we ask, for what values of x is the graph of f below (or equal to) the graph of g? The answer here is "Never!" The original statement is never true, and the solution set is empty, or $\varnothing$.

Solving a business application graphically can be handled effectively with a graphing calculator. This is illustrated in Example 4.

Example 4 Solving a Business Application Graphically

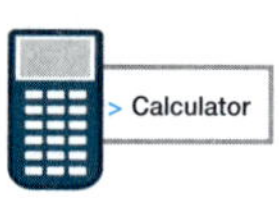

A company's cost per item is \$14.29, and its fixed costs per week total \$1,735. If the company charges \$25.98 per item, how many items must be manufactured and sold per week for the company to make a profit?

The company's cost function is

$C(x) = 14.29x + 1{,}735$

and the revenue function is

$R(x) = 25.98x$

where x is the number of items made and sold per week.

We want to know when revenue exceeds cost, so we need to solve the inequality

$$R(x) > C(x)$$

$$25.98x > 14.29x + 1{,}735$$

Using a graphical approach, we define functions in the calculator:

$Y_1 = 25.98x$

$Y_2 = 14.29x + 1{,}735$

To learn approximately where the two graphs cross, we explore using the TABLE utility:

X	Y_1	Y_2
0	0	1735
100	2598	3164
200	5196	4593

X=

Note that when $x = 100$, Y_1 (revenue) is lower than Y_2 (cost). But when $x = 200$, Y_1 is higher than Y_2. So the graphs must intersect between 100 and 200 on the x-axis, and a suitable viewing window could be $100 \le x \le 200$, $2{,}600 \le y \le 5{,}200$. Using this, we see (figure, below left):

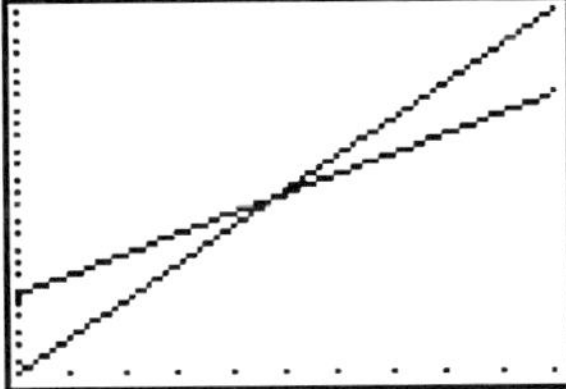

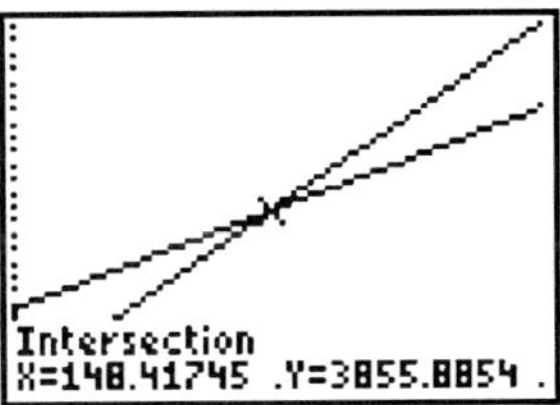

and we find the intersection (figure, above) at (148.41745, 3,855.8854). Since x needs to be a whole number, we conclude that the company makes a profit when x is at least 149, or when $x \ge 149$.

B.1 exercises

Boost *your* GRADE at ALEKS.com!

ALEKS®

- Practice Problems
- Self-Tests
- NetTutor
- e-Professors
- Videos

Name ______________________

Section __________ Date __________

Answers

1. ______________________
2. ______________________
3. ______________________
4. ______________________
5. ______________________
6. ______________________
7. ______________________
8. ______________________
9. ______________________
10. ______________________

< Objective 1 >

Solve each linear inequality graphically.

1. $2x < 8$

2. $-x < 4$

3. $\dfrac{x+3}{2} < -1$

4. $\dfrac{-3x+3}{4} > -3$

5. $7x - 7 < -2x + 2$ 

6. $7x + 2 > x - 4$

7. $6(1 + x) \geq 2(3x - 5)$

8. $2(x - 5) \geq 2x - 1$

9. $7x > \dfrac{9x - 5}{2}$ 

10. $-4x - 12 < x + 8$

Answers

1. Solution set $\{x \mid x < 4\}$

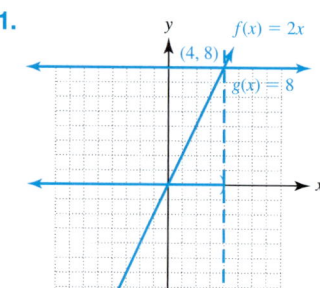

3. 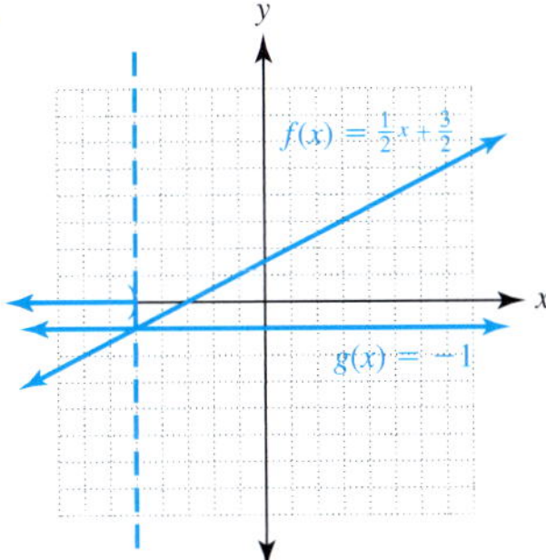

Solution set $\{x \mid x < -5\}$

5. 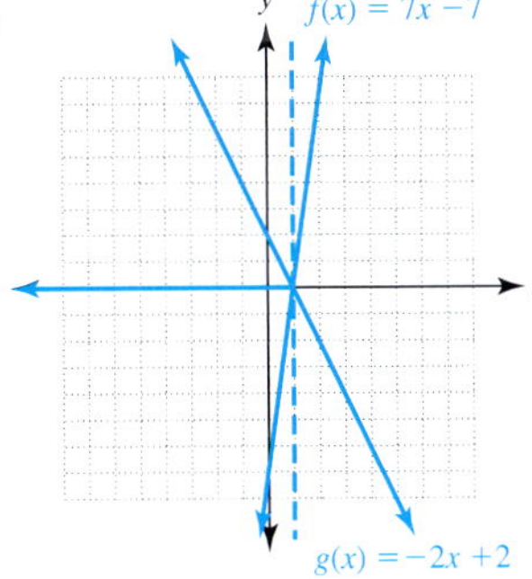

Solution set $\{x \mid x < 1\}$

7. 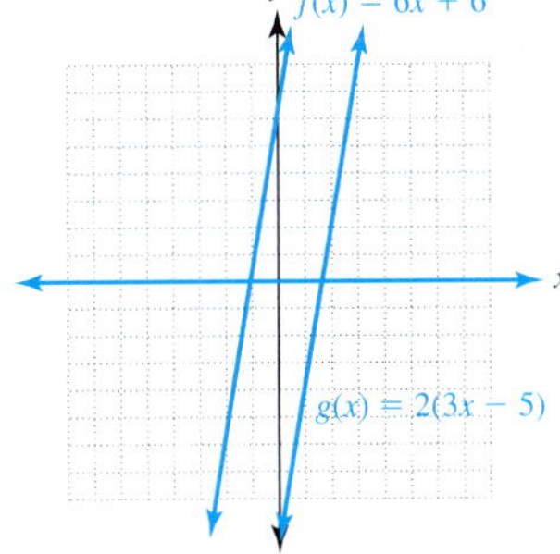

Solution set $\{x \mid x \in \mathbb{R}\}$

9. 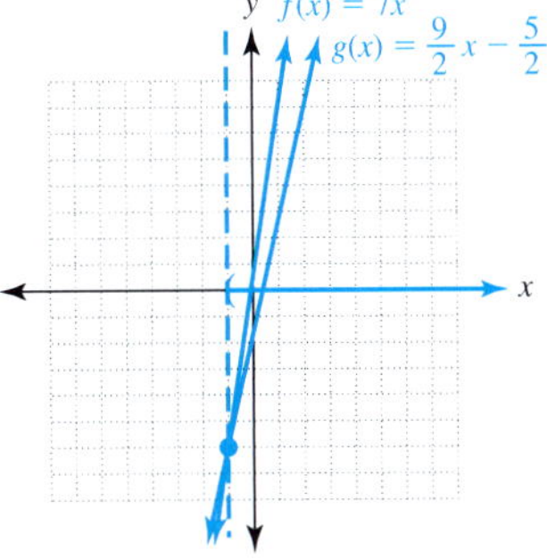

Solution set $\{x \mid x > -1\}$

B.2 Solving Absolute-Value Equations

< B.2 Objectives >

1 > Find the absolute value of an expression

2 > Solve an absolute-value equation

Equations may contain absolute-value notation in their statements. In this appendix, we look at algebraic solutions to statements that include absolute values. First, we review the concept of absolute value.

The **absolute value** of a real number is the distance from that real number to 0. Because absolute value is a distance, it is always positive. Formally, we say

Definition

Absolute Value

The **absolute value** of a number x is given by

$$|x| = \begin{cases} -x & \text{if } x < 0 \\ x & \text{if } x \geq 0 \end{cases}$$

Example 1 — **Finding the Absolute Value of a Number**

< Objective 1 >

Find the absolute value for each expression.

(a) $|-3|$ **(b)** $|7 - 2|$ **(c)** $|-7 - 2|$

(a) Because $-3 < 0$, $|-3| = -(-3) = 3$.

(b) $|7 - 2| = |5|$ Because $5 \geq 0$, $|5| = 5$.

(c) $|-7 - 2| = |-9|$ Because $-9 < 0$, $|-9| = -(-9) = 9$.

> CAUTION

The constant p, in the property box below, must be positive because an equation such as $|x| = -3$ has no solution. The absolute value of a quantity must always be equal to a nonnegative number.

Given an equation such as

$$|x| = 5$$

there are two possible solutions. The value of x could be 5 or -5. In either case, the absolute value is 5. This can be generalized with a property of absolute-value equations.

Property

Absolute-Value Equations—Property 1

For any positive number p, if

$$|x| = p$$

then

$$x = p \quad \text{or} \quad x = -p$$

We use this property in the next several examples.

Example 2 Solving an Absolute-Value Equation

< Objective 2 >

> CAUTION

A common mistake is to solve only the equation $x - 3 = 4$. You must solve *both* equations to find the two solutions.

Solve the equation

$$|x - 3| = 4$$

From Property 1, we know that the expression inside the absolute-value signs, $x - 3$, must equal either 4 or -4. We set up two equations and solve them both.

$(x - 3) = 4$ or $(x - 3) = -4$

$x - 3 = 4$ $\quad$ $x - 3 = -4$ $\quad$ Add 3 to both sides of the equation.

$x = 7$ $\quad$ $x = -1$

We arrive at the solution set, $\{-1, 7\}$.

We use Property 1 to solve subsequent examples in this appendix section.

Example 3 Solving an Absolute-Value Equation

Solve for x.

$$|3x - 2| = 4$$

From Property 1, we know that $|3x - 2| = 4$ is equivalent to the equations

$3x - 2 = 4$ or $3x - 2 = -4$ $\quad$ Add 2.

$3x = 6$ $\quad$ $3x = -2$ $\quad$ Divide by 3.

$x = 2$ $\quad$ $x = -\dfrac{2}{3}$

The solution set is $\left\{-\dfrac{2}{3}, 2\right\}$. These solutions are easily checked by replacing x with $-\dfrac{2}{3}$ and 2 in the original absolute-value equation.

An equation involving absolute value may have to be rewritten before you can apply Property 1. Consider Example 4.

Example 4 Solving an Absolute-Value Equation

Solve for x.

$$|2 - 3x| + 5 = 10$$

To use Property 1, we must first isolate the absolute value on the left side of the equation. This is easily done by subtracting 5 from both sides.

$$|2 - 3x| = 5$$

We can now proceed as before, by using Property 1.

$2 - 3x = 5$ or $2 - 3x = -5$ $\quad$ Subtract 2.

$-3x = 3$ $\quad$ $-3x = -7$ $\quad$ Divide by -3.

$x = -1$ $\quad$ $x = \dfrac{7}{3}$

The solution set is $\left\{-1, \dfrac{7}{3}\right\}$.

In some applications, there is more than one absolute value in an equation. Consider an equation of the form

$$|x| = |y|$$

Since the absolute values of x and y are equal, x and y are the same distance from 0, which means they are either *equal* or *opposite in sign*. This leads to a second general property of absolute-value equations.

Property

Absolute-Value Equations—Property 2

If $|x| = |y|$

then $x = y$ or $x = -y$

We look at an application of this second property in Example 5.

Example 5 **Solving Equations with Two Absolute-Value Expressions**

Solve for x.

$|3x - 4| = |x + 2|$

By Property 2, we can write

$$3x - 4 = x + 2 \quad \text{or} \quad 3x - 4 = -(x + 2)$$

$$3x - 4 = -x - 2 \qquad \text{Add 4 to both sides.}$$

$$3x = x + 6 \qquad 3x = -x + 2 \qquad \text{Isolate the } x\text{-term.}$$

$$2x = 6 \qquad 4x = 2 \qquad \text{Divide by 2.}$$

$$x = 3 \qquad x = \frac{1}{2}$$

The solution set is $\left\{\frac{1}{2}, 3\right\}$.

B.2 exercises

< Objective 1 >

Find the absolute value for each expression.

1. $|15|$

2. $|-18|$

3. $|8 - 3|$

4. $|-23 - 11|$

5. $|-12 - 19|$

6. $|-13| - |-12|$

7. $|-13| + |12|$

8. $|-13 - 12|$

9. $-|-13| - |-12|$

10. $-|(-13) - (-12)|$

< Objective 2 >

Solve the equations.

11. $|x| = 5$

12. $|x - 2| = 6$

13. $|2x - 1| = 6$

14. $2|3x - 5| = 12$

15. $|5x - 2| = -3$

16. $|x + 5| - 2 = 5$

17. $8 - |x - 4| = 5$ 

18. $|3x + 1| = |2x - 3|$

19. $|5x - 2| = |2x - 4|$

20. $\left|2 + \frac{5}{8}x\right| = \frac{3}{8}$ 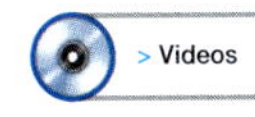

Answers

1. 15 **3.** 5 **5.** 31 **7.** 25 **9.** -25 **11.** $\{5, -5\}$

13. $\left\{-\frac{5}{2}, \frac{7}{2}\right\}$ **15.** No solution **17.** $\{7, 1\}$ **19.** $\left\{-\frac{2}{3}, \frac{6}{7}\right\}$

Name ______________

Section ________ Date ________

Answers

1. ________ 2. ________

3. ________ 4. ________

5. ________ 6. ________

7. ________ 8. ________

9. ________ 10. ________

11. ________ 12. ________

13. ________ 14. ________

15. ________

16. ________

17. ________

18. ________

19. ________

20. ________

B.3 Solving Absolute-Value Equations Graphically

< B.3 Objectives >

1 > Graph an absolute-value function

2 > Solve absolute-value equations in one variable graphically

In appendix section B.2, we learned to solve absolute-value equations algebraically. In appendix section B.3, we examine a graphical method for solving similar equations.

To demonstrate the graphical method, we first look at the graph of an absolute-value function. We start by looking at the graph of the function $f(x) = |x|$. All other graphs of absolute-value functions are variations of this graph.

The graph can be found using a graphing calculator (most graphing calculators use **abs** to represent the absolute value). We will develop the graph from a table of values.

NOTE

Graph the function $y = |x|$ as $Y_1 = \text{abs}(x)$

x	$f(x) = \|x\|$
−3	3
−2	2
−1	1
0	0
1	1
2	2

Plotting these ordered pairs, we see a pattern emerge. The graph is like a large V that has its vertex at the origin. The slope of the line to the right of 0 is 1, and the slope of the line to the left of 0 is −1.

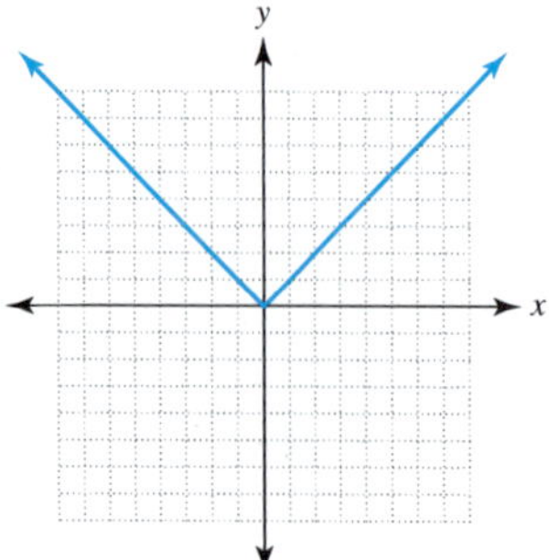

Let us now see what happens to the graph when we add or subtract some constant inside the absolute-value bars.

Example 1 — Graphing an Absolute-Value Function

< Objective 1 >

Graph each function.

(a) $f(x) = |x - 3|$

Again, we start with a table of values.

x	$f(x)$
−2	5
−1	4
0	3
1	2
2	1
3	0
4	1
5	2

NOTE

The equation

$f(x) = |x - 3|$

would be entered as

$Y_1 = \text{abs}(x - 3)$

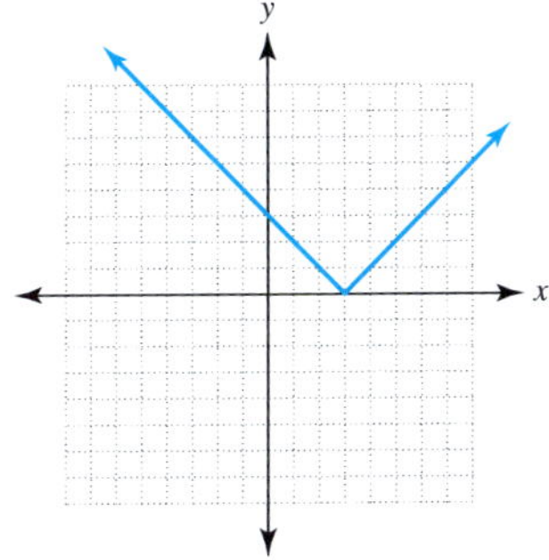

Then we plot the points associated with the set of ordered pairs. The graph is shown to the left.

The graph of the function $f(x) = |x - 3|$ is the same shape as the graph of the function $f(x) = |x|$; it has just shifted to the right 3 units.

(b) $f(x) = |x + 1|$

We begin with a table of values.

x	$f(x)$
−2	1
−1	0
0	1
1	2
2	3
3	4

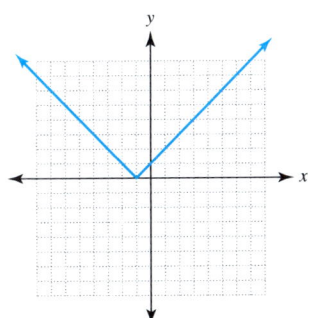

Again, you will find the graph in the margin.

Note that the graph of $f(x) = |x + 1|$ is the same shape as the graph of the function $f(x) = |x|$, except that it has shifted 1 unit to the left.

We can summarize what we have discovered about the horizontal shift of the graph of an absolute-value function.

Property

Horizontal Shifts of Absolute-Value Functions

The graph of the function $f(x) = |x - a|$ will be the same shape as the graph of $f(x) = |x|$ except that the graph will be shifted a units

To the right if a is positive

To the left if a is negative

If a is negative, $x - a$ will be x plus some positive number.

We now use these methods to solve equations that contain an absolute-value expression.

Example 2 Solving an Absolute-Value Equation Graphically

< Objective 2 >

Graphically find the solution set for the equation

$|x - 3| = 4$

We graph the function associated with each side of the equation.

$f(x) = |x - 3|$ and $g(x) = 4$

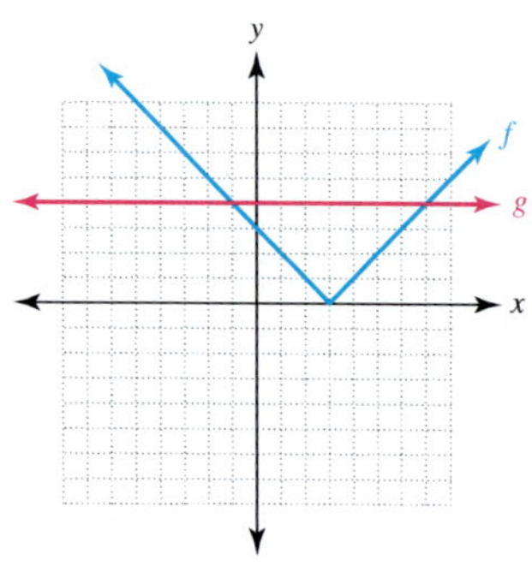

Then we draw a vertical line through each of the intersection points.

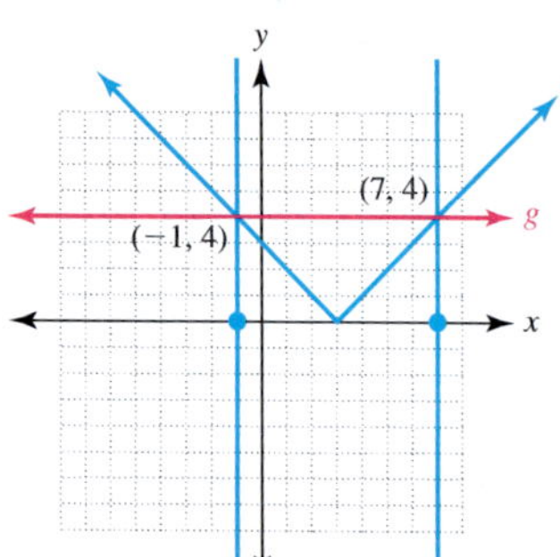

We ask the question: For what values of x do f and g coincide?

Looking at the x-values of the two vertical lines, we find the solutions to the original equation. There are two x-values that make the statement true: -1 and 7. The solution set is $\{-1, 7\}$.

Example 3 illustrates a case involving an absolute-value expression and a linear (but not simply constant) expression.

Example 3 Solving an Absolute-Value Equation Graphically

Solve the equation graphically.

$|x - 2| = 4 - \frac{1}{2}x$

Let $f(x) = |x - 2|$

and $g(x) = 4 - \frac{1}{2}x = -\frac{1}{2}x + 4$

We graph each function.

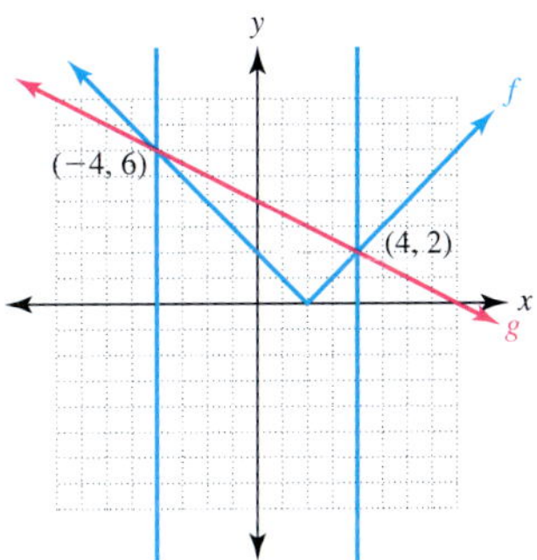

Draw a vertical line through each point of intersection. The vertical lines hit the x-axis when $x = -4$ and when $x = 4$. There are therefore two solutions, and the solution set is $\{-4, 4\}$.

In each of the examples of this appendix section, we have found two solutions. It is also quite possible for an absolute-value equation to have one solution, no solution, or even an infinite number of solutions. Fortunately, the situation will be clear if we are employing graphical methods.

Example 4 Solving an Absolute-Value Equation Graphically

Solve the equation graphically.

$|x + 2| = 2x + 7$

Let $f(x) = |x + 2|$

and $g(x) = 2x + 7$

Graph the functions.

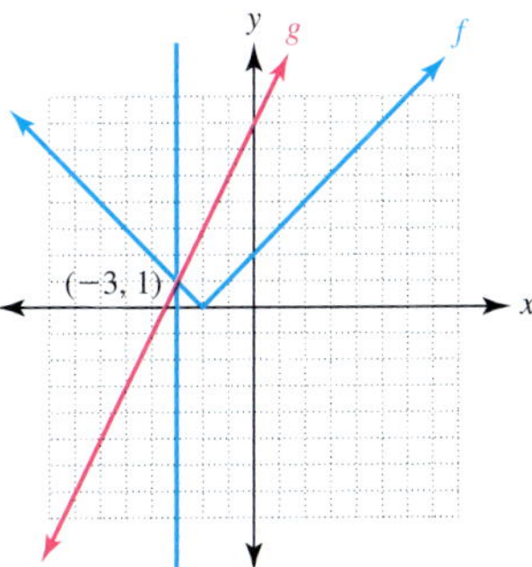

Since the slope of g is 2, we are convinced that the graph of g will meet the graph of f exactly once. The point of intersection is $(-3, 1)$. Drawing a vertical through this point, we note the x-value of -3. The solution set is $\{-3\}$.

If you are creating graphs by hand, it can be very difficult to locate the point(s) of intersection. But as you gain experience with a graphing calculator (changing the viewing window and finding intersection points), you will find that you can apply the graphical methods shown here to solve "messy" equations with great accuracy.

B.3 exercises

Boost your GRADE at ALEKS.com!

ALEKS®

- Practice Problems
- Self-Tests
- NetTutor
- e-Professors
- Videos

Name ____________________

Section ________ Date ________

Answers

1. ____________________
2. ____________________
3. ____________________
4. ____________________
5. ____________________
6. ____________________
7. ____________________
8. ____________________
9. ____________________
10. ____________________
11. ____________________
12. ____________________

< Objective 1 >

Graph each function.

1. $f(x) = |x - 3|$

2. $f(x) = |x + 2|$ 

3. $f(x) = |x + 3|$

4. $f(x) = |x - (-5)|$

< Objective 2 >

Solve the equations graphically.

5. $|x| = 3$

6. $|x| = 5$

7. $|x - 2| = 5$ 

8. $|x + 4| = 2$

9. $|x - 3| = 5 - \frac{1}{3}x$ 

10. $|x + 1| = \frac{1}{3}x + 5$

Determine the function represented by each graph.

11.

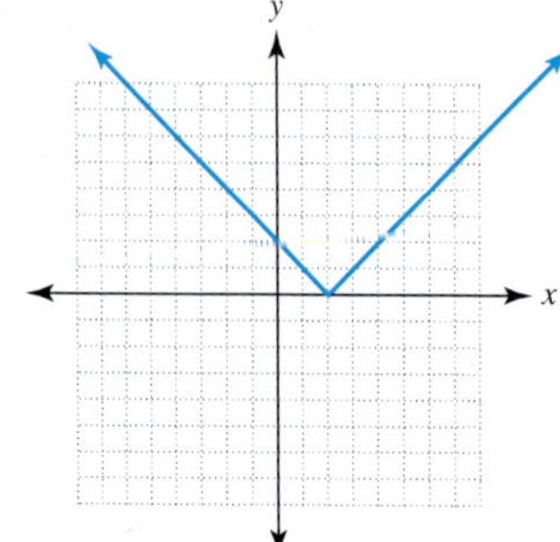

12.

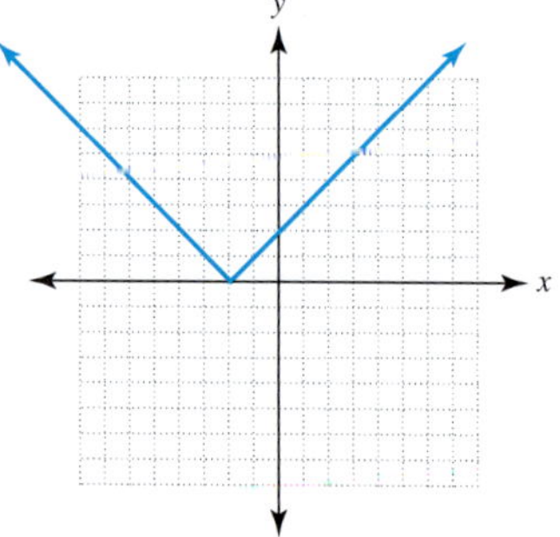

Answers

1.

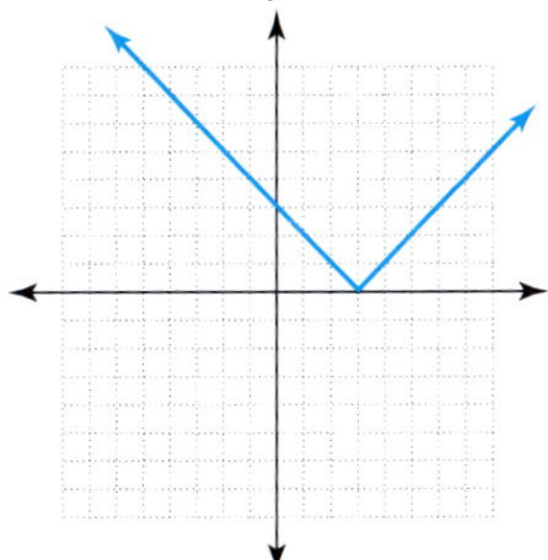

3.

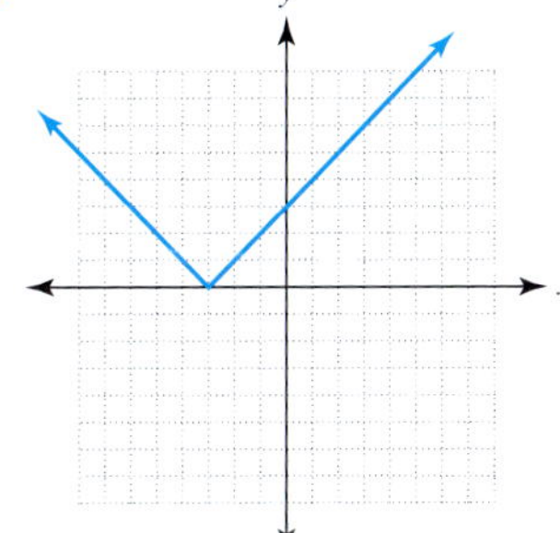

5.

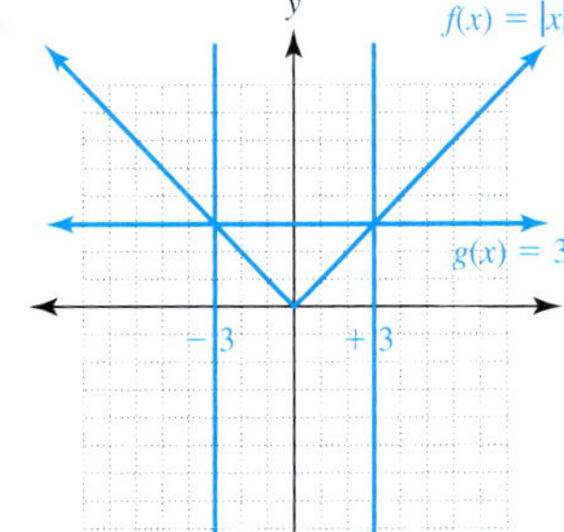

Solution set $\{-3, 3\}$

7.

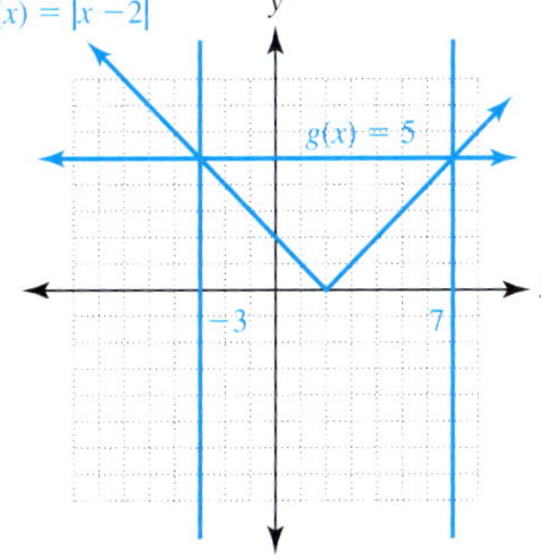

Solution set $\{-3, 7\}$

9.

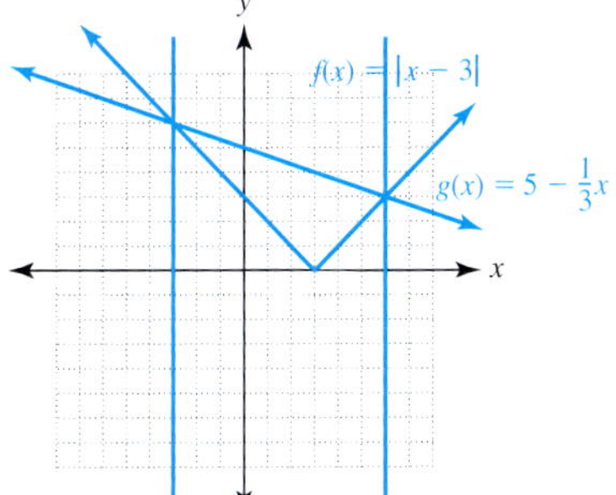

Solution set $\{-3, 6\}$

11. $f(x) = |x - 2|$

B.4 Solving Absolute-Value Inequalities

< B.4 Objectives >

1 > Solve a compound inequality

2 > Solve an absolute-value inequality

NOTE

Because there are two inequality signs in a single statement, these are sometimes called *double inequalities*.

In this appendix section B.4, we solve absolute-value inequalities.

First, we look at two types of inequality statements that arise frequently in mathematics. Consider a statement such as

$$-2 < x < 5$$

It is called a **compound inequality** because it combines the two inequalities

$$-2 < x \qquad \text{and} \qquad x < 5$$

When we begin with a compound inequality such as

$$-3 \le 2x + 1 \le 7$$

we find an equivalent statement in which the variable is isolated in the middle. Example 1 illustrates.

Example 1 Solving a Compound Inequality

< Objective 1 >

NOTES

We are really applying the addition property to each of the *two* inequalities that make up the double-inequality statement.

When we divide by a positive number, the direction of the inequality is preserved.

In interval notation, we write [−2, 3].

Solve and graph the compound inequality.

$$-3 \le 2x + 1 \le 7$$

First, we subtract 1 from each of the three members of the compound inequality.

$$-3 - 1 \le 2x + 1 - 1 \le 7 - 1$$

or

$$-4 \le 2x \le 6$$

We now divide by 2 to isolate the variable x.

$$-\frac{4}{2} \le \frac{2x}{2} \le \frac{6}{2}$$

$$-2 \le x \le 3$$

The solution set consists of all numbers between −2 and 3, including −2 and 3, and is written

$$\{x \mid -2 \le x \le 3\}$$

That set is graphed as shown here.

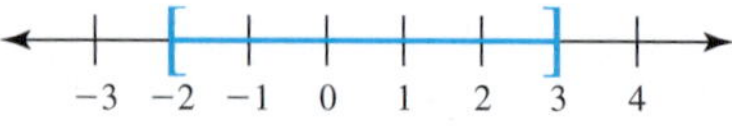

Note: Our solution set is equivalent to

$$\{x \mid x \ge -2 \text{ and } x \le 3\}$$

Look at the individual graphs.

$\{x \mid x \ge -2\}$

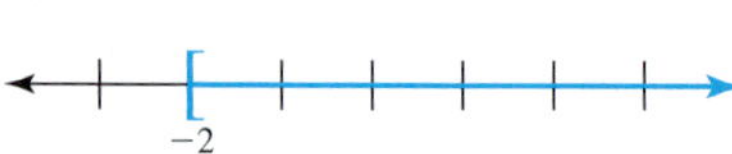

$\{x \mid x \le 3\}$

$\{x \mid x \ge -2 \text{ and } x \le 3\}$

NOTE

Using set-builder notation, we can write

$\{x \mid x \ge -2\} \cap \{x \mid x \le 3\}$

Because the connecting word is *and,* we want the *intersection* of the sets, that is, those numbers common to both sets.

A compound inequality may also consist of two inequality statements connected by the word *or.* Example 2 illustrates how to solve that type of compound inequality.

Example 2 Simplifying a Compound Inequality

Solve and graph the inequality

$2x - 3 < -5 \quad \text{or} \quad 2x - 3 > 5$

In this case, we must work with each of the inequalities *separately.*

$2x - 3 < -5 \quad \text{or} \quad 2x - 3 > 5$ Add 3.

$2x < -2 \qquad\qquad 2x > 8$ Divide by 2.

$x < -1 \qquad\qquad x > 4$

The graph of the solution set, $\{x \mid x < -1 \text{ or } x > 4\}$, is shown.

$\{x \mid x < -1 \text{ or } x > 4\}$

NOTES

In interval notation, we write $(-\infty, -1) \cup (4, \infty)$.

In set-builder notation, we can write

$\{x \mid x < -1\} \cup \{x \mid x > 4\}$

The connecting word is *or* in this case, so the solution set of the original inequality is the *union* of the two sets. That is, all of the numbers that belong to either or both of the sets.

Imagine that you receive a phone call from a friend who says that his car is stuck on the freeway, within 3 mi of mileage marker 255. Where is he? He must be somewhere between mileage marker 252 and marker 258. What you've actually solved here is an example of an **absolute-value inequality.** Absolute-value inequalities are in one of two forms:

$|x - a| < b \quad \text{or} \quad |x - a| > b$

The mileage marker example is of the first form. We could say

$|x - 255| < 3$

To solve an equation of this type, use the rule shown.

Property

Absolute-Value Inequalities—Property 1

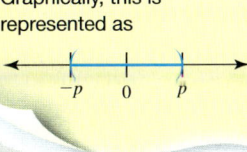

For any positive number p, if

$|x| < p$

then

$-p < x < p$

NOTE

Graphically, this is represented as

In our example, because $|x - 255| < 3$,

$-3 < x - 255 < 3$

Adding 255 to each member of the inequality, we get

$-3 + 255 < x - 255 + 255 < 3 + 255$

or $\quad 252 < x < 258$

Example 3 Solving an Absolute-Value Inequality

< Objective 2 >

Solve the inequality; then graph the solution set.

$|x - 5| < 9$

According to Property 1, we have

$|x - 5| < 9$

$-9 < x - 5 < 9$ Add 5 to each member.

$-9 + 5 < x < 9 + 5$

$-4 < x < 14$

NOTE

In interval notation, we write $(-4, 14)$.

Graphing the solution set, $\{x \mid -4 < x < 14\}$, we get

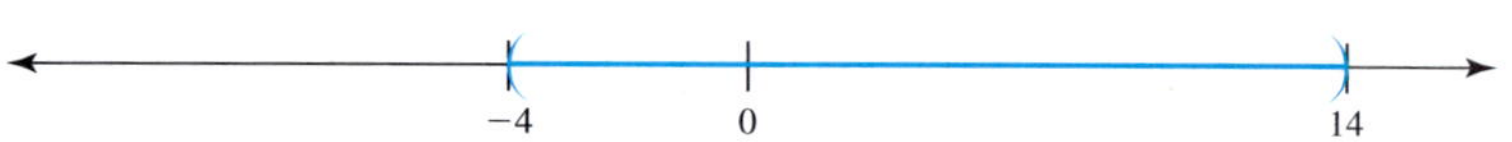

What about inequalities of the form $|x - a| > b$? The rule below applies.

Property

Absolute-Value Inequalities—Property 2

For any positive number p, if

$|x| > p$

then

$x < -p \quad \text{or} \quad x > p$

NOTE

Graphically, this is represented as

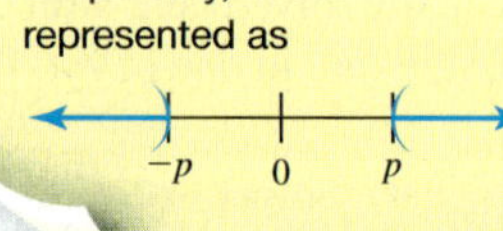

Example 4 Solving an Absolute-Value Inequality

Solve the inequality; then graph the solution set.

$|x - 7| > 19$

By Property 2,

$x - 7 > 19 \quad \text{or} \quad x - 7 < -19$

Solving each inequality by adding 7 to each side, we get

$x > 26 \quad \text{or} \quad x < -12$

Now we can graph the solution set.

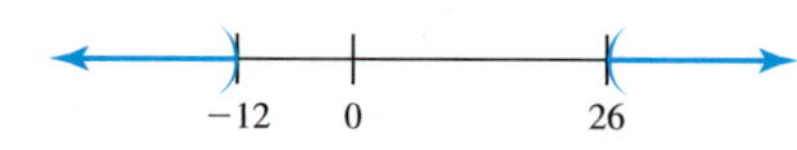

B.4 exercises

< Objective 1 >

Solve each inequality and then graph the solution set.

1. $4 \le x + 1 \le 7$

2. $-6 \le 3x \le 9$

3. $1 \le 2x - 3 \le 6$ > Videos

4. $-7 \le 3 + 2x \le 8$

5. $x - 1 < -3$ or $x - 1 > 3$

6. $2x + 5 < -3$ or $2x + 5 > 3$

< Objective 2 >

7. $|x| < 5$

8. $|x| \le 4$ > Videos

9. $|x + 6| \le 4$

10. $|5 - x| < 3$

11. $|3x + 4| \ge 5$

12. $|2x + 3| \le 9$

Answers

1. $\{x \mid 3 \le x \le 6\}$

3. $\left\{x \mid 2 \le x \le \frac{9}{2}\right\}$

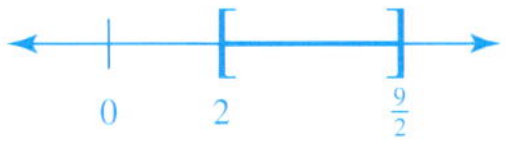

5. $\{x \mid x < -2 \text{ or } x > 4\}$

7. $\{x \mid -5 < x < 5\}$

9. $\{x \mid -10 \le x \le -2\}$

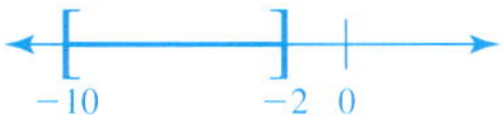

11. $\left\{x \mid x \le -3 \text{ or } x \ge \frac{1}{3}\right\}$

Name ____________

Section ________ Date ________

Answers

1. ____________

2. ____________

3. ____________

4. ____________

5. ____________

6. ____________

7. ____________

8. ____________

9. ____________

10. ____________

11. ____________

12. ____________

B.5 Solving Absolute-Value Inequalities Graphically

< B.5 Objectives >

1 > Solve absolute-value inequalities in one variable graphically

2 > Solve absolute-value inequalities in one variable algebraically

In Appendix B.3, we looked at a graphical method for solving an absolute-value equation. In appendix section B.5, we look at a graphical method for solving absolute-value inequalities.

Absolute-value inequalities in one variable x are obtained from absolute-value equations by replacing the symbol for equality ($=$) with one of the inequality symbols ($<, >, \leq, \geq$).

The general form for an absolute-value inequality in one variable is

$$|x - a| < b$$

where the symbol $<$ can be replaced with $>$, $\leq$, or $\geq$. Examples of absolute-value inequalities in one variable include

$$|x| < 6 \qquad |x - 4| \geq 2 \qquad |3x - 5| \leq 8$$

Example 1 Solving an Absolute-Value Inequality Graphically

< Objective 1 >

Graphically solve

$$|x| < 6$$

As we did in previous appendix sections, we begin by letting each side of the inequality represent a function. Here

$$f(x) = |x| \qquad \text{and} \qquad g(x) = 6$$

Now we graph both functions on the same set of axes.

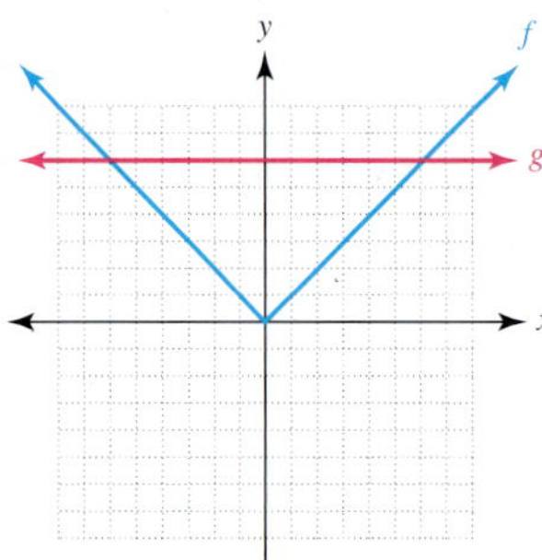

We now ask the question: For what values of x is f *below* g?

We next draw a vertical dotted line (equality is not included) through the points of intersection of the two graphs.

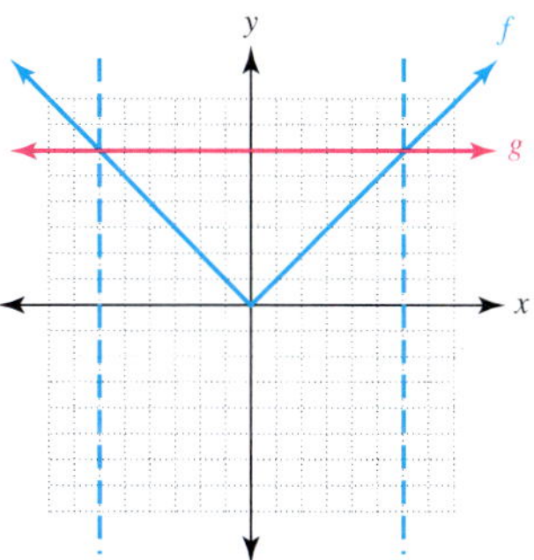

The solution set is any value of x for which the graph of $f(x)$ is below the graph of $g(x)$.

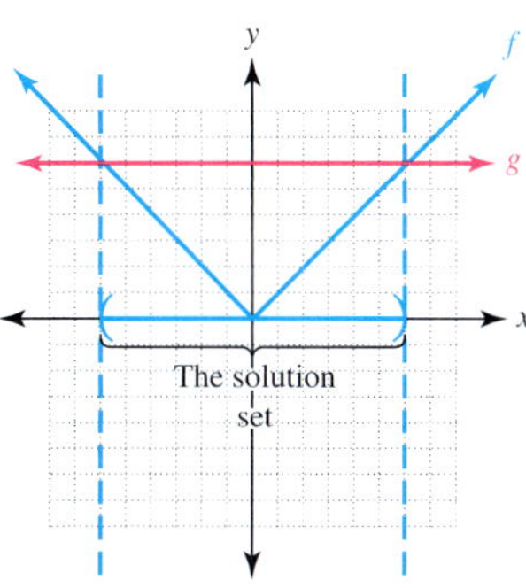

Keep in mind that we are searching for x-values that make the original statement true.

In set-builder notation, we write $\{x \mid -6 < x < 6\}$.

The graphical method of Example 1 relates to the general statement from Appendix B.4.

Property

Absolute-Value Inequalities

For any positive number p, if

$|x| < p$

then $-p < x < p$

Before we continue with a graphical approach, let's review the use of this property in solving an absolute-value inequality.

Example 2 **Solving an Absolute-Value Inequality Algebraically**

< Objective 2 >

NOTE

With this property we can *translate* an absolute-value inequality to a compound inequality *not* containing an absolute value, which can be solved by our earlier methods.

Solve the inequality algebraically and graph the solution set on a number line.

$|x - 3| < 5$

From this property, we know that the given absolute-value inequality is equivalent to the compound inequality

$-5 < x - 3 < 5$

Solve as before.

$-5 < x - 3 < 5$ Add 3 to all three parts.

$-2 < x < 8$

NOTE

The solution set is an open interval on the number line.

The solution set is

$\{x \mid -2 < x < 8\}$

The graph of the solution set is shown below.

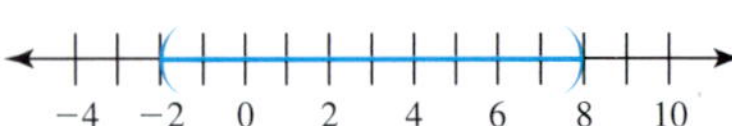

In Example 3, we look at a graphical method for solving the same inequality.

Example 3 Solving an Absolute-Value Inequality Graphically

Solve the inequality graphically, and graph the solution set on a number line.

$|x - 3| < 5$

Let $f(x) = |x - 3|$ and $g(x) = 5$, and graph both functions on the same set of axes.

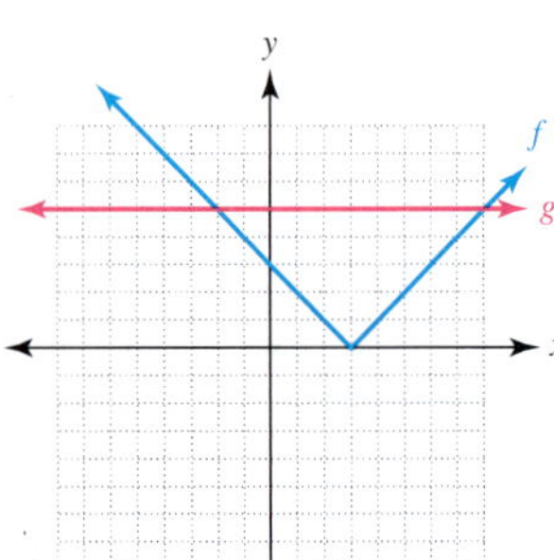

Here again, we ask: For what values of *x* is *f below g*?

Drawing a vertical dotted line through the intersection points, we find the set of x-values for which $f(x) < g(x)$.

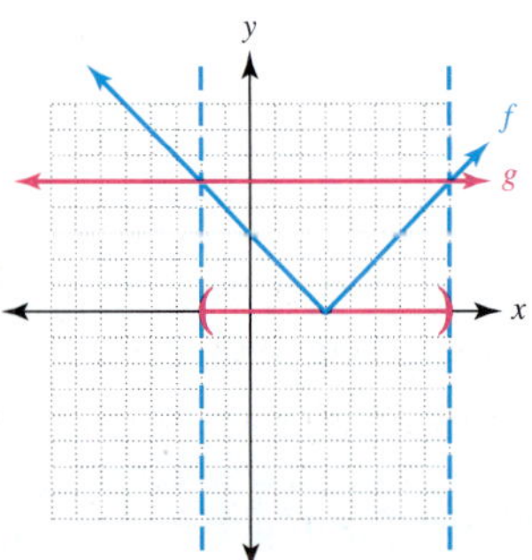

We see that the desired x-values are those that lie between -2 and 8. The solution set is $\{x \mid -2 < x < 8\}$. The graph of the solution set is

Note that the graph of the solution set shown here is precisely the portion of the x-axis that has been marked in the previous two-dimensional graph.

We have seen that the solution set for the statement $|x| < 6$ is the set of all numbers between -6 and 6. Now, how does the result change for the statement $|x| > 6$? Solving graphically will make this clear.

Example 4 Solving an Absolute-Value Inequality Graphically

Solve the inequality graphically, and graph the solution set on a number line.

$|x| > 6$

As before, we define

$f(x) = |x|$ and $g(x) = 6$

We graph both functions on the same set of axes, and we draw vertical dotted lines through the points of intersection.

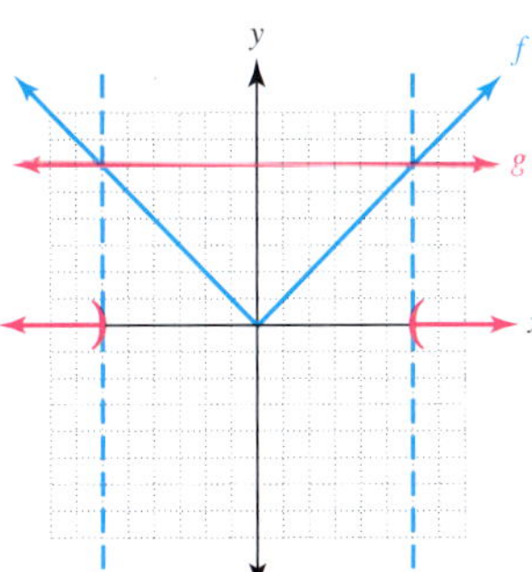

Now we ask: For what values of x is the graph of f *above* the graph of g?

We see that f is *above* g when $x < -6$ or when $x > 6$. The solution set is $\{x \mid x < -6$ or $x > 6\}$. The graph of the solution set is

−6 0 6

The solution set for the statement $|x| > 6$ is the set of all numbers that are greater than 6 or less than −6. We can describe these numbers with the compound inequality

$x < -6$ or $x > 6$

This relates to the general statement from Appendix B.4:

Property

Absolute-Value Inequalities

For any positive number p, if

$|x| > p$

then $x < -p$ or $x > p$

We now review the use of this property in solving an absolute-value inequality.

Example 5 Solving an Absolute-Value Inequality Algebraically

Solve the inequality algebraically, and graph the solution set on a number line.

$|2 - x| > 8$

NOTE

Again we *translate* the absolute-value inequality to the compound inequality *not* containing an absolute value.

From our second property, we know that the given absolute-value inequality is equivalent to the compound inequality

$2 - x < -8$ or $2 - x > 8$

Solving as before, we have

$$2 - x < -8 \quad \text{or} \quad 2 - x > 8$$
$$-x < -10 \qquad -x > 6$$
$$x > 10 \qquad x < -6$$

When we divide by a negative number, we reverse the direction of the inequality.

The solution set is $\{x \mid x < -6 \text{ or } x > 10\}$, and the graph of the solution set is shown here.

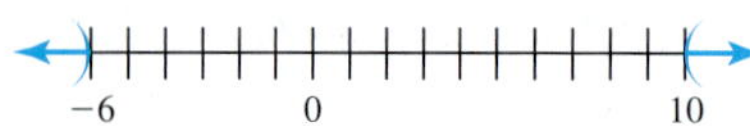

A property that can be useful in working with absolute values is given in the next box.

Property

Absolute-Value Expressions

For any real numbers *a* and *b*

$|a - b| = |b - a|$

Our final example looks at a graphical approach to solving the same inequality studied in Example 5.

Example 6 Solving an Absolute-Value Inequality Graphically

Solve the inequality graphically, and graph the solution set on a number line.

$|2 - x| > 8$

Let $f(x) = |2 - x|$ and $g(x) = 8$. Using Property 3, we know that $|2 - x| = |x - 2|$, so we define f as

$f(x) = |x - 2|$

Graphing f and g on the same set of axes, we have

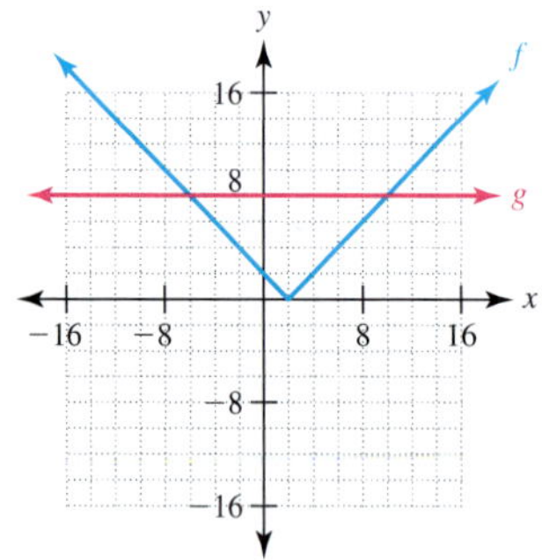

Drawing a vertical dotted line through each intersection point, we mark (on the x-axis) all x-values for which $f(x) > g(x)$.

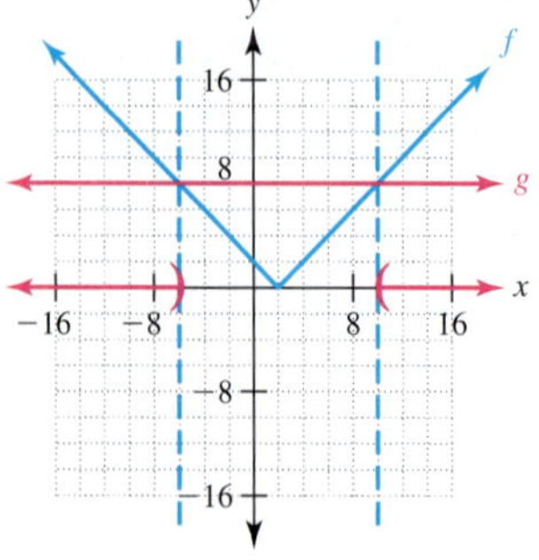

The solution set is therefore $\{x \mid x < -6 \text{ or } x > 10\}$, and the graph of the solution set is

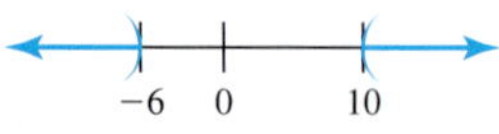

B.5 exercises

< Objective 2 >

Solve each inequality algebraically. Graph the solution set on a number line.

1. $|x| < 5$ 

2. $|x + 5| < 3$

3. $|x + 6| \leq 4$ 

4. $|x + 5| \geq 0$

5. $|3x + 4| \geq 5$ 

6. $|2x + 3| \leq 9$

< Objective 1 >

Solve each inequality graphically.

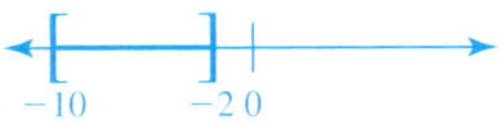

7. $|x| < 4$

8. $|x| \geq 2$

9. $|x - 3| < 4$

10. $|x + 2| > 4$

11. $|x + 1| \leq 5$

12. $|x - 4| > -1$

Name ____________

Section ______ Date ______

Answers

1. ____________
2. ____________
3. ____________
4. ____________
5. ____________
6. ____________
7. ____________
8. ____________
9. ____________
10. ____________
11. ____________
12. ____________

Answers

1. $\{x \mid -5 < x < 5\}$ (number line: −5, 0, 5)

3. $\{x \mid -10 \leq x \leq -2\}$ (number line: −10, −2 0)

5. $\left\{x \mid x \le -3 \text{ or } x \ge \dfrac{1}{3}\right\}$ −3 0 $\frac{1}{3}$

7. $f(x) = |x|$ y

$g(x) = 4$ x

Solution set $\{x \mid -4 < x < 4\}$

9.

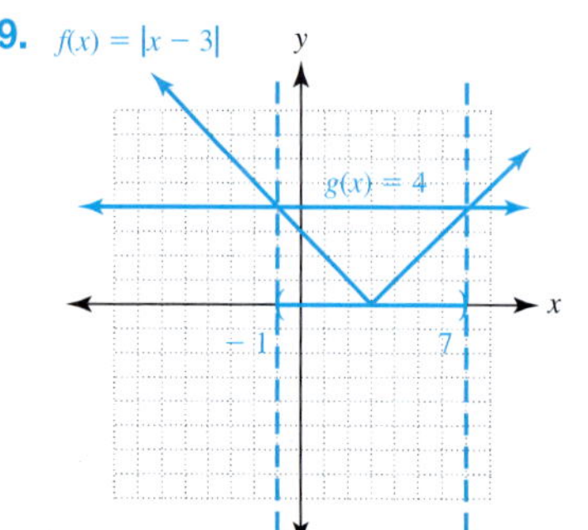

Solution set $\{x \mid -1 < x < 7\}$

11.

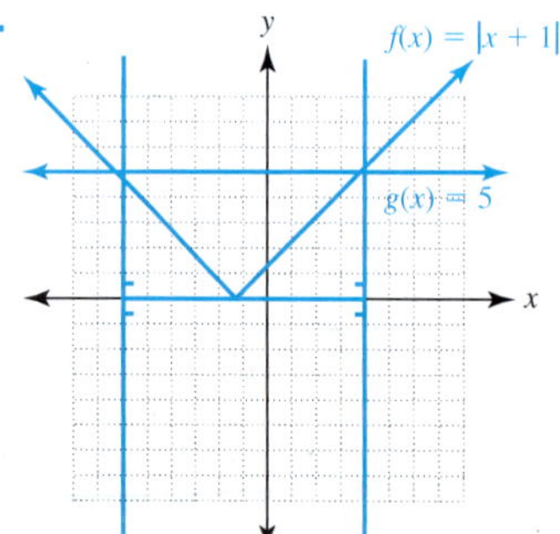

Solution set $\{x \mid -6 \le x \le 4\}$

Answers to Reading Your Text Exercises, Summary Exercises, Self-Tests, Cumulative Reviews, and Final Exam

Reading Your Text

Section 0.1 **(a)** natural; **(b)** one; **(c)** product; **(d)** Subtracting
Section 0.2 **(a)** whole; **(b)** negative; **(c)** integers; **(d)** magnitude
Section 0.3 **(a)** magnitudes; **(b)** order; **(c)** associative; **(d)** zero
Section 0.4 **(a)** Multiplication; **(b)** different; **(c)** positive; **(d)** area
Section 0.5 **(a)** factor; **(b)** grouping; **(c)** exponential; **(d)** exponent

Summary Exercises—Chapter 0

1. $\frac{10}{14}, \frac{20}{28}, \frac{50}{70}$ **3.** $\frac{8}{18}, \frac{16}{36}, \frac{24}{54}$ **5.** $\frac{1}{9}$ **7.** $\frac{2}{3}$ **9.** $\frac{3}{2}$
11. $\frac{7}{18}$ **13.** \$198 **15.** 48 lots **17.** 12 **19.** -3
21. -4 **23.** 16 **25.** $<$ **27.** $<$ **29.** 8 **31.** -11
33. $-\frac{31}{39}$ **35.** 4 **37.** -5 **39.** 5 **41.** $\frac{13}{2}$ **43.** 1
45. 0 **47.** \$467.66 **49.** 36 **51.** -15 **53.** 16
55. -360 **57.** -4 **59.** -24 **61.** -4 **63.** Undefined
65. $\frac{7}{6}$ **67.** -2 **69.** \$42,000 **71.** $3 \cdot 3 \cdot 3$
73. $2 \cdot 2 \cdot 2 \cdot 2 \cdot 2 \cdot 2$ **75.** 12 **77.** 48 **79.** 41
81. 20 **83.** 324 **85.** 11 **87.** 62 points

Self-Test—Chapter 0

1. $\frac{3}{11}$ **2.** $\frac{25}{16}$ **3.** $\frac{29}{30}$ **4.** $\frac{21}{80}$ **5.** -3 **6.** -12
7. -7 **8.** 4 **9.** -35 **10.** 54 **11.** -12 **12.** 11
13. 11 **14.** 113 **15.** $\frac{9}{44}$ **16.** $\frac{14}{5}$ **17.** $-7 < -5$
18. $8 + (-3)^2 > 8 - (-3)$ **19.** 32 lots **20.** \$1,775

Reading Your Text

Section 1.1 **(a)** variables; **(b)** sum; **(c)** multiplication; **(d)** expression
Section 1.2 **(a)** variable; **(b)** evaluating; **(c)** operations; **(d)** square
Section 1.3 **(a)** first; **(b)** term; **(c)** factors; **(d)** (numerical) coefficient
Section 1.4 **(a)** fresh; **(b)** expressions; **(c)** solution; **(d)** one
Section 1.5 **(a)** nonzero; **(b)** reciprocal; **(c)** original; **(d)** reciprocal
Section 1.6 **(a)** divide; **(b)** Multiplying; **(c)** substituting; **(d)** simplified
Section 1.7 **(a)** formula; **(b)** coefficient; **(c)** original; **(d)** distance
Section 1.8 **(a)** greater; **(b)** equivalent; **(c)** positive; **(d)** negative

Summary Exercises—Chapter 1

1. $y + 8$ **3.** $8a$ **5.** $x(x - 7)$ **7.** $\frac{a+2}{a-2}$ **9.** $\frac{9}{x}$
11. -7 **13.** 15 **15.** 25 **17.** 1 **19.** -20
21. $4a^3, -3a^2$ **23.** $5m^2, -4m^2, m^2$ **25.** $16x$ **27.** $3xy$
29. $19a + b$ **31.** $3x^3 + 9x^2$ **33.** $10a^3$ **35.** $(37 - x)$ ft
37. $(x + 4)$ m **39.** $x, 25 - x$ **41.** $(2x + 8)$ m **43.** Yes
45. Yes **47.** 2 **49.** -5 **51.** 1 **53.** -7 **55.** 7
57. -4 **59.** 27 **61.** -2 **63.** $\frac{7}{2}$ **65.** 18
67. $\frac{2}{5}$ **69.** 6 **71.** 6 **73.** 4
75. $\frac{1}{2}$ **77.** $\frac{5}{3}$ **79.** 12 **81.** 3 **83.** $W = \frac{V}{LH}$
85. $y = \frac{c - ax}{b}$ **87.** $t = \frac{A - P}{Pr}$ **89.** 6 **91.** 42, 43
93. 54 mi/h, 48 mi/h; 216 mi **95.** 150 pairs

97. -4 0

99. 0

101. -3 0

103. -15 0

105. $-\frac{4}{3}$ 0

107. -6 0

Self-Test—Chapter 1

1. $x + y$ **2.** $m - n$ **3.** ab **4.** $\frac{p}{q - 3}$ **5.** $c - 5$
6. $3(2x - 3y)$ **7.** $3(m - n)$ **8.** 4 **9.** 36 **10.** -4
11. 5 **12.** $3a - b$ **13.** $2x^2 + 8$ **14.** No
15. Yes **16.** 12 **17.** 30 **18.** 2 **19.** $\frac{35}{4}$
20. $B = \frac{3V}{h}$ **21.** 7 **22.** Juwan 6, Jan 12, Rick 17
23. 3:30 P.M.

24. 0 14

25. -4 0

Reading Your Text

Section 2.1 **(a)** elements; **(b)** empty; **(c)** set-builder; **(d)** interval
Section 2.2 **(a)** solution; **(b)** infinite; **(c)** ordered-pair; **(d)** dependent
Section 2.3 **(a)** x-axis; **(b)** y-axis; **(c)** origin; **(d)** quadrants
Section 2.4 **(a)** relation; **(b)** domain; **(c)** range; **(d)** independent
Section 2.5 **(a)** function; **(b)** relation; **(c)** domain; **(d)** outputs

Summary Exercises—Chapter 2

1. $\{1, 3\}$ **3.** $\{-1, 0, 1, 2, 3\}$ **5.** $\{1, 3\}$
7. $\{x \mid x > 9\}; (9, \infty)$ **9.** $\{x \mid x \le -5\}; (-\infty, -5]$

11. 0 1

13.

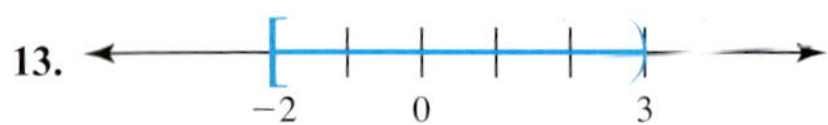

15. $\{x \mid x \le 3\}$; $(-\infty, 3]$ **17.** $\{x \mid -3 \le x < 2\}$; $[-3, 2)$
19. $\{x \mid -2 < x < 4\}$; $(-2, 4)$ **21.** $\{1, 2, 5, 7, 9, 11, 15\}$
23. $\{2, 5, 9, 11, 15\}$ **25.** $(6, 0), (3\ -3), (0, -6)$
27. $(4, 1)$ **29.** $(-1, -5)$

31–34.

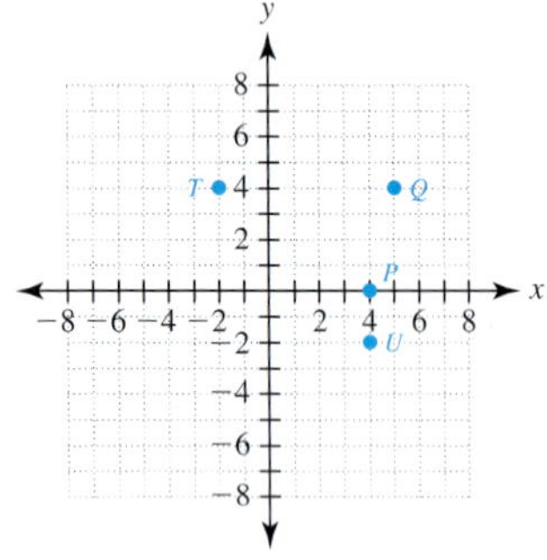

35. I **37.** III **39.** x-axis

41.

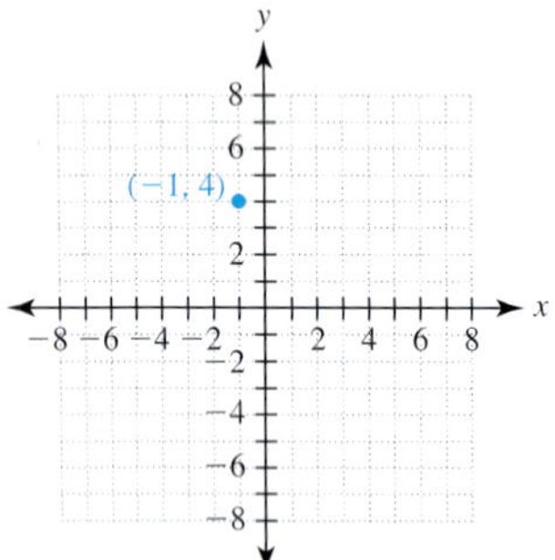

43.

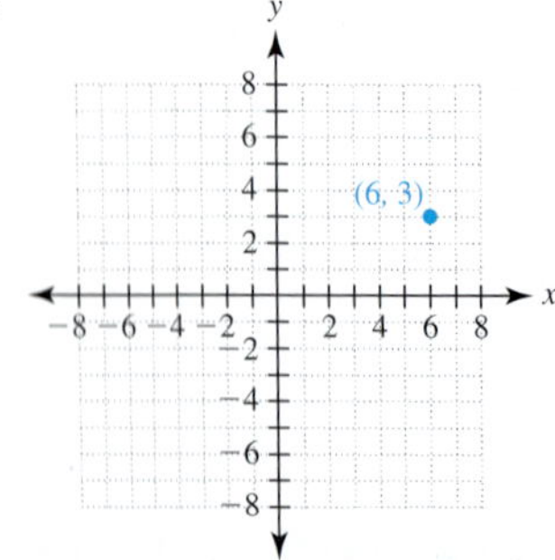

45. Domain: {Maine, Massachusetts, Vermont, Connecticut}; Range: {5, 13, 7, 11} **47.** Domain: {Dean Smith, John Wooden, Denny Crum, Bob Knight}; Range: {65, 47, 42, 41}
49. Domain: {1, 3, 4, 7, 8}; Range: {1, 2, 3, 5, 6}
51. Domain: {1}; Range: {3, 5, 7, 9, 10} **53.** Function
55. Not a function **57.** Function **59.** Not a function
1. (a) 5; **(b)** 9; **(c)** 3 **63. (a)** 5; **(b)** 5; **(c)** 5
. (a) 9; **(b)** 1; **(c)** 3 **67.** $f(x) = -2x + 5$

69. $f(x) = -\frac{2}{3}x + 2$ **71.** $f(x) = \frac{3}{4}x + 3$ **73.** $-\frac{3}{4}t + 2$

75. $-\frac{3}{4}x - \frac{3}{4}h + 2$ **77.** $3a - 2$ **79.** $3x + 3h - 2$

81.

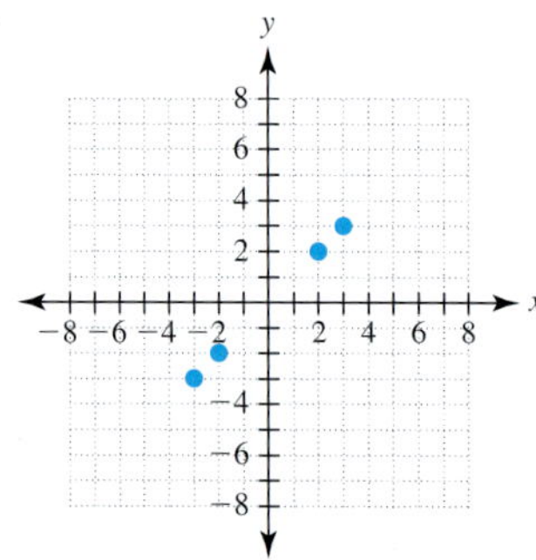

Function

83.

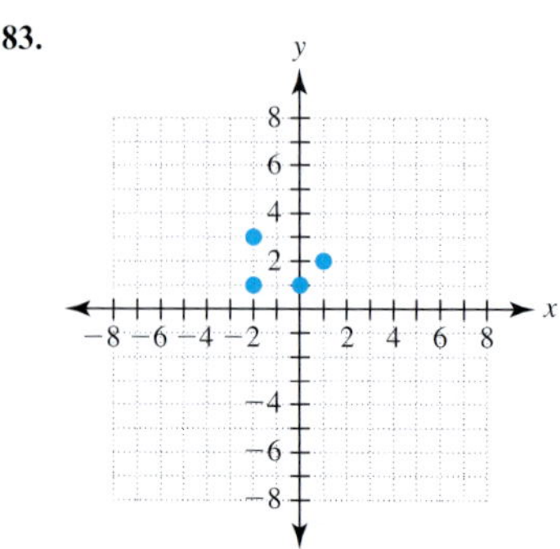

Not a function

85. Function; $D = \mathbb{R}$; $R = \{y \mid y \ge -4\}$
87. Not a function; $D = \{x \mid -6 \le x \le 6\}$; $R = \{y \mid -6 \le y \le 6\}$
89. (a) 0; **(b)** -3; **(c)** -5; **(d)** 5; **(e)** $-3.3, -2, 3.4$; **(f)** $-4, 1$
91. (a) -2; **(b)** 12; **(c)** $-5, 4$; **(d)** -2; **(e)** 8; **(f)** 12

Self-Test—Chapter 2

1. −5 0 3

2. (a) $\{x \mid -3 < x \le 2\}$; **(b)** $(-3, 2]$ **3.** $(4, 0), (5, 4)$
4. $(3, 6)$ **5.** $(4, -2)$ **6.** $(0, -7)$
7. (a) $D = \{-3, 1, 2, 3, 4\}$; $R = \{-2, 0, 1, 5, 6\}$;
(b) $D =$ {United States, Germany, Russia, China};
$R = \{50, 63, 65, 101\}$ **8. (a)** $D = \{-4, -1, 0, 2\}$; $R = \{2, 5, 6\}$;
(b) not a function; $D = \{-3, 0, 1, 2\}$; $R = \{0, 1, 2, 4, 7\}$;
9. Function

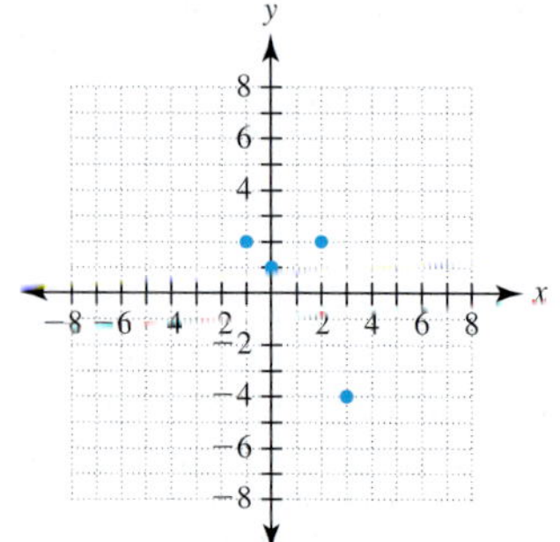

10. Function **11.** Not a function

12–14.

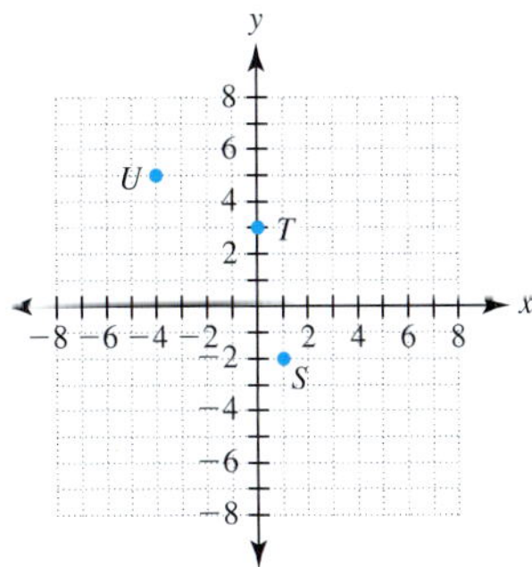

15. (3, 0), (0, 4), $\left(\frac{3}{4}, 3\right)$ **16.** **(a)** 6; **(b)** 12; **(c)** 2

17. **(a)** 2; **(b)** −5

18.

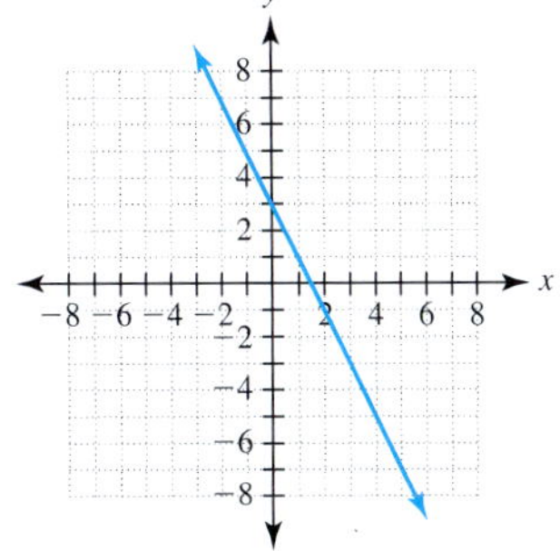

19. **(a)** {1, 2, 3, 5, 7}; **(b)** {5} **20.** **(a)** 3; **(b)** 2; **(c)** 1; **(d)** −5, 5

Cumulative Review—Chapters 0–2

1. $\frac{7}{11}$ **2.** $\frac{6}{5}$ **3.** 2 **4.** 80 **5.** 7 **6.** 7 **7.** −16
8. 4 **9.** 63 **10.** −16 **11.** 0 **12.** −5 **13.** 9
14. $\frac{1}{3}$ **15.** $\frac{19}{12}$ **16.** $\frac{7}{10}$ **17.** −9 **18.** 13 **19.** 3
20. 7 **21.** $15x - 9y$ **22.** $10x^2 - 13x + 2$ **23.** 4
24. 71 **25.** 4 **26.** $\{x \mid x \le -4\}$ **27.** $\{x \mid -3 \le x < 3\}$
28. $r = \frac{I}{Pt}$ **29.** $h = \frac{2A}{b}$ **30.** $y = \frac{c - ax}{b}$
31. $W = \frac{P}{2} - L$ or $W = \frac{P - 2L}{2}$ **32.** −5 **33.** −2
34. 5 **35.** 13; $4x - 7 = 45$ **36.** 42, 43; $x + (x + 1) = 85$
37. 7; $3x = (x + 2) + 12$ **38.** \$420; $x + (x + 120) = 720$
39. 5 cm, 17 cm; $2x + 2(3x + 2) = 44$
40. 8 in., 13 in., 16 in.; $x + (x + 5) + 2x = 37$

Reading Your Text

Section 3.1 **(a)** solutions; **(b)** vertical; **(c)** constant; **(d)** zero
Section 3.2 **(a)** slope; **(b)** horizontal; **(c)** rise; **(d)** negative
Section 3.3 **(a)** parallel; **(b)** perpendicular; **(c)** undefined; **(d)** zero
Section 3.4 **(a)** one; **(b)** slope; **(c)** zero; **(d)** line
Section 3.5 **(a)** half-plane; **(b)** test; **(c)** solid; **(d)** false

Summary Exercises—Chapter 3

1.

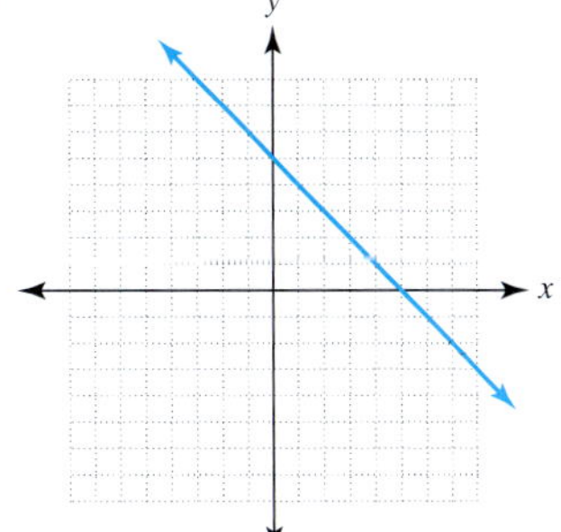

3.

5.

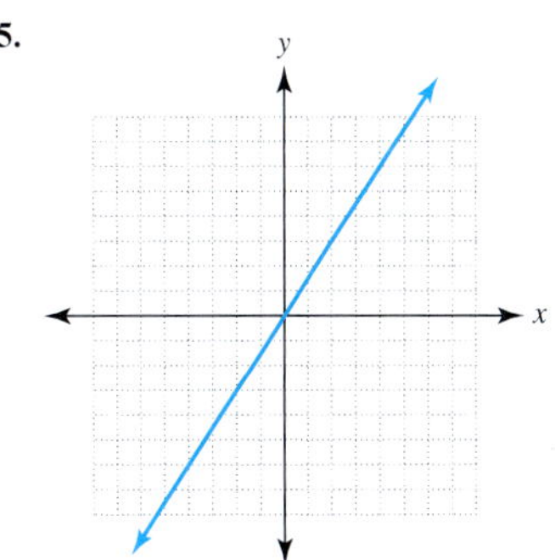

7.

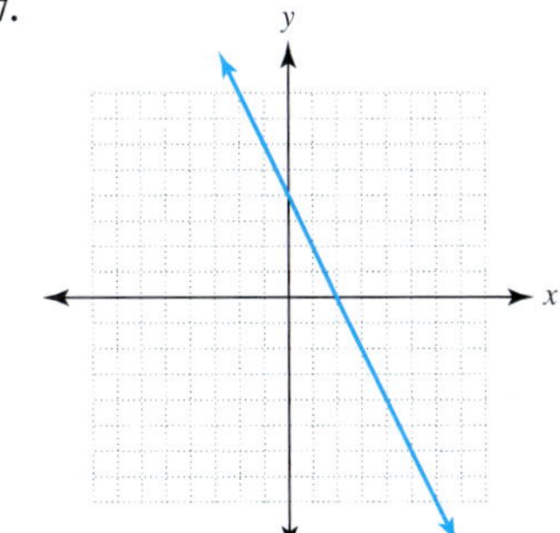

9.

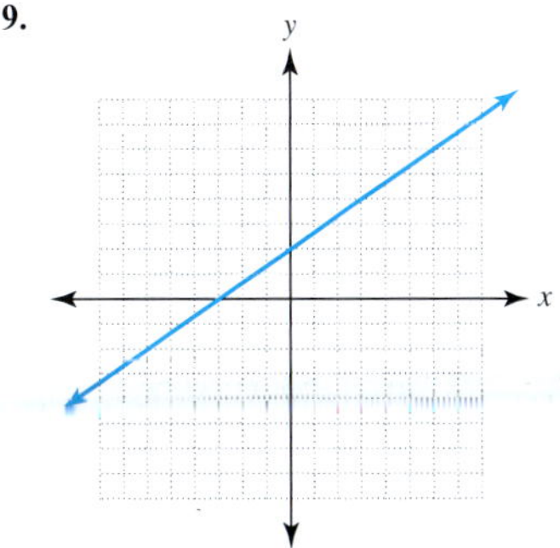

11.

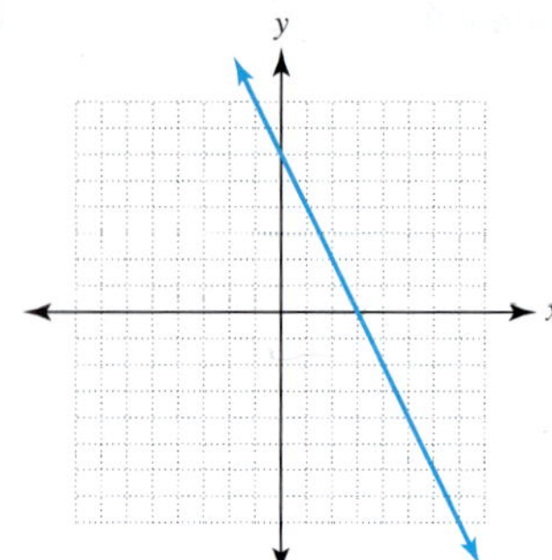

13.

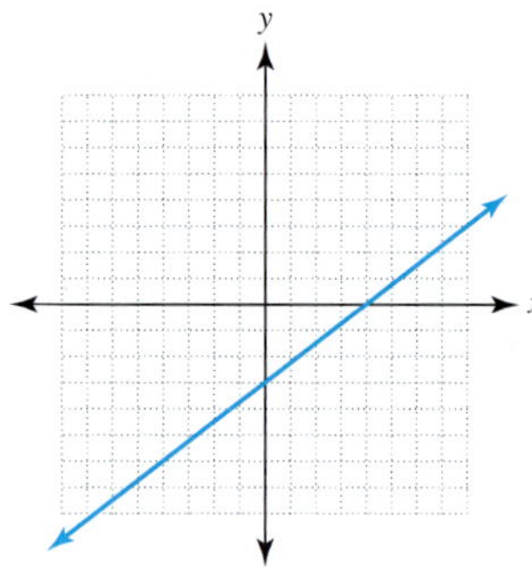

15.

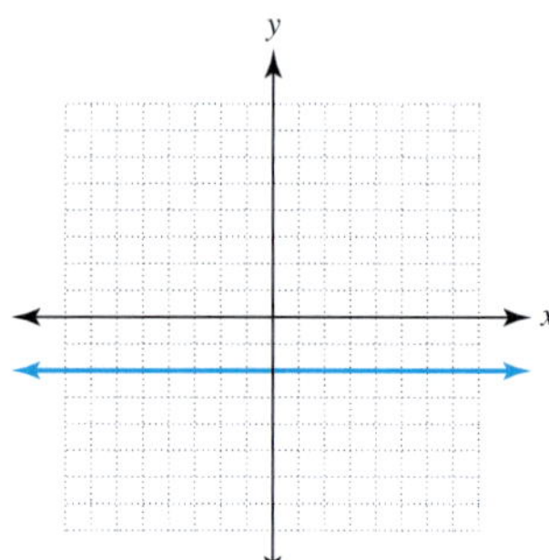

17.

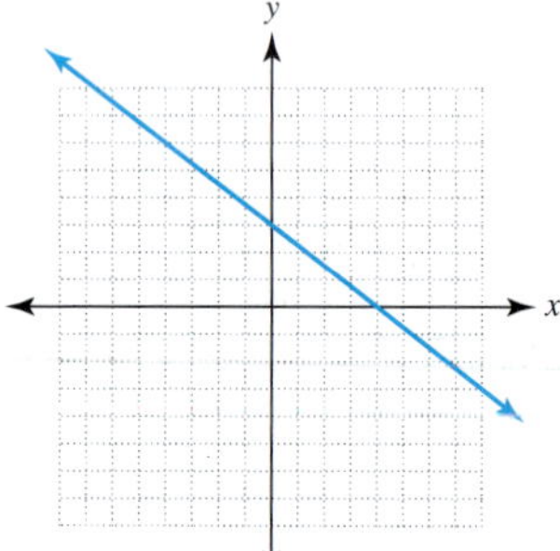

19.

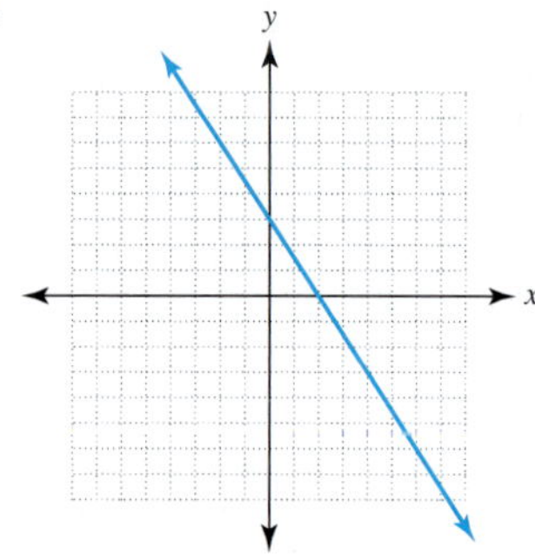

21. 2 **23.** $-\frac{1}{2}$ **25.** 0 **27.** $\frac{1}{2}$

29. Slope 2, y-intercept (0, 5) **31.** Slope $-\frac{3}{4}$, y-intercept (0, 0)

33. Slope $-\frac{2}{3}$, y-intercept (0, 2) **35.** Slope 0, y-intercept (0, −3)

37. $y = 2x + 3$ **39.** $y = -\frac{2}{3}x + 2$ **41.** Perpendicular

43. Parallel **45.** $y = -3$ **47.** $x = 4$ **49.** $y = -3$

51. $y = -\frac{4}{3}x - 2$ **53.** $x = -\frac{5}{2}$ **55.** $y = 2x - 1$

57. $y = -\frac{5}{4}x + 2$ **59.** $y = -\frac{3}{5}x - 1$ **61.** $y = \frac{4}{3}x + \frac{14}{3}$

63. \$91.25 **65.** $P(g) = 2.75g - 30$ **67.** 48 gyros

69. $\frac{2}{63}$ **71.** (0, 0)

73.

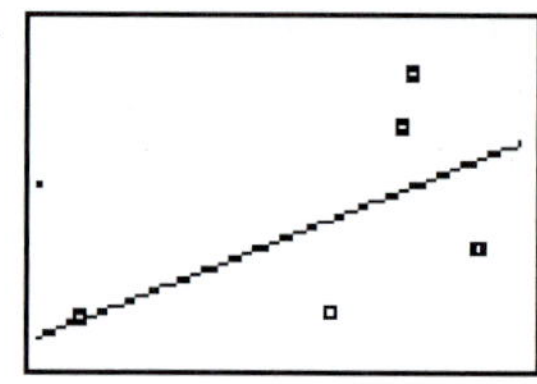

75. 5.73 **77.** 109 books

79.

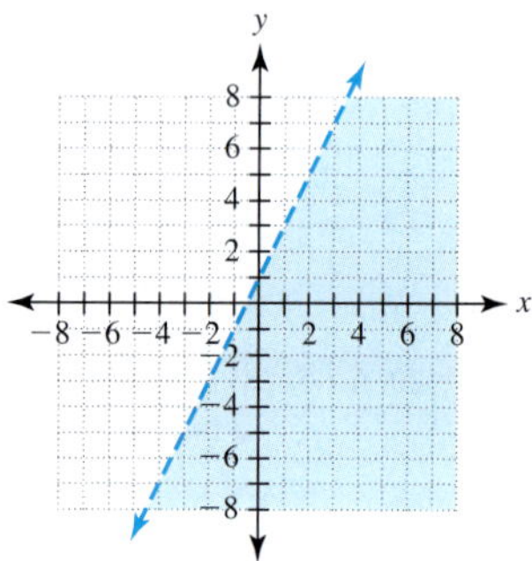

81.

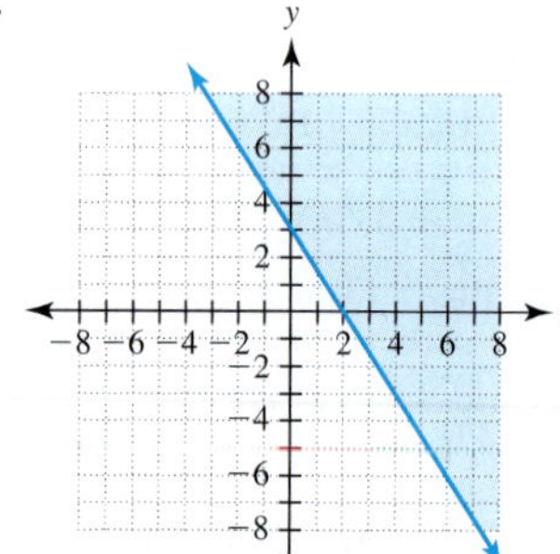

83.

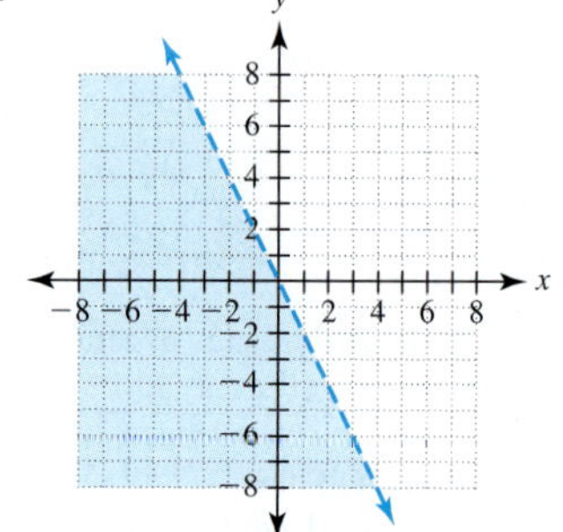

85.

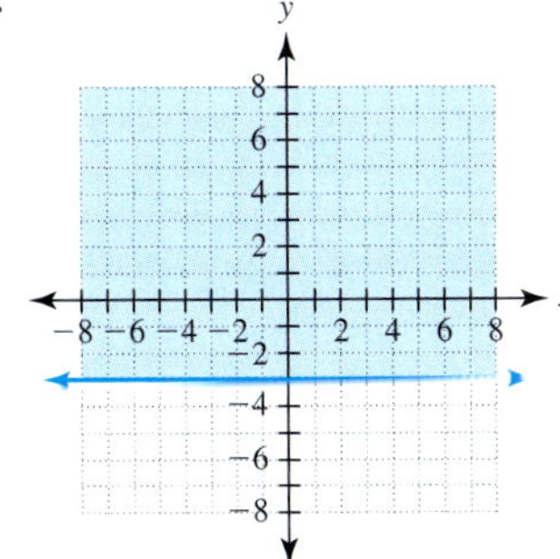

Self-Test—Chapter 3

1. 1 **2.** $\frac{3}{7}$

3. $y = -3x + 6$

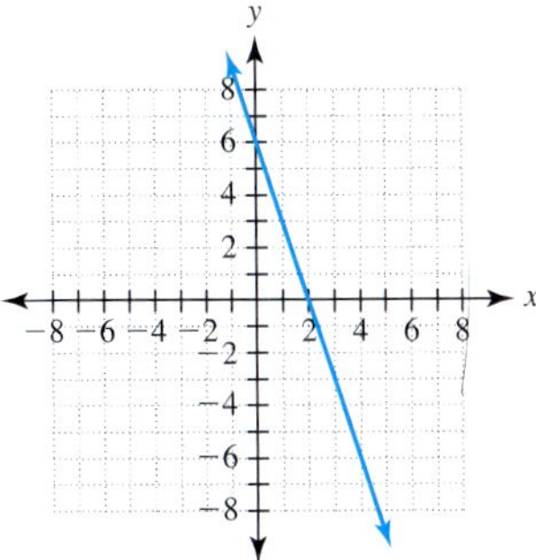

4. $y = \frac{2}{5}x - 3$

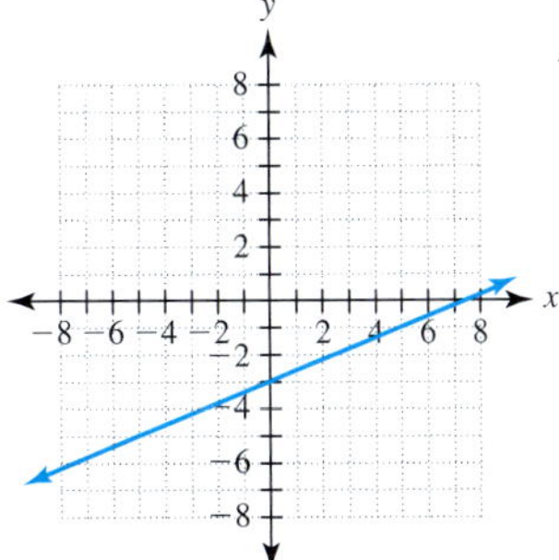

5. $y = \frac{1}{4}x + \frac{3}{2}$ **6.** $5\frac{1}{2}$ h

7. $x + y = 4$

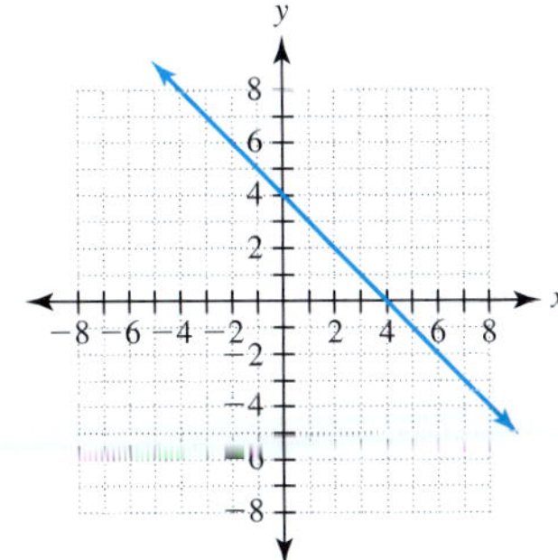

8. $y = 3x$

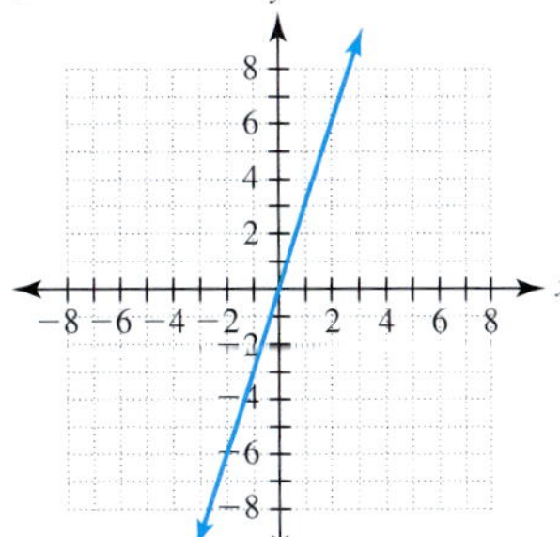

9. $y = \frac{3}{4}x - 4$

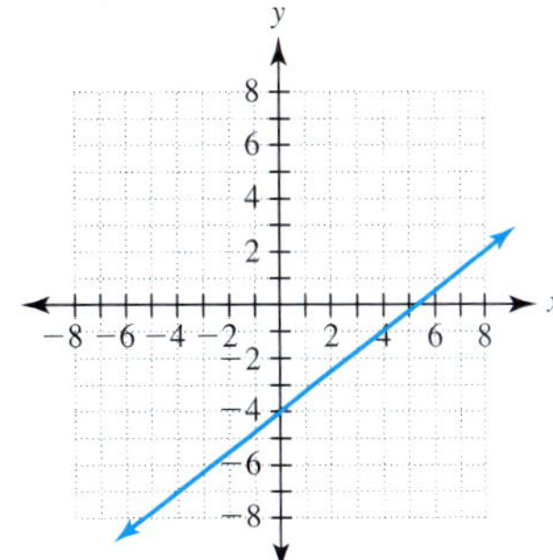

10. $x + 3y = 6$

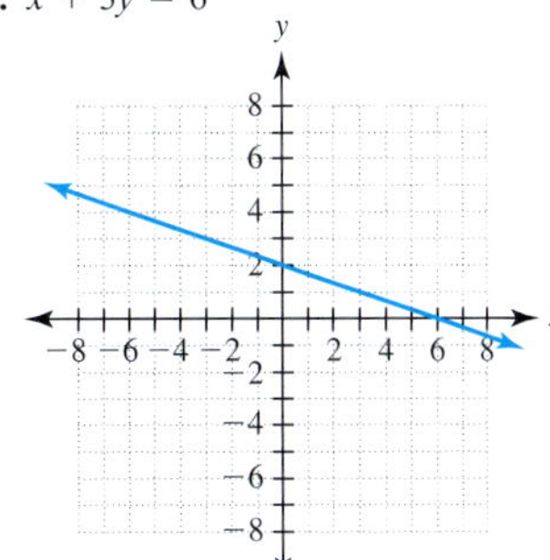

11. $2x + 5y = 10$

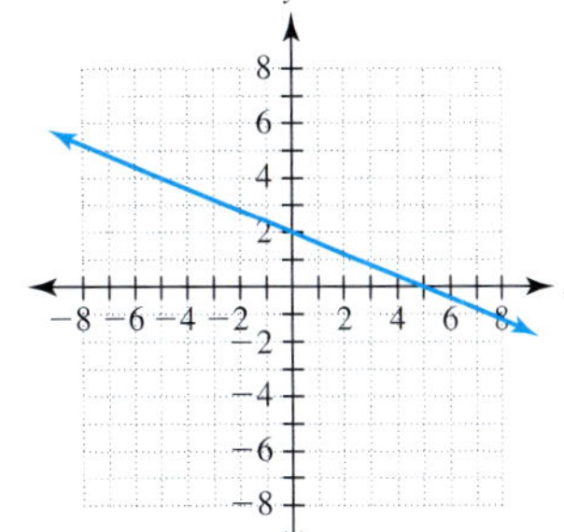

12. $y = -4$

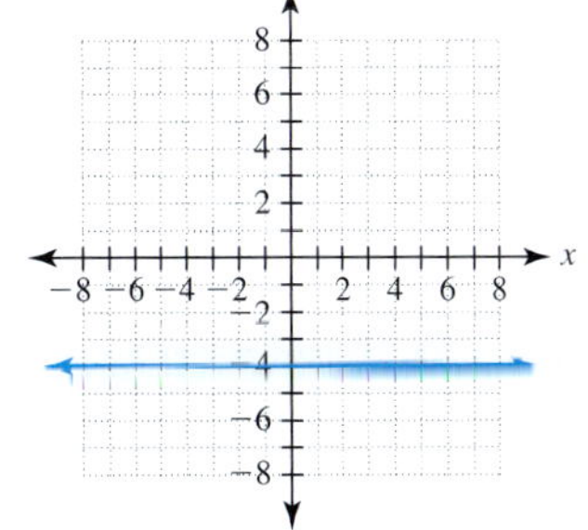

13. $5x + 6y \le 30$

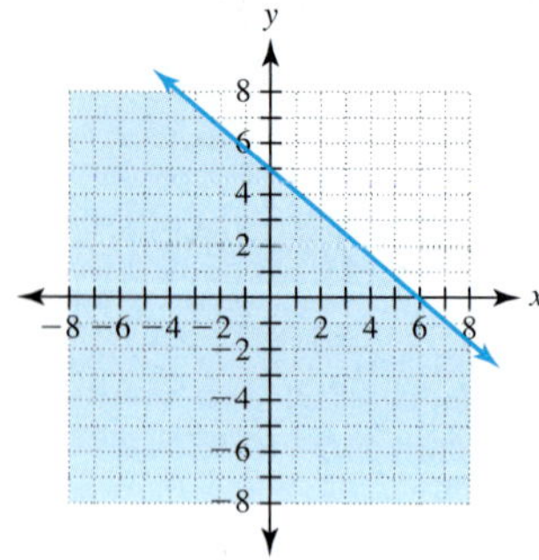

14. $x + 3y > 6$

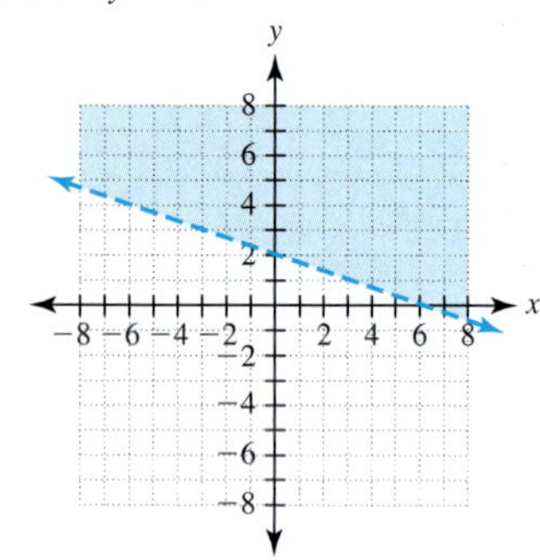

15. $4x - 8 \le 0$

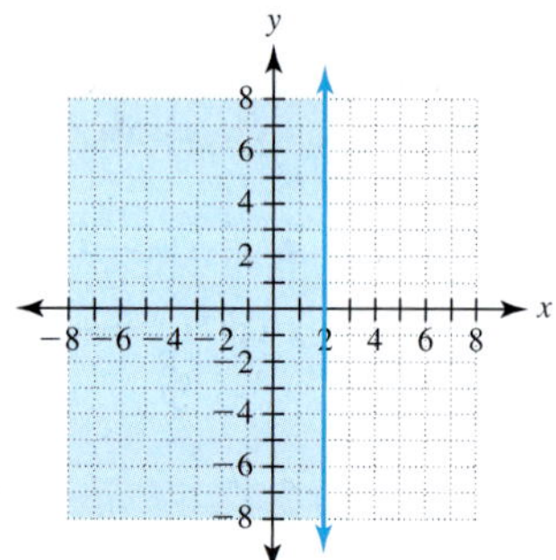

16. $2y + 4 > 0$

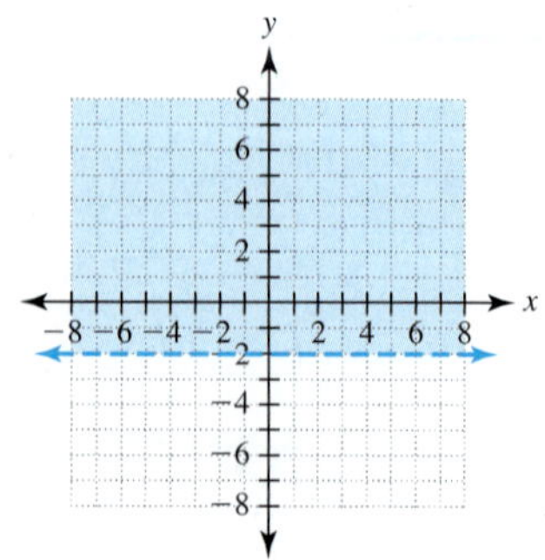

17. Slope -5; y-intercept $(0, -9)$ **18.** Slope $-\frac{6}{5}$; y-intercept $(0, 6)$

19. Slope 0; y-intercept $(0, 5)$ **20.** $y = 5x - 2$

21. $y = -4x - 16$ **22.** $y = 4x + 3$ **23.** $y = -\frac{5}{2}x - 17$

24.

25. $y = 0.15x + 67.41$

Cumulative Review—Chapters 0–3

1. $-\frac{1}{3}$ **2.** $\frac{2}{3}$ **3.** 17 **4.** 8 **5.** 1

6. $-\frac{33}{5}$ **7.** $6x + 3y$ **8.** $-x - 4$ **9.** $-6x^2 + x + 2$

10. $-8x^2 - 5x + 25$ **11.** -11 **12.** 100

13. $-\frac{7}{2}$ **14.** $-\frac{3}{2}$ **15.** $C = \frac{5}{9}(F - 32)$ **16.** $h = \frac{3V}{\pi r^2}$

17.

0 4

18.

-4 0

19.

0 5 9

20.

0 2 7

21. -2 **22.** 0 **23.** Slope -4; y-intercept $(0, 9)$

24. Slope $\frac{2}{5}$; y-intercept $(0, -2)$ **25.** Slope 0; y-intercept $(0, 9)$

26. Slope undefined; no y-intercept **27.** $y = 5x - 6$

28. $x + 10y = 86$ **29.** $y = -\frac{2}{3}x + 6$ **30.** $4y + 5x = 26$

31. $y = \frac{3}{2}x - 3$ **32.** $y = 3x + 10$ **33.** 9 **34.** 7

35. Coach \$450; first-class \$900 **36.** 12 m, 24 m, 28 m

37. $y < -\frac{2}{3}x + 2$

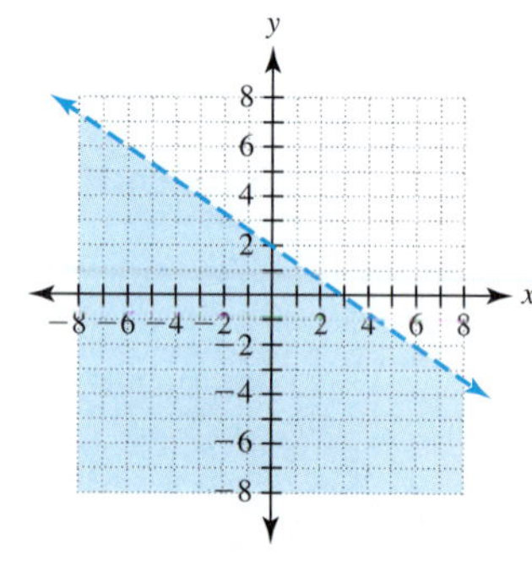

38. $f(x) = 3.95x + 30.77$ **39.** \$78.17 **40.** Each additional pound adds \$3.95 to the shipping costs.

Reading Your Text

Section 4.1 **(a)** related; **(b)** ordered pair; **(c)** consistent; **(d)** inconsistent

Section 4.2 **(a)** solution; **(b)** check; **(c)** intersect; **(d)** original

Section 4.3 **(a)** equivalent; **(b)** intersection; **(c)** no; **(d)** time

Section 4.4 **(a)** three; **(b)** three; **(c)** infinite; **(d)** consistent

Section 4.5 **(a)** satisfy; **(b)** inequalities; **(c)** boundary; **(d)** bounded

Summary Exercises—Chapter 4

1. $\{(6, 2)\}$

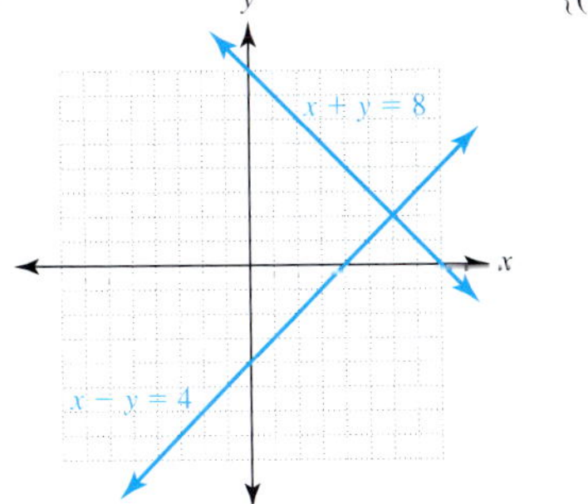

3. $\{(3, 2)\}$

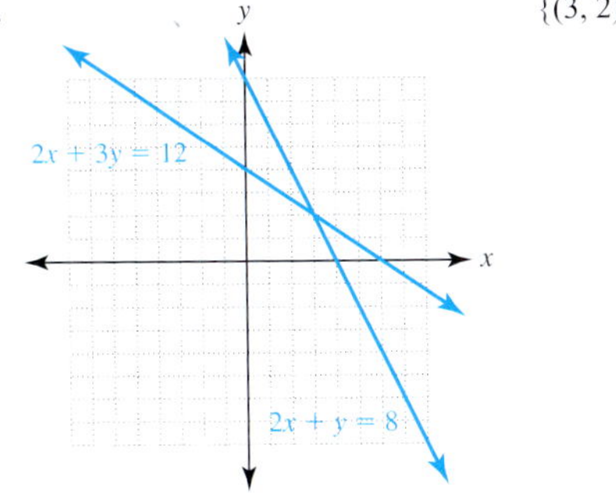

5. $\{(38.05, -15.98)\}$

7. $\{2\}$

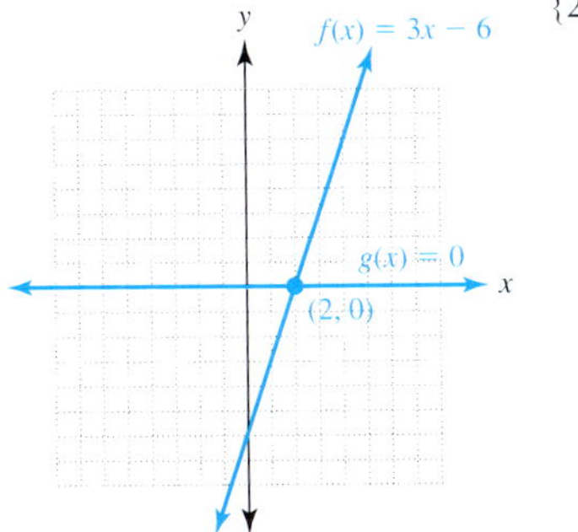

9. $\{1\}$

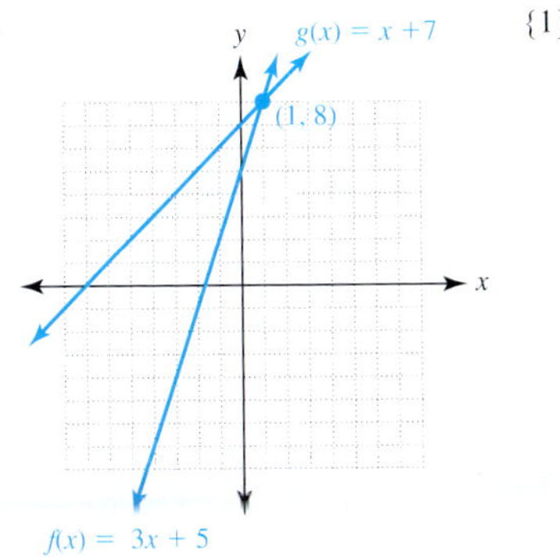

11. $\left\{-\dfrac{3}{2}\right\}$

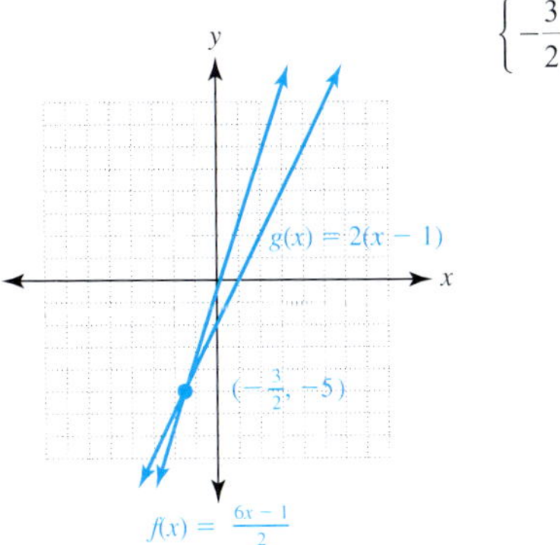

13. $\{4\}$

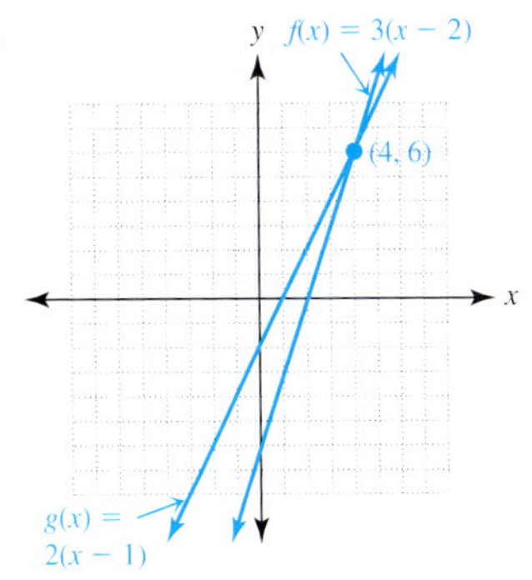

15. $\{(3, 2)\}$ **17.** $\{(0, -1)\}$ **19.** $\{(-4, 3)\}$ **21.** $\{(-3, 2)\}$ **23.** $\{(6, -4)\}$ **25.** $\{(9, 5)\}$ **27.** Inconsistent system, no solutions **29.** $\{(2, -1)\}$ **31.** $\left\{\left(-\dfrac{2}{5}, 5\right)\right\}$ **33.** 7, 23 **35.** 800 adult tickets, 400 student tickets **37.** 12 cm, 20 cm **39.** Savings: \$8,000; time deposit: \$9,000 **41.** Jet: 500 mi/h; wind: 50 mi/h **43.** (15, 195) **45.** $\left\{\left(3, \dfrac{8}{3}, -\dfrac{1}{3}\right)\right\}$ **47.** Inconsistent **49.** $\left\{\left(\dfrac{1}{2}, -\dfrac{3}{2}, 0\right)\right\}$ **51.** 2, 5, 8 **53.** 200 orchestra, 40 balcony, 120 box seats **55.** \$6,000 savings, \$2,000 stock, \$4,000 mutual fund

57.

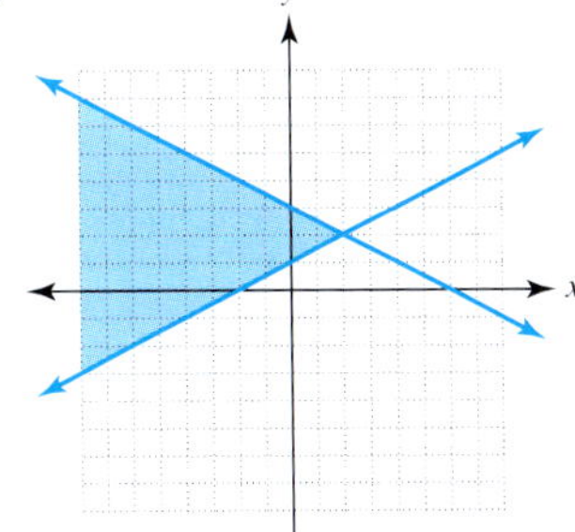

59.

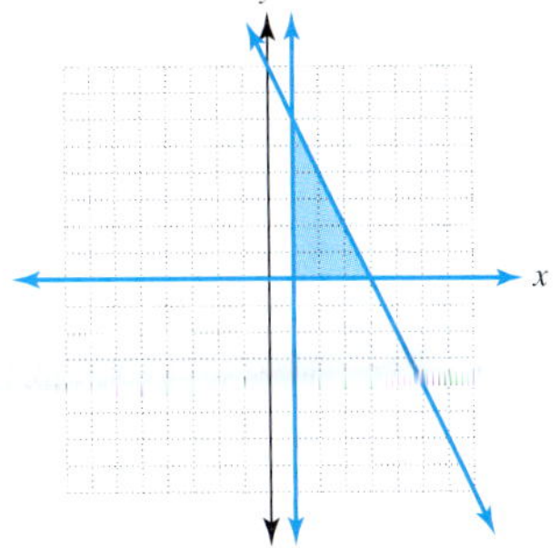

61.

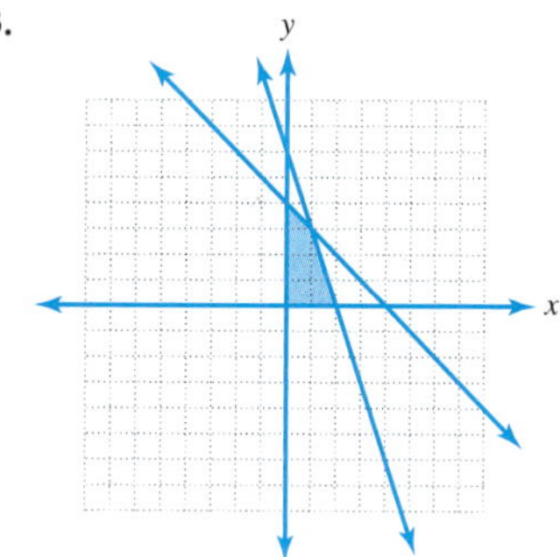

63.

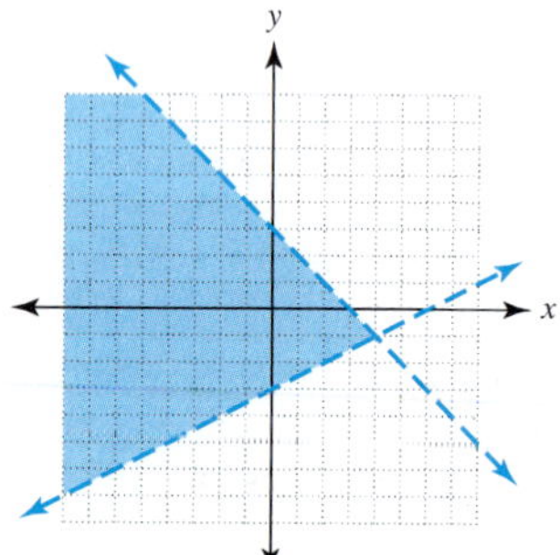

Self-Test—Chapter 4

1. $\{(-3, 4)\}$
2. Infinite number of solutions, dependent system
3. No solutions, inconsistent system 4. $\{(-2, -5)\}$
5. $\{(5, 0)\}$ 6. $\left\{\left(3, -\frac{5}{3}\right)\right\}$ 7. $\{(-1, 2, 4)\}$
8. $\left\{\left(2, -3, -\frac{1}{2}\right)\right\}$
9.

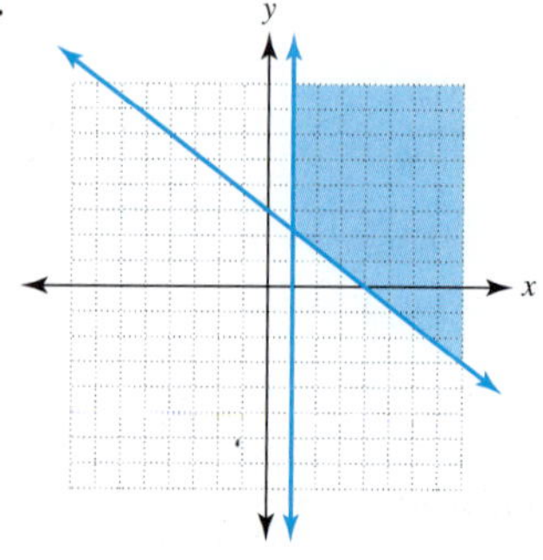

10.

11.

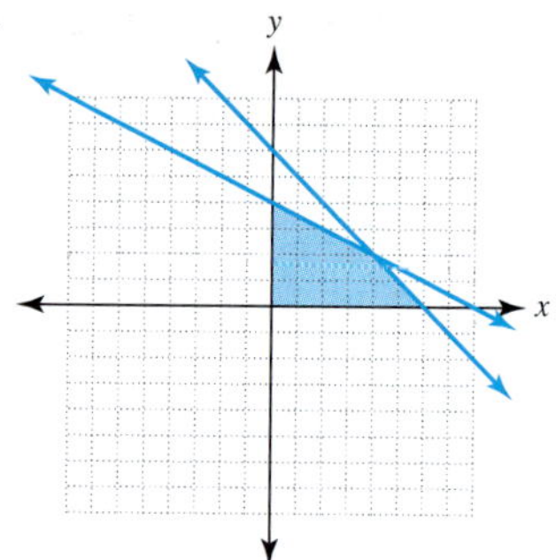

12. $\{3\}$

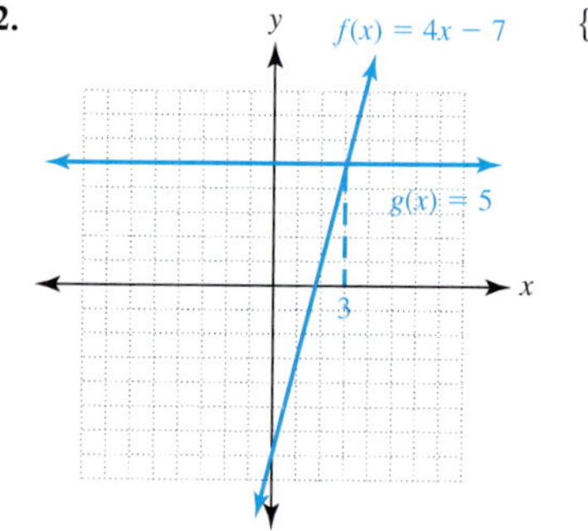

13. $\{2\}$

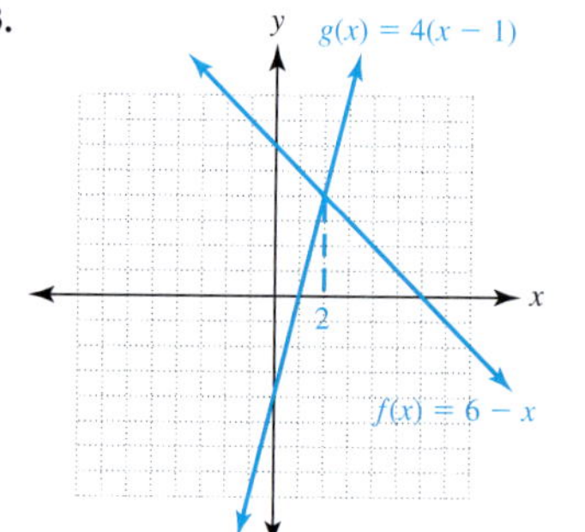

14. $\{2\}$

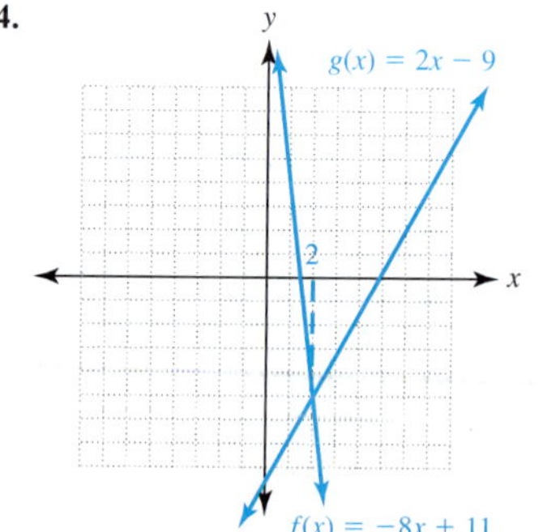

15. $\{2\}$

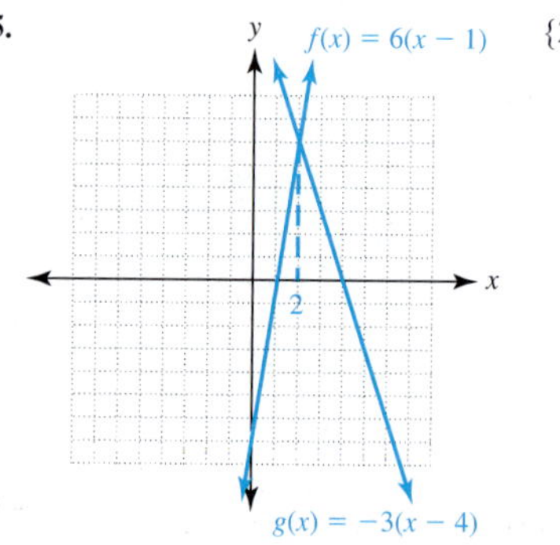

16. Disks \$2.50; ribbons \$6 17. 60 lb jawbreakers; 40 lb licorice
18. Four 5-in. sets; six 12-in. sets
19. \$3,000 savings; \$5,000 bond; \$8,000 mutual fund
20. 50 ft by 80 ft

Cumulative Review—Chapters 0–4

1. $\left\{\frac{11}{2}\right\}$ 2. $\{x \mid x > -2\}$ 3. $\left\{x \mid x \le \frac{11}{3}\right\}$ 4. $\{2\}$

5.

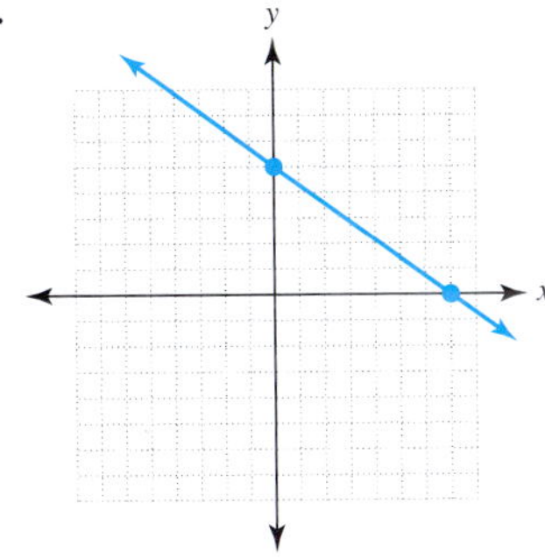

6.

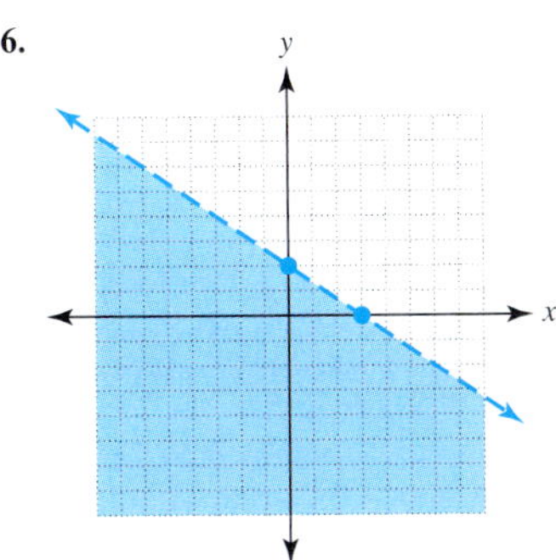

7. $R = \frac{P - P_0}{IT}$ 8. 7 9. $y = -x + 3$ 10. $y = \frac{4}{5}x - \frac{2}{5}$
11. 90 12. $\left\{\left(-10, \frac{26}{3}\right)\right\}$
13. $\{(2, 2, -1)\}$ 14. 8 cm by 19 cm 15. 73

Reading Your Text

Section 5.1 (a) multiplication; (b) exponential; (c) add; (d) denominator
Section 5.2 (a) add; (b) reciprocal; (c) one; (d) meters
Section 5.3 (a) term; (b) coefficient; (c) binomial; (d) tri-
Section 5.4 (a) plus; (b) sign; (c) distributive; (d) first
Section 5.5 (a) coefficients; (b) distributive; (c) binomials; (d) three
Section 5.6 (a) coefficients; (b) term; (c) degree; (d) zero

Summary Exercises—Chapter 5

1. r^{13} 3. $\frac{1}{8w^3}$ 5. $\frac{1}{y^3}$ 7. $\frac{1}{x^3}$ 9. $\frac{10}{b^3}$ 11. $\frac{m^3}{n^{21}}$
13. $\frac{y^6}{x^9}$ 15. $9a^8$ 17. 4.25×10^{-5} 19. Binomial
21. Trinomial 23. Binomial 25. $13x^2$; 2 27. $x + 5$; 1
29. $7x^6 + 9x^4 - 3x$; 6 31. $9x^2 - 3x - 7$ 33. $4x^2 + 8x$
35. $4x$[illegible] $- 4x - 3$ 37. $10a - 5$ 39. $3x^2 + 9x - 6$
41. $5x^2 - 5x + 5$ 43. $9m^2 - 8m$ 45. $6x^7$ 47. $-21a^5b^7$
49. $10a^2 - 6a$ 51. $21m^3n^2 - 14m^2n^3 + 35m^2n^2$

53. $w^2 - 19w + 90$ 55. $p^2 - 6pq + 9q^2$
57. $2b^2 - 13b - 24$ 59. $15r^2 - 24rs - 63s^2$
61. $b^3 - 2b^2 - 22b - 21$ 63. $m^4 + 4m^2 - 21$
65. $4a^3 - 24a^2b + 35ab^2$ 67. $a^2 - 14a + 49$
69. $9p^2 + 24p + 16$ 71. $64x^2 - 48xy + 9y^2$
73. $y^2 - 81$ 75. $16r^2 - 25$ 77. $49a^2 - 9b^2$
79. $3c^3 - 75cd^2$ 81. $3a^3$ 83. $3a - 2$
85. $-3rs + 6r^2$ 87. $x - 5$ 89. $x - 3 + \frac{2}{x - 5}$
91. $x^2 + 2x - 1 + \frac{-4}{6x + 2} = x^2 + 2x - 1 + \frac{-2}{3x + 1}$
93. $x^2 + x + 2 + \frac{1}{x + 2}$

Self-Test—Chapter 5

1. $-6x^3y^4$ 2. $\frac{16m^4n^{10}}{p^6}$ 3. x^8y^{10} 4. $\frac{c^2}{2d}$ 5. $108x^8y^7$
6. $\frac{9y^4}{4x}$ 7. 3 8. $10x^2 - 12x - 7$
9. $7a^3 + 11a^2 - 3a$ 10. $3x^2 + 11x - 12$
11. $7a^2 - 10a$ 12. $15a^3b^2 - 10a^2b^2 + 20a^2b^3$
13. $4x^2 + 7xy - 15y^2$ 14. $9m^2 + 12mn + 4n^2$
15. $4x^2 + 3x + 13 + \frac{17}{x - 2}$
16. $8x^4 - 3x - 7$; 8, 4; −3, 2; −7, 0; 4 17. Binomial
18. Trinomial 19. $4x^2 + 5x - 6$ 20. $2x^2 - 7x + 5$

Cumulative Review—Chapters 0–5

1. −10 2. $\frac{7}{5}$ 3. $\{7\}$ 4. $\{-36\}$
5. $\{4\}$ 6. $\left\{-\frac{2}{3}\right\}$ 7. $\{23\}$ 8. $\mathbb{R}$ 9. $\varnothing$ 10. $\left\{-\frac{13}{8}\right\}$
11. $\{x \mid x < -15\}$ 12. $\{x \mid x < 5\}$ 13. −13
14. $\left\{-\frac{8}{3}\right\}$ 15. $R(x) = 39.95x$ 16. \$679.15
17. slope: −4; y-intercept: (0, 9) 18. slope: 3
19. $y = -2x + 4$ 20. $y = \frac{4}{5}x - \frac{2}{5}$ 21. $y = 4x - 7$
22. $y = -2x$ 23. $3x^2 + 2x - 4$ 24. $x^2 - 3x - 18$
25. $12x^2 - 20x$ 26. $6x^2 + x - 40$
27. $x^3 - x^2 - x + 10$ 28. $4x^2 - 49$ 29. $9x^2 - 30x + 25$
30. $20x^3 - 100x^2 + 125x$ 31. $4xy - 2x^3 + 1$
32. $2x^2 + 6x + 3 + \frac{2}{x - 3}$ 33. $18x^7$ 34. $\frac{x^{16}}{y^{12}}$
35. $72x^4y^2$ 36. $\frac{2x^5}{y^{11}}$ 37. 2.1×10^{10} 38. 8
39. 65, 67 40. 4 cm by 24 cm

Final Exam—Chapters 0–5

1. 2 2. 10 3. $\frac{5}{8}$ 4. $\frac{4}{7}$ 5. 180 6. 118
7. −16 8. −285 9. −2 0 5
10. $\{x \mid -4 < x \le 2\}$ 11. $2x^2y + 3x + 2xy$
12. $x^2 + 9x$ 13. $\{11\}$ 14. $\{-33\}$ 15. $\left\{\frac{8}{5}\right\}$
16. $\{1\}$ 17. $\{18.5\}$ 18. $\{-6\}$ 19. $\left\{-\frac{31}{2}\right\}$

20. $\{x \mid x > -6\}$ **21.** $\{x \mid x > -6\}$ **22.** $\{x \mid -4 < x < 2\}$
23. $\{x \mid x < -3 \text{ or } x > 2\}$ **24.** $-3a - 5b$ **25.** $8mn^2 - 9m^2n$

26.

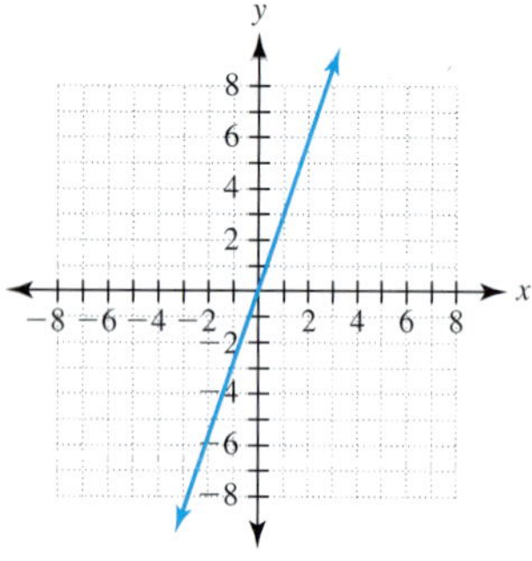

27.

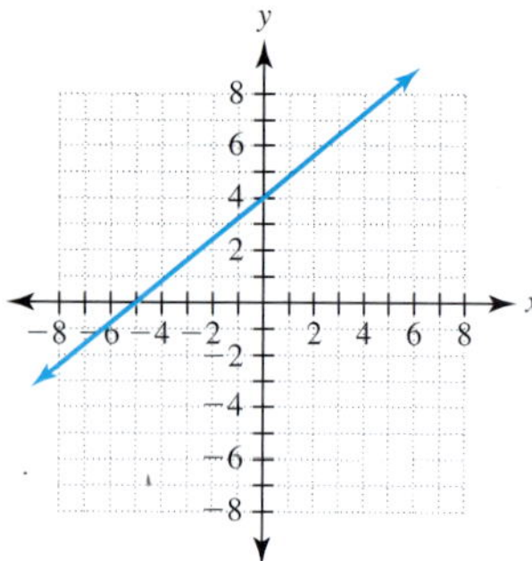

28.

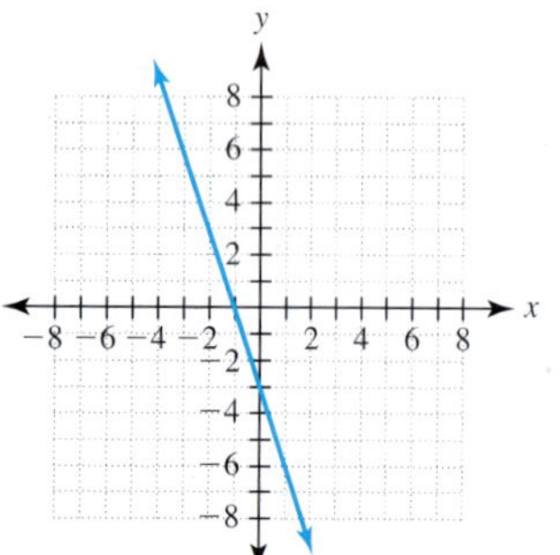

29.

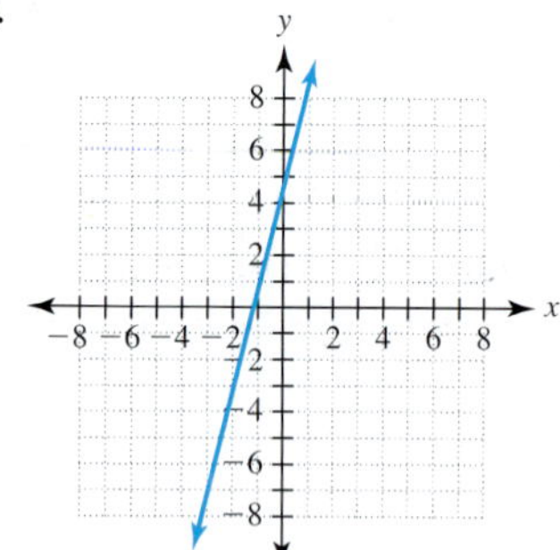

30. -2 **31.** Slope -4, y-intercept $(0, 9)$ **32.** $y = -3x - 1$
33. $y = 2x - 1$ **34.** Function **35.** Not a function
36. Not a function **37.** Function **38.** 160 watches
39. 37, 39, 41 **40.** $h = \dfrac{S - 2\pi r^2}{2\pi r}$ or $h = \dfrac{S}{2\pi r} - r$
41. $2m^3n^8$ **42.** $\dfrac{x^7}{y^6}$ **43.** $(2, 0), (3, -3), (-3, 15)$

44. $a^2 - 9b^2$ **45.** $x^2 - 4xy + 4y^2$ **46.** $x^2 + 3x - 10$
47. $a^2 + a - 12$ **48.** $3x + 2$ **49.** $3x + 3 + \dfrac{1}{x - 1}$
50. $(-4, 3)$ **51.** Dependent **52.** $\left(\dfrac{1}{3}, -\dfrac{5}{3}\right)$ **53.** $(5, -2)$
54. $\left(-\dfrac{3}{4}, \dfrac{1}{3}\right)$ **55.** Inconsistent **56.** $(2, -4, 5)$
57. $\left(\dfrac{1}{2}, -\dfrac{1}{3}, 3\right)$ **58.** Dependent

59.

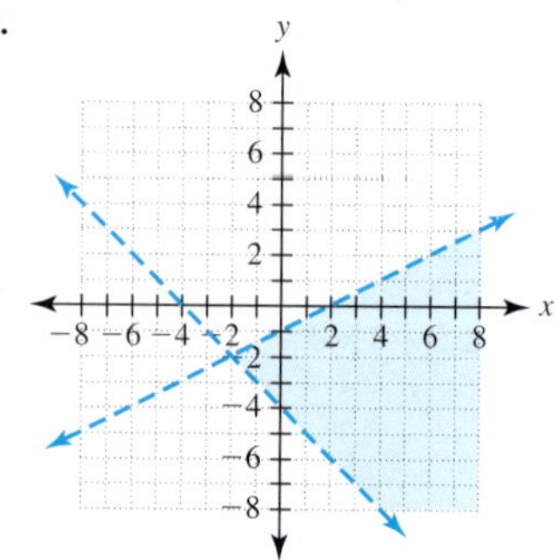

60.

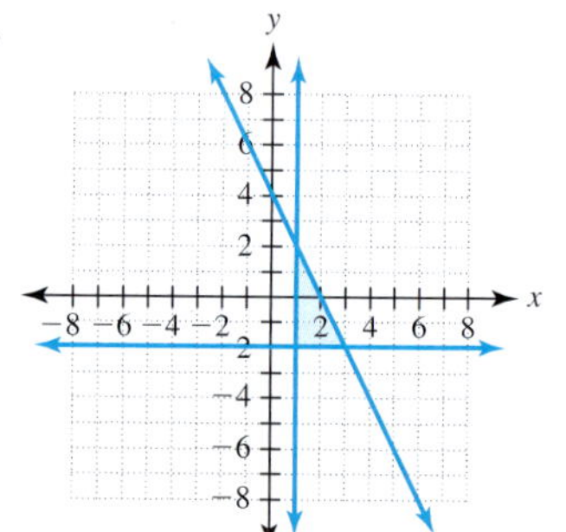

61. 30.5 in. by 31.5 in. **62.** Biology \$73, math \$68
63. \$43.50 **64.** 15%: 620 mL; 60%: 280 mL
65. 840 mi **66.** 48°, 60°, 72°
67. Bond \$3,300; time deposit \$5,200; savings \$500

Reading Your Text

Section 6.1 **(a)** grouping; **(b)** common; **(c)** multiplying; **(d)** distributive
Section 6.2 **(a)** perfect; **(b)** difference; **(c)** common; **(d)** cubes
Section 6.3 **(a)** binomials; **(b)** binomial; **(c)** opposite; **(d)** common
Section 6.4 **(a)** coefficients; **(b)** four; **(c)** factorable; **(d)** GCF
Section 6.5 **(a)** sum; **(b)** grouping; **(c)** polynomial; **(d)** *ac*
Section 6.6 **(a)** standard; **(b)** zero-product; **(c)** roots; **(d)** *x*-axis
Section 6.7 **(a)** factoring; **(b)** no; **(c)** consecutive; **(d)** negative

Summary Exercises—Chapter 6

1. $7(2a - 5)$ **3.** $8s^2(3t - 2)$ **5.** $9s^3(3s + 2)$
7. $9m^2n(2n - 3 + 4n^2)$ **9.** $8ab(a + 3 - 2b)$
11. $(3x + 4y)(2x - y)$ **13.** $(p + 7)(p - 7)$
15. $(3n + 5m)(3n - 5m)$ **17.** $(5 + z)(5 - z)$
19. $(5a + 6b)(5a - 6b)$ **21.** $3w(w + 2z)(w - 2z)$
23. $2(m + 6n^2)(m - 6n^2)$ **25.** $(x - 4)(x + 5)$
27. $(4x + 3)(2x - 5)$ **29.** $x(2x + 3)(3x - 2)$
31. $(y + 7)(y - 7)$ **33.** $(2x + 1)(2x - 1)$
35. $(x - 9)^2$ **37.** $(4m + 7n)(4m - 7n)$

39. $(a^2 + 4b^2)(a + 2b)(a - 2b)$ 41. $(2x + 1)(4x^2 - 2x + 1)$
43. $(5m + 4n)(25m^2 - 20mn + 16n^2)$ 45. $(x + 10)(x + 2)$
47. $(w - 6)(w - 9)$ 49. $(x - 12y)(x + 4y)$
51. $(7x - 3)(2x + 7)$ 53. $-1(x + 4)(x + 5)$
55. $(a - 4)(a - 3)$ 57. $(x + 8)^2$
59. $(b - 7c)(b + 3c)$ 61. $m(m + 7)(m - 5)$
63. $3y(y - 7)(y - 9)$ 65. $(3x + 5)(x + 1)$
67. $(2b - 3)(b - 3)$ 69. $(5x - 3)(2x - 1)$
71. $(4y - 5x)(4y + 3x)$ 73. $-4x(2x + 1)(x - 5)$
75. $3x(2x - 3)(x + 1)$ 77. Factorable; $m = -6, n = 5$
79. Factorable; $m = -3, n = -8$ 81. $5x^2(x - 7)(x + 6)$
83. $2y(6 - y)(6 + y)$ 85. $10x^3(x + 4)^2$
87. $(x - 5)(x + 5)(x + 2)$ 89. $\{-6, 1\}$ 91. $\{-10, 3\}$
93. $\{-5, 4\}$ 95. $\{0, 10\}$ 97. $\{-5, 5\}$
99. $\left\{-1, \frac{3}{2}\right\}$ 101. $\{-5, 2\}$ 103. $\{-4, 9\}$
105. $\{0, -3, 5\}$ 107. -6 and 9; -9 and 6
109. Width 12 ft, length 18 ft 111. 4 s 113. 50

Self-Test—Chapter 6

1. $2b(4a + 5b)(4a - 5b)$ 2. $(x + 2)(x - 5)$ 3. $7(b + 6)$
4. $(x^2 + 9)(x + 3)(x - 3)$ 5. $(3x - 2y)^2$ 6. $5(x^2 - 2x + 4)$
7. $(4y + 7x)(4y - 7x)$ 8. $(4x - 3y)(2x + y)$
9. $(3y - 2x)(9y^2 + 6xy + 4x^2)$ 10. $(3w + 7)(w + 1)$
11. $(3x + 2)(2x - 5)$ 12. $(a - 7)(a + 2)$
13. $-3x(2x + 5)(x - 2)$ 14. $(y + 10z)(y + 2z)$ 15. $\{5, 6\}$
16. $\{-5, -3\}$ 17. $\{-1, 3\}$ 18. $\left\{-\frac{1}{3}, \frac{3}{2}\right\}$ 19. 20 cm
20. 2 s

Cumulative Review—Chapters 0–6

1. 9 2. $\{11\}$ 3. $\{-33\}$ 4. $r = \frac{A - P}{Pt}$
5. $W = \frac{P - 2L}{2}$ 6. $\{x \mid x > -6\}$ 7. $\{x \mid x \geq -6\}$
8. $\{(0, -1)\}$ 9. $\left\{\left(4, -\frac{5}{3}\right)\right\}$ 10. Slope: $-\frac{2}{5}$; y-intercept: (0, 2)
11. $y = 2x + 5$ 12. $y = -3x - 2$ 13. $2x^2y + 5xy$
14. $-x^2 + 2x - 1$ 15. $x^2 + 9x - 2$ 16. $3x^3 - 6x^2 + 6x$
17. $2x^2 - 3x - 5$ 18. $x^3 + 27$ 19. $8x^2 + 14x - 15$
20. $a^2 - 9b^2$ 21. $x^2 - 4xy + 4y^2$ 22. $50x^3 - 18xy^2$
23. $-\frac{8y^9}{x^6}$ 24. $3a^4b^6$ 25. $\frac{y^3}{4x^9}$ 26. $\frac{2x^3z}{y^5}$
27. $4(3x + 5)$ 28. $(5x + 7y)(5x - 7y)$ 29. $(4x - 5)(3x + 2)$
30. $(x - 5)(2x - 3)$ 31. $\{-5, 3\}$ 32. $\{-3, 3\}$
33. $8xy^4 - 5x^3y + 3$ 34. $x + 4 + \frac{6}{x - 1}$
35. 17; $3x - 5 = 46$ 36. \$43; $x + (x - 5) = 81$
37. 60 mi/h 38. 10 or 50 items

Reading Your Text

Section 7.1 **(a)** square root; **(b)** two; **(c)** negative; **(d)** radius
Section 7.2 **(a)** product; **(b)** radical; **(c)** denominator; **(d)** perfect-square
Section 7.3 **(a)** polynomials; **(b)** radicands; **(c)** rationalizing; **(d)** conjugate
Section 7.4 **(a)** extraneous; **(b)** integer; **(c)** isolated; **(d)** solutions
Section 7.5 **(a)** real; **(b)** principal; **(c)** radicals; **(d)** exponent
Section 7.6 **(a)** imaginary; **(b)** complex; **(c)** imaginary; **(d)** conjugates

Summary Exercises—Chapter 7

1. 11 3. Not real 5. -4 7. $\frac{3}{4}$ 9. 8
11. a^2 13. $7w^2z^3$ 15. $-3b^2$ 17. $2xy^2$
19. $\sqrt{13}$ 21. $\sqrt{41}$ 23. Center (3, 0); radius 6

25.

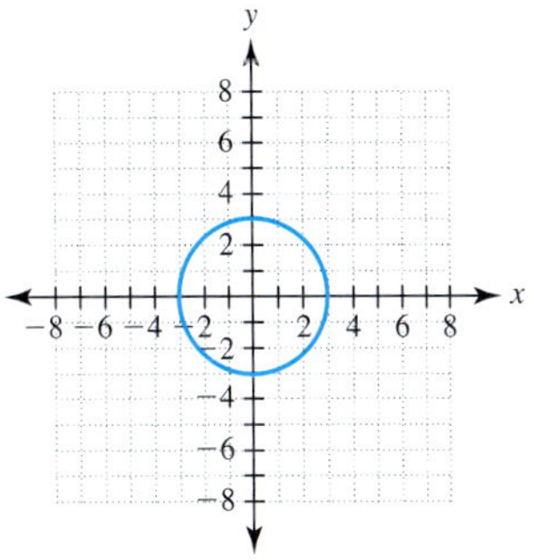

27.

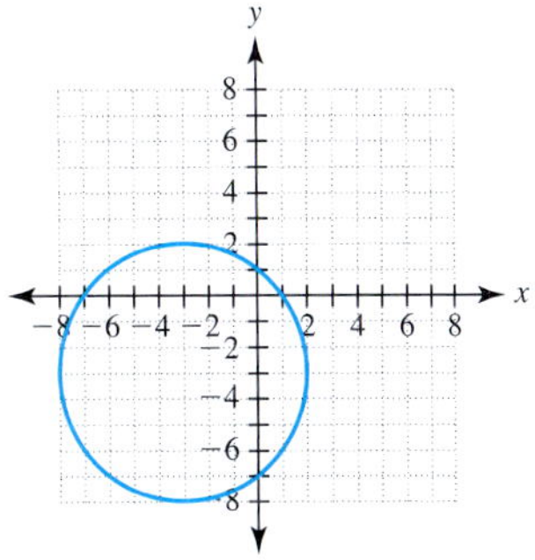

29. $-5\sqrt{3}$ 31. $6a\sqrt{3a}$ 33. $-2wz\sqrt[3]{10w}$ 35. $\frac{\sqrt{7}}{6}$
37. $\frac{\sqrt{2x}}{3}$ 39. $\frac{\sqrt[3]{5a^2}}{3}$ 41. $3\sqrt{3x}$ 43. $7\sqrt{10}$ 45. $\sqrt{6}$
47. $9\sqrt{5}$ 49. $w\sqrt{3w}$ 51. $\frac{13}{5}\sqrt{5}$ 53. $\frac{8a\sqrt[3]{3a}}{3}$
55. $\sqrt{21xy}$ 57. $ab\sqrt[3]{4}$ 59. $2\sqrt{3}$ 61. $-32 - 2\sqrt{3}$
63. 1 65. $7 + 4\sqrt{3}$ 67. $\frac{\sqrt{21}}{7}$ 69. $\frac{\sqrt{2ab}}{b}$
71. $\frac{2\sqrt[3]{9x^2}}{3x}$ 73. $\frac{3 - \sqrt{2}}{7}$ 75. $9 - 4\sqrt{5}$ 77. $\{21\}$
79. $\{9\}$ 81. $\{5\}$ 83. $\{9\}$ 85. 7 87. -3 89. 16
91. $\frac{8}{27}$ 93. $\frac{1}{27}$ 95. $b^{13/6}$ 97. $a^{3/4}$ 99. $\frac{1}{y^8}$
101. $8x^{1/4}y^{1/2}$ 103. $\frac{3xy}{z^2}$ 105. $\sqrt[5]{w^4z^2}$ 107. $\sqrt[3]{9a^2}$
109. $4w^2$ 111. $2a^2b^4$ 113. $i\sqrt{13}$ 115. $5 - 2i$
117. $3 - 8i$ 119. $8 + 28i$ 121. $-7 - 24i$ 123. $-3 - i$
125. $\frac{5}{13} - \frac{12}{13}i$

Self-Test—Chapter 7

1. $p^2q\sqrt[3]{9pq^2}$ 2. $\frac{\sqrt{10xy}}{4y}$ 3. $4x\sqrt{3}$ 4. $7a^2$ 5. $\frac{7x}{8y}$
6. $3 - 2\sqrt{2}$ 7. $64x^6$ 8. $\frac{32x^4}{y}$ 9. $-3w^2z^3$
10. $2x^3\sqrt[3]{4x}$ 11. 6.6 12. 1.4 13. $\{11\}$ 14. $\{4\}$
15. $\frac{\sqrt[3]{3x^2}}{x}$ 16. $x^3\sqrt{14x}$ 17. $5m\sqrt[3]{2m}$ 18. $3x\sqrt{3x}$

19. $a^2b\sqrt[5]{a^4b}$ **20.** $(125p^9q^6)^{1/3} = 5p^3q^2$ **21.** $22\sqrt{5}$
22. $4x^2y^3z^3\sqrt{2xy}$ **23.** $\dfrac{9m}{n^4}$ **24.** $\dfrac{8s^3}{r}$ **25.** $16 - 2\sqrt{55}$
26. $\sqrt{74}$; 8.602 **27.** $\sqrt{185}$; 13.601
28. Center $(4, -1)$; radius 7 **29.** $-7 + 51i$
30. $-\dfrac{10}{13} + \dfrac{24}{13}i$

Cumulative Review—Chapters 0–7

1. $\{-15\}$ **2.** 5 **3.** $y = \frac{3}{2}x - 6$ **4.** 17, 18, 19
5. $10x^2 - 2x$ **6.** $10x^2 - 39x - 27$ **7.** $x(x-1)(2x+3)$
8. $9(x^2 + 2y^2)(x^2 - 2y^2)$ **9.** $(x + 2y)(4x - 5)$
10. $(x + 4)(x - 4)(x^2 + 3)$ **11.** $-\dfrac{4}{3}$ **12.** $y = -\frac{1}{4}x + 1$
13. 20, 25 **14.** $2x^4y^3\sqrt{3y}$ **15.** $\sqrt{6} - 5\sqrt{2} + 3\sqrt{3} - 15$
16.

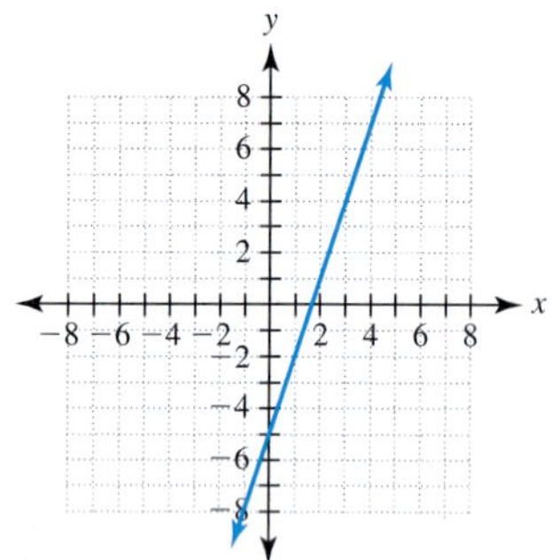

17.

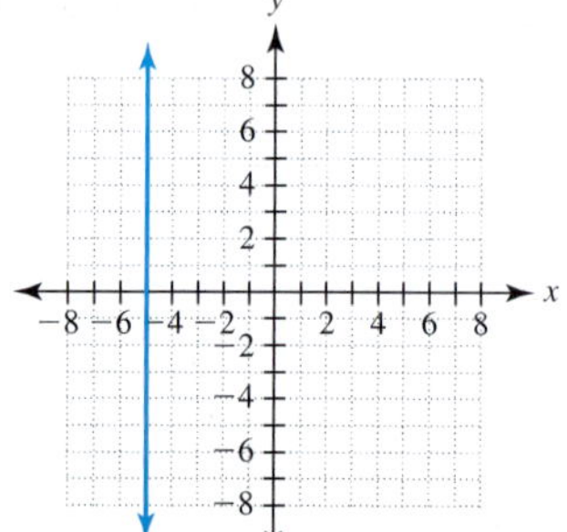

18.

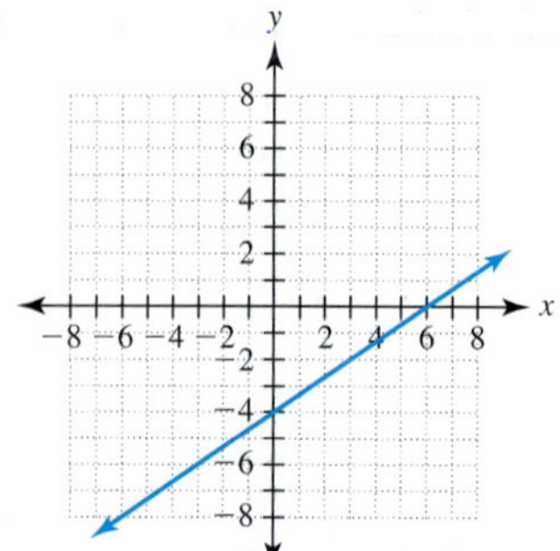

19. $\{(3, -1)\}$ **20.** $\left\{\left(5, \frac{3}{5}\right)\right\}$ **21.** x^6y^{14} **22.** $216x^4y^5$
23. $\dfrac{4xy^3}{3}$ **24.** $\{1\}$ **25.** $\{x \mid x \le -4\}$ **26.** $\frac{1}{2}, -5$
27. 17 in. by 31 in.

Reading Your Text

Section 8.1 **(a)** quadratic; **(b)** square-root; **(c)** completing; **(d)** constant
Section 8.2 **(a)** formula; **(b)** factoring; **(c)** discriminant; **(d)** no
Section 8.3 **(a)** downward; **(b)** minimum; **(c)** axis; **(d)** vertical
Section 8.4 **(a)** approximate; **(b)** lowest; **(c)** four; **(d)** quadratic formula

Summary Exercises—Chapter 8

1. $\{\pm 2\sqrt{3}\}$ **3.** $\{2 \pm 2\sqrt{5}\}$ **5.** 49 **7.** $\{-1, 5\}$
9. $\{5 \pm 2\sqrt{7}\}$ **11.** $\left\{\dfrac{3 \pm \sqrt{7}}{2}\right\}$ **13.** $\{-3, 8\}$
15. $\left\{\dfrac{5 \pm \sqrt{17}}{2}\right\}$ **17.** $\left\{\dfrac{-2 \pm \sqrt{7}}{3}\right\}$ **19.** $\{1 \pm \sqrt{29}\}$
21. $\left\{\dfrac{-1 \pm i\sqrt{14}}{3}\right\}$ nonreal **23.** None **25.** One **27.** 4, 8
29. 5 **31.** Width 5 cm; length 7 cm **33.** $30 \le x \le 50$
35. Width 4 cm; length 9 cm **37.** Width 8 ft; length 15 ft
39. $-1 + \sqrt{31}, 1 + \sqrt{31}$ or 4.6, 6.6 m **41.** $\{-1, 2\}$
43. $\left\{-2, -\dfrac{3}{2}\right\}$ **45.** $x = 0; (0, 0)$ **47.** $x = 0; (0, -5)$
49. $x = -2; (-2, 0)$ **51.** $x = -3; (-3, 1)$ **53.** $x = 5; (5, -2)$
55. $x = 1; (1, 1)$ **57.** $x = -\frac{1}{2}; \left(-\frac{1}{2}, \frac{25}{4}\right)$
59. $x = -3; (-3, 13)$ **61.** $\{-3.712, 1.212\}$ **63.** $\{0.807, 6.192\}$
65.

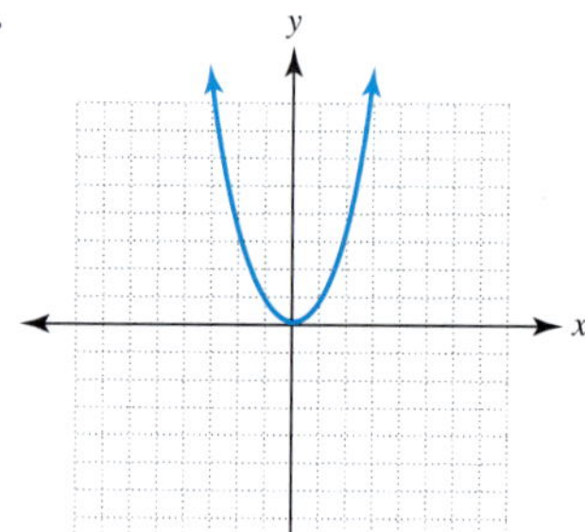

67.

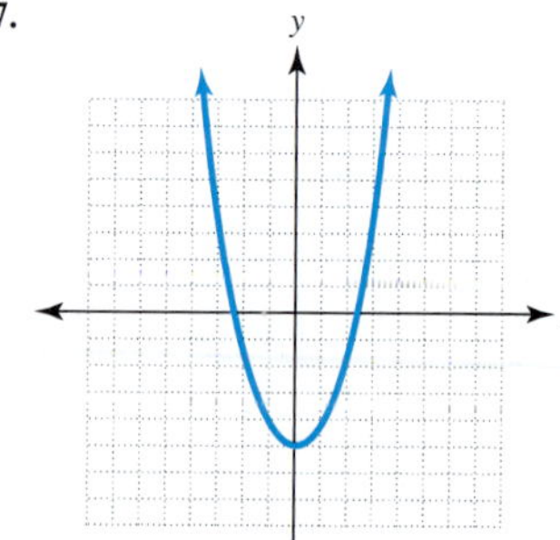

69.

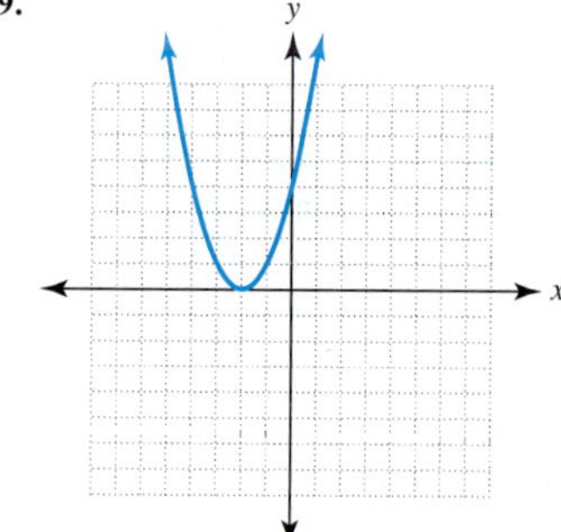

71.

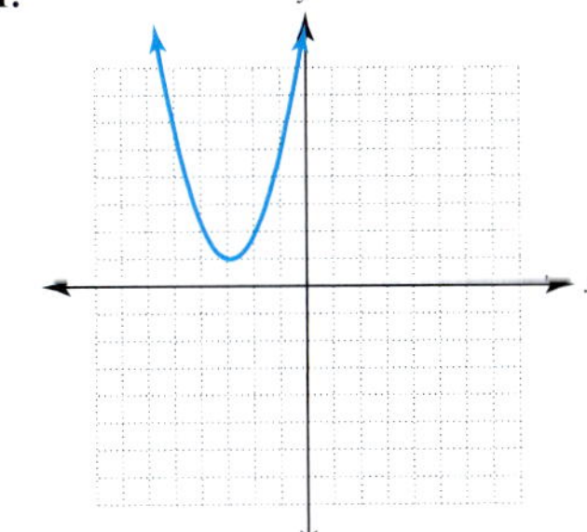

73.

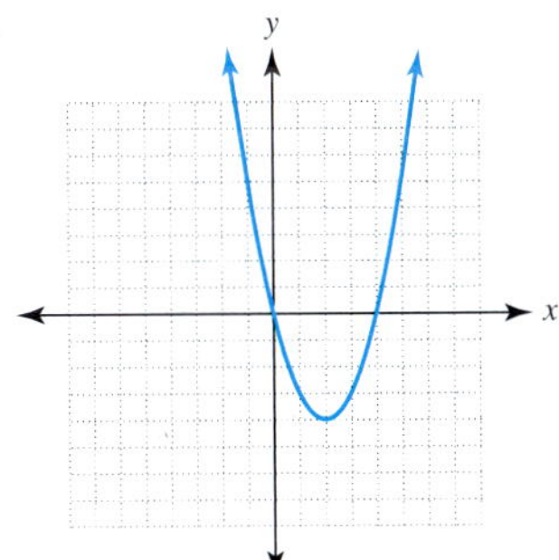

75.

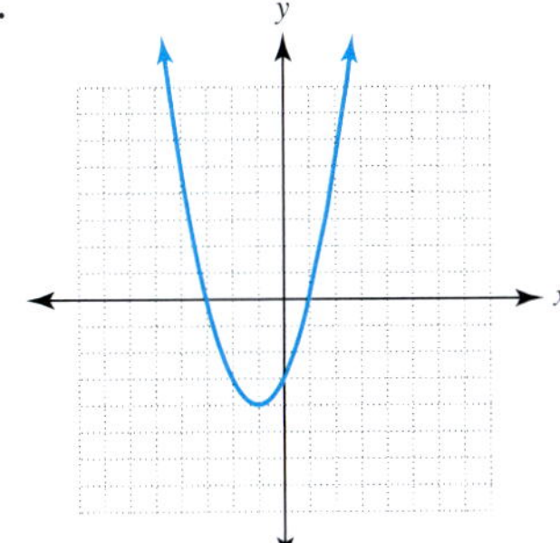

77.

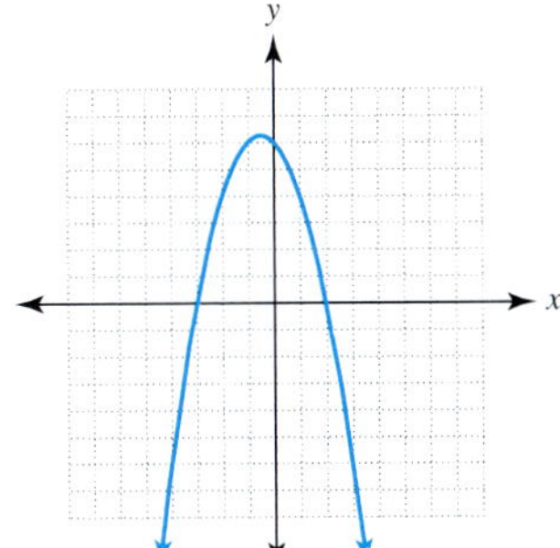

79.

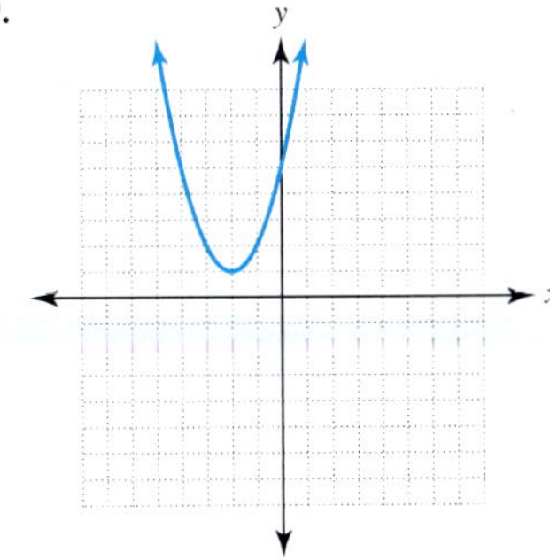

81.

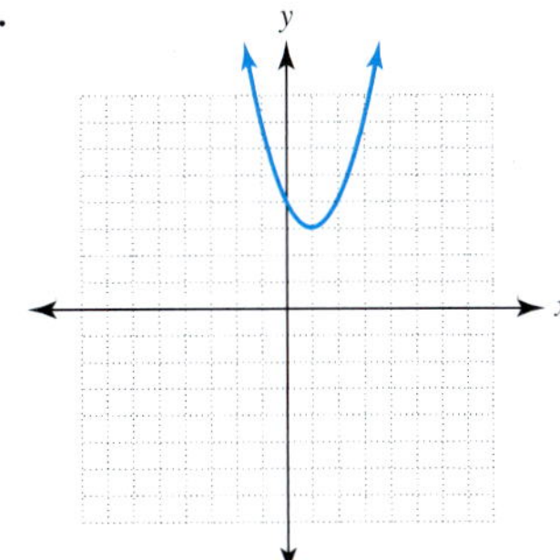

83.

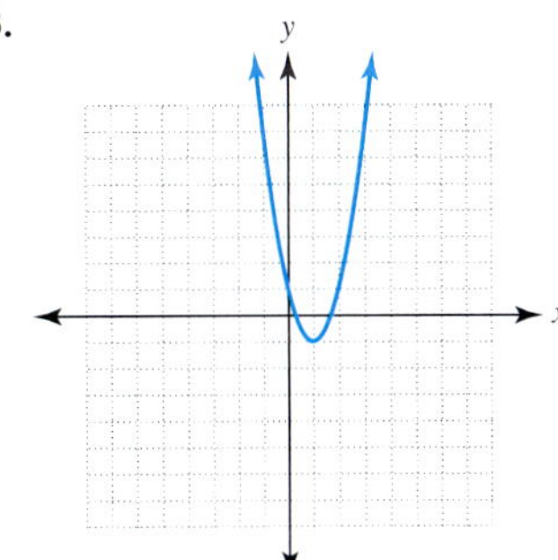

85. $\{\pm 2\sqrt{2}, \pm\sqrt{5}\}$ **87.** $\{\pm 3, \pm 2i\}$ **89.** $\{9, 16\}$
91. $\{\pm 2.713\}$

Self-Test—Chapter 8

1. $x = -2; (-2, 1)$ **2.** $x = 2; (2, -9)$ **3.** $x = \frac{3}{2}; \left(\frac{3}{2}, \frac{3}{2}\right)$

4. $x = 3; (3, -2)$ **5.** $x = 3; (3, -7)$ **6.** $\left\{\frac{-3 \pm \sqrt{13}}{2}\right\}$

7. $\left\{\frac{5 \pm \sqrt{19}}{2}\right\}$ **8.** $\left\{-\frac{2}{3}, 4\right\}$ **9.** $\{-4.541, 1.541\}$

10. $\{-1.396, 0.896\}$

11.

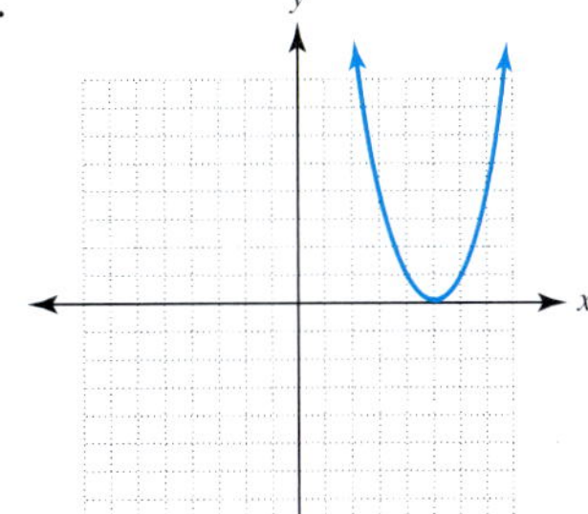

12.

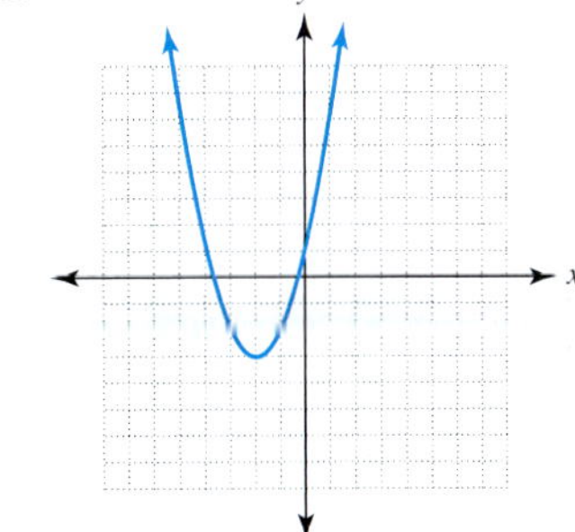

13.

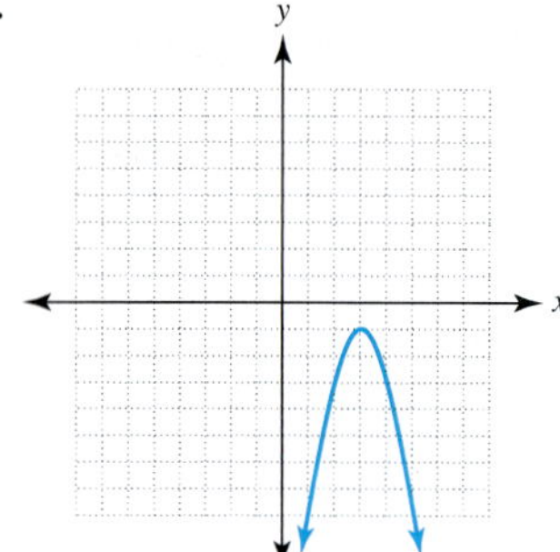

14.

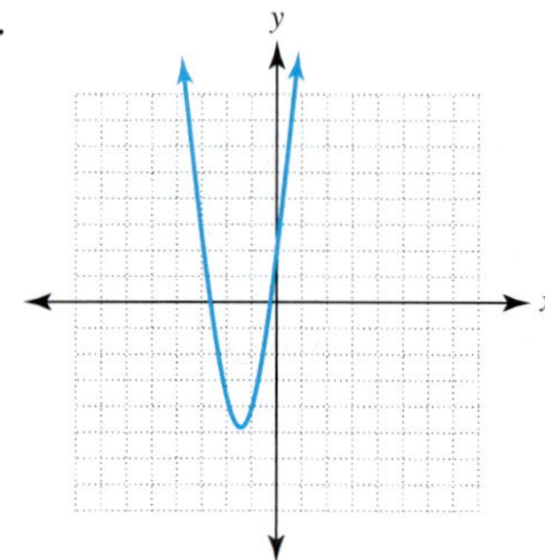

15. $\left\{-3, -\frac{1}{2}\right\}$ 16. $\left\{-\frac{5}{2}, \frac{2}{3}\right\}$ 17. $\left\{0, \frac{3}{2}, -\frac{3}{2}\right\}$

18. 7, 9 19. 2.7 s 20. $\left\{\frac{5 \pm \sqrt{37}}{2}\right\}$ 21. $\{-2 \pm \sqrt{11}\}$

22. $\left\{-\frac{2}{3}, \frac{4}{5}\right\}$ 23. $\left\{\frac{-1 \pm 3i}{2}\right\}$ nonreal 24. $\{\pm 2\sqrt{3}, \pm\sqrt{3}\}$

25. $\{\pm 2\sqrt{2}, \pm 2i\}$ two nonreal solutions 26. $\{25, 36\}$

27. $\{\pm\sqrt{5}\}$ 28. $\{1 \pm \sqrt{10}\}$ 29. $\left\{\frac{2 \pm \sqrt{23}}{2}\right\}$

30. $\{\pm 1.907\}$

Cumulative Review—Chapters 0–8

1.

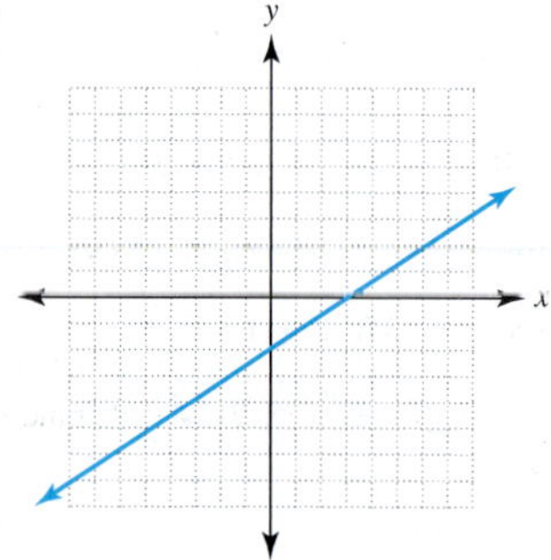

2.

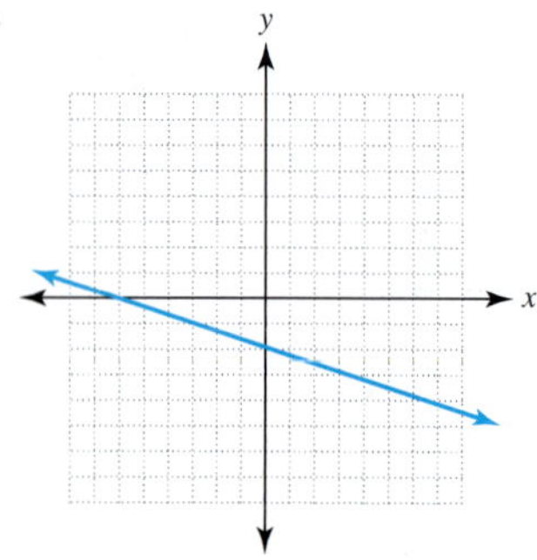

3.

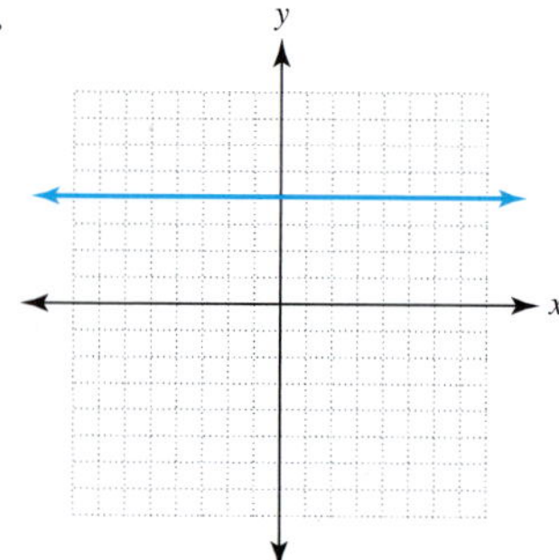

4. -3 5. $\frac{4}{3}$ 6. 35 7. $x^3 + 3x^2 - x - 3$

8. $x(x + 3)(x - 2)$ 9. $\frac{22\sqrt{6}}{3}$ 10. $6xy^2\sqrt{2xy}$

11. $\left\{\frac{7}{2}\right\}$ 12. $\{-4\}$ 13. $\{-5, 3\}$ 14. $\{1, 2\}$

15. $\{-10, 3\}$ 16. $\left\{\frac{3 \pm \sqrt{21}}{2}\right\}$ 17. $\{3 \pm \sqrt{5}\}$

18. $\{-3, 0, 5\}$ 19. $\left\{\frac{13}{6}\right\}$ 20. $\{1\}$ 21. $\{x \mid x \le 9\}$

22. $\sqrt{74}$ 23. $5x - 7 = 72, -13$

24. $x^2 + (x + 4)^2 = 28^2, -2 + 2\sqrt{97}, 2 + 2\sqrt{97}$ or 17.7 ft, 21.7 ft

25. 21 or 59 receivers

Reading Your Text

Section 9.1 **(a)** polynomials; **(b)** denominator; **(c)** integers; **(d)** factors

Section 9.2 **(a)** multiply; **(b)** factored; **(c)** reciprocal; **(d)** always

Section 9.3 **(a)** polynomial; **(b)** factored; **(c)** denominators; **(d)** exist

Section 9.4 **(a)** complex; **(b)** fundamental; **(c)** invert; **(d)** simplest

Section 9.5 **(a)** y-intercept; **(b)** denominator; **(c)** vertical; **(d)** horizontal

Section 9.6 **(a)** variable; **(b)** check; **(c)** similar; **(d)** rate

Summary Exercises—Chapter 9

1. Never undefined 3. $x \ne 5$ 5. $\frac{3x^2}{4}$ 7. 7

9. $-\frac{(x + 3)}{x + 5}$ 11. $\frac{2a - b}{3a - b}$ 13. $x - 4, x \ne -1$

15. $\frac{2x^3}{3}$ 17. $\frac{4}{3y}$ 19. $\frac{3}{2a}$ 21. $\frac{x + 2y}{x - 5y}$

23. **(a)** -8; **(b)** $x^2 + x - 20$; **(c)** -8 25. $\frac{3x^2 - 16x - 8}{(x + 4)(x - 4)}$

27. $\frac{x + 5}{x(x - 5)}$ 29. $-\frac{11}{6(m - 1)}$ 31. $\frac{4(x - 4)}{5(x + 1)(x - 1)}$

33. $-\frac{4}{s + 2}$ 35. $\frac{3x - 1}{x + 2}$ 37. **(a)** 8; **(b)** $\frac{3x^2 - 8x}{x^2 - 5x + 6}$; **(c)** (4, 8)

39. $\frac{2}{3x^2}$ 41. $\frac{b + a}{b - a}$ 43. $\frac{s + r}{rs}$ 45. $\frac{x - 4}{(x - 1)(x - 1)}$

47. $-y + 2$ 49. $\frac{x + 2}{(x - 1)(x + 4)}$ 51. $\{x \mid x \ne -4\}$

53. $\{x \mid x \ne 2\}$ 55. $\{x \mid x \ne -1\}$ 57. x-intercept: (3, 0), y-intercept: $\left(0, -\frac{3}{4}\right)$ 59. x-intercept: (0, 0), y-intercept: (0, 0)

61. x-intercept: (3, 0), y-intercept: (0, −6)

63. Vertical: $x = -4$, horizontal: $y = 1$

65. Vertical: $x = 2$, horizontal: $y = 8$

67. Vertical: $x = -1$, horizontal: $y = 2$

69.

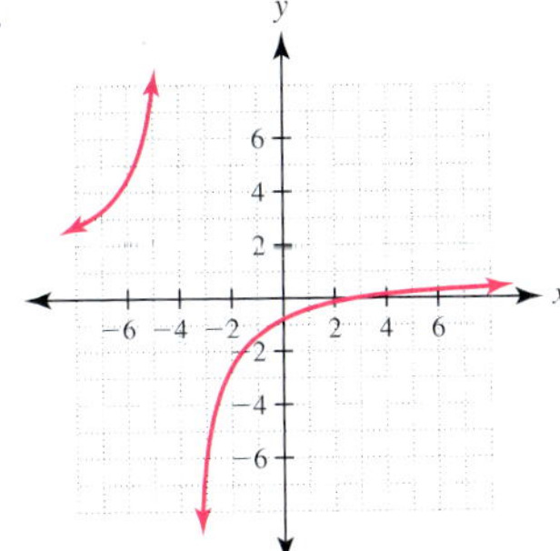

71.

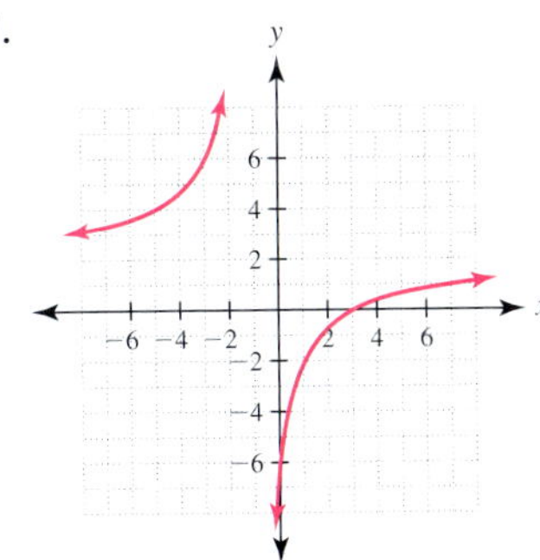

73. $\{5\}$ **75.** $\{6\}$ **77.** $\{3\}$ **79.** $\{-1\}$ **81.** $\{4\}$
83. $\{0, 7\}$ **85.** First: 4 and 11, second: 8 and 22
87. 5 and 6 **89.** 12 h

Self-Test—Chapter 9

1. $\frac{m-4}{4}$ **2.** $\frac{2}{x-3}$ **3.** $\frac{y}{3y+x}$ **4.** $\frac{4a^2}{7b}$

5. $\frac{8x+17}{(x-4)(x+1)(x+4)}$ **6.** $\frac{2}{x-1}$ **7.** $\frac{-5(3x+1)}{3x-1}$

8. $\frac{z-1}{2(z+3)}$ **9.** $\frac{2x+y}{x(x+3y)}$ **10.** $\frac{4}{x-2}$ **11.** $\frac{2(2x+1)}{x(x-2)}$

12. $\{2, 6\}$ **13.** $\frac{w+1}{w-2}$ **14.** $\frac{x+3}{x-2}$ **15.** $\frac{-3x^4}{4y^2}$ **16.** 3

17.

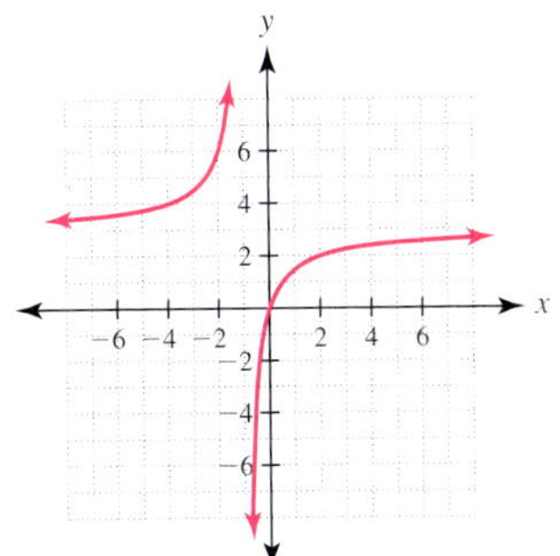

18.

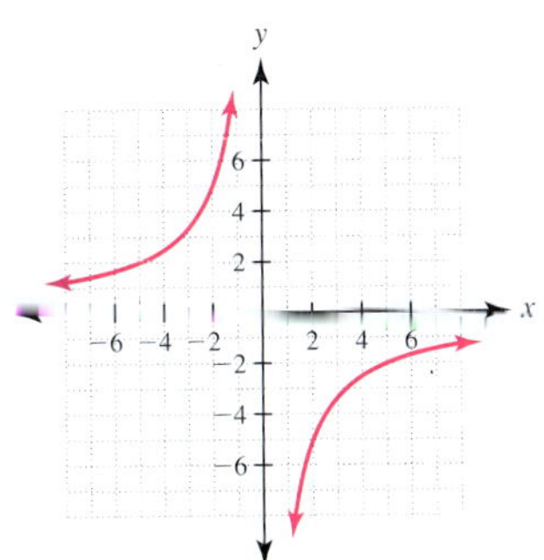

19. $f(x) = x - 1, x \neq 4$ **20.** **(a)** x-intercept: $\left(\frac{1}{4}, 0\right)$, y-intercept: $\left(0, -\frac{1}{2}\right)$ **(b)** vertical: $x = -\frac{2}{3}$, horizontal: $y = \frac{4}{3}$

Cumulative Review—Chapters 0–9

1. $\{12\}$ **2.** -14 **3.** $6x + 7y = -21$
4. x-intercept $(-6, 0)$; y-intercept $(0, 7)$
5. $f(x) = 6x^2 - 5x - 2$ **6.** $f(x) = 2x^3 + 5x^2 - 3x$
7. $\{x \mid x \neq 0\}$ **8.** 16 **9.** $x(2x + 3)(3x - 1)$
10. $(4x^8 + 3y^4)(4x^8 - 3y^4)$ **11.** $\frac{3}{x-1}$ **12.** $\frac{1}{(x+1)(x-1)}$
13. $x^2 - 3x$ **14.** 0.76 s and 2.86 s **15.** $\left\{\frac{7}{2}\right\}$
16. $\{\pm 4, \pm\sqrt{2}\}$ **17.** $\left\{-\frac{30}{17}\right\}$ **18.** $\{4\}$ **19.** $\{-5\}$
20. $\{x \mid x > -2\}$ **21.** 16.1 cm **22.** Vertical: $x = 5$, horizontal: $y = -2$ **23.** $\frac{b^6}{a^{10}}$ **21.** Don 6 h, Barry 3 h
25. Length 18 cm, width 10 cm

Reading Your Text

Section 10.1 **(a)** domain; **(b)** product; **(c)** quotient; **(d)** binomials
Section 10.2 **(a)** Composing; **(b)** g; **(c)** simpler; **(d)** operations
Section 10.3 **(a)** inverse; **(b)** symmetric; **(c)** one-to-one; **(d)** horizontal
Section 10.4 **(a)** exponential; **(b)** base; **(c)** greater; **(d)** Euler's
Section 10.5 **(a)** logarithm; **(b)** power; **(c)** y-intercept; **(d)** Loudness
Section 10.6 **(a)** exponent; **(b)** power; **(c)** common; **(d)** neutral
Section 10.7 **(a)** equation; **(b)** 10; **(c)** variable; **(d)** extraneous

Summary Exercises—Chapter 10

1. 2 **3.** Does not exist **5.** Does not exist **7.** $\{-2, 3, 7\}$
9. $2x^2 + 6x - 8$ **11.** $2x^4 + x^3 - x^2 + 6x + 5$
13. $5x^2 + 3x + 10$ **15.** $4x^2 + 8x$
17. $6x^2 - 10x$ **19.** $3x^3$ **21.** $\frac{2x}{x-3}$; $D = \{x \mid x \neq 3\}$
23. $\frac{3}{x}$; $D = \{x \mid x \neq 0\}$ **25.** 8 **27.** 2 **29.** -3
31. **(a)** -2; **(b)** -8; **(c)** -7; **(d)** $3x - 2$ **33.** **(a)** 25; **(b)** 49; **(c)** 4; **(d)** $x^2 - 5$ **35.** $(f \circ g)(x)$ **37.** $f^{-1}(x) = \frac{3-x}{2}$
39. $f^{-1}(x) = 4x + 3$ **41.** $\{(1, -3), (2, -1), (3, 2), (4, 3)\}$; function
43. $\{(5, 1), (7, 2), (9, 3)\}$; function
45. $\{(4, 2), (3, 4), (4, 6)\}$; not a function

47.

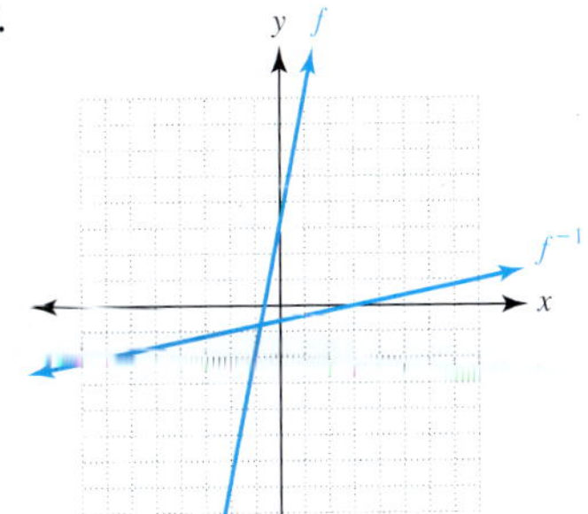

$f^{-1}(x) = \frac{x-3}{5}$

49.

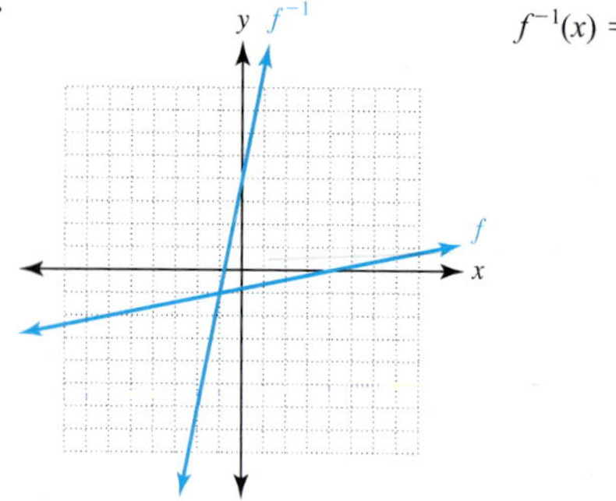

$f^{-1}(x) = 5x + 4$

51. One-to-one; f^{-1}: $\{(2, -1), (3, 1), (5, 2), (7, 4)\}$; function
53. One-to-one; f^{-1}: $\{(2, -1), (4, 3), (5, 4), (6, 5)\}$; function
55. Not one-to-one; f^{-1} is not a function **57.** 4
59. 8 **61.** x

63.

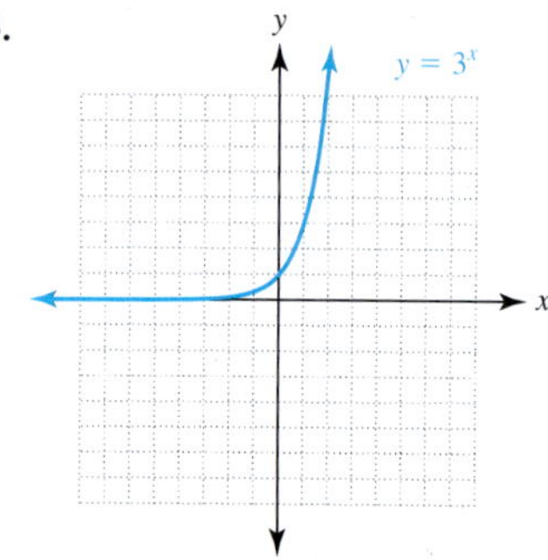

65. $\{3\}$ **67.** $\{-1\}$ **69.** 64,000

71.

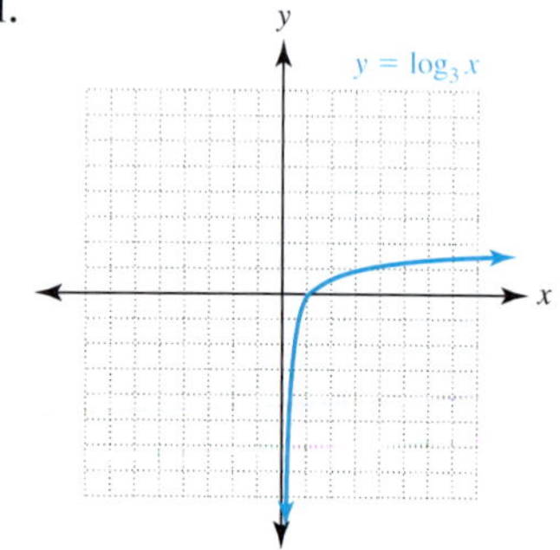

73. $\log_2 32 = 5$ **75.** $\log_5 1 = 0$ **77.** $\log_{25} 5 = \frac{1}{2}$
79. $4^3 = 64$ **81.** $81^{1/2} = 9$ **83.** $10^{-3} = 0.001$
85. $\{3\}$ **87.** $\{49\}$ **89.** $\{9\}$ **91.** $\left\{\frac{1}{3}\right\}$ **93.** 80 dB
95. 100 **97.** 10 **99.** $3 \log_4 y - \log_4 5$
101. $3 \log_5 x + \log_5 y + 2 \log_5 z$
103. $\frac{2}{3}\log_b x + \frac{1}{3}\log_b y - \frac{1}{3}\log_b z$ **105.** $\log_b \frac{x^3}{z^2}$
107. $\log_5 \frac{x^2}{y^3 z}$ **109.** $\log_b \sqrt[3]{\frac{x}{y^2}}$ **111.** 1.204 **113.** 0.239
115. 9.495, basic **117.** 6.3×10^{-11} **119.** 67

121.

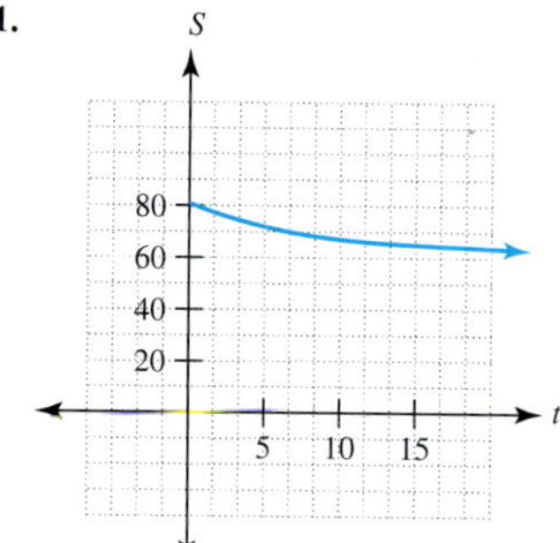

123. 1.969 **125.** $\{250\}$ **127.** $\{2\}$ **129.** $\{3\}$
131. $\{-2\}$ **133.** $\{2.5\}$ **135.** $\{4.419\}$ **137.** 8.21 yr
139. 166 yr **141.** 8.7 yr **143.** 6.6 mi

Self-Test—Chapter 10

1. $\log 10{,}000 = 4$ **2.** $\log_{27} 9 = \frac{2}{3}$ **3.** x
4. x **5.** $\frac{1}{2}(\log_5 x + 2\log_5 y - \log_5 z)$
6. $(f \cdot g)(x) = 3x^3 - 2x^2 - 3x + 2; D = \mathbb{R}$
7. $(g \div f)(x) = \frac{3x - 2}{x^2 - 1}; D = \{x | x \neq \pm 1\}$
8. $(f \circ g)(x) = 9x^2 - 12x + 3$ **9.** $(g \circ f)(x) = 3x^2 - 5$
10. $\{-2\}$ **11.** $\left\{\frac{5}{2}\right\}$ **12.** $5^3 = 125$ **13.** $10^{-2} = 0.01$
14. Not one-to-one; the inverse is not a function.
15. One-to-one; the inverse is a function. **16.** $\log_b \sqrt[3]{\frac{x}{z^2}}$
17. $\{0.262\}$ **18.** $\{2.151\}$ **19.** **(a)** $-3x^3 + 3x^2 + 5x - 9$; **(b)** $-3x^3 + 7x^2 - 9x - 5$

20.

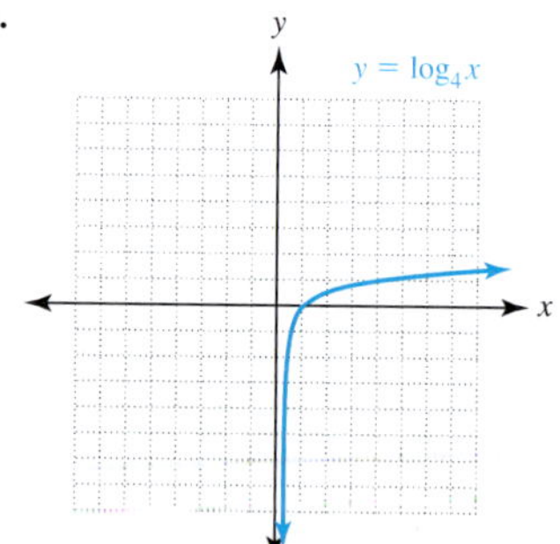

21. $\{8\}$ **22.** $\left\{\frac{11}{8}\right\}$ **23.** $f^{-1}(x) = 5x + 3$
24. $f^{-1} = \{(-2, -3), (2, 1), (5, 4), (6, 5)\}$
25. $f^{-1} = \{(1, -3), (2, 4), (-2, 5)\}$

26.

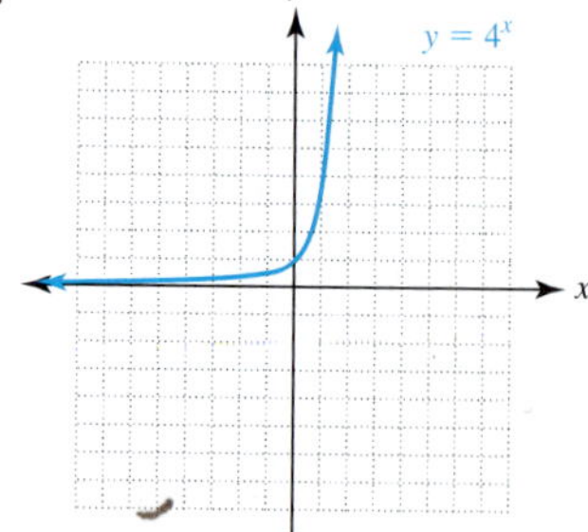

27.

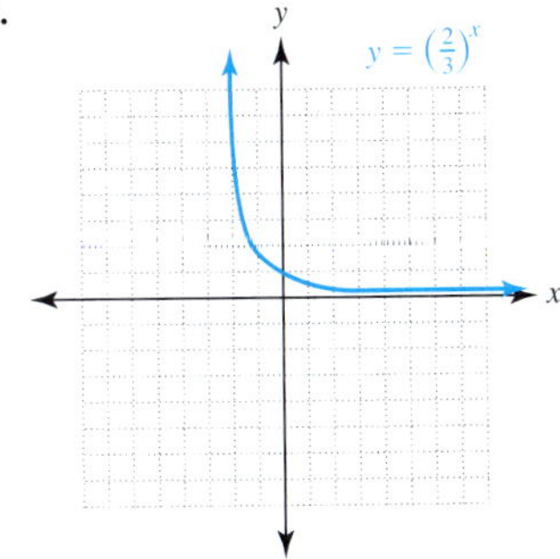

28. $\{6\}$ 29. $\{4\}$ 30. $\{5\}$

Cumulative Review—Chapters 0–10

1. $\{11\}$ 2. $\left\{\frac{5}{3}\right\}$ 3. $\left\{\frac{10}{9}\right\}$

4.

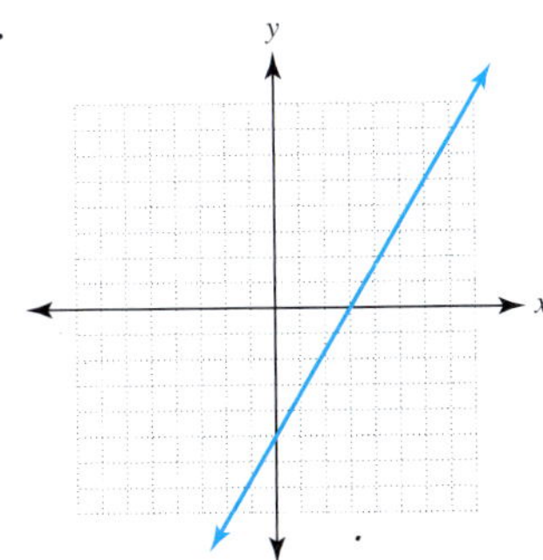

5.

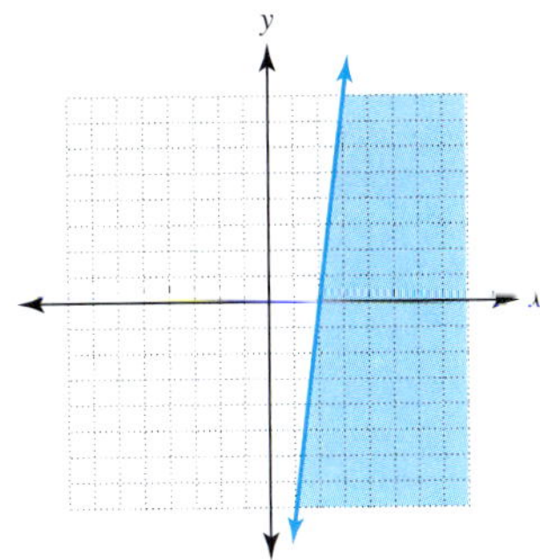

6. $6x + 5y = 7$ 7. $\{x \mid x \geq 10\}$ 8. $2x^2 - 6x + 1$
9. $6x^2 - 13x - 5$ 10. $(2x - 5)(x + 2)$
11. $x(5x + 4y)(5x - 4y)$ 12. $\dfrac{-x + 2}{(x - 4)(x - 5)}$ 13. $\dfrac{x + 2}{x - 2}$
14. $-4\sqrt{2}$ 15. $22 + 12\sqrt{2}$ 16. $\frac{5}{3}(\sqrt{5} + \sqrt{2})$
17. 77, 79, 81 18. $\{-2, 1\}$ 19. $\left\{\dfrac{3 \pm \sqrt{19}}{2}\right\}$
20. $R = \dfrac{R_1R_2}{R_1 + R_2}$ 21. $3x^3 - 4x^2 - 17x + 6$; $D = \mathbb{R}$
22. $\dfrac{3x - 1}{x^2 - x - 6}$; $D = \{x \mid x \neq -2, x \neq 3\}$
23. **(a)** $-6x + 3$; **(b)** -27 24. **(a)** $-6x + 5$; **(b)** -25
25. $7ab(2ab - 3a + 5b)$ 26. $(x - 3y)(x + 5)$
27. $(5c - 8d)(5c + 8d)$ 28. $(3x - 1)(9x^2 + 3x + 1)$
29. $2a(2a + b)(4a^2 - 2ab + b^2)$ 30. $(x - 8)(x - 6)$
31. $(5x - 2)(2x - 7)$ 32. $3x(x + 3)(2x - 5)$ 33. $3x^2$
34. $\dfrac{12}{3 - y}$ 35. $\dfrac{m - 13}{(m + 3)(m - 3)(m - 1)}$ 36. $\dfrac{10}{x - 5}$
37. 10, 24, and 26 38. $\{8\}$ 39. $\{5\}$
40. Center: $(-5, 2)$; radius 4

index